Satellite Altimetry over Oceans and Land Surfaces

Earth Observation of Global Changes

Series Editor
Chuvieco Emilio

Satellite Altimetry over Oceans and Land Surfaces
edited by Detlef Stammer and Anny Cazenave

Global Forest Monitoring from Earth Observation
edited by Frédéric Achard and Matthew C. Hansen

Earth Observation of Ecosystem Services
edited by Domingo Alcaraz-Segura, Carlos Marcelo Di Bella, and Julieta Veronica Straschnoy

Satellite Altimetry over Oceans and Land Surfaces

Detlef Stammer
Anny Cazenave

CRC Press
Taylor & Francis Group
Boca Raton London New York

CRC Press is an imprint of the
Taylor & Francis Group, an **informa** business

CRC Press
Taylor & Francis Group
6000 Broken Sound Parkway NW, Suite 300
Boca Raton, FL 33487-2742

First issued in paperback 2019

ISBN-13: 978-1-4987-4345-7 (hbk)
ISBN-13: 978-0-367-87484-1 (pbk)

Library of Congress Cataloging-in-Publication Data

Names: Stammer, Detlef, editor. | Cazenave, Anny, editor.
Title: Satellite altimetry over oceans and land surfaces / edited by Detlef
Stammer and Anny Cazenave.
Description: Boca Raton, FL : Taylor & Francis, 2017.
Identifiers: LCCN 2017013954 | ISBN 9781498743457 (hardback : alk. paper)
Subjects: LCSH: Oceanography--Remote sensing. | Satellite geodesy--Technique.
Classification: LCC GC10.4.R4 S26 2017 | DDC 551.48--dc23
LC record available at https://lccn.loc.gov/2017013954

Visit the Taylor & Francis Web site at
http://www.taylorandfrancis.com

and the CRC Press Web site at
http://www.crcpress.com

Contents

Preface

Dear Reader:

With this book, we aim to provide a state-of-the-art overview of the satellite altimetry techniques and related missions and to review recent applications with respect to ocean dynamics (large- and small-scale circulation, ocean tides, waves, El Niño-Southern Oscillation [ENSO], coastal processes, etc.) and to global and regional sea level change. The book also discusses related, auxiliary, space-based, and in situ observations (e.g., space gravimetry, Argo data, or tide gauges) as well as applications of satellite altimetry to the cryosphere, land surface waters, and seafloor topography.

In one sense, this book is written to update the expert community about recent results. Most of all, however, the book is designed as a textbook for students and researchers who are interested in getting acquainted with satellite altimetry for the study of the Earth as well as a reference for a graduate-level course of satellite altimetry. The style of each chapter provides the reader a broad exposure to the subject from the basics to the state-of-the-art. Following a similar first book on altimetry by Fu and Cazenave (2001), this new book provides an updated comprehensive description of the state-of-the-art technology, the historic developments of altimetry, and of ongoing scientific applications to a variety of topics in Earth sciences. By introducing the techniques and by providing a wide range of applications of satellite altimetry, this book also highlights the need for collecting long-term records of key processes acting in the Earth system.

The ocean is a turbulent fluid showing variability from basin scale down to a few meters and on time scales from a few minutes to climate time scales. Because of this, it requires the combination of all in situ observing components with satellite remote sensing to observe the full ocean variability with global coverage. It is especially the satellite remote sensing that provides an all-season full view of the surface ocean, globally and with a repeat cycle of just a few days. Because of this unique observing capability, several ocean parameters are routinely monitored by a variety of space sensors, such as sea surface temperature and salinity, surface winds and wave heights, ocean color, ocean mass changes, and sea surface topography. In this list, satellite altimetry is unique in that it is a technique that measures sea surface height and thereby the geometric shape of the sea surface relative to a reference system.

The concept of the satellite altimetry measurement is simple: The onboard radar altimeter transmits microwave radiation toward the sea surface, which partly reflects back to the satellite. Measurement of the round-trip travel time provides the height of the satellite above the instantaneous sea surface (called "range"). The principal quantity of interest is the geometry of the sea surface height above a reference fixed surface, typically a conventional reference ellipsoid. It is obtained by the difference between the altitude of the satellite above the reference (deduced from precise orbitography) and the measured range. The estimated sea surface height is further corrected for a variety of factors due to atmospheric delay and biases between the mean electromagnetic (EM) scattering surface and sea at the air–sea interface. Other environmental and geophysical corrections are also applied (Fu and Cazenave 2001).

Measurements of the topography of the sea surface are important for oceanography and geodesy alike: One contribution to the measured sea surface shape originates from a (nearly) time-invariable equipotential surface called the geoid (an equipotential surface of the Earth gravity field), consisting of highs and lows of 1–100 m amplitude. Superimposed to the marine geoid are contributions resulting from the time-mean and fluctuating ocean circulation. On spatial scales roughly larger than 10 km, the ocean is in geostrophic balance, implying a relationship between the sea surface slope relative to the geoid and ocean currents. The sea surface topography, therefore, is a dynamically unique parameter that provides unprecedented information about the upper ocean flow field, a long-standing goal of physical oceanographers for better understanding the complex full-depth ocean circulation and its role in climate. Through its unique capability to map the sea surface shape, even on an eddy and—in the future—a sub-eddy scale, it is an ideal system for studying the mesoscale and sub-mesoscale component of the ocean circulation and its changes in time.

Associated amplitudes in the measured sea surface deviation from the marine geoid are of the order of ± 1 m around the globe for the time-mean circulation, with the largest changes across permanent frontal structures such as boundary currents. The time-variable component related to the dynamics of the oceans (currents, tides, and so on) is even smaller and more in the range 0.1–0.5 m. It is thus obvious that highest accuracy is required for a system that can be used for quantitative ocean studies. It took the international community several decades to develop the altimetry system. With its nowadays approximately 1-cm accuracy of sea surface height measurements and a stability of approximately 1cm per decade, existing altimeter data time series have revolutionized the observation of the ocean. Measuring climate-related global mean sea level rise and associated regional variability is the most demanding application of satellite altimetry because of the low signal level. All sources of errors affecting the altimetry system need to be understood and reduced. At the time of the TOPEX/Poseidon satellite launch in 1992, it was considered that this objective could not be reachable. Yet subsequent improvements at all steps of altimetry data processing have allowed the production of accurate sea level time series, now considered as one of the best indicators of climate change and variability.

The first altimeter satellite, the Geodynamic Experimental Ocean Satellite (GEOS 3), was launched in 1975 by NASA. Since then, other U.S. altimetry missions have flown with the purpose of studying the oceans from space (e.g., the Seasat, 1978, and GEOSAT, 1985–1989). However, the uncertainty of the position of these satellites in space (the orbital error) was large (several decimeters), so that detection of ocean dynamics phenomena was very difficult. In the early 1990s, the era of high-precision altimetry began with the launch of the European satellite ERS-1 in 1991 and, especially, with the U.S./French TOPEX/Poseidon mission in 1992. The accuracy of individual sea surface height measurements from these missions and their successors (e.g., the Jason series, ERS-2, GEOSAT Follow-On (GFO), Envisat, Sentinel-3, and SARAL/AltiKa) has now reached the 1-cm level, which is required to obtain quantitative and—until then—unprecedented oceanographic information. This is the result of substantial technological developments and model improvements involving international collaborations of radar engineers, geodesists, geophysicists, and oceanographers as well as program managers. Since about the 1990s, such a system has provided routine and continuous time series of precise altimetry with data being used in a wide range of scientific and operational applications.

Today, satellite altimetry is a mature technique and a generally accepted element of the ocean and climate observing system with a long-planning horizon ahead through the European Sentinel-3 Series and continued long-term involvement from the European Organization for the Exploitation of Meteorological Satellites (EUMETSAT)—together with other space agencies—in sustaining the Jason series with the development of the future Jason-CS/Sentinel 6 mission, another component of the COPERNICUS European program. New technology such as the Synthetic Aperture Radar (SAR) interferometry already in orbit with the European CryoSat satellite and the wide-swath altimetric interferometry under development with the U.S./French Surface Water and Ocean Topography (SWOT) mission (to be launched in 2021) will even further expand and revolutionize the applications of altimetry over the ocean and land. Continuation and maintenance of significant data records from satellite altimetry for the benefit of science and society remains a pressing issue for the international science community as well as governmental organizations.

Ongoing applications of satellite altimetry cover a wider and ever-expanding range in oceanography, geodesy, and geophysics as well as in terrestrial hydrology and cryospheric research. In detail, satellite altimetry provides information about a variety of phenomena related to the internal structure of the Earth, seafloor bathymetry and plate tectonics, ocean tides, large-scale ocean circulation and eddies, waves, and sea level changes (the latter resulting from changes in ocean heat content and ocean mass changes). Although designed to observe the oceans, satellite altimetry is also now currently used to measure level variations of surface waters on land (lakes and rivers) and ice sheet elevation changes in Greenland and Antarctica, thus offering invaluable information on land hydrology and glaciology. The structure of the book is designed to cover all these subjects, with individual chapters dedicated to individual themes.

Chapters 1 and 2 provide information on existing and new altimeter technology. Chapter 1 presents the basic principle of satellite altimetry and discusses in great detail the satellite orbit and the various geophysical corrections that need to be applied to the altimetry measurement. It also provides some historical information on the various altimetry missions launched during the past few decades and on the evolution through time of the performances of the altimetry system. Chapter 2 reviews the concept of wide-swath altimetry, derives the measurement error budget for both ocean and surface water bodies, and also discusses science applications that require this new technology.

Chapters 3 and 4 describe auxiliary data sets required to process and analyze satellite altimetry. Chapter 3 deals with in situ data sets such as Argo profiles, tide gauge data, or velocity measurements; Chapter 4 deals with the Earth's gravity field, its theoretical representation and measurements. Some focus is directed toward the Gravity Recovery and Climate Experiment (GRACE) space gravimetry mission launched in 2002 for measuring temporal variations of the geoid.

Applications to global and regional problems are summarized in Chapters 5 through 9. Chapter 5 discusses the global mean sea level record derived from satellite altimetry and the various contributions (ocean thermal expansion, glacier and ice sheet mass balance, and land water storage) causing sea level variations. Chapters 6 through 9 then focus on specific regional aspects of satellite altimetry, notably the tropical oceans, the mid-latitude oceans, the Arctic, and the Southern Ocean. For each region, the importance of satellite altimetry for the study of dynamic and kinematic features is highlighted.

The study by altimetry of specific ocean phenomena is addressed in Chapters 10 through 13. Chapter 10 addresses the unique way ocean eddies can be observed by satellite altimetry and updates the reader on recent insights gained through the altimeter time series. Chapter 11 discusses problems in interpreting altimetry data in coastal zones arising from data inaccuracies in the proximity of land and how to improve those via the development of specific retracking of raw radar echoes and of new geophysical corrections. Chapter 12 summarizes the use of altimetry to monitor ocean waves and wind speeds and reviews several recent applications. Ocean tides and how they have been studied using altimetry are then addressed in Chapter 13.

Applications other than to the ocean are provided in Chapters 14 through 16. Chapter 14 describes the application of satellite altimetry to the monitoring of surface water levels on land. It discusses the difficulty in interpreting radar waveforms (echoes) over small lakes and rivers (because of land contamination) and the recent development of adapted retracking methodologies, allowing the calculation of long water level time series on lakes and rivers. Chapter 15 deals with altimetry on the Antarctic ice sheet. It describes in great detail how satellite altimetry has contributed to an improved understanding of the complex dynamics of the ice sheet and how it informs on the mass balance, one of the main contributions to the sea level budget. In Chapter 16, the global mapping of seafloor bathymetry is presented. The global bathymetric maps derived by combining data from different altimetry missions have many practical and scientific applications—in particular in marine geophysics.

Finally, altimetry in the context of ocean modeling and operational oceanography is addressed in Chapters 17 and 18. Both chapters provide background information on modeling and assimilation techniques and review applications with a focus on climate science in Chapter 17 and on operational oceanography in Chapter 18.

Compiling a book covering a broad range of topics as has been done here cannot be accomplished without the effort and involvement of a group of very dedicated experts. We would like to thank all the authors who made contributions to this book as well as the anonymous reviewers who helped by improving each chapter. Jointly, they have spent a great deal of effort in writing and revising passages as required by the peer-review process, making this a worthwhile book to read and to own. We are also indebted to a group of anonymous reviewers who greatly enhanced the quality of this book by evaluating each chapter. We would like to express our appreciation and thanks for their efforts.

Detlef Stammer
Anny Cazenave

REFERENCE

Fu, L. L., and A. Cazenave. 2001. Satellite altimetry and Earth sciences. In *A Handbook of Techniques and Applications*. International Geophysics Series, Volume 69, Academic Press, San Diego, CA, p. 463.

Editors

Detef Stammer is a professor of physical oceanography and Earth system remote sensing at the Universität Hamburg in Germany, where he is director of the Center für Erdsystemforschung und Nachhaltigkeit. His research interests include ocean and climate variability analyzed from ocean and climate data in combination with ocean and climate models (assimilation). The use of satellite altimetry and other satellite observations for studies of ocean circulation, ocean dynamics, and mixing and eddy transports are central to his work. One such aspect is the use of altimeter data for quantitative testing of general ocean circulation models. He has published more than 180 articles in international journals and was lead author in the Fifth Intergovernmental Panel on Climate Change Working Group I Assessment report. He has served on several national and international scientific committees and is fellow of the American Geophysical Union.

Anny Cazenave is senior scientist at the Laboratoire d'Etudes en Géophysique et Océanographie Spatiale, Centre National d'Etudes Spatiales, Toulouse, France, and director for Earth sciences at the International Space Science Institute, Bern, Switzerland. Her research deals with the applications of space techniques to geosciences (geodesy and solid Earth geophysics; sea level variations and the study of climatic causes; land hydrology from space). She has extensive experience in using satellite altimetry for studying the marine geoid, land surface waters, and sea level. She has contributed in several space missions in geodesy and oceanography and has served as lead author of the Intergovernmental Panel on Climate Change Working Group I (fourth and fifth assessment reports). She is fellow of the American Geophysical Union, a member of the French academy of science, and a foreign member of the U.S., Indian, and Belgian academies of science.

Contributors

Saleh Abdalla
ECMWF
Reading, UK

Michaël Ablain
CLS
Ramonville-Saint-Agne, France

Laïba Amarouche
CLS
Ramonville-Saint-Agne, France

Ole B. Andersen
Technical University of Denmark
Kongens Lyngby, Denmark

Sabine Arnault
Laboratoire LOCEAN
Paris, France

Jérôme Benveniste
ESA/ESRIN
Frascati, Italy

Srinivas Bettadpur
University of Texas
Austin, TX

Florence Birol
LEGOS
Toulouse, France

Stéphane Calmant
LEGOS
Toulouse, France

Loren Carrere
CLS
Ramonville-Saint-Agne, France

Anny Cazenave
LEGOS
Toulouse, France
and
ISSI
Bern, Switzerland

Don Chambers
University of South Florida
Tampa, FL

Curtis W.Chen
Jet Propulsion Laboratory
California Institute of Technology
Pasadena, CA

John Church
University of New South Wales
Sydney, Australia

Paolo Cipollini
National Oceanography Centre
Southampton, UK

Alexandre Couhert
CNES
Toulouse, France

Jean-François Cretaux
LEGOS
Toulouse, France

Jan-Willem De Bleser
Jet Propulsion Laboratory
California Institute of Technology
Pasadena, CA

Gérald Dibarboure
CNES
Toulouse, France
and
CLS
Ramonville-Saint-Agne, France

Shenfu Dong
Physical Oceanography Division
Atlantic Oceanographic and Meteorological Laboratory
Miami, FL

Joël Dorandeu
CLS
Ramonville-Saint-Agne, France

Pierre Dubois
CLS
Ramonville-Saint-Agne, France

Theodore Durland
Oregon State University
Corvallis, OR

Gary D. Egbert
Oregon State University
Corvallis, OR

Philippe Escudier
CNES
Toulouse, France

Daniel Esteban Fernandez
Jet Propulsion Laboratory
California Institute of Technology
Pasadena, CA

J. Thomas Farrar
Woods Hole Oceanographic Institution
Woods Hole, MA

M. Joana Fernandes
Universidade do Porto
Faculdade de Ciências
Porto, Portugal

Fréderic Frappart
LEGOS
Toulouse, France

Lee-Lueng Fu
JPL/NASA
California Institute of Technology
Pasadena, CA

Sarah T. Gille
Scripps Institution of Oceanography
University of California San Diego
La Jolla, CA

Stephen M. Griffies
NOAA/GFDL
and
Princeton University
Princeton, NJ

Weiqing Han
University of Colorado
Boulder, CO

Gregg Jacobs
NRL
Washington D.C., MS

Peter A. E. M. Janssen
ECMWF
Reading, UK

Svetlana Jevrejeva
National Oceanography Centre
Liverpool, UK

Johnny A. Johannessen
Nansen Environmental and Remote Sensing Center and University of Bergen
Bergen, Norway

Kathryn A. Kelly
Applied Physics Laboratory
University of Washington
Seattle, WA

Matthias Lankhorst
Scripps Institution of Oceanography
University of California San Diego
La Jolla, CA

Pierre-Yves Le Traon
Mercator Ocean
Ramonville St Agne, France
and
Ifremer
Pllouzané, France

Tong Lee
Jet Propulsion Laboratory
California Institute of Technology
Pasadena, CA

Eric Leuliette
NOAA
Silver Spring, MD

Alain Mallet
CNES
Toulouse, France

Matt Martin
Met Office
Exeter, UK

Kara J. Matthews
University of Oxford
Oxford, UK

Nikolai Maximenko
University of Hawaii at Manoa
Honolulu, HI

Anthony Memin
GEOAZUR
University of Nice
Sophia Antipolis, France

Flavien Mercier
CLS
Ramonville-Saint-Agne
France

Michael P. Meredith
British Antarctic Survey
Cambridge, UK

Benoit Meyssignac
LEGOS
Toulouse, France

Rosemary Morrow
LEGOS
Toulouse, France

R. Dietmar Müller
University of Sydney
Sydney, Australia

R. Steven Nerem
University of Colorado
Boulder, CO

Karina Nielsen
DTU
Kongens Lyngby, Denmark

Estelle Obligis
CLS
Ramonville-Saint- Agne
France

Fabrice Papa
LEGOS
Toulouse, France

Marcello Passaro
Deutsches Geodätisches Forschungsinstitut der Technischen Universität München
München, Germany

Eva Peral
Jet Propulsion Laboratory
California Institute of Technology
Pasadena, CA

Bruno Picard
CLS
Ramonville-Saint-Agne, France

Sarah Purkey
Scripps Institution of Oceanography
University of California
San Diego La Jolla, CA

Richard D. Ray
NASA Goddard Space Flight Center
Greenbelt, MD

Elisabeth Rémy
Mercator Ocean
Toulouse, France

Frédérique Remy
LEGOS
Toulouse, France

Gilles Reverdin
Laboratoire LOCEAN
Paris, France

Jacques Richard
TAS
Cannes, France

Marie-Hélène Rio
CLS
Ramonville-Saint-Agne, France

Ernesto Rodriguez
Jet Propulsion Laboratory
California Institute of Technology
Pasadena, CA

Dean Roemmich
Scripps Institution of Oceanography
University of California San Diego
La Jolla, CA

Reiner Rummel
Technical University Munich
Munich, Germany

David T. Sandwell
Scripps Institution of Oceanography
University of California San Diego
La Jolla, CA

Andreas Schiller
CSIRO
Hobart, Australia

Uwe Send
Scripps Institution of Oceanography
University of California San Diego
La Jolla, CA

Hyodae Seo
WHOI
Woods Hole, MA

Detlef Stammer
CEN
University of Hamburg
Hamburg, Germany

Nathalie Steunou
CNES
Toulouse, France

P. Ted Strub
Oregon State University
Corvallis, OR

Pierre Thibaut
CLS
Ramonville-Saint-Agne, France

LuAnne Thompson
School of Oceanography
University of Washington
Seattle, WA

Ngan Tran
CLS
Ramonville-Saint-Agne
France

Guillaume Valladeau
CLS
Ramonville-Saint-Agne
France

Isabella Velicogna
University of California
Irvine, CA

Stefano Vignudelli
Consiglio Nazionale Delle Ricerche
Pisa, Italy

David Wiese
Jet Propulsion Laboratory
California Institute of Technology
Pasadena, CA

John Wilkin
Rutgers University
New Brunswick, NJ

Brent Williams
Jet Propulsion Laboratory
California Institute of Technology
Pasadena, CA

Joshua K. Willis
Jet Propulsion Laboratory
California Institute of Technology
Pasadena, CA

Philip Woodworth
National Oceanography Centre
Liverpool, UK

1 Satellite Radar Altimetry
Principle, Accuracy, and Precision

Philippe Escudier, Alexandre Couhert, Flavien Mercier, Alain Mallet, Pierre Thibaut, Ngan Tran, Laïba Amarouche, Bruno Picard, Loren Carrere, Gérald Dibarboure, Michaël Ablain, Jacques Richard, Nathalie Steunou, Pierre Dubois, Marie-Hélène Rio, and Joël Dorandeu

1.1 INTRODUCTION

Radar altimetry was, very early in the development of space technology, identified as a key technique to provide essential information on solid Earth and ocean dynamics (see Williamstown report, Kaula 1969). This results from the fact that several important geophysical phenomena impacting the sea surface topography can be monitored using this measurement (see Chapters 4 to 11):

- *Earth gravity*. The geoid (equipotential surface of the Earth gravity field) is the largest signal in amplitude of topography undulations with respect to an ellipsoid (hundreds of meters). It includes large-scale signals related to Earth interior heterogeneity and short-scale signals related to bathymetry.
- *Ocean dynamics*. The ocean is a turbulent fluid the dynamics of which include multiple time and space scales (see Figure 1.1). Altimetry provides integral information on an ocean's physical state (current speed, temperature, and salinity) from surface to the bottom that is key to monitoring these dynamics.

Moreover, space altimetry techniques have proved to be efficient for non-ocean surfaces in monitoring such features as rivers, lakes, ice, snow, and so on.

In this chapter, we provide:

- An overall description of the measurement principles (Section 1.1.2)
- A detailed description of the measurement built up (Sections 1.2–1.6)
- A historical perspective of satellite radar altimetry (Section 1.1.1)
- An overall view of the performance requirements (Section 1.1.3)
- A detailed description of error budgets and sampling performance is given in Sections 1.7 and 1.8.

1.1.1 Satellite Altimetry Measurement Principle

Satellite altimetry calculation results from the combination of two measurements. The first one is the estimation of the satellite altitude with respect to an Earth reference (H), while the second is the measurement of the distance between the satellite and the targeted surface (D). By subtracting this distance to the satellite altitude, one obtains the required elevation of the targeted surface with respect to the reference (Sea Surface Height (SSH)):

$$SSH = H - D \tag{1.1}$$

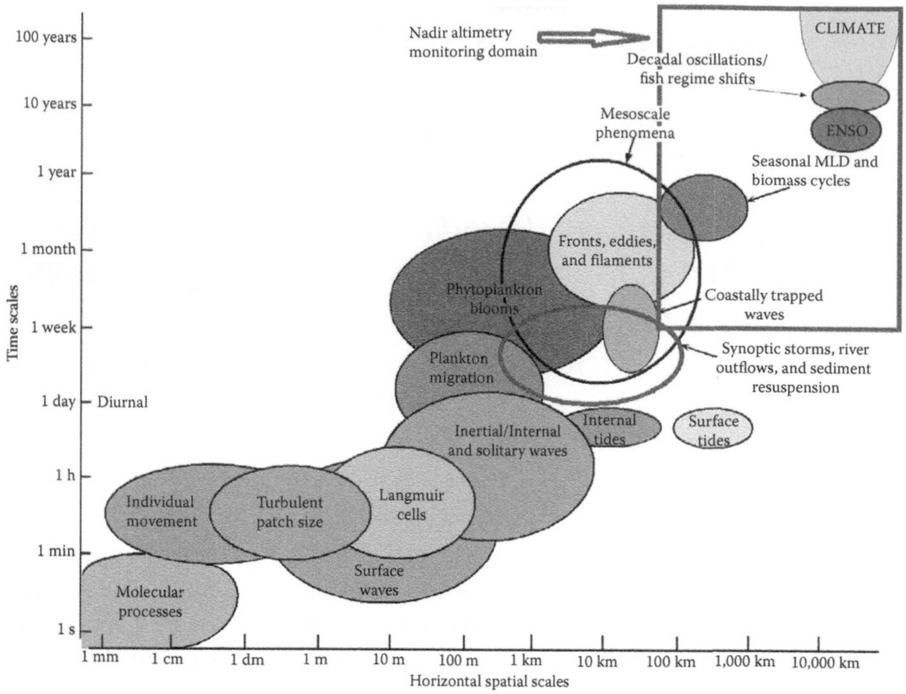

FIGURE 1.1 Typical spatial and timescales of ocean variability (Tommy 2003) and, superimposed in the rectangular shape located in the right upper corner, nadir altimetry monitoring domain.

Figure 1.2 shows that satellite radar altimetry is a composite measurement resulting from the combination of data provided by multiple sensors combined with modeling and external data.

The height measurement that is provided is an average of all elementary elevations over a zone the size and area of which depends upon the radar and antenna characteristics and the characteristics of the overflight surface (see Section 1.2). Over the ocean, the typical size of the zone encompassed in every radar measurement has an order of magnitude of several kilometers. It is larger when the significant wave heights (SWHs) are larger. Over land surfaces, the measured area is driven by the antenna aperture and the reflectiveness (backscatter coefficient) of the overflight zone. Water and ice reflectiveness is much larger than the reflectiveness of the surrounding soils; when lakes, rivers, ice, or snow are present, they drive the shape of the zone covered by the radar measurements.

When speaking about altitude, one needs to define the reference precisely. The primary reference, which is provided by the Precise Orbit Determination (POD) system, is an Earth ellipsoid, which is defined by its semi-major axis and eccentricity.

When this reference is used, the larger signal captured in the topography measurement is the geoid (i.e., the signature of the solid Earth signal). In theory, the dynamical parameter of interest to oceanographers would be a topography reference to the static geoid: This quantity is usually known as the absolute dynamic topography. However, despite major progress in its measurement (see Chapter 4, Chambers et al.), the geoid is not yet known at short scales with accuracy that is sufficient for such a direct use. That is why alternative references based on a mean topography of the ocean (also known as mean sea surface or MSS) computed over several years using historical satellite altimetry data are often used (see Section 1.6.1). The SSH is therefore often given with reference to the MSS and therefore expressed as a sea-level anomaly (SLA) or sea surface height anomaly (SSHA).

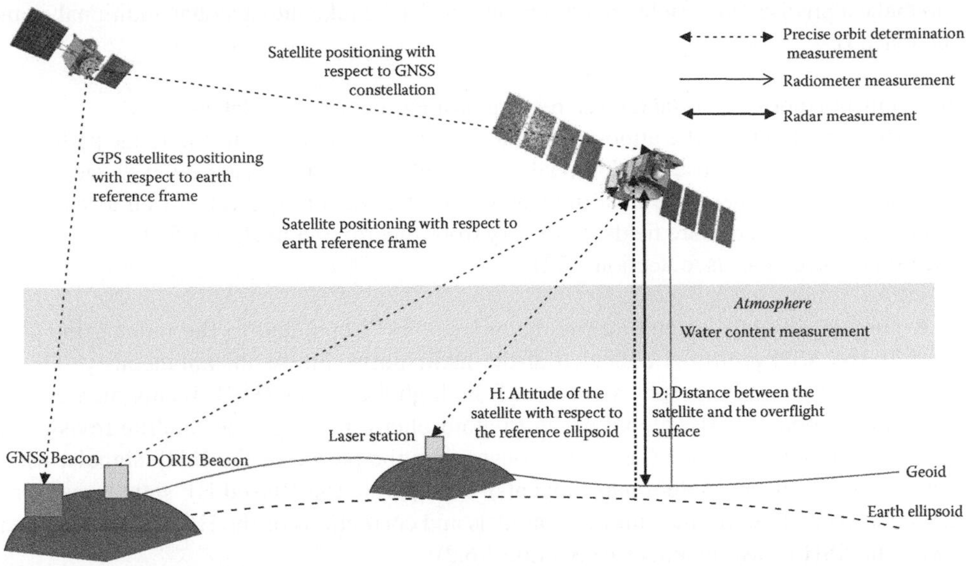

FIGURE 1.2 Satellite radar altimetry measurement principle: The main sensor used to compute the distance between the satellite and the targeted surface is a radar; however, to obtain the appropriate measurement accuracy, one needs a radiometer to measure the quantity of water that impacts the atmospheric propagation of the radar signal. To compute the altitude of the satellite with respect to an in situ network that constitutes the Earth reference, sensors on board the satellite are used in combination with modeling of satellite trajectory to perform the Precise Orbit Determination (POD) of the spacecraft (see Section 1.4).

The first element of the measurement equation (1.1) is the computation of the altitude. This is done through POD, which requires the combination of (see Section 1.4):

- An Earth reference system that will be the Earth base to compute the altitude
- A set of sensors on board the satellite to compute and validate the POD
- Modeling of the forces that act on the satellite to get an optimal estimate of the satellite trajectory

To compute the distance between the satellite and the overflight surface, the main instrument is the microwave radar, which emits an echo; this echo travels through the atmosphere and is reflected back to the radar by the overflight surface (see Section 1.2). Then the radar measures the time duration between emission and reception of the echo, which provides the distance. To make this computation, the intrinsic hardware resolution of the radar is limited, and one needs to make an analysis of this echo combining (see Section 1.3) the measured echo and a theoretical modeling of sea surface elevations over the overflight zone (also known as the Brown model for oceanic surfaces) to get an optimal estimate of the distance.

Over the ocean, this echo processing (also known as retracking) provides the following geophysical parameters:

- The distance between the satellite and the subsatellite point illuminated by the radar
- The SWH (or Hs)
- The backscatter coefficient (or sigma), which can be expressed as a sea surface wind speed modulus

Then, to make a precise topography measurement, one has to take into account additional elements (see Section 1.5):

- Instrumental corrections (antennas' relative positioning, internal delays, etc.).
- Propagation effects in the atmosphere that may introduce a bias in the radar path delay (PD)—this drives the use of additional equipment on board the satellite (e.g., radiometer for the wet troposphere and dual frequency radar for the ionosphere) as well as auxiliary data (e.g., surface pressure fields for the dry troposphere) (see Section 1.5.2).
- Sea state corrections (see Section 1.5.1).

Due to the velocity of the satellite (on the order of 6-7 km/s) and to the radar principle (see Section 1.2), the SSH profile is collected at the nadir point almost instantaneously. The SSH, therefore, also contains topography signatures from high-frequency (HF) signals such as tides or the barotropic response of the global ocean to atmospheric forcing. The satellite revisit time is on the order of 10 days or more (i.e., much longer than the period of these HF signals, which is therefore aliased in the altimeter measurements). To separate the aliased HF signatures from the ocean circulation and mesoscale, auxiliary models and corrections of the HF effects are provided along with the SSH measurements (see Section 1.6.2).

1.1.2 Satellite Radar Altimetry Historical Perspective

1.1.2.1 Satellite Altimetry Missions

1.1.2.1.1 Early Ages

Satellite altimetry potential value was identified and given a high priority in 1969 at the Williamstown Seminar, which gathered experts from solid Earth and ocean domains. Resulting from this analysis and the set of recommendations issued by this group (Kaula 1969), the 1970s and 1980s saw the development of increasingly accurate satellite altimeter systems.

Skylab (1973–1974) produced the first measurements of undulations in the marine geoid due to seafloor features. Then the Geodynamic Experimental Ocean Satellite (GEOS 3, 1975–1979) inherited radar technology from the Skylab and provided more accurate altimetric measurements (range measurement and satellite altitude determination using a laser retroreflector array on board the satellite). The coverage provided by this mission demonstrated the ability of such a system to measure the geoid and oceanographic parameters. In addition, the altimeter was shown to be capable of providing valid measurements over land and sea ice.

Seasat (1978) laid the foundations for a new generation of ocean satellites thanks to the improved quality of the data, which were widely and freely distributed to scientists throughout the world. Seasat included major innovation in its altimetric measurement design (Tapley et al. 1982):

- The radar altimeter was inherited from Skylab and GEOS 3 but included the first use of the full deramp technique for the radar altimeter (see Section 1.2), providing an improved resolution of the measurement. This became a "standard" for altimeters that followed.
- Radiometric measurements for water content correction in the radar measurement were available on-board (see Section 1.5).
- POD was performed using a specifically designed laser retroreflector array.

GEOSAT's (1985) primary mission was geodesy. The first part of the mission used a so-called geodetic orbit (very long repeat cycle), and data were classified by the U.S. Navy until 1995. During the second phase of the mission, the satellite was moved to a 17-day repeat orbit (Exact Repeat Mission, or ERM) similar to Seasat orbit and its data were distributed thanks to an agreement between the U.S. Navy and the National Atmospheric Oceanic and Atmospheric

Administration (NOAA) (Lillibridge et al. 2006). The GEOSAT radar altimeter was inherited from Seasat technology.

Seasat, followed by GEOSAT ERM, evidenced the ability to monitor ocean mesoscale using satellite altimetry, paving the way for "the maturity age" of altimetry.

1.1.2.1.2 Maturity Age

From these experiences, a new mission was designed and developed: TOPEX/Poseidon (T/P; 1992–2006). It was fully optimized to provide high-precision altimetry for mesoscale and large-scale monitoring (Fu and Lefebvre 1989). For that mission, two radar altimeter instruments were used: a dual-frequency (C and Ku bands) altimeter (ALT) and an experimental solid-state altimeter, (Poseidon-1). ALT was inherited from GEOSAT, and the addition of dual frequency measurement provided a correction of ionospheric propagation effects (see Section 1.5). Poseidon-1 shared the same antenna as ALT and was used in time-sharing to demonstrate the capabilities of the solid-state amplifier technology. A three-band microwave radiometer (TMR) was used to correct for wet tropospheric propagation effects (see Section 1.5). A specific effort was made to significantly improve POD techniques. This included the development of a new tracking system, Doppler Orbitography and Radiopositioning Integrated by Satellite (DORIS; with precise range rate measurements), an experimental global positioning system demonstration receiver (GPSDR), and the use of a laser retroreflector array (LRA). T/P used a specific non-sun-synchronous orbit, which was designed to allow good aliasing and restitution of tide signals and was higher than usual to minimize the impact of errors in Earth gravity and drag modeling on the POD accuracy.

This mission was concurrent with the European Remote Sensing (ERS-1; 1991–2000) satellite, which utilized a Ku-band radar altimeter (RA-1), a dual-frequency microwave water radiometer (MWR) for wet tropospheric correction (WTC), and precise range and range-rate equipment (PRARE), which failed shortly after launch, associated with a laser retroreflector array for POD (Francis et al. 1991). RA-1 also included a specific mode optimized for measurement over ice. ERS-1 flew on a sun-synchronous orbit that provided high latitude information and a sampling dedicated to mesoscale monitoring.

GEOSAT Follow On (GFO; 1998–2008) complemented these missions with an orbit inherited from GEOSAT, but GFO's payload was optimized for mesoscale monitoring (Barry 1993). GFO utilized a Ku-band radar altimeter combined with a dual-frequency radiometer, with both instruments sharing the same antenna. POD was performed using the laser retroreflector array combined with Doppler measurements due to a failure of the GPS receiver.

This set of combined missions provided the minimal time and space coverage of the ocean that could open a new era for oceanography: the sampling and accuracy of these altimeters allowed the development of operational oceanography as well as the beginning of a very long ocean climate record. Absolute accuracy was provided by a "reference" mission, T/P, while ERS-1, ERS-2, and GFO could provide adequate coverage thanks to intercalibration among these missions (Koblinsky et al. 1992).

T/P was then followed by the Jason series (Jason-1, 2001–2013; Jason-2, 2008– ; and Jason-3, 2016–), which maintained and even improved T/P accuracy while being optimized for an operational series. Jason-1 (Ménard et al. 2000) utilized Poseidon-2; a dual-frequency (Ku and C) radar altimeter derived from the Poseidon-1 altimeter of T/P. WTC was performed using a three-frequency radiometer (the Jason-1 Microwave Radiometer, or JMR) inherited from the TMR. For POD, DORIS was complemented by an experimental GPS receiver (Turbo Rogue Space Receiver (TRSR)-2) and a laser retroreflector array. The technology used for the radar made it possible to reduce the satellite mass by a factor of five with respect to the T/P spacecraft. The payload of Jason-2 and Jason-3 was derived from Jason-1. The radar altimeters (Poseidon-3 and Poseidon-3B) benefit from the addition of an open-loop mode, which uses high-precision DORIS immediate onboard orbit determination (DIODE) to perform improved tracking of non-ocean surfaces. WTC is provided through an enhanced version of JMR (an advanced microwave radiometer, or AMR). POD is computed using DORIS, GPS (TRSR-2), and laser retroreflector array.

In parallel, ERS-1 was followed by ERS-2 and then Envisat (2000–2012) (Benveniste et al. 2001). Envisat's radar altimeter (RA-2) was inherited from RA-1 with the addition of a second frequency measurement (Ku and S) for ionospheric correction. The dual-frequency radiometer (MWR) was inherited from ERS-1, and the satellite's POD was based on DORIS combined with laser retroreflectors.

HY-2 (2011–), cf. section 1.1.2.2, utilizes a dual-frequency altimeter, a five-frequency radiometer, and a DORIS receiver for POD.

1.1.2.1.3 New Technologies

CryoSat-2 (2010–) is an innovative radar altimetry mission, fully optimized for ice measurements. It utilizes a new radar altimeter concept (the Synthetic Aperture Radar Altimeter, or SRAL) designed for polar sea ice thickness measurement (see Chapter 15, Shepherd et al.), which is inherited from Poseidon-class altimeters. The SRAL includes a delay-Doppler mode (a Synthetic Aperture Radar, or SAR mode) (see Raney, 1998) as well as SAR-Interferometric modes so that the cross-track slopes complement the nadir topography. CryoSat-2 utilizes a DORIS receiver and a laser retroreflector array for POD. To provide coverage of the polar ice zone, the satellite is flying on a non-synchronous 92-degree inclination orbit. The repetition cycle is long (369 days) to provide higher space sampling of ice.

The new altimetry features of SRAL also proved to be useful for ocean monitoring and in particular for coastal applications. CryoSat-2 did not have any dual frequency on the radar altimeter nor did it have a radiometer; neither sensors were needed for its primary cryosphere objectives. For oceanography, this design was more limiting, and the effect was mitigated using a GPS-based model ionospheric correction, and a model-based WTC, as well as a cross-calibration with the Jason series. CryoSat-2 replaces CryoSat, which was lost after a launch failure in 2005.

From a different perspective, SARAL, which stands for "Satellite with Argos and AltiKa," (2013–) is another innovative radar altimetry mission that utilizes a new Ka-band altimeter concept (AltiKa). AltiKa benefits from the use of a higher frequency and wider bandwidth. It yields more precise measurements thanks to a better signal-to-noise ratio and the reduced beam width of the Ka-band antenna (Verron et al. 2010). The mission includes a dual-frequency radiometer sharing the same antenna as the radar for WTC and a DORIS system combined with a laser retroreflector array for POD. The use of the Ka band reduces the magnitude of ionospheric PDs on the topographic measurement so that dual frequency radar measurement is no longer necessary. SARAL used the ERS/Envisat orbit during the first 3 years of the mission to provide continuity of the measurements on this type of orbit while waiting for the launch of Sentinel-3. It is now flying on a drifting orbit providing geodesy measurements (high cross-track resolution) in addition to its other mission objectives (ocean, ice, and hydrology).

In the future, Surface Water and Ocean Topography (SWOT) will bring major innovation to the altimetric measurement history by demonstrating the potential of wide swath Ka-band altimeter radar interferometry. This new technology will provide high resolution and wide swath information as well as an unprecedented precision and resolution (see, e.g., Chapter 2, Rodriguez et al.).

1.1.2.1.4 Operational Missions

The experience gained from the combination of complementing missions provided the basis for the Ocean Surface Topography Virtual Constellation (OST-VC) concept (Escudier and Fellous 2008) formalized by the Committee on Earth Observing Satellites (CEOS). This committee ensures international coordination of civil space-based Earth observation programs. The OST measurement was the first parameter benefiting from this virtual constellation concept, which derives from the intrinsic space and time resolution of a single mission using nadir altimetry (see, e.g., Section 1.7). The higher space/time sampling provided by the constellation was leveraged by various systems and products, and notably the AVISO/DUACS multi-altimeter maps that have been providing a systematic and near-real-time monitoring of the ocean mesoscale over the last two decades (Pujol et al. 2016).

Sentinel-3 provides a long-term commitment for altimetry measurement in the context of the Copernicus program (Donlon 2011). The Sentinel-3 payload includes dual frequency (Ku and C) radar (SRAL) inherited from the Poseidon-3 altimeter, with the addition of a delay-Doppler capability inherited from CryoSat (sometimes called synthetic aperture radar mode or SARM), which provides high along-track resolution and reduced noise level. The WTC is computed from a dual-frequency radiometer (MWR) and POD from the combination of a Global Navigation Satellite System (GNSS) receiver, DORIS, and a retroreflector. The first spacecraft, Sentinel-3A, was launched in 2016 and will be followed in the short term by Sentinel-3B.

In the future, Sentinel-6 (also known as Jason-CS) will ensure continuation of Jason-3. Table 1.1 provides synthetic past and current altimetry mission characteristics.

1.1.2.2 Geographical Perspective and International Cooperation

The United States played a key role in the design of early altimetry techniques. Since then, it has stayed a major player in altimetry virtual constellation implementation.

On the European side, early involvement came from the European Space Agency (ESA), with the ERS missions, and from France (the T/P and Jason series), in cooperation with NASA.

Operational agencies from Europe and the United States (Eumetsat and NOAA) joined the Jason program later on (Jason-2 and Jason-3). Presently, European involvement strengthens in the framework of the Copernicus global Earth monitoring program.

In parallel, it is important to note other initiatives that have marked altimetry technique history:

- GEOSAT and later on GEOSAT Follow On were developed by the U.S. Navy.
- In Russia, the GEOIK program (1985–1995) was mainly dedicated to geodesy.
- HY-2A (2011–) is part of the *HaiYang* ("ocean," in Chinese) program, dedicated to altimetry (Lin et al. 2015).
- SARAL is due to cooperation between France and India that includes the AltiKa altimetry payload (discussed earlier this chapter).

An important factor of success was the fact that these programs were designed and developed in the framework of important international oceanography programs. One can note in particular:

- The World Ocean Circulation Experiment (WOCE) was a part of the World Climate Research Program (WCRP) that used resources from nearly 30 countries to make unprecedented in situ and satellite observations of the global ocean between 1990 and 1998 and to observe poorly understood but important physical processes (Bretherton and Woods 1986).
- Later on, the Global Ocean Data Assimilation Experiment (GODAE), proposed by Smith and Lefebvre in 1997 to support the development of national ocean prediction systems, was conceived as a 10-year demonstration of both the feasibility and the utility of high-resolution global-scale ocean predictions and led by an International GODAE Science Team (IGST) incorporating the key players in the teams developing the ocean prediction systems at the national level. GODAE gave a major boost to the establishment and improvement of operational ocean prediction systems in a number of countries and developed substantial capabilities for the robust, real-time collection and processing of measurements and the generation and dissemination of analyses and forecasts. It demonstrated that forecasting of open ocean mesoscale phenomena is feasible in many regions, and GODAE products showed real benefit for a number of applications (Smith and Lefebvre 1997).

TABLE 1.1
Past and Current Altimetry Mission Characteristics

Mission	Altitude/ Repetitivity	Inclination	Payload	Agencies in Charge	Launch/End of Mission	Primary Mission Objective
Skylab	440 km	50°	Ku band radar alt	NASA	1973/1974	Geodesy
Geos-3	830 km	114.98°	• Ku band radar • Laser retroreflector, Doppler system, Satellite to satellite tracking,	NASA	1975/1979	Geodesy
Seasat	800 km/17 days	108°	• Ku band radar, • Multichannel microwave radiometer • Laser retroreflector	NASA	1978/1978	Ocean
Geosat	Geodetic mission: 757 km Exact repeat mission: 800 km/17 days	108°	• Ku band radar (Seasat Heritage)	US Navy	1985/1990	Geodesy, Ocean
ERS 1	785 km/35 days, 3 days (ice), 168 days (geodetic)	Sun synchronous	• Ku band radar (includes an ice mode) • Dual freq. radiometer • Laser retroreflector, PRARE	ESA	1991/2000	Ocean, Ice, Geodesy
TOPEX/ Poseidon			• Ku and C band radar • 3 freq. radiometer • DORIS, laser retroreflector, GPS	NASA/CNES	1992/2006	Ocean
ERS 2	785 km/35 days (as ERS-1)	Sun synchronous	• Similar to ERS-1	ESA	1995/2011	Ocean, Ice
GFO	800 km/17 days, Same as Geosat	Same as Geosat	• Ku band radar • Bi frequency radiometer • Laser retroreflector array, GPS	US Navy	1998/2008	Ocean

(Continued)

TABLE 1.1 ***(Continued)***
Past and Current Altimetry Mission Characteristics

Mission	Altitude/ Repetitivity	Inclination	Payload	Agencies in Charge	Launch/End of Mission	Primary Mission Objective
Jason 1	Same as TOPEX/ Poseidon	Same as TOPEX/ Poseidon	• Ku & C band radar • 3 freq. radiometer • DORIS, laser retroreflector, GPS	NASA/CNES	2001/2013	Ocean
ENVISAT	782 km/35 days 799 km/30 days (end of life)	Sun synchronous	• ERS-1 heritage, Ku and S band radar alt • 2 freq. radiometer • DORIS, laser retroreflector	ESA	2002/2012	Ocean
Jason-2	Same as Jason-1	Same as Jason-1	Similar to Jason-1	NASA/CNES/ NOAA/ EUMETSAT	2008/-	Ocean
Cryosat-2	717 km/369 days	92°	• Ku nadir looking radar, SAR and interferometric SAR mode • - DORIS	ESA	2010/-	Ice
HY 2	971 km/14 days, first phase, 973 km/168 days geodetic, second phase	Sun synchronous	• Ku & C band radar alt • 5 freq radiometer • DORIS, laser retroreflector	CNSA/CNES	2011/-	Ocean, Geodesy
SARAL/ Altika	Same as ENVISAT	Sun synchronous	• Ka Band radar alt • Dual freq. radiometer • Doris, laser retroreflector	CNES/ISRO	2013/-	Ocean, Hydrology, Ice
Jason-3	Same as Jason-1	Same as Jason-1	• Similar to Jason-1	NASA/CNES/ NOAA/ EUMETSAT	2016/-	Ocean
Sentinel 3	815 km/27 days	Sun synchronous	• Ku and C band radar alt including an along track SAR mode • Dual freq. radiometer • GNSS, DORIS laser retroreflector	EC/ESA	2016/-	Ocean

1.1.2.3 Altimetry Products: History of Continuous Progress

The history of altimetry is not limited to spacecraft history (Figure 1.3). The products delivered to a wide range of users are derived from the sensor measurements, but they include major contributions from auxiliary information and modeling (see, e.g., Section 1.3, 1.4, 1.5, 1.6).

Among the major milestones in a multidecadal story of progress, one can note in particular:

- The story of POD progress (see, e.g., Section 1.4). Thanks to better auxiliary information (Earth gravity models in particular, see Chapter 4, Chambers et al.), progress in sensor performance, and progress in processing techniques, the accuracy of POD has been increased from metric-level accuracy in the early years to decimetric accuracy at the beginning of the T/P mission and centimetric accuracy in the present day (see Figure 1.4), This represents major progress; with a low level of POD accuracy, altimeter applications would be limited to local geoid and large mesoscale feature measurement, whereas a high level of POD accuracy is necessary to monitor large-scale and climate signals.
- Altimeter echo processing evolution (see, e.g., Section 1.3)—thanks to better modeling of the echo interaction with sea surface, better characterization of radar instrument and new geophysical parameters estimation algorithms, this progress has allowed a decrease in measured noise and biases.
- Specific analyses of the altimetry data set itself have allowed the determination of more and more precise modeling of EM bias, tides, and reference surfaces (see, e.g., Section 1.5), which in return has provided a mitigation of geophysical errors and more precise altimetry products.

The history of altimetry can be summarized as a continuous effort to improve measurement performance and to quantify and monitor this performance in order to meet new science and operational objectives at every step of this evolutive process. This is carried out through close interactions between engineers involved in space component design and ground processing and users. To that extent, the Ocean Surface Topography Science Team (OSTST) has been a key forum and exchange body between research and engineering (http://www.aviso.altimetry.fr/en/user-corner/science-teams/ostst-swt-science-team.html). OSTST is comprised of a worldwide team of investigators selected through International Announcements of Opportunity released every 4 years. Interaction between this group and space agency project teams has provided the key input for:

- Mission design:
  - Orbit selection at individual mission level or at virtual constellation level
  - Instrument design and improvements
  - Processing algorithms and auxiliary data and models
  - Product definition and adequation to new or evolving usages

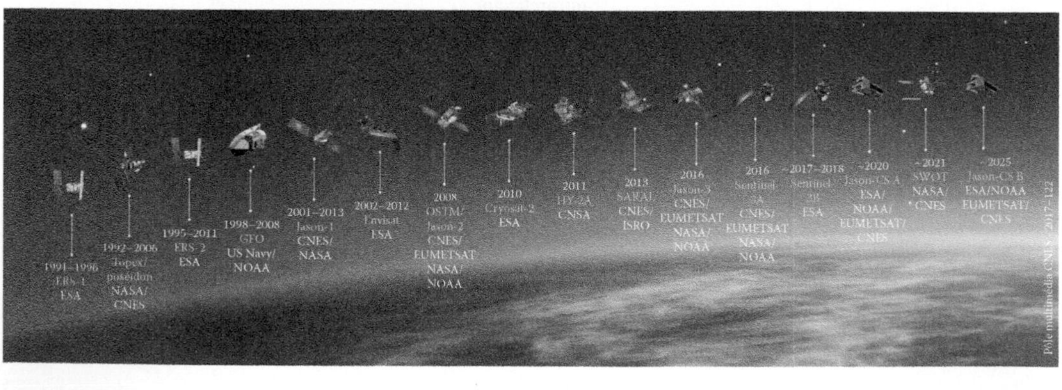

FIGURE 1.3 Radar altimetry missions' perspective.

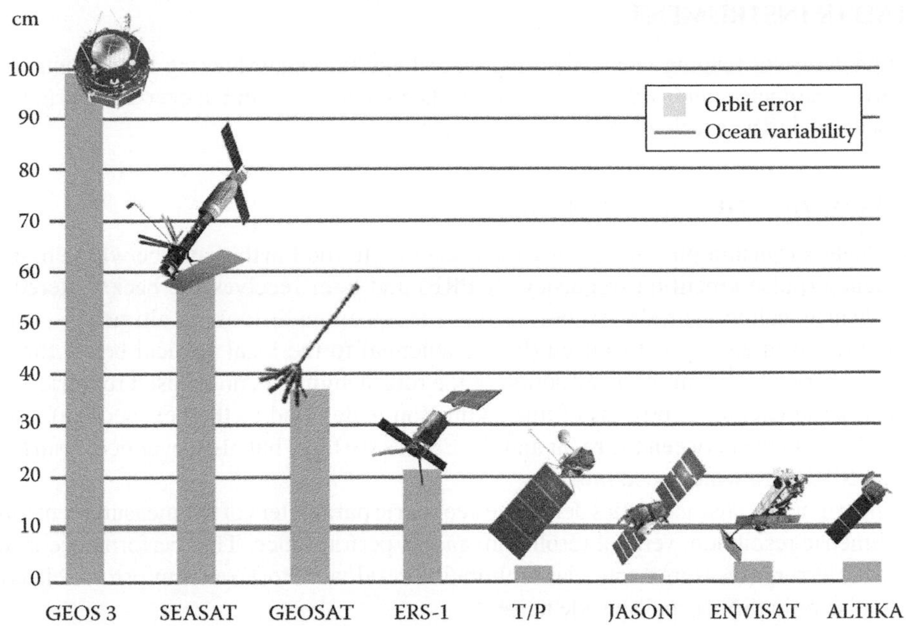

FIGURE 1.4 Evolution of POD accuracy from GEOS 3 to AltiKa.

- Calibration and validation
- Error budget consolidation
- Feedback and requests from new altimetry applications and research

Furthermore, the most demanding scientific applications carried out by those investigators have been the driver for this progress; as emblematic examples one can note:

- Mean sea level (MSL) monitoring as the driver for absolute accuracy and long-term stability
- Mesoscale monitoring as a driver for OST-VC design

1.1.3 Altimetry System Requirements

The ocean signals that are measured with altimetry (see Chapters 5 to 13) range from mesoscale activity to large-scale signals such as the MSL rise associated with climate change. A set of requirements shall be put on the altimetry system definition to ensure that the system is able to monitor most of these signals (see Section 1.8). Furthermore, beyond the error budget requirement for each elementary measurement, an important property of the altimetric system is that the requirements ensure its ability to adequately sample the signals of interest in time and space.

This point is critical because the very limited field of view of the radar altimeter provides only a one-dimensional subsatellite profile as opposed to imagery sensors (e.g., sea surface temperature [SST] or ocean color sensors can have hundreds of kilometers of fields of view). To that extent, along-track resolution is given by the spacecraft motion. For across-track information, one has to rely on the revisit of the satellite over several days. Ocean dynamics' signals have different time and space variation characteristics (see Figure 1.4). Then, to analyze the altimetry monitoring performance of a given signal, careful examination of the time and space sampling of the altimetry virtual constellation has to be performed. This is presented in Section 1.7.

1.2 RADAR INSTRUMENT

This section describes ranging radars, focusing on real-aperture nadir-looking instruments. The synthetic aperture altimeters and wide-swath instruments are addressed in Chapter 2 (Rodriguez et al.), which is dedicated to new technologies and applications.

1.2.1 Radar Altimeter Instrument Principles

Radar altimeters transmit pulsed signals from a satellite to the Earth's surface with a high repetition frequency (pulse repetition frequency, or PRF) and later receives the backscattered signals from the surface, called the radar echoes. Here we focus on nadir-looking altimeters—where the transmitted/received energy is focalized (by the antenna) to the local vertical below the satellite (nadir). The shape of the amplitude evolution of the return signal (echo) is used to retrieve surface geophysical parameters (the process of this estimation is detailed in the next section), especially the range (the distance between the radar and the Earth's surface) but also, over ocean surfaces, the subsatellite SWH and wind speed modulus.

These instruments' characteristics define the geometric parameters of the measurement (observed area, planimetric resolution, vertical resolution) and its performance. This performance is affected by systematic errors (measurements bias, calibration residual error), random errors related to the signal-to-noise ratio (SNR), and speckle noise.

1.2.2 Observation Geometry

Figure 1.5 illustrates the observation geometry over the ocean of a nadir altimeter on board a satellite. The antenna acts as a spatial filter that focalizes the energy (both transmitted and received) toward the nadir direction, which is defined as the local vertical from the instrument on the satellite

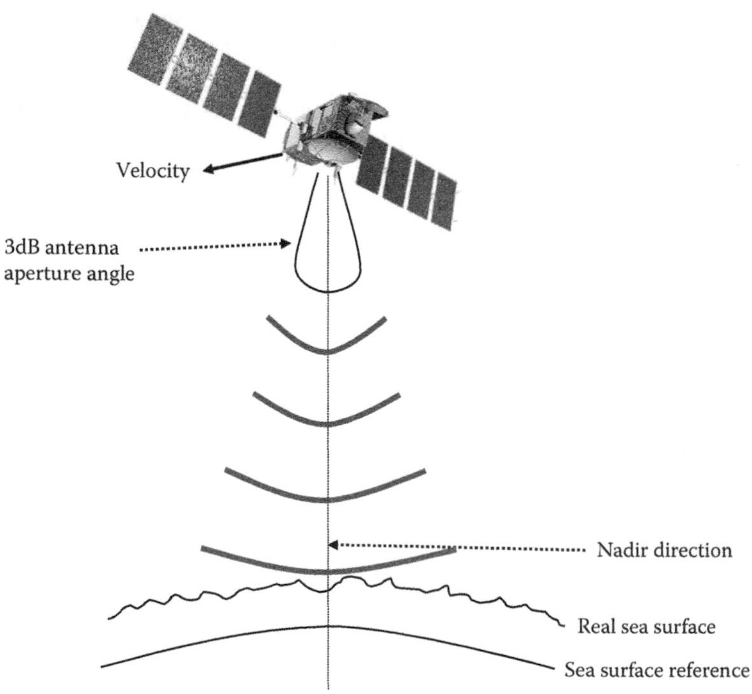

FIGURE 1.5 Observation geometry of an altimeter over the ocean.

to the Earth. This direction is approximately orthogonal to the velocity vector of the satellite (which is supposed to have a circular orbit) that references the azimuth direction.

Each individual height measurement is associated with a given surface over the observed area. It is conventionally defined as the projected 3-dB aperture angle of the antenna beam (the incidence angle where the antenna gain defers 3 dB from the maximum gain associated with the nadir direction).

1.2.3 Radar Operation

Radars emit short pulses ("transmit events," duration: τ) at a PRF rate. The PRF is chosen so that the return signals ("receive events," backscattered from the surface) fall down between two consecutive transmit events spaced out by the pulse repetition interval (PRI). For satellite altimetry operation, each received signal corresponds to a transmit event that occurred several PRIs earlier. This exact number of PRI is called the ambiguity rank of the radar (Figure 1.6).

1.2.4 Transmitted Waveform

The range radar principle is basically described by the following equation linking the distance (d) between the radar and the surface, the speed of light (c_0), and the time (Δt taken by the signal to travel back and forth).

$$d = \frac{c_0 . \Delta t}{2} \tag{1.2}$$

With a constant wave (CW = sinusoidal) modulated signal, the time resolution is equal to half of the pulse duration. The associated distance estimation uncertainty can then be expressed as follows:

$$\Delta d = \pm \frac{c_0 . \tau}{2} \tag{1.3}$$

To correctly design radar, a very short pulse duration is needed to achieve the required resolution (a range resolution of 50 cm corresponds to a pulse duration shorter than 2 ns). However,

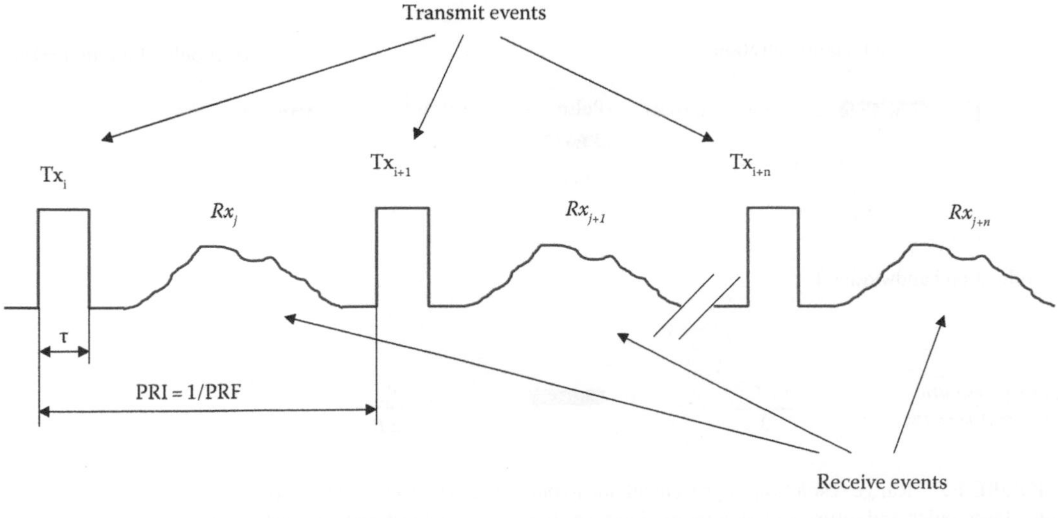

FIGURE 1.6 Radar principle chronogram.

in order to meet the link budget requirement, sufficient energy has to be transmitted, and the pulse duration has to be long enough. These two considerations are usually conflicting (called the range radar dilemma), which disqualifies CW signals for such types of operations. So, frequency modulated constant wave (FMCW) signals must be used in order to drastically improve both the vertical (range) resolution and the SNR. The principle is to spread the energy during the transmission on a pretty large pulse and to compress the return signal on a very short pulse that corresponds to the required time resolution. This pulse compression technique (that uses matched filtering) is well described in the literature (e.g., Ulaby and Long 2014). We simply recall hereafter its improvement in terms of time (or range, or vertical) resolution and SNR (Figure 1.7).

The transmit signal frequency is linearly modulated during the pulse duration. The associated resolution of such signals is inversely proportional to the extent of the frequency bandwidth (*B*) that has been emitted.

$$\text{Range resolution (after pulse compression)} = \frac{c_0}{2.B}$$

This offers two major advantages: First, the time resolution is no longer linked to the pulse duration, and the resolution is improved by enlarging the spectral content (a larger bandwidth, *B*, means a better range resolution) of the transmitted signal. The possibility of enlarging the pulse is also useful for managing the energy to be transmitted considering technological constraints, such as the maximum peak power of the high power amplifier. One of the main limits to enlarging the transmitted pulse is the chronogram constraint addressed earlier. This improvement is beneficial in terms of energy due to the SNR also being drastically improved because the useful signal is compressed by the matched filter whereas the noise keeps spreading (Figure 1.8).

The range compression gain characterizes the improvement offered by the pulse compression. It is defined as the product of the pulse duration and the frequency bandwidth, or *Range compression gain* = $B.\tau$. For typical satellite altimetry operation points, the range compression gain is around several tens of thousands.

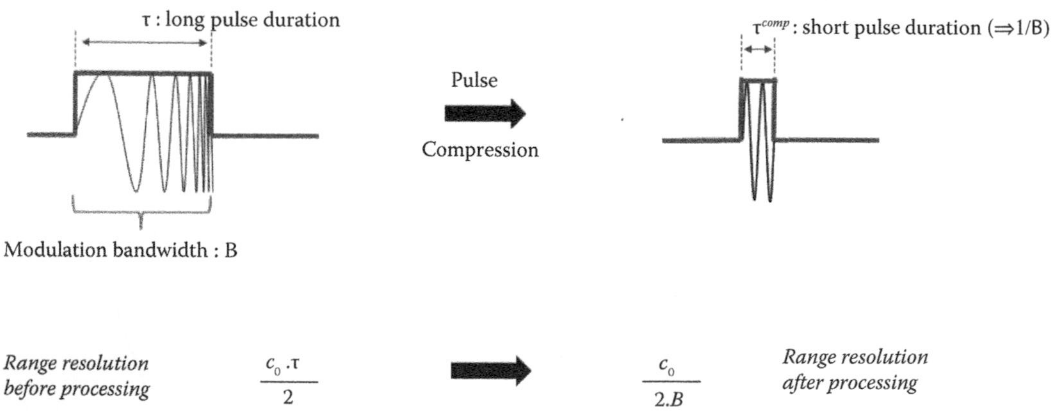

FIGURE 1.7 Range resolution improvement due to pulse compression—energy spread during the transmission on a large pulse and return signal compression on a short pulse that corresponds to the required time resolution.

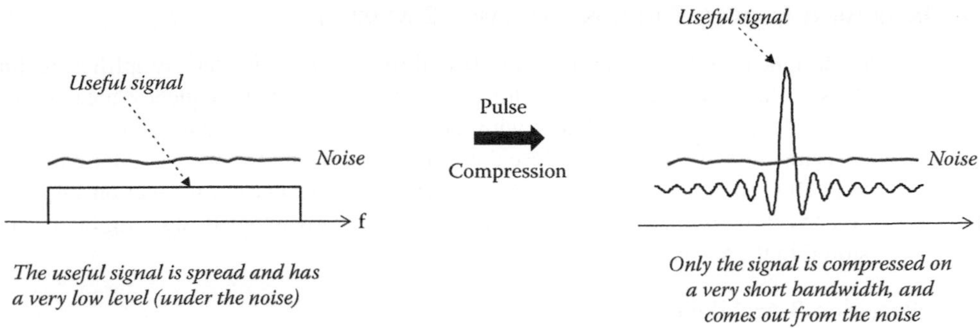

FIGURE 1.8 SNR improvement due to pulse compression.

1.2.5 Instrument Architecture

The symbolic architecture of an altimeter instrument is presented as follows. It typically consists of three main subsystems (Figure 1.9):

- The digital processing unit (DPU, the digital part of the instrument)
- The radio frequency unit (RFU, the analog part of the instrument)
- The antenna used to transmit and receive

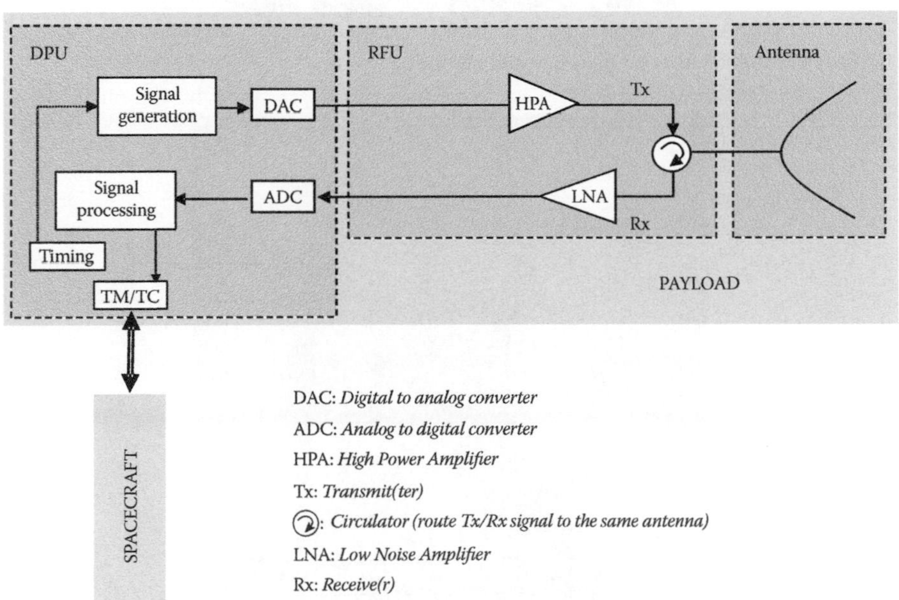

FIGURE 1.9 Altimeter symbolic architecture: First, the radar signal is digitally generated by the DPU (or by the processing and control unit, PCU); then it is converted to an analog signal through a digital to analog converter. This signal is also frequency converted to reach the radar microwave frequency before being amplified by the HPA. This transmit signal is emitted through the antenna, and the return signal is caught by the antenna. The LNA amplifies this very low signal (affected by the high transmission losses). It is then digitalized and processing is applied (the processing scheme is shared on board and on the ground). The HPA, LNA, and antenna are the three main technological contributors to the SNR of the instrument.

1.2.6 Instrument Example: Poseidon-3 of Jason-2 Mission

Poseidon-1 was designed to be compact radar, benefiting from solid-state amplifier technology. Poseidon-2 used the same basic design with the addition of a dual frequency measurement. Poseidon-3 inherited designs from Poseidon-2, with the addition of a new mode to provide optimal tracking of non-ocean surfaces. With that, Poseidon-3 receives (in real time) the satellite altitude computed by the DIODE navigation software in the DORIS instrument (see, e.g., Section 1.4) so that it can track the return echo coming from a desired target altitude in open loop using a digital elevation model (DEM) stored in the altimeter.

1.2.6.1 Poseidon-3 Architecture

Compared to the theoretical architecture presented earlier, the Poseidon-3 instrument (Figure 1.10) on board Jason-2 (Figure 1.11) includes specific characteristics:

- This instrument operates at two carrier frequencies, the C band and the Ku band, in order to provide ionospheric propagation correction capacity.

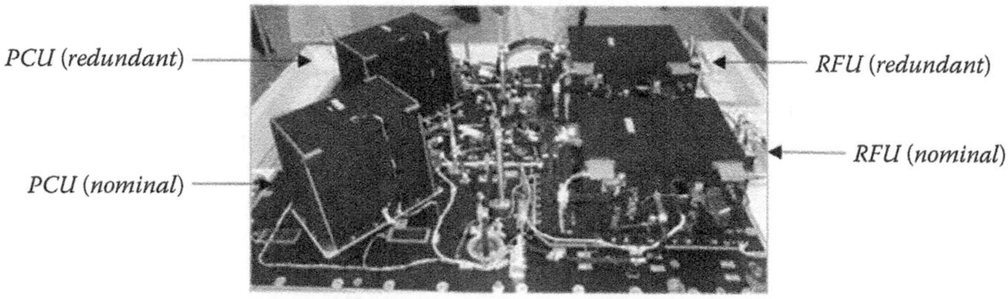

FIGURE 1.10 Poseidon-3 altimeter (nominal and redundant paths).

FIGURE 1.11 Jason-2 satellite.

- This instrument has frequency conversions realized with mixers and local oscillators.
- The altimeter uses an analog implementation of the compression technique (called the Deramp technique). At the estimated reception time of the return signal, a replica of the transmitted signal is generated and mixed with the received signal. The mixing product is a very low-frequency (LF) signal, the frequency of which is directly proportional to the time difference between the replica and this received signal. This equivalence among frequency/time/distance enables a very precise estimate of the range without having to digitalize all the emitted bandwidth. The scheme is completed with a Fourier transform to reconstruct the echo shape with optimum noise filtering.

1.2.6.2 Poseidon-3 Main Characteristics

The Poseidon-3 (Figure 1.11) main characteristics are given in Table 1.2. They are distinguished by the two carrier frequencies that are used (the Ku band, which is the main bandwidth, and the C band, which is used to correct for ionospheric delay).

- A 320-MHz bandwidth is used, after range compression, and then the range resolution is equal to 47 cm.
- Time slots are shared between the Ku band and the C band, with priority given to the Ku band.
- The overall consumption (70 W) can be considered as small.

1.2.7 Key Instrument Performance

Echoes are affected by thermal and speckle noise. The thermal noise is determined by the SNR equation

$$SNR = \frac{P_{rx}}{N} = K \cdot \frac{P_{tx} \cdot \lambda^2 \cdot G^2 \cdot \sigma}{R^4 \cdot k \cdot T \cdot B \cdot Nf} \tag{1.4}$$

where
P_{rx} = received power
N = noise level at the receiver
K = constant
P_{tx} = transmitted power

TABLE 1.2
Poseidon-3 Ku Band and C Band Radar Characteristics

	Poseidon 3	
	Ku-band	**C-band**
Central frequency	13.575 GHz	5.3 GHz
Chirp bandwidth	320 MHz	320 MHz or 100 MHz
Pulse length	105.6 μs	105.6 μs
Pulse repetition frequency	6Ku/1C: 1765 Hz 4Ku/1C: 1648 Hz	6Ku/1C: 294 Hz 4Ku/1C: 411 Hz
Antenna gain	42.2 dB	32.3 dB
Transmitted power	7 W	20 W
Number of accumulated pulses (in 50 ms packets)	6Ku/1C: 90 4Ku/1C: 84	6Ku/1C: 15 4Ku/1C: 21
Number of range samples	128	128
Mass	2 × 20 kg + 7 kg (antenna) + 4 kg (harness) = 51 kg	
Consumption	70 W	
Lifetime	5 years with continuous operation (data availability > 0.95)	

λ = wavelength (the inverse of the carrier frequency)
G = antenna gain
σ = radar cross section
R = range, the distance between the radar and the surface
K = Boltzmann constant
T = temperature of the receiver
B = bandwidth (reception bandwidth that determines the noise level)
N_f = noise figure

Besides altitude and the wavelength used for the radar, the driving instrument parameters are the transmitted power, the antenna gain, and the noise figure of the receiver (ratio of the output noise power to the portion thereof that is attributable to thermal noise).

The randomly distributed phases of the scatterers at the sea surface generate a modulation of the return power (sample by sample) called the speckle noise. This speckle noise is a multiplicative noise, and it is a major contributor to the overall random error budget. This is due to the coherent nature of the radar illumination. A large number of individual scatterers contribute (coherent addition) to the signal of a resolution cell. Then it is necessary to average (accumulate) several (typically, a hundred) individual echoes in order to decrease the speckle noise.

Systematic errors can be compensated in principle if they are known and characterized. A calibration process is activated on board on a regular basis to remove those systematic errors as often as possible. Residual systematic errors come from calibration errors, which are from the drift between two calibrations and a non-calibrated path. In addition to the internal instrument calibration, inter-calibration among missions and with other independent measurement systems is systematically used for each new satellite.

It is worth emphasizing that besides the tens of centimeters of resolution at the instrument level, the radar has to ensure long-term stability (better than a millimeter for the Poseidon's reference altimeters on Jason's mission). This long-term stability drives the full architecture of the instrument (the time reference given by the ultra-stable oscillator).

1.2.8 Echo Formation

While using the pulse compression technique, the vertical resolution of altimetry radars is typically on the order of tens of centimeters. So, to determine the distance between the radar and the surface at a centimeter level, additional processing techniques are used (described in the following). They are based on the use of the complete echo with an *a priori* knowledge of the waveform shape. Figure 1.12 illustrates geometrically how the waveform of the echoes is obtained from the shape of the received power amplitude with respect to time.

As long as the transmitted signal does not intercept the surface, the received signal (formally received later due to the propagation time) is almost zero (only thermal noise is present at the altimeter input). Due to the spherical shape of the wave front, the incidence pulse starts to intercept the surface at the middle of the observed area (the shortest distance between the radar and the surface). The received signal amplitude increases to a maximum. Then it tends to decrease accordingly to the shape of the antenna gain.

1.3 ECHO CHARACTERIZATION AND PROCESSING

The overall performance of an altimeter relies on the characteristics of the on-board sensor as described in Section 1.2 and on processing algorithms (covered in this section).

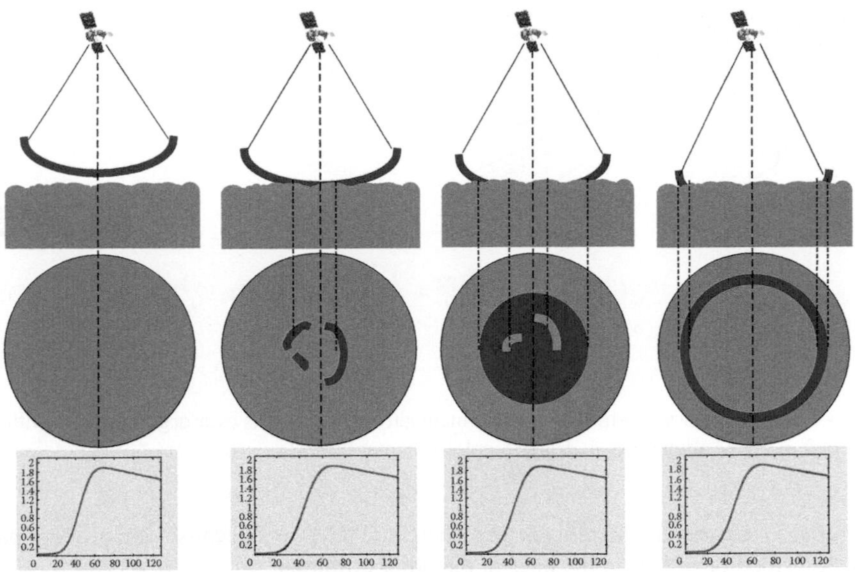

FIGURE 1.12 Altimeter echoes: This figure depicts the time evolution of radar altimeter signal reflection on the sea surface. The upper panel provides a side view, while the middle panel provides a top view of the phenomenon. The lower panel provides the amplitude of the backscattered received signal as a function of time (referred as waveform).The red part of the waveform curve corresponds to the time period of echo reflection depicted in the upper panels. (From http://www.altimetry.info.)

The list of parameters that have a direct impact on performance can be grouped into four main categories:

- Mission-related parameters: Height of the orbit, pointing of the antenna, and so on
- Parameters related to the on-board instrumental parameters: Altimeter mode (low resolution mode [LRM], delay-Doppler, interferometer), frequency (*f*), bandwidth, pulse length (*t*), number of integrated pulses (*N*), PRF, tracker mode ("closed loop" like a median tracker, sliding gate tracker, and early detectable point or "open loop" like a DIODE/DEM tracker), antenna pattern (σ_0), SNR, quantization of the waveform power, quantization of the telemetry, instrument stability with respect to variation of temperature along the orbit and ageing, and so on
- Parameters depending on the geophysical conditions of observation: Significant wave height, sea state (sea state bias [SSB], sigma naught bloom events, rain cells, etc.)
- Parameters linked to the ground processing: Focusing/stacking method for delay-Doppler mode, choice of the retracker (physical, empirical), width of the analysis windows (along track and across track), and—of course—all corrections, including instrumental and geophysical corrections such as tide, ionosphere, troposphere, Dynamic Atmospheric Correction (DAC), and SSB corrections (see Sections 1.5 and 1.6).

All these parameters influence the final performance of a particular altimeter mission at a large variety of spatiotemporal scales. The parameter selection is thus crucial—at the industrial level when building the altimeter, at the mission level when choosing the orbit characteristics, and at the ground level when designing ground processing.

The theoretical waveform shapes over the ocean are plotted in Figure 1.13 for the main LRM altimeters (as described in Section 1.2) and for a delay/Doppler altimeter (as described in Chapter 2,

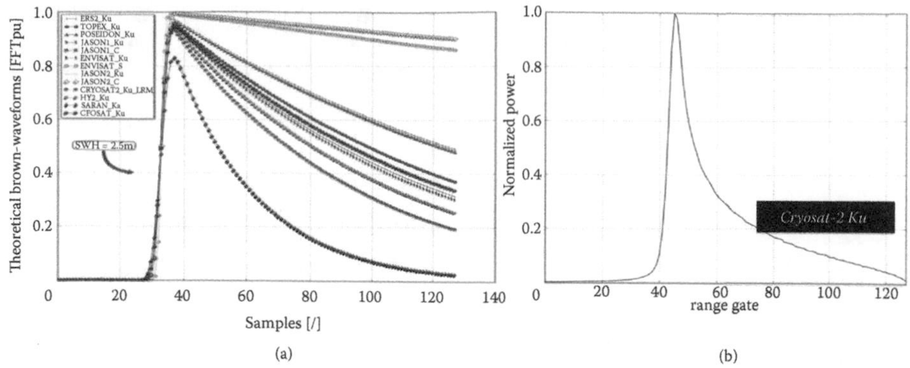

FIGURE 1.13 Theoretical waveform shapes of the main altimeter missions over ocean surfaces with significant wave height of 2 m, LRM altimeters (a), delay/Doppler altimeter (b).

Rodriguez et al.) for the same significant wave height. They vary because one or more parameters previously described depart from the ones of the other missions.

The significant wave height also impacts the echo shape (mainly the leading edge for conventional altimeters) as shown in Figure 1.14.

Echo modeling has evolved over time. First, it was shown by Moore and Williams (1957) that the backscattered power can be obtained by convoluting the emitted pulse with a function dependent on the surface backscattered coefficient. Then, Barrick (1972) showed that the measured power could be written as a double integral finally expressed as a double convolution among three terms (Barrick and Lipa 1985): the flat sea surface response (FSSR), the probability density function (PDF) of the heights of the specular scatterers, and the point target response (PTR) of the radar as follows (Brown 1977; Chelton 2001).

$$S(t) = \text{FSSR}(t) * \text{PTR}(t) * \text{PDF}(t) \tag{1.5}$$

where t is the two-way incremental ranging time.

More recently, Halimi et al. (2014) and Boy et al. (2016) used a similar formulation to describe the delay-Doppler measured power (e.g., CryoSat-2 or Sentinel-3) but introducing the Doppler frequency (f).

$$S(t,f) = \text{FSSR}(t,f) * \text{PTR}(t,f) * \text{PDF}(t) \tag{1.6}$$

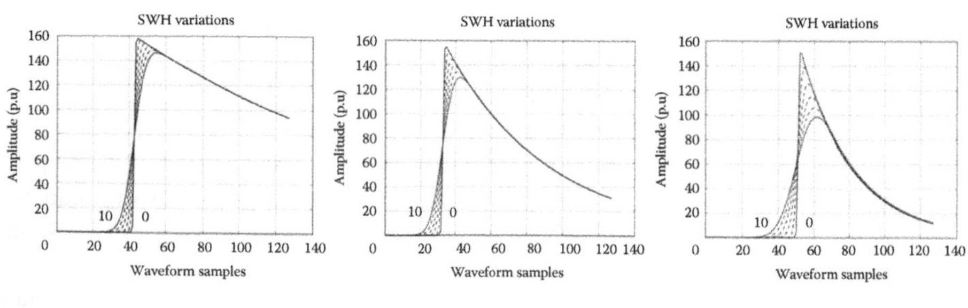

FIGURE 1.14 Impact of different SWHs (between 0 and 10 m) for Jason, CryoSat (LRM), and SARAL/AltiKa missions.

1.3.1 Speckle Noise

The theoretical waveforms cannot be observed in real measurements *W(t)* due to the multiplicative speckle noise *N(t)* that corrupts the altimeter returns:

$$W(t) = S(t).N(t) \tag{1.7}$$

The influence of the speckle noise is generally reduced by averaging a sequence of *L* consecutive echoes. This operation reduces the noise variance by *sqrt(L)* when assuming pulse-to-pulse statistical independence, and the resulting noise is generally assumed to be gamma distributed (LRM and SAR echoes). Examples of noisy 20-Hz waveforms (e.g., Jason-2) or 40-Hz waveforms (e.g., AltiKa) are shown in Figure 1.15.

1.3.2 Analytical and Numerical Models

For the mean power, *S(t)*, several analytical and numerical models have been proposed in the literature. The accuracy of the altimeter model depends on the three convoluted terms. This accuracy often comes at the price of a high computational complexity if no approximations are made. Choosing the appropriate model depends on accuracy requirements (which can depend on the application) and on ground processing facilities.

For conventional altimetry, the models by Hayne (1980) or Amarouche (2004) usually have been considered. They are based on the original formulation given by Brown (1977), accounting for higher order moments in the PDF function but with approximations in the FSSR and PTR functions.

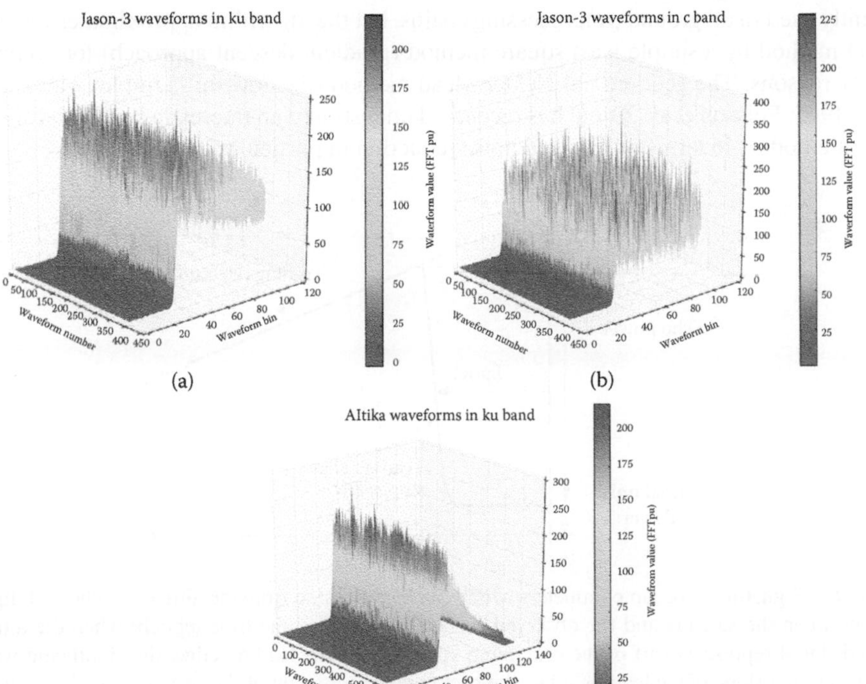

FIGURE 1.15 Examples of Jason-2 (a) Ku band, (b) C band, and (c) AltiKa band waveforms over ocean.

To process delay-Doppler multilooked echoes (CryoSat-2 and Sentinel-3), some analytical solutions were first developed by Halimi et al. (2014) and Ray et al. (2015). But, to be tractable, the analytical approach introduces some substantial approximations for the PTR and antenna gain profile. Phalippou and Enjolras (2007) and Boy et al. (2016) have presented numerical approaches without approximations. In the latter, the Doppler echo model is generated through simulations of the altimeter response (accounting for the real antenna pattern and PTR). A point-by-point radar response (FSSR) is computed by integrating echo returns from a gridded simulated surface. Each term of the double convolution is computed separately, and the overall convolution is numerically performed in order to obtain a set of models that are integrated (after range migration) to obtain the multilooked models.

1.3.3 Estimation Strategies

The parameter estimation algorithm (also called the retracking function) provides the relevant geophysical parameters by fitting a received power evolution model to each observed echo (see Figure 1.16).

Several strategies can be followed according to the considered waveform model, the criterion to minimize, and the optimization algorithms used. Performance can be evaluated in terms of bias (on simulated echoes, for example) and noise level at different spatial frequencies. The number of parameters determined by the retracking algorithm can be adjusted to the characteristics of the mission (occurrence of off-nadir pointing sequences, for example). Basically, the three main parameters (time, significant wave height, and power) are estimated, but one or two additional parameters can be also considered (slope of the trailing edge of the waveform, skewness, or mean square slope). A popular estimation strategy is based on a maximum likelihood estimation procedure (Rodriguez 1988, 1994; Challenor et al. 1990). This method maximizes the measurement statistics and assumes knowledge of the noise distribution corrupting the observed echoes. The Levenberg-Marquardt method (Bertsekas 1995) and the iterative Newton-Raphson method (maximum likelihood estimator—MLE3 or MLE4—algorithms (Dumont 1985; Amarouche et al. 2004; Thibaut et al., 2004, 2010) are currently used in all ground (re)processing chains, but the algorithm approximates the maximum likelihood method by a simple least square method (gradient descent approach) for computational complexity reasons. The geometrical Nelder-Mead method (the downhill simplex method) (Nelder and Mead 1965; Poisson et al. 2016a) has recently demonstrated an interesting potential with respect to former methods—in terms of bias and noise reduction in particular.

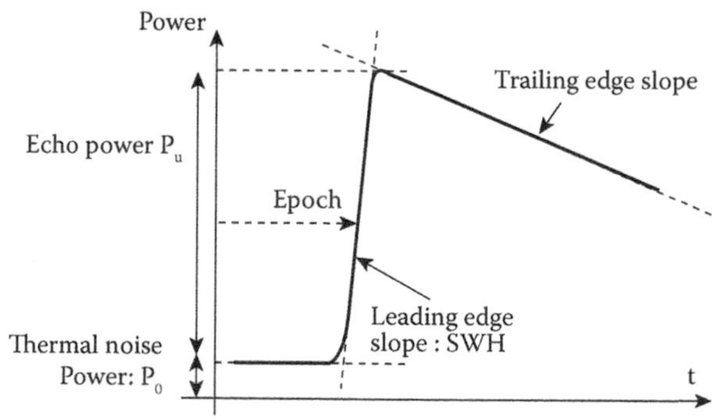

FIGURE 1.16 Signature of ocean parameters which can be estimated from the altimeter echoe (cf. figure 1.12): the range between the satellite and the observed surface is related to the time (epoch) when the amplitude of the received signal represents half of the maximum amplitude in the leading edge; the significant wave height is deduced from the slope of the leading edge; the backscatter coefficient of the ocean, related to wind speed, is estimated from the maximum amplitude received. The curvature of the leading edge is linked to waves skewness and the slope of the trailing edge provides information on the radar antenna mispointing with respect to nadir. (From http://www.altimetryinfo).

Finally, it is worth noting that most operational algorithms estimate each echo independently from the others. However, some algorithms use information about adjacent waveforms to improve estimation performance—for example, the weighted least squares algorithm proposed in Maus et al. (1998) for a group of waveforms, the Kalman filter-based algorithms (Jordi and Wang 2010) or the two-pass methods investigated by Sandwell (2005), Garcia (2014), and Amarouche et al. (2014).

1.3.4 New Altimeters

For the first missions (GEOSAT, ERS-1, etc.), the measurement frequency of the main altimeters (T/P, Envisat, ERS-2, Jason-1, Jason-2, CryoSat-2, Jason-3, and Sentinel-3) has always been the Ku band (13.6 GHz). Since February 2013, the SARAL mission has been using the AltiKa altimeter operating on the Ka band (35.75 GHz) to improve knowledge of the ocean mesoscale variability thanks to an increase in spatial and vertical resolutions. Its higher frequency allows improvement of the range accuracy, taking advantage of a smaller antenna beam width but making it more sensitive to moisture in the atmosphere (in particular rain cells) and to antenna mispointing. In the Ka band, the bandwidth is also increased with range gate of 30 cm instead of 47 cm in the Ku band (the range gate inversely proportional to the bandwidth). Moreover, the accuracy of the estimated parameters has increased because of a higher PRF (3600 Hz for SARAL/AltiKa instead of 2100 Hz for Jason), which permits a better along-track sampling of the surface (40 Hz for AltiKa; 20 Hz for Ku band missions). Then the overall assessment is highly positive. SARAL/AltiKa is based on the same principle as conventional altimetry, which means that the previous models and algorithms can be directly used.

The main hypothesis of the Brown model used to retrack waveforms is that the backscattering coefficient of the surface is constant in the waveform footprint (Brown 1977). This strong hypothesis is all the more satisfied because the footprint is small, which is the case for SARAL when compared to Jason or Envisat. Figure 1.17 compares waveform footprints of various altimeters. Differences in areas are important among Jason (300 km^2), SARAL (100 km^2), and missions like CryoSat-2 or Sentinel-3 operating in delay-Doppler mode (5 km^2). Of course, the smaller the waveform footprint, the closer to the coasts the instrument can provide valuable results. This is another very positive improvement brought about by the SARAL/AltiKa mission—especially for coastal applications.

In order to provide improved geophysical parameter estimates by increasing the number of observations (looks), the delay-Doppler altimeter (DDA) concept was proposed (Raney 1998). The CryoSat-2 and Sentinel-3 missions (SIRAL and SRAL instruments, respectively) have been the first satellites to exploit the DDA mode. The DDA requires a coherent correlation among pulses. This is obtained by transmitting pulses with a high PRF (PRF = 18,182 Hz for the SIRAL altimeter of CryoSat-2 [Wingham et al. 2006]). The second benefit of DDA is its ability to increase the along-track resolution; this is achieved by an appropriate processing that uses the Doppler information contained in the data to form a small and very thin strip-shaped synthetic aperture footprint (see Figure 1.18). Note that because the processing needed to obtain the 20-Hz waveforms differs from conventional altimetry, a new formulation of the model is required that is fitted to the characteristics of the new measurement geometry (Halimi et al. 2014; Ray et al. 2015; Boy et al. 2016).

1.3.5 Non-Ocean Surfaces

In addition to ocean measurements, most radar altimeters also acquire exploitable waveforms over other inland waters and ice surfaces. Useful geophysical parameters—such as topography, water level for rivers and lakes, sea ice parameters such as freeboard height or thickness, backscattering coefficient for different surfaces, and so on—can be derived from these atypical waveforms.

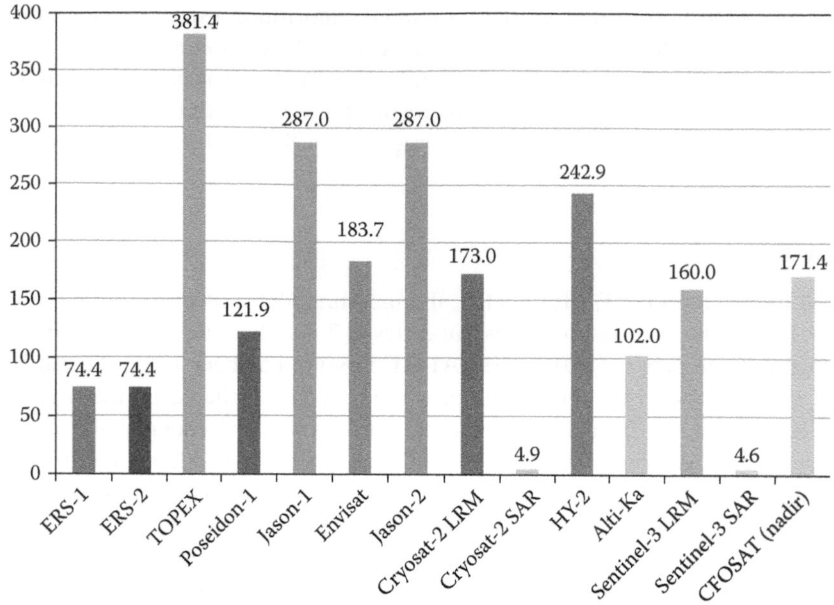

FIGURE 1.17 Waveform footprints (in km²) for the main altimeter missions considering all range gates transmitted on the ground. The number of range gates in the retracking process can be reduced for the price of an increase in the estimation noise for all geophysical parameters.

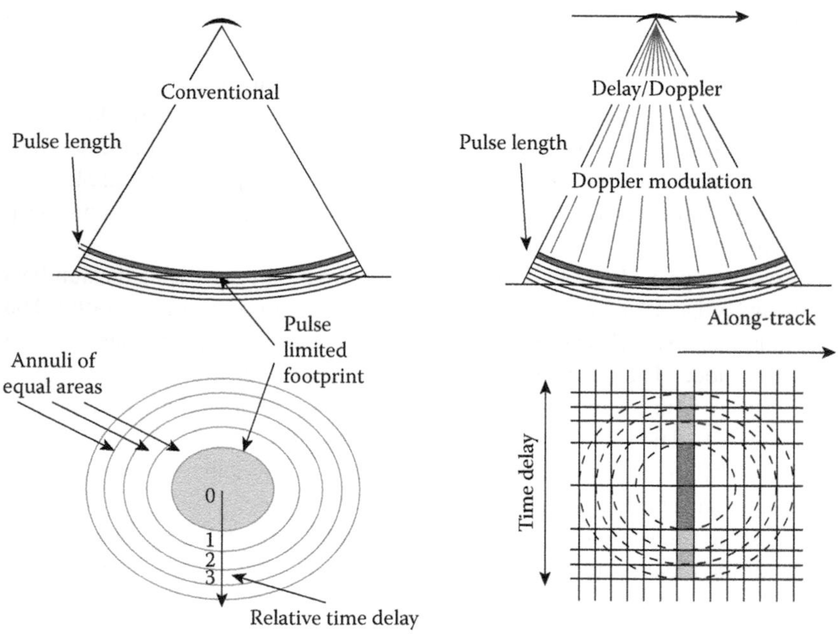

FIGURE 1.18 The Delay/Doppler technique consists of coherently processing the 64 pulses of each burst. Each received pulse contains the response from the sea surface over the circular footprint, as in conventional altimetry (left panels, side view and top view). Then, a Doppler processing is performed over the 64 pulses of each burst in the along-track dimension, generating 64 Doppler beams (right panels, side view and top view). The Doppler beams have a beam-limited illumination pattern in the along-track direction (around 300 m along track resolution for Cryosat-2, represented in green in the bottom right figure) , while maintaining the pulse-limited form in the across-track direction. From R. K. Raney, Johns Hopkins University, Applied Physics Laboratory

Although ocean waveforms are relatively stable over the ocean (except for speckle noise variation) and fit the Brown model well, waveforms from heterogeneous surfaces (coastal zones, inland waters, sea ice regions, and ice sheet areas) have chaotic shapes, introducing additional complexity in the process to derive geophysical information from the raw measurements (see Chapter 14 and Chapter 15).

Two examples are given in Figure 1.19: Jason-2 flying over an inland water region and SARAL/AltiKa over a sea ice region. The Figure 1.19a shows a transition from ocean waveforms to specular waveforms corresponding to returns from calm water surfaces and then a second transition over inland waters. The center of the image shows very specular waveforms generated by a small and isolated reflection from a single target, but the right-hand portion of the Figure 1.19a shows that the altimeter waveforms become very complex due to multiple reflections coming from small reflective surfaces of different origins. The Figure 1.19b shows similar transitions seen by SARAL/AltiKa, when the altimeter flies over open ocean (left part of the Figure 1.19b) to ice-covered regions (specular waveforms on the right-hand side of the Figure 1.19b).

Chapter 15 describes precisely the methodologies that can be used to analyze these non-ocean waveforms. Several strategies can be adopted using purely empirical methods such as threshold retrackers (Davis 1995, 1997; Lee et al. 2008), center of gravity retrackers (Wingham et al. 1986), or retrackers based on physics for open ocean waveforms. An overview of these algorithms is given in Vignudelli et al. (2011). Recent research has tried to define methods that can guarantee a good continuity of observation from the open ocean to coastal zones (Passaro et al. 2014) or from the open ocean to sea ice regions (Poisson et al. 2016b) as well as from coastal zones to inland waters.

1.4 PRECISE ORBIT DETERMINATION

For altimetry missions, knowledge of the satellite altitude is essential because it provides the reference for the nadir altimeter range measurement. Fortunately, the dynamics of a satellite in circular orbit implies that the average radial value is very stable due to the third Kepler law: a good along-track measurement implies that the semi-major axis is well known and so is the average radial component. To compute precise trajectories, one has to combine measurements with precise modeling of the dynamics of the spacecraft.

Two major measurement systems are used for POD of such satellites: DORIS and GPS. They have excellent characteristics for the along-track observation that, combined with trajectory modeling, allow precise monitoring of satellite altitude. In complement, laser ranging from laser retroreflectors

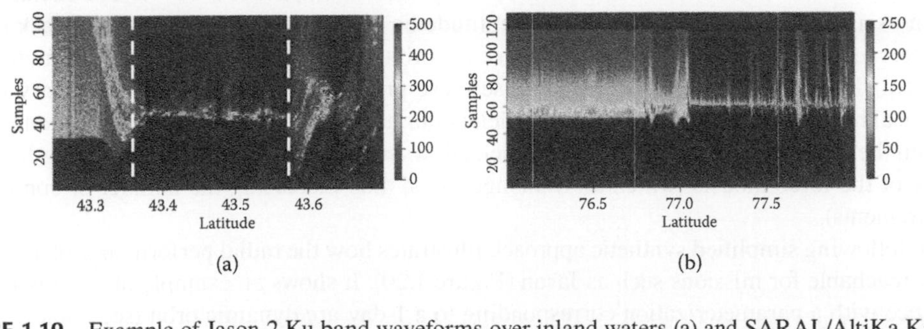

FIGURE 1.19 Example of Jason-2 Ku band waveforms over inland waters (a) and SARAL/AltiKa Ka band waveforms over sea ice (b). The images show a set of consecutive waveforms along the orbit at increasing latitudes (20 Hz rate for Jason, 40 Hz for SARAL/AltiKa). Each column represents a waveform of N samples (N = 104 for Jason and N = 128 for SARAL). Colors indicate the amplitude of the waveform range gates from dark blue for the noise level observed on the first samples to red, observed for the most energetic samples, 40 for Jason and 60 for SARAL.

TABLE 1.3
Orbital Characteristics (geodetic phases not considered here)

	T/P, Jason	Envisat	CryoSat-2	Altika	HY-2A	Sentinel-3
Altitude (km)	1336	782	717	800	971	814
Period (s)	6746	6036	5952	6036	6267	6060
Inclination (deg.)	66	98.6	92	98.6	99.3	98.6
Sun synchronous		x		x	x	x
Cycle (days)	9.9	35	369	35	14	27

on board the spacecraft have been used either as a primary measurement source (early missions) or as independent calibration and validation information.

Table 1.3 shows the different orbit characteristics of the altimetry satellites used in altimetry products since the beginning of the high-precision altimetry era.

1.4.1 Orbit Determination Technique

1.4.1.1 Performance Requirements

To be able to sample adequately the major ocean signals and in particular large-scale signals, the radial orbit performance is a key element. Since 2003, radial orbit performance of 1 cm root mean square (RMS) have been achieved. To monitor MSL trends at the regional level, further requirements regarding the long-term stability of the orbit are needed (see Chapter 5, Nerem et al.).

New altimetry techniques (interferometry) require short-term accuracy characteristics. The mission requirement is thus expressed using a spatial frequency spectrum (Fu et al. 2009). A new difficulty to be considered is that the frequencies of interest correspond to errors over very short periods of time (less than 100 s), where the characteristics of the satellite dynamical environment are not precisely known.

1.4.1.2 Radial Error Properties

For circular orbits, an error δa in the semi-major axis produces about 10 δa along-track error per orbit, and for 1 day, this corresponds to about 130 δa for the Jason orbit. If the dynamic model is correct for 1 day's duration, an along-track error variation of 10 cm/day will produce an error in the semi-major axis of only 0.7 mm. The average radial error is of this order of magnitude.

For the radial error, the other significant parameter is the eccentricity error δe. The corresponding along-track error is an oscillation at the orbital period of amplitude 2 δe * a. The radial error is also an oscillation at the orbital period of amplitude δe * a. If we suppose the along-track oscillation observed with an error noise of 10 cm on each orbit, uncorrelated, this produces a radial error of 1.4 cm (RMS) amplitude. To efficiently use these properties, it is of course necessary to have dynamical models that are precise over a sufficient duration (typically 1 day). This is a compromise between the quality of the measurements (which allows reducing the observation arc length) and the quality of the force models (which allows longer arcs, thus increasing the number of contributing measurements).

The following simplified synthetic approach illustrates how the radial performance of 1 cm RMS is now reachable for missions such as Jason (Figure 1.20). It shows an example of the possible performance with a parameterization corresponding to a 1-day arc dynamic orbit (see Table 1.5). The measurements are the along-track positions, with a 20-min sampling and 5-cm error noise. For this example, the radial bias is below 1 mm, and the RMS value is 7 mm. This is an ideal case; assuming a perfect model for the orbit dynamics (only a small along-track model error), the measurement errors and the dynamic errors are unbiased, but this shows how the radial performance of 1 cm RMS (now on Jason, for example) is achievable.

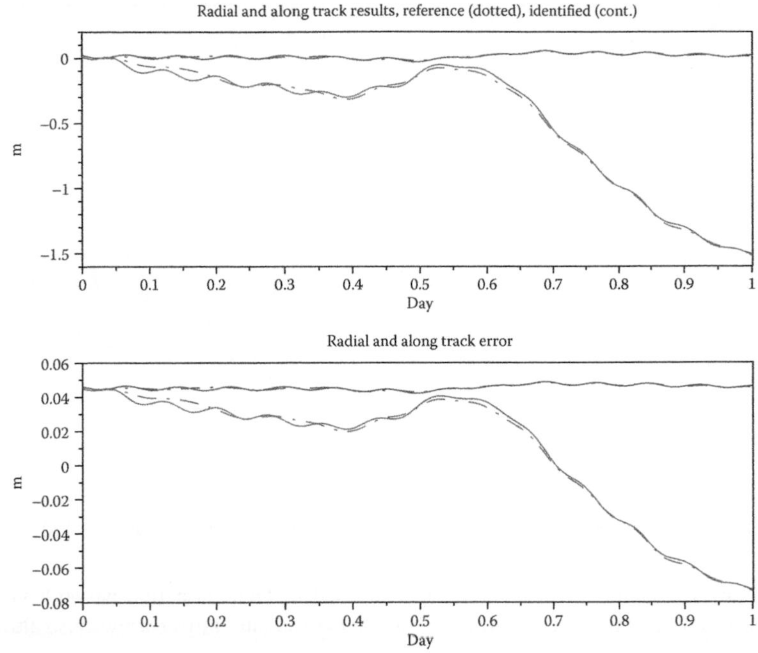

FIGURE 1.20 Orbit determination characteristics in the radial (blue, and along-track (green) directions. Example of achievable performance using a parameterization corresponding to a 1-day arc dynamic orbit (see Table 1.5).

1.4.2 Orbit Determination Measurement Systems

Different tracking systems are classically used for POD: satellite laser ranging (SLR), DORIS, or GPS. The observations are usually derived from the propagation time of a signal between the satellite and the Earth or another satellite. The signal can be in the visible domain (laser ranging (Pearlman et al. 2002)), radio frequency (DORIS [Willis et al. 2010], or GPS (Dow et al. 2009)).

For radio-frequency systems, ionospheric propagation effects are corrected; the models are not precise enough, and this is the reason for the use of dual-frequency measurement systems.

SLR was the first technology used by early altimetry missions (see Table 1.1). The measurement is a two-way propagation time: The laser impulse is emitted from the ground, reflected by the satellite retroreflector array, and received on the ground. The advantage of the two-way measurement is that it is not limited by clock errors such as for DORIS or GPS. The usual wavelength is 585 nm (green). The International Laser Ranging Service (ILRS; Pearlman et al. 2002) is coordinating the SLR activities for geodesy. The difficulty in using SLR is the ground network, which is not homogeneously distributed. Also, the station performance is not homogeneous. So, in order to have a stable orbit performance, SLR data are not directly used in current operational products. As a consequence, the specific models needed to achieve sub-centimeter SLR measurement modeling are not detailed further. But the SLR contribution is essential because it is used as an external and absolute validation of the radial performance.

The DORIS system was developed specifically for the orbit determination of satellites for orbits ranging from 500 to 2000 km (Tavernier et al. 2003). The system uses ground transmitters (ground network of 50 stations), and the measurement is the phase of the signal propagating from the ground to the onboard DORIS receiver (400 MHz and 2 GHz). An important characteristic of the DORIS system is the homogeneity of the ground station network; Figure 1.21 shows the network visibilities for Envisat.

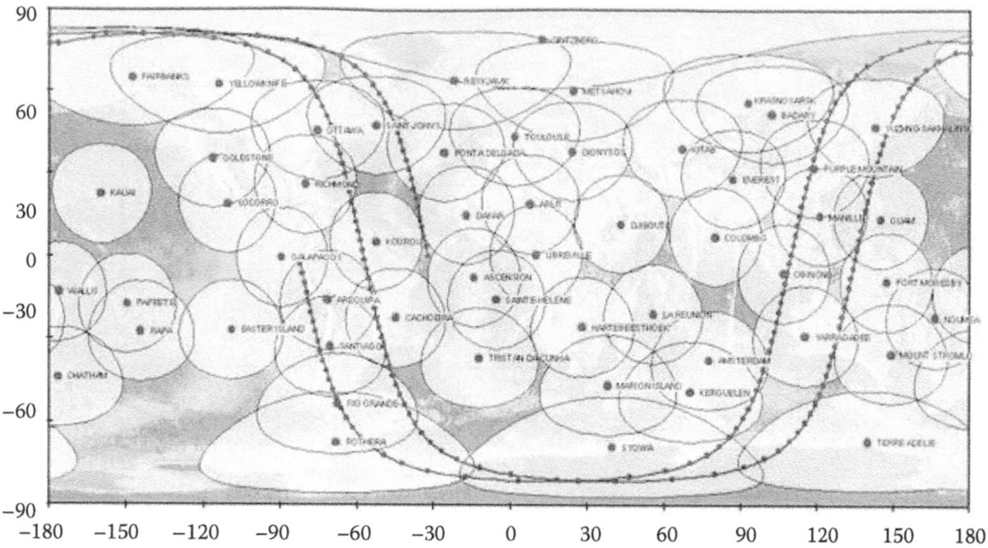

FIGURE 1.21 Envisat satellite ground track superimposed to DORIS station network visibility areas at ENVISAT altitude and 12° minimal elevation. (From Aviso website: https://www.aviso.altimetry.fr/en/my-aviso.html).

The phase measurement corresponds to a very precise propagation time (a few millimeters of noise error) but with an unknown bias. In addition, there are errors due to the transmitter and receiver clocks' stability. This is why the phase variations over 10 s are used as reference measurements (this eliminates the unknown biases). As for the ILRS, the International DORIS Service (Willis et al. 2010) provides the geodetic community with the necessary input (measurements, station coordinates). The DORIS measurement is very sensitive to along-track orbit error and less sensitive to transverse errors. DORIS-derived orbits have radial performance similar to GPS-based orbits when the parameterization is limited (reduced dynamics solutions are limited by the density of available measurements).

With GPS, the navigation systems (GNSS) are designed to provide precise positioning of mobiles on Earth or in the air. It has also proved to be efficient for satellite tracking. Up to now, the different altimetry satellites using GNSS systems have only used GPS. In the near future, the receivers will have the ability to track different GNSS constellations, such as Galileo or GLONASS. For precise positioning applications, the dual-frequency phase measurements are used (1.227 Mhz and 1.575 Mhz for GPS). Due to the geometry of the GPS system, there are always several visible GPS satellites, typically eight on average at the altitude of Jason (the Jason-3 receiver has 12 tracking channels). For GPS, the phase is used in absolute and not as phase variation over a given period, which imposes to solve for the unknown phase bias. Due to the important number of satellites simultaneously visible, the receiver clock bias can be solved epoch by epoch (a clock model is not necessary). The International GNSS Service (Dow et al. 2009) provides the geodetic community with the necessary input (Kouba et al. 2016), (i.e., measurements, stations coordinates, orbits, and clocks). The very dense GPS measurements along the orbit allow efficient reduced-dynamic solutions—for example, to improve the along-track model performance (drag effects).

To use DORIS and GPS tracking measurements, some unknown parameters must be estimated in every precision orbit determination process. This includes, especially, clock biases or atmospheric propagation parameters. Table 1.4 shows the parameters that have to be adjusted for the two systems.

TABLE 1.4
Measurement Parameters to Be Estimated

Name	Parameters
DORIS	Zenith troposphere delay, per pass
	Transmitter frequency bias, per pass
GPS	Ambiguities, per pass
	Receiver clock offset, at each epoch

1.4.3 Satellite Trajectory Modeling and Parameterization

The satellite position is not directly measured. It is necessary to combine the measurements with *a priori* information on the satellite trajectory to meet the required orbit accuracy. For that purpose, a parametric model describing the trajectory is constructed. This model has to take into account the various forces acting on the satellite: Earth's gravity and other gravity effects (planets, tides), atmospheric drag, solar radiation pressure, and so on. The amplitude of the perturbations induced by these forces varies according to the orbital characteristics (altitude, inclination, eccentricity). As an example, the T/P altitude was selected in order to minimize the unmodeled effects of Earth's gravity and of atmospheric drag.

To use these models to reconstruct the precise trajectory followed by the satellite, parameters need to be estimated. This estimation is done through an optimization process (least square adjustment, Kalman filtering, etc.) using tracking observations. This allows filtering out the tracking measurement noise and compensation for potential interruptions in those measurements. (The GPS system allows direct positioning of a mobile but with an accuracy limited to the 5-cm class, using dual-frequency, pseudo-range, and phase.)

The minimal set of parameters to be estimated is the initial position and velocity of the satellite (six parameters), but other components are not well known; the atmospheric density, as an example, is fluctuating with space and time, and the interactions of the satellite with the surrounding atmosphere are not precisely known. Thus, it is necessary to adjust model parameters to deal with these errors.

The dynamic system corresponding to a satellite flying on a circular orbit has a resonance frequency at the orbital period ω. Thus, if we consider a reference orbit, a very small periodic perturbation with a frequency close to ω will produce oscillations around the reference orbit with an important amplification. For example, for the drag effect or for the solar radiation pressure, the modeling precision is not sufficient to achieve a 1-cm radial error on 1 day. To avoid such errors at the orbital period, empirical accelerations at the frequency ω are adjusted (Colombo 1989). The tracking measurements must also be sufficiently dense along the orbit to allow a correct estimation of these unstable eigenvalues of the system.

Table 1.5 shows the set of parameters estimated for Jason POD. In the dynamic configuration, 16 parameters describing the satellite trajectory are estimated for 1 day. This parameterization is sufficient to achieve a centimeter radial performance. No constraint is applied on the parameters. To have a stable performance, the dynamic orbits are computed over a complete cycle (9.9 days).

However, thanks to the high quality of the tracking measurements, it is possible to increase the number of parameters to identify. These are called "reduced dynamic" solutions (Haines et al. 2003). The number of parameters raises to about 100 (Table 1.5); but there are some constraints in order to stabilize the parameter values, so this number of parameters is not directly comparable to the dynamic configuration. The reduced-dynamic parameterization is usually more constrained (less parameters or higher constraints) for DORIS-only than for GPS-based solutions. The constraints τ in Table 1.5 are relative constraints: The difference in the accelerations a_i and a_{i+1} between successive segments is constrained with a covariance τ. The acceleration global bias is not constrained;

TABLE 1.5
Parameterization Used for Present Products (Dynamic and Reduced-Dynamic Orbits) Jason GPS/DORIS Case (Couhert et al. 2015); Daily Number of Parameters in Brackets

Name	Dynamic	Reduced-Dynamic GPS
Initial conditions	Position, velocity (6)	Position, velocity (6)
Tangential const.	4 h segments (6)	30 mn segments (48), $\tau = 1.10^{-9}$ ms^{-2}
Tangential periodic ω	24 h segment (2)	112 mn segments (26), $\tau = 5.10^{-10}$ ms^{-2}
Normal periodic ω	24 h segment (2)	112 mn segments (26), $\tau = 2.10^{-9}$ ms^{-2}
DORIS weight	1.5 cm over 10 s	1.5 cm over 10 s
GPS weight (phase)	2 cm	2 cm
GPS weight (code)	20 m	20 m

this allows absorbing the same errors as in the dynamic solutions. Due to the important number of parameters, the orbit errors become uncorrelated after 1 day, so the reduced-dynamic orbits can be computed on a daily basis (instead of over several days for the dynamic case).

1.4.4 Major Modeling Evolution since the Beginning of the 1990s

The T/P mission pointed out the close relationship between successive gravity modeling improvements and associated leaps forward in radial orbit accuracy (nearly a factor of 10, from a 13-cm RMS mission goal to 2–2.5 cm achieved rapidly after launch) (Bertiger et al. 1994; Marshall et al. 1995). This is of primary interest in monitoring the change in the height of the ocean surface. For the successor missions, Jason-1 and Jason-2, orbits with a radial precision of 1-cm RMS (Luthcke et al. 2003; Choi et al. 2004; Haines et al. 2004) and 7-mm (Jalabert et al. 2015), respectively, have been achieved.

Since the launch of Jason-1 (the T/P follow-on mission), GDR-A1* orbits were computed at the Centre National d'Etudes Spatiales (CNES) Orbit Determination Department using ZOOM orbit estimation software. GDR-A stands for the precise orbits supplied for placement on the Geophysical Data Record's Version A that were released to the scientific community as of January 2002. At that time, the available dynamic models were initially inherited from those used for T/P. Since then, they have been progressively updated (see Table 1.6) based on the state-of-the-art set of International Earth Rotation and Reference Systems Service (IERS) geophysical standards (McCarthy 1996; McCarthy et al. 2004; Petit and Luzum 2010), the outcomes of the Gravity Recovery and Climate Experiment (GRACE) mission, and the successive versions of the International Terrestrial Reference System (ITRS) realizations. The drivers that usually lead to the definition of new POD standards combine significant improvements in the gravity field, solid Earth/ocean/pole tide models (to account for the tidal perturbation of the geopotential), reference frame definition, and tracking station coordinates. Reducing gravity field errors is of utmost importance because they produce fixed geographically correlated errors (geographically correlated errors, or GCE, are the same for repeated ascending and descending overflights of the same region), which result in a bias in estimated SSHs. Until the launches of Jason-1 (December 2001) and Envisat (March 2002), the new Earth gravity field models were used. GRIM5-S1 (Biancale et al. 2000) was developed jointly by a German-French team (GeoforschungsZentrum [GFZ] and CNES/Groupe de Recherche en Géodésie Spatiale [GRGS]).

* The Geophysical Data Record (GDR) is the science product of the altimetry missions. It includes every component of the altimetry measurement: distance measured by the radar and associated corrections to be applied, precise orbit determination, and so on. The algorithm used for data processing is documented, and various versions of this processing may have been applied: GDR-A, GDR-B, and so on.

TABLE 1.6
Progressive Improvements in Geopotential Models

GDR-A (2002)	GDR-B (2005)	GDR-C (2008)	GDR-D (2012)
Mean gravity field model GRIM5-S1	EIGEN-CG03C	EIGEN-GL04S-ANNUAL	EIGEN GRGS.RL02bis. MEAN-FIELD
Non-tidal TVG Drifts in $C_{2,0}$,$C_{3,0}$,$C_{4,0}$	Unchanged	Drifts in $C_{2,0}$,$C_{2,1}$,$S_{2,1}$,$C_{3,0}$,$C_{4,0}$ + annual and semi-annual terms up to deg./ord. 50	Annual, semi-annual and drift terms up to deg./ord.50
Solid Earth tides IERS Conventions (1996)	IERS Conventions (2003)	Unchanged	Unchanged
Ocean tides *FES 95.2*	FES2004	Unchanged	Unchanged
Atmospheric gravity None	Only tides from Haurwitz and Cowler model	6-h NCEP pressure fields + tides from Haurwitz and Cowley model	6-h NCEP pressure fields + tides from Biancale and Bode model
Pole tide None	Solid Earth from IERS Conventions (2003)	Solid Earth and ocean from IERS Conventions (2003)	Solid Earth from IERS Conventions (2010)
Terrestrial reference frame ITRF2000	Unchanged	ITRF2005	ITRF2008

As a satellite-only model, the solution was based on analysis of gravitational satellite orbit perturbations and tracking data from 21 satellites (including T/P). Also contributing to the GDR-A orbit standards, the finite element solution, FES95.2, ocean tides model (Le Provost 1998) was derived by assimilating T/P satellite altimetry data.

In 2005, the switch to the GDR-B standard enabled the use of the gravity field combination model EIGEN-CG03C (Förste et al. 2005), based on surface data (gravimetry) as well as on the Challenging Minisatellite Payload (CHAMP) and GRACE satellite data. Given the proximity with the launch of the latter mission (March 2002), only 376 days of the GRACE mission data were used in the preparation of EIGEN-CG03C. As a consequence, non-tidal time-variable gravity (TVG) modeling was still limited to the inclusion of the secular rates for the degree 2, 3, and 4 of zonal harmonics of the gravity field expansion derived from SLR measurements. The GDR-B orbit solutions also benefited from the updated version of the FES2004 (Lyard et al. 2006) and the modeling of the solid Earth rotational deformation due to the polar motion (pole tide). These solutions included both dynamic (geopotential) and geometric (station displacements) corrections. The tidal part of the atmospheric gravity effects was accounted for by Haurwitz and Cowley (1973) and updated from Biancale and Bode (2006).

When the GDR-C orbit standards became operational (2008), the static gravity field was no longer a significant source of POD error. The main non-tidal time-variable components in the hydrosphere and cryosphere captured by the GRACE observations (Tapley et al. 2004) were modeled for the first time in the EIGEN-GL04S-ANNUAL mean gravity field, as annual and semiannual variations and are currently available at: http://gravitegrace.get.obs-mip.fr/grgs.obsmip.fr/data/RL01/static/EIGEN_GL04S_ANNUAL.txt. For extrapolation purposes, the GDR-C standards omitted the trend terms in the model because they were determined over a limited 2-year time span (the GRACE-derived trend terms from the later EIGEN-GRGS.RL02bis.MEAN-FIELD were used starting with the GDR-D standards). Thus, in addition to the first zonal coefficients, only rates for the degree 2, order 1 terms of the geopotential (describing the drift in the position of the Earth's figure axis) were modeled, according to the IERS Conventions (2003). The non-tidal contribution from atmospheric gravity was also introduced based on the National Center for Environmental Prediction (NCEP) three-dimensional pressure field at 6-h intervals over land (inverted barometer hypothesis over the ocean).

1.4.5 Long-Term Orbit Error and Stability Budget

Although the orbit precision was constantly improved with the successive modeling upgrades, remaining errors over the lifetime of altimeter missions had to be periodically characterized and quantified. Identifying the principal sources of long-term errors (yearly to decadal timescales) affecting the orbit solutions at regional scales is of primary importance to prevent aliasing into calculations of regional MSL rate. These errors come from:

- Tracking systems
- Reference frame uncertainties
- Non-tidal TVG modeling issues

The spectral nature of these errors is rather complex and better described as geographically dependent patterns with seasonal, interannual, and secular variations, rather than in term of radial RMS values. Couhert et al. (2015) provided a long-term error budget of the 10-year Jason-1 and Ocean Surface Topography Mission/Jason-2 (OSTM/Jason-2) GDR-D orbit time series presented in Table 1.7. Since 2015, GDR orbits have been computed in the Version E standards. The models that were retained for the last GDR-E standards are summarized in Tables 1.8 and 1.9.

As stated in Couhert et al. (2015), inclusion of a seasonal non-tidal geocenter motion model in the GDR-E POD standards improved orbits consistency between GPS-based and SLR-DORIS-derived orbits computed by independent analysis centers. Indeed, the regional annual variation patterns, shown in Figure 1.22, were significantly reduced (about 2 mm) between the latest Standard 1504 DORIS+SLR dynamic solution provided by the NASA Goddard Space Flight Center (GSFC) and the GDR-E GPS+DORIS reduced-dynamic orbits, both of them using the SLR-derived annual geocenter model from Ries (2013). For comparison purposes, the Jet Propulsion Laboratory (JPL) release 16a GPS-based reduced-dynamic orbits do not include such a geocenter motion model.

Concerning interannual POD errors, the combined use of the CNES/GRGS mean geopotential model EIGEN-GRGS.RL03-v2.MEAN-FIELD (http://grgs.obs-mip.fr/grace/variable-models-grace-lageos/mean_fields, based on 12 years of GRACE and LAGEOS data and including time-variable terms—bias, drift, annual, semiannual—up to spherical harmonic degree and order 80) and a reduced-dynamic analysis strategy enables the reduction of TVG-induced radial regional drifts between the GDR-E solution and the external orbit series (GSFC and JPL) to a sub-millimeters per year (mm/y) level over the span of the Jason-2 mission (Figure 1.23).

TABLE 1.7
Upper Bound Estimates of GDR-D Radial Orbit Error Budget for the Jason Series

Error Source	Time Scale	Global	Regional	Rationale
Tracking Data Residual Consistency	Seasonal		3–8 mm	SLR v. GPS/DORIS orbits
	Interannual		3 mm/y	
	Decadal		2 mm/y	
Reference Frame	Seasonal		8 mm	GPS v. SLR+DORISITRF08 v.05
	Interannual	0.03 mm/y	1 mm/y	
	Decadal	0.05 mm/y	0.3 mm/y	
Time Variable Gravity	Seasonal		4 mm	Mean field v. 10-days series and external orbits

TABLE 1.8
Force Modeling Differences between the Current GDR-E Orbit Standards and the Previous GDR-D Orbit Standards

	GDR-D (2012)	GDR-E (2015)
	Geopotential	
Non Tidal TVG	EIGEN-GRGS.RL02bis.M-F annual, semi-annual, and drift up to deg/ord 50	EIGEN-GRGS.RL03-v2.M-F, annual, semi-annual, and drift terms per year up to deg/ord 80; $C2,1/S2,1$ modeled w.r.t. IERS Conventions
Solid Earth tides	IERS Conventions (2003)	Unchanged
Ocean tides	FES2004	FES2012
Atmospheric gravity	6-h NCEP pressure fields (20 × 20) + tides from Biancale and Bode model	6-h NCEP pressure fields (72 × 72) + tides from Biancale and Bode model
Pole tide	Solid Earth and ocean from IERS Conventions (2010)	Unchanged
Third bodies	Sun, Moon, Venus, Mars, and Jupiter	Unchanged
	Surface Forces	
Solar radiation	Thermo-optical coefficient from pre-launch box and, wing model with smoothed Earth shadow model	Calibrated semi-empirical solar radiation pressure model
Earth radiation	Knocke and Ries albedo and IR satellite model	Unchanged
Atmospheric drag	DTM-94 for Jason satellites, and MSIS-86 for the others	DTM-13 for Jason/HY-2A, and MSIS-86 for the others

TABLE 1.9
Measurement Modeling Differences between the Current GDR-E Orbit Standards and the Previous GDR-D Orbit Standards

	GDR-D (2012)	GDR-E (2015)
Displacement of reference points	IERS Conventions (2003)	Unchanged
Solid Earth tides		
Ocean loading	FES2004	FES2012
Pole tide	Solid Earth from IERS Conventions (2010)	Solid Earth and ocean from IERS Conventions (2010)
Reference GPS constellation	JPL solution at IGS (orbits and clocks) – fully consistent with IGS08	JPL solution in "native" format (orbits and clocks) consistent with IGS08
Geocenter variations		
Tidal	None	Ocean loading and S1-S2 atmospheric pressure loading
Non-tidal	None	Seasonal model from Ries
Terrestrial reference frame		
SLR	SLRF/ITRF2008	Unchanged
DORIS	DPOD2008	Unchanged
GPS	IGS08	Unchanged
Earth orientation	Consistent with ITRF2008 and IERS Conventions (2010)	Unchanged
Propagation delays		
SLR troposphere correction	Mendes and Pavlis model	Unchanged
Doris Troposphere correction	GPT/GMF model	Unchanged

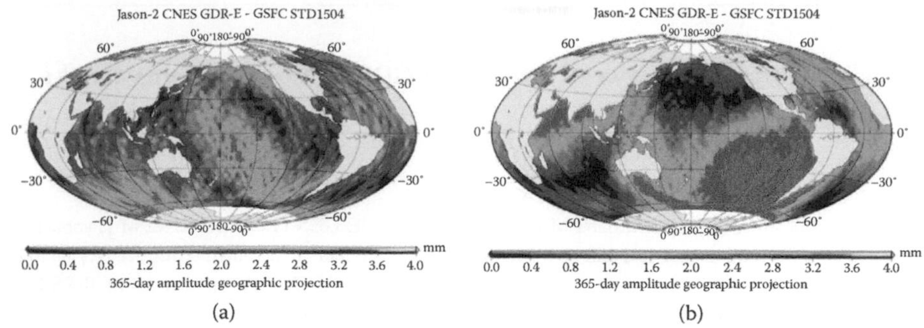

FIGURE 1.22 Jason-2 geographically correlated radial orbit difference 365-day signals of GSFC Standard 1504 (a) and JPL rlse16A (b) with respect to CNES GDR-E orbits.

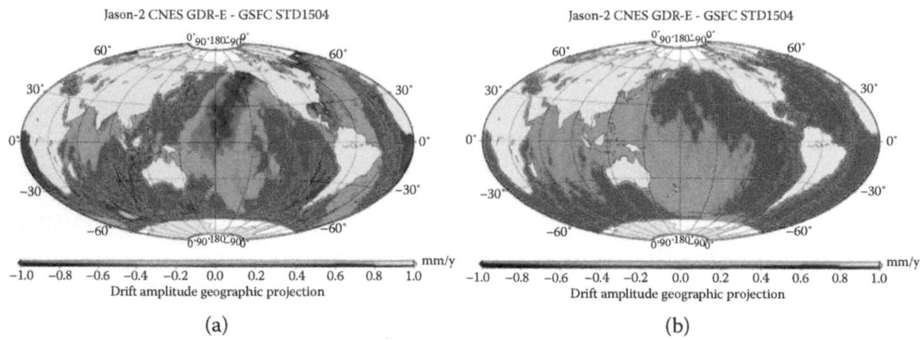

FIGURE 1.23 Jason-2 geographically correlated radial orbit difference drifts of GSFC Standard 1504 (a) and JPL rlse16A (b) with respect to CNES GDR-E orbits.

Thus, independent indicators, such as the ability to characterize orbit errors through regional comparisons of high-accuracy orbits determined from independent data (GPS, DORIS, and SLR) and different parameterization (the near-continuous tracking supplied by DORIS and GPS enable reduced-dynamic strategy), allow insight into model and geodetic technique error and help to validate improvements. Yet, there are still limitations in validating the orbit accuracy and stability as no direct measurement of absolute orbit accuracy at the 1 mm/y level exists. In situ oceanographic measurements (tide gauges and Argo networks, calibration and validation [CALVAL] sites), and statistics of the altimeter crossover residuals are currently used at the forefront of their measurement accuracy. Even SLR measurements from historically well-performing observatories can be subject to occasional biases or data gaps (e.g., Greenbelt and Yarragadee; see Couhert et al. 2015), making orbit regional drifts at an interannual timescale difficult to detect.

1.4.6 Foreseen Modeling Improvement

The GRACE mission also provides seasonal variations of the Earth's gravity field that are of potential interest for POD computation. Since the launch of the two formation flying spacecraft in 2002, four different processing centers have continuously released monthly gravity solutions:

The Center for Space Research (CSR) at the University of Texas, Austin
The GFZ in Potsdam, Germany
JPL
CNES/GRGS

Analyzing the differences between the low degree and order terms (below 20 by 20) of the Release-05 (RL05) GRACE monthly geopotential time series from the four data centers shows important discrepancies. Standard deviations plotted in Figure 1.24 exhibit that the dispersion among the four GRACE monthly gravity field estimates from CSR, GFZ, JPL, and CNES/GRGS is quite high for the degree 2/3 as well as for the sectorial harmonics (degree l = order m).

The radial orbit sensitivity of Jason-2 and CryoSat-2 (the highest and lowest altitude of the current altimetry missions, respectively) to individual variations of the spherical harmonics of the geopotential has been analyzed using the standard deviations derived from the dispersion in the GRACE analysis center solutions. Figure 1.25 shows that the projected errors from dispersion in GRACE TVG solutions affect the satellite orbits at specific orders and sets of coefficients of the geopotential expansion. Jason's orbit is most sensitive to GRACE gravity field errors in the degree 3, order 1 harmonic of this expansion, while CryoSat-2's orbit is more affected by inaccuracies in the degree 3, order 2/3 and resonant degree 14, order 14 harmonics. Such errors can affect the regional trends of the MSL.

The dynamics of a spacecraft flying in a circular orbit is such that its orbit will be centered at the center of gravity of the Earth; however, the use of tracking measurements from stations located in a given reference frame tends to center it at the origin of the reference system. Then, specific care should be given to the orbit centering to avoid biases and drifts in the POD. As such, orbit centering is a measure of the stability and accuracy of the computed reference frame. Orbit differences in X, Y, and Z (terrestrial reference frame axes) averaged over one cycle have been used as an index for orbit centering.

As can be seen in Figure 1.26, orbit-centering discrepancies among the different POD analysis centers are still visible and unexplained:

- Y bias of 2 to 3 mm between CNES GDR-E and GSFC Standard 1504
- Y annual signal from 1 to 2 mm between CNES GDR-E and JPL rlse16a

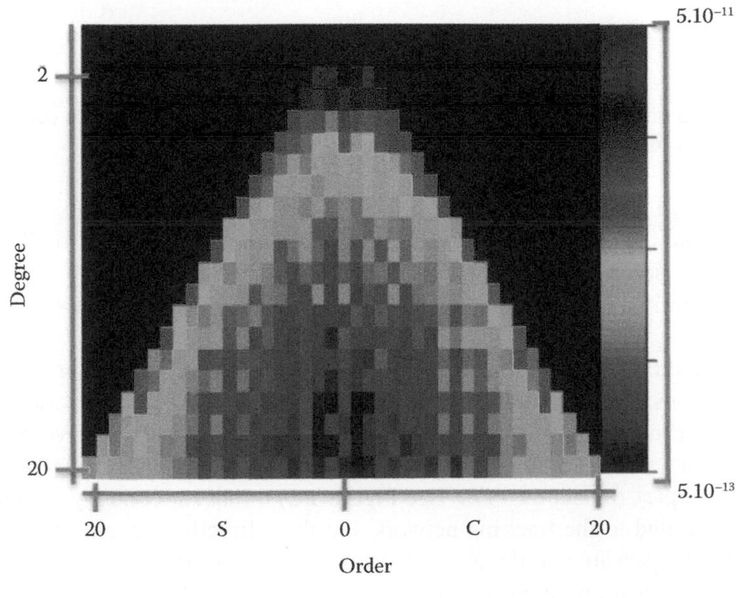

FIGURE 1.24 Standard deviation values of the low degree and order spherical harmonic coefficients (below 20 by 20) of the monthly geopotential solutions produced.

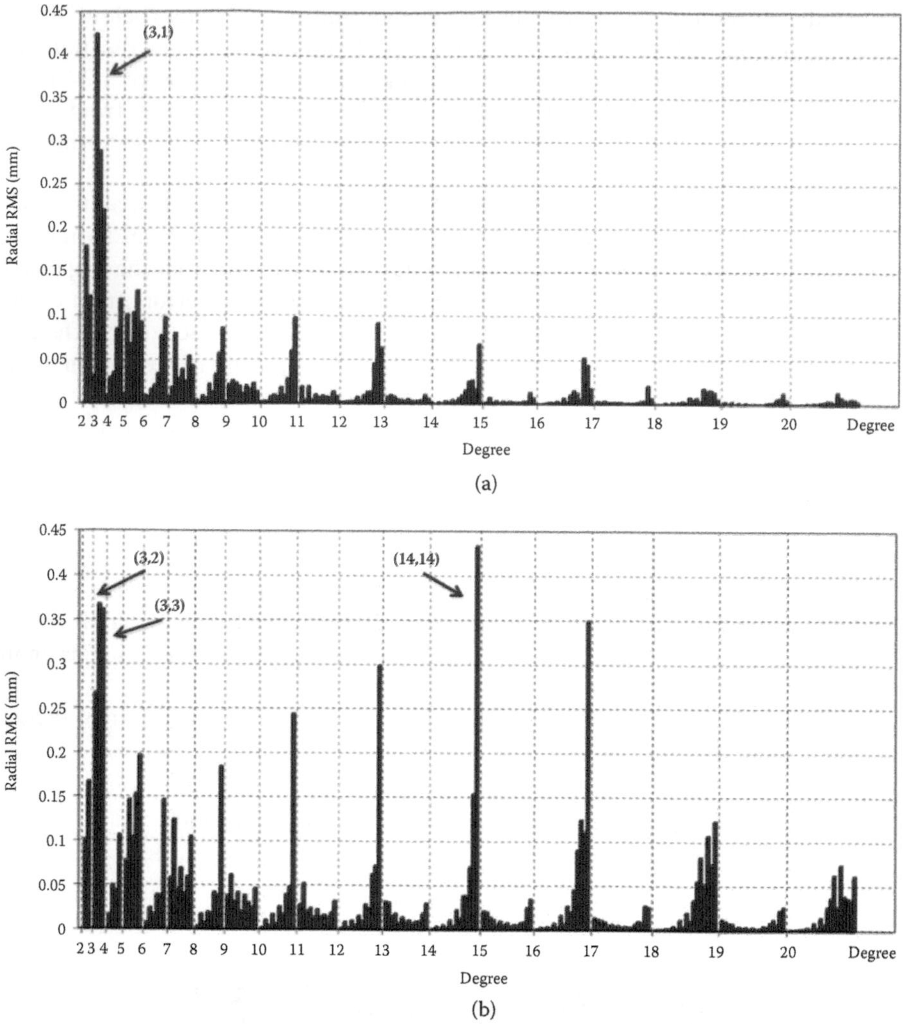

FIGURE 1.25 Radial orbit sensitivity to individual variations in spherical harmonics corresponding to their associated GRACE internal error estimates, for Jason-2 (a) and CryoSat-2 (b).

- Z annual signal and bias from 1 to 2 mm (opposite sign) with regard to GSFC Standard 1504 and JPL rlse16a
- Z drift of 0.5 mm/year between CNES GDR-E and GSFC Standard 1504

A partial explanation could be the uncertainty in modeling the geocenter motion. Geocenter motion determinations can not only rely on SLR measurements. Further works on independent GPS and DORIS estimates are necessary because geocenter-related errors give rise to geographically correlated orbit errors (Figure 1.22). Accounting for annual geocenter variations in the orbit standards has a complex impact on POD (see Figure 1.27). Indeed, contrary to reduced-dynamic solutions, which are tied to the tracking network and then directly impacted by geocenter errors, dynamic orbits are not sensitive to the X and Y (equatorial) network errors due to averaging of the errors because of the Earth's rotation (Figure 1.27a). Furthermore, this effect may not be a pure harmonic function as the geocenter variations that are linked to climatology parameters (i.e., change in water masses' locations) exhibit a nonstationary behavior as well as secular trends.

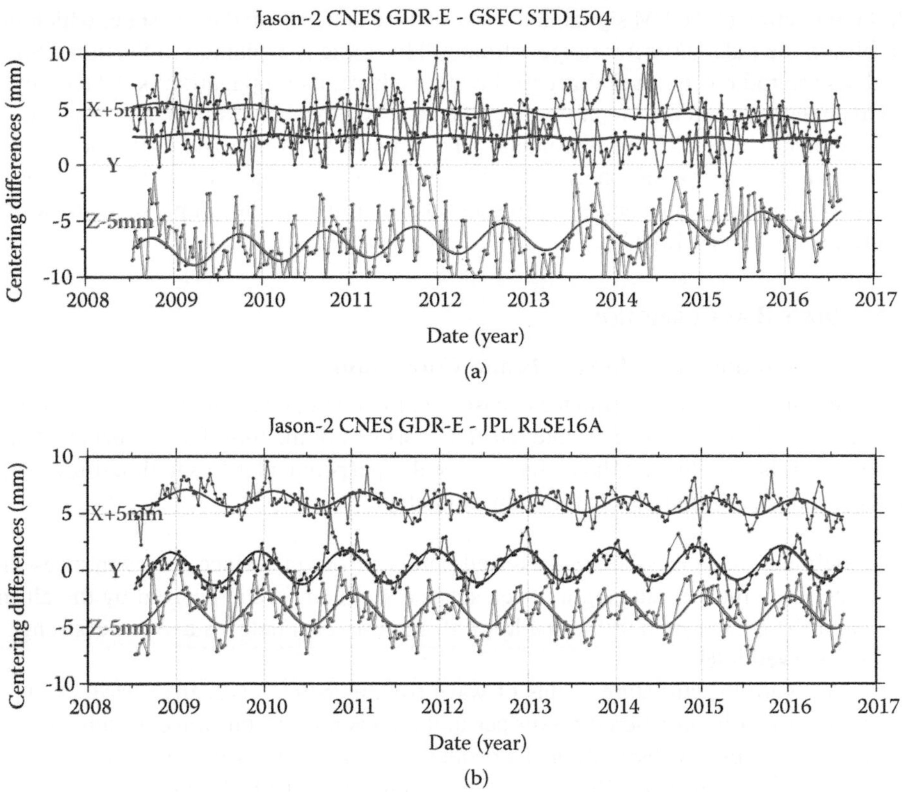

FIGURE 1.26 Mean X (blue), Y (red), and Z (green) orbit differences of GSFC Standard 1504 (a) and JPL rlse16A (b) with respect to CNES GDRE orbits (X offset by +5 mm, Z by −5 mm).

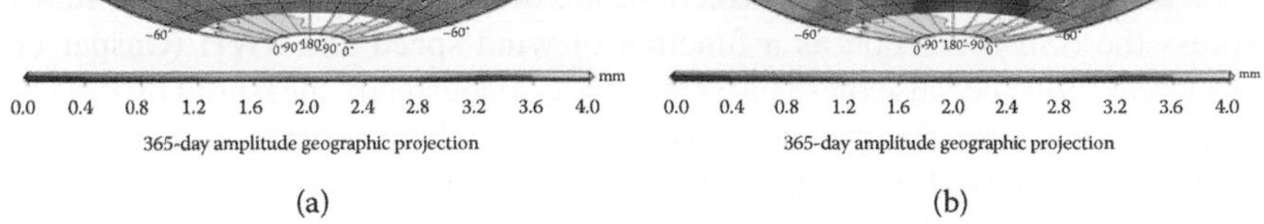

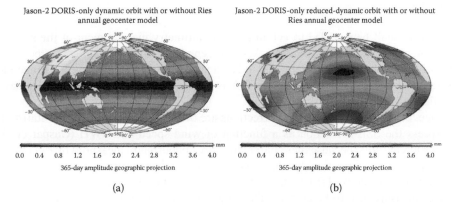

FIGURE 1.27 Jason-2 geographically correlated radial orbit difference 365-day signals of DORIS-only dynamic (a) and reduced-dynamic (b) orbits when including or not including Ries annual geocenter model.

1.5 GEOPHYSICAL CORRECTIONS

While on its path between the satellite and the Earth's surface, the radar signal is impacted by various geophysical phenomena that must be accounted for to provide a precise range measurement. The amplitude of these effects is large (from decimeters to meters). This section describes the methodologies used to correct the radar measurements impacted by these effects and the effect on the global range error budget.

First, the reflection of the EM signal at the sea surface depends on the sea state, which induces a potential bias in the radar altimeter range estimate. This topic is explained in Section 1.5.1.

Second, as the radar echo goes through the atmosphere, its propagation is delayed by various effects. This delay generates an apparent height error, so it must be minimized using a correction of the altimeter range or SSH measurement. The main PD effects include the presence of ionized particles in the upper atmosphere (ionosphere) as discussed in Section 1.5.2.1 as well as the composition of the lower atmosphere (troposphere) and, in particular, in presence of water vapor (WV; Sections 1.5.2.2 and 1.5.2.3).

1.5.1 Sea State Bias Correction

1.5.1.1 Origins of the Sea State Effects and Correction

The range measured by a radar altimeter is based on the reflections from the ocean surface of the EM signal generated by the altimeter antenna. The portions of the illuminated surface that re-emit the signal toward the satellite are those that are locally perpendicular to satellite direction, that is, the specular facets (horizontal facets for nadir altimetry).

- The nonlinear, non-Gaussian, and skewed nature of ocean surface waves induces differences between the median height of sea surface specular facets received by the altimeter echo and the mean height of the sea level of interest. This induces a skewness range bias in the retrieved SSH.
- Moreover, with the curvature radius of wave troughs being larger than wave crests' curvature, the radar backscattered power per unit area is larger from wave troughs than from wave crests, leading to observation of a mean scattering level lowered by as much as tens of centimeters below the true MSL. This effect is known as the EM bias and is theoretically a function of the EM pulse frequency and the sea state.
- The altimeter range is also affected by errors due to instrument processing and range computation methods (Section 1.3) also related to sea state, which are commonly referred as tracker biases.

These effects shall be modeled to get an accurate altimetry measurement of the sea level. All these components (EM, skewness, and tracker) are generally combined and corrected together as a global SSB correction. The SSB remains, today, a significant error affecting the range measurements of radar altimeters (see, e.g., Section 1.8) as it is not a white noise, and it may have space scales similar to ocean dynamic scales.

In practice, current operational SSB corrections are derived from empirical and statistical studies that express the SSB correction as a function of wind speed and SWH (Gaspar et al. 1994; Gaspar and Florens 1998) because these parameters are concurrently measured by radar altimeters. The SSB correction is a correction that depends on altimeter characteristics because of its empirical nature; it cannot yet be derived from a theoretical model let alone be computed independently from the altimeter measurement to be correct. Thus, the SSB correction must be empirically adjusted for each mission, and it must sometimes be updated after significant changes in the processing algorithms. However, good consistency among the SSB corrections of all Jason-class altimeters and similar missions has been reported due to the continuity of the design of the instruments and processing techniques.

1.5.1.2 Theoretical Solutions

Since the first experimental observations of the SSB bias by Yaplee et al. (1971), several theoretical studies (Jackson 1979; Barrick and Lipa 1985; Srokosz 1986) have attempted to model the sum of the EM and skewness biases. However, the most comprehensive studies (i.e., Elfouhaily et al. 2000, 2001) did not propose algorithms approaching the performance of early empirical models

(Gaspar et al. 1994) in terms of validated accuracy of the correction. This indicates that there may be room for improvements on the theoretical approach of this phenomenon or that the instrument processing-related component is not negligible and may vary significantly among altimeters. Theoretical activities on the SSB still proved to be useful in improving the understanding of the sea surface complex processes and provided the key inputs to tune the parameterization of the empirical models.

Potential improvements of theoretical modeling include the use of additional input parameters that are not directly measured by the altimeter (spectral peak period, spectral width parameter, peak amplitude, fetch, wave age, etc.). Another potential improvement could be to obtain additional information on the space and time characteristics of the nonlinear behavior of the sea surface at the wave scales that are involved in the SSB to better tune the modeling.

The SSB theoretical studies also produce, as a side effect, realistic end-to-end simulators (Amarouche 2001; Dubois 2011). These measurement simulators make it possible to anticipate the impact of sea state on the range measurements of new instrument concepts. This was used, for instance, to make preflight analysis of DDA (e.g., Sentinel-3) and wide swath interferometric altimeters (e.g., SWOT, see Chapter 2, Rodriguez et al.). This anticipated knowledge is necessary to gauge the ability to fulfill scientific requirements and to prepare the ground algorithms needed for launch.

1.5.1.3 Empirical Solutions

Since the launch of T/P, empirical, satellite-based SSB estimates have been delivered for operational purposes. They are built as a mission-specific look-up table that describes the SSB amplitude as a two-dimensional function of the SWH and the wind speed. Different aspects of the empirical determination have been analyzed and improved through the years: statistical methods for the SSB modeling, ways to extract the SSB signals from the SSH data, and improvement in the sea state description to better describe the SSB behaviors.

Multiple parametric formulations have been developed with SSB functional forms expressed as linear, polynomial, and quadratic functions of wind speed and SWH. For T/P, the BM4 formulation was proposed (Gaspar et al. 1994; Chambers et al. 2003), where four coefficients have to be estimated. Then, as more satellite data became available, Gaspar and Florens (1994) demonstrated that such parametric models based on SSH differences are not true least square approximations of the SSB due to the specification of a predefined fixed form. They developed and refined a nonparametric estimation technique based on kernel smoothing (Gaspar et al. 1998; 2002) that provides SSB solutions as a regular two-dimensional grid against wind speed and SWH. To perform the error minimizations, the data set used is usually based on SSH differences measured at the same locations at different times. This can be either differences at orbit crossover points or along collinear tracks from subsequent cycles (Chelton 1994; Gaspar et al. 1994; Labroue et al. 2004).

The use of SSH differences to compute the SSB model was necessary for the first altimetry missions in order to eliminate the poorly known geoid signal and to cope with relatively large orbit errors that dominated the altimeter error budget at that time. The pitfall of this methodology is that it recovers only a relative SSB effect not an absolute zero-offset correction. Nowadays, it becomes possible to derive SSB solutions directly from the SSH data. The MSS is now known with sufficient accuracy, and residual geoid signals that remain in the SLA appear to be negligible in the context of SSB algorithms. With state-of-the-art geophysical corrections and MSS, the SSB model can be built with a high level of accuracy (Vandemark et al. 2002; Labroue et al. 2004; Scharroo and Lillibridge 2004; Feng et al. 2010).

These empirical models solely use SWH and wind speed data as input (Tran et al. 2010). Non-satellite field studies (Millet et al. 2003; Melville et al. 2004) have shown that SSB uncertainty can be lowered if additional information on the instantaneous surface was used to better capture the full range of sea state effects. Wave age, the overall degree of wave development, and long wave slope variance are some of the variables that have been proposed. To obtain this information as additional input for the modeling, Kumar et al. (2003) and Millet et al. (2003) suggested using operational wave model output. Tran et al. (2006, 2010) showed that when using a simple additional parameter such as mean wave period (Tm) from a numerical wave model to build a three-dimensional global SSB model, significant improvement can be regionally obtained and can reduce the altimeter range error budget by approximately 7.5%. This also provides evidence that operational wave modeling can now support improved ocean altimetry. In the near future, the flight of new satellites dedicated to sea state monitoring—such as CFOSAT (Hauser et al. 2016), which will measure the directional wave spectrum—may further improve the performance of operational wave models.

Over the last decades, major improvements have been observed with regard to this geophysical correction, yet theoretical and empirical modeling both remain important challenges to further improvement of the altimeter error budget of past and present missions. This topic will also remain a major challenge in the context of new generations of altimeters (e.g., DDA) or wide-swath topography interferometers—for instance, due to the anisotropic nature of their measurements that might add another layer of complexity to the isotropic footprint of a traditional altimeter.

1.5.2 Atmospheric Propagation Effect Corrections

In order to reach the expected range measurement accuracy, the effects of the atmosphere on the propagation of the altimeter microwave radar pulse must be accounted for.

The consequent PDs, expressed as a function of the atmospheric refractivity, vary with the physical state and the composition of the atmosphere, temperature, pressure, WV density, cloud liquid water density, and ionospheric electron density (Fu and Cazenave 2000).

Three atmospheric corrections are estimated, one for each independent atmospheric source of delay:

- The ionospheric correction is related to the total electron content (TEC; Section 1.5.2.1).
- The dry tropospheric correction accounts for the impact of dry gases and is proportional to the sea-level pressure (Section 1.5.2.2).
- The WTC depends on the integrated WV content and, at a lesser level of magnitude, on the cloud liquid water content (CLWC), as per Section 1.5.2.3.

1.5.2.1 Ionospheric Correction

The presence of free electrons in the upper atmosphere between an altitude of 50 and 2000 km, the so-called ionosphere, makes this medium a dispersive one for altimetric radar signals. This PD can be up to 30 cm at solar maximum or during ionospheric storms due to disturbed solar and geomagnetic conditions. To that extent, the effect requires a precise correction to reach centimeter-level accuracy.

The radar altimeters flying on most missions (T/P, Jason-1, Envisat, Jason-2, and Jason-3) were operating on two different frequencies; they exploited the frequency dependence of the ionospheric refraction to directly retrieve the PD affecting the propagation of the radar pulse simultaneously with the range measurements (Chelton et al. 2001).

The necessary correction is computed from the difference between the measured ranges on the two frequencies. The range values have to be corrected for the SSB estimations because they are also frequency-dependent terms (see Section 1.5.1). For altimeters operating in the Ku band (approximately 13.6 GHz) with the secondary channel in the C band (5.3 GHz for TOPEX and Jason), the dual-frequency ionospheric correction is expressed as (e.g., Imel 1994):

$$\mathrm{Iono\_corr} = \frac{f_C^2}{f_{Ku}^2 - f_C^2} * [(R_C - SSB_C) - (R_{Ku} - SSB_{Ku})] \tag{1.8}$$

where f is the radar frequency. Because the range measurements at the secondary channel are much noisier than those measured in the main Ku band, the dual-frequency retrieved ionospheric correction is usually smoothed with the use of an along-track low-pass filter as first suggested by Imel (1994).

For single-frequency altimeters (i.e., GEOSAT, GFO, ERS-1, ERS-2, and CryoSat-2), it is necessary to use alternative means to remove this propagation delay imposed by the ionosphere. The range correction relies on the JPL GPS-derived global ionosphere maps (GIM) (Mannucci et al. 1998; Komjathy and Born 1999; Hernandez-Pajares et al. 2002). Indeed, since 1998, the JPL has developed the capability to monitor the global distribution of ionospheric TEC using dual-frequency observations from a worldwide network of ground GPS receivers. For altimeters older than 1998, a climatological global model such as the NOAA Ionosphere Climatology 2009 (NIC09) model (Scharroo and Smith 2010) is used instead. The NIC09 model adjusted their climatology on JPL GIM for the period covering 1998 to 2008 and extended the estimation of the ionospheric PDs prior to 1998.

The SARAL/AltiKa altimeter is a single-frequency altimeter but, due to its operation in the Ka band, the amplitude of the ionospheric PD is about seven times smaller than Ku-band altimeters so that this correction can almost be neglected. However, ionospheric effects are still corrected with the same process as a single-frequency Ku altimeter, even though the correction magnitude is smaller.

1.5.2.2 Dry Tropospheric Correction

The propagation of a radio pulse is delayed by dry gases and the quantity of WV in the Earth's troposphere. The dry gas contribution is nearly constant and produces height errors of approximately –2.3 m. This effect can be modeled. The refractive index depends on pressure and temperature. When a hydrostatic equilibrium and the ideal gas law are assumed, the vertically integrated range delay is a function of the surface pressure only (Chelton 2001). The dry meteorological tropospheric range correction is defined by the following equation:

$$\text{Dry\_Tropo} = -2.277 * P_{atm}\left(1 + 0.0026 * \cos(2 * \text{LAT})\right) \tag{1.9}$$

where P_{atm} is the surface atmospheric pressure in mbars, LAT is the latitude, and Dry_Tropo is the dry tropospheric correction in millimeters.

There is no straightforward way of measuring the nadir surface pressure from altimetry, but it can be approximated from a global atmospheric model analysis, such as the European Center for Medium-Range Weather Forecasts (ECMWF) operational analyses pressure field, with a 6-h resolution (AVISO 2014).

As this 6-h temporal sampling of the ECMWF operational analysis does not allow a proper estimate of diurnal and semidiurnal components of atmospheric pressure variations (Nyquist theory), specific processing is necessary for the dry tropospheric correction. The same method as that used for DAC (see, e.g., Section 1.6.2) is used to remove diurnal and semidiurnal signals from the ECMWF pressure field.

The ERA (ECMWF (European Center for Medium-range Weather Forecast) re-analysis) -interim global atmospheric reanalysis can also be used to estimate the dry atmospheric correction from model reanalyses; using this correction significantly reduces the discrepancy between ascending and descending tracks for older missions such as ERS and T/P because modern reanalysis perform better than historical operational analyses (see, e.g., Figure 1.28 from Carrere et al. [2016a]).

1.5.2.3 The Wet Tropospheric Correction

1.5.2.3.1 Order of Magnitude and Sources of Observations

The WTC, or PD, mainly depends on the integrated amount of WV and, to a lesser extent, on the integrated amount of liquid water:

$$\text{WTC} = PD_{vap} + PD_{liq} \tag{1.10}$$

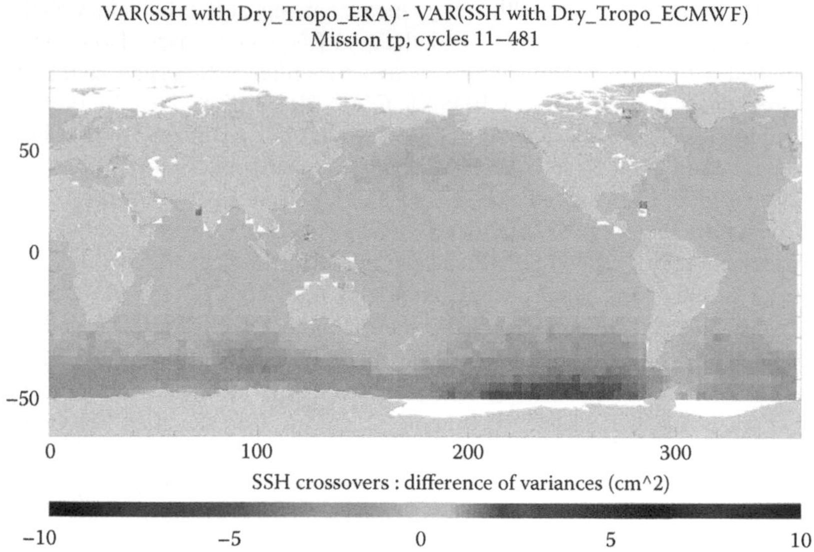

FIGURE 1.28 SSH crossover variance reduction (in blue) when using an ERA-interim-based correction instead of the operational dry tropospheric correction (cm²).

From the computation of the refractivity of the WV (Smith and Weintraub 1953; Liebe et al. 1993) and the cloud liquid water droplets (Resh 1984), and using thermodynamic laws, the PDs caused by WV PD_{vap} and by liquid water PD_{lip} over a distance R are expressed as:

$$PD_{vap} = \beta'_{vap} \int_0^R \frac{\rho_{vap}}{T(z)} dz \quad (1.11)$$

where z is the altitude, T is the temperature of the atmosphere (in K), ρ_{vap} is the WV density (in $g.cm^{-3}$), $\beta'_{vap} = 1720.6\ K.cm^3.g^{-1}$, and:

$$PD_{liq} = 1.6 \int_0^R \rho_{lip}\, dz = 1.6\ CLWC \quad (1.12)$$

where ρ_{lip} is the liquid droplet density, and CLWC is the integrated CLWC.

Note that typical values of PD_{lip} are less than 5 mm, that is, almost negligible compared to PD_{vap}.

Figure 1.29 shows the average (top right) and the standard deviation (top left) of SARAL/AltiKa radiometer WTC computed per grid cell of 2° × 2° over the entire year (2015).

The geographical distribution of the WTC is close to the distribution of WV and, therefore, at the first order related to the atmosphere and SST distribution (see Figure 1.30 for the SST). Over the ocean, WTC amplitude varies from a few millimeters at high latitudes up to 40 cm over the Intertropical Convergence Zone (ITCZ), with a global mean of about 15 cm. For latitudes higher than 60° (north and south), or various specific regions (e.g., west of South America and South Africa), the correction is stable with a small standard deviation.

Figure 1.31 shows the variations of WTC during the year at different latitudes. At global scale and over intertropical regions, the average WTC correction is about 25 cm, with seasonal variability. The largest seasonal cycle is observed at latitudes between 20°N and 40°N, with a difference of about 10 cm between the minimum during the northern winter (15 cm) and the maximum during the northern summer (25 cm). The amplitude is smaller between 20°S and 40°S (5 cm), with the same

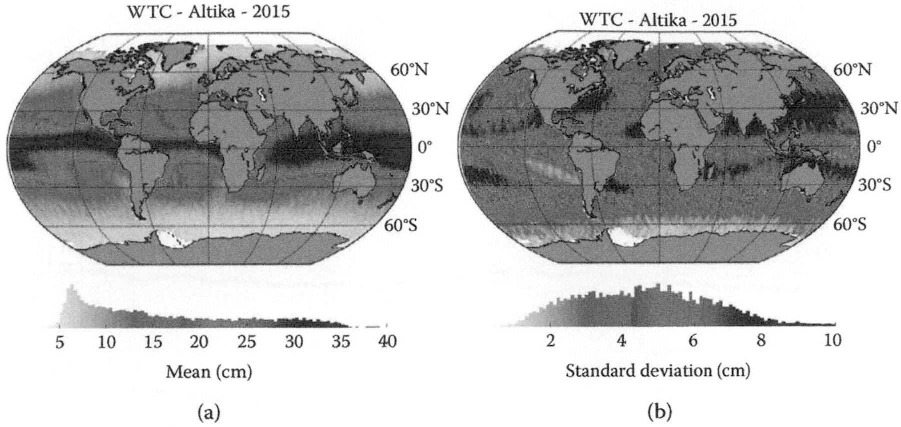

FIGURE 1.29 The WTC retrieved with the SARAL/AltiKa radiometer for the year 2015. (a) Gridded averaging over a 2° × 2° map. (b) Standard deviation over the same grid.

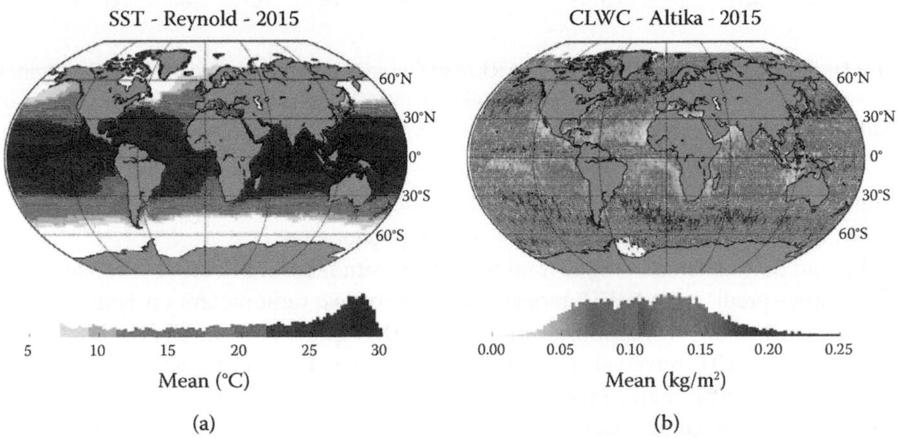

FIGURE 1.30 SST (a) (Reynolds et al. 2007) CLWC retrieved with the SARAL/AltiKa radiometer (b) for the year 2015. Gridded averaging over a 2° × 2° map.

minimum during the southern winter but a smaller maximum during the southern summer (20 cm). Above 40°N, the seasonal amplitude has a minimum of 8 cm and a maximum of 15 cm, while it is very low above 40°S (2 cm).

An approximate relationship between WV and WTC is (Fu and Cazenave 2000: Chapter 1, Section 3.1.2):

$$\text{WTC} \sim 6.4\ \text{WV} \tag{1.13}$$

where WTC is in cm and WV in $g.cm^{-2}$. Keeping in mind that 1 $g.cm^{-2}$ = 10 $kg.m^{-2}$, we also have 1 10^{-2} m (WTC) = 1.56 $kg.m^{-2}$ (WV) or 1 $kg.m^{-2}$ (WV) = 6.4 10^{-3} m.(WTC).

Taking into account the altimetry error budget (see Section 1.8), the constraints on the WTC retrieval error are very high if the goal is to reach 1 cm RMS. Considering the WTC amplitude combined with its high spatial and temporal variability, WTC plays a major role in the altimeter error budget, and efficient measurement principles and retrieval techniques are required. The typical correlation distance of the WTC ranges from 50 to 90 km, and its typical correlation time decorrelation

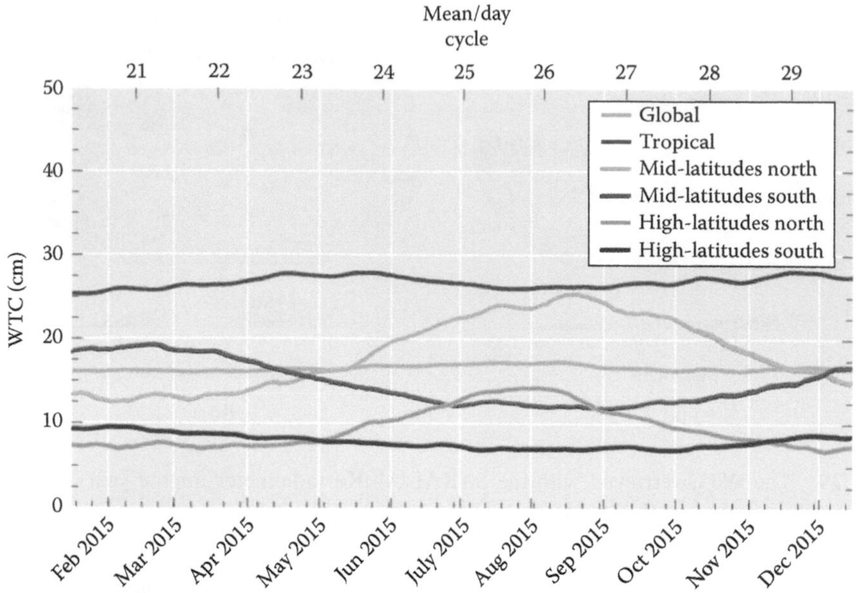

FIGURE 1.31 The WTC retrieved with the SARL/AltiKa radiometer for the year 2015. Top right: Gridded averaging over a 2° × 2° map. Top left: Standard deviation over the same grid. Bottom: Time series of a daily mean for different geographical selections.

ranges from 1.5 to 5 h (Stum et al. 2011), with particularly small values over the ITCZ that are dominated by short time and spatial scale convection phenomena.

The WTC can be estimated using several sources: in situ measurements (Fernandes et al. 2015), numerical weather prediction (NWP) models, and microwave radiometers on board altimetry satellites. As pointed out by the literature (Stum et al. 2011; Legeais et al. 2014), only a microwave radiometer providing WTC retrievals in coincidence with the altimetry range can meet the stringent requirements of modern altimetry missions.

Figure 1.32 shows the comparison of the WTC spectra retrieved from radiometer observations (SARAL/AltiKa in yellow, Jason-2 in red) with WTC computed from ECWMF analysis (light blue). This plot highlights the capability of microwave radiometers to catch smaller scales of WTC. Details on the computation of spectrum can be found in Ubelmann et al. (2014). Between 3000 and 200 km (the large scale associated with the zonal distribution of the WTC), the modeled and measured spectra of WTC are almost overlapping. Below 200 km, the WTC spectrum derived from ECMWF analysis departs from the linearly decreasing pattern standard deviation of SARAL/AltiKa and Jason-2. At these scales, the radiometers are still capable of restituting fine variations of WTC. With a spatial resolution of about 12 km, the WTC retrieved from the SARAL/AltiKa radiometer is linear up to the smallest scales observed by the spectral analysis (14 km). For Jason-2, the impact of the filtering process applied on brightness temperatures (BTs) is clearly seen below 50 km. In other words, the radiometer captures a geophysical process generally expressed as a power law of the wavelength involved down to a wavelength where the radiometer footprints act as a smoothing filter.

1.5.2.3.2 Measurement Principle and Implementation

The retrieval of the WTC uses the measured top of the atmosphere (BTs) realized at various frequencies. All instruments have a common 23.8-GHz channel that is centered on the WV absorption line. This channel captures the major part of the WTC information. A higher frequency channel is also used to capture the contribution of CLWC. This second channel ranges from 34 GHz to 37 GHz. The radiometers used by the Jason missions and HY-2A also use a third channel at

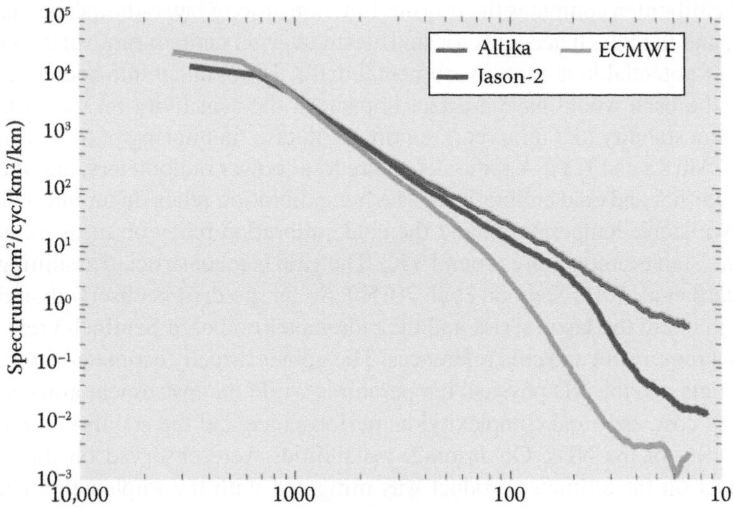

FIGURE 1.32 WTC spectrum as a function of the wavelength in km: From SARAL/AltiKa (yellow) and Jason-2 (red) radiometers as well as from the ECMWF model interpolated on the SARAL/AltiKa track (light blue).

18.7 GHz that, in addition to a good sensitivity to WV, brings additional information about the sea surface roughness. Lacking this third channel, other radiometers use the altimeter backscattering coefficient σ0 as a proxy for the surface roughness. The SST is also sometimes used as an additional input to the retrieval algorithms (Obligis et al. 2009; Picard et al. 2015; Thao et al. 2015). Adding the atmospheric temperature lapse rate (the slope of the temperature decrease from the surface to the top of atmosphere) improves the overall performance of the retrieval (Obligis et al. 2009).

Radiometer instruments can be categorized by the frequencies used and by their internal calibration scheme as per Table 1.10.

TABLE 1.10
Characteristics of In-Flight Microwave Radiometers

Mission/Instrument	Launch Date	Altitude	Architecture Internal Calibration	Frequencies Used	Spatial Resolutions
Jason-2 AMR	20/06/2008	1336 km	NIR	18.7 GHz 23.8 GHz 34 GHz	~ 25 km
HY-2D ACMR	15/08/2011	971 km	Total power, Internal hot load, Cold sky horn	18.7 GHz 23.8 GHz 37 GHz	24 km 19 km 10 km
SARAL/AltiKa MWR	25/02/2013	800 km	Total power, internal hot load, cold sky horn	23.8 GHz 37 GHz	12 km 8 km
Jason-3 AMR	17/01/2016	1336 km	NIR, cold sky via pitch manoeuvers	18.7 GHz 23.8 GHz 34 GHz	~ 25 km
Sentinel 3A	16/02/2016	814.5 km	Dicke NIR + cold sky horn	23.8 GHz 36.5 GHz	24 km 16 km

Note: AMR, advanced microwave radiometer; NIR, noise injection radiometer.

The internal calibration computes the transfer function among raw radiometric measurements, the antenna counts, and the BTs. It accounts for short-term internal temperature variations and allows for the monitoring of potential long-term instrumental drifts. If not taken into account, the temperature variation along the orbit would have a direct impact on the sensitivity on the instrument, and the required long-term stability (0.3 mm/year) requires a precise monitoring of any instrumental drift.

The SARAL/AltiKa and HY2-A radiometers are total power radiometers, and their calibration is ensured by regular hot and cold calibrations. The hot calibration relies on an internal stable thermistor (with a known stable temperature), and the cold calibration relies on measurements of the cold space continuum (stable temperature around 3 K). The gain is reconstructed assuming linearity of the instrument (Picard et al. 2015; Steunou et al. 2015b). So far, no drift is observed on the instrument.

The AMRs on board the Jason series and the radiometer on board Sentinel-3 rely on noise diodes (NDs) and do not require hot and cold references. The gain is directly estimated from the comparison between ND counts and the ND physical temperatures while the instrument aims at the Earth. This approach reduces cost, size, and complexity; its performance and the stability of the measurements rely on the stability of the NDs. On Jason-2, instabilities were observed on the 34-GHz channel ND. Their impact on the altimeter product was mitigated with the implementation of an external calibration scheme using calibrated targets (e.g., Amazonian forest and cold ocean; Brown et al. 2007; Brown 2013). This ensures the long-term stability of Jason-2 (Ruf 2000; Eymard et al. 2005). In order to improve the gain estimate and the stability monitoring, cold sky measurements have been introduced in Jason-3 and Sentinel-3 radiometer calibration processes.

1.5.2.3.3 Retrieval Approaches over the Open Ocean

The forward problem of the simulation of microwave BTs from atmospheric and surface conditions is commonly addressed by radiative transfer models (RTMs).

An RTM is applied to the simulation of LF channels (18.7 GHz, 23.8 GHz, and channels around 34 GHz). Principal constituents that are taken into account are: oxygen and WV (Liebe et al. 1993; Cruz-Pol et al. 1998), cloud liquid water absorption models (Benoit 1968), and seawater dielectric models (Stogryn 1971; Ellison et al. 2003). When altimeter backscattering is used for the retrieval in a two-channel configuration, the double-scale emissivity model developed by Guissard and Sobieski (1987), and improved by Lemaire (1998) and Boukabara et al. (2002), can be used. It is associated with the Elfouhaily spectrum (Elfouhaily et al. 1997) in describing sea surface roughness.

However, the inverse problem (i.e., the estimation of the WTC from the BTs) is complex and cannot be addressed by a purely physical approach. The retrieval uses a semi-empirical method: The RTM is used to build a database containing reference WTC and corresponding simulated BTs on which an empirical inversion is applied to establish the relation between BTs and WTC. This approach is qualified as semi-empirical because the physics of the impact of surface and atmospheric conditions appears in all its complexity through the simulations of the BTs using an RTM.

Different methods may exist in the building of the database, the empirical method used for the inversion, and the structure of the retrieval algorithm; the JPL method used for the Jason-class mission is detailed in Keihm et al. (1995), and the Collecte Localisation Satellites (CLS)/Institut Pierre Simon Laplace (IPSL) method used for Envisat is defined in Obligis et al. (2009). The main difference is in the structure of the retrieval algorithm. Although the CLS/IPSL approach is global, with a single neural network being trained on a global database, the JPL method uses a per-class of wind and WTC approach.

Another alternative is to define a purely empirical approach: The database is built from the measured BTs and the corresponding ECMWF WTC interpolated along the instrument timeline (Picard et al. 2015). This method is robust and simple (because it does not require an RTM) and is well adapted when simulations are complex, especially over heterogeneous surfaces. On a global scale, the pure empirical approach performs well over the ocean, better than a model WTC, but the inconsistency in the database limits its performance with respect to a semi-empirical approach. Moreover, this semi-empirical approach benefits from the constant improvement of the physics used by the radiative transfer modeling.

1.5.2.3.4 Performance over the Open Ocean

The WTC retrieved by microwave radiometers can be assessed over the ocean by comparison to the ECMWF WTC. Two metrics are used: the statistics of the difference on one hand and the difference in SSH variances at crossovers on the other. Then, the performance among different instruments is compared through these proxies.

Figure 1.33 shows the difference (radiometer WTC vs. ECMWF WTC) for SARAL/AltiKa and Jason-2 radiometers, computed over the whole year (2015) on a 2° × 2° grid. The bias of the difference usually ranges from −1 to +1 cm. Small biases among missions may be observed and explained by instrument calibration and cross-calibration schemes. The differences in WTC results from the combination of over and under estimation comparing radiometer and model WTC depend on the geophysical situation. The WTC retrieved from a three-channel radiometer (such as Jason-2) shows larger contrasts than the one retrieved with a two-channel radiometer (such as SARAL/AltiKa). In both cases, the WTC is systematically underestimated by the model on tropical regions and systematically overestimated over mid-latitude areas (between 30° and 60°). Jason-2's WTC is wetter than SARAL/AltiKa's WTC in tropical regions—particularly over cloud-free regions such as to the west of South America and South Africa or in the Arabic Sea (see the cloud liquid content map in Figure 1.30b). Over these regions with clear sky conditions, the 18.7 GHz on Jason-2 has a larger impact and may increase the accuracy of the retrieval.

Except for some evolutions of the model at known dates, the monitoring of the difference between radiometer and model WTC is stable through time and is consequently also used to detect unexpected instrumental events. For instance, as illustrated on Figure 1.34, a Satellite Safe Hold Mode occurred on SARAL/AltiKa (yellow solid line) on October 2014: Instruments were switched off then switched on 3 days later, and the internal thermal equilibrium of the radiometer environment has been modified. With the instrument calibration being changed (the statistics of the BTs are slightly modified), this induces a drop of a few millimeters on the WTC that is clearly seen in the time series. Likewise, a new calibration of the 18.7- and 23.8-GHz channels (following Autonomous Radiometer Calibration System (ARCS) long-term drift monitoring) is applied and clearly seen in December 2015.

A metric based on the difference in SSH variance at crossovers is used to discuss the relative performance among different versions of the same algorithm for a given mission. SSH differences between ascending and descending tracks at the same locations with a maximum temporal difference of 10 days are considered. Assuming that the ocean variability is small over the scale of a

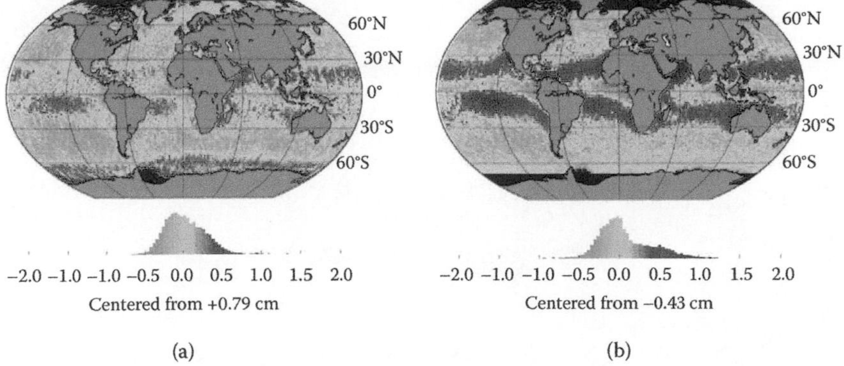

(a) (b)

FIGURE 1.33 Difference between radiometer and ECMWF WTC (radiometer versus ECMWF) for the year 2015: Gridded averaging over a 2° × 2° map. The difference is centered on the global mean. (a) SARAL/AltiKa WTC retrieved from a neural network semi-empirical algorithm. (b) Jason-2 WTC based on the JPL algorithm.

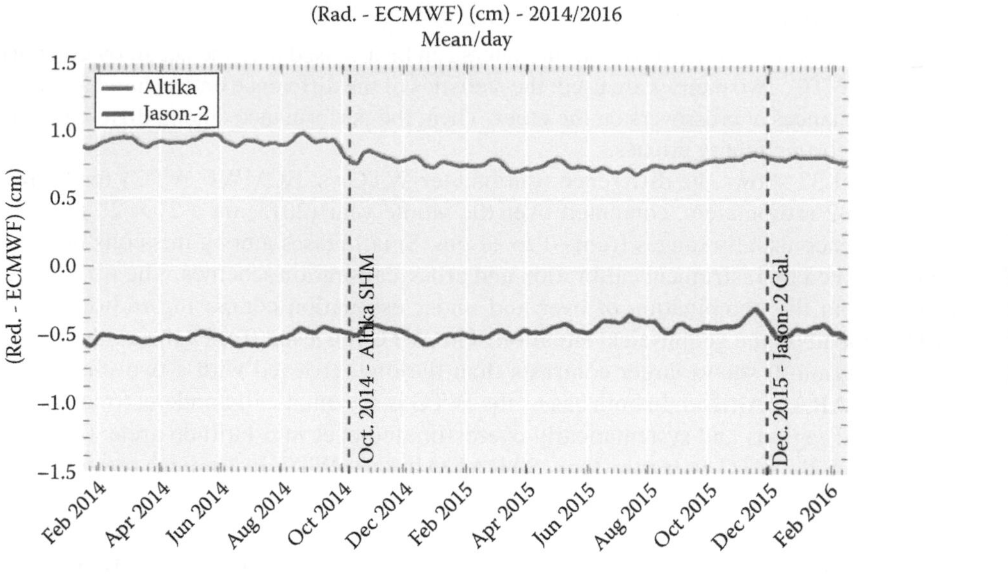

FIGURE 1.34 Monitoring of the difference between radiometer and ECMWF WTC from January 2014 to March 2016 for SARAL/AltiKa (yellow) and Jason-2 (red).

few days, the best WTC will then result in the smallest SSH variance at the crossover differences (see Legeais et al. 2014).

The improvement of the altimetric system performance using the radiometer WTC instead of the model WTC is global: A reduction of about −2 cm^2 of the SSH variance is obtained, which is quite constant in time (see Figure 1.35). It represents a reduction of between 10% and 20% of the SSH variance.

Figure 1.36 highlights the spatial distribution of the SSH variance reduction between an ECMWF-based WTC and the SARAL/AltiKa radiometer (the same distribution is observed for Jason-2). The larger improvements, between −3 and −4 cm^2 of SSH variance reduction, are observed over the ITCZ, where the WTC is most variable, spatially and temporarily. Over clear sky regions, which are more stable, the improvement is less important (west of South America). It is also the case at high latitudes where the reduction is about −1 cm^2. Over some specific areas, the use of the radiometer appears as a degradation of the altimetry system performance (red boxes in Figure 1.36). It is actually

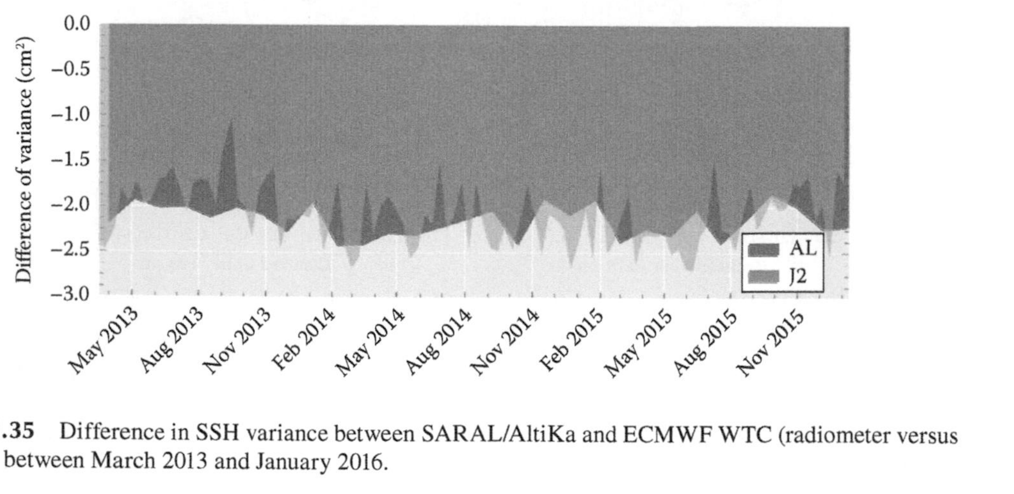

FIGURE 1.35 Difference in SSH variance between SARAL/AltiKa and ECMWF WTC (radiometer versus ECMWF) between March 2013 and January 2016.

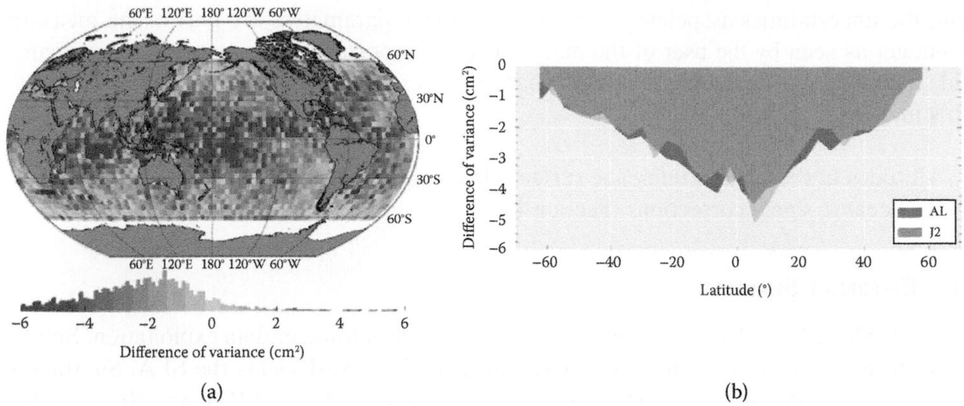

FIGURE 1.36 Difference in SSH variance between SARAL/AltiKa and ECMWF WTC (radiometer versus ECMWF) between March 2013 and January 2016. (a) Gridded averaging over a 2° × 2° map. (b) Zonal averaging.

difficult to draw conclusions about the significance of this apparent degradation because these patterns appear over regions of very high variability (e.g., Agulhas, Brazil, and Gulf Stream currents).

1.5.2.3.5 Outlook

Various features of new and future radar altimeter technology focus on improving the SSH precision or its spatial resolution down to a few kilometers or better (see Section 1.1.1). This drives the need for improved WTC retrievals at small scales (from 1 to 50 km).

From an instrumental perspective, this improvement requires the use of higher observation frequencies. Indeed, the Jason-CS radiometer is expected to add 90 GHz, 130 GHz, and 166 GHz channels to the current three LF channels. There are two benefits to this strategy: These additional channels will make it possible to capture smaller features, and they will improve the performance over coastal areas, in the detection of clouds, and in retrieval over non-oceanic surfaces (e.g., over inland waters).

Cloud detection is a potential improvement for the quality of the WTC itself and also for the radar altimeter performance because water droplets directly affect the shape of the altimeter waveforms. The accuracy of the WTC retrieval is driven by the capability of the inversion algorithm to remove the contribution from the surface and from the CLWC. The latter is challenging today due to the difficulty of taking into account clouds' contribution to the observations; convective phenomena may make a strong contribution to the measured BT but with a potential size of only a few kilometers (i.e., a small fraction of the instrument footprint). Higher measurement frequencies with smaller spatial resolution and higher sensitivity to clouds would better handle such cases.

The future of WTC estimation may also be improved with naturally flexible retrieval methods such as the one-dimensional variational (1D-VAR) approach (Desportes et al. 2010). Observed BTs are used as constraints in a converging system using ECMWF analysis profiles as first guess. The final iteration provides the atmospheric profiles that better explain the observations. Because 1D-VAR does not rely on a learning data set, it is flexible and the same approach can be applied on all surface types.

1.6 ALTIMETRY PRODUCT AUXILIARY INFORMATION: REFERENCE SURFACES, TIDES, AND HIGH-FREQUENCY SIGNAL

To allow adequate use of the altimetry measurement, one needs auxiliary information necessary to perform an efficient use of these data. Although not being a direct component of the measurement system, the uncertainties associated with these auxiliary parameters are part of the measurement error budget as seen by the user of the data. It is then important to characterize their accuracy to provide a global view on the measurement characteristics.

This includes:

- Altitudes of the various reference surfaces that can be used (Section 1.6.1)
- HF oceanic signal corrections (Section 1.6.2)

1.6.1 Reference Surfaces

Altimeter MSS and geoid are two key reference surfaces for altimeter data exploitation. Subtracting the MSS from instantaneous altimeter measurements of the SSH yields the SLA. Subtracting the geoid from the MSS yields the mean dynamic topography (MDT). MDT and SLA are the two contributions (the time-mean and the time-variable contributions, respectively) of the absolute dynamic topography from which ocean surface currents can be calculated through the geostrophic approximation.

Chapter 4 (Chambers et al.) details the contribution of altimetry to the improvement of the knowledge of these reference surfaces. In the 1990s, geoid models were based on the analysis of a satellite's orbit perturbation. Their accuracy at mesoscale exceeded one meter, which prevented any direct use of this information for ocean dynamics studies. In the 2000s, important technological and conceptual steps were made for computing the Earth's gravity field from space, which led to the launch of dedicated space gravity missions such as CHAMP (2000), GRACE (2002), and the Gravity Field and Steady-State Ocean Circulation Explorer GOCE mission (2009). The global accuracy of geoid models over the ocean at around a 100-km resolution decreased to half a meter with CHAMP data, to decimeter level with GRACE, and today reaches the centimeter level with GOCE (see Chapter 4 for further details on satellite gravity).

On the other hand, the accuracy of altimeter MSS has also greatly improved with time due to the increasing number of observations entering in the SSH average (the latest MSS are calculated from more than 20 years of data), the improvements made in POD (Section 1.4), retracking algorithms of the altimeter waveform (Section 1.2), and the various corrections applied on the altimeter SSHs (instrumental, environmental, geophysical, and surface state corrections, see Sections 1.3 and 1.5). Finally, the MSS mapping procedure has also beneficiated from numerous improvements in time (see Chapter 4). Improvements have been achieved for both the open ocean and the coastal ocean and also for partially ice-covered areas at high latitudes. Globally, the MSS accuracy in the open ocean (away from the coast and seasonally ice-covered areas) dropped from several tens of centimeters for the first solutions (which were dominated by orbit errors) to a few centimeters today.

The drastic improvements achieved for calculating both the MSS and the geoid reference surfaces had a direct impact on the resolution and accuracy of the ocean MDT (see Figure 4.10 from Chapter 4.4.2). Today, spatial scales of the ocean MDT greater than 100 km can be retrieved from the filtered difference between the MSS and the geoid with centimeter accuracy. Although this represents a true revolution compared to earlier altimeter missions, further improvement is still needed.

The 100-km resolution geoid is still insufficient for allowing direct computation of the ocean dynamic topography from along-track differences between altimeter SSH and geoid height (this would require knowing the geoid height with centimeter accuracy at a few kilometers of resolution, even less in the case of the SWOT mission, which will provide high resolution SSH measurements) or for estimating the full spectra of the MDT spatial scales. The calculation of the ocean MDT at scales smaller than 100 km is achieved by combining the altimeter and geoid data with in situ measurements of the ocean dynamic heights and surface velocities (Chapter 4.4.3). Recent global combined MDTs resolve spatial scales down to 25–50 km with 5–10 cm accuracy.

Improving the MDT resolution and accuracy will depend on the availability in the future of continuous in situ observation arrays (Argo floats, drifting buoys) together with the launch of challenging space gravity missions going beyond the performance of GRACE and GOCE.

Finally, while in the open ocean, the limiting factor for accurate high-resolution MDT calculation is the geoid; other areas would benefit from increased MSS accuracy (high latitudes, coastal zones).

1.6.2 Tides, High-Frequency Signals

The ocean is continuously changing under the effect of various physical processes involving many temporal and spatial scales (e.g., currents and tides, see Figure 1.4).

Following the Nyquist theory, the time sampling of altimeters (e.g., time revisit of 10 days for T/P-Jason altimeters) induces an aliasing of the HF ocean signal (noted as HF, which corresponds to periods less than 20 days for the T/P-Jason orbit) into the LF band (noted LF, which corresponds to periods larger than 20 days for the T/P-Jason orbit). These aliased signals include most tidal signals (short period waves in the tidal spectrum; see, e.g., Table 1.11) and the HF ocean response to atmospheric forcing. The altimetric measurements have to be corrected from these HF signals to avoid confusion between these aliased signals and the larger temporal scale signals and access to applications such as mesoscale and sub-mesoscale circulation studies, climate and long-term studies, or satellite calibration campaigns. Independent models at centimetric accuracy are necessary for those. This section presents a description of the main components of the HF corrections used in altimetry, the tide correction, and the DAC.

1.6.2.1 The Tide Correction

The tide correction for altimeter measurements includes several effects: the geocentric tide, the solid Earth tide, and the pole tide.

The solid Earth responds to external gravitational forces similarly to that of the oceans. The response of the Earth is fast enough that it can be considered to be in equilibrium with the tide-generating forces. The solid Earth tide correction is based on the tidal potential from Cartwright and Tayler (1971) and Cartwright and Edden (1973).

The pole tide is generated by small perturbations to the Earth's rotation axis primarily occurring at the Chandler wobble and annual periods. The pole tide correction has been improved recently to take into account a more accurate mean pole location and updated Love numbers (Desai et al. 2015).

TABLE 1.11
Aliasing Periods for the Main Tidal Waves for the T/P-Jason and the ERS-Envisat-SARAL/AltiKa, GFO, and Sentinel-3 Orbits

Satellite	T/T - Jason	ERS-1, 2, ENVISAT, SARAL/AltiKa	GFO	Sentinel-3
Cycle (days)	9.9156	35	17.0505	27
Darwin name	Aliasing (days)	Aliasing (days)	Aliasing (days)	Aliasing (days)
Q_1	69.4	132.8	74	229.6
O_1	45.7	75.1	113	277
P_1	88.9	365.2	4466.7	365.2
K_1^L	173.2	365.2	175.4	365.2
N_2	49.5	97.4	52.1	141
M_2	62.1	94.5	317.1	157.5
S_2	58.7	∞	168.8	∞
K_2^S	86.6	182.6	87.7	182.6

The geocentric tide correction includes the ocean tide signal forced by the lunisolar potential, and the loading tide component. It is a critical correction as tides represent a significant part of the ocean variability. The loading tide represents the ocean bottom displacement under the effect of the ocean tide loading, and it reaches up to 10% of the ocean tide signal.

Versions of global ocean tide models that can be used to perform this correction have been deeply improved since the beginning of altimetry as presented in Stammer et al. (2014). See, for example, Chapter 13 (Ray and Egbert) of this book. Models such as Godard Ocean Tide (GOT) (GOT4V10; Ray 2013) are purely empirical models fitted using altimeter measurements, while others such as Finite Element Solution (FES) (see, e.g., FES2014 [Carrere et al. 2016b; Lyard, F. personal communication, LEGOS, 2017]) are based on finite elements such as hydrodynamic modeling and data assimilation, both being barotropic models (Figure 1.37).

Loading tide solutions based on these oceanic models are available, including some nonlinear frequencies and long period components (i.e., FES2014, J. P. Boy, pers. comm., Ecole et Observatoire des Sciences de la Terre in Strasbourg France, 2015 [Carrere et al. 2016b]).

1.6.2.2 The High-Frequency Correction

The HF ocean signal forced by the atmosphere has a strong variability and is mostly located at high latitudes and in shallow water regions (Willebrand et al. 1980); it is predominately barotropic if considering large spatial scales (Vinogradova et al. 2007). As proposed by Stammer et al. (2000), an independent geophysical correction with centimetric accuracy can be used to remove this HF variability from altimeter measurements.

For that purpose, the DAC has been proposed in altimeter GDRs since 2004; it is a combination of the high frequencies of the 2D Gravity Waves mode (MOG2D)-G barotropic model forced by pressure and wind (Carrere and Lyard 2003) and the low frequencies of the inverted barometer (noted as IB), assuming a static response of the ocean to atmospheric forcing for low frequencies.

The MOG2D-G model uses finite element spatial discretization, which allows increasing resolution in areas of interest such as shallow waters and bathymetry sharp gradients. The global mesh (see, e.g., Figure 1.38) used for the operational correction has a minimum resolution of 10–15 km near the coasts to 150 km in some deep ocean regions; it also includes all marginal seas, such as the Weddell and Ross seas, Arctic Ocean, Hudson Bay, Black Sea, and Red Sea.

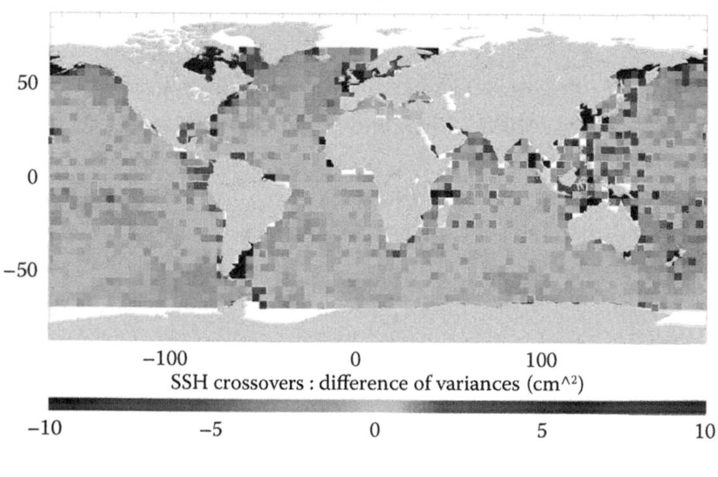

FIGURE 1.37 SSH variance differences at Jason-1 crossovers when successively using the FES2014 tide model and the GOT4v10 model in the SSH computation (cm^2).

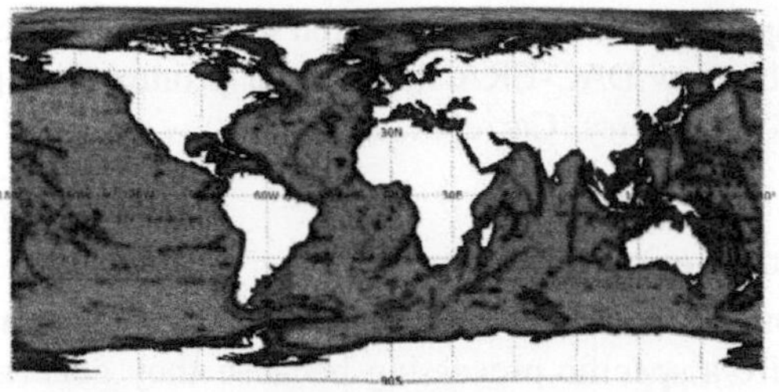

FIGURE 1.38 Finite element mesh used to compute the operational DAC.

The filtering wavelength is based on a T/P-Jason-1/Jason-2 Nyquist frequency of 20 days (twice a cycle length) because this correction is primarily a de-aliasing correction made for reference altimeter missions (Carrere and Lyard 2003).

$$DAC = MOG2D\text{-}G_{HF\,(T \le 20\text{ days})} + IB_{LF\,(T > 20\text{ days})} \tag{1.14}$$

For ERS, Envisat, and SARAL/AltiKa, the sampling Nyquist period is 70 days, which means that the DAC does not remove all atmospheric forced signals aliased in these data. For altimeter multi-mission products (AVISO 2014), remaining aliased signals are smoothed thanks to a long wavelength error correction; however, for mono-mission products such as GDRs, these signals remain aliased in lower frequency signals and can interfere with climate/seasonal variability (Ponte and Lyard 2002).

The operational Jason barotropic model is forced by the 6-h ECMWF operational analysis (sea-level pressure and 10-m winds). The DAC significantly reduced the discrepancy between ascending and descending tracks for the Jason-2 mission, as shown on Figure 1.39, and this result applies on all altimeter missions.

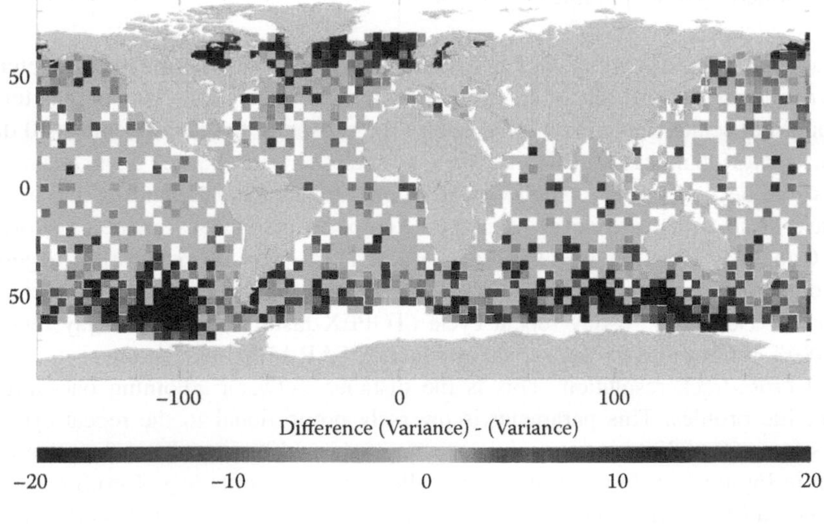

FIGURE 1.39 SSH crossover variance reduction (in red) for Cycle 301 of Jason-2 mission when using DAC instead of IB in the SSH computation (cm²).

For climate studies, a specific DAC-ERA correction has been produced using the ERA-Interim atmospheric reanalysis: This new DAC-ERA shows great promise compared to operational products from the first decade of altimetry (Carrere et al. 2016a).

1.6.2.3 S1 and S2 Atmospheric and Ocean Signals

As far as altimeter data correction is concerned, the diurnal (S1) and semidiurnal (S2) atmospheric tides (thermal tides) require a specific processing because they generate radiational tides in the ocean at the same frequencies as the diurnal and semidiurnal ocean tides. As the radiational and the gravitational components cannot be well separated from observations, both components are generally included in global ocean tide models, both FES2014 and GOT4V10. As a result, the radiational tides should not be included in the DAC correction to avoid redundancy when correcting altimetry data.

The methodology chosen to correct the operational DAC from S1 and S2 radiational tides makes it complementary with the ocean tide correction and is based on Ponte and Ray (2002) and Ray and Ponte (2003). It consists of removing S1 and S2 atmospheric pressure climatology from the DAC forcing; monthly climatology information computed from 11 years of operational ECMWF data (1993–2003) is used for the operational DAC.

As for dry troposphere correction, reanalysis such as the ERA-Interim may be used for climate studies as described in Carrere et al. (2016a).

1.7 ALTIMETRY TIME AND SPACE SAMPLING: ORBIT SELECTION AND VIRTUAL CONSTELLATION APPROACH

1.7.1 Sampling Properties of a Single Altimeter Orbit

In contrast with wide-swath imagers (e.g., SST or ocean color), measurements from radar altimeters are exceedingly anisotropic. SSHs from pulse-limited radar altimeters are one-dimensional profiles along the subsatellite track, with no SSH information in the cross-track direction. Figure 1.40 shows that, for a single altimeter flying on the TOPEX-Jason orbit, the along-track (white segment) resolution can be as small as 7 km (level 2 product, 1 Hz rate), whereas in the cross-track resolution (black segment), it can be as large as 300 km. The sampling properties of a radar altimeter are therefore controlled by three parameters:

- Repeat cycle or revisit time: This is the number of days needed for the altimeter profile to revisit the exact same location, typically within less than 1 km. This parameter defines the temporal scales that can be observed by the sensor. If the revisit time is 10 days, the Nyquist criterion makes it difficult to observe scales shorter than 20 days. It is not strictly impossible because of the complex mesh of one-dimensional ascending and descending profiles; at so-called crossover points, the temporal resolution can be as short as five days. Generally speaking, the temporal characteristics of an altimeter orbit are geographically dependent, as extensively discussed by Wunsch (1989). The largest data set collected so far is based on a 10-day repeat cycle (TOPEX-Jason series), a 17-day repeat cycle (GEOSAT series), or a 35-day repeat cycle (ERS/SARAL/AltiKa series).
- Spatial cross-track resolution: This is the distance between adjoining one-dimensional subsatellite profiles. This parameter is inversely proportional to the repeat cycle, and it ranges from 50 to 300 km for typical oceanographic orbits (e.g., Jason-3 or Sentinel-3) as shown in Figure 1.41. In contrast, the so-called geodetic orbits (e.g., CryoSat-2 or Jason-1 GM) have a very high spatial resolution on the order of 8 km due to their yearly repeat cycle.
- Inclination: This parameter defines the latitude band covered by the altimeter (i.e., how much of the global ocean can be observed).

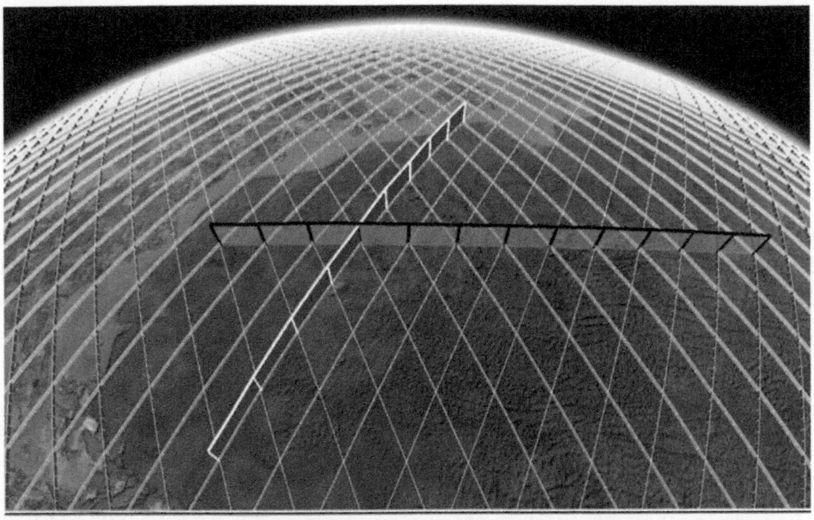

FIGURE 1.40 Ten-day sampling from an altimeter on the TOPEX-Jason orbit. The white segment highlights the along-track direction with one measurement every 7 km, and the black segment highlights the worst case configuration in the cross-track direction with one measurement every 315 km × cos(latitude). (From Dibarboure, G., et al., *J. Atmos. Ocean. Technol.*, 30, 1511–1526, 2013.)

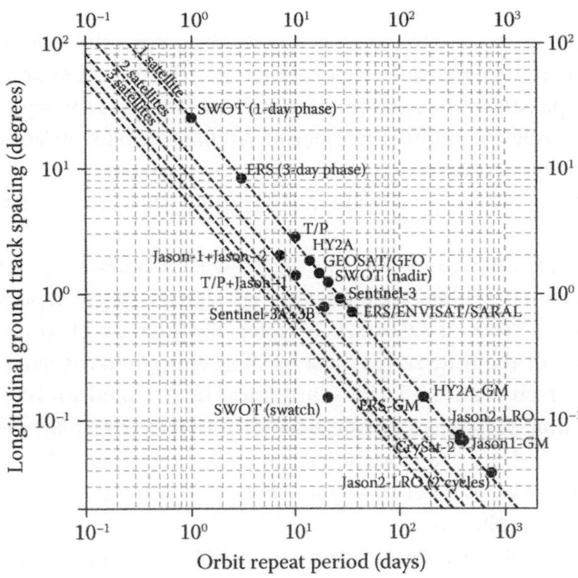

FIGURE 1.41 The relationship between orbit repeat period and the longitudinal separation of neighboring ground tracks for altimeters in exact repeat orbit configurations. The ground track spacing is displayed in degrees along the left axis and in kilometers at 40° latitude along the right axis. In log-log space, the choices of repeat period and ground track spacing for a single satellite fall approximately along the top straight line in the figure. The repeat periods and ground track separations of past and present altimeter missions are shown by the solid circles. The improvements in the resolution that would be obtained from multiple satellites in coordinated orbit configurations with evenly spaced ground tracks are shown for constellations of 2, 3, 4, and 5 satellites. Inspired from Chelton, D.B., Report of the High-Resolution Ocean Topography Science Working Group Meeting, 28-29 March 2001, College Park, MA. College of Oceanic and Atmospheric Sciences, Oregon State University, Corvallis, Oregon. 2001-4. Available at http://www-po.coas.oregonstate.edu/~poa/www-po/research/po/research/hotswg/

The sampling capability of an altimeter orbit greatly depends on the phenomena to be observed as explained by Escudier and Fellous (2008). Large and rapid atmospheric-related features can be captured by the altimetric SWH, and these measurements are assimilated in near real time by the operational wave model; they involve temporal scales shorter than 1 day and spatial scales generally larger than hundreds of kilometers. In contrast, geoid features are almost stationary in comparison with the lifespan of a satellite, so they accommodate with a yearly revisit; yet they require a spatial resolution less than 10 km. Among the intermediate space/time oceanic scales, one of the most commonly monitored is the large mesoscale. Although geographically varying, as shown by Jacobs et al. (2001), mesoscale features are on the order of 10–20 days and 100–300 km. These values explain why historical orbits have a repeat cycle on the order of 10–30 days.

1.7.2 Orbit Sub-Cycles and Sampling Properties

In theory, a single repeat cycle should not be able to accommodate such vast differences in terms of space/timescale requirements. In practice, the repeat-cycle grid (e.g., Figure 1.40) is generally not sampled linearly but through interleaved lower resolution grids. For the T/P and Jason orbit, the full grid is 300 km for a 10-day cycle, and it is composed of sub-cycles with a 3-day and 1000 km resolution (Figure 1.40). From one sub-cycle to the next, the low resolution grid is "moving" or "drifting." For T/P, the grid of the second 3-day subperiod mostly looks like a 300-km translation of the 1000-km grid of the first subperiod. In the case of very long repeat cycles (e.g., geodetic orbits), there can be four sub-cycles or more. This phenomenon is explained by Rees (1992), and practical consequences for oceanography are discussed by Dibarboure et al. (2012) or Dibarboure and Morrow (2016).

Thanks to the composition of sub-cycles, a single altimeter orbit can accommodate multiple space/timescales and, therefore, multiple applications. To illustrate, Dibarboure et al. (2012) have shown that it could be possible to use a geodetic orbit with a yearly repeat cycle and still have mesoscale-friendly sub-cycles to monitor the ocean circulation plus an additional short sub-cycle more suited for sea-state applications.

Yet most altimeter orbits use a repeat cycle ranging from 10 to 35 days for two reasons:

- Some altimetry applications require time series with a precise co-location. To illustrate this, hydrologists built so-called virtual hydrology stations at given points where the altimeter profile crosses a river or lake (see Chapter 14, Cretaux et al.). In that context, the temporal resolution of the time series must be short enough to resolve monthly to seasonal phenomena (e.g., Arsen et al. 2015). Similarly, most tide-related studies use harmonic analysis of the cyclic time series to separate tide constituents (e.g., Cartwright and Ray 1990; Stammer et al. 2014). Using a geodetic repeat cycle longer than 1 month would increase the spatial coverage (i.e., number of virtual stations) but make them irrelevant because of a very low temporal resolution.
- Some altimetry applications require a precise reference surface (sometimes called a mean profile) in order to get anomalies with respect to this surface. With a short revisit time, this reference can be built rapidly, such as for charted altimeter tracks (Jason or ERS series). In contrast, when the revisit time is very long, such as with CryoSat-2, it is not possible to build a local along-track mean profile. Instead an external and global gridded reference must be used, and this reference is not as accurate (see Chapter 18, Le Traon et al.).

To that extent, using an orbit with a short repeat cycle contributes to improving the precision of the altimetry topography. For instance, this is necessary to measure mesoscale SLAs (e.g., Dibarboure et al. 2012; Pujol et al. 2016) or for topography of rugged continental ice sheets (e.g., Remy et al. 2015).

One should note the recent and major improvements of mean surfaces built from CryoSat-2 or Jason-1 GM geodetic data sets (e.g., Sandwell et al. 2014; see also Chapter 16, Müller et al.). Thanks to upcoming and future geodetic data sets (CryoSat-2, SARAL/AltiKa Drifting Phase, HY-2A GM, tentative Jason-2 geodetic phase) one can expect to ultimately get the same order of accuracy for all altimetry orbits.

A recent and notable change observed with altimetry orbits is that when the platform is aging, the satellite is sometimes moved to a different orbit (sometimes called Extension of Life, or EoL) in order to protect the historical orbit and follow-on mission from potential collision with an old satellite that would have failed while flying this orbit (e.g., T/P-Jason series, ERS-Envisat-SARAL/AltiKa). This happened for Jason-1, Envisat, SARAL/AltiKa, and a preemptive plan has been defined for Jason-2. The consequence is that, in addition to a traditional coordinated set of altimeters, additional and older sensors might be able to complement the constellation from a new set of EoL orbits.

These so-called EoL orbits may sometimes be suboptimal in terms of coordination (e.g., as discussed by Dibarboure et al. 2012, for Jason-1) but still provide unique scientific benefits (e.g., Dibarboure and Morrow 2016). In this context, the SARAL/AltiKa Drifting Phase is unique because the satellite altitude is no longer maintained; it decreases slowly due to uncontrolled atmospheric drag. Still, the SARAL/AltiKa-DP altitude range was selected to guarantee an excellent mesoscale sampling capability for 5–7 years despite the unprecedented uncontrolled nature of this new orbit.

1.7.3 Altimeter Virtual Constellation and Phasing

The geometry of altimeter products (one-dimensional profile) sometimes makes it complex to analyze some features of the oceanic circulation. To that extent, more convenient two-dimensional gridded fields of SLAs are reconstructed from the global sampling of one-dimensional profiles (e.g., Leben et al. 2002; Dibarboure et al. 2011; Pujol 2016). The resolution of two-dimensional altimetry maps is, however, limited by the number of altimeters in operation because the anisotropy of SSH profiles is by far the most limiting factor of gridded SSH mesoscale fields (e.g., Dibarboure et al. 2013). Chelton and Schlax (2003) and Le Traon and Dibarboure (2002) have shown that mesoscale maps have a limited global resolution capability related to the altimeter constellation size and coordination.

Two coordinated altimeters is the minimum required to monitor the global large ocean circulation in delayed time (e.g., Le Traon and Dibarboure 2004; Chelton et al. 2011). For near real time application and operational oceanography, having four altimeters is very important because measurements "from the map's future" are not yet available (Pascual et al. 2008). Still, from the perspective of the operational forecasts, the key limitation at present is sufficient data to constrain features in the ocean models. Jacobs et al. (2014) report that with one data stream the general eddy positions are forecast, and skill measured by RMS SSH accuracy degrades very quickly after one data stream. However, with only one data stream, errors in the frontal positions of eddies are so large that there is no skill in determining the location of the eddy fronts. With three altimeters, operational models begin to see skill in predicting the location of eddy fronts, though errors are typically on the order of 50 to 100 km. There is linearly increasing skill in parameters related to frontal positions even out to four good altimeters.

For geodetic applications (see Chapter 16, Müller et al.), the sampling need is defined by the resolution achieved when all cumulated records are aggregated. CryoSat-2 and Jason-1 GM (Figure 1.41) have provided new non-repeat data. The wavelength resolution of the current gravity models is approximately 12 km (i.e., features on the order of 6 km). To significantly improve bathymetry fields, the spacing between geodetic tracks must now become smaller than 6 km. This is not entirely possible with CryoSat-2 alone due to its irregular repeat orbit (approximately 8-km resolution) and poor accuracy in the east–west direction nor with Jason-1 GM due to its wide track spacing. The only way to further reduce the MSS errors is to collect a denser and longer geodetic time series. The tentative geodetic phase of Jason-2 EoL and follow-on missions could be instrumental because they can provide a 1- to 4-km grid with good east–west accuracy after 2 years of geodetic phase (as opposed to the irregular tracks and poor east–west accuracy for CryoSat-2 and coarser accuracy for the SARAL/AltiKa Drifting Phase).

To a lower extent, the optimization (i.e., coordination) among concurrent altimeter satellites can substantially improve the sampling capability of a constellation for a given number of altimeters (Le Traon and Dibarboure 2002, 2004). The interleaved configuration between TOPEX and Jason-1 or more recently Jason-2 and Jason-3 is an example of excellent two-satellite coordination. The coordination between Sentinel-3A and Sentinel-3B discussed by Dibarboure and Lambin (2015) shows the challenge to optimize the tandem when three sensors (altimetry, SST, and ocean color) involve very different space/timescales (Figure 1.42).

1.8 ALTIMETRY ERROR BUDGET

This section discusses the global error budget of SSH derived from altimetry measurements. An elementary error budget is associated with individual measurements from the level-2 product that is delivered to users at a 1-Hz rate (which is equivalent to roughly 7-km spacing at the Earth's surface, varying slightly with satellite altitude). Depending on the space/timescales of interest, a user-oriented error budget can be estimated, taking into account the filtering that can be applied to smooth both unneeded scales and the associated elementary errors. Section 1.8.1 presents the error budget associated with mesoscale analysis; Section 1.8.2 discusses the error budget associated with large-scale MSL trend studies. A focus is also placed on altimetry errors at ocean scales smaller than 100 km (Section 1.8.3).

1.8.1 Error Budget for Mesoscale Oceanography

From 1993 onward, altimetry data provide very good monitoring of large mesoscale oceanic processes (Pujol et al. 2016) thanks to an adequate spatiotemporal data sampling with at least two altimetry missions in orbit simultaneously and very accurate altimetry measurements. In this section, we discuss the error budget of altimetry data for mesoscale oceanography with scales ranging from 100 to 300 km over a time period between 10 and 30 days.

The estimation of the budget error is carried out from different metrics. The most often used are: (1) spectral analyses of 1 Hz and 20 Hz measurements in order to evaluate the data noise of sea-level and other potential errors, SWH, or altimeter ionospheric correction; (2) sea-level crossover

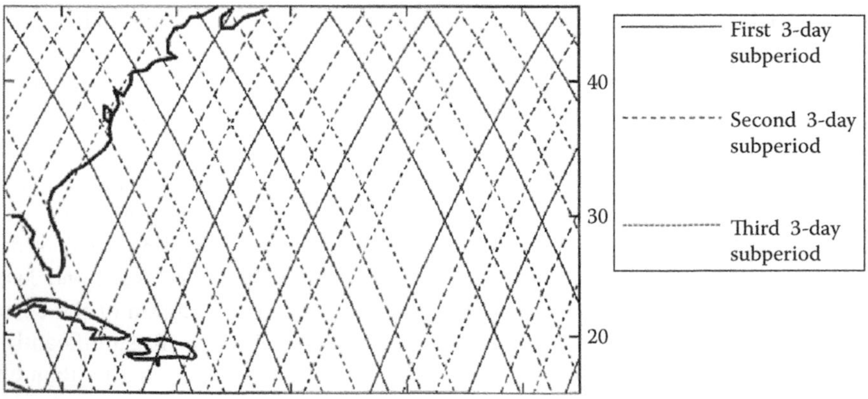

FIGURE 1.42 TOPEX-Jason sub-cycles. (From Greenslade, D. J. M., et al., *J. Atmos. Ocean. Tech.*, 14(8), 849–870, 1997).

analyses in order to evaluate the relative errors among different corrections for the same phenomenon (e.g., ocean tide) or orbit solutions applied in the SSH estimate. For the Jason missions, the operational phase begins in a tandem configuration between the satellite recently launched and its predecessor. These phases are very useful to cross-calibrate the two missions and evaluate the measurement uncertainties. During these specific phases, T/P and Jason-1 (2002), Jason-1 and Jason-2 (2008), and then Jason-2 and Jason-3 (2016) are spaced out by roughly 1 min, so that the same portions of the ocean are observed by both missions. The analysis of SSH differences allows the estimation of uncorrelated errors between both missions.

Table 1.12 presents the error budget estimated on an elementary SSH measurement for the Jason-2 missions at a global scale over the open ocean and for timescales shorter than 10 days (Jason-2 validation annual report 2013). The uncertainties for the altimeter range, the radial orbit component, and the main corrections used in the sea-level calculation are estimated separately.

For the altimeter range, the error mainly originates in random noise of about 1.6 cm to 1.7 cm for SWH of 2 m. This error increases with SWH (Zanife et al. 2003). It depends on the instrumental performance and also the measurement estimation (see Section 1.2). The noise variance for scales ranging from 100 to 300 km is generally small with respect to the amplitude of the oceanic signal. Other errors due to the footprint contamination (e.g., rain cells and sigma blooms) may impact the product in various regions (Dibarboure al. 2014). But their impact is most significant for scales shorter than 100 km (see Section 1.8.3).

Regarding geophysical correction, specific uncertainties have been reported: 0.2 cm for the altimeter ionosphere corrections after a 300-km smoothing and 0.4 cm for the SSB. Concerning the POD, the error of the radial orbit component at timescales less than 10 days is approximately 1.5 cm (Couhert et al. 2015). It is worth noting that this uncertainty increases up to 1.7 cm for near-real time orbit solutions. But the POD error essentially affects scales much larger than ocean mesoscale; to that extent, it essentially behaves as a bias for the altimeter track segment.

For the wet troposphere correction derived from microwave radiometers, the uncertainty is close to 0.2 cm on average (Thao et al. 2014), with larger errors in coastal areas. Using an atmospheric model instead, the WTC uncertainty increases to 1 cm on average, with higher errors in the tropical band (Legeais et al. 2014). The dry atmosphere and dynamical atmospherical corrections derived from atmospheric models have an uncertainty of approximately 0.7 cm (Carrere et al. 2016a), with larger errors at high latitudes where atmosphere variability (pressure, wind) is strong.

TABLE 1.12
Global Jason-2 Error Budget at Global Scale and for Short Timescales

Parameters		**Altimetry Uncertainties (cm)**
Parameters and correction for sea-surface height	Altimeter range	1.7 (noise)
	Filtered-out altimeter ionosphere correction	0.2
	Sea state bias	0.2
	Dry troposphere and dynamical atmospheric corrections	0.7
	Radiometer wet troposphere	0.2
	Ocean tide	1.0
	Orbit (radial component)	1.5
Sea surface height	Corrected with all the corrections	<3.5

Finally, the uncertainty induced by the ocean tide model is approximately 1 cm (Ray 2013; Stammer et al. 2014) in the open ocean, with larger errors in coastal areas and at high latitudes.

The total uncertainty on sea level at meso and larger scales is deduced from crossover analyses. The variance of sea-level differences between ascending and descending tracks yields an estimate of the error at a short timescale (less than 10 days). Assuming that the errors are the same on both ascending and descending tracks, the Jason-2 sea-level uncertainty has been estimated as 3.5 cm (see Table 1.12). Although this estimate does not account for correlated errors or systematic biases that are common to ascending and descending tracks, its value is usually considered as an upper-bound of the uncertainty estimation because a fraction of this number originates in the ocean variability (the true SSH changes over the crossover time difference).

A similar error budget at mesoscale was computed for other altimeters. The results are the same as Jason-2 for Jason-1 and Jason-3 (similar design). It slightly changes for missions in different orbits and with different instrumental characteristics (e.g., Envisat and SARAL/AltiKa). In all cases, the sea-level uncertainty ranges from 3.5 to 3.7 cm (Ollivier et al. 2012; Prandi et al. 2015).

1.8.2 Error Budget for Mean Sea Level Trend Monitoring

Altimetry data provide accurate climate data records from 1993 to 2016, essentially thanks to the so-called reference mission. The global MSL rise estimate ranges from 3.0 to 3.3 mm/year (see Chapter 5, Nerem et al.) whereas the regional trends range from −5 to 10 mm/year over this 24-year period. However, global and regional MSL time series are impacted by errors at different timescales.

An error budget dedicated to the main temporal scales (i.e., long-term: 5–10 years or more, interannual: less than 5 years, and seasonal) has been established by Ablain et al. (2015), and it is given in Table 1.13. Regarding the Global Mean Sea Level (GMSL) trend, an uncertainty of 0.5 mm/year was estimated over the whole altimetry era (1993 to 2015) within a confidence interval of 90%. On a regional scale, the regional trend uncertainty is on the order of 2–3 mm/year. A map of uncertainties on sea-level trend has also been estimated by Prandi et al. (2017), highlighting the regions where the regional sea-level trend estimations are higher than the uncertainty estimation. One should note that for both the GMSL trend and regional sea-level trends, uncertainties remain higher than the Global Climate Observing System requirements (GCOS 2011) of 0.3 mm/year for the GMSL trend and 0.5 mm/year for the regional MSL trend.

1.8.3 Error Budget for Sub-Mesoscale

The SSH measurement provided by most of the conventional altimeters in the so-called LRM mode (see Section 1.2) does not allow the observation of ocean features smaller than 80–100 km (Dibarboure et al. 2014; Dufau et al. 2016). This limitation is mainly due to the surface instrumental noise and an

TABLE 1.13
Mean Sea-Level Error Budget for the Main Climate Scales

Spatial Scales	Temporal Scales	Altimetry Uncertainties	User Requirements
GMSL	Long-term evolution (>10 years)	<0.5 mm/year	0.3 mm/year
	Interannual signals (<5 years)	<2 mm over 1 year	0.5 mm over 1 year
	Annual signals	<1 mm	Not defined
Regional sea-level	Long-term evolution (>10 years)	<3 mm/year	1 mm/year
	Annual signals	<1 cm	Not defined

Source: Ablain, M., et al., *Ocean Sci.*, 11, 67–82, 2015.

imperfect handling of heterogeneities in the altimeter footprint (e.g., rain, sigma blooms). The heterogeneities are captured in the radar echo but not modeled in the retracking process.

A usual approach to quantify the error budget at small scales (less than 100 km) is to perform spectral analyses of the corrected sea-level anomalies. Such spectra based on Fourier transform are plotted in Figure 1.43 for Jason-2 and SARAL/AltiKa. The noise level of each altimeter is easily deduced from the HF plateau of both missions. The white noise level is estimated at 7.3 cm for Jason-2 and 5.4 cm for SARAL/AltiKa for SWH close to 2.7 m on average. This difference between missions mainly results from AltiKa instrument characteristics (see Section 1.3.4). The noise level prevents the observation of small oceanic structures for length scales (on average) close to 50 km for SARAL/AltiKa and 60 km for Jason-2. This length scale is defined by the ratio between the oceanic signal slope and HF plateau equal to 1. This corresponds in Figure 1.43 to the X-axis of the intersection between the HF plateau and the oceanic slope (Dufau et al. 2016).

Furthermore, a large energy bump ranging from 5 to 100 km is observed on all LRM altimeters (e.g., Figure 1.43 for Jason-2 and AltiKa). Dibarboure et al. (2014) showed it is likely due to the model and retracker's inability to account for surface heterogeneities in the LRM footprint (e.g., Brown echo models assume a statistically homogeneous response of the sea surface). Comparison between observed and expected (i.e., derived from oceanic slope and noise level) SLA spectra (green and blue dash spectra in Figure 1.43) yield a description of the error: at 80 km, about 50% of additional energy is measured. Therefore, the addition of this error and the white noise prevents the observation of small oceanic scales lower than 80 to 100 km (on average) for all the LRM missions. Dufau et al. (2016) explored the spatial and seasonal variability of the observing capability for small scales.

Data processing algorithms have a strong impact on SSH errors at small scales (i.e., choice of retracking algorithms to remove spurious sea-level measurements). Delay-Doppler altimetry (or SAR mode) currently exhibits a good resilience to these effects. To that extent, the global SAR mode coverage provided by the Sentinel-3A mission (launched in February 2016) should significantly improve small-scale SSH monitoring.

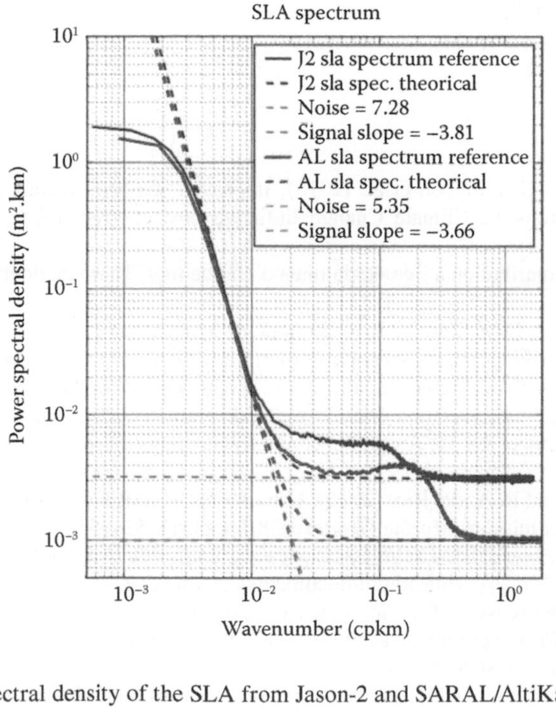

FIGURE 1.43 Power spectral density of the SLA from Jason-2 and SARAL/AltiKa (20 Hz and 40 Hz measurements, respectively).

GLOSSARY

DAC: dynamic atmospheric correction
ECMWF: European Center for Medium-Range Weather Forecasts
EM: electromagnetic
GDR: Geophysical Data Record
GNSS: Global Navigation Satellite System
GPS: Global Positioning System
HF: high frequency
LF: low frequency
LRA: laser retroreflector array
LRM: low resolution mode
MDT: mean dynamic topography
MSL: mean sea level
MSS: mean sea surface
PD: path delay
POD: Precise Orbit Determination
PRF: pulse repetition frequency
RMS: root mean square
SAR: Synthetic Aperture Radar
SLA: sea level anomaly
SLR: satellite laser ranging
SNR: signal-to-noise ratio
SSB: sea state bias
SSH: sea surface height
SST: sea surface temperature
SWH: significant wave height
T/P: TOPEX/Poseidon
TVG: time-variable gravity
WTC: wet tropospheric correction
WV: water vapor

REFERENCES

Ablain, M., A. Cazenave, G. Larnicol, et al. (2015). Improved sea level record over the satellite altimetry era (1993–2010) from the Climate Change Initiative project. *Ocean Sci.*, 11, 67–82. doi: 10.5194/os-11-67-2015.

Amarouche, L. (2001). Contribution à l'étude du biais d'état de mer. Thèse de doctorat, Paris 7, Paris, France, 2001.

Amarouche, L., P. Thibaut, O. Z. Zanife, J.-P. Dumont, P. Vincent, and N. Steunou. (2004). Improving the Jason-1 ground retracking to better account for attitude effects. *Marine Geodesy*, 27(1–2), 171–197.

Amarouche, L., L. Zawadzki, A. Vernier, et al. (2014). Reduction of the Sea Surface Height spectral hump using a new Retracker decorrelating ocean estimated parameters (DCORE). In: OSTST 2014, Constanz Lake, Germany. October 28-31, Available at: https://meetings.aviso.altimetry.fr/programs/program-by-abstract-type.html

Arsen, A., J. F. Crétaux, and R. A. del Rio. (2015). Use of SARAL/AltiKa over mountainous lakes, intercomparison with Envisat mission. *Marine Geodesy*, 38(Suppl. 1), 534–548.

AVISO. (2014). *A new version of SSALTO/DUACS products available in April 2014*. CNES Tech. Rep., 32 pp. https://www.aviso.altimetry.fr/fileadmin/documents/data/duacs/Duacs2014.pdf

Barrick, D. (1972). Remote sensing of the sea state by radar. In: V. E. Derr (ed.), *Remote sensing of the troposphere*. U.S. Govt. Printing Office, Washington, DC. https://www.researchgate.net/publication/3991543_Remote_sensing_of_sea_state_by_radar

Barrick, D. E., and B. J. Lipa. (1985). Analysis and interpretation of altimeter sea echo. *Adv. Geophys.*, 27, 61–100.

Barry, R. (1993). The GEOSAT Follow-On (GFO) program satellite. In: Space Programs and Technologies Conference and Exhibit, Huntsville, AL, September 21–23, AIAA-1993-4106. https://arc.aiaa.org/doi/abs/10.2514/6.1993-4106

Benoit, A. (1968). Signal attenuation due to neutral oxygen and water vapor, rain and clouds. *Microw. J.*, 11, 73–80.

Benveniste, J., M. Roca, G. Levrini, et al. (2001). *The radar altimetry mission: RA-2, MWR, DORIS, and LRR.* ESA Bulletin, No 106, p. 67. Available at: www.esa.int/esapub/bulletin/bullet106/bul106_5.pdf

Bertiger, W. I., Y. E. Bar-Sever, E. J. Christensen, et al. (1994). GPS precise tracking of TOPEX/Poseidon: Results and implications, *J. Geophys. Res. Oceans*, 99(C12), 24449–24464. doi: 10.1029/94JC01171.

Bertsekas, D. P. (1995). *Nonlinear programming*. Athena Scientific, Belmont, MA, 1995.

Biancale, R., G. Balmino, J.-M. Lemoine, et al. (2000). A new global Earth's gravity field model from satellite orbit perturbations: GRIM5-S1. *Geophys. Res. Lett.*, 27(22), 3611–3614. doi: 10.1029/2000GL011721.

Biancale, R., and A. Bode. (2006). *Mean annual and seasonal atmospheric tide models based on 3-hourly and 6-hourly ECMWF surface pressure data.* GFZ, Germany, Scientific Technical Report STR06/01.

Boy, F., Desjonquères, J-D., Picot, N., Moreau, T., Raynal, M. (2016). "*CRYOSAT-2 SAR Mode Over Oceans: Processing Methods, Global Assessment and Benefits*", IEEE Transaction Geoscience Remote Sensing, pp 1–11.

Bretherton and Woods, et al. (1986). *Scientific plan for the World Ocean Scientific Experiment.* WCRP Publication Series N° 6, WMO /TD-122.

Boukabara, S. A., L. Eymard, C. Guillou, D. Lemaire, P. Sobieski, and A. Guissard. (2002). Development of a modified two-scale electromagnetic model simulating both active and passive microwave measurements: Comparison to data remotely sensed over the ocean. *Radio Sci.*, 37, 1063. doi: 10.1029/1999RS002240.

Brown, G. (1977). The average impulse response of a rough surface and its applications. *IEEE Trans. Antenn. Propag.*, 25(1), 67–74.

Brown, S. (2013). Maintaining the long-term calibration of the Jason-2/OSTM advanced microwave radiometer through intersatellite calibration. *IEEE Trans. Geosci. Rem. Sens.*, 51(3), 1531–1543.

Brown, S. T., S. Dessai, W. Lu, and A. B. Tanner. (2007). On the long-term stability of microwave radiometers using noise diodes for calibration. *IEEE Trans. Geosci. Rem. Sens.*, 45(7), 1908–1920.

Callahan, P., and E. Rodriguez. (2004). Retracking of Jason-1 data. *Marine Geodesy*, 27(3), 391–407.

Carrere, L. (2005). *Rapport d'étude CNES/CLS.* Reference CLS-DOS-NT-05-007.

Carrere, L., Y. Faugère, and M. Ablain. (2016a). Major improvement of altimetry sea level estimations using pressure derived corrections based on ERA-interim atmospheric reanalysis. *Ocean Sci. Discuss.* doi: 10.5194/os-2015-112.

Carrere, L., and F. Lyard. (2003). Modeling the barotropic response of the global ocean to atmospheric wind and pressure forcing—Comparisons with observations. *Geophys. Res. Lett.*, 30. doi: 10.1029/2002GL016473. http://onlinelibrary.wiley.com/doi/10.1029/2002GL016473/abstract

Carrere, L., F. Lyard, M. Cancet, D. Allain, A. Guillot, and N. Picot. (2016b). Final version of the FES2014 global ocean tidal model including a new loading tide solution. In: OSTST 2016, La Rochelle, France. November 1-3, Available at: https://meetings.aviso.altimetry.fr/programs/program-by-abstract-type.htm

Cartwright, D., and A. C. Eden. (1973). Corrected table of tidal harmonics. *Geophys. J. Roy. Astron. Soc.*, 33, 253–264.

Cartwright, D.E., and R.D. Ray. (1990). Oceanic tides from GEOSAT altimetry. *J. Geophys. Res.*, 95, 3069–3090.

Cartwright, D. E., and R. J. Taylor. (1971). New computations of the tide-generating potential. *Geophys. J. Roy. Astron. Soc.*, 23, 45–74.

Challenor, P., M. Srokosz, and R. T. Tokmakian. (1990). Maximum likelihood estimation for radar altimetry. In: IEE Colloquium on Monitoring the Sea, Pages 10/1-10/3, December 18.

Chambers, D. P., S. A. Hayes, J. C. Ries, and T. J. Urban. (2003). New TOPEX sea state bias models and their effect on global mean sea level. *J. Geophys. Res.*, 108(10), 3305. doi: 10.1029/2003JC001839.

Chelton, D. B., "The sea state bias in altimeter estimates of sea level from collinear analysis of TOPEX data", *J. Geophys. Res*, 99, 24,995–25,008,1994.

Chelton, D. (ed.). (2001). *Report of the high-resolution ocean topography science working group meeting.* College of Oceanic and Atmospheric Sciences, Oregon State University. http://www-po.coas.oregonstate.edu/~poa/www-po/research/po/research/hotswg/

Chelton, D. B., J. C. Ries, B. J. Haines, L.-L. Fu, and P. S. Callahan. (2001). Satellite altimetry. In: L.-L. Fu and A. Cazenave (eds.), *Satellite altimetry and earth sciences: A handbook of techniques and applications*, vol. 69, pp. 1–131, International Geophysics Series. Academic Press. http://www-po.coas.oregonstate.edu/~poa/www-po/research/po/research/hotswg/

Chelton, D. B., and M. G. Schlax. (2003). The accuracies of smoothed sea surface height fields constructed from tandem altimeter datasets. *J. Atmos. Oceanic Technol.*, 20, 1276–1302.

Chelton, D. B., M. G. Schlax, and R. M. Samelson. (2011). Global observations of nonlinear mesoscale eddies. *Prog. Oceanogr.*, 91, 167–216.

Choi, K.-R., J. C. Ries, and B. D. Tapley. (2004). Jason-1 precision orbit determination by combining SLR and DORIS with GPS tracking data. *Marine Geodesy*, 27(12), 319–331. doi: 10.1080/01490410490465652.

Colombo, O. (1989). The dynamics of Global Positioning System orbits and the determination of precise ephemerides. *J. Geophys. Res. Atmos.*, 94(B7). doi: 10.1029/JB094iB07p09167. pp. 9167–9182.

Couhert, A., L. Cerri, and J.-F. Legeais. (2015). Toward the 1 mm/y stability of the radial orbit error at regional scales. *Adv. Space Res.*, 55(1), 2–23. doi: 10.1016/j.asr.2014.06.041.

Cruz-Pol, S., C. Ruf, and S. Keihm. (1998). Improved 20–32 GHz atmospheric absorption model. *Radio Sci.*, 33(5), 1319–1333.

Davis, C. (1995). Growth of the Greenland ice sheet: A performance assessment of altimeter retracking algorithms. *IEEE Trans. Geosci. Rem. Sens.*, 33(5), 1108–1116.

Davis, C. (1997). A robust threshold retracking algorithm for measuring ice-sheet surface elevation change from satellite radar altimeters. *IEEE Trans. Geosci. Rem. Sens.*, 35(4), 974–979.

Desai, S., J. Wahr, and B. Beckley. (2015). Revisiting the pole tide for and from satellite altimetry. *J. Geod.*, 89, 1233. doi: 10.1007/s00190-015-0848-7.

Desportes, C., E. Obligis, and L. Eymard. (2010). One-dimensional variational retrieval of the wet tropospheric correction for altimetry in coastal regions. *IEEE Trans. Geosci. Rem. Sens.*, 48(3 Pt 1), 1001–1008. doi: 10.1109/TGRS.2009.2031494.

Dibarboure, G., F. Boy, J. D. Desjonqueres, et al. (2014). Investigating short-wavelength correlated errors on low-resolution mode altimetry. *J. Atmos. Oceanic Technol.*, 31, 1337–1362. doi: 10.1175/JTECH-D-13-00081.1.

Dibarboure, G., and J. Lambin. (2015). Monitoring the ocean surface topography virtual constellation: Lessons learned from the contribution of SARAL/AltiKa. *Marine Geodesy*, 38(Suppl. 1), 684–703.

Dibarboure, G., P. Y. Le Traon, and N. Galin. (2013). Exploring the benefits of using CryoSat's cross-track interferometry to improve the resolution of multi-satellite mesoscale fields. *J. Atmos. Oceanic Technol.*, 30, 1511–1526.

Dibarboure, G., and R. Morrow. (2016). Value of the Jason-1 geodetic phase to study rapid oceanic changes and importance for defining a Jason-2 geodetic orbit. *J. Atmos. Ocean. Technol. https://doi.org/10.1175/JTECH-D-16-0015.1*

Dibarboure, G., M.-I. Pujol, F. Briol, et al. (2011). Jason-2 in DUACS: First tandem results and impact on processing and products. *Marine Geodesy*, 34(3–4), 214–241.

Dibarboure, G., Schaeffer, P., Escudier, P., et al. (2012). Finding desirable orbit options for the "extension of life" phase of jason-1. *Marine Geodesy*, 35(Suppl. 1), 363–399.

Donlon, C., B. Berruti, J. Frerick, et al. (2011). The sentinel-3 mission overview. In: *Proceedings of the 2011 EUMETSAT Meteorological Satellite Conference*, 5–9 September, Oslo, Norway. https://www.eumetsat.int/website/wcm/idc/idcplg?IdcService=GET_FILE&dDocName=ZIP_CONF_2011_PRESENTATIONS&RevisionSelectionMethod=LatestReleased&Rendition=Web

Dow, J. M., R. E. Neilan, and C. Rizos. (2009). The International GNSS Service (IGS) in a changing landscape of Global Navigation Satellite Systems. *J. Geodesy*, 83(3–4), 191.

Dubois, P. (2011). Impact de l'état de mer sur la mesure d'élévation des futurs instruments altimétriques. Thèse de doctorat de l'Université Paul Sabatier, Toulouse, France, 2011.

Dufau, C., M. Orsztynowicz, G. Dibarboure, R. Morrow, and P.-Y. Le Traon. (2016). Mesoscale resolution capability of altimetry: Present and future. *J. Geophys. Res. Oceans*, 121, doi: 10.1002/2015JC010904.

Dumont, J. P. (1985). Estimation optimale des paramètres altimétriques des signaux radar Poséidon. PhD thesis, Institut National Polytechnique de Toulouse, Toulouse, France.

Elfouhaily, T., B. Chapron, K. Katsaros, and D. Vandemark. (1997). A unified wave spectrum for long and short wind-driven waves. *J. Geophys. Res.*, 102, 15781–15796.

Elfouhaily, T., C. Thompson, C. Bertrand, and D. Vandemark. (2000). Improved electromagnetic bias theory. *J. Geophys. Res. Oceans*, 105(C1), 1299–1310.

Elfouhaily, T., C. Thompson, C. Bertrand, and D. Vandemark. (2001). Improved electromagnetic bias theory: Inclusion of hydrodynamic modulations. *J. Geophys. Res. Oceans*, 106(C3), 4655–4664.

Ellison, W. J., S. J. English, K. Lamkaouchi, et al. (2003). A comparison of ocean emissivity models using the Advanced Microwave Sounding Unit, the Special Sensor Microwave Imager, the TRMM Microwave Imager, and airborne radiometer observations. *J. Geophys. Res.*, 108, 4663. doi: 10.1029/2002JD003213.

Escudier, P., and J.-L. Fellous. (2008). *The next 15 years of satellite altimetry: Ocean surface topography constellation user requirements document.* Report of the Ocean Surface Topography Virtual Constellation (OST VC) at the 27th Committee on Earth Observation Satellites (CEOS) SIT Meeting. http://ceos.org/images/OST/SatelliteAltimetryReport_2009-10.pdf

Eymard, L., E. Obligis, N. Tran, F. Karbou, M. Dedieu, and A. Pilon. (2005). Long term stability of ERS-2 and TOPEX microwave radiometer in-flight calibration. *IEEE Trans. Geosci. Rem. Sens.*, 43, 1144–1158.

Feng, H., S. Yao, L. Li, N. Tran, D. Vandemark, and S. Labroue. (2010). Spline-based nonparametric estimation of the altimeter sea state bias correction. *IEEE Geosci. Rem. Sens. Lett.*, 7(3), 577–581. doi: 10.1109/LGRS.2010.2041894.

Fernandes, M. J., C. Lázaro, M. Ablain, and N. Pires. (2015). Improved wet path delays for all ESA and reference altimetric missions. *Rem. Sens. Environ.*, 169, 50–74.

Förste, C., F. Flechtner, R. Schmidt, et al. (2005). A new high resolution global gravity field model derived from combination of GRACE and CHAMP mission and altimetry/gravimetry surface gravity data. In: EGU General Assembly 2005, Vienna, Austria, 24–29 April. http://op.gfz-potsdam.de/grace/results/grav/g004_eigen-cg03c.html

Francis, C. R., G. Graf, P. G. Edwards, et al. (1991). *The ERS-1 spacecraft and its payload.* ESA Bulletin No 65, pp. 27–48. http://www.esa.int/About_Us/ESA_Publications/ESA_i_Bulletin_i_65_February_1991

Fu, L.-L., D. Alsdorf, E. Rodriguez, et al. (2009). The SWOT (Surface Water and Ocean Topography) mission: Spaceborne radar interferometry for oceanographic and hydrological applications. In: OCEANOBS'09 Conference. https://swot.oceansciences.org/docs/oceanobs09_swot.pdf

Fu, L.-L., and A. Cazenave. (2000). *Satellite altimetry and earth science*. International Geophysics Series, vol. 69. https://www.elsevier.com/books/satellite-altimetry-and-earth-sciences/fu/978-0-12-269545-2

Fu, L.-L., and M. Lefebvre. (1989). TOPEX/Poseidon: Precise measurement of sea level from space. *In: New satellite missions for solid earth missions, CSTG Bulletin No.* 11, pp. 51–54.

Garcia, E. S., D. T. Sandwell, and W. H. F. Smith. (2014). Retracking CryoSat-2, Envisat and Jason-1 radar altimetry waveforms for improved gravity field recovery. *Geophys. J. Int.*, 196(3), 1402–1422.

Gaspar, P., and J.-P. Florens. (1998). Estimation of the sea state bias in radar altimeter measurements of sea level: Results from a new nonparametric method. *J. Geophys. Res.*, 103(C8), 15803–15814.

Gaspar, P., S. Labroue, F. Ogor, G. Lafitte, L. Marchal, and M. Rafanel. (2002). Improving nonparametric estimates of the sea state bias in radar altimetry measurements of sea level. *J. Atmos. Ocean. Technol.*, 19, 1690–1707.

Gaspar, P., F. Ogor, P. Y. Le Traon, and O. Z. Zanife. (1994). Estimating the sea state bias of the TOPEX and POSEIDON altimeters from crossover differences. *J. Geophys. Res.*, 99. pp. 24981–24994.

GCOS. (2011). Systematic observation requirements for satellite-based data products for climate (2011 update)—Supplemental details to the satellite-based component of the "Implementation Plan for the Global Observing System for Climate in Support of the UNFCCC (2010 Update)." In: GCOS-154, WMO, Decemberhttps://www.wmo.int/pages/prog/gcos/Publications/gcos-154.pdf.

Greenslade, D. J. M., D. B. Chelton, and M. G. Schlax. (1997). The mid-latitude resolution capability of sea level fields constructed from single and multiple altimeter datasets. *J. Atmos. Ocean. Tech.*, 14(8), 849–870.

Guissard, A., and P. Sobieski. (1987). An approximate model for the microwave brightness temperature of the sea. *Int. J. Rem. Sens.*, 8, 1607–1627.

Haines, B., Y. Bar-Sever, W. Bertiger, et al. (2004). One-centimeter orbit determination for Jason-1: New GPS-based strategies. *Marine Geodesy*, 27(12), 299–318. doi: 10.1080/01490410490465300.

Haines, B., W. Bertiger, S. Desai, et al. (2003). Initial orbit determination for Jason 1: Toward a 1 cm Orbit. *Navigation*, 50, 171–180. doi: 10.1002/j.2161-4296.2003.tb00327.x.

Halimi, A., C. Mailhes, J. Y. Tourneret, P. Thibaut, and F. Boy. (2014). A semi-analytical model for delay/Doppler altimetry and its estimation algorithm. *IEEE Trans. Geosci. Rem. Sens.*, 52(7), 4248–4258.

Hauser D., C. Tison , T. Amiot , L. Delaye , A. Mouche , G. Guitton , L. Aouf , P. Castillan. (2016). CFOSAT: A new Chinese-French satellite for joint observations of ocean wind vector and directional spectra of ocean waves, Proc. SPIE 9878, Remote Sensing of the Oceans and Inland Waters: Techniques, Applications, and Challenges, 98780T. doi:10.1117/12.2225619. http://dx.doi.org/10.1117/12.2225619.

Haurwitz, B., and A. D. Cowley. (1973). The diurnal and semidiurnal barometric oscillations global distribution and annual variation. *Pure Appl. Geophys.*, 102(1), 193–222. doi: 10.1007/BF00876607.

Hayne, G. S., (1980). "Radar Altimeter Mean Return Waveforms from Near-Normal-Incidence Ocean surface scattering", *IEEE Trans. Antennas Propag.* AP-28, 5, 687–692.

Hernandez-Pajares, M., J. M. Juan, J. Sanz, and D. Bilitza. (2002). Combining GPS measurements and IRI model values for space weather specification. In: K. Rawer, D. Bilitza, B. W. Reinisch (eds.), *Modeling the topside ionosphere and plasmasphere*, Vol. 29, pp. 949–958. Pergamon-Elsevier Science Ltd., Oxford, UK.

Imel, D. A. (1994). Evaluation of the topex/poseidon dual-frequency ionosphere correction. *J. Geophys. Res. Oceans*, 99, 24895–24906.

Jackson, F. C. (1979). The reflexion of impulses from a nonlinear random sea. *J. Geophys. Res.*, 84(C8), 4939–4943.

Jacobs, G., C. Barron, R. Rhodes (2001): Mesoscale characteristics, Journal of Geophysical Research, VOL. 106, NO. C9, PP. 19, 581–19,595, 2001.

Jacobs, G. A., J. G. Richman, J. D. Doyle, et al. (2014). Simulating conditional deterministic predictability within ocean frontogenesis. *Ocean Modeling*, 78, 1–16.

Jalabert, E., A. Couhert, J. Moyard, et al. (2015). *Jason-2, Saral and CryoSat-2 status*. In: Ocean Surface Topography Science Team Meeting 2015, Reston, VA, October, 20–23. https://meetings.aviso.altimetry.fr/programs/program-by-abstract-type.html

Jason-2 pod status. J. Moyard, E. Jalabert, A. Couhert, S. Rios-Bergantinos, F. Mercier, S. Houry (2015). In: Ocean Surface Topography Science Team Meeting 2014, Lake Constance, Germany, October 28–31. https://meetings.aviso.altimetry.fr/programs/program-by-abstract-type.html

John, L., W. H. F. Smith, D. Sandwell, R. Scharroo, F. G. Lemoine, and N. P. Zelensky. (2006). 20 Years of improvements to GEOSAT altimetry. In: Symposium: 15 Years of Progress in Radar Altimetry, Venice, Italy, March 13–18.

Jordi, A. and D. P. Wang. (2010). "*Application of ensemble Kalman filter for satellite altimetry data assimilation in the Mediterranean sea*". In the 2nd International Workshop on Modeling the Ocean, Norfolk, VA, U.S.A, May 24–26 2010.

Kaula, W. (ed.). (1969). *NASA (Williamstown report): The terrestrial environment, solid-earth and ocean physics, application of space and astronomic techniques.* Report of a Study at Williamstown, MA., NASA. https://ilrs.cddis.eosdis.nasa.gov/docs/williamstown_1968.pdf

Keihm, S. J., M. A. Janssen, and C. S. Ruf. (1995). TOPEX/Poseidon microwave radiometer (TMR): Wet troposphere range correction algorithm and pre-launch error budget. *IEEE Trans. Geosci. Rem. Sens.*, 33(1). pp. 147–161.

Koblinsky, C. J., P. Gaspar, and G. Lagerloef. (1992). *The future of spaceborne altimetry: Oceans and climate changes.* Joint Oceanographic Institutions Inc. https://inis.iaea.org/search/search.aspx?orig_q=RN:24010750

Komjathy, A., and G. H. Born. (1999). GPS-based ionospheric corrections for single frequency radar altimetry. *J. Atmos. Solar Terrestr. Phys.*, 61(16), 1197–1203.

Kouba, J., J. Moyard, E. Jalabert, et al. (2016). *A guide to using the IGS products.* IGS. Available at: acc.igs.org/UsingIGSProductsVer21.pdf

Kumar, R., D. Stammer, W. K. Melville, and P. Janssen. (2003). Electromagnetic bias estimates based on TOPEX, buoy, and wave model data. *J. Geophys. Res.*, 108(C11), 3351. doi: 10.1029/2002JC001525.

Labroue, S., P. Gaspar, J. Dorandeu, et al. (2004). Non-parametric estimates of the sea state bias for Jason-1 radar altimeter. *Mar. Geod.*, 27, 453–481.

Leben, R., G. Born, and B. Engebreth. (2002). Operational altimeter data processing for mesoscale monitoring. *Marine Geodesy*, 25(1–2), 3–18.

Lee, H., C. Shum, Y. Yi, A. Braun, and C.-Y. Kuo. (2008). Laurentia crustal motion observed using Topex/Poseidon radar altimetry over land. *J. Geodyn.*, 46(3–5), 182–193.

Legeais, J.-F., M. Ablain, and S. Thao. (2014). Evaluation of wet troposphere path delays from atmospheric reanalyses and radiometers and their impact on the altimeter sea level. *Ocean Sci.*, 10, 893–905. doi: 10.5194/os-10-893-2014.

Lemaire, D. (1998). Non-fully developed sea state characteristics from real aperture radar remote sensing. Ph.D. thesis, l'Université Catholique de Louvain, 220 pp.

Le Provost, C., F. Lyard, J. M. Molines, et al. (1998). A hydrodynamic ocean tide model improved by assimilating a satellite altimeter-derived data set. *J. Res. Oceans*, 103(C3), 5513–5529. doi: 10.1029/97JC01733.

Le Traon, P. Y., and G. Dibarboure. (2002). Velocity mapping capabilities of present and future altimeter missions: The role of high-frequency signals. *J. Atmos. Ocean. Technol.*, 19, 2077–2087.

Le Traon, P. Y., and G. Dibarboure. (2004). Illustration of the contribution of the tandem mission to mesoscale studies. *Marine Geodesy*, 27, 3–13.

Liebe, H. J., G. A. Hufford, and M. G. Cotton. (1993). Propagation modeling of moist air and suspended water/ice particles at frequencies below 1000 GHz. In: *Proceedings of the NATO/AGARD Wave Propagation Panel 52nd Meeting*, May 17–23, Palma de Mallorca, Spain, AGARD, pp. 1–10.

Lin, M., Y. Jia, and Y. Zhang. (2015). Global assessment of HY-2 satellite. In: *Proceedings of the IGARSS (International Geoscience and Remote Sensing Symposium) 2015*, Milan, Italy, July 26–31.

Luthcke, S. B., N. P. Zelensky, D. D. Rowlands, et al. (2003). The 1-centimeter orbit: Jason-1 precise orbit determination using GPS, SLR, DORIS and altimeter data. *Marine Geodesy*, 26(3–4), 399–421. doi: 10.1080/714044529.

Lyard, F., F. Lefevre, T. Letellier, and O. Francis. (2006). Modeling the global ocean tides: Modern insights from FES2004. *Ocean Dynam.*, 56, 394–415. doi: 10.1007/s10236-006-0086-x.

Lyard, F. personal communication, LEGOS, 2017.

Mannucci, A. J., B. D. Wilson, D. N. Yuan, et al. (1998). A global mapping technique for GPS-derived ionospheric total electron content measurements. *Radio Sci.*, 33, 565–582.

Marshall, J. A., N. P. Zelensky, S. M. Klosko, et al. (1995). The temporal and spatial characteristics of TOPEX/Poseidon radial orbit error. *J. Geophys. Res. Oceans*, 100(C12), 25331–25352. doi: 10.1029/95JC01845.

Maus, S., C. M. Green, and J. D. Fairhead. (1998), "*Improved ocean-geoid resolution from retracked ERS-1 satellite altimeter waveforms*". Geophys. J. Int., 134(1): 243–253.

McCarthy, D. D. (1996). *IERS conventions (1996)*. IERS Technical Note 21, Central Bureau of IERS—Observatoire de Paris, Paris, France.

McCarthy, D. D., and G. Petit. (2004). *IERS conventions (2003)*. IERS Technical Note 32, Frankfurt am Main, Germany.

Melville, W. K., F. C. Felizardo, and P. Matusov. (2004). Wave slope and wave age effects in measurements of electromagnetic bias. *J. Geophys. Res.*, 109(C7), 7018. doi: 10.1029/2002JC001708.

Menard, Y., P. Escudier, L. Fu, and G. Kunstmann. (2000). Cruising the ocean from space with Jason-1. *EOS/AGU Trans.*, 81(34), 1, 390–391.

Millet, F. W., D. V. Arnold, K. F. Warnick, and J. Smith. (2003). Electromagnetic bias estimation using in situ and satellite data: 1. RMS wave slope. *J. Geophys. Res.*, 108(C2), 3040. doi: 10.1029/2001JC001095.

Moore, R., and C. Williams . (1957). Radar terrain return at near-vertical incidence. *Proc. IRE*, 45(2), 228–238.

Nelder, J.A., Mead. R., (1965). "*A Simplex Method for Function Minimization*", Comput J (1965) 7 (4): 308–313. https://doi.org/10.1093/comjnl/7.4.308

Obligis, E., A. Rahmani, L. Eymard, S. Labroue, and E. Bronner. (2009). An improved retrieval algorithm for water vapor retrieval: Application to the Envisat Microwave Radiometer. *IEEE Trans. Geosci. Rem. Sens.*, 47(9) pp. 3057–3064. .

Ollivier, A., G. Dibarboure, B. Picard, N. Picot, and N. Steunou. Spectral characterization of geophysical correction for nadir altimetry—Toward altimetry spectral Error budget. Accepted in IEEE TGRS.

Ollivier, A., Y. Faugere, N. Picot, M. Ablain, P. Femenias, and J. Benveniste. (2012). Envisat Ocean altimeter becoming relevant for mean sea level trend studies. *Marine Geodesy*, 35(Suppl. 1). pp. 118–136.

Pascual, A., C. Boone, G. Larnicol, and P. Y. Le Traon. (2008). On the quality of real time altimeter gridded fields: Comparison with in situ data. *J. Atmos. Ocean. Technol.* http://journals.ametsoc.org/doi/full/10.1175/2008JTECHO556.1

Passaro, M., P. Cipollini, S. Vignudelli, G. Quartly, and H. M. Snaith. (2014). ALES: A multi-mission adaptive subwaveform retracker for coastal and open ocean altimetry. *Rem. Sens. Environ.*, 145, 173–189.

Pearlman, M. R., J. J. Degnan, and J. M. Bosworth. (2002). The International Laser Ranging Service. *Adv. Space Res.*, 30(2), 135–143. doi: 10.1016/S0273-1177(02)00277-6.

Petit, G., and B. Luzum. (2010). *IERS conventions (2010)*. IERS Technical Note 36, Frankfurt am Main, Germany. https://www.iers.org/SharedDocs/Publikationen/EN/IERS/Publications/tn/TechnNote36/tn36.html

Picard, B., M.-L. Frery, E. Obligis, L. Eymard, N. Steunou, and N. Picot. (2015). SARAL/AltiKa wet tropospheric correction: In-flight calibration, retrieval strategies and performances. *Marine Geodesy*, 38(Suppl. 1), 277–296. doi: 10.1080/01490419.2015.1040903.

Phalippou, L., and V. Enjolras. (2007). Re-tracking of SAR altimeter ocean power waveforms and related accuracies of the retrieved sea surface height, significant wave height and wind speed In: *Proceedings of the IEEE International IGARSS*, July 23–28, pp. 3533–3536. http://ieeexplore.ieee.org/document/4423608/

Poisson, J. C., F. Piras, P. Thibaut, S. Le Gac, F. Boy, and N. Picot. (2016a). New powerful numerical retracker solution accounting for speckle noise statistics. In: OSTST 2016, November 1–4, La Rochelle. https://meetings.aviso.altimetry.fr/programs/program-by-abstract-type.html

Poisson J. C., F. Piras, P. Thibaut, S. Le Gac, F. Boy, N. Picot. (2016b). New Powerful Numerical Retracker Solution Accounting for Speckle Noise Statistics. In: 2016 OSTST (Ocean Surface Topography Science Team) Meeting, La Rochelle, France, November 1–4. The OSTST presentations can be downloaded from AVISO web site: https://meetings.aviso.altimetry.fr/programs/program-by-abstract-type.html

Ponte, R. M., and F. Lyard. (2002). Effect of unresolved high frequency signals in altimeter records inferred from tide gauge data. *J. Atmos. Ocean. Technol.*, 19, 534–539.

Ponte, R. M., and R. D. Ray. (2002). Atmospheric pressure corrections in Geodesy and oceanography: A strategy for handling air tides. *Geophys. Res. Lett.*, 29, 2253–2256. doi: 10.1029/2002GL016340.

Prandi, P., M. Ablain, L. Zawadzki, and B. Meyssignac. (2017). How reliable are local sea level trends?

Prandi, P., S. Philipps, V. Pignot, and N. Picot. (2015). SARAL/AltiKa global statistical assessment and cross-calibration with Jason-2. *Marine Geodesy*, 38(Suppl. 1). pp. 297–312.

Pujol, M.-I., Y. Faugère, G. Taburet, et al. (2016). DUACS DT2014: The new multi-mission altimeter data set reprocessed over 20 years. *Ocean Sci.*, 12, 1067–1090. doi: 10.5194/os-12-1067-2016.

Raney, R. K. (1998). The delay/Doppler radar altimeter. *IEEE Trans. Geosci. Rem. Sens.*, 36(5), 1578–1588.

Ray, C., C. Martin-Puig, M. P. Clarizia, et al. (2015). SAR altimeter backscattered waveform model. *IEEE Trans. Geosci. Rem. Sens.*, 53(2), 911–919.

Ray, R. D. (2013). Precise comparisons of bottom-pressure and altimetric ocean tides. *J. Geophys. Res. Oceans*, 118, 4570–4584. doi: 10.1002/jgrc.20336.

Rees, W. G. (1992). Orbital subcycles for earth remote sensing satellites. *Int. J. Rem. Sens.*, 13(5), 825–833.

Rémy, F., T. Flament, and A. Michel. (2015). Envisat and SARAL/AltiKa observations of the Antarctic ice sheet: A comparison between the Ku-band and Ka-band. *Marine Geodesy*, 38(Suppl. 1), 510–521.

Resch, G. Water-vapor radiometry in geodetic applications. In Geodetic Refraction; Brunner, F., Ed.; Springer-Verlag:New York, NY, USA, 1984; pp. 53–84.

Reynolds, R. W., T. M. Smith, C. Liu, D. B. Chelton, K. S. Casey, and M. G. Schlax. (2007). Daily high-resolution-blended analyses for sea surface temperature. *J. Clim.*, 20, 5473–5496. doi: 10.1175/JCLI-D-14-00293.1.

Ries, J. C. (2013). Seasonal geocenter motion from space geodesy and models. In: IERS Retreat 2013, Paris, France, May 23–24. https://www.iers.org/SharedDocs/Publikationen/EN/IERS/Workshops/Retreat2013/7_Ries.pdf?_blob=publicationFile&v=1

Rodriguez, E. (1998). Altimetry for non-Gaussian oceans: Height biases and estimation of parameters. *J. Geophys. Res.*, 93(C11), 14107–14120.

Rodriguez, E., and J.-M. Martin. (1994). Assessment of the TOPEX altimeter performance using waveform retracking. *J. Geophys. Res.*, 99(C12), 24957–24969.

Ruf, C. (2000). Detection of calibration drifts in spaceborne microwave radiometers using a vicarious cold reference. *IEEE Trans. Geosci. Rem. Sens.*, 38(1), 44–52.

Sandwell, D., R. D. Müller, W. Smith, E. Garcia, and R. Francis. (2014). New global marine gravity model from CryoSat-2 and Jason-1 reveals buried tectonic structure. *Science*, 346(6205), 65–67.

Sandwell, D. T., and W. H. F. Smith. (2005). Retracking ERS-1 altimeter waveforms for optimal gravity field recovery. *Geophys. J. Int.*, 163(1), 79–89.

Scharroo, R., and J. Lillibridge. (2005). Non-parametric sea state bias models and their relevance to sea level change studies. In: *Proceedings of the 2004 Envisat and ERS Symposium*, 6–10 September, Salzburg, Austria. Eds., H. Lacoste and L. Ouwehand. Published on CD-Rom., #39.1 available on http://adsabs.harvard.edu/full/2005ESASP.572E..39S

Scharroo, R., and W. H. F. Smith. (2010). A global positioning system-based climatology for the total electron content in the ionosphere. *J. Geophys. Res. Space Phys.* doi: 10.1029/2009JA014719. 115(A10), available at: http://onlinelibrary.wiley.com/doi/10.1029/2009JA014719/full

Smith, E. K., & Weintraub, S. (1953). The Constants in the Equation for Atmospheric Refractive Index at Radio Frequencies. Proceedings of the IRE, 41(8), 1035–1037. https://doi.org/10.1109/JRPROC.1953.274297

Smith, N., and M. Lefebvre. (1997). The Global Ocean Data Assimilation Experiment (GODAE). In: Monitoring the oceans in the 2000s: An integrated approach, International Symposium, Biarritz, October 15–17.

Srokosz, M. A. (1986). On the joint distribution of surface elevation and slopes for a nonlinear random sea, with an application to radar altimetry. *J. Geophy. Res.*, 91(C1), 995–1006.

Stammer, D., R. D. Ray, O. B. Andersen, et al. (2014). Accuracy assessment of global barotropic ocean tide models. *Rev. Geophys.*, 52(3), 243–282. doi: 10.1002/2014RG000450.

Stammer, D., C. Wunsch, and R. M. Ponte. (2000). De-aliasing of global high frequency barotropic motions in altimeter observations. *Geophys. Res. Lett.*, 27, 1175–1178.

Steunou, N., J. D. Desjonqueres, N. Picot, P. Sengenes, J. Noubel, and J. C. Poisson. (2015b). AltiKa altimeter: Instrument description and in flight performance. *Marine Geodesy*, 38(Suppl. 1): 22–42.

Steunou, N., N. Picot, P. Sengenes, J. Noubel, and M.-L. Frery. (2015b). AltiKa radiometer: Instrument description and in-flight performance. *Marine Geodesy*, 38(Suppl. 1), 43–61.

Stogryn, A. (1971). Equations for calculating the dielectric constant of saline water. *IEEE Trans. Microw. Theory Tech.*, MTT-(8), 733–736.

Stum, J., P. Sicard, L. Carrère, and J. Lambin. (2011). Using objective analysis of scanning radiometer measurements to compute the water vapor path delay for altimetry. *IEEE Trans. Geosci. Rem. Sens.*, 49(9) pp. 3211–3224. http://ieeexplore.ieee.org/stamp/stamp.jsp?arnumber=5710662.

Tapley, B. D., S. Bettadpur, M. Watkins, et al. (2004). The gravity recovery and climate experiment: Mission overview and early results. *Geophys. Res. Lett.*, 31(9), L09607. doi: 10.1029/2004GL019920.

Tapley, B. D., G. H. Born, and M. E. Parke. (1982). The SEASAT altimeter data and its accuracy assessment. *J. Geophys. Res. Ocean*, 87(C5), 3179–3188. doi: 10.1029/JC087iC05p03179. 87(C5), pp. 3179–3188.

Tavernier, G., J. P. Granier, C. Jayles, P. Sengenes, and F. Rozo. (2003). The last evolutions of the DORIS system. In: EGS-AGU-EUG Joint Assembly, Nice, France, April.

Thao, S., L. Eymard, E. Obligis, and B. Picard. (2014). Trend and variability of the atmospheric water vapor: A mean sea level issue. *J. Atmos. Ocean. Technol.*, 31, 1881–1901. doi: 10.1175/JTECH-D-13-00157.1.

Thao, S., L. Eymard, E. Obligis, and B. Picard. (2015). Comparison of regression algorithms for the retrieval of the wet tropospheric path. *IEEE J. Select. Topic. Appl. Earth Observ. Rem. Sens.*, 8, 9.

Thibaut, P., Amarouche, L., Zanife, O.Z., Steunou, N., Vincent, P., Raizonville, P. (2004) : "*Jason-1 Altimeter Ground Processing Look-Up Correction Tables*", Marine Geodesy, Special Issue on Jason-1 Calibration/ Validation, vol 27, Part III : 409-431–197

Thibaut, P., J. C. Poisson, E. Bronner, and N. Picot. (2010). Relative performance of the MLE3 and MLE4 retracking algorithms on Jason-2 altimeter waveforms. *Marine Geodesy*, 33(Suppl. 1), 317–335.

Tommy, D. (2003). Emerging ocean observations for interdisciplinary data assimilation systems. *J Marine Syst.*, 40–41, 5–48.

Tran, N., D. Vandemark, B. Chapron, S. Labroue, H. Feng, B. Beckley, and P. Vincent: "*New models for satellite altimeter sea state bias correction developed using global wave model data*", J. Geophys. Res., 111, C09009, doi: 10.1029/2005JC003406, 2006. Available at: http://onlinelibrary.wiley.com/doi/10.1029/2005JC003406/abstract

Tran, N., D. Vandemark, S. Labroue, et al. (2010). Sea state bias in altimeter sea level estimates determined by combining wave model and satellite data. *J. Geophys. Res.*, 115, C03020. doi: 10.1029/2009JC005534.

Ubelmann, C., L.-L. Fu, S. Brown, E. Peral, and D. Esteban-Fernandez. (2014). The effect of atmospheric water vapor content on the performance of future wide-swath Ocean. *J. Atmos. Ocean. Technol.* doi: 10.1175/JTECH-D-13-00179.1. http://journals.ametsoc.org/doi/abs/10.1175/JTECH-D-13-00179.1

Ulaby, F. T., and D. G. Long. 2014. *Microwave radar and radiometric remote sensing*. University of Michigan Press http://us.artechhouse.com/Microwave-Radar-And-Radiometric-Remote-Sensing-P1738.aspx.

Vandemark, D., N. Tran, B. D. Beckley, B. Chapron, and P. Gaspar. (2002). Direct estimation of sea state impacts on radar altimeter sea level measurements. *Geophys. Res. Lett.*, 29(24), 2148. doi: 10.1029/2002GL015776.

Verron, J., F. Ardhuin, S. Arnault, et al. (2010). *SARAL/AltiKa*—An altimetry mission in Ka-band. In: 2010 OSTST (Ocean Surface Topography Science Team) Meeting, Lisbon, Portugal, October 18–20. https://meetings.aviso.altimetry.fr/programs/program-by-abstract-type.html

Vignudelli, S., A. G. Kostianoy, P. Cipollini, and J. Benveniste. (2011). *Coastal altimetry*. Springer Verlag, Berlin.

Vinogradova, N. T., R. M. Ponte, and D. Stammer. (2007). Relation between sea level and bottom pressure and the vertical dependence of oceanic variability. *Geophys. Res. Lett.*, 34, L03608. doi: 10.1029/2006GL028588.

Willebrand, J., S. Philander, and R. Pacanowski. (1980). The oceanic response to large-scale atmospheric disturbances. *J. Phys. Oceanogr.*, 10, 411–429.

Willis, P., H. Fagard, P. Ferrage, et al. (2010). The International DORIS Service (IDS); Toward maturity. *Adv. Space Res.*, 45(12), 1408–1420. doi: 10.1016/j.asr.2009.11.018.

Wingham, D., C. R. Francis, S. Baker et al. (2006). CryoSat: A mission to determine the fluctuations in Earth's land and marine ice fields. *Adv. Space Res.*, 37(4), 841–871.

Wingham, D. J., C. G. Rapley, and H. Griths. (1986). New techniques in satellite altimeter tracking systems. In: *Proceedings of the IEEE International Conference Geoscience and Remote Sensing (IGARSS)*, Zurich, September, pp. 1339–1344 https://www.researchgate.net/publication/269518510_New_Techniques_in_Satellite_Altimeter_Tracking_Systems

Wunsch, C. (1989). Sampling characteristics of satellite orbits. *J. Atmos. Ocean. Technol.*, 6, 891–907.

Yaplee, B. S., A. Shapiro, D. L. Hammond, B. D. Au, and E. A. Uliana. (1971). Nanosecond radar observations of the ocean surface from a stable platform. *IEEE Trans. Geosci. Electron.*, 9(3), 170–174.

Zanife, O. Z., P. Vincent, L. Amarouche, J. P. Dumont, P. Thibaut, and S. Labroue. (2003). Comparison of the Ku band range noise level and the relative sea-state bias of the Jason-1, TOPEX, and Poseidon-1 radar altimeters special issue: Jason-1 calibration/validation. *Marine Geodesy*, 26(3–4), 201–238.

2 Wide-Swath Altimetry
A Review

Ernesto Rodriguez, Daniel Esteban Fernandez, Eva Peral, Curtis W. Chen, Jan-Willem De Bleser, and Brent Williams

2.1 INTRODUCTION

Conventional nadir radar altimetry is already in its third decade and has proven invaluable in understanding ocean sea surface height (SSH) from mesoscales to basin scales (see, e.g., Fu and Cazenave 2001, and other contributions in this volume). There is also a growing community of hydrologists using nadir altimetry data to monitor large lakes and rivers (see, e.g., Alsdorf et al. 2007). Significant advances have been made using small constellations of nadir altimeters to improve the coverage and sampling possible with a single altimeter instrument. However, limitations in nadir altimeter spatial resolution, or space–time sampling of the constellation, have not yet allowed the systematic mapping of small mesoscale and sub-mesoscale ocean features or smaller surface water bodies.

The mapping of SSH at high spatial resolution has been a goal for the ocean community. At the turn of the century, the first High Resolution Ocean Topography Science Working Group (HOTSWG; Chelton 2001) was formed to examine the science benefits of high resolution coverage and potential measurement methods to achieve it. In a parallel effort, thanks to encouraging results coming from nadir altimeters, the hydrology community formed a Surface Water Working Group (Alsdorf and Lettenmaier 2003; Alsdorf et al. 2003) to examine how altimetric measurements of surface water elevation (SWE) could be extended to obtain global coverage of large and small surface water bodies with temporal sampling sufficient for studying global water storage dynamics.

Over time, one of the leading candidates that has emerged, capable of meeting the needs of both communities, is *wide-swath altimetry*; namely, the use of radar interferometry at near-nadir incidence to achieve global measurements of water elevation for ocean and surface water bodies with centimeter-level accuracy. An initial concept that met some of the ocean science requirements, the Wide-Swath Ocean Altimeter (WSOA; Rodríguez and Pollard 2001; Pollard et al. 2002; Fu and Rodriguez 2004b), was examined in detail as a potential demonstration mission in the National Aeronautics and Space Administration/Centre National d'Etudes Spatiales (NASA/CNES) Ocean Surface Topography Mission/Jason-2. The concept was refined further in the Water Elevation and Recovery (WaTER) mission proposed to the European Space Agency (ESA) in 2005 (N. Mognard-Campbell, PI; Enjolras and Rodriguez 2009) and further in the Surface Water Ocean Topography (SWOT) mission that was recommended for implementation by the U.S. National Research Council Decadal Survey (National Research Council 2007). The SWOT mission is currently under development in a partnership among NASA, CNES, Canadian Space Agency (CSA), and United Kingdom Space Agency (UKSA), and is expected to be launched around 2021.

The purpose of this review is to present the measurement concept and evolution of wide-swath altimetry in a unified way because no such comprehensive description is available in the open literature. The emphasis of the review will be on the theory and capabilities of this measurement approach, while the many scientific applications that it enables are mainly addressed in the references. In Section 2.2, we give a brief description of the space–time characteristics and the magnitude of the water elevation signal for wide-swath altimetry's primary targets, together with the measurement requirements that have been endorsed by the ocean and hydrology communities (Rodríguez 2015). The measurement principles behind the concept and their evolution are then described in Section 2.3. Section 2.4

presents a derivation of all the components of the error budget for wide-swath altimeters, which contains many more terms than for the traditional nadir altimeter. Not only is the measurement principle different from traditional altimetry, but, due to the off-nadir viewing angles and the need to use Synthetic Aperture Radar (SAR) and interferometry, the measurement physics changes as well. These changes in measurement physics are reviewed in Section 2.5. The synthesis of these components into the design of a spaceborne mission is presented in Section 2.6, while the final section presents an overview of the current state of the measurement and possible avenues for future improvements.

2.2 OCEAN AND HYDROLOGY SAMPLING REQUIREMENTS

Figure 2.1 presents a graph of the space–time characteristics of various ocean phenomena whose SSH expression would be suitable for mapping with an altimeter. An overview of how high resolution

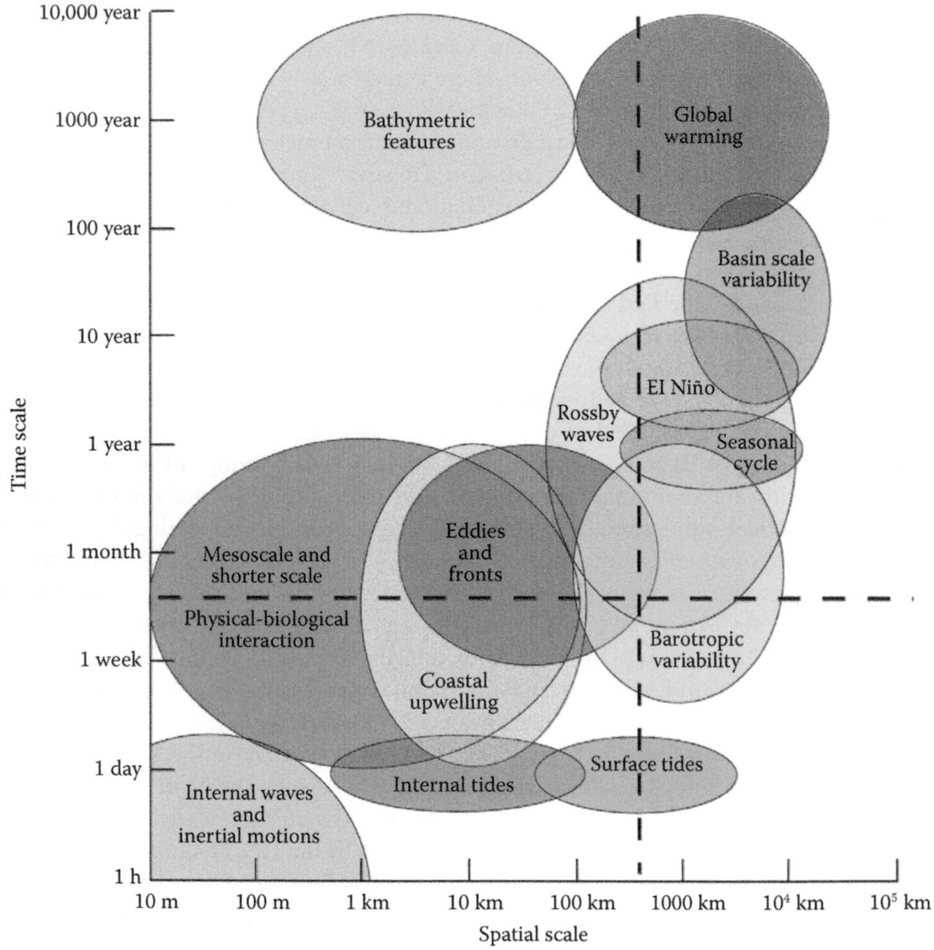

FIGURE 2.1 The approximate space and time scales of phenomena of interest that could be investigated from altimetric measurements of ocean topography with adequate spatial and temporal resolution. The dashed lines indicate the approximate lower bounds of the space and time scales that can be resolved in SSH fields constructed from measurements by a single altimeter in the T/P 10-day repeat orbit configuration. Processes with spatial scales to the left of the vertical dashed line and time scales below the horizontal dashed line require higher resolution measurements of ocean topography from a constellation of nadir-looking altimeters or a wide-swath altimeter. (From Chelton, D., *Report of the High-Resolution Ocean Topography Science Working Group Meeting*, 2001-4, College of Oceanic and Atmospheric Sciences Oregon State University, Corvallis, OR, 2001. With permission.)

SSH measurements might inform our understanding of the ocean mesoscale and sub-mesoscale circulation (Klein et al. 2016), tides (Arbic et al. 2016), coastal, and shelf processes (Ayoub et al. 2016; Laignel et al. 2016), as well as ocean bathymetry and sea ice, is given by Fu et al. (2012) and the previous references.

As expected with turbulent phenomena, shorter spatial scales are typically associated with shorter time scales, so that an increase in spatial resolution must be accompanied by a proportional improvement in temporal resolution. The space–time sampling improvements that can be obtained by using a constellation of nadir altimeters is illustrated in Figure 2.2, which shows that even a constellation of four to five altimeters will have problems providing appropriate sampling of small mesoscale features. However, as illustrated in Figure 2.2, even a single high spatial resolution swath instrument can provide significant improvements in sampling over a constellation of conventional altimeters, although temporal sampling of the fastest scales may still be challenging for just a single swath instrument.

Not only does the temporal scale decrease with decreasing spatial scale, but the magnitude of the SSH signal decreases as well, so that a high-resolution altimeter must not only meet the space–time

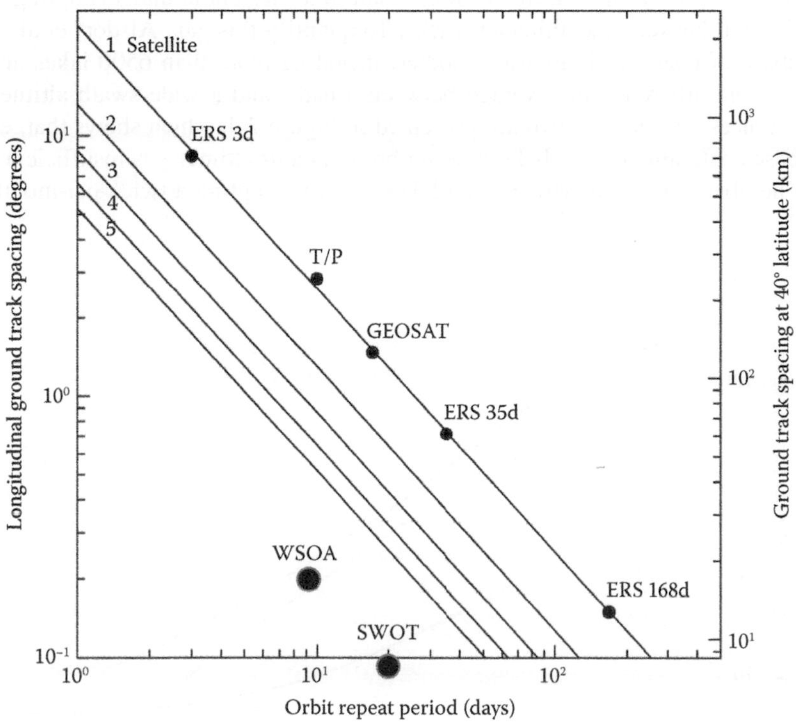

FIGURE 2.2 The relationship between orbit repeat period and the longitudinal separation of neighboring ground tracks for altimeters in exact repeat orbit configurations. The ground track spacing is displayed in degrees along the left axis and in kilometers at 40° latitude along the right axis. In log-log space, the choices of exact repeat period and ground track spacing for a single satellite fall approximately along the top straight line in the figure. The repeat periods and ground track separations of past and present altimeter missions are shown by the solid circles. The two sampling patterns shown for the ERS altimeter correspond to the Multi-Disciplinary Phase (35-day repeat) and the Geodetic Phase (168-day repeat) of the ERS-1 satellite. The improvements in the resolution that would be obtained from multiple satellites in coordinated orbit configurations with evenly spaced ground tracks are shown for constellations of 2, 3, 4 and 5 satellites. Also shown is the resolution expected to be achieved by WSOA and SWOT. (The figure without the WSOA and SWOT information is adapted from Chelton, D. B., *Report of the High-Resolution Ocean Topography Science Working Group Meeting*, College of Oceanic and Atmospheric Sciences Oregon State University, 2001.)

sampling requirements but must also have improved precision at smaller scales. To estimate how much the measurement precision must be improved, Fu and coworkers (Xu et al. 2011; Xu and Fu 2011; Xu and Fu 2012; Fu and Ubelmann 2014) have characterized the global power-law decay for the mesoscales observed by conventional altimetry and have extrapolated these signatures to smaller scales, as shown in Figure 2.3. Of course, this figure may be conservative since it does not include the potential contribution of non-mesoscale features, such as internal waves and tides. Given a desire to have a good signal-to-noise ratio through most of the observation range, Figure 2.3 also presents the SSH accuracy requirements that have been adopted by the SWOT science team as suitable goals for a high-resolution SSH mission (Rodríguez 2015).

Surface water stored in lakes, reservoirs, and rivers is a major component in the Earth's water cycle, and contains most of the potable water. Measuring how much water is stored in these bodies and how the storage changes over time and space is an important scientific question that can be addressed by wide-swath altimetry measurements of SWE (Alsdorf et al. 2007; Biancamaria et al. 2010; Durand et al. 2010a; Lee et al. 2010; Fu et al. 2012; Pavelsky et al. 2014). Given the inverse power law relation that exists between lake size and the number of lakes (Biancamaria et al. 2010; Lee et al. 2010), using only nadir altimetry may result in a substantial undersampling of the many small lakes that fall between the altimeter tracks. To quantify this gap, Alsdorf et al. (2007) used a global database of relatively large water bodies, including more than 6500 lakes and 3700 rivers, to examine the difference in coverage between a nadir and a wide-swath altimeter. Sample coverage differences between the two are presented in Figure 2.4, which shows that, even for this limited database containing relatively large water bodies, nadir altimetry is insufficient as it misses a substantial number of even moderate-sized lakes when attempting a twice-per-month sampling.

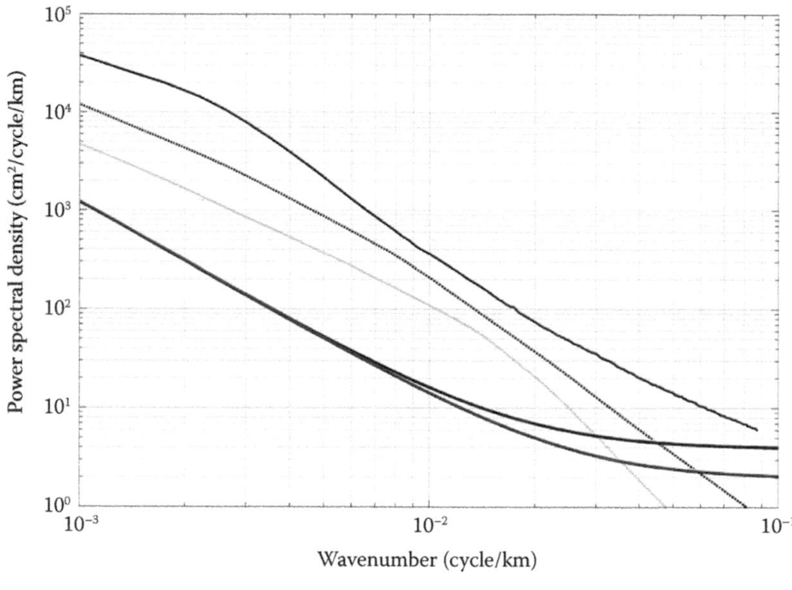

FIGURE 2.3 Global mean SSH spectrum estimated from the Jason-1 and Jason-2 observations (thick black line), the lower boundary of 68% of the spectral values (the upper gray dotted line), and the lower boundary of 95% of the spectral values (the lower gray dotted line) from the work of Fu and colleagues (Xu and Fu 2011, 2012; Fu and Ubelmann 2014). The red and blue lines represent the baseline (red) and threshold (blue) performance curves that would be desired as a measurement requirement. The intersections of the two dotted lines with the baseline spectrum at ~15 km (68%) and ~25 km (95%) determine the resolving capabilities of the SWOT measurement. The respective resolution for the threshold requirement is ~20 km (68%) and ~30 km (95%). (From Rodríguez, E., *Surface Water and Ocean Topography Mission (SWOT) Science Requirements Document*, Technical report, Jet Propulsion Laboratory, Pasadena, CA, 2015. With permission.)

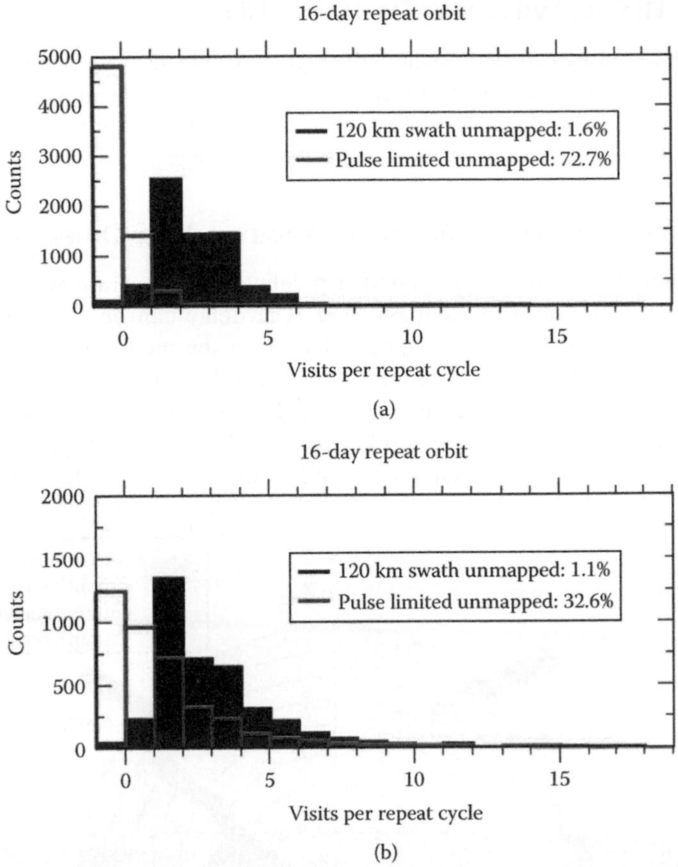

FIGURE 2.4 Histograms of the number of times lakes (a) and rivers (b) in a database of relatively large water bodies are imaged during a 16-day cycle by a nadir altimeter (red) and the WaTER KaRIn wide-swath instrument (black). Notice that the first bin corresponds to no observations during the repeat cycle. Even though 16 days is not sufficient to get complete global coverage with the wide-swath instrument only a small fraction of lakes and rivers are missed, and many are mapped multiple times. The nadir altimeter, on the other hand, misses a substantial number of water bodies altogether. (Based on data used in Alsdorf, D., et al., *Rev. Geophys.*, 45, 2007.)

Additional coverage could be obtained by lengthening the orbit repeat cycle but at the expense of degraded temporal sampling of monthly and seasonal dynamics. A wide-swath altimeter, on the other hand, can achieve good coverage on time scales of a few weeks, allowing the study of monthly (or better) variability of storage.

In addition to having the appropriate space–time coverage, high spatial resolution is needed to measure the inundation area required to calculate water storage. Various studies (e.g., Biancamaria et al. 2010; Pavelsky et al. 2014; Rodríguez 2015) have concluded that significant progress in the characterization of global storage change can be accomplished if a spaceborne mission maps all lakes with area greater than 100 m^2 and all rivers of width greater than 100 m (with a goal of 50 m). The SWE accuracy required to map these water bodies for storage change should be better than 10 cm.

In addition to storage change, SWEs can be used to calculate water slope, which, in conjunction with the width and elevation measurements, can be used to estimate river discharge via Manning's equation (see, e.g., Durand et al. 2010b). The SWOT science team has concluded that a slope accuracy of 1.7 cm/km (with a goal of 1 cm/km) is sufficient to estimate the discharge for the 100–50 m wide rivers targeted by the first dedicated global storage water change and discharge mission (Rodríguez 2015).

2.3 APPROACHES TO WIDE-SWATH ALTIMETRY

The need to extend altimetry to wide swath coverage was evident even before the launch of the TOPEX nadir altimeter mission, but technical challenges delayed the implementation of such a system. In the sections that follow, we outline these challenges and the way they were addressed through the use of radar interferometry and advanced technology.

2.3.1 From Nadir Altimetry to Wide-Swath Altimetry: Three-Dimensional Geolocation

The basic observable for a radar is the round-trip delay time of a transmitted signal scattered from a target and detected by the radar receiver. The delay can be converted to a distance, or range, by *a priori* knowledge of the speed of light in the medium through which the signal propagates. This one-dimensional range estimate can be converted into a measurement of topography only in the simplest circumstances, such as nadir altimetry from the ocean surface. Figure 2.5 illustrates the relationship between timing and the regions in the plane that contribute to signals arriving at the same time. When the radar pulse first strikes the surface,

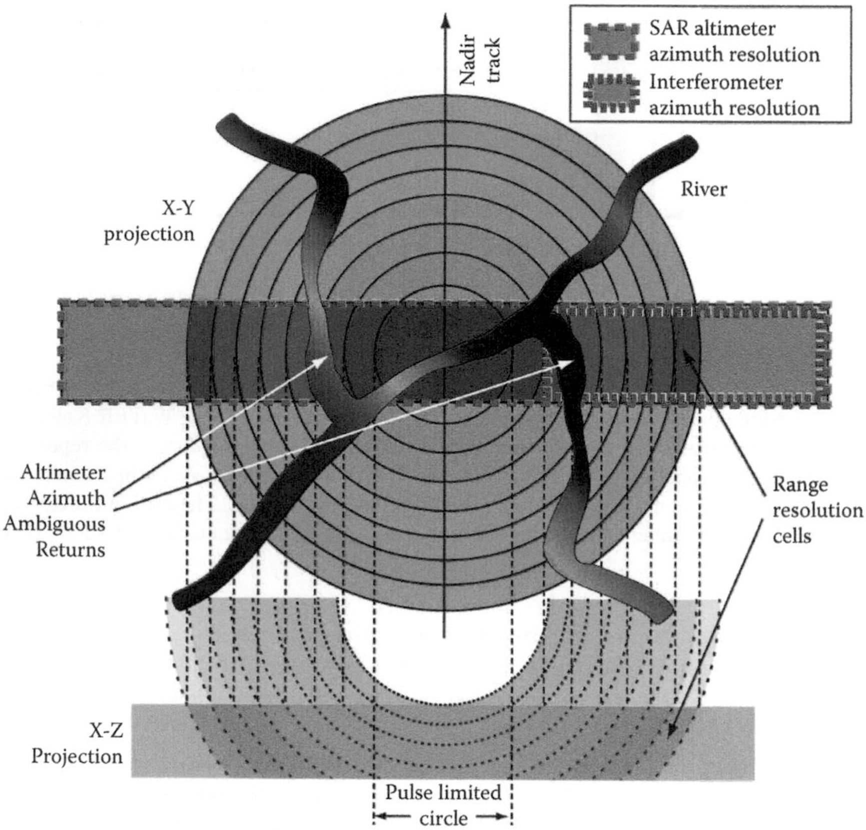

FIGURE 2.5 Illustration of the steps required to achieve geolocation in the plane. The upper figure represents the iso-range circles from a radar pulse impinging on a flat surface, in this case containing a river which one desires to image. The lower image illustrates the temporal progression of the radar pulse as it propagates in time. In the absence of azimuth resolution, all points inside an iso-range annulus arrive in the same range cell in the radar, and cannot be separated. By using SAR or Doppler processing, a narrow section (the green rectangle) perpendicular to the flight direction can be isolated, but symmetric points on the left and right hand side of the track still cannot be resolved. Finally, left and right can be discriminated by means of an illumination spot produced by the radar antenna pattern (yellow rectangle).

a well-defined locus, *the pulse limited circle*, is established. Tracking the average elevation of this pulse-limited circle through waveform retracking (Rodriguez 1988; Chelton et al. 1989; Rodriguez and Chapman 1989; Fu and Cazenave 2001) constitutes the basis for nadir altimetry. However, timing alone is insufficient for separating points that arrive from the same range annulus, and additional information is required to locate points on the plane.

Examining Figure 2.5, one sees that the range slices provide good geolocation in the cross-track direction but not in the along-track direction or for points located symmetrically to the left and right of the flight track. Two approaches have been used for narrowing the geolocation in the along-track, or azimuth, direction: the use of the antenna footprint and the use of SAR processing.

By pointing the radar antenna to one side of the nadir track, one can resolve the left-right ambiguity between points on either side of the track and in the along-track direction at the resolution of the antenna footprint. The limits on the physical size of the antenna place a significant restriction on how small the along-track resolution can be for real-aperture radar systems. For instance, for the Wide-Swath Ocean Altimeter concept (Rodríguez and Pollard 2001; Pollard et al. 2002; Fu and Rodriguez 2004a) described here, only a 15-km along-track spot could be achieved using a 3-m antenna at Ku band (2.2 cm wavelength).

To achieve a finer resolution in the along-track direction, SAR (Curlander and McDonough 1991) processing must be used. Using SAR, one can achieve an along-track resolution as small as half of the physical antenna size, which is more than sufficient to meet oceanography and hydrology needs. Fully focused processing for near-nadir incidence has been demonstrated recently for Synthetic Aperture Interferometric Radar Altimeter (SIRAL) on CryoSat-2 by Egido and Smith (2017).

SAR processing uses the fact that returns from points with different along-track locations will induce different Doppler shifts on the scattered radar signal. The Doppler shift, f_D, experienced by a point on the ground is given by

$$f_D = \frac{2\mathbf{v}_p \cdot \hat{\ell}}{\lambda} = \frac{2v_p}{\lambda} \sin\theta \sin\phi \tag{2.1}$$

where $\mathbf{v}_p$ is the platform velocity, $\hat{\ell}$ is the unit vector in the look direction, λ is the radar wavelength, θ is the look angle (see Figure 2.6), and ϕ is the azimuth angle measured relative to an axis perpendicular to the velocity vector (i.e., the interferometric baseline axis defined as follows). Examination of this equation shows that the iso-Doppler surface is given by a cone centered about the platform velocity vector. The iso-range surfaces are spheres centered at the radar location, and the intersection of these two surfaces isolates the returns along lines (but not points) in three-dimensional space.

Given a scattering plane at a given elevation, iso-range lines form circles, iso-Doppler lines form hyperbolas, and the two are roughly perpendicular, forming a two-dimensional geolocation grid for a given height. However, the selection of which of these height planes should be chosen to achieve full geolocation in three dimensions requires an additional item of information not contained in the delay and Doppler information. Next, we discuss two approaches that have been taken to resolve the height dimension: waveform tracking and radar interferometry.

Once the plane of incidence has been defined by the azimuth antenna pattern or Doppler constraints, the elevation above a reference plane and cross-track distance, x, of a point can be obtained if the platform height, H, the range, r_1, and the look angle, θ, are known (see Figure 2.6):

$$h = H - r_1 \cos\theta \tag{2.2}$$

$$x = r_1 \sin\theta \tag{2.3}$$

Radar and positioning systems can provide H and r, but the look angle θ is under-determined.

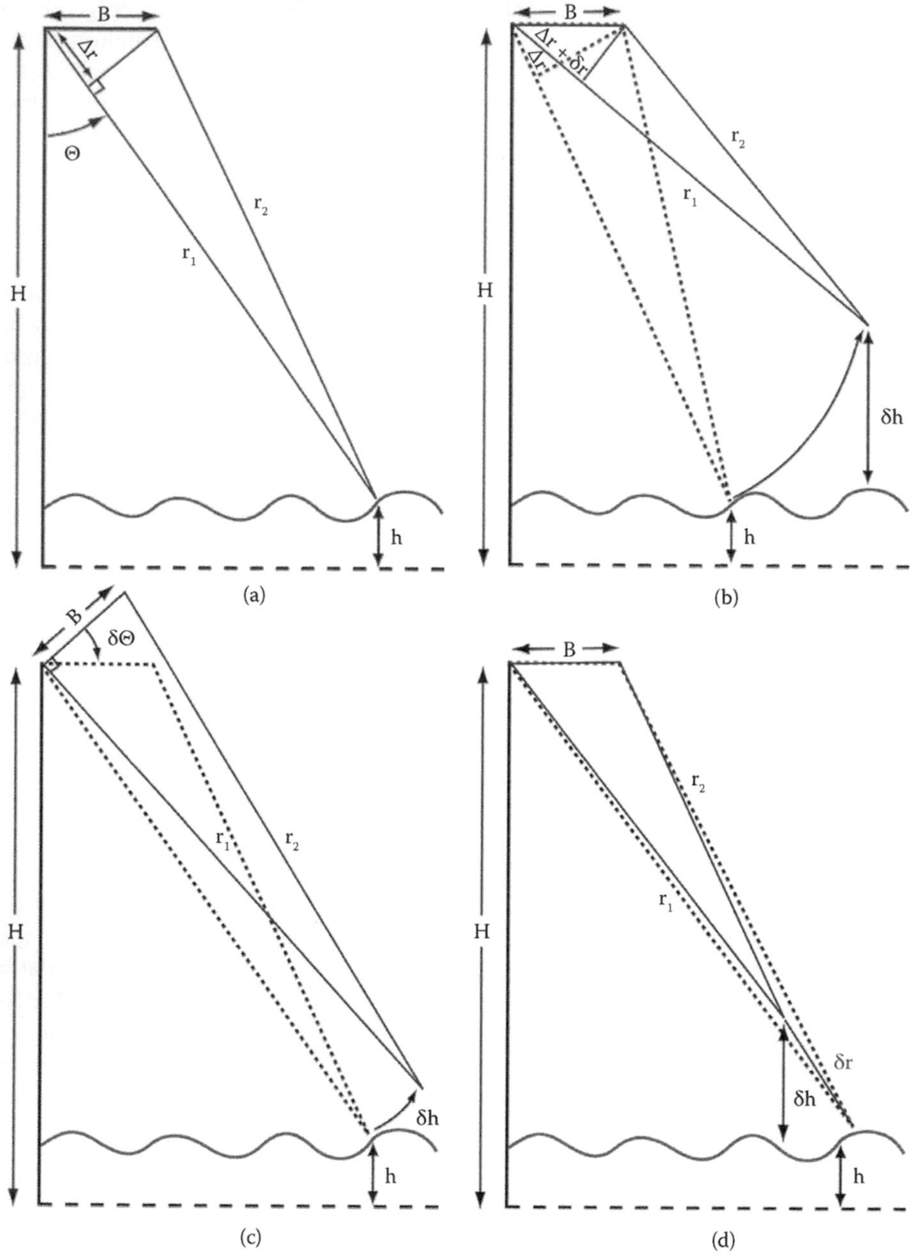

FIGURE 2.6 (a) Concept behind off-nadir altimetry and interferometry. Given the range, $r1$, the platform altitude, H, and the look angle, θ, the elevation, h, of a point in the ground can be obtained by $h = H - r1 \cos\theta$. Lidars and scanning altimeters determine θ by using a narrow beam. Interferometry uses Δr, the range difference between two antennas of know separation, B, to estimate θ: $\Delta r \approx B \sin\theta$. (b) Effect of an error in Δr (or, equivalently a phase error) on the height measurement: a positive error in Δr will cause the estimated θ to increase, resulting in a positive height error perpendicular to the look direction. (c) Effect of an error $\delta\theta$ in the baseline roll will cause a constant bias in the look angle, resulting on a surface tilted by an amount equal to the roll error. The error is perpendicular to the look direction and very similar to the phase error. (d) An error due to an error in the propagation speed will produce an error of geolocation that occurs in the look direction. This error is common with nadir altimetry, and perpendicular to the other interferometric errors.

2.3.2 Wide-Swath Altimetry Using Waveform Tracking

The first approach to estimating θ was based on using the antenna boresight as θ and estimating the range to the intersection of the boresight vector and the surface by retracking the radar waveform, as presented by Elachi et al. (1990). In this initial design, a SAR system used a large, electronically scanning antenna to achieve small enough spots on the ground to define θ. However, unlike nadir altimetry, off-nadir return waveforms do not exhibit a sharp leading edge that can be used for estimating the range. Rather, the waveforms have the shape of the antenna pattern, and some quantity—such as the peak or the energy centroid—must be used to estimate the range. Elachi et al. considered several alternatives and concluded that, given antenna sizes achievable on the Space Shuttle, a 5-m height error could be achieved globally by a scanning SAR altimeter system.

An alternate approach for ocean altimetry was proposed by Bush et al. (1984) and subsequently elaborated by Parsons and coworkers (Parsons and Walsh 1989; Parsons et al. 1994). In their approach, signals are collected from a pair of antennas separated by a baseline, *B*, and their normalized difference, computed on board, is used to form interferometric fringes, as in the famous Young's two-slit experiment. The angular separation among the fringes can be changed by modifying the baseline so that, in principle, extremely fine radar beams can be obtained in the cross-track direction, thus improving the accuracy of the range that could be assigned to the waveform for each interferometric fringe (i.e., power among subsequent nulls in the fringe pattern). The determination of which interferometric fringe the power is coming from can be achieved for flat surfaces such as the ocean by use of the measured range because consecutive fringes will be separated by a range difference greater than the range resolution of the instrument.

For scenes with constant radar cross section, this approach can be used to obtain height precisions that are compatible with SSH measurement needs. However, this method suffers from an insidious systematic source of error that is hard to resolve: Spatially varying wind and waves introduce changes in the return waveform shape, and the estimated height can be quite sensitive to these variations, which are unknown *a priori*. This puts a strong limitation on the accuracy that can be achieved by waveform tracking the normalized incoherently detected power of the complex difference between the signals from the two interferometer antennas.

2.3.3 Wide-Swath Altimetry Using Radar Interferometry

Given the sensitivity to scene backscatter inhomogeneity of the height estimates obtained by retracking, it is desirable to estimate the look angle independently of the brightness of the imaged scene. Graham (1974) was the first one to introduce the use of coherent radar interferometers, which utilize the relative phase between the signals from the two antennas to estimate the look angle to a point on the surface. Li and Goldstein (1990) demonstrated the technique from space using the Seasat SAR, while Rodríguez and Martin (1992) presented a systematic analysis of the errors, including volumetric decorrelation, and system design considerations. The Shuttle Radar Topography Mission (SRTM), the first spaceborne mission dedicated to radar interferometry (Farr et al. 2007), flew in the year 2000 and achieved meter-scale height accuracies. A good review of radar interferometry principles is given by Rosen et al. (2000).

The first spaceborne instrument to bring interferometry to near-nadir incidence was the SIRAL radar, which flew in the CryoSat-2 mission launched in 2010 (Wingham et al. 2006). SIRAL combines synthetic aperture altimetry with waveform tracking and interferometric phase capabilities in the very near-nadir region, limited by the antenna beamwidth to about 15 km in the cross-track direction. Initially, the interferometric phase measurements were foreseen as being useful for removing angle of arrival ambiguities (e.g., Armitage and Davidson 2014), but interferometric capabilities have also been used to map the cross-track ocean slopes (e.g., Galin et al. 2013) and ice-sheet topography (e.g., Foresta et al. 2016; Smith et al. 2016).

In the discussion that follows we simplify the full treatment, which is applicable to arbitrary surfaces, and concentrate on the mapping of relatively flat water bodies from space. This limitation achieves a significant simplification in the mathematics without significant loss in accuracy. The major assumptions are that the interferometric baseline is taken to be nearly orthogonal to both the velocity vector and the local vertical direction; that the azimuth beamwidth is narrow, so the scattering can be viewed has happening mostly in the plane defined by the interferometric baseline and the antenna boresight; and that the angle subtended by the interferometric baseline as seen from the ground is very small, so that the received fields at both antennas can be regarded as plane waves, without the need to take into account wave front curvature.

Given two antennas imaging in the same plane and separated by a baseline **B** (see Figure 2.6), simple geometry shows that the range difference is given

$$r_1 - r_2 = \Delta r \approx \mathbf{B} \cdot \hat{\ell} = B \sin\theta$$

and the look angle can be derived from knowledge of the baseline and the range difference. Incoherent mapping techniques that use two receivers, such as radar stereo imaging, use cross-correlation of the return power from each channel to estimate the range difference, but this is not sufficient for estimating SSH accurately because the accuracy of the range difference is then limited by the system range resolution (or bandwidth), which—given current technology—can produce signals with a range resolution on the order $\mathcal{O}$ (0.5 m), which is orders of magnitude too coarse for the required accuracy. On the other hand, the returned fields at each antenna from a point (x, y) on the ground (assuming we transmit from the first antenna and receive in both) is given, up to an unimportant constant, by

$$E_1(t, y_0) = \chi_a(x, y - y_0)\,\chi_r(t - 2r_1/c)\,s(x, y)\exp[2ikr_1]$$

$$E_2(t, y_0) = \chi_a(x, y - y_0)\,\chi_r(t - (r_1 - r_2)/c)\,s(x, y)\exp[ik\,(r_1 + r_2)]$$

where $k = 2\pi/\lambda$ is the electromagnetic wavenumber; χ_a is the normalized azimuth point target response, which includes the combined effect of the product of the antenna illumination and the SAR azimuth point target response, which isolates the scattering area around the antenna or Doppler azimuth, y_0; $s(x, y)$ is the scattering strength at the point; χ_r is the range point target response, a normalized function that peaks around 0 and has a width on the order of the range resolution of the system; and the final exponential terms represent the propagation phase. If we were to co-register these two signals in time (which involves a resampling), the complex product between the two of them would be given by

$$E_1E_2^* = \left|\chi_a(y - y_0)\,\chi_r(t - 2r_1/c)s\right|^2 \exp\,[ik(r_1 - r_2)]$$

$$= |A|^2 \exp\,[i\Phi]$$

(i.e., by the product of a real amplitude and a phase term related to the antenna range difference by $\Phi = k\Delta r$). Because the relative phase between two signals can be measured to a small fraction of the wavelength, the coherent interferometric technique offers the possibility of improvement over radar stereo by orders of magnitude in accuracy.

In practice, the range and azimuth point target responses are such that the interferometric or scattering phases can vary over the region they illuminate, and the spatial variability must be accounted for by integrating over all points contributing to the return on the surface. In addition, one must take into account that the surface return "speckles" (Goodman 1985) due to the separation among different scattering points on the surface, which can many wavelengths, so that the power from the surface must be treated statistically, and we assume, consistent with the deep phase approximation (Tsang et al. 1985) that $\langle s(x,y)s^*(x', y')\rangle = \delta(x-x')\,\delta(y-y')\sigma_0$, where σ_0 is the normalized radar cross section. The final result for the expected interferometric return is given by

$$\begin{aligned} \langle E_1E_2^*\rangle &= \int dx\,dy\,\left|\chi_a(y-y_o)\chi_r(t-2r_1/c)\right|^2\sigma_0(x,y)\exp\left[i\Phi(x,y)\right] \\ I &= \langle E_1E_2^*\rangle = P\gamma\exp\left[i\Phi_0\right] \end{aligned} \tag{2.4}$$

where I is the so-called *radar interferogram*; $P=\sqrt{|E_1|^2|E_2|^2}$ is the return power; $0\le\gamma\le 1$ is the so-called interferometric correlation, which is different from 1 due to changes in the interferometric phase inside the range/azimuth footprint; and Φ_0 is the observed interferometric phase that can be used for the final interferometric height equation

$$\Phi_0 = \mathrm{kB}\cdot\hat{\ell} = \mathrm{kB}\sin\theta\cos\phi_0 \tag{2.5}$$

where ϕ_0 is the azimuth angle with respect to the baseline direction, which is assumed to be known through the estimates of the Doppler (Equation 2.1) or the pointing of the antenna boresight. If the antenna pattern or the SAR azimuth resolution is small enough, then one can take $\phi_0\approx 0$. However, if there is significant variability of the phase within the radar footprint, or if the Doppler or antenna boresight is in error, then cos ϕ_0 term must be taken into account as it will introduce both height biases and increased height noise, as shown in the following.

Equation 2.5 shows that iso-phase surfaces are cones centered about the baseline vector (see Figure 2.7). Since we have assumed that the baseline vector and velocity vectors are orthogonal, it can be shown that intersection of the range sphere, the Doppler cone (Equation 2.1), and the iso-phase cone (Equation 2.5) occurs only at two points located symmetrically to the left and right of the satellite nadir track. This last ambiguity is resolved by the direction to which the antenna pattern is pointed, thus fully resolving the problem of locating the centroid of the scattering pixel in three-dimensional space.

At first sight, it is not evident how the interferometric equation is related to the heights on the surface. To make the relationship more obvious, one can multiply the observed interferometric phase by the phase that one would expect from a known reference surface, a process called *phase flattening* (see, e.g., Ferretti et al. 2007). If the reference surface is sufficiently close to the true surface, it is simple to show that the height obtained from the interferometric measurements can be written in the intuitive form (Peral et al. 2015)

$$\begin{aligned} h_M &= \int dx\,dy\,\chi_a^2(y-y_0)\chi_r^2(t-2r(x,y)/c)h(x,y)\frac{\sigma_0(x,y)}{\bar{\sigma}_0} \\ \bar{\sigma}_0 &= \int dx\,dy\,\chi_a^2(y-y_0)\chi_r^2(t-2r(x,y)/c)\sigma_0(x,y) \end{aligned} \tag{2.6}$$

This shows that the measured height, h_M, is the true height not only averaged over the areas defined by the point target responses in range and azimuth but also weighted by the surface backscatter cross section. For stationary flat surfaces, the product of the range and azimuth point target responses results in a well-defined grid, as depicted in Figure 2.5. However, in the presence of ocean surface waves, distortion can be introduced by the range point target response, motion of the surface, or brightness variations, as will be discussed in Section 2.5.2.

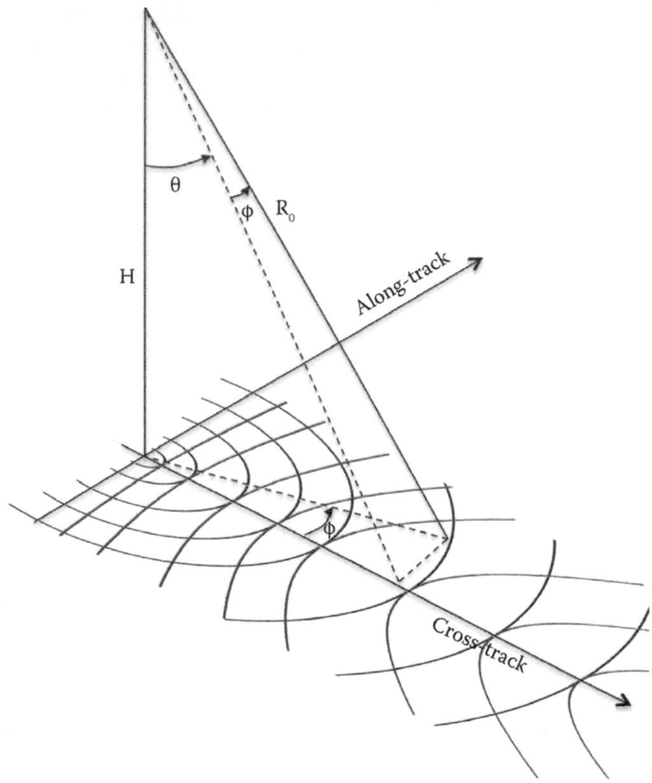

FIGURE 2.7 Intersection of iso-phase lines (red), which are hyperbolas with an axis defined by the baseline vector, and iso-range lines, which are circles centered around the nadir point. In the plane of incidence defined by the baseline vector and the local normal, the tangents between the two sets of lines are in the along-track direction. Away from the baseline plane, the two lines diverge. Since radar data are sampled along iso-range lines, sampling away from the baseline plane will result in phase biases and decorrelation.

2.4 THE INTERFEROMETRIC ERROR BUDGET

The SAR interferometry concept has been demonstrated from space in the SRTM (Rodriguez et al. 2006; Farr et al. 2007), which demonstrated global mapping on 11 days with a 90% height accuracy better than 5 m. The validation effort conducted during that mission (Rodriguez et al. 2006), as well as from airborne sensors, has allowed the validation of many of the error sources in the interferometric technique. In the following sections, we give a detailed account of the major error contributors as they apply to wide-swath interferometry for water bodies.

The components of the interferometric error budget can be obtained by differentiating the interferometric equations (2.2, 2.3, and 2.5). In the following, we examine each of the error terms in order.

2.4.1 Roll Errors

Roll errors (i.e., errors in the estimation of θ) are the dominant error sources for all swath-altimetry techniques. The effect of a roll error is to rotate the entire swath (see Figure 2.6c), as can be seen directly by differentiating Equation 2.2

$$\delta h = x\delta\theta \tag{2.7}$$

$$\delta x = (H - h)\,\delta\theta \tag{2.8}$$

The height error is linear in cross-track distance, while the cross-track position error is independent of it. This linear increase of height error with cross-track implies that height error can be minimized by choosing the cross-track distance and, hence, the incidence angle, to be as small as possible within the constraints of achieving a given swath; this approach has been taken in the WSOA and SWOT interferometers, described here.

To achieve a 1-cm error at a cross-track distance of 100 km, a knowledge of 10^1 μrad is required. This level of absolute knowledge for spacecraft pointing is beyond current capabilities. Even if it were possible to estimate this level of spacecraft roll, a very tight control of the antenna phase centers is required (e.g., for a 10-m baseline, the displacement of one of the antenna phase centers must be controlled with an accuracy of 1 μm, which is again quite challenging).

In practice, rather than imposing a requirement on the absolute knowledge of the attitude, one requires that the spacecraft attitude be measured to high precision, with drift slow enough so that the errors are concentrated in the longer wavelengths where alternate methods, such as the crossover calibration described as follows, can be applied to mitigate the error. This is possible using modern-day gyroscopes, which have good measurement precision but suffer from long-wavelength drift caused by integration of the white noise in the angular acceleration.

The roll introduced by distortions in flexing of the spacecraft or distortions of the baseline or the antennas is difficult to measure and is controlled by using very stiff mechanical assemblies. In that case, there will be a static roll bias, to be calibrated using crossover calibration, and the stiffness requirement is on the how slowly the components must deform (from thermal expansion, for instance).

If one is only interested in small spatial scale phenomena, it is sufficient to impose a requirement on the time rate of change of $\delta\theta$ such that the spectral contribution of $x\delta\theta(t)$ lies below the measurement requirement lines shown in Figure 2.3. This is the approach that has been taken for the SWOT mission (Rodríguez 2015) for the study of the ocean mesoscales and sub-mesoscales.

2.4.2 Phase Errors

The effect of a phase error, $\delta\Phi$, will be to move the estimated height and cross-track distance approximately along an iso-range line, as shown in Figure 2.6a. By taking the derivative with respect to phase, the induced height error and position are given by

$$\delta h = \frac{x}{kB\cos\theta}\delta\Phi \tag{2.9}$$

$$\delta x = \frac{H-h}{kB\cos\theta}\delta\Phi \tag{2.10}$$

For near-nadir incidence radar, as with WSOA or SWOT, $\cos\theta \approx 1$, and the error induced by the phase is completely equivalent to a roll error $\delta\theta_\Phi = \delta\Phi/kB$. Since $kB \gg 1$, the phase error requirement is typically orders of magnitude smaller than the roll requirement for the same level of error.

The phase error can be separated into a random component, which is independent from pixel to pixel and thus results in high-frequency height noise, and systematic phase errors, which are common for pixels in the same observation plane. Causes for the systematic phase errors are unmatched changes between the two channels in the signal path-length in the system or by temperature dependent delays in the components. To calibrate this source of error, systems typically include a calibration loop for each channel that shares many of the same delays of that channel. This can be done with high accuracy but not for the entire transmit and receive path because usually it is not possible to include the antennas in the calibration loop. The design of all components is such that the phase difference drifts slowly and can be corrected at the same time as the roll drifts discussed earlier.

Another cause for systematic phase biases is the interaction between the interferometer antennas and the spacecraft or baseline structures. These biases result in systematic differential phase ripples in the cross-track direction, which have been called the *interferometric phase screen*. It is impossible to calibrate the phase screen *a priori* before flying, and it must be corrected from the data itself. A possible approach to make this correction is to use the interferometer crossovers, as described here.

The cause of random phase noise is the presence of random signal components that are different between the two channels. The most easily understood unshared signal is due to thermal noise, which is independent from channel to channel. Notice that a drop in signal-to-noise ratio can also be due to applying a non-ideal range or azimuth compression function, which will drop the coherent system gain. Such a loss could be due to very fast antenna or mast motions during the aperture formation but is typically much smaller than the thermal noise component. The second source of difference between the two channels is the fact that the *speckle* (Goodman 1985) at each channel is slightly different due changes in geometry. To understand this phenomenon, one can model the scattering from the surface as consisting of the sum of phasors from many scatterers, with the phase of each phasor being proportional to the electromagnetic (EM) wave propagation phase. If the scattering points are randomly separated by many wavelengths, the resulting sum follows circular Gaussian statistics, with a uniformly distributed phase. Equation 2.4 illustrates the source of this loss of correlation between the channels. For small baselines, the interferometric phase varies little over the pixel defined by the range and azimuth point target responses. However, as the baseline increases, the phase variability within the pixel also increases so that points are no longer adding coherently, and significant cancellation will occur when the phase variability approaches one cycle (also called an interferometric fringe). The result of this phase variability is that γ, the channel-to-channel correlation, or coherence, will decrease from its maximum value of 1. Because the range point target response samples along a sphere of iso-range (not just along a circle in the plane), the phase variability can be due to either points located on the same plane (in which case, one speaks about *geometric decorrelation*) or at different heights (in which case, one speaks about *volumetric decorrelation*). A third source of channel-to-channel differences is quantization noise, which is a multiplicative noise source proportional to the signal strength. For a single channel, the signal-to-quantization-noise ratio (SQNR) is a complicated nonlinear function* that roughly increases by 6 dB for every bit added, and the quantization noise can be lumped with the thermal noise. However, for high signal-to-noise ratios, the quantization noise will be highly correlated between channels and the resulting error can show correlations among non-neighboring pixels. In practice, this is a minor effect, and thermal and quantization noise levels are often lumped as uncorrelated noise.

It can be shown (Rodríguez and Martin 1992) that the phase noise variance can be predicted given γ:

$$\sigma_{\Phi}^{2} = \frac{1-\gamma^{2}}{2N_{L}\gamma^{2}} \tag{2.11}$$

where N is the number of independent samples (also called *looks*) used in estimating the correlation, typically by averaging interferograms from neighboring pixels and extracting the resulting phase. It can also be shown (Rodríguez and Martin 1992) that, for homogeneous surfaces, the total correlation coefficient can be broken up as a product of noise, geometric, and volumetric correlations: $\gamma = \gamma_N \gamma_G \gamma_V$. The noise correlation is given by

$$\gamma_N = \frac{\text{SNR}}{\text{SNR}+1} \tag{2.12}$$

The geometric correlation can be shown to be proportional to the Fourier transform of the range point-target response and is typically close to 1 for WSOA and SWOT. It has been shown by

* See, for instance, https://en.wikipedia.org/wiki/Signal-to-quantization-noise_ratio

Gatelli et al. (1994) that, by appropriate signal filtering to ensure that the projected wavenumbers on the plane are the same for each channel, $\gamma_G \rightarrow 1$.

The decorrelation due to a homogeneous distribution of scatterers in the vertical direction can be shown to be proportional to the Fourier transform of the normalized distribution of backscatter cross section (Rodríguez and Martin 1992):

$$\gamma_V = \frac{\int dz \sigma_0(z) e^{ik_z z}}{\int dz \sigma_0(z)} \approx \exp\left[-\frac{1}{2}(\kappa_z \sigma_h)^2\right] \tag{2.13}$$

where $\kappa_z = kB \cos \theta / r \sin \theta_i$ is the interferometric fringe wavenumber in the vertical direction, r is the range, and θ_i is the local incidence angle. The last equality applies to Gaussian-distributed scatterers, with height standard deviation σ_h. To lowest order, this is a good approximation for ocean scattering when the pixel size in the range direction is large compared to the wavelength, in which case the distribution of points inside the resolution cell approaches the statistical distribution. It is a fair predictor of the decorrelation due to waves, even for smaller resolutions, but not a good predictor of how the noise is distributed in wavenumber space, as will be seen in Section 2.5.2, where wave effects are examined in greater detail. Trees will also have a volumetric decorrelation, as will be discussed in Section 2.5.3, but in that case, the scatterers cannot be considered as a homogeneous medium.

The relationship between height and geolocation error is given by $\delta h = \tan \theta \delta x$, so that even small height errors can lead to large geolocation errors for small angles of incidence. Although the random geolocation errors predicted by Equation 2.10 are not very significant for kilometer-scale ocean imaging, they can be significant for the delineation of small water bodies. In that case, the phase noise must be reduced to a magnitude consistent with the body being imaged. One way to do this is to average the phase within the water body so that the phase noise in Equation 2.11 is reduced to an appropriate level. Desroches et al. (2016) have used the geometry of the SAR image to identify points that should be in the same neighborhood in physical space, so that they can be averaged to produce images of rivers and lakes that are significantly less impacted by random phase errors, yielding a better geometrical representation of the water body.

2.4.3 Range Errors

The impact of range errors due to unknown media or system delays is given by

$$\delta h = -\cos \theta \delta r \tag{2.14}$$

$$\delta x = \sin \theta \delta r \tag{2.15}$$

For near-nadir incidence, $\cos \theta \rightarrow 1$, and $\delta h \approx \delta r$ and $\delta x \approx 0$. Similarly, errors due to uncertainties in the platform height, H, are given by $\delta h = \delta H$. Both of these equations are identical to those that apply to conventional nadir altimetry. The reader is referred to the literature (e.g., Fu and Cazenave 2001) for a discussion of delays due to the dry and wet troposphere, the ionosphere, or orbit errors. A thorough discussion for the SWOT mission is given by Esteban-Fernandez (2013).

2.4.4 Baseline Errors

An error in the length of the baseline will result in a height error given by

$$\delta h = -\frac{x^2}{H-h}\frac{\delta B}{B} \tag{2.16}$$

Thus the error varies quadratically with cross-track distance and, for near-nadir incidence, is much less sensitive than the roll and phase errors due to the inverse dependence of the platform height. As an example, for a platform height of 10^3 km, a cross-track distance of 50 km, and a baseline of 10 m, the requirement to achieve a 1-cm error is for the baseline length to be known to 40 μm, which is an order of magnitude less stringent than that required to control the baseline roll given earlier. Modern materials are very stable given a benign thermal environment, and changes in baseline will be slow and can be calibrated using the crossover technique described as follows.

2.4.5 Finite Azimuth Footprint Biases

In the previous discussion, we have been assuming that the azimuth resolution is fine enough so that the variation in phase over the azimuth direction can be neglected, which corresponds to setting $\phi \approx 0$ in Equation 2.5. This is not always the case, and the mismatch between range and differential phase iso-lines shown in Figure 2.7 will result in phase biases and increased noise due to decorrelation. The magnitude of the effect will depend on the shape of the azimuth point target response and, potentially, on cross-section variations in the azimuth direction, if the azimuth footprint is large enough. As an example, we will assume constant back scatter cross section and that the azimuth footprint size is small enough in the azimuth angle such that $\epsilon = kB \sin \theta (\cos \phi - \cos \phi_0) << 1$, where ϕ_0 is the nominal azimuth pointing angle for the azimuth footprint. Assuming that the range footprint is narrow, and expanding about ϕ_0 to second order in $(\phi - \phi_0)$, one finds that the phase error predicted by integrating Equation 2.4 is given by

$$\delta\Phi \approx -\Phi_0(\tan\phi_0\mu_1 + \frac{1}{2}\mu_2)$$

$$\mu_n \equiv \frac{\int d\phi \chi_a^2(\phi - \phi_0|\theta)(\phi - \phi_0)^n}{\int d\phi \chi_a^2(\phi - \phi_0|\theta)} \tag{2.17}$$

where Φ_0 is the true phase difference for a point with azimuth ϕ_0; $\chi_a^2(\phi - \phi_0|\theta)$ is the azimuth impulse response evaluated at constant look angle; and μ_n is the n th moment of the azimuth angle about the antenna boresight. The resulting height error can be obtained by Equation 2.9.

$$\delta h = -x \tan\theta\,[\sin\phi_0\,\mu_1 + \cos\phi_0\,\mu_2]$$

In practice, the azimuth pattern is symmetric enough about ϕ_0 and the μ_1 term is small. The width in azimuth of the antenna pattern is on the order of $\sqrt{\mu_2} \approx \alpha / \sin\theta$, where α is the antenna beamwidth standard deviation in the azimuth direction, and one can approximate

$$\delta h \approx -(H - h)\cos\phi_0\alpha^2$$

Due to the large $(H - h)$, this term can be sizable if the azimuth resolution angle is large enough, as when using real aperture radar or unfocused SAR processing, and it must accounted for by performing the integral in Equation 2.4 exactly. The remaining error will then be proportional to the unaccounted for variations of σ_0 or errors in the antenna pattern.

2.4.6 Radial Velocity Errors

When using SAR, one must rely on knowledge of the radial (i.e., along the look direction) component of the relative velocity between the platform and the surface for geolocation in the azimuth direction.

In this section, we deal with the effect of a constant radial velocity error and defer until Section 2.5.2 the impact of spatially varying motion, such as the one introduced by ocean waves.

From Equation 2.1, assuming the range footprint is narrow enough so that the look angle can be viewed as constant, an error in the knowledge of the radial velocity, $\delta\upsilon_r$, will induce a shift in the apparent azimuth of the target along an iso-range line (see Figure 2.8) given by

$$\delta\phi = \frac{\delta\upsilon_r}{\upsilon_p}\frac{1}{\sin\theta\cos\phi_0}$$

This error in azimuth angle will result in an error in position along the iso-range circle given by

$$\delta\rho = r\frac{\delta\upsilon_r}{\upsilon_p}\frac{1}{\cos\phi_0}$$

where ρ is the radial distance along the surface from the nadir point to the scatterer. In the nominal case when the viewing plane is along zero-Doppler, $\delta\rho = \delta y = r\delta\upsilon_r/\upsilon_p$, which is the well-known azimuth shift in SAR wave imaging.

The reconstruction of the look angle θ using the interferometric equation (2.5) assumes that the scattering occurs in the ϕ_0 direction, so that errors in the estimate of ϕ will induce errors in the estimated phase

$$\begin{aligned}\delta\Phi &= -kB\sin\theta\sin\phi_0\delta\phi \\ &= -kB\tan\phi_0\frac{\delta\upsilon_r}{\upsilon_p}\end{aligned}$$

and a corresponding height error of

$$\delta h = -x\frac{\tan\phi_0}{\cos\theta_0}\frac{\delta\upsilon_r}{\upsilon_p}$$

Thus, to first order, for zero-Doppler imaging, one will have a geolocation error but not a height error. This will not be the case if ϕ_0 is substantial as can occur for near-nadir angles if the

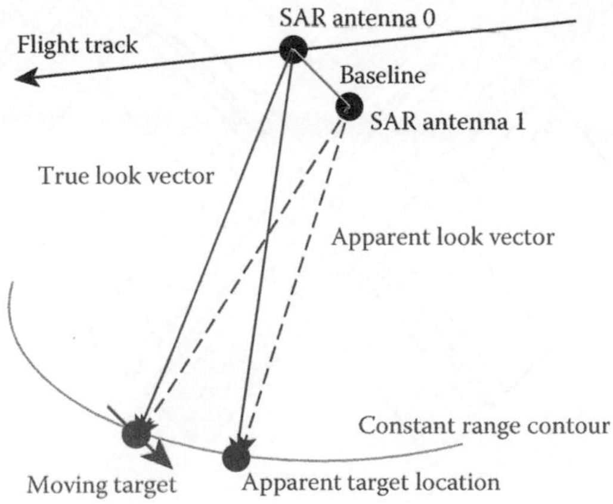

FIGURE 2.8 The effect of a constant error in radial motion will be to shift the apparent target azimuth location along a constant range line. This will result in an error in the look vector azimuth component, and hence in the translation of interferometric phase into an elevation angle, so that a both location and height errors are induced.

platform pitch angle, β, is significant because $\phi_0 \approx \beta/\sin\theta$. In practice, this puts a severe constraint on the pitch control of the spacecraft because the radial velocity due to spatially varying waves (see Section 2.5.2) is unknown *a priori*.

2.4.7 Calibration Methods

Given present day capabilities, it is not possible to measure directly either the interferometric phase screen or the varying roll/phase values. Fu and Rodriguez (2004a) introduced two possible solutions for obtaining roll and phase (but not phase screen) values: One can either use a preexisting constellation of nadir altimeters or an altimeter collocated with the interferometer to provide independent calibration, or one can use the crossovers provided by the satellite orbit to estimate these quantities. Subsequently, the use of the altimeter constellations, crossovers, and partial overlaps has been examined for the SWOT mission by Dibarboure et al. (2012) and Dibarboure and Ubelmann (2014). The idea of using crossovers to calibrate the phase screen was introduced subsequently by Rodríguez and Chen (2015, personal communication).

Figure 2.9 illustrates the geometry of observations at an orbit crossover. Two types of differences can occur at each crossover: altimeter–interferometer and interferometer–interferometer. Because the roll/phase or phase-screen errors are only a function of cross-track distance, and assuming that the rate of change of the parameters is small enough that they can be regarded as constant

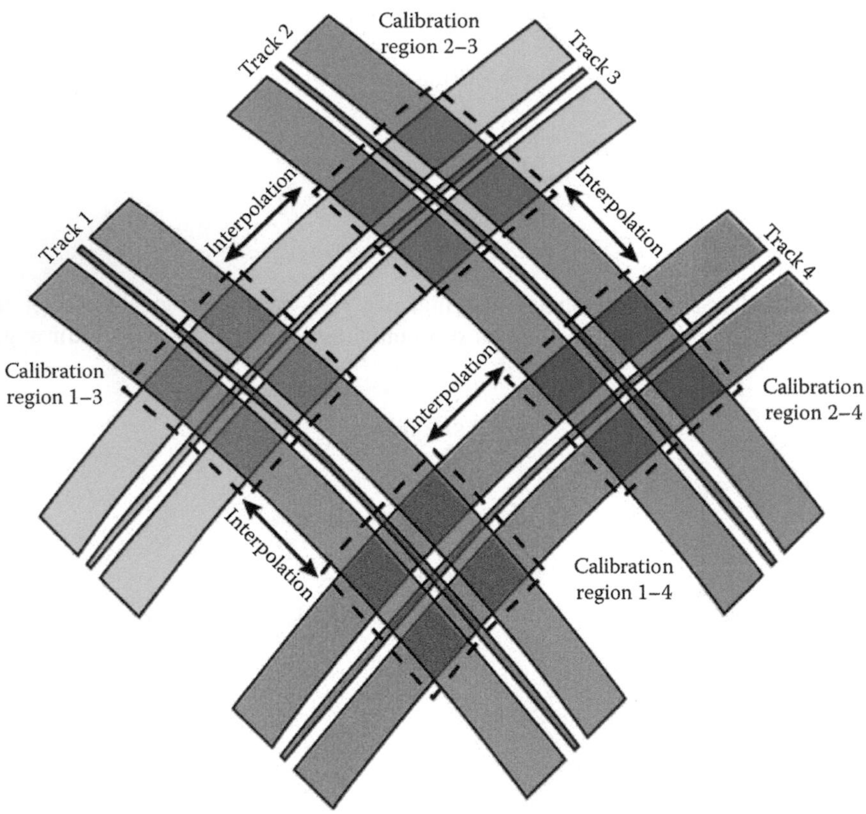

FIGURE 2.9 Geometry of observations for four orbits (blue, green, orange, and purple) intersecting at four cross-over points. The measurements for each orbit consist of wide swath measurements on either side of a nadir altimeter track. Two types of cross-over differences exist: altimeter-interferometer and interferometer-interferometer. Baseline roll/phase and range drift are estimated at each cross-over and the estimated values are optimally interpolated between cross-overs.

over that crossover, one can estimate in principle the calibration parameters if the surface variation between the two orbit passes at the crossover is small enough.

We model the height measurement at a crossover location, and at a time t, as

$$h_i(t) = h_{mss} + h_{sla}(t) + \delta h_G(t) + \delta h_I(t) + \alpha_i x_i + p_i + n_i$$

where the index i labels the cross-track position; h_{mss} is the stationary mean sea surface; $h_{sla}(t)$ is the SSH anomaly; δh_G are uncompensated geophysical delays (wet-tropo, EM bias residuals); δh_I is the range bias (which may drift); α_i is the uncompensated phase/tilt bias, assumed to be constant over the crossover; x_i is the cross-track distance; p_i is the phase screen value for the ith cross-track index position; and n_i is the interferometer random noise. The range bias can be controlled quite accurately but may drift over very long times. We assume that it can be calibrated by altimeter-interferometer crossovers or nearby altimeter- interferometer measurements. Roll biases at every crossover can be estimated using the procedure described as follows, and a mean roll bias subtracted prior to estimation, so that only the fluctuating roll components are estimated. This roll bias subtraction is a convenience but not a necessary component of the calibration and error propagation process outlined in the following text.

We proceed to estimate the αs at the crossover by taking the best estimate of the phase screen at a given iteration and performing a least-square inversion, assuming that the geophysical slope variability and residual phase screen are uncorrelated with the interferometer roll/phase errors. After estimation of the αs at each crossover, the values are optimally interpolated using the *a priori* knowledge of the drift rates of gyroscope roll and instrument phase drift. An example of the resulting error map obtained using the SWOT orbit and realistic contributions for the geophysical fields and only interferometer-interferometer crossovers is given in Figure 2.10. As can be seen, the residual ocean slope is about 0.7 μrad, but the residual errors occurring over land can be a factor of 2 worse (or greater, in the worst cases) due to the distance that the error has to be propagated away from the crossovers.

Improving roll corrections over land, given sufficient signal-to-noise ratio over land targets, can be implemented by using preexisting digital elevation models, as advocated by Enjolras and Rodríguez (2009), or by using large interior lakes or crossovers over suitably flat places, to avoid layover issues.

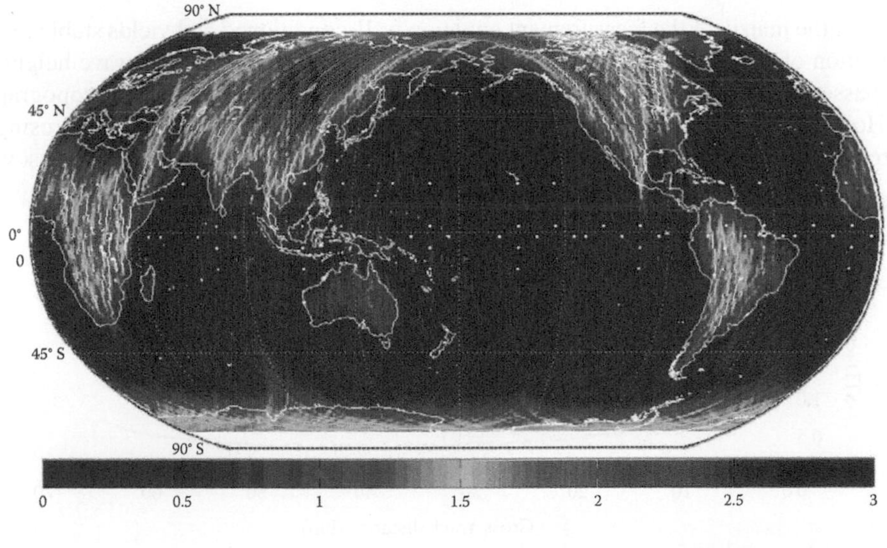

FIGURE 2.10 Slope errors (in μradians) after cross-over adjustments for roll/phase errors at cross-overs and optimal propagation away from the cross-overs. Notice that errors grow substantially inside the continents, where no water cross-overs exist. (From Esteban-Fernandez, D., *SWOT Mission Performance and Error Budget*, JPL D-79084, 2013. With permission.)

To estimate the phase screen at a crossover, we first remove the best estimates for roll and bias, as noted earlier. Assuming that the left and right swath phase screens are independent, we examine retrievals only using homogeneous (right/right or left/left) crossovers and assume that the inhomogeneous crossovers can be used for independent validation of the estimates. After bias and tilt removal, there will be N_p^2 crossover points (N_p is the number of phase screen values retrieved per sub-swath), which result in $N(N-1)/2$ equations of the form

$$h_i^A - h_j^D = p_i - p_j + \left(n_i - n_j\right) \text{for } i > j \tag{2.18}$$

where A/D stand for ascending/descending, and $N(N-1)/2$ equations of the form

$$\tilde{h}_i^A - \tilde{h}_j^D = -\left(p_i - p_j\right) + (\tilde{n}_i - \tilde{n}_j) \text{ for } i < j \tag{2.19}$$

where the tilde implies that the differences are collected at different positions (i.e., at the transpose of the grid) in the crossover diamond. Although every crossover will have N_p^2 phase screen overlaps, the size of the overlap area (which determines the measurement noise) will vary with the overlap angle. To bound the worst case error, we assume the worst case situation (in terms of overlap area), which occurs when cross-over tracks intersect at right angles, and the noise at the overlaps is given by the 1-km noise. Aside from the noise, the second set of equations is identical (up to a multiplicative factor) to the first set, so in reality we only have $N_p(N_p - 1)/2$ independent equations to derive the N_p unknown phase screen values. Examination of the equations shows that if the phase screen value is changed by an arbitrary constant, the resulting equations are identical. This makes the inversion of these equations highly singular and unstable. This instability can be fixed by imposing the requirement that the mean value of the phase screen across the swath is zero because it has been assumed that biases and tilts are not included in the phase screen. This additional condition leads to the additional equation

$$\sum_{i=1}^{N} p_i = 0 \tag{2.20}$$

which makes the matrix of the measurement equation well-conditioned and yields stable inversions. The estimation of the phase screen is most affected by changes in significant wave height (SWH) between passes because both the measurement noise characteristics and the actual topography will change. However, as shown in Figure 2.11, quite accurate estimates can be obtained using only a single crossover—even for significant values of the SWH, while using multiple crossovers will improve the estimate by the square root of the number of crossovers.

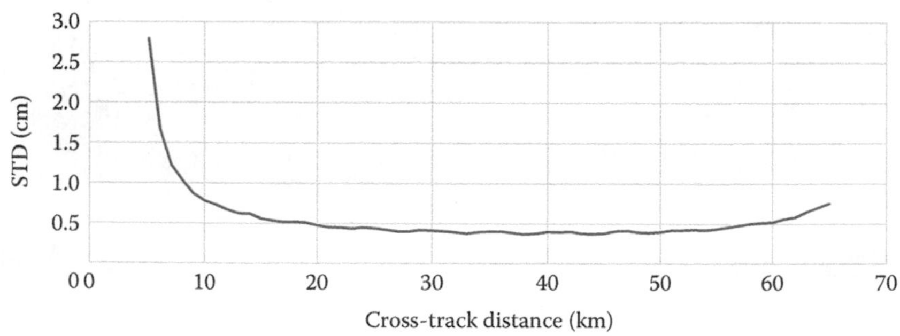

FIGURE 2.11 Standard deviation of the residual height error after correcting for phase screen estimates, given 4,000 Monte Carlo samples, as a function of cross-track position, for a white-noise phase screen. The cross-over averaging time is assumed to be 1 day and the SWH 6 m.

2.5 WIDE-SWATH ALTIMETRY PHENOMENOLOGY

Thus far, we have treated the measurement problem as if it were due to independent point targets alone. In this section, we study the impact of representing the scatterers as geophysical surfaces that may move or have variability in the backscatter, which may vary in the vertical and horizontal directions. We limit ourselves to the near-nadir direction and to the Ka band because this is the band selected for the upcoming SWOT mission, and it has been explored more thoroughly for near-nadir interferometry than other frequencies.

2.5.1 Water Brightness

Nadir altimetry has demonstrated that the ocean nadir backscatter cross section is a function of wind speed. Vandemark et al. (2004) review the nadir backscatter at the Ku and Ka bands and show that the nadir ocean radar backscatter is described well by a geometrical optics (GO) scattering model (Tsang et al. 1985) with a mean sea slope that depends on the wind logarithmically. Furthermore, the relationship at the Ku and Ka bands is quite similar, with the Ka band being slightly darker. Extrapolating this model from nadir to near-nadir angles (Tsang et al. 1985) results in

$$\sigma_0(\theta) \approx \frac{|R|^2}{s^2 \cos^4(\theta)} \exp\left[-\left(\frac{\theta}{s}\right)^2\right] \tag{2.21}$$

$$s^2 = 0.004 + 0.0093\, In\, (U_{10})$$

where U_{10} is the 10-m wind speed in m/s, and s is measured in radians.

Thanks to the recently launched Global Precipitation Mission (GPM) Ka band channel of the Dual-Frequency Precipitation Radar (DPR) instrument (Satoh et al. 2004), it is possible to validate Equation 2.21 for the large water bodies resolved by the DPR. The results are shown in Figure 2.12a, and they agree fairly well with the previous predictions.

In addition to global validation, validation of σ_0 has been done using radar data collected using airborne platforms (the Jet Propulsion Laboratory's [JPL] AirSWOT [https://swot.jpl.nasa.gov/airswot/], ONERA's BUSARD [Fjortoft et al. 2014]), using bridge mounted radars (Moller and Esteban-Fernandez 2014; Fjortoft et al. 2014), or radar data collected during wave tank experiments (Boisot et al. 2015b). In general, these observations are in agreement with the geometrical optics model, although Boisot et al. (2015b) find that the refined GO4 model (Boisot et al. 2015a) provides even better agreement as the incidence angle increases.

A significant difference is observed between the backscatter characteristics for small inland water bodies and those of ocean or large inland water bodies (see Figure 2.12b). In general, small water bodies are darker at off-nadir angles and sometimes exhibit a strong specular peak and very little off-nadir return (see Figure 2.12 for quantitative statistics). This difference is attributed to the fact that wind speeds over land are generally smaller, with a mode around 3 m/s, while ocean speeds have a mode between 7 and 8 m/s (Yin 2000; Archer and Jacobson 2003, 2005). In addition, it has been observed that for low enough wind speeds, somewhere around 2 m/s, light winds may fail to excite the small waves necessary for off-nadir scattering, so that small lakes may exhibit a completely reflective surface (Donelan and Pierson 1987; Plant 2000; Shankaranarayanan and Donelan 2001). In spite of this, the results shown in Figure 2.12 show that the backscatter is still generally bright enough to allow the retrieval of elevations from small lakes most of the time (although wind shadowing may cause certain parts of the lakes to be dark).

2.5.2 Wave Effects

The combination of the near-nadir incidence of ocean interferometers and moving surface waves leads to several effects not captured by the point target error budget discussed here.

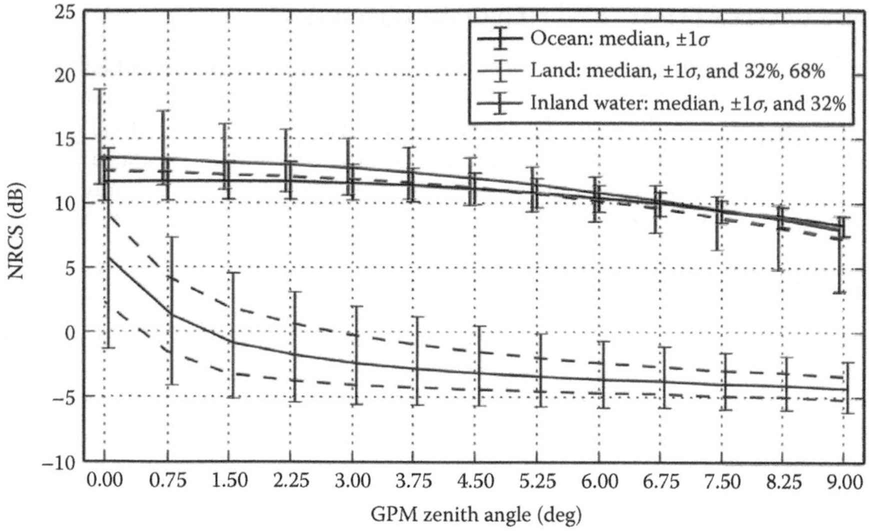

(a)

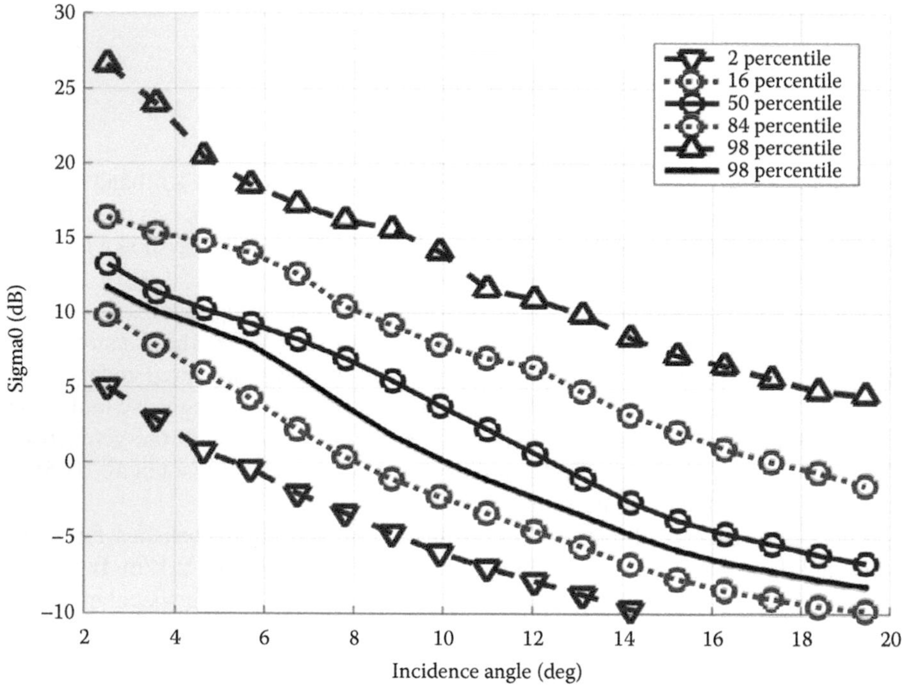

(b)

FIGURE 2.12 (a) Ocean, inland water and land measured σ_0 from the GPM mission. Note that the inland water measurements from GPM apply only for large water bodies. (b) Percentiles of σ_0 measured by the AirSWOT instrument over smaller water bodies.

2.5.2.1 The "Surfboard Effect"

As shown in Equation 2.6 and Figure 2.13, for near-nadir observations, sampling of the wave field occurs along iso-range planes ("surfboards") that have an angle relative to the local horizontal direction equal to the incidence angle. Even though the slope of the wave field is small, for small incidence angles, the same iso-range plane may sample the wave field many times

(Figure 2.13, upper panel) so that the measured wave field, which averages the interferogram among the multiply sampled points, can have significant distortions both in its height and spatial distribution (Figure 2.13, lower panel). Peral et al. (2015) examined this effect in detail and found that the main effect of this surfboard sampling effect was a distortion of the surface wave spectra that could leak into lower frequencies and measurement noise that depends on the dominant wavelength and direction of the surface waves (see Figure 2.14). These effects are maximized when the waves are traveling in the cross-track direction. In the long wavelength limit, the spectral distortion is given to the lowest order by a term proportional to the convolution of the surface height spectrum with itself, leading to spectral leakages at both higher and lower frequencies. This spectral distortion effect limits the usefulness of near-nadir interferometers for studying surface wave effects and may contaminate sub-mesoscale signatures to some extent for high SWH conditions.

2.5.2.2 Temporal Correlation Effects

For SAR interferometry, wave motion causes a change in the interferometric speckle pattern such that after a certain time the backscatter field at the beginning of the formation of the synthetic aperture is no longer correlated to the field at later points in the aperture (see, e.g.,

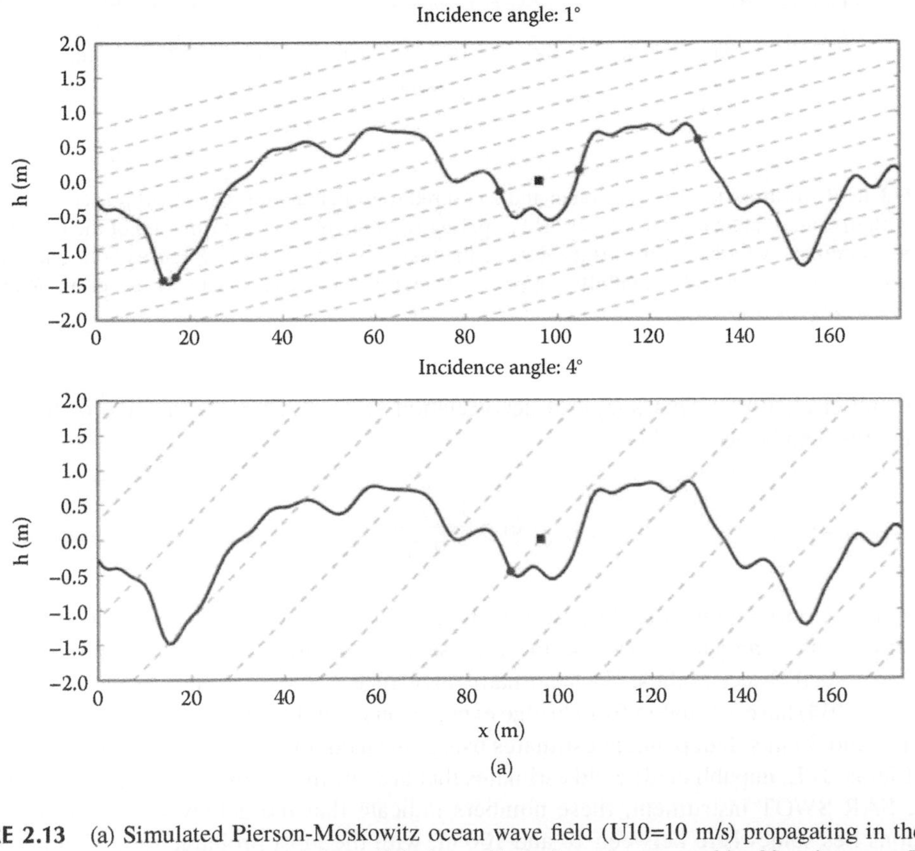

FIGURE 2.13 (a) Simulated Pierson-Moskowitz ocean wave field (U10=10 m/s) propagating in the range direction (blue) and sampled by a set of parallel planes (dashed gray) separated by 20 m in range. The top figure represents the sampling when the local incidence angle is 1°, while the bottom figure shows the same wave field when the incidence angle is 4°. The intersection of a sample plane with the mean level is shown as a black square, while the sampled heights are shown as red circles. The radar is on the left and the radar pulses propagate from left to right. *(Continued)*

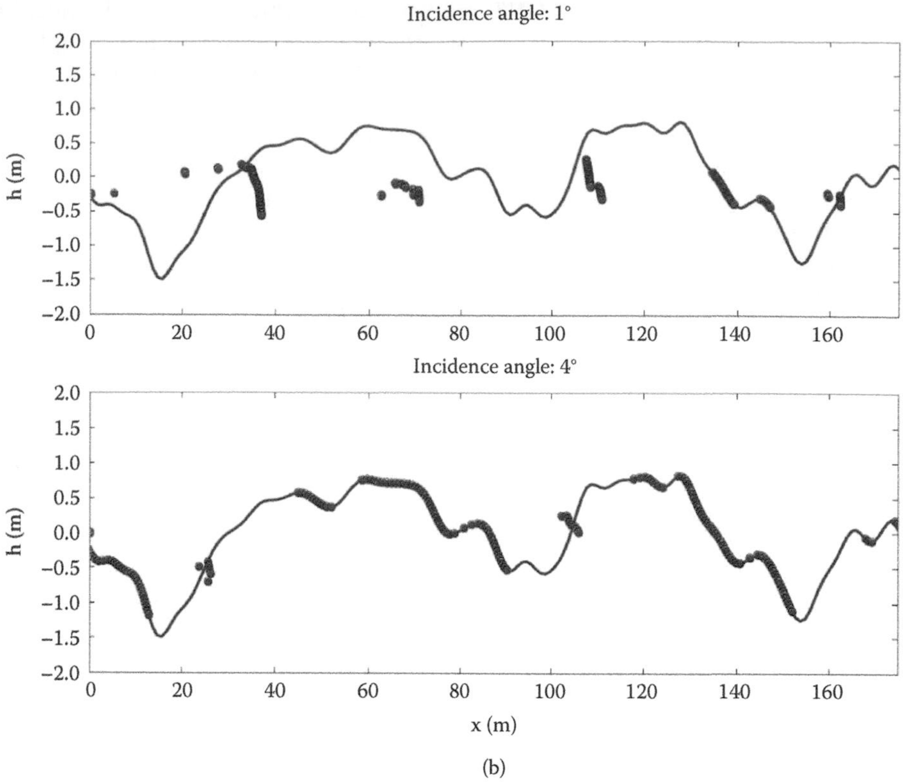

FIGURE 2.13 (Continued) (b) True (blue) and measured (red) heights for wave field presented in (a) for a system with finite resolution bandwidth (surface impulse response), plotted as a function of the intersection of the range coordinate with the mean surface. Notice the many-to-one resampling causes both uneven spatial coverage of the wave field and potentially significant height distortion. (From Peral, E., et al., *Rem. Sens.*, 7(11), 14509–14529, 2015.)

Hasselmann et al. 1985). This temporal decorrelation sets a limit as to the size of the azimuth resolution that can be achieved to

$$\delta y \approx \frac{1}{2}\frac{r\lambda}{\upsilon_p \tau_c} \tag{2.22}$$

where τ_c is the correlation time. In practice, the ocean correlation time has little effect for the measurement of ocean sub-mesoscale and mesoscale phenomena, but it may impact the measurement of small water bodies. For near-nadir incidence at the Ka band, Moller and Esteban-Fernandez (2014) have estimated from bridge experiments that the correlation time varied roughly between 5 and 50 m/s. Independent estimates using smearing of the AirSWOT azimuth response (B. Williams, JPL, unpublished) yield estimates that are toward the middle range of this interval. For the SAR SWOT instrument, these numbers indicate that the achievable range of azimuth resolutions lies somewhere between 10 and 100 m, with the most probable range being between 20 and 50 m. This degradation in resolution has little impact on SWOT's ocean goals but limits the size of rivers that can be resolved by the instrument. Note that for real aperture systems, such as WSOA, the temporal correlation does not impact the resolution because only one pulse at a time is used for aperture formation.

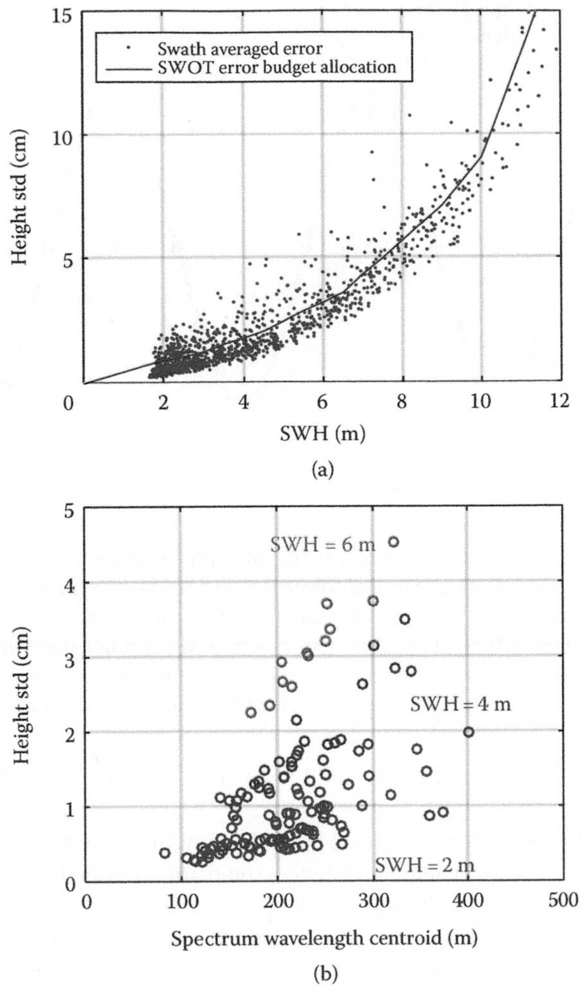

FIGURE 2.14 (a) Swath average height error for 1 × 1 km pixels as a function of SWH for multiple WAVEWATCH-III spectra. The red curve is the height error as a function of SWH that is used to project SWOT performance. (b) Swath average height error for 1 × 1 km pixels as a function of the spectrum wavelength centroid. (From Peral, E., et al., *Rem. Sens.*, 7(11), 14509–14529, 2015.)

2.5.2.3 Wave Bunching

As discussed in Section 2.4.6, a knowledge error in radial velocity will lead to a horizontal displacement for a moving target and, given a pitch in the spacecraft, a height error. In the case of a moving surface, when scatterers can have radial velocities that differ in magnitude and sign, this will result in a nonlinear resampling of the surface, as illustrated in Figure 2.15. The shifts in position can be substantial compared to ocean wavelengths; for instance, a 1-m/s wave orbital velocity will result in shifts near 150 m for the SWOT instrument. This surface resampling is well known for SAR imagery (e.g., Hasselmann et al. 1985), and its effect for the imaging of surface waves has been studied by Schulz-Stellenfleth and Lehner (2001) and Schulz-Stellenfleth et al. (2001), who characterized the distortion of the wave height spectra using Monte-Carlo simulations.

Previous literature concentrated on measuring surface waves rather than on assessing the impact on SSH at scales appropriate for sub-mesoscale and mesoscale phenomena. Several of the authors of this review have recently investigated this impact on longer wavelengths and near-nadir incidence

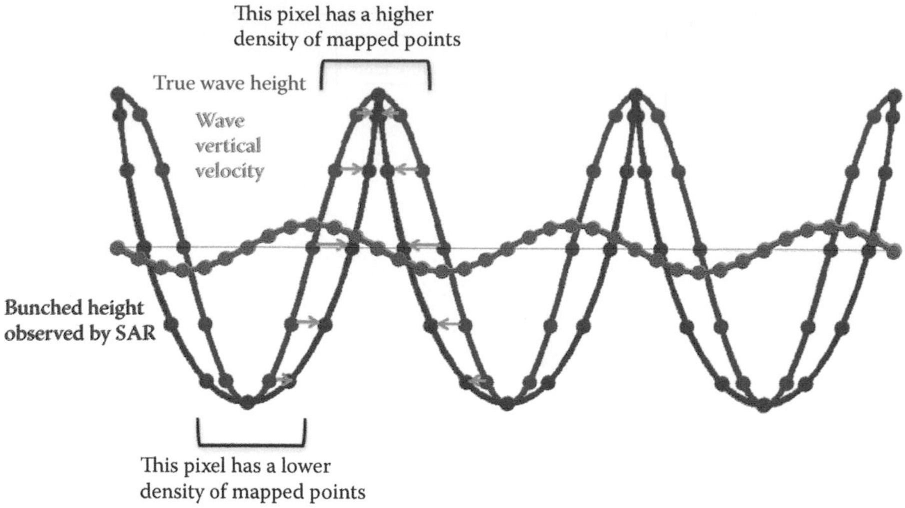

FIGURE 2.15 Cartoon illustrating the height wave-bunching effect for a sinusoidal moving surface, such as long-wavelength swell, viewed from the near-nadir direction. The blue line represents the true surface, and the green line the orbital vertical velocity. The radial motion induces the distortions shown by the arrows, leading to the surface points to be shifted into the red line, which shows significant skewness, with flatter troughs and peaky crests. Finite pixel resolution (pixel size indicated by brackets), will result in more points being contained in trough than in crest pixels, leading to brightness variations in the imaged field.

and have found several effects that, though small for surface wave imaging, must be considered for SSH studies.

The long-wavelength issues brought about by wave bunching can be understood qualitatively by examining Figure 2.15. At near-nadir incidence, the primary contributor to the radial velocity error is the vertical component of the wave orbital velocity, so that the direction of wave travel is not a significant factor in the net effect. The nonlinear mapping of scatterers will tend to preferentially map troughs and crests in opposite directions, leading to pixels that have an excess or a deficit of scatterers mapped into them, as well as an apparent skewness in the wave field and spectral distortions. If the spatial resolution was infinitely precise, averaging the heights of all the scatterers remapped in azimuth would yield the same mean value as before the remapping. However, if scatterers are binned into finite pixels, without taking into account the density of scatterers (i.e., brightness), and the height average is performed, then a height bias may appear, as well as long-wavelength leakage from beat patterns in the brightness. Therefore, when performing averages, the surface brightness must be taken into account: This can be done to some degree by flattening the interferogram (i.e., removing the phase contribution for a flat surface, which has a ramp in range) and averaging the resulting flattened complex interferogram, which includes the brightness variations.

Even if one averages the flattened interferogram, the nonlinear shifts will distort the wave spectrum. Similarly to the surfboard effect, the lowest order effect can be shown to lead to the measured spectrum having additional terms that are proportional to convolutions of the true spectrum with itself, resulting in shifting energy to both lower and higher frequencies.

To account for all of the effects (interferogram averaging, finite pixel size, surfboard effect, and large azimuth shifts) to all orders, it is necessary to simulate the return signal. The results for a swell-like wave spectrum are presented in Figure 2.16, where, in addition to the interferogram to 250-m resolution, we average the cross-track elevations with a 7.5-km window, sufficient for meeting the SWOT requirement of measuring 15-km sub-mesoscale features. The results for using high-resolution heights without significant interferogram averaging over scales

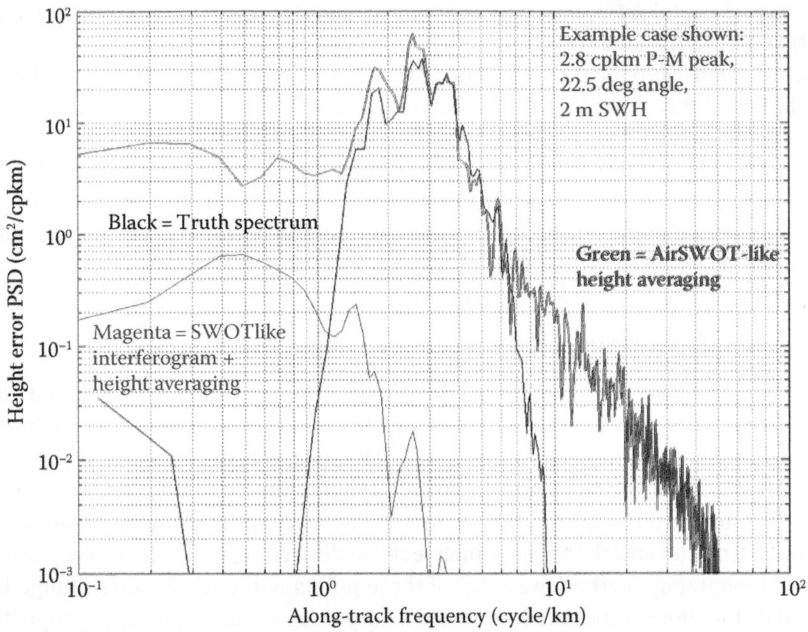

FIGURE 2.16 Effect of wave bunching on along-track height spectra for a simple long wavelength swell-like ocean spectrum (black line). The green line is the resulting spectra for high-resolution imaging, where heights are averaged without taking into account surface brightness or without significant cross-track interferogram averaging. The magenta line represents the results of averaging the interferogram, which includes power weighted averaging, followed by averaging in the cross-track direction by 7.5 km, corresponding to the SWOT goal of resolving 15 km sub mesoscale features.

similar to the wave period (green line) show that wave bunching can have significant impact on the retrieved spectrum—not only at small scales but especially at long wavelengths, introducing distortions that are on the same order of magnitude as the expected levels shown in Figure 2.3. On the other hand, if interferogram averaging is performed to scales similar to the peak wavelength, followed by the same level of cross-track averaging that SWOT assumes (Esteban-Fernandez 2013), then the low-frequency contributions are significantly lower than the expected sub-mesoscale spectral levels.

We conclude that for ocean imaging, or imaging of large water bodies with significant SWH, interferogram averaging must be performed to suppress wave bunching leakage into the sub-mesoscale. The direction of the wave propagation is of secondary importance to this along-track spectral distortion for near-nadir incidence, where vertical orbital velocities dominate. This consideration does not apply to rivers and smaller water bodies, where waves are much smaller and of shorter wavelength, and so contribute substantially less to the radial velocity.

2.5.2.4 The EM Bias

The EM bias is introduced by modulations in surface brightness that have a net height dependence, leading to a lower mean level of estimated SSH. The cause of the brightness modulation is due to both the modulation of large-scale waves slopes (flatter at the trough than the peaks, with a dependence on significant slope) and hydrodynamic modulation of small wave scatterers by the wind and surface currents (see, e.g., Walsh et al. 1989, 1998; Rodriguez et al. 1992; Elfouhaily et al. 2000, 2001; Vandemark et al. 2005 for a sampling of the extensive and still evolving literature). The nadir incidence EM bias is also frequency dependent, with generally larger biases at lower frequencies, although the experimental difference between the Ku band and the Ka band has been reported to be large (Walsh et al. 1989) and relatively small (Vandemark et al. 2005).

There are no extensive published measurements of the angular dependence of the EM bias for general conditions at this time. However, Peral et al. (2015) note that, for off-nadir incidence, the surfboard effect can modify the values of the nadir EM bias. Figure 2.17 shows the results of detailed simulations that include realistic sampling and processing for a large ensemble of WAVEWATCH-III spectra, when the input EM bias is set to be a fixed 3% of SWH. As can be seen, the resulting EM bias is modified at the incidence angles where the surfboard layover effect dominates, and the resulting EM bias is less than the theoretical expectation.

2.5.3 Layover and Vegetation Effects

In the previous section, we saw that, due to the tilted sampling implicit in the radar range resolution, returns from different waves at different heights could arrive simultaneously at the radar, causing distortion and location and height errors. This is a general phenomenon in radar remote sensing that is called *layover* because the imaged topography appears "laid over" in the image. This situation is illustrated in Figure 2.18, where it is shown how, for imaging of surface water bodies, an iso-range line can simultaneously intersect the desired water surface, vegetation along the riverbank, and topography further away. All of these points will have the same range but different look angles and, therefore, different interferometric phases because phase is proportional to the look angle. In general, all points on a line having a slope equal to the angle incidence will arrive at the same time at the radar. For near-nadir incidence, this slope is small, and the likelihood of having layover increases.

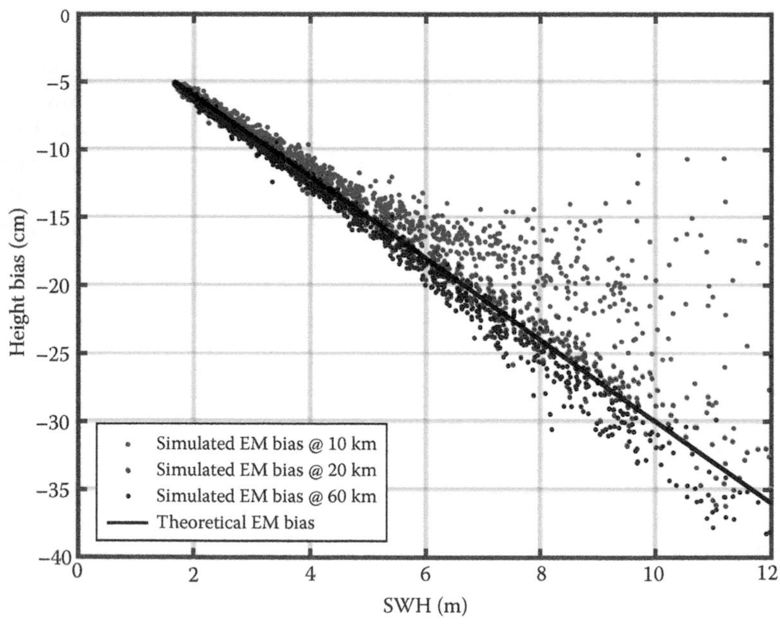

FIGURE 2.17 Electromagnetic bias from simulation including the surfboard effect at 10 km (blue), 20 km (green) and 60 km (red) as a function of SWH, compared with input nadir EM bias in black. The results are for an ensemble of 300 globally sampled WAVEWATCH-III spectra. The EM bias for larger incidence angle follows closely the input EM bias, but deviates for smaller incidence angles where the surfboard effect is larger. (From Peral, E., et al., *Rem. Sens.*, 7(11), 14509–14529, 2015.)

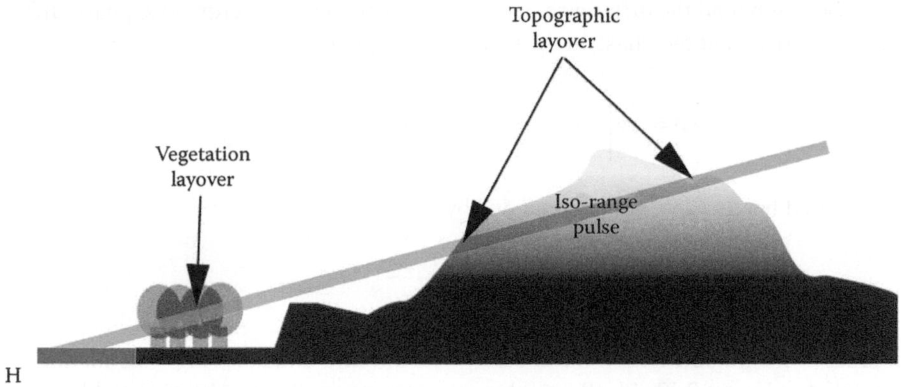

FIGURE 2.18 Cartoon illustrating how a radar interferogram can have contributions from the desired water body (shown in blue), stream-side vegetation, and topography, since the scatterers at each of these locations have the same range relative to the radar. The red line indicates an iso-range line for a radar located to the left and up from the edge of the image. (From Fu, L., et al., *SWOT: The Surface Water and Ocean Topography Mission: Wide-Swath Altimetric Measurement of Water Elevation on Earth*, JPL-Publication 12-05, Jet Propulsion Laboratory, Pasadena, CA, 2012. With permission.)

To assess the impact of layover on measured heights, we can separate the total interferogram in Equation 2.4 into the contributions from the water target and layover points to obtain

$$\langle E_1 E_2^* \rangle = P_\omega \gamma_\omega \exp[-i\Phi_\omega]\left[1+\frac{P_L\gamma_L}{P_\omega\gamma_\omega}\exp[-i(\Phi_L-\Phi_W)]\right] \tag{2.23}$$

where the power, correlation, and phase contributions from water and layover have been explicitly separated, as follows:

$$P_\omega \gamma_\omega \exp[-i\Phi_\omega] = A\int_{-\Delta x/2}^{+\Delta x/2} dx\,dy\; \chi_a^2(y,x)\chi_r^2\left(t-2r(x)/c\right)\exp[-i\Phi(x)]\sigma_{0_\omega}(x)$$

$$P_\omega = A\int_{-\Delta x/2}^{+\Delta x/2} dx\,dy\; \chi_a^2(y,x)\chi_r^2\left(t-2r(x)/c\right)\sigma_{0_\omega}(x)$$

$$P_L\gamma_L \exp[-i\Phi_L] = A\int dx\,dy\,dz\; \chi_a^2(y,x)\chi_r^2\left(t-2r(x,z)/c\right)\exp[-i\Phi(x,z)]\sigma_{0L}(x)$$

$$P_L = A\int dx\,dy\,dz\; \chi_a^2(y,x)\chi_r^2\left(t-2r(x,z)/c\right)\sigma_{0L}(x)$$

The first two integrals are taken over the water areas, while the last two are taken over all space excluding the surface contribution from the water. We identify P_w and P_L as the return powers from water and layover, respectively; the real quantities γ_w and γ_L are the interferometric correlations from water and land; finally, $\Phi_w$and Φ_L are the effective interferometric phases from water and land, respectively.

The second term inside the square brackets in Equation 2.23 is the contribution due to the layover, while the factors outside represent the return from the water surface in the absence of layover. When we are in the near-nadir regime, water will be significantly brighter than land and, due to the greater angular extent of the layover, the water correlation will be greater than the land correlation.

In this case, we can bound the difference between the expected interferometric phase difference in the absence of layover and the phase in the presence of layover as

$$\delta\Phi = \arg\left[1 + \frac{P_L \gamma_L}{P_\omega \gamma_\omega} \exp[-i(\Phi_L - \Phi_W)]\right] \leq \frac{P_L \gamma_L}{P_\omega \gamma_\omega}$$

and the associated height error will be bounded by

$$\delta h \leq \frac{x_0}{kB} \frac{P_L \gamma_L}{P_\omega \gamma_\omega}$$

As an order of magnitude, for SWOT at the $x_0 = 35$ km mid-swath, $x_0/kB \approx 4.5$ m, and, given the results presented in Figure 2.12, $PL/Pw \leq 0.1$, so the maximum effect of layover will be on the order of 0.5 m. However, as we will see in the following, the scattering areas for the layover, as well as the greater decorrelation exhibited by the layover, can reduce significantly the error associated with layover.

To quantify the layover problem, we separate the layover by topographic variations due to the presence of vegetation surrounding the water body. To assess the effect of topographic layover in greater detail, one must use a high-resolution digital elevation model (DEM) to integrate the interferogram equations numerically. Simulations over different river basins have been conducted using a variety of DEMs. We show representative results in Figure 2.19 for the Ohio River Basin using the SRTM DEM. In this simulation, we assume, conservatively, that the water normalized cross section is only 10 dB brighter than the land cross section. The pixels containing some water are geolocated using the interferometric equations, and the elevation errors relative to the known water surface are calculated. In Figure 2.19, results are presented for approximately 4° and less than 1° incidence angles. To illustrate the layover impact, a color is associated with each point, depending on whether it has no layover (blue) or layover with a small height error (less than 5 cm), or it is contaminated by

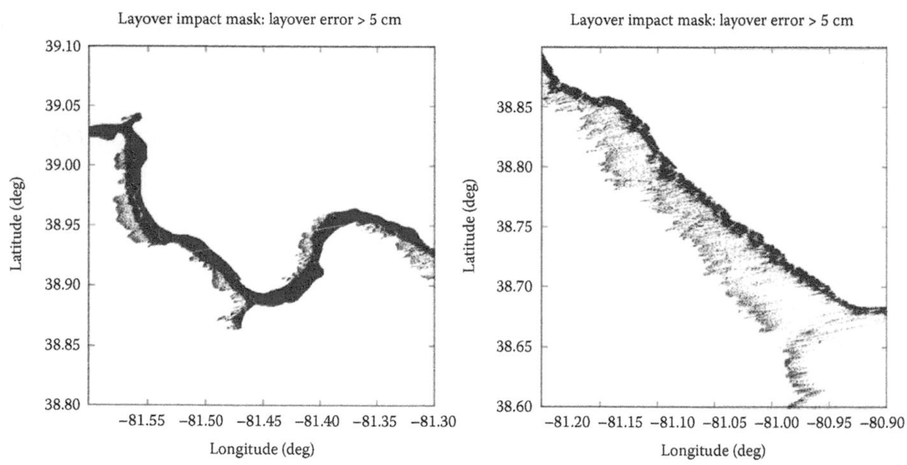

FIGURE 2.19 (a) Mid-swath mask of all points containing some water return from the Ohio River simulation. Blue points are pure water, while red points are a combination of water and layover. The points have been geolocated according to their interferometric phase, so points with large phase errors will lie outside the river. (b) Same as (a), but for a portion of the river in the near range, where layover is prevalent. In the right image, all pixels with a height error smaller than 5 cm have been colored blue, as this is an acceptable height error (otherwise, most near-range pixels would be colored red, as most all exhibit some level of layover). (From Fu, L., et al., *SWOT: The Surface Water and Ocean Topography Mission: Wide-Swath Altimetric Measurement of Water Elevation on Earth*, JPL-Publication 12-05, Jet Propulsion Laboratory, Pasadena, CA, 2012. With permission.)

layover and has a greater height error (red). (If the height error threshold had not been used, most pixels for the 1o incidence would have been affected by layover.)

The results shown in the figure show the two main features of topographically induced layover. First, when the land brightness dominates, a large phase error occurs, and this is associated with large geolocation errors, as predicted by Equation 2.10. Many points will lie wide of the river and could potentially be filtered as outliers. The second characteristic is clumping in areas where the viewing geometry leads to significant layover from topography near the river. If the land topography and brightness were known *a priori*, these areas could be identified and excluded from the estimates of surface elevation, but due to the relationship between height and geolocation errors, this level of *a priori* knowledge may not be available everywhere. Additional factors, such as surface distortions and drops in correlation, can be used to identify layover areas but, at this point, the automated identification and exclusion of areas impacted significantly by topographic layover remains an area of active research.

Vegetation present at the fringes of a river, lake, or wetland, or contained within the water body for flooded areas, can distort the signal from the water surface in two ways: First, if it happens to lie in the path between the antenna and the water surface, it will attenuate the return from the desired target. The attenuated signal will have a lower signal-to-noise ratio and will therefore present a higher level of random noise. However, this random noise will be unbiased and will not distort the estimation of the mean water level. Due to the fact that the return is darker, water that lies below trees may be misclassified as land. If μ is the ratio between the vegetation σ_0 and the open water σ_0, one will not be able to differentiate between water and land if the vegetation attenuation term (which, for a homogeneous canopy, will exhibit exponential attenuation $\exp[-\beta\,(T-z)]$, where z is the height above the surface, and T is the tree height) is approximately equal to μ (Tsang et al. 1985). Experimental evidence suggests that for vegetated areas, μ ranges between –20 and –10 dB, with a likely value of about –15 dB. This implies that β must range from 4.6 to 2.3, with a most likely value of 3.5 for water to be separable from land.

The second distortion due to vegetation is height bias that occurs when layover occurs. Typically, when layover from a vegetation patch occurs, there will be a range of distances, $[x_1, x_2]$, from the water scatterer (and corresponding set of heights, $[z_1, z_2]$, $0 < z_1 < z_2 < T$) that will be laid over. For each layover height, the vegetation will introduce a phase error proportional to the vegetation brightness at that height (including attenuation) and the intercepted volume. Integrating the interferogram equation, the total phase error will be

$$\delta\Phi = \arg\left[1+\frac{\mu\upsilon}{\beta+ik_zT}\left(\exp\left[ik_z(T-z_1)\right]e^{-\beta z_1/T}-\exp\left[ik_z(T-z_2)\right]e^{-\beta z_2/T}\right)\right]$$

$$k_z=\frac{kBT}{(H-h)\sin\theta}$$

$$v=\frac{\beta}{1-e^{-\beta}}$$

This equation predicts that the phase (and height) errors depend on the range of distances from the water that are laid over along the plane of incidence. To quantify this effect for realistic vegetation distributions, a detailed simulation of the interferometric errors for the water bodies in the Amazon Basin was conducted based on the vegetation/water mask from Hess et al. (2003). We assume that the vegetation is present everywhere with a height of 20 m and that penetration into the canopy occurs with the observed penetration of other radar interferometric data that typically reports a tree height of about 60% to 70% of the true height (Rosen et al. 2000). (Note that there is only limited Ka band experimental data at near-nadir, so this assumption may need revisiting as additional data becomes available.) Sample results of the simulation are presented in Figure 2.20 for three areas with different river and vegetation morphologies. The root mean

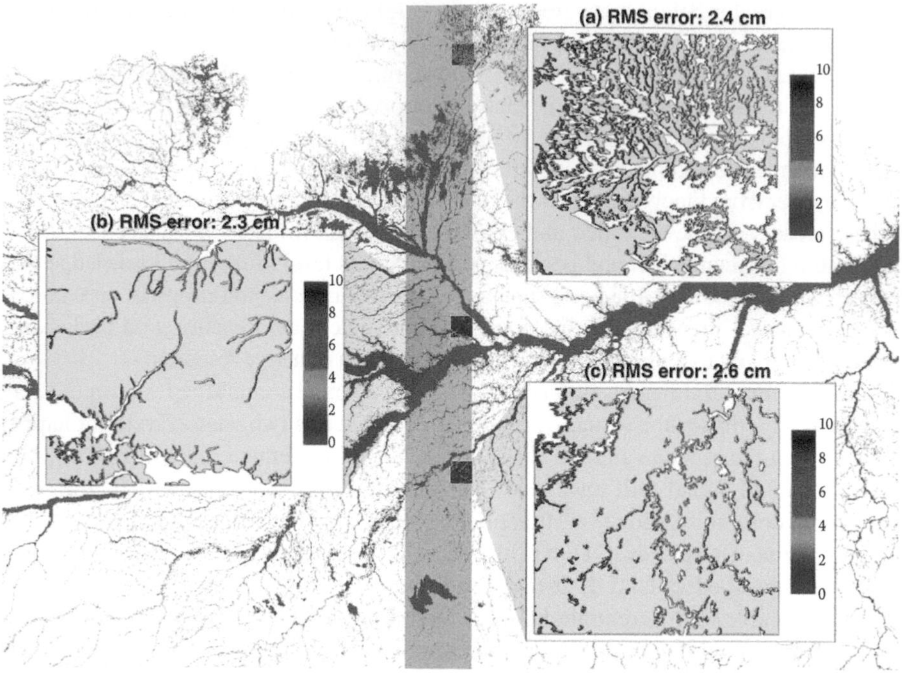

FIGURE 2.20 The area in green shows the region used to simulate the effect of vegetation on the interferometric heights for a region in the Amazon basin. The water mask used is shown in blue, and we magnify three 60 km² areas containing small rivers with different river and wetland morphology. The instrument was looking from the left (west), and one can see the effects of layover errors on the right (east) bank of the water bodies. In the far range (east side of the swath), the errors can be larger than in the near (west) side, due to higher volumetric correlation of the vegetation returns. The root mean squared error for the entire simulation is 2.8 cm, assuming that the water was 10 times brighter than the vegetation. (Based on data used in Alsdorf, D., et al., *Rev. Geophys.*, 45, 2007.)

squared height error for the entire data set was about 2.8 cm, with a more pronounced error for smaller water bodies and for the larger incidence angles. Although in a few extreme cases the errors can be as large as 30 cm, the performance over the scene meets the desired accuracy for the mapping water bodies. Errors can be reduced by excluding the areas on the affected banks at the expense of reducing coverage.

In summary, due to the phase variation within the vegetation layer and the associated decorrelation, vegetation effects tend to be significantly lower than the effects due to topographic layover for simple vegetation models. This conclusion should be validated by means of direct measurements, such as the ones that will be collected by AirSWOT, and greater sophistication in the vegetation models.

2.6 WIDE-SWATH ALTIMETRY MISSION DESIGN

The design of a wide-swath mission is driven by two needs: global coverage and temporal revisit time. From Figures 2.1 and 2.2, it is clear that the maximum desirable revisit time for ocean phenomena must be on the order of 10–20 days, or shorter, to capture short mesoscale variability. Hydrology applications at basin scale would also prefer sampling at better than monthly scales. In both cases, there will be fast changing phenomena (sub-mesoscales and flood waves) that will be under-sampled even if 10-day sampling is achieved, but intermittent observations will still be useful for studying these processes. To achieve global sampling at these time scales, one can see from Figures 2.1 and 2.2 that a swath width on the order of 100–200 km is required.

An additional driving restriction on the instrument design is the requirement to achieve centimetric precision through the entire swath. Given the extreme sensitivity to roll and phase errors discussed earlier, achieving this level of precision, even after using crossover adjustments, requires that the incidence angles be as close to nadir-looking as possible. Even for near-nadir incidence, the stability requirements in both the length and roll distortions of the interferometric baseline, which are at the micron level, require that the mechanical structure for the baseline be as short and stiff as possible, which reduces phase sensitivity. To overcome this need for short mechanical baselines, one can reduce the electromagnetic wavelength because the total phase sensitivity is proportional to the ratio B/λ.

In addition to these driving requirements, additional requirements may arise for specific applications. Ocean tides will be under-sampled by any altimeter mission, and the orbit inclination must be selected so that the principal tidal components do not alias into the phenomena of interest. For hydrology applications, it is required that the spatial resolution be commensurate with small water bodies, leading to the requirement for SAR imaging. Finally, it is desirable to be able to fill in the nadir gap, where interferometry does not work, to provide water vapor corrections and independent instrument validation, so including a conventional two-frequency altimeter and water vapor radiometer (WVR) enhances the measurement suite significantly.

These design considerations have led to the proposal of two wide-swath altimeter instruments that have very similar general architecture. The WSOA (Rodríguez and Pollard 2001; Pollard et al. 2002; Fu and Rodriguez 2004a; Enjolras et al. 2006), pictured in Figure 2.21, a demonstration instrument with the NASA/CNES Ocean Surface Topography Mission (OSTM)/Jason-2 but ultimately descoped, pioneered the basic architecture for a wide-swath interferometer. Subsequently, the WaTER concept (Enjolras and Rodriguez 2009) and the NASA/CNES/CSA SWOT mission (Durand et al. 2010a), used the Ka band Radar Interferometer (KaRIn), together with a Jason or AltiKa class altimeter/WVR, to achieve greater precision and spatial resolution. A comparison in the basic parameters for each mission concept is presented in Table 2.1.

The measurement scheme, illustrated in Figure 2.22, deploys two antennas on either side of a conventional nadir altimeter and WVR instrument. One of the antennas is capable of alternately illuminating swaths on either side of the nadir track, and an interferometric pair for each swath is formed by combining the backscattered signal received by both antennas simultaneously. The returns from the left and right swaths are separated from each other by using orthogonal polarizations for each

FIGURE 2.21 The Wide-Swath Ocean Altimeter (WSOA) mission concept consisted of a Ku-band real-aperture interferometer using a 5.4 m deployable baseline, working in conjunction with the Ocean Surface Topography Mission/Jason-2, to fill in the nadir gap and provide tropospheric ionospheric corrections at nadir.

TABLE 2.1
Comparison of WSOA, WaTER and SWOT Parameters

Parameter	WSOA	WaTER	SWOT
Orbit height	1334 km	834 km	981 km
Orbit inclination	66°	93:7°	77:6°
Orbit repeat period	10 days	16 days	20.86 days
Average revisit period	5 days	8 days	11 days
Total swath	200 km	120km	120 km
Spatial resolution	15 km	500 m (ocean) ~ 50 m (land)	500 m (ocean) ~ 50 m (land)
Wavelength	2.2 cm	0.8 cm	0.8 cm
Baseline length	6.4 m	10 m	10 m
Peak transmit power	100W	1.5 kW	1.5 kW
Altimeter	Ku/C Jason	AltiKa	Ku/C Jason
Radiometer	Nadir WVR	Nadir WVR	2-beam WVR

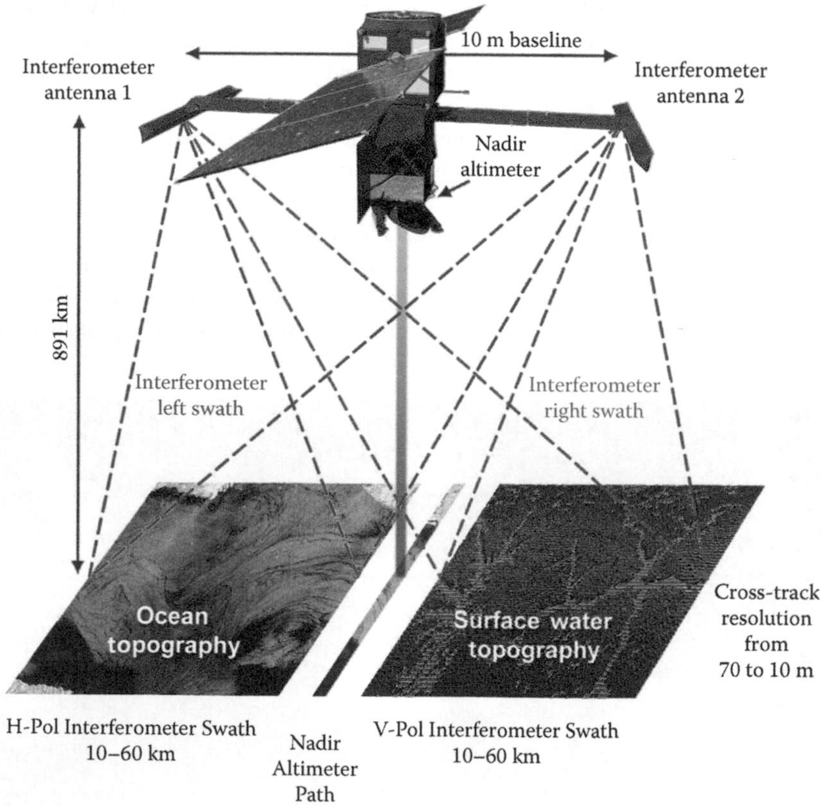

FIGURE 2.22 The forthcoming Surface Water Ocean Topography (SWOT) mission uses a Ka-band SAR interferometer with a 10-m rigid interferometric baseline. A 120-km swath is achieved by imaging with separate beams on either side of the nadir track. The nadir gap is filled in by a Jason class real-aperture Ku/C-band altimeter. A two-beam water vapor radiometer provides wet-tropospheric corrections at the center of each swath. The spatial resolution over land is on the order of 50 m, while onboard processing provides 500 m ocean resolution.

swath and by ensuring that the antenna cross-polarization isolation is sufficient to exclude ambiguous signals that arrive from the opposite swath.

The WSOA design, conceived as a demonstration mission, used real aperture Ku-band radar to achieve 15-km height postings. Due to technology limitations, only a 5.4-m baseline could be deployed in the relatively small OSTM bus, while a peak power of only 100 W was available. Within these restrictions, the measurement random height error for 15-km height posting (i.e., sufficient to resolve a 30-km wavelength) was on the order of 2 cm throughout the swath, as shown in the upper panel of Figure 2.23. Translated into the spectral domain, this level of noise is sufficient to resolve

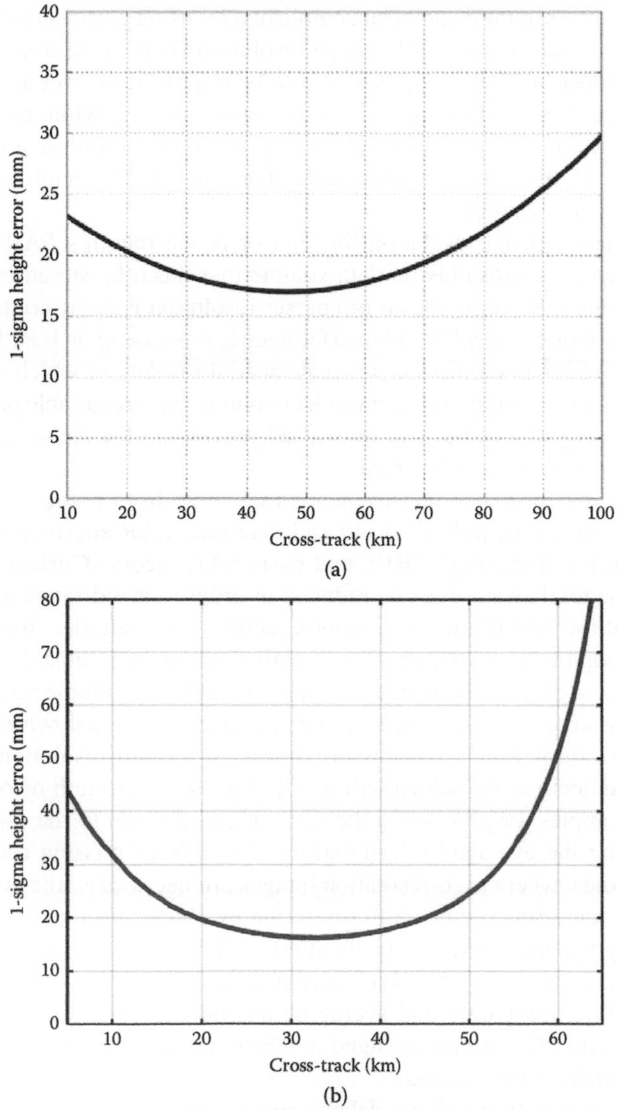

FIGURE 2.23 (a) WSOA 1 σ ocean height precision for a posting of 15 × 15 km (From Fu, L., and Rodriguez, E., High-resolution measurement of ocean surface topography by radar interferometry for oceanographic and geophysical applications. In *The State of the Planet: Frontiers and Challenges in Geophysics, IUGG Geophysical Monograph*, vol. 19, International Union of Geodesy and Geophysics and the American Geophysical Union, pp. 209–224, 2004.). (b) SWOT 1 σ ocean height precision for a posting of 1 × 1 km in along and across-track directions. (From Esteban-Fernandez, D., *SWOT Mission Performance and Error Budget*, JPL D-79084, 2013. With permission). The noise level at 15 km posting will be a factor of 15 smaller than shown in the figure.

mesoscale signals at scales on the order of 100 km, according to Figure 2.3, but it is not really sufficient for sub-mesoscale applications. In addition, due to the real aperture processing, it is also not sufficient for surface water applications.

To improve the performance and resolution, the WaTER and SWOT KaRIn design introduces several major improvements that were enabled by new technology. The baseline is now made of solid articulated pieces, so that a 10-m baseline is now realizable within the desired stability and control constraints. In addition, by moving from the Ku band (2.2 cm) to the Ka band (0.8 cm), the combined effect of baseline and wavelength changes results in a factor of four improvement in phase sensitivity (Equation 2.9) for the same level of phase noise. Finally, improvements in resolution and signal-to-noise ratio can be accomplished by using a transmitter with a peak power output 15 times higher, and improving the range resolution from 5.6 to 0.75 m allows the signal-to-noise level and number of independent samples to be improved by over an order of magnitude, as displayed in the lower panel of Figure 2.23, which shows similar white noise levels, but with a factor of 15 improvement in the spatial resolution, so that the spectral noise levels achieved at high frequencies are about 15 times smaller for KaRIn than for WSOA, enabling mapping of 15-km wavelength sub-mesoscale features.

Achieving appropriate spatial resolution for land or ocean requires SAR processing, but the desired resolution imposes constraints on data volume that much be stored in the spacecraft and transmitted to the ground. Because 250-m or coarser resolution is appropriate for the ocean, one can contemplate using unfocused SAR and interferometric processing on board. On the other hand, appropriate sampling of rivers requires meter-scale spatial resolution, which is not yet achievable with commercially available storage and downlink or commercially available processors. Therefore, SWOT implements a hybrid scheme with unfocused processing for the ocean and fully focused processing for land, as will be described next.

Although the noise performance improvements that come from going to SAR processing are significant, they come at a cost in both swath and data rate. SAR aperture formation requires a very fast pulse repetition frequency (PRF), and basic SAR theory (Curlander and McDonough 1991) dictates that the total swath must be reduced in order to avoid range ambiguities. In practice, this implies that the SWOT swath is about a factor of two smaller than for WSOA, with a consequent degradation of the revisit time by a similar factor (see Table 2.1). Another practical disadvantage of going to SAR mode is that the data rate and data volume required for SAR aperture formation is very large and cannot be stored in the spacecraft and downlinked globally at this time. To overcome this limitation, it is necessary to process the full-resolution SAR data on board by either filtering and decimating (which will result in a loss of azimuth resolution and the number of independent samples) or processing the data on board down to the average interferogram level and downloading the averaged interferogram. The SWOT mission uses both approaches. For surface water bodies, where high-resolution images are necessary, filtering and decimation is used: The penalty is degrading the azimuth resolution by about a factor of two to approximately 5 m, while increasing the random noise by about 40%. For the ocean, where spatial resolution is not critical but height accuracy is, a novel onboard processor (OBP) that will perform multi-beam interferogram formation, flattening, and averaging has been designed (Peral 2016). The output from the OBP will be interferograms averaged to 500-m resolution and 250-m posting that keep the full accuracy available from the data.

The correction of the wet tropospheric delay is another area where conventional altimetry and wide-swath interferometry differ. Using a single-beam WVR will correct heights along the nadir track over the ocean but will not be able to correct for cross-track variations in water vapor, which can be several centimeters over the entire swath. To mitigate this cross-track variation, the SWOT mission uses a two-beam WVR to provide water vapor corrections mid-swath for each of the swaths, while corrections at other swath locations are obtained by interpolation. Jason-class water vapor radiometers cannot estimate the wet tropospheric delay over land. If uncorrected, this delay can be as large as approximately 30 cm, with a variability on the order of 5–10 cm, depending on location.

TABLE 2.2
Representative Hydrology Error Budget for the SWOT KaRIN Instrument

Hydrology Error Component	Height Error (cm)	Slope Error (μrad)	Comments
Ionosphere signal	0.8	0.1	RMS of the full signal for maximum solar activity (100 TECU), using IONEX model
Dry troposphere signal	0.7	0.1	RMS after correction with weather models, based on Jason heritage.
Wet troposphere signal	1.0	1.5	Model-based correction
Orbit radial component	1.6	0.02	Orbit error RMS
KaRIn random & systematic errors	6.9	8.5	KaRIn roll-up, after cross-over correction.
KaRIn random	2.4	8.32	See Section 4.2
KaRIn systematic cross-track errors after cross-over correction	6.33	1.62	See Section 4.7
KaRIn systematic along-track height bias error	1.5	0.08	Uncalibrated bias
High frequency errors	0.035	<0.01	Mechanical motion aliasing
Motion errors	0.4	0.8	See Section 4.6
Total (RSS)	7.2	8.7	Current best estimate
Requirement	10	17	(Rodríguez, 2015)

Note: The bottom line represents the science requirement, while the lines above represent the expected performance, as of summer 2016 (which may change, not to exceed the requirements, as the mission evolves)

Fortunately, much of this effect can now be reduced by applying seasonal means or forecasts from numerical weather prediction models (that assimilate *in situ* data), and it is now estimated that the residual wet tropospheric delay over land will be on the order of approximately 1 cm by the time SWOT launches (S. Brown, personal communication).

A complete summary of the error budget for a wide-swath mission involves specific assessment of the performance of mechanical and electrical hardware and is beyond the scope of this review. A detailed accounting of the various components of the error budget has been presented in Fu et al. (2012) and Esteban-Fernandez (2013). As a summary, we present a high-level breakdown of the hydrology error budget in Table 2.2. For the ocean, the current total mission error budget is below the requirement (blue line) in Figure 2.3, so it is expected that sub-mesoscale features of wavelengths greater than 15 km will be resolved.

2.7 SUMMARY AND PROSPECTS

Wide-swath altimetry using radar interferometry has had a long maturation period since its original conception in the late 1990s. Radar interferometry for land topographic applications, such as the SRTM mission, helped mature and validate the measurement concept and error models. Additional modeling, experimental near-nadir instruments (AirSWOT and BUSARD), as well as spaceborne sensors (GPM and AltiKa) helped the understanding of additional phenomenological issues, such as the variations of the backscatter cross section with incidence angle and the impact of layover due to waves, topography, and vegetation. The understanding of the processing of the data is mature and has been defined to the level that the SWOT mission is about to enter its implementation phase, with an expected launch within a few years. The maturation process has improved the expected performance and the spatial resolution of wide-swath systems by orders of magnitude relative to what could be envisioned at the turn of the century.

However, in the absence of a global mission or additional experimental data, it cannot yet be claimed that wide-swath altimetry's capability to contribute to better understanding of the ocean sub-mesoscale or global surface water storage change has been realized. Significant work remains in trying to get a global assessment of the measurement physics, with outstanding areas in the EM bias and other wave effects and in the effects of topographic and vegetation layover. Some of this work is ongoing, while some of it may have to wait for the SWOT mission to fly.

In terms of technology, the SWOT instrument will probably represent a baseline that will be hard to improve with a single instrument. The development of extremely stable radio frequency and mechanical systems is probably at the stage where these components are not the limitation for future wide-swath missions. Algorithms to process the data to high accuracy are also maturing quickly and will be in place by the launch of SWOT.

However, for both oceanography and hydrology, SWOT can only be considered as a first step. SWOT solves the spatial mapping problem for the oceans, but the very high spatial resolution is not consistent with the temporal resolution to map fast-changing sub-mesoscales, internal waves, or incoherent tides.

For hydrology, the spatial resolution is sufficient to map the largest contributors to the surface water budgets, but future measurements will need to improve the spatial resolution and also the height accuracy over surface water to reach the next level of understanding global discharge and storage change. The temporal resolution for SWOT is sufficient for mapping changes at monthly scales but will often miss important intermittent effects, such as floods and flood waves.

The natural place to push wide-swath interferometry is in trying to improve the temporal coverage. One possible way to do this is to try to increase swath. Unfortunately, this will not be an easy thing to do. The radar cross section drops rapidly with incidence angle, so that increases in transmit power by an order of magnitude may be required. In addition, SAR data collection puts very severe constraints on how large the swath can be due to both data volume and radar ambiguity constraints that would dictate that the antenna length grow by a factor proportional to the increase of the swath, which will be technologically challenging for any significant increase in the swath.

A more realistic approach for improving the temporal coverage to match the spatial resolution will be to fly a constellation of moderately improved (higher power and bandwidth—to improve spatial resolution and height noise) SWOT descendants. The temporal improvement of a two-member constellation would bring the average revisit time to around 5 days, while a four-member constellation would reduce the average revisit time to around 2 days, consistent with many of the phenomena one wishes to study.

A preview of the possibilities of properly sampled data will be obtained by the SWOT mission on a regional basis because it will start its operations in a 1-day repeat orbit to perform its calibration. This phase of the mission, called the fast-sampling phase, could last as long as 90 days, potentially yielding important new observations of the temporal variability of the ocean and surface water bodies.

ACKNOWLEDGMENTS

The research presented in the paper was carried out at the Jet Propulsion Laboratory, California Institute of Technology, under contract with NASA. Support from the SWOT Project for some of the authors is acknowledged.

REFERENCES

Alsdorf, D., and D. Lettenmaier, 2003. Tracking fresh water from space. *Science*, **301**, 1485–1488.

Alsdorf, D., D. Lettenmaier, C. Vörösmarty, and the NASA Surface Water Working Group, 2003. The need for global, satellite-based observations of terrestrial surface waters. *EOS Transactions of AGU*, **84 (269)**, 275–276.

Alsdorf, D., E. Rodríguez, and D. Lettenmaier, 2007. Measuring surface water from space. *Reviews of Geophysics*, **45**. doi:10.1029/2006RG000197.

Arbic, B. K., F. Lyard, A. Ponte, R. D. Ray, J. G. Richman, J. F. Shriver, E. D. Zaron, and Z. Zhao, 2016. *Tides and the SWOT mission: Transition from science definition team to science team*. Technical report, Jet Propulsion Laboratory. http://swot.jpl.nasa.gov/files/swot/Tides_and_the_SWOT_ mission_Transition_from_Science_Definition_Team_to_Science_Team. pdf.

Archer, C. L., and M. Z. Jacobson, 2003. Spatial and temporal distributions of us winds and wind power at 80 m derived from measurements. *Journal of Geophysical Research*, **108 (D9)**, 4289.

Archer, C. L., and M. Z. Jacobson, 2005. Evaluation of global wind power. *Journal of Geophysical Research*, **110 (D12)**, D12110.

Armitage, T. W., and M. W. Davidson, 2014. Using the interferometric capabilities of the ESA cryosat-2 mission to improve the accuracy of sea ice freeboard retrievals. *IEEE Transactions on Geoscience and Remote Sensing*, **52 (1)**, 529–536.

Ayoub, N., et al., 2016. *Coastal and estuaries white paper. Part 2: Coastal seas and shelf processes*. Technical report, Jet Propulsion Laboratory. http://swot.jpl.nasa.gov/files/swot/Coastal_and_Estuaries_White_Paper_Part%202_Coastal_seas_and_shelf_processes.pdf .

Biancamaria, S., et al., 2010. Preliminary characterization of SWOT hydrology error budget and global capabilities. *IEEE Journal f Selected Topics in Applied Earth Observations and Remote Sensing*, **3 (1)**, 6–19. doi:10.1109/JSTARS.2009.2034614.

Boisot, O., F. Nouguier, B. Chapron, and C.-A. Guerin, 2015a: The go4 model in near-nadir microwave scattering from the sea surface. *IEEE Transactions on Geoscience and Remote Sensing*, **53 (11)**, 5889–5900. doi:10.1109/tgrs. 2015.2424714.

Boisot, O., S. Pioch, C. Fatras, G. Caulliez, A. Bringer, P. Borderies, J. Lalaurie, and C.-A. Guérin, 2015b: Ka-band backscattering from water surface at small incidence: A wind-wave tank study. *Journal of Geophysical Research: Oceans*, **120 (5)**, 3261–3285. doi:10.1002/2014JC010338.

Bush, G. B., E. B. Dobson, R. Matyskiela, C. C. Kilgus, and E. J. Walsh, 1984. An analysis of a satellite multibeam altimeter. *Marine Geodesy* **8 (1–4)**, 345–384.

Chelton, D., Ed., 2001: *Report of the high-resolution ocean topography science working group meeting*, 2001-4, College of Oceanic and Atmospheric Sciences Oregon State University, Corvallis, OR. http://www-po.coas.oregonstate.edu/research/po/research/ hotswg/HOTSWG_report.html.

Chelton, D., E. J. Walsh, and J. MacArthur, 1989. Pulse compression and sea level tracking in satellite altimetry. *Journal of Atmospheric and Oceanic Technology*, **6**, 407–438.

Curlander, J., and R. McDonough, 1991. *Synthetic Aperture Radar: Systems and Signal Processing*, Wiley-Interscience.

Desroches, D., R. Fjørtoft, D. Massonnet, J. Duro, J.-M. Gaudin, and N. Pourthie, 2016. Inland water height estimation without ground control points for near-nadir InSAR data. *IEEE Geoscience and Remote Sensing Letters*, **13 (9)**, 1354–1358.

Dibarboure, G., S. Labroue, M. Ablain, R. Fjortoft, A. Mallet, J. Lambin, and J.-C. Souyris, 2012. Empirical cross-calibration of coherent SWOT errors using external references and the altimetry constellation. *IEEE Transactions on Geoscience and Remote Sensing*, **50 (6)**, 2325–2344.

Dibarboure, G., and C. Ubelmann, 2014. Investigating the performance of four empirical cross-calibration methods for the proposed SWOT mission. *Remote Sensing*, **6 (6)**, 4831–4869.

Donelan, M., and W. Pierson, Jr., 1987. Radar scattering and equilibrium ranges in wind-generated waves with application to scatterometry. *Journal of Geophysical Research*, **92**, 4971–5029.

Durand, M., L. L. Fu, D. P. Lettenmaier, D. E. Alsdorf, E. Rodriguez, and D. Esteban-Fernandez, 2010a: The surface water and ocean topography mission: Observing terrestrial surface water and oceanic submesoscale eddies. *Proceedings of the IEEE*, **98 (5)**, 766–779. doi:10.1109/JPROC.2010.2043031.

Durand, M., E. Rodriguez, D. E. Alsdorf, and M. Trigg, 2010b: Estimating river depth from remote sensing swath interferometry measurements of river height, slope, and width. *IEEE Journal of Selected Topics in Applied Earth Observations and Remote Sensing*, **3 (1)**, 20–31. doi:10.1109/JSTARS.2009. 2033453.

Egido, A., and W. H. Smith, 2017. Fully focused SAR altimetry: Theory and applications. *IEEE Transactions on Geoscience and Remote Sensing*, **55 (1)**, 392–406.

Elachi, C., K. E. Im, F. Li, and E. Rodriguez, 1990. Global digital topography mapping with a synthetic aperture scanning radar altimeter. **11 (4)**, 585–601.

Elfouhaily, T., D. Thompson, B. Chapron, and D. Vandemark, 2000. Improved electromagnetic bias theory. *Journal of Geophysical Research: Oceans (1978–2012)*, **105 (C1)**, 1299–1310.

Elfouhaily, T., D. Thompson, B. Chapron, and D. Vandemark, 2001. Improved electromagnetic bias theory: Inclusion of hydrodynamic modulations. *Journal of Geophysical Research: Oceans (1978–2012)*, **106 (C3)**, 4655–4664.

Enjolras, V., and E. Rodriguez, 2009. An assessment of a ka-band radar interferometer mission accuracy over Eurasian rivers. *IEEE Transactions on Geoscience and Remote Sensing*, **47 (6)**, 1752–1765.

Enjolras, V., P. Vincent, J. Souyris, E. Rodriguez, L. Phalippou, and A. Cazenave, 2006. Performances study of interferometric radar altimeters: From the instrument to the global mission definition. *Sensors*, **6 (3)**, 164.

Esteban-Fernandez, D., 2013. SWOT mission performance and error budget. *JPL D-79084*. http://swot.jpl.nasa.gov/files/SWOT_D-79084_v5h6_SDT. pdf .

Farr, T., et al., 2007. The shuttle radar topography mission. *Reviews of Geophysics*, **45**. doi:10.1029/2005RG000183.

Ferretti, A., A. Monti-Guarnieri, C. Prati, F. Rocca, and D. Massonnet, 2007. *InSAR principles: Guidelines for SAR interferometry: Processing and interpretation*, vol. TM-19. European Space Agency.

Fjortoft, R., et al., 2014. Karin on SWOT: Characteristics of near-nadir ka-band interferometric SAR imagery. *IEEE Transactions on Geoscience and Remote Sensing*, **52 (4)**, 2172–2185. doi:10.1109/TGRS.2013.2258402.

Foresta, L., N. Gourmelen, F. Pálsson, P. Nienow, H. Björnsson, and A. Shepherd, 2016. Surface elevation change and mass balance of icelandic ice caps derived from swath mode cryosat-2 altimetry. *Geophysical Research Letters*, **43 (23)**.

Fu, L., D. Alsdorf, R. Morrow, E. Rodríguez, and N. Mognard, Eds., 2012. *SWOT: The Surface Water and Ocean Topography Mission: Wide-Swath Altimetric Measurement of Water Elevation on Earth*, JPL-Publication 12-05, Jet Propulsion Laboratory, Pasadena, CA. http://swot.jpl.nasa. gov/files/SWOT_MSD_final-3-26-12.pdf.

Fu, L., and A. Cazenave, Eds., 2001. *Satellite Altimetry and Earth Sciences*, Academic Press. San Diego, California.

Fu, L.-L. and Rodriguez, E. (2004) High-Resolution Measurement of Ocean Surface Topography by Radar Interferometry for Oceanographic and Geophysical Applications, in The State of the Planet: Frontiers and Challenges in Geophysics (eds R.S.J. Sparks and C.J. Hawkesworth), American Geophysical Union, Washington, D. C.. doi: 10.1029/150GM17

Fu, L.-L., and C. Ubelmann, 2014. On the transition from profile altimeter to swath altimeter for observing global ocean surface topography. *Journal of Atmospheric and Oceanic Technology*. **31 (2)**. doi:10.1175/jtech-d-13-00109.1. pp. 560–568.

Galin, N., D. J. Wingham, R. Cullen, M. Fornari, W. H. Smith, and S. Abdalla, 2013. Calibration of the cryosat-2 interferometer and measurement of across-track ocean slope. *IEEE Transactions on Geoscience and Remote Sensing*, **51 (1)**, 57–72.

Gatelli, F., A. Monti-Guarnieri, F. Parizzi, P. Pasquali, C. Prati, and F. Rocca, 1994. The wave-number shift in SAR interferometry, *IEEE Transactions on Geoscience and Remote Sensing*, **32 (4)**, 855–865.

Goodman, J., 1985. *Statistical Optics*, Wiley-Interscience, New York.

Graham, L., 1974. Synthetic interferometer radar for topographic mapping. *Proceedings of the IEEE*, **62 (6)**, 763–768.

Hasselmann, K., R. Raney, W. Plant, W. Alpers, R. Shuchman, D. Lyzenga, C. Rufenach, and M. Tucker, 1985. Theory of synthetic aperture radar ocean imaging: A marsen view. *Journal of Geophysical Research: Oceans (1978–2012)*, **90 (C3)**, 4659–4686. doi:10.1029/JC090iC03p04659.

Hess, L. L., J. M. Melack, E. M. Novo, C. C. Barbosa, and M. Gastil, 2003. Dual-season mapping of wetland inundation and vegetation for the central amazon basin. *Remote Sensing of Environment*, **87 (4)**, 404–428.

Klein, P., et al., 2016. *Mesoscale/sub-mesoscale dynamics in the upper ocean. Technical report, Jet Propulsion Laboratory*. http://swot.jpl.nasa.gov/files/swot/Mesoscale_sub-mesoscale_dynamics_in_the_upper_ocean.pdf.

Laignel, B., et al., 2016. *Issues and SWOT contribution in the coastal zones and estuaries. Technical report, Jet Propulsion Laboratory*. http://swot.jpl.nasa.gov/files/swot/Issues%20_and_SWOT_ contribution_in_the_coastal_zones_and_estuaries_Part_1.pdf.

Lee, H., M. Durand, H. Jung, D. Alsdorf, C. Shum, and Y. Sheng, 2010. Characterization of surface water storage changes in arctic lakes using simulated SWOT measurements. *International Journal of Remote Sensing*, **31 (14)**, 3931–3953.

Li, Fuk K., and Richard M. Goldstein. "Studies of multibaseline spaceborne interferometric synthetic aperture radars." *IEEE Transactions on Geoscience and Remote Sensing*, **28** (1) (1990): 88–97.

Moller, D., and D. Esteban-Fernandez, 2014. Near-nadir ka-band field observations of freshwater bodies. *Remote Sensing of the Terrestrial Water Cycle*, **206**, 143.

National Research Council, 2007. *Earth Science and Applications from Space: National Imperatives for the Next Decade and Beyond*, The National Academies Press, Washington, DC.

Parsons, C.L., and E. J. Walsh, 1989. Off-nadir radar altimetry. *IEEE Transactions on Geoscience and Remote sensing* **27 (2)**, 215–224.

Parsons, C. L., E. J. Walsh, and D. C. Vandemark, 1994. Topographic mapping using a multibeam radar altimeter. *IEEE Transactions on Geoscience and Remote sensing*, **32 (6)**, 1170–1178.
Pavelsky, T. M., M. T. Durand, K. M. Andreadis, E. R. Beighley, R. Paiva, G. H. Allen, and Z. F. Miller, 2014. Assessing the potential global extent of SWOT river discharge observations. *Journal of Hydrology*, 519, 1516–1525. doi:10.1016/j.jhydrol.2014.08.044.
Peral, E., 2016. *KarIn: Ka-band radar interferometer on-board processor (OBP) algorithm theoretical basis document (ATBD)*. Technical report, JPL D-79130, Jet Propulsion Laboratory.
Peral, E., E. Rodríguez, and D. Esteban-Fernández, 2015. Impact of surface waves on SWOT's projected ocean accuracy. *Remote Sensing*, **7 (11)**, 14509–14529.
Plant, W. J., 2000. Effects of wind variability on scatterometry at low wind speeds. *Journal of Geophysical Research-Oceans*, **105**, 16899–16910.
Pollard, Brian D., et al. "The wide swath ocean altimeter: radar interferometry for global ocean mapping with centimetric accuracy." *Aerospace Conference Proceedings*, 2002. IEEE. Vol. 2. IEEE, 2002.
Rodriguez, E., 1988. Altimetry for non-Gaussian oceans—Height biases and estimation of parameters. *Journal of Geophysical Research*, **93 (C11)**, 14107–14120.
Rodríguez, E., 2015. *Surface water and ocean topography mission (SWOT) science requirements document.* Technical report, Jet Propulsion Laboratory, Pasadena, CA. http://swot.jpl.nasa.gov/files/SWOT_science_reqs_final.pdf.
Rodriguez, E., and B. Chapman, 1989. Extracting ocean surface information from altimeter returns—The deconvolution method. *Journal of Geophysical Research*, **94 (C7)**, 9761–9778.
Rodriguez, E., Y. J. Kim, and J. M. Martin, 1992. The effect of small-wave modulation on the electromagnetic bias. *Journal of Geophysical Research*, **97 (C2)**, 2379–2389.
Rodríguez, E., and J. M. Martin, 1992. Theory and design of interferometric synthetic aperture radars. *IEE Proceedings F*, **139 (2)**, 147–159.
Rodriguez, E., C. S. Morris, and J. Belz, 2006. A global assessment of the SRTM performance. *Photogrammetric Engineering & Remote Sensing,* **72 (3)**, 249–260.
Rodríguez, E., and B. Pollard, 2001. The measurement capabilities of wide-swath ocean altimeters. *Report of the High-Resolution Ocean Topography Science Working Group Meeting*, D. Chelton, Ed., College of Oceanic and Atmospheric Sciences Oregon State University, Corvallis, OR, 190–216, 2001-4.
Rodriguez, E., and C. W. Chen,' Non-Parametric Phase Screen Estimation at Cross-Overs", SWOT Technical Memorandum, Jet Propulsion Laboratory, California Institute of Technology, December, 2015.
Rosen, P., S. Hensley, I. Joughin, F. Li, S. Madsen, E. Rodriguez, and R. Goldstein, 2000. Synthetic aperture radar interferometry—Invited paper. *Proceedings of the IEEE*, **88 (3)**, 333–382.
Satoh, S., et al., 2004. Development of spaceborne dual-frequency precipitation radar for the global precipitation measurement. *24th International Symposium on Space Technology and Science, ISTS.*
Schulz-Stellenfleth, J., J. Horstmann, S. Lehner, and W. Rosenthal, 2001. Sea surface imaging with an across-track interferometric synthetic aperture radar: The sinewave experiment. *IEEE Transactions on Geoscience and Remote Sensing*, **39 (9)**, 2017–2028.
Schulz-Stellenfleth, J., and S. Lehner, 2001. Ocean wave imaging using an airborne single pass across-track interferometric SAR. *IEEE Transactions on Geoscience and Remote Sensing,* **39 (1)**, 38–45.
Shankaranarayanan, K., and M. Donelan, 2001. A probabilistic approach to scatterometer model function verification. *Journal of Geophysical Research*, **106 (C9)**, 19969–19990.
Smith, B. E., N. Gourmelen, A. Huth, and I. Joughin, 2016. Connected subglacial lake drainage beneath Thwaites Glacier, West Antarctica. *The Cryosphere Discussion,* **10**, 841–871.
Tsang, L., J. Kong, and R. Shin, 1985. *Theory of Microwave Remote Sensing*, Wiley-Interscience, New York.
Vandemark, D., B. Chapron, T. Elfouhaily, and J. W. Campbell, 2005. Impact of high-frequency waves on the ocean altimeter range bias. *Journal of Geophysical Research: Oceans (1978–2012),* **110 (C11)**. doi:10.1029/2005jc002979. http://onlinelibrary.wiley.com/doi/10.1029/2005JC002979/full
Vandemark, D., B. Chapron, J. Sun, G. Crescenti, and H. Graber, 2004. Ocean wave slope observations using radar backscatter and laser altimeters. *Journal of Physical Oceanography*, **34**, 2825–2842.
Walsh, E., F. Jackson, E. Uliana, and R. Swift, 1989. Observations on electromagnetic bias in radar altimeter sea-surface measurements. *Journal of Geophysical Research* **94 (C10)**, 14575–14584.
Walsh, E., D. Vandemark, C. Friehe, S. Burns, D. Khelif, R. Swift, and J. Scott, 1998. Measuring sea surface mean square slope with a 36-ghz scanning radar altimeter. *Journal of Geophysical Research*, **103 (C6)**, 12587–12601.
Wingham, D., et al., 2006. Cryosat: A mission to determine the fluctuations in earth's land and marine ice fields. *Advances in Space Research*, **37 (4)**, 841–871.
Xu, Y., and L. Fu, 2011. Global variability of the wavenumber spectrum of oceanic mesoscale turbulence. *Journal of Physical Oceanography*, **41 (4)**, 802–809.

Xu, Y., and L.-L. Fu, 2012. The effects of altimeter instrument noise on the estimation of the wavenumber spectrum of sea surface height. *Journal of Physical Oceanography*, **42 (12)**, 2229–2233. doi:10.1175/jpo-d-12-0106.1.

Xu, Y., L. Fu, and R. Tulloch, 2011. The global characteristics of the wavenumber spectrum of ocean surface wind. *Journal of Physical Oceanography*, **41 (8)**, 1576–1582.

Yin, X., 2000. Surface wind speed over land: A global view. *Journal of Applied Meteorology*, **39 (11)**, 1861–1865.

3 *In Situ* Observations Needed to Complement, Validate, and Interpret Satellite Altimetry

Dean Roemmich, Philip Woodworth, Svetlana Jevrejeva, Sarah Purkey, Matthias Lankhorst, Uwe Send, and Nikolai Maximenko

3.1 INTRODUCTION

Most satellite-observed ocean variables, such as sea surface temperature, sea surface salinity, and ocean color, are specific to the skin of the ocean and not directly related to the state of the subsurface or deep ocean. Altimetric sea surface height (SSH) is an exception. The height of the sea surface depends on variables spanning the full ocean depth, and this is part of what makes SSH especially valuable in oceanography and climate science. In particular, the height of the sea surface above the sea bottom (H) is equal to the ocean mass per unit area divided by the mean density of the water column ($\bar{\rho}$) and is proportional to the difference between the ocean bottom pressure (p_b) and atmospheric pressure at sea level (p_a). To see this, the hydrostatic balance, $dp = -\rho g\, dz$, is integrated from the sea bottom to the surface:

$$H = \frac{1}{\bar{\rho}} * \text{Mass per unit area} = \frac{p_b - p_a}{\bar{\rho} g} \tag{3.1}$$

Time variability in H thus can result either from changes in the mass or in the mean density of the water column. The time variability of H is equivalent to that in SSH from differencing the satellite orbit (height above the Earth's center) and its height above the sea surface (Chapter 1, refer to Chapter 1, Figure 1.2).

A broad range of *in situ* observations is needed for calibration and validation of altimetric SSH and, in combination with SSH, for understanding the spatial, temporal, and depth characteristics of subsurface ocean variability. In this chapter, the *in situ* measurement systems are described, including a brief history of their implementation and examples of their uses together with satellite altimetry. A goal is to show how satellite and *in situ* observations combine to form an integrated ocean observing system. Importantly, the limitations of the observing system are also illustrated. Spatial coverage remains very sparse on the global scale, leaving boundary currents, mesoscale, and submesoscale variability unresolved. Full-ocean depth coverage is also a challenge, and climate signals including global ocean heat gain extend to the ocean bottom. New technologies will continue to make progress against these limitations. The modern ocean observing system has made dramatic progress against the lack of systematic global observations in the pre-satellite era. Auxiliary measurement systems described in this chapter include the following:

- Sea level gauges at islands or continental margins provide a time-series record of changes in ocean height relative to the land. Additional measurements of the height of the land at the gauge location, relative to the center of the Earth, combine to produce a pointwise record of SSH. Sea level gauge observations are valuable for validation of the satellite SSH record and for connecting SSH across the dynamically complex regions separating the open ocean and the coast (Section 3.2).

- Bottom pressure recorders (SBE 2015) set on the sea floor provide a record of the variability in ocean mass (Equation 3.1). This is of interest not only for closing the time-varying regional sea level budget but also for estimation of the mass added to the global ocean by melting ice. Bottom pressure recorders are another pointwise measurement having a satellite areal counterpart, namely satellite gravity measurements (Section 3.5).
- Profile measurements of ocean temperature and salinity versus pressure are used to calculate seawater density through the equation of state. Historically, profile measurements have been obtained by research vessels using lowered sensors, ships of opportunity using expendable sensors, and more recently by autonomous instrumentation. Density variability is vertically integrated (steric height) to reveal the contribution of temperature and salinity to changes in SSH (Sections 3.3 and 3.4).

In addition to the hydrostatic relationship (Equation 3.1) among ocean mass, pressure, and density, horizontal gradients of SSH and of subsurface density are fundamental to the geostrophic dynamics of large-scale ocean circulation. For large-scale flows in geostrophic balance, the horizontal velocity is proportional and at right angles to the horizontal gradient in pressure:

$$v = \frac{1}{\rho f}\frac{\partial p}{\partial x},\ u = -\frac{1}{\rho f}\frac{\partial p}{\partial y} \tag{3.2}$$

Here, f is the Coriolis parameter, equal to 2Ω sin φ, where Ω is the Earth's rotation rate, and φ is the latitude. The coordinates x and y are positive eastward and northward, respectively, and u and v are the eastward and northward components of velocity. The vertical height of observations in the ocean is not known well enough to enable direct measurement of the horizontal pressure gradient (Equation 3.2).

For geostrophic flow on the sea surface, the geostrophic balance can be expressed in terms of the time-varying slope in SSH, as:

$$v' = \frac{g}{f}\frac{\partial(SSH')}{\partial x},\ u' = -\frac{g}{f}\frac{\partial(SSH')}{\partial y} \tag{3.3}$$

where u', v', and SSH′ are deviations from a common time mean. Combining the geostrophic equation (Equation 3.2) and the hydrostatic balance (Equation 3.1) and integrating from depth z_0 to z,

$$v(z) = -\frac{g}{\rho f}\int_{z_0}^{z}\frac{\partial \rho}{\partial x}\,dz + v_0,\ u(z) = \frac{g}{\rho f}\int_{z_0}^{z}\frac{\partial \rho}{\partial y}\,dz + u_0 \tag{3.4}$$

Thus, the velocity at any depth can be calculated from the velocity (u_0, v_0) at a reference depth plus the depth integral of the horizontal density gradient. Historically, reference velocity estimation has been approached in a number of different ways. These include assuming velocity is so negligibly small at great depth (Sverdrup et al. 1942) that it can be calculated from higher order dynamics (Stommel and Schott 1977) or through conservation of mass (Hidaka 1940; Wunsch 1978) or based on a combination of these approaches (Reid 1986). These estimates all have limitations, so the modern ocean observing system takes the direct approach with systematic observations of velocity at the sea surface and at intermediate depth. Thus, additional requirements for *in situ* observations needed to interpret satellite SSH in the context of geostrophic velocity and large-scale ocean circulation include:

- The profile measurements of temperature and salinity versus pressure mentioned earlier, in addition to providing the density-related (steric) component of SSH, also provide the horizontal gradients of density for estimating geostrophic shear (Equation 3.4) (Section 3.3).
- Surface velocity is measured by satellite-tracked drifting buoys. Decomposition of surface velocity into geostrophic and frictional ageostrophic components completes the surface balance (Equation 3.3) (Section 3.6).
- Velocity at mid-depth is measured by Argo floats, typically with 10-day trajectories at 1000 m. The mid-depth 10-day velocity is in approximate geostrophic balance, so it can be applied as (u_0, v_0) in (Equation 3.4) (Section 3.3).

The combination of SSH with *in situ* measurements of density, velocity, and bottom pressure is complete in the sense of providing the terms in the hydrostatic balance to separate mass and density contributions to SSH and the terms in the thermal wind equation (Equation 3.4) to calculate geostrophic velocity at any depth. Thus, the static and the dynamic variability in SSH can be understood if the satellite coverage is matched by complementary *in situ* observing systems.

3.2 SEA SURFACE HEIGHTS OBTAINED FROM TIDE GAUGE/GNSS NETWORKS

3.2.1 Sea Level Measurements before the Altimeter Era

Tide gauges (or sea level recorders) have been used since ancient times to measure the rise and fall of the tide at the coast (Cartwright 1999). One notable development occurred with the publication by Robert Moray of an article in the first edition of *Philosophical Transactions of the Royal Society* (Moray 1665), in which he suggested the use of a stilling well for measuring sea level accurately. A stilling well is a vertical tube, with a small hole open at its bottom so that sea level inside the tube is, on average, the same as that outside. The hole acts as a low-pass filter that removes the high-frequency motions due to waves inside the well (i.e., it dampens or "stills" the water) enabling "still water level" to be measured.

Almost one and half centuries went by until the 1830s when automatic (or "self-registering") tide gauges were developed. These instruments consisted of a stilling well inside of which was a float connected by a wire run over pulleys to a pen that moved up and down as the tide rose and fell, thereby drawing a tidal curve on a drum of paper that rotated at a speed set by an accurate clock. The resulting continuous heights and times of "relative sea level" could then be expressed relative to the height of a benchmark on the nearby land.

The first such tide gauges are often credited to those installed in the Thames Estuary, England (Palmer 1831; Matthäus 1972). The first self-registering tide gauge in the United States was made by Joseph Saxton for the U.S. Coast Survey in 1851. By the end of the nineteenth century, similar instruments had been installed at many ports around the world, and the same technology was employed until the second half of the twentieth century, thereby providing the set of historical sea level measurements still used by many researchers today (Woodworth 2015). The operation of float and stilling well gauges in modern times is described in manuals (IOC 2015) and textbooks (Pugh and Woodworth 2014; Simon 2015; Woodworth et al. 2015). In the last few decades, many such gauges have been replaced by pressure, acoustic, or radar devices that do not require a stilling well.

3.2.2 Tide Gauge and Altimeter Data Complementarity

It is important to avoid the impression that tide gauges and altimetry provide alternative sea level monitoring systems that are somehow in competition. In fact, there is only one monitoring system with two components that have complementary strengths. The most obvious use of the two techniques in combination is in the calibration of altimetry by tide gauges discussed as follows.

Altimetry provides a synoptic coverage of the deep ocean within latitudes determined by the inclination of the satellite. However, its temporal–spatial sampling is coarse and especially unsuited to sea level measurements near the coast (Chapter 12 this volume). The important attributes of tide gauges are, therefore, that they are located at the coast itself; they provide as rapid a sampling as required (even 1 Hz if necessary) so they can provide information on extreme as well as mean sea levels; they can be located to monitor flows through straits (e.g., Gibraltar or Skagerrak) that altimetry cannot measure adequately because of the altimeter and radiometer contamination by land; they provide the sea level monitoring at high latitudes above the satellite inclination and in ice-covered areas; they provide detailed tidal information from which, in combination with altimetry, advanced tide models have been developed (Stammer et al. 2014; Chapter 14 this volume); and, perhaps most important as explained earlier, they provide the historical sea level record on which much research in geophysics, ocean circulation, and climate change is based.

Altimetry has traditionally been validated by comparison of its sea level record to that of a tide gauge or sets of gauges (e.g., Harangozo et al. 1993). When the most energetic spatial scales of sea level variability are large (e.g., in the tropical Pacific during El Niño events), then a high correlation can be expected between the two sea level time series. However, at other locations, there may be sufficient short spatial-scale variability that significant differences can be expected over the approximately 10 s or approximately 100 km between an island or continental coastal tide gauge and the nearest altimeter overpass, resulting in a weaker correlation. Examples of such short scale variability have been demonstrated by many authors (e.g., Mitchum 1995; Vinogradov and Ponte 2011) and explained using ocean models (e.g., Williams and Hughes 2013). In addition, there may be local variability in the tide gauge data due to harbor seiches or wave setup, although these signals can probably be low-pass filtered prior to comparison to the altimetry. All these considerations enter into selection of tide gauges to use for altimeter calibration, discussed later in this chapter.

An interesting question concerns the complementarity between altimetry and tide gauges on continental coastlines at long timescales, and whether the two systems measure the same long-term sea level trends. There could be differences in detail at any one location depending on long-term fluctuations in the nearby ocean circulation (e.g., along the U.S. coast due to long-term changes in the Gulf Stream; Blaha 1984). However, although some early investigations suggested a generally faster sea level rise near the coast than in the deep ocean during the 1990s (Holgate and Woodworth 2004; White et al. 2005), this interpretation was later discounted when considering the global coastline overall (Prandi et al. 2009). In general, tide gauges and altimetry have been found to measure approximately the same trends when adjusted appropriately for land movement contributions (e.g., Watson et al. 2015).

3.2.3 Tide Gauges Used for Altimeter Calibration

An important role for tide gauges is in the determination of altimeter height bias, which is the systematic error in the measurement of SSH by the altimeter radar system. In these exercises, attempts are made to compare measurements of the height of the same area of ocean by both the altimeter and a tide gauge. The height of the satellite is known in geocentric coordinates by some combination of laser, Global Navigation Satellite System (GNSS), or Doppler tracking; there is a range measurement by the altimeter between the satellite and the sea surface; and meanwhile there is an independent measurement of the SSH by a tide gauge equipped with a GNSS receiver. Any error in the closure among these three sets of information provides an estimate of the altimeter bias.

In one of the first altimeter calibration campaigns for the Seasat altimeter, data from a tide gauge on the island of Bermuda were used, with due allowance for the fact that the altimeter could not measure exactly at the tide gauge itself, but some distance off-shore, requiring assumptions on variation in the geoid across the small island (Kolenkiewicz and Martin 1982; Tapley et al. 1982). In the later TOPEX/Poseidon mission, tide gauges were instead used at the Harvest oil platform off the coast of California, which allowed for direct overflight of the satellite and no need for geoid interpolation (Haines et al. 2010). This activity has subsequently been maintained in the Jason-1 and Jason-2/ Ocean Surface Topography Mission (OSTM) era. Additional long-term calibration sites have been established for the TOPEX series of missions, and also for other altimeter satellites, at specially instrumented coastal tide gauge sites (e.g., Chelton et al. 2001; Bonnefond et al. 2011; Mertikas et al. 2015).

The problem with a small number of special sites is that, given the long repeat period of altimeter satellite orbits, the amount of comparison data obtained from each site is limited, and there will be uncertainties in the sea level difference between altimeter and tide gauge that can be reduced to an acceptable level only by repeated measurements. Therefore, it is highly desirable to make use of as large a distributed subset of the global tide gauge network as possible in order to monitor the temporal stability of an individual altimeter's measurement system; within the Intergovernmental Oceanographic Commission (IOC) Global Sea Level Observing System (GLOSS) this subset is sometimes called GLOSS-ALT. Such comparisons of altimeter and tide gauge data from the global network led to an identification of a major error in the range measurements of TOPEX/Poseidon, which had been responsible for incorrect findings of sea-level rise in the early years of the mission. This method, described by Mitchum (2000) and Leuliette et al. (2004), is used only for routine validation and not for a "hard calibration" of the altimeter data. In fact, a "harder calibration," in which the altimeter SSHs from different missions are adjusted for time-dependent biases with the use of data from more than 100 tide gauges and incorporating GNSS information at tide gauges, has resulted in significant changes in estimated rates of global sea level change in recent decades that are more consistent with expectations (Watson et al. 2015). However, this procedure is not employed operationally.

3.2.4 Tide Gauge and Altimeter Data in Combination in Studies of Long-Term Sea Level Change

Before the era of precise satellite altimetry (i.e., before the early 1990s), and since then in some cases, secular sea level changes were estimated by combining individual tide gauge records into global and regional averages in various ways (e.g., Barnett 1984; Douglas 1991; Holgate and Woodworth 2004). Such combinations of records yielded composite time series starting at the end of the nineteenth century or the beginning of the twentieth, which is when most of the float and stilling well gauges mentioned earlier were first installed. Most of our knowledge of sea level change prior to the nineteenth century comes from geological information (e.g., Kemp et al. 2015) rather than instrumental data. (A very small number of historical sea level records extend back to the seventeenth century; Woodworth et al. 2011.)

These combinations of individual tide gauge records inevitably could not take into consideration the sea level changes that must have occurred over the vast areas of ocean between the coastlines where the tide gauges were located. In addition, the spatial biases in the global data set, with relatively fewer records in the Southern Hemisphere, meant that the "global" combined time series were primarily Northern Hemisphere ones (Holgate et al. 2013). Nevertheless, in spite of the many reservations about the data set, the various analyses provided conclusive evidence that the sea level had been rising throughout the twentieth century.

The scientific literature contains a vast number of papers concerned with such tide gauge studies of long-term sea level change, including changes in extreme sea levels as well as mean levels. Syntheses of research findings include the scientific assessments of the Intergovernmental Panel on Climate Change (e.g., Church et al. 2014), studies by expert teams (e.g., Lowe et al. 2010; Mitchum et al. 2010), and occasional book reviews (e.g., Pugh and Woodworth 2014).

As the altimeter record has lengthened, attempts have been made to combine tide gauge and altimeter data in optimal ways. Altimeter data from the last two and half decades have been used to learn about the main spatial and temporal modes of sea level variability (Chambers et al. 2002). This insight has then been applied to the sparser, but much longer, tide gauge data set in order to obtain more rigorous estimates of regional and global change over an extended period that is much longer than the altimeter era. These exercises are called "sea level reconstructions" and have provided "reconstructed time series" of sea level change, also back to the late-nineteenth century.

Such reconstruction exercises have employed empirical orthogonal functions (EOFs) to parameterize the modes of ocean variability from the altimetry (Church and White 2011; Ray and Douglas 2011). Alternatively, EOFs have been derived from ocean circulation models, with model runs performed over many decades and, therefore, in principle, capable of representing lower-frequency sea-level changes more reliably than the EOFs based on altimetry (Llovel et al. 2009). Other investigations have involved the use of cyclo-stationary EOFs to represent progressive motions in sea-level variations instead of the standing waves of conventional EOFs (Hamlington et al. 2011). Some analyses have made use of tide gauge data within neural networks (Wenzel and Schröter 2010), while sophisticated "virtual station" techniques for averaging individual records in a region, or globally, without consideration of particular modes of variability have been developed (Jevrejeva et al. 2006). Each of these methods has its own technical drawbacks that are inevitable when using altimetry in combination with a sparse data set (see discussions in Christiansen et al. 2010; Ray and Douglas 2011; Calafat et al. 2014).

3.2.5 GNSS Equipment at Tide Gauges

A requirement for all tide gauges in the GLOSS network of the IOC, specified by the program's 2012 Implementation Plan, is that they be equipped with a nearby GNSS receiver (IOC 2012). The same requirement also applies as far as possible to sites with long tide gauge records. The main form of GNSS to date has been the U.S. Global Positioning System (GPS), but many receivers can now also accept signals from the Russian GLONASS, European Galileo, and Chinese Beidou systems. GLOSS requires the GNSS antenna to be installed as near as possible to the tide gauge and, ideally, on it. A further requirement is that all GNSS data from GLOSS, and other contributing, tide gauge stations are made available to international data banks, including especially the GLOSS center for GNSS data at the University of La Rochelle called SONEL (Systéme d'Observation du Niveau des Eaux Littorales, www.sonel.org).

Figure 3.1 summarizes the status of GNSS installations in the GLOSS network and the flow of their data to SONEL, while Figure 3.2 shows an ideal installation with the GNSS located as close as possible to the tide gauge. An up-to-date version of Figure 3.1, and other maps of the GLOSS Core Network showing its status using other criteria, such as overall sea level data availability at the Permanent Service for Mean Sea Level (PSMSL), can be obtained from www.psmsl.org. Advice is available from GLOSS on how to monument and operate GNSS systems at tide gauges (Bevis et al. 2002; Foster 2015; IOC 2015) and on the priorities for new installations (King 2014).

There are two main scientific applications for such GNSS data. One application is to monitor vertical crustal movement at the site that can be employed in studies of the reasons for the relative sea level change observed by the tide gauge. Such estimates of rates of change of vertical crustal movement are also required in altimeter calibration (Mitchum 2000; Leuliette et al. 2004). A second application is so that the sea levels recorded by the gauge can be expressed as ellipsoidal heights for direct comparison to those obtained from space by satellite altimetry and so enable an inter-calibration as described earlier.

However, there are several other practical benefits of GNSS to tide gauge operations other than the aforementioned scientific ones. One is that GNSS provides an accurate time standard that can be used in control of other clocks in the tide gauge hardware (inaccurate timing has plagued tide gauge measurement throughout its history). The precise positional information from GNSS can be added to the station's metadata, enabling the tide gauge's position to be shown accurately on maps. GNSS also leads to wider application of the tide gauge data. For example, sea levels

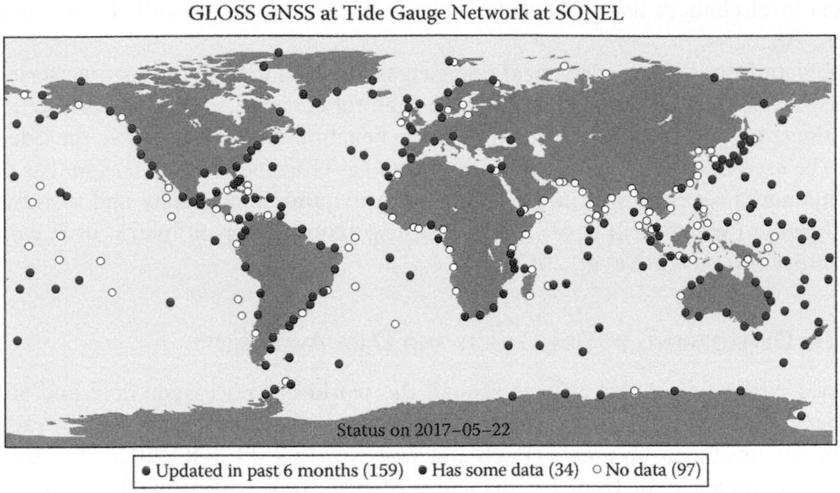

FIGURE 3.1 Locations of the approximately 300 stations in the Global Sea Level Observing System (GLOSS) core network of the Intergovernmental Oceanographic Commission (IOC 2012). Green dots indicate sites that are equipped with a continuous Global Navigation Satellite System (GNSS) near to the GLOSS tide gauge and that have reported recent GNSS data to the SONEL data center; brown dots indicate that only historical data are available from the site; and white indicates no GNSS data is available at SONEL at all. Brown and white suggest that the installation of new GNSS equipment may be required. (Courtesy of Permanent Service for Mean Sea Level, Liverpool, UK.)

FIGURE 3.2 An acoustic tide gauge at Burnie in northern Tasmania, Australia, and, to its right, a special pillar with a continuous Global Navigation Satellite System (GNSS) receiver on top. (Courtesy of © Commonwealth of Australia (Geoscience Australia) 2016. This product is released under the Creative Commons Attribution 4.0 International License.)

expressed in geocentric coordinates can be applied to modern methods of hydrographic chart production. At some locations, if the GNSS is operated at a high sampling rate and connected to high-bandwidth telemetry, the receivers can contribute seismic information to regional warning centers for determination of earthquake magnitudes and dissemination of near real-time tsunami alerts (Blewitt et al. 2006). In addition, GNSS can also be used in some locations to measure both land and sea level changes using the same equipment in a technique called reflectometry (e.g., Larson et al. 2013).

Although data from GNSS receivers at tide gauges are used extensively as time series, an often neglected aspect of their operations concerns the need for accurate, and often repeated, ties (leveling connections) between the GNSS and tide gauge benchmarks. These enable the tide gauge sea level data to be used in an absolute sense (i.e., expressed as ellipsoidal heights) as well as time series. Such information is required within studies of mean dynamic topography and worldwide height system unification (Woodworth et al. 2012). It is also required for altimetric drift calibration as discussed earlier (Woodworth et al. 2017).

3.2.6 New Developments in Tide Gauges and Data Availability

As mentioned earlier, most tide gauges around the world are no longer float and stilling well devices. They have been replaced by systems that use water pressure or acoustic or radar ranging to derive the sea level measurements (Pugh and Woodworth 2014; IOC 2015). The main reasons for these developments stem from the difficulty of installing stilling wells at some locations, which requires local engineering infrastructure, and is also due to the need for careful maintenance by trained staff. In fact, where good float and stilling well gauges have operated for many years, there is no overriding urgency in replacing them if they are operating satisfactorily as they can be updated to serve modern requirements for real-time reporting by using digital shaft encoders.

Nevertheless, radar gauges in particular are replacing the earlier technologies, primarily because of their low cost, ease of installation, and low maintenance. These tide gauges are also completely electronic and so can be readily interfaced with satellite transmission systems and other forms of telemetry. This has resulted in a great increase in the availability of real-time data that are accessible via national Web sites (see http://www.psmsl.org/links/) and from the Intergovernmental Oceanographic Commission Sea Level Monitoring Facility (http://ioc-sealevelmonitoring.org). Of course, these data are of great importance to operational oceanography, including flood and tsunami warning. However, a drawback with these new data sets is that many of them are uncalibrated (e.g., their sea level time series have unknown offsets), and their large quantity means that only a subset of the data can be subjected to the rigorous quality control (QC) needed for scientific research and for inclusion into international sea level databanks such as the PSMSL (http://www.psmsl.org; Holgate et al. 2013) or the University of Hawaii Sea Level Center (UHSLC; http://uhslc.soest.hawaii.edu/). At the same time, some national agencies no longer undertake a routine QC of their data sets, as they did previously. As a result, unless sophisticated, automated methods for calibration, documentation, and QC can be developed in the future, there is a danger that, although the overall quantity of tide gauge data may increase, the amount of data suitable for research may actually decrease. For example, already some of the long-time high-quality records (used, e.g., by Douglas 1991) are no longer maintained.

3.2.7 Tide Gauges and Altimetry in the Future

As mentioned previously, tide gauges are becoming cheaper and easier to install and maintain. Therefore, there is no reason to expect a reduction in data flow overall from coastal sites in the future, although it will always be the case that data will be easier to obtain from some countries than others and that inhomogeneous data coverage is inevitable. At the same time, more altimeter satellites are now operational, and altimetric monitoring of the deep ocean looks assured for the next decade.

An important remaining problem concerns bridging the gap in information between the monitoring of sea level provided by conventional nadir pointing altimetry over the deep ocean (up to typically 10 s of km from the coast) and that at the coast itself by tide gauges. This part of the coastal ocean is most important from economic and environmental perspectives, which is why coastal altimetry has emerged as a research topic in its own right (Vignudelli et al. 2011, and Chapter 12 this volume).

Modern altimeters such as those of the CryoSat-2 or Sentinel-3 missions, and the development of wide-swath altimetry, will sample much closer to the coast than previous missions. However, the coastal gap will always remain to some extent. One could imagine the gap being bridged by other sea level technologies such as GNSS buoys (e.g., Testut et al. 2010) or GNSS reflectometry (e.g., Santamaría-Gómez et al. 2015). However, these techniques have their own problems, and they are unlikely to provide information on a quasi-global basis in the near future. The IOC GLOSS global network of high-quality tide gauges, based on 300 stations worldwide, is still only two-thirds complete. Therefore, the priority must be to ensure that the global network is completed, with tide gauges accompanied by GNSS equipment, that data flow is ensured by international agreements, and that methods are established for timely quality control of their data for both operational and research users.

3.3 UPPER-OCEAN (0 TO 2000 DECIBARS) STERIC VARIABILITY: THE XBT AND ARGO NETWORKS

3.3.1 The Relationship of SSH Variability with Subsurface T and S—Steric Height

In Section 3.1, it was noted that SSH changes in response to changes in the mass of water per unit area and to changes in density (Equation 3.1). There are some processes in which the mass change is the larger of the two. For example, about two-thirds of the multidecadal increase in global sea level in the past decade is due to mass added to the oceans by melting ice (Llovel et al. 2014). In other cases, such as seasonal and interannual variability of SSH in the tropical oceans, the density contribution dominates SSH changes.

The density-dependent change in SSH, known as steric height, can be written as

$$\Delta\,\text{steric height} = \frac{1}{g}\int_{p=p_{ref}}^{0} \Delta\alpha\, dp, \tag{3.5}$$

where α is the specific volume ($1/\rho$), a function of temperature, salinity, and pressure, and Δ represents the change. Thus, if temperature and salinity versus pressure profiles, and their time variability, are measured over part of the water column, the steric height change can be calculated for that pressure range. Variability in density tends to be largest at the sea surface, where air–sea heat exchange causes warming or cooling of the mixed layer and where rainfall and evaporation cause changes in salinity. Conversely, variability is smallest in the deep ocean, far from sources of heat and freshwater, where the advection of properties is slow and property gradients are minimal.

Figure 3.3 shows examples of the annual cycle in upper-ocean steric height (0/500 decibars; Roemmich and Gilson 2009) and SSH (Ducet et al. 2000) for dynamically different ocean locations, to illustrate their similarities and differences. The annual cycle in steric height is mainly due to air–sea exchanges of heat and water (Gill and Niiler 1973). The contribution to steric height from temperature alone is also illustrated. At mid-latitudes, the thermosteric contribution dominates, with larger proportions due to salinity at low- and high-latitude locations having large rainfall. Substantial differences between SSH and upper-ocean steric height annual cycles indicate that ocean dynamical processes are responsible for deep steric or mass-related variability at these locations. In the tropics, the oceans respond to wind-forced annual variability with vertical displacement of density that extends below the surface layer (Donguy and Meyers 1996). At high latitudes, the seasonal warming and cooling is smaller, and deep winter mixed layers spread the cooling over a greater depth range.

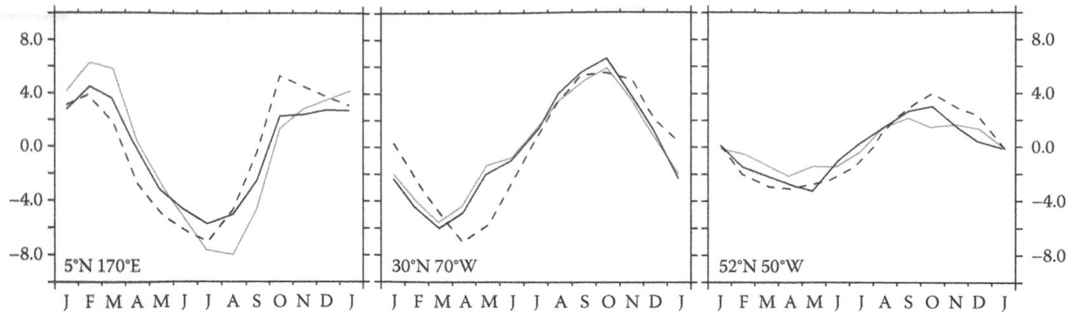

FIGURE 3.3 Annual cycle, 2005 to 2014, of steric height (cm, 0/500 decibars, thick solid lines), steric height component due to temperature (thin solid lines), and SSH (dashed lines), for example, locations at low, middle, and high latitudes as labeled. (Based on data described in Roemmich, D., and Gilson, J., *Prog. Oceanogr.*, 82, 81–100, 2009; Ducet et al., *J. Geophys. Res.*, 105(C8), 19477–19498, 2000.)

The typical steric height annual cycle amplitudes shown in Figure 3.3, amounting to about 5 cm for thermosteric and 1–2 cm for halosteric (salinity) variability, are equivalent to heat content anomalies of 8×10^8 J m^{-2} (50 W m^{-2} for 6 months) and freshwater anomalies of 0.5 to 1 m, respectively. The close relationship among SSH and subsurface temperature and salinity, including surface layer buoyancy exchanges and full water-column ocean dynamics, defines the importance of having a global ocean observing system that includes both satellite and *in situ* components.

3.3.2 A Brief History of Systematic Ocean Sampling by the XBT and Argo Networks

Exploration of subsurface ocean properties began in the nineteenth century, and global scale observations were initiated by the Challenger Expedition of 1873–1876 (Wyville Thomson 1873). HMS *Challenger* obtained about 300 deep ocean temperature profiles in a globe-circling voyage. Measurements were made by suspending pressure-protected thermometers on a cable lowered by a steam-powered winch. Due to the requirements for specialized sampling equipment, subsurface measurements remained in the province of research vessels until the late 1960s. Moreover, since research vessels capable of working on the high seas have always been small in number, the data set collected in this way was very sparse. During the 1990s, the total number of temperature and salinity versus depth profiles collected to a depth greater than 1000 m averaged 5000 profiles per year, as seen in Figure 3.4. Moreover, they were concentrated in the Northern Hemisphere, near the coasts of continents, and during summer. The pinnacle of research vessel hydrographic expeditions was the World Ocean Circulation Experiment of 1991–1997 (WOCE, Woods 1985). WOCE carried out a global survey of top-to-bottom temperature/salinity profiles and geochemical tracers. WOCE provided a compelling justification for the TOPEX altimetry mission, which was needed to provide the global and continuous temporal context for interpreting the spatially sparse WOCE data set.

Prior to WOCE and TOPEX, the necessity for research vessels to make subsurface observations was mitigated by the invention of the expandable bathythermograph (XBT). The XBT probe (Figure 3.5) includes a thermistor mounted in a streamlined plastic housing with a heavy zinc nose. Spools of very thin insulated copper wire in the hollow tail of the XBT and in a shipboard canister link the probe electrically to a shipboard recording unit. Wire is paid out from both spools as the probe falls through the water, with the recorder logging temperature and converting time to depth. Early XBTs went to 450 m and later models to 750 m or even deeper. With the advent of the XBT, systematic and regular observations of upper ocean temperature were possible. But what made the XBT unique was being deployed from any type of ship, including commercial ships while underway. Thus, the XBT technology made it practical to collect temperature

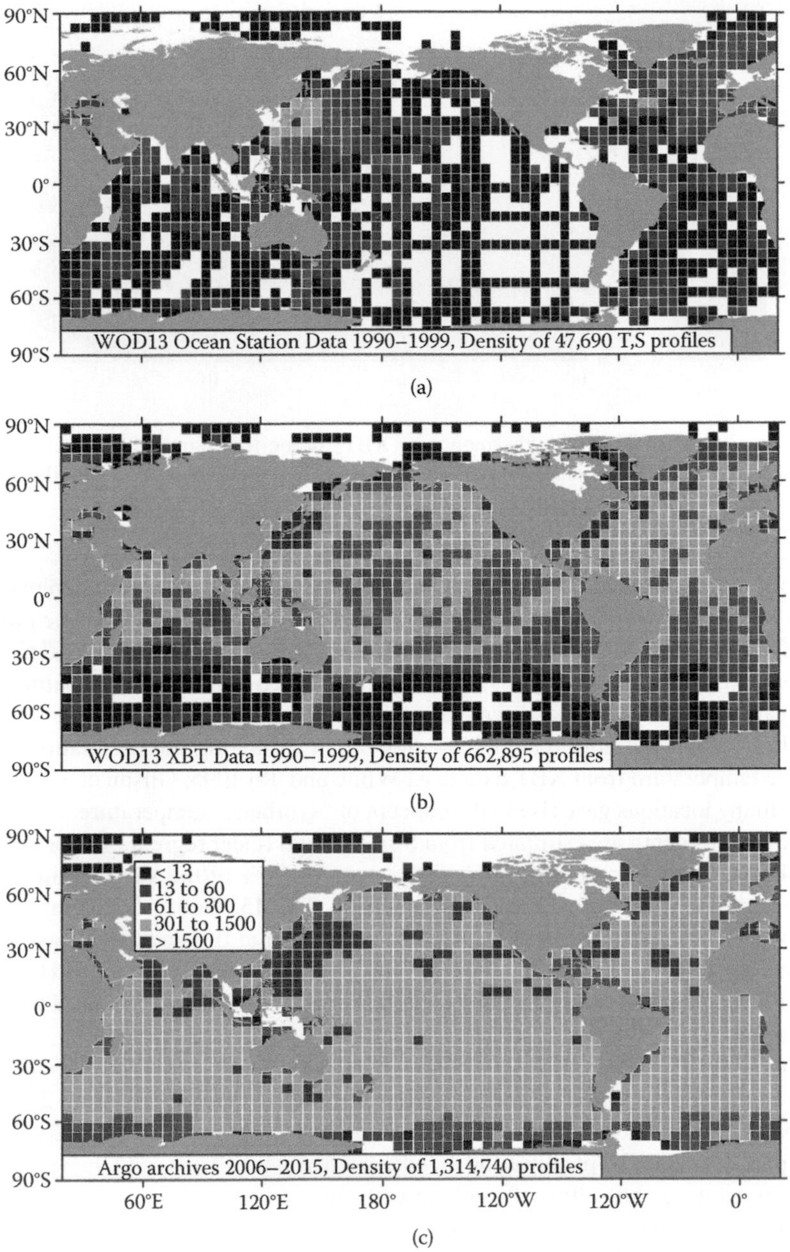

FIGURE 3.4 Accumulation of stations in 5° × 5° boxes over one decade of sampling for (a) ocean station data (47,690 temperature/salinity profiles to at least 1000 m) 1990 to 1999; (b) 662,895 XBT temperature profiles 1990 to 1999; and (c) 1,314,740 Argo temperature/salinity profiles 2006 to 2015. (World Ocean Database (http://doi.org/10.7289/V5NZ85MT) and Argo Global Data Assembly Centers (http://doi.org/10.17882/42182).)

profiles wherever there is ship traffic. Network design studies (Bernstein and White 1974; White and Bernstein 1979; Meyers et al. 1991; Sprintall and Meyers 1991) determined the sampling requirements for large-scale ocean climate variability. XBT networks were implemented initially around "broadscale" coverage and later to resolve boundary currents and mesoscale features with repeating "high resolution" transects. Limitations of the XBT are its depth range, lack of salinity, and inaccuracy in the time-to-depth conversion.

(a) (b)

FIGURE 3.5 (a) An expandable bathythermograph (XBT) probe is deployed from a container ship by a Scripps XBT Autolauncher. (G. Pezzoli).(b) An Argo float shortly before its recovery by the Japanese coast-guard vessel *Takuyo*. (http://www.argo.ucsd.edu/pictures.html)

As seen in Figure 3.4, during the 1990s, about 70,000 XBTs per year were deployed worldwide (i.e., an order of magnitude more profiles than the research vessel station data). The global distribution of XBT profiles was also an improvement as many ocean-crossing commercial shipping routes were instrumented. However, there remained large unsampled regions in the Southern Ocean and sparser sampling throughout the Southern Hemisphere than the Northern.

Early in the TOPEX mission, scientific studies began exploiting the correlation between SSH and subsurface temperature from XBT data (e.g., White and Tai 1995; Gilson et al. 1998). The high correlation in many locations gave rise to the concept of "synthetic" temperature, in which temperature anomalies versus depth are estimated from SSH through linear regression. For example, on the Equator at 140°W, a time series of SSH anomaly captures 90% of the variance in temperature at 100 m, on timescales longer than 20 days (Gasparin et al. 2015). Subsurface temperature anomalies have a long vertical scale, and a 10-cm anomaly in SSH at this location is associated with a downward displacement of the thermocline by 30 m. The close relationship of SSH and subsurface temperature allows the latter to be viewed at the much greater spatial and temporal resolution of the satellite observations.

WOCE not only surveyed the global ocean, it also developed new technologies for the future of ocean observation, notably including the autonomous profiling float (Davis et al. 2001) that would later be the basis of the Argo Program. The modern Argo float (Figure 3.5) descends from the sea surface to a "parking depth," typically 1000 m, where it drifts for 10 days and then descends to 2000 m to profile temperature and salinity from here up to the sea surface. Using Iridium communications, the float's data are transmitted in 15 min, after which a new cycle begins. Modern lithium–ion battery technology enables the 20-kg float to cycle for more than 5 years. The hundreds of floats deployed during WOCE provided a demonstration not only of the data quality and high scientific value produced by this transformational technology but also its global reach.

At the conclusion of WOCE in 1997, a global array of 3200 profiling floats, one every three degrees of latitude and longitude, was proposed (Argo Science Team 1998). The name Argo was chosen for this array to underline the strong scientific linkage between Argo floats and the Jason series of altimeters. A multinational effort was mounted, and deployment of Argo floats began in late 1999. By 2007, the Argo array included more than 3000 instruments, and that number increased to 3800 at the end of 2016.

Figure 3.4 shows that the distribution of Argo data is much more homogeneous than was possible with XBTs or other ship-mounted technologies. Moreover, profiling float measurement errors are an order of magnitude less than for XBTs; the measurements include salinity as well as temperature, and they extend much deeper into the ocean than XBTs. Most dramatically, the figure shows that the Northern Hemisphere bias of earlier observations has been greatly reduced. Argo has provided more than 1.5 million temperature/salinity profiles, with the data used in more than 2500 research papers, demonstrating the feasibility and high value of systematic global observations of the subsurface oceans.

3.3.3 Ocean Heat Content and Steric Sea Level

More than 90% of the heat increase in the Earth's climate system is absorbed by the oceans (Rhein et al. 2013). Because of this, the planetary energy imbalance—with more downward than upward radiation—can be measured more accurately through the increasing temperature of the oceans than by any other means, including radiation measurements at the top of the atmosphere. During the past decade, with good Argo global coverage, about half of the global heat gain has gone into the 0–500 decibars upper ocean and half into the 500–2000 decibars mid-range. Nearly all of the heat gain is in the Southern Hemisphere oceans (Roemmich et al. 2015). The increasing temperature of the oceans causes an expansion of the water column (Equation 3.5), and the globally averaged increase in the steric height of the sea surface is 1.1 cm/decade for the period 2006–2015, as seen in Figure 3.6. This corresponds to a heat gain by the upper and mid-depth ocean, 0–2000 decibars, of 1×10^{23} J over the past decade, or 0.6 W/m^2 averaged over the total surface area of the Earth (Wijffels et al. 2016). On a regional basis, the heat gain and steric height increase can be an order of magnitude greater than the global mean (or similarly large and negative). Decadal variability in wind-forcing is among the major causes of regional variability in SSH (e.g., Merrifield and Maltrud 2011). Thermal expansion is a large component in the global and regional sea level budgets (Chapter 6), and observations of steric variability and water column heat content are prominent among the objectives for *in situ* observations in relation to satellite altimetry.

3.3.4 The Global Pattern of SSH and Upper-Ocean Steric Height

Figure 3.7 shows that the spatial patterns of the magnitude of time variability in SSH and in upper-ocean steric height are remarkably similar. Both exhibit prominent maxima in variability in each

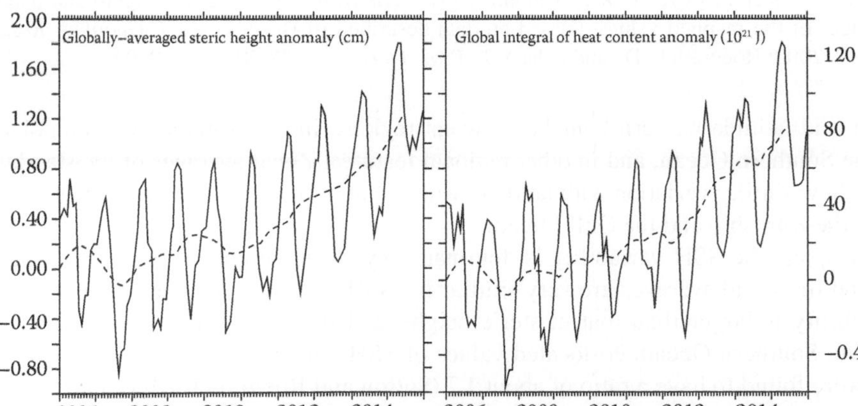

FIGURE 3.6 (a) Globally averaged steric height anomaly (cm, 0/2000 decibars) based on Argo data (Roemmich and Gilson 2009) for monthly (solid) and 12-month running mean (dashed) values. (b) Global ocean heat content anomaly (10^{21} J, 0 to 2000 decibars), same data set. (Adapted from Roemmich, D., and Gilson, J., *Prog. Oceanogr.*, 82, 81–100, 2009.)

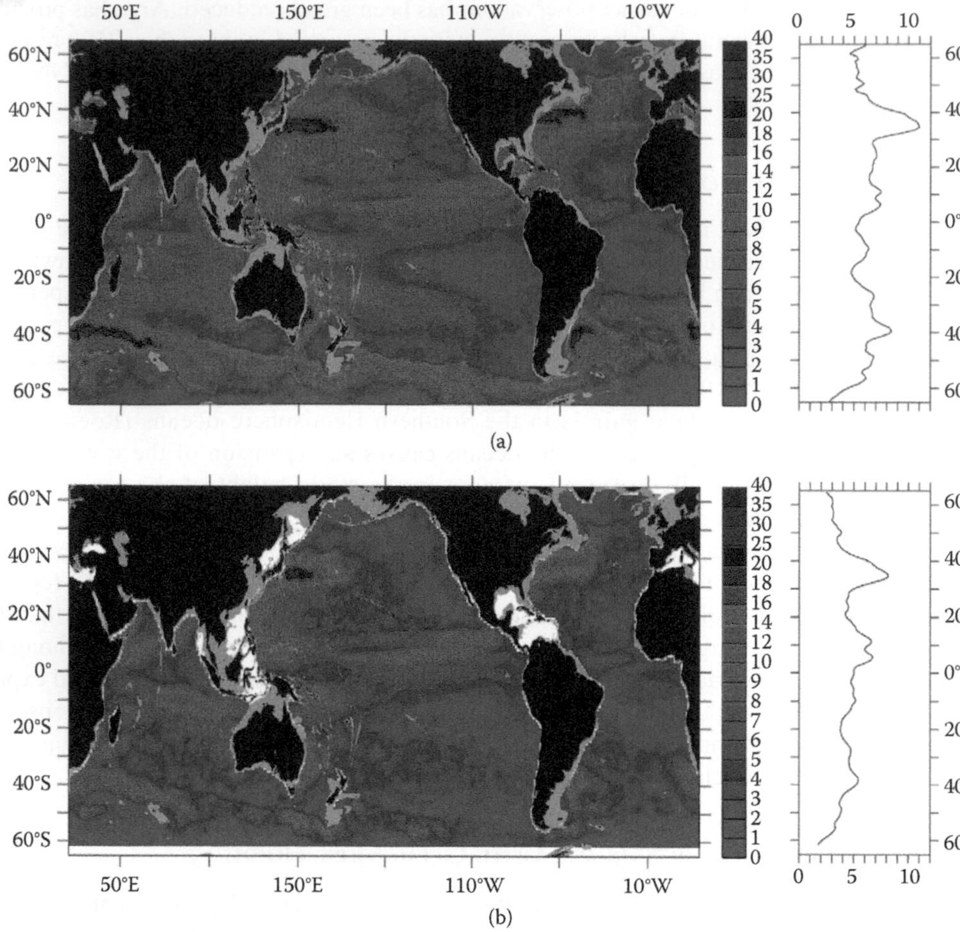

FIGURE 3.7 Standard deviation (cm) of altimetric sea surface height (a) and of Argo steric height ((b) 0/2000 decibars), averaged over 1° × 1° × monthly grid cells from 2006 to 2015. Right-hand panels (b) are zonal averages of the mapped values. (Based on data described in Ducet et al., *J. Geophys. Res.*, 105(C8), 19477–19498, 2000; Roemmich, D., and Gilson, J., *Prog. Oceanogr.*, 82, 81–100, 2009.)

of the five mid-latitude western boundary current regions, in the Antarctic Circumpolar Current circling the Southern Ocean, and in other regional features. Zonal averages of the standard deviations also show similar variation with latitude, with the maximum in both cases at about 37°N, the latitude of the Kuroshio and the Gulf Stream.

In the tropics, the SSH variability is dominated by steric height. In Figure 3.7, the regional features and the zonal averages are very similar for SSH and steric height. In the subtropics, the SSH variability is larger than that of steric height, and this ratio increases further toward the poles. In the Southern Ocean, co-located values of SSH anomaly and upper-ocean steric height anomaly were found to have a ratio of about 1.7 (Sutton and Roemmich 2011). Factors that affect the latitude dependence of the SSH-to-upper ocean steric height ratio include the reduced vertical stratification of the upper ocean at high latitude and the bottom pressure variability.

3.3.5 Geostrophic Ocean Circulation

The ocean, like the atmosphere, is not static. Heat and freshwater exchanged across the sea surface do not remain in place but are redistributed horizontally and vertically by the ocean circulation and mixing. Patterns of temperature and salinity anomaly contain evidence of the past histories of air–sea exchange, modified by flow and mixing from source regions into the ocean interior. Understanding such patterns requires observations of the transport of ocean properties as well as their local changes.

On large spatial scales and below the ocean surface layer, horizontal velocity is governed by geostrophic dynamics (Equations 3.2–3.4). The observing system needed for ocean circulation must include the horizontal gradient in density for geostrophic shear (Equation 3.4) and the reference velocity (u_0, v_0) measured at some level. For altimetry, this would be the surface velocity. In the case of the Argo Program, in addition to profiles of temperature and salinity, a trajectory velocity observation is collected during each 10-day cycle. The velocity is a 10-day mean at the float's parking depth, usually 1000 m.

Argo's collection of more than a million velocity observations in the past decade has been used to create global maps of the mean velocity at 1000 m (Gray and Riser 2014; Lebedev et al, 2007; Ollitrault and Colin de Verdiere 2014; Figure 3.8) and of absolute geostrophic transport in the upper 2000 m. The combined satellite and *in situ* data sets make it possible to explore the circulation of the global oceans, to calculate the transport and renewal of water masses in the Meridional Overturning Circulations, to test dynamical underpinnings of ocean circulation, and to initialize ocean and coupled models for better description and understanding of the climate system. Ocean circulation is thus another valuable point of intersection between satellite altimetry and subsurface ocean observations.

3.3.6 Horizontal Scales of Variability in the Ocean: The Challenge of Resolution

The Argo array of 3800 profiling floats is a revolutionary achievement in global ocean observation, but its limitations must none the less be recognized. Free-drifting Argo floats have typical separation of several 100 km, while property and circulation variability in the ocean extends to much shorter scales. The Argo concept (Argo Steering Team 1998) was to observe variability on spatial scales of a 1000 km and longer, and timescales of months and longer, by averaging multiple realizations in such large-scale regions. Many important features such as mesoscale eddies, fronts, and boundary currents are not resolved. A 10° x 10° x 60-day bin of Argo data includes about 60 profiles, and these measurements can be averaged to reduce noise from the unresolved sources of variability.

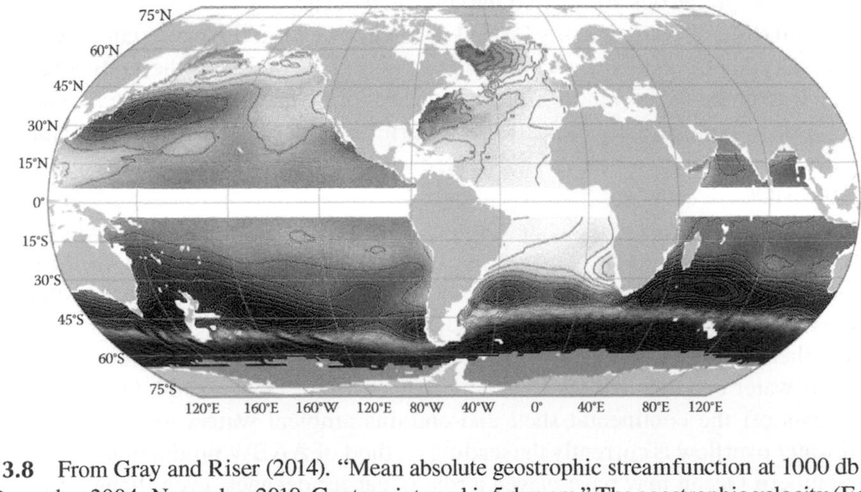

FIGURE 3.8 From Gray and Riser (2014). "Mean absolute geostrophic streamfunction at 1000 db from Argo data for December 2004–November 2010. Contour interval is 5 dyn cm." The geostrophic velocity (Equation 3.2) is parallel to streamfunction contours (clockwise around a high in the Northern Hemisphere, counterclockwise in the Southern Hemisphere) with magnitude equal to the horizontal gradient of the streamfunction divided by the Coriolis parameter.

The resolution of satellite altimetry is finer than that of Argo, particularly in the along-track direction but also by making use of multiple overlapping altimetry missions. Altimetry can be used to amplify the resolution of *in situ* observations where they are well correlated through the "synthetic" profile concept (e.g., Ridgway et al. 2008).

The Argo array is designed for observing the large-scale ocean interior. Other elements of the ocean observing system based on different technologies can provide better resolution for boundary current regions. For example, gliders (Sherman et al. 2001) are a technology similar to profiling floats in their ability to adjust their buoyancy, but they are able to control their direction and navigate between fixed waypoints. High Resolution XBT (HRX) transects provide repeated crossings of boundary currents with approximately 10-km spacing between temperature profiles. Satellite altimetry, Argo, gliders, and HRX transects can each mitigate sampling limitations of the other systems and in combination provide comprehensive boundary current transport time series (Ridgway et al. 2008; Zilberman et al. 2016).

3.4 DEEP-OCEAN (GREATER THAN 2000 M) STERIC VARIABILITY: REPEAT HYDROGRAPHY AND DEEP ARGO

The previous section introduced the relationship between SSH and subsurface density and highlighted how density variability in the upper ocean is monitored. The deep ocean, however, can also experience small changes in density over very large volumes, which causes an important contribution to steric SSH when integrated over space. Here, we will highlight the variability of density in the deep ocean, how it is measured, and how it contributes to total SSH.

3.4.1 Ventilating the Deep Ocean: Deep Water Production and the Global MOC

The deep ocean is filled with cold, dense waters that are ventilated at high latitudes and slowly circulated throughout the globe through the buoyancy-driven Meridional Overturning Circulation (MOC) (e.g., Lumpkin and Speer 2007; Johnson 2008). Two distinct varieties of deep water exist: North Atlantic Deep Water (NADW), formed in the North Atlantic, and Antarctic Bottom Water (AABW), formed primarily along the continental shelf around Antarctica. At these high latitudes where surface conditions allow for dense water formation, the deep ocean's isopycnals shoal and unique surface water properties are carried into the deep ocean.

NADW is a mixture of water masses formed in the Labrador and Norwegian seas (LeBel et al. 2008). There, salty subtropical surface waters, advected northward by the Gulf Stream and the North Atlantic Current, lose buoyancy through surface cooling and sink to create a cold, salty water mass. NADW flows southward along the deep western boundary and fills the deep North Atlantic before it is displaced from below by AABW and, eventually, upwelled along the ACC to form the deep cell of the MOC.

AABW is formed primarily through complex ocean-ice-atmosphere interactions along the Antarctic continental shelf. This process occurs in select locations where cold, dense shelf waters are found. Shelf waters form when polynyas open over shallow continental shelves and allow surface waters to cool and become saltier through brine rejection. The surface waters sink and circulate under the ice, thereby allowing interaction with the overlying ice shelf and formation of a fresh, super cold shelf water end member (Foster and Carmack 1976; Jacobs 2004; Orsi et al. 1999). The shelf water flows off the continental shelf and entrains ambient waters to form AABW (Jacobs 2004). Shelf water overflow is currently the leading method of AABW production, but open convection in the Southern Ocean may have played a role in earlier decades, even though it has not been observed to be a significant source of deep water since the Weddell Polynya closed in 1976 (Gordon 1978). AABW fills the Southern Ocean, mixes with overlying circumpolar waters in the ACC, and travels north along the bottom. Eventually, it upwells and mixes with overlying NADW before returning south to complete the bottom limb of the MOC (Schmitz 1995; Sloyan and Rintoul 2001; Lumpkin and Speer 2007).

3.4.2 Monitoring Deep Steric Variability through Repeat Hydrography

The WOCE conducted the first ever top-to-bottom, ship-based global hydrographic survey, thus effectively taking a "snap shot" of the ocean in the early 1990s, and provided extremely accurate measurements of deep ocean density. Through the WOCE hydrography program, thousands of ocean profiles were taken nominally every 55 km along coast-to-coast ocean transects. Each top-to-bottom WOCE profile consists of continuous measurements of temperature, salinity, and pressure obtained with a conductivity-temperature-depth (CTD) sensor. The CTD data set is of the highest accuracy with standards of 0.002 PSS-78, 0.002 °C, and 3 dbar for salinity, temperature, and pressure, respectively.

A key subset of the WOCE one-time transects has been reoccupied through the Climate and Ocean: Variability, Predictability, and Change (CLIVAR) project and carbon cycle science programs, which are now coordinated by the international Global Ocean Ship-Based Hydrographic Investigation Program (GO-SHIP). The repeated sections all achieve WOCE accuracy standards or better and are exact repeats where time, weather, and ice conditions permit. This allows for evaluation of any temporal changes in deep ocean density and, hence, the deep ocean's contribution to steric sea level between the 1990s and the present.

3.4.3 The Deep Ocean Contribution to Steric Sea Level

The contribution of the abyssal ocean to the steric height of the sea surface is derived from the integrated change in density below the surface (Equation 3.5). Conceptually, density changes in the deep ocean can come about in two ways. The first mechanism for these changes occurs if source conditions in water mass formation regions change in a way that alters the average density of new abyssal waters but not the overall rate of their formation. These density changes in bottom waters are then carried by the ocean circulation far from the source regions into the ocean interior. Except for regions of the ocean directly downstream from formation sites, most of the deep ocean has been isolated from the atmosphere for hundreds to thousands of years. The second mechanism for the deep ocean's contribution to surface steric height variability is through changes in the rate of deep water formation. When a greater or smaller amount of bottom water is produced, isopycnals above the newly formed waters must rise or sink proportionately. For example, if the ventilation rate of the deep water decreases, a global-scale deepening of the mean depth of the dense AABW isopycnal will occur, causing a change in the local vertical density profile through the water column and an integrated change in sea level height (Kawase 1987).

By comparing re-occupations of hydrographic sections, it is seen that both advective changes at high latitudes and isopycnal heave have contributed to variability and change in steric sea level over the past three decades. NADW properties have shown significant interannual and decadal variations. For example, the salinity, temperature, and production rates of NADW have been tied to the North Atlantic Oscillation (NAO). Colder, fresher varieties formed during the positive NAO phase between the mid-1960s and mid-1990s, and warmer, saltier varieties have been produced more recently (Dickson et al. 2002; Hatun et al. 2005; Desbruyères et al. 2014).

Decadal warming trends in AABW observed around the globe have also contributed to steric sea level rise. This heave-derived warming is strongest in the Southern Ocean but extends north to most basins filled with AABW, thereby contributing 0.15 mm year^{-1} to the global sea level rise budget between the 1990s and 2000s (Purkey and Johnson 2010). In addition, AABW found in the Pacific and Indian sectors of the Southern Ocean has freshened, likely owing to the increase in glacial melt rates along West Antarctica (Jacobs and Giulivi 2010; Swift and

Orsi 2012), contributing as much as 0.4 mm year^{-1} to local steric sea level rise due to halosteric expansion (Purkey and Johnson 2013).

3.4.4 Future of Deep Observing: Deep Argo

Although repeat hydrography has started to paint a picture of decadal variability in the deep ocean and its contribution to steric sea level, the ship-based hydrography observations are limited in space and time. In addition, ship-based work is biased toward summer months when oceanographic conditions are milder and ship-based exploration is easier. As a result, current deep observations only capture a small fraction of the observations needed to fully monitor ocean steric variability.

To solve this observational shortcoming, new Argo floats have been developed that are capable of measuring temperature and salinity from the surface to depths as great as 6000 m. Regional pilot arrays of these new instruments have been deployed in the Atlantic, Pacific, Indian, and Southern oceans to demonstrate their capabilities and scientific value. A global array of 1200 Deep Argo floats spaced roughly every five by five degrees will be employed to monitor the full water column (Johnson et al. 2015) and to complement the present Argo Program. Similar to the existing 2000-m Argo array, this deep observing array will revolutionize our ability to monitor and understand deep ocean steric variability.

3.5 GEOSTROPHIC TRANSPORTS AND BOTTOM PRESSURE OBSERVATIONS

3.5.1 Complementarity among Altimetry, Water Column Density, and Bottom Pressure

As explained in Section 3.1, the SSH fluctuates due to changes in interior density (or steric height, see Equation 3.5) and due to pressure changes at the reference level or at the sea floor p_{bot} (changes in mass). It was made clear there that this is a closed system, such that one quantity can be inferred from measurements of the other two. In the present section, we will show how this complementarity can be exploited by different combinations of density measurements, satellite altimetry, and bottom pressure.

Sufficiently accurate measurements of all three terms (altimetry for SSH, *in situ* temperature and salinity for density, and seafloor pressure) are now possible to validate one from the other two or to choose the most suitable combination of them based on a given situation. Figure 3.9 shows an example of an SSH time series observed with satellite altimetry, compared against fixed-point time series of SSH anomaly derived from *in situ* measurements of density and seafloor pressure. It is clear that the satellite and the *in situ* data measure the same signals, but some disagreements exist. The reasons for the latter will include sensor uncertainties, calibration errors, and lack of true co-location of the observations in space and time. The following subsections include discussions of measurement accuracies for the aforementioned terms. Figure 3.9 also shows the typical size of the oceanic SSH signals illustrated elsewhere in this book (order of a few centimeters), which provides a basic requirement for the measurement accuracy of the three terms under consideration. In the following, we will assume that the altimetry data already include a time mean of the dynamic ocean topography (Niiler et al. 2003a; or the current version of "ADT" data from AVISO, SSALTO/DUACS User Handbook 2015) as opposed to only sea level anomalies against an unknown mean and/or geoid.

3.5.2 Volume Transports from End Points, and Accuracy Requirements

The earlier relations (Equation 3.2) and (Equation 3.4) for the geostrophic currents can also be expressed via the geopotential height ϕ of a pressure surface (same as steric or dynamic height, Equation 3.5, multiplied by g) $\phi(p) = \int_{pref}^{p} \frac{1}{\rho} dp + \phi_{pref}$ from which the geostrophic velocity at the

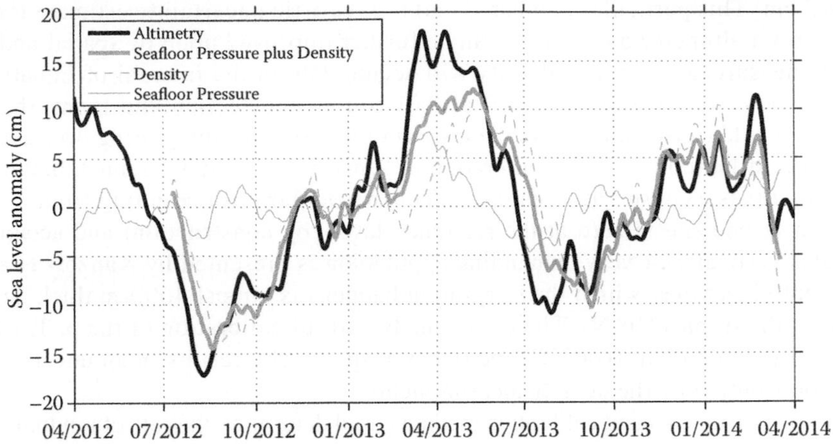

FIGURE 3.9 Time series of altimetry-derived sea level anomaly versus that derived from in situ measurements of seafloor pressure and density. Seafloor pressure was measured with a Pressure Inverted Echo Sounder (PIES) instrument, and density comes from a mooring equipped with temperature/salinity sensors. Altimetry data are from a gridded product generated by Archiving, Validation and Interpretation of Satellite Oceanographic data (AVISO). All data have been filtered with a 10-day low-pass; PIES data were first de-tided and de-trended as per Watts and Kontoyiannis (1990). The location of the site is near Gizo, Solomon Islands. (A. Anutaliya, Scripps Institution of Oceanography.)

pressure level p becomes $u(p) = -\frac{1}{f}\frac{\partial \phi}{\partial x}$; $v(p) = \frac{1}{f}\frac{\partial \phi}{\partial x}$. The volume transport of an ocean current then is obtained by vertical integration over the depth layer of interest and by horizontal integration over the geographic section across which the transport is sought (assuming here, for simplicity, that the section is aligned in the x direction):

$$T = \int_{x1}^{x2}\int_{z1}^{z2} \frac{1}{f}\frac{d\phi}{dx} dz\, dx = \frac{1}{f}\left[\int_{z1}^{z2} \phi_{x2}\, dz - \int_{z1}^{z2} \phi_{x1}\, dz\right] \tag{3.6}$$

Note that the second equality is exact if variations in 1/f are negligible, which means that for purely geostrophic flow the total horizontally integrated transport can be obtained from observations of the vertical density profiles simply at the end points of the section.

This implies that time series of density profile information from two fixed locations may be sufficient to capture volume transports through straits, boundary currents, or entire ocean basins (Hirschi et al. 2003; Kanzow et al. 2007). Such information can, for example, be collected via moorings, station-keeping gliders, or bottom-mounted instruments that capture the main modes of density profile variability. The reference geopotential height ϕ_{pref} in the aforementioned relations can be that of the sea surface or of a deep pressure level, in which cases the vertical integration is either from the surface down or from a deep level up (see examples that follow).

An SSH difference of 10 cm over a section at mid-latitudes (45°) and integrated over a 1000-m layer would result in a geostrophic volume transport of 10 Sv across the section. This is independent of the length of the section; only the pressure difference between the end points matters for the transport. The same sea level or pressure difference at 20° latitude would cause approximately double the transport, highlighting the sensitivity of the measurements at low values of *f*.

Inversely, we can use the previous relations to obtain accuracy requirements for the satellite altimeter, the density, or bottom pressure measurements. For example, to determine an upper-layer (0–1000 m) transport to within 3 Sv at mid-latitude, the altimetry data need to be accurate to 3 cm. This particular number (3 cm) is a realistic uncertainty estimate for instantaneous along-orbit altimetry data, which can be further improved upon by spatial and temporal averages. A measurement error in density will accumulate in the integral of Equation 3.6: If, for example, there were a salinity offset of 0.01 psu at one section end point, the resulting density offset would cause an erroneous transport of 0.7 Sv after integrating over a depth layer of 1000 m at mid-latitude. Careful calibration of sensors can achieve a much better accuracy of salinity, so this is a conservative estimate. This particular error resulting from a systematic density offset at one site is zero at the reference level (by construction) and accumulates as one integrates vertically. A very demanding application is presented by Kanzow et al. (2006), where the depth layer over which the integration happens is almost 4000-m thick and the latitude is within the tropics (16°N). There, extremely careful calibration of the p, T, and S measurements (targeting 5 dbar, 0.002°C, and 0.003, respectively) results in an uncertainty of 1.6 Sv for the contribution by the density measurements.

The types of instruments and calibration methods used determine the role of random versus systematic errors in the density measurements. Random errors resulting from a vertically distributed set of discrete sensors on a mooring tend to cancel out in the vertical integration of Equation 3.6, while systematic errors from, for example, a single profiling sensor will accumulate. However, systematic sensor biases cancel out when the instruments at both end points of the section are subject to the same offsets due to the horizontal difference in Equation 3.6. This leads to better transport accuracy if all instrumentation is calibrated in one batch and against the same reference standard (e.g., when preparation, calibration, and deployment of the sensors on both end points occurs during the same ship expedition against CTD or water sampling references rather than during independent operations).

3.5.3 Upper-Layer Transports

An illustration of how to compute the volume transport using Equation 3.6 from a mooring pair (for density profiles) together with satellite altimetry (for ϕ_{pref}) is provided in Figure 3.10: Panel A shows the locations of two moorings that span the California Current (Lankhorst and Send 2016). There were 15 to 16 instruments on each mooring, measuring temperature and salinity to the same target accuracies as those by Kanzow et al. (2006). Panels B–D show profiles of density and geopotential height anomaly and how the geopotential profiles are adjusted using altimetry-derived SSH as a starting point for the vertical integration. Horizontal differencing then yields a transport-per-unit-depth profile that represents the section-average geostrophic flow across the section. Here, the California Current stands out as a strong signal in the upper 300 m. Integrating the profile in the vertical results in a net transport of around 4 Sv by the California Current. The referencing to the altimetry data shows that the California Current extends to a 300-m depth (depth of the zero crossing). The information contributed by altimetry is the surface velocity, and the information from the moorings quantifies how this flow decays with depth. Again, this approach requires that the altimetry data be augmented with knowledge of the mean dynamic topography, implying knowledge of the geoid.

Many cases of surface currents can be approximated by a one-and-a-half-layer model in which there is a flow at the surface overlying a deep layer that is at rest. Assuming the flow is geostrophic, this will result in a SSH signal that describes the surface flow, while the interface depth between the two layers will be an amplified, mirror-reverse image of the SSH (e.g., Figure 3.3 by Tomczak and Godfrey 1994). In principle, all of the information about the system is available in the sea surface signal once the interior structure is known; thus, the altimeter alone would be able to observe the corresponding changes in absolute transport after suitable calibration. An example is shown in Figure 3.11, where mooring data referenced to bottom pressure were used to compute the geostrophic transport of the California Current (same instrumentation as used in Figure 3.10, plus

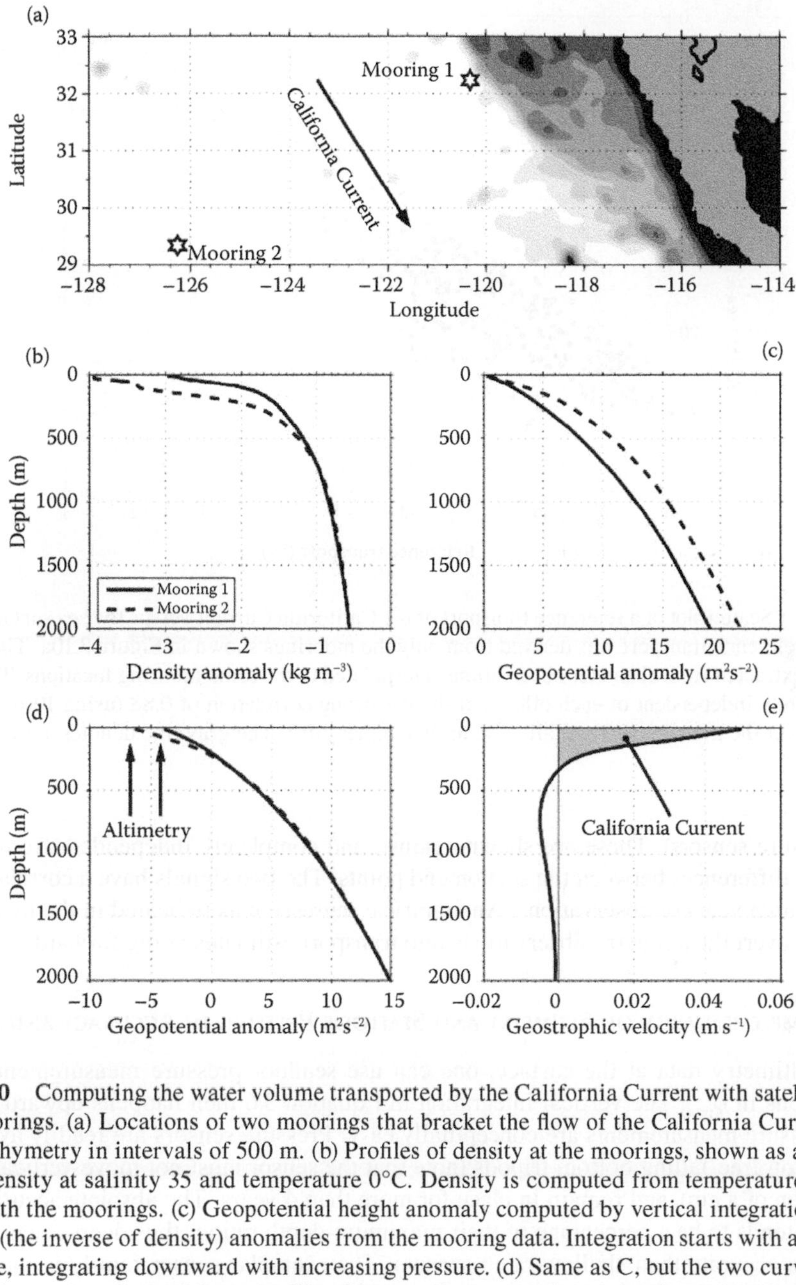

FIGURE 3.10 Computing the water volume transported by the California Current with satellite altimetry and two moorings. (a) Locations of two moorings that bracket the flow of the California Current. Shading indicates bathymetry in intervals of 500 m. (b) Profiles of density at the moorings, shown as anomaly relative to the density at salinity 35 and temperature 0°C. Density is computed from temperature and salinity measured with the moorings. (c) Geopotential height anomaly computed by vertical integration of the specific volume (the inverse of density) anomalies from the mooring data. Integration starts with a value of zero at the surface, integrating downward with increasing pressure. (d) Same as C, but the two curves have been offset by surface values derived from satellite altimetry. To convert the sea level data (an elevation given in meters) to geopotential height, one multiplies with the gravitational acceleration g. (e) Geostrophic velocity derived from the difference between the two curves from (d). This is the mean flow across the section defined by the two mooring locations. The signal of the California Current in the upper 300 m is shaded. Vertical integration of these velocities, and multiplication with the section length (ca. 650 km), results in the volume transport. All data are 1-month averages from June 2009. (Based on data from Lankhorst and Send, 2016.)

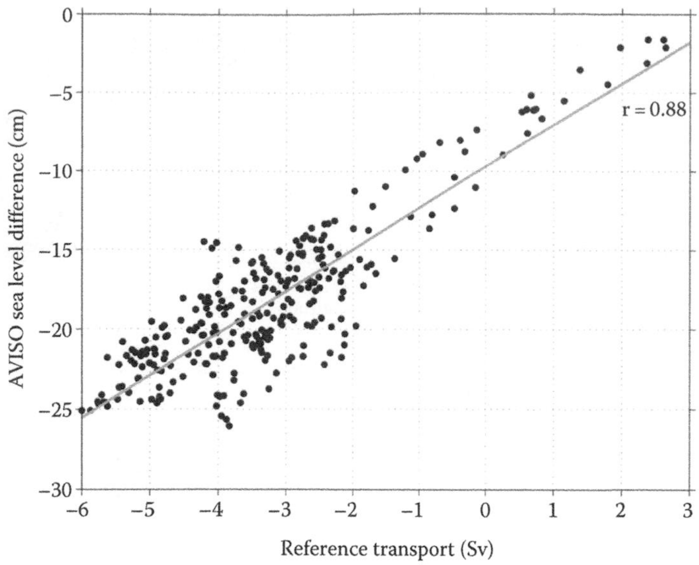

FIGURE 3.11 Scatterplot of a reference transport of the California Current versus the sea surface height difference. The reference transport was derived from only the moorings shown in Figure 3.10a. The sea surface heights were extracted from altimeter observations interpolated to the same mooring locations. The two properties are entirely independent of each other and show a strong correlation of 0.88 (using 10-day low passed data). Data cover the time range from fall 2008 until summer 2012. The gray line denotes a linear regression fit to the data.

bottom pressure sensors). These are shown against, and completely independent from, altimetry-derived SSH differences between the section end points. The two signals have a correlation of 0.88 over more than 3 years of observations. An empirical regression as indicated in the figure can then be used to convert the altimetry observations into transport estimates going forward.

3.5.4 Complementarity of Altimetry and Seafloor Pressure in Accuracy and Timescales

Instead of altimetry data at the surface, one can use seafloor pressure measurements to obtain the fluctuations in ϕ_{pref}. The vertical integration in Equation 3.6 then happens upward from there. Seafloor pressure measurements are conceptually easy: Pressure sensors are readily available, can be deployed on free-falling bottom tripods (note that the sensor must not move vertically by more than a fraction of a cm), and remain in place for more than 4 years. The absolute accuracy of pressure sensors tends to be a percentage of their maximum-depth rating; thus deep-ocean sensors are typically less accurate than shallow-rated sensors. The exemplar sensor family is the Digiquartz series (see Paroscientific, Inc., 2014) which has an accuracy of 0.01% of its depth rating. For a deep-ocean version, this would amount to 70 cm, which is not sufficient to capture the signals of interest. However, the resolution and repeatability of the measurements are substantially better, and the measurements used in oceanic applications such as Kanzow et al. (2006) are based on these rather than absolute accuracy. Concurrent deployments have shown (Kanzow et al. 2006, their Figure 10) that on timescales of a few days to months, agreement among separate deep-rated sensors can be 1 cm or better, despite a nominal accuracy of 70 cm (and even better for shallower applications). This offers a unique complementarity between satellite altimetry and seafloor pressure sensors, in that the pressure sensors are particularly good at short (hourly to intra-annual) timescales, whereas altimetry-derived data products excel at longer (monthly to interannual) timescales, including secular trends, due to the averaging then possible. Note that referencing the vertical integrals to the sea surface for the long timescales and for the mean also resolves the problems of sensor drift and unknown absolute depth for the bottom pressure sensors.

To merge the three types of measurements, altimetry data can provide the low-passed geopotential reference, and the seafloor pressure is the high-passed (periods less than 2 years for 4-year long deployments) reference for the vertical integration of the interior density data. In the overlapping period band, the comparison provides cross-validation and redundancy.

3.5.5 Constraining Transports in Two-Mode Systems with Altimetry and Bottom Pressure

Even without concurrent *in situ* density profiles, satellite altimetry data and seafloor pressure data can sometimes still be combined in a way that may provide information about the interior structure of the ocean. Because there are only two data points in such a situation (altimeter and seafloor pressure), the system has to be reduced to two descriptors that capture the status of the system, for example, by decomposing the profiles of transport or ϕ into vertical EOFs. In many areas of the oceans, the first two EOFs will capture most of the variability. In the case of Figure 3.10 (Panel c or d), this situation would mean that the surface and seafloor values of the geopotential height profiles are known, but the shape of the curves in between is not. Projecting the two values onto the first two EOFs approximates the internal structure of the geopotential height profile or of the transport.

An application of this is shown in Figure 3.12 for the flow across the Solomon Sea as a function of depth and time as measured by moorings (density) and Pressure Inverted Echo Sounder (PIES; Inverted Echo Sounder, 2008; seafloor pressure) on the left side and a reconstruction using two vertical modes based on altimetry and seafloor pressure alone on the right. The two methods produce nearly identical results, showing the vertical structure of the flow as well as its evolution in time.

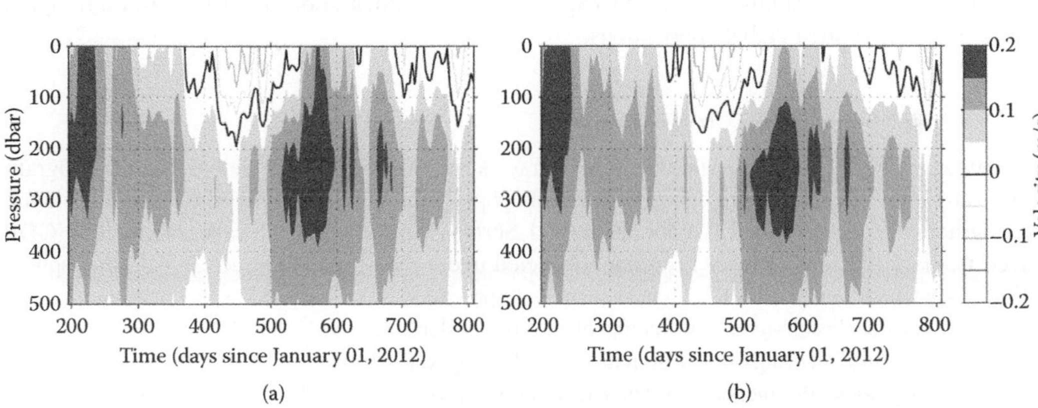

FIGURE 3.12 Depth-time contour plots of velocity through the Solomon Sea. Data are based on observations at two sites, each of which is equipped with a seafloor pressure sensor (Pressure Inverted Echo Sounder, or PIES) and a mooring measuring temperature and salinity in the water column. One site is off Misima, Papua New Guinea, and the other near Gizo, Solomon Islands. Panel (a) shows the geostrophic currents computed from the vertical density profiles based on mooring data, with the PIES pressure as the reference level. Panel (b) shows a reconstruction that uses the PIES pressure data and satellite altimetry. In this case, the vertical structure comes from empirical orthogonal functions (EOF) that were determined from the mooring data. These vertical modes were then utilized in a linear combination with altimetry and PIES data to obtain the velocity profiles. (The analyses shown here are preliminary and were made available by A. Anutaliya [Scripps Institution of Oceanography]; they are meant to illustrate the technique rather than present scientific results.)

3.6 DYNAMIC TOPOGRAPHY AND SURFACE VELOCITY

Ocean circulation reveals itself through its dynamics in the dynamic topography. In the geostrophic current, fluid flows along the lines of constant dynamic topography (Equation 3.3). Satellite altimeters measure the shape of the ocean surface in which the dynamic topography signal is dominated (by nearly two orders of magnitude on the global scale) by geographic variations in gravity. To extract dynamic topography from altimeter data, accurate geoid models have been developed using results from recent satellite gravity missions (see Chapter 5). An alternative method, used for many applications, derives dynamic topography from direct measurements of the ocean current velocity with various instruments. At present, these two methods provide the most accurate dynamic topography on the ocean mesoscale.

3.6.1 Eulerian Velocity Measurements

A classic example of the dynamic topography estimate is the dynamic height, calculated from the density profiles and referenced to an assumed "no-motion" level (Section 3.1). This "dynamical method" was used for many decades, giving satisfactory descriptions of the ocean circulation. However, the growing need for accurate oceanographic data and better understanding of the significance of subsurface motions has dictated the development of more advanced methods. One of the first experiments that successfully combined the data of current meters, measurements of velocity at intermediate and abyssal depths, and hydrographic surveys to reference the altimetry from then-flying TOPEX/Poseidon was the Affiliated Surveys of the Kuroshio off Cape Ashizuri (ASUKA) line (Imawaki et al. 2001), a project aligned with one of the satellite tracks south of Japan. The study revealed a very high correlation between the velocities measured by current meters extending to the ocean surface using CTD-surveys and the geostrophic velocities derived from the altimeter data. This correlation allowed calibrating the altimetry and continued monitoring of the Kuroshio transport after the end of the field campaign by using only a remotely sensed sea level anomaly.

Although Eulerian and ship-borne velocity measurements provided satellite altimetry with an important reference in a number of regional experiments, implementation of the array of Lagrangian drifting buoys facilitated global applications.

3.6.2 Lagrangian Velocity Measurements

Approximately 90% of the operational drifter array is maintained by the Global Drifter Program (GDP) (http://www.aoml.noaa.gov/phod/dac/dacdata.php), jointly managed by the NOAA Atlantic Oceanographic & Meteorological Laboratory and Scripps Institution of Oceanography. The GDP evolved from the Surface Velocity Program conducted under the Tropical Ocean Global Atmosphere (TOGA) experiment and the WOCE that, in turn, were preceded by many local experiments with drifting buoys of various designs (c.f., Maximenko et al. 2013, Niiler 2001 and Niiler 2003). After many comparisons, a "standard" drifting buoy design has been developed (Sybrandy and Niiler 1991) that includes a 6-m tall "holey sock" drogue, centered at a 15-m depth, providing a 40:1 drag area ratio. The surface float is equipped with a satellite transmitter, drogue sensor, and sea surface temperature sensor as well as optional wind, pressure, salinity, and other sensors. Though the design of the GDP drifter and drogue sensor evolved over nearly three decades, its dynamic characteristics have remained nearly unchanged. Currently, with approximately 1400 active drifters, the GDP archive includes data for nearly 20,000 historic buoy trajectories with the 15-m drogue of a total length exceeding 10,000 years. A comparable volume of data has been collected from drifters after losing their drogues by making simple calculations that reduce the differences between velocity measurements of drogued and undrogued buoys.

Ocean velocity is derived from time series of coordinates of drifters, which move in response to the full spectrum of ocean processes, ranging from the Stokes drift in wind waves to large-scale circulation. In addition, the direct wind force acting on the surface of the buoys affects the drifters' motion relative to the water (Poulain et al 2009). For drogued drifters, this downwind slip is as small as 0.7 cm/s under a 10 m/s wind (Niiler et al. 1995) but increases to 8 cm/s for undrogued buoys (Pazan and Niiler 2001). Nominal interpretation of drifter velocity as a surface current at the drogue center depth or as an average through the depth range occupied by the drogue may not be accurate, especially under rough ocean conditions. However, the array of drifters, having a uniform design, continues to provide an excellent "index" of ocean surface currents that, combined with other observations, allows characterization of ocean motions of different scales and different physics, from tides and inertial oscillations (Elipot et al. 2016) to Ekman currents (e.g., Ekman 1905), to mesoscale eddies (Lumpkin 2016) to the seasonal cycle (Lumpkin and Johnson 2013) and the large-scale circulation. In Figure 3.13, the combined trajectories of historic drifters outline all main ocean currents and connect them into an amazingly complex global system.

3.6.3 Geostrophic Currents and Mean Dynamic Topography

Kinetic energy of the ocean motions peaks on the mesoscale (see Chapter. 11). On this scale, the dynamic is essentially geostrophic (Section 3.1). Outside of relatively small regions of very low kinetic energy and away from narrow zones along the coastlines and the equator, low-pass filtered drifter velocities (after elimination of inertial and higher frequency components) correlate at 0.7 to 0.9 with collocated geostrophic velocities derived from the sea level anomaly (e.g., Niiler et al. 2003a). Relatively simple methods parameterizing Ekman currents (e.g., Rio and Hernandez 2003) can help to better isolate the geostrophic part of drifter velocity and further increase correlation with altimetry.

After filtering and subtracting Ekman velocity, drifter data provide an estimate of the instantaneous gradient of the dynamic topography along the trajectory. The combination of drifter data with collocated (in space and in time) satellite sea level anomaly gradients produces time series and time averages of the dynamic topography gradients. The ensemble of even a relatively small number of active drifters provides excellent coverage over the drifters' lifetime. Unlike Eulerian observations limited to particular points or lines, drifters travel around the ocean covering it with a dense network of historical data and effectively outlining its circulation pattern, as seen in Figure 3.13.

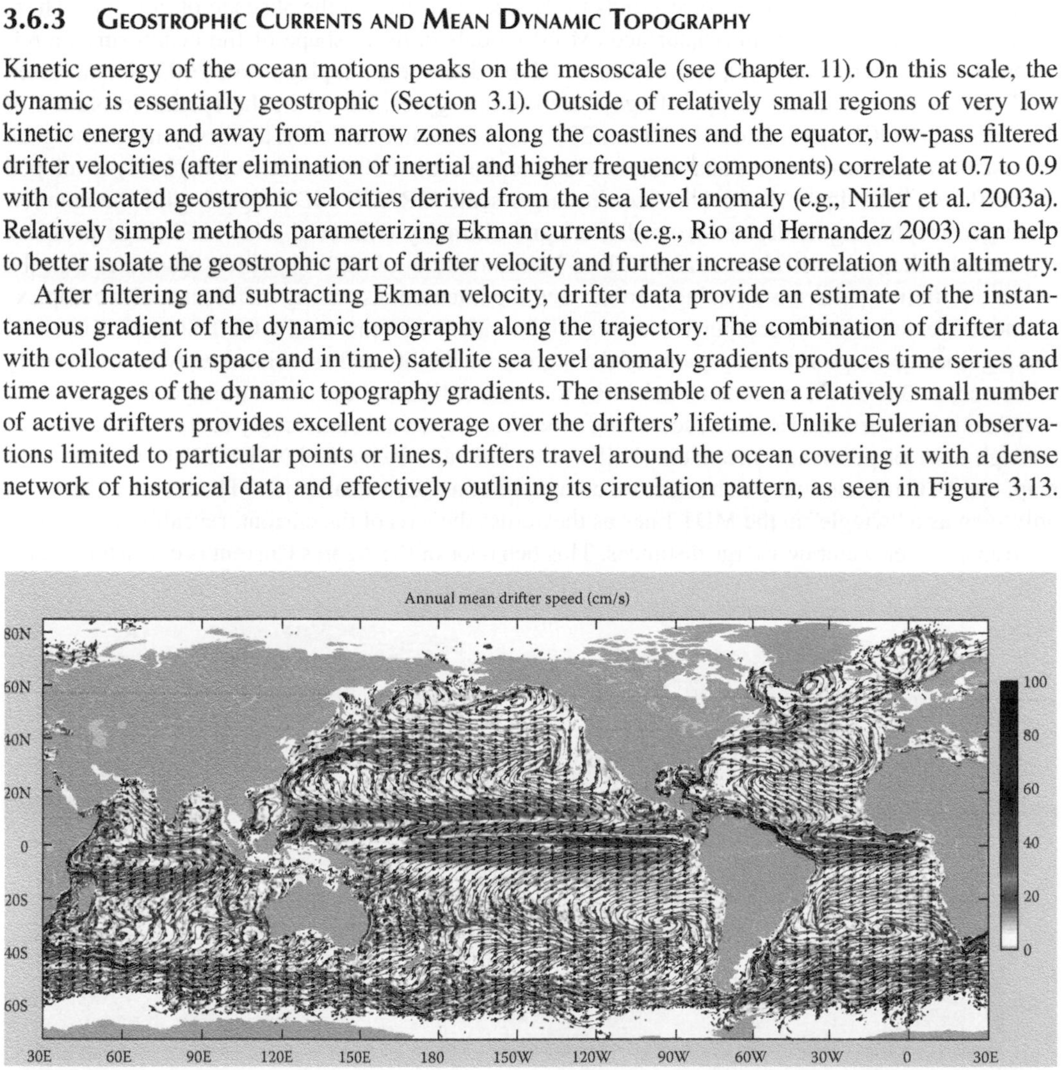

FIGURE 3.13 Mean current speeds (colors, in cm s^{-1}) from near-surface drifter data with streamlines (black lines). Streamlines are calculated from spatially smoothed currents to indicate flow direction and qualitatively illustrate large-scale circulation features, including surface divergence. Light gray areas have less than 10 drifter days per bin (0.8 per square degree). In addition, only bins with mean current speeds statistically different from zero at one standard error of the mean are shaded. (Updated from Lumpkin, R., and Johnson, G., *J. Geophys. Res. Oceans*, 118(6), 2992–3006, 2013. With permission.)

Over the years, most locations are visited by many different drifters, improving estimates of the time-mean dynamic topography gradient.

Integration over the large scale of the gradient of dynamic topography, estimated locally from drifters and geostrophic velocities, accumulates significant errors (Niiler et al. 2003a). These errors are due to ageostrophic ocean motions and can be partly reduced by parameterizing local wind-driven currents. Early mean dynamic topography (MDT) products (Rio and Hernandez, 2004) have radically improved with observations from such satellite gravity missions as NASA's Gravity Recovery and Climate Experiment (GRACE, since 2002, Tapley et al. 2004) and ESA's Gravity Field and Steady-State Ocean Circulation Explorer (http://www.esa.int/Our_Activities/Observing_the_Earth/The_Living_Planet_Programme/Earth_Explorers/GOCE; GOCE, 2009–2013). These resulted in greatly improved models of the Earth's geoid (see Chapter 5), the equipotential surface referenced to the ellipsoid that corresponds to the ocean surface in the absence of currents. These models, combined with mean sea surface (MSS) models (a mean shape of the ocean surface referenced to the same ellipsoid), resulted in MDT products (e.g., Tapley et al. 2003; Johannessen et al. 2014) that accurately mapped large-scale oceanic gyres. However, the spatial resolution of the GRACE/GOCE fields (around 400 km/200 km) is/was not sufficient to survey most energetic mesoscale ocean currents (jets and eddies). Maximenko et al. (2009) and Rio et al. (2011) developed two different techniques that synthesize the larger-scale GRACE MDT signal with the smaller-scale geostrophic signal derived from drifters and altimetry (Rio et al. also used Argo profiles). Figure 3.14 shows the 1992 to 2012 MDT of Maximenko et al. (2009), which illustrates that the ocean circulation pattern is amazingly robust and the interactions among different scales complex.

Indeed, extensions of such persistent western boundary currents as the Gulf Stream, Kuroshio, and Agulhas correspond to the regions of the highest variability and associated eddy kinetic energy in the ocean. One would expect this variability to mix and diffuse these jets in multiyear averages into broad and slow currents. However, according to MDTs, they remain amazingly narrow while flowing through a set of permanent meanders (Niiler et al. 2003a,b; Pazan and Niiler 2004). On the same map, the ACC reveals a very rich structure composed of a dozen smaller jets. The Azores Current is only seen as a "wiggle" in the MDT lines as they cross the axis of the current, revealing that it does not transport sea water over large distances. This behavior of the Azores Current is consistent with a

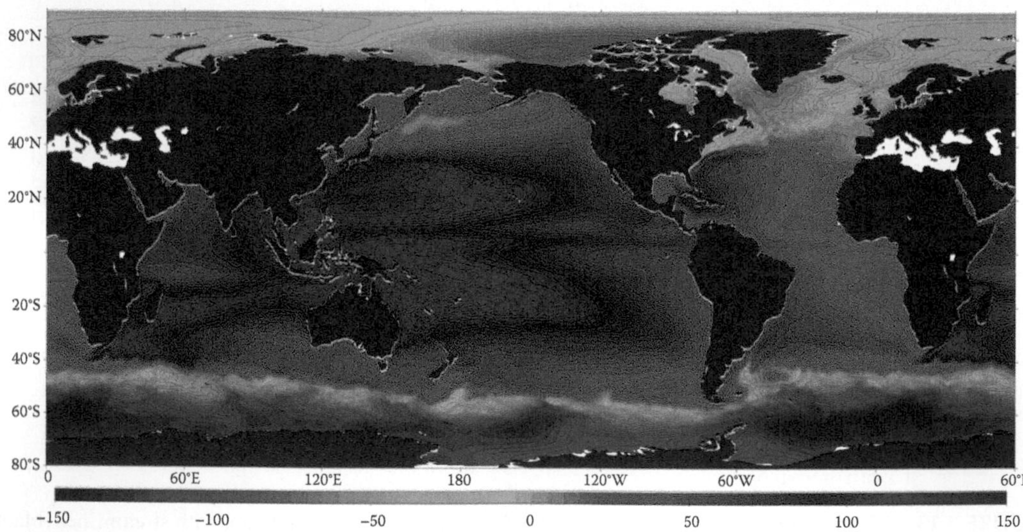

FIGURE 3.14 The 1992 to 2012 mean dynamic topography, updated from Maximenko et al. (2009) by adding the data of the Gravity Field and Steady-State Ocean Circulation Explorer (GOCE) gravity mission. Contour interval is 5 cm. (Modified from Maximenko, N. A., et al., *J. Atmos. Oceanic Tech.*, 26(9), 1910–1919, 2009. With permission.)

beta-plume dynamic (Kida et al. 2007), which could also explain the multiple stationary anisotropic jet-like features (striations) found in various regions (Maximenko et al. 2008) of the ocean.

3.6.4 Ageostrophic Motions

Ageostrophic motions are described by the terms of the full momentum equations, excluding the geostrophic balance (Equation 3.3); they do not always produce a signal in the sea level. Understanding ageostrophic processes is important for correct interpretation and joint use of sea level and velocity data.

For example, for a water parcel spinning in an eddy, nonlinear terms describe a "centripetal" acceleration pointing toward the eddy center. A corresponding "centrifugal" force, an inertia force "working" in the direction opposite to centripetal acceleration, is aligned with the Coriolis force in cyclonic eddies, yielding an additional sea level anomaly to support the "cyclostrophic" balance. Respectively, in anticyclones, the centrifugal force is against the Coriolis force, resulting in a small sea level anomaly to maintain the balance. Therefore, the same rotational velocity anticyclonic eddies have weaker signatures in the sea level than cyclonic eddies. Although the relative value of nonlinear terms is small under the small Rossby numbers common for the oceanic mesoscale, the effect of the centripetal acceleration is visible in the statistics of drifter and altimeter velocities. Fratantoni (2001) showed that drifter-derived eddy kinetic energy (EKE) slightly exceeds the EKE estimated from the altimetry south of the Gulf Stream Extension, and the difference changes sign north of the Gulf Stream. Maximenko and Niiler (2006) showed that southern and northern flanks of the Gulf Stream are dominated by cyclonic and anticyclonic eddies, respectively, and that the pattern of the difference in drifter and altimeter EKE is consistent with the effect of the centripetal acceleration.

Motions on periods shorter than the local inertial period are also ageostrophic. Shorter timescales correspond as a rule to smaller space scales that make them hard to detect in today's observing systems. Inertial oscillations are a special case of ageostrophic motions because they reflect an adjustment process that restores geostrophic balance or modifies it depending upon external conditions. Oscillations on periods close to the local inertial period are usually triggered by sudden changes in the wind (passing storms and fronts). Induced circular motions are distinct in the data of drifting buoys (Elipot et al. 2016); they are controlled by the Coriolis force and do not necessarily have any signature in the sea level.

Under a steady wind, vertically sheared currents develop a profile known as an Ekman spiral (Ekman 1905). These motions are maintained by the balance among the wind stress, Coriolis force, and vertical mixing. At the sea surface, the Ekman current flows at an angle to the right from the wind direction and rotates further to the right and decays with depth in the Northern Hemisphere (to the left in the Southern Hemisphere). In the full formulation of the mixed layer dynamics, including buoyancy flux, diurnal cycle, Stokes drift, breaking waves, and so on, Ekman currents become difficult to describe or simulate numerically. Rio and Hernandez (2003) extracted ageostrophic velocities from drifter data and analyzed their correspondence with the local wind. In all cases, they confirmed the predicted sign of the angle between current and wind, but the degree of the correspondence depended on many parameters, including the latitude, season, and frequency. Elipot and Gille (2009) also pointed out that the ocean response depends upon the rotational polarity of the wind.

Uchida and Imawaki (2003) introduced a model of surface circulation in the North Pacific as a function of both geostrophic and Ekman currents determined from a simple parameterization. Bonjean and Lagerloef (2002) built the Ocean Surface Currents Analysis Real-Time (OSCAR) model that includes the equatorial regions, where simple geostrophic balance does not hold. Maximenko and Hafner (2010) developed a similar global diagnostic model (Surface Currents from Diagnostic, or SCUD), in which currents are fitted to local historical drifter velocities. They demonstrated the utility of this approach through simulating the drift of marine debris (e.g., Maximenko et al. 2015). A surface velocity product is also available from Rio and Etienne (2017).

Scientific interest and technological progress point to future investigations of the sub-mesoscale ocean dynamics, in which the relation between sea level and velocity is indirect and demands the development of new models, such as "surface quasi-geostrophic models" (Klein et al. 2010) (see Chapter 11).

3.7 THE TECHNOLOGY REVOLUTION AND THE FUTURE OF OCEAN OBSERVATIONS

This chapter has shown how technology advances are revolutionizing the capabilities of *in situ* ocean observing systems in relation to satellite altimetry. Autonomous arrays such as the Argo Program and the Global Drifter Network now have the same global reach as satellite altimetry. Thus, the observations needed to understand satellite altimetry, by separating the ocean density and mass contributions to SSH and by direct measurements of velocity, are now available on space and timescales spanning much of the variability of the ocean interiors. The physical processes of the global oceans, including water mass formation and spreading, meridional overturning circulations, air–sea exchanges, and transports of heat, mass, and freshwater, and mesoscale eddy generation and propagation, are now observed by *in situ* as well as satellite observing systems. Even tide gauges, with their centuries of history, can now be installed and maintained in greater numbers and with increased utility through advances including radar and pressure-measuring gauges and GNSS observations of land motion.

The technology revolution in ocean observing is only beginning. Oceanic boundary currents are a critical element of the ocean circulation, and these are now accessible to gliders and advanced mooring technologies. Regional pilot experiments have demonstrated the feasibility of sustained observation of boundary current velocity and transports globally. Similarly, the deep oceans below the Argo Program's present 2000-m limit are beginning to be explored with Deep Argo floats having 6000-m capability. New autonomous sensors for biogeochemical observations including pH, dissolved oxygen, and others, are opening multidisciplinary dimensions in ocean observations. These extensions of the ocean observing system—into the swift and narrow boundary currents, into the deep oceans, and through new sensors—are essential to complete inventories and transports for closing global budgets to document climate variability and change. Today's ocean observing system is merging satellite and *in situ* data streams, for coverage and understanding of the global ocean, to a greater extent than seemed possible a few years ago.

3.8 OUTLOOK

Sustained observing systems for particular current systems can exploit the complementarity described earlier, especially as sensor and platform technologies evolve. Fixed end-point installations using moorings, bottom pressure, or inverted echo sounders can provide volume transport with high temporal resolution (inertial timescales and up) and can work well in strong currents with easy access to either side (e.g., straits and passages). Availability of good altimetry data across the section reduces the degrees of freedom required from the fixed installations. Gliders can be used either in station-keeping mode to provide density profiles or can do less frequent crossings of the section with full x–z resolution, in which case combination with fixed end-point sensors provides volume transport constraints and the high-frequency variability. The problem of bottom pressure sensor drift on long timescales may be correctable with GRACE satellite bottom pressure observations that work well on longer space and timescales. Depending on requirements, suitable sensor and platform combinations, including altimetry, now exist to design efficient transport observing systems.

ACKNOWLEDGMENTS

Section 3.3 and the integration and editing of all sections in this chapter by M. Scanderbeg were supported by NASA Ocean Surface Topography Science Team Grant No. NNX13AE82G to the

University of California, San Diego, and Section 3.6 by Grant No. NNX13AK35G to the University of Hawaii. Section 3.5 was funded in part by the Climate Observation Division, Climate Program Office, NOAA, U.S. Department of Commerce. Some of the altimeter products were produced by Ssalto/Duacs and distributed by Aviso, with support from Cnes (http://www.aviso.altimetry.fr/duacs/).

REFERENCES

Argo Science Team. 1998. *On the Design and Implementation of Argo—An Initial Plan for the Global Array of Profiling Floats*. International CLIVAR Project Office Report, 21(GODAE Report No. 5). Qingdao, China.

AVISO data reference: SSALTO/DUACS User Handbook: (M)SLA and (M)ADT Near-Real Time and Delayed Time Products. Reference: CLS-DOS-NT-06-034. Nomenclature: SALP-MU-P-EA-21065-CLS. Issue 4.4, date 2015/06/30, accessed online 2016/04/05. https://www.aviso.altimetry.fr/fileadmin/documents/data/tools/hdbk_duacs.pdf

Barnett, T. P. 1984. The estimation of "global" sea level change: A problem of uniqueness. *Journal of Geophysical Research*, 89(C5), 7980–7988. doi:10.1029/JC089iC05p07980.

Bernstein, R., and W. White. 1974. Time and length scales of baroclinic eddies in the central North Pacific Ocean. *Journal of Physical Oceanography*, 4, 613–624.

Bevis, M., W. Scherer, and M. Merrifield. 2002. Technical issues and recommendations related to the installation of continuous GPS stations at tide gauges. *Marine Geodesy*, 25, 87–99. doi:10.1080/01490410275351 6750.

Blaha, J. P. 1984. Fluctuations of monthly sea level as related to the intensity of the Gulf Stream from Key West to Norfolk. *Journal of Geophysical Research*, 89(C5), 8033–8042. doi:10.1029/JC089iC05p08033.

Blewitt, G., C. Kreemer, W. C. Hammond, H.-P. Plag, S. Stein, and E. Okal. 2006. Rapid determination of earthquake magnitude using GPS for tsunami warning systems. *Geophysical Research Letters*, 33, L11309. doi:10.1029/2006GL026145.

Bonjean, F., and G. S. E. Lagerloef. 2002. Diagnostic model and analysis of the surface currents in the tropical Pacific Ocean. *Journal of Physical Oceanography*, 32, 2938–2954.

Bonnefond, P., B. Haines, and C. Watson. 2011. In situ calibration and validation: A link from coastal to open-ocean altimetry. In *Coastal Altimetry* (eds. S. Vignudelli, A. Kostianoy, P. Cipollini, and J. Benveniste). Springer, Berlin, pp. 259–296.

Calafat, F.M., D. P. Chambers, and M. N. Tsimplis. 2014. On the ability of global sea level reconstructions to determine trends and variability. *Journal of Geophysical Research Oceans*, 119, 1572–1592. doi:10.1002/2013JC009298.

Cartwright, D. E. 1999. *Tides: A Scientific History*. Cambridge University Press: Cambridge. 292 p.

Chambers, D. P., C. A. Mehlhaff, T. J. Urban, D. Fujii, and R. S. Nerem. 2002. Low-frequency variations in global mean sea level: 1950–2000. *Journal of Geophysical Research*, 107, 3026. doi:10.1029/2001JC001089.

Chelton, D. B., J. C. Ries, B. J. Haines, L.-L. Fu, and P. S. Callahan. 2001. Satellite altimetry. In *Satellite Altimetry and Earth Sciences. A Handbook of Techniques and Applications* (eds. L.-L. Fu, and A. Cazenave). Academic Press, San Diego, CA, pp. 1–131.

Church, J. A., and N. J. White. 2011. Sea-level rise from the late 19th to the early 21st century. *Surveys in Geophysics*, 32, 585–602. doi:10.1007/s10712-011-9119-1.

Christiansen, B., T. Schmith, and P. Thejll. 2010. A surrogate ensemble study of sea level reconstructions. *Journal of Climate*, 23, 4306–4326. doi:10.1175/2010JCLI3014.1.

Church, J. A., et al. 2014. Sea level change. In *Climate Change 2013: The Physical Science Basis*. Working Group I Report of the Fifth Scientific Assessment of the Intergovernmental Panel on Climate Change. Cambridge University Press, Cambridge. pp. 1137–1216.

Davis, R. E., J. T. Sherman, and J. Dufour. 2001. Profiling ALACEs and other advances in autonomous subsurface floats. *The Journal of Atmospheric and Oceanic Technology*, 18(6), 982–993.

Desbruyères, D. G., E. L. McDonagh, B. A. King, F. K. Garry, A. T. Blaker, B. I. Moat, and H. Mercier. 2014. Full-depth temperature trends in the northeastern Atlantic through the early 21st century. Geophys. Res. Lett., 41, 7971–7979, doi:10.1002/2014GL061844.

Dickson, B., I. Yashayaev, J. Meincke, B. Turrell, S. Dye, and J. Holfort. 2002. Rapid freshening of the deep North Atlantic Ocean over the past four decades. *Nature*, 416, 832–837. doi:10.1038/416832a.

Digiquartz® Broadband Pressure Transducers and Depth Sensors with Frequency Outputs. *User Manual. Paroscientific, Inc.*, 4500 148th Ave. NE, Redmond, WA 98052, USA, Document No. G8203, Rev. P, February 2014.

Donguy, J.-R., and G. Meyers. 1996. Mean annual variation of transport of major currents in the tropical Pacific Ocean. *Deep-Sea Research Part I*, 43, 1105–1122.

Douglas, B. C. 1991. Global sea level rise. *Journal of Geophysical Research*, 96, C4. doi:10.1029/91JC00064.

Ducet, N., P.-Y. Le Traon, and G. Reverdin. 2000. Global high resolution mapping of ocean circulation from TOPEX/Poseidon and ERS-1 and -2. *Journal of Geophysical Research*, 105(C8), 19477–19498.

Ekman, V. W. 1905. On the influence of the Earth's rotation on ocean currents. *Arkiv för matematik, astronomi och fysik*, 2(11), 1–36.

Elipot, S., and S. T. Gille. 2009. Estimates of wind energy input to the Ekman layer in the Southern Ocean from surface drifter data. *Journal of Geophysical Research*, 114, C06003. doi:10.1029/2008JC005170.

Elipot, S., R. Lumpkin, R. C. Perez, J. M. Lilly, J. J. Early, and A. M. Sykulski. 2016. A global surface drifter data set at hourly resolution. *Journal Geophysical Research: Oceans*, 121, 2937–2966.

Foster, J. 2015. GPS and surveying. In *Handbook of Sea-Level Research* (eds. I. Shennan, A.J. Long, and B.P. Horton), Wiley, Chichester, pp. 157–170.

Foster, T. D., and E. C. Carmack. 1976. Frontal zone mixing and Antarctic bottom water formation in the southern Weddell Sea. *Deep Sea Research and Oceanographic Abstracts*, 23, 301–317. doi:10.1016/0011-(76)90872-X.

Fratantoni, D. M. 2001. North Atlantic surface circulation during the 1990's observed with satellite-tracked drifters. *Journal of Geophysical Research*, 106(C10), 22067–22093.

Gasparin, F., D. Roemmich, J. Gilson, and B. Cornuelle. 2015. Assessment of the upper-ocean observing system in the Equatorial Pacific: The role of Argo in resolving intraseasonal to interannual variability. *Journal of Atmospheric and Oceanic Technology*, 32, 1668–1688. doi:10.1175/JTECH-D-14-00218.s1.

Gill, A., and P. Niiler. 1973. The theory of the seasonal variability in the ocean. *Deep-Sea Research*, 20, 141–177.

Gilson, J., D. Roemmich, B. Cornuelle, and L.-L. Fu. 1998. Relationship of TOPEX/Poseidon altimetric height to steric height and circulation in the North Pacific. *Journal of Geophysical Research*, 103(C12), 27947–27965. doi:10.1029/98JC01680.

Gordon, A. L. 1978. Deep Antarctic convection west of Maud Rise. *Journal of Physical Oceanography*, 8, 600–613.

Gray, A. R., and S. C. Riser. 2015. A method for multiscale optimal analysis with application to Argo data. *Journal of Geophysical Research: Oceans*, 120(6), 4340–4356. doi:10.1002/2014JC010208.

Haines, B. J., S. D. Desai, and G. H. Born. 2010. The Harvest Experiment: Calibration of the climate data record from TOPEX/Poseidon, Jason-1 and the Ocean Surface Topography Mission. *Marine Geodesy*, 33 Supplement 1, 91–113. doi:10.1080/01490419.2010.491028.

Hamlington, B. D., R. R. Leben, R. S. Nerem, W. Han, and K.-Y. Kim. 2011. Reconstructing sea level using cyclostationary empirical orthogonal functions. *Journal of Geophysical Research*, 116, C12015. doi:10.1029/2011JC007529.

Harangozo, S. A., P. L. Woodworth, R. H. Rapp, and Y. M. Wang. 1993. A comparison of monthly mean sea level variability determined from Geosat altimetry and a global tide gauge data set. *International Journal of Remote Sensing*, 14(4), 789–795. doi:10.1080/01431169308904377.

Hatun, H., A. B. Sandø, H. Drange, and B. Hansen. 2005. Influence of the Atlantic subpolar gyre on the thermohaline circulation. *Science*, **309**, 1841, doi:10.1126/science.1114777.

Hidaka, K. (1940). Absolute evaluation of ocean currents in dynamical calculations. Proceedings of the Imperial Academy, 16(8), 391–393.

Hirschi, J., J. Baehr, J. Marotzke, J. Stark, S. Cunningham, and J. O. Beismann. 2003. A monitoring design for the Atlantic meridional overturning circulation. *Geophysical Research Letters*, 30(7), 1413. doi:10.1016/0198-0149(83)90096-1.

Holgate, S. J., and P. L. Woodworth. 2004. Evidence for enhanced coastal sea level rise during the 1990s. *Geophysical Research Letters*, 31, L07305. doi:10.1029/2004GL019626.

Imawaki, S., H. Uchida, H. Ichikawa, M. Fukasawa, S.-I. Umatani, and the ASUKA Group. 2001. Satellite altimeter monitoring the Kuroshio transport south of Japan. *Geophysical Research Letters*, 28(1), 17–20.

Inverted Echo Sounder User's Manual, IES Model 6.2B. University of Rhode Island, Graduate School of Oceanography, 215 South Ferry Road, Narragansett, RI 02882, USA. Revision August 2008.

IOC. 2012. *The Global Sea Level Observing System (GLOSS) Implementation Plan—2012*. UNESCO/Intergovernmental Oceanographic Commission. 37 p. (IOC Technical Series No. 100). Available from ioc.unesco.org (accessed July 10, 2017).

IOC. 2015. *Manual on Sea Level Measurement and Interpretation. Volume V: Radar Gauges. Manuals and Guides 14.* Intergovernmental Oceanographic Commission (in preparation). The earlier Volumes I-IV may be obtained from http://www.psmsl.org/train_and_info/training/manuals/ (accessed July 10, 2017).

Jacobs, S. S. 2004. Bottom water production and its links with the thermohaline circulation. *Antarctic Science*, 16, 427–437. doi:10.1017/S095410200400224X.

Jacobs, S. S., and C. F. Giulivi. 2010. Large multidecadal salinity trends near the Pacific–Antarctic continental margin. *Journal of Climate*, 23, 4508–4524. doi:10.1175/2010JCLI3284.1.

Jevrejeva, S., A. Grinsted, J. C. Moore, S. J. and Holgate. 2006. Nonlinear trends and multiyear cycles in sea level records. *Journal of Geophysical Research*, 111, C09012. doi:10.1029/2005JC003229.

Johannessen, J. A., et al. 2014. Toward improved estimation of the dynamic topography and ocean circulation in the high latitude and Arctic Ocean: The importance of GOCE. *Surveys in Geophysics*, 35(3), 661–679. doi:10.1007/s10712-013-9270-y.

Johnson, G. C. 2008. Quantifying Antarctic bottom water and North Atlantic deep water volumes. *Journal of Geophysical Research*, 113, C05027. doi:10.1029/2007JC004477.

Johnson, G. C., J. M. Lyman, and S. G. Purkey. 2015. Informing deep Argo array design using Argo and full-depth hydrographic section data. *Journal of Atmospheric and Oceanic Technology*, 32, 2187–2198. doi:10.1175/JTECH-D-15-0139.1.

Kanzow, T., S. A. Cunningham, D. Rayner, J. J.-M. Hirschi, W. E. Johns, M. O. Baringer, H. L. Bryden, L. M. Beal, C. S. Meinen, and J. Marotzke. 2007. Observed flow compensation associated with the MOC at 26.5 N in the Atlantic. *Science*, 317(5840), 938–941.

Kanzow, T., U. Send, W. Zenk, A. D. Chave, and M. Rhein. 2006. Monitoring the integrated deep meridional flow in the tropical North Atlantic: Long-term performance of a geostrophic array. *Deep Sea Research Part I: Oceanographic Research Papers*, 53(3), 528–546. doi:10.1016/j.dsr.2005.12.007.

Kawase, M. 1987. Establishment of deep ocean circulation driven by deep-water production. *Journal of Physical Oceanography*, 17, 2294–2317. doi:10.1175/1520-0485(1987)017<2294:EODOCD>2.0.CO;2.

Kemp, A. C., A. D. Hawkes, J. P. Donnelly, C. H. Vane, B. P. Horton, T. D. Hill, S. C. Anisfeld, A. C. Parnell, and N. Cahill. 2015. Relative sea-level change in Connecticut (USA) during the last 2200 yrs. *Earth and Planetary Science Letters*, 428, 217–229. doi:10.1016/j.epsl.2015.07.034.

Kida, S., J. F. Price, and Y. Jiayan. 2007. The upper-oceanic response to overflows: A mechanism for the Azores current. *Journal of Physical Oceanography*, 38, 880–895.

King, M. 2014. *Priorities for Installation of Continuous Global Navigation Satellite System (GNSS) Near to Tide Gauges.* Report by Professor Matt King, University of Tasmania to the Global Sea Level Observing System. 20 p.

Klein, P., G. Lapeyre, G. Roullet, S. Le Gentil, and H. Sasaki. 2011. Ocean turbulence at meso and submesoscales: connection between surface and interior dynamics. *Geophysical and Astrophysical Fluid Dynamics*, 105(4–5), 421–437. doi:10.1080/03091929.2010.532498.

Kolenkiewicz, R., and C. F. Martin. 1982. SEASAT altimeter height calibration. *Journal of Geophysical Research*, 87(C5), 3189–3197. doi:10.1029/JC087iC05p03189.

Lankhorst, M., and U. Send. 2016. *Seawater temperature and salinity observed from the CORC1 and CORC2 moorings in the southern California Current (NE Pacific) from 2008-09-20 to 2012-11-14 (NCEI Accession 0137858). Version 1.1.* NOAA National Centers for Environmental Information. Dataset. doi:10.7289/V51834J4

Larson, K. M., J. Löfgren, and R. Haas. 2013. Coastal sea level measurements using a single geodetic GPS receiver. *Advances in Space Research*, 51(8), 1301–1310. doi:10.1016/j.asr.2012.04.017.

Lebedev, K. V., H. Yoshinari, N. A. Maximenko, and P. W. Hacker. 2007. *YoMaHa'07: Velocity Data Assessed from Trajectories of Argo Floats at Parking Level and at the Sea Surface.* IPRC Technical Note No. 4(2), June 12, 2007, 16 p. Available from http://apdrc.soest.hawaii.edu/projects/Argo/data/Documentation/YoMaHa070612.pdf (accessed July 9, 2017).

LeBel, D. A., et al. 2008. The formation rate of North Atlantic deep water and eighteen degree water calculated from CFC-11 inventories observed during WOCE. *Deep Sea Research Part I: Oceanographic Research Papers*, 55, 891–910. doi:10.1016/j.dsr.2008.03.009.

Leuliette, E. W., R. S. Nerem, G. T. and Mitchum. 2004. Calibration of TOPEX/Poseidon and Jason altimeter data to construct a continuous record of mean sea level change. *Marine Geodesy*, 27, 79–94. doi:10.1080/01490410490465193.

Llovel, W., A. Cazenave, P. Rogel, A. Lombard, and M. B. Nguyen. 2009. Two-dimensional reconstruction of past sea level (1950-2003) from tide gauge data and an Ocean General Circulation Model. *Climate of the Past*, 5, 217–227. doi:www.clim-past.net/5/217/2009/.

Llovel, W., J. Willis, F. Landerer, and I. Fukumori. 2014. Deep-ocean contribution to sea level and energy budget not detectable over the past decade. *Nature Climate Change*, 4, 1031–1035. doi:10.1038/nclimate2387.

Lowe, J. A., et al. 2010. Past and future changes in extreme sea levels and waves. In *Understanding Sea-Level Rise and Variability* (eds. J.A. Church, P.L. Woodworth, T. Aarup, and W.S. Wilson). Wiley-Blackwell, London, pp. 326–375.

Lumpkin, R. 2016. Global characteristics of coherent vortices from surface drifter trajectories. *Journal of Geophysical Research: Oceans*, 121, 1306–1321. doi:10.1002/2015JC011435.

Lumpkin, R., S. Grodsky, M.-H. Rio, L. Centurioni, J. Carton, and D. Lee. 2013. Removing spurious low-frequency variability in surface drifter velocities. *Journal of Atmospheric and Oceanic Technology*, 30(2), 353–360. doi:10.1175/JTECH-D-12-00139.1.

Lumpkin, R., and G. Johnson. 2013. Global ocean surface velocities from drifters: Mean, variance, ENSO response, and seasonal cycle. *Journal of Geophysical Research: Oceans*, 118(6), 2992–3006. doi:10.1002/jgrc.20210.

Lumpkin, R., and K. Speer. 2007. Global ocean meridional overturning. *Journal of Physical Oceanography*, 37, 2550–2562. doi:10.1175/JPO3130.1.

Matthäus, W. 1972. On the history of recording tide gauges. *Proceedings of the Royal Society of Edinburgh B*, 73, 26–34. doi:10.1017/S0080455X00002083.

Maximenko, N.A., A. MacFadyen, and M. Kamachi. 2015. Modeling drift of marine debris from the Great Tohoku tsunami. *PICES Press*, Summer 2015, 23(2), 32–36.

Maximenko, N. A, and J. Hafner. 2010. *SCUD: Surface Currents from Diagnostic model.* IPRC Technical Note No. 5, 17 p. Available from http://apdrc.soest.hawaii.edu/projects/SCUD/SCUD_manual_02_17.pdf (accessed July 9, 2017).

Maximenko, N. A., R. Lumpkin, and L. Centurioni. 2013. Ocean surface circulation. In *Ocean Circulation and Climate*(eds. G. Siedler, S.M. Griffies, J. Gould, and J.A. Church), 2nd Edition. A 21st century perspective, International Geophysics Series, Volume 103, Academic Press, Amsterdam, The Netherlands. pp. 283–300. doi:1.1016/B978-0-12-391851-2.000012.X.

Maximenko, N. A., O. V. Melnichenko, P. P. Niiler, and H. Sasaki. 2008. Stationary mesoscale jet-like features in the ocean. *Geophysical Research Letters*, 35, L08603. doi:10.1029/2008GL033267.

Maximenko, N. A., and P. Niiler. 2006. Mean surface circulation of the global ocean inferred from satellite altimeter and drifter data. Symposium "15 Years of Progress in Radar Altimetry," March 13–18, 2006, Venice, Italy.

Maximenko, N. A., P. Niiler, M.-H. Rio, O. Melnichenko, L. Centurioni, D. Chambers, V. Zlotnicki, and B. Galperin. 2009. Mean dynamic topography of the ocean derived from satellite and drifting buoy data using three different techniques. *Journal of Atmospheric and Oceanic Technology*, 26(9), 1910–1919.

Merrifield, M. A., & Maltrud, M. E. (2011). Regional sea level trends due to a Pacific trade wind intensification. Geophysical Research Letters, 38(21).

Mertikas, S. P., A. Daskalakis, I. N. Tziavos, G. Vergos, X. Fratzis, and A. Tripolitsiotis. 2015. First calibration results for the SARAL/AltiKa Altimetric mission using the Gavdos permanent facilities. *Marine Geodesy*, 38, Supplement 1, 249–259. doi:10.1080/01490419.2015.1030052.

Meyers, G., H. Phillips, N. Smith, and J. Sprintall. 1991. Space and time scales for optimal interpolation of temperature-tropical Pacific Ocean. *Progress in Oceanography*, 28, 198–218.

Mitchum, G. T. 1995. The source of 90-day oscillations at Wake Island. *Journal of Geophysical Research*, 100(C2), 2459–2475. doi:10.1029/94JC02923.

Mitchum, G. T. 2000. An improved calibration of satellite altimetric heights using tide gauge sea levels with adjustment for land motion. *Marine Geodesy*, 23, 145–166. doi:10.1080/01490410050128591.

Mitchum, G. T., R. S. Nerem, M. A. Merrifield, and W. R. Gehrels. 2010. Modern sea-level-change estimates. In *Understanding Sea-Level Rise and Variability* (eds. J.A. Church, P.L. Woodworth, T. Aarup and W. S. Wilson). Wiley-Blackwell, London, pp. 122–142.

Moray, R. 1665. Considerations and enquiries concerning tides, by Sir Robert Moray; Likewise for a further search into Dr. Wallis's Newly Publish't Hypothesis. *Philosophical Transactions of the Royal Society of London*, 1, 298–301. doi:10.1098/rstl.1665.0113.

Niiler, P. P. 2001. The world ocean surface circulation. In *Ocean circulation and climate* (eds. G. Siedler, J. Church, and J. Gould), International Geophysics Series, Volume 77, Academic Press, San Diego, CA, pp. 193–204.

Niiler, P. P. 2003. *A brief history of drifter technology. In: Autonomous and Lagrangian Platforms and Sensors Workshop.* Scripps Institution of Oceanography, La Jolla, CA.

Niiler, P. P., and J. D. Paduan. 1995. Wind-driven motions in the northeast Pacific as measured by Lagrangian drifters. *Journal of Physical Oceanography*, 25, 2819–2830.

Niiler, P. P., N. A. Maximenko, and J. C. McWilliams. 2003a. Dynamically balanced absolute sea level of the global ocean derived from near-surface velocity observations. *Geophysical Research Letters*, 30(22), 2164. doi:10.1029/2003GL018628, 2003.

Niiler, P. P., N. A. Maximenko, G. G. Panteleev, T. Yamagata, and D. B. Olson. 2003b. Near-surface dynamical structure of the Kuroshio extension. *Journal of Geophysical Research*, 108, 3193. doi:10.1029/2002JC001461.

Niiler, P. P., A. Sybrandy, K. Bi, P. Poulain, and D. Bitterman. 1995. Measurements of the water-following capability of holey-sock and TRISTAR drifters. *Deep Sea Research*, 42, 1951–1964.

Ollitrault, M., and A. Colin de Verdière. 2014. The ocean general circulation near 1000-m depth. *Journal of Physical Oceanography*, 44(1), 384–409. doi:10.1175/JPO-D-13-030.1

Orsi, A. H., G. C. Johnson, and J. L. Bullister. 1999. Circulation, mixing, and production of Antarctic Bottom Water. *Progress in Oceanography*, 43, 55E109.

Palmer, H. R. 1831. Description of a graphical register of tides and winds. *Philosophical Transactions of the Royal Society of London*, 121, 209–213. doi:10.1098/rstl.1831.0013.

Pazan, S. E., and P. P. Niiler. 2001. Recovery of near-surface velocity from undrogued drifters. *Journal of Atmospheric and Oceanic Technology*, 18, 476–489.

Pazan, S. E., and P. P. Niiler. 2004. Ocean sciences: New global drifter data set available. *Eos, Transactions American Geophysical Union*, 85, 17.

Poulain, P.-M., R. Gerin, E. Mauri, and R. Pennel. 2009. Wind effects on drogued and undrogued drifters in the eastern Mediterranean. *Journal of Atmospheric and Oceanic Technology*, 26, 1144–1156. doi:10.1175/2008JTECHO618.1.

Prandi, P., A. Cazenave, and M. Becker. 2009. Is coastal mean sea level rising faster than the global mean? A comparison between tide gauges and satellite altimetry over 1993-2007. *Geophysical Research Letters*, 36, L05602. doi:10.1029/2008GL036564.

Pugh, D. T., and P. L. Woodworth. 2014. *Sea-Level Science: Understanding Tides, Surges, Tsunamis and Mean Sea-Level Changes*. Cambridge University Press, Cambridge, 408 p.

Purkey, S. G., and G. C. Johnson. 2010. Warming of global abyssal and deep Southern Ocean waters between the 1990s and 2000s: Contributions to global heat and sea level rise budgets. *Journal of Climate*, 23, 6336–6351. doi:10.1175/2010JCLI3682.1.

Purkey, S. G., and G. C. Johnson. 2013. Antarctic Bottom Water warming and freshening: Contributions to sea level rise, ocean freshwater budgets, and global heat gain. *Journal of Climate*, 26, 6105–6122. doi:10.1175/JCLI-D-12-00834.1.

Ray, R. D., and B. C. Douglas. 2011. Experiments in reconstructing twentieth-century sea levels. *Progress in Oceanography*, 91, 496–515. doi:10.1016/j.pocean.2011.07.021.

Reid, J. L. (1986). On the total geostrophic circulation of the South Pacific Ocean: Flow patterns, tracers and transports. Progress in Oceanography, 16(1), 1–61.

Rhein, M., et al. 2013. Observations: Ocean. In: *Climate Change 2013: The Physical Science Basis. Contribution of Working Group I to the Fifth Assessment Report of the Intergovernmental Panel on Climate Change* (eds. T.F. Stocker, D. Qin, G.-K. Plattner, M. Tignor, S.K. Allen, J. Boschung, A. Nauels, Y. Xia, V. Bex, and P.M. Midgley). Cambridge University Press, Cambridge. pp. 255–316.

Ridgway, K. R., R. C. Coleman, R. J. Bailey, and P. Sutton. 2008. Decadal variability of East Australian Current transport inferred from repeated high-density XBT transects, a CTD survey and satellite altimetry. *Journal of Geophysical Research*, 113, C08039. doi:10.1029/2007JC004664.

Rio, M.-H., and H. Etienne. 2017. Global Ocean delayed mode in situ observations of ocean surface currents. *SEANOE*. doi:10.17882/41334. http://www.seanoe.org/data/00302/41334/

Rio, M. H., S. Guinehut, and G. Larnicol. 2011. New CNES-CLS09 global mean dynamic topography computed from the combination of GRACE data, altimetry and in situ measurements. *Journal of Geophysical Research*, 116, C07018. doi:10.1029/2010JCOO6505.

Rio, M.-H., and F. Hernandez. 2003. High-frequency response of wind-driven currents measured by drifting buoys and altimetry over the world ocean. *Journal of Geophysical Research*, 108 (C8), 3283. doi:10.1029/2002JC001655.

Rio, M.-H., and F. Hernandez. 2004. A mean dynamic topography computed over the world ocean from altimetry, in situ measurements, and a geoid model. *Journal of Geophysical Research*, 109, C12032. doi:10.1029/2003JC002226.

Roemmich, D., J. Church, J. Gilson, D. Monselesan, P. Sutton, and S. Wijffels. 2015. Unabated planetary warming and its ocean structure since 2006, *Nature Climate Change*, 5(3), 240–245. doi:10.1038/nclimate2513.

Roemmich, D., and J. Gilson. 2009. The 2004–2008 mean and annual cycle of temperature, salinity, and steric height in the global ocean from the Argo Program. *Progress in Oceanography*, 82, 81–100.

Santamaría-Gómez, A., C. Watson, M. Gravelle, M. King, and G. Wöppelmann. 2015. Levelling co-located GNSS and tide gauge stations using GNSS reflectometry. *Journal of Geodesy*, 89, 241–258. doi:10.1007/s00190-014-0784-y.

SBE 53 BPR Bottom Pressure Recorder. *User Manual. Sea-Bird Electronics*, 13431 NE 20th Street, Bellevue, WA 98005, USA. Manual version 013, 18 February 2015.

Schmitz, W. J. 1995. On the interbasin-scale thermohaline circulation. *Reviews of Geophysics*, 33, 151–173. doi:10.1029/95RG00879.

Sherman, J., R. E. Davis, W. B. Owens, and J. Valdes. 2001. The autonomous underwater glider "Spray." *IEEE Journal of Oceanic Engineering*, 26, 437–446.

Simon, B. 2015. *Coastal Tides*. (Translated by D. Manley). Published by the Service Hydrographique et Océanographique de la Marine (SHOM), Brest, 409 p. Available from refmar.shom.fr (accessed July 10, 2017).

Sloyan, B. M., and S. R. Rintoul. 2001. The Southern Ocean limb of the global deep overturning circulation. *Journal of Physical Oceanography*, 31, 143–173. doi:10.1175/1520-0485(2001)031<0143:TSOLOT>2.0.CO;2.

Sprintall, J., and G. Meyers. 1991. An optimal XBT sampling network for the eastern Pacific Ocean. *Journal of Geophysical Research*, 96, 10539–10552.

Stammer, D., et al. 2014. Accuracy assessment of global barotropic ocean tide models. *Reviews of Geophysics*, 52, 243–282. doi:10.1002/2014RG000450.

Stommel, H., & Schott, F. (1977). The beta spiral and the determination of the absolute velocity field from hydrographic station data. Deep Sea Research, 24(3), 325–329.

Sutton, P., and D. Roemmich. 2011. Decadal steric and sea surface height changes in the Southern Hemisphere. *Geophysical Research Letters*, 38, L08604. doi:10.1029/2011GL046802.

Sverdrup, Harald Ulrik, Martin W. Johnson, and Richard H. Fleming. The Oceans: Their physics, chemistry, and general biology. Vol. 7. New York: Prentice-Hall, 1942.

Swift, J., and A. H. Orsi. 2012. Sixty-four days of hydrography and storms: RVIB Nathaniel B. Palmer's 2011 S04P Cruise. *Oceanography*, 25, 54–55. doi:10.5670/oceanog.2012.74.

Sybrandy, A. L., and P. P. Niiler. 1991. *WOCE/TOGA Lagrangian Drifter Construction Manual*. WOCE Rep. 63, SIO Ref. 91/6. Scripps Institution of Oceanography, La Jolla, CA, 58 p.

Tapley, B. D., S. Bettadpur, M. M. Watkins, and C. Reigber. 2004. The gravity recovery and climate experiment: Mission overview and early results. *Geophysical Research Letters*, 31, L09607. doi:10.1029/2004GL019920.

Tapley, B. D., G. H. Born, and M. E. Parke. 1982. The SEASAT altimeter data and its accuracy assessment. *Journal of Geophysical Research*, 87(C5), 3179–3188. doi:10.1029/JC087iC05p03179.

Tapley, B. D., D. P. Chambers, S. Bettadpur, and J. C. Ries. 2003. Large scale ocean circulation from the GRACE GGM01 Geoid. *Geophysical Research Letters*, 30(22), 2163. doi:10.1029/2003GL018622.

Testut, L., B. Martín Míguez, G. Woppelmann, P. Tiphaneau, N. Pouvreau, and M. Karpytchev. 2010. Sea level at Saint Paul Island, southern Indian Ocean, from 1874 to the present. *Journal of Geophysical Research*, 115, C12028. doi:10.1029/2010JC006404.

Tomczak, M., and J. S. Godfrey. 1994. *Regional Oceanography: An Introduction*. Pergamon, New York.

Uchida, H., and S. Imawaki. 2003. Eulerian mean surface velocity field derived by combining drifter and satellite altimeter data. *Geophysical Research Letters*, 30(5), 1229. doi:10.1029/2002GL016445.

Vignudelli, S., A. Kostianoy, P. Cipollini, and J. Benveniste (eds). 2011. *Coastal Altimetry*. Springer, Berlin, 578 p.

Vinogradov, S. V., and R. M. Ponte. 2011. Low-frequency variability in coastal sea level from tide gauges and altimetry. *Journal of Geophysical Research*, 116, C07006. doi:10.1029/2011JC007034.

Watson, C. S., N. J. White, J. A. Church, M. A. King, R. J. Burgette, and B. Legresy. 2015. Unabated global mean sea-level rise over the satellite altimeter era. *Nature Climate Change*, 5, 565–568. doi:10.1038/NCLIMATE2635.

Watts, D. R., and H. Kontoyiannis. 1990. Deep-ocean bottom pressure measurement: Drift removal and performance. *Journal of Atmospheric and Oceanic Technology*, 7(2), 296–306. doi:10.1175/1520-0426(1990)007<0296:DOBPMD>2.0.CO;2.

Wenzel, M., and J. Schröter. 2010. Reconstruction of regional mean sea level anomalies from tide gauges using neural networks. *Journal of Geophysical Research*, 115, C08013. doi:10.1029/2009JC005630.

White, N. J., J. A. Church, and J. M. Gregory. 2005. Coastal and global averaged sea level rise for 1950 to 2000. *Geophysical Research Letters*, 32, L01601. doi:10.1029/2004GL021391.

White, W., and R. L. Bernstein. 1979. Design of an oceanographic network in the mid-latitude North Pacific. *Journal of Physical Oceanography*, 9, 592–606.
White, W., and C.-K. Tai. 1995. Inferring interannual changes in global upper ocean heat storage from TOPEX altimetry. *Journal of Geophysical Research*, 100, 24943–24954. doi:10.1029/95JC02332.
Wijffels, S., D. Roemmich, D. Monselesan, J. Church, and J. Gilson. 2016. Ocean temperatures chronicle the ongoing warming of Earth. *Nature Climate Change*, 6, 116–118. doi:10.1038/nclimate2924.
Williams, J., and C. W. Hughes. 2013. The coherence of small island sea level with the wider ocean: A model study. *Ocean Science*, 9, 111–119. doi:10.5194/os-9-111-2013.
Woods, J. D. 1985. The world ocean circulation experiment. *Nature*, 314, 501–511. doi:10.1038/314501a0.
Woodworth, P. L. 2015. Tidal measurement. In *The History of Cartography: Cartography in the Twentieth Century* (ed. M. Monmonier). 1st Edition. Vol. 6. University of Chicago, Chicago, IL, pp. 1525–1528.
Woodworth, P. L., C. W. Hughes, R. J. Bingham, and T. Gruber. 2012. Toward worldwide height system unification using ocean information. *Journal of Geodetic Science* 2(4), 302–318. doi:10.2478/v10156-012-004-8.
Woodworth, P. L., M. Menéndez, and W. R. Gehrel. 2011. Evidence for century-timescale acceleration in mean sea levels and for recent changes in extreme sea levels. *Surveys in Geophysics*, 32(4–5), 603–618 (erratum page 619). doi:10.1007/s10712-011-9112-8.
Woodworth, P. L., D. T. Pugh, and A. J. Plater. 2015. Sea-level measurements from tide gauges. In *Handbook of Sea-Level Research* (eds. I. Shennan, A.J. Long, and B.P. Horton). Wiley, Chichester, pp. 557–574.
Woodworth, P. L., G. Wöppelmann, M. Marcos, M. Gravelle, and R. M. Bingley. 2017. Why we must tie satellite positioning to tide gauge data. *Eos, Transactions of the American Geophysical Union*, 98. doi:10.1029/2017EO064037. Available from https://eos.org/opinions/why-we-must-tie-satellite-positioning-to-tide-gauge-data.
Wunsch, C. (1978). The North Atlantic general circulation west of 50 W determined by inverse methods. Reviews of Geophysics, 16(4), 583–620.
Wyville Thomson, C. 1873. The Challenger expedition. *Nature*, 7, 385–388.
Zilberman, N., D. Roemmich, and S. Gille. 2016. Estimating the velocity and transport of Western Boundary Current systems: A case study of the East Australian Current System near Brisbane. In preparation.

4 Auxiliary Space-Based Systems for Interpreting Satellite Altimetry

Satellite Gravity

Don Chambers, Ole B. Andersen, Srinivas Bettadpur, Marie-Hélène Rio, Reiner Rummel, and David Wiese

4.1 INTRODUCTION

Satellite altimeter measurements are intrinsically linked with gravity. More than 99% of the sea surface height (SSH) measured by satellite altimeters is explained by the Earth's geoid, an equipotential surface of Earth's mean gravity field. Until recently, the largest error source in precise orbits needed to make measurements of SSH has been uncertainty in the Earth's gravity field (Chapter 1). Gravity can also be used along with altimetry data to obtain two important measurements: surface geostrophic currents (Section 4.4) and the mass component of global mean sea level change (Section 4.5).

The use of satellites to make detailed measurements of the gravity field was first proposed in the 1960s (e.g., Kaula 1966), and numerous space mission concepts were developed in the 1970s and 1980s to measure the longest wavelengths of the Earth's mean gravity field (e.g., NASA 1987; NRC 1979, 1982). Although a geodetic satellite (LAGEOS 1) was launched into a medium Earth orbit in 1976 partly in order to measure the Earth's oblateness (J_2) using satellite laser ranging tracking, the data were insufficient to measure the other dominant wavelengths of the Earth's gravity field. The portion of the gravity field between 1000 and 20,000 km wavelength had to be deduced from tracking to multiple satellites in low- and medium-Earth orbits and had varying accuracies (e.g., Balmino et al. 1978; Lerch et al. 1981; Marsh et al. 1990; Nerem et al. 1994a; Tapley et al. 1996). Although these gravity fields were accurate enough to improve precise orbits for satellite altimeters (Tapley et al. 1994; Chapter 1) they were not sufficient to determine the geoid accurately enough to resolve the absolute geostrophic surface circulation (Nerem et al. 1994b; Stammer and Wunsch 1994; Tapley et al. 1994).

By the mid-1990s, it became clear that to make progress in using altimetry to measure the absolute geostrophic surface circulation and to further improve orbit determination, a dedicated satellite gravity mission was required. Numerous technologies and configurations had been previously proposed (e.g., NRC 1979; Reigber et al. 1987), and the U.S. National Academies of Science funded a report to study the options (NRC 1997). That report for the first time quantified how time-variable gravity, if it could be measured accurately enough, could be used to measure important climate variables, including ice mass loss from the Greenland and Antarctic ice sheets, changes in land water storage, and, most importantly for applications with altimetry, ocean bottom pressure and global mean ocean mass variability.

Following the NRC report, NASA selected a proposal that would, in partnership with the German space agency (Deutsches Zentrum für Luft und Raumfahrt, DLR), fly a pair of satellites in low Earth orbit with an advanced microwave satellite-to-satellite ranging system with precise accelerometers.

The goal of the mission, the Gravity Recovery and Climate Experiment (GRACE), was to measure the time-mean and time-variable gravity field monthly with a spatial resolution of 200–300 km at the equator (Tapley et al. 2004b).

During GRACE development, the German Challenging Minisatellite Payload (CHAMP) mission was launched. Its low-altitude, near polar orbit, accelerometers, and high-precision Global Positioning System (GPS) receiver allowed for the determination of a more accurate long-wavelength gravity field than had been derived over the previous 25 years using multiple satellites (Reigber et al. 2002).

The GRACE satellites were successfully launched in March 2002. As of this writing (August 2016), the GRACE satellites are still operational and returning useful science data, with only some minor degradation in the last 6 years of the mission.

Seven years later, in March 2009, the European Space Agency (ESA) launched the Gravity Field and Steady-State Ocean Circulation Explorer (GOCE), using very different technology than GRACE (Rummel and Gruber 2010). GOCE was a single satellite flying at a much lower altitude than GRACE and housing an advanced gravity gradiometer. This allowed GOCE to measure aspects of the time-mean gravity field at much smaller spatial scales than GRACE, down to 100 km at the equator. GOCE reentered the Earth's atmosphere in November 2013 after it ran out of propellant.

Both GRACE and GOCE have revolutionized our knowledge of the Earth's gravity field. Fully documenting their contributions to geodesy and climate science would require a book itself. In this chapter, we limit ourselves to results that are most directly useful to satellite altimetry: the mean geoid and the mass component of sea level, or ocean bottom pressure variability. Section 4.2 will discuss how the mean gravity field is determined from GRACE and GOCE, and how it can be combined with surface gravity from marine and airborne gravity, satellite altimetry, and terrestrial gravity to obtain even higher resolution geoids. Section 4.3 will describe how time-variable gravity is estimated from GRACE, how it is converted to mass density, and how the relatively small signal of ocean mass variability is extracted from the noisy measurement. Section 4.4 will describe how the mean geoid and altimetry can be combined to determine the mean dynamic topography, from which the surface geostrophic currents can be derived. Section 4.5 will discuss how ocean bottom pressure variations are related to sea level measured by an altimeter and to changes in the deep ocean circulation. Finally, Section 4.6 will conclude with an overview of future gravity missions.

4.2 MEASUREMENTS: MEAN GEOID AND SEA SURFACE

The Earth's geoid is the equipotential surface that the ocean's surface would follow if there were no tides, currents, or other dynamic adjustments (Gill 1982). The geoid is commonly expressed in terms of geoid heights relative to a best-fitting Earth ellipsoid that approximates the mean radius of the Earth (approximately 6378 km) and the Earth's oblateness (the equatorial radius is larger than the polar radius by about 21 km). The root mean square (RMS) of geoid heights over the ocean is about 30 m with maximum values of −105 m South of India and +85 m around New Guinea (Figure 4.1).

Before dedicated gravity satellite missions were launched, our knowledge of the global gravity field and geoid was still poor, and satellite altimetry played an important role in mapping of the ocean geoid from space because 99% of the SSH measurement is the geoid. The remaining 1% is comprised of dynamic signals—such as tides, geostrophic currents, inverted barometer effects driven by atmospheric pressure changes, or other geophysical signals, many of which are modeled and removed in processing the reported SSH (Chapter 1). For computing the geoid from altimetry, the primary issue is the residual dynamic topography required to balance surface geostrophic currents (Section 4.4). The time-mean dynamic topography (MDT) has variations of order ± 1 m

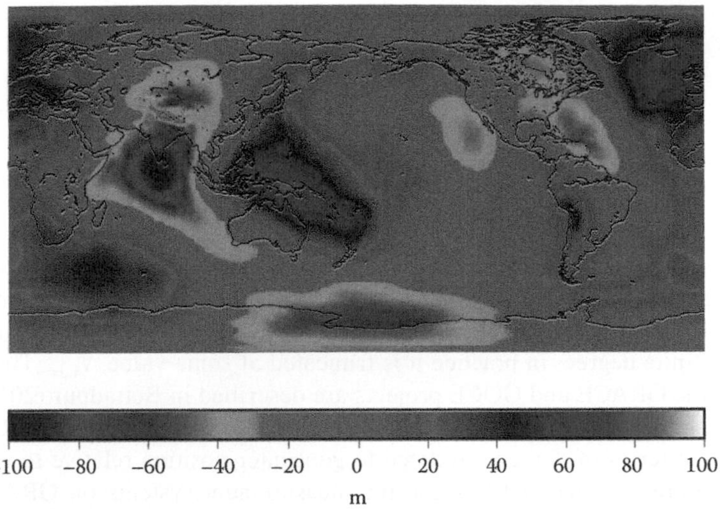

FIGURE 4.1 Global geoid map based on GRACE and GOCE data. (Based on GGM05G model.)

over wavelengths of a few tens of kilometers to order ±0.5 m over several thousand kilometers (i.e., two orders of magnitude smaller than the geoid).

Even moderately accurate MDT models, such as those estimated from *in situ* data (Section 4.4), could be applied to the altimeter SSH measurements to obtain geoid heights with an accuracy of better than 10 cm at the longest wavelengths. On the other hand, geoid models estimated from combinations of satellites were not sufficient to allow a direct computation of MDT from the difference of the altimeter mean sea surface (MSS) and satellite-based geoid heights (Nerem et al. 1994b; Stammer and Wunsch 1994; Tapley et al. 1994).

The situation changed profoundly with the launch of the gravimetric satellites CHAMP in 2000, GRACE in 2002, and GOCE in 2009. Although CHAMP was not designed to provide as accurate measures of the gravity field as GRACE and GOCE, its low orbit, onboard accelerometers, and precise GPS receiver did allow improved measurements of the mean and time-variable gravity field for the 2 years before GRACE was launched (Reigber et al. 2002).

Since the launch of GRACE, and especially after adding observations from GOCE, global geoid models are now available with accuracy comparable to the MSS as derived from satellite altimetry at scales of order 100 km. For the first time, MDT can be derived at these scales directly as a difference of the MSS and geoid, purely from geodetic space methods. In the remainder of this section, we will describe how the mean geoid and gravity field are commonly parameterized (Section 4.2.1), the two main satellite missions (GRACE and GOCE) that have improved our knowledge of the mean geoid and their different measurement strategies (Section 4.2.2), how MSS models are obtained from altimetry (Section 4.2.3), and how the satellite gravity data can be combined with *in situ* data (and MSS data) to estimate a combined geoid that has higher spatial resolution (Section 4.2.4). Section 4.4 will focus on how one can combine the MSS and geoid to estimate the MDT (and hence the mean surface geostrophic currents), and how the combination with oceanographic *in situ* data allows achieving higher spatial resolution.

4.2.1 Parameterizing Gravity and the Geoid

In geodesy, the Earth's gravitational field potential (V) at time t and at any point exterior to the Earth system is typically expressed as an infinite spherical harmonic series that is a function of the Earth's gravitational constant (μ), the mean equatorial radius (a_E), the geocentric radius to the point (R), as well as the geocentric latitude (ϕ) and longitude (λ) (e.g., Heiskanen and Moritz 1967)

$$V(R,\phi,\lambda,t)=\frac{\mu}{R}+\frac{\mu}{R}\sum_{n=2}^{N\max}\left(\frac{R_{eq}}{R}\right)^{n}\sum_{m=0}^{n}P_{nm}(\sin\phi)\{C_{nm}(t)\cos m\lambda+S_{nm}(t)\sin m\lambda\} \tag{4.1}$$

where $P_{nm}(\sin\phi)$ are the fully normalized associated Legendre polynomials of degree n and order m. $C_{nm}(t)$ and $S_{nm}(t)$ are the fully normalized, time-variable spherical harmonic geopotential Stokes coefficients. Although theoretically the series is exact only if the summation is carried out to an infinite degree, in practice it is truncated at some value N_{max}. The normalization factors used by the GRACE and GOCE projects are described in Bettadpur (2012a) and Gruber et al. (2014a, 2014b), respectively. Although these are commonly used normalizations, others do exist. Note that terms of degree 1 (related to geocenter position relative to the center of the reference frame) are not included because the measurement systems on GRACE and GOCE are insensitive to this gravity coefficient. Thus, the gravitational field potential is relative to the instantaneous center of mass of the Earth—not the center of the reference frame. Over very long times, it is expected that the center of the frame and center of mass should correspond, but for shorter periods they do not, and estimates of the time-variable degree 1 coefficients become more important (see Section 4.3).

Thus, we can determine the gravity potential (and, hence, force on a satellite) at any time and location from the center of the Earth defined by R, ϕ, and λ given the Stokes coefficients $C_{nm}(t)$ and $S_{nm}(t)$ that describe the time-variable gravity field. Problems arise because we do not know the exact values of $C_{nm}(t)$ and $S_{nm}(t)$, only approximate values, and we cannot compute the full expansion but can only do so to some maximum degree ($N_{\max}$). However, in the process of determining the precise orbits of a satellite, one can also allow all or certain of the gravity coefficients to be estimated at the same time. The quality of the recovered coefficients is determined by the quality and quantity of the tracking data, its type, other measurements of gravitational or nongravitational forces on the satellites, and the height of the satellite above the surface of the Earth, among others. This is, of course, a very simplistic overview of the gravity estimation problem. A more rigorous derivation is beyond the scope of this chapter; curious readers are directed to the book on precise orbit determination by Tapley et al. (2004a).

The best representation of the mean Earth gravity field is the time average of the variable $C_{nm}(t)$ and $S_{nm}(t)$ geopotential harmonics estimated from satellites and other data. The mean Earth models are derived by a combination of information equations from the least-squares adjustment analyses for the time-variable gravity field over a chosen time span, as further described in Section 4.2.2. The mean Earth gravity field model is represented by using Stokes coefficients without the time argument (t).

Assuming that one has estimated a set of accurate Stokes coefficients, how does one estimate the geoid? First, the gravity potential associated with the geoid is typically defined by adding a second term to Equation 4.1 to account for centrifugal potential due to the Earth's rotation (e.g., Hughes and Bingham 2008). This geopotential (W) is then

$$W(R,\phi,\lambda)=V(R,\phi,\lambda)+\frac{\Omega^2R^2\cos^2\phi}{2} \tag{4.2}$$

where Ω is the rotation rate of the Earth in radians s^{-1}. The geoid at a location (ϕ, λ), then, is a solution to Equation 4.2 for R given a constant equipotential value that represents the value of the MSS at rest (W_0), which is typically calculated for a reference ellipsoid. Clearly, Equation 4.2 is nonlinear, and there is no direct solution for R_{geoid} given values of W_0, ϕ, λ, the Stokes coefficients using this equation. However, iterative methods have been devised that converge rapidly (Marsh and Vincent 1974; Shum 1982).

The value R_{geoid}, measured from the Earth's center, is not typically shown, however, because it is dominated by the nearly constant radius of the Earth. Instead, one uses a geoid undulation (N), which measures the departure of the geoid from a best-fit ellipsoid along the normal to the ellipsoid.

In common oceanographic usage, the word "undulation" is dropped, and N is most often simply referred to as the *geoid*. In all maps of the geoid (e.g., Figure 4.1), it is N that is displayed. Note that the reference ellipsoid is only a function of latitude—it has the same shape for any longitude. It can be defined by two parameters. The most common are the radius at the equator (R_{eq}) and the inverse flattening ($1/f$). There are several ellipsoids that have been defined in the literature (Table 4.1). It is important to note that the geoid can change substantially (even using the same Stokes coefficients) if different reference ellipsoids are used. One has to be extremely careful with this when computing mean dynamic topography (Section 4.4).

There is another common method to compute the geoid, which is faster than using Equation 4.2 and the iterative method. It is a close approximation, if done properly. The approximation first derives a set of Stokes coefficients for the reference ellipsoid assuming a value of W_0. These are then subtracted from the full Stokes coefficients to form anomalous coefficients relative to the ellipsoid $\left(\bar{C}_{nm}, \bar{S}_{nm}\right)$. The geoid anomaly is then computed as:

$$N(\phi,\lambda) = R_{eq}\sum_{n=2}^{n_{\max}}\left(\bar{C}_{nm}\cos m\lambda + \bar{S}_{nm}\sin m\lambda\right)P_{nm}(\sin\phi) \tag{4.3}$$

The two methods are nearly identical if: (1) R_{eq} is the same in each, (2) the reference ellipsoids are the same, and (3) $\bar{C}_{nm}$ and $\bar{S}_{nm}$ in Equation 4.3 include the Earth's rotational effect (Equation 4.2). If any one of these is not valid, then the two geoids may differ substantially.

There are two different types of errors one has to consider in the geoid model. The first is the *commission error*, which represents the error in the geoid due to uncertainty in the gravity coefficients (σ_{Cnm}, σ_{Snm}) up to the truncation of n_{max}:

$$\sigma_n^2(COM) = R_{eq}^2\sum_{n=2}^{n_{\max}}\sum_{m=0}^{n}\left(\sigma_{Cnm}^2 + \sigma_{Snm}^2\right) \tag{4.4}$$

Quantifying σ_{Cnm}, σ_{Snm} is not trivial. Although it is often derived from the covariance matrix that is part of the estimation process (e.g., Tapley et al. 2004a), these errors tend to be optimistic. Calibration based on the differences among internal sub-solutions (Lerch et al. 1993; Pail et al. 2010;

TABLE 4.1
Different Reference Ellipsoids and Their Characteristics

Ellipsoid Name	R_{eq} (m)	$1/f$
"GRIM"	6378136.46	298.25765
"TOPEX"	6378136.3	298.257
"GRS80"	6378137.	298.257222101
"WGS84"	6378137.	298.257223563

Tapley et al. 2005) tends to increase the errors slightly and is generally used as a more conservative estimate.

The second error source is the *omission error*, which is due to the neglected part of the expansion from n_{max} +1 to infinity. To compute this, one needs to know the approximate size of that signal. Because the coefficients $\bar{C}_{nm}$ and $\bar{S}_{nm}$ become progressively smaller at higher degrees, this has been approximated by a geoid degree variance $\left(\sigma_n^2\right)$ (Kaula 1966):

$$\sigma_n^2 = R_{eq}^2 \sum_{m=0}^{n} \left(\bar{C}_{nm}^2 + \bar{S}_{nm}^2\right) \approx 6494.10^{-12}/n^3 \tag{4.5}$$

where the degree variances are in units of m^2, noting that we have converted from the dimensionless units used by Kaula (1966). From this, one can compute the omission error as:

$$\sigma_n^2(OM) = \sum_{n=n_{\max}+1}^{\infty} \sigma_n^2 \tag{4.6}$$

Its approximate size as a function of n_{max} is given in Table 4.2, although we note that the actual omission error depends on the roughness of the geoid (= signal RMS) in a particular area. The full geoid error will then be the root sum square (RSS) of the commission error plus the omission error.

Finally, the gravity field is also often shown in terms of the gravity anomaly (Figure 4.2). This is simply a map of the acceleration due to gravity at the geoid surface (determined from a gradient of the equipotential, Equation 4.1) minus the normal gravity on the ellipsoid (i.e., *g*). Units are generally expressed in milliGals (mGal = 10^{-5} m s^{-2}, or 1 millionth of *g*), meaning a +10 mGal gravity anomaly reflects a local gravitational acceleration that is 10^{-4} m s^{-2} higher than the mean value. Gravity anomaly maps are useful as they more clearly show the local gravity hotspots of mountains, spreading ridges, seamounts, and so on.

4.2.2 GRACE and GOCE

The GRACE mission, a collaboration between NASA and the German space agency DLR, was launched in March 2002. It consists of two co-planar satellites in a 500-km altitude polar orbit, separated nominally by an approximately 200-km mean distance. Each satellite travels a path

TABLE 4.2
Omission Error in Geoid Models for Various Maximum Degree Expansions, Based on Kaula's Rule of Thumb (Equation 4.7)

Maximum Degree/ Order of Geoid	Geoid Omission Error (cm)
200	30–50
300	20–30
500	10–20
1000	7–8
1500	3
2000	0.1

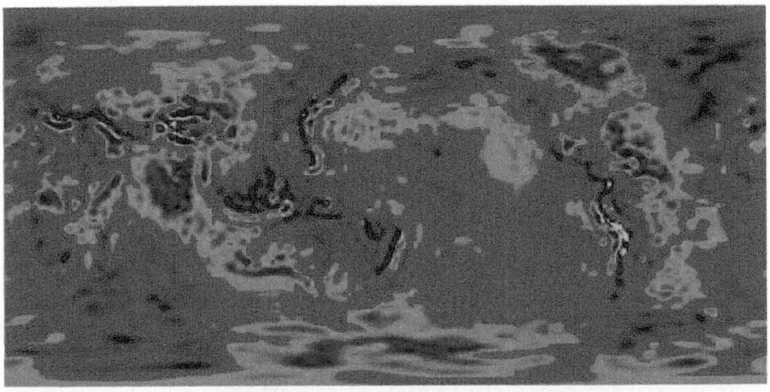

(a)

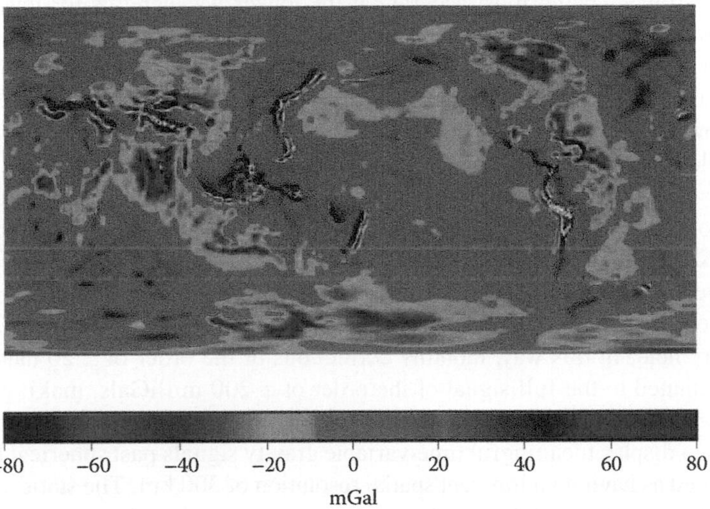

(b)

FIGURE 4.2 Gravity anomalies for (a) the JGM-3 gravity model. (Data from Tapley, B.D., et al., *J. Geophys. Res.*, 101(B12), 28029–28049, 1996.) Gravity anomalies for (b) the GGM05G gravity model that incorporates GRACE and GOCE data.

governed by the Earth's gravity field and by the nongravitational forces, but each satellite experiences slightly different orbital perturbations. The relative motion, or the first difference of the orbital perturbations, is precisely measured interferometrically using a dual-frequency, K/Ka band microwave ranging system (Dunn et al. 2004). Relative range-rate variations of the order of a few meters/second are measured to a precision of less than a micron/second. This relative motion contains information on the difference of the accelerations acting on the two satellites. The nongravitational contributions to the inter-satellite relative acceleration are measured and then removed in data analysis using a high-accuracy, three-axis electrostatic accelerometer (Touboul et al. 2004). The inertial orientation of the satellites is determined using twin star cameras, and the geocentric position and relative time-offset of the satellites are determined using the Blackjack GPS receiver (Dunn et al. 2004). The dual-one-way-ranging (DOWR; Thomas 1999) concept was derived from a synthesis of modern GPS phase-lock-loop tracking techniques with previous satellite-to-satellite tracking techniques (Pisacane et al. 1982). Combined with relative time synchronization using GPS, this gave rise to an economical satellite-to-satellite tracking concept that allowed continuous monitoring of relative motion between the twin GRACE spacecraft using available technologies.

The ability of GRACE to make precise and continuous measurements of satellite orbit perturbations allowed the estimation of the Earth's gravity field parameters based on their short-period orbit perturbation effects. In the language of Kaula's exposition of the orbit perturbation due to the geopotential (Kaula 1966), it was possible to derive the gravity field parameters using the "m-dailies" and the "n-cpr" (cpr = cycles per revolution) perturbations at time scales of minutes to a few hours. Although these are very small amplitude effects, there is a very rich spectrum of these perturbations (e.g., Cheng 2002), leading to a well-determined system for the estimation for the gravity field parameters from GRACE data. In particular, with GRACE, the gravity field was no longer captive to the large amplitude, low-frequency error spectra that accompany signals collected from any other tracking technique. The influence of these considerations on GRACE mission design is more completely discussed in Bettadpur (2008).

The time-integrated nature of the GRACE inter-satellite range-change measurement results in the GRACE estimates being most accurate at the longer wavelengths (or lower degree Stokes coefficients). GRACE mission data are consolidated over multi-day spans, ranging from 1 week to 30 days, so that each gravity field estimate represents a time-mean estimate of the Earth gravity field for that span. The GRACE gravity field estimate is derived as a time span averaged correction to the prior knowledge of the Earth's gravity field. Total gravity field variations due to the static and time-variable components (e.g., due to lunisolar tides in the solid Earth and ocean) as well as the atmospheric and non-tidal oceanic variability are modeled *a priori* in GRACE data processing. Variations of the orders of meter/second in inter-satellite range-rate can be *a priori* modeled to a few microns/second, representing residual signals due to the geophysical processes not known *a priori*. These represent the corrections to the mean Earth gravity field and total gravitational time-variations due to the terrestrial water cycle, ice-sheet changes, and errors in the *a priori* models of Earth's gravity field. In this way, monthly corrections of the order of ± 20 nanoGals of gravity anomaly are estimated to the full signal of the order of ± 200 milliGals, making GRACE truly a "parts-per-billion" quality experiment. Current-best unconstrained, monthly gravity field solutions from GRACE data display meaningful time-variable gravity signals past spherical harmonic degree 30 and are regarded as having an inherent spatial resolution of 300 km. The static, or time averaged, Earth gravity field models from GRACE, however, contain considerably greater new information to degree 120 or beyond, due to the increased resolution from average GRACE mission data for more than 12 years. Access to mean geoid information at 100-km wavelengths relevant to oceanography, however, requires the simultaneous use of GOCE mission data.

GOCE was an ESA mission approved in 1999 and launched on March 17, 2009. The mission ended on November 11, 2013. Its main objectives were the measurement of the geoid with an accuracy of 1–2 cm (gravity anomaly better than 1 milliGal) and with a spatial resolution of about 100 km half wavelength at the equator, corresponding to a spherical harmonic series complete up to degree and order (d/o) 200. For this purpose, GOCE was equipped with a gravity gradiometer instrument, the first of its kind to be flown in space. It was composed of three one-axis gradiometers, each 50 cm long and mounted perpendicular to each other. The gradiometer measured acceleration differences using two ultra-precise three-dimensional accelerometers on the ends of each gradiometer axis. The accelerometer kept a 320-g rhodium platinum test mass (4 cm × 4 cm × 1 cm) in a levitated position via an electrostatic feedback system. The feedback signal was proportional to the gravity gradients plus angular effects due to the rotation of the spacecraft. Each accelerometer had two very precise axes $\left(10^{-12} m/s^2 \sqrt{Hz}\right)$ and one less accurate axis. The chosen arrangement of the accelerometers allowed very precise determination of the gravity gradient components V_{xx}, V_{yy}, V_{zz}, and V_{xz} with x pointing in flight direction, y orthogonal to the orbit plane, and z radially downward. The attained noise level in the measurement band between $5 \cdot 10^{-3}$ and 0.1 Hz was $10 \frac{mE}{\sqrt{Hz}}$ for V_{xx} and V_{yy} and $20 \frac{mE}{\sqrt{Hz}}$ for V_{zz} and V_{xz} ($1\ E = 10^{-9} s^{-2}$). Gravity gradients had to be isolated from the effects of angular motion. This was achieved from a combination of orientation changes of the satellite as measured with a set of three star trackers and certain combinations of the measured acceleration differences. In order to maximize gravity sensitivity, the spacecraft flew in an extremely low orbit of 265 km, which required drag compensation using a set of ion-thrusters. During the final phase of the mission, the orbit was lowered in several steps to 224 km, in a test to further enhance gravity sensitivity. The orbit was sun-synchronous with an inclination of 96.7°. Thus, two circular caps at the poles were left without measurements. GOCE collected its data in cycles of 61 days.

Because of the sensitivity of the gradiometers on GOCE, it was far more sensitive to the shorter-wavelength portion of the gravity field than GRACE and had lower errors as a result (Figure 4.3). However, the GOCE gradiometers alone were not sensitive to the long-wavelength part of the gravity field, to which GRACE was sensitive (Figure 4.3). Thus, the two missions complement each other nicely to obtain a more complete picture of the geoid for both the long and short wavelengths. Although some GOCE-only gravity field solutions are available, their long-wavelength accuracy is limited by the use of the GOCE GPS data in the solution (Brockmann et al. 2014; Bruinsma et al. 2014). GRACE-only gravity models are also available (e.g., Tapley et al. 2005), and although their long-wavelength accuracy is good, errors are much higher than GOCE geoids for wavelengths smaller than 400 km or so.

The best satellite-based gravity/geoid models come from a combination of GRACE and GOCE data. The geoid commission errors for contemporary GRACE/GOCE combined gravity field models are listed in Table 4.3 for three maximum degree/orders: 200, 220, and 280. They are derived from the error variances of the spherical harmonic coefficients following Equation 4.5 and using the GOCE-only model TIM-5 (Brockmann et al. 2014), based on a sophisticated error calibration procedure, and the GRACE/GOCE model DIR-5 (Bruinsma et al. 2014). They are given as global average and as error standard deviation of a chosen validation area at mid-latitude. Note the increasing error for maximum degree and order greater than 200. This indicates growing error at the shortest wavelengths, even with a combination of GRACE/GOCE.

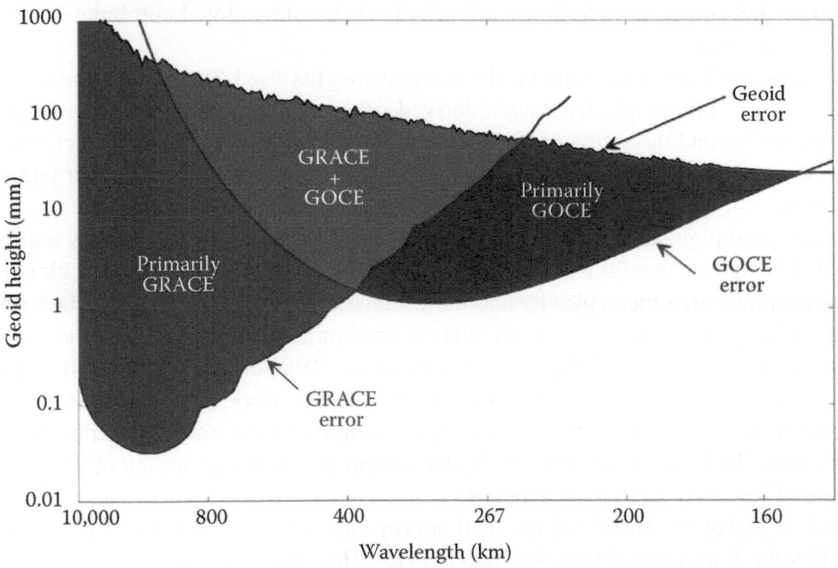

FIGURE 4.3 Accuracy of geoid (mm) versus resolution (wavelength) for gravity fields computed from GRACE satellite-to-satellite tracking data only or GOCE gravimeter data only. Shaded areas indicate where a combined gravity field would rely primarily on GRACE, GOCE, or a combination. The full geoid signal is also shown. (Based on calibrated error covariances provided by John Ries, University of Texas at Austin, Center for Space Research.).

TABLE 4.3
Geoid Commission Errors for Several Expansions of the Model and for a GOCE-Only Solution versus a GRACE/GOCE Combination, Based on Covariances

Maximum Degree/ Order and Area	GOCE-Only (cm)	GRACE/GOCE (cm)
200 Global	3.4	0.8
200 Germany	1.4	0.5
220 Global	4.3	1.2
220 Germany	2.4	0.8
280 Global	12.5	4.9
280 Germany	10.0	4.5

4.2.3 Surface Gravity Data and Combination Geoids

Although the gravitational information obtained from satellite observations such as GRACE and GOCE is global and highly accurate at long wavelengths, it lacks short wavelength details due to the attenuation of the gravitational signal with altitude. Terrestrial, airborne, marine, and altimetric gravity (hereinafter surface gravity), however, will measure the shortwave wavelength character of the gravity (down to a few kilometers along the ship, aircraft, or satellite tracks) but will be irregularly distributed on the Earth's surface and will be inaccurate at long wavelengths (greater than 100 km). Combination geoids such as EGM08 and recently EIGEN-6C or GOCO5C are based on an optimal least squares combination of the two sources of gravitational information (long-wavelength satellite gravity and short-wavelength surface gravity) to create a global combination grid with the advantage of both sources.

The optimal combination depends on the relative weights used for the two sources of gravitational information, with the relative weight derived using the associated error estimates. The specific methods are beyond the scope of this chapter, but interested readers are directed to the paper describing the development of the EGM08 models (Pavlis et al. 2012) or the earlier work by Tapley and Kim (2001).

In preparation for high-resolution combination geoids, comparable mean gravity anomalies must be compiled. This means that for a combination model to degree and order 2160, a global, complete file of five arc-minute area-mean gravity anomalies must be compiled from all surface data sources and subsequently used to compute a set of surface harmonic coefficients. Over the world's ocean, satellite altimetric gravity (i.e., DTU13 [Andersen et al. 2010] or SS13 [Sandwell et al. 2013]) is typically used as surface gravity for the oceans (70% of the world's surface). At a few locations (i.e., high latitudes), marine and airborne gravity data are merged with the satellite altimetry using methods discussed by Pavlis and Rapp (1990), due to limitations in the amount of altimeter observations because of ice cover (see also Chapter 9).

Over land, a global patchwork of regional surveys are compiled through international efforts coordinated by the International Gravity Field Service. However, many regions are still unsurveyed and have no surface gravity measurements. These are generally mountainous areas of the Earth, such as the Himalayas and the Andes. To create global surface gravity data, these missing areas will be "filled-in" based on gravity data estimated from topography information.

For the Arctic region, an international effort led to the ArcGP dataset (Kenyon and Forsberg 2008), which resulted in the most complete set of surface gravity measurements of the Arctic to date. Antarctica, however, is still the least surveyed part of the Earth. For EGM2008, the five arc-minute area-mean gravity anomalies were synthesized from the ITG-GRACE03S (Mayer-Gürr et al. 2010) satellite-only model and thus were not fully reflective of the short-wavelength surface gravity. Over the past several years, however, a large international effort has been undertaken to obtain more direct surface gravity measurements across Antarctica (Scheinert et al. 2016) (Figure 4.4).

4.2.4 Mean Sea Surface Models

The MSS is a key parameter in geodesy and physical oceanography. It is the time-averaged physical height of the ocean's surface, and it is frequently used as the reference to derive sea level anomalies. In principle, a complete separation of the ocean's mean and variable part requires uninterrupted infinite sampling in both time and space. The challenge in MSS mapping is to achieve the most accurate filtering of the temporal sea surface variability with a limited time span while simultaneously obtaining the highest spatial resolution. The MSS is normally expressed relative to a best-fitting Earth ellipsoid and given in a mean tide system (Section 4.4.1). The MSS resembles the geoid to a few meters accuracy as seen in Figure 4.5.

High-resolution MSS models are derived by merging several decades of repeated observations from the TOPEX/Poseidon (T/P), Jason-1, and Jason-2 10-day repeat ground tracks along with denser, non-repeating data from the geodetic missions such as ERS-1, CryoSat-2, and Jason-1 (end-of-life mission). Although older MSS models used GEOSAT geodetic mission data, these have been shown to be of poorer quality than the more modern CryoSat-2 and Jason-1 data, and thus are no longer used for MSS modeling.

The MSS is generally derived by direct interpolation of the averaged SSH observations or the along-track SSH gradients using various sophisticated interpolation techniques (e.g., Andersen and Knudsen 1998). When interpolating along-track SSH observations, it is important to ensure consistency between ascending and descending tracks and between different satellites and missions. Insufficient removal of these inconsistencies will lead to ground track-related striation in the resulting MSS (Knudsen 1993).

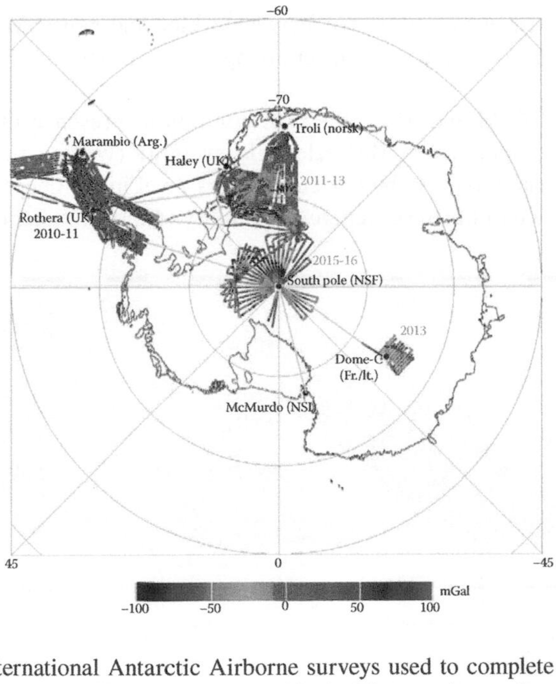

FIGURE 4.4 Recent international Antarctic Airborne surveys used to complete the surface gravity map. (Data from Forsberg, R., et al., GOCE polar gaps now filled—First gravity results of the ESA PolarGap project, in ESA Living Planet Symposium, SP, 2016.)

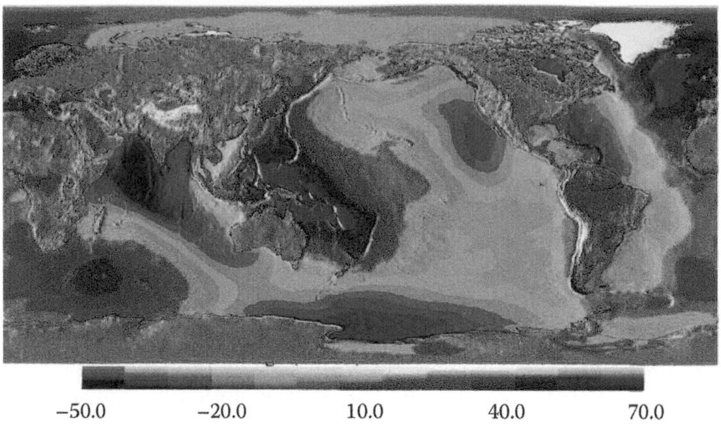

FIGURE 4.5 The global DTU13 mean sea surface (in meters of sea surface relative to the reference ellipsoid) from multiple satellite altimeters. (Data from Andersen, O.B., et al., *The DTU13 MSS (Mean Sea Surface) and MDT (Mean Dynamic Topography) from 20 Years of Satellite Altimetry*, IAG Symposia Series, Springer Verlag, 2015.)

Although it is preferable to use the longest temporal averaging period possible in order to better average out interannual and interdecadal sea level fluctuations, for various reasons either a 7-year or a 20-year period have been used for MSS computation. Older MSS models such as CLS01/DNSC08/MSS_CNES_CLS11 (Schaeffer et al. 2012) are referenced over a 7-year period (1993–1999) whereas newer MSS models such as DTU13/DTU15 (Andersen et al. 2015) are averaged over a 20-year period (1993–2012).

The center time for the older 7-year MSS models will be in 1996, whereas the center time for the newer 20-year MSS will be in 2003. Global long-term sea level change of 3 mm year^{-1} (Chapter 5) will contribute nearly 2 cm mean difference between older and newer MSS due to the different averaging periods. However, regional sea level changes will cause differences as large as 6 cm (Figure 4.6).

Within the bounds of 65°S and 65°N, the near-continuous presence of TOPEX, Jason-1, and Jason-2 data makes the computation of MSS relatively straightforward (e.g., Andersen and Knudsen 2009). The error is estimated to be at the few centimeters level in the deep ocean. Poleward of 65°, the presence of seasonal ice limits the coverage and quality of the altimeter observations and makes

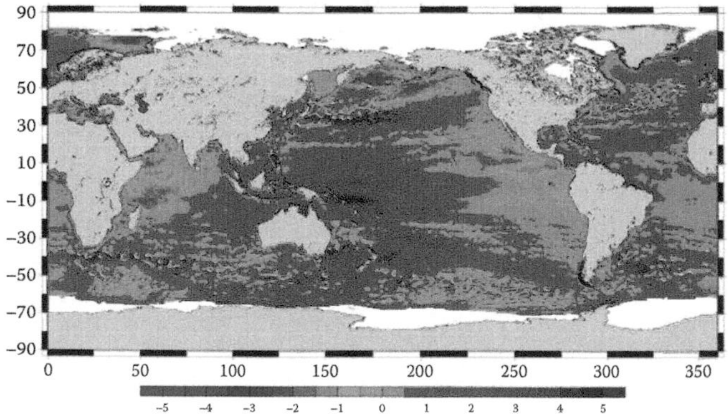

FIGURE 4.6 Difference between 20- and 7-year averaged mean profiles from TOPEX/Poseidon and Jason-1/Jason-2. The color scale is in cm.

determining the MSS far more difficult. Hence, the derived MSS models will be more inaccurate in polar regions. Between 65° and 82°, the MSS computation relies on mean profiles from ERS-1, ERS-2, HY-2, SARAL/AltiKa, and Envisat, whereas only CryoSat-2 data can be used to determine the MSS between 82° and 88°. As CryoSat-2 was not launched until 2010, the averaging period in the Arctic above 82°N will be far shorter and different than that for the rest of the world, but it is the best that can be done with the available data. All high-latitude satellites are also sun-synchronous satellites, which means that they pass over the same location on the Earth at the same local solar time. Although this is convenient to ensure constant sunlight to provide power to instruments, it also means the satellites will measure the second-largest tidal constituent (S2) at the same phase, causing errors in the S2 tide in polar regions to show up as an erroneous bias in polar MSS (Andersen and Scharroo 2011).

4.3 MEASUREMENTS: TIME-VARIABLE GRAVITY

The general principles of the determination of the Earth's time-variable gravity field from GRACE were outlined in Section 4.2.2. In particular, the time-variable Earth gravity field is represented by piece-wise constant adjustments to the *a priori* knowledge of the Earth's gravity field. The time average of such piecewise constant estimates leads to the mean Earth gravity field model, as described in Section 4.2. The remainder of the time span mean estimates relative to the long-term mean represent the time-variable gravity field of the Earth. These will be referred to as "monthly fields," though it is understood that the time span of averaging can be variable, depending on application.

There are two general methods used to convert the monthly mean gravity field Stokes coefficients ($\Delta C_{l,m}(t)$, $\Delta S_{l,m}(t)$) into maps of water mass density: the direct approach (e.g., Wahr et al. 1998), or the mass concentration (mascon) approach (e.g., Watkins et al. 2015).

The direct approach has been the most widely used as it is based on a simple algorithm:

$$\Delta\sigma(\phi,\lambda,t) = \frac{a_E \rho_E}{3} \sum_{l=0}^{N\max} \sum_{m=0}^{l} \frac{(2l+1)}{(1+k_l)} P_{lm}(\sin\phi)\{\Delta C_{lm}(t)\cos m\lambda + \Delta S_{lm}(t)\sin m\lambda\} \tag{4.7}$$

where ϕ is the latitude, λ is the longitude, $P_{lm}(\sin\phi)$ are the fully normalized associated Legendre polynomials of degree l and order m, and $Nmax$ is the maximum degree of the spherical harmonic expansion; $\Delta\sigma$ is the surface mass density in kg m^{-2}, ρ_E is the average density of the Earth (5517 kg m^{-3}), and k_l are load Love numbers of degree l. Over the ocean, this represents the change in mass of the entire water column above the location assuming the gravity signal is caused by water mass fluctuations and we have removed atmospheric mass variability, which is done in the processing. Multiplying this by g (the acceleration due to gravity) gives us the time-variable pressure at the bottom of the ocean ($\Delta P_b(\phi,\lambda,t)$) due to the total mass change of the water column above it. This is often converted to an equivalent sea level perturbation by dividing by g and a constant ocean density ρ_0 (typically 1027 kg m^{-3}). Using a mean density of ocean water compared to local values will introduce errors of less than 1%.

A major problem of the direct approach is how to treat random and correlated errors in the Stokes coefficients. Random errors in the Stokes coefficients increase with degree and become significantly larger than the signal at about degree 15 (e.g., Wahr et al. 2004). Without any smoothing to reduce the random error, maps of bottom pressure from the direct approach are too noisy to be usable (Figure 4.7). However, a straightforward Gaussian smoothing of the gravity coefficients (e.g., Chambers 2006a; Wahr et al. 1998) can reduce the noise significantly (Figure 4.7).

One known problem with Gaussian filters is that they also attenuate some of the longer wavelength signals in addition to the shorter-wavelength random noise. This is more of a problem over the land and ice sheets, though, where there are appreciable real gravity signals at wavelengths attenuated by the filter (e.g., Landerer and Swenson 2012). Over the deep ocean, mass variations are

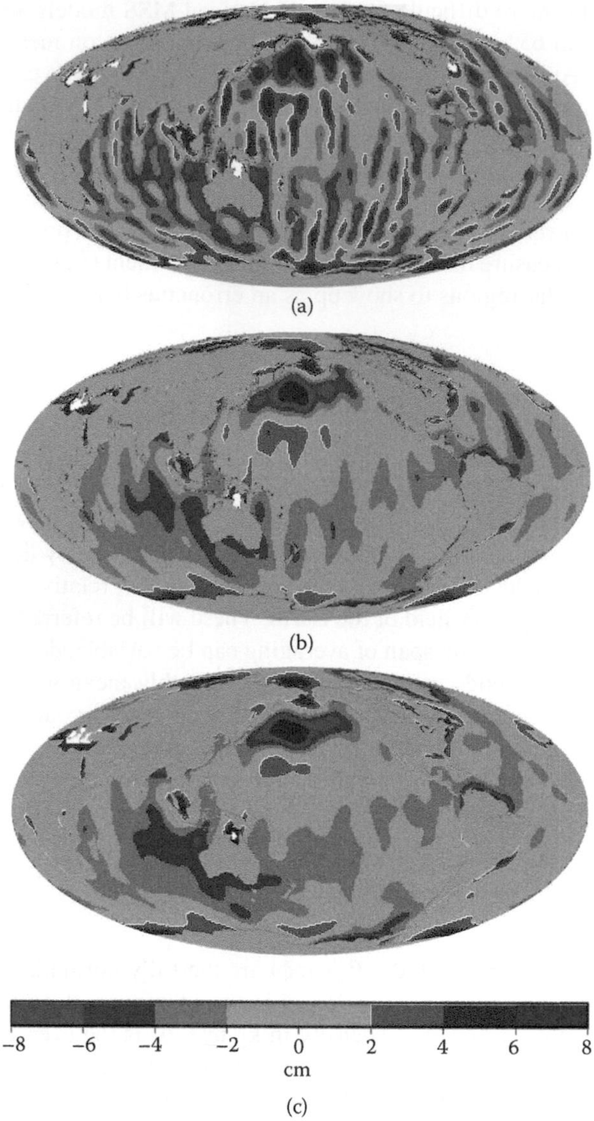

FIGURE 4.7 Maps of ocean bottom pressure from GRACE for January 2010 (a) without smoothing, (b) with 500-km Gaussian smoothing, and (c) with a de-striping algorithm and 500-km Gaussian smoothing.

typically correlated over a longer wavelength than the Gaussian filter, so the attenuation effect is minimal (Chambers and Bonin 2012).

There is an additional correlated error in the released Stokes coefficients (Swenson and Wahr 2006), which causes north-south stripes in the mapped data that cannot be removed with a Gaussian filter alone (Figure 4.7). The north-south stripes are primarily caused by the observation geometry of GRACE. Because the satellites are in a near-polar orbit, there is little sensitivity to east-west gravity gradients. The limited east-west sensitivity results in correlations between Stokes coefficients of a fixed order and the same parity of degree (Swenson and Wahr 2006). This correlated error manifests spatially as longitudinal stripes in the estimated gravity fields. Numerous algorithms have been devised to reduce the size of this correlated errors in $\Delta\sigma$ based on the direct gridding method (Davis et al. 2008; Kusche 2007; Swenson and Wahr 2006; Wouters and Schrama 2007), but only one has been optimized for the ocean (Chambers 2006b; Chambers and Bonin 2012). All of these have become known as "de-striping" algorithms.

Another significant problem with the direct gridded method caused by the de-striping and Gaussian filters is leakage of the much larger land hydrology and cryosphere signals into the ocean grids near land (Figure 4.8). These land-based gravity fluctuations are much larger than the ocean-based variations, leading to much larger errors within about 300 km of land. A simple method has been devised to estimate the leakage error using GRACE estimates over land and iterating (Chambers and Bonin 2012; Wahr et al. 1998). Although this does reduce the errors around coastlines significantly, it does not remove all of them (Figure 4.8).

The mascon approach, although more difficult to implement, does a superior job of reducing both random and correlated error, and more importantly, of partitioning the signal between

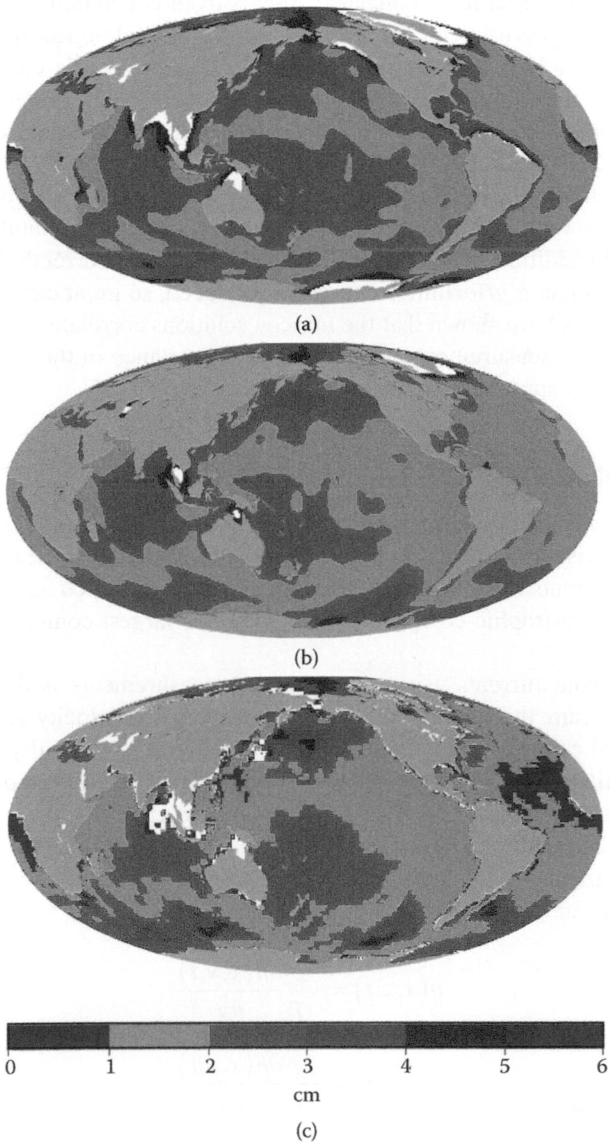

FIGURE 4.8 Root mean square of ocean bottom pressure grids (August 2002 to December 2015): (a) without land correction, (b) with land correction, and (c) for mascon approach.

ocean and land grids, resulting in less leakage of signal into the ocean (Figure 4.8). In this approach, rather than parameterizing the gravity field in terms of spherical harmonic coefficients, for which each C_{lm} and S_{lm} coefficient has a global expression, the gravity field is instead expressed in terms of regional mass elements, or *mascons* for short (Lemoine et al. 2007; Muller and Sjogren 1968; Watkins et al. 2015). The JPL RL05M GRACE mascon solution uses 4551 equal-area mascons (approximately 3° in diameter) to parameterize the gravity field (Watkins et al. 2015). The key advantage to this approach is that each mascon is expressed regionally, unlike the spherical harmonic coefficients, which are expressed globally. This allows for a convenient framework to introduce external *a priori* information into the filter when solving for the gravity field. Each mascon can be assigned an expected value along with an expected variance, both of which can be derived from a combination of global models and auxiliary observations. Combining these expected values and variances with the inter-satellite range-rate measurements enables the filter to better separate geophysical signals from correlated and random errors, and the result is an estimated gravity field that is largely devoid of the longitudinal stripes that plague the spherical harmonic solutions. In addition, land and ocean signals can be better separated with this approach as the majority of mascons are uniquely placed over land or ocean. For mascons that span both land and ocean, filters have been derived to separate the land and ocean signals (Wiese et al. 2016).

The primary difficulty with the mascon approach is that it must be implemented during the gravity processing, thus requiring significantly more expertise and computational resources than the direct approach. In addition, there does exist the potential to incorrectly bias the gravity field estimate if grossly incorrect *a priori* information is introduced, so great care must be taken in this process. However, studies have shown that the mascon solutions correlate better with *in situ* ocean bottom pressure recorder measurements and explain more variance in these *in situ* measurements than the traditional direct approach allows (Watkins et al. 2015).

4.4 APPLICATIONS: DYNAMIC OCEAN TOPOGRAPHY

The ocean's circulation can be broken into two components: the *geostrophic* and the *ageostrophic.* Geostrophic currents arise from a balance between pressure gradients in the ocean and the Coriolis force. Ageostrophic currents arise primarily from frictional forces, such as the Ekman currents in the upper ocean. The geostrophic component tends to be the largest component, especially in the time-mean flow.

Measuring geostrophic currents using ground-based measurements is difficult. Hydrographic data can be used to measure the geostrophic currents relative to the velocity at some reference level, but because the actual current at that depth is rarely known, it is generally assumed to be zero. This can lead to significant errors. Hydrographic sections are also limited due to the expense of ship-borne measurements. However, SSH data from altimetry represents the surface pressure of the ocean. Wunsch and Gaposchkin (1980) showed that if a geoid (N) was subtracted from the SSH to form a dynamic topography, h, the gradient of h was directly related to the absolute surface geostrophic currents (u,v) via:

$$
\begin{aligned}
u(x,y,t) &= -\frac{g}{f}\frac{\partial h(x,y,t)}{\partial y} \\
v(x,y,t) &= -\frac{g}{f}\frac{\partial h(x,y,t)}{\partial x}
\end{aligned}
\tag{4.8}
$$

where x is the east-west position (+ to east), y is the north-south position (+ to north), and f is the Coriolis parameter. The importance of this is that a reference velocity at depth is not needed because the geoid, being an equipotential surface, defines a level of no geostrophic motion in the ocean.

Typically, however, h is not computed at each time, t. Instead, it is broken into a time-mean component ($\bar{\eta}$) that reflects the average over some period, P, and a time-variable part ($\Delta\eta$)

$$\begin{aligned} h(x,y) &= \bar{\eta}(x,y) + \Delta\eta(x,y,t) \\ \Delta\eta(x,y,t) &= SSH(x,y,t) - MSS(x,y) \\ \bar{\eta}(x,y) &= MSS(x,y) - N(x,y) \end{aligned} \tag{4.9}$$

The period over which the MDT is computed significantly affects its characteristics. Ideally, many decades of the SSH are desired in order to average out climate-related signals and obtain the "true" mean surface over the period. This is not the case, however, and the MDT is affected by these time-variable signals. A mean over at least 5 years is required to correctly filter out the turbulent mesoscale eddy signals (Le Provost and Bremond 2003), which is the source of highest variance in ocean variability. An MDT calculated over a shorter time period may, therefore, include the signature of turbulent mesoscale structures. As of today, the altimeter time series length from T/P and Jason-1 exceeds 20 years, and altimeter sea level anomalies along these ground tracks can therefore be computed relative to a 20-year time period. It should be noted, however, that other satellite altimeters (GEOSAT, ERS-1, ERS-2, Envisat, CryoSat, etc.) have observed different regions of the ocean over periods less than 10 years in length. Thus, a consistent time period among the multiple altimeters should be used to reduce aliasing of climatic signals. For example, the 10-day ground track of T/P and Jason and 35-day ground track of ERS-2 and Envisat have a common period of 1995 to 2012.

Averaging over this greater-than-10-year period, we expect the turbulent signal to be mostly smoothed out, leaving "steady" features maintained by a geostrophic balance between the pressure gradient and Coriolis force. The expected spatial scales of an MDT calculated over such a long period can therefore be approximated by considering the so-called Rossby radius of deformation (Gill 1982; Pond and Pickard 1983), which is about 20–40 km at mid-latitudes, with larger values at low latitudes (100–200 km) and decreasing to zero at the poles (Chelton et al. 1998). In the Mediterranean Sea, it is of the order of only 10–15 km (Robinson et al. 2001).

This chapter will only discuss the time-mean dynamic topography (MDT, or $\bar{\eta}$) and its use to compute the mean geostrophic currents. Discussions of time-variable geostrophic currents can be found in Chapters 6 through 10.

4.4.1 Importance of Consistency between Geoid and MSS

The use of altimeter data to derive absolute geostrophic currents depends on both the accuracy of the geoid, N, and the MSS. More critically, one has to ensure the geoid and MSS are consistent (Hughes and Bingham 2008). For example, MSS and N should use the same ellipsoid, the same tide system, and have the same corrections for solid, ocean, and pole tides. However, the use of a consistent ellipsoid and tide system are the most dominant, provided the solid, ocean, and pole tide models are of recent heritage. Differences in either the ellipsoid or tide system can lead to large errors in the MDT, even if the MSS and geoid were perfectly known.

The reference ellipsoid was discussed in Section 4.2.1, and several common ellipsoid parameters are shown in Table 4.1. MSS models are most commonly computed relative to the TOPEX ellipsoid, but the GRACE/GOCE geoid models are computed relative to different ellipsoids, depending on the producer. However, it is a straightforward calculation to replace the reference ellipsoid, once one knows which ones were used, in order to make them consistent.

Geoid heights and the MSS can also differ depending on what system is implemented to deal with the permanent tide effects. Although most tidal signals are averaged out (or modeled), there is a residual permanent deformation in the ocean surface, a bulge at the equator, that will appear in both the SSH from altimetry and the gravitational potential measured by gravimetric satellites. There are multiple ways to handle this effect (Hughes and Bingham 2008; Rapp 1989). For oceanographic applications, it is preferred to use the mean tide (MT) system. This system includes the bulge in both the geoid and the MSS, so that the geoid defines an equipotential surface in the oceanographic sense—the ocean at rest. However, other systems exist. If the geoid is created to represent only the Earth's gravitational forces, the tidal bulge is removed and the zero tide (ZT) system is used. Another system is the tide-free (TF) system, which is a purely theoretical construct as it allows the Earth's gravitational bulge to relax from being free of tidal forces although this relaxation is not known and has to be estimated based on several assumptions (Hughes and Bingham 2008).

If the geoid is not in the MT system, it is straightforward to correct it from the FT or ZT system using (Hughes and Bingham 2008; Rapp 1989):

$$\begin{aligned} N_{MT} - N_{ZT} &= 9.9 - 29.6\sin^2\phi[\text{cm}] \\ N_{MT} - N_{TF} &= (1+k)\left(9.9 - 29.6\sin^2\phi\right)[\text{cm}] \end{aligned} \tag{4.10}$$

where k is the Love number of deformation used for the permanent mean tide deformation of the Earth. Unfortunately, this is not really known and is one reason the ZT system is theoretical. For the conversion, one must know which value has been adopted for k.

Finally, several corrections need to be applied to both MSS and the geoid consistently. These are solid Earth tides, ocean tides, and solid Earth and ocean pole tides (Petit and Luzum 2010). In particular, because the MSS is computed from multiple altimeters (Section 4.2.3), one must ensure they each use the same models. This is especially true of the sun-synchronous altimeter satellites as certain constituents of the tide model alias into biases (Chapter 1). Errors in these constituents will go directly into the MSS, and one should use consistent, accurate models to reduce the amount of error in the MSS. Because these corrections are usually already applied to both altimetry and gravimetry as part of the corresponding processing chain, one should ensure they are identical—or as similar as possible.

Even if the previous consistency checks have been done, the raw difference between an MSS and a geoid model will differ from the true ocean MDT due to errors in the gridded MSS (Section 4.2.3) and by both commission and omission errors in the geoid model (Section 4.2.2). For example, Figure 4.9 shows the raw difference between an MSS model (CNES-CLS11; Schaeffer et al. 2012) and the GOCE-DIR5 geoid model (Bruinsma et al. 2014), which combines GRACE and GOCE data. In strong subduction areas, the signal is dominated by the geoid omission error (i.e., the neglected geoid signal above degree n_{max}, which is contained in the MSS, but not the geoid model). These large unmodeled geoid errors, such as in the Tonga-Kermadec trench north of New Zealand or in the Izu-Ogasawara trench south of Japan, would imply unrealistic ocean currents if used without appropriate filtering.

There are two alternative strategies: either filtering in the space domain or in the spectral domain. A comparison of the two approaches has been done by Bingham et al. (2008), and the reader is directed there for details. Here, we merely summarize the methods and their relative benefits/deficiencies.

Filtering in the space domain means applying an appropriate spatial filter either to the MSS and N separately or, in one step, to their difference (i.e., the "raw" MDT as in Figure 4.9). Several filters are proposed in the literature, such as the Hamming (Jayne 2006) or Gaussian filters (Jekeli 1981; Wahr et al. 1998). Gaussian filtering has the advantage that it has a nice symmetry between space and spherical harmonic spectral domain, but its isotropy results in significant signal loss along slopes and coastal zones. Alternatively, more complex filters may be used. Vianna et al. (2009) developed an adaptive filter, based on principal component analysis techniques, in order to extract as much signal as possible while also minimizing attenuation. Bingham (2010), Bingham et al. (2011, 2014), and Siegismund et al. (2013) use nonlinear diffusive filtering in order to filter preferentially along

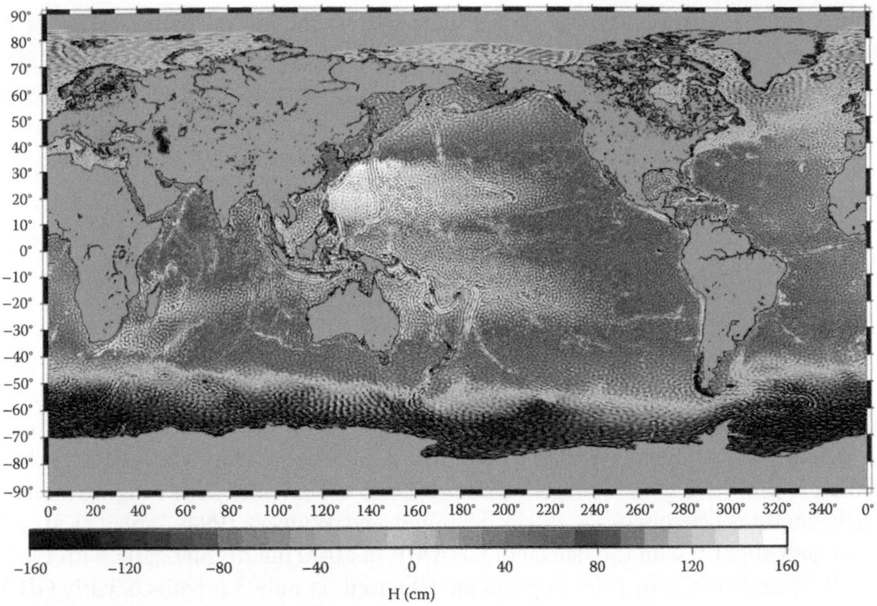

FIGURE 4.9 Raw difference (cm) between the CLS11 mean sea surface model (Schaeffer et al. 2012) and the GOCE-DIR5 geoid model (Bruinsma et al. 2014).

rather than across steep MDT gradients. They obtain an optimal filter length by minimizing the RMS difference between the currents derived from the filtered geodetic MDT and an independent reference (either surface drifter velocities or an independent MDT). Rio et al. (2011, 2014a) use an approach based on the objective analysis method (Bretherton et al. 1976) in which the output field (the filtered MDT) is a weighted average of the observations (the raw MDT). The weights depend on an estimate of the MDT errors and an estimate of the *a priori* variance and correlation scale of the true MDT signal. The method allows filtering out of the noise while preserving the strong MDT gradients. As a penalty for using these methods, however, one has to accept the loss of a simple relationship between space and spectral domain. Also, at ocean-land transitions, where no complete data coverage is available, filtering in the space domain creates significant issues for coastal MDT.

The alternative is to filter in the spherical harmonic spectral domain. As altimeter data are confined to ocean areas, the MSS has to be complemented by some geoid model over land. It is recommended to use the same geoid model as the one chosen for the ocean areas (Bingham et al. 2008; Haines et al. 2011) for this purpose. It is also necessary to merge the geoid and MSS data to create a smooth transition from ocean to land (Albertella and Rummel 2009) in order to avoid unwanted Gibbs effects when transferring back to the space domain. In order to prevent unnecessary signal loss due to filtering, it is recommended to artificially extend the chosen gravity model above n_{max} by some high resolution combined gravity model (Section 4.2.5), such as EGM2008 (Pavlis et al. 2012). Even if the two do not match perfectly, it will allow one to better confine filtering to spatial scales corresponding to N_{max}. This same principle can be applied in the spatial as well as in the spectral domain.

Alternatively, another approach is described in Becker et al. (2012, 2014). They use a least-squares adjustment to the gridded MSS and the set of spherical harmonic coefficients $\overline{C}_{nm}$ and $\overline{S}_{nm}$ of the selected gravity model. *In situ* ocean observations may be added as well. The observations are accompanied with an error variance matrix for the MSS set and the full error variance covariance matrix of the gravity model. The unknowns are: (1) the MDT values at a grid point, expressed as local polynomial base functions, and (2) a series of spherical harmonic (SH) coefficients. The unknown

SH-coefficient set is divided into three parts: (1) the coefficients up to degree N_{max}^{geoid} (for which the geoid is believed to be known with accuracy to less than 1 cm); (2) the coefficients from N_{max}^{geoid} to the truncation degree of the MSS, N_{max}^{MSS} (which are down-weighted); and (3) a set for coefficients above N_{max}^{MSS}. For the latter, some regularization of the Kaula rule of thumb (Equation 4.5) is introduced in the loss function. The result is an estimate of MDT values with realistic errors and an optimal down-weighting of the geoid signal beyond N_{max}^{geoid}, which is contained in the MSS.

4.4.2 Improvements in MDT with GRACE and GOCE Geoids

As mentioned in the introduction, at the beginning of the altimetry era, geoid models were not sufficient to calculate MDT for determination of geostrophic currents (Nerem et al. 1994b; Stammer and Wunsch 1994; Tapley et al. 1994). Even considering only the long-wavelength (greater than 1000 km wavelength) MDT, velocity errors were of the order of 10 to 20 cm s^{-1} RMS, roughly the same size as many of the mean currents. Zonal currents in the low latitudes could not be resolved in any of the pre-2000 era geoid models based primarily on satellite laser ranging (SLR) tracking to low-Earth satellites (Nerem et al. 1994b; Stammer and Wunsch 1994; Tapley et al. 1994). The situation changed slightly with the launch of CHAMP in 2000 but dramatically with the launch of GRACE in 2002 and GOCE in 2009. A geoid model based on only 3 months of early GRACE data was sufficient to estimate currents at scales of greater than 500 km with an accuracy approaching 3 cm s^{-1} RMS (Tapley et al. 2003). With nearly a decade of GRACE data and the higher resolution GOCE data, a similar or slightly better accuracy can now be achieved at scales of around 100 km (Bingham et al. 2011; Bruinsma et al. 2014; Knudsen et al. 2011; Mulet et al. 2012).

This is highlighted in Figure 4.10, which shows the geodetic MDT obtained using the geoid models representative of the four major eras: (1) Pre-2000 (GRIM5S1), which is based on primarily SLR tracking (Biancale et al. 2000); (2) CHAMP3S (Reigber et al. 2004), based only on CHAMP GPS and accelerometer data; (3) ITG-GRACE2010s (Mayer-Gürr et al. 2010), based only on GRACE satellite-satellite tracking and accelerometer data; and (4) GOCE-DIR5 (Bruinsma et al. 2014), which is a combination of GRACE data with GOCE gravimeter and accelerometer data, plus some SLR data for the degree 2.

The geodetic MDT shows the general features one would expect to recover: (1) high topography in the middle of the oceans at mid-latitudes, intensified near the western boundaries, which is caused by the subtropical gyres and western boundary currents; (2) a deep trough around 10°N in the Pacific associated with the equatorial counter current; and (3) the sharp gradient from low topography to high topography in the Southern Ocean caused by the Antarctic Circumpolar Current (ACC).

Pre-GRACE geoid models could only distinguish the Southern Ocean topography gradient and the high topography between 40°N and 40°S (Figure 4.10a,b), primarily because these are related mostly to the long-wavelength geoid that these models could recover adequately. However, GRIM5S1 had so much noise at higher degrees/orders that it could not be used to resolve any of the gyre circulations. CHAMP-only geoids, although clearly better than GRIM5S1, still had sufficient noise that low-latitude zonal currents could not be resolved. Although the Gulf Stream is apparent, the gyres are still difficult to observe with spatial smoothing of only 300 km.

The improvement with GRACE data is very apparent (Figure 4.10c), although there are still obvious errors at the 100-km scale, primarily around seamounts/trenches, and also from meridional stripes that are caused by a weakness of GRACE to adequately determine sectorial and tesseral gravity coefficients (Tapley et al. 2005). With the addition of almost 4 years of GOCE data, however, these artifacts are completely eliminated at 100-km spatial smoothing (Figure 4.10d).

The improvement is even more apparent if one computes geostrophic velocities (Equation 4.10) and compares the speed (sqrt($u^2 + v^2$)) (Figure 4.11a and b). Also shown are the surface speeds computed from 15-m drogued drifter buoys distributed by the Surface Drifter Data Assembly

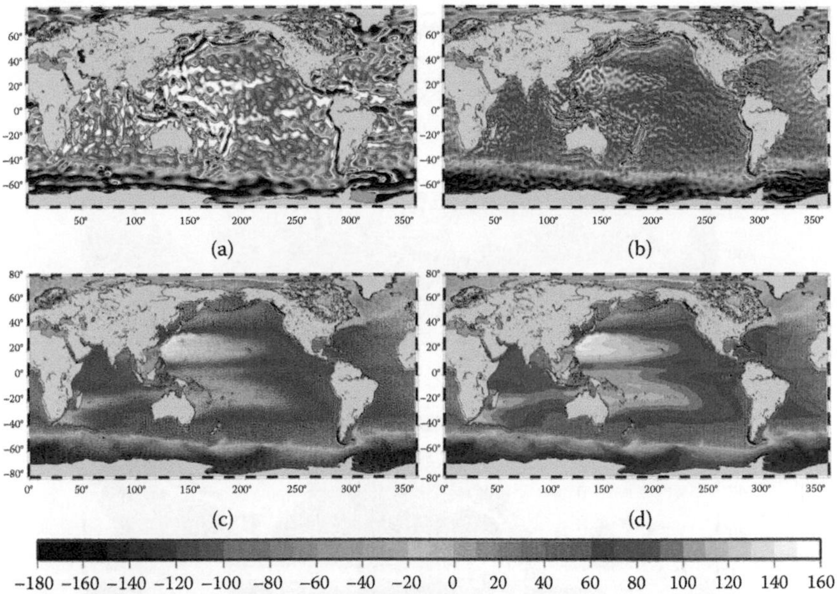

FIGURE 4.10 Geodetic mean dynamic topography obtained from the CLS11 mean sea surface model (Schaeffer et al. 2012) and different geoid models: (a) GRIM5S1, based on tracking to low-Earth satellites. (Data from Biancale, R., et al., *Geophys. Res. Lett.*, 27(22), 3611–3614, 2000.); (b) CHAMP3S, based on only CHAMP data. (Data from Reigber, C., et al., Earth gravity field and seasonal variability from CHAMP. In: Reigber, C., et al., (eds.), *Earth Observation with CHAMP—Results from Three Years in Orbit.* Springer, Berlin, pp. 25–30, 2004.); (c) ITG-GRACE2010s. (Data from Mayer-Gürr, T., et al., *ITG-Grace2010 Gravity Field Model*, http://www.igg.uni-bonn.de/apmg/index.php?id=itg-grace2010, 2010.), based only on GRACE data; and (d) GOCE-DIR5, based on a combination GRACE and GOCE. (Data from Bruinsma, S. L., et al., *Geophys. Res. Lett.*, 2014.). In (a) and (b), the filtering length used is 300 km, while in (c) and (d) it is 100 km.

Center (SD-DAC) (Figure 4.11c). It must be noted that the velocities deduced from the drifting buoy trajectories result from the full ocean dynamics and therefore include both the geostrophic component and ageostrophic components (Ekman currents, inertial currents, Stokes drift, and tidal currents). The geostrophic surface currents have been extracted from the total buoy velocity by first removing an estimate of the Ekman current (Maximenko et al. 2009; Niiler et al. 2003; Rio et al. 2004, 2011, 2014b) and further applying a 3-day low-pass filter to remove inertial and tidal currents as well as residual high-frequency geostrophic currents. The temporal variability of the geostrophic current derived from altimetry is also removed to obtain an estimate of the mean geostrophic current.

The improvement from a GRACE-only geoid to a GRACE/GOCE geoid is clear (Figure 4.11). Noise is reduced, especially in the low-latitude currents, but also around seamounts/trenches, and the GRACE/GOCE geodetic currents are qualitatively similar to the *in situ* currents in terms of location and magnitude. The *in situ* currents are stronger because they have not been spatially smoothed with a 100-km filter. In order to compare the *in situ* and geodetic mean geostrophic currents in a more quantitative way, RMS differences between the two types may be calculated using different smoothing filters (Bruinsma et al. 2014; Mulet et al. 2012) (Figure 4.12). There was significant improvement obtained at scales shorter than 150 km by incorporating even just 6 months of GOCE data into geoid models compared to a GRACE-only geoid (GOCE-DIR2) dropping errors from order 18 cm s^{-1} at spatial scales of 100 km to about 7 cm s^{-1}. GOCE-DIR5, which includes almost 4 years of GOCE data and also benefited from data acquired at a lower altitude than the nominal 255-km altitude during the last year of the mission, results in even smaller errors (order 4.5 cm s^{-1}

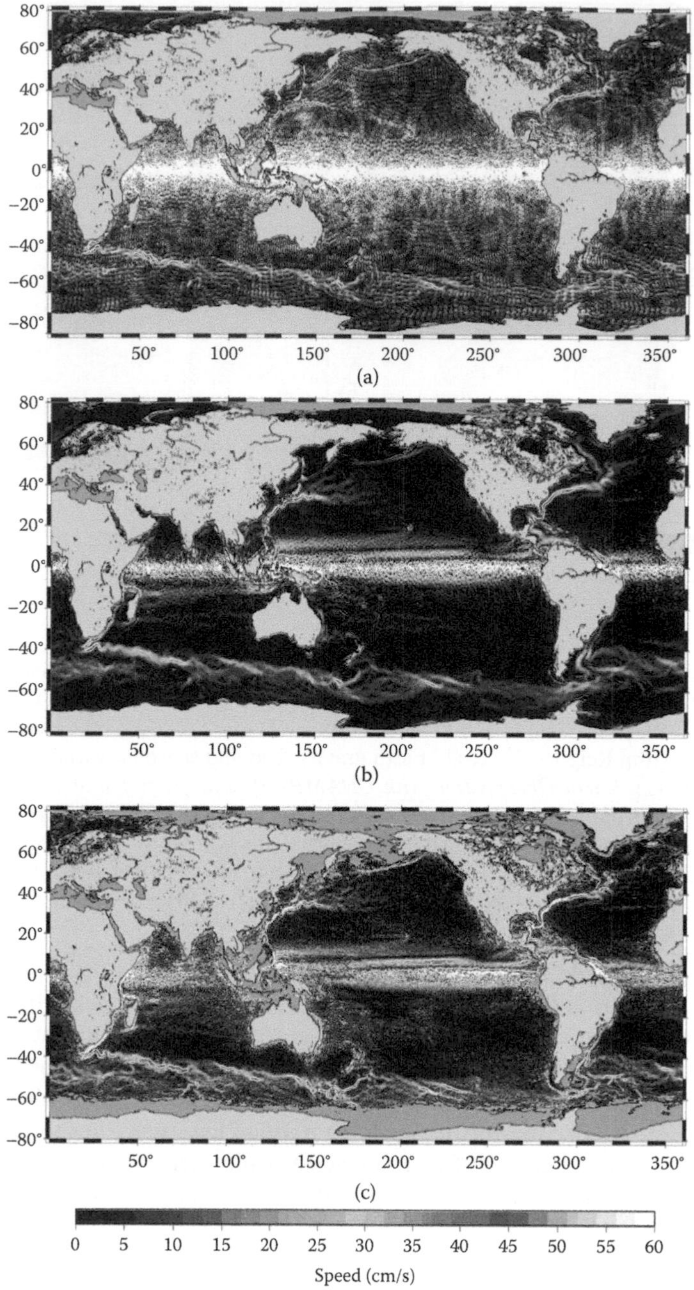

FIGURE 4.11 Mean geostrophic speeds derived from (a) the 100-km filtered geodetic mean dynamic topography based on ITG-GRACE2010s; (b) the 100-km filtered geodetic mean dynamic topography based on GOCE-DIR5; and (c) in situ velocity measurements.

at 100 km). These errors are below the standard deviation of *in situ* velocities at these spatial scales, indicating a significant signal-to-noise ratio (Figure 4.12). One should note that the error in zonal velocities is still substantially below the signal threshold even for resolutions lower than 100 km, although this is not true of the meridional component.

Although the comparison with filtered drifter velocities provides an upper bound of the MDT gradient commission errors, an indication of both omission and commission errors is obtained

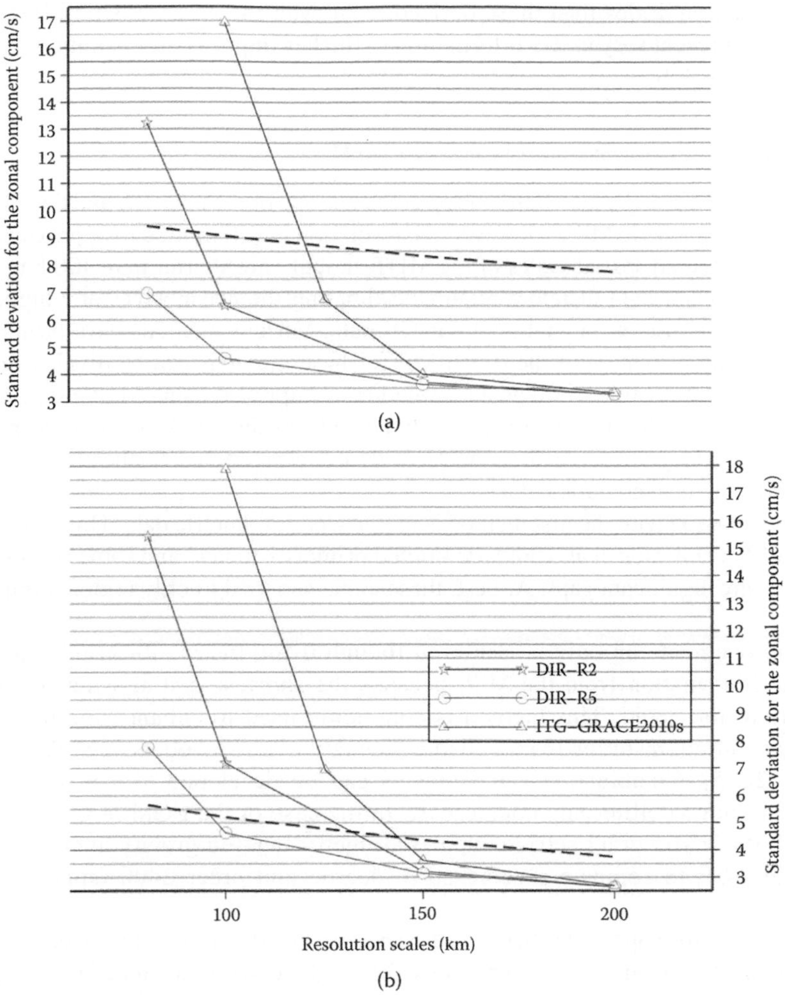

FIGURE 4.12 Zonal (a) and meridional (b) root mean square differences between in situ and geodetic mean geostrophic velocities at varying spatial scales (from 80 to 200 km). Geodetic mean geostrophic velocities are derived from different geodetic mean dynamic topography (blue: DIR2 to DIR4; pink: DIR5; and red: ITG-GRACE2010s). The dashed black line gives the in situ mean geostrophic velocities' standard deviation. The geodetic and in situ mean geostrophic currents were both filtered consistently using lengths ranging from 80 to 200 km.

using unfiltered *in situ* mean velocities. The minimum RMS difference between a GOCE DIR5 geodetic MDT and unfiltered velocities is reached between 100 and 125 km (Bruinsma et al. 2014, Figure S4). It is 7.0 cm s^{-1} for the zonal component of the velocity and 7.7 cm s^{-1} for the meridional component. These values increase to 9.3 (9.5) cm s^{-1} for the zonal (meridional) component of the velocity at 80-km resolution. A thorough analysis of the expected error level on recent GOCE-based geodetic MDTs can be found in Bingham et al. (2014).

Despite the remarkable improvement found with new geoids based on combinations of GRACE and GOCE data, further improvement is necessary and possible. First, the accuracy of geodetic MDT is still insufficient in coastal areas and semi-enclosed seas such as the Mediterranean, which are characterized by numerous straits and short spatial scales. This is due to limited accuracy both in the MSS (since altimetry data is less accurate close to the coasts) and the geoid. It is also clear from Figures 4.11 and 4.12 that even in the open ocean the *in situ* mean currents resolve shorter spatial scales than those derived from even the best geodetic MDT. The drifter data, therefore, represent very useful information to further improve the resolution of the geodetic MDT (see Section 4.4.3).

4.4.3 Toward a Higher Spatial Resolution MDT

The previous section has highlighted the strong contributions of the GRACE and GOCE missions to improvement of the geodetic MDT. However, the resolution is still limited to approximately 100 km and longer. The expected spatial scales of the MDT, though, are shorter than 100 km, and optimal exploitation of altimeter data therefore requires estimating the ocean MDT at a higher resolution than permitted by the use of only GRACE/GOCE data. Various methods have been developed in order to increase the resolution of the ocean MDT. These methods can be divided into two main categories. In the first approach, the geoid resolution is improved using *in situ* or aircraft gravity data (Section 4.2.4), which is then combined with the high-resolution MSS to obtain an MDT. In the second approach, a large-scale geodetic MDT is first computed and further improved at shorter scales using external oceanographic data.

One significant issue with the first method is that *in situ* gravimetric data (Hunegnaw et al. 2009; Pavlis et al. 2012; Thompson et al. 2009) are limited primarily to land and some coastal areas—and even then there are significant gaps. At best, these alone can lead to only regional improvement of the geoid.

Global improvement can only be achieved though using the *in situ* gravimeter data along with geodetic information from the MSS because the shortest spatial scales of the MSS are primarily due to the geoid. This method is commonly used to enhance the resolution of the satellite-only solution, resulting in so-called combination geoid models (e.g., Pavlis et al. 2012; Tapley et al. 2005). Combination geoid models can be developed to a much higher degree and order than their satellite-only counterpart. In the case of the EGM08 geoid model (Pavlis et al. 2012), both *in situ* gravimetric data and altimetry-derived gravity anomalies were combined with GRACE to compute a geoid up to degree and order 2400 (approximately 8-km resolution).

The drawback of this approach is that because an *a priori* MDT is used to correct the MSS before its conversion into gravity anomaly observations, the geodetic MDT calculated using a combination geoid model will not be fully independent of the choice of the *a priori* MDT.

The second way of obtaining higher resolution estimates of the MDT is to enhance the larger-scale geodetic MDT using *in situ* oceanographic measurements. This approach has used only drifter velocity data (Maximenko et al. 2009; Niiler et al. 2003) or drifter buoy velocity information and *in situ* measurements of dynamic height from conductivity-temperature-depth (CTD) casts and Argo floats (Rio and Hernandez 2004; Rio et al. 2007, 2011, 2014a, 2014b). As mentioned previously, mean geostrophic current estimates are extracted from the drifting buoy velocities by removing the ageostrophic components and the temporal velocity variability as measured by altimetry. Similarly, the *in situ* dynamic heights are processed and the altimeter temporal variability is removed to recover estimates of the mean dynamic topography. Mean geostrophic velocities and mean dynamic topography values, together with their relative error estimates, are then used to improve the large-scale geodetic MDT through a multivariate objective analysis (Bretherton et al. 1976) based on the *a priori* knowledge of the MDT variance and spatial correlation lengths.

The most recent MDT computed from this method, the CNES-CLS13 MDT (Rio et al. 2014a, 2014b), is on one-quarter-degree grid (Figure 4.13). It is based on the CNES-CLS11 MSS (Schaeffer et al. 2012) and the GOCE_DIR-4 geoid model (Bruinsma et al. 2013) together with 20 years (1993 to 2012) of *in situ* measurements of surface velocities, dynamic heights, and altimeter sea level anomalies. The velocities from this new MDT (Figure 4.13b) are significantly stronger than those from the comparable geodetic MDT (Figure 4.11b).

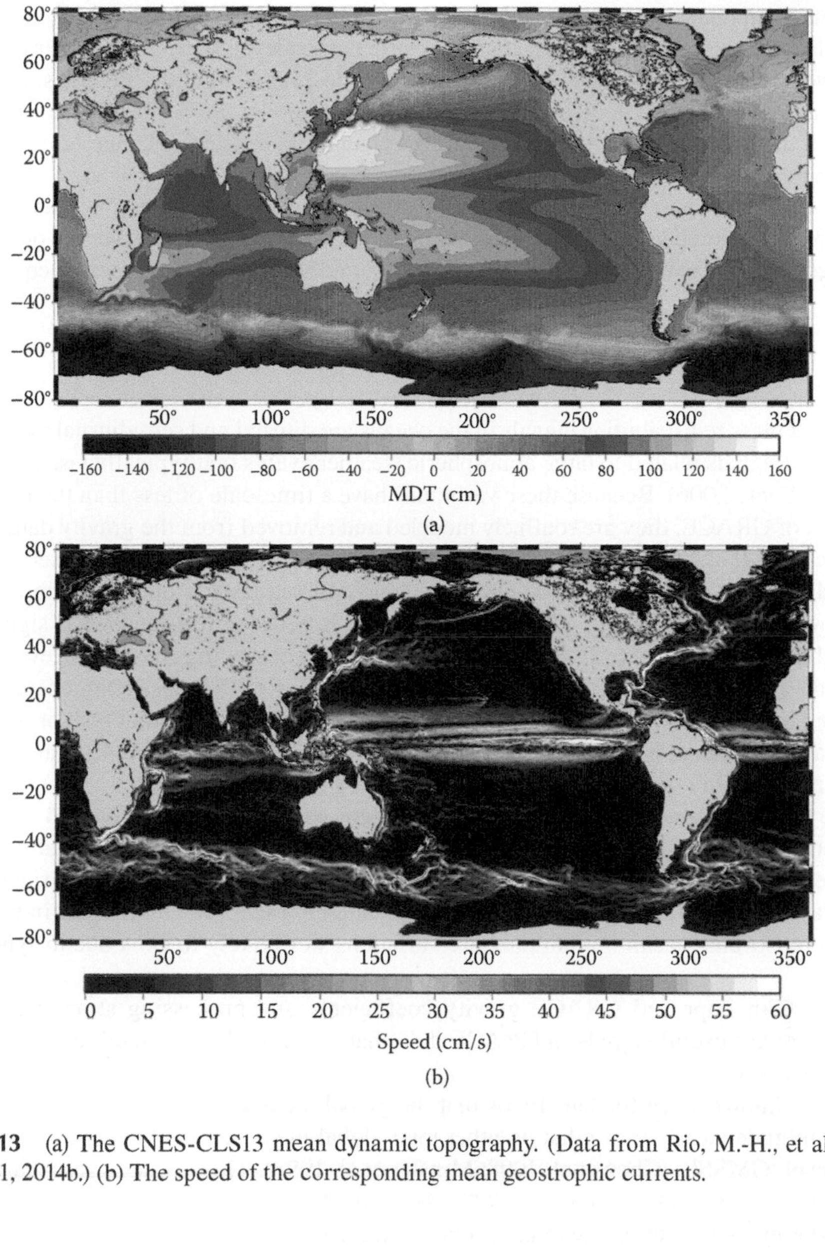

FIGURE 4.13 (a) The CNES-CLS13 mean dynamic topography. (Data from Rio, M.-H., et al., *Geophys. Res. Lett.*, 41, 2014b.) (b) The speed of the corresponding mean geostrophic currents.

4.5 APPLICATIONS: GLOBAL AND REGIONAL OCEAN MASS VARIATIONS

Satellite altimetry measures SSH at specific locations (ϕ,λ) and times (t). An MSS (Section 4.2) is generally removed to estimate sea level anomalies ($\Delta\eta$), which are often mapped or averaged over large areas to estimate time-variable sea level. There are many physical processes that cause sea level to change over time, from fast processes such as tides to slower ones such as climatic-scale changes in the winds and ocean circulation.

Here, we consider only applications of using time-variable gravity that are most directly related to sea level measured by altimetry. Sea level can be broken into two components, the *mass* component and the *steric* component:

$$\Delta\eta(\phi,\lambda,t) = \Delta\eta_{mass}(\phi,\lambda,t) + \Delta\eta_{steric}(\phi,\lambda,t) \tag{4.11}$$

The steric component is caused by temperature/salinity changes within the water column. These result in small changes in ocean density that will cause the sea level to expand or contract (this chapter). However, because the sea level expands as density decreases (volume increases) and contracts as density increases (volume decreases), the integrated pressure at the bottom of the ocean (i.e., the total mass of the water above the seafloor) does not change. Thus, steric changes are not detectable in time-variable gravity data.

Any process that causes a redistribution of mass, however, will cause a change in gravity (Section 4.3) and theoretically be observable by missions such as GRACE. Over the ocean, this can be estimated from the mass density from Equation 4.8 and the hydrostatic pressure equation:

$$\Delta\eta_{mass}(\phi,\lambda,t)=\frac{\Delta\sigma(\phi,\lambda,t)}{\rho_0} \tag{4.12}$$

The largest mass redistribution signals in the oceans are diurnal and semidiurnal tides, as well as fast barotropic signals related to large atmospheric weather events that typically last a period of less than a week (Ponte 2006). Because these variations have a timescale of less than the monthly averaging period of GRACE, they are routinely modeled and removed from the gravity data (Bettadpur 2012a; Flechtner et al. 2015) so that, in an ideal situation, GRACE would only observe variability with periods longer than a month. The mass component of sea level is quite large in coastal areas with shallow bathymetry (Figure 4.14) due to interactions between the deep steric signals and the shelf (Bingham and Hughes 2012). Because of problems with observing coastal sea level with altimetry (Chapter 12) and with GRACE (Figure 4.8), most studies of the mass component of sea level have focused on the deep ocean away from the shelves. Here, the mass component of sea level is an order of magnitude smaller than steric variations in the deep ocean (Figure 4.14) and close to the expected uncertainty in the mass estimate.

Because the mass signal in the ocean is so small, many early studies actually used it as a measure of the uncertainty in the GRACE mass density fields (Kanzow et al. 2005; Wahr et al. 2004). These early studies concluded that the signal-to-noise ratio in the early GRACE time-variable data was likely too large to make them useful for ocean applications, except in small areas in high-latitude regions where extreme ocean bottom pressure variations were likely to exist due to synoptic-scale wind events.

However, with improved GRACE gravity coefficients and processing strategies, it became clear that there are useful signals in GRACE data that can complement satellite altimetry. Here, we discuss just a few.

It has been known since the late 1990s that the global mean ocean mass (GMOM) has a large seasonal amplitude, nearly twice that of either total global mean sea level (GMSL) or global mean steric sea level (GMSSL) (Chen et al. 1998; Minster et al. 1998). This is due to seasonal changes in cycling of water over land (Chen et al. 1998). The global steric variation, which is dominated by warming/cooling of the larger ocean area south of the equator, peaks 6 months out of phase with GMOM and hence cancels approximately half the signal in GMSL.

Chambers et al. (2004) demonstrated that GRACE could observe this seasonal signal even in the earliest releases of data. Subsequent analysis by numerous scientists (e.g., Boening et al. 2012; Chambers et al. 2017; Fasullo et al. 2013; Leuliette and Miller 2009; Willis et al. 2008) revealed significant non-seasonal variations in mean ocean mass (Figure 4.15). It was expected that a significant fraction of the trend in GMSL would be from increasing ocean mass due to addition of water from melting of ice on the continents and ice sheets. The size of interannual variations was more surprising. Although several studies based on steric-corrected altimetry (Chambers et al. 2000; Willis et al. 2004) and hydrology models (Ngo-Duc et al. 2005) had suggested interannual global ocean mass fluctuations could be as large as seasonal ones, their results were inconsistent. Moreover, other studies argued that the large interannual fluctuations in GMSL (such as the rise during the 1997–1998 El Niño) were related to ocean heat (i.e., steric) changes, which some models supported (Nerem et al. 1999).

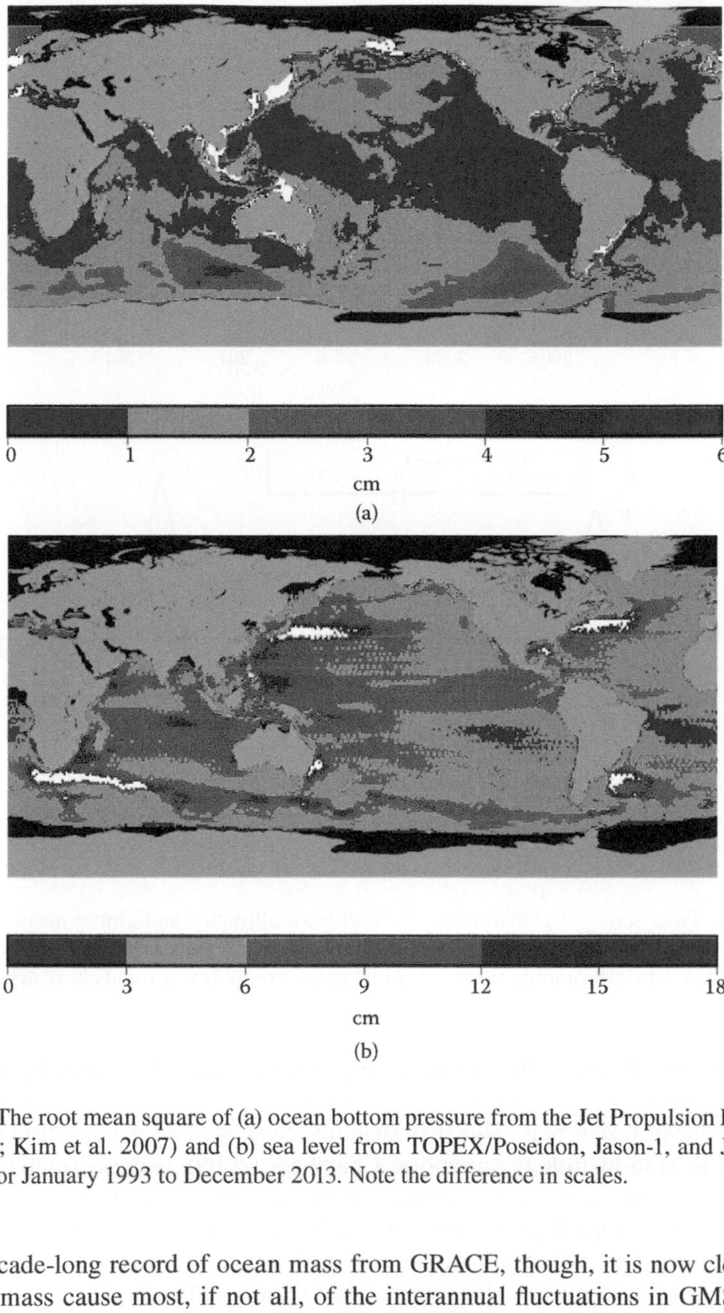

FIGURE 4.14 The root mean square of (a) ocean bottom pressure from the Jet Propulsion Lab ECCO model (Fukumori 2002; Kim et al. 2007) and (b) sea level from TOPEX/Poseidon, Jason-1, and Jason-2 altimetry, both computed for January 1993 to December 2013. Note the difference in scales.

With the decade-long record of ocean mass from GRACE, though, it is now clear that fluctuations in ocean mass cause most, if not all, of the interannual fluctuations in GMSL. These arise from the changes in precipitation and water transport between the oceans and land (e.g., Fasullo et al. 2013). The large drop in 2011 has been well documented, for instance, and the mechanism tied primarily to an anomalous precipitation over Australia due to a convergence of storms caused by a La Niña, a strong negative phase of the Indian Ocean Dipole, and a strong Southern Annual Mode (Fasullo et al. 2013). These storms dropped most of the rainfall over the interior of Australia, which has no natural outflow to the ocean. The rainfall filled a large, normally dry lake called Lake Eyre, resulting in the appearance of a shallow inland sea for several months until it evaporated, returning the water mass eventually back to the ocean.

Chapter 5 will discuss many more aspects of global mean sea level and global ocean mass, so a curious reader is directed there. However, before continuing with a discussion of regional

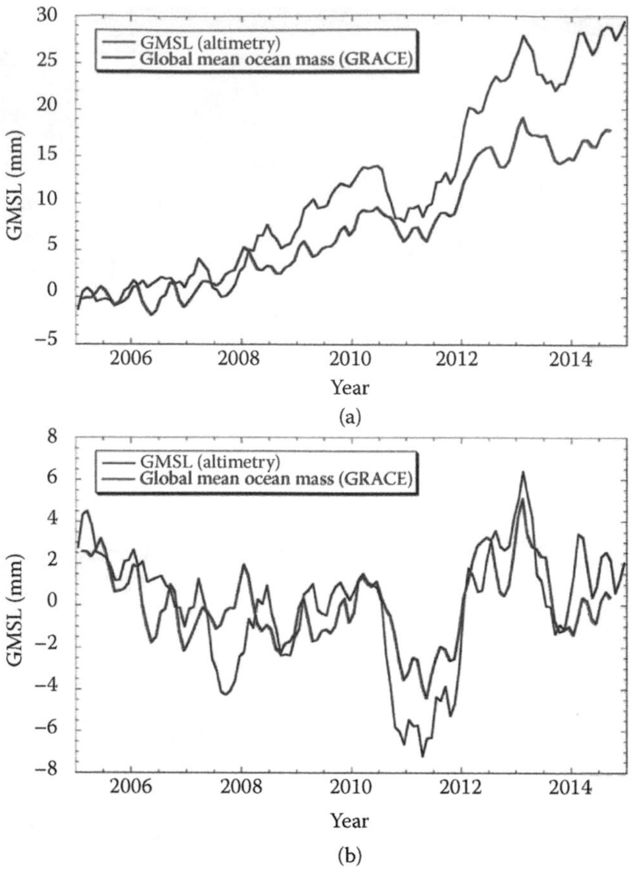

FIGURE 4.15 (a) Time series of global mean sea level from altimetry and global mean ocean mass from GRACE, January 2005 to December 2014. Both time series have had seasonal variations removed and have been smoothed with a 3-month running mean. (b) Same as (a), except linear trends have also been removed.

mass changes, we would like to discuss some important aspects of computing the global mean ocean mass time series.

Although the mean ocean mass signal is spatially correlated over long distances, it is quite small compared to land hydrology variations. Chambers (2009), using a simulated data set for hydrology variations and ice mass loss, found that the leakage of these two larger signals could artificially bias the ocean mass calculation when using an averaging kernel method. However, if one masked the ocean data within 300 km of land, the leakage was minimal. Although this introduces a small error from loss when observing coastal regions, Chambers (2009) showed this was much smaller than the error introduced from leaked hydrology. Johnson and Chambers (2013), furthermore, showed that the effect of the smoothing and de-striping algorithms used to produce mapped bottom pressure also aliases hydrology and cryosphere variations into the ocean, causing significant errors in the mean ocean mass; but this was minimal if an averaging kernel was used with unsmoothed/unfiltered coefficients. Gridded data produced with the mascon method do not have this issue, so they can be averaged over the ocean to the coastline to compute time-variable mean ocean mass (Watkins et al. 2015).

Although regional ocean mass variations are much smaller than those in total sea level (Figure 4.14), they are not completely insignificant. Moreover, with the more-than-14-year record from GRACE now available, it has become increasingly clear that there are changes in ocean mass redistribution with periods from 1 to 5 years that explain a significant fraction of total sea level in many areas of the ocean (e.g., Chambers 2011; Piecuch et al. 2013; Ponte and Piecuch 2014). Most of these locations are poleward of 45° (Figure 4.16), although one notable exception is in the eastern Mediterranean Sea (e.g., Calafat et al. 2012; Fenoglio-Marc 2012; Fukumori et al. 2007; Tsimplis et al. 2013). The mass variations explain anywhere from 50% to 75% of the interannual sea level fluctuations in these areas. Studies have linked these fluctuations to low-frequency changes in local wind-stress curl in the North Pacific (Chambers 2011; Cheng et al. 2013) and to both wind-stress curl and local topographic effects in the south Indian Ocean (Ponte and Piecuch 2014). Although studies have not verified the exact forcing mechanisms in the other regions, low-frequency wind variations are the most likely reason.

There are also decadal-scale changes in ocean mass redistribution, reflected as trends over the GRACE period (Figure 4.17). Although the trends around coastlines are suspect due to leakage of hydrology, the trends in the North Pacific, Southern Ocean, and Arctic are large and, more importantly, have been confirmed with other observations. The trend in the North Pacific has been validated with steric-corrected altimetry (Chambers 2011; Chambers and Willis 2008) and has been linked to decadal changes in the zonal wind stress north and south of 40°N (Chambers 2011; Cheng et al. 2013). These wind stress changes are most likely related to the Pacific Decadal Oscillation. The Arctic mass trend has been validated via a combination of ICESat altimetry, hydrographic data, and bottom pressure recorders (Armitage et al. 2016; Morison et al. 2007, 2012). It has been linked to a change in the circulation due to a change in the Arctic Oscillation (Armitage et al. 2016; Morison et al. 2012; Peralta-Ferriz et al. 2014).

The mass trends in the Southern Ocean (Figure 4.17) are interesting in that they are increasing in the South Atlantic and South Indian Ocean, but they are decreasing in the South Pacific. Johnson and Chambers (2013) argued that the trends in the Indo-Atlantic Oceans were unlikely to be caused

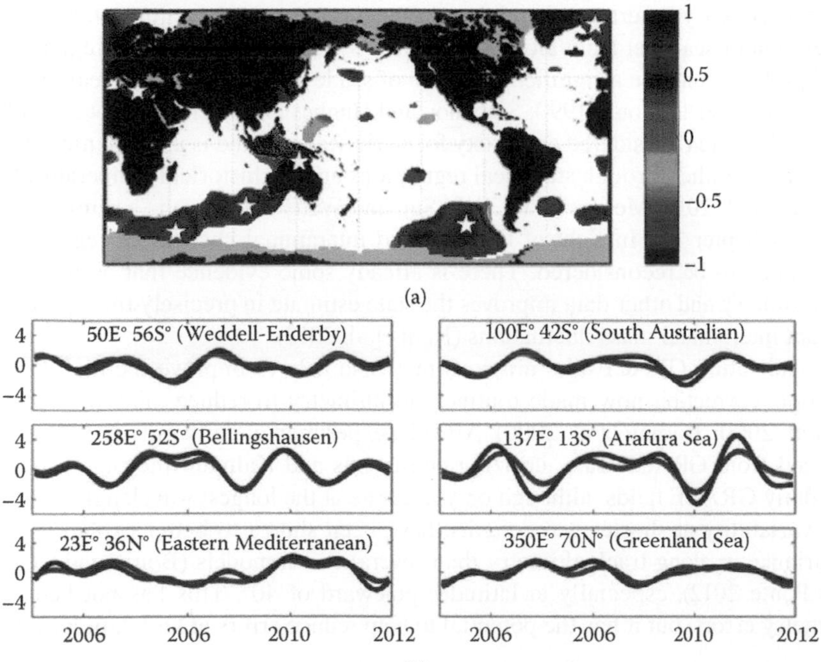

FIGURE 4.16 (a) Fractional variance of low-frequency (detrended) sea level explained by bottom pressure and (b) low-frequency sea level and ocean bottom pressure (in centimeters of water thickness) at locations of stars indicated in (a). (Courtesy of C. Piecuch. Updated from Piecuch, C. G., et al., *Geophys. Res. Lett.*, 40, 2013.)

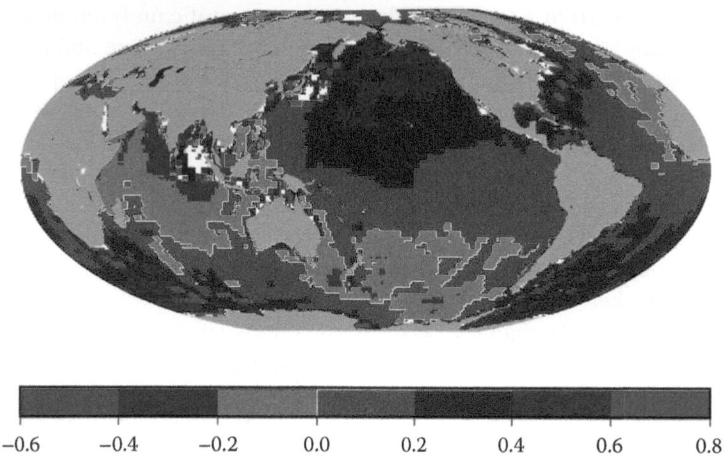

FIGURE 4.17 Ocean bottom pressure trends, January 2003 to December 2015, computed from Jet Propulsion Lab mascons, version 2.

by leakage of ice melting from Antarctica or systematic errors in GRACE but felt the trends in the South Pacific could result from leakage. Subsequently, however, Purkey et al. (2014) have found evidence of similar decadal trends in ocean mass over the South Pacific area based on steric-corrected altimetry, where the steric estimate is based on full depth hydrography. Makowski et al. (2015) have also found this level of mass trend in a high-resolution state estimate and showed that zonal wind stress in the South Pacific has decreased considerably, which is consistent with the bottom pressure trends.

These low-frequency (interannual and decadal) mass variations have important implications for the interpretation of sea level from altimetry. Previously, it was assumed that regional mass variations would probably only be a significant portion of sea level at periods of 1 year or less, based on analysis of ocean models (Ponte 1990; Stepanov and Hughes 2006; Zlotnicki et al. 2007). Because of this, sea level is often considered as a proxy for steric sea level and is used to infer the sub-surface density structure—either through statistical regressions against historical temperature/salinity profiles (Guinehut et al. 2012; Meijers et al. 2010; Sun and Watts 2001) or via assimilation into a state estimate (e.g., Chapter 18). In light of the observed interannual changes in regional mass, these assumptions need to be reconsidered. There is already some evidence that assimilating GRACE along with altimetry and other data improves the state estimate in precisely the regions where there are significant interannual mass fluctuations (Köhl et al. 2012).

GRACE and future GRACE-like missions may also help to improve the high-frequency non-tidal barotropic correction now made routinely to altimetry to reduce aliased errors (Chapter 1; Stammer et al. 2000; Tierney et al. 2000). Already, experimental daily gravity field estimates have been produced from GRACE data, utilizing constraints and Kalman filtering (Kurtenbach et al. 2012). The daily GRACE fields, although only accurate at the longest wavelengths and constrained to have no variability at shorter wavelengths, have been shown to better explain high-frequency sea level variance in along-track altimetry than several ocean models (Bonin and Chambers 2011; Quinn and Ponte 2012), especially at latitudes poleward of 40°. This has not been exploited to reduce altimetry errors, but it has the potential to help reduce errors at high latitudes in altimetry.

4.6 CONCLUSIONS AND FUTURE PROSPECTS

In preparation for future altimetric missions such as the Surface Water and Ocean Topography (SWOT) Mission (Chapter 3), an MSS with higher resolution and accuracy than currently available will be required. This will require either a geodetic mission (GM) with systematic ground-track spacing better than what is available today or many years of SWOT data. Today, the ground-track spacing is 8 km for all GM mission data (ERS-1, Jason-1 end-of-life, and CryoSat-2), which corresponds to roughly a 1-year geodetic mission. Multi multiple-year geodetic missions would be required to bring the cross-track spacing down to 4 km (2 years) or 2 km (4 years). Decreasing the cross-track sampling will also require increasing the standard along-track sampling from 1 Hz (6–7 km) to 2 Hz (3 km) or even 4 Hz (1.5 km).

The future of satellite gravimetry lies in improving both the spatial and temporal resolution of the gravity field estimates as well as their accuracy. The spatial resolution is fundamentally limited by the satellite's altitude, and realistically it will be difficult to ever achieve spatial resolutions better than approximately 100 km with satellite data alone. Such improved spatial resolution, however, could, for instance, enable the better detection of ocean bottom currents, especially those for the fronts of the ACC. The temporal resolution of the gravity field is limited by how quickly it is sampled. Thus, increasing the number of satellites is the only way to improve the temporal resolution. The accuracy of the gravity fields can be primarily improved by reducing the magnitude of the correlated error in the gravity solution, as has been demonstrated with the mascon solutions described earlier.

GRACE Follow-On (GRACE-FO) is expected to demonstrate improved ranging precision with a laser interferometer instrument, as well as slight improvements in the precision of the on-board accelerometer, which is used to measure nonconservative forces acting on the spacecraft. GOCE has already demonstrated the ability to measure the nonconservative forces at a level that exceeds what will be done with GRACE-FO. Accounting for these technological improvements, it is now expected that the largest source of error for future missions will come from mismodeling the high-frequency mass variations (i.e., errors in ocean tide models and models of high-frequency atmosphere and ocean non-tidal mass variations, as well as unmodeled high-frequency hydrology signals) during the gravity processing (Wiese et al. 2011a) (Figure 4.18). The only way to reduce

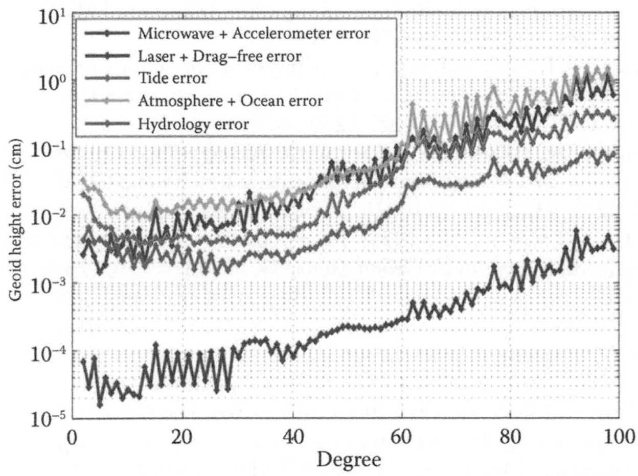

FIGURE 4.18 Error in determining the height of the geoid versus spherical harmonic degree of the gravity field showing individual sources of error for a satellite gravimetry mission consisting of a single pair of polar orbiting satellites at 500-km altitude. The plot shows: (1) Instrument errors from a microwave ranging instrument with accelerometers similar to GRACE (blue); (2) instrument errors from a hypothetical laser-ranging instrument (similar to the one on GRACE Follow-On) and a GOCE drag-free system (red); (3) error introduced from mismodeling ocean tides (green); (4) error introduced from mismodeling high-frequency atmospheric and oceanic mass variations (cyan); and (5) error introduced from undersampling continental hydrology mass variations (magenta).

these errors is to either: (1) improve the models or (2) sample the gravity field more frequently such that these high-frequency mass variations are more directly observed, such as has already been demonstrated for long-wavelength high-frequency ocean signals (Bonin and Chambers 2011; Quinn and Ponte 2012).

To put it simply, we have now come to a point where the inability to sample the gravity field frequently enough is expected to be the largest source of error for future missions, limiting both the spatial resolution and accuracy of the derived gravity fields. Although a constellation of satellites is an extremely attractive option for increasing the temporal sampling of the gravity field, this option is currently considered to be cost-prohibitive. A more likely scenario would involve a mission architecture consisting of two pairs of satellites (Bender et al. 2008; Wiese et al. 2012), with one pair in a polar orbit and the second pair in a moderately inclined orbit. Such an architecture has been shown to increase both the temporal and spatial resolution while also improving the accuracy of the estimated gravity fields with respect to an architecture consisting of a single pair of satellites. It has been shown that such a pair would enable the estimation of daily gravity fields at spatial scales of 1000 km (Wiese et al. 2011a), which could potentially aid in reducing high-frequency ocean aliasing errors in satellite altimetry, while also reducing errors in recovering ocean bottom pressure signals by up to 80% (Wiese et al. 2011b).

REFERENCES

Albertella, A., and R. Rummel. (2009). On the spectral consistency of the altimetric ocean and geoid surface: A one-dimensional example. *J. Geod.*, *83*, 805–815. doi: 10.1007/s00190-008-0299-5.

Andersen, O.B., and P. Knudsen. (1998). Global marine gravity field from the ERS-1 and Geosat geodetic mission altimetry. *J. Geophys. Res.*, *103*, 8129–8137.

Andersen, O.B., and Knudsen, P. (2009). DNSC08 mean sea surface and mean dynamic topography models. *J. Geophys. Res.*, *114*, C11001. doi: 10.1029/2008JC005179.

Andersen, O.B., P. Knudsen, and P.A.M. Berry. (2010). The DNSC08GRA global marine gravity field from double retracked satellite altimetry. *J. Geod.*, *84*, 191–199. doi: 10.1007/s00190-009-0355-9.

Andersen, O.B., P. Knudsen, and L. Stenseng. (2015). *The DTU13 MSS (Mean Sea Surface) and MDT (Mean Dynamic Topography) from 20 Years of Satellite Altimetry*. IAG Symposia Series, Springer Verlag. Switzerland. doi: 10.1007/1345_2015_182.

Andersen, O.B., and Scharroo, R. (2011). Range and geophysical corrections in coastal regions: And implications for mean sea surface determination. In: Vignudelli, S., et al. (eds.), *Coastal Altimetry*, pp. 103–145. Springer, New York. doi: 10.1007/978-3-642-12796-0_5.

Armitage, T.W.K., S. Bacon, A.L. Ridout, S.F. Thomas, Y. Aksenov, and D.J. Wingham. (2016). Arctic sea surface height variability and change from satellite radar altimetry and GRACE, 2003–2014. *J. Geophys. Res. Oceans, 121*, 4303–4322. doi: 10.1002/ 2015JC011579.

Balmino, G., Reigber, C., and Moynot, B. (1978). Le modèle de potentiel gravitationnel terrestre GRIM2: Détermination, evaluation. *Annales de Géophysique, 34*, 55–78.

Becker, S., J.M. Brockmann, and W.-D. Schuh. (2014). Mean dynamic topography estimates purely based on GOCE gravity models and altimetry. *Geophys. Res. Lett.*, **41**, 1–7. doi: 10.1002/2014GL059510.

Becker, S., G. Freiwald, M. Losch, and W.-D. Schuh. (2012). Rigorous fusion of gravity field into stationary ocean models. *J. Geodyn.*, 59–60, 99–110. doi: 10.1016/j.jog.2011.07.006.

Bender, P. L., D. N. Wiese, and R. S. Nerem. (2008). A possible dual-GRACE mission with 90 degree and 63 degree inclination orbits. In: *Proceedings of the Third International Symposium on Formation Flying, Missions and Technologies*, 23–35 April 2008, ESA/ESTEC, Noordwijk, The Netherlands, 1–6.

Bettadpur, S. (2008). Orbital mechanics, perturbations and GRACE science and mission design. In: AAS Paper 08-179, AAS Spaceflight Mechanics Conference, Galveston, TX, January 29. pp. 1219–1233.

Bettadpur, S. (2012a). *Level-2 Gravity Field Product User Handbook*. GRACE 327-734, CSR Publ. CSR-GR-03-01, Rev 3.0. University of Texas, Austin, TX, 19 p, May 29.

Bettadpur, S. (2012b). *UTCSR Level-2 Processing Standards Document*. GRACE 327-742, CSR Pub. CSR-GR-12-xx, Rev 4.0 (For Level-2 Product Release 0005). University of Texas, Austin, TX, 17 p, May 29.

Biancale, R., Balmino, G., Lemoine, J.-M., Marty, J.-C., Moynot, B., Barlier, F., Exertier, P., et al. (2000). A new global earth's gravity field model from satellite orbit perturbations: GRIM5-S1. *Geophys. Res. Lett., 27*(22), 3611–3614.

Bingham, R.J. (2010). Nonlinear anisotropic diffusive filtering applied to the ocean's mean dynamic topography. *Rem. Sens. Lett., 1*(4), 205–212. doi: 10.1080/01431161003743165.

Bingham, R.J., K. Haines, and C. Hughes. (2008). Calculating the ocean's mean dynamic topography from a mean sea surface and a geoid. *J. Atmos. Oceanic Tech., 25*, 1808–1822.

Bingham, R.J., K. Haines, and D.J. Lea. (2014). How well can we measure the ocean's mean dynamic topography from space? *J. Geophys. Res. Oceans, 119*, 3336–3356. doi: 10.1002/2013JC009354.

Bingham, R.J., and C.W. Hughes. (2012). Local diagnostics to estimate density-induced sea level variations over topography and along coastlines. *J. Geophys. Res., 117*, C01013. doi: 10.1029/2011JC007276.

Bingham, R.J., P. Knudsen, O. Andersen, and R. Pail. (2011). An initial estimate of the North Atlantic steady-state geostrophic circulation from GOCE. *Geophys. Res. Lett., 38*, L01606. doi: 10.1029/2010GL045633

Boening, C., J.K. Willis, F.W. Landerer, R.S. Nerem, and J. Fasullo. (2012). The 2011 La Niña: So strong, the oceans fell. *Geophys. Res. Lett., 39*, L19602. doi: 10.1029/2012GL053055.

Bonin, J., and D.P. Chambers. (2011). Evaluation of high-frequency oceanographic signal in GRACE data: Implications for de-aliasing. *Geophys. Res. Lett., 38*, L17608. doi: 10.1029/2011GL048881.

Bretherton, F.P., R.E. Davis, and C.B. Fandry. (1976). A technique for objective analysis and design of oceanographic experiments applied to MODE-73. *Deep Sea Res., 23*, 559–582.

Brockmann, J.M., N. Zehentner, E. Höck, R. Pail, I. Loth, T. Mayer-Gürr, and W.-D Schuh. (2014). EGM TIM RL05: An independent geoid with centimeter accuracy purely based on the GOCE mission. *Geophy. Res. Lett.* Vol. 41, pp. 8089–8099, doi: 10.1002/2014GL061904.

Bruinsma, S.L., C. Föerste, O. Abrikosov, J.-M. Lemoine, J.-C. Marty, S. Mulet, M.-H. Rio, et al. (2014). ESA's satellite-only gravity field model via the direct approach based on all GOCE data. *Geophys. Res. Lett.* Vol. 41, pp. 7508–7514, doi: 10.1002/2014GL062045.

Bruinsma, S.L., C. Foerste, O. Abrikosov, J.C. Marty, M.H. Rio, S. Mulet, and S. Bonvalot. (2013). The new ESA satellite-only gravity field model via the direct approach. *Geophys. Res. Lett., 40*, 3607–3612. doi: 10.1002/grl.50716.

Calafat, F.M., D.P. Chambers, and M.N. Tsimplis. (2012). Mechanisms of decadal sea level variability in the Eastern North Atlantic and the Mediterranean Sea. *J. Geophys. Res. 117*, C09022, doi: 10.1029/2012JC008285.

Chambers, D.P. (2006a). Observing seasonal steric sea level variations with GRACE and satellite altimetry. *J. Geophys. Res.*, 111(C3), C03010. doi: 10.1029/2005JC002914.

Chambers, D.P. (2006b). Evaluation of new GRACE time-variable gravity data over the ocean. *Geophys. Res. Lett., 33*, L17603. doi: 10.1029/2006GL027296.

Chambers, D.P. (2009). Calculating trends from GRACE in the presence of large changes in continental ice storage and ocean mass. *Geophys. J. Int., 176*, 415–419. doi: 10.1111/j.1365-246X.2008.04012.x.

Chambers, D.P. (2011). ENSO-correlated fluctuations in ocean bottom pressure and wind-stress curl in the North Pacific. *Ocean Sci., 7*, 685–692. doi: 10.5194/os-7-685-2011.

Chambers, D.P., and J.A. Bonin. (2012). Evaluation of release-05 GRACE time-variable gravity coefficients over the ocean. *Ocean Sci., 8*, 1–10. doi: 10.5194/os-8-1-2012.

Chambers, D.P., A. Cazenave, N. Champollion, H. Dieng, W. Llovel, R. Forsberg, K. von Schuckmann, et al. (2017). Evaluation of the global mean sea level budget between 1993 and 2014. *Surv. Geophys.* Vol 38, pp. 309–327, doi: 10.1007/s10712-016-9381-3.

Chambers, D.P., J.L. Chen, R.S. Nerem, and B.D. Tapley. (2000). Interannual sea level change and the earth's water mass budget. *Geophys. Res. Lett., 27*, 3073–3076.

Chambers, D.P., J. Wahr, and R.S. Nerem. (2004). Preliminary observations of global ocean mass variations with GRACE. *Geophys. Res. Lett., 31*, L13310. doi: 10.1029/2004GL020461.

Chambers, D.P., and J.K. Willis. (2008). Analysis of large-scale ocean bottom pressure variability in the North Pacific. *J. Geophys. Res., 113*, C11003. doi: 10.1029/2008JC004930.

Chelton, D.B., R.A. de Szoeke, M.G. Schlax, K. El Naggar, and N. Siwertz. (1998). Geographical variability of the first-baroclinic Rossby radius of deformation, *J. Phys. Oceanogr., 28*, 433–460.

Chen, J.L., C.R. Wilson, D.P. Chambers, R.S. Nerem, and B.D. Tapley. (1998). Seasonal global water mass balance and mean sea level variations. *Geophys. Res. Lett., 25*, 3555–3558.

Cheng, M.K. (2002). Gravitational perturbation theory for intersatellite tracking. *J. Geodesy., 76*, 169–185. doi: 10.1007/s00190-001-0233-6.

Cheng, X., L. Li, Y. Du, J. Wang, and R.-X. Huang. (2013). Mass-induced sea level change in the northwestern North Pacific and its contribution to total sea level change. *Geophys. Res. Lett., 40*, 3975–3980, doi: 10.1002/grl.50748.

Davis, J.L., M.E. Tamisiea, P. Elósegui, J.X. Mitrovica, and E.M. Hill. (2008). A statistical filtering approach for Gravity Recovery and Climate Experiment (GRACE) gravity data. *J. Geophys. Res.*, *113*, B04410. doi: 10.1029/2007JB005043.

Dunn, C.E., W. Bertiger, Y. Bar-Sever, S. Desai, B. Haines, D. Kuang, G. Franklin, et al. (2003). Instrument of GRACE: GPS augments gravity measurements. *GPS World*, *14*(2), 16.

Fasullo, J.T., C. Boening, F.W. Landerer, and R.S. Nerem. (2013). Australia's unique influence on global sea level in 2010–2011. *Geophys. Res. Lett.*, *40*, 4368–4373. doi: 10.1002/grl.50834.

Fenoglio-Marc, L., M. Becker, R. Rietbroeck, J. Kusche, S. Grayek, and E. Stanev. (2012). Water mass variation in Mediterranean and Black Sea. *J. Geodyn.* Vol 59–60, pp. 168–182, doi: 10.1016/j.jog.2012.04.001.

Flechtner, F., H. Dobslaw, and E. Fagiolini. (2015). *AOD1B Product Description Document for Product Release 05.* GRACE 327-750, GFZ Publ. GR-GFZ-AOD-0001 Rev 4.3. GFZ German Research Centre for Geosciences, 34 p. Potsdam, Germany.

Forsberg, R., A.V. Olesen, F. Ferraccioli, T. Jordan, K. Matsuoka, and J. Kohler. (2016). *GOCE polar gaps now filled—First gravity results of the ESA PolarGap project.* In: ESA Living Planet Symposium, SP. 9–13 May 2016, Prague, Czech Republic.

Fukumori, I. (2002). A partitioned Kalman filter and smoother. *Mon. Weather Rev.*, *130*, 1370–1383.

Fukumori, I., D. Menemenlis, and T. Lee. (2007). A near-uniform basin-wide sea level fluctuation of the Mediterranean Sea. *J. Phys. Ocean.*, *37*, 338–358. doi: 10.1175/JPO3016.1.

Gill, A.E. (1982). *Atmosphere-Ocean Dynamics*. Academic Press, London, 662 pp.

Gruber, T., O. Abrikosov, and U. Hugentobler. (2014b). *GOCE Standards.* GO-TN-HPF-GS-0111, Issue 4. The European GOCE Gravity Consortium.

Gruber, T., R. Rummel, O. Abrikosov, and R. van Hees. (2014a). *GOCE Level-2 Data Handbook.* EGG-C, GO-MA-HPF-GS-0110, Issue 5. The European GOCE Gravity Consortium.

Guinehut, S., A.-L. Dhomps, G. Larnicol, and P.-Y. Le Traon. (2012). High resolution 3D temperature and salinity fields derived from in situ and satellite observations. *Ocean Sci.*, *8*, 845–857. doi: 10.5194/os-8-845-2012.

Haines, K., J.A. Johannessen, P. Knudsen, D. Lea, M.-H. Rio, L. Bertino, F. Davidson, and F. Hernandez. (2011). An ocean modelling and assimilation guide to using GOCE geoid products. *Ocean Sci.*, *7*, 151–164. doi: 10.5194/os-7-151-2011.

Heiskanen, W.A., and H. Moritz. (1967). *Physical Geodesy*. W. H. Freeman and Co., San Francisco, CA.

Hughes, C.W., and R.J. Bingham. (2008). An oceanographer's guide to GOCE and the geoid. *Ocean Sci.*, *4*, 15–29.

Hughes, C.W., Woodworth, P.L., Meredith, M.P., Stepanov, V., Whitworth, T., and Pyne, A.R. (2003). Coherence of Antarctic sea levels, Southern Hemisphere Annular Mode, and flow through Drake Passage. *Geophys. Res. Lett.*, *30*(9), 1464. doi: 10.1029/2003GL017240.

Hunegnaw, A., F. Siegismund, R. Hipkin, and K.A. Mork. (2009). Absolute flow field estimation for the Nordic seas from combined gravimetric, altimetric, and in situ data. *J. Geophys. Res.*, *114*, C02022. doi: 10.1029/2008JC004797.

Jayne, S.R. (2006). Circulation of the North Atlantic Ocean from altimetry and the Gravity Recovery and Climate Experiment geoid. *J. Geophys. Res.*, *111*, C03005. doi: 10.1029/2005JC003128.

Jekeli, C. (1981). *Alternative Methods to Smooth the Earth's Gravity Field.* Rep. 327, Dep. of Geod. Sci. and Surv., Ohio State University, Columbus, OH.

Johnson, G.F., and D.P. Chambers. (2013). Ocean bottom pressure seasonal cycles and decadal trends from GRACE release-05: Ocean circulation implications. *J. Geophys. Res. Oceans, 118*, doi: 10.1002/jgrc.20307.

Kanzow, T., Flechtner, F., Chave, A., Schmidt, R., Schwintzer, P., and Send, U. (2005). Seasonal variation of ocean bottom pressure derived from Gravity Recovery and Climate Experiment (GRACE): Local validation and global patterns. *J. Geophys. Res.*, *110*, C09001. doi: 10.1029/2004JC002772.

Kaula, W.M. (1966). *Theory of Satellite Geodesy*. Blaisdell Publishing Co., Waltham, MA. Republished by Dover, New York City, NY in 2000.

Kim, S.-B., T. Lee, and I. Fukumori. (2007). Mechanisms controlling the interannual variation of mixed layer temperature averaged over the Niño-3 region. *J. Clim.*, *20*(15), 3822–3843. doi: 10.1175/JCLI4206.1.

Kenyon, S., and R. Forsberg. (2008). New gravity field for the Arctic. *EOS,* Transactions American Geophysical Union, *89*(32), 289–290.

Knudsen, P. (1993). Satellite altimetry for geodesy and geophysics. In: Kakkuri, J. (ed.), *Lecture Notes for NKG Autumn School.* Korpilampi, Finland, pp. 87–126.

Knudsen, P., R. Bingham, O. Andersen, and M.-H. Rio. (2011). A global mean dynamic topography and ocean circulation estimation using a preliminary GOCE gravity model. *J. Geod.*, *85*, 861–879. doi: 10.1007/s00190-011-0485-8.

Köhl, A., F. Siegismund, and D. Stammer. (2012). Impact of assimilating bottom pressure anomalies from GRACE on ocean circulation estimates. *J. Geophys. Res., 117*, C04032. doi: 10.1029/2011JC007623.

Kurtenbach, E., A. Eicker, T. Mayer-Gürr, M. Holschneider, M. Hayn, M. Fuhrmann, and J. Kusche. (2012). Improved daily GRACE gravity field solutions using a Kalman smoother. *J. Geodyn., 59*, 39–48.

Kusche, J. (2007). Approximate decorrelation and non-isotropic smoothing of time-variable GRACE-type gravity field models. *J. Geodesy, 81*, 733–749. doi: 10.1007/s00190-007-0143-3.

Landerer, F.W., and S.C. Swenson. (2012). Accuracy of scaled GRACE terrestrial water storage estimates. *Water Resour. Res., 48*, W04531. doi: 10.1029/2011WR011453.

Lemoine, F.G., S.B. Luthcke, D.D. Rowlands, D.S. Chinn, S.M. Klosko, and C.M. Cox. (2007). The use of mascons to resolve time-variable gravity from GRACE. In: Tregoning, P. and Rizos, C. (eds.), *Dynamic Planet: Monitoring and Understanding a Dynamic Planet with Geodetic and Oceanographic Tools, Intl. Assoc. of Geodesy Symposia, Vol. 130*, Springer, Berlin, pp. 231–236.

Le Provost, C. and M. Brémond. (2003). Resolution needed for an adequate determination of the mean ocean circulation from altimetry and an improved geoid. *Space Sci. Rev., 108*, 163–178.

Leuliette E.W., and L. Miller. (2009). Closing the sea level rise budget with altimetry, Argo, and GRACE. *Geophys. Res. Lett., 36*, L04608. doi: 10.1029/2008GL036010.

Lerch, F., R. Nerem, D. Chinn, J. Chan, G. Patel, and S. Klosko. (1993). New error calibration tests for gravity models using subset solutions with independent data: Applied to GEM-T3. *Geophys. Res. Lett., 20*, 249–252.

Lerch, F.J., B.H. Putney, C.A. Wagner, and S.M. Klosko. (1981). Goddard earth models for oceanographic applications (GEM-10B and GEM-10C). *Mar. Geod., 5*, 145–187.

Makowski, J.K., D.P. Chambers, and J.A. Bonin. (2015). Using Ocean Bottom Pressure from the Gravity Recovery and Climate Experiment (GRACE) to estimate transport variability in the Southern Indian Ocean. *J. Geophys. Res. Oceans,* 120. 4245–4259, doi: 10.1002/2014JC010575.

Marsh, J.G., F.J. Lerch, B.H. Putney, T.L. Felsentreger, B.V. Sanchez, S.M. Klosko, G.B. Patel, et al. (1990). *J. Geophys. Res. Solid Earth, 95*, 22043–22071.

Marsh, J.G., and S. Vincent. (1974). Global detailed geoid computation and model analysis. *Geophys. Surv.,* 1–4, 481–511.

Maximenko, N.P., P. Niiler, M.-H. Rio, O. Melnichenko, L. Centurioni, D. Chambers, V. Zlotnicki, et al. (2009). Mean dynamic topography of the ocean derived from satellite and drifting buoy data using three different techniques. *J. Atmos. Oceanic Tech., 26*(9), 1910–1919.

Mayer-Gürr, T., Kurtenbach, E., and Eicker, A. (2010). *ITG-Grace2010 Gravity Field Model.* http://www.igg.uni-bonn.de/apmg/index.php?id=itg-grace2010

Meijers, A.J.S., N.L. Bindoff, and S.R. Rintoul. (2010). Estimating the four-dimensional structure of the Southern Ocean using satellite altimetry. *J. Atmos. Oceanic Technol., 28*, 548–568. doi: 10.1175/2010JTECHO790.1.

Minster, J.F., A. Cazenave, Y.V. Serafini, F. Mercier, M.C. Gennero, and P. Rogel. (1999). Annual cycle in mean sea level from Topex-Poseidon and ERS-1: Inference on the global hydrological cycle. *Global and Planetary Change 20*, 57–66.

Morison, J., R. Kwok, C. Peralta-Ferriz, M. Alkire, I. Rigor, R. Andersen, and M. Steele. (2012). Changing Arctic Ocean freshwater pathways. *Nature, 481*, 66–70.

Morison, J., J. Wahr, R. Kwok, and C. Peralta-Ferriz. (2007). Recent trends in Arctic Ocean mass distribution revealed by GRACE. *Geophys. Res. Lett., 34*, L07602. doi: 10.1029/2006GL029016.

Mulet, S., M.-H. Rio, and S. Bruinsma. (2012). Accuracy of the preliminary GOCE GEOID models from an oceanographic perspective. *Marine Geodesy, 35*, 314–336. doi: 10.1080/01490419.2012.718230.

Muller, P.M., and W.L. Sjogren. (1968). Mascons: Lunar mass concentrations. *Science, 161*, 680–684.

NASA. (1987). *Geophysical and Geodetic Requirements for Global Gravity Field Measurements, 1987–2000.* Report of a Gravity Workshop, Colorado Springs, CO. NASA Geodynamics Branch, Division of Earth Science and Applications, Washington, DC, 45 pp.

Niiler, P.P., N.A. Maximenko, and J.C. McWilliams. (2003). Dynamically balanced absolute sea level of the global ocean derived from near-surface velocity observations. *Geophys. Res. Lett., 30*(22), 2164. doi: 10.1029/2003GRL018628.

Ngo-Duc, T., K. Laval, J. Polcher, A. Lombard, and A. Cazenave. (2005). Effects of land water storage on global mean sea level over the past half century. *Geophys. Res. Lett., 32*, L09704. doi: 10.1029/2005GL022719.

Nerem, R.S., D.P. Chambers, E. Leuliette, G.T. Mitchum, and B.S. Giese. (1999). Variations in global mean sea level during the 1997–98 ENSO event. *Geophys. Res. Lett., 26*, 3005–3008.

Nerem, R.S., F.J. Lerch, J.A. Marshall, E.C. Pavlis, B.H. Putney, B.D. Tapley, R.J. Eanes, et al. (1994a). Gravity model development for TOPEX/POSEIDON: Joint Gravity Model-1 and 2. *J. Geophys. Res., 99*, 24421–24447.

Nerem, R.S., E.J. Schrama, C.J. Koblinsky, and B.D. Beckley. (1994b). A preliminary evaluation of ocean topography from the TOPEX/POSEIDON mission. *J. Geophys. Res.*, *99*, 24565–24583.

NRC. (1979). *Applications of a Dedicated Gravitational Satellite Mission*. National Academy Press, Washington, DC, 53 pp.

NRC. (1982). *A Strategy for Earth Science from Space in the 1980s: Part 1: Solid Earth and Oceans*. National Academy Press, Washington, DC, 99 pp.

NRC. (1997). *Satellite Gravity and the Geosphere*. National Academy Press, Washington, DC, 112 pp.

Pail, R., H. Goiginger, W.-D. Schuh, E. Höck, J.M. Brockmann, T. Fecher, T. Gruber, et al. (2010). Combined satellite gravity field model GOCO01S derived from GOCE and GRACE. *Geophys. Res Lett.*, *37*, L20314. doi: 10.1029/2010GL044906.

Pavlis, N.K., S.A. Holmes, S.C. Kenyon, and J.K. Factor. (2012). The development and evaluation of the Earth Gravitational Model 2008 (EGM2008). *J. Geophys. Res.*, *117*, B04406. doi: 10.1029/2011JB008916.

Pavlis, N.K., and R.H. Rapp. (1990). The development of an isostatic gravitational model to degree 360 and its use in global gravity modeling. *Geophys. J. Int.*, *100*, 369–378, doi: 10.1111/j.1365-246X.1990.tb00691.x.

Peralta-Ferriz, C., J.H. Morison, J.M. Wallace, J.A. Bonin, and J. Zhang. (2014). Arctic Ocean circulation patterns revealed by GRACE. *J. Clim.*, *27*, 1445–1468. doi: 10.1175/JCLI-D-13-00013.1.

Petit, G., and B. Luzum. (2010). *IERS Conventions*. IERS Technical Note 36. Bundesamt fur Kartographie und Geodasie, Frankfurt, Germany.

Piecuch, C.G., K.J. Quinn, and R.M. Ponte. (2013). Satellite-derived interannual ocean bottom pressure variability and its relation to sea level. *Geophys. Res. Lett.*, 40. 3106–3110, doi: 10.1002/grl.50549.

Pisacane, V.L., J.C. Ray, J.L. MacArthur, and S.E. Bergeson-Willis. (1982). Description of the dedicated gravitational satellite mission (GRAVSAT). *IEEE Trans. Geosci. Rem. Sens.*, *GE*-20(3), 315–321.

Pond, S., and G. L. Pickard. (1983). *Introductory Dynamical Oceanography*. 2nd edition. Pergammon Press, Oxford, England, 329 pp.

Ponte, R. M. (1990), Barotropic motions and the exchange of angular momentum between the oceans and solid Earth, J. Geophys. Res., *95*, 11,369–11,374.

Ponte, R.M. (2006). Oceanic response to surface loading effects neglected in volume conserving models. *J. Phys. Oceanogr.*, *36*, 426–434. doi: 10.1175/JPO2843.1.

Ponte, R.M., and C.G. Piechuch. (2014). Interannual bottom pressure signals in the Australian–Antarctic and Bellingshausen Basins. *J. Phys. Ocean.*, *44*, 1456–1465. doi: 10.1175/JPO-D-13-0223.1.

Purkey, S.G., G.C. Johnson, and D.P. Chambers. (2014). Relative contributions of ocean mass and deep steric changes to sea level rise between 1993 and 2013. *J. Geophys. Res. Oceans*, *119*, 7509–7522, doi: 10.1002/2014JC010180.

Quinn, K.J., and R.M. Ponte. (2012). High frequency barotropic ocean variability observed by GRACE and satellite altimetry. *Geophys. Res. Lett.*, *39*, L07603. doi: 10.1029/2012GL051301.

Rapp, R.H. (1989). The treatment of permanent tidal effects in the analysis of satellite altimeter data for sea surface topography. *Manuscripta Geodaetica*, *14*, 368–372.

Reigber, C., H. Jochmann, J. Wünsch, S. Petrovic, P. Schwintzer, F. Barthelmes, K.H. Neumayer, et al. (2004). Earth gravity field and seasonal variability from CHAMP. In: Reigber, C., Lühr, H., Schwintzer, P., and Wickert, J. (eds.), *Earth Observation with CHAMP—Results from Three Years in Orbit*. Springer, Berlin, pp. 25–30.

Reigber, C., H. Lühr, and P. Schwintzer. (2002). CHAMP mission status. *Adv. Space Res.*, *30*, 129–134.

Reigber, C., P. Schwintzer, P. Hartl, K.H. Ilk, R. Rummel, M. van Gelderen, E.J.O. Schrama, et al. (1987). *Study of a Satellite-to-Satellite Tracking Gravity Mission: Final Contract Report to the European Space Agency*. ESTEC/Contract No. 6557/85/NL/PP(SC). ESA.

Rio, M.H., S. Guinehut, and G. Larnicol. (2011). New CNES-CLS09 global mean dynamic topography computed from the combination of GRACE data, altimetry, and in situ measurements. *J. Geophys. Res.*, 116, C07018. doi: 10.1029/2010JC006505.

Rio, M.-H., and F. Hernandez. (2004). A mean dynamic topography computed over the world ocean from altimetry, in-situ measurements and a geoid model. *J. Geophys. Res.*, *109*, C12032.

Rio, M.-H., S. Mulet, and N. Picot. (2014b). Beyond GOCE for the ocean circulation estimate: Synergetic use of altimetry, gravimetry, and in situ data provides new insight into geostrophic and Ekman currents. *Geophys. Res. Lett.*, 41. 8918–8925, doi: 10.1002/2014GL061773.

Rio, M.-H., P.-M. Poulain, A. Pascal, E. Mauri, G. Larnicol, and R. Santoleri. (2007). A mean dynamic topography of the Mediterranean Sea computed from altimetric data, in-situ measurements and a general circulation model. *J. Marine Syst.*, *65*, 484–508.

Rio, M.-R., A. Pascual, P.-M. Poulain, M. Menna, B. Barceló, and J. Tintoré. (2014a). Computation of a new mean dynamic topography for the Mediterranean Sea from model outputs, altimeter measurements and oceanographic in situ data. *Ocean Sci., 10,* 731–744. doi: 10.5194/os-10-731-2014.

Robinson, A.R., G.L. Wayne, A. Theocharis, and A. Lascaratos. (2001). Mediterranean Sea circulation. In: *Encyclopedia of Ocean Sciences*, Eds., Steve A. Thorpe and Karl K. Turekian, Academic Press, London, pp. 1689–1705. doi: 10.1006/rwos.2001.0376, 2001.

Rummel, R., and T. Gruber. (2010). Gravity and steady-state ocean circulation explorer. In: Flechtner, F., et al. (eds.) *System Earth via Geodetic-Geophysical Space Techniques*. Springer-Verlag, Berlin, pp. 203–212.

Sandwell, D.T., E. Garcia, K. Soofi, P. Wessel, M. Chandler, and W.H.F. Smith. (2013). *Towards 1-mGal Accuracy in Global Marine Gravity from Cryosat-2, Envisat and Jason-1, The Leading Edge*. SEG, Houston, pp. 892–898.

Schaeffer, P., Y. Faugére, J.F. Legeais, A. Ollivier, T. Guinle, and N. Picot. (2012). The CNES_CLS11 Global Mean Sea surface computed from 16 years of satellite altimeter data. *Marine Geodesy*, 35(Suppl. 1), 3–19. doi: 10.1080/01490419.2012.718231.

Scheinert, M., F. Ferraccioli, J. Schwabe, R. Bell, M. Studinger, D. Damaske, W. Jokat, et al. (2016). New Antarctic gravity anomaly grid for enhanced geodetic and geophysical studies in Antarctica. *Geophys. Res. Lett., 43*, 600–610. doi: 10.1002/2015GL067439.

Shum, C.K. (1982). *Altimeter Methods in Satellite Geodesy*. University of Texas, Austin, TX, 392 pp.

Siegismund, F. (2013). Assessment of optimally filtered recent geodetic mean dynamic topographies. *J. Geophys. Res. Oceans, 118*, 108–117. doi: 10.1029/2012JC008149.

Stammer, D., and C. Wunsch. (1994). Preliminary assessment of the accuracy and precision of the TOPEX/POSEIDON altimeter with respect to the large scale ocean circulation. *J. Geophys. Res., 99*, 24584–24604.

Stammer, D., C. Wunsch, and R.M. Ponte. (2000). De-aliasing of global high frequency barotropic motions in altimeter measurements. *Geophys. Res. Lett., 27*, 1175–1178.

Stepanov, V.N., and C.W. Hughes. (2006). Propagation of signals in basin-scale ocean bottom pressure from a barotropic model. *J. Geophys. Res., 111*, C12002. doi: 10.1029/2005JC003450.

Sun, C., and D.R. Watts. (2001). A circumpolar gravest empirical mode for the Southern Ocean hydrography. *J. Geophys. Res.,* 106(C2), 2833–2855.

Swenson, S.C., and J. Wahr. (2006). Post-processing removal of correlated errors in GRACE data. *Geophys. Res. Lett., 33*, L08402. doi: 10.1029/2005GL025285.

Tapley, B.D., S. Bettadpur, M. Watkins, and C. Riegber. (2004b). The gravity recovery and climate experiment: Mission overview and early results. *Geophys. Res. Lett., 31*, L09607, 1–4.

Tapley, B.D., D.P. Chambers, S. Bettadpur, and J.C. Ries. (2003). Large scale ocean circulation from the GRACE GGM01 geoid. *Geophys. Res. Lett.*, 30, 2163, doi: 10.1029/2003GL018622.

Tapley, B.D., D.P. Chambers, C.K. Shum, R.J. Eanes, J.C. Ries, and R.H. Stewart. (1994). Accuracy assessment of the large-scale dynamic ocean topography from TOPEX/POSEIDON altimetry. *J. Geophys. Res., 99*, 24605–24618.

Tapley, B.D., and M.C. Kim. (2001). Application to geodesy. In: Fu, L.-L., and Cazanave, A. (eds.), *Satellite Altimetry and Earth Science. International Geophysics Series*, Vol. 69. Academic Press, pp. 371–406.

Tapley B.D., B.E. Schutz, and G.H. Born. (2004a). *Statistical Orbit Determination*. Elsevier Academic Press, Burlington, MA, 547 pp.

Tapley, B., J. Ries, S. Bettadpur, D. Chambers, M. Cheng, F. Condi, B. Gunter, et al. (2005). GGM02—An improved earth gravity field model from GRACE. *J. Geod.* 467–478, doi: 10.1007/s00190-005-0480-z.

Tapley, B.D., M.M. Watkins, J.C. Ries, G.W. Davis, R.J. Eanes, S.R. Poole, H.J. Rim, et al. (1996). The JGM-3 geopotential model. *J. Geophys. Res.,* 101(B12), 28029–28049.

Thomas, J. B. (1999) An analysis of gravity-field estimation based on intersatellite dual-1-way biased ranging, JPL Publication 98–15, Jet propulsion Laboratory, Pasadena, CA, 195pp.

Thompson, K. R., J. Huang, M. Ve´ronneau, D. G. Wright, and Y. Lu (2009), Mean surface topography of the northwest Atlantic: Comparison of estimates based on satellite, terrestrial gravity, and oceanographic observations, J. Geophys. Res., 114, C07015, doi:10.1029/2008JC004859.

Tierney, C., J. Wahr, F. Bryan, and V. Zlotniki. (2000). Short-period oceanic circulation: Implications for satellite altimetry. *Geophys. Res. Lett., 27*, 1255–1258.

Touboul, P., B. Foulon, M. Rodrigues, and J.P. Marque. (2004). In orbit nano-g measurements, lessons for the future. *Aerospace Sci. Technol.*, 8, 431–441. doi: 10.1016/j.ast.2004.01.006.

Tsimplis, M.N., F.M. Calafat, M. Marcos, G. Jorda, D. Gomis, L. Fenoglio-Marc, M.V. Struglia, et al. (2013). The effect of the NAO on sea level and on mass changes in the Mediterranean Sea. *J. Geophys. Res. Oceans*, 118, 944–952. doi: 10.1002/jgrc.20078.

Vianna, M.L., V.V. Menezes, and D.P. Chambers. (2007). A high resolution satellite-only GRACE-based mean dynamic topography of the South Atlantic Ocean. *Geophys. Res. Lett., 34*, doi: 10.1029/2007GL031912, L24604.

Wahr, J., M. Molenaar, and F. Bryan. (1998). Time-variability of the earth's gravity field: Hydrological and oceanic effects and their possible detection using GRACE. *J. Geophys. Res., 103*, 32205–30229.

Wahr, J., S. Swenson, V. Zlotnicki, and I. Velicogna. (2004). Time-variable gravity from GRACE: First results. *Geophys. Res. Lett., 31*, L11501. doi: 10.1029/2004GL019779.

Watkins, M.M., D.N. Wiese, D.-N. Yuan, C. Boening, and F.W. Landerer. (2015). Improved methods for observing earth's time variable mass distribution with GRACE. *JGR Solid Earth.* 2648–2671, doi: 10.1002/2014JB011547.

Wiese, D. N., F. W. Landerer, and M. M. Watkins (2016), Quantifying and reducing leakage errors in the JPL RL05M GRA CE mascon solution, Water Resour. Res., *52*, 7490–7502, doi:10.1002/2016WR019344.

Wiese, D.N., R.S. Nerem, and S.-C. Han. (2011b). Expected improvements in determining continental hydrology, ice mass variations, ocean bottom pressure signals, and earthquakes using two pairs of dedicated satellites for temporal gravity recovery. *J. Geophys. Res., 116*, B11405. doi: 10.1029/2011JB008375.

Wiese, D.N., R.S. Nerem, and F.G. Lemoine. (2012). Design considerations for a dedicated gravity recovery satellite mission consisting of two pairs of satellites. *J. Geodesy, 86*(2), 81–98. doi: 10.1007/s00190-011-0493-8.

Wiese, D.N., P. Visser, and R.S. Nerem. (2011a). Estimating low resolution gravity fields at short time intervals to reduce temporal aliasing errors. *Advances in Space Research, 48*, 1094–1107. doi: 10.1016/j.asr.2011.05.027.

Willis, J.K., D.P. Chambers, and R.S. Nerem. (2008). Assessing the globally averaged sea level budget on seasonal to interannual time scales. *J. Geophys. Res., 113*, C06015. doi: 10.1029/2007JC004517.

Willis, J.K., D. Roemmich, and B. Cornuelle. (2004). Interannual variability in upper ocean heat content, temperature, and thermosteric expansion on global scales. *J. Geophys. Res., 109*, C12036. doi: 10.1029/2003JC002260.

Wouters, B., and E.J.O. Schrama. (2007). Improved accuracy of GRACE gravity solutions through empirical orthogonal function filtering of spherical harmonics. *Geophys. Res. Lett., 34*, L23711. doi: 10.1029/2007GL032098.

Wunsch, C., and E.M. Gaposchkin. (1980). On using satellite altimetry to determine the general circulation of the oceans with application to geoid improvement. *Rev. Geophys., 18*, 725–745.

Zlotnicki, V., J. Wahr, I. Fukumori, and Y.T. Song. (2007). Antarctic circumpolar current transport variability during 2003–05 from GRACE. *J. Phys. Ocean., 37*, 230–244.

5 A 25-Year Satellite Altimetry-Based Global Mean Sea Level Record

Closure of the Sea Level Budget and Missing Components

R. Steven Nerem, Michaël Ablain, Anny Cazenave, John Church, and Eric Leuliette

5.1 INTRODUCTION

Global mean sea level (GMSL)—the height of the oceans averaged over the globe (or over a large portion of the globe) is a crucial barometer of climate change because it reflects both the amount of heat being added to the oceans and the mass loss of global ice reservoirs (Greenland, Antarctica, glaciers, etc.). As such, it is important to precisely measure changes in GMSL and determine what is causing these changes. GMSL can also be used as a metric for assessing the impacts of sea level rise on coastal infrastructure and populations, and so knowledge of changes in GMSL is important for communicating climate change impacts to policymakers, stakeholders, and the public.

For many years, measurements of changes in GMSL were made primarily by averaging measurements from a global network of tide gauges (e.g., Douglas 1991). However, these measurements are limited by poor spatial distribution over the global oceans, and they can be contaminated by vertical land movement. The launch of TOPEX/Poseidon (T/P) in 1992 ushered in a new paradigm for measuring sea level, providing for the first time precise globally distributed sea level measurements at 10-day intervals. At the time of the launch of T/P, the measurements were not expected to have sufficient accuracy for measuring GMSL. However, as the radial orbit accuracy was improved from approximately 10 cm at launch to approximately 1 cm presently, it was determined that physical variations in GMSL could in fact be detected (e.g., Nerem 1995). The T/P measurements were seamlessly continued with the launch of Jason-1 (2001), Jason-2 (2008), and Jason-3 (2016) (see Chapter 1 of this volume for more details). By design, each of these missions has an overlap period with the next during which the measurements can be intercompared and instrument biases determined (e.g., Nerem et al. 2010; Ablain et al. 2015) (see Chapter 1). With one exception, this has allowed the construction of a continuous uninterrupted time series of GMSL that is currently 25 years in length (Figure 5.1). Only when the TOPEX altimeter was switched from Side A to Side B of its electronics on February 10, 1999, was there a potential discontinuity. In addition to GMSL, the measurements provide the spatial variability of sea level rise between ± 66° latitude (Figure 5.2), so that the regional variability of GMSL can be mapped.

Over these 25 years, a number of discoveries have been made using these measurements. Most importantly, GMSL has been observed to be rising at an average rate of approximately 3 mm/year. Interannual variability has been shown to have a link to El Niño-Southern Oscillation (ENSO) variations (e.g., Nerem et al. 1999).

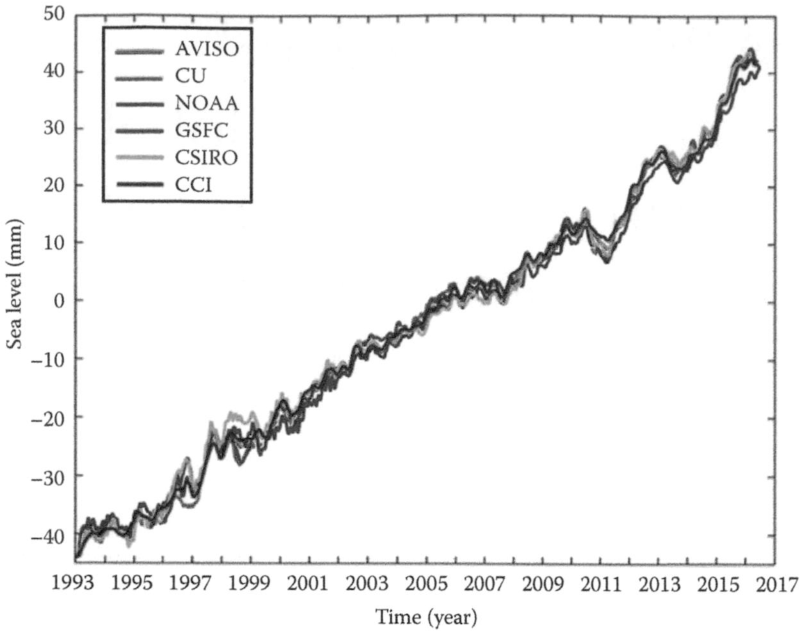

FIGURE 5.1 Ten-day estimates of global mean sea level (GMSL) change (1993–2016) from a 25-year record of measurements from the TOPEX/Poseidon, Jason-1, and Jason-2 missions. Results are from AVISO, University of Colorado (CU), NOAA, NASA Goddard Space Flight Center (GSFC), CSIRO, and the ESA CCI.

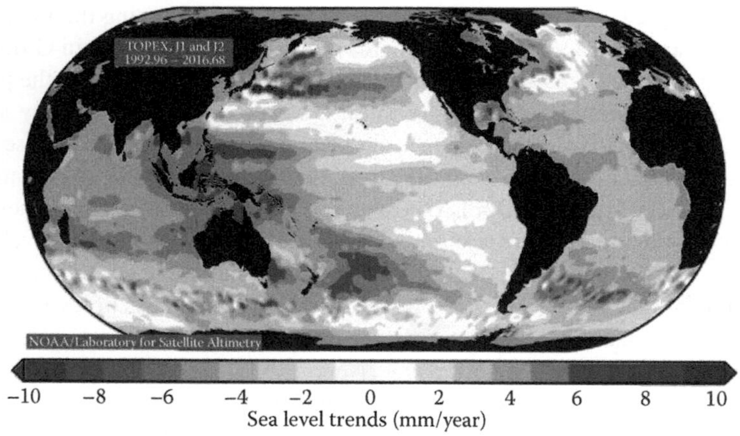

FIGURE 5.2 Spatial variations of sea level rise over the 25-year altimeter record (NOAA analysis). The average of this map is 3.4 mm/year.

The launch of the Gravity Recovery and Climate Experiment (GRACE) mission in 2002 gave us another tool to interpret GMSL variations as GRACE can provide estimates of changes in global ocean mass (Chambers et al. 2004, 2010) and the sources (Greenland and Antarctica, e.g., Velicogna and Wahr 2013; land-water storage, e.g., Llovel et al. 2011). These measurements have shown that interannual variability in GMSL is dominated by changes in ocean mass (e.g., Cazenave et al. 2014), and these changes are driven by land-water storage variations related to changes in land-ocean precipitation (Boening et al. 2012; Fasullo et al. 2013).

The altimeter data have also been combined with tide gauge data to construct long (more than 75 years) records of GMSL variations (e.g., Church and White 2011). These types of analyses together with purely tide gauge-based estimates showed that GMSL was rising more slowly over the last century (in the range 1.2 to 1.9 mm/year; e.g., Church and White 2011; Jevrejeva et al. 2014; Hay et al. 2015) than over the altimeter era. As shown by Hamlington and Thompson (2015), the tide gauge selection has a strong impact on the estimated twentieth-century GMSL rate. A reconstructed record of GMSL variations has been used to characterize decadal variability in GMSL (e.g., Hamlington et al. 2013).

The development of the Argo network of profiling floats has allowed thermosteric sea level change from 0 to 2000 m to be precisely mapped for the first time (e.g., Willis et al. 2008; Roemmich et al. 2012; see also Chapter 3 of this volume). When combined with GRACE and altimetry, this has allowed a detailed assessment of the closure of the sea level budget since 2005 (Willis et al. 2008; Llovel et al. 2014; Dieng et al. 2015a, 2015b; Leuliette 2015; Chambers et al. 2016) and has shown the excellent accuracy of the observing systems over the last decade.

5.2 THE ALTIMETER MEAN SEA LEVEL RECORD

5.2.1 Computing Global and Regional Mean Sea Level Time Series

The processing to provide the GMSL record mainly depends on the geophysical corrections that are applied to the sea surface height (SSH) measurements (hereafter called "altimeter standards"), on the global and regional biases applied to accurately link the altimeter missions together, and on the gridding process applied to average the along-track measurements.

5.2.2 Altimeter Missions

All the groups producing estimates of GMSL (University of Colorado, Centre National d'Etudes Spatiales (CNES)/Archiving, Validation and Interpretation of Satellite Oceanographic data (AVISO), Climate Change Initiative (CCI)/European Space Agency (ESA), CSIRO, National Oceanic and Atmospheric Administration (NOAA), National Aeronautics and Space Administration (NASA)/Goddard Space Flight Center (GSFC)) use 1-Hz altimetry measurements derived from TOPEX/Poseidon, Jason-1, Jason-2, and very recently Jason-3 as reference missions in order to compute GMSL. These missions provide the most accurate long-term stability at global and regional scales (Ablain et al. 2009, 2016), and they are all on the same historical TOPEX ground track allowing the construction of a seamless record of GMSL change from 1993 to the present. In addition, complementary missions (ERS (European Remote Sensing)-1, ERS-2, Envisat, GEOSAT Follow-on, CryoSat-2, SARAL/AltiKa, and Sentinel-3A) are also very useful for increasing the spatial resolution of mean sea-level grids and for providing relevant cross-comparison of the GMSL evolution (Ollivier et al. 2012).

5.2.3 Altimeter Corrections

Satellite altimeters measure the distance between the satellite and the sea surface. The satellite altitude minus this distance gives the SSH. However, numerous perturbations have to be accounted for, and corrections need to be applied to take into account various physical phenomena: propagation corrections (the altimeter radar wave is delayed during travel through the atmosphere—ionospheric correction, wet tropospheric correction, and dry tropospheric correction), corrections for the sea state that directly affects the radar wave (electromagnetic bias), geophysical corrections for the tides (ocean, solid earth, and polar tides as well as loading effects), and atmospheric corrections for the ocean's response to atmospheric dynamics (inverse barometer correction at low frequency, atmospheric dynamics correction at high frequency). These corrections together with satellite orbit determination are discussed in detail in Chapter 1. However, it should be noted that GMSL can be very sensitive to small errors in these corrections when they do not average to zero globally. For example, errors in the inverted barometer (IB) correction and the wet troposphere correction can be particularly problematic for the computation of GMSL.

SSH is calculated for each altimetric measurement considered valid according to certain criteria (e.g., threshold, spline, and statistics on the ground track) applied either to the main altimetric parameters, the geophysical corrections, or to the SSH directly. These criteria may vary from one mission to another depending on the altimeter's instrumental characteristics. The precise references for the corrections and orbits used when calculating the GMSL depends on each group (University of Colorado, CMEMS/AVISO/CNES, CSIRO, NOAA, NASA, and CCI/ESA) (e.g., Masters et al. 2012). Chapter 1 provides information about the temporal evolution of accuracy of SSH estimates since the early altimetry missions.

5.2.4 Intermission Biases

An important step for accurately computing the GMSL time series is the calculation of the instrument biases between the altimeter missions, both at global and regional scales. Thanks to the verification phases between two consecutive missions (TOPEX/Jason-1, Jason-1/Jason-2, and Jason-2/Jason-3) when both satellites are on the same ground track following each other by a few seconds, the global intermission biases are estimated with an accuracy of approximately 0.5 to 1 mm (Zawadzki and Ablain 2016). The absolute value of this bias depends on the corrections selected in the SSH calculation. It is worth noting that the absolute GMSL bias is often arbitrarily set to 0 (determined in 1993 by convention). A similar method is also used to derive regional biases generated by systematic effects such as geographically correlated errors in the orbits (Couhert et al. 2014). One particularly problematic bias is between Side A and Side B of TOPEX as there is by definition no overlap between the measurements, unlike the boundaries between different satellite missions. Essentially, one of two different techniques can be used: (1) The tide gauge validation can be used to solve for the bias directly or (2) one can constrain GMSL to be smooth across the boundary. Although the evidence is that this bias is small (less than 1 mm), there is still considerable uncertainty in this estimate.

5.2.5 Averaging Process

The second main step is to average the along-track altimeter data in order to derive regional and GMSL time series. Two different approaches can be considered. The first approach developed by Dibarboure et al. (2011) and applied to derive the sea-level products of the Climate Change Initiative (CCI) project of the European Space Agency (ESA) (hereinafter called SL_cci), consists of using the reference missions (TOPEX/Poseidon, Jason-1, and Jason-2) to provide the long-term evolution and the complementary missions (ERS-1, ERS-2, Envisat, GEOSAT Follow-on, CryoSat-2, and SARAL/AltiKa) to increase the spatial resolution and also to better cover high latitudes (beyond 66°N/S). Basically, after removing spurious along-track altimeter data (e.g., impacted by rain cells and sea ice) and applying GMSL biases among all the missions, long wavelength errors (e.g., orbit errors) are removed among all the missions through a global minimization of the cross-over differences observed for the reference mission, and between reference and complementary missions (Dibarboure et al. 2011). Then the calculation of sea-level grids combining all the missions together by an objective analysis approach (Ducet et al. 2000; Le Traon et al. 2003) is performed. This method allows monthly sea-level grids to be derived, with a spatial resolution of 0.25°. The GMSL time series is easily deduced from the sea-level grids after averaging with area weighting over the oceanic latitudes observed by the altimetry data (from 82°S to 82°N).

The second approach, applied by other groups (University of Colorado, CNES/AVISO, CSIRO, NOAA, and NASA), consists of only using the reference missions (TOPEX/Poseidon, Jason-1, and Jason-2) to provide GMSL time series. In this case, the method is simpler because the altimetry data are not merged but just linked together. Then, a simple averaging on a cycle-by-cycle basis for each mission is applied. This averaging can either be applied to the along-track data directly (e.g., Nerem 1995) or gridded fields can be computed (e.g., 1° along the latitudinal axis, 3° along the longitudinal axis for CNES/AVISO), and then averaging done from the grids. The main advantages of this approach are the reduction of the computing time (fewer altimeter missions and no objective analysis) and the uniformity of the data coverage (the additional satellites do not provide uniform temporal coverage). The GMSL time series is only estimated between 66°S and 66°N, and the regional sea-level variations are not as well represented. A disadvantage is that errors in the altimeter measurements such as long wavelength orbit errors or oceanic tide errors are less well averaged out than in the multisatellite solutions and impact the mean sea-level estimate by up to 1–2 mm at each cycle. However, differences among gridding processes have little impact on the GMSL trend or the interannual signals. Studies (e.g., Henry et al. 2014) have shown differences in trend estimates less than 0.05 mm/year over the whole altimetry period of 1–2 mm at interannual time scales (periods in the range of 1–3 years).

5.2.6 Validation of the GMSL Record with Tide Gauge Measurements

One of the main objectives of comparing altimeter and tide gauge measurements is to ensure that the altimeter instruments are not drifting over time (Mitchum 1998, 2000). Basically, the approach consists of using a global network of tide gauges, which are accurate enough and provide a large enough ensemble to detect regional or global biases between altimeter instruments and tide gauges. As the differences between absolute sea level measured by altimetry and relative sea level measured by tide gauges could also arise from vertical motion of the land in which the tide gauge is grounded, stations to be used need to be carefully selected and also corrected for vertical land motion if known (Fenoglio-Marc et al. 2004; Santamaría-Gomez et al. 2014).

All comparison methods rely on similar processing, which is briefly described here. More complete description of the comparison method can be found in Mitchum (2000), Valladeau et al. (2012), and Wöppelmann and Marcos (2016). Relative SSH measurements from tide gauges are first corrected for various effects (e.g., tides and vertical land motions) so that the physical content is comparable to absolute SSH measurements from altimetry. Then, the altimeter data are collocated to the tide gauge stations and a time series of altimetry minus tide gauge differences are extracted at each *in situ* station. Eventually, a global average is estimated from the ensemble mean of the individual difference time series, assuming that the instrument behavior has no geographic variations (Figure 5.3).

As tide gauges unevenly sample the global coastlines and comparisons do not cover the deep open ocean, the drift estimated may contain local biases, and caution is required before applying the estimated drift to correct any global scale altimetry drift. Furthermore, the uncertainty of the method has been estimated as 0.7 mm/year (Mitchum 1998, 2000; Valladeau et al. 2012). One important error source is the correction for vertical land motion at the tide gauge sites. Without precise Global Navigation Satellite System (GNSS) positioning at the tide gauges, corrections often include only a single process, the Glacial Isostatic Adjustment (GIA) of the viscoelastic solid Earth to the last deglaciation. GIA is due to the rebound of the Earth's crust in response to the removal of

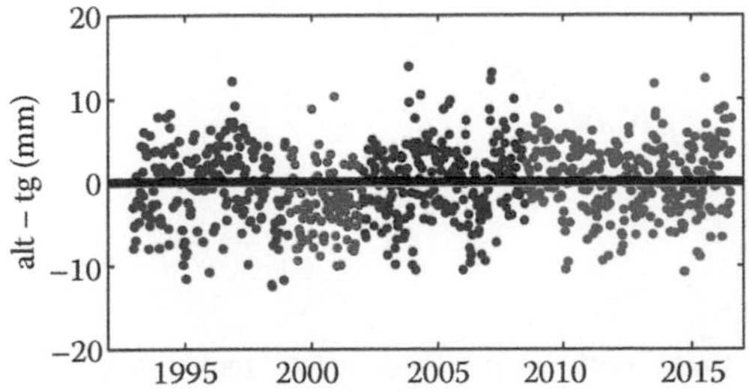

FIGURE 5.3 Altimeter—tide gauge sea level differences (TOPEX-A blue, TOPEX-B green, Jason-1—red, Jason-2—pink), averaged globally, to monitor the performance of the different instruments over the 25 years of the altimeter era (After Mitchum 2000).

the ice load at the end of the last glacial cycle. Vertical land motions due to GIA can be estimated from models (e.g., Peltier 2004) and thus can be calculated everywhere but with limited accuracy. In addition, correcting tide gauge measurements for only GIA may lead to ignoring contemporary vertical land motion sources (present-day ice melt, surface loading, groundwater extraction, etc.) unless these are separately estimated (Watson et al. 2015). Using this approach of correcting for land motion, Watson et al. (2015) estimated an uncertainty of 0.4 mm/year in the tide gauge validation. Wöppelmann and Marcos (2016) quantified vertical land movements not linked to the GIA process and showed they may reach up to 10 mm/year, although for a global average they cancel out (0.01 ± 0.27 mm/year) if the number of tide gauge stations is large enough.

Recent results from global comparisons between altimetry and tide gauges (Watson et al. 2015) suggest that there is no significant drift detected in either the Jason-1 or Jason-2 instrument (lower than 0.5 mm/year and not statistically significant). However, a statistically significant drift is observed with the TOPEX instrument, mainly confined to the TOPEX-A record (January 1993 to February 1999). This drift is estimated at approximately 1 mm/year for TOPEX-A (Watson et al. 2015). This means that the current GMSL rise (approximately 3 mm/year) would have been overestimated during the TOPEX period, and the rate over the entire altimeter era (1993–2016) would be reduced by approximately 0.5 mm/year (Watson et al. 2015).

5.2.7 Mean Sea Level Variation and Uncertainties

5.2.7.1 Global Scale Uncertainty

GMSL increased by approximately 3 mm/year over the 1993–2016 period (Figure 5.4), according to the different groups (University of Colorado, CNES/AVISO, CSIRO, NOAA, NASA/GSFC, and ESA/SL_cci project). There is the possibility of systematic drifts in the measurements, but this can be bounded by comparisons to tide gauges (see 5.2.7). Although the global evolution is nearly linear (formal error of the linear fit is about 0.02 mm/year), interannual variations are also observed. Removing the trend from the global MSL time series highlights these variations over a 1- to 3-year time period. Their magnitudes depend on the period (+ 3 mm in 1998–1999, −5 mm in 2011–2012, and + 10 mm in 2015–2016) and are well correlated in time with El Niño and La Niña events. The comparison of the Multivariate ENSO Index (MEI) with the detrended global MSL time series

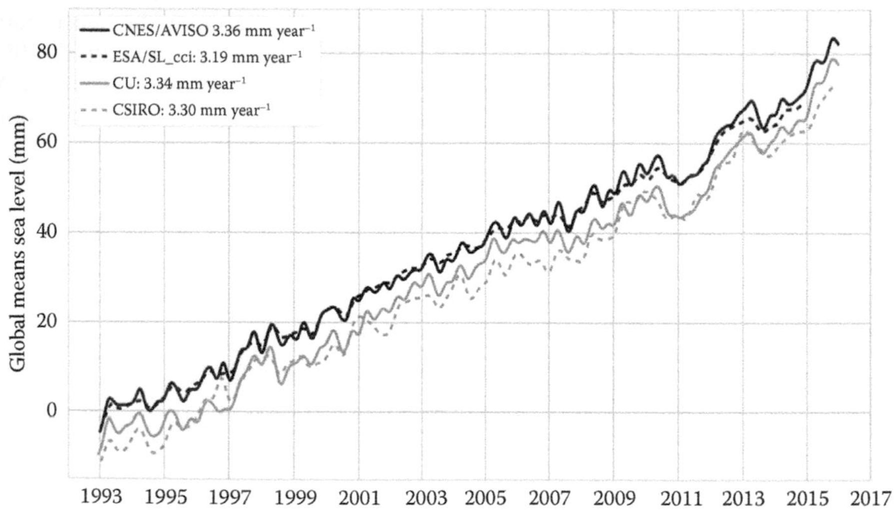

FIGURE 5.4 Global mean sea level evolution from 1993 to 2015 included for the following MSL groups: AVISO/CNES, CCI/ESA (until 2014), University of Colorado, CSIRO. Periodic annual signals have been removed and signals lower than 6 months have been filtered out, and an offset has been applied to the curves.

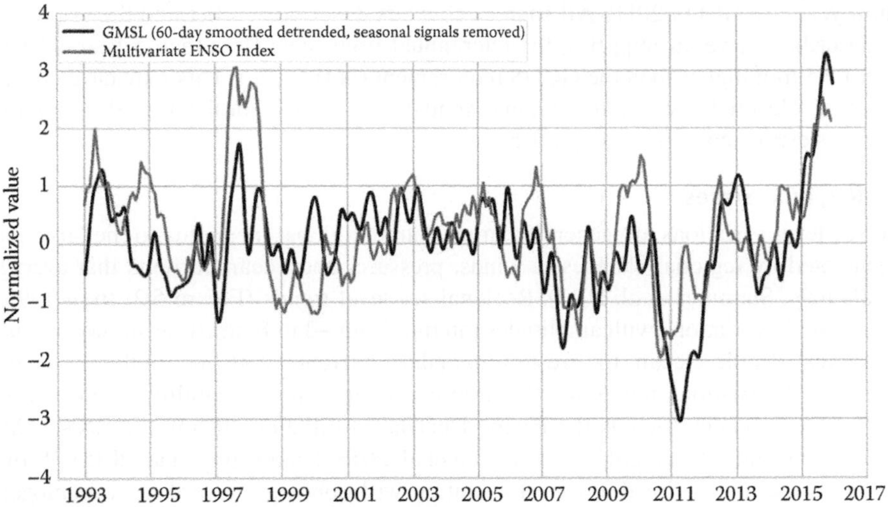

FIGURE 5.5 Comparison of MEI index and the global mean sea level time series (from CNES/AVISO global MSL time series) after removing the global trend.

highlights the temporal correlation (Figure 5.5) very well, although there is no strict linear relationship between MEI and GMSL.

The error budget for global and regional sea level variations is shown in Table 5.1. For the GMSL trend, an uncertainty of 0.5 mm/year is estimated over the whole altimetry era (1993–2015) within a confidence interval of 90%. The main source of error is the radiometer wet tropospheric correction with a drift uncertainty in the range of 0.2–0.3 mm/year (Legeais et al. 2014). To a lesser extent, the orbit error (Couhert et al. 2014) and the altimeter parameter (range, sigma-0, and significant wave height) instabilities (Ablain et al. 2012) also contribute to the GMSL trend uncertainty at the level of 0.1 mm/year. It is worth noting that, for these two corrections, the uncertainties are higher in the first altimetry decade (1993–2002) where TOPEX/Poseidon, ERS-1, and ERS-2 measurements display greater errors at climate scales. For instance, the orbit solutions contain larger uncertainties due to the gravity fields calculated without GRACE data. Furthermore, imperfect links between TOPEX-A and TOPEX-B, TOPEX-B and Jason-1 (April 2003), and Jason-1 and Jason-2 (October 2008) lead to uncertainties of 2, 1, and 0.5 mm, respectively (Ablain et al. 2009; Zawadzki and Ablain 2016). This causes a GMSL trend uncertainty of about 0.1 mm/year over the 1993 to 2015 period. It is relevant to note that the remaining uncertainty of 0.5 mm/year on the GMSL trend remains higher than the Global Climate Observing System (GCOS) requirements

TABLE 5.1
Mean Sea Level Error Budget at Different Time Scales

Spatial Scales	Temporal Scales	Altimetry Errors	GCOS Requirements
GMSL	Long-term evolution (>10 years)	<0.5 mm/year	0.3 mm/year
	Interannual signals (<5 years)	<2 mm over 1 year	0.5 mm over 1 year
	Annual signals	<1 mm	Not defined
Regional sea level	Long-term evolution (>10 years)	<3 mm/year	1 mm/year
	Annual signals	<1 cm	Not defined

Source: Ablain, M., et al., *Surveys Geophys.*, 2016.

(of 0.3 mm/year, see GCOS 2011). All sources of errors described earlier and the gridding process (already described) have an impact at the interannual time scale (less than 5 years). The level of error is still 1.5 mm higher than the GCOS requirement (of 0.5 mm). This may have consequences on the sea level closure budget studies on interannual time scales. For the annual signal, the amplitude error was estimated to be less than 1 mm.

5.2.7.2 Regional Scales

Regional sea level variations are generally larger than the global mean due to the large local variability generated by regional changes in winds, pressure, and ocean currents that average out at global scale (e.g., Stammer et al. 2013). Regional sea level trends (Figure 5.2) from 1993 to 2015 exhibit large-scale variations with amplitudes ranging from –3 to 8 mm/year in regions such as the western tropical Pacific Ocean, the western boundary currents, and the Southern Ocean. A part of these regional variations represents the unforced interannual variability of the ocean related to internal climate modes such as the Pacific Decadal Oscillation (PDO) and ENSO (Merrifield et al. 2011; Zhang and Church 2012; Hamlington et al. 2013; Palanisamy et al. 2015). With a longer observation period, the regional sea level evolution will converge toward more homogenous sea level trends.

The regional trend uncertainty due to altimeter errors is of the order of 2–3 mm/year depending on the region (Prandi et al. 2016). The orbit solutions are the main source of errors in the range of 1–2 mm/year (Couhert et al. 2014) with large spatial patterns at hemispheric scale. The effect of Earth gravity field errors on orbit determination explains an important part of these uncertainties (Rudenko et al. 2014). Furthermore, errors are higher in the first decade (1993 to 2002) where the Earth gravity field models are less accurate due to the unavailability of the GRACE data before 2002. Additional errors are still observed (e.g., for the radiometer-based wet tropospheric correction in tropical areas, other atmospheric corrections in high latitudes, and high frequency corrections in coastal areas). The combined errors suggest an uncertainty of 0.5–1.5 mm/year. Finally, the 2–3 mm/year uncertainty on regional sea level trends remains a significant metric compared to the 1 mm/year GCOS requirement.

5.3 INTERPRETING THE ALTIMETER GMSL RECORD

The physical processes causing global mean and regional sea level changes are not identical. In terms of global mean, the primary mechanisms leading to current GMSL rise consist of (1) ocean volume changes due to thermal expansion of the ocean water and (2) ocean mass changes due to ice melting and discharge from ice sheets, glaciers, and ice caps; changes in land water storage; and in atmospheric water content. This is usually expressed by the sea level budget equation:

$$GMSL(t) = GSSL(t) + M_{ocean}(t) \tag{5.1}$$

where $GSSL(t)$ refers to the global steric sea level change (i.e., the contributions of ocean temperature and salinity changes) and $M_{ocean}(t)$ refers to the change in mass of the oceans; t is time. Note that when global averages are considered, salinity changes have small effect and therefore $GSSL(t)$ largely reflects ocean thermal expansion.

Because of mass conservation in the climate system, ocean mass change (i.e., $M_{ocean}(t)$) is more complicated, so that:

$$M_{Ocn}(t) + M_{Gla}(t) + M_{Gre}(t) + M_{Ant}(t) + M_{LWS}(t) + M_{Atm}(t) + M_{Snow}(t) + \text{missing terms} = 0 \tag{5.2}$$

where $M_{Gla}(t)$, $M_{Gre}(t)$, $M_{Ant}(t)$, $M_{LWS}(t)$, $M_{Atm}(t)$, and $M_{Snow}(t)$ represent temporal changes in mass of glaciers, Greenland and Antarctica ice sheets, land water storage (LWS), atmospheric water vapor, and snow mass changes, respectively. Missing terms include, for example, permafrost melting.

At regional scales, the regional sea level (RSL) budget can be written as:

$$RSL(t) = SSL(t) + RM_{Ocn}(t) + atmospheric\ loading + static\ terms, \tag{5.3}$$

where $SSL(t)$ and $RM_{Ocn}(t)$ are regional steric sea level and ocean mass variations. The so-called "static terms" (Stammer et al. 2013) consist of GIA and other solid Earth and gravitational effects due to present-day mass redistributions (Milne et al. 2009; Tamisiea and Mitrovica 2011).

In this chapter, we do not address regional variations and focus on GMSL changes. The following subsections successively discuss the steric and (main) ocean mass contributions to GMSL variations during the altimetry era.

5.3.1 Steric Sea Level Contribution

As discussed in Chapter 3 of this volume, steric sea level variations result from temperature and salinity-related density changes of water between two depth levels Z_1 and Z_2. Sea water density is a nonlinear function of temperature T, salinity S, and pressure. However, it is common to consider a first-order expansion of the inverse of density around a reference T_0/S_0 profile, usually from a climatology. Hence, the total steric effect H can be separated into two terms: a (temperature-related) thermosteric component H_T, and a (salinity-related) halosteric component H_S, so that:

$$H = H_T + H_S = \int \alpha(T - T_0)dz + \int \beta(S - S_0)dz \tag{5.4}$$

where z is depth (integration between Z_1 and Z_2); α and β are coefficients of thermal and haline expansion (note that a temperature/salinity increase causes an increase/decrease in steric sea level). In terms of the global mean, H_S is almost zero to first-order (Bindoff et al. 2007), and so it is often better to ignore this term rather than introduce biases due to incomplete salinity data coverage.

Chapter 3 provides a history of temperature and salinity measurements and presents the various devices used for that purpose, in particular expandable bathythermographs (XBTs) and Argo profiling floats. Beginning in the 1970s, temperature measurements of the upper ocean were collected down to an approximately 700-m depth by XBTs along commercial shipping routes. Although the coverage was improved compared to previous decades, large regions remain unsampled, particularly the Southern Hemisphere and in the Arctic. In addition, salinity measurements were very few, preventing estimation of the halosteric component.

Since the beginning of the 2000s, temperature/salinity (T/S) profiles have been collected by the autonomous Argo floats (Roemmich et al. 2012). Deployment of 3000 active Argo floats was achieved around 2005. At the present, approximately 3800 floats provide systematic temperature and salinity data, with quasi-global coverage down to 2000 m in depth.

Throughout the altimetry era, several groups have provided gridded time series of temperature data for different depth levels based on XBTs, with additional data from mechanical bathythermographs (MBTs) and conductivity-temperature-depth (CTD) devices, as well as Argo, from which it is easy to compute the thermosteric component. The halosteric component is mostly based on Argo.

For the period 1993–2005, the temperature data sets are mostly based on XBTs (plus a few MBTs and CTDs) and cover the upper ocean only (down to 700 m or 1000 m). The main data sets that are regularly extended and updated are from Ishii and Kimoto (2009), Levitus et al. (2012), Good et al. (2013), and Domingues et al. (2008). These data sets differ because of different strategies adopted for data editing, temporal and spatial data gap filling, mapping methods, baseline climatology, and instrument bias corrections (in particular, the time-to-depth correction for XBT data). Abraham et al. (2013) review the impact of these details on the data uncertainties (see also Boyer et al. 2014; Lyman and Johnson 2014).

The steric component can also be estimated from ocean reanalyses (e.g., Ocean Reanalysis System (ORAS)4, Balmaseda et al. 2013 and GECCO, Köhl and Stammer 2008). These provide steric sea level estimates integrated from the surface to the ocean floor over the whole oceanic domain. Figure 5.6 shows the thermosteric sea level (upper 700 m) since 1993 estimated from four T/S data sets (latest versions from Ishii and Kimoto 2009) and NOAA; EN4 and ORAS4. The Domingues et al. (2008) data are not superimposed because unlike the other data sets, they do not contain high-frequency fluctuations due to the 3-year smoothing filter applied.

For the Argo era (2005 to present), the main data sets available come from the International Pacific Research Center (IPRC), the Japan Agency for Marine-Earth Science and Technology (JAMSTEC), and the Scripps Institution of Oceanography (Roemmich and Gilson 2009). Another data set based on Argo T/S measurements gathered at the French Research Institute for Exploitation of the Sea (IFREMER) (CORIOLIS data base) is provided by von Schuckmann et al. (2014). These data and associated uncertainties are available for January 2005 to present and provide temperature and salinity data down to 2000 m.

Figure 5.7 shows Argo-based steric sea levels (thermosteric plus halosteric components) for the four data sets over January 2005 to December 2014 and associated uncertainty (except for IPRC and Scripps data for which no errors are provided).

Tables 5.2 and 5.3 gather trend estimates and associated uncertainties for the two groups of data sets and two periods.

For the altimetry era (starting in 1993), as noted previously (e.g., Lyman et al. 2010), large discrepancies are visible among the different data sets. In terms of trend, the EN4 data lead to a very low steric trend compared to other data sets. The trend difference between EN4 and other data is not reflected by the associated uncertainty. There is better agreement among the other data sets. The mean steric trend (0–700 m) over 1993–2012 agrees well with the Intergovernmental Panel on Climate Change (IPCC) Fifth Assessment Report (AR5; Church et al. 2013) estimates for 1993–2010 (amounting to 0.8 ± 0.3 mm/year for the same depth range) and those provided in Chambers et al. (2016) for January 1992 to December 2013 (i.e., roughly the same time span) based on a combination of NOAA data and an update of Domingues et al. (2008) (amounting to 0.85 ± 0.2 mm/year). Neglecting the EN4 data, the mean steric trend (0.84 ± 0.15 mm/year) is even closer

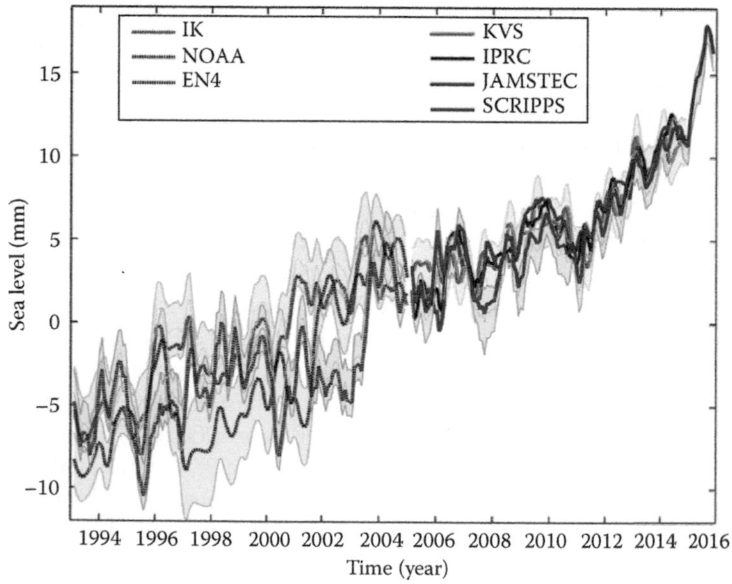

FIGURE 5.6 Global mean steric sea level time series over January 1993–December 2012 based on the latest T/S version of Ishii and Kimoto (2009)—noted IK- and NOAA, as well as EN4 and ORAS4 reanalysis. Shaded areas represent the uncertainty range (only for IK and EN4; no errors are provided for the other data sets).

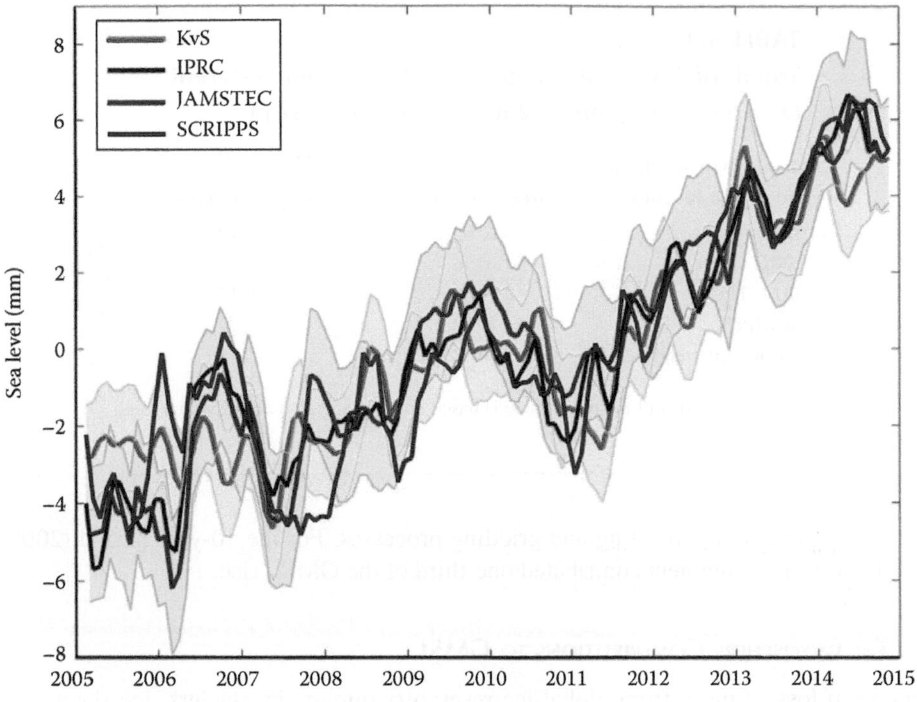

FIGURE 5.7 Global mean steric sea level time series over January 2005–December 2014 (Argo era) from four different data sets (KvS, IPRC, JAMSTEC, and SCRIPPS). Shaded areas represent uncertainty range (no errors are available for the IPRC data).

TABLE 5.2
Trends of Global Mean Steric Sea Level from Different Data Sets over January 1992 to December 2012

Steric Sea Level Trends	January 1993–December 2012	
(0–700 m)	Trend (mm/year)	Error (1-sigma) (mm/year)
IK	0.67	0.1
NOAA	0.86	
EN4.0.2	0.37	0.1
ORAS4	1.0	
Mean steric trend	0.725	0.2

Note: The error of the mean value is based on the dispersion of individual trend values.

to Chambers et al.'s value. In Chambers et al. (2016), the steric contribution of the 700–2000 m layer is estimated to 0.24 ± 0.1 for 1992–2014. The deep ocean contribution (below 2000 m) was estimated to approximately 0.1 mm/year by Purkey and Johnson (2010).

There is much better agreement among the different steric sea level time series during the Argo era (see Figure 5.7 and Table 5.3). This is a consequence of the better global data coverage, leading

TABLE 5.3
Trends of Global Mean Steric Sea Level from Different Data Sets over January 2005 to December 2014

Steric Sea Level Trends from Argo (0–2000 m)	January 2005–December 2014 Trend (mm/year)	1-sigma (mm/year)
KvS	0.82	0.12
IPRC	0.96	
JAMSTEC	1.02	0.14
SCRIPPS	0.90	
Mean steric trend	0.92	0.15

Note: The error of the mean value is based on the dispersion of individual trend values.

to reduced differences in gap filling and gridding processes. For the 10-year period (2005–2014), the 0–2000 m steric component contributed one-third of the GMSL rise.

5.3.2 The Cryosphere Contributions to GMSL

The long-term loss of mass from global ice reservoirs (mountain glaciers, ice sheets, ice caps, etc.) contributes significantly to long-term GMSL variations. Associated mass redistribution also contributes to regional variations in sea level through the "ice sheet fingerprints" (e.g., Mitrovica et al. 2001; Milne et al. 2009), although it is not at present the dominant source of regional variability (Stammer et al. 2013), unlike the steric effects (Church et al. 2013; Han et al. 2016; Meyssignac et al. 2016). In the future, melting of Greenland and Antarctica is expected to accelerate and 1 day will be the dominate driver of global sea level change and its regional variation. In fact, Greenland and Antarctica are of the most concern for the future because of the potential multi-meter contributions to sea level change (e.g., DeConto and Pollard 2016). Therefore, understanding the cryosphere contributions to present-day sea level change is one of the top objectives of sea level science.

Unfortunately, direct observations of changes in the cryosphere are less plentiful than direct observations of sea level change. Only a fraction of the Earth's mountain glaciers are precisely monitored and so global estimates of the contribution of mountain glaciers to GMSL must be extrapolated from these sparse observations (e.g., Gardner et al. 2013; Leclercq et al. 2014; Marzeion et al. 2016). A recent consensus estimate of the contribution of mountain glaciers and small ice caps to GMSL is from Gardner et al. (2013), who found a value of −259 ± 28 Gt/year over 2003 to 2009, which is equivalent to 0.7 mm/year (see also the update by Marzeion et al. 2016).

Similarly, estimates of the contributions of Greenland and Antarctica to GMSL are also highly uncertain prior to the 1990s. However, with the launch of GRACE in 2002; the Ice, Cloud, and Land Elevation Satellite (ICESat) in 2004, and various Interferometric Synthetic Aperture Radar (InSAR) and altimeter satellites, estimates of the ice sheet contribution to GMSL have greatly improved (there is abundant literature on the ice sheet mass balances during the altimetry era; the most recent publications include Jacob et al. 2012; Velicogna and Wahr 2013; Velicogna et al. 2014; and Forsberg et al. 2016). Shepherd et al. (2012) provided consensus estimates of mass loss over 1992–2011 of −142 ± 49 Gt/year for Greenland, +14 ± 43 Gt/year for East Antarctica, −65 ± 26 Gt/year for West Antarctica, and −20 ± 14 Gt/year for the Antarctic Peninsula. Mass loss has accelerated since that time (e.g., Velicogna et al. 2014). Recent GRACE estimates of mass loss over 2002–2016 are −281 ± 24 Gt/year for Greenland and −118 ± 44 Gt/year for Antarctica based on the

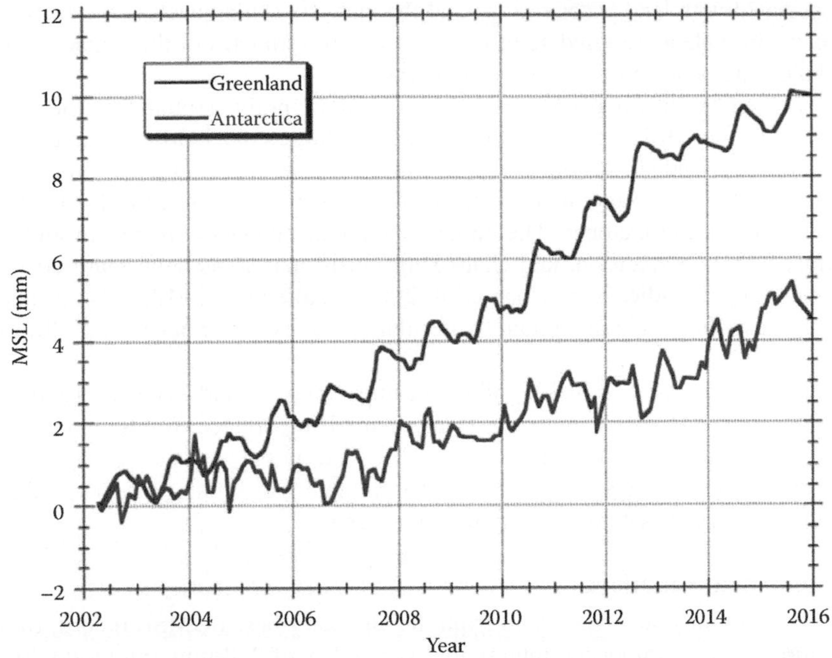

FIGURE 5.8 GRACE estimates of Greenland mass loss, Antarctic mass loss, in units of contribution to GMSL, estimated from the JPL mascon solutions. (Watkins, M. M., et al., *J. Geophys. Res. [Solid Earth]*, 120(4), 2648–2671, 2015.)

Jet Propulsion Laboratory (JPL) mascon solutions (Watkins et al. 2015) (Figure 5.8), which translates to a combined GMSL contribution of 1.1 mm/year. The errors in the GRACE estimates are largely driven by GIA uncertainty (Shepherd et al. 2012), in particular over the Antarctica ice sheet, and measurement errors. Forsberg et al. (2016) provide an overview of the most recent estimates of the Greenland and Antarctica mass balance over the altimetry era while Chapter 15 of this volume discusses altimetry-based mass balance measurements for Antarctica.

When taken together (and ignoring the different time frames of the estimates), the contribution of the cryosphere to the GMSL trend over the altimetry era is approximately 1.8 ± 0.16 mm/year. This is similar to GRACE estimates of changes in global ocean mass (1.8 ± 0.4 mm/year) from the JPL mascon solutions (Watkins et al. 2015) as well as GRACE spherical harmonic solutions (Chambers et al. 2004) over the last decade only.

5.3.3 The Land Water Storage Contributions to GMSL

Water is exchanged between the ocean and the terrestrial environment, directly affecting sea level. Land water storage includes rivers, lakes, reservoirs, swamps, and surface snow pack (in addition to the glacier and ice sheet regions). Subsurface water storage includes soil water, the "vadose zone" (from the land surface to the top of the underground water table), and groundwater.

The volume of water in each of these reservoirs varies as a result of climate fluctuation such as the ENSO phenomenon and other climatic variations, including anthropogenic climate change. In addition, direct anthropogenic interferences in the hydrological cycle affect sea level through the building of dams with the subsequent storage of water in artificial reservoirs (Chao et al. 2008), the depletion of groundwater for agriculture and human consumption, and through other land surface changes (Wada et al. 2012). Milly et al. (2010) provide an excellent introduction to the range of issues associated with land water storage and the analysis techniques used to understand these issues. Here, we provide a brief update of recent results in estimates of the impact on sea level of changes in land water storage for a range of time scales.

Direct measurement of changes in ocean mass using monthly estimates of the Earth's gravitational field from the GRACE satellites have shown a large contribution of land water storage to the annual GMSL budget (Chambers et al. 2004). These measurements indicate a seasonal exchange between the ocean and the terrestrial environment with a magnitude of 8.5 mm with a maximum in early to mid-October. The annual-cycle mass-budget closure (through changes in ocean volume, thermal expansion, and changes of ocean mass) has been featured in a number of more recent budget studies (e.g., Chen et al. 2005; Leuliette and Miller 2009) and has been extended to address changes in individual (particularly tropical) river basins (e.g., Ramillien et al. 2008; Llovel et al. 2011).

The clear results from these studies highlight the importance of land water storage in understanding GMSL variations and trends and raises the potential importance of interannual and decadal variations in land water storage for understanding longer-term sea level change. The closure of the annual budget also demonstrates the power of the multiple global data sets, particularly satellite altimetry and time variable satellite gravity measurements.

5.3.3.1 Interannual Variations

If we remove the 3 mm/year linear trend from the altimetry-based GMSL record, the detrended GMSL variations display important interannual variability of 2–4 mm amplitude, highly correlated with various ENSO indices (Nerem et al. 1999). During El Niño/La Niña, the GMSL curve exhibits positive/negative anomalies of approximately a 2-year duration (Figure 5.5). Several studies have shown that these mainly reflect temporary changes in ocean mass linked to excess or deficit precipitation over the tropical Pacific (e.g., Landerer et al. 2008; Cazenave et al. 2014; Dieng et al. 2014).

For example, during 2010 to 2011 La Niña event, Boening et al. (2012) demonstrated (using GRACE data) that the fall in GMSL was dominated by increased terrestrial water storage, particularly in floods in Australia, northern South America, and south and southeast Asia, with a much smaller component from changes in ocean thermal expansion (Figure 5.1). Ngo-duc et al. (2005) and Llovel et al. (2011) emphasized the role of tropical river basins in the interannual variability, but Fasullo et al. (2013) demonstrated that it was Australia's endorheic basins, which largely trapped the runoff of the flood-waters until they evaporated, that was responsible for the persistence of the signal from 2010 to 2011 La Niña event.

Although changes in snow pack are important for the annual cycle, they make only a small contribution to longer time scales (Biancamaria et al. 2011).

Changes in land water storage are not only related to ENSO, but direct quantitative linkages to other climate indices have not yet been made. Using a continental water balance model driven by climate variability, Milly et al. (2003) estimated a trend of 0.12 mm/year over in GMSL from 1981 to 1998, with substantial interannual fluctuations; for example, 1993 to 1998 trend was 0.25 mm/year. However, on decadal time scales, they argued the steric changes were larger than the mass changes. After removing estimates of glacier mass loss and direct anthropogenic influence in the hydrological cycle (see following) from GRACE observations of terrestrial water storage, Reager et al. (2016) estimated an even larger contribution to sea level from natural climate variability of -0.7 ± 0.2 mm/year from 2002 to 2014 (sea level falling; including the large 2010 to 2011 La Niña event). As this estimate is a residual difference among several other terms, it necessarily contains biases in these terms. Dieng et al. (2015c) also used a residual approach for estimating total land water storage from the global sea level budget, arriving at a somewhat different estimate to that of Reager et al. These differences demonstrate the need for improved observations and understanding of land water storage changes.

5.3.3.2 Long-Term Variations

Although the natural fluctuations are substantial, the limited evidence available suggests that the trends over multidecadal time scales (Ngo-Duc et al. 2005; Llovel et al. 2011) and longer (Milly et al. 2010) are small. However, during the twentieth century, direct anthropogenic interference in the hydrological cycle has had a significant impact on sea level (Sahagian et al. 1994; Gornitz et al. 1997). The two largest anthropogenic contributions to sea level change are from the impoundment of water in reservoirs and the depletion of groundwater.

Chao et al. (2008) used a comprehensive record of the reservoirs constructed since 1900 to estimate the total water storage. Most of the capacity was constructed since 1950, with a peak in growth in about 1980 and significantly reduced construction after 2000. Storage capacity continues to grow (e.g., with the Three Gorges Dam and other recent dams), but may be completely or partially offset by sedimentation (Lettenmaier and Milly 2009). Chow et al. allowed for seepage into the surrounding ground, assumed that on average the dams were filled to 85% capacity, and did not include an allowance for sedimentation. Ignoring the seepage, the total increased impoundment was estimated to be about 23 mm by 2007.

Groundwater depletion has gradually increased from 1900. Wada et al. (2010), using a global hydrological model, estimated that groundwater depletion was 0.8 mm/year in 2000, which was reduced to 0.57 mm/year in a subsequent study (Wada et al. 2012). However, this estimate does not adequately represent groundwater recharge processes and was substantially larger than the estimate of groundwater depletion based on direct (but incomplete) observations of 0.41 mm/year (Konikow 2011). A substantially larger estimate by Pokhrel et al. (2012) is likely unrealistic (Konikow 2013; Wada et al. 2016). Allowing for recharge, Wada et al. (2016) estimated only 80% of the groundwater depletion ends up in the ocean. The revised estimate of 0.3 mm/year for 2000–2010 is close to the recent estimate from Döll et al. (2014) of 0.31 mm/year for 2000–2009. When combined with updated reservoir storage estimates and small estimated contributions from deforestation, wetland loss and endorheic basin storage loss, the net twentieth century contribution to sea level rise is negligible and averages 0.12 ± 0.04 mm/year for 1993–2010 (smaller than used in the budget of Church et al. 2011, and the IPCC AR5; Church et al. 2013).

5.4 CLOSING THE SEA LEVEL BUDGET AND UNCERTAINTIES

Instrumental to interpreting the 25-year-long satellite altimetry-based global sea level record is the concept of an observational budget. GMSL is a reflection of incremental increases or decreases of the volume of water in the global ocean. On climate time scales, the causes of volume change in the ocean are primarily from two processes, density (steric) changes and the water exchange between the ocean and continents (Leuliette and Willis 2011). Variations in ocean temperature and salinity change the density of seawater, which is inversely proportional to volume changes. When expressed as a component of mean sea level, this effect is termed steric sea level (see Section 5.3). The exchange of water between the ocean and reservoirs of continental hydrology (ice sheets, glaciers, ice caps, groundwater, and inland surface waters; see Section 5.3) causes variations in the ocean's total mass. This component is sometimes called barystatic sea level (Gregory et al. 2013).

With sufficient observations of sea level from altimetry or tide gauges, ocean temperatures and salinity from hydrographic profiles such as Argo, and either land reservoirs or ocean mass from gravity missions such as GRACE (described in the following), the sea level budget can in principle be closed. Expressed in terms of globally averaged height, contributions to the total budget of GMSL are given in Equation 5.1. Improvements in the global ocean observing system, including GRACE and an Argo network with more than 3500 active floats have allowed the budget to be closed (e.g., Willis et al. 2008; Leuliette and Miller 2009; Llovel et al. 2014). Closure of the sea level budget both demonstrates the integrity of sea level monitoring for climate studies and reveals the relative contributions of sea level change.

5.4.1 Glacial Isostatic Adjustment

Changes in the volume and net mass of the ocean are inferred by observing variations in globally averaged sea level and by modeling the changes in the shape of the ocean basin that are caused by geophysical processes. These processes include tectonic deformations from earthquakes and plate motion, subsidence from water or gas/oil extraction, and the short- and long-term responses to the removal of ice and liquid water loads. The responses to load changes are particularly important when estimating the sea level budget. One is the nearly instantaneous, elastic response of Earth due to contemporary load changes (e.g., melting ice sheets). The second response is the time-delayed, viscous response due to load changes from the last ice age known as GIA. After some discussion about the appropriate methodology to compute the GIA effect on ocean mass estimates from GRACE (Peltier 2009; Chambers et al. 2010, 2012; Peltier et al. 2012), model predictions of the GIA effects on tide gauges (relative sea level), altimetry (geocentric sea level or the sea surface), and the time-varying geoid as measured by GRACE are available for sea level budget studies (Tamisiea and Mitrovica 2011; Tamisiea et al. 2014).

In the global mean, GIA produces a small net sinking of the ocean floor relative to the Earth's center, which, if the sea surface was static, would imply a small secular increase in volume (Douglas and Peltier 2002). Geophysical modeling suggests this volume change is 75–180 km^3/year, depending mainly on assumptions about Earth's lower mantle viscosity. Thus, the rate of GMSL observed by altimeters is higher by +0.2 to +0.5 mm/year when GMSL is expressed as a change in ocean volume (Tamisiea 2011). Similarly, GIA effects on mass must be removed from GRACE observations of changes in the geoid in order to isolate ocean mass variations (Chambers et al. 2010). This GIA "correction" for ocean mass variations accounts for mass redistribution from crustal motion, which produces significantly larger apparent changes in terms of equivalent water height, roughly 1 mm/year.

5.4.2 Ocean Mass/Barystatic Sea Level from GRACE

Without a global array of ocean bottom pressure gauges available to directly measure ocean mass changes, several approaches have been used to infer the contribution of ocean mass/barystatic sea level to the sea level budget. In one approach, all individual changes in continental water storage are estimated and combined (e.g., Church et al. 2011), including changes in the mass balance of the Greenland and Antarctic ice sheets, mountain glaciers and ice caps, changes in dam retention, the depletion of groundwater, and natural variability of terrestrial surface waters (Syed et al. 2010; Jensen et al. 2013). The net change in the sum of these reservoirs is assumed to reflect an exchange of freshwater with the ocean. In another approach, long-term trends of water transferred from continents to the ocean are inferred from freshening of the ocean (i.e., a decrease in global mean salinity), assuming that the contribution from melting of sea ice, which affects freshening but not sea level, can be estimated (Munk 2003; Wadhams and Munk 2004), but the errors are large. A more recent method of inferred ocean mass exploits "sea level fingerprints," regional patterns in sea level change that result from the deformation of the solid Earth and the changes in the Earth's gravitational field in response to ice sheets gaining or losing mass (Rietbroek et al. 2011). In theory, sea level measurements from tide gauges can be used to solve for "fingerprints" and infer the individual contributions to sea level (Hay et al. 2012).

More important for sea level budget studies, tracking water mass movements on land and in the ocean at unprecedented spatial scales and precision has been possible since the launch in 2002 of the GRACE mission (see Chapter 4, Section 4.5). GRACE provides observations of terrestrial water storage, the exchange of water between the ocean and the continents, and the redistribution of mass within the ocean. The largest signals in the gravity field result from the seasonal variations of water stored on the continents as snow, ice, and water in rivers, reservoirs, and underground. For regions near coasts, it is impossible to distinguish whether gravity variations observed by GRACE were influenced by water changes on land or in the ocean. Therefore, when analyzing the gravity field for mass variations in the ocean, various techniques are used to limit large land hydrology signals from being misinterpreted as changes in barystatic sea level (see Chapter 5). These include masking of the ocean near coasts, using special averaging functions, mass concentration (mascon) methods, and removing a hydrology model from the fields. Based on these leakage errors and GRACE instrument errors, the error bounds for monthly changes in the global mean of ocean mass anomalies are roughly 2 mm (Willis et al. 2008).

5.4.3 Closure and Missing Components

On a monthly basis, the sum of the steric component estimated from Argo (Roemmich and Gilson 2009) and the barystatic component from GRACE (Johnson and Chambers 2014) agree with total sea level from Jason-1 and Jason-2 (Leuliette and Scharroo 2010) within the estimated uncertainties (Figure 5.9), with the residual difference having a root mean square (RMS) of 1.6 mm for the period January 2005 to December 2015 after 2-month boxcar smoothing has been applied (Leuliette and Nerem 2016). Direct measurements of ocean warming above 2000 m in depth explain about one-third (1.1 mm/year) of the observed rate of global mean sea level rise (3.4 mm/year). The rate of GRACE-derived global barystatic sea level (2.3 mm/year) accounts for the remaining two-thirds of sea level rise.

Until the planned deployment of Deep Argo, the current ocean observing system continues to provide sparse sampling of temperature observations below 2000 dB. Direct observations suggest that the deep ocean (below 2000 db) contributes on the order of approximately 0.1 mm/year to global sea level rise (Purkey and Johnson 2010; Llovel et al. 2014; Purkey et al. 2014; Dieng et al. 2015a).

To close the sea level budget, all measurements must be made in a common terrestrial reference frame (see Chapter 1). The International Terrestrial Reference System (ITRS) uses station data from the global networks for the four major precise positioning systems (satellite laser ranging, GNSS, Very Long Baseline Interferometry [VLBI], and Doppler Orbitography and

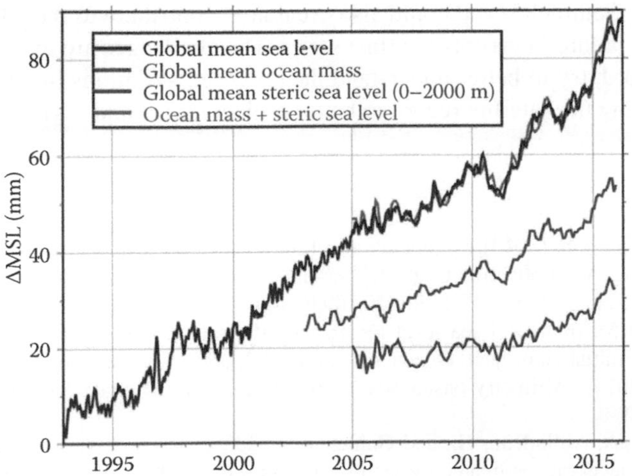

FIGURE 5.9 Comparisons of global mean sea level from TOPEX/Jason-1/Jason-2 (Leuliette and Scharroo 2010), global mean ocean mass from GRACE (Johnson and Chambers 2013), and steric (density) sea level from Argo (Roemmich and Gilson 2009), with seasonal variations removed and 2-month smoothing applied.

Radiopositioning Integrated by Satellite [DORIS]) that form the backbone of the global ocean observing system. The resulting International Terrestrial Reference Frame (ITRF) has translation rate and scale rate errors, which in terms of GMSL are on the order of approximately 0.1 mm/year (Collilieux and Wöppelmann 2011).

5.5 HOW ALTIMETRY INFORMS US ABOUT THE FUTURE

A major reason for monitoring changes in GMSL with satellite altimeter measurements is to help us understand how GMSL and components are changing with time. This leads to improved process understanding and thus improved climate change modeling and hopefully more realistic projections of future sea level change. GMSL is rising at a faster rate in the satellite record than has been observed in the longer tide gauge record (Church and White 2011; Hay et al. 2015). Thermosteric sea level rise, the loss of mass from mountain glaciers, and Greenland and Antarctica are all contributing roughly equally to the present-day rise of sea level. However, during the last decade, land ice melt has accelerated while thermal expansion has slightly reduced. Land ice melt contributed to nearly 60% of the recent years' GMSL rise (Chambers et al. 2016; Dieng et al. 2017).

The tide gauge record of sea level change shows that there has been a slight acceleration of sea level rise over the last 75 years. However, the detection of an acceleration in the satellite record has been more challenging—in fact, the nominal satellite record shows the rate of sea level rise over 1993–2003 is greater than over 2003–2013, although there is some evidence that there may be errors in the TOPEX part of the altimeter record (Watson et al. 2015). A careful reprocessing of the TOPEX data will hopefully resolve these uncertainties. There is also evidence that the eruption of Mount Pinatubo just prior to the launch of TOPEX caused a higher rate of sea level rise in the first half of the satellite GMSL record (Fasullo et al. 2016). Finally, there is evidence of other sources of decadal variability in the satellite GMSL record that might be obscuring an acceleration of GMSL (Hamlington et al. 2016).

We have seen a remarkable string of successful precision altimetry missions with the success of TOPEX/Poseidon, Jason-1, Jason-2, and Jason-3—all with overlap between the missions to allow intercalibration of the measurements and to continue a nearly uninterrupted climate data record (CDR). Barring any failures, Jason-CS/Sentinel-6 is poised to continue this record along the same ground track first flown by TOPEX but using an advanced delay-Doppler altimeter. Other altimeter missions flying different ground tracks are approaching the performance of the TOPEX/Jason missions (e.g., Sentinel-3) and could also eventually contribute to the CDR and be critical if a launch or satellite failure interrupts the time series. The challenges are to continue to extend the satellite GMSL record, try to better understand the observed variations in GMSL, and to develop new methods for using the satellite record to better project global and regional sea level change.

REFERENCES

A, G., Wahr, J., and Zhong, S. (2013). Computations of the viscoelastic response of a 3-D compressible earth to surface loading: An application to glacial isostatic adjustment in Antarctica and Canada. Geophys. J. Roy. Astronom. Soc., 192, 557–572. doi: 10.1093/gji/ggs030.

Ablain, M., et al. (2015). Improved sea level record over the satellite altimetry era (1993–2010) from the Climate Change Initiative project. *Ocean Sci.*, 11, 67–82, doi: 10.5194/os-11-67-2015.

Ablain, M., et al. (2016). Altimetry-based sea level, global and regional. *Surv Geophy.* doi: 10.1007/s10712-016-9389-8.

Ablain, M., Cazenave, A., Valladeau, G., and Guinehut, S. (2009). A new assessment of the error budget of global mean sea level rate estimated by satellite altimetry over 1993-2008. *Ocean Sci.*, 5, 193–201.

Ablain, M., Philipps, S., Urvoy, M., Tran, N., and Picot, N. (2012). Detection of long-term instabilities on altimeter backscattering coefficient thanks to wind speed data comparisons from altimeters and models. *Marine Geodesy*, 35(S1), 42–60. doi: 10.1080/01490419.2012.718675.

Abraham, J.P., et al. (2013). A review of global ocean temperature observations: Implications for ocean heat content estimates and climate change. *Rev. Geophys.*, 51, 450–483, RG000432.

Balmaseda, M.A., Mogensen, K., and Weaver, A. (2013b). Evaluation of the ECMWF ocean reanalysis ORAS4. *Q. J. R. Meteorol. Soc.*, doi: 10.1002/qj.2063.

Biancamaria, S., et al. (2011). Satellite-based high latitude snow volume trend, variability and contribution to sea level over 1989/2006. *Glob. Planet. Change.*, 75(3–4), 99–107.

Bindoff, N., et al. (2007). Observations: Oceanic climate and sea level. In: *Climate Change 2007: The Physical Science Basis. Contribution of Working Group I to the Fourth Assessment Report of the Intergovernmental Panel on Climate Change* [Solomon, S., et al. (eds.)]. Cambridge University Press, Cambridge, UK.

Boening, C., Willis, J.K., Landerer, F.W., Nerem, R.S. and Fasullo, J. (2012). The 2011 La Niña: So strong, the oceans fell. *Geophys. Res. Lett.*, 39(19), L19602.

Boyer, T.P., et al. (2014). 2013 World ocean atlas aids high-resolution climate studies. *Eos Trans. Am. Geophys. Union*, 95(41), 369–370. doi: 10.1002/2014EO410002.

Cazenave, A., Dieng, H.B., Meyssignac, B., von Schuckmann, K., Decharme, B. and Berthier, E. (2014). The rate of sea-level rise. *Nat. Clim. Change*, 4(5), 358–361.

Chao, B.F., et al. (2008). Impact of artificial reservoir water impoundment on global sea level. *Science*, 320(5873), 212–214.

Chen, J.L., et al. (2005). Low degree spherical harmonic influences in Gravity Recovery and Climate Experiment (GRACE) water storage estimates. *Geophys. Res. Lett.*, 32, L14405. doi: 14410.11029/12005GL022964.

Chambers, D.P., et al. (2016). Evaluation of the Global Mean Sea level budget between 1993 and 2014. *Surv. Geophys.*, 1–19. doi: 10.1007/s10712-016-9381-3.

Chambers, D.P., Wahr, J., and Nerem, R.S. (2004). Preliminary observations of global ocean mass variations with GRACE. *Geophys. Res. Lett.*, 31(13). doi: 10.1029/2004GL020461.

Chambers, D.P., Wahr, J., Tamisiea, M.E., and Nerem, R.S. (2010). Ocean mass from GRACE and glacial isostatic adjustment. *J. Geophys. Res.*, 115, B11415. doi: 10.1029/2010JB00753.

Chambers, D.P., Wahr, J., Tamisiea, M.E., and Nerem, R.S. (2012). Reply to comment by W. R. Peltier et al. on "Ocean mass from GRACE and glacial isostatic adjustment." *J. Geophys. Res.*, 117, B11404. doi: 10.1029/2012JB009441.

Church, J.A., et al. (2011). Revisiting the earth's sea-level and energy budgets from 1961 to 2008. *Geophys. Res. Lett.*, 38(18), L18601. doi: 10.1029/2011GL048794.

Church, J.A., et al. (2013). Sea level change. In: *Climate Change 2013: The Physical Science Basis. Contribution of Working Group I to the Fifth Assessment Report of the Intergovernmental Panel on Climate Change* [Stocker, T.F., et al. (eds.)]. Cambridge University Press, Cambridge, United Kingdom.

Church, J.A., and White, N.J. (2011). Sea-level rise from the late 19th to the early 21st century. *Surv. Geophys.*, 32(4–5), 585–602.

Collilieux, X., and Wöppelmann, G. (2011). Global sea-level rise and its relation to the terrestrial reference frame. *J. Geod.*, 85(1), 9–22. doi: 10.1007/s00190-010-0412-4.

Couhert, A., et al. (2015). Toward the 1 mm/y stability of the radial orbit error at regional scales. *Adv. Space Res.*, 55, 2–23.

DeConto, R.M., and Pollard, D. (2016). Contribution of Antarctica to past and future sea-level rise. *Nature*, 531, 591.

Dibarboure, G., et al. (2011). Jason-2 in DUACS: Updated system description, first tandem results and impact on processing and products. *Marine Geodesy*, 34(3–4), 214–241.

Dieng, H.B., Cazenave, A., Messignac, B., Henry, O., von Schuckmann, K., and Lemoine, J.M. (2014). Effect of La Niña on the global mean sea level and north Pacific ocean mass over 2005–2011. *J. Geodetic. Sci.*, 4, 19–27.

Dieng, H. B., A. Cazenave, B. Meyssignac, and M. Ablain (2017), New estimate of the current rate of sea level rise from a sea level budget approach, *Geophys. Res. Lett.*, 2017GL073308.

Dieng, H.B., Cazenave, A., von Schuckmann, K., Ablain, M., and Meyssignac, B. (2015b). Sea level budget over 2005–2013: Missing contributions and data errors. *Ocean Sci.*, 11, 789–802. doi: 10.5194/os-11-789-2015.

Dieng, H.B., Champollion, N., Cazenave, A., Wada, Y., Schrama, E., and Meyssignac, B. (2015c). Total land water storage change over 2003–2013 estimated from a global mass budget approach. *Environ. Res. Lett.*, 10, 124010. doi: 10.1088/1748-9326/10/12/124010.

Dieng, H.B., Palanisamy, H., Cazenave, A., Meyssignac, B., and von Schuckmann, K. (2015a). The sea level budget since 2003: Inference on the deep ocean heat content. *Surveys Geophys.*, 36, 1, doi: 10.1007/s10712-015-9314-6.

Döll, P., et al. (2014). Global-scale assessment of groundwater depletion and related groundwater abstractions: Combining hydrological modeling with information from well observations and GRACE satellites. *Water Resour. Res.*, 50(7), 5698–5720.

Domingues, C.M., et al. (2008). Improved estimates of upper ocean warming and multi decadal sea level rise. *Nature*, 453, 1090–1993.

Douglas, B.C. (1991). Global sea level rise. *J. Geophys. Res. C Oceans*, 96(C4), 6981–6992.

Douglas, B.C., and Peltier, W. (2002). The puzzle of global sea-level rise. *Physics Today*, 55, 35–40. doi: 10.1063/1.1472392.

Ducet, N., Le Traon, P.Y., and Reverdin, G. (2000). Global high resolution mapping of ocean circulation from the combination of TOPEX/POSEIDON and ERS-1/2. *J. Geophys. Res.*, 105(C8), 19477–19498.

Fasullo, J.T., et al. (2013). Australia's unique influence on global sea level in 2010–2011. *Geophys. Res. Lett.*, 40(16), 4368–4373.

Fasullo, J. T., R. S. Nerem, and B. Hamlington (2016), Is the detection of accelerated sea level rise imminent?, *Sci. Rep.*, 6, 31245.

Fenoglio-Marc, L., Groten, E., and Dietz, C. (2004). Vertical land motion in the Mediterranean Sea from altimetry and tide gauge stations. *Marine Geodesy*, 27(3–4), 683–701.

Forsberg, R., et al. (2016). Greenland and Antarctica ice sheet mass changes and effects on global sea level. *Surveys Geophys.*

Gardner, A.S., Moholdt, G., Cogley, J.G., and Wouters, B. (2013). A reconciled estimate of glacier contributions to sea level rise: 2003 to 2009. *Science.*

Good, S.A., Martin, M.J., and Rayner, N.A. (2013). EN4: Quality controlled ocean temperature and salinity profiles and monthly objective analyses with uncertainty estimates. *J. Geophys. Res. Oceans*, 118, 6704–6716, doi: 10.1002/2013JC009067.

Gornitz, V., et al. (1997). Effects of anthropogenic intervention in the land hydrological cycle on global sea level rise. *Glob. Planetary Change*, 14, 147–161.

Gregory, J.M., et al. (2013). Twentieth-century global-mean sea level rise: Is the whole greater than the sum of the parts? *J Clim.*, 26, 4476–4499. doi: 10.1175/JCLI-D-12-00319.1.

Hamlington, B.D., Leben, R.R., Strassburg, M.W., Nerem, R.S., and Kim, K.-Y. (2013). Contribution of the Pacific decadal oscillation to global mean sea level trends. *Geophys. Res. Lett.*, 40(19), 5171–5175.

Hamlington, B.D., and Thompson, P.R. (2015). Considerations for estimating the 20th century trend in global mean sea level. *Geophys. Res. Lett.* doi: 10.1002/2015GL064177.

Hamlington, B. D., J. T. Reager, M.-H. Lo, K. B. Karnauskas, and R. R. Leben (2017), Separating decadal global water cycle variability from sea level rise, *Sci. Rep., 7*(1), 995.

Han, W., et al. (2016). Spatial patterns of sea level; climate modes Decadal sea level variability regional sea level change. *Surveys Geophys.*

Hay, C.C., Morrow, E., Kopp, R.E., and Mitrovica, J.X. (2012). Estimating the sources of global sea level rise with data assimilation techniques. *Proc. Natl. Acad. Sci. U. S. A.* doi: 10.1073/pnas.1117683109.

Hay, C.C., Morrow, E., Kopp, R.E., and Mitrovica, J.X. (2015). Probabilistic reanalysis of twentieth-century sea-level rise. *Nature.* doi: 10.1038/nature14093.

Henry, O., et al. (2014). Effect of the processing methodology on satellite altimetry-based global mean sea level rise over the Jason-1 operating period. *J. Geodesy*, 88, 351–361. doi: 10.1007/s00190-013-0687-3.

Ishii, M., and Kimoto, M. (2009). Reevaluation of historical ocean heat content variations with time-varying XBT and MBT depth bias corrections. *J. Oceanogr.* 65(3), 287–299. doi: 10.1007/s10872-009-0027-7.

Jacob, T., Wahr, J., Pfeffer, W.T., and Swenson, S. (2012). Recent contributions of glaciers and ice caps to sea level rise. *Nature*, 482(7386), 514–518.

Jensen, L., Rietbroek, R., and Kusche, J. (2013). Land water contribution to sea level from GRACE and Jason-1measurements. *J Geophys. Res. Oceans.*, 118, 212–226. doi: 10.1002/jgrc.20058.

Jevrejeva, S., et al. (2014). Trends and acceleration in global and regional sea levels since 1807. *Glob. Planet. Change*, 113, 11–22.

Johnson, G.C., and Chambers, D.P. (2013). Ocean bottom pressure seasonal cycles and decadal trends from GRACE release-05: Ocean circulation implications. *J. Geophys. Res. Oceans*, 118(9), 4228–4240. doi: 10.1002/jgrc.20307.

Kohl, A., and Stammer, D. (2008). Decadal sea level changes in the 50-year GECCO ocean synthesis. *J Clim.*, 21(9), 1876–1890. doi: 10.1175/2007JCLI2081.1.

Konikow, L.F. (2011). Contribution of global groundwater depletion since 1900 to sea-level rise. *Geophys. Res. Lett.*, 38(17), L17401.

Konikow, L.F. (2013). Overestimated water storage. *Nat. Geosci.*, 6(1), 3–3.

Landerer, F.W., et al. (2008). El Niño-Southern Oscillation signals in sea level, surface mass redistribution, and degree-two geoid coefficients. *J. Geophys. Res.*, 113, C08014. doi: 08010.01029/02008JC004767.

Leclercq P., et al. (2014). A data set of worldwide glacier length fluctuations. *Cryosphere*, 8, 659–672.

Legeais, J.-F., Ablain, M., and Thao, S. (2014). Evaluation of wet troposphere path delays from atmospheric reanalyses and radiometers and their impact on the altimeter sea level. *Ocean Sci.*, 10, 893–905. doi: 10.5194/os-10-893-2014.

Le Traon, P.Y., Faugère, Y., Hernandez, F., Dorandeu, J., Mertz, F., and Ablain, M. (2003). Can we merge GEOSAT follow-on with TOPEX/Poseidon and ERS-2 for an improved description of the ocean circulation? *J. Atmos. Ocean. Technol*, 20, 889–895, doi: 10.1175/1520-0426(2003)020<0889:CWMGFW>2.0.CO;2.

Lettenmaier, D.P., and Milly, P.C.D. (2009). Land waters and sea level. *Nat. Geosci.*, 2(7), 452–454.

Leuliette, E.W. (2015). The balancing of the sea-level budget. *Curr Clim Change Rep*, 1(3), 185–191. doi: 10.1007/s40641-015-0012-8.

Leuliette, E.W., and Miller, L. (2009). Closing the sea level rise budget with altimetry, Argo, and GRACE. *Geophys. Res. Lett.*, 36(4), L04608. doi: 10.1029/2008GL036010.

Leuliette, E.W., and Nerem, R.S. (2016). Contributions of Greenland and Antarctica to global and regional sea level change. *Oceanography,* 29(4), 154–159. doi: 10.5670/oceanog.2016.107.

Leuliette, E.W., and Scharroo, R. (2010). Integrating Jason-2 into a multiple-altimeter climate data record. *Marine Geodesy*, 33(1), 504–517. doi: 10.1080/01490419.2010.487795.

Leuliette, E., and Willis, J. (2011). Balancing the sea level budget. *Oceanography*, 24, 122–129. doi: 10.5670/oceanog.2011.32.

Levitus, S., et al. (2012). World ocean heat content and thermosteric sea level change (0–2000 m), 1955–2010. *Geophys. Res. Lett.*, 39, L10603. doi: 10.1019/2012GL051106.

Llovel, W., et al. (2011). Terrestrial waters and sea level variations on interannual time scale. *Glob. Planet. Change*, 75(1–2), 76–82.

Llovel, W., Willis, J.K., Landerer, F.W., and Fukumori, I. (2014). Deep-ocean contribution to sea level and energy budget not detectable over the past decade. *Nat. Clim. Change*. doi: 10.1038/nclimate2387.

Lyman, J.M., et al. (2010). Robust warming of the global upper ocean. *Nature*, 465, 334–337. doi: 10.1038/nature09043.

Lyman, J.M., and Johnson, G.C. (2014). Estimating global ocean heat content changes in the upper 1800 m since 1950 and the influence of climatology choice. *J. Clim.*, 27(5), 1945–1957. doi: 10.1175/JCLIM-D-12-00752.1.

Masters, D., et al. (2012). Comparison of global mean sea level time series from TOPEX/Poseidon, Jason-1, and Jason-2. *Mar. Geod.*, 35, 20–41.

Marzeion, B., et al. (2016). Observation of glacier mass changes on the global scale and its contribution to sea level change. *Surveys Geophys.*

Merrifield, M.A., and Maltrud, M.E. (2011). Regional sea level trends due to a Pacific trade wind intensification. *Geophys. Res. Lett.*, 38. doi: 10.1029/2011GL049576.

Mertz, F., Mercier, F., Labroue, S., Tran, N., and Dorandeu, J. (2005). *ERS-2 OPR data quality assessment; long-term monitoring – Particular investigation.* CLS.DOS.NT-06.001. http://www.aviso.altimetry.fr/fileadmin/documents/calval/validation_report/E2/annual_report_e2_2005.pdf (accessed 11 May 2016).

Meyssignac, B., et al. (2016). Causes of the regional variability in observed sea level, sea surface temperature and ocean colour. *Surveys Geophys*. doi: 10.1007/s10712-016-9383-1.

Milly, P.C.D., et al. (2003). Contribution of climate-driven change in continental water storage to recent sea-level rise. *Proc. Natl. Acad. Sci.*, 100(23), 13158–13161.

Milly, P.C.D., et al. (2010). Terrestrial water-storage contributions to sea-level rise and variability. In *Understanding Sea-Level Rise and Variability* [Church, J.A., Woodworth, P.L., Aarup, T., and Wilson, W. S. (eds.)]. Blackwell Publishing Ltd, London.

Milne, G.A., et al. (2009). Identifying the causes of sea-level change. *Nat. Geosci.*, 2, 471–478.

Mitchum, G.T. (1998). Monitoring the stability of satellite altimeters with tide gauges. *J. Atmos. Ocean. Technol.*, 15(3), 721–730.

Mitchum, G. T. (2000), An Improved Calibration of Satellite Altimetric Heights Using Tide Gauge Sea Levels with Adjustment for Land Motion, *Mar. Geod., 23*(3), 145–166.

Mitrovica, J.X., Tamisiea, M.E., Davis, J.L., and Milne, G.A. (2001). Polar ice mass variations and the geometry of global sea level change. *Nature*, 409, 1026–1029. doi: 10.1038/35059054.

Munk, W. (2003). Ocean freshening, sea level rising. *Science*, 300, 2041–2043. doi: 10.1126/science.1085534.

Nerem, R.S. (1995). Global mean sea level variations from TOPEX/POSEIDON altimeter data. *Science*, 268(5211), 708–710.

Nerem, R.S., Chambers, D.P., Choe, C., and Mitchum, G.T. (2010). Estimating mean sea level change from the TOPEX and Jason altimeter missions. *Mar. Geodesy*, 33(1), 435–446.

Nerem, R.S., Chambers, D.P., Leuliette, E.W., Mitchum, G.T., and Giese, B.S. (1999). Variations in global mea sea level associated with the 1997–98 ENSO event. *Geophys. Res. Lett.*, 26(19), 3005–3008.

Ngo-Duc, T., et al. (2005). Effects of land water storage on global mean sea level over the past half century. *Geophys. Res. Lett.*, 32, 9704–9707.

Ollivier, A., Faugère, Y., Picot, N., Ablain, M., Femenias P., and Benveniste, J. (2012). Envisat Ocean altimeter becoming relevant for mean sea level trend studies. *Marine Geodesy*, 35(Suppl. 1), 118–136.

Palanisamy, H. Cazenave, A., Delcroix, T., and Meyssignac, B. (2015). Spatial trend patterns in Pacific Ocean sea level during the altimetry era: The contribution of thermocline depth change and internal climate variability. *Ocean Dyn.* doi: 10.1007/s10236-014-0805-7.

Peltier, W. R. (2004), GLOBAL GLACIAL ISOSTASY AND THE SURFACE OF THE ICE-AGE EARTH: The ICE-5G (VM2) Model and GRACE, *Annu. Rev. Earth Planet. Sci., 32*(1), 111–149.

Peltier, W.R. (2009). Closure of the budget of global sea level rise over the GRACE era: The importance and magnitudes of the required corrections for global glacial isostatic adjustment. *Quarter. Sci. Rev.*, 1–17. doi: 10.1016/j.quascirev.2009.04.004.

Peltier, W.R., Drummond, R., and Roy, K. (2012). Comment on "Ocean mass from GRACE and glacial isostatic adjustment" by D. *P. Chambers et al. J. Geophys. Res.*, 117, B11403. doi: 10.1029/2011JB008967.

Pokhrel, Y.N., et al. (2012). Model estimates of sea-level change due to anthropogenic impacts on terrestrial water storage. *Nat. Geosci.*, 5(6), 389–392.

Prandi, P., Ablain, M., Zawadzki, L., and Meyssignac, B. (2016). *How reliable are local sea level trends*?

Purkey, S.G., and Johnson, G.C. (2010). Warming of global abyssal and deep Southern Ocean waters between the 1990s and 2000s: Contributions to global heat and sea level rise budgets. *J Clim.*, 23, 6336–6351. doi: 10.1175/2010JCLI3682.1.

Purkey, S.G., Johnson, G.C., and Chambers, D.P. (2014). Relative contributions of ocean mass and deep steric changes to sea level rise between 1993 and 2013. *J. Geophys. Res. Oceans*, 119, 7509–7522. doi: 10.1002/2014JC010180.

Ramillien, G., et al. (2008). Detection of continental hydrology and glaciology signals from GRACE: A review. *Surveys Geophys.*, 29(4–5), 361–374.

Reager, J.T., et al. (2016). A decade of sea level rise slowed by climate-driven hydrology. *Science*, 351(6274), 699–703.

Rietbroek, R., Brunnabend, S.-E., Kusche, J., and Schröter, J. (2011). Resolving sea level contributions by identifying fingerprints in time-variable gravity and altimetry. *J. Geodyn.*, 1–10. doi: 10.1016/j.jog.2011.06.007.

Roemmich, D., and Gilson, J. (2009). The 2004–2008 mean and annual cycle of temperature, salinity, and steric height in the global ocean from the Argo Program. *Prog. Oceanogr.*, 82, 81–100. doi: 10.1016/j.pocean.2009.03.004.

Roemmich, D., Gould, W.J., and Gilson, J. (2012). 135 years of global ocean warming between the challenger expedition and the Argo Programme. *Nat. Clim. Change*, 2(6), 425–428. doi: 10.1038/nclimate1461.

Rudenko, S., et al. (2014). Influence of time variable geopotential models on precise orbits of altimetry satellites, global and regional mean sea level trends. *Adv. Space Res.*, 54(1), 92–118. doi: 10.1016/j.asr.2014.03.010.

Sahagian, D.L., et al. (1994). Direct anthropogenic contributions to sea level rise in the twentieth century. *Nature*, 367, 54–57.

Santamaría-Gómez, A., Gravelle, M., and Wöppelmann, G. (2014). Long-term vertical land motion from double-differenced tide gauge and satellite altimetry data. *J Geod.*, 88, 207–222. doi: 10.1007/s00190-013-0677-5.

Shepherd, A., et al. (2012). A reconciled estimate of ice-sheet mass balance. *Science*, 338(6111), 1183–1189.

Stammer, D., Cazenave, A., Ponte, R.M., and Tamisiea, M.E. (2013). Causes for contemporary regional sea level changes. *Annu. Rev. Mar. Sci.*, 5. doi: 10.1146/annurev-marine-121211-172406.

Syed, T.H., et al. (2010). Satellite-based global-ocean mass balance estimates of interannual variability and emerging trends in continental freshwater discharge. *Proc. Natl. Acad. Sci.* doi: 10.1073/pnas.1003292107.

Tamisiea, M.E. (2011). Ongoing glacial isostatic contributions to observations of sea level change. *Geophys. J. Int.*, 186(3), 1036–1044. doi: 10.1111/j.1365-246X.2011.05116.x.

Tamisiea, M.E., Hughes, C.W., Williams, S.D.P., and Bingley, R.M. (2014). Sea level: Measuring the bounding surfaces of the ocean. *Philos. Trans. Roy. Soc. A Math. Phys. Eng. Sci.*, 372, 20130336–20130336. doi: 10.1098/rsta.2013.0336.

Tamisiea, M.E., and Mitrovica, J. (2011). The moving boundaries of sea level change: Understanding the origins of geographic variability. *Oceanography*, 24, 24–39. doi: 10.5670/oceanog.2011.25.

Valladeau, G., Legeais, J.F., Ablain, M., Guinehut, S., and Picot, N. (2012). Comparing altimetry with tide gauges and argo profiling floats for data quality assessment and mean sea level studies. *Marine Geodesy*, 35(Suppl. 1), 42–60.

Velicogna, I., Sutterley, T.C., and van den Broeke, M.R. (2014). Regional acceleration in ice mass loss from Greenland and Antarctica using grace time variable gravity data. *Geophys. Res. Lett.*, 41, 8130–8137.

Velicogna, I., and Wahr, J. (2013). Time-variable gravity observations of ice sheet mass balance: Precision and limitations of the GRACE satellite data. *Geophys. Res. Lett.*

Von Schuckmann, K., et al. (2014). Consistency of the current global ocean observing systems from an Argo perspective. *Ocean Sci.*, 10, 547–557. doi: 10.5194/os-10-547-2014.

Wada, Y., et al. (2010). Global depletion of groundwater resources. *Geophys. Res. Lett.*, 37(20), L20402.

Wada, Y., et al. (2012). Past and future contribution of global groundwater depletion to sea-level rise. *Geophys. Res. Lett.*, 39(9), L09402.

Wada, Y., et al. (2016). Fate of water pumped from underground and contributions to sea-level rise. *Nat. Clim. Change*, 6(8), 777–780.

Wadhams, P., and Munk, W. (2004). Ocean freshening, sea level rising, sea ice melting. *Geophys. Res. Lett.*, 31, L11311. doi: 10.1029/2004GL020039.

Watkins, M.M., Wiese, D.N., Yuan, D.-N., Boening, C., and Landerer, F.W. (2015). Improved methods for observing earth's time variable mass distribution with GRACE using spherical cap mascons. *J. Geophys. Res. [Solid Earth]*, 120(4), 2648–2671.

Watson, C.S., White, N.J., Church, J.A., King, M.A., Burgette, R.J., and Legresy, B. (2015). Unabated global mean sea-level rise over the satellite altimeter era. *Nat. Clim. Change*. doi: 10.1038/nclimate2635.

Willis, J.K., Chambers, D.P., and Nerem, R.S. (2008). Assessing the globally averaged sea level budget on seasonal to interannual timescales. *J. Geophys. Res*, 113(C6), C06015. doi: 10.1029/2007JC004517.

Wöppelmann, G., and Marcos, M. (2016). Vertical land motion as a key to understanding sea level change and variability. *Rev. Geophys.*, 54. doi: 10.1002/2015RG000502.

Zhang, X., and Church, J.A. (2012). Sea level trends, interannual and decadal variability in the Pacific Ocean. *Geophys. Res. Lett.*, 39. doi: 10.1029/2012GL053240.

Zawadzki, L., and Ablain, M. (2016). Accuracy of the mean sea level continuous record with future altimetric missions: Jason-3 vs. Sentinel-3a. *Ocean Sci.*, 12, 9–18, doi: 10.5194/os-12-9-2016.

6 Monitoring and Interpreting Mid-Latitude Oceans by Satellite Altimetry

Kathryn A. Kelly, Joshua K. Willis, Gilles Reverdin, Shenfu Dong, and LuAnne Thompson

6.1 INTRODUCTION: ROLE OF MID-LATITUDE OCEANS

Large-scale shipboard observations of the global oceans as part of the World Ocean Circulation Experiment (WOCE) allowed a comprehensive description of the mean ocean circulation throughout the world's oceans and direct estimates of the heat transported by the ocean (Ganachaud and Wunsch, 2000). But until the advent of satellite altimetry, the time-varying ocean circulation remained elusive. Satellite altimetry has revolutionized studies of the variability of ocean circulation and the dynamics that control that variability by providing continuous observations of the time-varying sea level and thereby of dynamical features in the ocean. Within its first decade, radar altimetry allowed the first global observation of Rossby waves (Chelton and Schlax 1996), leading to a series of studies examining how wind stress drives variations in ocean circulation. Variability in western boundary current extensions were also quantified using satellite sea level, allowing a more detailed view of that variability than is possible with either repeat shipboard transects or moored observations, beginning with Kelly and Gille (1990). The altimeter also allowed the first estimate of the spatial structure of eddy kinetic energy, creating an important metric for assessing the fidelity of mid-latitude eddy-resolving ocean models (Ducet et al. 2000).

Understanding the variability of mid-latitude ocean circulation and its contributions to ocean heat and freshwater storage and transport has become essential for accurate predictions of climate change (Meehl et al. 2014). The mid-latitude ocean basins are composed of the subtropical gyres and subpolar gyres in both hemispheres. The subtropical gyres are characterized by strong poleward-flowing western boundary currents with weak equatorward return flows. These boundary currents carry heat from the tropics to higher latitudes, where much of the heat is fluxed to the atmosphere. The maximum meridional (poleward) heat transport of about 2 PW occurs near 20°S and 20°N (Trenberth and Caron 2001). The western boundary currents separate near the maximum in the westerly surface winds at around 35 to 40 degrees latitude to become western boundary current extensions. Heat transport decreases dramatically poleward of the boundary current separation and is reflected in the large flux of heat to the atmosphere in this region (see, e.g., Yu and Weller 2007).

Over the past decade, the scope of scientific topics that have been addressed using satellite altimeter data has increased dramatically. In the mid-latitudes, early work focused on topics such as the structure and propagation of eddies and Rossby waves, changes in ocean heat content, and large-scale changes in ocean circulation (see Fu and Cazenave 2001, for a comprehensive review). As researchers have continued to exploit the accuracy, resolution, and availability of ocean altimeter observations, more comprehensive aspects of ocean circulation have been explored. In recent years, the longer satellite altimetry record has allowed examination of the interactions of eddies with evolving western boundary current extensions and the relationship of western boundary extension path variability with wind forcing (Frankignoul et al. 2001). The combination of satellite altimetry with other satellite-derived fields as well as *in situ* observations, including Argo temperature and salinity profiles, has allowed insight into basin-scale measures of ocean variability such as the Atlantic Meridional Overturning Circulation (Willis 2010; Frajka-Williams 2015) and meridional heat transport (Kelly et al. 2014). In addition, satellite altimetry has allowed new insight in modes of ocean climate variability such as the North Pacific Gyre Oscillation (Di Lorenzo et al. 2008) and the examination of sea surface height (SSH) and heat content variability in the northern North Atlantic (Häkkinen et al. 2013).

Here, we review some examples of new insights that satellite altimetry has brought to the understanding of mid-latitude ocean dynamics, the interaction of the mid-latitude ocean with other ocean basins and with the atmosphere, and the role of the ocean in climate. We highlight only some of the topics that have been advanced through the use of satellite altimetry in the study of the mid-latitude oceans. Progress has been made in characterizing and understanding interannual variability in the paths of western boundary current extensions. The heat budgets in the vicinity of the boundary current extensions are now better quantified by combining satellite altimetry with *in situ* profiles of temperature and salinity. Variability in basin-wide heat, freshwater, and volume budgets can now be constructed, enabled by altimetry-derived proxies for meridional heat transport at fixed latitudes. We also highlight studies of decadal scale changes in water mass properties, heat and freshwater uptake, and large-scale structure and circulation of subpolar and subtropical gyres throughout the globe. The linkages between the observed changes in mid-latitude ocean circulation and wind-stress changes through shifts in atmospheric modes of variability are highlighted throughout this chapter. For more contributions of the altimeter to understanding mid-latitude ocean variability, see the chapters on eddy variability and on data assimilation.

The linkages between the observed changes in mid-latitude ocean circulation and wind-stress changes through shifts in atmospheric and climate modes of variability are highlighted throughout this chapter. Although the dominant mode of global climate variability, the El Nino Southern Oscillation (Holton and Dmowska, 1989), originates in the tropical ocean, its effects are also observed in the mid and high latitudes. The modes of variability that are associated with large-scale changes in the mid-latitude ocean include the Pacific Decadal Oscillation, the dominant mode of variability of SST in the Pacific north of 20°N (Mantua and Hare, 2002); the North Atlantic Oscillation or Northern Annular Mode, the dominant mode of sea level pressure variability in the Northern Hemisphere in winter (Hurrell et al. 2003); and the similarly defined Southern Annular Mode in the Southern Hemisphere (Thompson and Wallace, 2000). All of these modes are associated with both surface heat and freshwater flux and with wind-stress anomalies that drive changes in ocean circulation, temperature, and salinity.

6.2 WESTERN BOUNDARY CURRENTS

Western boundary currents (WBCs) are a common feature of all mid-latitude ocean basins, characterized by highly energetic currents that flow poleward along continents in the subtropics (between about 20°N and 40°N) and equatorward in the subpolar regions (poleward of about 40°) and then eastward into the interior ocean, shedding energetic eddies along the way. Here we focus on the subtropical WBCs, where the contrast between the warm waters that originate in the tropics and the cooler overlying air temperatures gives rise to complex air-sea interaction and local heat storage in unstratified water masses known as "mode waters." An extensive discussion of the dynamics of WBCs overall and specifically for each of the major current systems is given in Imawaki et al. (2013). Many early altimeter studies focused on the WBCs in particular because the anomalies in sea level are relatively large.

The initial few years of radar altimetry introduced oceanographers to a wealth of new phenomena, some of which were described in the first altimeter book (Fu and Cazenave 2001), and researchers inferred WBC characteristics and possible dynamics from short data records. The subsequent two decades of radar altimeter data have allowed a more quantitative examination of the statistics and dynamics of the WBCs and their role in large-scale ocean circulation and climate.

The Northern Hemisphere subtropical Atlantic and Pacific WBCs, the Gulf Stream (GS) and the Kuroshio Extension (KE), respectively, have been extensively studied. A robust feature of these WBCs is an alternation between types of paths of the current cores, labeled by Qiu and Chen (2005) as a "stable" or an "unstable" mode with time scales of several years, based on studies of the KE. The stable mode is characterized by large transports, a lack of meandering, and a poleward shift of path latitude, with the opposite true for the unstable mode. The Gulf Stream has similar path stability modes, although with smaller anomalies compared with the KE (see Kelly et al. 2010 for a comparison) but with the same relationship between transport and meandering. Changes in the stability of the KE are associated with the westward propagation of sea level anomalies by Rossby waves, forced by large-scale winds over the Pacific Ocean and associated with the Pacific Decadal Oscillation (PDO) as shown in Figure 6.1. Qiu and Chen (2005) demonstrated predictability of the modes using a simple wind-forced ocean model. Large-scale shifts in GS path latitude, determined both by altimetry and hydrographic data, were shown to be correlated with the North Atlantic Oscillation (NAO) climate index (Joyce et al. 2000; Kelly et al. 2010). However, the dynamics of the GS are not as readily modeled as those of the KE as the GS circulation is embedded in the large-scale circulation known at the Atlantic Meridional Overturning Circulation (AMOC), which is discussed in Section 6.3.

The subtropical WBC of the southern Indian Ocean is the Agulhas Current, which is not confined by a continental boundary poleward of the southern end of the African continent at the relatively low latitude of 35°S. Consequently, the Agulhas Current flows westward around the tip of Africa

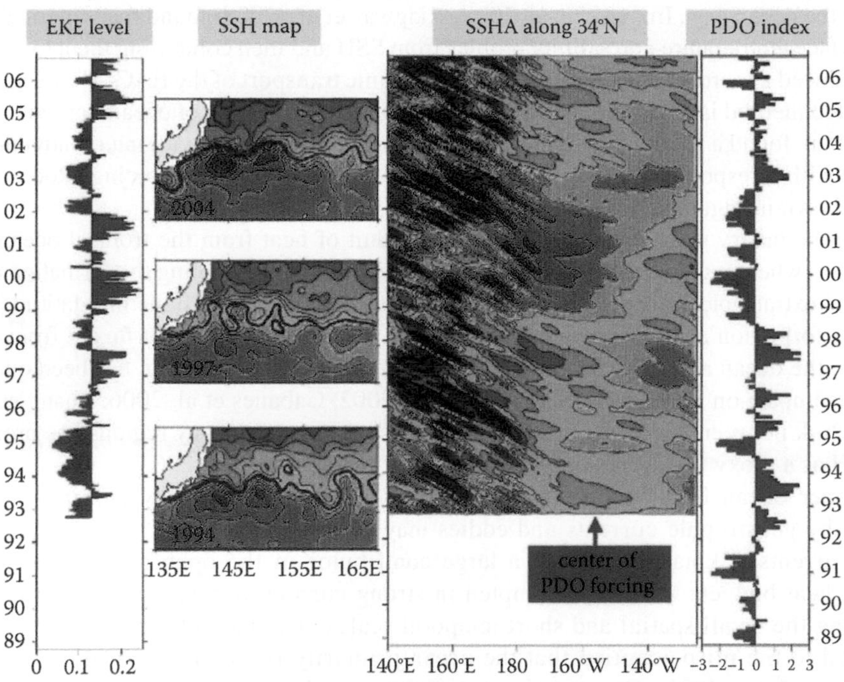

FIGURE 6.1 Dynamics of the Kuroshio Extension region. (a) Eddy kinetic energy (left) in the region shown by the maps. Sea surface height (right) for stable paths in 2004 and 1994 (top and bottom, respectively) and for an unstable path in 1997 (middle). Stable paths correspond to periods of low EKE (red regions). (b) Time-longitude plot of SSH showing westward propagation and the center of action for wind forcing associated with. (c) The Pacific Decadal Oscillation index. (From Qiu, B., and Chen, S., Deep-Sea Research II 57: 1098–1110, 2010a. With permission.)

and retroflects to flow eastward, shedding warm rings and leaking heat and salt into the Atlantic Ocean. The large warm and salty rings are visible in altimeter data as they travel westward toward South America. Assuming the rings to be the dominant contributor to the Indian–Atlantic Ocean exchange, estimates of the volume of water transported by the rings and the northward heat transport across the ring paths were made by Souza et al. (2011) by combining altimetric estimates of ring diameters with temperature profiles from Argo profiling floats (http://www.argo.net). However, a subsequent model study by Loveday (2015) found that the rings contributed less than half of the leakage from the Indian to the Atlantic Ocean.

Altimeter measurements have largely improved our understanding of the variability of the two western boundary currents of the South Atlantic, the warm poleward flowing Brazil Current and the cold equatorward flowing Malvinas Current, which combine to form the Brazil-Malvinas Confluence (BMC). Combining a 6-year altimeter record and a two-layer reduced gravity ocean model, Goni and Wainer (2001) found that meridional excursions of the Brazil Current are dominated by annual and semiannual signals, with southward frontal movement corresponding to a stronger current. Vivier and coauthors (2001) investigated the dynamics of the Malvinas Current transport using estimates derived from a 5-year altimeter record in combination with current meter measurements at the BMC, moorings in the Drake Passage, and wind stress fields. The authors concluded that the annual latitudinal excursions of the BMC are driven by the annual excursion of the maximum in wind stress and that anomalies at scales of several months are likely in a Sverdrup balance with South Atlantic winds.

The western boundary current of the subtropical South Pacific is the East Australian Current (EAC). Although direct estimates of ocean transport using SSH have been successful in some regions of the ocean (e.g., Imawaki et al. 2001), Ridgway et al. (2008) found that separately estimating subsurface temperature and salinity profiles from SSH and then combining them to obtain steric height produced a more reliable estimate of the baroclinic transport of the EAC. They suggested that this two-step method is better able to account for eddies and the effect of ocean bottom topography. Ridgway et al. found a factor of three variability in the EAC transports with interannual fluctuations consistent with a response to the South Pacific gyre spin-up from wind forcing (Roemmich et al. 2007) as shown in Figure 6.2.

Western boundary currents are the principal conduit of heat from the tropical oceans into the mid-latitudes where much of the heat is fluxed into the atmosphere, fueling the formation and intensification of extratropical storms; some heat continues poleward through the mid-latitudes to affect deep water formation and the cryosphere, and some is stored locally. Heat fluxed from the atmosphere into the ocean also contributes to local storage and that contribution has been quantified in terms of its impact on SSH anomalies (Vivier et al. 2002; Cabanes et al. 2006; Zhang et al. 2016). The close link between SSH anomalies and ocean heat content in many regions has prompted the use of SSH as a proxy for heat content (Gilson et al. 1998; Willis et al. 2003).

The upper ocean heat budget becomes complicated near boundary currents, where heat advection by geostrophic currents and eddies may exceed advection of heat by locally wind-induced currents (Ekman transport), a large contributor in the open ocean. Before satellite altimetry, heat budgets were not attempted in strong current systems, owing to the difficulty of resolving the small spatial and short temporal scales. The lack of advection measurements contributed to the misperception that the ocean primarily responds passively to atmospheric heat forcing (Cayan 1992). Direct estimates of the contributions of advection to the heat budget were made possible using the altimeter for geostrophic currents and the scatterometer for improved Ekman transport estimates. These estimates, combined with simple models, quantified the importance of WBCs to the heat budget (e.g., Qiu and Kelly 1993).

These early studies of the heat budget used mixed-layer models of the upper ocean forced by specified heat fluxes with geostrophic currents derived from SSH. Later models were augmented by a temporally and spatially varying subsurface layer. Heat content from these models compares favorably with SSH anomalies. Vivier et al. (2002) demonstrated the importance of advection below the mixed

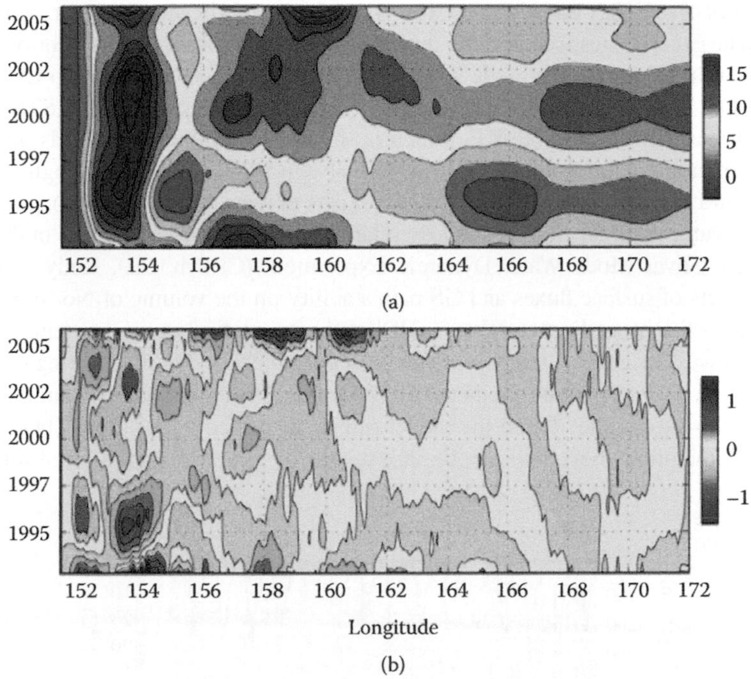

FIGURE 6.2 Geostrophic transport of the East Australian Current. (a) Time series of poleward transport in Sv (10^6 m^3/s) accumulated from the coast at left to the given longitude along a measurement line, showing decadal variations in magnitude. (b) Transport anomalies per unit of distance from the coast that contribute to the cumulative transport. (From Ridgway, K. R., et al., *J. Geophys. Res.*, 113, 1–18, 2008. With permission.)

layer on interannual time scales: Although horizontal advection is comparable in magnitude to surface fluxes in the mixed layer, it dominates the budget of the KE region in the upper 400 m. They further linked the heat budget with the state of the KE, showing that the warm subsurface waters during a stable state make vertical entrainment less efficient at cooling the upper ocean. Dong and Kelly (2004) found that an increase in ocean heat storage near a WBC is likely caused by advection by the warm currents rather than being the result of increased fluxes of heat from the atmosphere, as in the open ocean (Cayan 1992). This increased heat is then fluxed to the atmosphere, resulting in a negative correlation between ocean heat content and air–sea heat fluxes. Similarly, Roemmich et al. (2005) used altimetric SSH to help close the heat budget in a region east of Australia that encloses the EAC, with much of the ocean data provided by expandable bathythermograph (XBT) surveys along the perimeter. SSH was used to estimate heat storage using the method of Willis et al. (2003) and also for estimating surface geostrophic currents in conjunction with currents derived from XBT data. They found evidence for oceanic heat transport convergence causing ocean heat fluxes to the atmosphere as found in the KE and GS regions, as well as ocean heating by surface fluxes, particularly by shortwave fluxes.

The altimeter has been used in combination with other measurements to examine the evolution and dynamics of the subtropical mode waters (STMWs) that occur in the recirculation gyres equatorward of WBCs. STMW is a thick layer of unstratified water that has a distinct SSH signature reflecting the vertical structure. In the North Atlantic, the mode water has a relatively low heat content, creating a negative SSH anomaly, whereas in the North Pacific, the STMW is associated with a deepening of the thermocline, giving a positive SSH anomaly (see Kelly et al. 2010 for a comparison of the two basins). These surface signatures have been used to give qualitative estimates of STMW location, volume, and evolution. Quantitative estimates of mode water volume have come primarily from Argo profiling floats and other hydrographic data, but the altimeter has been critical in understanding the interaction between the WBC and mode water (Dong et al. 2007).

The stability of the WBC path is related to the volume of the mode water, with stable paths associated with larger volumes, because the meandering of the path is an indicator of lateral mixing across the WBC that decreases the mode water volume. As part of the Kuroshio Extension System Study (KESS), Qiu and Chen (2006) showed that the volume of North Pacific STMW is substantially smaller in years of an unstable KE, which is shown by comparing periods of high potential vorticity (smaller mode water volumes) in Figure 6.3b with the larger path length in Figure 6.3c. Further, they found that anomalies in air–sea heat fluxes (shown in Figure 6.3d), generally thought to control mode water volume, were not clearly related to the observed volumes in the KE region.

As part of the Clivar Mode Water Dynamic Experiment (CLIMODE), Kelly and Dong (2013) examined the effects of surface fluxes and GS path stability on the volume of North Atlantic STMW, which is known as "Eighteen-Degree Water" (EDW) for its relatively uniform temperature. Using a simple model that was forced by surface fluxes and that parameterized lateral mixing using path length (Figure 6.4a and b), they compared predicted EDW volumes with observed volumes (black line in Figure 6.4c) for two cases. They found the model run with surface fluxes alone (blue line) is able to predict EDW volume anomalies, unlike in the KE region; however, the addition of mixing (red line)

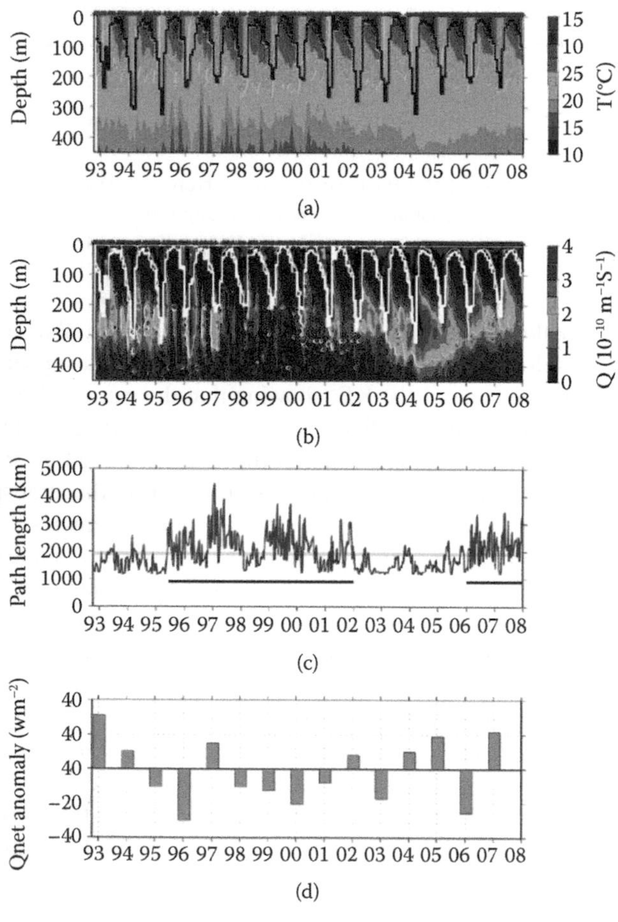

FIGURE 6.3 Subtropical mode water relationships near the Kuroshio Extension. (a) Time series of temperature profiles (color) with mixed layer depth in black, (b) potential vorticity (low values correspond to thicker STMW layers), (c) KE path length derived from the altimeter, and (d) surface flux anomalies. Thicker STMW layers coincide with periods of decreased path length, but not to anomalies in surface heat flux. (From Qiu, B., and Chen, S., *J. Phys. Oceanogr.*, 36(7), 1365–1380, 2006. With permission.)

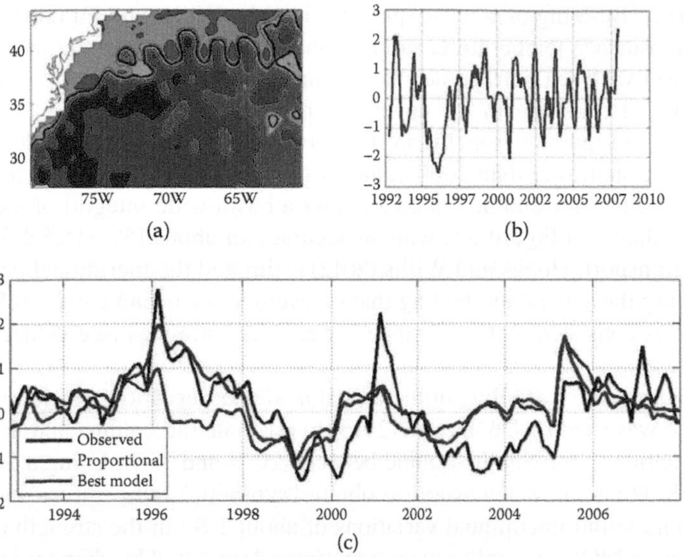

FIGURE 6.4 Subtropical mode water relationships near the Gulf Stream. (a) Example of SSH anomaly map for an unstable path, (b) time series of index of path length, and (c) time series of mode water volume (black) and simulations using surface fluxes only (blue) and fluxes plus effect of path length (red). (From Kelly, K. A., and Dong, S., *Deep Sea Res. II*, 2013. With permission.)

improves the volume prediction, giving a reduction during periods of more meandering, as is the case for the KE.

A fundamental difference between the two basins that can explain the difference in dynamics is the balance between wind and buoyancy forcing. In the North Pacific, thermocline deepening in response to wind forcing lowers vertical stratification and facilitates mode water formation (Oka and Qiu 2012). In contrast, the North Atlantic thermocline is well below the depth of the mode water layer and exerts no clear influence on EDW volume. Therefore, surface buoyancy forcing dominates interannual variability, as shown by Billheimer and Talley (2013) who combined Argo, SSH, and surface forcing fields in a study of interannual variations in EDW. In the study of the South Pacific heat budget near the EAC, Roemmich et al. (2005) found that surface heat fluxes and oceanic heat advection contributed equally to the interannual STMW volume anomalies, confirming earlier work by Sprintall et al. (1995). Larger-scale circulation may also contribute to the heat budget and the formation of mode water, in particular the AMOC, which is discussed in Section 6.3.

6.3 MERIDIONAL CIRCULATION AND INTERBASIN EXCHANGES

The existence of a large-scale, slow-moving, overturning circulation that carries warm water to the north near the surface and cold water to the south at depth in the Atlantic Ocean has been well-known for decades (see Buckley and Marshall 2016, for a recent review). However, measuring the strength of the Meridional Overturning Circulation (MOC) involves teasing out small imbalances between large northward and southward parts of the circulation near the ocean's surface across the width of an entire basin and a good knowledge of the subsurface temperature, salinity, and density structure. For this reason, efforts to characterize and study the MOC using satellite altimetry did not really begin until complementary observing systems, particularly the Argo array of profiling floats, became widely available.

A direct estimate of the volume transport of the AMOC and its variability during the mid-2000s was generated by Willis (2010) using a combination of satellite altimeter observations and data from the Argo array. Building on a technique described by Willis and Fu (2008) to combine Argo observations of subsurface temperature, salinity, and float displacement (a measure of integrated, subsurface velocity) with sea surface height measurements from altimetry, Willis found that the basin-wide integral of transport in the upper ocean could be calculated seasonally at 41°N, just north of where the GS separates from the western boundary. At this latitude, despite the absence of Argo data in regions shallower than 2000 m and within a few hundred kilometers of the coastline, the combination of altimetry and *in situ* data allows a basin-wide integral of the northward upper ocean velocity, as shown in Figure 6.5, with an accuracy of about 15% (15.5 ± 2.4 Sv). In addition to mean volume transport, Hobbs and Willis (2012) estimated the meridional ocean heat transport at this latitude using the same data, finding that on average the ocean carries 0.5 ± 0.1 PW of heat northward, which is consistent with the range of previous estimates (see Ganachaud and Wunsch 2000, for estimates).

A similar combination of satellite altimeter and *in situ* observations (including data from Argo and other sources) was used by Dong et al. (2015) to estimate the volume and heat transport of the overturning circulation in the South Atlantic between 20°S and 35°S. Using a technique to project the estimates of SSH anomaly downward to obtain "synthetic" temperature and salinity profiles, Dong and coauthors found interannual variations of about 1 Sv in the strength of the MOC in the South Atlantic. These MOC anomalies were sometimes dominated by changes in the Ekman transport and sometimes by the geostrophic flow in the upper ocean, as shown in Figure 6.6. In addition, they found that the time-mean overturning varied with latitude between 17 Sv and 20 Sv, and heat transport varied between 1.2 and 0.5 PW, with the strongest volume transport and the smallest heat transport at the more southern latitudes of 30°S and 34.5°S.

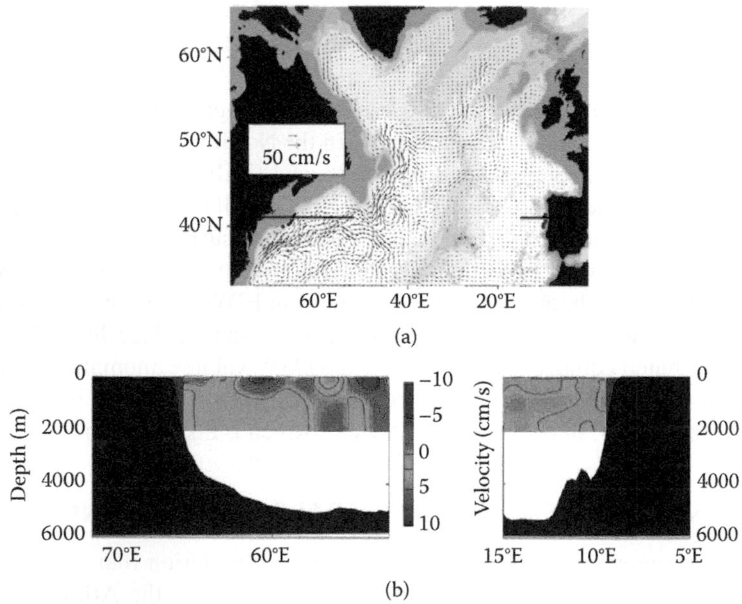

FIGURE 6.5 (a) 2004 through 2006 time-mean of surface geostrophic currents based on absolute dynamic height estimated using Argo and altimeter data in the study region. Arrows are shown every 1° of longitude and ½° of latitude for clarity. Orange vectors have been scaled down by a factor of two and generally lie within the Gulf Stream. The red dots show the basin boundaries marked by the 2000 m isobath between 40°N and 41.5°N, the latitudes at which the overturning is estimated. From light to dark, shading indicates bathymetry <4000 m, <3000 m, <2000 m, <1000 m, and < 500 m. (b) Black lines in (a) indicate transects at 41°N along which northward velocity is plotted. Colors indicate velocity and contours are as indicated in the color scale. The red areas mark the shallow boundary regions that are not sampled by Argo. (From Willis, J. K., *Geophys. Res. Lett.*, 37, L06602, 2010. With permission.)

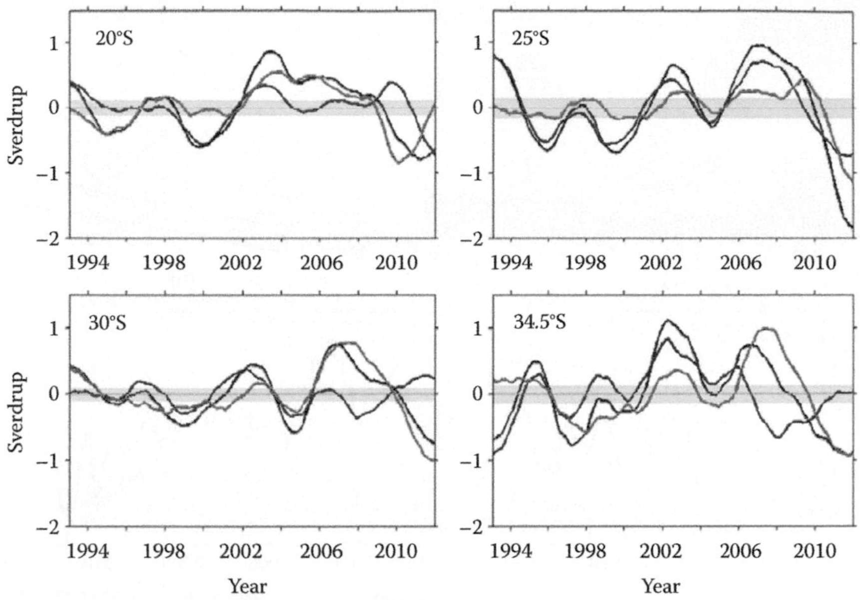

FIGURE 6.6 Interannual variations of the MOC (black) and contributions from the geostrophic (red) and Ekman (green) components at 20°S, 25°S, 30°S, and 34.5°S, respectively. The gray shading denotes the range where anomalies are not significantly different from zero. (From Dong, S., et al., *Geophys. Res. Lett.*, 42, 7655–7662, 2015. With permission.)

A more subtle analysis of the heat and freshwater transports of the AMOC was recently performed by Kelly et al. (2016). Building on previous work with a box model that demonstrated coherent anomalies in the meridional heat transport between 35°S and 40°N (Kelly et al. 2014), these authors used four basin-spanning boxes between 67°N and 35°S to expand the analysis to include freshwater transports. Using altimetric SSH in combination with ocean mass changes from the Gravity Recovery and Climate Experiment (GRACE) satellite mission and estimates of heat and freshwater content from Argo, Kelly, and coauthors estimated the convergence of heat and freshwater in each box. Meridional heat and freshwater transports by the ocean were computed as residuals by integrating the convergence estimates and using a transport estimate at a single latitude to supply the integration constant. The resulting estimates of heat transport across 35°S were then compared with estimates of the leakage of warm water between the Indian and Atlantic Oceans that is known to occur in the region where the warm Agulhas Current overshoots the southern tip of Africa into the South Atlantic before retroflecting back to the east around 40°S. Agulhas Leakage was calculated using satellite altimeter observations to estimate the transport of the Agulhas that does not return east after retroflection (Le Bars et al. 2014). As shown in Figure 6.7c, the phasing and magnitude of Agulhas Leakage are consistent with the AMOC heat transport at 35°S after the availability of GRACE (2002) and Argo data (2004). Since such signals are coherent between 35°S and 40°N in the Atlantic, as shown by Kelly et al. (2014), this is evidence for the idea that the slowdown in the AMOC has a southern origin.

The theory of forcing of AMOC anomalies by the Agulhas Leakage, postulated by Biastoch et al. (2008), is supported by a large literature on how the Southern Ocean affects the overturning circulation (see Beal et al., 2011, for a review). This runs contrary to the more traditional assumption that AMOC variations are driven primarily by changes in deep water formation rates in the polar North Atlantic. Recent direct observations of the Agulhas Current, extrapolated using an altimeter proxy, show that the slowdown of the current is accompanied by an increase in eddy energy and

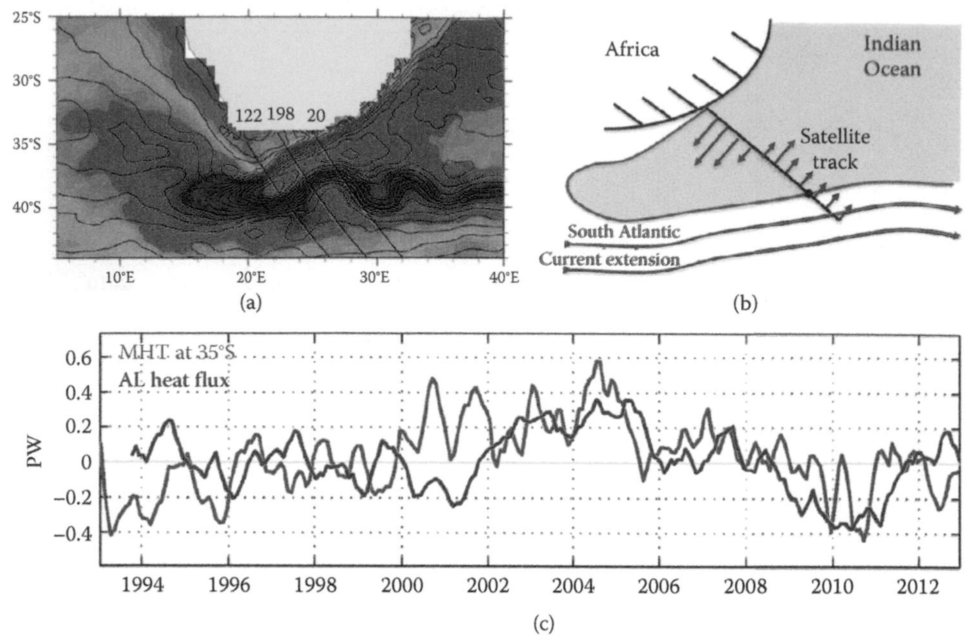

FIGURE 6.7 Meridional Heat Transport (MHT) and Agulhas Leakage. (a) Variability of SSH for the Agulhas Current and altimeter tracks, (b) Schematic of net leakage across an altimeter track (c) baseline MHT (blue) at 35°S and inferred Agulhas Leakage heat transport (red). (From Kelly, K. A., et al., *Geophys. Res. Lett.* 43, 2016. With permission.)

more heat loss to the atmosphere (Beal and Elipot, 2016), reinforcing the close connection between changes in ocean circulation and atmosphere–ocean coupling.

Interestingly, Biastoch et al. (2008) used westward propagation of large-scale SSH anomalies across the South Atlantic at 30°S derived from satellite altimeter data to help validate their ocean model, as shown in Figure 6.8. The origin of the propagating anomalies in the region near the Agulhas retroflection is an indicator of the importance of Agulhas Leakage in the model ocean. This simple use of altimeter data to support model results illustrates the rather impressive evolution of the use of altimeter data for understanding and explaining changes in the MOC. More recently, Biastoch et al. (2015) used a comparison between altimeter data and observations of SST in the Agulhas region to estimate changes in leakage from the late 1800s to the present. In the process, they drew a stronger link between wind forcing in the southern hemisphere and the MOC strength throughout the Atlantic. As the planet warms due to human interference with the climate, this mechanism may prove an important driver of MOC-related climate change in the North Atlantic. And as the MOC evolves, satellite altimetry is sure to play an increasingly important role in observing, understanding, and ultimately predicting these effects.

Although the altimeter has made enormous contributions to our understanding of ocean circulation, the basic measurement, SSH, is ambiguous in that it combines the effects of changes in the density of the ocean and changes in the mass of the ocean. The GRACE satellite measures gravity, and those measurements can be converted into an estimate of the bottom pressure or, equivalently, the mass of the ocean. Ocean circulation can be decomposed into depth-averaged and depth-varying components, which correspond to changes in bottom pressure and changes in density, respectively. Recent studies by Piecuch and coauthors (Piecuch et al., 2013; Piecuch and Ponte, 2014) have highlighted the contributions of GRACE to understanding the importance of depth-averaged circulation in deep extratropical regions, such as the Southern Ocean, and in shallow seas. Further, they have found a bottom pressure signal in the North Atlantic along the boundary between the subtropical and subpolar gyres.

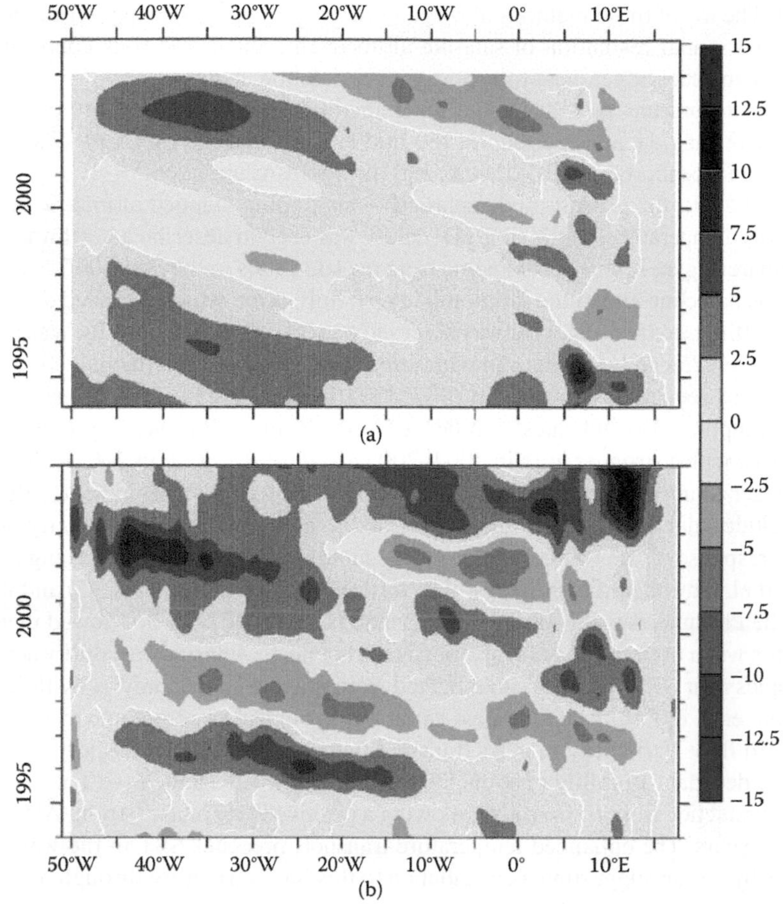

FIGURE 6.8 Decadal variability of the sea surface height at 30°S. (a) Hovmoeller diagram of the 23-month-long low-pass filtered SSH anomalies (in cm) at 30°S from a global model incorporating a high-resolution Agulhas component and (b) relative SSH from Aviso satellite data. (From Biastoch, A., et al., *Nature*, 456(27), 2008. With permission.)

The bottom pressure variations in this region correspond to changes in the wind stress associated with the NAO consistent with an analysis of North Atlantic SSH by Zhang et al. (2016).

6.4 CLIMATE CHANGE

Altimetric fields of sea level are now long enough (more than two decades) to tackle interannual-to-decadal variability in the upper ocean circulation. The combination of the altimetric SSH with subsurface sampling of temperature and salinity, in particular by Argo profilers, yields upper-ocean or mid-ocean internal steric changes and thus contributes to interpreting the variability in terms of dynamical processes, water mass changes, and heat or freshwater uptake. Although the current record is inadequate to reliably document and interpret decadal or multidecadal variability, the altimeter record provides some suggestions on the causes of that variability.

At mid-latitudes outside of western boundary currents, several methods have been developed to convert sea level from altimetry into synthetic baroclinic profiles, taking advantage of the large contribution of the upper ocean density to large-scale sea level variability. Examples of these synthetic profile methods, or "gravest-modes," are given by Swart et al. (2010) for the Antarctic Circumpolar Current (ACC), by Guinehut et al. (2012) for global products, and by Dong et al. (2015), as discussed in Section 6.3. The use of these methods allows an inference of upper ocean internal properties with the temporal and spatial resolution of satellite altimetric products and thus contributes to understanding the ocean heat budget, heat transport, and circulation changes.

The altimeter record has provided input for numerous recent studies on gyre-scale variability at near-decadal or multidecadal periods in the mid-latitude oceans: the North Pacific, the South Pacific, the Indian Ocean, the North Atlantic, and the South Atlantic.

In the North Pacific, the gravest-mode approach—combining mapped altimetric sea level and *in situ* (conductivity-temperature-depth, or CTD) data—was used to describe a northward shrinkage of the western subarctic gyre in the North Pacific from the late 1990s to the mid-2000s associated with a deepening of the halocline and of the minimum temperature layer, which covaried during that period (Nagano et al., 2016a), confirming what was observed on some repeat stations by Wakita (2013). This shrinkage was likely forced by changes in wind stress curl. Interestingly, the halocline and the minimum temperature layer ceased to covary in 2007–2009, indicating some limitations on the approach. Recent Argo sampling also indicates significant North Pacific subtropical gyre changes resulting from wind stress, with a stronger gyre in 2004–2005 and a weaker gyre in 2008–2009 (Giglio et al. 2012). The longer altimetric record, which is consistent with Argo, also indicates the presence of decadal to multidecadal variability in the gyre structure and transports particularly near its southern edge. The response of the southern edge of the subtropical gyre to wind forcing on interannual time scales had already been diagnosed for the North Equatorial Current (NEC) and its bifurcation latitude near the Philippines using altimetry by Qiu and Chen (2010b), who showed that higher NEC transport occurs when the latitude of bifurcation is farther north. Internal variability across the North Pacific subtropics near 30°N was also investigated from gravest-mode analysis (with XBT and Argo data) by Nagano et al. (2016b) who identified variations in current structure and average temperature of the southward flow. Variability in mass transport dominated the heat transport anomalies, which presented quasi-decadal variability of about 5%, with peaks in 1998 and 2007. The peaks are related to changes in subduction from the surface following a positive PDO phase (strong Aleutian low) with a lag of up to 4 years. The enhanced temperature transport precedes SST in the warm pool east of the Philippines by 1 year, suggesting a contribution to this SST variability through modulation of the entrainment in the surface mixed layer.

A trend of increasing circulation in the eastern part of the South Pacific subtropical gyre has been found for the altimetric period of more than 22 years and confirmed with Argo data (2005–2014) by Roemmich et al. (2016). During the last decade (2005–2014), the 5-Sv northward transport increase from 160°W to South America was countered by a southward transport anomaly between the date line and 160°E, as illustrated in Figure 6.9c. In the previous decade, maximum SSH changes were found farther west by Roemmich et al. (2007), indicating increasing northward transport throughout most of the subtropical gyre. A large SSH trend of 0.08 m/decade from 1993 to 2014 was found at 35°S, 160°E, close to where sea level pressure and SST trends since the early 1980s have been the largest. These multidecadal trends support the interpretation by Roemmich et al. (2015) of a 40°S maximum in global ocean heat gain as the result of anomalous wind forcing and Ekman convergence that likely has spanned at least three decades. Other interannual variability in the interior transport may be related to La Niña events. Integrated mass transports across 32°S from Australia to South America indicate interannual variability related to the Southern Annular Mode (SAM), with an increase in northward transport during the positive phase of SAM and a decrease during the negative phase, as noted by Zilberman et al. (2014). Westward-propagating SSH anomalies at 32°S reflect El Niño-Southern Oscillation (ENSO) forcing time scales and baroclinic Rossby wave time scales of several years; the SSH anomalies are associated with large zonal shifts in the meridional transports of the deep and abyssal circulation but without a net transport change (Hernández-Guerra and Talley, 2016).

In the Indian Ocean, Lee and McPhaden (2008) found large decadal trends in sea level and heat content with a change of sign between 1993 and 2000 and 2000 and 2007. The spatial pattern corresponding to the trends has large anomalies of one sign near 10°S–15°S in the central to western Indian Ocean with anomalies of the opposite sign in the eastern Indian Ocean, in particular east of

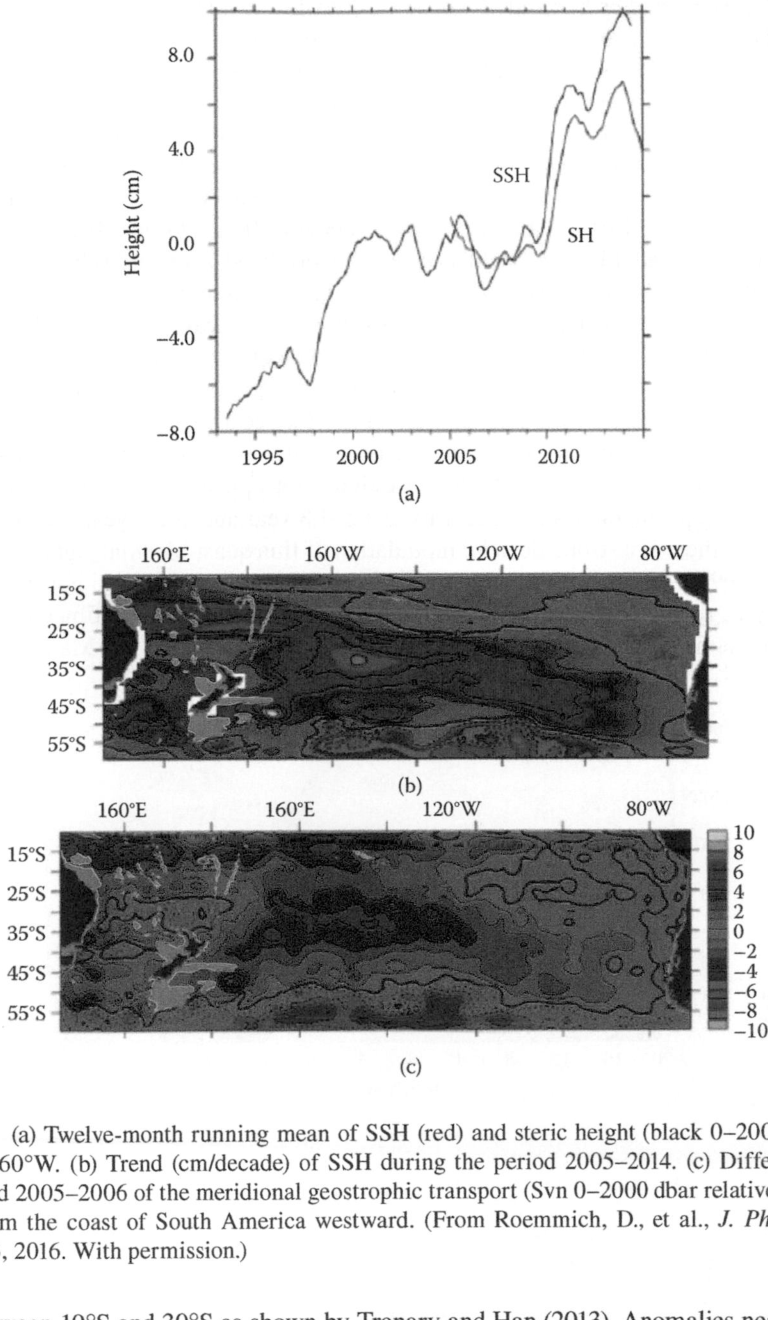

FIGURE 6.9 (a) Twelve-month running mean of SSH (red) and steric height (black 0–2000 dbar) anomaly at 35°S, 160°W. (b) Trend (cm/decade) of SSH during the period 2005–2014. (c) Difference between 2013–2014 and 2005–2006 of the meridional geostrophic transport (Svn 0–2000 dbar relative to 2000 dbar) integrated from the coast of South America westward. (From Roemmich, D., et al., *J. Phys. Oceanogr.*, 46, 1871–1883, 2016. With permission.)

Australia between 10°S and 30°S as shown by Trenary and Han (2013). Anomalies near and west of Madagascar tend to have the same sign as near 10°S–15°S. South of 25°S, the trends tend to be positive for both periods and are known to have lasted over a longer period. The trends in SSH are associated with deepening of the thermocline in the tropics and throughout the water column farther south, resulting in warming. Near 10°S–15°S, which is in the northern part of the subtropical gyre near and south of the Mascarene ridge, the slow shoaling of the thermocline in 1993–2000 is associated with negative SST trends. Through a series of numerical experiments, Trenary and Han (2013) found that most of the sea level and thermocline depth changes are related to local Indian Ocean wind forcing, except near Australia where decadal variability of sea level since the 1990s is also strongly influenced by the Indonesian Throughflow (ITF, the currents that bring water from the western Pacific to the

Indian Ocean) as suggested by Lee and McPhaden (2008). Llovel and Lee (2015) found a strong influence of the ITF on upper ocean salinity in the southeastern Indian Ocean where salinity decreased greatly from 2005 to 2013, as shown in terms of the salinity (halosteric) contribution to sea level in Figure 6.10b, right panel. Decreasing salinity, which occurred despite the absence of local changes in freshwater fluxes, was largely responsible for the sea level rise illustrated in Figure 6.10, top panel. These different results all relate to the variability after 1990, and one can wonder how representative it is of earlier or future decadal variability. Interestingly, a change in decadal SST variability was found in the Indian Ocean in the 1980s, with the dominant modes of Indian Ocean SST for 1900–1980 correlated with the Interdecadal Pacific Oscillation but anticorrelated afterward (Han et al. 2014). There is also the possibility that part of the decadal variability in the southern Indian Ocean south of 20°S is of internal ocean origin. A possible connection between Agulhas Leakage and variability of southern Indian Ocean properties was suggested, with some evidence that Agulhas Leakage has increased in recent decades (Beal et al. 2011), but this was contradicted by later analyses showing a weakening Agulhas by Le Bars et al. (2014) and Beal and Elipot (2016). See also Figure 6.7 (c).

In addition to decadal and longer trends, quasi-biennial variability has been identified from altimetry in the eastward currents of the South Indian subtropical gyre (20°S–30°S) (Menezes et al. 2016) with two separate modes of variability at 1.5–1.8 year and at 2.5 year (this one only found west of 80°E). As there is a strong decadal modulation of this quasi-biennial signal (e.g., high variance found in 1999–2003), the origin of these two modes of variability is still not fully interpreted. The variability of the northernmost eastward currents observed with the altimetric data could be associated, for example, with variations in the northern recirculation cell off Madagascar.

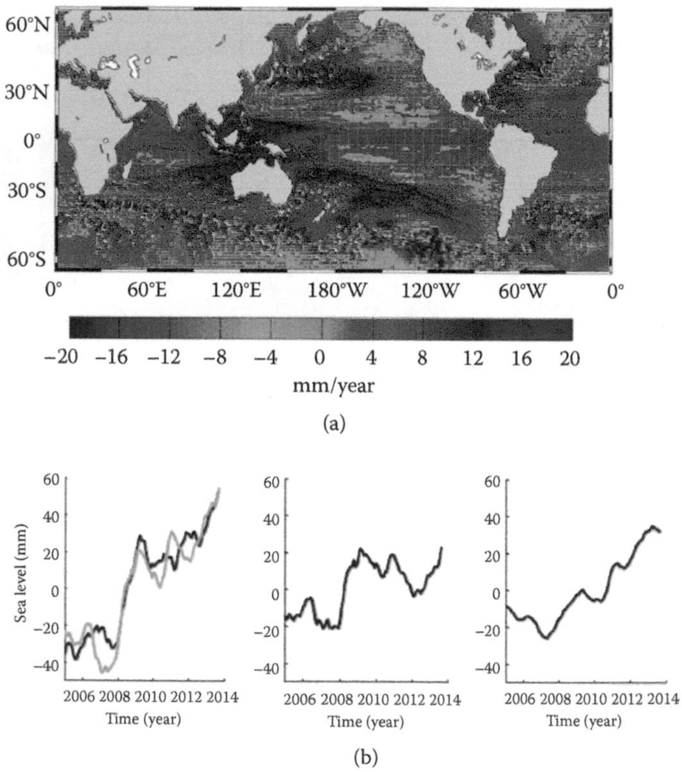

FIGURE 6.10 (a) Sea level trend over 2005–2013 by satellite altimetry. (b) Averages over the Southeastern Indian Ocean (roughly, the area with large positive trend on [a]) of (left) sea level (black curve) and steric component (grey); middle thermosteric component, right, halosteric component. (From Llovel, W., and Lee, T., *Geophys. Res. Lett.*, 42, 1148–1157, 2015. With permission.)

The North Atlantic has witnessed large decadal and multidecadal climate variability in the last 50 years, as noted—for example—by Liu (2012). Recently, Häkkinen et al. (2013) found covarying trends in sea level and upper ocean heat content with large increases in the core of the Labrador and Irminger Seas from 1993 to 2011 and decreases over the GS region, as shown in Figure 6.11. This "gyre mode" of variability seems largely driven by the wind stress and brings subtropical waters into the subpolar gyre on decadal to multidecadal time scales (Häkkinen et al. 2011). The gyre mode was previously identified by Hatun et al. (2005) as a driver for current and water mass variability in the northeast subpolar Atlantic. The mode projects strongly onto a second mode of Atlantic wind stress curl that is similar to the East Atlantic Pattern (EAP; the second most prominent mode of atmospheric variability in the North Atlantic, after the NAO), which is linked to atmospheric blocking over the Atlantic (Msadek and Frankignoul 2009). The gyre mode also has a connection with multidecadal surface temperature variability in the North Atlantic.

Focusing on the poleward currents in the North Atlantic subpolar gyre, Chafik et al. (2014) found an anticorrelation between currents in the Irminger Sea branch and in the Iceland Sea/Rockall Channel branches. However, the combined poleward currents show an overall decrease of 10% in northward transport from 1998 to 2010 (–1.7 Sv). The anticorrelated signal is multidecadal and is associated with the NAO, whereas the overall decrease is mostly associated with the Atlantic Multidecadal Oscillation (AMO), according to Häkkinen et al. (2013). Chafik et al. (2014) present evidence for topographic steering of the barotropic currents during positive phases of the NAO (see, e.g., in the Faroe-Shetland Channel) and a contraction of the subpolar gyre with negative NAO. A role for the EAP in subpolar gyre strength is also identified on decadal time scales. An investigation of the Azores Current in the eastern Atlantic by Volkov and Fu (2011) shows that surface current eddy kinetic energy, averaged over large spatial scales (30°N–40°N) is related to wind stress curl (25°N–45°N). They further identify a link between the Azores Current and the NAO, with strongest Azores Current and largest eddy energy lagging a positive NAO.

The influence of the EAP or blocking mode is also seen in subtropical gyre dynamics and sea level variability (Volkov and Fu, 2011). Sea level shows a strong decadal or multidecadal component to some extent forced by EAP (e.g., a large contrast between high sea level in the 1960s and low sea level in the 1970s). To place this in a larger context requires other analyses than *in situ* data. For instance, large bidecadal variability of sea level was identified in Simple Ocean Data Assimilation (SODA) ocean reanalyses (1908–2008) by Vianna and Menezes (2013). They identify sea level variability in this spectral band in all subtropical gyres, including the South Atlantic subtropical gyre. The bidecadal variability in the different gyres of the Atlantic Ocean is not found to be phase-locked through the whole reanalysis period, thus putting into question the nature of this variability.

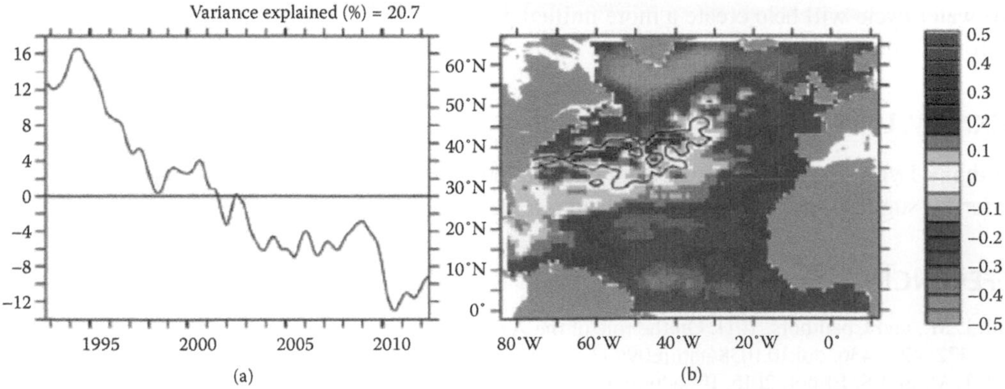

FIGURE 6.11 Sea surface height PC1 ([a] in cm) and EOF1 ([b], dimensionless) of North Atlantic SSH. (From Häkkinen, S., et al., *J. Geophys. Res.*, 118, 3670–3678, 2013. With permission.)

Measurements from satellite altimetry also revealed variability in the South Atlantic Ocean on interannual to longer time scales. The southern edge of the subtropical gyre, in particular the Brazil Current front where it separates from the coast, has been observed to shift southward by 1.5° latitude, but the transports of neither the Brazil Current nor the Malvinas Current show a trend (Goni et al. 2011; Lumpkin and Garzoli 2011). These changes were linked to concurrent southward shifts in the latitude of the maximum wind stress curl (Goni et al. 2011), indicating that the governing mechanisms are different from the seasonal variability discussed in Section 6.2. Analysis of combined drifter–altimeter data by Lumpkin and Garzoli (2011) showed that low-frequency BMC anomalies are negatively correlated with sea surface temperature (SST) along the westward paths of the Agulhas rings, which they suggested may influence the latitude excursions of the BMC. A proxy for the confluence latitude derived from wind stress curl for the pre-altimeter time period suggested that the recent trend may be part of a longer-term oscillation. Upper ocean heat transport at 35°S, based mostly on XBT sampling, is observed to have largely varied on near-decadal time scales (from short records starting in 2002), with maxima in 2005 and in 2013, with contributions from western boundary, eastern boundary, and interior transport variability (time series in Blunden and Arndt, 2015).

6.5 SUMMARY AND FUTURE RESEARCH

Much has been learned from altimeter data about the mid-latitude ocean currents and gyres and their forcing mechanisms at the seasonal-to-interannual time scales. The two decades-long altimetry record has prompted examinations into longer time scale variability and possible trends related to climate change. Although the current record is too short for statistical reliability, the need to understand how the ocean will respond to increases in greenhouse gases is great. It is too early to make pronouncements about trends, but the altimeter has helped to frame the questions that future measurements need to answer. For example, by demonstrating the importance of ocean advection in the heat budget and its impact on air–sea heat fluxes, altimeter measurements spurred research into the ocean's role in anchoring storm paths and in intensifying extratropical storms (Nakamura et al. 2004; Joyce et al. 2009; Kwon et al. 2010).

Although much was learned in the first decade of the altimetry mission, synergy with subsequent data streams, such as the Argo profiling floats and the GRACE mission, accelerated the pace of ocean understanding. Continuing records of all of these measurements will yield more conclusive information about the ocean's response to changes in greenhouse forcing and the distinction between decadal-scale ocean variability and long-term trends. The advent of the Surface Water and Ocean Topography (SWOT) mission (expected in 2021) will improve the resolution of boundary currents and allow a more detailed examination of the relationship between ocean eddies and the larger-scale variability. Satellite missions that measure changes in the cryosphere and in the terrestrial water cycle will help create a more unified picture of the interactions in the Earth system and improve climate models and predictions.

ACKNOWLEDGMENTS

We thank Lynne Talley for a review and helpful comments on a draft of this chapter and Frederic Vivier for suggestions on research to include in the discussion.

REFERENCES

Beal, L.M., and Coauthors. 2011. On the role of the Agulhas system in ocean circulation and climate. *Nature* 472: 429–436. doi:10.1038/nature09983.

Beal, L.M. and S. Elipot, 2016. Broadening not strengthening of the Agulhas Current since the early 1990s, *Nature* 540, doi:10.1038/nature19853.

Biastoch, A., C.W. Boning, and J.R.E. Lutjeharms. 2008. Agulhas leakage dynamics aspects decadal variability in Atlantic overturning circulation. *Nature* 456 (27). doi:198 10.1038/nature07426.

Biastoch, A., J.V. Durgadoo, A.K. Morrison, E. van Sebille, W. Weijer, and S.M. Griffies. 2015. Atlantic multidecadal oscillation covaries with Agulhas leakage. *Nat. Commun.* 6: 10082. doi:10.1038/ncomms10082.
Billheimer, S. and L. D. Talley, 2013. Near-cessation of Eighteen Degree Water renewal in the western North Atlantic in the warm winter of 2011-2012. *J. Geophys. Res. Oceans* 118: 1-16. doi:10.1002/2013JC009024.
Buckley, M.W., and J. Marshall. 2016. Observations, inferences, and mechanisms of Atlantic meridional overturning circulation variability: A review, *Rev. Geophys.* 54. doi: 10.1002/2015RG000493.
Cabanes, C., Huck, T. and C. de Verdière. 2006. Contributions of wind forcing and surface heating to interannual sea level variations in the Atlantic Ocean. *J. Phys. Oceanogr.* 36 (9): 1739–1750.
Cayan, D.R. 1992. Latent and sensible heat flux anomalies over the northern oceans: The connection to monthly atmospheric circulation. *J. Climate* 5(4): 354–369.
Chafik, L., T. Rossby and C. Schrum. 2014. On the spatial structure and temporal variability of poleward transport between Scotland and Greenland. *J. Geophys. Res.* 119. doi:10.1002/2013JC009287.
Chelton, D.B., and M.G. Schlax. 1996. Global observations of oceanic Rossby waves. *Science*. 272(5259): 234.
Di Lorenzo, E., N. Schneider, K.M. Cobb, P.J. Franks, K. Chhak, A.J. Miller, J.C. McWilliams, S.J. Bograd, H. Arango, E., Curchitser and T.M. Powell. 2008. North Pacific Gyre Oscillation links ocean climate and ecosystem change. *Geophys. Res. Lett.* 35 (8).
Dong, S. and K.A. Kelly. 2004. Heat budget in the Gulf Stream region: the importance of heat advection and storage, *J. Phys. Oceanogr.* 34: 1214–1231.
Dong, S., S. L. Hautala, and K. A. Kelly. 2007. Interannual variations in upper-ocean heat content and heat transport convergence in the western North Atlantic, *J. Phys. Oceanogr.* 37: 2682–2697.
Dong, S., G. Goni, and F. Bringas. 2015. Temporal variability of the South Atlantic Meridional Overturning Circulation between 20°S *and* 35°S. *Geophys. Res. Lett.* 42: 7655–7662. doi:10.1002/2015GL065603.
Ducet, N., P.Y. Le Traon, and G. Reverdin. 2000, Global high-resolution mapping of ocean circulation from TOPEX/Poseidon and ERS-1 and -2. *J. Geophys. Res.* 105(C8): 19477–19498, doi:10.1029/2000JC900063.
Frajka-Williams, E. 2015. Estimating the Atlantic overturning at 26°N using satellite altimetry and cable measurements. *Geophys. Res. Lett.* 42 (9): 3458–3464.
Frankignoul C., G. de Coëtlogon, T.M. Joyce and S. Dong. 2001. Gulf Stream variability and ocean-atmosphere interactions. *J. Phys. Oceanogr.* 31 (12): 3516–3529.
Fu, L.L. and Cazenave A. 2001. *Satellite Altimetry and Earth Sciences: A handbook of Techniques and Applications. International Geophysics Series*, Vol. 69. Academic Press. San Diego, 463 pp.
Ganachaud, A. and C. Wunsch. 2000. Improved estimates of global ocean circulation, heat transport and mixing from hydrographic data. *Nature*, 408 (6811): 453–457.
Giglio, D., D. Roemmich, and S.T. Gille. 2012. Wind-driven variability of the subtropical North Pacific Ocean. *J. Phys. Oceanogr.* 42(12): 2089–2100.
Gilson, J., D. Roemmich, B., Cornuelle and L.L. Fu. 1998. Relationship of TOPEX/Poseidon altimetric height to steric height and circulation in the North Pacific, *J. Geophys. Res.* 103(C12): 27947–27965.
Goni, G.J., F. Bringas, and P.N. diNezio. 2011. Observed low frequency variability of the Brazil Current front. *J. Geophys. Res.* 116 doi:10.1029/2011JC007198.
Goni, G.J. and I. Wainer. 2001. Investigation of the Brazil Current front variability from altimeter data. *J. Geophys. Res.* 106: 31117–31128.
Guinehut, S., A.-L. Dhomps, G. Larnicol, and P.-Y. Le Traon. 2012. High resolution 3D temperature and salinity fields derived from in situ and satellite observations. *Ocean Sci.*, 8: 845–857.
Häkkinen, S., P.B., Rhines and D.L. Worthen. 2011. Warm and saline events embedded in the meridional circulation of the northern North Atlantic, *J. Geophys. Res.* 116 (C03006) doi:10.1029/2010JC006275.
Häkkinen, S., P.B. Rhines, and D.L. Worthen. 2013. Northern North Atlantic sea-surface height and ocean heat content variability, *J. Geophys. Res.* 118: 3670–3678, doi:10.1002/jgrc.20268.
Han, W., J. Vialard, M.J. McPhaden, T. Lee, Y. Masumoto, M. Feng, and W.P. de Ruijter. 2014. Indian Ocean decadal variability: A review. *Bull. Amer. Meteor. Soc.* 95: 1679–1703 doi:10.1175/BAMSD1300028.1.
Hatun, H., B. Hansen, A.B. Sando, H. Drange, and H. Valdimarsson. 2005. De-stabilization of the North Atlantic thermohaline circulation by the gyre mode. *Science* 309: 1841–1844.
Hernández-Guerra, A. and L. D. Talley. 2016. Meridional overturning transports at 30°S in the Indian and Pacific Oceans in 2002-2003 and 2009. *Prog Oceanogr.* 146: 89–120.
Hobbs, W.R., and J. K. Willis. 2012. Midlatitude North Atlantic heat transport: A time series based on satellite and drifter data, *J. Geophys. Res. 117* (C01008) doi:10.1029/2011JC007039.
Holton J.R., and R. Dmowska. (C01008) doi:10.1029/2011JC007039. S.G. Philander, editor. Academic press; 1989 Dec 14.
Hurrell, J. W., Y. Kushnir, G. Ottersen and M. Visbeck. An overview of the North Atlantic oscillation. American Geophysical Union; 2003 Mar 19.

Imawaki, S., H. Uchida, H. Ichikawa, M., Fukasawa and S.I. Umatani. 2001. Satellite altimeter monitoring the Kuroshio transport south of Japan. *Geophys. Res. Lett.* 28 (1): 17–20.

Imawaki, S., A.S. Bower, L. Beal and B. Qiu. 2013. Western boundary currents. *Ocean circulation and climate—a 21st century perspective.* 2nd edn. Academic Press, London, 305–338.

Joyce, T.M., C. Deser and M. A. Spall. 2000. The relation between decadal variability of subtropical mode water and the North Atlantic Oscillation. *J. Climate*, 13 (14): 2550–2569.

Joyce, T., Y.-O. Kwon, and L. Yu. 2009. On the relationship between synoptic wintertime atmospheric variability and path shifts in the Gulf Stream and the Kuroshio Extension. *J. Climate* 22: 3177–3192.

Kelly, K.A., and S. T. Gille. 1990. Gulf Stream surface transport and statistics at 69oW from the Geosat altimeter. *J. Geophys. Res.* 95 (C3): 3149–3161.

Kelly, K.A., R. Justin Small, R.M. Samelson, B. Qiu, T. Joyce, M. Cronin, and Y.-O. Kwon. 2010. Western boundary currents and frontal air-sea interaction: Gulf Stream and Kuroshio Extension, *J. Climate* 23: 5644–5667.

Kelly, K.A., and S. Dong. 2013. The contributions of atmosphere and ocean to North Atlantic Subtropical Mode Water volume anomalies, *Deep-Sea Res. II http://dx.doi.org/*10.1016/j.dsr2.2013.02.020i.

Kelly, K.A., L. Thompson, and J. Lyman. 2014. The coherence and impact of meridional heat transport in the Atlantic Ocean inferred from observations, *J. Climate* 27 (4): 1469–1487.

Kelly, K.A., K. Drushka, L. Thompson, D. Le Bars and E.L. McDonagh. 2016. Impact of AMOC slowdown on heat and freshwater transports, *Geophys. Res. Lett.* 43 doi:10.1002/2016GL069789.

Kwon, Y.O., Alexander, M.A., Bond, N.A., Frankignoul, C., Nakamura, H., Qiu, B. and Thompson, L.A., 2010. Role of the Gulf Stream and Kuroshio-Oyashio systems in large-scale atmosphere-ocean interaction: A review. *J. Climate*, 23 (12): 3249–3281.

Le Bars, D., J.V. Durgadoo, H.A. Dijkstra, A. Biastoch and W.P.M. de Ruijter. 2014. An observed 20-year time series of Agulhas leakage, *Ocean Sci.* 10 doi:10.5194/os-10-601-2014.

Lee, T., and M.J. McPhaden. 2008. Decadal phase change in large-scale sea level and winds in the Indo-Pacific region at the end of the 20th century. *Geophys. Res. Lett.* 35 doi:10.1029/2007GL032419.

Liu, Z. 2012. Dynamics of interdecadal climate variability: A historical perspective. *J. Climate* 25: 1963–1995.

Llovel, W., and T. Lee. 2015. Importance and origin of halosteric contribution to sea level change in the southeast Indian Ocean during 2005–2013. *Geophys. Res. Lett* 42: 1148–1157 doi:10.1002/2014GL062611.

Loveday, B.R., P. Penven and C.J.C. Reason. 2015. Southern Annular Mode and westerly-wind-driven changes in Indian-Atlantic exchange mechanisms. *Geophys. Res. Lett.* 42: 4912–4921. doi:10.1002/2015GL064256.

Lumpkin, R., and S. Garzoli. 2011. Interannual to decadal changes in the southwestern Atlantic's surface circulation. *J. Geophys. Res.* 116 (C01014) doi:10.1029/2010JC006285.

Mantua, N.J. and Hare, S.R. 2002. The Pacific decadal oscillation. *Journal of oceanography*, 58(1): 35-44.

Meehl, G.A., L. Goddard, G. Boer, R. Burgman, G. Branstator, C. Cassou, S. Corti, G. Danabasoglu, F. Doblas-Reyes, E. Hawkins and A. Karspeck. 2014. Decadal climate prediction: an update from the trenches. *Bull. Am. Meteorol. Soc.* 95(2): 243–267.

Nakamura, H., Sampe, T., Tanimoto, Y. and Shimpo, A., 2004. Observed associations among storm tracks, jet streams and midlatitude oceanic fronts. *Earth's Climate* 329–345.

Menezes, V.V., H.E. Phillips, M.L. Vianna, and N.L. Bindoff. 2016. Interannual variability of the South Indian Countercurrent. *J. Geophys. Res.* 121*:* 3465–3487 doi:10.1002/2015JC011417.

Msadek, R. and C. Frankignoul. 2009. Atlantic multidecadal variability and its influence on the atmosphere on a climate model. *Clim. Dyn.* 33: 45–62 doi:10.1007/s00382-008-0452-0.

Nagano, A., M. Wakita, and S. Watanabe. 2016a. Dichothermal layer deepening in relation with halocline depth change associated with northward shrinkage of North Pacific western subarctic gyre in early 2000s. *Ocean Dyn.,* 66: 163–172 doi:10.1007/s10236-015-0917-8.

Nagano, A., S. Kizu, K. Hanawa and D. Roemmich. 2016b. Heat transport variations due to change of North Pacific subtropical gyre interior flow during 1993-2012. *Ocean Dynam.*, 66, 1637–1649, doi: 10.1007/s10236-016-1007-2.

Oka, E. and B. Qiu. 2012. Progress of North Pacific mode water research in the past decade. *J. Oceanogr.* 68(1): 5–20.

Piecuch, C.G., K.J. Quinn and R.M. Ponte. 2013. Satellite-derived interannual ocean bottom pressure variability and its relation to sea level, *Geophys. Res. Lett.* 40. doi:10.1002/grl.50549.

Piecuch, C.G., and R.M. Ponte. 2014. Nonseasonal mass fluctuations in the midlatitude North Atlantic Ocean, *Geophys. Res. Lett.* 41, 4261-4269, doi:10.1002/2014GL060248.

Qiu, B., and K.A. Kelly. 1993. Upper-ocean heat balance in the Kuroshio Extension region. *J. Phys. Oceanogr.* 23: 2027–2041.

Qiu, B. and S. Chen. 2005. Variability of the Kuroshio Extension Jet, Recirculation Gyre, and Mesoscale Eddies on Decadal Time Scales, *J. Phys. Oceanogr.* 35: 2090–2103, doi: 10.1175/JPO2807.1.

Qiu, B. and S. Chen. 2006. Decadal variability in the formation of the North Pacific Subtropical Mode Water: Oceanic versus atmospheric control. *J. Phys. Oceanogr.* 36 (7): 1365–1380.

Qiu, B., and S. Chen. 2010a. Eddy-mean flow interaction in the decadally modulating Kuroshio Extension system. Deep-Sea Research II 57: 1098–1110.

Qiu, B., and S. Chen. 2010b. Interannual-to-decadal variability in the bifurcation of the North Equatorial Current off the Philippines. *J. Phys. Oceanogr.* 40: 2525–2538.

Ridgway, K.R., R.C. Coleman, R.J. Bailey and P. Sutton. 2008. Decadal variability of East Australian Current transport inferred from repeated high density XBT transects, a CTD survey and satellite altimetry. *J. Geophys. Res.* 113: 1–18.

Roemmich, D., J. Gilson, J. Willis, P. Sutton and K. Ridgway. 2005. Closing the time-varying mass and heat budgets for large ocean areas: The Tasman Box, *J. Climate* 18: 2330–2343.

Roemmich, D., J. Gilson, R. Davis, P. Sutton, S. Wijffels and S. Riser. 2007. Decadal spin-up of the South Pacific subtropical gyre, *J. Phys. Oceanogr.* 37: 162–173.

Roemmich, D., J. Church, J. Gilson, D. Monselesan, P. Sutton and S. Wijffels. 2015. Unabated planetary warming and its ocean structure since 2006. *Nat. Clim. Change* 5(3): 240–245.

Roemmich, D., J. Gilson, P. Sutton and N. Zilberman. . . 2016. Multidecadal change of the South Pacific gyre circulation. *J. Phys. Oceanogr.* 46: 1871–1883.

Souza, J.M.A.C., C. de Boyer Montégut, C. Cabanes and P. Klein. 2011. Estimation of the Agulhas ring impacts on meridional heat fluxes and transport using ARGO floats and satellite data. *Geophys. Res. Lettr.* 38: doi:10.1029/2011GL049359.

Sprintall, J., D. Roemmich, B. Stanton and R. Bailey. 1995. Regional climate variability and ocean heat transport in the southwest Pacific Ocean. *J. Geophys. Res.* 100: 15,865–15,871.

Swart, S., S. Speich, I.J. Ansorge and J.E. Lutjeharms. 2010. An altimetry-based gravest empirical mode south of Africa. 1. Development and validation. *J. Geophys. Res.* 115 (C03002) doi:10.1029/2009JC005299.

Thompson, D.W. and J.M. Wallace. 2000. Annular modes in the extratropical circulation. *Part I: Month-to-month variability. Journal of climate, 13*(5), pp.1000–1016.

Trenary, L.L., and W. Han. 2013. Local and remote forcing of decadal sea level and thermocline depth variability. *J. Geophys. Res.* 118: 381–398 doi:10.029/2012JC008317.

Trenberth, K.E., and J.M. Caron. 2001. Estimates of meridional atmosphere and ocean heat transports. *J. Clim.* 14(16): 3433–3443.

Vivier, F., C. Provost and M.P. Meredith. 2001. Remote and Local Forcing in the Brazil-Malvinas Region. *J. Phys. Oceanogr.* 31(4).

Vivier, F., K.A. Kelly and L. Thompson. 2002. Heat budget in the Kuroshio Extension region: 1993–1999, *J. Phys. Oceanogr.* 32: 3436–3454.

Vianna, M., and V.V. Menezes. 2013. Bidecadal sea level modes in the North and South Atlantic Oceans. *Geophys. Res. Lettr.* 40: 5926–5931 doi:10.1002/2013GL058162

Volkov, D.L., and L.L. Fu. 2011. Interannual variability of the Azores Current strength and eddy energy in relation to atmospheric forcing, *J. Geophys. Res.* 116, C11011, doi:10.1029/2011JC007271.

Wakita, M., et al. 2013. Ocean acidification from 1997 to 2011 in the subarctic western North Pacific Ocean. *Biogeosciences* 10: 7817–7827 doi:10.5194/bg-10-7817-2013.

Willis, J.K. 2010. Can in situ floats and satellite altimeters detect long term changes in Atlantic Ocean overturning? *Geophys. Res. Lett., 37*: L06602 doi:10.1029/2010GL042372.

Willis, J.K., D. Roemmich and B. Cornuelle. 2003. Combining altimetric height with broadscale profile data to estimate steric height, heat storage, subsurface temperature, and sea-surface temperature variability. *J. Geophys. Res.* 108: doi:10.1002/2015JC011492.

Willis, J. K., and L.-L. Fu. 2008. Combining altimeter and subsurface float data to estimate the time-averaged circulation in the upper ocean. *J. Geophys. Res., 113*, C12017 doi:10.1029/2007JC004690.

Yu, L, and R. A. Weller 2007. Objectively analyzed air–sea heat fluxes for the global ice-free oceans (1981–2005). *Bulletin of the American Meteorological Society* 88.4: 527–539.

Zhang, J., K.A. Kelly and L. Thompson. 2016. The role of heating, winds and topography on sea level changes in the North Atlantic. *J. Geophys. Res.* 121: 2887–2900 doi:10.1002/2015JC011492.

Zilberman, N.V., D.H. Roemmich and S.T. Gille. 2014. Meridional volume transport in the South Pacific: Mean and SAM-related variability. *J. Geophys. Res.* 119: 2658–2678 doi:10.1002/2013JC009688.

7 Monitoring and Interpreting the Tropical Oceans by Satellite Altimetry

Tong Lee, J. Thomas Farrar, Sabine Arnault, Benoit Meyssignac, Weiqing Han, and Theodore Durland

7.1 INTRODUCTION

The tropical oceans, which are responsible for meridional and zonal heat transports that are comparable to or larger than those carried by the tropical atmosphere, have significant influence on the Earth's climate. The interactions between the tropical oceans and atmosphere and their impact on intraseasonal to multidecadal timescales result in climate variability that has regional to global impact. This variability includes the impact of the Madden-Julian Oscillations (MJO) on intraseasonal timescales, the monsoons on seasonal timescales, El Niño-Southern Oscillation (ENSO), Indian Ocean Zonal/Dipole Mode (IOZDM), and Tropical Atlantic Niño on interannual timescales, and the Pacific Decadal Oscillation (PDO). Variability associated with tropical-ocean currents influences the geographical distribution of large-scale upper-ocean heat content. Under the influence of global warming, the climate of the tropical region will probably undergo significant changes. The tropical Pacific easterly trade winds are expected to weaken under the global warming scenario, leading to a slower equatorial oceanic circulation (including equatorial upwelling) (Vecchi and Soden 2007). In response, the surface ocean temperatures are expected to warm fastest near the equator and more slowly farther away, the equatorial thermocline is expected to shoal, and the temperature gradients across the thermocline are expected to become steeper (e.g., Collins et al. 2010). The climate modes such as ENSO and IOZDM, which are controlled by a delicate balance of amplifying and damping feedback, depend on these thermocline features, and thus they will probably be modified under global warming. Understanding how they will change, however, remains a challenge.

The stratification of the tropical oceans is characterized by a relatively shallow pycnocline with an averaged depth ranging from 50 to 200 m depending on longitudes, and most of the variability occurs above the pycnocline (Hastenrath and Merle 1987; Durand and Delcroix 2000; Schott et al. 2009). A two-layer system is a reasonably good approximation of the tropical oceans where the response to wind forcing is dominated overall by the first baroclinic mode. Consequently, sea surface height (SSH) variability provides a good proxy for changes of pycnocline geostrophic flow and upper-ocean heat content. Altimeter-derived SSH anomalies thus have been providing critical measurements for the study of tropical ocean variability and the related ocean-atmosphere interaction.

The equatorial oceans provide a waveguide that hosts a suite of large-scale oceanic waves that are important to ocean dynamics and ocean-atmosphere interactions, such as the oceanic equatorial Kelvin and Rossby waves and tropical instability waves (TIWs). In particular, Kelvin and Rossby waves transmit the oceanic response to wind forcing zonally across the equatorial oceans, thereby modulating sea surface temperature (SST) and the feedback to the atmosphere (e.g., through the so-called Bjerknes feedback) and providing the oceanic memory associated with seasonal-to-interannual climate variability such as ENSO. Prior to the early 1990s, the limited observational capability to monitor these waves relied primarily on the sparsely distributed tide gauges and moorings (Knox and Halpern 1982; Lukas et al. 1984). Furthermore, in the Atlantic and Indian oceans, there was no equivalent to the

Tropical Atmosphere Ocean (TAO) mooring array of the Pacific back then. Altimeter-derived SSH measurements have greatly enhanced our ability to observe the variability of the tropical oceans, and they have been fundamental to the study of tropical ocean dynamics and air-sea interaction.

Following the proof-of-concept missions such as Skylab, GEOS 3, and Seasat, in the 1980s, GEOSAT brought a noticeable improvement in terms of altimeter accuracy (Tai 1988; Cheney et al. 1989). Tropical ocean studies using GEOSAT SSH data were first dedicated to comparisons with *in situ* data or models on basin scales (Arnault et al. 1989; Carton 1989; Arnault et al. 1990; Picaut et al. 1990; Arnault and LeProvost 1997) or on more regional scales (Carton and Katz 1990; Cheney and Miller 1990; Forristall et al. 1990; Wyrtki and Mitchum 1990; Arnault et al. 1992; Carton et al. 1993; Didden and Schott 1993; Arnault and Cheney 1994; Minster et al. 1995; Verstraete and Park 1995; Arnault et al. 1999). These comparisons showed that altimetric observations were mature enough to depict the seasonal cycle and to detect the interannual variability of Tropical Ocean SSH. Applications to study oceanographic physical processes were emerging (e.g., Miller and Cheney 1990; Périgaud 1990; Périgaud and Delecluse 1992) especially those concerning wave propagation associated with ENSO (Delcroix et al. 1991). The famous picture from Miller et al. (1988) showing, for the first time, propagating Kelwin and Rossby waves along the whole Pacific basin during 1987 to 1988 El Niño was a remarkable milestone in altimetry history.

TOPEX/Poseidon (T/P), launched in 1992, represented a major improvement in high-precision satellite altimetry. Jason-1, Jason-2, and Jason-3 missions continue the precision altimeter records. Besides the CalVal studies inherent to each new satellite mission (Busalacchi et al. 1994; Giese et al. 1994; Mitchum 1994; Picaut et al. 1995; Verstraete and Park 1995; Mayer et al. 2003; Arnault et al. 2004; Provost et al. 2004; Arnault et al. 2011; Liu et al. 2014), altimetry began to be used as an independent source of information to improve knowledge of tropical dynamics, allowing new insights into mesoscale activity, wave propagation, and low-frequency variability from seasonal to decadal timescales.

Much of the efforts describing the utility of altimetry measurements to study intraseasonal to interannual tropical ocean variability and the related climate variability prior to 2000 has been described in Chapter 4, "Tropical Ocean Variability" (authored by J. Picaut and A. Busalacchi), of the *Altimetry and Earth Sciences: A Handbook of Techniques and Applications* (Fu and Cazenave 2000). Since then, significant progress has been made toward understanding and predicting such variability. Moreover, the nearly two-and-half decades of precision satellite altimetry measurements have enabled the studies of decadal variability of the tropical ocean and coupled ocean-atmosphere system. This chapter mostly focuses on the contributions of altimetry-derived SSH measurements in these areas of studies since 2000. However, some related studies in the late 1990s that have not been discussed by Picaut and Busalacchi (2000) in the previous altimetry book are also highlighted here to provide a more complete scope and context.

Section 7.2 of this chapter focuses on the Atlantic Ocean variability. Variability of the tropical Pacific and Indian Ocean as well as their linkages (e.g., the Indonesian Throughflow) are described in Sections 7.3, 7.4, and 7.5, respectively. The discussions in these sections are generally organized by subsections that focus on intraseasonal, seasonal, interannual, and decadal timescales. Hereafter, SSH measurements generally refer to altimetry-derived SSH anomalies from various missions since 1992 (e.g., T/P, Jason-1 and Jason-2), including the SSH products derived by merging data from two or more altimeter missions such as the Archiving, Validation, and Interpretation of Satellite Oceanographic (AVISO) gridded data products (see Pujol et al. 2016 and references therein).

7.2 TROPICAL ATLANTIC OCEAN

7.2.1 Intraseasonal and Eddy Activities

7.2.1.1 Eddy Structures

Eddy activities have been identified from altimetry along the African coast, in the Mauritania-Sénégal and the Benguela upwelling systems (Chaigneau et al. 2009), and in the Gulf of Guinea

(Djakoure et al. 2014). However, the mesoscale eddy activity in the tropical Atlantic is mostly located in the western basin associated with the North Brazil Current (NBC) and its retroflection into the North Equatorial Countercurrent (NECC) (e.g., Goni and Johns 2001) (Figure 2.1). These eddies may account for more than 25%–50% of the inter-hemispheric transports of mass and heat within the Atlantic meridional overturning cell (Didden and Schott 1993; Frantantoni et al. 1995; Goni and Johns 2001; Garzoli et al. 2003). Following previous work by Didden and Schott (1993), Nystuen and Andrade (1993), and Fratantoni et al. (1995) based on GEOSAT SSH data, Pauluhn and Chao (1999) tracked 16–20 anticyclonic eddies using T/P SSH data between October 1992 and June 1997. Goni and Johns (2001) depicted approximately six rings per year between 1993 and 1998, nearly twice that previously thought from *in situ* observations. Extending the series up to 2001, Goni and Johns (2003) found a weak tendency for rings to form during the first half of the year, suggesting that a formation mechanism other than the NBC retroflection also exists, with important year-to-year variability. Castelao and Johns (2011) questioned the altimetric underestimation of the rings' intensity due to the inexact crossing of the ring centers during individual satellite passes.

7.2.1.2 Tropical Instability Waves

TIWs play a role in climate variability as they can influence, for instance, the phase of the seasonal cycle, the position of the equatorial cold tongue and of the Intertropical Convergence Zone (ITCZ), the oxygen and salinity front along the equator, and the plankton and nutrient distribution (Menkes et al. 2002; Jochum and Malanotte-Rizzoli 2003; Lee et al. 2012). Away from the equator (3°N–5°N), where Kelvin and Rossby waves play a key role (see the following), SSH and thermocline variations in the 40- to 60-day band in the tropical Atlantic are caused largely by TIWs (Katz 1997; Han et al. 2008). Strong seasonal variations of tropical Atlantic TIW activities have been identified in SSH as well as in SST and SSS associated with seasonal variations of the equatorial current system (e.g., Lee et al. 2014) (Figure 2.2). This is in contrast to tropical Pacific TIWs that have stronger interannual than seasonal variations.

Von Schuckmann et al. (2008) used altimetry SSH data to validate a numerical eddy-resolving model and investigated the spatial and temporal distributions of these TIWs and the role of baroclinic versus barotropic instabilities to generate these waves. Grodsky et al. (2005) used temperature, salinity, velocity, and wind from a mooring at 0°N, 23°W along with satellite data for SST and SSH to examine the contribution of TIWs to the energy and heat balance of the equatorial Atlantic mixed layer. The TIW intensification occurs in phase with strengthening of the southeasterly trade winds and the seasonal appearance of the equatorial cold tongue. The waves contribute to a net warming of the mixed layer by 0.35°C during the summer months, and they are maintained by both barotropic and baroclinic conversions that are of comparable size. They also showed that salinity fluctuations, previously neglected, increase the magnitude of baroclinic energy conversion, a result expanded on by Lee et al. (2014) in an analysis of satellite measurements including SSH, SSS, and SST. Perez et al. (2012) developed metrics based on SST and SSH fluctuations to examine interannual TIW variability in the Atlantic. They found that in contrast to the near equatorial region, years with low TIW variance along the off-equatorial latitude bands are associated with anomalously warm SSTs in the cold tongue region, weak wind stress divergence and curl in the equatorial region, and weak zonal current shear in the NECC/South Equatorial Current (SEC) region.

7.2.2 The Seasonal Cycle

The seasonal cycle is the dominant source of variability in the tropical Atlantic Ocean. Its main features—including the annual strengthening/weakening of the surface currents, the retroflection of the NBC, the zonal pressure gradient along the equator, and the annual and semiannual upwelling and downwelling cycle in the Gulf of Guinea—were apparent even in the early satellite altimetry missions. They suggested that the annual cycle accounts for at least 60% of the total SSH variance in the Tropical Atlantic. However, the limited accuracy of these early missions prevented a precise quantification of this seasonal cycle.

Two decades of high-precision altimetry from T/P and subsequent missions have confirmed these early findings (Schouten et al. 2005; Aman et al. 2007; Ding et al. 2010). They offer the opportunity to obtain a detailed picture of the tropical Atlantic equatorial upwelling (Schlundt et al. 2014) and surface circulation (Arnault et al. 1999; Arnault and Kestenaere 2004; Fonseca et al. 2004; Stramma et al. 2005; Artamonov 2006; Urbano et al. 2008; Goes et al. 2013). For instance, Fonseca et al. (2004) showed that the NBC retroflection exhibits a mean position of 6.6°N ± 2.0°N, with a strong annual signal that follows the meridional migration of the ITCZ and the strength of the wind stress curl over the equatorial Atlantic. Yang and Joyce (2005) found that the seasonal variation of the NECC geostrophic transport in the tropical Atlantic Ocean is dominated by changes in the southern flank of the current and hypothesized that the wind stress forcing along the equator is the leading driver for the seasonal cycle of the NECC transport. The wind stress curl in the NECC region is an important but smaller contributor.

7.2.3 Equatorial Waves

Equatorially trapped waves are an important component of tropical ocean variability because they are efficient carriers of information along the equator and can lead to remote effects on ocean-atmosphere interaction. Prior to the altimeter era, the existence of Kelvin and Rossby waves in the tropical Atlantic was questioned on the basis of narrow width of the basin, the typical length scales of the wind forcing, and the high propagation speed of these waves. Katz (1997) was one of the first to identify in the 1993–1995 T/P data eastward propagations as first-mode Kelvin waves. Franca et al. (2003) explained much of the variance of the T/P data during 1992–1999 as a combination of the Kelvin and first two Rossby mode waves. Schouten et al. (2005) found that the seasonal adjustment of the tropical Atlantic to the wind leads to a cycle of consecutive Kelvin and Rossby waves, with the African coastal signal partially forced by the equatorial waves (Figure 2.3). However, Bunge and Clarke (2009) argued that the phase speed of the propagation found by Schouten et al. (2005) is too slow. They concluded that this annual eastward propagating signal in SSH at the equator is not the result of waves but rather the superposition of two independent modes of variability: variations in the warm water volume above the 20°C isotherm and a tilt in the thermocline. But Polo et al. (2008a) presented an intraseasonal climatology of T/P data that reveals regular boreal autumn-winter equator to coast propagations with phase speeds varying from 1.5 to 2 m s^{-1} in good agreement with the theory. A remote forcing effect of the Kelvin waves was evidenced over coastal regions as far as 10–15 degrees latitude and have no counterpart in the hypothesis put forth by Bunge and Clarke (2009). Furthermore, the modeling results of Ding et al. (2009), validated using SSH data, show that Kelvin and Rossby waves are both important at the equator, with the Kelvin wave being most important. Additionally, there is an important contribution of boundary reflections equal to that of the wind-forced response.

More studies on the role of Kelvin and Rossby waves during particular events in the tropical Atlantic Ocean have been conducted. Illig et al. (2006) detailed the 1996 warm event from a numerical investigation associated with SSH data and argued for an essential role of equatorial waves triggered by distinct wind events. A possible teleconnection with the tropical Pacific Ocean through the modification of the Walker Circulation has been suggested. Han et al. (2008) found that the 40- to 60-day Sea Surface Height Anomaly (SSHA) was dominated in 2002–2003 by the equatorially symmetric Kelvin waves driven by intraseasonal winds within the 3°S–3°N equatorial belt. Interestingly, if the 10- to 40-day periods are dominated by TIWs west of 10°W, east of it, SSHA variations result almost entirely from wind-driven equatorial waves. During the boreal spring of 2002, when TIWs were weak, Kelvin waves dominated the SSHA across the equatorial basin (2°S–2°N). Hormann and Brandt (2009) continued analysis of the 2002 warm event, comparing it to the 2005 strong cold event using moored and satellite observations. Prior to the warm event in 2002, equatorial easterly winds in the western to central Atlantic, covarying with the equatorial

Kelvin wave mode, relax. In response, the thermocline shoals in the west and the basin-wide adjustment of the equatorial thermocline slope via downwelling Kelvin waves results in a deeper thermocline in the east. In contrast, such a zonal wind mechanism could not be established for the strong cold event in 2005. Hormann et al. (2012), using surface currents derived from altimetry, suggested that alongshore winds in the northeastern tropical Atlantic generate Rossby waves that propagate westward from the eastern boundary and are responsible for a large part of the NECC interannual variability. These studies dedicated to specific years showed that besides the intense annual cycle, the tropical Atlantic also presents interannual variations but of comparatively small amplitude.

7.2.4 Interannual Variability

Interannual variations in the early 1990s have been identified through their signature on SST (Servain 1991; Houghton and Tourre 1992). An attempt using GEOSAT altimetry was conducted (Arnault and Cheney 1994), but the analysis of their signature on SSH began in earnest in the 2000s, enabled by the longer record of precise satellite altimetry (Arnault and Kestenaere 2004; Schouten et al. 2005; Handoh et al. 2006a, 2006b; Foltz and McPhaden 2010; Arruda et al. 2011; Arnault and Mélice 2012). These studies revealed that the interannual variability of the tropical Atlantic basin-scale SSH is dominated by two modes of variability: (1) an equatorial mode akin to ENSO but with a weaker magnitude and (2) a meridional mode varying between north (10°N–20°N) and south (along 20°S) of the equator.

The equatorial mode is characterized by changes in the east-west SSH slope along the equator in response primarily to changes in the zonal wind stress (Provost et al. 2006; Ding et al. 2010; Foltz and McPhaden 2010; Arnault and Mélice 2012) or by the emergence of a cold tongue in the Gulf of Guinea (such as in 1997 or 2002). The positive feedback among SST, zonal wind, and thermocline depth or SSH (i.e., Bjerknes feedback) tends to prolong this equatorial mode (Keenlyside and Latif 2007; Ding et al. 2010).

The meridional mode variability has also been attributed to variability in the trade winds. In the northern and southern tropics, the variations of SSH are in phase with the variations in SST because both are a response to the thermosteric variations generated by the wind variability: As trade winds intensify, SSH and SST decrease (Ruiz-Barradas et al. 2000; Handoh et al. 2006a, 2006b; Arnault and Mélice 2012). Foltz and McPhaden (2010) analyzed SST and SSH observations to conclude that the meridional and equatorial modes of the tropical Atlantic are linked through anomalous changes in equatorial winds.

The possible teleconnection of the Atlantic tropical modes of variability with the Pacific ENSO mode is intensely debated. Several studies have shown an influence of ENSO on the northern tropical Atlantic SST with warming in the North Atlantic occurring 4–5 months later than in the Pacific (Enfield and Mayer 1997; Huang et al. 2005; Arnault and Mélice 2012) (Figure 2.4). This ENSO-related warming seems to have a signature on SSH in the Gulf of Guinea as well, with a longer lag of the latter (Andrew et al. 2006; Arnault and Mélice 2012). Other climate modes such as the IOZDM and the North Atlantic Oscillation (NAO) do not seem to influence the tropical Atlantic sea level variability (Andrew et al. 2006).

The role of equatorial Kelvin and Rossby waves in interannual variability has also been suggested in the Angola/Benguela area, south of 15°S. Shannon et al. (1986) suggested that wind stress anomalies to the north, and possibly in the equatorial Atlantic, are the dominant factor of SST variations in that area. Polo et al. (2008a, 2008b) point out that the influence of Kelvin waves triggered by equatorial wind stress anomalies might not reach further poleward than 12°S, while other studies have articulated the importance of remote equatorial influences (Florenchie et al. 2003, 2004; Grodsky et al. 2006; Huang and Hu 2007; Rouault et al. 2007). Richter et al. (2010) argued that meridional wind anomalies along the southwest African coast contribute substantially. These wind anomalies form part of a basin-scale weakening of the subtropical anticyclone that extends to the equator. Results also indicate that the close correlation between Benguela and Atlantic Niños in observations might result from the large spatially coherent wind stress anomalies associated with the weakened anticyclone.

Very recently, the tropical Atlantic has raised special interest in the climate community because the particularly low interannual variability in SSH in this region makes it an area where the SSH signal due to greenhouse gas (GHG) emissions might be detected over the signals generated by the natural climate variability. Indeed, recent detection and attribution studies based on satellite altimetry and atmosphere-ocean general circulation models or Earth system models suggest that the SSH signal forced by the GHG emissions may be detectable above the level of unforced internal variability in the tropical Atlantic within the next decade (Jordà 2014; Richter et al. 2014; Carson et al. 2015; Fernandez Bilbao et al. 2015; Lyu et al. 2015).

7.3 TROPICAL INDO-PACIFIC OCEAN

7.3.1 Tropical Pacific

SSH measurements by satellite altimeters have proven especially useful for allowing insight into the dynamics of the tropical Pacific Ocean. Unlike variability seen in SST or SSS, the SSH anomalies detected from altimetry directly reflect the upper-ocean pressure fluctuations associated with ocean dynamical variations, and, because of a strong correlation of SSH and thermocline depth in the tropical Pacific (Rebert et al. 1985), SSH anomalies provide important information about thermocline depth and ocean heat content. The enormous size of the tropical Pacific, combined with a dearth of shipping lanes south of the equator, has made it a challenge to adequately resolve the broad spectrum of spatiotemporal variability of the tropical Pacific with *in situ* measurements. Although much was learned about tropical Pacific dynamics in the pre-satellite years, satellite altimetry has given us the first opportunity to observe the entire region in great detail, to confirm or refute previous conjectures, and to find new aspects of the equatorial dynamics that were previously unavailable to us.

7.3.1.1 Intraseasonal Variability

Several distinct types of variability with intraseasonal periods (20 to 100 days) coexist in the tropical Pacific Ocean, including eastward propagating Kelvin waves and shorter-wavelength, westward-propagating variability, such as TIWs.

TIWs were first discovered in satellite infrared SST measurements in the Pacific (Legeckis 1977). They arise from instabilities of the equatorial current system (Philander 1976; Luther and Johnson 1990; Lyman et al. 2005) and are clearly visible in satellite SST measurements as a meandering of the northern side of the equatorial cold tongue near 2°N. The discovery of TIWs in the late 1970s stimulated a great deal of subsequent theoretical and observational work to understand their generation mechanisms, properties, and consequences.

A relatively clear separation of TIWs and intraseasonal Kelvin waves can be achieved by examining the spectrum of SSH variability in the zonal-wavenumber-frequency domain, which allows separation of the eastward-propagating Kelvin waves from the westward-propagating TIWs. Examination of the zonal-wavenumber/frequency spectrum is also helpful because the most common theoretical approaches to Kelvin waves and TIWs are carried out in the zonal-wavenumber/frequency domain (e.g., Matsuno 1966; Moore and Philander 1977; Philander 1978).

The zonal-wavenumber/frequency spectrum of SSH in the equatorial Pacific exhibits two broad regions of elevated SSH variance, one corresponding to eastward propagating Kelvin waves at small, positive zonal wavenumbers (wavelengths exceeding 50° longitude) and one corresponding to TIWs and other westward-propagating variability with zonal wavelengths of about 9°–30° (Figure 7.1; Perigaud 1990; Zang et al. 2002; Wakata 2007; Farrar 2008; Shinoda et al. 2009; Farrar 2011). The SSH spectrum shown in Figure 7.1 (after Farrar 2011) was estimated using data from the DT2014 AVISO gridded altimetry product (Pujol et al. 2016) over the period 1993–2014 and almost the full width of the equatorial Pacific (149°E–88°W) and has been averaged over 7°S–7°N.

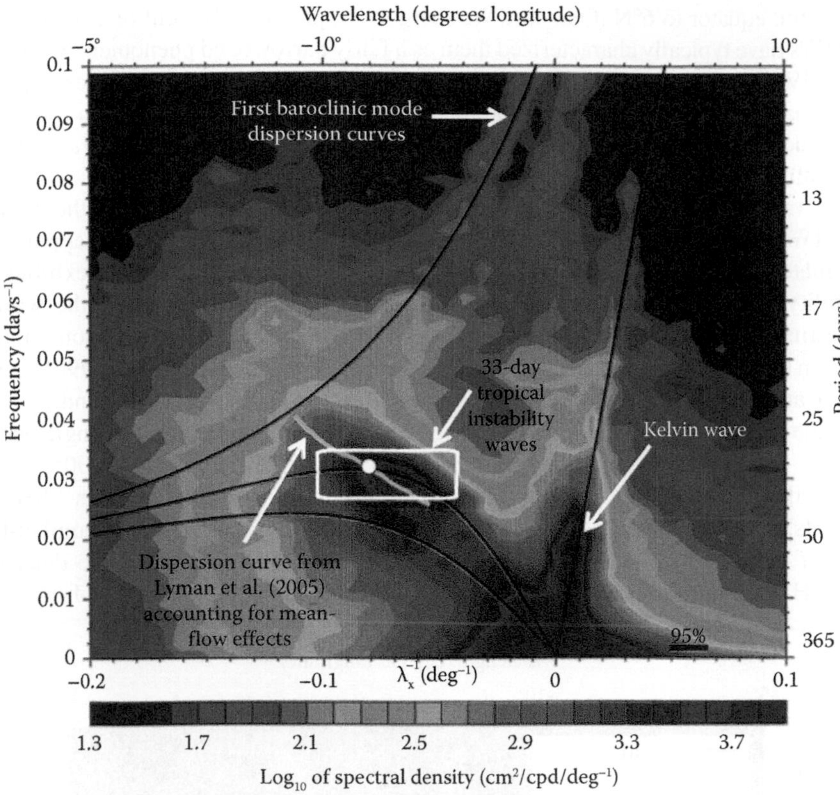

FIGURE 7.1 Zonal-wavenumber/frequency spectrum of SSH, averaged over 7°S–7°N (after Farrar, J. T., *J. Phys. Oceanogr.*, 41, 1160–1181, 2011. With permission), using data from 1993–2014 and almost the full width of the Pacific (149°E–88°W). At periods of 20–100 days, there are two broad regions of elevated SSH variance, one corresponding to eastward propagating Kelvin waves at small, positive wavenumbers (wavelengths exceeding 50°), and one corresponding to TIWs and other westward-propagating variability having zonal wavelengths of about 9°–30°. The TIW spectral peak in SSH near 33-day periods (white box) has wavelengths of 12°–17°. The four black curves are the theoretical dispersion curves for the first-baroclinic-mode Kelvin, Rossby, and mixed Rossby-gravity waves. The grey line depicts a theoretical dispersion curve for the unstable TIW mode, and the white circle indicates the fastest growing wavelength (from Lyman, J., et al., *J. Phys. Oceanogr.*, 35, 232–254, 2005. With permission). The 95% confidence interval should be measured against the color scale; a difference of two contour intervals is significant at 95% confidence.

There is a well-defined spectral peak in the SSH wavenumber-frequency spectrum near periods of 33 days (frequency of 0.03 days^{-1}) and wavelengths of 12°–16° (zonal wavenumbers of 0.06–0.08 degrees^{-1}). This is close to the period and wavelength (33 days and 12.4°) found for the most unstable mode of the equatorial current system in a linear stability analysis by Lyman et al. (2005). The ridge of high power in the 25- to 40-day period band roughly follows the dispersion curve of this most unstable mode (Farrar 2008, 2011; Figure 7.1). The SSH variability in this wavenumber-frequency band also has a spatial structure that closely resembles the structure of the most unstable mode predicted by the linear stability analysis (Farrar 2008, 2011). However, the linear stability analysis of Lyman et al. (2005) predicted many unstable modes that grew more slowly than the most unstable mode, and the SSH spectrum in Figure 7.1 suggests there is in fact a broad spectrum of instabilities.

Descriptions of the properties of TIWs have varied widely: TIWs have been reported to occur at periods of 14–50 days, at zonal wavelengths of 7° to 25°, and to have maximum amplitude at locations ranging from the equator to 6°N (Qiao and Weisberg 1995, provide a helpful review). Observational studies of TIWs have typically characterized them as a fairly narrowband phenomenon, but this apparently conflicts with the broad range of wavenumbers and frequencies reported. For example, Halpern et al. (1988) characterized the TIW signal in meridional velocity measurements on the equator as a narrowband fluctuation with a period of 20 days, while Lyman et al. (2005) and Farrar (2008) found a clear maximum in SSH variability at periods of about 33 days.

When the various wavenumbers and frequencies that have been reported or theoretically predicted for TIWs are plotted over the spectrum shown in Figure 7.1, it becomes apparent that the various estimates collectively span the range of wavenumber-frequency space that exhibits energetic SSH variability in the 7°S–7°N latitude band (Figure 7.2). Figure 7.2 includes the wavenumber-frequency estimates summarized in Table 1 of Qiao and Weisberg (1995) and some more recent estimates from Chelton et al. (2000), Donohue and Wimbush (1998), McPhaden (1996), Lyman et al. (2005, 2007), and Lee et al. (2012). The figure shows boxes, lines, and points depending on whether a study provided a range of wavenumbers and/or frequencies or simply stated a single wavenumber and frequency. Studies using SSH measurements (Perigaud 1990; Lyman et al. 2005; Farrar 2008, 2011) identified TIW wavenumbers and frequencies around the spectral peak in SSH seen near 33-day periods, consistent with the well-defined spectral peak in the SSH wavenumber-frequency spectrum near periods of 33 days. Those studies and Shinoda et al. (2009) also determined the peak TIW SSH variability to occur near 5°N. At the other extreme of the reported frequency range,

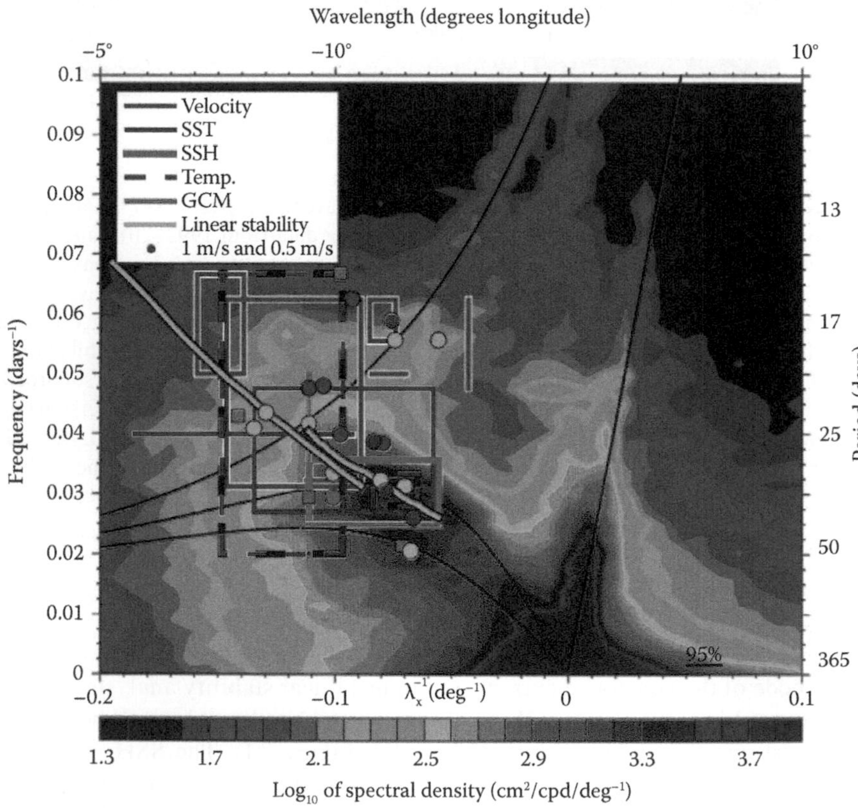

FIGURE 7.2 Previous estimates of the wavenumbers and frequencies of tropical instability waves (TIWs). (The contoured field is the zonal-wavenumber/frequency spectrum of SSH from Figure 7.1). A box indicates that a particular study provided a range of wavenumbers and frequencies, a line indicates a range of frequencies or wavenumbers, and a point indicates that a single wavenumber-frequency value was given. "SST" and "SSH" are from satellite observations, and "Velocity" and "Temp." are from moored in situ measurements.

in situ equatorial velocity measurements have yielded shorter periods, near 20 days, with the strongest signals on the equator being seen in meridional velocity (e.g., Halpern et al. 1988). Satellite SST and *in situ* temperature measurements have tended to yield estimates in between these two extremes and have also tended to identify maximum variability as occurring near 2°N.

One interpretation of these disparate observations is that some of the TIW variability resembles mixed Rossby-gravity waves, which have a relatively weak SSH signal and a strong signal in equatorial meridional velocity, and other TIW variability resembles first-meridional-mode (and perhaps second-meridional-mode) equatorial Rossby waves, with a stronger off-equatorial SSH signal (Lyman et al. 2007). In SSH, the TIW signal is largest near 5°N, where the SSH signal of the Rossby-wave-like theoretical most unstable mode is largest (Lyman et al. 2005; Farrar 2008). Equatorial current meter measurements tend to emphasize the unstable modes that resemble mixed Rossby-gravity waves, which have a maximum meridional velocity signal at the equator.

This interpretation is supported by the analysis of Lee et al. (2012), who used SSS and SST to estimate propagation speeds of about 1 m/s for TIWs near the equator and of about 0.5 m/s near 5°N, which roughly spans the range of propagation speeds that have been reported in the literature and that dominate the westward-propagating part of the SSH spectrum (Figure 7.2). TIWs produce signals in variables like SSS and SST largely through advection, with meridional velocity fluctuations deforming the sharp meridional gradients of these quantities. SSS has its strongest propagating signature of tropical Pacific TIWs near the equator where the salty South Pacific waters meet the fresher waters under the ITCZ, forming a relatively large meridional SSS gradient. In contrast, the meridional gradient of SST is weaker at the equator and is the strongest near the northern edge of the cold tongue near 2°N, where TIW signature in SST is strongest. If the meridional velocities contributing to meridional advection of SST near 2°N are from a combination of the Rossby-wave-like and mixed Rossby-gravity-wave-like unstable modes, this could explain the fact that a broader range of frequencies of TIWs have been reported from temperature observations.

Kelvin waves are the other major form of intraseasonal variability in the tropical Pacific Ocean. These waves are visible in Figures 7.1 and 7.2 as a low-wavenumber band of high variance along the first-baroclinic-mode Kelvin wave dispersion curve. They dominate the variance of SSH and other fields at 40- to 90-day periods and have received considerable attention (e.g., Enfield 1987; McPhaden and Taft 1988; Johnson 1993; Johnson and McPhaden 1993; Kessler et al. 1995; Kessler and McPhaden 1995; Hendon et al. 1998; Kutsuwada and McPhaden 2002; Zang et al. 2002; Cravatte et al. 2003; Roundy and Kiladis 2006; Farrar 2008). The waves are forced by intraseasonal wind fluctuations in the western and central Pacific (Kessler et al. 1995; Hendon et al. 1998). Interestingly, altimetry observations and numerical simulations indicate that intraseasonal Kelvin waves affect the propagation and energetics of TIWs by modulating the mean flow that produces the TIWs (Holmes and Thomas 2016).

7.3.1.2 Seasonal Variability

Before satellite altimetry, investigations into the seasonal cycle of thermocline depth or dynamic topography in the equatorial Pacific were mostly based on hydrographic and current measurements that were sparsely and irregularly distributed in time and space; lack of good data coverage (primarily south of the equator) was invoked more than once as a possible explanation for puzzling results. The completion of the TAO/TRITON mooring array at about the same time that satellite altimetry was becoming available greatly improved the subsurface sampling characteristics, although the two- to three-degree meridional spacing of the moorings is barely able to resolve the relatively short meridional scales of equatorially trapped phenomena, and the 15-degree zonal spacing barely resolves important features of the seasonal cycle near the western and eastern boundaries.

Following is a review of major features of the seasonal cycle as understood before the modern satellite altimetry record, based primarily on Wang et al. (2000), Yu and McPhaden (1999), and Johnson et al. (2002). This is followed by an overview of papers that have advanced our knowledge through the use of satellite altimetry, focusing on those published since 2000.

7.3.1.2.1 Background

On the northern and southern flanks of the mean ITCZ and the South Pacific Convergence Zone (SPCZ), high annual thermocline variability is coincident with zones of elevated annual Ekman pumping produced by the annual march of the convergence zones. A basin-wide strip of high variability between 10°N and 15°N is intensified in the east where the annual march of the ITCZ is greatest, but it is non-propagating. Within 10 degrees of the equator, the maximum oceanic variability is collocated with the Ekman pumping maximum on the southern flank of the ITCZ in the east-central Pacific (170°W–110°W); with a peak near 4°N, 140°W, and extending across the equator with a smaller amplitude. The annual minimum and maximum thermocline depth propagate westward at both 5°N and 6°S with a phase speed close to the theoretical speed for a first baroclinic, first meridional mode Rossby wave. Mitchum and Lukas (1990) concluded that this wave is resonantly forced by a westward-propagating easterly wind pattern, which spans the wave guide but is obscured in climatologies by stronger stationary winds everywhere except near the equator.

Near the equator, thermocline anomalies propagate eastward in the central and eastern Pacific, and westward with a strong semiannual signal in the western Pacific (Yu and McPhaden 1999). Wang et al. (2000) found the eastward propagation of the maximum thermocline depth to proceed at close to the first baroclinic mode Kelvin wave speed, with a timing that suggested excitation by the Western Pacific Monsoon westerlies. This monsoon also contributes to elevated annual variability centered at 5°N near the western boundary. The eastward propagation of the minimum thermocline depth on the equator was found to be quite a bit slower, and Yu and McPhaden concluded that the annual Kelvin wave was a mix of first and second baroclinic modes.

In the central equatorial Pacific, Wang et al. (2000) found the seasonal cycle to be unimodal, with the annual harmonic explaining more than 90% of the variance. Near the western and eastern boundaries, and along the mean positions of the convergence zones, they found the seasonal cycle to be bimodal and explained this as an overlap between two distinct annual forcing regimes.

The far eastern equatorial Pacific is a dynamically complex place, and even a cursory description is beyond the scope of this work. An excellent review of this region and its seasonal cycle is given in Kessler (2006), including discussion of altimetry studies that have helped to define it.

Johnson et al. (2002) estimated the annual cycle of the equatorial Pacific currents from 15 years of hydrographic and ADCP sections, and we present here a summary of their conclusions for the major surface geostrophic currents. The NECC flows eastward between about 3°N and 10°N, with a maximum in August in the eastern Pacific, October in the central Pacific, and December in the western Pacific. Between the NECC and the equator, the northern branch of the westward-flowing South Equatorial Current (or NSEC) peaks in June to October in the east and in December in the central and western Pacific. South of the equator, the southern branch of the South Equatorial Current (or SSEC) peaks in October to December in the east and in February to April in the west, with a weak seasonal cycle in the central Pacific. The westward propagation of peak currents is consistent with the dominant westward propagation of thermocline depth noted earlier. Near the equator, the SEC in the eastern Pacific reverses direction in late spring and early summer, and this reversal propagates westward in concert with the annual Rossby wave but in contradiction to the eastward thermocline propagation. Yu and McPhaden (1999) showed that this unusual pattern could result from a superposition of the aforementioned Rossby and Kelvin waves.

Johnson et al. (2002) showed that La Niña conditions tend to strengthen all of the surface currents and push the NECC northward in the central and western Pacific. El Niño conditions tend to weaken the SEC and the NECC in the eastern Pacific.

7.3.1.2.2 Contributions from Altimetry: The Seasonal SSH Anomaly Field

Work done with altimetry data prior to 2000 is well covered in Fu and Cazenave (2000), and we focus on subsequent work. As a sole exception, we note that Stammer (1997) analyzed the global annual cycle of SSH anomalies from 3 years of T/P measurements (October 1992 to

November 1995). The results were plotted on a global scale, but even at that scale, some of the features described here for the equatorial Pacific were visible in the signal of SSH minus steric height. Stammer (1997) noted that the amplitude of interannual variability was similar to that of the annual cycle in the Pacific, "rendering a climatological annual cycle nearly nonexistent." This is worth keeping in mind, but we note that the Niño 3.4 index was positive throughout all but a few months of Stammer's 3-year measurement period and was above the El Niño threshold during two of the years (Figure 7.3), a period during the prolonged ENSO event (1990 to 1995) (Trenberth and Hoar 1996). This was not a representative period for climatology of the equatorial Pacific but was part of the 1990 to 1995 prolonged ENSO that was noted by Trenberth and Hoar (1996) to be the "longest on record."

Wakata and Kitaya (2002) examined the annual cycle in Pacific SSH anomalies for the time period January 1992 to December 1997—the period of Stammer's analysis plus the two following years. They thus encountered one weak La Niña but also the buildup and peak of the strong 1997 to 1998 El Niño, so even this 5-year period would be biased heavily toward El Niño conditions (Figure 7.3). Their Figure 1 is the first good view of the SSH annual cycle in the equatorial Pacific, and it is reproduced in our Figure 7.4.

The equatorial features in Figure 7.4 are consistent with the description of Wang et al. (2000), with some exceptions. The SSH anomaly maximum near 4°N peaks near 120°W, roughly 20° further east than the peak reported by Wang et al. The local maximum reported by Wang et al. near 5°N at the western boundary is not evident in the results of Wakata and Kitaya. The eastward phase propagation on the equator reported by Wang et al. and Yu and McPhaden (1999) is also not evident in Figure 7.4, although it is possible that latitudinal smoothing has obscured this feature.

Qu et al. (2008) analyzed 13 years of merged satellite altimetry (1993 to 2005) and demonstrated that the bimodal seasonal signal noted by Wang et al. (2000) near the western boundary is in part due to a semiannual complement to the annual Rossby wave. The amplitude and phase of their annual and semiannual harmonics are shown in Figure 7.5. The maximum of SSH anomalies near 5°N is further west than the maximum of Wakata and Kitaya (2002) and more in line with

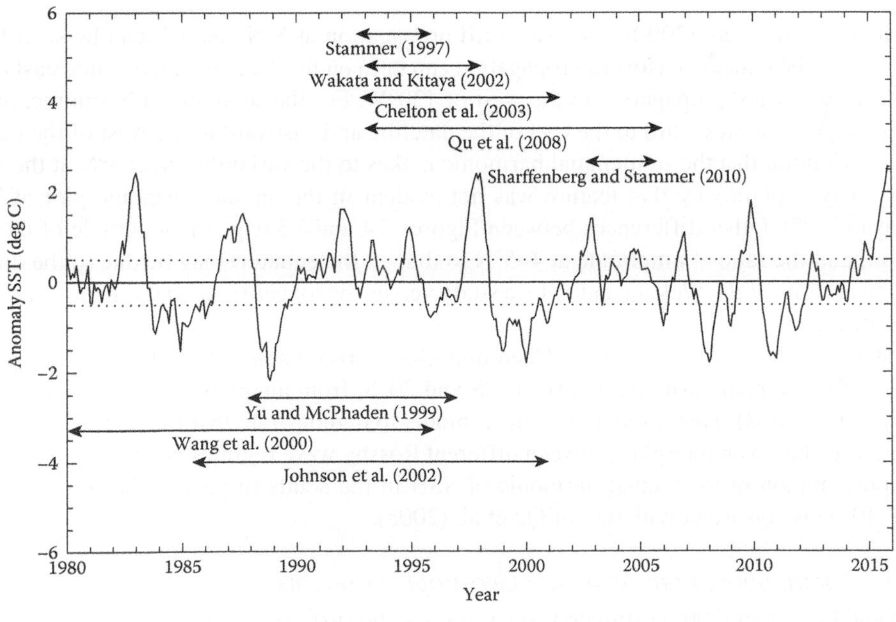

FIGURE 7.3 Niño 3.4 index with measurement spans of cited analyses. Studies shown below the index curve analyzed subsurface thermal data. Studies above the curve analyzed satellite altimetry data. Dotted lines are the ± 0.5C El Niño and La Niña thresholds.

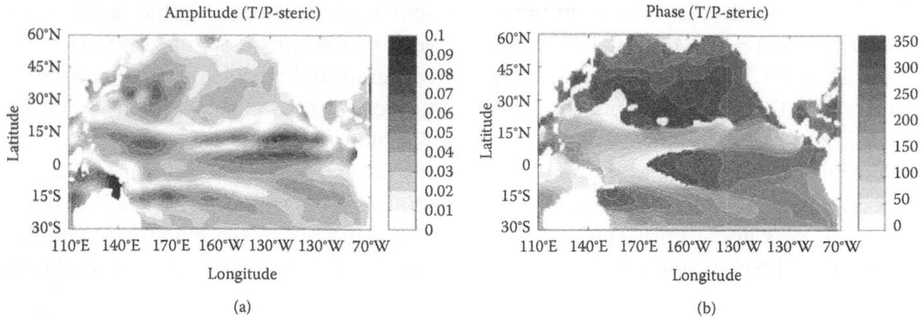

FIGURE 7.4 Amplitude (a: cm) and phase (b: year day) of annual harmonic from 5 years of T/P SSHa measurements (1/1993–12/1997). (From Wakata, Y., and Kitaya, S., *J. Oceanogr.*, 58, 439–450, 2002. With permission.)

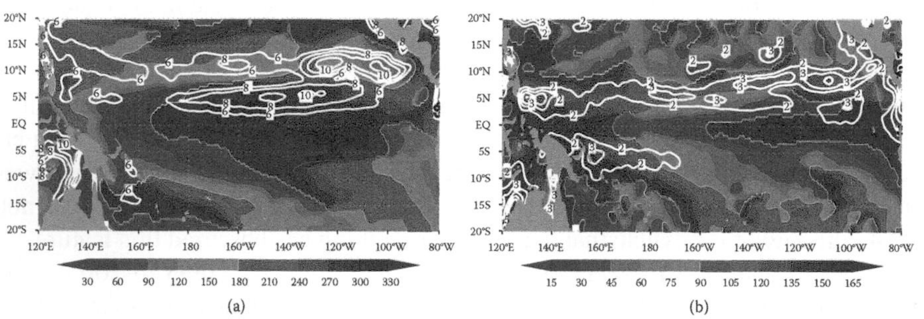

FIGURE 7.5 Amplitude (contours: cm) and phase (colors: year day) of the annual (a) and semiannual (b) harmonics from analysis of 13 years of merged satellite altimetry data sets (1/1993–12/2005) by Qu et al. (2008). From their Figure 1.

the results of Wang et al. (2000). The westward propagation at 5°N and 5°S can be seen for both harmonics. On the equator, eastward propagation can be seen for the annual harmonic east of about 130°W, with westward propagation to the west of 130°W. For the semiannual harmonic, propagation on the equator is westward to the east of the dateline and eastward to the west of the dateline.

The contribution that the semiannual harmonic makes to the variability near 5°N at the western boundary may explain why that feature was not evident in the annual harmonic plot of Wakata and Kitaya (2002). Other differences between Figures 7.4 and 7.5 (e.g., the longitude of the 4–5°N maximum, and the ratio of amplitude at 4–5°N to that on the equator) may be due to the dominant El Niño conditions of the Wakata and Kitaya study versus the longer time series incorporated in the Qu et al. study.

In a study of the SECC near 10°S, Chen and Qiu (2004) displayed the basin-wide amplitude and phase of the annual harmonic between 4°S and 20°S, from an analysis of 10 years of satellite altimetry (1992–2003). They used a dynamical model to demonstrate that the annual variability of the SECC was due to an interplay between different Rossby wave regimes north and south of 10°S. Their representation of the annual harmonic of SSH in the southern part of the equatorial waveguide (4–10°S) is consistent with that of Qu et al. (2008).

7.3.1.2.3 Contributions from Altimetry: Geostrophic Currents

Bonjean and Lagerloef (2002) estimated equatorial Pacific surface currents using satellite measurements of SSHA (T/P), winds (SSM/I speed and QuikSCAT wind vectors), and SST (AVHRR). Their current anomalies included geostrophic, Ekman, Stommel shear, and thermal wind shear components, averaged over the upper 30 m. The annual cycle, as represented by climatologies for the months of January, April, July, and October, over the years 1993–1996 and 1999 (i.e., without the 1997 to 1998 El Niño period) compared well with coincident cycles of 15-m drifter data and TAO array current measurements. From this effort, sprang the Ocean Surface Current Analyses Real-Time (OSCAR) product that has been used in subsequent analyses of the annual cycle.

Scharffenberg and Stammer (2010) used data from the 3-year tandem T/P and Jason-1 mission (September 2002 to September 2005) to calculate geostrophic velocity anomalies everywhere outside ±1° from the equator. They analyzed global distributions of the annual harmonics of both zonal and meridional current anomalies. Their findings for the equatorial Pacific were in general agreement with those of Johnson et al. (2002). Westward propagation is evident (in their Figure 9) north of the equator (NECC and NSEC) but not south of the equator (SSEC). There is a hint of phase propagation on the equator in the far eastern Pacific, but with results plotted from 65°S to 65°N, it is difficult to say anything about this feature. Scarffenberg and Stammer noted only minor discrepancies between their annual cycle phases and those of Johnson et al. (with 165°W in their paragraph 45 apparently intended to be 165°E). They noted one major discrepancy between their phase and that reported by Yu and McPhaden (1999) at 5°N in the eastern Pacific. The significance of this is unclear, but it may be simply a phase difference between the currents analyzed by Scharffenberg and Stammer and the thermocline anomalies reported by Yu and McPhaden. The 3-year period of the tandem mission examined by Scharffenberg and Stammer (2010) was another that contained 2 years during which the Niño 3.4 index exceeded the El Niño threshold, and the index was positive for all but a few months. Once again, these results would be biased toward El Niño conditions, and this may also have contributed to phase discrepancies in the eastern Pacific.

Hsin and Qiu (2012) used 18 years (October 1992 to September 2010) of OSCAR surface currents to construct monthly climatologies of zonal current anomalies north of the equator in order to define the seasonal cycle of the NECC. Although the OSCAR currents contain Ekman currents and thermal wind shear in addition to the geostrophic component, Hsin and Qiu showed that the OSCAR-derived climatologies compared well with those constructed from only geostrophic current anomalies calculated from AVISO SSHA during the same time span. Their analysis was able to show that the annually varying transport of the NECC does not propagate; it remains in phase across the width of the basin. What does propagate westward is the meridional march of the current centerline. Complementing the observations with a process model, they showed that the seasonal cycle of the NECC east of the dateline is controlled by both Ekman pumping and the annual Rossby wave, while west of the dateline it is controlled primarily by the latter.

7.3.1.2.4 The Annual Rossby Wave

Historical studies of the annual Rossby wave in the equatorial Pacific (e.g., Lukas and Firing 1985; Kessler and McCreary 1993) noted a meridional asymmetry in the thermal signature that is in contrast to classical equatorial theory (Moore and Philander 1977) and that they could not satisfactorily explain. Chelton et al. (2003) showed that the meridional asymmetry (as revealed by analysis of 8.5 years of satellite altimetry, November 1992 to May 2001) is consistent with solutions of the shallow water equations linearized about the mean equatorial currents reported by Johnson et al. (2002). In the eigenvalue solutions, the current system slows the phase speed (consistent with observations) through eastward advection by the Equatorial Undercurrent and a decrease in the background potential vorticity gradient at the peaks of the NSEC and SSEC, which coincide with the latitudes where the Rossby wave's meridional velocity has extrema. The meridional structure of the wave appears to deform in a manner that reduces changes to the dispersion relation (Durland et al. 2011).

The studies cited here are representative of a larger group that we have not had room to cover but that have advanced our knowledge of the equatorial Pacific's seasonal cycle through analyses of satellite altimetry. The excellent spatial and temporal coverage of the satellite measurements has allowed investigation of details that were inaccessible to the pre-satellite studies, and it is apparent that the long altimetric SSH record we now have presents continued opportunities for improving our understanding of the seasonal cycle in this dynamically important region.

7.3.1.3 Interannual and Decadal Variability

Since the mid-1990s, SSH- and SSH-derived surface geostrophic currents have been used routinely to monitor equatorial Kelvin waves and Rossby waves, large-scale zonal slope of the thermocline, equatorial upper-ocean heat content, and surface currents in the tropical Pacific associated with ENSO as well as to test and improve ENSO forecast models for studies. Picaut and Busalacchi (2000) provided an overview of the related studies prior to 2000. Here we focus on major areas of studies of interannual variability of the tropical Pacific Ocean in relation to studies since the 2000s, especially regarding the additional understanding of the roles of Kelvin and Rossby waves in ENSO cycle, the related physics in association with ENSO theories, and ENSO diversity that was not discussed by Picaut and Busalacchi (2000).

7.3.1.3.1 ENSO-Related SSH Variability

Dominant SSH fluctuations associated with classical El Niño/La Niña reflect the variations in the zonal tilt of the thermocline associated with the variation in the strength of the trade wind, with increasing (decreasing) SSH in the eastern (or western) equatorial Pacific during El Niño (e.g., McPhaden et al. 1998) vice versa during La Niña. During the peak of the 1997–1998 El Niño, the positive SSH in the eastern equatorial Pacific exceeded 30 cm, with negative SSH anomalies in the western equatorial Pacific being about half this magnitude (Picaut et al. 2002) (Figure 7.6). The SSH signature associated with this strong El Niño, fortuitously captured within the first few years of the T/P altimeter, provided researchers with significant insights about processes associated with El Niño development and challenged models to reproduce the observed changes.

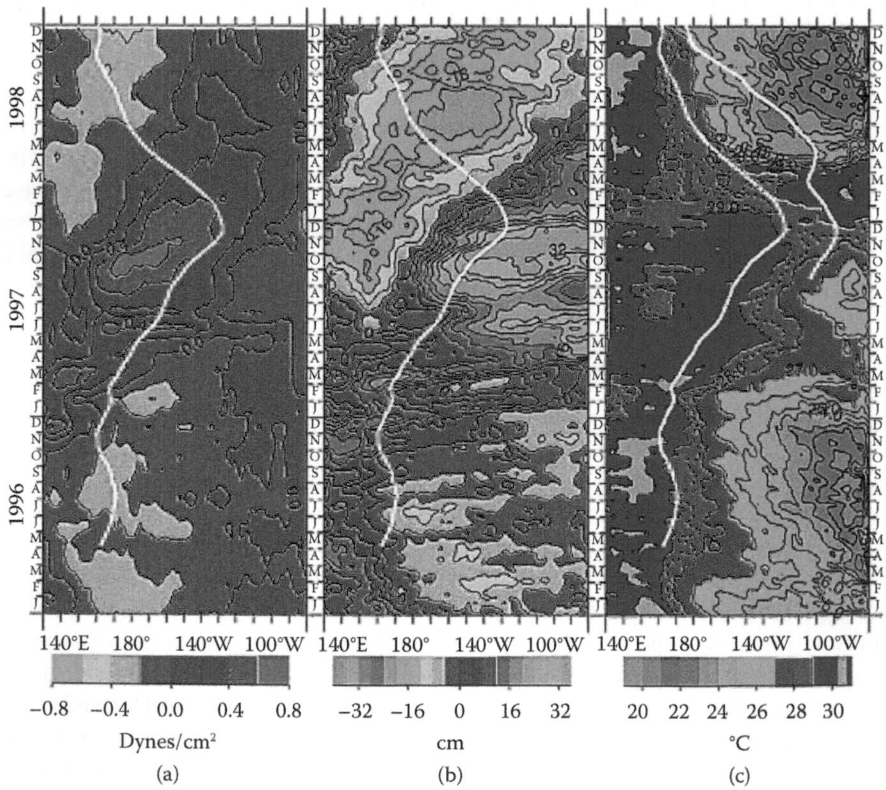

FIGURE 7.6 Longitude-time distribution of the following parameters averaged within 2_N–2_S: (a) SSM/I zonal wind stress anomaly, (b) TOPEX/Poseidon sea level anomaly, and (c) sea surface temperature. The thick lines represent the trajectory of a hypothetical drifter moved by the h2_N–2_Si zonal total geostrophic current.

In particular, the SSH observations together with other measurements illustrated that the delayed action oscillator mechanism was active during the onset of the 1997 El Niño, while both delayed oscillator and the recharge/discharge oscillator mechanisms were at work during the transition to the 1998 La Niña.

Altimeter-derived SSH has significantly advanced our knowledge of how the combined effects of wind-forced equatorial Pacific Kelvin and Rossby waves and their reflections at the eastern and western boundaries regulate the zonal displacement of the eastern edge of the western Pacific warm pool and thermocline depth fluctuations that are important to the development and decay of El Niño and La Niña events (e.g., Delcroix et al. 2000). The reflection of Rossby waves at western boundaries is key to the termination of El Niño in the Delayed Oscillator theory. Although such reflection was found to have 80%–90% efficiency, based on altimeter SSH, the resultant upwelling Kelvin waves during the peak of the 1997–1998 El Niño was insufficient to terminate the event without the upwelling Kelvin waves generated by easterly wind anomalies during the peak of that event (Boulanger et al. 2003, 2004). It suggests that reflected Rossby waves and wind-forced Kelvin waves are both important to the ENSO cycle. The equatorial Kelvin and Rossby waves also influence the stretching of upper-ocean isopycnals and barrier layer thickness that have implications in air-sea interactions (Bosc et al. 2009).

ENSO-related SSH variations are not limited to the zonal seesaw of SSH about the dateline associated with the Delayed Oscillation theory but also are associated with a meridional fluctuation of SSH (or warm water volume, WWV) with a fulcrum near 5°N that is dominated by geostrophic advection across 5°N (Alory and Delcroix 2002). The Recharge/Discharge Oscillator theory that describes the fluctuations of tropical Pacific WWV cannot explain the hemispheric asymmetry of the observed meridional SSH changes. The tropical Pacific WWV changes were attributed to the residual of opposing effects of meridional Ekman and geostrophic transports (Bosc and Delcroix 2008).

In the tropical Pacific, the OSCAR current estimates described previously have greatly facilitated diagnostic analysis of ENSO events. In the equatorial Pacific, surface current anomalies typically lead SST anomalies about 3 months with a magnitude that scales with the SST anomaly magnitude (Figure 7.7, Lumpkin et al. 2013). These features suggest the importance of surface current in regulating ENSO SST. Near the equator, accurate representation of ocean surface currents based on altimeter and scatterometer data (e.g., OSCAR) remains challenging because both the geostrophic and Ekman theories become invalid as the Coriolis parameter approaches zero.

El Niño characteristics experience changes on multidecadal timescales. Since 2000, there has been more frequent occurrence of the so-called El Niño Modoki (Ashok et al. 2007) or central-Pacific El Niño (Kao and Yu 2009). Different from the classical eastern-Pacific El Niño where large positive SSH anomalies developed in the eastern-equatorial Pacific, El Niño Modoki events are generally associated with positive SSH anomalies and surface current convergence in the central-equatorial Pacific due to the anomalous zonal wind convergence associated with El Niño Modoki (e.g., Ashok et al. 2007; Singh et al. 2011). The recharge/discharge of WWV as inferred from SSH is also different between the two types of El Niño events (Singh and Delcroix 2013). The more frequent occurrence of El Niño Modoki in the early to mid-2000s resulted in a decadal trend of increasing SSH in the central-equatorial Pacific region during the 1998–2007 period (Behera and Yamagata 2010).

7.3.1.3.2 Decadal and Multidecadal SSH Changes

Large decadal fluctuations in SSH and ocean surface winds have been observed in the Indo-Pacific region since the early 1990s (Lee and McPhaden 2008). Large-scale trends of the SSH and wind fields during 1993–2000 were generally opposite to those during 2000 to 2006 over much of the Indo-Pacific domain, although the trends in the earlier period were generally larger. The 1993–2000 and 2000–2006 SSH trends for the tropical Indo-Pacific domain (Figure 7.8c and d) were associated with opposite trends of zonal wind stress between these two periods both in the tropical Pacific and Indian oceans (Figure 7.8a and b). The coherent decadal changes of SSH and wind fields in the

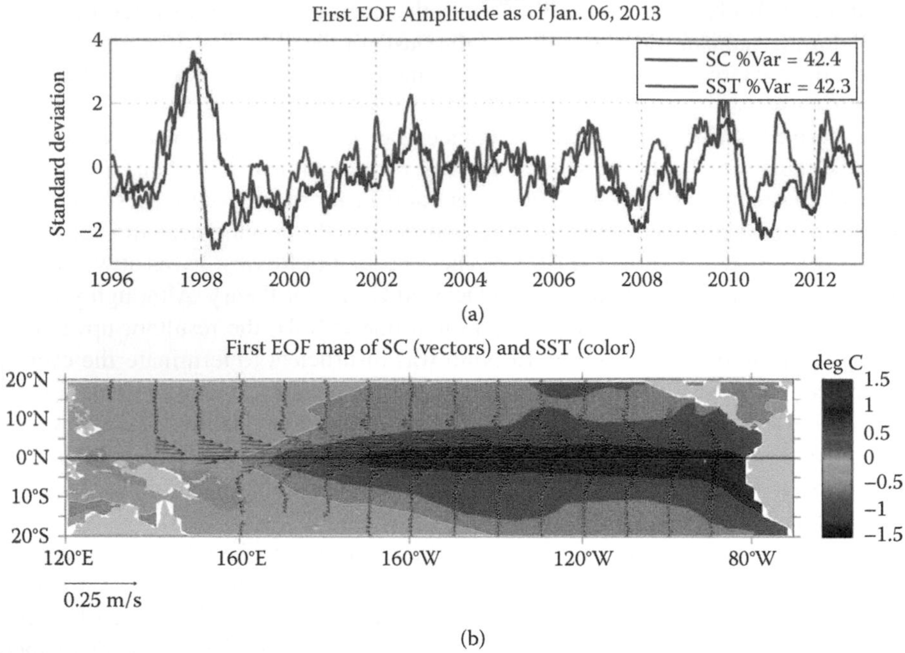

FIGURE 7.7 The time series (a) and spatial structure (b) of the first EOF of non-seasonal ocean surface current and SST in the tropical Pacific, showing the 2–3-month lead time of the surface current over SST. (After Lumpkin, R., et al., *Bull. Am. Meteorol. Soc.*, 94, S62–S65, 2013. With permission.)

Pacific and Indian oceans reflect oceanic and atmospheric linkages between the two basins caused by decadal oscillation of the Walker Circulation in the tropical Pacific and Indian sectors. These linkages are further discussed in the next subsection. The maximum net increase of SSH in the northwest tropical Pacific (as seen from Figure 7.8c) during 1993–2000 was 20 cm–25 cm, rivaling the magnitude of SSH changes in the eastern equatorial Pacific associated with strong El Niño events. It has been shown that the Interdecadal Pacific Oscillation (IPO), which is highly correlated with decadal variability of ENSO, can have a large influence on decadal variability of SSH patterns over the tropical Pacific (Frankcombe et al. 2015).

The SSH trend observed by altimetry during 1993 to 2009 in the Pacific was examined in the context of longer tide gauge records and wind stress patterns (Merrifield 2011). The dominant regional trends are also found to be associated with SSH rise in the western tropical Pacific (WTP), accompanied by weak SSH rise or fall in the eastern Pacific basin. This basin-scale SSH pattern is shown to be associated with the IPO (or decadal variability of ENSO; see Section 2.1 of Han et al. 2017 for a review and references therein). The rate of SSH rise in the WTP, however, has increased significantly since the early 1990s relative to the preceding 40 years, as revealed by tide gauge records. The IPO cannot explain this intensification because its strength did not intensify. Modeling studies suggest that warming of the tropical Indian Ocean enhances surface easterly trade winds and thus contributes to the intensified WTP SSH rise in the past few decades (Luo et al. 2012; Hamlington et al. 2014; Han et al. 2014). Also through numerical model experiments, it has been shown that warming of the Atlantic strengthens the equatorial easterlies and intensified SSH rise in the WTP since the early 1990s and plays an important role in causing the global warming "hiatus" since the early 2000s—a period when global mean surface temperature does not increase (McGregor et al. 2014). During this period, altimeter data however show a persistent rising trend of global mean SSH (Church et al. 2013). Because SSH represents an integral heating effect from both surface and subsurface ocean,

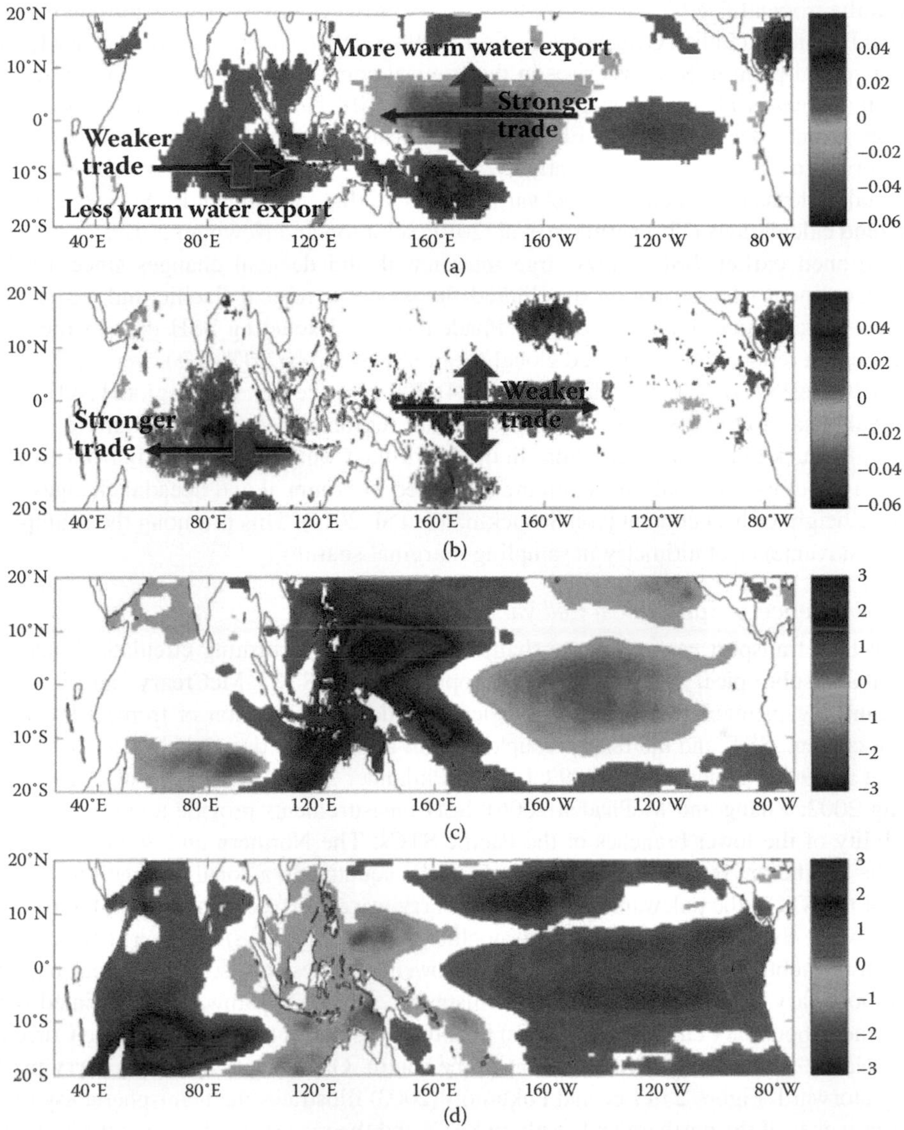

FIGURE 7.8 Trends of zonal wind stress during 1993–2000 estimated from ERS scatterometer data (a) and those during 2000–2006 estimated from QuikScat scatterometer data. SSH trends for the 1993–2000 (a) and 2000–2006 (b) periods estimated from altimeter data. These observations were also used to infer changes in the upper and lower branches of the shallow overturning circulations in the Pacific and Indian Oceans and their linkages as described in Indo-Pacific linkage subsection. (a) Trend of zonal wind stress: 1993–2000 (dyn/cm^2/yr), (b) Trend of zonal wind stress: 2000–2006 (dyn/cm^2/yr), (c) Trend of SSH: 1993–2000 (cm/yr), and (d) Trens of SSH: 2000–2006 (cm/yr). (After Lee, T., et al., *Dyn. Atmos. Oceans.*, 2010; adapted from Lee, T., and McPhaden, M. J., *Geophys. Res. Lett.*, 35, L01605, 2008.)

the SSH rise indicates that the excessive heat received by the Earth's surface has gone into the subsurface ocean.

Robust intensification also has been detected for decadal (10 to 20 years) SSH variability in the WTP since the early 1990s. This intensification results from the "out of phase" relationship of SST anomalies between the Indian Ocean and IPO since 1985, which produces "in phase" effects on the WTP SSH variability (Han et al. 2014). These results suggest that decadal and multidecadal SSH changes in the tropical Pacific also involve the linkage of the coupled ocean-atmosphere systems in the tropical Pacific and Indian Ocean sectors. From 1998 to 2007, the decadal trend of sea level shows a different spatial pattern: Sea level rises in the tropical central Pacific flanked by sea level fall in the western and eastern basins (Behera and Yamagata 2010). This abnormal condition is due to the frequent occurrence of central Pacific El Niño events during 2000 to 2004, which are associated with wind convergence to the dateline. Evidently, decadal changes in ENSO behavior will induce changes in the spatial patterns of decadal sea level variations. Note that interannual ENSO events may affect decadal trend calculations (Timmermann et al. 2010; Solomon and Newman 2012).

As mentioned earlier, SSH shows large interannual and decadal changes since 1992 in the Maritime Continent oceanic region that linked the western tropical Pacific and eastern tropical Indian Ocean warm pool (e.g., Lee and McPhaden 2008). Excluding SSH data in the Maritime Continent region reduced the estimated global mean SSH trend by 20% (7%) over the 2005 to 2010 (1992 to 2010) period (von Schuckmann et al. 2014). Because the trends of regional SSH over much of the global ocean to the first order resemble those of regional thermosteric height (e.g., Stammer et al. 2013), the lack of *in situ* observations in the Maritime Continent (especially from Argo) raises the question about its potential impact on the estimated interannual and decadal changes of global ocean steric height or heat content (von Schuckmann et al. 2014). This is among the examples highlighting the advantages of altimetry in sampling marginal seas.

7.3.1.3.3 Inference of Subtropical Cell Variability Using SSH

Meridional heat transport carried by the shallow meridional overturning circulations that connect the tropical and subtropical Pacific (i.e., the subtropical cells or STCs; McCreary and Lu 1994) plays an important role in regulating the climatological and decadal variation of tropical Pacific upper-ocean heat content, SST, and the related coupled ocean-atmosphere decadal climate variability (e.g., Kleeman et al. 1999; Schneider et al. 1999; Liu and Philander 2000; Hazeleger et al. 2001, McPhaden and Zhang 2002; Zhang and McPhaden 2006). SSH measurements provide a good proxy to study the variability of the lower branches of the Pacific STCs. The Northern and Southern Hemisphere STCs in the Pacific are somewhat symmetric about the equator. In a zonal average sense, the upper branches of the STCs (the poleward Ekman flows) carry warm surface waters from the tropics to the subtropics, while the lower branches (the pycnocline geostrophic flows) transport colder subsurface waters from the subtropics toward the tropics. The lower branches consist of an interior pathway and a western-boundary (WB) pathway in both hemispheres. The WB pathway is associated with low-latitude western-boundary currents (LLWBCs) in both hemispheres (i.e., the Mindanao Current in the north and New Guinea Coastal Undercurrent in the south). On average, they both carry pycnocline waters equatorward. Figure 2 in Lee and Fukumori (2003) illustrates the hemispheric asymmetry of the lower branches of the northern and southern STCs and the respective WB and interior pathways.

The SSH difference between the western and eastern boundaries of the tropical Pacific Ocean is found to be anticorrelated with that of thermocline depth difference and thus provides a good proxy for the variation of net meridional geostrophic transport in the pycnocline (i.e., the strength of the lower branches of the STCs; Lee and Fukumori 2003). On the other hand, meridional Ekman transport inferred from zonal wind stress is indicative of the strength of the upper branches of the STCs. Therefore, west-east SSH difference derived from altimeter measurements and zonally averaged zonal wind stress from scatterometer observations together are complementary to monitoring the lower and upper branches of the STCs on interannual and decadal timescales (e.g., Lee and Fukumori 2003; Lee and McPhaden 2008).

The zonal slope of the SSH anomalies across the WB and that across the interior (from the eastern side of the LLWBCs to the eastern boundary) also provide a good proxy for the variations of the WB and interior pycnocline transports (Lee and Fukumori 2003). Because of this, SSH measurements have offered insight into the zonal structure of the interannual and decadal variation of interior and WB pycnocline flows associated with the STCs. In contrast to the time mean picture where the interior and WB pycnocline flows reenforce each other (both being equatorward), the interannual and decadal anomalies of the interior and WB pycnocline transports counteract each other (Lee and Fukumori 2003). Therefore, they play opposite roles in regulating the tropical Pacific upper-ocean heat content. However, the variation of the interior pycnocline transport anomalies is more dominant than that of the WB transport anomalies. The compensation of the interior pycnocline transport by WB transport in the Southern Hemisphere is more significant (approximately 60%) than that in the Northern Hemisphere (approximately 30%). These features are robustly reproduced by various ocean models and data assimilation products (e.g., Hazeleger et al. 2004, Capotondi et al. 2005; Schott et al. 2007). The anticorrelated variability of WB transport and interior pycnocline transport is the result of the oscillation of the tropical gyres in the western Pacific caused by off-equatorial wind stress curl associated with the trade wind variations (Lee and Fukumori 2003).

The counteracting WB transport and interior pycnocline transport have important implications for the monitoring of the lower branch of the STC. Previous studies of decadal variation of the lower branch of the Pacific STC were mostly based on interior expandable bathythermograph (XBT) and conductivity-temperature-depth (CTD) observations and assumed that the WB transports do not change significantly (e.g., McPhaden and Zhang 2002, 2004). While broad-scale *in situ* observations in the interior ocean (e.g., from XBT, CTD, and Argo) provide observations to estimate the interior pycnocline transport, they do not have sufficient coverage near the LLWBCs to estimate their transports. Process-oriented experiments such as the Northwestern Pacific Ocean Climate Experiment (NPOCE) and the Southwest Pacific Ocean Circulation and Climate Experiment (SPICE) have developed the capability to study the LLWBCs based on measurements from gliders, moorings, and research vessels. However, the capability to develop sustained monitoring systems for the LLWBCs based on *in situ* instruments has not been established. SSH measurements from satellites provide a means to fill this observational gap, especially with sustained multiple altimeter missions operating simultaneously and with the upcoming Surface Waters Ocean Topography (SWOT) mission that will enhance the temporal and spatial samplings important to resolve the mesoscale variability associated with the LLWBCs. The contribution of satellite SSH in filling this observational gap is important to closing the volume, heat, and freshwater budget of the tropical Pacific Ocean. Altimeter data along with high-resolution ocean models have provided further insights about the variability of the South Pacific LLWBC in the Solomon Sea—in particular, the regional distributions of SSH anomalies and eddy kinetic energy and their relationships to ENSO (e.g., Melet et al. 2010, 2011).

7.3.2 Tropical Indian Ocean

7.3.2.1 Intraseasonal Variability

Over the tropical Indian and Pacific oceans, the large amplitude intraseasonal variability (20–90 days) in the troposphere is dominated by the MJO (Madden and Julian 1971, 1972), which exhibits strong spectral peaks at 30- to 60-day periods. The MJO propagates eastward during boreal winter and both eastward and poleward during boreal summer (see reviews by Zhang 2005; Lau and Waliser 2012). The summertime MJOs, together with the quasi-biweekly mode, are often referred to as the monsoon IntraSeasonal Oscillations (ISOs; Goswami 2012). The ISOs constantly interact with the underlying ocean and influence many weather and climate systems over the globe.

Compared to the MJO, oceanic intraseasonal variability was detected a decade later. Earlier *in situ* observations revealed 30- to 60-day variability in zonal currents at several locations in the equatorial Indian Ocean (Luyten and Roemmich 1982; McPhaden 1982; Schott et al. 1994; Reppin et al. 1999). Spectral peaks at 30- to 60-day and 26-day periods of meridional currents were observed in the western basin, but the maximum peak shifted to 12–15 days in the central and eastern basin (Mertz and Mysak 1984; Reverdin and Luyten 1986; Reppin et al. 1999; *review of* Schott and McCreary 2001). Numerical modeling studies suggested that the 30- to 60-day currents were forced by the 30- to 60-day oscillations of winds associated with the MJO (Moore and McCreary 1990), and the 26-day meridional current in the western basin resulted from oceanic instabilities (Kindle and Thompsen 1989; Woodberry et al. 1989; Tsai et al. 1992). Oceanic instabilities are not forced by the corresponding atmospheric variability (e.g., wind); rather, they arise from nonlinearity of the oceanic system. The sporadic *in situ* observations, however, could not depict the basin-scale structure and evolution associated with the intraseasonal currents.

Since the beginning of the twenty-first century, rapid advancement has been made in understanding the Indian Ocean intraseasonal variability, and satellite altimetry has played a vital role in this rapid development. Han et al. (2001) performed spectral analysis using the T/P altimeter data with 10-day resolution and model zonal surface current from 1993 to 1999 and showed that the strongest spectral peaks of SSH and zonal current occur at the 90-day period and secondary peaks occur at 40–70 days across the equatorial Indian Ocean (Figure 7.9), even though the MJO winds that force them peak at the 30- to 60-day periods. Similar spectral peaks are found using T/P and Jason data (Fu 2007), weekly AVISO SSH, and OSCAR zonal surface currents for various temporal periods (Iskandar and McPhaden 2011; Nagura and McPhaden 2012). Model experiments suggested that while the 40- to 70-day peaks are directly driven by the strong 30- to 60-day winds, the large 90-day peak is associated with the excitation of the basin resonance mode for the second baroclinic mode at the 90-day period in the equatorial Indian Ocean, which is established by the constructive interference between the eastward-propagating Kelvin wave directly forced by winds and the westward-propagating Rossby waves reflected from the eastern boundary (Han et al. 2001; Han 2005; Nagura and McPhaden 2012). While the T/P SSH showed an eastern basin concentration of the 90-day resonance (Han 2005; Fu 2007), the multiple-satellite merged AVISO SSH with finer spatial resolution showed a clear 90-day basin mode structure as theoretically predicted, with a significantly larger amplitude in the western basin (Han et al. 2011).

The strong intraseasonal variability of zonal current and SSH can significantly impact the spring and fall Wyrtki Jets (Masumoto et al. 2005), and can induce eastward monsoon jets during summer ISOs (Senan et al. 2003). The intraseasonal signals observed in the equatorial basin do not just stay there; they exert remote influence on intraseasonal variability in the Bay of Bengal and the Arabian Sea via coastal Kelvin waves that propagate around the perimeter of the North Indian Ocean (with the coasts to their right in the Northern Hemisphere) and through Rossby waves that radiate westward, as revealed by the AVISO SSH data (Girish Kumar et al. 2013; Suresh et al. 2013). They can also have large influence on the Indonesian Seas via eastward-propagating equatorial Kelvin waves, coastal Kelvin waves that propagate with the coast to their left in the Southern Hemisphere (Sprintall et al. 2000; Schouten et al. 2002; Iskandar et al. 2005, 2006; Zhou and Murtugudde 2010), and subsequently westward-radiating Rossby waves from the eastern basin, affecting the east Indian Ocean upwelling (Chen et al. 2015), impacting the IOZDM (Rao and Yamagata 2004; Han et al. 2006) and affecting the Indonesian Throughflow (Qiu et al. 1999; Drushka et al. 2010; Schiller et al. 2010; Pujiana et al. 2013; Shinoda et al. 2016).

In the central and western interior basin, analyses of the AVISO SSH together with *in situ* and other satellite observations showed that the arrival of downwelling intraseasonal Rossby waves might play an important role in triggering the primary MJO events, which were not immediately preceded by other MJO activities (Webber et al. 2012). Indeed, a downwelling intraseasonal Rossby wave reflected from the eastern Indian Ocean boundary was observed by SSH during the CINDY/DYNAMO field campaign period (Yoneyama et al. 2013; Zhang et al. 2013). Together with the analysis of *in situ* data from the Research Moored Array for African-Asian-Australian Monsoon Analysis and Prediction (RAMA) array (McPhaden et al. 2009) and from the CINDY/DANAMO field campaign, it was shown that this Rossby wave significantly influenced the upper ocean structure in the thermocline ridge region, where many MJO events were initiated (Shinoda et al. 2013).

In addition to the equatorial Kelvin and Rossby waves, satellite altimeter data were also used to identify Yanai waves at 20- to 30-day periods in the western equatorial basin and at approximately 14-day periods in the central and eastern basin, even though the weekly AVISO data appeared to

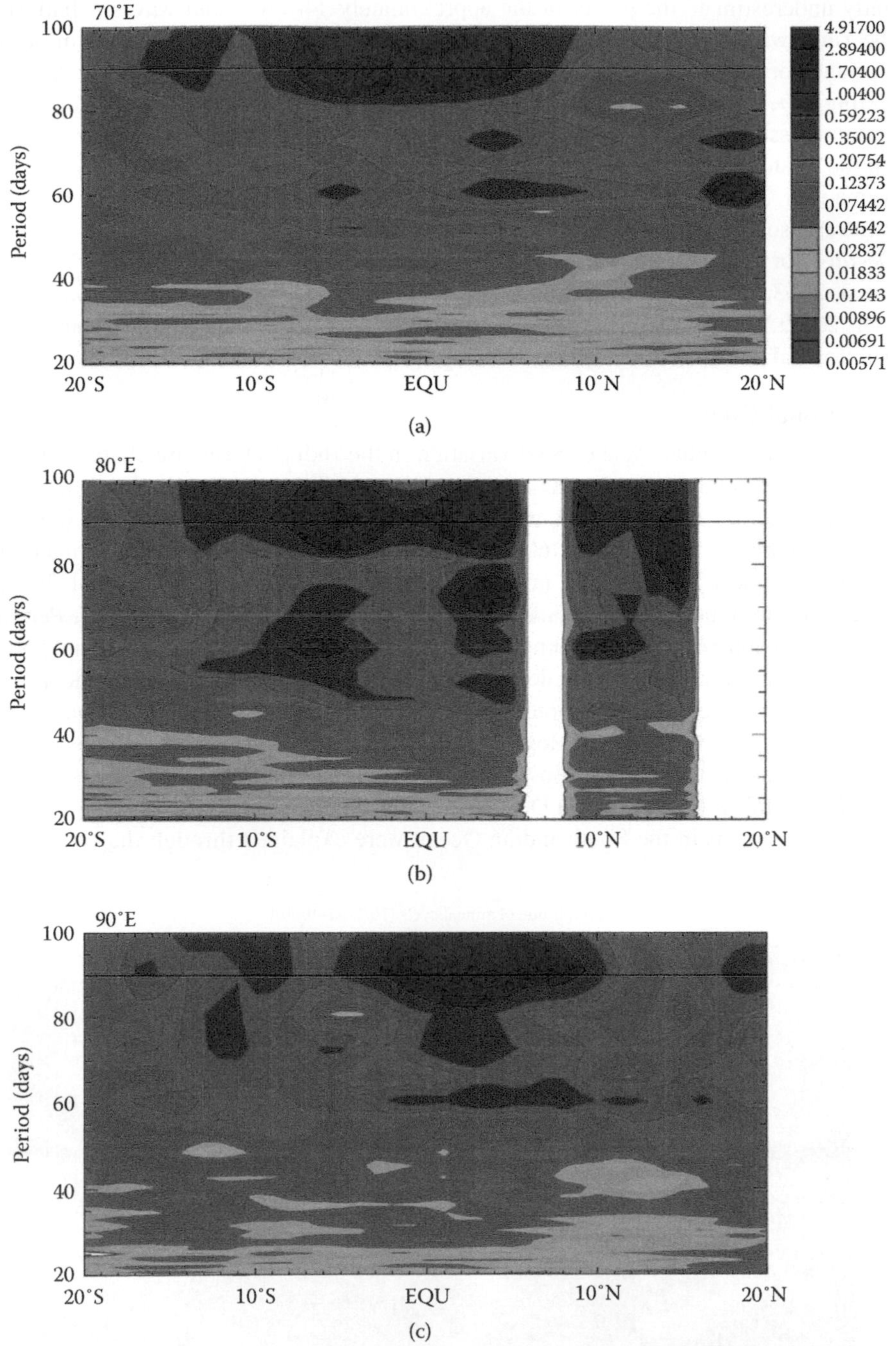

FIGURE 7.9 Meridional section for variance spectra (cm^2) of TOPEX/Poseidon SSH anomaly along (a) 70°E, (b) 80°E, and (c) 90°E of the Indian Ocean, based on an 8-year record (1993–2000) with a 10-day resolution. The 90-day and most 30- to 60-day spectral peaks are above 95% significance. (From Han, W., *J. Phys. Oceanogr.*, 35, 708–728, 2005.)

significantly underestimate the power of the approximately 14-day Yanai waves (Chatterjee et al. 2013). The Yanai waves (also referred to as "mixed Rossby-gravity waves") are antisymmetric about the equator and correspond to the *in situ* observed meridional currents at the equator discussed earlier. Modeling experiments suggested that the Yanai waves at the approximately 14-day period were forced by winds associated with the 10- to 20-day mode of monsoon ISOs, and those at 20- to 30-day periods in the western basin were influenced by western boundary reflection and oceanic instabilities (Miyama et al. 2006; Chatterjee et al. 2013). Furthermore, satellite altimeter data were also used to detect intraseasonal variability that resulted from oceanic instabilities in various regions of the Indian Ocean—for example, the western equatorial basin; the Somali, Omani, and Yemeni coasts (Brandt et al. 2003; Han 2005); the South Equatorial Current in the southeast tropical basin (Feng and Wijffels 2002; Yu and Potemra 2006; Ogata and Masumoto 2010, 2011; Trenary and Han 2012); and the subtropical south Indian Ocean (Palastanga et al. 2007).

7.3.2.2 Seasonal Cycle

The amplitudes of the annual cycle of SSH variation in the Indian Ocean are characterized by two local maxima in the South Indian Ocean between 5°S and 15°S as well as by local maxima in the Arabian Sea and the Bay of Bengal (Fu and Smith 1996) (Figure 7.10). In the South Indian Ocean, the two local maxima are centered near 60°E and 95°E (with amplitudes of approximately 9 and 12 cm) and a minimum near 75°E (with an amplitude of approximately 2 cm) (Wang et al. 2001). These observed features are reproduced reasonably well by ocean model simulations (e.g., Périgaud and Delecluse 1992; Fu and Smith 1996). Although these studies were based on just a couple of years of altimeter data, annual variations of the depth of the 20C isotherm based on over a decade of XBT data show a similar structure (Masumoto and Meyers 1998). These annual SSH variations were attributed to local wind-forced annual Rossby waves in the interior ocean superimposed on Rossby waves forced by winds in the east and those radiated from the eastern boundary near the exits of the Indonesian Throughflow (Périgaud and Delecluse 1992). The two local maxima and the minimum of annual SSH variations in the South Indian Ocean were explained through the constructive and

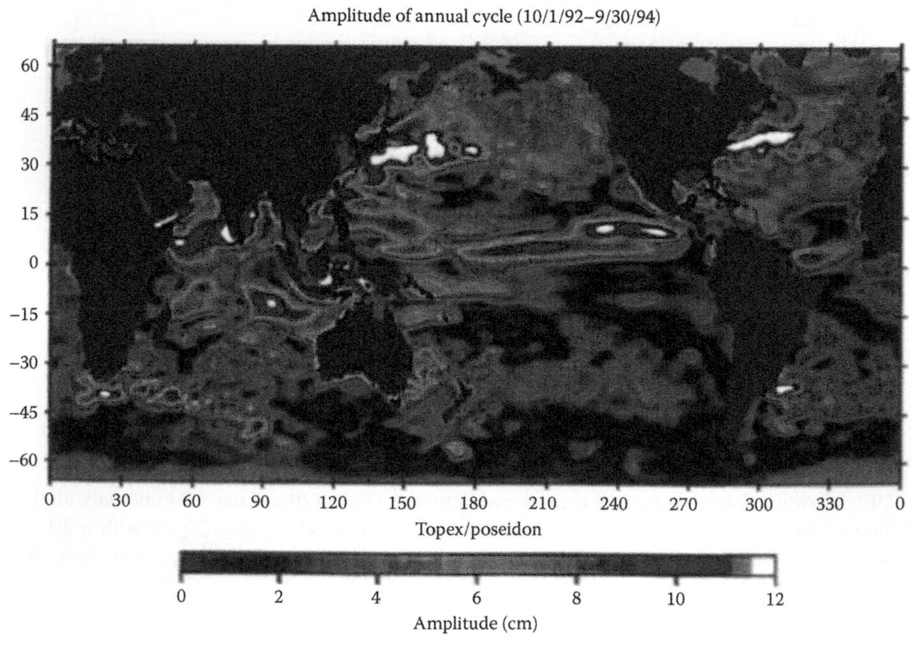

FIGURE 7.10 Amplitude of the annual cycle of SSH estimated from TOPEX/Poseidon altimeter data. (After Fu, L., and Smith, R., *Bull. Am. Meteorol. Soc.*, 77, 2625–2636, 1996. With permission.)

destructive interferences of the local response to Ekman pumping and Rossby wave propagation, respectively (Wang et al. 2001). Radiation of the Rossby waves from the ITF exit has some effects on annual SSH variations in the eastern part but not the western part of the basin (Wang et al. 2001; Trenary and Han 2012).

In the Arabian Sea, strong seasonal fluctuations of SSH (with an annual-harmonic amplitude of 10–15 cm) were observed between 6°N and 10°N and were associated with westward propagating annual Rossby waves radiated from the western side of the Indian subcontinent and continuously forced by the wind-stress curl over the central Arabian Sea (Brandt et al. 2002). These waves are related to a time-dependent meridional overturning cell that sloshes water northward and southward. Between 58°E and 68°E in the central Arabian Sea, Rossby waves induced up to 10 Sv of transport in the upper 500 m (southward in August 1993 and northward in January 1998) and a few Sv below 2000 m, as shown by hydrographic measurements. The annual SSH variations can be explained by the first- and second-mode annual Rossby waves. The reflection of annual Rossby waves also affects the western boundary currents.

In the Bay of Bengal, large-scale annual SSH variations reflect the oceanic response to seasonal wind stress curl over the bay (Schott and McCreary 2001). However, there are smaller-scale features in the bay, especially in the southwestern part. SSH-derived surface geostrophic currents suggest that during the peak of the summer monsoon, there is a western-boundary confluence near 10°N, with the East India Coastal Current (EICC) flowing northward to the north of 10°N and southward to the south (Eigenheer and Quadfasel 2000). This confluence is supplied mostly by the South Monsoon Current across 6°N that circulates cyclonically around an area of low SSH (i.e., the Sri Lanka Dome) (Vinayachandran et al. 1999). After the summer monsoon in late September, the EICC begins to flow southward, and by November, southward flow is seen everywhere along the east coasts of India and Sri Lanka. During this time, SSH suggests a slow cyclonic circulation around the Bay of Bengal, which breaks up into several cells as the monsoon season develops (Eigenheer and Quadfasel 2000).

7.3.2.3 Interannual Variability

A dominant mode of interannual climate variability of the coupled ocean-atmosphere system in the Indian Ocean sector is the IOZDM (Saji et al. 1999). During a positive IOZDM event, anomalous southeasterly winds in the equatorial Indian Ocean raise the thermocline depth and depress SSH in the southeastern equatorial Indian Ocean, reducing SST and triggering a coupled ocean-atmosphere interaction across the equatorial Indian Ocean that is akin to the Bjerknes feedback process associated with ENSO in the Pacific sector. The strongest IOZDM event occurred during 1997 to 1998 and associated with the El Niño event in the Pacific. During November 1997 to May 1998, negative anomalies of SSH in the southeastern equatorial Indian Ocean reached –30 cm, associated with approximately –2°C of SST anomaly (Webster et al. 1999; Yu and Reinecker 1999) (Figure 7.11). During this period, the SSH in the western equatorial Indian Ocean was higher than normal, but the magnitude of the anomaly was smaller than that in the east by a factor of three. Negative IOZDM events (e.g., in 1996 and 1998) are associated with opposite changes of SSH. The effects of the IOZDM on SSH are not limited to the equatorial band but extend northward into the western part of the Bay of Bengal and southward to the region off the west coast of Australia and to the southwest tropical Indian Ocean (SWTIO) (Han and Webster 2002; Rao et al. 2002).

The SWTIO region (5°S–15°S, 50°E–70°E), referred to as the Seychelles-Chagos thermocline ridge (Hermes and Reason 2008), is associated with a shallow thermocline and a unique region with open-ocean upwelling as evidenced by enhanced phytoplankton concentration (Murtugudde et al. 1999; Schott and McCreary 2001). The shallowness of the thermocline makes the SST in this region easily affected by the fluctuations of thermocline depth. This is well characterized by the highly correlated SSH and SST measurements (e.g., Xie et al. 2002; Rao and Behera 2005). Local Ekman pumping associated with the variability of IOZDM- or ENSO-related wind fields causes Rossby wave

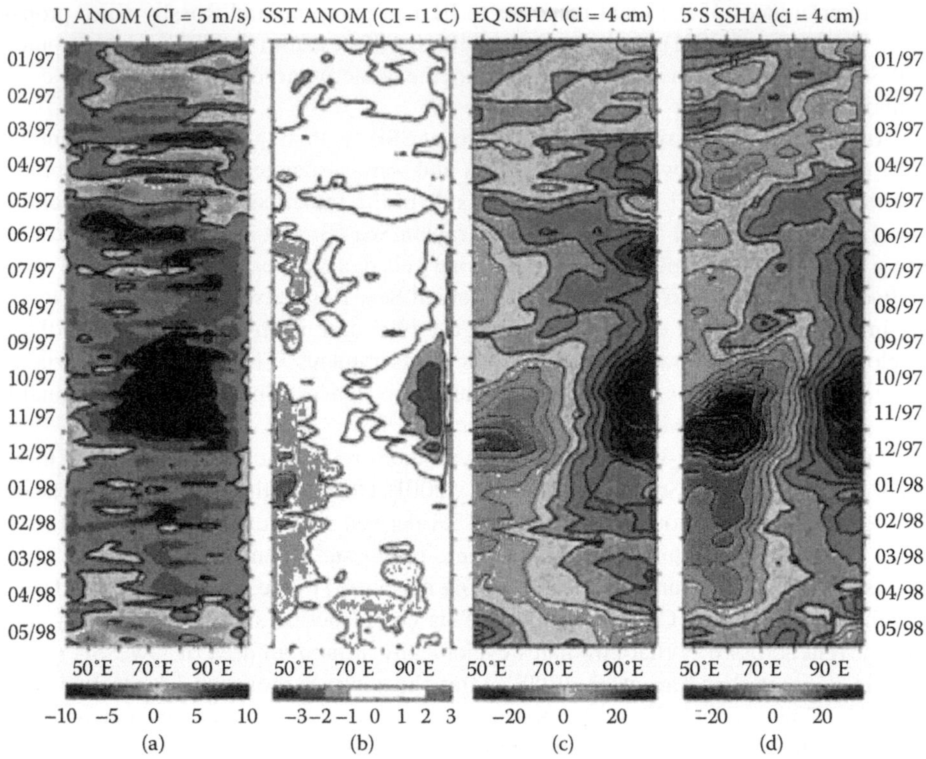

FIGURE 7.11 Time-longitude sections (an average over 2°S–2°N) of anomalous (a) zonal winds (5-day means) from NCEP reanalyses and (b) weekly-mean SST from Reynolds analyses; and time-longitude sections of anomalous SSH from TOPEX/Poseidon altimeter (10-day average) along (c) the equator and (d) 5°S. The anomalies are all derived from the 1981–1996 base period monthly means except the SSH anomalies which are deviations from the 1993–96 base period monthly means. (After Yu, L., and Rienecker, M. M., *Geophys. Res. Lett.*, 26, 735–738, 1999. With permission.)

propagation into this region that influences the thermocline depth (and thus SSH), thereby inducing feedback to the atmosphere (including precipitation) through its effect on SST (Xie et al. 2002). These processes, referred to as coupled ocean-atmosphere Rossby waves (Xie et al. 2002), offer potential predictability for SST and tropical cyclones in the western Indian Ocean.

In contrast to the IOZDM-related interannual maxima of SSH variations in the eastern and western equatorial Indian Ocean, the Seychelles-Chagos thermocline ridge, and the western Bay of Bengal, there are local minima of interannual SSH variations in the central equatorial Indian Ocean, Arabian Sea, and the eastern Bay of Bengal off the east coast of India and Sri Lanka. The minima (maxima) of interannual variability have been explained by the destructive (constructive) interference of the direct wind-forced response and reflected Rossby waves from the eastern boundaries (Shankar et al. 2010).

Shankar et al. (2010) also concluded that, on interannual timescales, the SSH adjustment to wind forcing in the Indian Ocean is less than that of the forcing, leading to a quasi-steady balance. This distinguishes the interannual minima from those at annual and semiannual timescales—for instance, the local minimum of annual Rossby wave amplitude in the middle of the southern tropical Indian Ocean (Wang et al. 2001).

McPhaden and Nagura (2014) examined whether SSH in the equatorial Indian Ocean could be used as a predictor of IOZDM development as it is for ENSO in the Pacific sector. They used SSH measurements and an analytical linear equatorial wave model to interpret the evolution of the

IOZDM in the context of the recharge oscillator theory. They found that, as in the Pacific, there are zonally coherent changes in SSH (heat content) along the equator prior to the onset of IOZDM events, an indication of a recharge oscillator being at work. These SSH changes are modulated by wind-forced westward propagating Rossby waves at 5°S–10°S, which at the western boundary reflect into Kelvin waves trapped at the equator. The biennial character of the IOZDM is affected by this cycling of wave energy between 5°S and 10°S and the equator. However, SSH changes are a weaker leading indicator of IOZDM-related SST anomalies than they are for ENSO. This is because IOZDM is also affected by ENSO through atmospheric teleconnection in addition to the recharge oscillator process in the Indian Ocean.

7.3.2.4 Decadal and Multidecadal Changes

The South Indian Ocean experienced large decadal changes from early 1993–2006 (Lee 2004; Lee and McPhaden 2008). During 1993–2000, SSH over much of the interior of the southern tropical Indian Ocean (8°S–20°S, 50°E–100°E) dropped by 10–20 cm while that in the east from the equator to 30°S (i.e., off the coasts of southern Sumatra and Java, near the ITF exit, and off the west coast of Australia) increased by similar amounts (Figure 7.8c). This pattern reversed during 2000–2006 (Figure 7.8d). These decadal changes of SSH were attributed to two factors. The SSH changes in the interior of the southern tropical Indian Ocean were caused by the wind stress curl forcing associated with decadal changes of the South Indian Ocean trade winds as well as the westward propagation of Rossby waves. In the east, the decadal SSH changes are largely due to the transmission of the SSH signals from the tropical Pacific through the Indonesian Seas via coastal Kelvin waves. The decadal SSH changes in the South Indian Ocean were associated with regional SSH changes in the entire Pacific Ocean that also had opposite trends between 1993 and 2000 and 2000 and 2006, as discussed in Section 7.3.2.2. The associated changes in SSH differences between the western and eastern boundaries have strong implications for the changes in the lower branches of the Indian and Pacific Ocean STCs. The linkages of the lower branches of the Pacific and Indian Ocean STCs through an oceanic tunnel and the relationship of the upper branches of the Pacific and Indian Ocean STCs through an atmospheric bridge are discussed in Section 7.5. In the equatorial and North Indian Ocean, SSH experienced a basin-wide decrease from 1993 to 2003 but a sharp increase from 2004 to 2013 (Srinivasu et al. 2017).

Both observational analyses and OGCM experiments suggest that winds over the Indian Ocean are the primary forcing for the basin-wide decadal sea level patterns, with the ITF having a significant contribution primarily in the eastern basin (e.g., Lee and McPhaden 2008; Trenary and Han 2012; Nidheesh et al. 2013, Zhuang et al. 2013; Wang et al. 2015). The observed North Indian Ocean basin-wide decadal reversal of SSH trends around 2003–2004 resulted from the combined effect of changing surface turbulent heat flux and cross-equatorial heat transport, both being associated with decadal changes of surface winds over the Indian Ocean (Srinivasu et al. 2017). Thermosteric effect is the primary contributor to the spatial patterns of decadal SSH variability (e.g., Fukumori and Wang 2013; Nidheesh et al. 2013; Srinivasu et al. 2017), while the halosteric effect is significant only in certain regions (Nidheesh et al. 2013). In particular, nearly two-thirds of the SSH rise in the southeastern-south-central tropical Indian Ocean in the past decade was due to a halosteric effect (Llovel and Lee 2015). In fact, this is the only region in the global ocean between approximately 66°S and 66°N (the domain sampled by most altimeters and Argo floats) where large-scale decadal SSH changes are primarily due to a halosteric instead of a thermosteric effect. The large halosteric effect in this region was associated with a freshening in this region in the upper 200–300 m as observed by Argo data. Possible causes for this freshening in the past decade include the relatively abrupt strengthening of the ITF since 2007 as observed by mooring observations in the Makassar Strait (the main branch of the ITF inflow from the Pacific; Gordon et al. 2012) and enhanced precipitation in the Maritime Continent region (Llovel and Lee 2015; Hu and Sprintall 2016).

The forcing that caused decadal SSH variability in the Indian Ocean is associated with various climate modes, including decadal changes in the IOZDM, the Decadal Indian Ocean Basin

(DIOB) mode, and the PDO, as well as decadal changes in ENSO. In particular, the basin-wide decadal SSH patterns over the tropical Indian Ocean (north of 20°S) are primarily forced by wind stress associated with climate modes, with the maximum amplitude occurring in the SCTR in the southwest tropical Indian ocean. Further studies are necessary to better decipher the relative contributions of different climate modes in forcing decadal SSH changes in the Indian Ocean. Given the dominant roles of wind forcing, the high correlation of decadal SSH variations near Mumbai with Indian monsoon rainfall (Shankar and Shetye 1999) may result from monsoon wind influence (Li and Han 2015) rather than the salinity effect associated with monsoon rainfall as suggested previously (Shankar and Shetye 1999). In addition to the effects of surface forcing, ocean internal variability also has large amplitudes near the Somali coast, western Bay of Bengal, and subtropical south Indian Ocean.

7.3.3 Indo-Pacific Linkage and Indonesian Throughflow

As mentioned in Sections 7.3.2.2 and 7.4.4, the variations of SSH differences between the western and eastern boundaries in the northern and southern tropical Pacific and southern tropical Indian Oceans reflect the variations of net meridional pycnocline transports associated with the lower branches of the STCs in these oceans. The rise (fall) of SSH at the western boundaries of the northern and southern tropical Pacific Ocean during 1993–2000 (2000–2006) (Figure 7.8c and d in Section 7.3.2.2) and the lack of significant SSH changes at the eastern boundaries implied an increased (reduced) equatorward convergence of pycnocline waters into the tropical Pacific (Lee and McPhaden 2008). The SSH signals at the western boundary of the tropical Pacific are transmitted through the Indonesian Seas via coastal Kelvin waves, affecting the SSH at the ITF exits and the eastern boundary of the southern tropical Indian Ocean off the west coast of western Australia. Therefore, the increase (decrease) of SSH at the western boundary of the northern tropical Pacific during 1993 to 2000 (2000 to 2006) caused similar SSH changes in the eastern Indian Ocean. This mechanism has in fact been at work throughout the entire period of the altimeter record since 1993 (Wang et al. 2015). Because the SSH at the western boundary of the southern tropical Indian Ocean did not experience significant changes, the SSH differences between the western and eastern boundary of the southern tropical Indian Ocean implied strengthening (weakening) of the northward pycnocline flow associated with the lower branch of the Indian Ocean STC. Therefore, the decadal anomalies of pycnocline transports associated with the Pacific and Indian Ocean STCs are anticorrelated, with the ITF region providing the oceanic tunnel that links the lower branches of the Pacific and Indian Ocean STCs.

Wind stress measurements from ERS and QuikSCAT scatterometers implied similarly opposite decadal changes of the upper branches of the Pacific and Indian Ocean STCs that were driven by anticorrelated decadal changes of the tropical Pacific and South Indian Ocean trade winds (Lee and McPhaden 2008). The anticorrelated changes in the Pacific and Indian Ocean trade winds were attributed to the decadal oscillation of the Walker Circulation, providing an atmospheric bridge that links the upper branches of the Pacific and Indian Ocean STCs. Therefore, the oscillation of the Walker circulation and the SSH signals transmitted through the Indonesian Seas provide an atmospheric bridge and oceanic tunnel that connect the Pacific and Indian Ocean STCs. As a result of these two mechanisms, the Pacific and Indian Ocean STCs play opposite roles in regulating the upper-ocean heat content of the Indo-Pacific domain.

Sustained, direct measurements of ITF transport are extremely challenging in large part because of the complicated geometry with many inflow and outflow passages (Sprintall et al. 2014). Systematic *in situ* measurements of transport at the main ITF inflow (Makassar Strait) and three main outflow passages were undertaken by the international program INSTANT but only for 3 years (2004 to 2006) (Sprintall et al. 2009). Only the Makassar Strait has mooring measurements that have encompassed decadal timescales (1996–1998 and 2004 onward) (Gordon et al. 2012). The ITF is known to be forced by the positive pressure difference between the tropical Pacific and tropical Indian oceans (Wyrtki 1987). Early attempts to infer ITF transport variations using SSH measurements from tide gauge stations in the islands of the tropical Pacific and Indian oceans or altimeter-derived SSH measurements showed that the resultant estimates were sensitive to the locations of reference points in the two oceans (e.g., Wyrtki 1987; Clarke and Liu 1994; Potemra et al. 1997; Potemra 2005). The lack of *in situ* measurements of ITF transport in the past has also made it difficult to validate the proxy estimates of ITF transport derived from SSH measurements. The extended record of satellite altimetry and gravimetry, along with estimates of upper-700 m ITF transport based on XBT data have provided an opportunity to identify the optimal locations in the tropical Pacific and Indian oceans where SSH from altimetry and ocean bottom pressure (OBP) can be used as a proxy to estimate ITF transport. The SSH and OBP differences between 11°N, 162°E in the Pacific and 0°E, 80°E in the Indian Ocean were found to provide optimal proxy estimates of the ITF transport when these measurements are used in a theoretical calculation that combined the "geostrophic control" theory (Garrett and Toulany 1982) and "hydraulic control" theory (Whitehead 1989) as proposed by Song (2006) and Qu and Song (2009). The derived proxy time series of ITF transport compared reasonably well (both for time mean and interannual variations) with estimated ITF transport derived from *in situ* measurements in the Makassar Strait (the main branch of the ITF inflow from the Pacific) during 1996–1998 and 2004–2011.

The large increase of SSH in the western tropical Pacific during 1993–2000 associated with the strengthening of the tropical Pacific trade winds implied a strengthening of the ITF transport. Although no direct measurements of ITF transports were available for this period, ITF transport estimates derived from a suite of ocean data assimilation products consistently depicted the strengthening of the ITF transport during this period (Figure 12 in Lee et al. 2010), which is also evident in the estimates derived from SSH by Susanto and Song (2015). This further supports the notation that SSH measurements can be used as a proxy to infer interannual and decadal changes of the ITF transport.

7.4 SUMMARY

Following the success of the T/P mission, sustained measurements of SSH from precision altimetry into the twenty-first century with the Jason series satellites have significantly advanced the understanding far beyond what was learned in the 1990s in terms of tropical ocean variability, its relationship with various climate variability, and inter-basin linkage, from intraseasonal to multi-decadal timescales.

On intraseasonal timescales, SSH measurements have allowed in-depth understanding of the wavenumber-frequency characteristics associated with different behaviors of TIWs (e.g., mixed Rossby-gravity wave versus meridional-mode Rossby wave nature). The measurements have also enabled the detection of coastal Kelvin waves in various basins and marginal seas, including those transmitting through the Indonesian archipelago. On seasonal timescales, SSH measurements help identify and improve the understanding of annual Rossby waves and basin modes that have been identified in various basins as well as the dynamics of the Wyrtki Jet in the Indian Ocean.

The sustained altimeter measurements also have been essential to improving the understanding of interannual ocean variability and its relationships with interannual climate modes such as ENSO, IOZDM, and Tropical Atlantic Interannual climate variability. In particular, these measurements illustrate the importance of the combined effects of wind-forced equatorial Kelvin waves and the reflections of equatorial Rossby waves at boundaries to regulate the zonal movement of the eastern edge of the tropical Pacific warm pool and the thermocline depth fluctuations as well as the influence on the ENSO cycle in the context of the Delayed Oscillator theory. Altimeter measurements also characterize the meridional recharge/discharge of tropical Pacific warm water volume on ENSO timescales and reveal the hemispheric asymmetry of the recharge/discharge that was not described by the Recharge Oscillator theory. The continuing altimeter record has made it possible to examine the diversity of ENSO events such as the eastern- versus central-equatorial Pacific El Niño in terms of their distinct SSH and surface current signatures, which have significant implications for the interaction between the thermocline and the mixed-layer and thus the associated air-sea interaction.

Significant new knowledge about decadal to multidecadal SSH variability has been gained from the analysis of SSH measurements and the related ocean modeling effort. Studies of the nearly two-and-one-half decade SSH record derived from precision altimetry in the context of the longer record of the sparsely distributed tide gauge data help in the understanding of the processes responsible for enhanced decadal variability during the altimeter record and shed light on the relative contributions of natural decadal variability versus potential climate change forcing.

Satellite-derived SSH and wind measurements have been used to infer the variations in the strengths of the shallow meridional overturning cells that are the primary mechanisms for tropical-subtropical exchanges both in the Pacific and Indian oceans (i.e., the STCs). They also elucidated the linkages of the STCs in the Pacific and Indian oceans, including the oceanic tunnel of the Indonesia Seas (coastal Kelvin wave propagation and the Indonesian Throughflow) and the atmospheric bridge associated with the two branches of the Walker Circulation over the tropical Pacific and Indian oceans. SSH measurements have also revealed the zonal structure of the lower branch (pycnocline flow) of the STC in the Pacific Ocean, characterized by anticorrelated variability of the LLWBC transport and that of the interior pycnocline flow and a substantial compensation of the latter by the former. This has important implications for the closure of the volume, heat, and freshwater budgets of the tropical Pacific Ocean and supports the need for sustained observing systems in the LLWBC regions together with broad-scale observations in the ocean interior to close these budgets.

Finally, it is important to note that sustaining the climate data record of satellite SSH to monitor tropical ocean variability is critical to address the knowledge gaps in terms of the interplay among multidecadal variation of interannual climate modes such as ENSO, intrinsic multidecadal variability that may be linked to the extratropics, and climate change signals. Sustaining the SSH climate data record is also important to the continued improvement of ENSO forecast models, especially in light of the apparent degradation of ENSO forecast skills after the turn of the century. As the length of the SSH climate data record continues to increase, the measurements become increasingly useful for evaluating climate models. The current generation of altimeters still has limited capability for characterizing SSH variations in coastal regions and for sub-mesoscale variability (e.g., in the Solomon Sea and the Maritime Continent region). The upcoming SWOT mission will significantly improve this capability.

The ocean and climate research community has recognized the importance of sustaining and enhancing altimeter measurements of the world's ocean (including tropical oceans) to advance research and applications in the aforementioned areas. This was well described by several community whitepapers of the OceanObs'09 Conference (http://www.oceanobs09.net/proceedings/cwp/) as well as the Tropical Pacific Observing System 2020 (TPOS2020) First Report released recently (http://tpos2020.org/). Sustaining and enhancing satellite altimetry will continue to improve our understanding of tropical ocean and climate variability and predictions.

ACKNOWLEDGMENTS

The research by T. Lee was carried out at the Jet Propulsion Laboratory (JPL), California Institute of Technology, under a contract with NASA, supported by the NASA Physical Oceanography Program. J. T. Farrar and T. Durland acknowledge support from NASA grants NNX13AE46G and NNX14AM71G. W. Han is supported by NASA NNH16ZDA001N-OSTST.

REFERENCES

Alory, G., and T. Delcroix, (2002), Interannual sea level changes and associated mass transports in the tropical Pacific from TOPEX-Poseidon data and linear model results (1964–1999). *J. Geophys. Res.*, 107, C10, 3153, doi:101029/2001JC00106.

Aman, A., L. Testut, P. Woodworth, T. Aarup, and D. Dixon, 2007. Seasonal sea level variability in the Gulf of Guinea from altimetry and tide gauge, *Rev. Ivoir. Sci. Technol.*, 09 (2007) 105–118.

Andrew, J. A. M., H. Leach, and P. L. Woodworth, (2006), The relationships between tropical Atlantic sea level variability and major climate indices. *Ocean Dyn.*, 56, 452–463, doi:10.1007/s10236-006-0068-z.

Arnault, S., B. Bourlès, Y. Gouriou, and R. Chuchla, (1999), Intercomparison of the upper layer circulation of the western equatorial Atlantic ocean: In situ and satellite data, *J. Geophys. Res.*, 104, C9, 21171–21194, doi:10.1029/1999JC900124.

Arnault, S., and R.E. Cheney, (1994), Tropical Atlantic sea level variability from GEOSAT (1985–1989), *J. Geophys. Res.*, 99, C9, 18207–18223, doi:10.1029/94JC01301.

Arnault, S., G. Eldin, B. Bourlès, Y. Dupenhoat, Y. Gouriou, A. Aman, R. Chuchla, F. Gallois, E. Kestenaere, and F. Marin, (2004), In situ and satellite data in the tropical Atlantic ocean during the EQUALANT99 experiment, *Int. J. Remote Sens.*, vol. 25, n°7-8, 1291–1296.

Arnault, S., L. Gourdeau, and Y. Ménard, (1992), Comparison of the altimetric signal with in situ measurements in the Tropical Atlantic Ocean, *Deep Sea Res.*, 39, 3/4, 481–499.

Arnault, S., and E. Kestenare, (2004), Tropical Atlantic surface current variability from 10 years of TOPEX/Poséïdon altimetry, *Geophys. Res. Lett.*, 31, L03308, doi:10.1029/2003GL019210.

Arnault, S., and C. LeProvost, (1997), Regional identification in the tropical Atlantic ocean of residual tide errors from an empirical orthogonal function analysis of the TOPEX/Poséïdon altimetric data, *J. Geophys. Res.*, 102, C9, 21011–21036, doi:10.1029/97JC00983.

Arnault, S., and J. L. Mélice, (2012), Investigation of the interannual variability of the tropical Atlantic Ocean from satellite data, *Mar. Geod.*, vol. 35, n°1, 151–174, doi:10.1080/01490419.2012.718212.

Arnault, S., Y. Ménard, and J. Merle, (1990), Observing the Tropical Atlantic Ocean in 86-87 from altimetry, *J. Geophys. Res.*, 95, C10, 17921–17945, doi:10.1029/JC095iC10p17921.

Arnault, S., Y. Ménard, and M. C. Roucquet, (1989), Variability of the Tropical Atlantic in 1986–1987 as observed by GEOSAT and in situ data. *Adv. In Space Res.*, 9, n°7, 7,383–7,386, doi:10.1016/0273-1177(89)90189-0.

Arnault, S., I. Pujol, and J. L. Mélice, (2011), In situ validation of Jason-1 and Jason-2 altimetry missions in the tropical Atlantic Ocean, *Mar. Geod.*, 34, 319–339, doi:10.1080/01490419.2011.584833.

Arruda, W. Z., and E. J. Campos, Sea level anomaly trends in the South Atlantic around 10°S, Anais XV Simpósio Brasileiro de Sensoriamento Remoto - SBSR, Curitiba, PR, Brasil, 30 de abril a 05 de maio de 2011, INPE p.7200.

Artamonov, Y. V., (2006), Seasonal variability of geostrophic currents in the Atlantic ocean according to the altimetry data. *Phys. Oceanogr.*, 16, 3, 60–71.

Ashok, K., S. K., Behera, S. A. Rao, H. Weng, and T. Yamagata, (2007), El Niño Modoki and its possible teleconnection, *J. Geophys. Res.*, 112, C11007, doi:10.1029/2006JC003798.

Behera, S., and T. Yamagata, (2010), Imprint of the El Niño Modoki on decadal sea level change. *Geophys. Res. Lett.*, 37, L23702, doi:10.1029/2010GL045936.

Bonjean, F., and G. S. E. Lagerloef, (2002), Diagnostic model and analysis of the surface currents in the tropical Pacific Ocean, *J. Phys. Oceanogr.*, 32(10), 2938–2954.

Bosc, C., and T. Delcroix, (2008), Observed equatorial Rossby waves and ENSO-related warm water volume changes in the equatorial Pacific Ocean. *J. Geophys. Res.*, 113, C06003, doi:10.1029/2007JC004613.

Bosc, C., T. Delcroix, and C. Maes, (2009), Barrier layer variability in the western Pacific warm pool from 2000–2007. *J. Geophys. Res.*, 114, C06023, doi:10.1029/2008JC005187.

Boulanger, J.-P., S. Cravatte, and C. Menkes, (2003), Reflected and locally wind-forced interannual equatorial Kelvin waves in the western Pacific Ocean. *J. Geophys. Res.*, 108, 3311, doi:10.1029/2002JC001760.

Boulanger, J.-P., C. Menkes, and M. Lengaigne, (2004), Role of high- and low-frequency winds and wave reflection in the onset, growth and termination of the 1997–1998 El Niño. *Clim. Dyn.*, 2, 267–280, doi:10.1007/s00382-003-0383-8.

Brandt, P., M. Dengler, A. Rubino, D. Quadfasel, and R. Schott, (2003), Intraseasonal variability in the southwestern Arabian Sea and its relation to the seasonal circulation, *Deep Sea Res. II*, 50, 2129–2141.

Brandt, P., L. Stramma, F. Schott, J. Fischer, M. Dengler, and D. Quadfasel, (2002). Annual Rossby waves in the Arabian Sea from TOPEX/POSEIDON altimeter and in situ data. *Deep-Sea Res. II*, 49, 1197–1210.

Bunge, L., and A. J. Clarke, (2009), Seasonal propagation of sea level along the Equator in the Atlantic, L., *J. Phys. Oceanogr.*, 39, 1069–1074. doi:10.1175/2008.4003.1.

Busalacchi, A.J., M. J. McPhaden, J. Picaut, (1994), Variability in the equatorial Pacific sea surface topography during the verification phase of the TOPEX/Poséïdon mission. *J. Geophys. Res.*, 99, C12, 24725–24738.

Capotondi, A., M. A. Alexander, C. Deser, et al., (2005), Anatomy and decadal evolution of the Pacific Subtropical-Tropical Cells (STCs). *J. Clim.*, 18, 3739–3758.

Carson, M., A. Köhl, and D. Stammer, (2015), The impact of regional multidecadal and century-scale internal climate variability on sea level trends in CMIP5 Models. *J. Clim.*, 28, 853–861, doi:10.1175/JCLI-D-14-00359.1.

Carton, J.A., (1989), Estimates of Sea Level in the Tropical Atlantic Ocean Using GEOSAT Altimetry, *J. Geophys. Res.*, 94, C6, 8,029–8,039, doi:10.1029/JC094iC06p08029.

Carton, J. A., and E. Katz, (1990), Estimates of the zonal slope and seasonal transport of the Atlantic North Equatorial Countercurrent, *J. Geophys. Res.*, 95, C3, 3,091–3,100, doi:10.1029/JC095iC03p03091.

Carton, J. A., G. A. Chepurin, G. K. Korotaev, and T. Zhu, (1993), Comparison of dynamic height variations in the tropical Atlantic during 1987–1989 as viewed in Sections hydrography and GEOSAT altimeter. *J. Geophys. Res.*, 98, 14369–14377, doi:10.1029/93JC01505.

Castelão, G. P., and W. E. Johns, (2011), Sea surface structure of North Brazil Current rings derived from shipboard and moored acoustic Doppler current profiler observations, *J. Geophys. Res.*, 116, C01010, doi:10.1029/2010JC006575.

Chaigneau, A., G. Eldin, and B. Dewitte, (2009), Eddy activity in the 4 major upwelling systems from satellite altimetry (1992–2007), *Prog. In Oceanogr.*, 83, 117–123, doi:10.1016/J.pocean.2009.07.012.

Chatterjee, A., D. Shankar, J. P. McCreary Jr., and P. N. Vinayachandran, (2013), Yanai waves in the western equatorial Indian Ocean, *J. Geophys. Res. Oceans*, 118, 1556–1570, doi:10.1002/jgrc.20121.

Chelton, D. B., M. G. Schlax, J. M. Lyman and G. C. Johnson, (2003), Equatorially trapped Rossby waves in the presence of meridionally sheared baroclinic flow in the Pacific Ocean. *Progr. Oceanogr.*, 56, 323–380.

Chelton, D. B., F. Wentz, C. Gentemann, R. deSzoeke, and M. Schlax, (2000), Satellite microwave SST observations of transequatorial tropical instability waves. *Geophys. Res. Lett.*, 27, 1239–1242.

Chen, S., and B. Qiu, (2004), Seasonal variability of the South Equatorial Countercurrent. *J. Geophys. Res.*, 109, C08003, doi:10.1029/2003JC002243.

Chen, G., W. Han, Y. Li, D. Wang, and T. Shinoda, 2015: Intraseasonal Variability of Upwelling in the Equatorial Eastern Indian Ocean. J. Geophys. Res.-Oceans, 120, 7598–7615, DOI: 10.1002/2015JC011223.

Cheney, R. E., B. C. Douglas, and L. Miller, (1989), Evaluation of GEOSAT altimeter data with applications to tropical Pacific sea level variability. *J. Geophys. Res.*, 94, C4, 4737–4747.

Cheney, R. E., and L. Miller, (1990), Recovery of the sea level signal in the western tropical Pacific from GEOSAT altimetry. *J. Gophys. Res.*, 95, C3, 2977–2984.

Church, J. A., P. U. Clark, A. Cazenave, et al., (2013), Sea Level Change. In: *Climate Change 2013: The Physical Science Basis. Contribution of Working Group I to the Fifth Assessment Report of the Intergovernmental Panel on Climate Change*, Stocker, T. F., D. Qin, G.-K. Plattner, M. Tignor, S. K. Allen, J. Boschung, A. Nauels, Y. Xia, V. Bex and P. M. Midgley (eds.). Cambridge University Press, Cambridge, United Kingdom and New York, NY.

Clarke, A. J., and X. Liu, (1994), Interannual sea level in the northern and eastern Indian Ocean, *J. Phys. Oceanogr.*, 24, 1224–1235, doi:10. 1175/1520-0485(1994)024<1224:ISLITN>2.0.CO;2.

Collins, M., S. An, W. Cai, et al., (2010), The impact of global warming on the tropical Pacific Ocean and ElNiño. *Nat. Geosci.*, 3, 391–397.

Cravatte, S., J. Picaut, and G. Eldin, (2003), Second and first baroclinic Kelvin modes in the equatorial Pacific at intraseasonal timescales. *J. Geophys. Res.*, 108, 3266, doi:10.1029/ 2002JC001511.

Delcroix, T., B. DwWitte, Y. duPenhaot, et al., (2000), Equatorial waves and warm pool displacements during the 1992–1998 El Niño Southern Oscillation events: Observation and modeling. *J. Geophys. Res.*, 105, 26045–26062.

Delcroix, T., J. Picaut, and G. Eldin, (1991), Equatorial Kelvin and Rossby waves evidenced in the Pacific ocean through GEOSAT sea level and surface current anomalies. *J. Geophys. Res.*, 96, 3249–3262.

Didden, N., and F. Schott, (1993), Eddies in the North Brazil Current retroflection region observed by GEOSAT altimetry. *J. Geophys. Res.*, 98, c11, 20121–20131.

Ding, H., N. S. Keenlyside, and M. Latif, (2009), Seasonal cycle in the upper equatorial Atlantic Ocean, *Geophys. Res. Lett.*, 114, C09016, doi:10.1029/2009JC005418.

Ding, H., N. S. Keenlyside, and M. Latif, (2010), Equatorial Atlantic interannual variability: Role of heat content, *J. Geophys. Res.*, 115, C09020, doi:10.1029/2010JC006304.

Djakoure, S., P. Penven, B. Bourles, J. Veitch, and V. Kone, (2014), Coastally trapped eddies in the north of the Gulf of Guinea, *J. Geophys. Res.*, 119, 6805–6819, doi:10.1002/2014JC010243.

Donohue, K., and M. Wimbush, (1998), Model results of flow instabilities in the tropical Pacific Ocean, *J. Geophys. Res.*, 103, 21401–21412.

Drushka, K., J. Sprintall, S. T. Gille, and I. Brodjonegoro, (2010), Vertical structure of Kelvin waves in the Indonesian throughflow exit passages, *J. Phys. Oceanogr.*, 40(9), 1965–1987. doi:10.1175/2010JPO4380.1.

Durand, F., and T. Delcroix, (2000), On the variability of the Tropical Pacific Thermal Structure during the 1979–96 period, as Deduced from XBT Sections. *J. Phys. Oceanogr.*, 30, 3261–3269.

Durland, T. S., R. M. Samelson, D. B. Chelton, and R. A. deSzoeke, (2011), Modification of long Rossby wave phase speeds by zonal currents. *J. Phys. Oceanogr.*, 41, 1077–1101.

Eigenheer, A., and D. Quadfasel, (2000), Seasonal variability of the Bay of Bengal circulation inferred from TOPEX/POSEIDON altimetry. *J Geophys. Res.*, 105, 3243–3252.

Enfield, D., (1987), The intraseasonal oscillation in eastern Pacific sea levels: How is it forced? *J. Phys. Oceanogr.,* 17, 1860–1876.

Enfield, D. B., and D. A. Mayer, (1997), Tropical Atlantic sea surface temperature variability and its relation to El Niño-Southern Oscillation. *J. Geophys. Res.: Oceans*, 102(C1), 929–945.

Farrar, J. T., (2008), Observations of the dispersion characteristics and meridional sea-level structure of equatorial waves in the Pacific Ocean. *J. Phys. Oceanogr.*, 38, 1669–1689.

Farrar, J. T., (2011), Barotropic Rossby waves radiating from tropical instability waves in the Pacific Ocean. *J. Phys. Oceanogr.*, 41, 1160–1181.

Feng, M., and S. Wijffels, (2002), Intraseasonal variability in the south equatorial current of the East Indian Ocean. *J. Phys. Oceanogr.*, 32 (2002), 265–277.

Fernandez Bilbao, R. A. F., J. M. Gregory, and N. Bouttes, (2015), Analysis of the regional pattern of sea level change due to ocean dynamics and density change for 1993–2099 in observations and CMIP5 AOGCMs. *Clim. Dyn.*, 45, 2647–2666, doi:10.1007/s00382-015-2499-z.

Florenchie, P., J. R. E. Lutjeharms, C. J. C. Reason, S. Masson, and M. Rouault, (2003), The source of Bengula Niños in the South Atlantic Ocean, *Geophys. Res. Lett.*, 30(10), 1505, doi:10.1029/2003GL017172.

Florenchie, P., C. J. C. Reason, J. R. E. Lutjeharms, M. Rouault, C. Roy, and S. Masson, (2004), Evolution of interannual warm and cold events in the southeast Atlantic Ocean, *J. Clim.*, 17, 2318–2334, doi:10.1175/1520-0442(2004)017<2318:EOIWAC>2.0.CO;2.

Foltz, G.R., and M. J. McPhaden, (2010), Interaction between the Atlantic meridional and Niño modes, *Geophys. Res. Lett.*, 37, L18604, doi:10.1029/2010GL044001.

Fonseca, G. A., G. J. Goni, W. E. Johns, and E. J. D. Campos, (2004), Investigation of the North Brazil Current retroflection and North Equatorial Countercurrent variability, *Geophys. Res. Lett.*, vol. 31, L21304, doi:10.1029/2004GL020054.

Forristall, G. Z., K. J. Schaudt, and J. Calman, (1990), Verification of GEOSAT altimetry for operational use in the Gulf of Mexico. *J. Gophys. Res.*, 95, C3, 2985–2989.

Franca, C. A. A., I. Wainer, and A. R. de Mesquita, (2003), Planetary equatorial trapped waves in the Atlantic ocean from Topex/Poseidon altimetry, In: Gustavo Jorge Goni, Paola Malanotte-Rizzoli. (org.). Interhemispheric water exchange in the Atlantic ocean. Amsterdam: Elsevier Science, 2003, v. 68, p. 213–232.

Frankcombe, L. M., S. McGregor, and M. H. England, (2015), Robustness of the modes of Indo-Pacific sea level variability, *Clim. Dyn.*, 45, 1281–1298, doi:10.1007/s00382-014-2377-0.

Fratantoni, D. M., W. E. Johns, and T. L. Townsend, (1995), Rings of the North Brazil Current: Their structure and behavior inferred from observations and a numerical simulation. *J. Geophys. Res.*, 100, C6, 10635–10654, doi:10.1029/95JC00925.

Fu, L., and R. Smith, (1996), Global ocean circulation from satellite altimetry and high-resolution computer simulation. *Bull. Am. Meteorol. Soc.*, 77, 2625–2636.

Fu, L.-L. and A. Cazanave (2000), Satellite Altimetry and Earth Science: a Handbook for Techniques and Applications, Academic Press, San Diego, California, USA. 463 pp.

Fu, L.-L., (2007), Intraseasonal variability of the equatorial Indian Ocean observed from sea surface height, wind, and temperature data. *J. Phys. Oceanogr.*, 37, 188–202, doi:10.1175/JPO3006.1.

Fukumori, I., and O. Wang, (2013), Origins of heat and freshwater anomalies underlying regional decadal sea level trends, *Geophys. Res. Lett.*, 40, 563–567, doi:10.1002/grl.50164.

Garrett, C. J. R., and B. Toulany (1982), Sea level variability due to meteorological forcing in the northeast of St. Lawrence, J. Geophys. Res., 87, 1968–1978.

Garzoli, S. L., A. Field, and Q. Yao, (2003), North Brazil current rings and the variability in the latitude of retroflection. In: *Interhemispheric Water Exchange in the Atlantic Ocean*, Goni G. J. and P. Malanotte-Rizzoli (Ed.), Elsevier Oceanography Series, Elsevier B.V., Amsterdam, The Netherlands. 68, pp. 357–373.

Giese, B. S., J. A. Carton, and L. J. Holl, (1994), Sea level variability in the eastern tropical Pacific as observed by TOPEX and Tropical Ocean Global Atmosphere, Tropical Atmosphere Ocean experiment. *J. Geophys. Res.*, 99, C12, 24739–24748.

Girish Kumar, M. S., M. Ravichandran, and W. Han, (2013), Observed intraseasonal thermocline variability in the Bay of Bengal. *JGR-Oceans*, 118, 3336–3349, doi:10.1002/jgrc.20245.

Goes, M., G. Goni, V. Hormann, and R. C. Perez, (2013), Variability of the Atlantic off-equatorial eastward currents during 1993–2010 using a synthetic method, *J. Geophys. Res.*, 118, 1–20, doi:10.1002/jgrc.20186.

Goni, G. J, and W.E. Johns, (2001), A census if North Brazil Current rings observed from Topex/Poseidon altimetry: 1993–1998, *Geophys. Res. Lett.*, 28, 1, 1–4.
Goni, G. J., and W. E. Johns, (2003), Synoptic study of warm rings in the North Brazil Current retroflection region using satellite altimetry, In: *Interhemispheric Water Exchange in the Atlantic Ocean, Elsevier Oceanogr. Ser.*, vol. 68, G. J. Goni, and P. Malanotte-Rizzoli, (Ed.), chap. 13, pp. 335–356, Elsevier, Amsterdam.
Gordon, A. L., B. A. Huber, E. J. Metzger, R. D. Susanto, H. E. Hurlburt, and T. R. Adi, (2012), South China Sea throughflow impact on the Indonesian throughflow, *Geophys. Res. Lett.*, 39, L11602, doi:10.1029/2012GL052021.
Goswami, B. N., (2012), South Asian monsoon. In Intraseasonal Variability in the Atmosphere-Ocean Climate System, Lau W. and D. Waliser (Eds.), Springer Fachmedien, Wiesbaden, 2012, ISBN 978 3 642 13913 0.
Grodsky, S. A., J. A. Carton, and F. M. Bingham, (2006), Low frequency variation of sea surface salinity in the tropical Atlantic, *Geophys. Res. Lett.*, 33, L14604, doi:10.1029/2006GL026426.
Grodsky, S. A., J. A. Carton, C. Provost, J. Servain, J. A. Lorenzzetti, and M. J. McPhaden, (2005), Tropical instability waves at 0°N, 23°W in the Atlantic: A case study using Pilot Research Moored Array in the Tropical Atlantic (PIRATA) mooring data, *J. Geophys. Res.*, 110, C08010, doi:10.1029/2005JC002941.
Halpern, D., R. A. Knox, and D. S. Luther, 1988: Observations of 20-day period meridional current oscillations in the upper ocean along the Pacific equator. J. Phys. Oceanogr., 18, 1514–1534.
Hamlington, B. D., M. W. Strassburg, R. Leben, W. Han, R. S. Nerem, and K. Y. Kim, (2014), Uncovering the anthropogenic warming-induced sea level rise signal in the Pacific Ocean, *Nat. Clim. Change*, *4*, 782–785.
Han, W., (2005), Origins and dynamics of the 90-day and 30–60-day variations in the equatorial Indian Ocean. *J. Phys. Oceanogr.*, 35, 708–728.
Han, W., D. M. Lawrence, and P. J. Webster, (2001), Dynamical response of the equatorial Indian Ocean to intraseasonal winds: Zonal flow. *Geophys. Res. Lett.*, 28, 4215–4218.
Han, W., J. P. McCreary, Y. Masumoto, J. Vialard, and B. Duncan, (2011), Basin resonances in the equatorial Indian Ocean. *J. Phys. Oceanogr.*, 41, 1252–1270.
Han, W., G. Meehl, D. Stammer, A. Hu, B. Hamlington, J. Kenigson, H. Palanisamy, and P. Thompson (2017), Spatial Patterns of Sea Level Variability Associated With Natural Internal Climate Modes. Surveys in Geophysics, 38(1), 217–250, DOI:10.1007/s10712-016-9386-y.
Han, W., G. Meehl, D. Stammer, A. Hu, B. Hamlington, J. Kenigson, H. Palanisamy, and P. Thompson, (2017), Spatial patterns of sea level variability associated with natural internal climate modes, *Surv. Geophys.*, 38(1), 217–250, doi:10.1007/s10712-016-9386-y.
Han, W., T. Shinoda, L.-L. Fu, and J. P. McCreary, (2006), Impact of atmospheric intraseasonal oscillations on the Indian Ocean dipole during the 1990s. *J. Phys. Oceanogr.*, 36, 670–690.
Han, W., J. Vialard, M. McPhaden, T. Lee, Y. Masumoto, M. Feng, and W. de Ruijter, (2014), Indian Ocean decadal variability: A review. *Bull. Amer. Meteorol. Soc.*, 95, 1679–1703. 10.1175/BAMS-D-13-00028.1.
Han, W., and P. J. Webster, (2002), Forcing mechanisms of sea-level interannual variability in the Bay of Bengal, *J. Phys. Oceanogr.*, 32, 216–239.
Han, W., P. J. Webster, J. L. Lin, W. T. Liu, R. Fu, D. Yuan, and A. Hu, (2008), Dynamics of intraseasonal sea level and thermocline variability in the equatorial Atlantic during 2002–03, *J. Phys. Oceanogr.*, 38, doi:10.1175/2008.3854.1.
Handoh, I. C., J. Matthews, G. R. Bigg, and D. P. Stevens, (2006a), Interannual variability of the tropical Atlantic independent of and associated with ENSO: Part 1. The north tropical Atlantic. *Int. J. Climatol.*, 26, 14, 1937–1956, doi:10.1002/joc.1343.
Handoh, I. C., J. Matthews, G. R. Bigg, and D. P. Stevens, (2006b), Interannual variability of the tropical Atlantic independent of and associated with ENSO: Part 2. The south tropical Atlantic. *Int. J. Climatol.*, 26, 14, 1957–1976, doi:10.1002/joc.1342.
Hastenrath, S., and J. Merle, (1987), Annual cycle of subsurface thermal structure in the Tropical Atlantic Ocean. *J. Phys. Oceanogr.*, 17, 1518–1537.
Hazeleger, W., R. Seager, M. A. Cane, and N. H. Naik, (2004), How can tropical Pacific ocean heat transport vary? *J. Phys. Oceanogr.*, 34, 320–333.
Hazeleger, W., M. Visbeck, M. Cane, A. Karspeck, and N. Naik, (2001), Decadal upper ocean temperature variability in the tropical Pacific. *J. Geophys. Res.*, 106, 8971–8988.
Hendon, H., B. Liebmann, and J. Glick, (1998), Oceanic Kelvin waves and the Madden–Julian oscillation. *J. Atmos. Sci.*, 55, 88–101.

Hermes, J. C., and C. J. C. Reason, (2008), Annual cycle of the south Indian Ocean (Seychelles-Chagos) thermocline ridge in a regional ocean model. *J. Geophys. Res.*, 113, C04035, doi:10.1029/ 2007JC004363.

Holmes, R. M., and L. N. Thomas (2016), Modulation of tropical instability wave intensity by equatorial Kelvin waves, J. Phys. Oceanogr., 46, 2623–2643.

Hormann, V., and P. Brandt, (2009), Upper equatorial Atlantic variability during 2002 and 2005 associated with equatorial Kelvin waves, *J. Geophys. Res.*, 114, C03007, doi:10.1029/2008JC005101.

Hormann, V., R. Lumpkins, and G. R. Foltz, (2012), Interannual North Equatorial Countercurrent variability and its relation to Tropical Atlantic climatic modes. *J. Geophys. Res.*, 117, c04035, doi:10:1029/2011jc007697.

Houghton, R. W., and Y. M. Tourre, (1992), Characteristics of low-frequency sea surface temperature fluctuations in the tropical Atlantic, *J. Clim.*, 5, 765–771.

Hsin, Y.-C., and B. Qiu, (2012), Seasonal fluctuations of the North Equatorial Countercurrent. *J. Geophys. Res.*, 117, C06001, doi:10.1029/2011JC007794.

Hu, S., and J. Sprintall, (2016), Interannual variability of the Indonesian Throughflow: The salinity effect. *J. Geophys. Res.*, 121, 2596–2615, doi:10.1002/2015JC011495.

Huang, B., and Z.-Z. Hu, (2007), Cloud-SST feedback in southeastern tropical Atlantic anomalous events, *J. Geophys. Res.*, 112, C03015, doi:10.1029/2006JC003626.

Illig, S., D. Gushchina, B. Dewitte, N. Ayoub, and Y. du Penhoat, (2006), The 1996 equatorial Atlantic warm event: Origin and mechanisms, *Geophys. Res. Lett.*, 33, L09701, doi:10.1029/2005GL025632.

Iskandar, I., W. Mardiansyah, Y. Masumoto, and T. Yamagata, (2005), Intraseasonal Kelvin waves along the southern coast of Sumatra and Java. *J. Geophys. Res.*, 110, C04013, doi:10.1029/ 2004JC002508.

Iskandar, I., and M. J. McPhaden, (2011), Dynamics of wind-forced intraseasonal zonal current variations in the equatorial Indian Ocean, *J. Geophys. Res.*, 116, C06019, doi:10.1029/2010JC006864.

Iskandar, I., T. Tozuka, H. Sasaki, Y. Masumoto, and T. Yamagata, (2006), Intraseasonal variations of surface and subsurface currents off Java as simulated in a high-resolution ocean general circulation model. *J. Geophys. Res.*, 111, C12015, doi:10.1029/2006JC003486.

Jochum, M., and P. Malanotte-Rizzoli, (2003), The flow of AAIW along the equator. In: *Interhemispheric Water Exchange in the Atlantic Ocean*, Goni G.J. and P. Malanotte-Rizzoli (Eds.), Elsevier Oceanography Series, Elsevier B.V., Amsterdam, The Netherlands. 68, 193–212.

Johnson, E., (1993), Effects of a mean three-dimensional flow on intraseasonal Kelvin waves in the equatorial Pacific Ocean. *J. Geophys. Res.,* 98, 10185–10194.

Johnson, E., and M. McPhaden, (1993), Structure of intraseasonal Kelvin waves in the equatorial Pacific Ocean. *J. Phys. Oceanogr.*, 23, 608–625.

Johnson, G. C., B. M. Sloyan, W. S. Kessler, and K. E. McTaggart, (2002), Direct measurements of upper ocean currents and water properties across the tropical Pacific during the 1990s. *Progr. Oceanogr.*, 52, 31–61.

Jordà, G., (2014), Detection time for global and regional sea level trends and accelerations. *J. Geophys. Res. Oceans,* 119, 7164–7174. doi:10.1002/2014JC010005.

Kao, H.-Y., and J.-Y. Yu, (2009), Contrasting Eastern-Pacific and Central-Pacific types of ENSO. *J. Clim.*, 22, 615–632.

Katz, E. J., (1997), Waves along the Equator, *J. Phys. Oceanogr.*, 27, 2536–2544.

Keenlyside, N. S., and M. Latif, (2007), Understanding equatorial Atlantic interannual variability, *J. Clim.*, 20,131–142, doi:10.1175/JCLI3992.1.

Kessler, W. S., (2006), The circulation of the eastern tropical Pacific: A review. *Progr. Oceanogr.*, 69, 181–217.

Kessler, W. S., and J. P. McCreary, (1993), The annual wind-driven Rossby wave in the sub-thermocline equatorial Pacific. *J. Phys. Oceanogr.*, 23, 1192–1207.

Kessler, W. S., and M. McPhaden, (1995), Oceanic Kelvin waves and the 1991–93 El Niño. *J. Clim.*, 8, 1757–1774.

Kessler, W. S., M. McPhaden, and K. Weickmann, (1995), Forcing of intraseasonal Kelvin waves in the equatorial Pacific. *J. Geophys. Res.*, 100, 10613–10631.

Kindle, J. C., and J. D. Thompson, (1989), The 26- and 50-day oscillations in the western Indian Ocean: Model results. *J. Geophys. Res.*, 94, 4722–4735.

Kleeman, R., J. P. McCreary, and B. Klinger, (1999), A mechanism for generating ENSO decadal variability. *Geophys. Res. Lett.*, 26, 1743–1746.

Knox, R. A., and D. Halpern, (1982), Long range Kelvin wave propagation of transport variations in Pacific Ocean equatorial currents. *J. Mar. Res.*, 40 Suppl., 329–339.

Kutsuwada, K., and M. McPhaden (2002), Intraseasonal variations in the upper equatorial Pacific Ocean prior to and during the 1997–98 El Niño. J. Phys. Oceanogr., 32, 1133–1149.

Lau, W. K. M., and D. E. Waliser (Editors), (2012), *Intraseasonal variability in the Atmosphere–Ocean Climate System* (Second Edition). Springer-Verlag, Berlin.

Lee, T., (2004), Decadal weakening of the shallow overturning circulation in the South Indian Ocean, *Geophys. Res. Lett.*, 31, L18305, doi:10.1029/2004GL020884.

Lee, T., T. Awaji, M. Balmaseda, et al., (2010), Consistency and fidelity of Indonesian-throughflow total volume transport estimated by 14 ocean data assimilation products. *Dyn. Atmos. Oceans*. 50, 201–223, doi:10.1016/j.dynatmoce.2009.12.004.

Lee, T., and I. Fukumori, (2003), Interannual to decadal variation of tropical-subtropical exchange in the Pacific Ocean: Boundary versus interior pycnocline transports. *J. Clim.,* 16, 4022–4042.

Lee, T., G. Lagerloef, H. Y. Kao, J. Mcphaden, J. Willis, and M. M. Gierach, (2014), The influence of salinity on tropical Atlantic instability waves, *J. Geophys. Res.*, 119, 8375–8394, doi:10.1002/2014jc010100.

Lee, T., and M. J. McPhaden, (2008), Decadal phase change in large-scale sea level and winds in the Indo-Pacific region at the end of the 20th century. *Geophys. Res. Lett.*, 35, L01605, doi:10.1029/2007GL032419.

Lee, T., B. Qiu, S. Hakkinen, et al., (2010), Satellite observations of ocean circulation changes associated with climate variability. *TOS, Oceanogr.*, 23(4), 70–81.

Legeckis, R. (1977), Long waves in the eastern equatorial Pacific Ocean: a view from a geostationary satellite. Science. 197, 1179–1181.

Li, Y., and W. Han, (2015), Decadal sea level variations in the Indian Ocean investigated with HYCOM: Roles of climate modes, Ocean internal variability and stochastic wind forcing, *J. Clim.*, 28, 9143–9165, doi:10.1175/JCLI-D-15-0252.1.

Liu, Z., and S. G. H. Philander, (2000), Tropical-extratropical oceanic exchange. Chapter 4.4 *of Ocean Circulation and Climate: Observing and Modeling the Global Ocean*, G. Siedler, J. Church, and J. Gould, (Eds.), AP International Geophys. Series, Academic Press, London, UK and San Diego, California. 77, pp. 247–257.

Liu, Y., R. H. Weisberg, S. Vignudelli, and G. T. Mitchum, (2014), Evaluation of altimetry-derived surface current products using Lagrangian drifter trajectories in the eastern Gulf of Mexico, *J. Geophys. Res.*, 119, doi:10.1002/2013JC009710.

Llovel, W., and T. Lee, (2015), Importance and origin of halosteric contribution to sea level change in the southeast Indian Ocean during 2005–2013, *Geophys. Res. Lett.*, 42, 1148–1157, doi:10.1002/2014GL062611.

Lukas, R., and E. Firing, (1985), The annual Rossby wave in the central equatorial Pacific Ocean. *J. Phys. Oceanogr.*, 11, 55–67.

Lukas, R., S. P. Hayes, and K. Wyrtki, (1984), Equatorial sea-level response during the 1982–1983 El Niño. *J. Geophys. Res.*, 89, 10425–10430.

Lumpkin, R., G. Goni, and K. Dohan, (2013), Surface currents, in "State of the Climate in 2012". *Bull. Amer. Meteor. Soc.*, 94, S62–S65.

Luo, J. J., W. Sasaki, and Y. Masumoto, (2012), Indian Ocean warming modulates Pacific climate change, *Proc. Natl. Acad. Sci.,* 109, 18701–18706.

Luther, D., and E. Johnson (1990), Eddy energetics in the upper equatorial Pacific during the Hawaii-to-Tahiti Shuttle Experiment. J. Phys. Oceanogr., 20, 913–944.

Luyten, J. R., and D. H. Roemmich, (1982), Equatorial currents at semiannual period in the Indian Ocean, *J. Phys. Oceanogr.*, 12, 406–413, doi:10.1175/1520-0485(1982)012<0406:ECASAP>2.0.CO;2.

Lyman, J., D. Chelton, R. deSzoeke, and R. Samelson, (2005), Tropical instability waves as a resonance between equatorial Rossby waves. *J. Phys. Oceanogr.,* 35, 232–254.

Lyman, J., G. Johnson, and W. Kessler, (2007), Distinct 17-day and 33- day tropical instability waves in subsurface observations. *J. Phys. Oceanogr.*, 37, 855–872.

Lyu, K., X. Zhang, J. A. Church, and J. Hu, (2015), Quantifying internally generated and externally forced climate signals at regional scales in CMIP5 models, *Geophys. Res. Lett.*, 42, 9394–9403, doi:10.1002/2015GL065508.

Madden, R. A., and P. R. Julian, (1971), Detection of a 40–50 day oscillation in the zonal wind of the tropical Pacific. *J. Atmos. Sci.,* 28, 702–708.

Madden, R. A., and P. R. Julian, (1972), Description of global-scale circulation cells in the tropics with a 40–50 day period. *J. Atmos. Sci.,* 29, 1109–1123.

Masumoto, Y., H. Hase, Y. Kuroda, H. Matsuura, and K. Takeuchi, (2005), Intraseasonal variability in the upper-layer currents observed in the eastern equatorial Indian Ocean, *Geophys. Res. Lett.*, 32, L02607, doi:10.1029/2004GL021896.

Masumoto, Y., and G. Meyers, (1998), Forced Rossby waves in the southern tropical Indian Ocean, *J. Geophys. Res.,* 103, 12287–12293.

Matsuno, T. (1966), Quasi-geostrophic motions in the equatorial area. J. Meteor. Soc. Japan, 44, 25–43.

Mayer, D. A., M. O. Baringer and G. J. Goni, (2003), Comparison of hydrographic and altimeter based estimates of sea level height variability in the Atlantic Ocean, In: Gustavo Jorge Goni, Paola Malanotte-Rizzoli. (org.). Interhemispheric water exchange in the Atlantic ocean. Amsterdam: Elsevier Science, 2003, v. 68, pp. 213–232.

McCreary, J. P., and P. Lu, (1994), On the interaction between the subtropical and the equatorial oceans: The subtropical cell. *J. Phys. Oceanogr.*, 24, 466–497.

McPhaden, M. J., (1982), Variability in the central equatorial Indian Ocean. Part 1: Ocean dynamics, *J. Mar. Res.*, 40, 157–176.

McPhaden, M. J., (1996), Monthly period oscillations in the Pacific North Equatorial Countercurrent. *J. Geophys. Res.*, 101, 6337–6359.

McPhaden, M. J., A. J. Busalacchi, R. Cheney, et al., (1998), The Tropical Ocean-Global Atmosphere observing system: A decade of progress. *J. Geophys. Res.*, 103, 14169–14240.

McPhaden, M. J., G. Meyers, K. Ando, Y. Masumoto, V. S. N. Murty, M. Ravichandran, F. Syamsudin, J. Vialard, L. Yu, and W. Yu, (2009), RAMA: The research moored array for African-Asian-Australian monsoon analysis and prediction. *Bull. Amer. Meteorol. Soc.*, 90, 459–635.

McPhaden, M. J., and M. Nagura, (2014), Indian Ocean Dipole interpreted in terms of Recharge Oscillator theory, *Clim. Dyn.*, 42, 1569–1586. doi:10.1007/s00382-013-1765-1.

McPhaden, M. J., and B. Taft, (1988), Dynamics of seasonal and intraseasonal variability in the eastern equatorial Pacific. *J. Phys. Oceanogr.*, 18, 1713–1732.

McPhaden, M. J., and D. Zhang, (2002), Slowdown of the meridional overturning circulation in the upper Pacific Ocean. *Nature*, 415, 603–608.

McPhaden, M. J., and D. Zhang, (2004), Pacific Ocean circulation rebounds, *Geophys. Res. Lett.*, 31, L18301, doi:10.1029/2004GL020727.

Melet, A., L. Gourdeau, and J. Verron, (2011), Variability in the Solomon Sea circulation derived from altimeter sea level data, *Ocean Dyn.*, 60, 833–900. doi:10.1007/s10236-010-0302-6.

Melet, A., J. Verron, L. Gourdeau, et al., (2011), Equatorward pathways of Solomon Sea water masses and their modifications. *J. Phys. Oceanogr.*, 41, 810–826. doi:10.1175/2010JPO4559.1.

Menkes, C. E., S. C. Kennan, P. Flament, et al., (2002), A whirling ecosystem in the equatorial Atlantic, *Geophys. Res. Lett.*, 29(11), 1553, doi:10.1029/2001GL014576.

Merrifield, M. A., (2011), A shift in Western Tropical Pacific Sea Level Trends during the 1990s. *J. Clim.*, 24, 4126–4136, doi:10.1175/2011JCLI3932.1.

Mertz, G. L., and L. A. Mysak, (1984), Evidence for a 40–60 day oscillation over the western Indian Ocean during 1976 and 1979, *Mon. Weather Rev.*, 112, 383–386.

Miller, L., and R. E. Cheney, (1990), Large scale meridional transport in the tropical Pacific Ocean during 1986–1987 El Niño from GEOSAT. *J. Gophys. Res.*, 95, C10, 17905–17919.

Miller, L., R. E. Cheney, and B. C. Douglas, (1988), GEOSAT Altimeter Observations of Kelvin Waves and the 1986–87 El Niño, *Science*, 239, 52–54.

Minster, J. F., M. l. Genco, and C. Brossier, (1995), Variations of the sea level in the Amazon estuary, *Continental Shelf Res.*, 15(10), 1287–1302.

Mitchum, G., (1994), Comparison of TOPEX Sea Surface Heights and tide gauges levels. *J. Geophys. Res.*, 99, C12, 24541–24553.

Mitchum, G. T., and R. Lukas, (1990), Westward propagation of annual sea level and wind signals in the western Pacific Ocean. *J. Clim.*, 3, 1102–1110.

Miyama, T., J. P. McCreary, D. Sengupta, and R. Senan, (2006), Dynamics of biweekly oscillations in the equatorial Indian Ocean. *J. Phys. Oceanogr.*, 36(5), 827–846. doi:10.1175/JPO2897.1.

Moore, D. W., and J. P. McCreary, (1990), Excitation of intermediate-frequency equatorial waves at a western ocean boundary: With application to observations from the Indian Ocean, *J. Geophys. Res.*, 95, 5219–5231.

Moore, D. W., and S. G. H. Philander, (1977), Modeling of the tropical ocean circulation, *The Sea*, 6, 319–361. New York, John Wiley and Sons.

Murtugudde, R., S. Signorini, J. Christian, A. Busalacchi, C. McClain, and J. Picaut, (1999), Ocean color variability of the tropical Indo-Pacific basin observed by SeaWiFS during 1997–98. *J. Geophys. Res.*, 104, 18351–18366.

Nagura, M., and M. J. McPhaden, (2012), The dynamics of wind-driven intraseasonal variability in the equatorial Indian Ocean. *J. Geophys. Res.*, 115, C07009, doi:10.1029/2011JC007405.

Nidheesh, A. G., M. Lengaigne, J. Vialard, A. S. Unnikrishnan, and H. Dayan, (2013), Decadal and long-term sea level variability in the tropical Indo-Pacific Ocean, *Clim. Dyn.*, 41, 381–402, doi:10.1007/s00382-012-1463-4.

Nystuen, J. A., and C. A. Andrade, (1993), Tracking mesoscale ocean features in the Caribbean sea using GEOSAT altimetry. *J. Geophys. Res.*, 98, C5, 8389–8394.

Ogata, T., and Y. Masumoto, (2010), Interactions between mesoscale eddy variability and Indian Ocean dipole events in the south-eastern tropical Indian Ocean—Case studies for 1994 and 1997/ 1998, *Ocean Dyn.*, 60, 717–730, doi:10.1007/s10236-010-0304-4.

Ogata, T., and Y. Masumoto, (2011), Interannual modulation and its dynamics of the mesoscale eddy variability in the southeastern tropical Indian Ocean. *J. Geophys. Res.*, 116, C05005, doi:10.1029/2010JC006490.

Palastanga, V., P. J. van Leeuwen, M. W. Schouten, and W. P. M. de Ruijter, (2007), Flow structure and variability in the subtropical Indian Ocean: Instability of the South Indian Ocean Countercurrent, *J. Geophys. Res.*, 112, C01001, doi:10.1029/2005JC003395.

Pauluhn, A., and Y. Chao, (1999), Tracking Eddies in the Subtropical North-Western Atlantic Ocean, *Phys. Chem. Earth (A)*, 24(4), 415–421.

Perez, R. C., R. Lumpkin, W. E. Johns, G. R. Foltz, and V. Hormann, (2012), Interannual variations of Atlantic tropical instability waves, *J. Geophys. Res.*, 117, C03011, doi:10.1029/2011JC007584.

Périgaud, C., (1990), Sea level oscillations observed with GEOSAT along the 2 shear fronts of the Pacific North Equatorial Countercurrent. *J. Gophys. Res.*, 95, C5, 7239–7248.

Périgaud, C., and P. Delecluse, (1992), Annual sea level variations in the southern tropical Indian Ocean from GEOSAT and shallow water simulations. *J. Geophys. Res.*, 97, 20169–20178.

Philander, S. G. H. (1976), Instabilities of zonal equatorial currents, J. Geophys. Res., 81, 3725–3735.

Philander, S. G. H. (1978), Instabilities of zonal equatorial currents, 2. J. Geophys. Res., 83, 3679–3682.

Picaut, J., A. J. Busalacchi, M. J. McPhaden, B. Camusat, (1990), Validation of the geostrophic method for estimating zonal currents at the equator from GEOSAT altimeter data. *J. Gophys. Res.*, 95, C3, 3015–3024.

Picaut, J., A. J. Busalacchi, M. J. McPhaden, M. McPhaden, L. Gourdeau, F. I. Gonzales, and E. C. Hackert, (1995), Open-ocean validation of TOPEX/Poseidon sea level in the western equatorial Pacific. *J. Geophys. Res.*, 100, 25109–25127. doi:10.1029/95JC02128.

Picaut, J. and A.J. Busalacchi (2000), Tropical ocean variability, Chpt. 4, 217-236 in "Satellite Altimetry and Earth Science: a Handbook for Techniques and Applications", L.-L. Fu and A. Cazanave (Eds), Academic Press, San Diego, California, USA. 463 pp.

Picaut, J., E. Hackert, A. J. Busalacchi, et al., (2002), Mechanisms of the 1997–1998 El Niño La Niña, as inferred from space-based observations. *J. Geophys. Res. – Oceans.*107, 5-1-5-18, doi:10.1029/2001JC000850.

Polo, I., A. Lazar, B. Rodriguez-Fronseca, S. Arnault, and G. Mainsaint, (2008a), Ocean Kelvin waves and tropical Atlantic intraseasonal variability: Part I: Kelvin wave characterization, *J. Geophys. Res*, 113(7), C07009.

Polo, I., B. Rodríguez-Fonseca, T. Losada, and J. García-Serrano, (2008b), Tropical Atlantic variability modes (1979–2002). Part I: time-evolving SST modes related to West African rainfall, *J. Clim.*, 21, 6457–6475, doi:10.1175/2008JCLI2607.1.

Potemra, J., (2005), Indonesian throughflow transport variability estimated from satellite altimetry, *Oceanography*, 18(4), 98–107, doi:10.5670/oceanog.2005.10.

Potemra, J. T., R. Lukas, and G. T. Michum, (1997), Large-scale estimation of transport from the Pacific to the Indian Ocean, *J. Geophys. Res.*, 102, 27795–27812, doi:10.1029/97JC01719.

Provost, C., S. Arnault, L. Bunge, N. Chouaib, and E. Sultan, (2006), Interannual variability of the zonal sea surface slope in the tropical Atlantic ocean in the 1990s, *Adv. Space Res.*, 4, 823–831.

Provost, C., S. Arnault, N. Chouaib, A. Kartavsteff, and L. Bunge, (2004), Equatorial pressure gradient in the Atlantic in 2002: TOPEX/Poséïdon and Jason versus the first PIRATA current measurements, *Mar. Geod.*, 27, n°1–2, 31–47.

Pujiana, K., A. L. Gordon, and J. Sprintall, (2013), Intraseasonal Kelvin Wave in Makassar Strait, *J. Geophys. Res.*, 118, 2023–2034 doi:10.1002/jgrc.20069.

Pujol, M.-I., Faugère, Y., Taburet, G., Dupuy, S., Pelloquin, C., Ablain, M., and Picot, N., (2016), DUACS DT2014: The new multi-mission altimeter data set reprocessed over 20 years, *Ocean Sci.*, 12, 1067–1090, doi:10.5194/os-12-1067-2016.

Qiao, L., and R. Weisberg, (1995), Tropical instability wave kinematics: Observations from the Tropical Instability Wave Experiment. *J. Geophys. Res.,* 100, 8677–8693.

Qiu, B., M. Mao, and Y. Kashino, (1999), Intraseasonal variability in the Indo-Pacific throughflow and regions surrounding the Indonesian seas, *J. Phys. Oceanogr.*, 29, 1599–1618.

Qu, T., J. Gan, A. Ishida, Y. Kashino and T. Tozuka, (2008), Semiannual variation in the western tropical Pacific Ocean. *Geophys. Res. Lett.*, L16602, 35, doi:10.1029/2008GL035058.

Qu, T., and Y. T. Song, (2009), Mindoro Strait and Sibutu Passage transports estimated from satellite data, *Geophys. Res. Lett.*, 36, L09601, doi:10.1029/2009GL037314.

Rao, S. A., and S. K. Behera, (2005), Subsurface influence on SST in the tropical Indian Ocean: Structure and interannual variability. *Dyn. Atmos. Oceans*, 39, 103–139.

Rao, S. A., S. K. Behera, Y. Masumoto, and T. Yamagata, (2002), Interannual variability in the subsurface tropical Indian Ocean, *Deep-Sea Res. II*, 49, 1549–1572.

Rao, S. A., and T. Yamagata, (2004), Abrupt termination of Indian Ocean dipole events in response to intraseasonal oscillations. *Geophys. Res. Lett.*, 31, L19306, doi:10.1029/2004GL020842.

Rebert, J. P., J. R. Donguy, G. Eldin, and K. Wyrtki, (1985), Relations between sea level, thermocline depth, heat content, and dynamic height in the tropical Pacific Ocean. *J. Geophys. Res.*, 90, 11719–11725.

Reppin, J., F. A. Schott, J. Fischer, and D. Quadfasel, (1999), Equatorial currents and transports in the upper central Indian Ocean: Annual cycle and interannual variability, *J. Geophys. Res.*, 104, 15495–15514.

Reverdin, G., and J. Luyten, (1986), Near-surface meanders in the equatorial Indian Ocean. *J. Phys. Oceanogr.*, 17, 903–927.

Richter, I., S. K. Behera, Y. Masumoto, B. Taguchi, N. Komori, and T. Yamagata, (2010), On the triggering of Benguela Niños: Remote equatorial versus local influences, *Geophys. Res. Lett.*, 37, L20604, doi:10.1029/2010GL044461.

Richter, K., and B. Marzeion, (2014), Earliest local emergence of forced dynamic and steric sea-level trends in climate models. *Environ. Res. Lett.*, 9, 114009, doi:10.1088/1748-9326/9/11/114009.

Rouault, M., S. Illig, C. Bartholomae, C. J. C. Reason, and A. Bentamy, (2007), Propagation and origin of warm anomalies in the Angola Benguela upwelling system in 2001, *J. Mar. Syst.*, 68, 473–488, doi:10.1016/j.jmarsys.2006.11.010.

Roundy, P., and G. Kiladis, (2006), Observed relationships between oceanic Kelvin waves and atmospheric forcing. *J. Clim.*, 19, 5253–5272.

Ruiz-Barradas, A., J. A. Carton, and S. Nigam, (2000), Structure of interannual-to-decadal climate variability in the tropical Atlantic sector. *J. Clim.*, 13, 3285–3297.

Saji, N. H., B. N. Goswami, P. N. Vinayachandran, and T. Yamagata, (1999), A dipole mode in the tropical Indian Ocean. *Nature*, 401, 360–363.

Servain, J., (1991), Simple climatic indices for the tropical Atlantic Ocean and some applications, *J. Geophys. Res.*, 96, C8, 15137–15146, doi:10.1029/91JC01046.

Scharffenberg, M. G., and D. Stammer, (2010), Seasonal variations of the large-scale geostrophic flow field and eddy kinetic energy inferred from the TOPEX/Poseidon and Jason-1 tandem mission data. *J. Geophys. Res.*, 115, C02008, doi:10.1029/2008JC005242.

Schiller, A., S. E. Wijffels, J. Sprintall, R. Molcard, and P. R. Oke, (2010), Pathways of intraseasonal variability in the Indonesian Throughflow region. *Dyn. Atmos. Oceans*, 50, 174–200, doi:10.1016/j.dynatmoce.2010.02.003.

Schlundt, M., P. Brandt, M. Dengler, R. Hummels, T. Fischer, K. Bumke, G. Krahmann, and J. Karstensen, (2014), Mixed layer heat and salinity budgets during the onset of the 2011 Atlantic cold tongue, *J. Geophys. Res.*, 119, 7882–7910. doi:10.1002/2014JC010021.

Schneider, N., A. J. Miller, M. A. Alexander, and C. Deser, (1999), Subduction of decadal North Pacific temperature anomalies: Observations and dynamics. *J. Phys. Oceanogr.*, 29, 1056–1070.

Schott, F., J. Reppin, J. Fischer, and D. Quadfasel, (1994), Currents and transports of the monsoon current south of Sri Lanka. *J. Geophys. Res.*, 99(C12), 25127–25141.

Schott, F., and J.P. McCreary (2001), The monsoon circulation of the Indian Ocean. Prog. Oceanogr., 51, 1–123.

Schott, F. A., W. Wang, D. Stammer, (2007), Variability of Pacific subtropical cells in the 50-year ECCO assimilation. *Geophys. Res. Lett.*, 34, L05604, doi:10.1029/2006GL028478.

Schott, F. A., S. P. Xie, and J. P. McCreary, (2009), Indian Ocean circulation and climate variability. *Rev. of Geophys.*, 47, RG1002, doi:10.1029/2007RG000245.

Schouten, M. W., W. P. M. de Ruijter, P. J. van Leeuwen, and H. A. Dijkstra, (2002), An oceanic teleconnection between the equatorial and southern Indian Ocean. *Geophys. Res. Lett.*, 29, 1812, doi:10.1029/2001GL014542.

Schouten, M. W., R. P. Matano, and T. P. Strub, (2005), A description of the seasonal cycle of the equatorial Atlantic from altimeter data. *Deep-Sea Res. I*, 52, 477–493.

Senan, R., D. Sengupta, and B. N. Goswami, (2003), Intraseasonal "monsoon jets" in the equatorial Indian Ocean, *Geophys. Res. Lett.*, 30(14), 1750, doi:10.1029/ 2003GL017583.

Shankar, D., S. G. Aparna, J. P. McCreary, I. Suresh, S. Neetu, F. Durand, S. S. C. Shenoi, M. A. Al Saafani, (2010), Minima of interannual sea-level variability in the Indian Ocean, *Prog. In Oceanogr.*, 84, 225–241.

Shankar, D., and S. R. Shetye, (1999), Are interdecadal sea level 1120 changes along the Indian coast influenced by variability of monsoon rainfall? *J. Geophys. Res.*, 104, 26031–26042.

Shannon, L. V., A. J. Boyd, G. B. Bundrit, and J. Taunton-Clark, (1986), On the existence of an El Niño–type phenomenon in the Benguela system, *J. Mar. Sci.*, 44, 495–520.

Shinoda, T., G. N. Kiladis, and P. E. Roundy, 2009: Statistical representation of equatorial waves and tropical instability waves in the Pacific Ocean. Atmos. Res., 94, 37–44.

Shinoda, T. J., M. Flatau, S. Chen, W. Han, and C. Wang, (2013), Large-scale oceanic variability during the CINDY/DYNAMO field campaign from satellite observations. *Remote Sensing*, 5, 2072–2092.

Shinoda, T., W. Han, T. Jensen, L. Zamudio, E. J. Metzger, and R.-C. Lien, (2016), Impact of the Madden-Julian Oscillation on the Indonesian Throughflow in Makassar Strait during the CINDY/DYNAMO field campaign. *J. Clim.* (DYNAMO/CINDY/AMIE/LASP special collection), 29, 6085–6108. doi:10.1175/JCLI-D-15-0711.1

Singh, A., and T. Delcroix, (2013), Eastern and central Pacific ENSO and their relationships to the recharge/discharge oscillator paradigm. *Deep Sea Res.*, 82, 32–43.

Singh, A., T. Delcroix, and S. Cravatte, (2011), Contrasting the flavors of El Niño Southern Oscillation using sea surface salinity observations. *J. Geophys. Res.*, 116, C06016, doi:10.1029/2010JC006862.

Solomon, A., and M. Newman, (2012), Reconciling disparate twentieth-century Indo-Pacific ocean temperature trends in the instrumental record. *Nat. Clim. Change*, 2, 691–699.

Song, Y. T., (2006), Estimation of interbasin transport using ocean bottom pressure: Theory and model for Asian marginal seas, *J. Geophys. Res.*, 111, C11S19, doi:10.1029/2005JC003189.

Sprintall, J., A. L. Gordon, T. Lee, J. T. Potemra, K. Pujiana, and S. E. Wijffels, (2014), The Indonesian seas and their role in the couple ocean climate system, *Nat. Geosci.*, 7, 487–492, doi:10.1038/ngeo2188.

Sprintall, J., A. L. Gordon, R. Murtugudde, and R. D. Susanto, (2000), An Indian Ocean Kelvin wave observed in the Indonesian seas in May 1997. *J. Geophys. Res.*, 105, 17217–17230.

Sprintall, J., S. E. Wijffels, R. Molcard, and I. Jaya, (2009), Direct estimates of the Indonesian throughflow entering the Indian Ocean: 2004–2006, *J. Geophys. Res.*, 114, C07001, doi:10.1029/2008JC005257.

Stammer, D., (1997), Steric and wind-induced changes in TOPEX/POSEIDON large-scale sea surface topography observations. *J. Geophys. Res.,* 102, No. C9, 20987–21009.

Stammer, D., A. Cazenave, R. M. Ponte, and M. E. Tamisiea, (2013), Causes for contemporary regional sea level changes, *Annu. Rev. Mar. Sci.*, 5, 21–46, doi:10.1146/annurev-marine-121211-172406.

Stramma, L., S. Huttl, and J. Schafstall, (2005), Water masses and currents in the upper tropical northeast Atlantic off northwest Africa, *J. Geophys. Res.*, 110, C12006, doi:10.1029/2005JC002939.

Srinivasu, U., Ravichandran, M., Han, W. et al. Clim Dyn (2017), Causes for decadal reversal of North Indian Ocean sea level in recent two decades, *Clim. Dyn.*, in revision. doi:10.1007/s00382-017-3551-y. https://link.springer.com/article/10.1007/s00382-017-3551-y

Suresh, I., J. Vialard, M. Lengaigne, W. Han, J. McCreary, F. Durand, P. M. Muraleedharan, (2013), Origins of wind-driven intraseasonal sea level variations in the North Indian Ocean coastal waveguide. *Geophys. Res. Lett.*, 40, 5740–5744, doi:10.1002/2013GL058312.

Susanto, R. D., and Y. T. Song, (2015), Indonesian throughflow proxy from satellite altimeters and gravimeters, *J. Geophys. Res. Oceans*, 120, 2844–2855, doi:10.1002/2014JC010382.

Tai, C. K., (1988), GEOSAT crossover analysis in the tropical pacific 1: Constrained sinusoidal crossover adjustment. *J. Geophys. Res.*, 93, 10621–10629.

Timmermann, A., S. McGregor, and F. F. Jin, (2010), Wind effects on past and future regional sea level trends in the southern Indo-Pacific, *J. Clim.*, 23, 4429–4437.

Trenary, L., and W. Han, (2012), Intraseasonal-to-interannual variability of South Indian Ocean sea level and thermocline: Remote versus local forcing. *J. Phys. Oceanogr.*, 42, 602–627.

Trenberth, K. E., and T. J. Hoar, (1996), The 1990-1995 El Niño-Southern Oscillation event: Longest on record. *Geophys. Res. Lett.*, 23(1), 57–60.

Tsai, P. T. H., J. J. O'Brien, and M. E. Luther, (1992), The 26-day oscillation observed in the satellite sea surface temperature measurements in the equatorial western Indian Ocean. *J. Geophys. Res.*, 97, 9605–9618.

Urbano, D. F., R. A. F. de Almeida, and P. Nobre, (2008), Equatorial undercurrent and North Equatorial Countercurrent at 38°W: A new perspective from direct velocity data, *J. Geophys. Res.*, 113, c04041, doi:10.1029/2007jc004215.

Vecchi, G.A., and B. J. Soden, (2007), Global warming and the weakening of the tropical circulation. *J. Clim.*, 20, 4316–4340. doi:10.1175/JCLI4258.1.

Verstraete, J. M., and Y. H. Park, (1995), Comparison of Topex Poseidon altimetry and in situ sea level data at Sao Tome island, Gulf of Guinea, *J. Geophys. Res.*, 100(12), 25129–25134, doi:10.1029/95JC01960.

Vinayachandran, P. N., Y. Masamoto, T. Mikawa, and T. Yamagata, (1999), Intrusion of the Southwest Monsoon Current into the Bay of Bengal. *J. Geophys. Res.*, 104, 11077–11085.

Von Schuckmann, K., P. Brandt, and C. Eden, (2008), Generation of tropical instability waves in the Atlantic Ocean, *J. Geophys. Res.*, 113, C08034, doi:10.1029/2007JC004712.

Von Schuckmann, K., J. B. Sallée, D. Chambers, et al., (2014), Consistency of the global ocean observing systems from an Argo perspective. *Ocean Sci.*, 10, 547–557. doi:10.5194/os-10-547-2014.

Wakata, Y., and S. Kitaya, (2002), Annual variability of sea surface height and upper layer thickness in the Pacific Ocean. *J. Oceanogr.*, 58, 439–450.

Wakata, Y., 2007: Frequency-wavenumber spectra of equatorial waves detected from satellite altimeter data. J. Oceanogr., 63, 483–490.
Wang, T. Y., Y. Dy, W. Zhuang, and J. B. Wang, (2015), Connection of sea level variability between the tropical western Pacific and the southern Indian Ocean during recent two decades. *Science China-Earth Sci.*, 58(8), 1387–1396. doi:10.1007/s11430-014-5048-4.
Wang, L.-P., C. J. Koblinksky, and S. Howden, (2001), Annual Rossby waves in the Southern Indian Ocean: Why does it appear to break down in the middle ocean? *J. Phys. Oceanogr.*, 31, 54–74.
Wang, B., R. Wu, and R. Lukas, (2000), Annual adjustment of the thermocline in the tropical Pacific Ocean. *J. Clim.*, 13, 596–616.
Webber, B. G., A. J. Matthews, K. J. Heywood, and D. P. Stevens, (2012), Ocean Rossby waves as a triggering mechanism for primary Madden–Julian events. *Q. J. R. Meteorol. Soc.*, 138, 514–527.
Webster, P. J., A. M. Moore, J. P. Loschnigg, and R. R. Leben, (1999), Coupled oceanic–atmospheric dynamics in the Indian Ocean during 1997–98. *Nature*, 401, 356–360.
Whitehead, J. A., (1989), Internal hydraulic control in rotating fluids: Applications to oceans, *Geophys. Astrophys. Fluid Dyn.*, 48, 169–192, doi:10.1080/03091928908219532.
Woodbery, K. E., M. E. Luther, and J. J. O'Brien, (1989), The wind-driven seasonal circulation in the southern tropical Indian Ocean. *J. Geophys. Res.*, 94, 17985–18002.
Wyrtki, K., (1987), Indonesian throughflow and the associated pressure gradient, *J. Geophys. Res.*, 92, 12941–12946, doi:10.1029/JC092iC12p12941.
Wyrtki, K., and G. Mitchum, (1990), Interannual differences of GEOSAT altimeter heights and sea level. *The importance of a datum. J. Geophys. Res.*, 95, C3, 2969–2975.
Xie, S.-P., H. Annamalai, F. A. Schott, and J. P. McCreary, (2002), Structure and mechanisms of south Indian Ocean climate variability. *J. Clim.*, 15, 864–878.
Yang, J., and T. M. Joyce, (2005), Local and Equatorial forcing of seasonal variations of the North equatorial Countercurrent in the Atlantic Ocean., *J. Phys. Oceanogr.*, 36, 238–254.
Yoneyama, K., C. Zhang, C. N. Long, (2013), Tracking pulses of the Madden-Julian oscillation. *Bull. Amer. Meteorol. Soc.*, 94, 1871–1891. doi:10.1175/BAMS-D-12-00157.1.
Yu, X., and M. J. McPhaden, (1999), Seasonal variability in the equatorial Pacific. *J. Phys. Oceanogr.*, 29, 925–947.
Yu, Z., and J. Potemra, (2006), Generation mechanism for the intraseasonal variability in the Indo-Australian basin. *J. Geophys. Res.-Oceans*, 111, C01013, doi:10.1029/2005JC003023.
Yu, L., and M. M. Rienecker, (1999), Mechanisms for the Indian Ocean warming during the 1997–98 El Niño. *Geophys. Res. Lett.*, 26, 735–738, 10.1029/1999GL900072.
Zang, X., L.-L. Fu, and C. Wunsch, (2002), Observed reflectivity of the western boundary of the equatorial Pacific Ocean. *J. Geophys. Res.*, 107, 3150, doi:10.1029/2000JC000719.
Zhang, C., (2005), Madden-Julian Oscillation, *Rev. Geophys.*, 43, RG2003, doi:10.1029/2004RG000158.
Zhang, C., J. Gottschalck, E. D. Maloney, M. W. Moncrieff, F. Vitart, D. E. Waliser, B. Wang, and M. C. Wheeler, (2013), Cracking the MJO nut, *Geophys. Res. Lett.*, 40, 1223–1230, doi:1210.1002/grl.50244.
Zhang, D., and M. J. McPhaden, (2006), Decadal variability of the shallow Pacific meridional overturning circulation: Relation to tropical sea surface temperatures in observations and climate change models. *Ocean Model.*, 15, 250–273.
Zhou, L., and R. Murtugudde, (2010), Influences of Madden-Julian oscillations on the eastern Indian Ocean and the maritime continent. *Dyn. Atmos. Oceans*, 50, 257–274.
Zhuang, W., M. Feng, Y. Du, A. Schiller, and D. Wang, (2013), Low-frequency sea level variability in the southern Indian Ocean and its impacts on the oceanic meridional transports, *J. Geophys. Res.*, 118, 1302–1315, doi:10.1002/jgrc.20129.

8 The High Latitude Seas and Arctic Ocean

Johnny A. Johannessen and Ole B. Andersen

8.1 INTRODUCTION

Changes in the dynamic topography and ocean circulation between the North Atlantic and the Arctic Ocean result from variations in the atmospheric forcing field and convective overturning combined with changes in freshwater runoff and their pathways, mean sea level, sea ice extent and deformation, and water mass transformation. The ocean circulation in this region has been subject to investigations since Helland-Hansen and Nansen (1909). In general, it can be characterized by four regional circulation regimes and cross-regional exchanges and volume transports, namely the Northeast Atlantic, the Labrador Sea and Canadian archipelago, the Nordic and Barents Seas, and the Arctic Ocean, as illustrated in Figure 8.1.

Accurate knowledge of the ocean transport variability together with understanding of the water mass transformations within and across these regions is greatly needed to quantify changes in the overturning circulation with acceptable uncertainty. The Atlantic meridional overturning circulation is, among other factors, influenced by: variations in the upper ocean and sea ice interaction; ice sheet mass changes and their effect on the regional sea-level change; changes in freshwater fluxes and pathways; and variability in the large-scale atmospheric pressure field. For instance, changes in the pathways of the freshwater from the Eurasian runoff forced by shifts in the Arctic Oscillation (AO) can lead to increased trapping of freshwater in the Arctic Ocean, as presented by Morison et al. (2012), which, in turn, may alter the thermohaline circulation in the sub-Arctic Seas.

The Arctic Ocean is the smallest and shallowest of the world's five major oceans. It covers roughly 4.2% of the world's ocean and, as the average depth of the Arctic Ocean is just 1300 m, it holds only about 1.2% of the total volume of ocean water. The International Hydrographic Organization (IHO) recognizes the Arctic Ocean as an ocean, although some oceanographers call it the Arctic Mediterranean with limited pathways and water exchange with the larger sub-Arctic oceans. The four major pathways include: the Fram Strait between Greenland and Svalbard; the Bearing Strait between Russia and the United States; the Canadian Archipelago region and Nares Strait; and the northeastern Barents Sea.

Besides these open boundaries, the Arctic Ocean has significant input of freshwater from rivers and melting ice caps. Eight of the nine largest rivers contributing freshwater to the Arctic Ocean are located in the Russian sector with the Siberian rivers Yenisei, Ob, and Lena each providing up to 600 km^3 of water per year, whereas the Canadian-sector Mackenzie River provides of the order of 340 km^3 per year (Aagaard and Carmack 1989). In addition, the melting of the Greenland Ice Sheet and glaciers adds between 200 and 400 km^3 freshwater per year (Khan et al. 2015).

The Arctic Ocean currently stores around 84,000 km^3 of freshwater in its surface layer (Serreze et al. 2006), and this number seems to be increasing. Measurements from ships and moorings have shown that the deep Arctic basins, in particular the Canada Basin, accumulated up to 10,000 km^3 of freshwater during the 1990s and 2000s (Proshutinsky et al. 2009; Krishfield et al. 2014; Rabe et al. 2014). This was confirmed from satellite altimetry by Giles et al. (2012) who

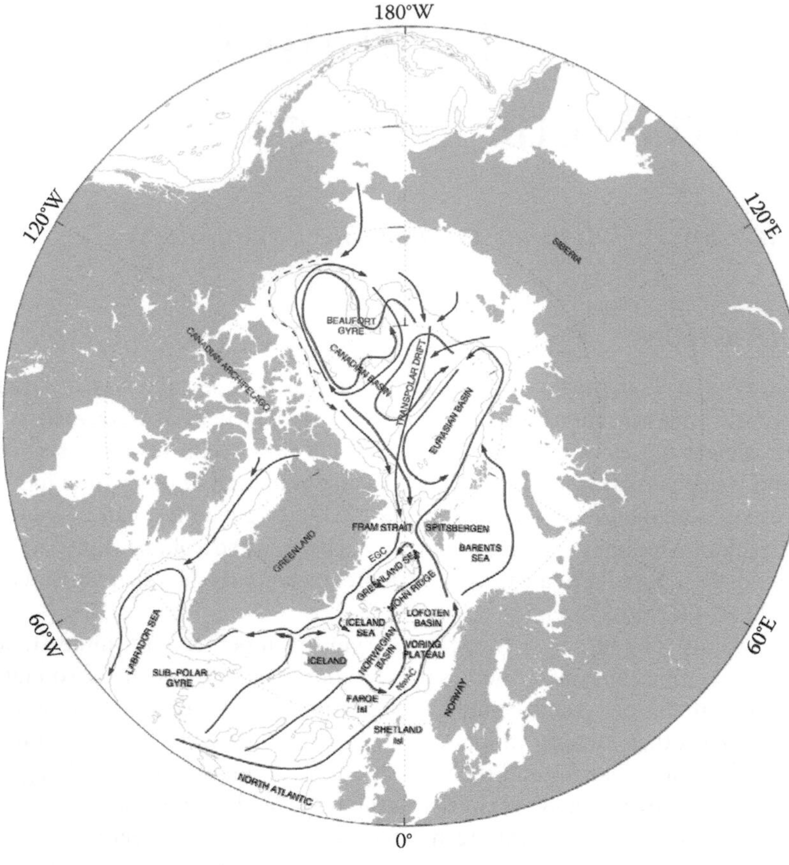

FIGURE 8.1 General circulation in the upper ocean of the Arctic Ocean, Nordic Seas, and North Atlantic from Furevik and Nilsen (2005). Red arrows represent the warmer Atlantic waters, which reside in the surface in the Nordic Seas and are submerged in the Arctic Ocean. Blue arrows represent polar water, residing in the surface. Bottom contours marked by 1000 and 3000 m are outlining the shelves and basins.

demonstrated that the Beaufort Gyre accumulated 800 ± 200 km^3 of freshwater per year during the 2000s. The freshwater storage in the Arctic Ocean is highly important to monitor as the enhancement of the freshwater outflow, mainly occurring through the Fram and Nares Straits, may be able to disrupt the North Atlantic Meridional Overturning Circulation (Manabe and Stouffer 1995). Such enhanced freshwater outflow has previously been linked to the North Atlantic "Great Salinity Anomalies" of the 1970s, 1980s, and 1990s (Dickson et al. 1988; Belkin 2004). The "Great Salinity Anomaly" of the 1970s consisted of only approximately 2000 km^3 excess liquid and solid freshwater exported to the North Atlantic (Häkkinen 1993). Hence, monitoring the balance and/or sea level budget of the Arctic Ocean freshwater fluxes is climatically important.

There is a growing concern regarding the Arctic region's response to climate change. There is evidence that the polar ocean is undergoing rapid climate change resulting from processes including a reduction of sea ice extent, thickness, and volume (e.g., Kwok and Rothrock 2009; Stroeve et al. 2012; IPCC 2013; Kern et al. 2014, Tilling et al. 2015). Sea level observations in the Arctic Ocean have traditionally been based on tide gauges. However, the coverage is inhomogeneous and in some regions sparse in time and space. A reasonable amount of tide gauge data (more than 100) has been available along the Norwegian and Russian coasts since 1950. Unfortunately, a substantial part of the Russian gauges were discontinued with the collapse of the Soviet Union in the early 1990s. A challenge with many of the Arctic tide gauges is also their location in the vicinity of river mouths and even inside rivers. For instance, the tide gauge Antipaiuta (69°N, 76°E) in the Permanent Service for Mean Sea Level (PSMSL) database (Holgate et al. 2013) is located inside the Ob River estuary nearly 900 km from the open Arctic Ocean. Consequently, a very careful editing of the tide gauges is required to isolate the sea level signals that are of oceanographic origin (Proshutinsky et al. 2015; Svendsen et al. 2016). Figure 8.2 shows the 66 tide gauges carefully edited by Henry et al. 2012.

8.1.1 Satellite Altimetry in the High Latitude and Arctic Ocean

Sea surface height (SSH) is an essential climate variable and global indicator (IPCC 2013; GCOS 2016), and satellite altimetry is a key component to retrieve this variable. Despite the fact that satellite altimetry is a mature and precise technique for global mean sea level (GMSL) monitoring, it suffers from increasing uncertainties in the Arctic Ocean region due to a number of factors such as:

- The inclination of the Centre National d'Etudes Spatiales/National Aeronautics and Space Administration (CNES/NASA) TOPEX/Poseidon and Jason satellites limits their coverage to 66°N, which means that these satellites do not cover the Arctic Ocean with regular routine sea level observations.
- The European Space Agency (ESA) satellites European Remote Sensing (ERS)-1, ERS-2, and Envisat provided continuous altimetry observations up to nearly 82°N from 1991 to 2012. Covering 78% of the Arctic Ocean, they have therefore been fundamental in deriving the long altimeter-based SSH record for the Arctic Ocean (Prandi et al. 2012).
- Standard processing of satellite radar altimetry is faced with severe difficulties in the presence of sea ice and as such they do not provide regular routine monitoring of the Arctic sea level. Sea ice affects the returned radar echo or waveform recorded by the satellite or even prevents it from reflecting off the sea surface. The waveform can become highly complex with multiple peaks resulting from scattering within the sea ice-covered footprint. The waveform can also be very specular if returned by water within leads. In any case, the waveform does not resemble a normal open ocean Brown waveform

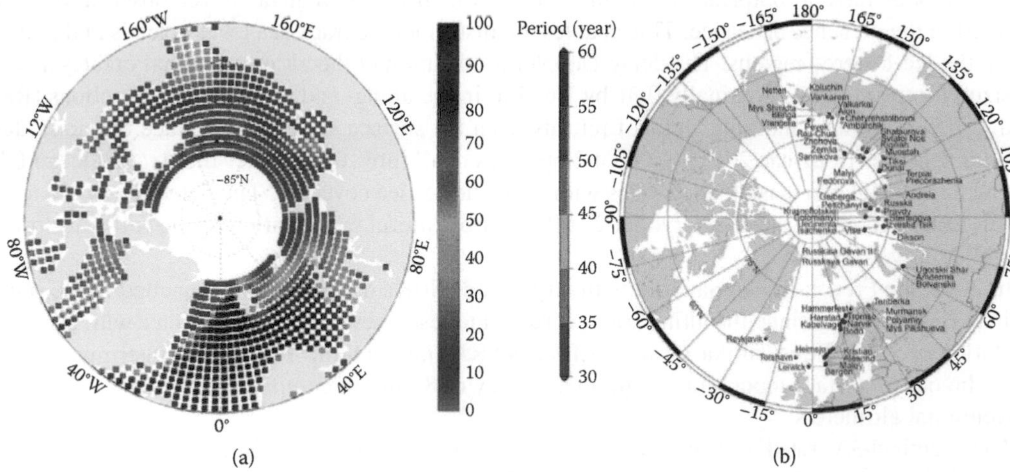

FIGURE 8.2 (a) Percentage of available data along the ERS-1/ERS-2/Envisat reference tracks in the radar altimeter database system (RADS) standard edited sea level anomaly (SLA) data set during the 1993 to 2010 period (From Cheng et al. 2015). (b) Locations of the 66 tide gauge stations selected for the sea level study by Henry et al. (2012).

(Brown 1977) for conventional altimeters and is subsequently often discarded by the ESA processing schemes. Moreover, the sea ice contaminates the radiometer observations and hence the ability to provide accurate range corrections. Finally, several range corrections are missing and/or are less accurate in the Arctic Ocean. For instance, the tides are less accurate in the Arctic Ocean than elsewhere (Stammer et al. 2015; Cheng and Andersen, 2011).

- In the northern part of the Nordic Seas and in the Barents Sea, nearly 100% of conventional altimetric data are available for the 1991 to 2010 period. However, this number rapidly decreases when moving into the interior of the Arctic Ocean, where an average of only between 5% and 10% of data are available (Figure 8.2) during the 1993 to 2010 period.

In order to increase the coverage and quality of SSH observations in ice-affected regions, several new altimeter satellite missions have been launched and operated since 2003, such as ICESat (Ice, Cloud, and Land Elevation Satellite), CryoSat-2, Sentinel-3A, and SARAL/AltiKa. ICESat provided 17 monthly epochs of laser altimetry up to 86°N between 2003 and 2009 and will likely be continued by ICESat-2 in 2018. The higher resolution (footprint: 50–70 m) and precision of the lidar (shot-to-shot repeatability of approximately 2–3 cm) onboard ICESat (Zwally et al. 2002) allowed unambiguous identification of open water between ice floes and therefore importantly supplied a new source of SSH observations in the Arctic Ocean.

CryoSat-2 (Wingham et al. 2006), a new generation Earth Explorer mission launched by the ESA, acquires the SSH by the single frequency (Ku) Synthetic Aperture Radar (SAR) Interferometric Radar Atimeter (SIRAL). With an inclination of 92°, the altimeter covers 95% of the Arctic Ocean while operating in three different modes. These are conventional or low resolution mode (LRM); the SAR mode, in which a delay-Doppler modulation partitions the radar footprint into a number of along-track slices; and finally, the interferometric Synthetic Aperture Radar Mode (SAR-in) in which two receiving antenna chains onboard CryoSat-2 are enabling the detection of the cross-track angle to the prime scatter in the footprint. This SAR-in mode has proven to be particularly useful in the Arctic Ocean (Laxon et al. 2013) as well as for coastal regions (i.e., Abulaitijiang and Andersen 2015) where the satellite can detect coastal sea level even when the satellite is flying over land close to the coast.

Whereas conventional or LRM observations have a footprint of roughly 100 km^2, the SAR and SAR-in modes have a footprint of roughly 4 km^2, which means that far fewer observations are affected in the presence of sea ice. This enables the instrument to track sea level in leads and polynyas in the ice-covered regions. The delay-Doppler modulation of the altimeter signal creates a synthesized footprint that is nominally 0.31 by 1.67 km in the along- and across-track directions (Ray et al. 2015). Similarly, multiple looks of returns from the surface are used to reduce the noise due to radar speckle (Bouzinac 2013). CryoSat-2 operates in all three modes over the Arctic Ocean following a sophisticated mask that changes with time and sea ice coverage. The detection of sea level within leads and polynyas from CryoSat-2 still requires careful examination of the full waveforms (so-called Level-1B data).

In 2013, the French-Indian Satellite with Argos and AltiKa (SARAL) was launched in the same orbit as Envisat to continue the altimetric record and to test a new Ka-band altimeter with a spatial resolution of 2 km (https://altika-saral.cnes.fr/en/SARAL/altika.htm). The AltiKa instrument measures the ocean surface topography with an accuracy of 8 mm, in comparison to about 2.5 cm for conventional altimeters.

The Sentinel-3A satellite that was successfully launched in 2016 carries a dual-frequency (Ku- and C-band) SRAL instrument. SRAL is similar to the SIRAL onboard CryoSat-2 but does not provide the interferometric capabilities as for CryoSat-2. On the other hand, it has a microwave radiometer for atmospheric corrections. This satellite is the first-ever satellite to operate in high-resolution SAR mode everywhere including the Arctic Ocean up to 81°N. Altogether the continuity of altimetry satellites in principle provides a data record for the Arctic Ocean that by 2016 reached 25 years.

Returning to the conventional altimeter satellites ERS-1/ERS-2 and Envisat, there are basically two approaches to increase the number of reliable SSH observations in the Arctic Ocean. One is the use of a more robust radar waveform retracker tailored to the conditions in the Arctic Ocean; the other is a reevaluation of existing data sets consisting of fine-tuning/tailoring the editing and reprocessing of the data within the Arctic Ocean. The drawback of using robust retrackers (Gommenginger et al. 2011) is that they do not provide estimates of significant wave height.

Retracking of the ERS-2 data was applied by Peacock and Laxon (2004) who developed a robust empirical retracker to extract the SSH for the 1995 to 2004 period. More recently, Armitage et al. (2016) retracked the Envisat data using the same method developed for ERS-2, and recently the combined reprocessed ERS-1 and ERS-2 REAPER (REprocessing of Altimeter Products for ERS) data set has been made available. Within the ESA climate change initiative (http://cci.esa.int/), the Sea Level - Climate change initiative (SL-CCI) has issued the study of a more sophisticated retracking system of Envisat (Ablain et al. 2015). Here, Collecte Location de Satellite (CLS), in corporation with Plymouth Marine Laboratories (PML), has developed a system in which the data are first classified by ocean surface type in order to separate frozen ocean areas from open water corresponding to leads and polynyas and subsequently using new retracking algorithms for each class. Cheng et al. (2015) created an Arctic SSH data set without retracking but by reprocessing the data with a combination of fine-tuning editing criteria for the Arctic Ocean and replacing range and geophysical corrections. Thereby, they retrieved between four and ten times as many sea level anomaly (SLA) observations in large part of the interior of the Arctic Ocean without degrading the quality of the data. In turn, they could derive a time series of Arctic Ocean SSH anomalies for 20 years. This has recently been updated to a 25-year time series called Technical University of Denmark (DTU)-SSH (Andersen and Piccioni 2016) taking into account five additional years of retracked CryoSat-2 SAR altimetry data. This data set was used in the study of Arctic Sea level changes recently published by Carret et al. (2016).

8.2 MAPPING THE SEA ICE THICKNESS IN THE ARCTIC OCEAN

The key approach to derive sea ice thickness estimation in the Arctic Ocean is by satellite altimetry. The caveat is that the uncertainty estimation is challenging and that the existence of *in situ* observations of snow depth, snow density, and sea ice thickness are sparse. The first comprehensive estimate of changes in sea ice thickness from altimetry was published by Laxon et al. (2003). Exploring radar altimeter measurements from ERS-1, ERS-2, and Envisat from the 1990s, they found a strong interannual variability in the observed sea ice freeboard height and, in turn, in the sea ice thickness. Moreover, they encountered circumpolar thinning of the Arctic sea ice. In comparison, Kwok et al. (2009) found a decline in Arctic sea ice thickness of 0.18 m/year between 2003 and 2008 based on analyses of laser altimeter measurements from NASA's ICESat. At the end of the ICESat period in 2008, a winter thickness of 1.89 m was reported, being 1.75 m lower than the mean sea ice thickness from the 1980s based on reported submarine data in the central Arctic (Kwok and Untersteiner 2011). In comparison, Hendricks et al. (2013) derived a mean sea ice thickness of 1.87 m in the central Arctic in the winter of 2012–2013 (October to March) using CryoSat-2 data.

The freeboard, the part of the ice above the water level, is obtained by using the elevation over leads as the instantaneous SSH and then calculating the difference between the SSH and ice floes (Zwally et al. 2002; Kwok et al. 2007; Hendricks et al. 2013). The elevation measurements from leads and ice floes are distinguished by the shape of the waveform based on the pulse peakiness structure (see Figure 8.3). Specular echoes occur when the radar burst is reflected from a smooth, mirror-like surface such as a lead or very thin ice. In these cases, the power in the range window rises and falls again very rapidly, creating an echo that looks like a spike (Figure 8.3a). Diffuse echoes occur when the radar burst is reflected from a rougher surface such as an ice floe. In these cases, the power in the range window rises rapidly but gently decays, creating an echo that looks like a step (Figure 8.3b). After retracking the range and applying necessary corrections (e.g. Doppler range, the

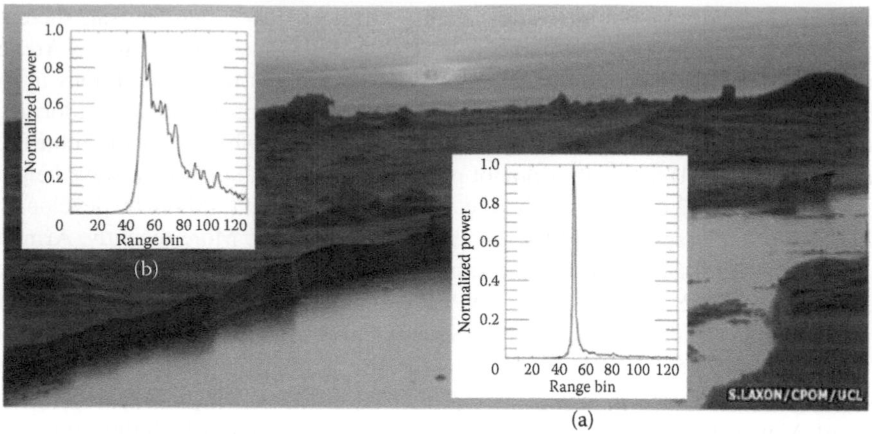

FIGURE 8.3 Typical radar altimeter waveforms representing leads (a) and sea ice (b). Courtesy S. Laxon, UCL.

ionospheric, the dry tropospheric and the modeled wet tropospheric, ocean tide, long-period tide, loading tide, earth tide, pole tide, and inverse barometer corrections), and filters (removal of complex waveforms, failed re-tracking, and echoes that yielded elevations more than 2 m from the mean dynamic SSH), the local sea level at ice floe locations is interpolated from nearby lead elevations. The freeboard is then calculated as the difference of radar altimetry measured ice floe elevation and the local sea level. As the freeboard measurement is known to be noisy, it is necessary to average several measurements. Although the radar altimeter signal is assumed to be reflected from the snow and ice interface (Beaven et al. 1995), thus providing the ice freeboard, the laser signal is reflected from the air-snow interface and hence provides the snow plus the ice freeboard. Assuming hydrostatic equilibrium, the freeboard can be converted into an estimate of sea ice thickness based on given knowledge of the sea ice density as well as the snow depth and snow density.

During the last three decades, the sea ice area in the Arctic Ocean has shown a distinct decline (Stroeve et al. 2014). The largest reductions are found for the month of September when the annual minimum sea ice area is reached. Declines in sea ice area and thickness also result in a reduction of sea ice volume. Based on data from the laser altimeter on board ICESat, Kwok et al. (2009) found a net loss of 5400 km^3 in October to November and 3500 km^3 in February to March during the ICESat period from 2003 to 2008. Recent results, exploring new data from CryoSat-2, report a further decline in Arctic sea ice volume (Laxon et al. 2013; Tilling et al. 2015). The average sea ice volume in October to November for 2010 and 2011 was estimated to be 7560 km^3 (i.e., 64% of the 2003–2008 mean value estimated from ICESat) (Kwok et al. 2009). However, all these findings are associated with large uncertainties. According to Tilling et al. (2015), the key contributor to the uncertainty in the sea ice thickness and volume estimates is associated with the snow load (depth and density) that invokes an error of around 15% to the total monthly estimated sea ice volume.

8.3 SEA LEVEL CHANGE

Most of the Arctic Ocean is covered with sea ice that varies in extent and thickness on seasonal to interannual timescales with a distinct decrease in ice coverage and sea ice thickness during the recent decades as mentioned in Section 8.2 and reported by Laxon et al. (2013), Kwok et al. (2009), and Tilling et al. (2015). Armitage et al. (2016) performed an empirical orthogonal function (EOF) analysis of monthly retracked Envisat data to inspect the dominant modes of seasonal and non-seasonal Artic SSH variability. They first computed the EOF on the full SSH time series and concluded that the leading two modes account for 62.6% of the variance. They next removed the mean seasonal cycle in the time series and repeated the EOF analysis on the non-seasonal SSH variability. The result is shown in Figure 8.4.

Armitage et al. (2016) concluded that the Arctic SSH variability is dominated by the seasonal cycle and that the seasonal EOF1 captures 38.7% of the total SSH variance. In comparison, the non-seasonal EOF1 analysis is dominated by secular changes capturing 33.5% of the variance. The second EOF2 is dominated by wind stress, being largest along the Siberian shelf seas and accounting for 21.9% of the non-seasonal SSH variance. Peralta-Ferriz et al. (2014), moreover,

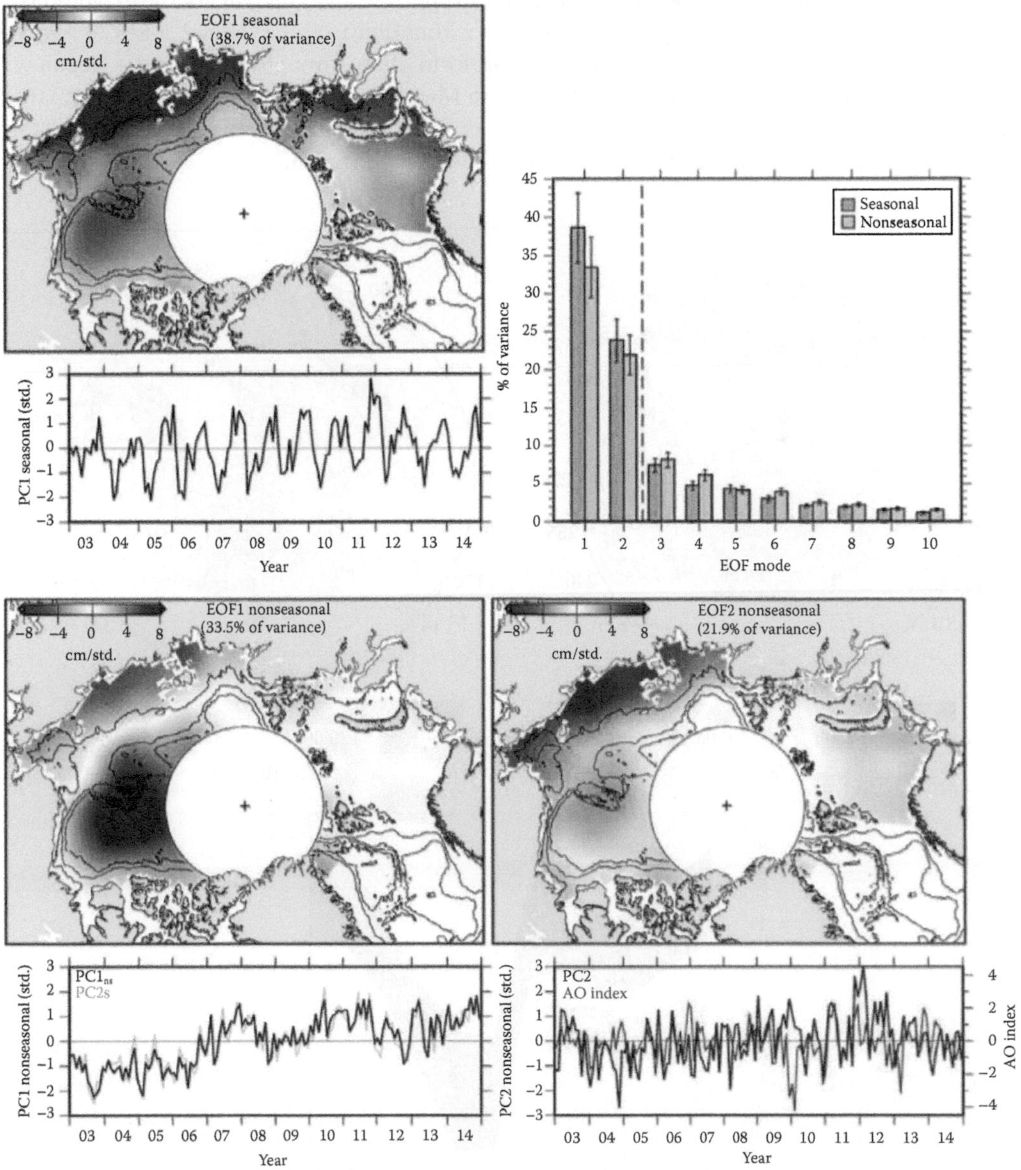

FIGURE 8.4 The dominant modes of seasonal and non-seasonal Arctic SSH variability from Envisat. The leading seasonal EOF is shown in the upper left figure. The upper right figure shows the variance explained by the first seasonal and first nonseasonal mode. The leading two EOFs of the non-seasonal SSH field are shown in the lower two figures. The AO is superimposed on the second non-seasonal EOF. Also, contours of bathymetry are shown in the pictures.

found that this mode of variability was significantly correlated with the AO index (www.cpc.noaa.gov). A positive AO index indicates low atmospheric pressure over the central Arctic Ocean, which is responsible for driving the eastward alongshore wind anomalies in the Siberian Arctic (Peralta-Ferriz et al. 2014). Variations in sea level pressure and winds are also largely responsible for sea ice drift in the Arctic.

8.3.1 The Seasonal Cycle

The seasonal cycle detected by Armitage et al. (2016) using EOF analyses can also be explored from the 20-year DTU-SSH data set augmented with the estimation from CryoSat-2 north of 82°N as shown in Figure 8.5. The seasonal cycle has a near-uniform phase throughout the Arctic Ocean with maximum in October to December and a minimum in May to June (Figure 8.5a and b). The largest amplitudes, reaching 10 cm, are found on the Siberian shelf. This amplitude clearly decreases to a few centimeters pole ward toward the central area of the Arctic Ocean.

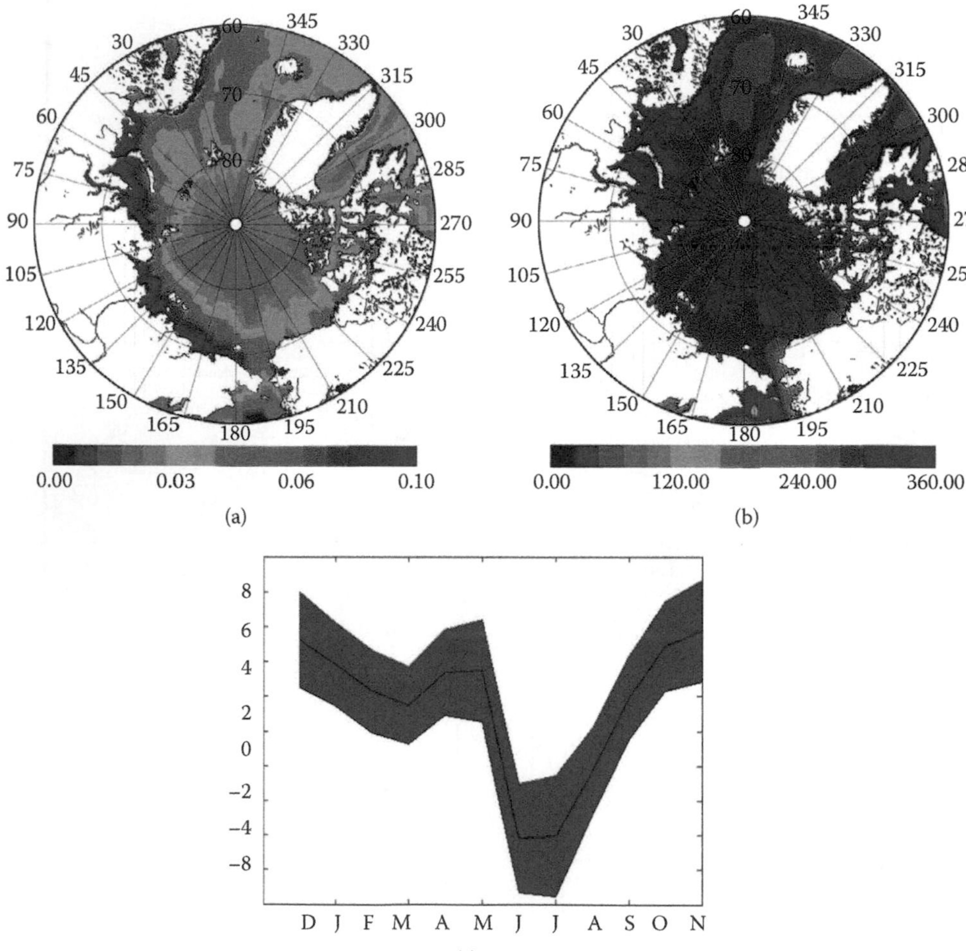

FIGURE 8.5 The seasonal SSH signal in the Arctic Ocean derived from satellite altimetry with the amplitude (a), phase (b), and basin average with the associated uncertainty estimate (c). The amplitude in (a) is given in meters and in centimeters in (c). The amplitude in (b) is given in degrees.

The basin-averaged mean Arctic SSH seasonal cycle shows a broad maximum of 4–5 cm between October and January and a minimum of −4 cm in June and July. A secondary maximum of nearly 4 cm is also seen in April and May. Smaller intermediate peaks noticed in ocean mass data have previously been linked to the annual cycle of river runoff (Peralta-Ferriz and Morison 2010).

The large annual signal in the Arctic Ocean provides a challenge for tidal modeling using satellite altimetry. With only sun-synchronous satellites being available in the Arctic, the problem is that the diurnal constituents K_1 and P_1 have alias periods of exactly 1 year (365 days) when observed by the altimeter satellites. As such, they are inseparable from the seasonal cycle. Similarly, the semidiurnal K_2 tidal constituent has an alias period of 183 days, which makes it inseparable from the semiannual signal (S_{sa}). Although this signal generally has less amplitude in the Arctic Ocean, it modifies the annual variation in e.g., freshwater fluxes in and out of the Arctic Ocean.

The phase of the SSH seasonal cycle from altimetry is fairly similar to that observed directly by tide gauges (Proshutinsky 2004; Richter et al. 2012). However, the amplitude is somewhat smaller. This is partly explained because the tide gauges measure SSH variations closer to the coast (compared to the altimeters) where the seasonal cycle is largest and where the persistent seasonal ice coverage will limit the satellite altimetry.

Another explanation for the differences in the observed amplitude is the fact that the tide gauges observe the real variations in the sea level, including the variations caused by the atmospheric pressure effect. Satellite altimetry, in contrast, is corrected for the atmospheric pressure effect via the inverse barometer effects. This inverse barometer correction (Wunsch and Stammer 1997)—also frequently called Dynamic Atmosphere Correction—corrects for the hydrostatic response to surface pressure variations. The altimeter SSH signals examined without applying the inverse barometer effect are therefore yielding a more direct comparison to tide gauges' data. This is shown in Figure 8.6 for the seasonal SSH amplitude (Figure 8.6a) and phase (Figure 8.6b). As expected, the structures reveal distinct differences from the patterns shown in Figure 8.5. The sea level pressure is, therefore, clearly contributing to the seasonal to interannual variations in SSH. This is further documented by the averaged Arctic Ocean sea level pressure changes derived over several decades and reported by Thorndyke (1982). In Figure 8.6c, significant variations in SSH are displayed throughout the Arctic Ocean. On average, the amplitude of the sea level pressure is 4 mb, which roughly corresponds to 4 cm in sea level.

Throughout the ocean, the amplitude of the seasonal cycle is increased and, again, the largest amplitudes are found along the Siberian shelves. The analysis also indicates that the sea level pressure shifts the peak of the annual signal in sea level slightly later in the season (toward November) in the northern Atlantic Ocean and slightly earlier in the interior of the Arctic Ocean. The latter shift inside the Arctic Ocean should be considered with caution as the non-inverted barometer data were obtained from the radar altimeter database system (RADS), which has very few data here.

8.3.2 Secular and Long-Term Sea Level Changes

Several studies have dealt with the linear sea level trend over the altimetry era. Prandi et al. (2012) reported 3.6 ± 1.3 mm/year trend from reprocessed altimetry and 2.2 ± 1.3 mm/year using standard Archiving, Validation and Interpretation of Satellite Oceanographic (AVISO) data set over the 1993–2009 period, while Scharroo et al. (2006, personal communication) estimated mean sea level (MSL) drop in the Arctic Ocean over 1996–2003 of 2 mm/year. Figure 8.7 shows the averaged sea level variations over the last 23 years (blue) with a 1-year moving average superimposed in order to remove the annual sea level variations. The regional sea level trend computed over the period 1993–2015 indicates an increase of 2.2 ± 1.1 mm/year, which is relatively consistent with the results obtained by Svendsen (2015) and recently by Carret et al. (2016). In the latter, the DTU altimetry-based sea-level trend in the Arctic Ocean (north of 66°N) was estimated to 2.10 ± 0.63 mm/year over the period from 1992 to 2014. In contrast, the averaged steric contribution to the sea level trend

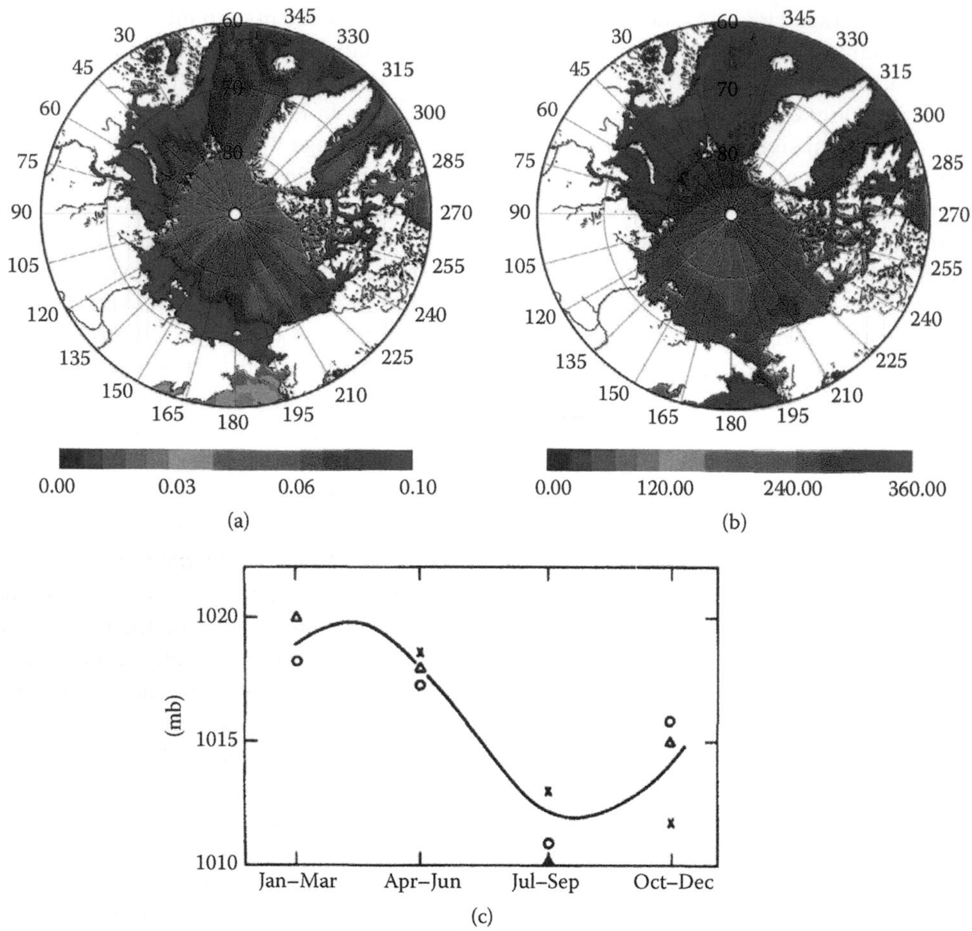

FIGURE 8.6 The seasonal SSH signal in the Arctic Ocean derived from satellite altimetry without applying for the inverse barometer correction. The amplitude in (a) is given in meters. The amplitude in (b) is given in degrees. (c) The Arctic-wide averaged sea level pressure from Thorndyke (1982) is shown. Data from various years (triangles, crosses, and circles) are shown.

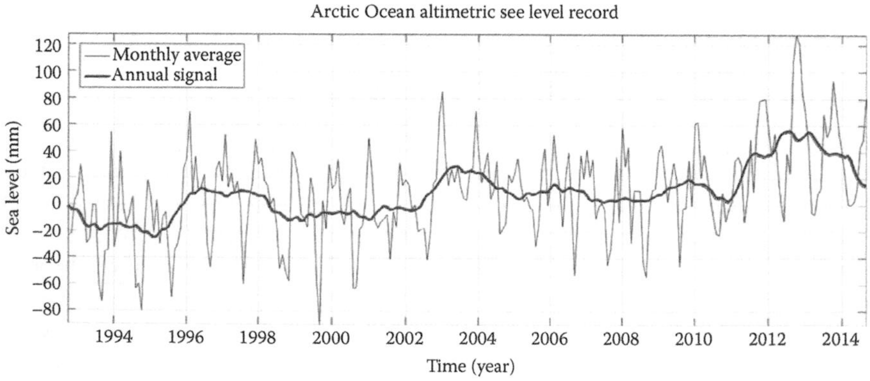

FIGURE 8.7 Regional sea level variations over 1993–2015. Monthly values (blue curve) are averaged with a 13-month moving mean (red curve) (Andersen and Piccioni, 2016).

derived from the Ocean Reanalysis Pilot 5 (ORAP5) was shown to be minor, although a slight increase is noticed from 2008.

Large interannual variations in sea level variability are seen in the Arctic Ocean linked to the AO and sea level pressure as revealed by a study on the Arctic Ocean Model Intercomparison project reported by Proshutinsky et al. (2007). The averaged Arctic sea level change also exhibits large interannual variation and the shorter averaging period reveals a period of sea level increase and sea level drop. Interesting enough, the sea level drop found by Scharroo et al. (2006) for the 1995–2003 period is also found here with a similar negative trend of 1.8 mm/year. A detailed view of the spatial pattern of the linear sea level trend for the period 1993–2015 is presented in Figure 8.8a with a spatial resolution of 1° × 1°.

Changes in SSH reflect changes in volume and mainly increase or decrease in storage of freshwater in the Arctic. The increased sea level confirms the findings from moorings, ships, and Ice-Tethered Profilers (ITPs) that the Arctic accumulated up to 10,000 km^3 of freshwater during the 1990s and 2000s (Proshutinsky et al. 2009; Krishfield et al. 2014; Rabe et al. 2014). The SSH trend pattern is dominated by a significant positive trend in the area of the Beaufort Sea, where an increase of almost 15 mm/year is registered (Figure 8.8b). This is due to the Beaufort Gyre, a wind-driven phenomenon that leads to freshwater accumulation (Rabe et al. 2011). Giles et al. (2012) estimated that the Beaufort Gyre accumulated 8000 ± 2000 km^3 of freshwater in the 2000s, and Bulczak et al. (2015) associated that with a change in sea level of roughly 2 mm/year.

In the northern part of the Nordic Seas and the Barents Sea, we observe regional sea level trends of 3–5 mm/year, which is comparable to what is seen by Nerem et al. (2010). Very close to the east coast of Greenland, the high sea level trend is questionable and can be attributed to the fact that limited data existed during the ERS-1/ERS-2/Envisat period due to heavy sea ice coverage, whereas in the same area, CryoSat-2 provided a very narrow strip of SAR-in data.

In order to extend the studies of sea level change in the Arctic Ocean beyond the altimetry era to longer periods, it is important to use a careful selection of tide gauges. Proshutinsky (2004) used tide gauges to estimate a secular sea level change in the Siberian Arctic of 1.85 mm/year between 1954 and 1989, while Richter et al. (2012) estimate trends of 1.3 to 2.3 mm/year along the Norwegian coast between 1960 and 2010. Reconstructing historical Arctic Ocean sea level change is also highly challenging due to the relatively small amount of usable data. In a recent paper by Svendsen et al. (2016), the datum-fit sea level reconstruction method (Ray and Douglas 2011) produces a very stable Arctic linear sea level trend of around 1.5 ± 0.3 mm/year for the period

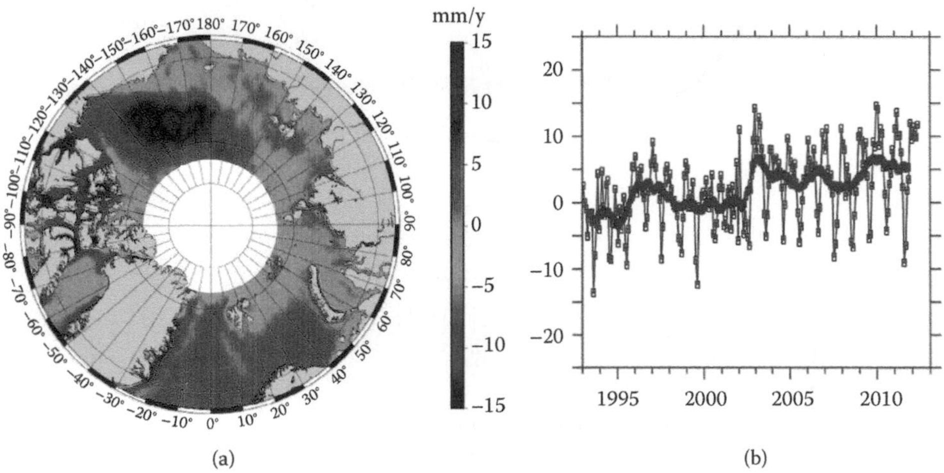

FIGURE 8.8 (a) Spatial pattern of linear sea level trend for the 1993 to 2015 period. (b) Shows the sea level averaged over the Beaufort Gyre bounded by longitudes 150°E and 240°E.

1950–2010 between 68°N and 82°N (see Figure 8.9). This value is also in good agreement with the global mean trend of 1.8 ± 0.3 mm/year for 1950–2000, as reported by Church et al. (2004).

In Henry et al. (2012), the MSL was studied along the Norwegian and Russian coasts using high-quality tide gauge data from 62 stations over the 1950–2009 time period. The coastal sea level (after correction for the Glacial Isostatic Adjustment and the inverse barometer effects) did not rise significantly from 1950 to 1980. In this period, the mean coastal sea level fluctuations closely follow the AO index. Since 1995, on the other hand, the coastal MSL presents an increasing trend of approximately 4 mm/year, with fluctuations out of phase with the AO index. Moreover, Henry et al. (2012) found that spatial trend patterns of the observed altimetry-based sea level from 1993 to 2009 are largely explained by observed steric patterns. However, residual spatial trends suggest that regional ocean mass changes cannot be ignored. In particular, along the Norwegian coast, they found that the mass component partly explains the observed sea level rise of approximately 4 mm/year over the altimetry era.

8.3.3 Arctic Sea Level Budget

The sea level budget equation in its most simple form reads:

$$\Delta S_{sl} = \Delta S_{mass} + \Delta S_{steric}$$

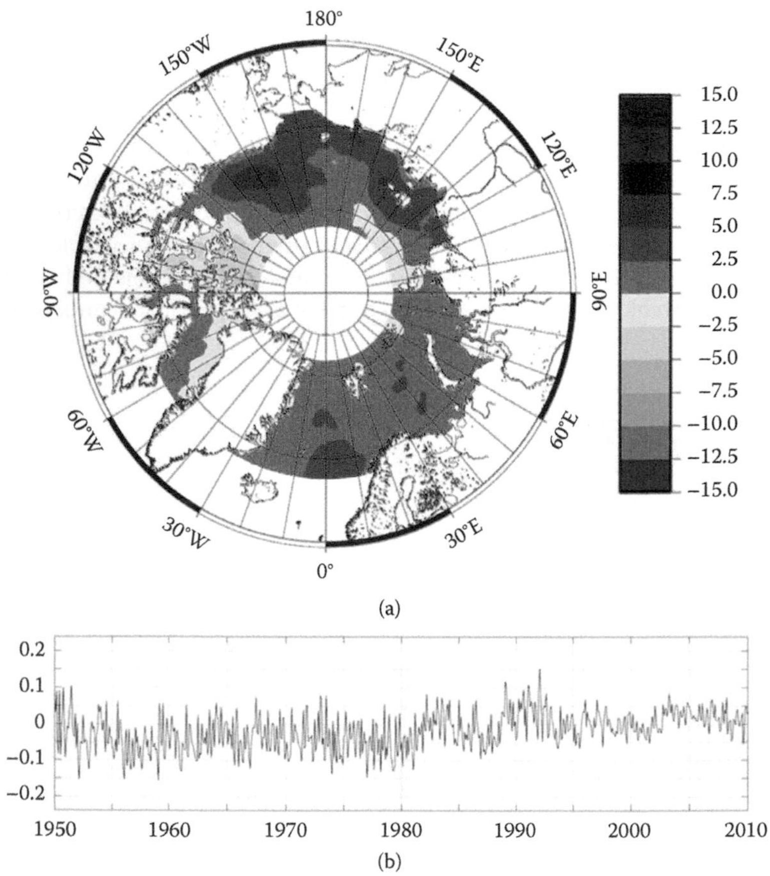

FIGURE 8.9 Reconstructed linear sea level trend for the 1950–2009 period. The values in (a) are given in mm/year, whereas the monthly averaged sea level for the period in (b) is given in meters.

where ΔS_{sl} is the observed sea level, ΔS_{mass} is the ocean mass variation, and ΔS_{steric} is the steric component associated with the subsurface temperature (thermosteric) and salinity (halosteric) structures of the water masses. Smaller contributions due to inflow and outflow from the Arctic Ocean, as well as the sea level pressure variations, should be accounted for in the overall sea level budget. However, as a first estimate, these contributions can be neglected in a first approximation. The Arctic budget closure can be evaluated by comparing the updated sea level record with the Gravity Recovery and Climate Experiment (GRACE) equivalent water thickness (EWT) ocean mass solution (Watkins et al. 2015; Volkov and Landerer, 2013) and thermosteric and halosteric trends from either *in situ* data (sparse) or from reanalysis data.

Carret et al. (2016) used temperature and salinity data obtained from the ORAP5 reanalysis (Zuo et al. 2015) that can be downloaded from https://reanalyses.org/ocean/overview-current-reanalyses2015. They reported significant regional differences in terms of the trends of the halosteric (+ 0.84 mm/year) and thermosteric (+ 0.59 mm/year) components, as revealed in Figure 8.10. In general, these regional trends suggest a decrease in salinity and an increase in temperature. The former is consistent with the increase in freshwater input to the Beaufort Gyre as described in Morison et al. (2012), whereas the thermosteric trend found in the Norwegian Sea may correspond to the inflow of warm water by the Norwegian-North Atlantic Current. Altogether, Carret et al. (2016) concluded that the sea level budget is dominated by the mass contribution in the Greenland, Norwegian, and Barents seas sector. In contrast, the steric contribution due to the strong decrease in salinity is more important in the Beaufort Gyre, displaying values of the same order as the mass component.

8.3.4 The Polar Gap and Accuracy Estimates

Since as much as 27% of the Arctic Ocean is not covered by the ERS-1, ERS-2, and Envisat altimeters, the estimates should be considered with significant uncertainty. Similarly, large regions of the Arctic Ocean will have large voids in the temporal resolution— even after retracking. Temporal voids in the altimeter record will cluster during the winter months for particularly conventional satellites, biasing the estimate of the annual variation. With CryoSat-2, a significant step forward has been made in terms of temporal and spatial sampling of the Arctic Ocean. This satellite now covers

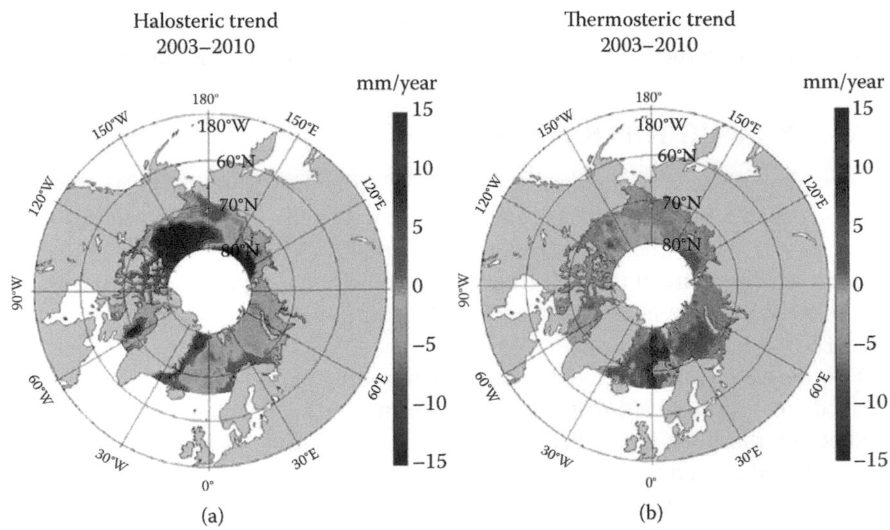

FIGURE 8.10 Spatial trend patterns of halosteric (a) and thermosteric (b) sea levels from 2003 to 2010 from Ocean Reanalysis Pilot (ORAP) 5. All values are in mm/year. (Courtesy of Carret, A., et al., *Surv. Geophys.*, 38, 251–275, 2016. With permission.)

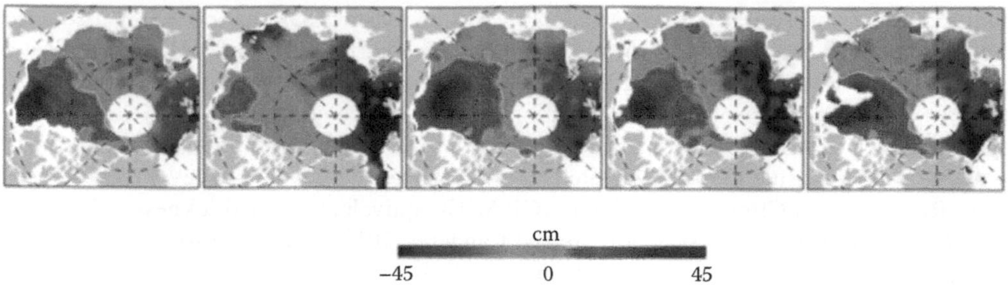

FIGURE 8.11 Mean dynamic topography from ICESat for each of the winters 2004 to 2008 from Kwok and Morrison (2012) illustrating how the Beaufort Gyre anomaly extends outside the coverage of the ERS/Envisat satellites.

more than 96% of the entire Arctic Ocean, and the temporal sampling is hugely increased with significantly more data from leads in the ice during winter and in regions with near-permanent ice coverage. However, since CryoSat-2 was only launched in 2010, observations of long-term changes north of 82°N are incomplete.

Recently, Armitage et al. (2016) investigated the effect of the polar gap using CryoSat-2 data in which monthly mean SSH was calculated using (1) all CryoSat-2 data south of 88°N and (2) just data south of 81.5°N. In so doing, the impact of the inclusion of data north of 81.5°N on the basin mean estimates could be determined. They achieved very good correlation between month-to-month variations north and south of 81.5°N, giving confidence to the fact that temporal variations in mean SSH south of 81.5°N are representative of variations across the whole basin at monthly to seasonal timescales. For long-term changes, on the other hand, the situation is more challenging. The net accumulation across the Arctic Basin will be partly missed by the polar gap in satellite altimetry. Whereas the Beaufort Gyre is mostly within the coverage of the ERS-1, ERS-2, and Envisat satellites, there are periods when the Gyre extends north of 82°N. For instance, ERS-2 and Envisat both captured the changes in SSH and freshwater content in the Canada Basin between 2004 and 2008. However, the satellite altimeters did not observe the changes in the SSH north of 81.5°N as reported by Kwok and Morison (2011). They used the ICESat data to compute the mean dynamic topographies for each of the ICESat winter campaigns during the same period and detected significant variations associated with the Beaufort Gyre north of 82°N as shown in Figure 8.11.

8.4 MEAN DYNAMIC TOPOGRAPHY

The dynamic topography can be interpreted in terms of influences from the water mass properties, ocean currents, ocean-atmosphere fluxes, and near-surface winds. The time mean of the dynamic ocean topography is called the mean dynamic topography (MDT). This reflects the long-term dynamically driven departure of the SSH from the geoid, which is represented by the mean sea surface (MSS). In turn, the time-mean ocean geostrophic current is estimated from the corresponding slope of the MDT (= MSS − h_{geoid}).

The MDT is derived with respect to the temporal averaging period over which the corresponding MSS is derived from satellite altimetry. The time-mean used to calculate the surface geostrophic currents and ocean transports is particularly sensitive to geoid residuals because accurate models of the gravitational field are required to separate the marine geoid and oceanographic signals.

The main features of the Arctic MDT (DTU15MDT, Andersen et al. 2015) derived from the new DTU15MSS and the EIGEN6-C4 (Figure 8.12) are, as expected, consistent with Johannessen et al. (2014) and clearly reveal the presence of a high (greater than 0.3 m) in the Beaufort Sea associated with the anticyclonic Beaufort Gyre; a large-scale slope (approximately 0.6 m/1300 km) in topography from the Amerasian Basin to the Eurasian Basin associated with both the Beaufort Gyre and

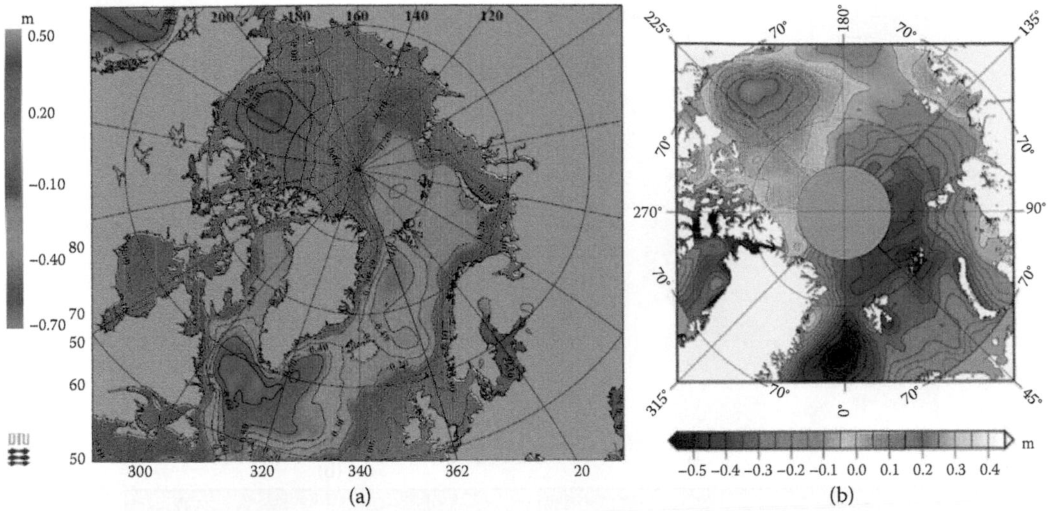

FIGURE 8.12 Two independent mean dynamic topography models for the Arctic Ocean. (a) The DTU15MDT derived from the DTU15MSS and the EIGEN6-C4 geoid model. (b) The MDT by Farrell et al. (2012) derived from ICESat. The gray sector centered at the pole marks lack of coverage.

transpolar current; a low (less than –0.4 m) in the Norwegian and Greenland seas associated with the cyclonic circulation; the expression of a sloping MDT (approximately 0.4 m/500 km) in the northeast Atlantic consistent with the North Atlantic Current and its extension to the Norwegian North Atlantic Current; and a distinct low (less than –0.6 m) in the sub-polar gyre connected with the circulation in the North Atlantic and Labrador Sea.

In the recent work by Kwok and Morison (2015), the time MDT from ICESat (Kwok and Morison 2011), CS-2 (Kwok and Morison 2015), DOT2008A (Andersen and Knudsen 2009; Pavlis et al. 2012), and DTU13MDT (Andersen et al. 2015) are compared primarily for the Arctic Ocean (Figure 8.13). The MDTs are smoothed with a 250-km Gaussian averaging kernel to reduce the noise in the SSH measurements and the contribution of residual geoid errors at shorter wavelengths.

Although the ICESat and CryoSat-2 MDTs are from different epochs with potential biases between the two instruments, their spatial patterns are comparable and in agreement with Johannessen et al. (2014) with a well-defined dome in the Canada Basin located to the Beaufort Sea and an east-west gradient across the Amerasian and Eurasian basins. Some rather distinct changes are seen in comparison with the DOT200A, which was the MDT used for the development of EGM2008. This MDT covered the period 1993 to 2007 and was pre CryoSat-2. It only used a few monthly solutions of ICESat-2. These findings and results agree qualitatively well with previous and recent results derived from satellite altimetry (Kwok and Morison 2011; Farrell et al. 2012; Giles et al. 2012; Kwok and Morison 2015) as well as with those from ocean models (e.g., Koldunov et al. 2014; Proshutinsky et al. 2015). In the Nordic Seas, moreover, the spatial pattern in the MDT also agrees well with the spatial pattern in the mean steric height derived from hydrographic data (Nilsen et al. 2008) for the period 1950 to 2010 as shown by Johannessen et al. (2014).

8.5 OCEAN CIRCULATION AND VOLUME TRANSPORT

8.5.1 Surface Circulation

There are two main pathways by which the Arctic Ocean connects with the global ocean circulation (Figure 8.1), notably the Pacific-Arctic Ocean and the Atlantic-Arctic Ocean gateways (Rudels and Friedrich 2000). The Pacific-Arctic Ocean gateway is the narrow (approximately

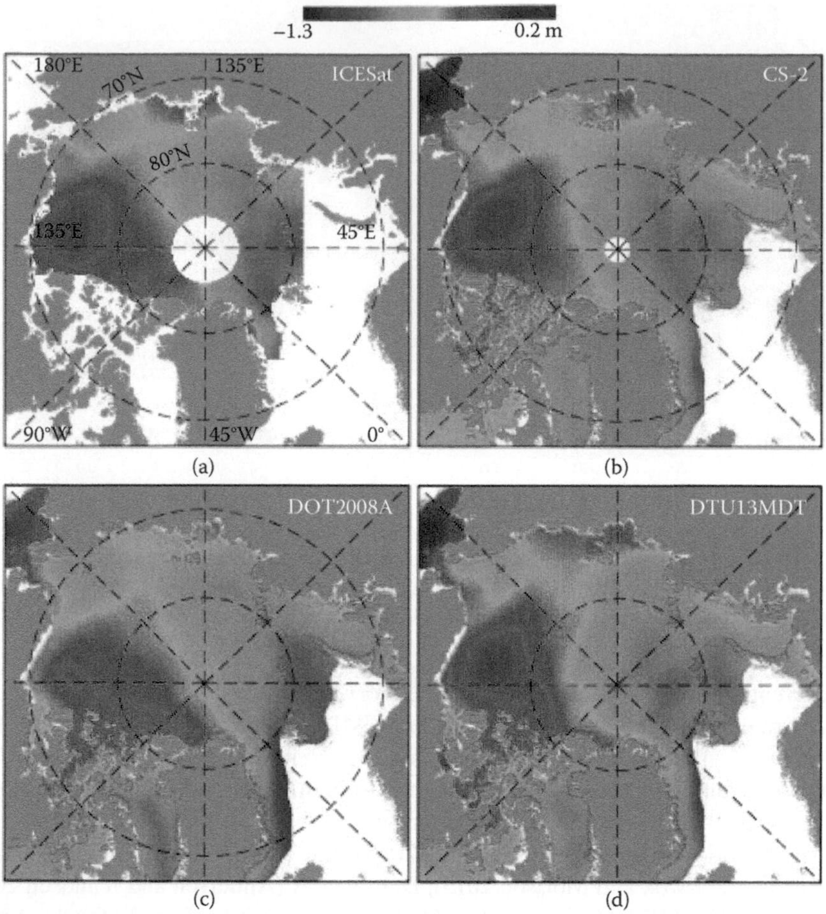

FIGURE 8.13 Mean dynamic topography of the ice-covered Arctic Ocean. (a) ICESat (mean of February and March, 2003 to 2008). (From Kwok, R., and Morison, J., *J. Geophys. Res.*, 38, L02501, 2011. With permission.); (b) CryoSat-2 (2011 to 2014); (c) DOT2008A; and (d) DTU13MDT. The fields have been smoothed with a 250-km Gaussian kernel. (Courtesy of Kwok, R., and Morison, J., *J. Geophys. Res. Oceans*, 121, 674–692, 2015. With permission.)

85-km wide), shallow (approximately 50-m deep) Bering Strait, through which about 0.8 Sv (1 Sv = 10^6 m^3/s) of water enters the Arctic. Properties of this inflow display significant seasonal variability, from about 0.4 Sv, −1.9°C, and 33 psu in winter to about 1.2 Sv, greater than 2°C, and less than 31.9 psu in summer (Woodgate et al. 2005a). The Atlantic–Arctic Ocean gateway is through both the Fram Strait (approximately 350-km wide, approximately 2700-m deep) and the Barents Sea (mostly via St. Anna Trough, approximately 200-km wide, approximately 600-m deep). The Atlantic inflow is generally saltier (greater than 34 psu), warmer (greater than 0°C), and about 10 times greater in volume than the Pacific-Arctic Ocean inflow (Beszczynska-Möller et al. 2012). The Fram Strait inflow is about 7 Sv and varies seasonally (Fahrbach et al. 2001), although complex recirculations in the Fram Strait return around half of the inflow (Rudls et al. 2000). The Barents Sea inflow is around 1 Sv in summer and 3 Sv in winter and is substantially modified during the transit across the Barents Sea (Schauer et al. 2002).

The other inputs to the Arctic Ocean are volumetrically small: Eurasian and Russian rivers (approximately 0.1 Sv) and precipitation minus evaporation (approximately 0.06 Sv). However, together, they contribute roughly two-thirds of the freshwater entering the Arctic Ocean, with the remaining third coming from the Pacific inflow (Aagaard and Carmack 1989; Serreze et al. 2006).

The outflows from the Arctic Ocean, on the other hand, are all to the North Atlantic Ocean, either through the western side of the Fram Strait (approximately 9 Sv, Fahrbach et al. 2001) or via the complex channels of the Canadian Archipelago (approximately 1–2 Sv, Melling et al. 2008). All these inflow and outflow estimates are approximate, with uncertainties typically ranging around 25% as pointed out in the review paper by Beszczynska-Möller et al. (2012).

Within the Arctic Ocean, the circulation is characterized by the eastward-flowing Atlantic water in the Eurasian basin, the transpolar drift from the Siberian shelf region to the Fram Strait, and the clockwise circulation in the Beaufort Gyre in the Canadian Basin, as illustrated in Figure 8.1. The distinct development and presence of the dome in the Beaufort Gyre from 2003 to 2014 is primarily influenced by the atmospheric wind field and ocean circulation in the Arctic Ocean. For the period 2005 to 2009, Koldunov et al. (2014) suggest that an anomalous transport (referenced to the 1970 to 2000 mean transport) in the upper ocean driven by dominant negative anomalies in the Ekman pumping directed fresh Siberian, Alaskan, and Canadian Archipelago shelf and upper slope waters into the Beaufort Gyre, as shown in Figure 8.14. This led to a convergence of low-salinity water and a subsequent increase of the steric height and hence formation of the dome as expressed in the MDTs depicted in Figures 8.10–8.13.

The Gravity Field and Steady-State Ocean Circulation Explorer (GOCE) mission (Johannessen et al. 2003) was launched and operated by ESA from 2009 to 2011. GOCE derived unique models of the Earth's gravity field and of its equipotential surface, as represented by the geoid, on a global scale with high spatial resolution (100 km) and to very high accuracies (1 mGal and 1 cm). Within the Greenland-Island-Norwegian (GIN) seas, the mean surface geostrophic velocities (Figure 8.15) are computed from the GOCE-derived MDT (Johannessen et al. 2014) for the

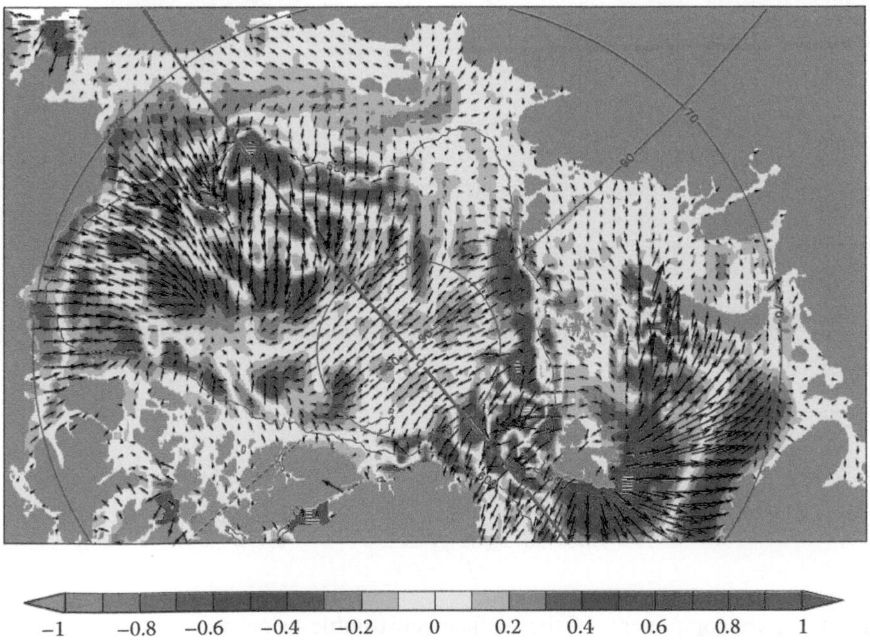

FIGURE 8.14 Ekman pumping anomaly (2005 to 2009 average minus 1970 to 2000 average) with the respective anomaly of Ekman transports superimposed (largest vector = $0.2\ m^2\ s^{-1}$) computed from the ATL12 wind stress. The isobath shown by the green line corresponds to 600-m depth.

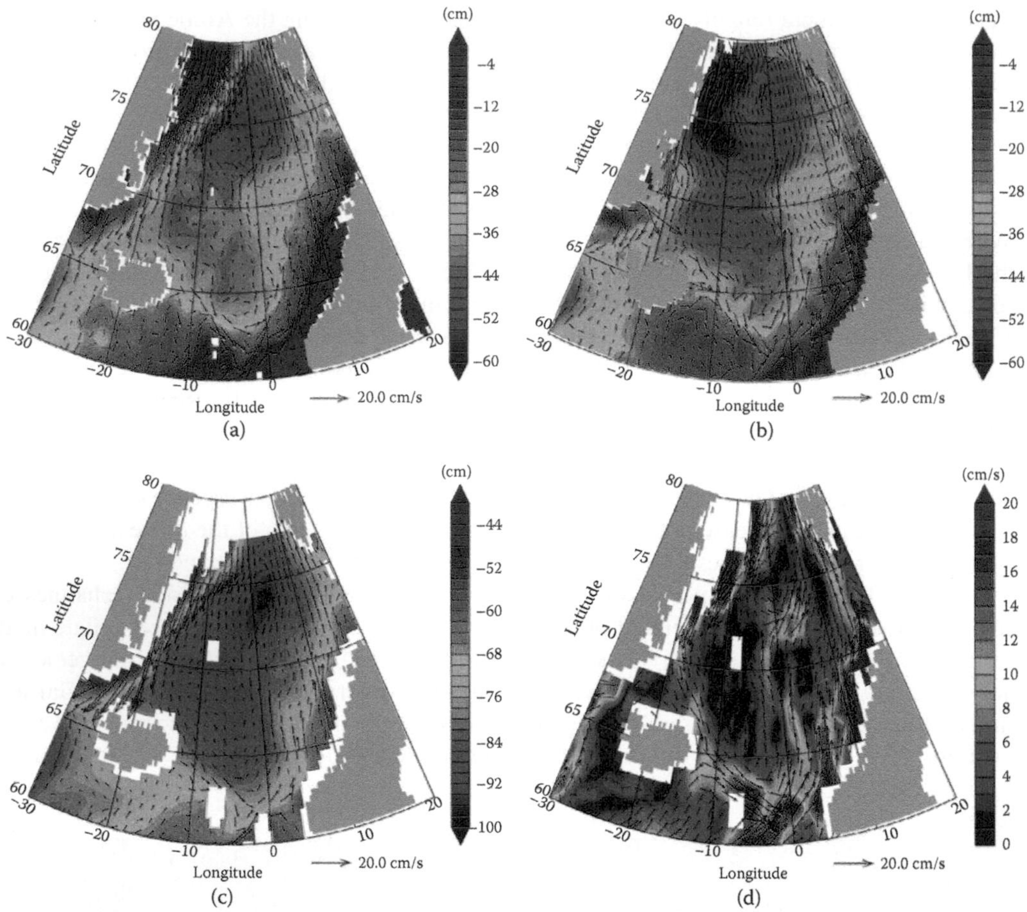

FIGURE 8.15 Mean surface geostrophic velocity vectors superimposed on the mean dynamic topography (MDT) derived from (a) GOCE, (b) CNES_CLS09, (c) Maximenko et al. (2009), and (d) mean surface velocity vectors derived from the climatology of the global surface drifter data. Color scale indicates the MDT in centimeters for (a) to (c) and speed in cm/s for (d). Current vector scale shown in the lower right corner. (Courtesy of Johannessen, J. A., et al., *Surv. Geophys.*, 35, 661–679, 2014. With permission.)

period 1993 to 2009 in consistence with the period for the construction of the DTU MSS data (e.g., MDT = MSS − h_{geoid}), whereby:

$$u_s = \frac{-g}{f} \cdot \frac{\partial MDT}{\partial y}$$

$$v_s = \frac{g}{f} \cdot \frac{\partial MDT}{\partial x}$$

Here, u_s and v_s are components of the surface geostrophic velocity, g is the acceleration due to gravity, f is the Coriolis parameter, and x and y are the zonal and meridional directions. The large-scale cyclonic surface circulation regime is well reproduced with the inflowing Atlantic water to the Norwegian Sea, reaching nearly 0.2 m/s. The broadening of the Norwegian Atlantic Current (NwAC) over the Vøring Plateau and in the Lofoten Basin is also noticed, as is the northward flowing West Spitsbergen Current (WSC) and the strong southbound East Greenland Current (EGC).

When the altimeter-based SLA is added to the MDT, these branches appear strongest in wintertime in consistence with the *in situ*-based (Steele et al. 2001, WOA01) observations reported by Mork and Skagseth (2005).

8.5.2 Volume Transport

By combining the GOCE-derived MDT and altimetric SLAs with the comprehensive hydrographic database, Johannessen et al. (2014) estimated the mean and variable transport of Atlantic water (salinity greater than 35) entering the Nordic seas for the period 1993 to 2011 at a spatial resolution of 100 km. Using 44 CTD sections for the Island-Faroe Ridge (IFR), 84 CTD sections for the Faroe-Shetland Channel (FSC), and 76 CTD sections taken along the Svinøy section, the baroclinic velocity structures in the Atlantic water were estimated across these sections. Combined with the barotropic velocity values, the absolute velocities are then retrieved and, when multiplied by the area covered by the Atlantic water, estimates of the corresponding volume transports of Atlantic water across the three sections are obtained and compared to previous observations and simulation as given in Table 8.1.

The mean inflows of Atlantic water across the IFR and through the FSC are estimated as 3.5 Sv and 4.1 Sv, respectively (1 Sv = 10^6 m^3s^{-1}). Moreover, the mean transport of the two branches of Atlantic water crossing the Svinøy section (e.g., the Norwegian Atlantic Slope Current [NwASC] and the Norwegian Atlantic Front Current [NwAFC]) are 3.0 Sv and 3.9 Sv, respectively. These estimates compare reasonably well with the earlier reported observed and simulated transport values despite the different integration periods. However, discrepancies exist, including the total combined GOCE-, altimeter-, and hydrographic-based transport estimates across the Svinøy section that are about 20% lower than the simulated transports (e.g., 6.9 Sv vs. 8.5 and 8.2 Sv). According to Johannessen et al. (2014), this is partly related to the definition and choice of layers for the

TABLE 8.1
Comparison of volume transport estimates in unit of sverdrup (Sv) from combined GOCE, altimetry, and *in situ* data to previous studies as well as estimates from simulation models for the Island-Faroe ridge (IFR), Faroe-Shetland channel (FSC), NwAFC, NwASC in the Svinøy Section, and the Total Svinøy Section

					Svinøy		
Source	**Data**	**Period**	**IFR**	**FSC**	**NwAFC**	**NwASC**	**Total**
Johannessen et al. (2014)	GOCE + Altim + hydr.	1993–2011	3.5	4.1	3.0	3.9	6.9
Mork and Skagseth (2010)	Altim + hydr.	1993–2009			1.7	3.4	5.1
Skagseth et al. (2008)	Current meters	1995–2006				4.3	
Orvik and Skagseth (2005)	Current meters	1995–1999				4.2	
Orvik and Skagseth (2003)	Current meters	1998–2000				4.4	
Orvik et al. (2001)	Current meters + ADCP + hydr.	1995–1999			3.4	4.2	7.6
Berx et al. (2013)	Altim + ADCP + hydr.	1995–2009	3.5				
Østerhus et al. (2005)	ADCP + hydr.	1999–2001	3.8	3.8			
Hansen et al. (2010)	ADCP + hydr.	1997–2008	3.5				
Hansen et al. (2013)	ADCP + hydr.	1997–2001	3.5				
Sandø et al. (2012)	MICOM model	1994–2007	4.7*	4.7			
Johannessen et al. (2014)	MICOM model	1993–2007	3.5	6.9	3.5	5.0	8.5
Johannessen et al. (2014)	ATL model	1993–2007	3.5	4.2	3.5	4.7	8.2

*Only from 1997 to 2007.

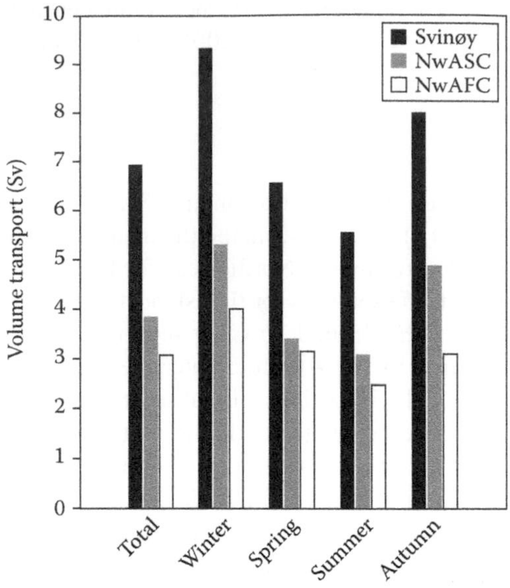

FIGURE 8.16 Mean annual and mean seasonal total volume transport estimates for the Svinøy section (black) compared to the mean transports of the NwASC (gray) and the NwAFC (white) for the period 1993 to 2011 based on combined use of GOCE, altimetry, and *in situ* hydrography data.

Atlantic water transport estimations. These disagreements in transport estimates imply significant differences in the mean northward advection of heat and salt to the Nordic Seas and Arctic Ocean. This, in turn, may affect both the evaporation-precipitation fluxes and convective overturning in the Norwegian and Greenland seas.

Taking advantage of the temporal variability observed in the altimeter SLA and the hydrographic data, the mean and seasonal cycle in the transport of the inflowing Atlantic water for the period 1993 to 2009 can also be estimated and intercompared as shown in Figure 8.16 for the Svinøy section. The mean seasonal variability reveals a pattern with the largest transports (9.3 Sv) in winter compared to the summer transport minimum (5.4 Sv). Moreover, the mean seasonal NwASC transport always exceeds the mean seasonal NwAFC transport, while the latter displays a narrower range of seasonal variability in the volume transport. This suggests that the seasonal changes of the transport across the Svinøy section are predominantly controlled by seasonal changes in the transport of the NwASC.

Altogether, these GOCE-based estimates combined with altimeter-based SLA and *in situ* hydrographic data are providing new and promising abilities to examine the seasonal transport variability (total as well as barotropic and baroclinic components) across key selected regions. As such, it is also providing an important tool for validations of models and simulation of transports between the northeast Atlantic Ocean, the Nordic Seas, and Arctic Ocean.

8.6 SUMMARY AND OUTLOOK

Satellite altimetry is fundamental for monitoring sea level and its changes on various timescales in the Arctic Ocean. Availability of data sources such as *in situ* observations, satellite altimetry, and GOCE- and GRACE-based gravity fields enables the derivation of MDTs that put valuable constrains on both the use of satellite altimetry and ocean modeling.

The success of ICESat has paved the way for the use of laser altimetry for sea ice thickness retrievals in the Arctic Ocean, and soon a continuation of ICESat-2 will enable ultra-high resolution surface mapping. Moreover, the availability of SAR altimetry from CryoSat-2 has advanced the multidisciplinary studies of the Arctic Ocean dynamics and MSL, variability in the sea ice thickness and volume, and the distribution of leads.

In addition, the success of the CryoSat-2 SAR altimetry enabled the important decision to maintain continuity of SAR altimetry observations in the Arctic Ocean from the Sentinel-3 (A/B) satellites for the next decade at an inclination of 98.65°. Moreover, within a few years, the SWOT satellite (http://swot.jpl.nasa.gov/) will be launched and provide high resolution (15 km) SAR-in in the Arctic Ocean. However, the inclination of the SWOT mission will only be 78°. Hence, Sentinel-3 and SWOT will lack between 27% and 35% coverage of the northernmost sector of the Arctic Ocean.

However, for studies of the Arctic Ocean, sea ice thickness, and sea level as well as monitoring of freshwater storage in the Arctic Ocean, a continuation of the CryoSat-2 mission is fundamentally important—such as, eventually, with a dual-frequency all interferometric mission called CryoSat-3. By measuring in both Ku and Ka bands, such a mission concept will also be able to map the snow depth on the sea ice in the Arctic Ocean. In so doing, a major contributor to the uncertainty in sea ice freeboard and sea ice thickness retrievals would be diminished.

ACKNOWLEDGMENT

This work was supported by the ESA Sea Level CCI project funded under the ESRIN contract no. 4000109872/13/I-NB.

REFERENCES

Aagaard, K., and E. Carmack (1989), The role of sea ice and other fresh water in the Arctic circulation, J. *Geophys. Res., 94,* 14,485–14,498,doi:10.1029/JC094iC10p14485.

Abulaitijiang, A., O. B. Andersen, and L. Stenseng (2015), Coastal sea level from inland CryoSat-2 interferometric SAR altimetry. *Geophys. Res. Lett.*, 42, 1841–1847. doi: 10.1002/2015GL063131.

Ablain, M., A. Cazenave, G. Larnicol, M. Balmaseda, P. Cipollini, Y. Faugère, M. J. Fernandes, O. Henry, J. A. Johannessen, P. Knudsen, O. Andersen, J. Legeais, B. Meyssignac, N. Picot, M. Roca, S. Rudenko, M. G. Scharffenberg, D. Stammer, G. Timms, and J. Benveniste (2015), Improved Sea Level record over the satellite altimetry era (1993-2010) from the Climate Change Initiative project, *Ocean Sci., 11,* 67–82, doi:10.5194/os-11-67-2015.

Andersen, O. B., and Knudsen, P. (2009) The DNSC08 mean sea surface and mean dynamic topography models, *J. Geophys. Res.*, doi:10.1029/2008JC005179.

Andersen, O. B. and G. Piccioni (2016) Recent Arctic Sea Level Variations from Satellites, Frontiers in Marine Science, 3, journal.frontiersin.org/article/10.3389/fmars.2016.00076, DOI=10.3389/fmars.2016.00076.

Andersen, O., B., P. Knudsen, and L. Stenseng (2015), The DTU13 MSS (Mean Sea Surface) and MDT (Mean Dynamic Topography) From 20 Years of Satellite Altimetry, pp. 1–10, Springer, Berlin, doi:10.1007/1345_2015_182.

Armitage, T. W. K., S. Bacon, A. L. Ridout, S. F. Thomas, Y. Aksenov, and D. J. Wingham (2016), Arctic sea surface height variability and change from satellite radar altimetry and GRACE, 2003–2014, *J. Geophys. Res. Oceans*, 121, 4303–4322, doi:10.1002/2015JC011579.

Beaven, S. G., Lockhart, G. L., Gogineni, S. P., Hosseinmostafa, A. R., Jezek, K., Gow, A. J., Perovich, D. K., Fung, A. K., and Tjuatja, S.: Laboratory measurements of radar backscatter from bare and snow-covered saline ice sheets, Int. *J. Remote Sens.*, 16(5), 851–876, 1995.

Belkin, I. (2004), Propagation of the "Great Salinity Anomaly" of the 1990s around the northern North Atlantic, *Geophys. Res. Lett.*, 31, L08306, doi:10.1029/2003GL019334. Prog. Oceanogr., 20, 103–151, doi:10.1016/0079-6611(88)90049-3.

Berx B, Hansen B, Østerhus S, Larsen KM, Sherwin T, Jochumsen K (2013) Combining in situ measurements and altimetry to estimate volume, heat and salt transport variability through the Faroe Shetland Channel. *Ocean Sci* 9. doi:10.5194/os-9-639-2013.

Beszczynska-Möller, A., Fahrbach, E., Schauer, U., and Hansen, E.: Variability in Atlantic water temperature and transport at the entrance to the Arctic Ocean, 1997–2000, *ICES J. Mar. Sci.*, doi:10.1093/icesjms/fss056, 2012.

Bouzinac, C. (2013), CryoSat Product Handbook, April 2013. ESRIN/ESA and Mullard Space Science Laboratory, University College London, London, U. K.

Brown, G.S. (1977), The average impulse response of a rough surface and its applications, IEEE Trans. Antennas Propag. 25 (1), Jan. 1977, pp. 67–74.

Bulczak, Anna I., Sheldon Bacon, Alberto C. Naveira Garabato, Andrew Ridout, Maike J. P. Sonnewald and Seymour W. Laxon (2015) Seasonal variability of sea surface height in the coastal waters and deep basins of the Nordic Seas, *Geophys. Res. Lett.*, 42, 113–120, doi:10.1002/2014GL061796.

Carret, A., Johannessen, J.A., Andersen, O.B. et al. Surv Geophys (2016) Arctic Sea Level During the Satellite Altimetry Era. *Surveys in Geophysics,* doi:10.1007/s10712-016-9390-2.

Cheng, Y., and Andersen, O. B. (2011) Multi-mission empirical ocean tide modeling for shallow waters and polar seas. *J. of Geophys. Res.,* 116(C11), 1–11, doi: 10.1029/2011JC007172.

Cheng Y., Andersen O. B., Knudsen P. (2015). An Improved 20- Year Arctic Ocean Altimetric Sea Level Data Record, *Marine Geodesy*, 38:2, 146–162, DOI: 10.1080/01490419.2014.954087

Church, J.A., N. J. White, R. Coleman, K. Lambeck, and J.X. Mitrovica (2004), Estimates of the Regional Distribution of Sea Level Rise over the 1950–2000 Period, Journal of Climate, July 2004, Vol. 17, No. 13.

Dickson, R. R., J. Meincke, S.-A. Malmberg, and A. J. Lee (1988), The "Great Salinity Anomaly" in the northern North Atlantic 1968–1982, Prog. *Oceanogr.*, 20, 103–151, doi:10.1016/0079-6611(88)90049-3. Prog. Oceanogr., 20, 103–151, doi:10.1016/0079-6611(88)90049-3.

Fahrbach, E., J. Meincke, S. Østerhus, G. Rohardt, U. Schauer, V. Tverberg, J. Verduin (2001), Direct measurements of volume transports through Fram Strait, *Polar Research*, DOI: 10.1111/j.1751-8369.2001.tb00059.x.

Farrell, S. L., D. C. McAdoo, S. W. Laxon, H. J. Zwally, D. Li, A. Ridout, and K. Giles (2012), Mean dynamic topography of the Arctic Ocean, *Geophys. Res. Lett.*, 39, L01601, doi:10.1029/2011GL050052.

Furevik, T. and J.E.Ø. Nilsen (2005) Large-scale atmospheric circulation variability and its impacts on the Nordic Seas ocean climate a review. In: The Nordic Seas: an integrated perspective. *AGU Geophysical Monograph Series,* vol 158. pp 105–136.

GCOS (2016), The Global Observing System for Climate: Implementation Needs, *GCOS-200.* World Meteorological Organization, 2016.

Giles, K. A., S. W. Laxon, A. L. Ridout, D. J. Wingham, and S. Bacon (2012), Western Arctic Ocean freshwater storage increased by wind-driven spin-up of the Beaufort Gyre, *Nat. Geosci.*, 5, 194–197, doi:10.1038/ngeo1379.

Gommenginger, C., Thibaut, P., Fenoglio-Marc, L., Quarty, G., Deng, X., 2011. Retracking altimeter waveforms near the coasts. In: Chapter 4, Coastal Altimetry. *Springer,* ISBN 978-3-642-12795-3.

Hansen B, Hatun H, Kristiansen R, Olsen SM, Østerhus S (2010) Stability and forcing of the Iceland-Faroe inflow of water, heat, and salt to the Arctic. Ocean Sci 6:1013–1026.

Helland-Hansen B and Nansen F (1909) The Norwegian Sea: its physical oceanography based upon the Norwegian Researches 1900–1904, Report on Norwegian Fishery and Marine Investigation, vol. II. The Royal Department of Trade, Navigation and Industries, Mallingske, Kristiania, pp 390.

Henry, O., P. Prandi, W. Llovel, A. Cazenave, S. Jevrejeva, D. Stammer, B. Meyssignac, and N. Koldunov (2012) Tide gauge based Sea level variations since 1950 along the Norwegian and Russian coasts of the Arctic Ocean; contribution of the steric component, *J. Geophys. Res.*, 117, C06023.

Holgate, S. J., A. Matthews, P. L. Woolworth, L. J. Rickards, M. E. Tamisiea, E. Bradshaw, P. R. Foden, K. M. Gordon, S. Jevrejeva, and J. Pugh (2013), New data systems and products at the Permanent Service for Mean Sea Level, *J. Coastal Res.*, 29, 493–504, doi:10.2112/JCOASTRES-D-12-00175.1.

IPCC (2013), Climate Change 2013: The Physical Science Basis, 1535 pp., Cambridge Univ. Press, New York, NY, 1535 p.

Johannessen, J.A., G. Balmino, C. Le Provost, R. Rummel, R. Sabadini, H. Sünkel, C.C. Tscherning, P. Visser, P. Woodworth, C. W. Hughes, P. LeGrand, N. Sneeuw, F. Perosanz, M. Aguirre-Martinez, H. Rebhan, and M. Drinkwater (2003), The European Gravity Field and Steady-State Ocean Circulation Explorer Satellite Mission: Impact in Geophysics, *Survey in Geophysics,* 24, 339-386, 2003.

Johannessen, J. A., R. P. Raj, J. E. Ø. Nilsen, T. Pripp, P. Knudsen, F. Counillon, D. Stammer, L. Bertino, O. B. Andersen, N. Serra and N. Koldunov (2014) Toward Improved Estimation of the Dynamic Topography and Ocean Circulation in the High Latitude and Arctic Ocean: The Importance of GOCE, *Survey in Geophysics, Springer,* DOI 10.1007/s10712-013-9270-y.

Kern, S., K. Khvorostovsky, H. Skourup, E. Rinne, Z. S. Parsakhoo, V. Djepa, P. Wadhams and S. Sandven (2015), About uncertainties in sea ice thickness retrieval from satellite radar altimetry: results from the ESA-CCI Sea Ice ECV Project Round Robin Exercise, *The Cryosphere Discuss.,* 8, 1–44, 2014, doi:10.5194/tcd-8-1-2014.

Khan, S. A., Aschwanden, A., Bjørk, A. A., Wahr, J., Kjeldsen, K. K., Kjær, K. H. (2015). Greenland ice sheet mass balance: a review, *Rep. Prog. Phys.*

Koldunov, N. V., N. Serra, A. Kohl, D. Stammer, O. Henry, A. Cazenave, P. Prandi, P. Knudsen, O. B. Andersen, Y. Gao and J. A. Johannessen (2014) Multi-model Simulations of Arctic Ocean Sea Surface Height Variability in the Period 1970–2009, *Journal of Geophys. Res., Oceans 2014*; Volume 119. (12)s. 8936–8954, doi:10.1002/2014JC010170.

Krishfield, R. A., A. Proshutinsky, K. Tateyama, W. J. Williams, E. C. Carmack, F. A. McLaughlin, and M.-L. Timmermans (2014), Deterioration of perennial sea ice in the Beaufort Gyre from 2003 to 2012 and its impact on the oceanic freshwater cycle, *J. Geophys. Res. Oceans,* 119,1271–1305, doi:10.1002/2013JC008 999.

Kwok R. and N. Untersteiner (2011) The thinning of Arctic sea ice, *Physics Today*, 2011.

Kwok, R., G. F. Cunningham, H. J. Zwally, and D. Yi (2007), Ice, Cloud, and land Elevation Satellite (ICESat) over Arctic sea ice: Retrieval of freeboard, *Journal of Geophysical Research-Oceans,* 112 (C12), C12013.

Kwok, R. and D. A. Rothrock (2009), Decline in Arctic sea ice thickness from submarine and ICESat records: 1958–2008, *Geophysical Research Letters,* 36 (15), L15501.

Kwok, R., and J. Morison (2011), Dynamic topography of the icecovered Arctic Ocean from ICESat, *J. Geophys. Res.,* 38, L02501, doi:10.1029/2010GL046063.

Kwok, R., and J. Morison (2015), Sea surface height and dynamic topography of the ice-covered oceans from CryoSat-2: 2011–2014, *J. Geophys. Res. Oceans,* 121, 674–692, doi:10.1002/2015JC011357.

Laxon, S., N. Peacock, and D. Smith (2003), High interannual variability of sea ice thickness in the Arctic region, *Nature*, 425 (6961), 947–950.

Laxon, S., K.A. Giles, A. Ridout, D. Wingham, R.C. Willatt, R. Cullen, R. Kwok, A. Schweiger, J. Zhang (2013), CryoSat-2 estimates of Arctic sea ice thickness and volume, *Geophysical Research Letters*, 40 (4), 732–737.

Manabe, S. and R. J. Stouffer (2012), Simulation of abrupt climate change induced by freshwater input to the North Atlantic Ocean, *Nature*, 378, 165 -167 (09 November 1995); doi:10.1038/378165a0.

Maximenko N, P. Niiler, M.-H. Rio, O. Melnichenko, L. Centurioni, D. Chambers, V. Zlotnicki, B. Galperin (2009) Mean dynamic topography of the ocean derived from satellite and drifting buoy data using three different techniques. *J. Atmos. Ocean Tech.* 26(9):1910–1919.

Melling, H., T.A. Agnew, K.K. Falkner, D.A. Greenberg, M.L. Craig, A. Münchow, B. Petrie, S.J. Prinsenberg, R.M. Samuelson, R.A. Woodgate (2008), Arctic-Subarctic ocean fluxes, *Springer*, pp 193–247.

Morison, J., R. Kwok, C. Peralta-Ferriz, M. Alkire, I. Rigor, R. Andersen and M. Stele (2012), Changing Arctic Ocean freshwater pathways. *Nature* 481, 66–70, doi:10.1038/nature10705.

Mork KA, and Skagseth Ø (2005) Annual sea surface height variability in the Nordic Seas, in The Nordic Seas: An Integrated Perspective, *Geophys Monogr Ser*, vol. *158*, edited by H. Drange et al. pp. 51–64, AGU, Washington, DC.

Mork KA, Skagseth Ø (2010) A quantitative description of the Norwegian Atlantic current by combining altimetry and hydrography, *Ocean Sci* 6:901–911. doi:10.5194/os-6-901-2010.

Nerem, R. S., D. P. Chambers, C. Choe, and G. T. Mitchum (2010), Estimating mean sea level change from the TOPEX and Jason altimeter missions, *Mar. Geod.*, 33, 435–446, doi:10.1080/01490419.2010.491031.

Nilsen JEØ, Hatun H, Mork KA and Valdimarsson H (2008), The NISE Data Set. Technical Report 08-01, Faroese Fisheries Laboratory, Box 3051, Torshavn, Faroe Islands.

Orvik KA, Skagseth Ø (2003) Monitoring the Norwegian Atlantic slope current using a single moored current meter. *Cont Shelf Res* 23:159–176.

Orvik KA, Skagseth Ø (2005) Heat flux variations in the eastern Norwegian Atlantic current toward the Arctic from moored instruments, 1995–2005. *Geophys Res Lett* 32:L14610. doi:10.1029/ 2005GL023487.

Orvik KA, Skagseth Ø, Mork M (2001) Atlantic inflow to the Nordic Seas: current structure and volume fluxes from moored current meters, VM-ADCP and SeaSoar-CTD observations, 1995–1999. Deep-Sea Res I:48. doi:10.1016/S0967-0637(00)00038-8.

Østerhus S, Turrell WR, Jo´nsson S, Hansen B (2005) Measured volume, heat, and salt fluxes from the Atlantic to the Arctic Mediterranean. *Geophys. Res. Lett* 32:L07603, doi:10.1029/2004GL022188.

Pavlis, N. K., S. A. Holmes, S. C. Kenyon, and J. K. Factor (2012), The development and evaluation of the Earth Gravitational Model 2008 (EGM2008), *J. Geophys. Res., 117*, B04406, doi:10.1029/2011JB008916.

Peacock, N. R., and S. W. Laxon (2004). Sea surface height determination in the Arctic Ocean from ERS altimetry, *J. Geophys. Res.*, 109, C07001, doi:10.1029/2001JC001026.

Peralta-Ferriz, C., J. H. Morison, J. M. Wallace, J. A. Bonin, and J. Zhang (2014), Arctic Ocean circulation patterns revealed by GRACE, *J. Clim.*, 27, 1445–1468, doi:10.1175/JCLI-D-13-00013.1

Prandi, P., M. Ablain, A. Cazenave, and N. Picot (2012). A new estimation of mean sea level in the Arctic Ocean from satellite altimetry, *Mar. Geod.,* 35, sup1, 61–81.

Proshutinsky, A. (2004) Secular sea level change in the Russian sector of the Arctic Ocean, *J. Geophys. Res.,* 109(C3), 1–19, doi:10.1029/2003JC002007.

Proshutinsky, A., R. Krishfield, M.-L. Timmermans, J. Toole, E. Carmack, F. McLaughlin, W. J. Williams, S. Zimmermann, M. Itoh, and K. Shimada (2009) Beaufort Gyre freshwater reservoir: State and variability from observations, *J. Geophys. Res.,* 114, C00A10, doi:10.1029/2008JC005104.

Proshutinsky, A, Ashik, I., Häkkinen, S (2007). Sea level variability in the Arctic Ocean from AOMIP models, *J. Geophys. Res.*, 112(C4), 1–25. doi:10.1029/2006JC003916.

Proshutinsky, A., D. Dukhovskoy, M.-L. Timmermans, R. Krishfield, J. L. Bamber (2015), Arctic circulation regimes, *Phil. Trans. R. Soc.* doi: 10.1098/rsta.2014.0160.

Rabe, B., M. Karcher, U. Schauer, J. M. Toole, R. A. Krishfield, S. Pisarev, F. Kauker, R. Gerdes, and T. Kikuchi (2011) An assessment of Arctic Ocean freshwater content changes from the 1990s to the 2006–2008 period. Deep Sea Research Part I: Oceanographic Research Papers 58(2):173–185. doi: 10.1016/j.dsr.2010.12.002.

Rabe, B., M. Karcher, F. Kauker, U. Schauer, J. M. Toole, R. A. Krishfield, S. Pisarev, T. Kikuchi, and J. Su (2014), Arctic Ocean basin liquid freshwater storage trend 1992–2012, *Geophys. Res. Lett.*, 41, 961–968, doi:10.1002/2013GL058121.

Ray, R. D. and Douglas, B. C. (2011). Experiments in reconstructing twentieth century sea levels, *Progress in* Oceanography*,* 91(4):496–515.

Ray, C., C. Martin-Puig, M. P. Clarizia, G. Ruffini, S. Dinardo, C. Gommenginger, J. Benveniste (2015) SAR Altimeter Backscattered Waveform Model, *IEEE Trans. Geosci. Remote Sens.* 53, 2, 911–919, doi: 10.1109/TGRS.2014.2330423.

Richter, K., J. E. Ø. Nilsen, and H. Drange (2012) Contributions to sea level variability along the Norwegian coast for 1960-2010, *J. Geophys. Res*, 117(C5), 1-12. doi:10.1029/2011JC007826.

Rudels, B. and H.J. Friedrich (2000), The transformation of Atlantic Water in the Arctic Ocean and their significance for the Freshwater budget. In: E.L. Lewis et al. (ed), *The Freshwater Budget of the Arctic Ocean*, Kluwer Academic Publishers, Dordrecht, 70, 503–532, 2000.

Rudels, B., Meyer, R., Fahrbach, E., Ivanov, V. V., Østerhus, S., Quadfasel, D., Schauer, U., Tverberg, V., and Woodgate, R. A.: Water mass distribution in Fram Strait and over the Yermak Plateau in summer 1997, *Ann. Geophys.*, 18, 687–705, doi:10.1007/s00585-000-0687-5, 2000.

Sandø A.B, Nilsen J.E.Ø., Eldevik T, Bentsen M (2012) Mechanisms for variable North Atlantic–Nordic seas exchanges. *J Geophys Res* 117:C12006. doi:10.1029/2012JC008177.

Scharroo, R., Smith, W. H. F., and Lillibridge, J. L.: Reply to Comment on "Satellite Altimetry and the Intensification of Hurricane Katrina", *Eos*, 87(8), 89 pp. 2006.

Schauer, U., Loeng, H., Rudels, B., Ozhigin, V. K., and Dieck, W. (2002): Atlantic Water flow through the Barents and Kara Seas, *Deep-Sea Research I,* 49, 2281–2298, doi:10.1016/S0967-0637(02)00125-5, 2002.

Serreze, M. C., Barrett, A. P., Slater, A. G., Woodgate, R. A., Aagaard, K., Lammers, R. B., ... & Lee, C. M. (2006). The large-scale freshwater cycle of the Arctic. *J. Geophys. Res. Oceans (1978–2012), 111*(C11).

Skagseth Ø, Furevik T, Ingvaldsen R, Loeng H, Mork KA, Orvik KA, Ozhigin V (2008) Volume and Heat Transports to the Arctic Ocean via the Norwegian and Barents Seas. In: Dickson (ed) Arctic-Subarctic Ocean Fluxes (ASOF): Defining the Role of the Northern Seas in Climate. Springer, Berlin, pp 45–64.

Stammer, D., R.D. Ray, O. B. Andersen, B. K. Arbic, W. Bosch, L. Carrère, Y. Cheng, D. S. Chinn, B. D. Dushaw, G. D. Egbert, S.Y. Erofeeva, H.S. Fok, J.A.M. Green, S. Griffiths, M. A. King, V. Lapin, F. G. Lemoine, S. B. Luthcke, F. Lyard, J. Morison, M. Müller, L. Padman, J. G. Richman, J. F. Shriver, C.K. Shum, E. Taguchi and Y. Yi (2014) Accuracy assessment of global barotropic ocean tide models, *Review of Geophysics*, doi:10.1002/2014RG000450.

Steele, Michael, Rebecca Morley, and Wendy Ermold (2001) PHC: A Global Ocean Hydrography with a High-Quality Arctic Ocean, *Journal of Climate,* https://doi.org/10.1175/1520-0442(2001)014<2079:PAGOHW>2.0.CO;2.

Stroeve, J. C., V. Kattsov, A. Barrett, M. Serreze, T. Pavlova, M. Holland, and W. N. Meier (2012), Trends in Arctic sea ice extent from CMIP5, CMIP3 and observations, *Geophys. Res. Lett.*, 39, L16502, doi:10.1029/2012GL052676.

Stroeve, J. C., T. Markus, L. Boisvert, J. Miller, and A. Barrett (2014), Changes in Arctic melt season and implications for sea ice loss, *Geophys. Res. Lett.*, 41 (4), 1216–1225.

Svendsen, P. L. (2015). Arctic Sea Level Reconstruction, PhD diss., Technical University of Denmark, Lyngby, Denmark.

Svendsen, P. L., Andersen, O. B., and Nielsen, A. A. (2016). Stable reconstruction of Arctic sea level for the 1950- 2010 period. *J. Geophys. Res. Oceans* 121(8), 5697–5710. DOI: 10.1002/2016JC011685.

Tilling, R. L., A. Ridout, A. Shepherd, and D. J. Wingham (2015), Increased Arctic sea ice volume after anomalously low melting in 2013, *Nature Geoscience*, 8, 643–646.

Thorndyke A. S. (1982). Statistical properties of the pressure field over the Arctic Ocean, *J. Atmospheric Sciences,* 39, 229–234.

Volkov, D., and F. W. Landerer (2013), Nons seasonal fluctuations of the Arctic Ocean mass observed by the GRACE satellites, *J. Geophys. Res.*, 6451–6460, doi:10.1002/2013JC009341.

Watkins, M. M., Wiese, D. N., Yuan, D.-N., Boening, C., Landerer, F. W. (2015). Improved methods for observing Earth's time variable mass distribution with GRACE using spherical cap mascons, *J. Geophys. Res. Solid Earth,* 120, doi:10.1002/2014JB011547.

Wingham, D. J., et al. (2006), CryoSat: A mission to determine the fluctuations in the Earth's land and marine ice fields, *Adv. Space Res.*, 37,841–871.

Woodgate, R. A., K. Aagaard, and T. J. Weingar tner (2005), Monthly temperature, salinity, and transport variability of the Bering Strait through flow, *Geophys. Res. Lett.,* 32, L04601, doi:10.1029/2004GL021880.

Wunsch and Stammer (1997). Atmospheric loading and the oceanic "inverted barometer" effect, *Rev. Geophys.*, 35(1), 79–107, doi:10.1029/96RG03037.

Zuo H, Balmaseda MA, Mogensen K (2015) The new eddy-permitting ORAP5 ocean reanalysis: description, evaluation and uncertainties in climate signals. *Clim Dyn.*, pp 1–21, doi:10.1007/s00382-015-2675-1.

Zwally, H. J., B. Schutz, W. Abdalati, J. Abshire, C. Bentley, A. Brenner, J. Bufton, J. Dezio, D. Hancock, D. Harding, T. Herring, B. Minster, K. Quinn, S. Palm, J. Spinhirne, R. Thomas (2002), ICESat's laser measurements of polar ice, atmosphere, ocean, and land, *J. Geodyn.*, 34, 405–445.

9 The Southern Ocean

Sarah T. Gille and Michael P. Meredith

9.1 INTRODUCTION

The Southern Ocean is the expanse of ocean encircling the Antarctic continent. Often it is defined as all ocean areas south of 35°S (e.g., Chelton et al. 1990) or 40°S (e.g., Mestas-Nuñez et al. 1992). Because of its vast size, encompassing nearly 30% of the global ocean, because of its notoriously strong wind and wave conditions, and because it is nearly unbroken by land, the Southern Ocean is difficult to monitor from ships. Sea surface height (SSH) measurements from satellite altimetry have proved crucial in characterizing both the time-mean features and the variability of the region.

The major current of the Southern Ocean, the Antarctic Circumpolar Current (ACC, Figure 9.1), carries approximately 150 Sv (where 1 Sv = 10^6 m^3 s^{-1}) water eastward (Cunningham et al. 2003; Mazloff et al. 2010), or clockwise, around the Antarctic continent. The ACC constitutes the major link between the Atlantic, Indian, and Pacific oceans and carries significant quantities of heat, carbon, and other climatically and ecologically important tracers (Rintoul et al. 2001). The ACC is comprised of multiple narrow jets, coinciding with sharp gradients in temperature or salinity that are referred to as "fronts." Traditionally, these fronts are identified from north to south as the Subantarctic Front (SAF), the Polar Front (PF), and the Southern ACC Front (SACCF) (see Figure 9.1, Orsi et al. 1995; Sokolov and Rintoul 2009). Although ACC fronts are constrained by topography in some locations (e.g., Gordon et al. 1978; Dong et al. 2006), at most longitudes of the Southern Ocean, time series of altimetric SSH indicate substantial variability (e.g., Chelton et al. 1990), which is a signature of the meridional displacement of the fronts (e.g., Gille 1994; Gille and Kelly 1996). This temporal variability allows the fronts to be detected via satellite altimetry. In addition to the ACC fronts, the Southern Ocean contains numerous other time-varying eddy features (Figure 9.1b), and altimetry has provided a valuable window for studying eddies and their energetics as well.

The objective of this chapter is to highlight how satellite altimetry has helped to illuminate the circulation and dynamics of the Southern Ocean. In Section 2, we show how altimetry has been used to identify the position of the ACC fronts. In Section 3, we expand in the vertical to describe the use of altimetry to characterize the four-dimensional temperature and salinity structure of the Southern Ocean, and in Section 4, we use this four-dimensional perspective to evaluate transport variability and changes in the strength of the current. In Section 5, we consider the dynamics governing the Southern Ocean circulation through the lens of eddy kinetic energy and its temporal changes, as seen from altimetry. Finally, in Section 6, we summarize the key contributions from altimetry to Southern Ocean research and consider where future research might lead.

9.2 CHARACTERIZING SPATIAL VARIABILITY OF THE ANTARCTIC CIRCUMPOLAR CURRENT FROM ALTIMETRY

Given the global importance of the ACC, and given its strong surface expression marked by more than a 1-m dynamic height difference across the full width of the current, the ACC has been a focal point of satellite altimeter studies, which have sought better understanding of its mean

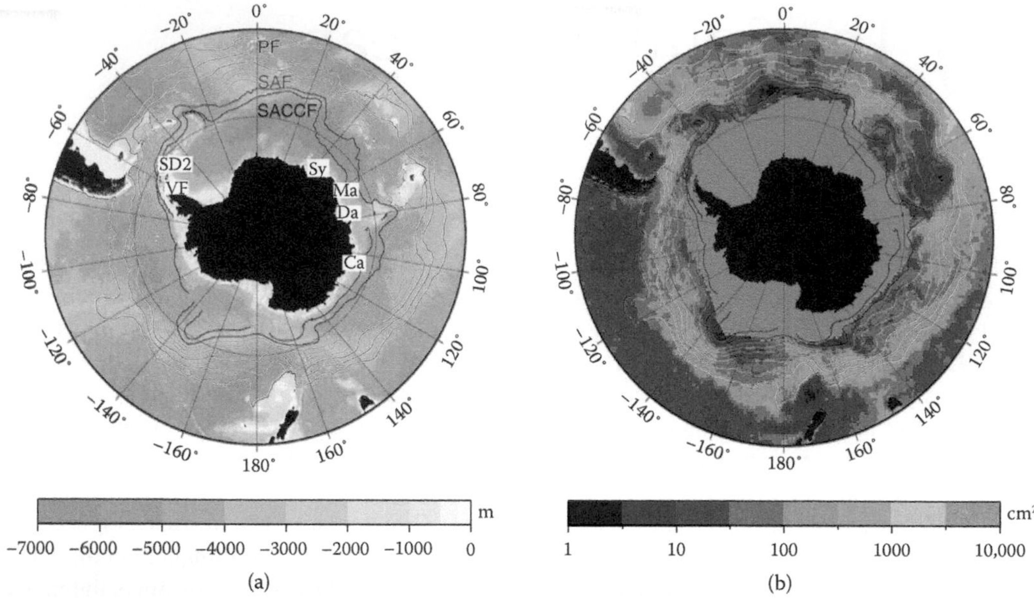

FIGURE 9.1 Schematic location of Antarctic Circumpolar Current (ACC) fronts superimposed on bathymetry (a) and on altimeter-derived eddy kinetic energy (b). Frontal positions here are from Sokolov and Rintoul (2009), who used gradients in dynamic topography to identify quasi-permanent frontal positions. The northern three lines (in red) represent the northern, central, and southern positions of the Subantarctic Front (SAF). The next three lines (in magenta) represent the northern, central, and southern positions of the Polar Front (PF). The southern two lines (in black) are the northern and southern positions of the Southern ACC Front (SACCF). (a) Letters on white backgrounds indicate the locations of tide gauges presented in Figure 9.6: (Casey, Ca; Davis, Da; Mawson, Ma; Syowa, Sy; Vernadsky/Faraday, VF; South Drake Passage, SD2).

flow, its transport variability, and the dynamics that govern its response to changing forcing. Through the geostrophic relationship, eastward flowing geostrophic velocities (u) are proportional to north-south, or meridional, gradients in SSH ($u = -g/f\ \partial\eta/dy$), where g is gravity, f is the Coriolis parameter, η is dynamic height, and y is distance in the meridional direction. Thus, strong gradients in SSH are co-located with strong surface velocities. Time-varying, meandering currents have large SSH anomalies, and altimeter data show that high mesoscale variability aligns with the mean path of the ACC (Chelton et al. 1990). Because ridges and plateaus on the sea floor can steer the currents, both the pathways of mean currents and the regions of maximum variability are strongly controlled by bathymetry (Sandwell and Zhang 1989; Chelton et al. 1990; Wunsch and Stammer 1995).

Coarse resolution climate models have suggested that ACC fronts, the geostrophic jets that define the current, could be displaced poleward as a result of a poleward displacement of the winds that drive the ACC (e.g., Fyfe and Saenko 2006; Cai 2006; Cai and Cowan 2007). The nearly 25-year-long record of SSH from altimetry offers one of the most extensive long-term time series for inferring possible frontal shifts, and a series of studies have examined strategies for inferring frontal displacement statistics from the altimeter record.

The frontal positions of the ACC can be challenging to identify from altimetry, which measures SSH variability but does not directly measure the time-mean dynamic topography—the component of SSH that depends on the time-invariant geostrophic dynamics of the ocean currents. Spatial variations in the Earth's geoid determine roughly 99% of the time-averaged spatial structure of the absolute sea surface measured by altimetry (e.g., Smith and Sandwell 1997). Because the geoid dominates the time-averaged absolute SSH, the dynamic topography associated with time-mean ocean currents is difficult to determine and requires ancillary data for gravity or ocean circulation (e.g., Maximenko and Niiler 2005; Tapley et al. 2005; Pavlis et al. 2008; Rio et al. 2009). Transport constraints have not been incorporated into geoid or mean dynamic topography products, and as a result, the best estimates of the mean dynamic topography have not been consistent with the notion that the ACC is a continuous system, transporting roughly the same volume of water at all longitudes around Antarctica (e.g., Griesel et al. 2012).

One strategy for inferring ACC frontal variability from altimetry has worked from the assumption that the biases in the mean dynamic topography products are small and that a useful time-varying dynamical topography can be inferred from the sum of the altimeter measurements of time-varying SSH plus the mean dynamic topography. Sokolov and Rintoul (2007, 2009) used gradients in dynamic topography to identify likely frontal positions, showing the ACC to be more complicated than previous studies had inferred, with multiple quasi-stable positions for each of the major fronts of the ACC. For example, their results (indicated in Figure 9.1) show that the SAF and PF both have a northern, middle, and southern position, and the SACCF has a northern and southern position. Each of these frontal positions is characterized by a specific gradient in dynamic topography and a specific dynamic height contour. These results provided a strategy for them to infer a long-term southward shift of the dynamic height contours identified with each of the ACC fronts from TOPEX/Poseidon and Jason altimetry (Sokolov and Rintoul 2009). Sallée et al. (2008) similarly used a set of fixed dynamic height contours to examine ACC frontal variability on a regional level, finding local correlations with both the Southern Annular Mode (SAM) and with an El Niño-Southern Oscillation (ENSO) index. Kim and Orsi (2014) also explored such an approach using 19 years of altimetry, finding substantial longitudinal variations in long-term trends. One caveat to identifying changes in ACC frontal positions from SSH stems from the fact that seawater expands as it warms, meaning that SSH contours can shift southward as a result of oceanic warming on seasonal to decadal timescales, without a corresponding shift in the position of the SSH gradient. Using an eddy-resolving climate model, Graham et al. (2012) noted that when used by themselves, SSH contours do not provide reliable measures of frontal variability, and they recommended supplementing SSH contours with additional information, such as SSH gradients, to monitor frontal variability.

Several studies have explored alternative methods for identifying temporal variations in ACC frontal positions without relying on fixed height contours. Gille (1994) used a meandering Gaussian jet model applied to SSH anomalies in GEOSAT altimetry to trace the mean position of the ACC fronts around the Southern Ocean, highlighting strong steering of the fronts by topography. For front identification in numerical model output, Chapman (2014) proposed a Wavelet Higher Order Statistics Enhancement approach to distinguish jets from eddies, a powerful method that does not appear to have been tested with altimeter data yet. Gille (2014) noted that if the ACC consists of many distinct fronts, then frontal position may be less important than the mean latitude of all eastward transport within the ACC. Her altimeter-based index for transport latitude showed no long-term trends and no significant correlation with SAM or ENSO indices. Shao et al. (2015) used the transition from positive to negative skewness in altimeter data (Thompson and Demirov 2006) to identify frontal positions in the ACC, also finding no long-term trend in frontal latitudes and no systematic dependence on SAM or ENSO.

A further question about ACC variability focuses on how jets shift among adjacent topographic constraints. At three specific points where the ACC is topographically constrained, Chapman and Morrow (2014) showed evidence for "jet jumping," computed as a lateral shift in the position of surface transport (computed as $g/f\ \Delta\eta$, where $\Delta\eta$ is the SSH jump between specified end points.)

9.3 MAPPING THE TIME-VARYING, THREE-DIMENSIONAL STRUCTURE OF THE SOUTHERN OCEAN

One fundamental challenge in the Southern Ocean has been to map not just the time-evolving horizontal structure of the Southern Ocean but also the vertical structure of water mass properties, which is critical, for example, for detecting and attributing climatic changes in the interior of the

Southern Ocean. Satellite altimetry can observe large-scale, time-varying spatial structures but nothing below the ocean surface, so satellites are not always the obvious tool for reconstructing three-dimensional oceanic fields. However, because of the low stratification of the Southern Ocean, surface and subsurface fields are strongly correlated (Sokolov and Rintoul 2003), and the ACC has sometimes been described as "equivalent barotropic" (e.g., Gordon et al. 1978; Killworth 1992), meaning that surface velocities are aligned with deep velocities, with an exponential decay with depth from strong surface velocities to weaker subsurface velocities (e.g., Gille 2003), and variability is conceived to be correlated throughout the water column, suggesting that ACC frontal positions derived from altimetry could also be used to infer top-to-bottom structure. Detailed investigation of the equivalent barotropic framework for the ACC has revealed that the system is not separable, meaning that vertical length-scales vary with latitude (Firing et al. 2011), and instantaneous *in situ* observations show velocities that rotate with depth (Phillips and Bindoff 2014). Nonetheless, if a degree of time averaging is applied, the strong correlation between surface and subsurface variability provides a useful framework for inferring the structure of temperature, salinity, and density fields in the sparsely sampled Southern Ocean.

Meijers et al. (2011a) adopted the Gravest Empirical Mode (GEM) approach (e.g., Meinen and Watts 2000; Meinen 2001; Sun and Watts 2001) to use satellite altimetry to investigate the changing internal structure of the Southern Ocean (see also Swart et al. 2010). This approach, referred to as satGEM, provides a statistical methodology to reconstruct a time-evolving, three-dimensional temperature/salinity field. GEM approaches can work well in regions dominated by strong horizontal gradients in bulk water column structure, and GEM was adopted previously for interpretation of inverted echo sounder data from the ACC chokepoints during the World Ocean Circulation Experiment (WOCE; Meredith et al. 1997; Watts et al. 2001). Instead of using acoustic travel time as the input variable, Meijers et al. (2011a) used altimeter-derived SSH, the spatial and temporal coverage of which enabled complete circumpolar fields of internal structure to be estimated. Fields of vertically varying (baroclinic) velocity were derived, with sufficient resolution that mesoscale features could be identified. As an example, Figure 9.2 shows reconstructed current speed on four neutral density surfaces for 1 day. Further, Meijers et al. (2011b) used this satGEM product to show that adiabatic changes (i.e., changes that require no extra heat or freshwater to be input, such as would be associated with movements of the ACC fronts) are insufficient to explain the long-term trends in Southern Ocean temperature and salinity. Accordingly, diabatic changes (i.e., changes in the heat and freshwater content of the Southern Ocean water masses) must also have occurred over recent decades.

Dynamical constraints are not part of the satGEM approach, but dynamics can be incorporated into data analysis using an assimilating model. Data assimilation optimally combines observations (e.g., satellite altimetry and observed profile information) in order to constrain a primitive equation numerical model, taking dynamical processes into consideration to obtain best estimates of ocean temperature/salinity properties and circulation (e.g., Wunsch and Heimbach 2007; see also Chapter 17 of this book). The Southern Ocean State Estimate (SOSE, Mazloff et al. 2010) is a 1/6° resolution regional version of the Estimating Circulation and Climate of the Ocean (ECCO) model that employs four-dimensional variational assimilation—a method that constrains the model to observations in time and space in a way that is consistent with the model physics, without introducing artificial forcing in the interior of the domain. SOSE focuses on the sector of the global ocean south of 20°S, and thus SOSE development efforts emphasize issues critical to the Southern Ocean, including improving the sea ice model and maintaining sufficient resolution to represent as much as possible of the small eddies that typify the Southern Ocean. Figure 9.3 shows time mean (a) and instantaneous (b) SOSE SSHs. SOSE output fields have been exploited for a broad range of purposes, including studies of mean dynamic topography (Griesel et al. 2012), eddy processes (e.g., Sallée and Rintoul 2011; Klocker et al. 2012), and air–sea–ice exchange (e.g., Cerovecki et al. 2011; Mazloff 2012; Abernathey et al. 2016).

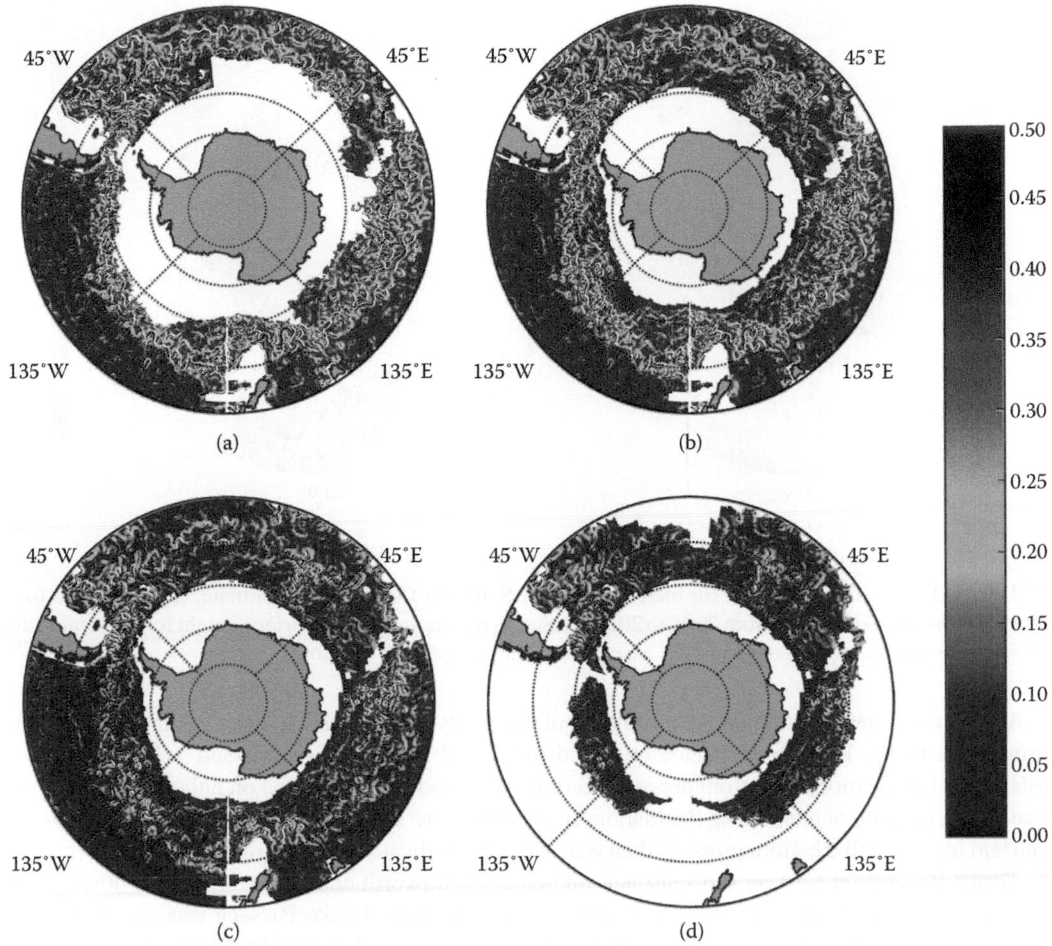

FIGURE 9.2 Speed of ocean currents on May 31, 1995 on four neutral density surfaces, estimated from satGEM. The four panels show speed at neutral densities of (a) 26.85 kg m^{-3}, roughly corresponding to Subantarctic Mode Water; (b) 27.4 kg m^{-3}, roughly corresponding to Antarctic Intermediate Water; (c) 28 kg m^{-3}, roughly corresponding to Circumpolar Deep Water; and (d) 28.2 kg m^{-3}, roughly corresponding to Antarctic Bottom Water. Units are cm s^{-1}. White regions indicate locations that are ice covered or where the density class is not found. (Figure courtesy of Andrew Meijers, adapted from a depth based map by Meijers, A., et al., *J. Atmos. Oceanic Technol.*, 28, 548–568, 2011a.)

9.4 ACC TRANSPORT FROM ALTIMETRY

In situ studies of ACC transport have often focused on the chokepoint of the current, with the assumption that ACC transport should be roughly consistent at all longitudes around Antarctica (e.g., Georgi and Toole 1982). The Drake Passage, the region between South America and the Antarctic Peninsula, is the narrowest chokepoint through which the full ACC must flow. It captures the full width of the ACC without complications relating to the presence of adjacent recirculating subpolar and subtropical gyres, making it one of the cleanest locations for transport studies. It is also a location where comparatively comprehensive deployments of *in situ* instrumentation have occurred, thus offering the possibility of integrating the remotely sensed data with *in situ* data in order to test and verify the results obtained and to maximize the utility of both approaches.

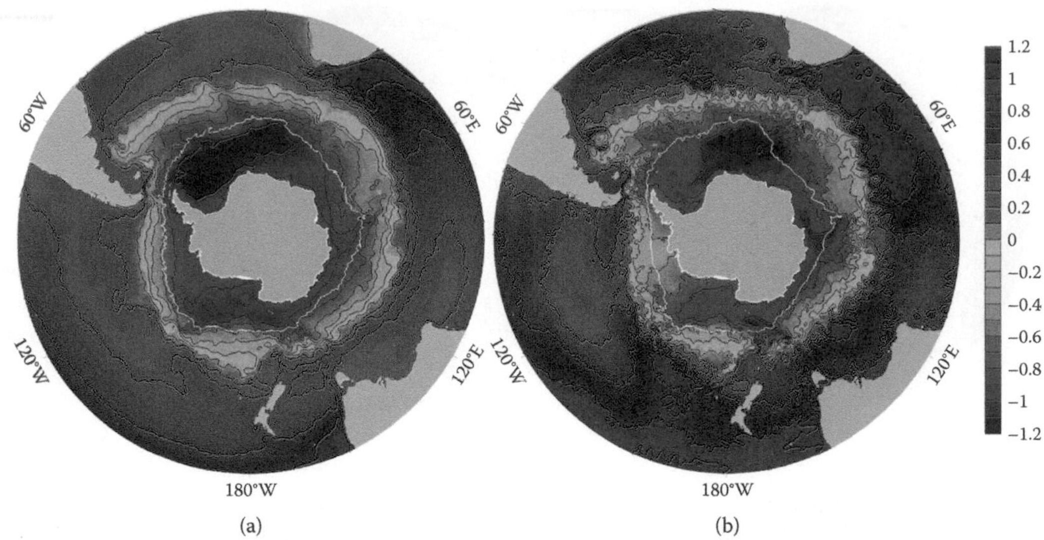

FIGURE 9.3 Sea surface height (in meters) from the Southern Ocean State Estimate. (a) Shows the time mean SSH for a 6-year period from 2005 to 2010. (b) Shows instantaneous sea surface height for September 26, 2007. White lines indicate maximum sea ice extent. (Courtesy of Matt Mazloff.)

Among these investigations was an early test of the ability of altimeters to detect changes in the flow through the Drake Passage conducted by Woodworth et al. (1996), who extracted TOPEX/Poseidon SSHs from the locations of bottom pressure recorders that had been deployed on either side of the passage for the purpose of monitoring fluctuations in the ACC flow. Previous studies that used purely *in situ* data had indicated that bottom pressure data were a reliable indicator of ACC transport changes through the Drake Passage, at least on sub-seasonal timescales (Whitworth and Peterson 1985). Although the agreement between SSH and bottom pressure at each side of the Drake Passage was reasonable, the presence of locally strong mesoscale variability and baroclinic signals led to the determination of transport changes being problematic. It was concluded that satellite altimetry alone was not capable of determining the Drake Passage transport changes with the required accuracy and that *in situ* instrumentation would be needed for the indefinite future as part of the ongoing monitoring effort.

Challenor et al. (1996) provided a separate example of the contemporaneous use of satellite altimetry and *in situ* data, using European Remote Sensing (ERS) 1 altimetry along a ground track that was collocated with the Drake Passage repeat section occupied as part of WOCE. *In situ* data from a towed undulator enabled surface geostrophic currents to be determined and used as absolute references for the altimetry-derived surface velocities. This enabled a time series of absolute surface current velocity to be derived with the temporal resolution of the satellite repeat (35 days). Distinct jets with peak velocities of 0.75–1 ms^{-1} were resolved, separated by more quiescent zones of water, consistent with previous conceptions based on sparse ship sections (e.g., Nowlin and Clifford 1982).

The Challenor et al. (1996) study was updated subsequently by Cunningham and Pavic (2007). Several full-depth hydrographic sections along the same transect were used in combination with the much longer series of altimetry available (1992–2004) to produce longer series of surface geostrophic currents. These surface currents were converted to transport anomalies by imposing an assumption that the velocity anomalies were independent of depth. Resulting 7-day transport anomalies had a standard deviation of 17 Sv, and interannual transport fluctuations had fluctuations of approximately 8 Sv (Cunningham and Pavic 2007).

Comparisons of the surface currents and transports with the SAM index indicated only weak relationships, and this was interpreted as indicating a dynamical decoupling of Southern Ocean winds and ACC transport. However, Meredith and Hughes (2005) demonstrated that SSH data along a single altimeter ground track provide insufficient information to construct reliable estimates of transport variability. This critical shortcoming stems from aliasing of higher-frequency fluctuations, due to the altimeter resampling time interval. To advance our understanding of the implications of this constraint, Meredith and Hughes (2005) studied in detail the altimetric sampling requirements for reliably detecting seasonal and interannual changes in ACC transport, using sea level data from tide gauges and models, which had been shown separately to reliably depict the changes in transport to an acceptable level of accuracy (Meredith et al. 2004). They determined the largest sampling interval that would still enable reliable calculations of annual mean ACC transport through the Drake Passage and found that sampling at a rate faster than once per week was required, even if the transport could be determined perfectly on each realization (i.e., that there was zero measurement error). In reality, given any reasonable expectation of measurement error, the sampling rate required would be much faster. This indicates that ACC transport variability estimates derived from a single altimeter ground track alone are indeed aliased, and that *in situ* instrumentation and/or the inclusion of satellite altimetry from multiple ground tracks is required.

Recently, interest has been rekindled in using altimetry from the Drake Passage to determine ACC transport changes, due in no small part to the deployment of new arrays of *in situ* instrumentation that enable further comparison and inclusion of subsurface information in the calculations. Notably, Ferrari et al. (2012) and Koenig et al. (2014) used velocity profile time series from moored acoustic Doppler current profilers (ADCPs) to provide accurate vertical structure estimates in the upper 350 m of the water column, supplementing the altimetric-derived surface velocities to obtain time-varying transport estimates of the full water column transport and its components, as illustrated in Figure 9.4. Koenig et al. (2014) obtained a mean top-to-bottom Drake Passage transport of 141 Sv, with a standard error of 2.7 Sv, a standard deviation of 13 Sv, and a range of 110 Sv. Yearly means varied from 133.6 Sv in 2011 to 150 Sv in 1993 and standard deviations from 8.8 Sv in 2009 to 17.9 Sv in 1995 (see Figure 9.4). These values compare well with the canonical values obtained during the early International Southern Ocean Studies (ISOS) program at the Drake Passage (mean 133.8 Sv and standard deviation 11.2 Sv; Whitworth 1983) and with the Drake Passage transport estimates from SOSE of 147 ± 5 Sv (e.g., Griesel et al. 2012). They also compare well with more

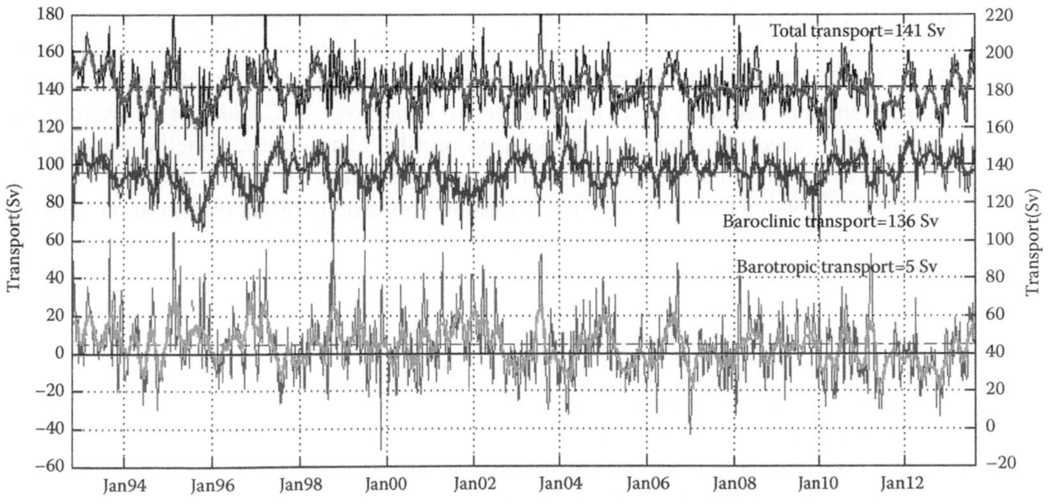

FIGURE 9.4 Drake Passage transport from Koenig et al. (2014) showing total top-to-bottom transport (in black), baroclinic transport (in red, transport relative to the bottom), and barotropic transport (in magenta, the component of transport associated with non-zero flow at the bottom). Time series smoothed with a 1-month running mean are in green, blue, and light blue, respectively. Total mean: 141 Sv, std: 13 Sv Barotropic mean: 5 Sv, std: 16 Sv. Baroclinic mean: 136 Sv; std: 11 Sv. The red right axis refers to the baroclinic transport only.

recent independent baroclinic estimates obtained from large arrays of current- and pressure-recording inverted echo sounders (127.7 ± 1.0 Sv; standard deviation 8.1 Sv) (Chidichimo et al. 2014).

Numerous studies have applied satellite altimetry techniques outside the Drake Passage in order to better constrain the transport and variability of the ACC in other parts of the Southern Ocean. For example, Swart et al. (2008) used gridded altimeter data in combination with expendable bathythermograph (XBT) and hydrographic section data to constrain the baroclinic variability of the ACC transport south of South Africa. The altimetry-derived mean baroclinic transport of the ACC, relative to 2500 dbar, was 84.7 ± 3.0 Sv, with the bulk of the transport (88%) residing in the northernmost two fronts (the SAF and PF). This work extended and developed the methods used by Rintoul et al. (2002), who utilized a similar data set south of Australia to investigate the mean ACC flow and its variability there, and who determined a standard deviation of net transport above and relative to 2500 m of 4.3 Sv. The variability in net transport was found to be largest (2.7 Sv) in the quasi-annual band (4 months to 1.5 years), slightly smaller (2.3 Sv) in the mesoscale band (less than 4 months), and smallest in the interannual band (greater than 1.5 years, 1.5 Sv).

In Fawn Trough, which carries less than a third of the net ACC transport on a pathway south of Kerguelen, Vivier et al. (2014) found that altimetry and current meter measurements were not uniformly correlated but that by regressing altimeter measurements against mooring measurements from the northern side of Fawn Trough, they developed an *ad hoc* method to infer transport from altimetry. Their results implied a gradual increase in the transport through Fawn Trough, with significant interannual variability. Although the observed transport variations suggested possible localized SAM or ENSO responses, neither was significant at a 95% level, implying a need for further study on wind forcing impacts on Fawn Trough transport.

At Macquarie Ridge, Rintoul et al. (2014) used altimeter data to evaluate whether the major transport of the SAF was passing through the two gaps in the ridge in which they had deployed a mooring array. Their results showed that the middle core of the SAF passed through an instrumented gap 96% of the time, and the northern core of the SAF passed through an instrumented gap 76% of the time. Thus, they could adjust their mooring-based transport estimates to account for possible underestimation when either of the frontal expressions meandered out of the instrumented array.

Although ACC transport is generally assumed to be consistent at all longitudes around Antarctica (Georgi and Toole 1982), detailed studies often show that ACC transport and variability differ at different locations. For studies using altimeter data, transport inconsistencies may stem partially from difficulties in determining the time-mean dynamic ocean topography or geoid. As noted earlier, Griesel et al. (2012) found that although numerical models do conserve transport along the length of the ACC, none of the four geoid products that they tested conserved ACC transport. Kosempa and Chambers (2014) fine-tuned the approach, using Argo profiles along with altimetry, and reached similar conclusions. These results suggest limitations in current absolute ACC transport estimates but imply that further work on the geoid and mean dynamic topography might lead to improvements.

A possible explanation for the spatially varying ACC transport variability is that the unusual bathymetry of the Southern Ocean permits different modes of transport variability to exist. Most quantifications of transport that span the major fronts of the ACC show little variability, subject to adequate sampling. The reasons for this low variability are becoming increasingly well understood. However, Hughes et al. (1999) demonstrated the presence of a "southern mode" around Antarctica—closely linked contours of planetary vorticity divided by depth that nearly connect around the continent. They argued that this mode is a dominant mode of circumpolar transport variability, explaining around half the total variance in transport in the Drake Passage. Using data from coastal tide gauges, Hughes et al. (2003) subsequently demonstrated that this mode has strong circumpolar coherence in transport changes. The spatial pattern of this mode, although strongly tied to bathymetric contours close to Antarctica around a significant part of its path, extends outward into the Southern Ocean in some locations. Figure 9.5, from Hughes et al. (2003), illustrates this mode by showing the significant correlation between negative altimetric SSH along the Antarctic margins and positive Drake Passage transport from the Ocean Circulation and Climate Advanced

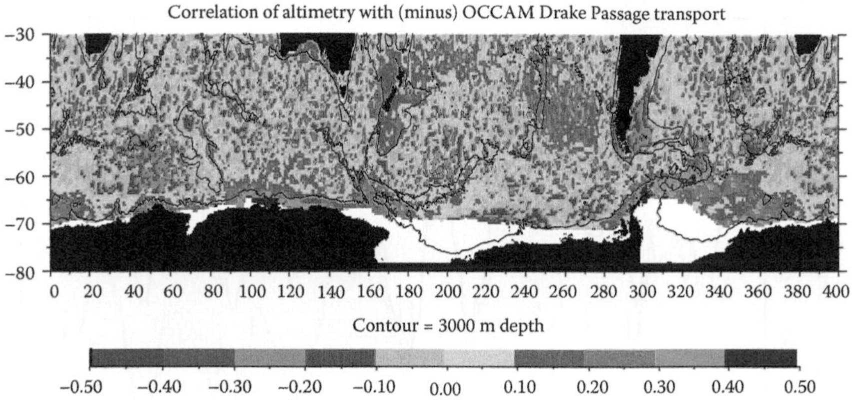

FIGURE 9.5 Correlation of altimeter-derived sea level with modeled Drake Passage transport from the Ocean Circulation and Climate Advanced Modeling (OCCAM) project forced with realistic forcing. White areas are those influenced by sea ice; correlations are plotted inverted. The strong positive correlation close to Antarctica and extending across the Weddell Sea, indicates that strong Drake Passage transport coincides with depressed sea level around most of Antarctica, implying that sea level fluctuations at the southern boundary of the Antarctic Circumpolar Current (ACC; sometimes termed the "southern mode") can be a useful signature of transport changes. The 3000-m contour is marked. (From Hughes, C. W., et al., *Geophys. Res. Lett.*, 30(9), 1464, 2003. With permission.)

Modeling (OCCAM) project. A number of other studies have also demonstrated that satellite altimeter-derived SSH from across this region can provide a valuable index of the changes in transport associated with this mode on sub-annual timescales (Hughes et al. 2003; Vivier et al. 2005; Hughes and Meredith 2006). In Section 2, we noted that fluctuations in the SAM are not correlated with the latitude of ACC fronts, but here we see that transport differs markedly from latitude: Figure 9.6

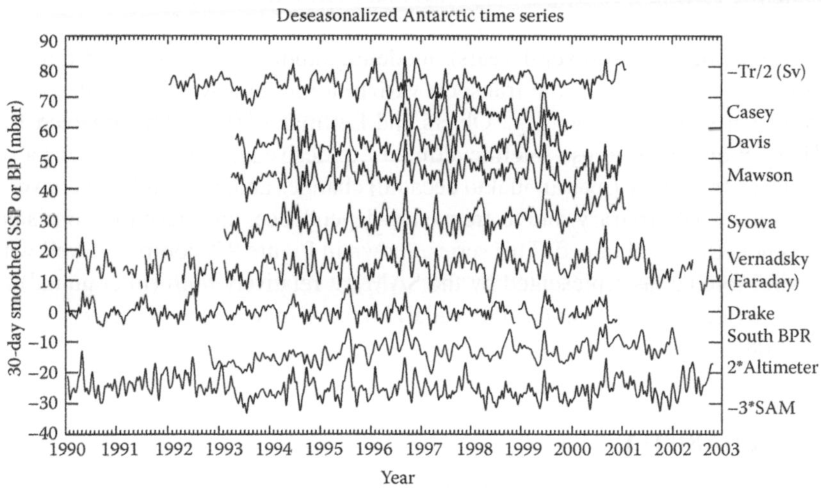

FIGURE 9.6 Deseasonalized time series of transport (top), sea level from tide gauges around Antarctica (Casey, Ca; Davis, Da; Mawson, Ma; Syowa, Sy; Vernadsky/Faraday, VF; South Drake Passage, SD2), sea level from altimetry (Alt), and the Southern Annular Mode (SAM). The altimetric sea level is obtained from the spatial region indicated in Figure 9.5. The tide gauge locations are marked on Figure 9.1a. Note the strong similarity of the tide gauge sea level data, the altimeter-derived sea level, the modeled Antarctic Circumpolar Current (ACC) transport, and the SAM, collectively illustrating the rapid response of the ACC to changing wind forcing on sub-seasonal timescales, and the robust representation of those transport changes in circumpolar SSH. (From Hughes, C. W., et al., *Geophys. Res. Lett.*, 30(9), 1464, 2003. With permission.)

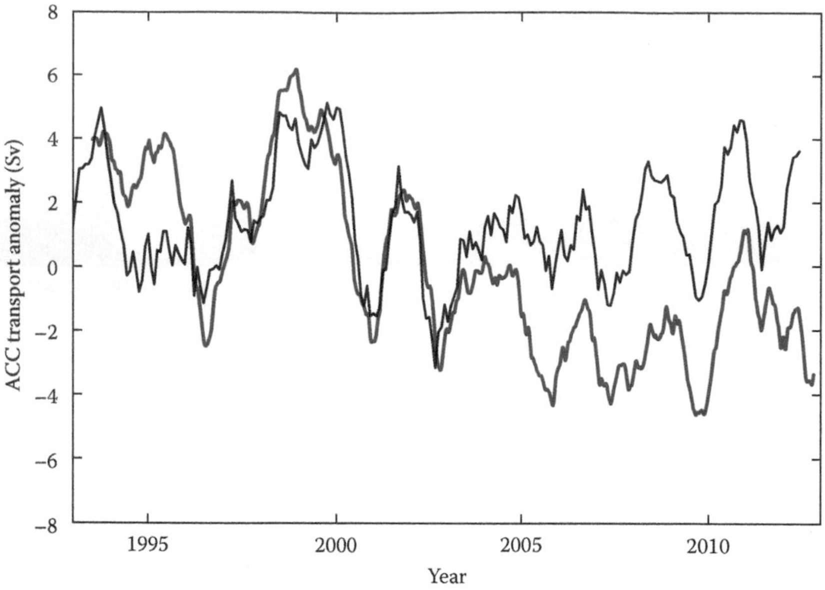

FIGURE 9.7 Antarctic Circumpolar Current (ACC) transport anomaly (magenta) estimated from circum-Antarctic sea level obtained from altimetry. Also shown is the 1-year running mean SAM index (black). Note the coherence of ACC transport with SAM changes on interannual timescales. (From Hogg, A. M., et al., *J. Geophys. Res.*, 120, 257–267, 2015. With permission.)

indicates a strong correlation on seasonal-to-interannual timescales between the transport of this mode and changing forcing from the zonal winds that overlie the Southern Ocean (as quantified by the SAM).

On longer timescales (up to several years), modeling studies have indicated that this southern mode remains a significant part of the transport variability around Antarctica, despite the increasing relevance of baroclinic changes (e.g., Olbers and Lettmann 2007). This enabled Hughes et al. (2014) and Hogg et al. (2015) to use satellite altimetry to derive an estimate of transport variability including the lower-frequency (interannual to decadal) changes and, although the conversion of SSH to transport is somewhat arbitrary due to frequency-dependence, an internally consistent transport time series was nonetheless obtained. This series, shown in Figure 9.7, compared well with estimates of changing wind forcing (as represented by the SAM) at relatively high (interannual) frequencies, but there was no evidence for a long-term (decadal) increase in transport (Hogg et al. 2015; see also Böning et al. 2008).

9.5 EDDY VARIABILITY

The Southern Ocean is rich in eddies with SSH variance maxima aligned along the path of the ACC (e.g., Zlotnicki et al. 1989; Chelton et al. 1990; Gille 1994). Figure 9.1b shows an intensification of eddy kinetic energy (EKE, computed from geostrophic velocity anomalies as the time average of $[u'^2+v'^2]/2$) along the ACC. Southern Ocean eddies are hypothesized to govern the meridional transport of momentum and heat across the ACC (e.g., de Szoeke and Levine 1981) and to be linked to the zonal transport (e.g., Johnson and Bryden 1989). Altimetry has proved a valuable resource for diagnosing the impact of eddies on horizontal stirring and mixing (e.g., Marshall et al. 2006; Abernathey and Marshall 2013).

Eddies are particularly important in efforts to understand ACC transport variability both from *in situ*- and satellite altimeter-based studies. At comparatively long timescales (interannual and longer), winds are expected to drive transport changes, but transport variability is proportionally small (less than a 5% transport change) compared with the observed approximately 20% strengthening of Southern Ocean wind stress from 1980 to 2010 (e.g., Swart and Fyfe 2012; Gent 2015). Further, although the SAM has shown a marked upward trend in recent decades, due at least partially to anthropogenic influences, as Figure 9.7 indicates, there is no evidence of a detectable acceleration in ACC transport in response (e.g., Meredith et al. 2011).

One putative reason for this is given by theoretical arguments that the baroclinic response time of the ACC transport to a change in forcing is long (centuries; Allison et al. 2011); however, internal ACC dynamics are also believed to play a significant role in modulating transport changes. As noted earlier, the Southern Ocean contains some regions of markedly high EKE (see Figure 9.1b), associated in particular with meanderings of the ACC current cores and the generation of mesoscale eddies. Dynamical and modeling studies have suggested that a regime can exist within which extra energy imparted to the ACC from winds is cascaded to this eddy field, with little or no acceleration in ACC transport as a result (Straub 1993; Hallberg and Gnanadesikan 2001).

Meredith and Hogg (2006) used satellite altimeter-derived quantifications of EKE in the Southern Ocean to look for the existence of this "eddy-saturated" regime. They found that, in response to the strong positive SAM event in the late 1990s, there was an enhancement of EKE across the Southern Ocean around 2–3 years later (Figure 9.8). This enhancement was marked by regions of high mean EKE having anomalously high values during the period of interest rather than a general rise in background levels (Figure 9.9b). The Pacific and Indian Ocean sectors were noted to be most sensitive and to show the greatest proportional changes in EKE; this was expanded on subsequently by Morrow et al. (2010), who noted that the ENSO phenomenon also modulates the wind forcing in these regions. The lag of 2–3 years was attributed to the timescale of eddy interactions with the bottom topography to drain the extra energy imparted to the system.

Thompson and Garabato (2014) recently examined the EKE changes in the Southern Ocean, focusing on local regions of high EKE linked to standing meanders in the ACC flow. They

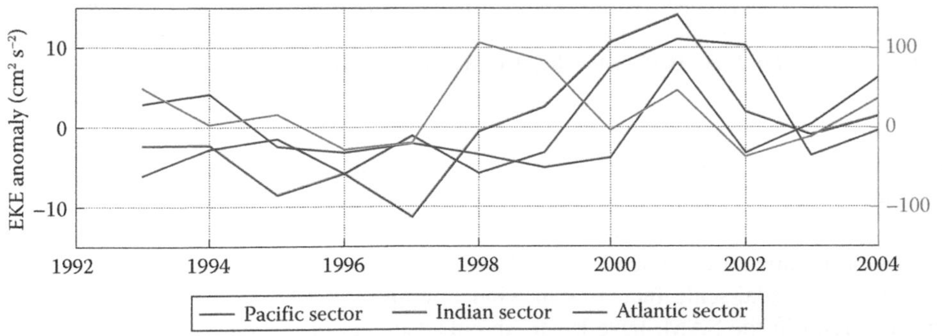

FIGURE 9.8 Annual means of eddy kinetic energy (EKE) from the Pacific (dark blue), Indian (green), and Atlantic (red) sectors of the Southern Ocean. Also shown are annual means of the Southern Annular Mode (SAM) index (light blue). Note the strong peak in SAM in the late 1990s, followed 2–3 years later by peaks in EKE. EKE is averaged over the boxes shown in Figure 9.9a. (From Meredith, M. P., and Hogg, A. M., *Geophys. Res. Lett.*, 33(16), L16608, 2006. With permission.)

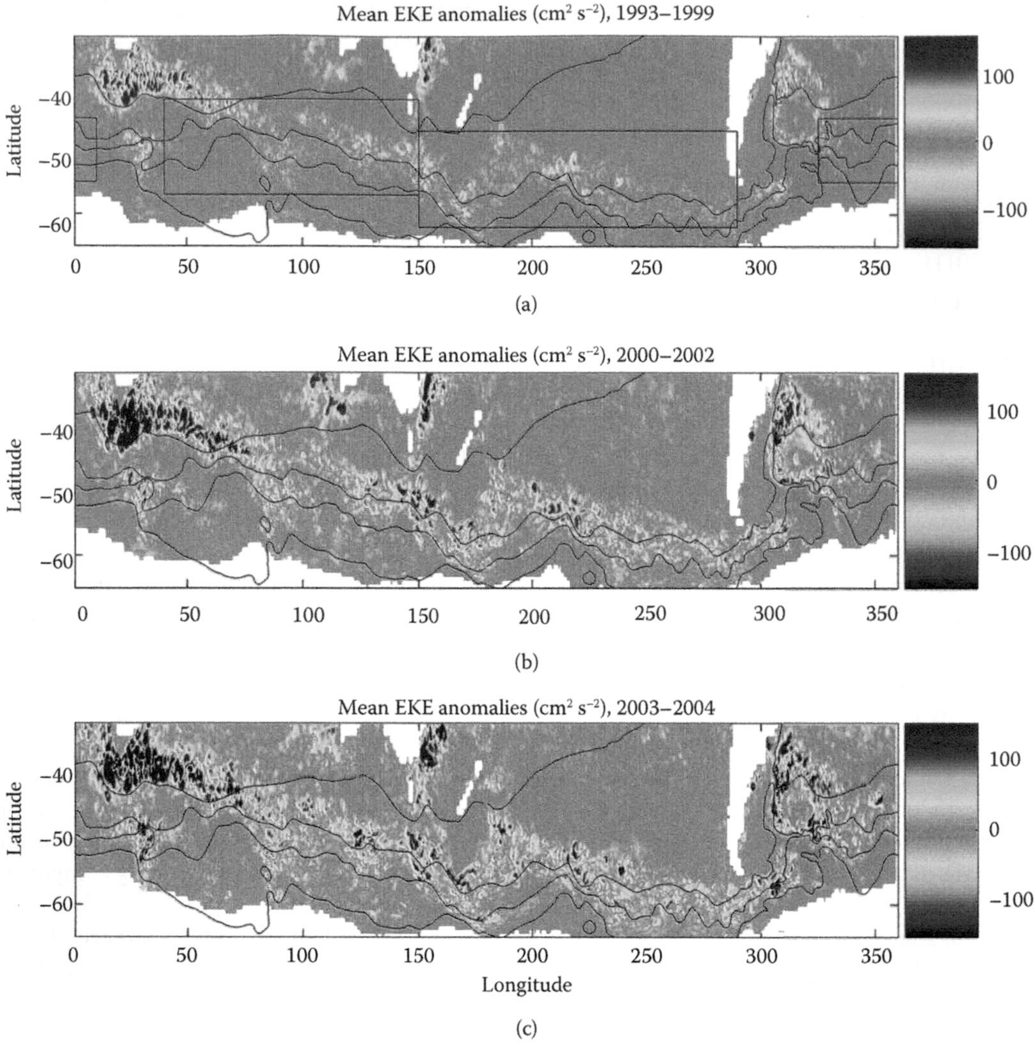

FIGURE 9.9 Spatial patterns of eddy kinetic energy (EKE) anomalies during (a) 1993–1999, (b) 2000–2002, and (c) 2003–2004. (b) Corresponds to the period of elevated EKE seen in Figure 9.8. (From Meredith, M. P., and Hogg, A. M., *Geophys. Res. Lett.*, 33(16), L16608, 2006. With permission.)

argued that the flexing of these meanders was an effective mechanism for reducing the sensitivity of the baroclinicity of the ACC to changes in forcing and hence effecting an eddy saturated regime.

Given the observational evidence for an (almost) eddy-saturated ACC, an obvious question is thus whether the decadal-scale increases in wind forcing over the Southern Ocean, as depicted in an upward trend in the SAM, have led to an enhancement of the EKE field. This was investigated by Hogg et al. (2015), who quantified basin-scale EKE using the same regional boxes as Meredith and Hogg (2006). They found strong, statistically significant rising trends in the Indian and Pacific Ocean sectors and a smaller (non-significant) rising trend in the Atlantic (Figure 9.10). They noted that this has profound implications for controls on the overturning circulation in the Southern Ocean, and hence on the drawdown of carbon and heat from the atmosphere, in addition to suppressing the transport change of the ACC.

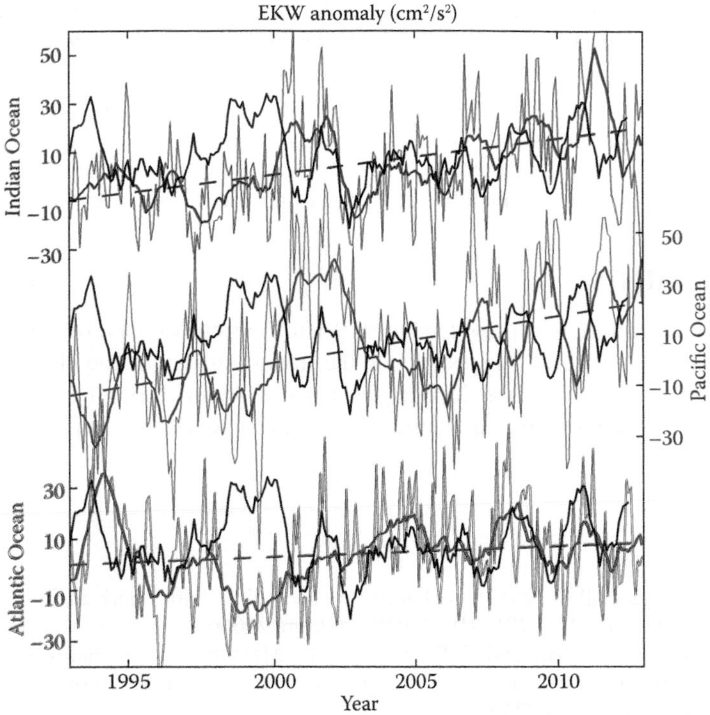

FIGURE 9.10 Time series of eddy kinetic energy (EKE) averaged over the boxes shown in Figure 9.9. Upward trends exist in each of the EKE series and are statistically significant in the Indian and Pacific Ocean sectors. Also shown (black) is the Southern Annular Mode (SAM) for the period of interest. (From Hogg, A. M., et al., *J. Geophys. Res.*, 120, 257–267, 2015. With permission.)

9.6 SUMMARY AND CONCLUSIONS

Satellite altimetry has proved valuable for characterizing the large-scale time-varying structure of the system, particularly because the Southern Ocean is difficult to measure from ships. Altimetry both provides metrics of frontal migration and transport changes over time and space and also provides statistical means to infer the time-varying, three-dimensional structure of the system (through satGEM or assimilation). Although the long-term shifts in the SAM have led to a hypothesis that we might expect equivalent long-term shifts in ACC position or transport, recent altimeter analyses have led to a nuanced view of this hypothesis: Although evidence of large-scale frontal displacements is inconclusive and there is no strong evidence for a decadal change in ACC transport, altimetry shows evidence for significant transport variability on interannual timescales and for multiyear shifts in the level of eddy kinetic energy within the Southern Ocean.

Southern Ocean eddies have spatial scales of just a few kilometers across much of the Southern Ocean (e.g., Chelton et al. 1998), and hence are not well resolved in Ku-band satellite altimeter products from recent-generation missions such as TOPEX/Poseidon, Jason-1, Jason-2, Jason-3, or Envisat. New types of altimeters, such as AltiKa, which uses a Ka-band radar and Sentinel-3, Jason-CS, or CryoSat-2, which use a Synthetic Aperture Radar (SAR) mode altimeter, offer the promise of much higher resolution measurements and the possibility of probing mesoscale or sub-mesoscale motions along the satellite nadir track (e.g., Dibarboure et al. 2014). The Surface Water Ocean Topography (SWOT) altimeter, due to launch in 2021, will expand further on this capability by measuring SSH along a 100-km wide swath on either side of the satellite nadir beam, offering the possibility to examine mesoscale or sub-mesoscale motions in two dimensions

(Fu et al. 2009, 2012). Existing altimeter measurements have exposed the details of balanced, geostrophic eddy-scale features in the ocean, but preliminary evidence from ADCP data in the Drake Passage and from a 1/48° model with tidal forcing suggest that, at least in the Southern Ocean, at scales smaller than 50–70 km, the dominant motions measured by altimetry may no longer be in geostrophic balance but instead may carry the signature of rapidly moving internal waves (Rocha et al. 2016). This new era in high-resolution satellite altimetry offers the potential for oceanographers to explore an entirely new range of physical processes in the ocean.

ACKNOWLEDGMENTS

Many thanks to Matt Mazloff, who provided Figure 9.3, to Andrew Meijers, who provided Figure 9.2, and to Steve Rintoul who offered suggestions on the initial outline for the chapter. STG was supported by the NASA Ocean Surface Topography Science Team through award NNX13AE44G. MPM acknowledges funding from the Natural Environment Research Council, most recently via award NE/N018095/1.

REFERENCES

Abernathey, R. P., and Marshall, J. (2013). Global surface eddy diffusivities derived from satellite altimetry. *J. Geophys. Res. Oceans*, 118, 901–916. doi:10.1002/jgrc.20066.

Abernathey, R. P., Cerovecki, I., Holland, P. R., Newsom, E., Mazloff, M., and Talley, L. D. (2016). Water-mass transformation by sea ice in the upper branch of the Southern Ocean overturning. *Nat. Geosci.*, 9, 596–601. doi:10.1038/ngeo2749.

Allison, L. C., Johnson, H. L., Marshall, D. P. (2011). Spin-up and adjustment of the Antarctic Circumpolar Current and global pycnocline. *J. Mar. Res.*, 69, 167–189.

Böning, C., Dispert, A., Visbeck, M., Rintoul, S. R., and Schwarzkopf, F. U. (2008). The response of the Antarctic Circumpolar Current to recent climate change. *Nat. Geosci.*, 1, 864–869. doi:10.1038/ngeo362.

Cai, W. (2006). Antarctic ozone depletion causes an intensification of the Southern Ocean super-gyre circulation. *Geophys. Res. Lett.*, 33, L03712. doi:10.1029/2005GL024911.

Cai, W., and Cowan, T. (2007). Trends in Southern Hemisphere circulation in IPCC AR4 models over 1950–99: Ozone depletion versus greenhouse forcing. *J. Clim.*, 20, 681–693.

Cerovecki, I., Talley, L. D., and Mazloff, M. R. (2011). A comparison of Southern Ocean air–sea buoyancy flux from an ocean state estimate with five other products. *J. Clim.*, 24(24), 6283–6306.

Challenor, P. G., Read, J. F., Pollard, R. T., and Tokmakien, R. T. (1996). Measuring surface currents in Drake Passage from altimetry and hydrography. *J. Phys. Oceanogr.*, 26, 2748–2759. doi:10.1175/1520-0485(1996)026%3C2748:MSCIDP%3E2.0.CO;2.

Chapman, C. C. (2014). Southern Ocean jets and how to find them: Improving and comparing common jet detection methods. *J. Geophys. Res. Oceans*, 119, 4318–4339. doi:10.1002/2014JC009810.

Chapman, C. C., and Morrow, R. (2014). Variability of Southern Ocean jets near topography. *J. Phys. Oceanogr.*, 44(2), 676–693.

Chelton, D. B., de Szoeke, R. A., Schlax, M. G., El Naggar, K., and Siwertz, N. (1998). Geographical variability of the first baroclinic Rossby radius of deformation. *J. Phys. Oceanogr.*, 28, 433–460.

Chelton, D. B., Schlax, M. G., Witter, D. L., and Richman, J. G. (1990). GEOSAT altimeter observations of the surface circulation of the Southern Ocean. *J. Geophys. Res.*, 95(C10), 17877–17903. doi:10.1029/JC095iC10p17877.

Chidichimo, M. P., Donohue, K. A., Watts, D. R., and Tracey, K. L. (2014). Baroclinic transport time series of the Antarctic Circumpolar Current measured in Drake Passage. *J. Phys. Oceanogr.*, 44, 1829–1853. doi:10.1175/JPO-D-13-071.1.

Cunningham, S. A., Alderson, S. G., King, B. A., and Brandon, M. A. (2003). Transport and variability of the Antarctic Circumpolar Current in Drake Passage, *J. Geophys. Res.*, 108(C5), 8084. doi:10.1029/2001JC001147.

Cunningham, S. A., and Pavic, M. (2007). Surface geostrophic currents across the Antarctic Circumpolar Current in Drake Passage from 1992 to 2004. *Prog. Oceanogr.*, 73(3–4), 296–310. doi:10.1016/j.pocean.2006.07.010.

de Szoeke, R. A., and Levine, M. D. (1981). The advective flux of heat by mean geostrophic motions in the Southern Ocean. *Deep Sea Res., Part A*, 28, 1057–1085.

Dibarboure, G., Boy, F., Desjonqueres, J. D., Labroue, S., Lasne, Y., Picot, N., Poisson, J. C., and Thibaut, P. (2014). Investigating short-wavelength correlated errors on low-resolution mode altimetry. *J. Atmos. Ocean. Tech.*, 31(6), 1337–1362.

Dong, S., J. Sprintall, and S. T. Gille, 2006. Location of the Polar Front from AMSR-E satellite sea surface temperature measurements, *J. Phys. Oceanogr.*, 36, 2075–2089.

Ferrari, R., Provost, C., Renault, A., Sennéchael, N., Barré, N., Park, Y.-H., and Lee, J. H. (2012). Circulation in Drake Passage revisited using new current time series and satellite altimetry: 1. The Yaghan Basin. *J. Geophys. Res.*, 117, C12024. doi:10.1029/2012JC008264.

Firing, Y. L., Chereskin, T. K., and Mazloff, M. R. (2011). Vertical structure and transport of the Antarctic Circumpolar Current in Drake Passage from direct velocity observations. *J. Geophys. Res.*, 116, C08015. doi:10.1029/2011JC006999.

Fu, L.-L., Alsdorf, D., Rodriguez, E., Morrow, R., and Mognard, N. (2012). *SWOT: The Surface Water and Ocean Topography Mission Wide-Swath Altimetric Measurement of Water Elevation on Earth.* JPL Publication 12-05. California Institute of Technology, Pasadena, CA.

Fu, L-L., Alsdorf, D., Rodríguez, E., Morrow, R., Mognard, N., Lambin, J., Vaze, P., and Lafon, T. (2009). The SWOT (Surface Water and Ocean Topography) Mission: Spaceborne Radar Interferometry for Oceanographic and Hydrological Applications, Proceedings of OceanObs'09: Sustained Ocean Observations and Information for Society (Vol. 2), Venice, Italy, 21-25 September 2009, Hall, J., Harrison, D.E. & Stammer, D., Eds., ESA Publication WPP-306, doi: 10.5270/OceanObs09.cwp.33.

Fyfe, J. C., and Saenko, O. A. (2006). Simulated changes in the extratropical Southern Hemisphere winds and currents. *Geophys. Res. Lett.*, 33, L06701. doi:10.1029/2005GL025332.

Gent, P. R. (2015). Effects of Southern Hemisphere wind changes on the meridional overturning circulation in ocean models. *Annu. Rev. Mar. Sci.*, 8, 79–94.

Georgi, S. T., and Toole, J. M. (1982). The Antarctic Circumpolar Current and the oceanic heat and freshwater budgets. *J. Mar. Res.*, 40, 183–197.

Gille, S. T. (1994). Mean sea surface height of the Antarctic circumpolar current from GEOSAT data: Methods and application. *J. Geophys. Res.*, 99, 18255–18273.

Gille, S. T., 2003. Float observations of the Southern Ocean: Part 1, Estimating mean fields, bottom velocities, and topographic steering, *J. Phys. Oceanogr*, 33, 1167–1181.

Gille, S. T. (2014). Meridional displacement of the Antarctic Circumpolar Current. *Phil. Trans. R. Soc. A.*, 372, 20130273. doi:10.1098/rsta.2013.0273.

Gille, S. T., and K. A. Kelly, 1996. Scales of spatial and temporal variability in the Southern Ocean, J. Geophys. Res., 101, 8759–8773.

Gordon, A. L., Molinelli, E., and Baker, T. (1978). Large-scale relative dynamic topography of the Southern Ocean, *J. Geophys. Res.*, 83(C6), 3023–3032. doi:10.1029/JC083iC06p03023.

Graham, R. M., deBoer, A. M., Heywood, K. J., Chapman, M. R., and Stevens, D. P. (2012). Southern Ocean fronts: Controlled by wind or topography? *J. Geophys. Res.*, 117, C08018. doi:10.1029/2012JC007887.

Griesel, A., Mazloff, M. R., and Gille, S. T. (2012). Mean dynamic topography in the Southern Ocean: Evaluating Antarctic Circumpolar Current transport. *J. Geophys. Res.*, 117, C01020. doi:10.1029/2011JC007573.

Hallberg, R., and Gnanadesikan, A. (2001). An exploration of the role of transient eddies in determining the transport of a zonally reentrant current. *J. Phys. Oceanogr.*, 31, 3312–3330.

Hogg, A. M., Meredith, M. P., Chambers, D. P., Abrahamsen, E. P., Hughes, C. W., and Morrison, A. K. (2015). Recent trends in the Southern Ocean eddy field. *J. Geophys. Res.*, 120, 257–267. doi:10.1002/2014JC010470.

Hughes, C. W., and Meredith, M. P. (2006). Coherent sea-level fluctuations along the global continental slope. *Philos. Trans. R. Soc. A*, 364, 885–901.

Hughes, C. W., Meredith, M. P., and Heywood, K. J. (1999). Wind-forced transport fluctuations at Drake Passage: A southern mode. *J. Phys. Oceanogr.*, 29(8), 1971–1992.

Hughes, C. W., Williams, J., Coward, A. C., and de Cuevas, B. A. (2014). Antarctic circumpolar transport and the southern mode: A model investigation of interannual to decadal timescales. *Ocean Sci.*, 10, 215–225. doi:10.5194/os-10-215-2014.

Hughes, C. W., Woodworth, P. L., Meredith, M. P., Stepanov, V., Whitworth T. III, Pyne, A. (2003). Coherence of Antarctic sea levels, Southern Hemisphere annular mode, and flow through Drake Passage. *Geophys. Res. Lett.*, 30(9), 1464. doi:10.1029/2003GL017240.

Johnson, G. C., and Bryden, H. L. (1989). On the size of the Antarctic Circumpolar Current. *Deep Sea Res. Part A*, 36, 39–53.

Killworth, P. (1992). An equivalent-barotropic mode in the fine resolution Antarctic model. *J. Phys. Oceanogr.*, 22, 1379–1387.

Kim, Y. S., and Orsi, A. H. (2014). On the variability of Antarctic Circumpolar Current fronts inferred from 1992–2011 altimetry. *J. Phys. Oceanogr.*, 44, 3054–3071. doi:10.1175/JPO-D-13-0217.1.

Klocker, A., Ferrari, R., and LaCasce, J. H. (2012). Estimating suppression of eddy mixing by mean flows, *J. Phys. Oceanogr.*, 42, 1566–1576.

Koenig, Z., Provost, C., Ferrari, R., Sennechael, N., and Rio, M.-H. (2014). Volume transport of the Antarctic Circumpolar Current: Production and validation of a 20 year long time series obtained from in situ and satellite observations. *J. Geophys. Res.*, 119, 5407–5433. doi:10.1002/2014JC009966.

Kosempa, M., and Chambers, D. P. (2014). Southern Ocean velocity and geostrophic transport fields estimated by combining Jason altimetry and Argo data. *J. Geophys. Res. Oceans*, 119, 4761–4776. doi:10.1002/2014JC009853.

Marshall, J., Shuckburgh, E., Jones, H., and Hill, C. (2006). Estimates and implications of surface eddy diffusivity in the southern ocean derived from tracer transport. *J. Phys. Oceanogr.*, 36, 1806–1821.

Maximenko, N. A., and Niiler, P. P. (2005). Hybrid Decade-Mean Global Sea Level with Mesoscale Resolution. In: Saxena N (ed.), *Recent Advances in Marine Science and Technology*, PACON Int., Honolulu, HI, pp. 55–59.

Mazloff, M. R. (2012). On the sensitivity of the Drake Passage transport to air–sea momentum flux. *J. Clim.*, 25, 2279–2290.

Mazloff, M. R., Heimbach, P., and Wunsch, C. (2010). An eddy-permitting Southern Ocean state estimate. *J. Phys. Oceanogr.*, 40, 880–899. doi:10.1175/2009JPO4236.1.

Meijers, A., Bindoff, N., and Rintoul, S. (2011a). Estimating the four-dimensional structure of the Southern Ocean using satellite altimetry. *J. Atmos. Oceanic Technol.*, 28, 548–568.

Meijers, A. J. S., Bindoff, N. L., and Rintoul, S. R. (2011b). Frontal movements and property fluxes: Contributions to heat and freshwater trends in the Southern Ocean. *J. Geophys. Res.*, 116, C08024. doi:10.1029/2010JC006832.

Meinen, C. (2001). Structure of the North Atlantic current in stream-coordinates and the circulation in the Newfoundland basin. *Deep Sea Res. Part I*, 48, 1553–1580.

Meinen, C., and Watts, D. (2000). Vertical structure and transport on a transect across the North Atlantic Current near 42°N. *J. Geophys. Res.*, 105(C9), 21869–21891.

Meredith, M. P., and Hogg, A. M. (2006). Circumpolar response of Southern Ocean eddy activity to changes in the southern annular mode. *Geophys. Res. Lett.*, 33(16), L16608. doi:10.1029/2006GL026499.

Meredith, M. P., and Hughes, C. W. (2005). On the sampling timescale required to reliably monitor interannual variability in the Antarctic circumpolar transport. *Geophys. Res. Lett.*, 32(3), L03609. doi:10.1029/2004GL022086.

Meredith, M. P., Vassie, J. M. Spencer, R., and Heywood, K. J., (1997). The processing and application of inverted echo sounder data from Drake Passage. *J. Atmos. Oceanic Technol.*, 14(4), 871–882.

Meredith, M. P., Woodworth, P. L., Chereskin, T. K., Marshall, D. P., Allison, L. C., Bigg, G. R., Donohue, K., et al. (2011). Sustained monitoring of the Southern Ocean at Drake Passage: Past achievements and future priorities. *Rev. Geophys.*, 49, RG4005. doi:10.1029/2010RG000348.

Meredith, M. P., Woodworth, P. L., Hughes, C. W., and Stepanov, V. (2004). Changes in the ocean transport through Drake Passage during the 1980s and 1990s, forced by changes in the Southern Annular Mode. *Geophys. Res. Lett.*, 31(21), L21305. doi:10.1029/2004GL021169.

Mestas-Nuñez, A. M., Chelton, D. B., and deSzoeke, R. A. (1992). Evidence of time-dependent Sverdrup circulation in the South Pacific from the Seasat scatterometer and altimeter. *J. Phys. Oceanogr.*, 22, 934–943.

Morrow, R., Ward, M. L., Hogg, A. M., and Pasquet, S. (2010). Eddy response to Southern Ocean climate modes. *J. Geophys. Res.*, 115, C10030. doi:10.1029/2009JC005894.

Nowlin, W. D., and Clifford, M. (1982). The kinematic and thermohaline zonation of the Antarctic Circumpolar Current at Drake Passage. *J. Mar. Res.*, 40, 481–507.

Olbers, D., and Lettmann, K. (2007). Barotropic and baroclinic processes in the transport variability of the Antarctic Circumpolar Current. *Ocean Dyn.*, 57, 559–578. doi:10.1007/s10236-007-0126-1.

Orsi, A. H., Whitworth, T., and Nowlin, W. D. (1995). On the meridional extent and fronts of the Antarctic Circumpolar Current. *Deep Sea Res. I*, 42(5), 641–673.

Pavlis, N. K., Holmes, S. A., Kenyon, S. C., and Factor, J. K. (2008). An earth gravitational model to degree 2160: EGM2008. *Geophys. Res. Abstr.*, 10, Abstract EGU2008-A-01891.

Phillips, H. E., and Bindoff, N. L. (2014). On the nonequivalent barotropic structure of the Antarctic Circumpolar Current: An observational perspective. *J. Geophys. Res. Oceans*, 119, 5221–5243. doi:10.1002/2013JC009516.

Rintoul, S. R., Hughes, C., and Olbers, D. (2001). The Antarctic Circumpolar System. In: Sielder G, Church J, Gould J (eds.), *Ocean Circulation and Climate*, Academic Press, San Diego. pp. 271–302.

Rintoul, S. R., Sokolov, S., and Church, J. (2002). A 6 year record of baroclinic transport variability of the Antarctic Circumpolar Current at 140E derived from expendable bathythermograph and altimeter measurements. *J. Geophys. Res.*, 107(C10), 3155. doi:10.1029/2001JC000787.

Rintoul, S. R., Sokolov, S., Williams, M. J. M., Peña Molino, B., Rosenberg, M., and Bindoff, N. L. (2014). Antarctic Circumpolar Current transport and barotropic transition at Macquarie Ridge. *Geophys. Res. Lett.*, 41, 7254–7261. doi:10.1002/2014GL061880.

Rio, M.-H., Schaeffer, P., Moreaux, G., Lemoine, J.-M., and Bronner, E. (2009). *A new mean dynamic topography computed over the global ocean from GRACE data, altimetry and in-situ measurements, Paper presented at OceanObs09 Symposium*. Eur. Space Agency, Venice, Italy, 21–15 Sept.

Rocha, C., Chereskin, T. K., Gille, S. T., and Menemenlis, D. (2016). Mesoscale to submesoscale wavenumber spectra in Drake Passage. *J. Phys. Oceanogr.*, 46, 601–620.

Sallée, J.-B., and Rintoul, S. (2011). Parameterization of eddy-induced subduction in the Southern Ocean surface-layer. *Ocean Model.*, 39, 146–153.

Sallée, J.-B., Speer, K., and Morrow, R. (2008). Response of the Antarctic Circumpolar Current to atmospheric variability. *J. Clim.*, 21(12), 3020–3039.

Sandwell, D. T., and Zhang, B. (1989). Global mesoscale variability from the GEOSAT Exact Repeat Mission: Correlation with ocean depth. *J. Geophys. Res.*, 94. 17971–17984, doi:10.1029/JC094iC12p17971.

Shao, A., Gille, S. T., Mecking, S., and Thompson, L. (2015). Properties of the Subantarctic Front and Polar Front from the skewness of sea level anomaly. *J. Geophys. Res. Oceans*, 120, 5179–5193.

Smith, W. H. F., and Sandwell, D. T. (1997). Global seafloor topography from satellite altimetry and ship depth soundings. *Science*, 277, 1957–1962.

Sokolov, S., and Rintoul, S. (2003). Subsurface structure of interannual temperature anomalies in the Australian sector of the Southern Ocean. *J. Geophys. Res.*, 108(C9), 3285. doi:10.1029/2002JC001494.

Sokolov, S., and Rintoul, S. R. (2007). Multiple jets of the Antarctic Circumpolar Current south of Australia. *J. Phys. Oceanogr.*, 37(5), 1394–1412.

Sokolov, S., and Rintoul, S. R. (2009). Circumpolar structure and distribution of the Antarctic Circumpolar Current fronts: 1. Mean circumpolar paths. *J. Geophys. Res.*, 114, C11018. doi:10.1029/2008JC005108.

Straub, D. N. (1993). On the transport and angular momentum balance of channel models of the Antarctic Circumpolar Current. *J. Phys. Oceanogr.*, 23, 776–782.

Sun, C., and Watts, D. (2001). A circumpolar gravest empirical mode for the Southern Ocean hydrography. *J. Geophys. Res.*, 106(C2), 2833–2855.

Swart, N. C., and Fyfe, J. C. (2012). Observed and simulated changes in the Southern Hemisphere surface westerly wind-stress. *Geophys. Res. Lett.*, 39, L16711. doi:10.1029/2012GL052810.

Swart, S., Speich, S., Ansorge, I. J., Goni, G. J., Gladyshev, S., and Lutjeharms, J. R. E. (2008). Transport and variability of the Antarctic Circumpolar Current south of Africa. *J. Geophys. Res.*, 113(C9), C09014. doi:10.1029/2007JC004223.

Swart, S., Speich, S., Ansorge, I. J., and Lutjeharms, R. (2010). An altimetry-based gravest empirical mode south of Africa: 1. Development and validation, *J. Geophys. Res.*, 115, C03002. doi:10.1029/2009JC005299.

Tapley, B. D., Ries, J., Bettadpur, S., Chambers, D., Cheng, M., Condi, F., Gunter, B., et al. (2005). GGM02—An improved Earth gravity field model from GRACE, *J. Geod.*, 79, 467–478.

Thompson, A. F., and Garabato, A. C. N. (2014). Equilibration of the Antarctic Circumpolar Current by standing meanders. *J. Phys. Oceanogr.*, 44(7), 1811–1828. doi:10.1175/JPO-D-13-0163.1.

Thompson, K. R., and Demirov, E. (2006). Skewness of sea level variability of the world's oceans. *J. Geophys. Res.*, 111, C05005. doi:10.1029/2004JC002839.

Vivier, F., Kelly, K. A., and Harismendy, M. (2005). Causes of large-scale sea level variations in the Southern Ocean: Analyses of sea level and a barotropic model. *J. Geophys. Res.*, 110(C9), C09014.

Vivier, F., Park, Y.-H., Sekma, H., and Le Sommer, J. (2014) Variability of the Antarctic Circumpolar Current transport through the Fawn Trough, Kerguelen Plateau. *Deep Sea Res. II*, 114, 12–26. doi: 10.1016/j.dsr2.2014.01.017i.

Watts, D., Sun, C., and Rintoul, S., 2001. A two-dimensional gravest empirical mode determined from hydrographic observations in the Subantarctic Front. *J. Phys. Oceanogr.*, 31, 2186–2209.

Whitworth, T. (1983). Monitoring the transport of the Antarctic Circumpolar Current at Drake Passage. *J. Phys. Oceanogr.*, 13(11), 2045–2057.

Whitworth, T., and Peterson, R. G. (1985). Volume transport of the Antarctic Circumpolar Current from bottom pressure measurements. *J. Phys. Oceanogr.*, 15(6), 810–816.

Woodworth, P. L., Vassie, J. M., Hughes, C. W., and Meredith, M. P. (1996). A test of the ability of TOPEX/POSEIDON to monitor flows through the Drake Passage. *J. Geophys. Res.*, 101(C5), 11935–911947.

Wunsch, C., and Heimbach, P. (2007). Practical global oceanic state estimation. *Physica D*, 230, 197–208. doi:10.1016/j.physd.2006.09.040.

Wunsch, C., and Stammer, D. (1995). The global frequency-wavenumber spectrum of oceanic variability estimated from TOPEX/POSEIDON altimetric measurements. *J. Geophys. Res.*, 100(C12), 24895–24910. doi:10.1029/95JC01783.

Zlotnicki, V., Fu, L.-L., and Patzert, W. (1989). Seasonal variability in global sea level observed with GEOSAT altimetry. *J. Geophys. Res.*, 94(C12), 17959–17969. doi:10.1029/JC094iC12p17959.

10 Ocean Eddies and Mesoscale Variability

Rosemary Morrow, Lee-Lueng Fu, J. Thomas Farrar, Hyodae Seo, and Pierre-Yves Le Traon

10.1 INTRODUCTION

The ocean, like the atmosphere, is a fundamentally turbulent system. When we observe animations of the ocean's sea level evolution, what is most striking is that all of the world's oceans are full of small-scale eddies and meanders and at all latitude bands. The circulation is indeed dominated by mesoscale variability, due to ocean eddies or isolated vortices, meandering currents or fronts. Mesoscale variability generally refers to ocean signals with spatial scales of 30 to 1000 km and timescales of one to several months (Wunsch and Stammer 1998).

Over the last three decades, satellite altimetry has provided global, high-resolution, regular monitoring of sea level and ocean circulation variations. The early satellite altimetry missions such as Seasat (1978) and GEOSAT (1986–1989) gave our first global view of ocean mesoscale sea level variations. Eddy variability maps and along-track sea level spectra were calculated from the 3 months of Seasat data, allowing the first view of the energy cascade over the mesoscale band (Fu 1983). GEOSAT data were extensively used for studying the ocean mesoscale dynamics, including characterizing the eddy space and timescales (Le Traon 1991), the position and strength of meandering jets (e.g., Kelly and Gille 1990), and eddy-mean flow interactions (Morrow et al. 1994).

In the 1990s, altimetry observations of ocean eddies improved, due to the presence of at least two altimeter missions in orbit at the same time, providing precise measurements and the complementary sampling that is needed for ocean mesoscale variability monitoring (see Koblinsky et al. 1992; Le Traon and Morrow 2001; Fu et al. 2010). This multi-mission period was underpinned by the very precise TOPEX/Poseidon (T/P) mission, launched in 1992 on a 10-day repeat orbit, serving as the reference mission for the altimeter constellation. T/P was followed by Jason-1 (J1) launched in 2001, Jason-2 (J2) in 2008, and recently Jason-3 (J3) in 2016. At the same time, European Remote Sensing (ERS) 1 (1991–1995), ERS-2 (1995–2005), Envisat (2002–2012), and SARAL/AltiKa (launched in 2013) provided higher-latitude and finer spatial coverage on a 35-day repeat orbit. Other missions provided complementary altimetric coverage (e.g., GEOSAT Follow-On [GFO] launched in 1998 on a 17-day repeat orbit, CryoSat-2 launched in 2010, and HY-2 launched in 2011).

There have been a number of review papers covering the early mesoscale studies with satellite altimetry by Le Traon and Morrow (2001), Fu et al. (2010), Morrow and Le Traon (2012), or Le Traon (2013). These review papers covered topics such as the seasonal and interannual variations in eddy kinetic energy used as a benchmark to validate the performance and realism of eddy-resolving numerical models. The 20-year time series of sea level anomaly (SLA) maps allowed detailed investigations into the generation, propagation, and termination of ocean eddies, both in regional and global analyses (e.g., Chelton et al. 2011b). These eddies can propagate for months or years across the ocean basins and, when combined with other tracer data, allow us to estimate the role of mesoscale eddies in the lateral transport of heat, salt, carbon, or other tracers (e.g., Stammer 1997; Roemmich and Gilson 2001; Morrow et al. 2003; Qiu and Chen 2005). The long time series also revealed the ubiquitous presence of alternating jet-like structures when time-averages are made in the global oceans, which may be associated with meandering jets and fronts (e.g., Sokolov and Rintoul 2007; Maximenko et al. 2008). Analyses of satellite observations of sea surface height (SSH) in conjunction with concurrent sea surface temperature (SST) and wind stress measurements have revealed persistent mesoscale oceanic influence on the lower atmosphere and surface fluxes (Small et al. 2008; Chelton and Xie 2010). This mesoscale air-sea coupling is known to affect not only the energetics of ocean circulation and biological response but also broader-scale climate. These topics will be revisited here with an update on more recent studies.

Most of the major advances in the altimetric observation of ocean mesoscale dynamics have been derived from mapped altimeter data, which resolve only the larger, more energetic mesoscale structures, with scales larger than 150 to 200 km wavelength (Chelton et al. 2011b; Dibarboure et al. 2011). Over the last few years, there has been great progress in improving the signal to noise ratio of the along-track altimeter data that can potentially allow us to observe smaller scale dynamics. This is possible due to the reprocessing of conventional altimeters, such as Jason, but also due to the new technology from the SARAL/AltiKa Ka-band mission or the CryoSat-2 or Sentinel-3 Synthetic Aperture Radar (SAR) mode missions. Improvement in the quality of all of the altimetric corrections has been aided by European Space Agency's (ESA) Climate Change Initiative (CCI) to obtain precise global mean sea level observations (Ablain et al. 2015). These improved corrections also feed into the different altimeter products for mesoscale studies. New mapping techniques that take into account the local scales of motion are also available, allowing a better representation of the regional eddy field and in coastal regimes where finer-scale dynamics occur.

In this chapter, we will present a review of the advances in observing the ocean eddy field with satellite altimetry over the last 10 years. We will particularly address the techniques being used to study the finer-scale ocean dynamics. Section 10.2 gives an overview of the recent reprocessing of along-track data, both from conventional altimetry and the new technology missions, and looks at the improvements in mapping the multi-mission data for mesoscale studies. We will then review various scientific applications of the fine-scale ocean eddies. These include recent analyses of mesoscale eddies and jets in the global ocean and regional seas (Section 10.3) and analyses of along-track spectra from different altimetric missions and their relation with instability regimes in the ocean (Section 10.4). Section 10.5 covers the potential and limits of resolving higher-order dynamical processes from the mapped data, such as eddy kinetic energy fluxes or using Lagrangian techniques to derive finer-scale dynamics from the temporal evolution of the mapped altimeter data. Section 10.6 addresses the recent understanding of the vertical structure of the ocean eddies, from combinations of altimetry with *in situ* data. Section 10.7 considers progress in understanding coupled mesoscale processes using altimetry combined with satellite wind, SST, or ocean color data. Section 10.8 deals with the new challenges in separating the internal wave signal from the smaller mesoscale SSH signals. The conclusions and perspectives for the future altimeter missions are given in Section 10.9.

10.2 IMPROVEMENTS IN ALONG-TRACK DATA AND MAPPING CAPABILITIES

Over the last decade, there have been important advances in the processing and reprocessing of the along-track altimeter data from all missions and also in the quality of the multi-mission mapping. Most mesoscale studies use the DUACS (Developing Use of Altimetry for Climate Studies) products from AVISO (Archiving Validation and Interpretation of Satellite Oceanographic), now distributed by CMEMS (Copernicus Marine and Environment Monitoring Service). Although the mesoscale band is the most energetic component of sea level, small-scale errors and noise are among the factors limiting our observation of the smaller mesoscales. The choice of editing, filtering, or mapping can also impact on the larger mesoscale signals we can observe. Here, we give a brief overview of the recent improvements in along-track data processing and multi-mission mapping.

10.2.1 Reprocessing of Along-Track Data

Some mesoscale studies use the along-track SLA data directly—for example, for calculating along-track wavenumber spectra (Section 4). There have been many improvements in the quality of these along-track data described in detail by Dibarboure et al. (2011) and aided by the CCI effort (Ablain et al. 2015). The most recent altimeter data sets have been reprocessed using the latest standards for climate studies applied homogeneously to all missions. The radar waveforms have been reprocessed, rigorous editing and selection processes have been applied, especially important in coastal and high-latitude regions, and the most up-to-date standards applied for orbits and atmospheric forcing (Ablain et al. 2015). This has improved the signal-to-noise ratio in the mesoscale band for each mission.

The next step is to cross-calibrate the different altimeter missions within the constellation. This is done first among the reference missions (T/P, J1, J2, and J3) to reduce the long wavelength, geographically correlated errors. A second cross-calibration is then performed to reduce the long wavelength errors between the reference missions and the other altimeter missions. The details are described in Dibarboure et al. (2011) and Pujol et al. (2016). The long wavelength reduction removes orbit errors but also errors in the tidal corrections, atmospheric forcing corrections, or in the mean sea surface for non-repeat altimeter missions. This can potentially improve the quality of the mapped mesoscale signals.

Once the missions are cross-calibrated, sea level anomalies are calculated by removing a reference surface—either a precise time-mean along-track profile (for the missions on a long-term repeating track) or a gridded two-dimensional mean sea surface product (for the new missions or non-repeating missions). There has been recent progress in improving the time-mean along-track profile calculation over a 20-year period, especially in coastal and high-latitude regions (Pujol et al. 2016), and a new generation of two-dimensional mean sea surface products are being developed using more geodetic altimeter data from CryoSat-2 and Jason-1 (DTU2015; O. Anderson; CNES-CLS-2015; P. Schaeffer; pers. commun.). These improvements in the mean sea surface have a direct impact on the mesoscale sea level anomalies, as shown by Pujol et al. (2016) and Dufau et al. (2016).

Finally, due to the limited temporal sampling of the altimeters, rapid barotropic signals are not well resolved. In the mapped and along-track SLA data, their high-frequency component less than 20 days is removed using a Dynamical Atmospheric Correction (DAC), based on a barotropic model forced by wind and atmospheric pressure (Carrère and Lyard 2003). This is a relatively large-scale correction and should not impact on the barotropic component of the mesoscale dynamics, except in the coastal zone where the DAC correction has higher resolution and resolves smaller scales. In general, the corrected altimetric data represents both barotropic and baroclinic motions at mesoscales.

10.2.2 Multi-Mission Mapping

Since 1992, the altimeter constellation has varied over time, having two to four altimeter missions available that can be combined for better spatial-temporal resolution of the mesoscale field (Figure 10.1). Two versions of the multi-mission merged SLA maps are used for mesoscale studies. A two-satellite or "reference" version has consistent sampling over the entire period from 1992 to 2015, using data from T/P-J1-J2 and ERS-Envisat-SARAL (CryoSat-2 data are used in the small gap between Envisat and SARAL). This allows consistent sampling for long-term studies of interannual variability in mesoscale processes. A second all-satellite version includes data from all available missions, with increased spatiotemporal sampling. This is available from 2000 onward and provides better anisotropic structure for mesoscale studies (Pascual et al. 2006) and improved sea level and velocity variances used to evaluate the realism of ocean circulation models (Le Traon and Dibarboure 2002). These global products, available weekly on a 1/3° Mercator grid up to 66°N

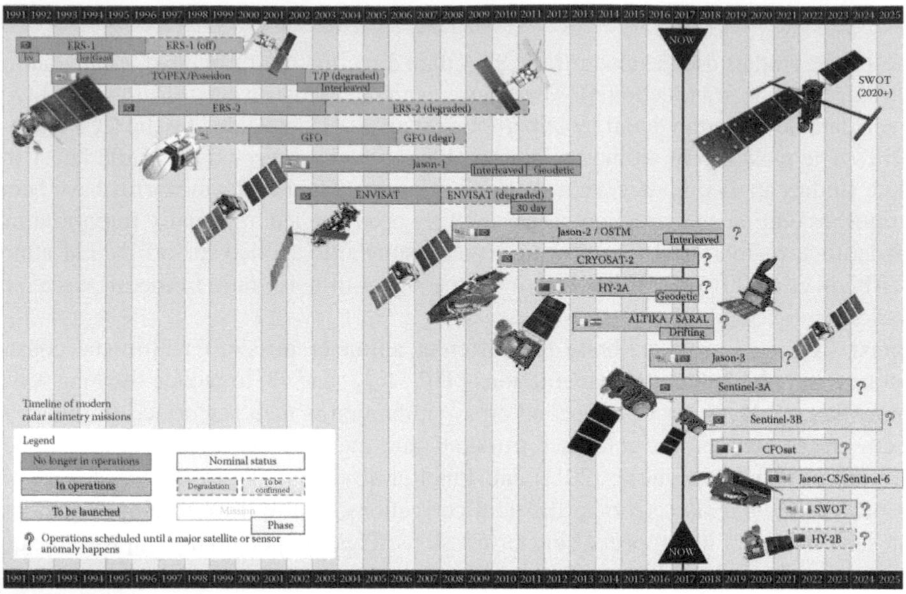

FIGURE 10.1 Timeline of the altimeter mission constellation used in the DUACS/AVISO gridded maps from 1992 up to 2016. Historical missions no longer in operation are in red; missions in operation in 2016 are in orange. Upcoming missions are in yellow (derived from Dibarboure and Lambin, 2015).

and S, have been widely used in mesoscale studies. An improved gridded multi-mission product is now available (DUACS DT2014, described in the following).

The improved coverage in recent years has allowed a number of regional gridded products to be developed in the Mediterranean Sea (Pascual et al. 2005; Pujol and Larnicol 2005), the Black Sea, the Mozambique region, and the Arctic Ocean. Regional studies have created gridded altimeter products with correlation scales that are better tuned to the local dynamics, allowing smaller-scale processes in the coastal zone (Dussurget et al. 2011) or including anisotropic scales associated with bathymetry (Escudier et al. 2012). As an alternative to the statistical optimal interpolation, Ubelmann et al. (2015) have recently developed a simple dynamical interpolation method to retrieve SSH in the temporal gap between two SSH fields separated by up to 20 days. This experimental method, based on the conservation of potential vorticity and assuming that most of the SSH variability is carried by the first baroclinic mode, was able to reconstruct gridded maps with wavelengths smaller than 150 km.

The latest version of the DUACS DT2014 gridded products has a number of advantages for mesoscale applications (Pujol et al. 2016). The data are referenced to a 20-year period, which provides better interannual variability; the editing has been improved and homogeneous standards applied to all missions; the along-track data input in the system have less filtering, maintaining scales down to 65-km wavelength (older versions had along-track filtering to 250 km in the tropics, decreasing to higher latitudes); the gridded data are available daily on a 1/4° Cartesian grid, reducing interpolation errors for users converting from the original 1/3° Mercator grid; correlation scales vary regionally (correlation scale maps are given in Dibarboure et al. 2011); and the error budget varies regionally and between missions depending on data quality. These processing changes allow additional mesoscale signals at wavelengths less than 250 km, and the global SLA variance has increased by 5%, the eddy kinetic energy (EKE variance) by 15%. This has particular benefits for mesoscale studies in the boundary region, as noted by Capet et al. (2014) for the eastern boundary upwelling systems.

The mean wavelength resolution of these gridded products remains 2°, that is, 200 km at mid-latitudes (Chelton et al. 2011b). A comparison with the along-track data shows that in the

65 to 300 km mesoscale band, around 40% of the mesoscale variability is missing in these gridded products, mainly associated with the smaller mesoscale signals (this depends very much on latitude) (Pujol et al. 2016). A second consequence of the reduced along-track filtering is that additional signals are now observed, associated with residual internal tide signals in both the along-track and mapped products (Ray and Zaron 2015; Dufau et al. 2016). As we resolve more smaller-scale structures, the separation and identification of internal tides and internal waves becomes more critical, which will be discussed in Section 8. Despite the progress in improving the along-track signal-to-noise ratio, mesoscale studies today are still limited by this along-track noise and the altimeter ground track sampling. The future Surface Water and Ocean Topography (SWOT) mission, to be launched in 2021, will provide the first two-dimensional swath sampling of SSH with low noise allowing us to resolve SSH variability down to wavelength scales of 15- to 20-km wavelength, revealing the anisotropic structure of smaller mesoscale eddies at all latitudes up to 78°N/S.

10.3 OBSERVED MESOSCALE EDDIES AND JETS

Animations of the mapped sea level anomalies are striking and show the surface signature of many dynamical processes. Large mesoscale eddies are clearly evident in the extratropics, and they propagate across the ocean in an organized way. Early studies suggested they are generated mainly by baroclinic or barotropic instabilities of the mean circulation (Stammer and Wunsch 1999; Qiu and Chen 2004). The large, long-duration eddies captured by the altimetry maps are in quasi-geostrophic balance and are influenced by the Earth's rotation. The sea-level anomalies associated with these mesoscale eddies include both the steric, baroclinic component and a non-steric, barotropic component.

A major altimetric discovery from the 1990s was that sea level anomalies in the mid-latitudes propagated westward with a phase speed nearly twice as fast as that predicted from linear Rossby wave theory (Chelton and Schlax 1996). This observational result led to many revised theories on Rossby wave propagation, including the effects of vertical shear on the background flow, bathymetric effects, and so on (see Fu and Chelton 2001; Tailleux 2003; Hunt et al. 2012). More recently, Chelton et al. (2007, 2011b) highlighted that the mapped sea level anomalies in the extra-tropics have nondispersive characteristics, closer to nonlinear vortices than Rossby waves. Nonlinear vortices can transport heat, salt, carbon, and other tracers in their core waters, with consequences for the mid-latitude tracer budgets (e.g., Dong et al. 2014; Zhang et al. 2014). In the last decade, the vertical structure associated with the cyclones and anticyclones has also been extensively studied by combining altimetric SLA maps with *in situ* vertical profiles (Section 6).

Different automatic eddy tracking techniques have been developed to study the propagation pathways of eddies at mid to high latitudes using the mapped altimetric data. These are based on the Okubu-Weiss parameter (e.g., Isern-Fontanet et al. 2003), the skewness of the relative vorticity (Niiler et al. 2003), criteria based on sea level (Fang and Morrow 2003; Chaigneau and Pizarro 2005; Chelton et al. 2011b), wavelet decomposition of the SLA (e.g., Lilly et al. 2003), or a geometric criteria using the winding angle approach (e.g., Souza et al. 2011). Depending on the technique used, there can be differences in the number of eddies detected, their duration, and their propagation velocities (Souza et al. 2011). A global analysis of cyclones and anticyclones detected from 16 years of altimetry maps was performed by Chelton et al. (2011b), and reviewed by Fu et al. (2010). The Chelton et al. (2011b) study established a statistical inventory of these large eddies, mapping their generation and termination locations, their average amplitude (8 cm), radius (90 km), lifetimes (32 weeks), and propagation distances (550 km). During their lifetimes, these eddies exhibit clear phases of rapid growth, a steady mature phase then a rapid decay that are remarkably robust and symmetric over time, when normalized by their duration. This has been demonstrated regionally (Liu et al. 2012; Pegliasco et al. 2015) and globally (Samelson et al. 2014). Samelson et al. (2016) also discuss the difficulty in reconciling this behavior with 2D turbulence theory.

Chelton et al. (2011b) confirmed that, outside the tropical band from 20°N to 20°S, eddies showed characteristics of nonlinear vortices. The propagation pathways for eddies lasting longer than 18 months are shown in Figure 10.2. Away from bathymetry or strong mean currents, these large, long-lived eddies tend to propagate westward, except in the strong eastward currents such as the Gulf Stream or the Circumpolar Current. Over long periods, the pathways of the cyclones and anticyclones diverge, with (cold) cyclones drifting poleward and (warm) anticyclones drifting equatorward. This can induce a net equatorward heat flux from these long-lived eddies at mid-latitudes that opposes the poleward heat fluxes by the western boundary currents (Morrow et al. 2004a; Dong et al. 2014). This meridional divergence in the large eddy propagation is due to the β-effect and the conservation of potential vorticity (Cushman-Roisin et al. 1990). Locally, eddy interactions with bathymetric features or strong currents can override this slight meridional deviation. This divergence has been noted in many regions, such as in the Agulhas (e.g., Boebel et al. 2003) and in the eastern subtropical basins (Morrow et al. 2004; Chaigneau and Pizarro 2005) and has been confirmed by global analyses (Chelton et al. 2011b; Dong et al. 2014).

Eddies have been tracked along "corridors" in regions where local forcing systematically generates instabilities due to merging currents, currents interacting with bathymetry, wind interacting with orography, or seasonal forcing. For example, vortex dipoles regularly form south of Madagascar where the East Madagascar Current separates from the shelf. The dipole strength varies with the main tropical climate mode forcing upstream (the Indian Ocean Dipole and El Niño-Southern Oscillation [ENSO]); then as the dipoles propagate downstream, they can trigger variations in the Agulhas retroflection and the associated Agulhas ring shedding (Ridderinkhof et al. 2013). Similar regular eddy generation occurs in many regions, such as near Hawaii (Yoshida et al. 2010), in the Gulf of Mexico (Ohlmann et al. 2001), around the Canary Islands (Sangra et al. 2009), in the California Current (Kurian et al. 2011), and from seasonal forcing of mid-ocean currents (Qiu 1999; Farrar and Weller 2006). In the southeast Indian Ocean, the systematic formation of eddies near capes or canyons at the coast are then steered by bathymetric features into zonal corridors offshore (Fang and Morrow 2003).

Altimetry has also revealed the symbiotic relation between these propagating ocean eddies and fine-scale meandering jets and fronts. In two-dimensional turbulence on a β-plane, a long-term free evolution of an eddy field should result in the formation of alternating zonal jets. Using merged

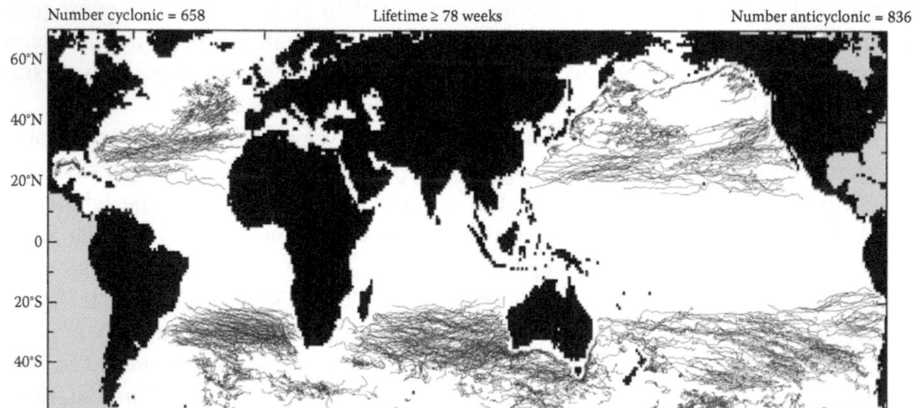

FIGURE 10.2 The trajectories of cyclonic (blue lines) and anticyclonic (red lines) eddies over the 16-year period October 1992–December 2008 for eddies with lifetimes longer than 78 weeks (18 months) for only those eddies for which the net displacement was westward. The numbers of eddies of each polarity are labeled at the top of each panel. (After Chelton et al. 2011b)

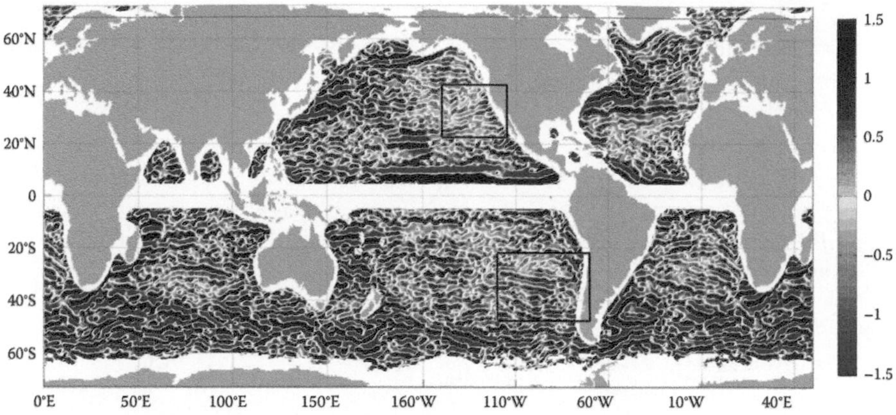

FIGURE 10.3 The 1993–2002 mean zonal surface geostrophic velocity calculated from a hybrid mean dynamic ocean topography product based on altimetry and surface drifters. (After Maximenko, N. A., et al., *Geophys. Res. Lett.*, 35, L08603, 2008. © American Geophysical Union.)

altimetric maps, Maximenko et al. (2008) have shown that alternating jet-like structures are seen in long-term averaged geostrophic velocity anomalies (Figure 10.3) and in geostrophic vorticity in every part of the global ocean. These striations are oriented nearly zonally, and *in situ* data comparisons show they are coherent vertically at least through 700 m in depth. These alternating jets are also observed in annual mean model currents at 1000 m in depth (Nakano and Hasumi 2005). In the Southern Ocean, these multiple meandering jets can be associated with absolute sea level contours and *in situ* tracer fronts (Sokolov and Rintoul 2007; Sallée et al. 2008a). Although the polar fronts detected from altimetry maps can split and reform, strengthen and weaken, their mean position remains close to the mean frontal locations derived from climatological *in situ* data.

10.4 SPECTRAL ANALYSES OF ALONG-TRACK SSH

Satellite altimetry has provided a comprehensive framework for estimating the frequency-wavenumber spectrum of oceanic variability on temporal and spatial scales not accessible by most other means. The global and decadal coverage of the observations from a series of satellite altimetry missions has made the altimetric record "the backbone of the resulting spectral model, jointly covering horizontal wavenumber and frequency," according to Wortham and Wunsch (2013).

In particular, the information content of altimetry in the wavenumber domain is unique, owing to the spatial extent and resolution of the along-track measurement. The wavenumber spectrum of SSH is generally "red" (e.g., Figure 10.4), with the largest SSH fluctuations on the largest spatial scales and increasingly smaller fluctuations at smaller scales. This decrease of SSH signal with decreasing spatial scale means that the smallest spatial scales that can be resolved are limited by the noise floor of the altimeters. The random error for the Jason series altimeters can be estimated directly by examination of data from the Jason-1/Jason-2 tandem mission, when the two satellites were in the same orbit sampling the same scenes 1 min apart (Figure 10.4). Assuming that the two instruments have equal levels of random instrument error, the along-track spectrum of their difference is twice the spectrum of their random errors. This exercise yields a noise spectrum that is white with a level of about 100 cm^2/cycle/km (e.g., Fu and Ubelmann 2014), which largely prevents us from examining ocean signals at wavelengths smaller than about 70–80 km for Jason-class missions (Dufau et al. 2016).

The dense along-track sampling every 7 km allows us to resolve the most energetic mesoscale energy-containing scales of a few hundred kilometers down to the 70- to 80-km scales above the noise limit. This wavenumber range covers the power-law regime of the energy cascade of geostrophic turbulence theory (Charney 1971; Blumen 1978). Geostrophic turbulence theory predicts

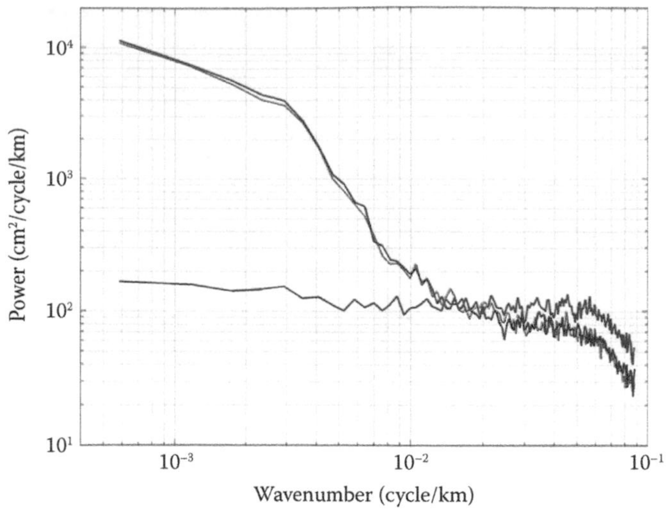

FIGURE 10.4 Along-track SSH wavenumber spectra of Jason-1 (green line) and Jason-2 (red line) along a pass in the eastern Pacific (Pass 132) during a time when the two satellites sampled the same ground track with a separation of one minute. The blue line shows the spectrum of their difference. (From Fu, L.-L., and Ubelmann, C., *J. Ocean. Atm. Tech.*, 31(2), 560–568, 2014. With permission.)

a particular slope of the power-law dependence of the spectrum on wavenumber in a log-log plot, being –5 in SSH and –3 in surface velocity. Fu (1983) explored the SSH wavenumber spectrum of the oceanic mesoscale variability from the limited Seasat data, suggesting a wide range of spectral slopes from –1 to –5. The steepness of the spectral slope increased with the variance level of the geographic location. Although the –5 slope in SSH is consistent with the geostrophic turbulence theory, the –1 slope is puzzling. The subject has been revisited many times with the ever-increasing altimetry database. Using 2 years of the GEOSAT altimeter data, Le Traon et al. (1990) analyzed the SSH wavenumber spectrum in the North Atlantic. They found the spectral slope was close to –4 in high eddy-energy regions and close to –2 in the less energetic regions. Stammer (1997) analyzed T/P data and, by fitting an analytical model to the data in the global extratropical oceans, obtained a uniform spectral slope closer to geostrophic turbulence with a value of –4.6. With a longer time series, Le Traon et al. (2008) presented evidence that the SSH wavenumber spectral slope in the high eddy-energy regions of the world's ocean is closer to the prediction of the theory of surface quasigeostrophic (SQG) turbulence (Blumen 1978; Held et al. 1995), –11/3 (–3.67), as opposed to –5 predicted by the quasigeostrophic (QG) turbulence theory of Charney (1971).

The wavenumber spectral slope estimate is sensitive to the level of instrument noise, which varies geographically and seasonally (Dufau et al. 2016). Xu and Fu (2011, 2012) have shown that this instrument noise can be reduced by removing an empirical fit to the white noise level; this procedure has a significant effect on the spectral slope estimate. A global map of the resulting annual mean spectral slope is displayed in Figure 10.5 (from Xu and Fu 2012). Regions shaded in dark orange have slopes close to –11/3 (–3.67) at the 95% confidence level, consistent with SQG theory. The larger regions shaded in red have slopes steeper than –11/3 but less than –5, including the high eddy-energy regions of the Gulf Stream, the Kuroshio, and the Antarctic Circumpolar Current. Using an empirical orthogonal function approach to estimating the instrument noise and high-frequency signals, Zhou et al. (2015) obtained results similar to Xu and Fu (2012) but with steeper slopes, closer to –5 in the core of the high eddy-energy regions. Only some regions in the tropics exhibit slopes flatter than –2.

The argument for the –5 slope to match the QG turbulence theory may not be justified everywhere because the theory was meant to be applicable to regions away from any boundary influence

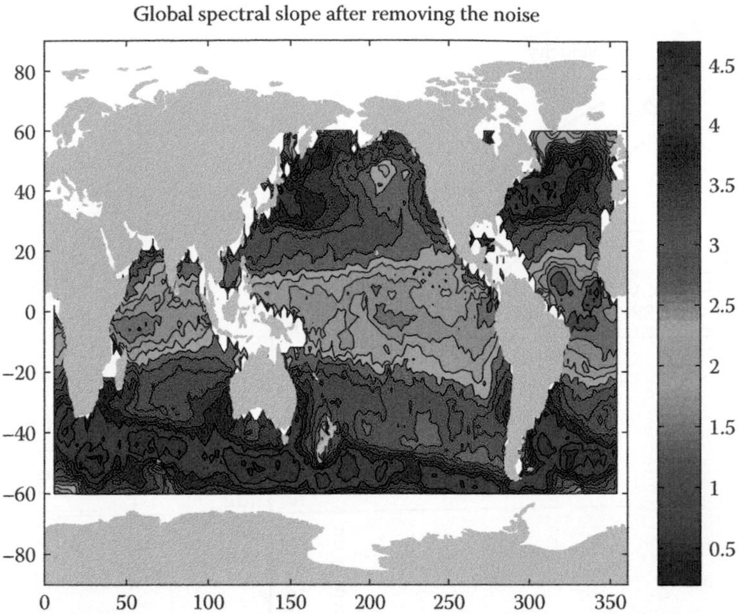

FIGURE 10.5 The global distribution of the spectral slopes of SSH wavenumber spectrum in the wavelength band of 70–250 km estimated from the Jason-1 altimeter measurements after removing the noise. The sign of the slopes was reversed to make the values positive. (From Xu, Y., and Fu, L.-L., *J. Phys. Oceanogr.*, 42, 2229–2233, 2012. With permission.)

(Charney 1971). In theory, the SSH spectrum should follow the SQG theory, as demonstrated by numerical simulations (Capet et al. 2008a; Klein et al. 2008). Although the altimetry observations in Figure 10.5 are closer to −11/3 than −5, they do not show a persistent −11/3 slope in the SSH spectrum. Some clues to this observed difference have been found from recent studies suggesting a prominent seasonal cycle in the sub-mesoscale variability, affecting the wavenumber spectral slope. Owing to the extra energy in the sub-mesoscale from winter mixed-layer instabilities, the simulation of an ocean general circulation model revealed flatter kinetic energy spectral slope close to −2 (−4 in SSH) in winter and −3 (−5 in SSH) in summer (Figure 10.6a, from Sasaki et al. 2014). Such seasonal variability has also been observed in altimetry observations (Qiu et al. 2014; Dufau et al. 2016) as well as *in situ* observations (Figure 10.6b, from Callies et al. 2015). The mixture of winter and summer observations has probably contributed to the annual mean SSH spectral slopes in the range of −4 and −4.5 in the high eddy-energy regions (see Figure 10.5), where the seasonal cycle is strong.

Such close scrutiny of the altimetry SSH wavenumber spectrum has motivated a number of studies to compare the SSH wavenumber spectral properties with *in situ* data. Wang et al. (2010) analyzed near-surface ocean velocity observations by acoustic Doppler current profiler (ADCP) and concluded that the slope of the surface velocity spectrum was indeed close to that of geostrophic turbulence (−3 in surface velocity) in the Gulf Stream region. They also reported that the energy level of the ADCP velocity spectrum was lower than that estimated from altimetry SSH via geostrophy and that the altimetry-derived velocity spectral slope was flatter: −2 as opposed to −3 from the ADCP observations. They attributed the discrepancies between the two types of observations to the instrument error of altimetry. Results from Ponte et al. (2013) suggest, however, that part of the discrepancies could be due to non-geostrophic velocities. Furthermore, since the altimetry spectrum only observes the cross-track component of the geostrophic velocity field, Callies and Ferrari (2013) also argued that it should be compared to the cross-track component of the ADCP kinetic

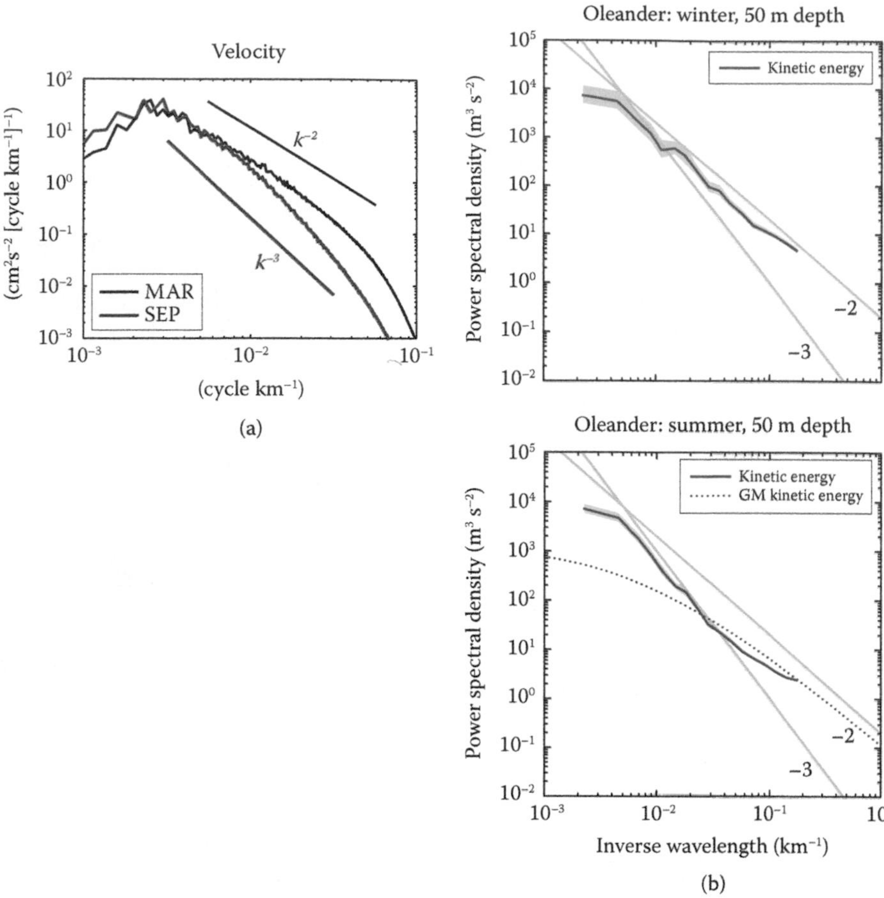

FIGURE 10.6 Seasonal variations in wavenumber velocity spectra from (a) model simulations and (b) *in situ* ADCP data. (a) OGCM simulations of the surface velocity spectra in the North Pacific in March (black) and September (red). The analyzed domain is 148°E–168°W and 20°N–43°N. (From Sasaki, H., et al., *Nat. Commun.*, 5, 5636, 2014. With permission.) (b) Kinetic energy spectra at 50 m from the Oleander ADCP observations in the Gulf Stream region in winter (upper panel) and summer (lower panel). (From Callies, J., et al., *Nat. Commun.*, 6, 6862, 2015. With permission.)

energy spectrum. When this is performed, the altimetry-ADCP comparisons were closer than suggested by Wang et al. (2010).

In the low eddy-energy regions, the altimetric spectral slopes are not consistent with the various QG theories. Richman et al. (2012; their Figure 10) compared simulations from an Ocean General Circulation Model (OGCM) that includes internal waves to altimetric observations in the tropical South Pacific. Their result indicated that the internal-wave induced SSH variance was comparable to the total SSH variance, suggesting significant influence of internal waves on the SSH spectrum in lower eddy energy regions. Rocha et al. (2016) analyzed the simulations of a different OGCM that resolved internal waves at a higher resolution in the region of Drake Passage. Their Figure 9 shows −2 spectral slopes from the contributions of internal waves to SSH, raising questions on the interpretation of the spectra in the vast areas with slopes close to −2 as shown in Figure 10.5. They also analyzed ADCP velocity observations, which suggested that the kinetic energy of the internal waves was comparable to that of the geostrophic motions at wavelengths shorter than 100 km. These results have suggest that internal waves can introduce significant flattening of the SSH spectrum at wavelengths shorter than 100 km (see also Section 8).

10.5 RESOLVING HIGHER-ORDER DYNAMICAL PROCESSES

10.5.1 Toward Two-Dimensional Spectral Energy Fluxes

The ocean community has great interest in observing the energy fluxes across spatial scales as energy is transferred from the large scales of the forcing to the small scales at which energy is dissipated. The nonlinear kinetic energy cascade across wavenumbers can be estimated from along-track spectra assuming isotropic motion, but the full two-dimensional spectral energy fluxes rely on the gridded altimetry maps. Spectral energy fluxes calculated from the gridded altimetry maps have been compared to model simulations by Scott and Wang (2005), Scott and Arbic (2007), Capet et al. (2008), and Arbic et al. (2013). The inverse kinetic energy cascade from the mesoscales up to the large scales is well-represented by altimetry and models (Arbic et al. 2013). However, the so-called "forward" cascade down to the smaller dissipative scales that has been estimated from spectral fluxes derived from altimetry maps (Scott and Wang 2005) is more difficult to interpret. Arbic et al. (2013) showed that a forward kinetic energy cascade across wavenumbers that has been observed with mapped altimetry can be introduced by either progressive smoothing of QG model outputs or variations in viscosity (representing the missing sub-mesoscale dynamics and internal waves). Therefore, it is difficult to confirm whether the forward spectral kinetic energy fluxes in the Scott and Wang (2005) altimeter-based calculations are physical or an artifact of the smoothing inherent in the AVISO gridded altimeter products. The same difficulty holds for the detection of the cascade of kinetic energy across frequencies (Arbic et al. 2012, 2014). In the future, the two-dimensional wide-swath observations from the SWOT mission should allow us to study the global variations of spectral kinetic energy fluxes across wavenumbers with much less spatial smoothing. Spectral fluxes across frequencies will remain difficult to resolve with SWOT, due to the 10- to 21-day repeat times between swaths. However, some local studies across frequencies and wavenumbers may be possible within the first few months of SWOT, during the 1-day fast-repeat orbit phase.

10.5.2 Lagrangian Fine-Scale Ocean Dynamics From Altimetry

Even though the mapped altimetric geostrophic currents only resolve the larger Eulerian mesoscale field at wavelengths larger than 150–200 km, the temporal evolution of these two-dimensional currents can generate smaller fronts and filamentation (Waugh and Abraham 2008; d'Ovidio et al. 2009). Indeed, eddies drive fluid particles in a complex evolution, straining and distorting the flow. Lagrangian particle trajectories can be created from the time series of gridded altimetry maps, and a number of studies have calculated higher-order dynamical properties from these Lagrangian trajectories.

Eddy diffusivity has been studied using dispersion statistics obtained from surface drifters or simulated particle trajectories based on surface currents derived from altimetry. Based on the long time series and regular spatial-temporal coverage of the altimetric surface currents, different statistical techniques have been applied to estimate eddy diffusivity, albeit driven by the larger eddies. Eddy diffusivities and mixing rates have been derived in the Southern Ocean (e.g., Marshall et al. 2006; Sallée et al. 2008b) and globally (Abernathey and Marshall 2013), revealing very high mixing rates (approximately 10^4 m^2s^{-1}) in western boundary current regions, high mixing rates in the tropics (approximately 3 to 5×10^3 m^2s^{-1}), and weak meridional mixing rates in the Southern Ocean (less than 1000 m^2s^{-1}). These eddy mixing rates are strongly modified by the presence of the mean flow. These altimeter-derived estimates can be used to test and develop new model parameterizations of eddy diffusivity.

Other higher-order diagnostics can be calculated from the separation rate of these altimetric-derived particle trajectories, such as finite-time or finite-size Lyapunov exponents (FTLE: Waugh and Abraham 2008; FSLE: d'Ovidio et al. 2009). These Lyapunov exponents highlight the transport barriers that control the horizontal exchange of water in and out of eddy cores. Large values of these Lyapunov exponents identify regions where the stretching induced

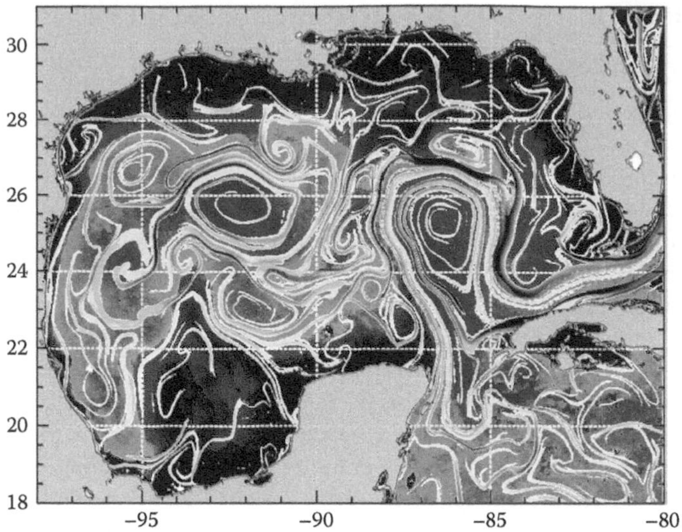

FIGURE 10.7 Example of FSLEs contours over the Gulf of Mexico calculated from altimetry maps, overlaid on an ocean color image for one date. Credits *CNES/CLS/*LOCEAN/*CTOH.*

by altimetric mesoscale velocities is strong, and they are often in good agreement with tracer filaments stirred by horizontal advection. Figure 10.7 highlights the very good agreement between the dynamical transport barriers identified by the Lyapunov exponents advected by the mapped altimetric currents and the satellite ocean color observations in the Gulf of Mexico. Global FSLE data are now made available by AVISO+ and are regularly updated based on the most recent multi-mission maps.

When horizontal stirring is applied to any large-scale, regionally contrasted tracer pattern, it acts to cascade the tracer field into finer-scale structures. Filaments generated by this mesoscale stirring not only create a distinctive pattern of tracer concentrations—well visible from satellite SST and ocean color data—they also structure the different phytoplankton communities (Abraham et al. 2000; d'Ovidio et al. 2010). These dynamical barriers modulate the spatiotemporal biodiversity of primary producers in the open ocean (De Monte et al. 2013) and impact the repartition of higher trophic levels up to top predators (Cotte et al. 2015). Lagrangian stirring of large-scale SST and Sea Surface Salinity (SSS) fields by altimetric surface geostrophic currents also shows promising results in different mid- to high-latitude regions (Desprès et al. 2011; Dencausse et al. 2014; Reverdin et al. 2015; Rogé et al. 2015) in reconstructing finer-scale tracer patterns in conditions where lateral advection dominates the tracer evolution.

10.6 UNDERSTANDING THREE-DIMENSIONAL VERTICAL STRUCTURE OF EDDY AND SUB-MESOSCALE PROCESSES

The vertical structures of the mesoscale eddies detected with altimetry are more difficult to observe because they rely on collocating *in situ* observations at depth with the transient, moving eddy. Most studies have explored individual eddies with dedicated *in situ* campaigns in restricted areas—for example, in the Agulhas Current region (de Ruijter et al. 1999), the California Current (Kurian et al. 2011), the Canary Current (Sangra et al. 2009), the Leeuwin Current (Morrow et al. 2004a), and in the Programme Ocean Multidisciplinary Meso scale (POMME) experiment in the northeast Atlantic (Assenbaum and Reverdin 2005).

Regular ocean monitoring from high-resolution expandable bathythermograph (XBT) transects is not designed to observe eddies. However, statistical composites can be made from long time series of XBT transects and altimetric data in order to estimate the vertical eddy structure and eddy heat transports. Roemmich and Gilson (2001) used collocated XBT and altimetry data in the North Pacific to determine composite warm and cold-core eddy structures, which slant westward with decreasing depth. This tilt produces a depth-varying velocity/temperature correlation and hence a vertical-meridional overturning circulation. On average, 3.9 Sv of thermocline waters are carried equatorward by the eddy field over the basin, balanced by the northward transport of surface waters. In the Southern Ocean south of Tasmania, Morrow et al. (2004b) used XBT and altimeter data to show that cyclonic eddies were instrumental in transporting cool, freshwater northward in the upper 1000 m into the Subantarctic Zone where mode waters form. The amount of eddy heat and salt transported was estimated to be of the same order as that carried by the northward Ekman transport in the surface layer. Similar analyses have been performed in the Drake Passage (Sprintall 2003).

The global coverage from Argo profiles has allowed a more widespread analysis of the typical vertical structure of cyclones and anticyclones in different regions. Chaigneau et al. (2011) reconstructed the mean thermohaline vertical structure of composite eddies in the Peru-Chile upwelling system from Argo floats collocated with eddies detected from altimetry. In this region, warm and salty anticyclonic eddies are mainly subsurface intensified, whereas cold and fresh cyclonic eddies are surface-intensified (Figure 10.8). This offset was attributed to the mechanisms involved in the eddy generation because cyclonic eddies are formed by instabilities in the equatorward near-coastal currents whereas sub-thermocline anticyclonic eddies are likely shed by the subsurface poleward Peru-Chile Undercurrent (Chaigneau et al. 2011; Colas et al. 2012). Pegliasco et al. (2015) extended this study to four eastern boundary upwelling systems and used a clustering analysis to refine the eddy types detected in each boundary system and their geographical distribution. Similar techniques of collocating Argo profiles within eddies detected from altimetric SSH or surface currents have also been used to characterize eddies in the subtropical North Pacific (Liu et al. 2012; Yang et al. 2013) and the Gulf Stream and the South Atlantic Bight (Castelao 2014).

The eddy-induced mass transport in the global ocean has been estimated from a combination of altimetry (to track eddies) and Argo floats (to estimate the volume of trapped fluid in a propagating eddy). Zhang et al. (2014) have shown that the dominant westward propagation of eddies in the tropics and subtropics leads to a large eddy-induced zonal mass transport of 30 to 40 Sv, which is of similar magnitude to the large-scale wind-driven or thermohaline circulation. The meridional eddy mass transport is only a few Sv but with a systematical structure, being poleward in the tropics and equatorward in the subtropics. Altimetry and Argo analyses have also shown that the divergent pathways of cyclones and anticyclones contribute to a global upgradient equatorward eddy-induced heat transport in the tropics and subtropics (Dong et al. 2014). This eddy-induced heat transport opposes the mean poleward heat transport carried by the mean circulation and is around 20% to 30% of the total heat flux. The eddy-induced freshwater flux also opposes the mean freshwater fluxes, leading to a moderate (20–30%) freshening of the evaporative subtropical zone and depleting of the precipitative equatorial and subpolar regions. These global studies depend on a limited number of collocated Argo floats but are generally consistent with high-resolution modeling studies if the eddy tracking algorithms are applied to both the mapped altimetry and modeled fields. However, the magnitude of this eddy-induced (trapped) transport is often much larger than the calculation of transient eddy fluxes across model sections as shown in modeling studies of the Cape Cauldron region south of Africa by Treguier et al. (2003). These studies highlight the important role of mesoscale eddies in the ocean's adjustment and for global mass, heat, and salt budgets—processes that are not explicitly included in today's generation of climate models.

Ocean gliders have also been flown along altimetric ground tracks across coastal currents and mesoscale eddies in different regions. The along-track altimeter data can resolve smaller-scale structures than the mapped data, reaching scales of 50- to 70-km wavelength depending on the satellite mission (Saral with lower noise levels can detect smaller-scale structures, although this varies regionally and seasonally with the surface roughness conditions). In the Alboran Sea in the

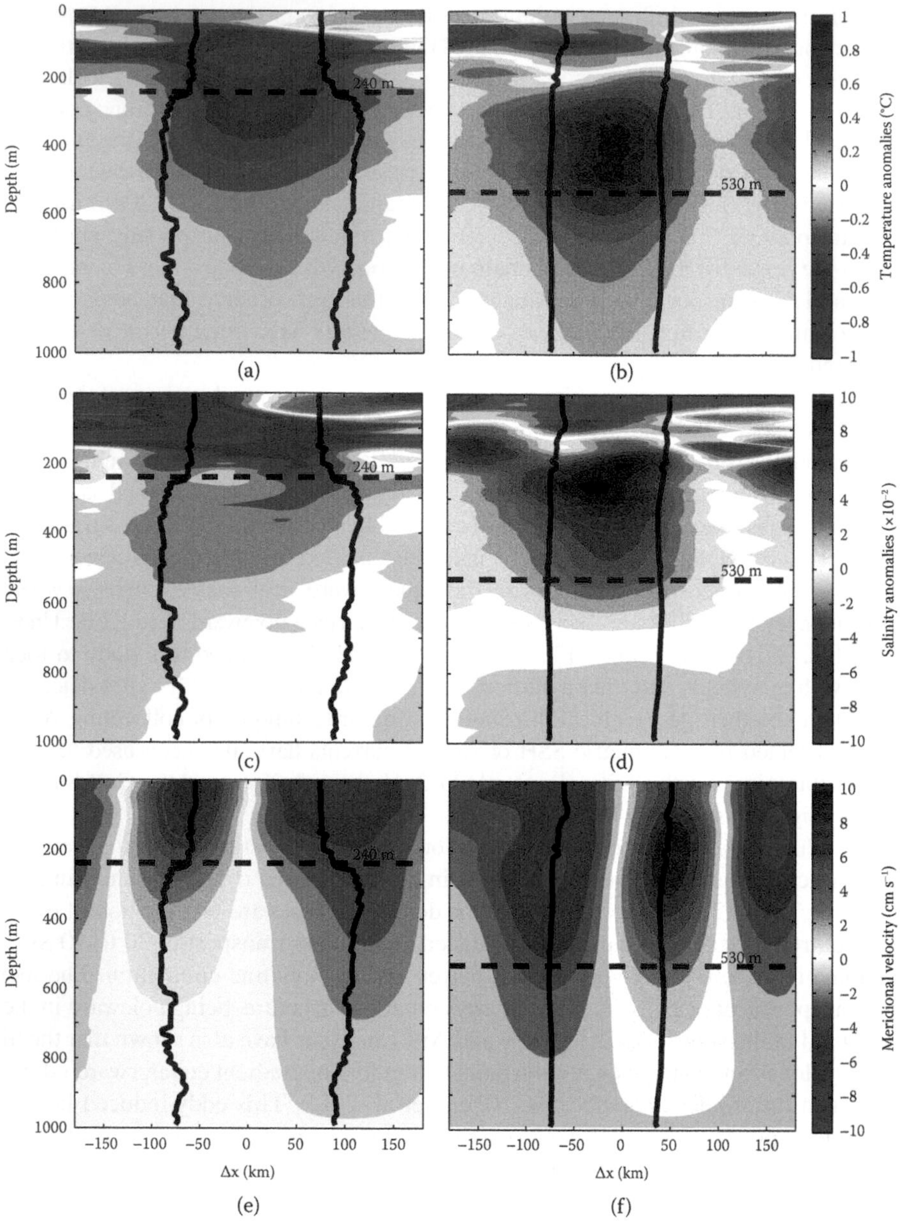

FIGURE 10.8 Vertical section across the composite cyclonic and anticyclonic eddies in the Peru-Chile Upwelling system. (From Chaigneau, A., et al., *J. Geophys. Res.*, 116, C11025, 2011. With permission.) (a, b) Potential temperature anomaly (°C); (c, d) salinity anomaly ($\times 10^{-2}$); and (e, f) meridional (cross section) geostrophic speed (cm s^{-1}) relative to 1000 m, indicating the clockwise (anticlockwise, respectively) rotation for the composite cyclonic (anticyclonic) eddy. Eddy edges are denoted by black lines, whereas horizontal dashed lines indicate the trapping depths.

western Mediterranean, gliders launched along a Jason ground track revealed sharp dynamical height gradients across strong fronts and eddies (Bouffard et al. 2010). Three-dimensional maps of dynamic height based on the glider and mapped altimeter data were also constructed and, using the omega-equation, large-scale vertical velocities (approximately 1 m/day) were diagnosed (Ruiz et al. 2009), which could help explain the subduction of a high chlorophyll tongue.

Many studies have explored the synergy between SSH and SST measurements to reconstruct the finer-scale surface circulation. As an example, Gonzalez-Haro and Isern-Fontanet (2014) combined the phase of SST, giving the relative position of ocean structures, with the amplitude of SSH measurements, for the distribution of energy across spectral scales. Their technique works well in regions where the two fields are correlated, especially for deep enough mixed layers, but it is less effective in situations of strong surface stratification (e.g., in summer). Combined SSH and SST observations have also been used to help identify whether individual vortices are intensified at the surface or subsurface (Assassi et al. 2016). However, modeling studies show that the surface detection becomes complicated when double-core eddies occur (when surface and subsurface eddies are aligned).

New techniques are being developed to reconstruct the three-dimensional dynamics in the ocean interior using satellite altimetry, SST, and *in situ* climatological fields to set the vertical stratification. The vertical reconstruction of the upper ocean three-dimensional circulation has been explored theoretically with surface QG dynamics (Lapeyre and Klein 2006). Although today's gridded altimetry maps do not have the resolution to reconstruct the finer-scale dynamics, satellite SST has been used as a proxy of the surface buoyancy and combined with SQG formalism to reconstruct the upper ocean circulation (Isern-Fontanet et al. 2008). This works well in moderately energetic regions. Wang et al. (2013) also proposed to extrapolate subsurface velocity and density fields from sea surface density and SSH via the SQG formalism. In their extended technique, the subsurface field was augmented by the addition of the barotropic and first baroclinic modes, with amplitudes derived from the SSH field (pressure), after subtracting the SQG contribution. These new techniques have mainly been explored with high-resolution models, but in the future, with the high-resolution two-dimensional SWOT SSH fields, reconstruction of the upper ocean circulation may be achieved with satellite SST/SSH data alone. Assimilation of altimetry and other satellite and *in situ* data into models is another way to reconstruct the four-dimensional dynamics in the ocean interior—this is addressed in Chapters 17 and 18.

10.7 UNDERSTANDING COUPLED MESOSCALE PROCESSES

The larger mesoscale eddy field detected from mapped altimetry has mean SST anomalies of 0.3°C to 0.5°C and interior geostrophic surface current speeds of greater than 20 cm/s (Gaube et al. 2015). These anomalies are large and persistent enough to influence the wind stress, a process referred to as eddy-wind interaction (Small et al. 2008; Chelton and Xie 2010).

The surface current-wind interaction can be represented in the simple form of a bulk aerodynamic formula of wind stress (ignoring the wave effects on currents) as $\underline{\tau} = \rho_a C_D (\underline{W} - \underline{U})|\underline{W} - \underline{U}|$, where $\underline{\tau}$ is the wind stress vector, ρ_a the air density, C_D the drag coefficient, $\underline{W}$ is the 10-m wind velocity, and $\underline{U}$ is the ocean surface current velocity. The wind stress τ on the ocean mesoscale is determined by the eddy-induced SST influencing $\underline{W}$ through the marine atmospheric boundary layer (MABL) dynamics and by the eddy-induced surface currents ($\underline{U}$) creating the velocity shear across the air-sea interface. The wind stress depends on the wind relative to the sea surface, so the influence of ocean currents $\underline{U}$ on the stress is sometimes called the relative wind effect.

Eddy SSTs alter the turbulent mixing of momentum (Wallace et al. 1989) and the pressure anomaly (Lindzen and Nigam 1987) in the MABL, resulting in enhanced surface wind and wind stress over a warm mesoscale SST anomaly. The opposite is true for a cold SST anomaly. Recent high-resolution, all-weather satellite observations have revealed the strong positive correlation between SST and wind speed at the oceanic mesoscale over the global oceans (Figure 10.9, see also Xie 2004), which represents the mesoscale oceanic forcing of the atmospheric boundary layer. This is in stark contrast to the predominant negative correlation that is commonly found on the large scale, where SSTs passively respond to changes in wind speed (Xie, 2004). Chelton et al. (2004) empirically identified a robust relation between the wind stress curl and the crosswind SST gradient (see also O'Neill et al., 2010; 2012). The SST-driven Ekman pumping velocity derived from this

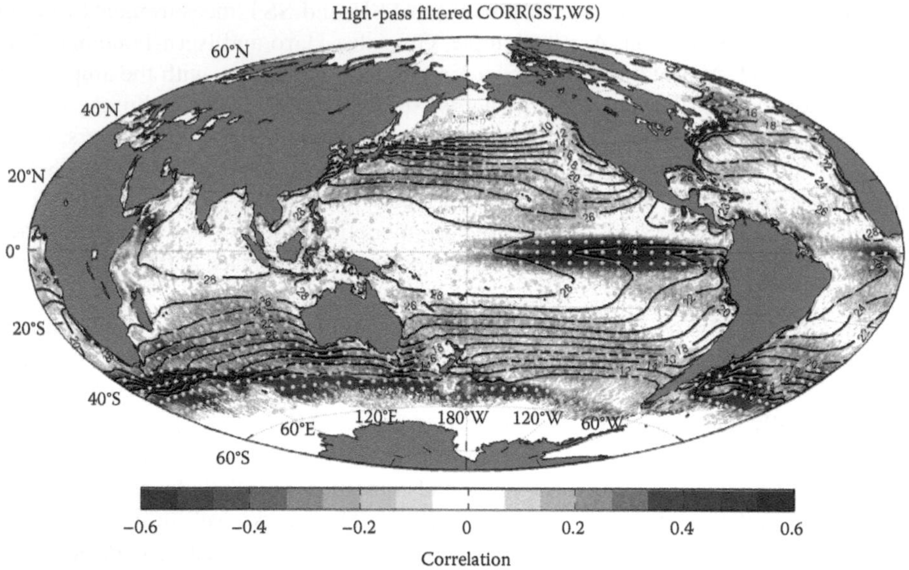

FIGURE 10.9 Map of correlation coefficient between the daily NOAA OI SST and the daily QuikSCAT wind speed for the period of 2001–2009, both of which are zonally high-pass filtered (10° longitude) to highlight the coupling at oceanic mesoscale. Gray dots denote the significant correlation at 95% confidence level. (From Seo, H., *J. Clim.*, 2017. With permission.)

empirical relationship produces a dipolar upwelling and downwelling that are in quadrature with the SSH of the eddy (Figure 10.10b). The resulting feedback affects the evolution and propagation of the eddy fields (Chelton et al. 2013; Gaube et al. 2015).

Mesoscale eddy surface currents (U) derived from altimetry can alter the surface wind stress by creating a velocity shear across the air-sea interface (i.e., the relative wind effect). Where the surface current of the mesoscale eddy is aligned with the wind, the relative wind effect will lower the wind stress, and where the wind and currents are in the opposite direction, the wind stress will be enhanced. Regardless of the local alignment between the wind and current, the relative wind effect results in a reduction of the total wind work imparted on the surface because of the reduced momentum input and the enhanced surface drag (Pacanowski 1987). Numerical studies have demonstrated a substantial reduction of the energy of the currents and eddy fields by the relative wind effect (Duhaut and Straub 2006; Seo et al. 2007; Small et al. 2009; Eden and Dietze 2009; Hutchinson et al. 2010). Some recent studies suggest that this damping effect is particularly enhanced at the oceanic mesoscale (e.g., Anderson et al. 2011; Seo et al. 2016). The eddy-wind interaction through the eddy surface currents and vorticity also makes a significant contribution to the Ekman pumping velocity (Stern 1965; Dewar and Flierl, 1985; Gaube et al. 2015). The current-induced Ekman pumping velocity tends to act to reduce the surface vorticity anomaly of the eddy, thus acting to weaken the amplitude of mesoscale eddy fields (McGillicuddy et al. 2007), while the vorticity-gradient-induced Ekman pumping produces dipoles of Ekman upwelling and downwelling in instantaneous maps (Figure 10.10). The dipolar structure in Figure 10c is evident in composite averages only in regions where the wind direction is relatively steady; elsewhere, the dipolar structure averages to near zero and the total Ekman pumping is dominated by the relative wind effects of surface currents shown in Figure 10b (Gaube et al. 2015).

In addition to the mesoscale eddies, the global oceans feature smaller scale (1–10 km) yet highly energetic and transient sub-mesoscale variability and mixed layer fronts. As the Rossby number (a nondimensional number measuring the relative importance of nonlinear advection and rotation) nears O(1), the sub-mesoscale fields are increasingly governed by nonlinear dynamics and

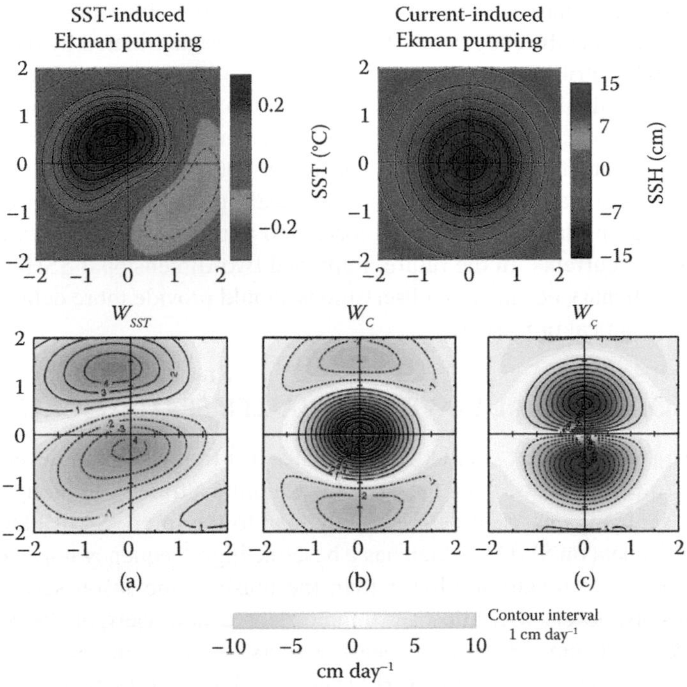

FIGURE 10.10 The various components of Ekman pumping for uniform eastward winds of 7 ms^{-1} over an idealized anticyclonic eddy with an axisymmetric Gaussian SSH profile with an amplitude of 14.5 cm (top right panel) and an SST structure equal to that of the composite-averaged SST fields in northern hemisphere anticyclones with a maximum SST anomaly of 0.3 deg C (top left panel). (a) The SST-induced Ekman pumping arising from crosswind SST gradients; (b) The surface current-induced Ekman pumping from the relative wind effects; and (c) the vorticity-gradient-induced Ekman pumping arising from the contribution of relative vorticity to the total potential vorticity (Stern, 1965). The x and y axes in all panels have been normalized by the radius of maximum rotational speed for a Gaussian eddy. (From Chelton, D. B., *Nat. Geosci.*, 6, 594–595, 2013. With permission.)

ageostrophic processes. We currently do not know whether the same air-sea coupling that has been observed at the ocean mesoscale should exist on the sub-mesoscales and the smaller end of the mesoscale range. The empirical relationship between mesoscale SST and wind anomalies may be altered as the horizontal length scale of the eddy fields approaches the vertical extent of the MABL (approximately 2 km). On the one hand, we can expect the finite response time of the MABL to become more important as the spatial scales of SST and surface current anomalies become smaller. On the other hand, it is possible that the strong gradients and lateral variability at the sub-mesoscale influences the MABL and air-sea interaction even more strongly than the mesoscale fields. The next-generation satellite altimeter measurements from SWOT, together with ongoing endeavors toward global high-resolution SST (e.g., GHRSST) and wind stress data sets, would help to understand the fundamental dynamics of the sub-mesoscale air-sea interactions.

Mesoscale eddies have a large influence on phytoplankton distribution, biogeochemical cycles, and pelagic ocean ecosystems. Identification of mesoscale eddies with satellite altimetry, and their joint analysis with ocean color maps, has led to great progress in understanding the relative roles of lateral stirring and mixing versus the vertical advection of nutrients in structuring the biogeochemical patterns. Using composites formed from thousands of eddies, Gaube et al. (2014) revealed that in most boundary current regions, cyclonic eddies exhibit higher chlorophyll anomalies in their interiors, and anticyclonic eddies contain lower chlorophyll. However, in the interior of the South Indian Ocean, the opposite occurs, due to the particular dynamics and chlorophyll distribution at the eastern boundary. At timescales longer than 2–3 weeks, the chlorophyll patterns evolve into dipoles with extrema outside the eddy cores. These patterns are dominated by the lateral stirring of the surface chlorophyll distribution by the rotational effects of mesoscale eddies (Siegel et al. 2008; Chelton et al. 2011a). In addition, modeling studies suggest that frontogenesis, occurring in the sub-mesoscale filaments surrounding these eddies, can trigger intense vertical velocities that enhance or reduce vertical nutrient supplies (e.g., Lévy and Klein 2004). Although altimetry today cannot resolve the dynamics associated with these sub-mesoscale processes, Lehahn et al. (2007) have shown that phytoplankton filaments were well correlated with time-evolving Lyapunov exponents derived from altimetric currents. In the future, improved two-dimensional SSH maps from SWOT combined with geostationary ocean color observations should provide more detail on these coupled physical and biogeochemical processes.

10.8 EFFECTS OF INTERNAL WAVES AT SMALLER SPATIAL SCALES

Interest in the ocean variability at smaller spatial scales has motivated the development of higher precision altimetry measurements, from the SARAL Ka-band altimetry to along-track SAR altimetry (CryoSat-2 in some regions, Sentinel-3 globally) and to the future SAR-interferometric wide-swath altimeter instrument on SWOT, which has a baseline high-frequency noise level (2 cm^2/km^{-2}) that is about two orders of magnitude lower than the noise of the Jason series altimeters. This great reduction in noise levels will allow an unprecedented new view of the SSH variability at wavelengths below 70–80 km. These new measurements, with a progressive increase in spatial resolution, are expected to reveal oceanic SSH variability that has remained below the noise floor of the previous altimeter missions.

In addition to the new view of mesoscale and sub-mesoscale variability that will be afforded by 2D SWOT measurements, the 20- to 200-km wavelength band will also contain SSH variability from internal waves. Internal waves can be generated in a stratified ocean by the interaction of tides or currents with bathymetry or by localized atmospheric forcing. Existing along-track SSH measurements are too noisy to allow direct estimation of the SSH wavenumber spectrum in the 10- to 80-km wavelength range where internal waves and sub-mesoscale variability coexist.

Numerical simulations are beginning to represent more realistic internal-wave fields (e.g., Richman et al. 2012; Muller et al. 2015; Rocha et al. 2016) that can be used to analyze the relative contributions of internal waves and lower-frequency variability. Figure 10.11 shows wavenumber spectra of SSH in the Drake Passage region from a 1/48° ocean general circulation model that represents internal waves and geostrophic variability down to scales as small as 5 km (figure modified from Rocha et al. 2016). The model output was from the Massachusetts Institute of Technology general circulation model (MITgcm) forced by realistic high-frequency wind and tidal forcing (see Rocha et al. 2016). By filtering the model output in time, one can crudely separate the variability at internal wave frequencies (minutes up to the inertial period, which is about 14 h in Drake Passage) from the lower frequency variability that is expected to be in geostrophic balance. Comparison of the SSH wavenumber spectra of hourly model output and daily averaged output gives an indication of the contribution of variability at periods of 2 h to 2 days, which is expected to be largely due to internal waves (Figure 10.11). On wavelengths larger than 50 km, the low-frequency variability dominates the SSH variability; at scales smaller than 40 km, the high frequency variability dominates. Figure 10.11 also shows the estimated noise levels for Jason (100 cm^2/cycle/km, e.g., Fu and Ubelmann 2014) and SWOT (from the SWOT Science Requirements)—the dominance of internal waves begins to occur at the wavelengths that will be newly accessible with the much lower noise levels of SWOT.

This example highlights the possible internal wave signal that should be expected in the 20- to 100-km wavelength band. We expect both internal-wave variability and sub-mesoscale variability to be apparent in the along-track SAR and two-dimensional SWOT data, and disentangling the two to fully exploit the information in the SWOT measurements will be an important and interesting

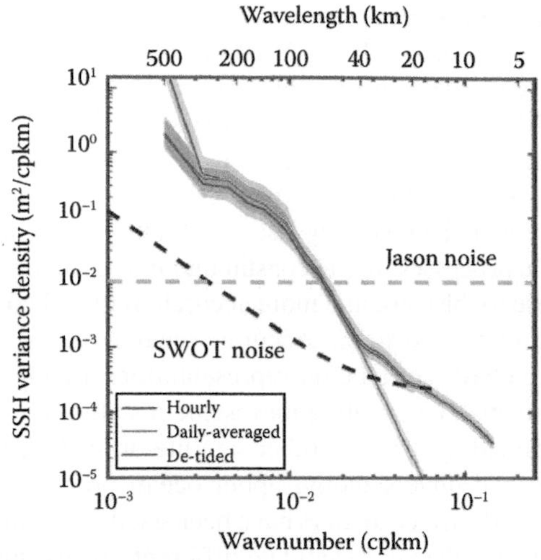

FIGURE 10.11 Wavenumber spectra of SSH in the Drake Passage region from a 1/48° ocean general circulation model that represents internal waves and geostrophic variability down to scales as small as 5 km. The difference between SSH wavenumber spectra of hourly model output (red line) and daily averaged output (blue line) gives an indication of the contribution of variability at periods of 2 h to 2 days, which is expected to be largely due to internal waves. At wavelengths larger than 50 km, the low-frequency variability dominates the SSH variability (and hence the blue curve and the red curve are almost the same); at scales smaller than 40 km, the high-frequency variability from internal waves becomes larger than the low-frequency variability. (Modified from Rocha et al. 2016.)

research challenge in the coming years. This example is from a single location, and many parallel efforts are underway to characterize and understand the space–time variability of different contributions to SSH variability at scales smaller than 100 km.

10.9 SUMMARY AND PERSPECTIVES

This chapter has highlighted the great progress that has been made in the last decade in our understanding of ocean mesoscale dynamics from altimeter observations. Most of the studies presented here are based on the mapped altimetric data and therefore represent the larger, mesoscale eddy field. Twenty-five years of altimetric SSH and velocity maps have given us great insight into the generation, propagation, and termination of these larger eddies. The coupling between larger mesoscale eddies and the atmospheric boundary layer has been revealed. However, we are still lacking in our observation of the smaller-scale eddies (wavelengths of 10–200 km). Smaller eddy dynamics will dominate in the regions of eddy generation and decay and have a strong role in ocean mixing. As we move to smaller scales, the dynamics of the ocean-atmospheric coupling may also change, which needs to be further explored with finer-scale observations of SSH, SST, and wind, and coupled air-sea interaction models.

Ocean dynamics at the sub-mesoscales (2–50 km) in the form of fronts and filaments, are most effective in driving the vertical exchanges between the surface layer and the deeper ocean (Lapeyre and Klein 2006). The next generation of satellite altimeters aims to better observe the smaller mesoscale eddy field and its anisotropic structure and may allow us to observe any balanced SSH signal resulting from finer-scale sub-mesoscale processes. Even at higher resolution, altimetric SSH cannot detect the unbalanced, ageostrophic component of frontal processes. However, if the future high-resolution altimetry is combined with SST and models, the ensemble should provide a better observation of the interactions between fronts and the smaller balanced mesoscale anomalies.

Over the next 10 years, advances in improving the along-track signal-to-noise ratio will continue for the conventional altimetric missions (Jason, SARAL) but also with the new nadir SAR missions (Sentinel-3 and then Jason-CS). This will be followed by the SWOT wide-swath altimetry using SAR-interferometry in 2020, providing two-dimensional images of SSH resolving scales down to 20-km wavelength. These new missions aim to provide accurate global observations of the smaller mesoscale signals, from 20- to 200-km wavelength. As we move forward with more precise observations, a number of altimetric data processing issues still need to be improved to correctly observe the full spectra of mesoscale processes. In the coastal and marginal seas, mesoscale studies require improved data coverage close to the coast and more accurate regional corrections for tides, mean sea surface, tropospheric corrections, and so on. Accurate mean dynamic topographies and mean sea surfaces are also required globally for a better representation of eddy-mean flow interactions and the role of bathymetric gradients in generating mesoscale instabilities.

Combining high-resolution altimetry with other satellite or *in situ* data provides complementary information for monitoring and understanding upper ocean processes. *In situ* data, such as tide gauges, gliders, Argo floats, and surface drifters have been used in the data processing stages to identify inaccuracies in the altimetry data or to validate different processing techniques. Altimetry and tide gauges have been combined to give better sea level coverage in coastal regions (e.g., Saraceno et al. 2008). Combining altimetry and ocean color data has given us enormous insight into the biological coupling associated with mesoscale eddies and fronts. Argo floats have revolutionized our vision of the time-variable vertical structure of the ocean, especially in the hard-to-reach regions of the Southern Hemisphere and the center of ocean gyres. The 10-day sampling of the Argo profiling floats makes them the ideal partner for altimetry to study the three-dimensional evolution of the larger mesoscale field (e.g., Pegliasco et al. 2015).

At smaller scales, the colocation of *in situ* profiles with small, rapidly moving eddy processes is more challenging (Morrow et al. 2017). Pegliasco et al. (2015) revealed the sampling bias in the colocation of 10 years of Argo profiles with eddies detected from altimetric maps, with the Argo profiles being located preferentially in the larger eddies. So building a statistical composite of the smaller-scale eddies' vertical structure is difficult with our present *in situ* array. The rapid evolution of the smaller-scale structures will require different *in situ* techniques, such as colocated underway CTDs, ADCP, Moving Vessel Profilers, and so on. Coastal radars that extend out over the coastal mesoscale field are also promising for their rapid sampling and two-dimensional coverage, albeit with the additional ageostrophic surface current component.

A number of theoretical studies are advancing, aiming to reconstruct the fine-scale upper three-dimensional ocean circulation from satellite data alone. These studies rely on the mesoscale space–time coverage of altimetry and SST, with *in situ* data providing the vertical stratification. Many of these studies use SST data as a proxy for the surface buoyancy field, which is adequate when SSH and SST are well correlated in the mixed layer. However, improved surface buoyancy fields to be combined with SSH depend on the availability of high-resolution surface salinity data. The satellite missions of Soil Moisture and Ocean Salinity (SMOS) and Aquarius, linked with Argo measurements, have greatly improved our monitoring of the large-scale salinity changes. New techniques are being developed to access the finer-scale salinity signal, including the horizontal stirring of the large-scale salinity field by altimetric currents (e.g., Desprès et al. 2011). Understanding the interplay between the buoyancy field in the surface mixed layer and the deeper dynamics revealed by altimetric SSH is an important ongoing question.

One of the key issues to be addressed in the coming years concerns the high-frequency ocean dynamics and how they impact on SSH at these shorter scales (20–200 km). The along-track altimetric SSH wavenumber spectra are based on basin-scale rapid "snapshots" collected in a few minutes. Although the along-track SSH data are corrected for large-scale high-frequency barotropic motions (from tides and atmospheric forcing), they can contain errors from these modeled fields as well as the surface signature of internal waves. The internal tide signature is at similar wavelengths to the mesoscale, and its coherent part has been estimated in the along-track data (Ray and Zaron 2015) and in the future may be predicted and removed. Recent modeling studies suggest that the internal tides interacting with major currents may generate a substantial incoherent signal that cannot be separated from the smaller mesoscales (Ponte and Klein 2015). The internal wave signature from different sources is greatly reduced in daily averaged model outputs or in the altimetric maps with a temporal smoothing over 15 days. However, analysis of high-resolution hourly output models (e.g., Rocha et al. 2016) and the wavenumber projection of mooring data (Farrar and Durland 2012) show that internal waves may have a substantial signature at the 20- to 200-km spatial scales. As we progress to resolving these smaller scales, a great challenge will be to improve our understanding of the internal wave SSH signature and its relative amplitude compared to internal ocean dynamics resulting from different dynamical regimes (QG, SQG, mixed-layer instabilities, etc.).

ACKNOWLEDGMENTS

The authors would like to acknowledge the outstanding work by the Ocean Surface Topography Science team, which brings together both scientific and technical teams, working to improve altimetric data and products for mesoscale studies. Much of the data analyzed by the different authors was made available by CNES/AVISO and via the DUACS project. RM and the CTOH receive finance from INSU, the Université Toulouse III, and the CNES TOSCA program. JTF acknowledges support from NASA grants NNX13AE46G and NNX16AH76G. HS receives support from the Office of Naval Research (N00014-15-1-2588).

REFERENCES

Abernathey, R. P., and Marshall, J., 2013, Global surface eddy diffusivities derived from satellite altimetry, *J. Geophys. Res. Oceans* 118:901–916. doi:10.1002/jgrc.20066.

Ablain, M., Cazenave, A., Larnicol, G., Balmaseda, M., Cipollini, P., Faugère, Y., Fernandes, M. J., et al., 2015, Improved sea level record over the satellite altimetry era (1993–2010) from the Climate Change Initiative project, *Ocean Sci.* 11:67–82. doi:10.5194/os-11-67-2015.

Abraham, E. R., Law, C. S., Boyd, P. W., Lavender, S. J., Maldonado, M. T., and Bowie, A., 2000, Importance of stirring in the development of an iron-fertilized phytoplankton bloom, *Nature* 407:727–730.

Anderson, L., McGillicuddy, D., Maltrud, M., Lima, I., Doney, S., 2011, Impact of eddy-wind interaction on eddy demographics and phytoplankton community structure in a model of the north Atlantic ocean, *Dyn. Atmos. Oceans* 52:80–94.

Arbic, B. K., Müller, M., Richman, J. G., Shriver, J. F., Morten, A. J., Scott, R. B., Sérazin, G., et al., 2014, Geostrophic turbulence in the frequency-wavenumber domain: Eddy-driven low-frequency variability, *J. Phys. Oceanogr.* 44(8):2050–2069.

Arbic, B. K., Polzin, K. L., Scott, R. B., Richman, J. G., Shriver, J. F., 2013, On eddy viscosity, energy cascades, and the horizontal resolution of gridded satellite altimeter products, *J. Phys. Oceanogr.* 43(2):283–300. doi:10.1175/JPO-D-11-0240.1.

Arbic, B. K., Scott, R. B., Flierl, G. R., Morten, A. J., Richman, J. G., Shriver, J. F., 2012, Nonlinear cascades of surface oceanic geostrophic kinetic energy in the frequency domain, *J. Phys. Oceanogr.* 42(9):1577–1600.

Assassi, C., Morel, Y., Vandermeirsch, F., Chaigneau, A., Pegliasco, C., Morrow, R., Colas, F., et al., 2016, An index to distinguish surface and subsurface intensified vortices from surface observations, *J. Phys. Oceanogr*. 46, 2529–2552, doi: 10.1175/JPO-D-15-0122.1

Assenbaum, M., and Reverdin, G., 2005, Near-real time analyses of the mesoscale circulation during the POMME experiment, *Deep-Sea Res. I* 52:1345–1373.

Blumen, W., 1978, Uniform potential vorticity flow: Part I. Theory of wave interactions and two-dimensional turbulence, *J. Atmos. Sci.* 35:774–783.

Boebel, O., Lutjeharms, J., Schmid, C., Zenk, W., Rossby, T., and Barron, C., 2003, The Cape Cauldron: A regime of turbulent inter-ocean exchange, *Deep Sea Res. II* 50:57–86.

Bouffard, J., Pascual, A., Ruiz, S., Faugère, Y., and Tintoré J., 2010, Coastal and mesoscale dynamics characterization using altimetry and gliders: A case study in the Balearic Sea, *J. Geophys. Res. C*, 115(10), C10029, doi:10.1029/2009JC006087

Callies, J., and Ferrari, R., 2013, Interpreting energy and tracer spectra of upper-ocean turbulence in the sub-mesoscale range (1–200 km), *J. Phys. Oceanogr.* 43:2456–2474. doi:10.1175/JPO-D-13-063.1.

Callies, J., Ferrari, R., Klymak, J. M., and Gula, J., 2015, Seasonality in sub-mesoscale turbulence, *Nat. Commun.* 6:6862. doi:10.1038/ncomms7862.

Capet, X., Klein, P., Hua, B., Lapeyre, G., and McWilliams, J. C., 2008a, Surface kinetic and potential energy transfer in SQG dynamics, *J. Fluid Mech.* 604:165–174.

Capet, A., Mason, E., Ross, V., Troupin, C., Faugere, Y., Pujol, M.-I., and Pascual, A., 2014, Implications of a refined description of mesoscale activity in the eastern boundary upwelling systems, *Geophys. Res. Lett.* 41, 7602–7610, doi:10.1002/2014GL061770.

Capet, X., McWilliams, J. C., Molemaker, M. J., and Shchepetkin, A. F., 2008b, Mesoscale to sub-mesoscale transition in the California current system. Part III: Energy balance and flux, *J. Phys. Oceanogr.* 38:2256–2269.

Carrère L., Lyard F., 2003, Modeling the barotropic response of the global ocean to atmospheric wind and pressure forcing – comparisons with observations. *Geophys Res Lett* 30(6):1275. doi:10.1029/2002GL016473

Castelao, R. M., 2014, Mesoscale eddies in the South Atlantic Bight and the Gulf Stream Recirculation region: Vertical structure, *J. Geophys. Res. Oceans*, 119, 2048–2065. doi:10.1002/2014JC009796.

Chaigneau, A., Le Texier, M., Eldin, G., Grados, C., and Pizarro, O., 2011, Vertical structure of mesoscale eddies in the eastern South Pacific Ocean: A composite analysis from altimetry and argo profiling floats, *J. Geophys. Res.* 116:C11025. doi:10.1029/2011JC007134.

Chaigneau, A., and Pizarro, O., 2005, Eddy characteristics in the eastern South Pacific, *J. Geophys. Res.* 110. doi:10.1029/2004JC002815, 2005.

Charney, J.G., 1971, Geostrophic turbulence, *J. Atmos. Sci.* 28:1087–1095.

Chelton, D. B., 2013, Ocean-atmosphere coupling: Mesoscale eddy effects, *Nat. Geosci.* 6:594–595.

Chelton, D.B., Gaube, P., Schlax, M. G., Early, J. J., and Samelson, R. M., 2011a. The influence of nonlinear mesoscale eddies on near-surface chlorophyll, *Science*, 334, 328–332, doi:10.1126/science.1208897.

Chelton, D. B., and Schlax, M. G., 1996, Global observations of oceanic Rossby waves, *Science* 272:234–238.

Chelton, D. B., Schlax, M. G., Freilich, M. H., and Milliff, R. F., 2004, Satellite measurements reveal persistent small-scale features in ocean winds, *Science* 303:978–983.

Chelton, D. B., Schlax, M. G., Samelson, R. M., and de Szoeke, R. A., 2007, Global observations of large oceanic eddies, *Geophys. Res. Lett.* 34: L15606. doi:10.1029/2007GL030812.

Chelton, D. B., Schlax, M. G., and Samelson, R. M., 2011b, Global observations of nonlinear mesoscale eddies, *Prog. Oceanogr.* 91(2):167–216. doi:10.1016/j.pocean.2011.01.002.

Chelton, D. B., and Xie, S.-P., 2010, Coupled ocean-atmosphere interaction at oceanic mesoscales, *Oceanogr.* 23:52–69.

Colas, F., McWilliams, J. C., Capet, X., and Kurian, J., 2012, Heat balance and eddies in the Peru-Chile current system, *Clim. Dyn.* 39(1–2):509–529. doi:10.1007/s00382-011-1170-6.

Cotte, C., d'Ovidio, F., Dragon, A.-C., Guinet, C., and Levy, M., 2015, Flexible preference of southern elephant seals for distinct mesoscale features within the Antarctic Circumpolar Current, Progress Oceanogr. 131, 46–58

Cushman-Roisin, B. E., Chassignet, P., and Tang, B., 1990, Westward motion of mesoscale eddies, *J. Phys. Oceanogr.* 20:97–113.

De Monte, S., Soccodato, A., Alvain, S., and d'Ovidio, F. (2013). Can we detect oceanic biodiversity hotspots from space ? *The ISME journal*, 7(10), 2054–2056.

Dencausse, G., Morrow, R., Rogé, M., and Fleury, S., 2014, Lateral stirring of large-scale tracer fields by altimetry, *Ocean Dynamics* 64:61–78. doi: 10.1007/s10236-013-0671-8.

de Ruijter, W. P. M., Biastoch, A., Drijfhout, S. S., Lutjeharms, J. R. E., Matano, R. P., Pichevin, T., van Leeuwen, P. J., et al., 1999, Indian-Atlantic inter-ocean exchange: Dynamics, estimation and impact, *J. Geophys. Res.* 104:20885–20910.

Desprès, A., Reverdin, G., and d'Ovidio, F., 2011, Mechanisms and spatial variability of meso scale frontogenesis in the northwestern subpolar gyre, *Ocean Modeling* 39:97–113. doi:10.1016/j.ocemod.2010.12.005

Dewar, W., and Flierl, G., 1987, Some effects of the wind on rings, *J. Phys. Oceanogr.* 17:1653–1667.

Dibarboure, G. and J. Lambin, 2015. Monitoring the ocean surface topography virtual constellation: Lessons learned from the contribution of SARAL/AltiKa. *Marine Geodesy* 38(suppl 1), 684–703.

Dibarboure, G., Pujol, M.-I., Briol, F., Le Traon, P.-Y., Larnicol, G., Picot, N., Mertz, F., et al., 2011, Jason-2 in DUACS: First tandem results and impact on processing and products, *Marine Geodesy* 34:214–241. doi:10.1080/01490419.2011.

d'Ovidio, F., Isern-Fontanet, J., López, C., Hernández-García, E., and García-Ladona, E., 2009, Comparison between Eulerian diagnostics and finite-size Lyapunov exponents computed from altimetry in the Algerian basin, *Deep Sea Res.* 56(1):15–31.

d'Ovidio, F., De Monte, S., Alvain, S., Danonneau, Y. and Lévy, M., 2010, Fluid dynamical niches of phytoplankton types, PNAS, 107, 18366-18370 doi: 10.1073/pnas.1004620107

Dong, C., McWilliams, J. C., Liu, Y., and Chen, D., 2014, Global heat and salt transports by eddy movement, *Nat. Commun.* 5. doi:10.1038/ncomms4294.

Dufau, C., Orsztynowicz, M., Dibarboure, G., Morrow, R., and Letraon, P-Y., 2016, Mesoscale resolution capability of altimetry: Present and future, submitted. *J. Geophys. Res.* Oceans, 121, 4910–4927, doi:10.1002/2015JC010904.

Duhaut, T. H. A., and Straub, D. N., 2006, Wind stress dependence on ocean surface velocity: Implications for mechanical energy input to ocean circulation, *J. Phys. Oceanogr.* 36:202–211.

Dussurget, R., Birol, F., Morrow, R., and De Mey, P., 2011, Fine resolution altimetry data for a regional application in the Bay of Biscay, *Marine Geodesy.* 34:3–4, 447–476.

Eden, C., and H. Dietze, 2009, Effects of mesoscale eddy/wind interactions on biological new production and eddy kinetic energy, *J. Geophys. Res.* 114:C05023.

Escudier, R., Bouffard, J., Pascual, A., Poulain, P.-M., Pujol, M.-I., 2012, Improvement of coastal and mesoscale observation from space: Application to the Northwestern Mediterranean Sea, *Geophys. Res. Lett.* 40(10):2148–2153.

Fang, F., and Morrow, R., 2003, Warm-core eddy propagation in the southeast Indian Ocean, *Deep Sea Res. II* 50:2245–2261.

Farrar, J. T., and Durland, T. S., 2012, Wavenumber-frequency spectra of inertia-gravity and mixed Rossby-gravity waves in the equatorial Pacific Ocean, *J. Phys. Oceanogr.* 42:1859–1881.

Farrar, J. T., and Weller, R.A., 2006, Intraseasonal variability near 10°N in the eastern tropical Pacific Ocean, *J. Geophys. Res.* 111:C05015. doi:10.1029/2005JC002989.

Fu, L. L., 1983, On the wave number spectrum of oceanic mesoscale variability observed by the SEASAT altimeter, *J. Geophys. Res.* 88:4331–4341.

Fu, L.-L., and Chelton, D. B., 2001, Large-scale ocean circulation, in *Satellite Altimetry and Earth Sciences. A Handbook of Techniques and Applications,* ed. L.-L. Fu and A. Cazenave, Eds., Academic Press, San Diego, 133–169.

Fu, L. L, Chelton, D. B., Le Traon, P.-Y., Morrow, R., 2010, Eddy dynamics from satellite altimetry, *Oceanography Magazine* 23(4): 14–25, http://dx.doi.org/10.5670/oceanog.2010.02.

Fu, L-L., and Ubelmann, C., 2014, On the transition from profile altimeter to swath altimeter for observing global ocean surface topography, *J. Ocean. Atm. Tech.* 31(2):560–568. doi:10.1175/JTECH-D-13-00109.1.

Gaube, P., Chelton, D. B., Samelson, R. M., Schlax, M. G., O'Neill, L. W., 2015, Satellite observations of mesoscale eddy-induced Ekman pumping, *J. Phys. Oceanogr.* 45:104–132.

Gaube, P., McGillicuddy, D. J., Jr., Chelton, D. B., Behrenfeld, M. J., and Strutton, P. G., 2014, Regional variations in the influence of mesoscale eddies on near-surface chlorophyll, *J. Geophys. Res.* 119, 8195–8220, doi:10.1002/2014JC010111.

González-Haro, C., and Isern-Fontanet, J., 2014, Global ocean current reconstruction from altimetric and microwave SST measurements, *J. Geophys. Res. Oceans* 119(6):3378–3391. doi:10.1002/2013jc009728.

Held, I. M., Pierrehumbert, R. T., Garner, S. T., and Swanson, K. L., 1995, Surface quasi-geostrophic dynamics, *J. Fluid Mech.* 282:1–20.

Hunt, F.K., Tailleux, R., and Hirsch, J.J.-M., 2012, The vertical structure of oceanic Rossby waves: A comparison of high-resolution model data to theoretical vertical structures, *Ocean Sci.* 8:19–35.

Hutchinson, D. K., Hogg, A. M. C., and Blundell, J. R., 2010, Southern ocean response to relative velocity wind stress forcing, *J. Phys. Oceanogr.* 40:326–339.

Isern-Fontanet, J., Garcia-Ladona, E., and Font, J., 2003, Identification of marine eddies from altimeter maps, *J. Atmos. Oceanic Tech.* 20:772–778.

Isern-Fontanet, J., Lapeyre, G., Klein, P., Chapron, B., and Hecht, M. W., 2008, Three-dimensional reconstruction of oceanic mesoscale currents from surface information, *J. Geophys. Res.* 113:C09005. doi:10.1029/2007JC004692.

Kelly, K. A., and Gille, S. T., 1990, Gulf stream surface transport and statistics at 69° W from the GEOSAT altimeter, *J. Geophys. Res.* 95:3149–3161.

Klein, P., Hua, B., Lapeyre, G., Capet, X., Gentil, S. L., and Sasaki, H., 2008, Upper ocean turbulence from high 3-D resolution simulations, *J. Phys. Oceanogr.* 38:1748–1763.

Koblinsky, C.J., Gaspar, P., and Lagerloef, G., 1992, *The future of spaceborne altimetry: Oceans and climate change, joint oceanographic institutions incorporated,* Washington, DC, 75 pp.

Kurian, J., Colas, F., Capet, X., McWilliams, J. C., and Chelton, D. B., 2011, Eddy properties in the California Current system, *J. Geophys. Res.* 116:C08027. doi:10.1029/2010JC006895.

Lapeyre, G., and Klein, P., 2006, Dynamics of the upper oceanic layers in terms of surface quasigeostrophy theory, *J. Phys. Oceanogr.* 36:165–176.

Lehahn, Y., d'Ovidio, F., Lévy, M., and Heitzel, E., 2007, Stirring of the Northeast Atlantic spring bloom: A Lagrangian analysis based on multi-satellite data, *J. Geophys. Res.* 112:C08005. doi:10.1029/2006JC003927.

Le Traon, P.Y., 1991, Timescales of mesoscale variability and their relationship with space scales in the North Atlantic, *J. Mar. Res.* 49:467–492, 1991.

Le Traon, P. Y., 2013, From satellite altimetry to argo and operational oceanography: Three revolutions in oceanography, *Ocean Science* 9(5):901–915. doi:10.5194/os-9-901-2013.

Le Traon, P.Y., and Dibarboure, G., 2002, Velocity mapping capabilities of present and future altimeter missions: The role of high frequency signals, *J. Atm. Ocean Tech.* 19:2077–2088, 2002.

Le Traon, P. Y., Klein, P., Hua, B. L., and Dibarboure, G., 2008, Do altimeter wavenumber spectra agree with the interior or surface quasi-geostrophic theory? *J. Phys. Oceanogr.* 38:1137–1142. doi:10.1175/2007JPO3806.1.

Le Traon, P-Y., and Morrow, R., 2001, Ocean currents and eddies, in *Satellite altimetry and earth sciences: A handbook for techniques and applications*, edited by L.-L. Fu and A. Cazenave, pp. 171–210, Academic Press, San Diego, CA, 423 pp.

Le Traon, P. Y., Rouquet, M. C., Boissier, C., 1990, Spatial scales of mesoscale variability in the North Atlantic as deduced from GEOSAT data, *J. Geophys. Res.* 95:20267–20285.

Lévy, M., and Klein, P., 2004, Does the low frequency variability of mesoscale dynamics explain a part of the phytoplankton and zooplankton spectral variability? *Proc. Royal Soc. Lon.* 460(2046):1673–1683. doi:10.1098/rspa.2003.1219.

Lilly, J.M., Rhines, P.B., Schott, F., Lavender, K., Lazier, J., Send, U., and D'Asaro, E., 2003, Observations of the Labrador Sea eddy field, *Progress in Oceanography* 59(1):75–176. doi:10.1016/j.pocean.2003.08.013, 2003.

Lindzen, R., and Nigam, S., 1987, On the role of sea surface temperature gradients in forcing low-level winds and convergence in the tropics, *J. Atmos. Sci.* 44:2418–2436.

Liu, Y., Dong, C., Guan, Y., Chen, D., McWilliams, J., and Nencioli, F., 2012, Eddy analysis in the subtropical zonal band of the North Pacific Ocean, *Deep Sea Res., Part I* 68:54–67. doi:10.1016/j.dsr.2012.06.001.

Marshall, J., Shuckburgh, E., Jones, H., and Hill, C., 2006, Estimates and implications of surface eddy diffusivity in the Southern Ocean derived from tracer transport, *J. Phys. Oceanogr.* 36:1806–1821. doi:10.1175/JPO2949.1.

Maximenko, N. A., Melnichenko, O. V., Niiler, P. P., and Sasaki, H., 2008, Stationary mesoscale jet-like features in the ocean, *Geophys. Res. Lett.* 35:L08603. doi:10.1029/2008GL033267.

McGillicuddy, D., Anderson, L., Bates, N., Bibby, T., Buesseler, K., Carlson, C., Davis, C., et al., 2007, Eddy/wind interactions stimulate extraordinary mid-ocean plankton blooms, *Science* 316:1021–1026.

Morrow, R., Birol, F., Griffin, D., Sudre, J., 2004a, Divergent pathways of anticyclonic and cyclonic eddies, *Geophys. Res. Lett.* 31:L24311. doi:10.1029/2004GL020974.

Morrow, R., Carret, A., Birol, F., Nino, F., Valladeau, G., Boy, F., Bachelier, C., et al., 2017, Observability of fine-scale ocean dynamics in the Northwest Mediterranean Sea, *Ocean Sci.* 13:13–29. doi:10.5194/os-13-13-2017.

Morrow, R. A., Coleman, R., Church, J. A., and Chelton, D. B., 1994, Surface eddy momentum flux and velocity variances in the Southern Ocean from GEOSAT altimetry, *J. Phys. Oceanogr.* 24:2050–2071.

Morrow, R., Donguy, J. R., Chaigneau, A., and Rintoul, S., 2004b, Cold core anomalies at the Subantarctic Front, south of Tasmania, *Deep-Sea Res. I* 51:1417–1440.

Morrow, R. A., Fang, F., Fieux, M., and Molcard, R., 2003, Anatomy of three warm-core Leeuwin Current eddies, *Deep Sea Res. II* 50:2229–2243.

Morrow, R., and Le Traon, P.-Y., 2012, Recent advances in observing mesoscale ocean dynamics with satellite altimetry, *Adv. Space Res.* 50(8):1062–1076. doi:10.1016/j.asr.2011.09.033.

Muller, M., Arbic, B. K., Richman, J. G., Shriver, J. F., Kunze, E. L., Scott, R. B., Wallcraft, A. J., and Zamudio, L., 2015, Toward an internal gravity wave spectrum in global ocean models, *Geophys. Res. Lett.* 42:3474–3481. doi:10.1002/2015GL063365.

Nakano, H., and Hasumi, H., 2005, A series of zonal jets embedded in the broad zonal flows in the Pacific obtained in eddy-permitting ocean general circulation models, *J. Phys. Oceagr.* 35(4):474–488.

Niiler, P. P., Maximenko, N. A., Panteleev, G. G., Yamagata, T., and Olson, D. B., 2003, Near-surface dynamical structure of the Kuroshio Extension, *J. Geophys. Res.* 108(C6):3193. doi:10.1029/2002JC001461, 2003.

Ohlmann, J. C., Niiler, P. P., Fox, C. A., and Leben, R. R., 2001, Eddy energy and shelf interactions in the Gulf of Mexico, *J. Geophys. Res.* 106(C2):2605–2620. doi:10.1029/1999JC000162.

Pacanowski, R. C., 1987, Effect of equatorial currents on surface stress, *J. Phys. Oceanogr.* 17, 833–838.

Pascual, A., Faugere, Y., Larnicol, G., and Le Traon, P.Y., 2006, Improved description of the ocean mesoscale variability by combining four satellite altimeters, *Geophys. Res. Letters* 33(2):L02611.

Pascual, A., Pujol, M. I., Larnicol, G., Le Traon, P. Y., and Rio, M. H., 2005, Mesoscale mapping capabilities of multisatellite altimeter missions: First results with real data in the Mediterranean Sea, *J. Mar. Systems* 65(2007):190–211. doi:10.1016/j.jmarsys.2004.12.004.

Pegliasco, C., Chaigneau, A., and Morrow, R., 2015, Main eddy vertical structures observed in the four major Eastern Boundary Upwelling Systems, *J. Geophys. Res. Oceans* 120:6008–6033. doi:10.1002/2015JC010950.

Ponte, A. L., and Klein, P., 2015, Incoherent signature of internal tides on sea level in idealized numerical simulations, *Geophys. Res. Lett.* 42:1520–1526. doi:10.1002/2014GL062583.

Ponte, A., Klein, P., Capet, X., Le Traon, P. Y., Chapron, B., and Lherminier, P., 2013, Diagnosing surface mixed layer dynamics from high-resolution satellite observations: Numerical insights, *J. Phys. Oceanogr.* 43(7):1345–1355. doi:10.1175/JPO-D-12-0136.1.

Pujol, M.-I., Faugère, Y., Taburet, G., Dupuy, S., Pelloquin, C., Ablain, M., and Picot, N., 2016, DUACS DT2014: The new multi-mission altimeter data set reprocessed over 20 years, *Ocean Sci. Discuss.* 12, 1067–1090, doi:10.5194/os-2015-110.

Pujol, M. I., and Larnicol, G., 2005, Mediterranean sea eddy kinetic energy variability from 11 years of altimetric data, *J. Mar. Systems* 58:121–142.

Qiu, B., 1999, Seasonal eddy field modulation of the North Pacific subtropical countercurrent: TOPEX/Poseidon observations and theory, *J. Phys. Oceanogr.* 29, 2471–2486.

Qiu, B. and S. Chen, 2004: Seasonal Modulations in the Eddy Field of the South Pacific Ocean. *J. Phys. Oceanogr.*, 34, 1515–1527, doi:10.1175/1520-0485(2004)034

Qiu, B., and Chen, S., 2005, Eddy-induced heat transport in the subtropical North Pacific from Argo, TMI and altimetry measurements, *J. Phys. Oceanogr.* 35:458–473.

Qiu, B., Chen, S., Klein, P., Sasaki, H., and Sasai, Y., 2014, Seasonal mesoscale and sub-mesoscale eddy variability along the North Pacific subtropical countercurrent, *J. Phys. Oceanogr.* 44:3079–3098. doi:10.1175/JPO-D-14-0071.1.

Ray, R. D., and Zaron, E. D., 2015, M2 internal tides and their observed wavenumber spectra from satellite altimetry, *J. Phys. Oceanogr.* 46, 3–22, doi:10.1175/JPO-D-15-0065.1.

Reverdin, G., Morisset, S., Marié, L., Bourras, D., Sutherland, G., Ward, B., Salvador, J., et al., 2015, Surface salinity in the North Atlantic subtropical gyre during the STRASSE/SPURS summer 2012 cruise, *Oceanography* 28(1):114–123. doi:10.5670/oceanog.2015.09.

Richman, J. G., Arbic, B. K., Shriver, J. F., Metzger, E. J., and Wallcraft, A. J., 2012, Inferring dynamics from the wavenumber spectra of an eddying global ocean model with embedded tides, *J. Geophys. Res.* 117:C12012. doi:10.1029/2012JC008364.

Ridderinkhof, W., Le Bars, D., von der Heydt, A. S., and de Ruijter, W. P. M., 2013, Dipoles of the South East Madagascar Current, *Geophys. Res. Lett.* 40:558–562. doi:10.1002/grl.50157.

Rocha, C. B., Chereskin, T. K., Gille, S. T., and Menemenlis, D., 2016, Mesoscale to sub-mesoscale wavenumber spectra in Drake Passage, *J. Phys. Oceanogr.* 46(2):601–620. doi:10.1175/JPO-D-15-0087.1.

Roemmich, D., and Gilson, J., 2001, Eddy transport of heat and thermocline waters in the North Pacific: A key to interannual/decadal climate variability? *J Phys. Ocean* 31:675–688.

Rogé, M., Morrow, R., Dencausse, G., 2015, Altimetric Lagrangian advection to reconstruct Pacific Ocean fine scale surface tracer fields, *Ocean Dynamics* 65:1249–1268. doi:10.1007/s10236-015-0872-4.

Ruiz, S., Pascual, A., Garau, B., Pujol, I., and Tintoré, J., 2009, Vertical motion in the upper ocean from glider and altimetry data, *Geo. Res. Let.* 36(14), L14607, doi:10.1029/2009GL038569.

Sallée, J. B., Speer, K., and Morrow, R., 2008a, Southern Ocean fronts and their variability to climate modes, *J. Climate* 21(12):3020–3039.

Sallée, J. B., Speer, K., Morrow, R., and Lumpkin, R., 2008b, An estimate of Lagrangian eddy statistics and diffusion in the mixed layer of the Southern Ocean, *J. Marine Res.* 66:441–463. doi:10.1357/002224008787157458.

Samelson, R. M., Schlax, M. G., and Chelton, D. B., 2014, Randomness, symmetry, and scaling of mesoscale eddy life cycles, *J. Phys. Oceanogr.* 44(3):1012–1029. doi:10.1175/JPO-D-13–0161.1.

Sangra, P., Pascual, A., Rodríguez-Santana, A., Machín, F., Mason, E., McWilliams, J. C., Pelegrí, J. L., et al., 2009, The Canary Eddy Corridor: A major pathway for long-lived eddies in the subtropical North Atlantic, *Deep Sea Res., Part I* 56(12):2100–2114. doi:10.1016/j.dsr.2009.08.008.

Saraceno, M., P. T. Strub, and P. M. Kosro, 2008, Estimates of sea surface height and near-surface alongshore coastal currents from combinations of altimeters and tide gauges, *J. Geophys. Res.* 113, C11013. doi:10.1029/2008JC004756.

Sasaki, H., Klein, P., Qiu, B., and Sasai, Y., 2014, Impact of oceanic-scale interactions on the seasonal modulation of ocean dynamics by the atmosphere, *Nature Comm.* 5:5636. doi:10.1038/ncomms5636.

Seo, H., 2017, Distinct influence of air-sea interactions mediated by mesoscale sea surface temperature and surface current in the Arabian Sea, *J. Climate.* https://doi.org/10.1175/JCLI-D-16-0834.1

Seo, H., Jochum, M., Murtugudde, R., Miller, A. J., and Roads, J. O., 2007, Feedback of tropical instability wave—Induced atmospheric variability onto the ocean. *J. Climate* 20:5842–5855.

Seo, H., Miller, A. J., and Norris, J. R., 2016, Eddy-wind interaction in the California Current System: Dynamics and impacts, *J. Phys. Oceanogr.* 46:439–459.

Scott, R. B., and Arbic, B. K., 2007, Spectral energy fluxes in geostrophic turbulence: Implications for ocean energetics, *J. Phys. Oceanogr.* 37:673–688.

Scott, R. B., and Wang, F., 2005, Direct evidence of an oceanic inverse kinetic energy cascade from satellite altimetry, *J. Phys. Oceanogr.* 35:1650–1666.

Siegel, D. A., Court, D. B., Menzies, D. W., Peterson, P., Maritorena, S., and Nelson, N.B., 2008, Satellite and in situ observations of the bio-optical signatures of two mesoscale eddies in the Sargasso Sea, *Deep-Sea Res. II* 55:1218–1230.

Small, R. J., de Szoeke, S., Xie, S. P., O'Neill, L., Seo, H., Song, Q., Cornillon, P., et al., 2008, Air-sea interaction over ocean fronts and eddies, *Dyn. Atmos. Oceans* 45:274–319.

Small, R. J., Richards, K. J., Xie, S.-P., Dutrieux, P., and Miyama, T., 2009, Damping of Tropical Instability Waves caused by the action of surface currents on stress, *J. Geophys. Res.* 114:C04009.

Sokolov, S., and Rintoul, S. R., 2007, Multiple jets of the Antarctic Circumpolar Current south of Australia, *J. Phys. Oceanogr.* 37:1394–1412.

Souza, J. M. A. C., de Boyer Montegut, C., and Le Traon, P. Y., 2011, Mesoscale eddies in the South Atlantic. *Ocean Sci. Discuss.* 8:483–531. doi:10.5194/osd-8-483-2011.

Sprintall, J., 2003, Seasonal to interannual upper-ocean variability in the Drake Passage, *J. Mar. Res.* 61(1):27–57. doi:10.1357/002224003321586408.

Stammer, D., 1997, Global characteristics of ocean variability estimated from regional TOPEX/Poseidon altimeter measurements, *J. Phys. Oceanogr.* 27:1743–1769.

Stammer, D., and Wunsch, C., 1999, Temporal changes in eddy energy of the oceans, *Deep Sea Res.* 46:77–108.

Stern, M., 1965, Interaction of a uniform wind stress with a geostrophic vortex, *Deep Sea Res. Oceanogr. Abstr.* 12:355–367.

Tailleux, R., 2003, Comments on 'the effect of bottom topography on the speed of long Extratropical planetary waves', *J. Phys. Oceanogr.* 33:1536–1541.

Treguier, A., Boebel, O., Barnier, B., and Madec, G., 2003, Agulhas eddy fluxes in a 1/6°Atlantic model, *Deep Sea Res. II* 50:119–139.

Ubelmann, C., Klein, P., Fu, L.-L., 2015, Dynamical interpolation of sea surface height and potential applications for future high-resolution altimetry mapping, *J. Atmos. Ocean. Tech.* 32:177–184. doi:10.1175/JTECH-D-14-00152.1.

Wang, D.-P., Flagg, C. N., Donohue, K., and Rossby, H. T., 2010, Wavenumber spectrum in the Gulf Stream from shipboard ADCP observations and comparison with altimetry measurements, *J. Phys. Oceanogr.* 40:840–844. doi:10.1175/2009JPO4330.1.

Wang, J., Flierl, G. R., LaCasce, J. H., McClean, J. L., and Mahadevan, A., 2013, Reconstructing the ocean's interior from surface data, *J. Phys. Oceanogr.* 43(8):1611–1626.

Wallace, J., Mitchell, T., and Deser, C., 1989, The influence of sea-surface temperature on surface wind in the eastern equatorial Pacific: Seasonal and interannual variability, *J. Climate* 2:1492–1499.

Waugh, D. W., and Abraham, E. R., 2008, Stirring in the global surface ocean, *Geophys. Res. Lett.* 35:L20605. doi:10.1029/2008GL035526.

Wortham, C., and Wunsch, C., 2013, A multidimensional spectral description of ocean variability, *J. Phys. Oceanogr.* 44:944–966. doi:10.1175/JPO-D-13-0113.1.

Wunsch, C. and Stammer, D., 1998. Satellite altimetry, the marine geoid and ocean general circulation. *Ann. Rev. Earth Planet. Sci.* 26, 219–254.

Xie, S.-P., 2004, Satellite observations of cool ocean–atmosphere interaction, *Bull. Amer. Meteor. Soc.* 85:195–209.

Xu, Y., and Fu, L.-L., 2011, Global variability of the wavenumber spectrum of oceanic mesoscale turbulence, *J. Phys. Oceanogr.* 41:802–809.

Xu, Y., and Fu, L.-L., 2012, The effects of altimeter instrument noise on the estimation of the wavenumber spectrum of sea surface height, *J. Phys. Oceanogr.* 42:2229–2233. doi:10.1175/JPO-D-12-0106.1.

Yang, G., Wang, F., Li, Y., and Lin, P., 2013, Mesoscale eddies in the northwestern subtropical Pacific Ocean: Statistical characteristics and three-dimensional structures, *J. Geophys. Res. Oceans* 118:1906–1925. doi:10.1002/jgrc.20164.

Yoshida, S., Qiu, B., and Hacker, P., 2010, Wind-generated eddy characteristics in the lee of the island of Hawaii, *J. Geophys. Res.* 115, C03019, doi:10.1029/2009JC005417.

Zhang, Z., Wang, W. and Qiu, B., 2014, Oceanic mass transport by mesoscale eddies, *Science* 345:322–324.

Zhou, X.-H., Wang, D.-P., and Chen, D., 2015, Global wavenumber spectrum with corrections for altimeter high-frequency noise, *J. Phys. Oceanogr.* 45(2):495–503. doi:10.1175/JPO-D-14-0144.1.

11 Satellite Altimetry in Coastal Regions

Paolo Cipollini, Jérôme Benveniste, Florence Birol, M. Joana Fernandes, Estelle Obligis, Marcello Passaro, P. Ted Strub, Guillaume Valladeau, Stefano Vignudelli, and John Wilkin

GLOSSARY

1DVAR:	One-Dimensional VARiational scheme
ACCRA:	Atmospheric Correction for Coastal Radar Altimeters
ALADIN:	Aire Limitée Adaptation dynamique Développement InterNational (model)
ALBICOCCA:	ALtimeter-Based Investigations in COrsica, Capraia and Contiguous Area
ALES:	Adaptive Leading-Edge Subwaveform retracker
ARCOM:	Altimetry for Regional and Coastal Ocean Modeling
AVISO:	Archiving, Validation and Interpretation of Satellite Oceanographic Data
CAW:	Coastal Altimetry Workshop
CCI:	Climate Change Initiative
CNES:	Centre National d'Études Spatiales
COASTALT:	ESA COASTal ALTtimetry - Development of radar altimetry data processing in the oceanic coastal zone
COSS-TT:	Coastal Ocean and Shelf Seas Task Team of GODAE,
COSTA:	COastal Sea level Tailored ALES
CSIRO:	Commonwealth Scientific and Industrial Research Organisation
CTOH:	Center for Topographic studies of the Ocean and Hydrosphere
DAC:	Dynamic Atmosphere Correction
DLM:	Dynamically Linked Model
DLR:	Deutsches Zentrum für Luft- und Raumfahrt (German Aerospace Center)
DMSP:	Defense Meteorological Satellite Program
DTC:	Dry Tropospheric Correction
ECMWF:	European Center for Medium-range Weather Forecasts
EnOI:	Ensemble Optimal Interpolation
ERA:	ECMWF ReAnalysis
ERS-1:	European Remote Sensing satellite-1
ERS-2:	European Remote Sensing satellite-2
ESA:	European Space Agency
FASTEM:	FAST microwave Emissivity Model
FOV:	Field of View
GDR:	Geophysical Data Records
GFO:	GEOSAT Follow-On
GLORYS:	Global Ocean ReanalYsis and Simulation
GNSS:	Global Navigation Satellite System
GODAE:	Global Ocean Data Assimilation Experiment
GPD:	GNSS derived Path Delay
G-POD:	ESA Grid Processing On Demand

GPS:	Global Positioning System
GPT:	Global Pressure and Temperature
IB:	Inverse Barometer
JPL:	Jet Propulsion Laboratory
LCA:	Land Contamination Algorithm
LEGOS:	Laboratoire d'Études en Géophysique et Océanographie Spatiales
LRM:	Low-Resolution Mode
MDT:	Mean Dynamic Topography
MHS:	Microwave Humidity Sounder
MLE3:	Maximum-Likelihood Estimation of 3 parameters
MOG2D:	Modèle aux Ondes de Gravité - 2 dimensions
MPA:	Mixed-Pixel Algorithm
MWR:	Microwave Radiometers
NASA:	National Aeronautics and Space Administration
NCEP:	National Centers for Environmental Prediction
NOAA:	National Oceanic and Atmospheric Administration
NPC:	NOAA Prediction Center
NWM:	Numerical Weather Models
NWS:	National Weather Service
ODES:	Online Data Extraction Service
PEACHI:	Prototype for Expertise on AltiKa for Coastal, Hydrology and Ice
PISTACH:	Prototype Innovant de Système de Traitement pour les Applications Côtières et l'Hydrologie
PLRM:	Pseudo-Low Resolution Mode
POD:	Precise Orbit Determination
PO.DAAC:	Physical Oceanography Distributed Active Archive Center
RADS:	Radar Altimetry Database System
RMS:	Root Mean Square
SAR:	Synthetic Aperture Radar
SARAL:	Satellite with ARgos and ALtiKa
SGDR:	Sensor Geophysical Data Record
SLA:	Sea Level Anomaly
SLP:	Sea Level Pressure
SSB:	Sea State Bias
SSHA:	Sea Surface Height Anomaly
SSM/I:	Special Sensor Microwave Imager
SST:	Sea Surface Temperature
SWH:	Significant Wave Height
SWOT:	Surface Water Ocean Topography
T-UGO:	Toulouse Unstructured Grid Ocean model
TB:	Brightness Temperature
TG:	Tide Gauge
T/P:	TOPEX/Poseidon
WTC:	Wet Tropospheric Correction

11.1 INTRODUCTION AND RATIONALE

The ocean plays a major role in the climate system and its evolution, not only as a regulator but also as an indicator of changes. These changes are mostly felt in the coastal zone, where they affect local populations, biodiversity, accessible marine resources, and coastal engineering and management. The coastal zone is where changing circulation, sea level, and sea state have by far the largest impact on human society. Knowledge of ocean dynamics is essential, but near the coast the processes are much more complex than in the open ocean and require dedicated observing tools. Coastal observations are also required for marine meteorology forecasting and climate predicting models.

The success of altimetry over the open ocean, which is illustrated in detail in the previous and following chapters of this book, leads most naturally to consider the potential of this remote sensing technique in the coastal zone, the natural interface between the oceans and humans.

In the last 15 years, increasing international efforts have been put toward the improvement and exploitation of altimetry in the coastal zone, a broad field of research, development, and applications that is now routinely called *coastal altimetry* and was first reviewed in a book with the same name in 2011 (Vignudelli et al. 2011). In this chapter, we describe both the improvements in coastal altimetry on the technical side and the ensuing applications.

The technical improvements derive from the combination of three approaches that have been developed in coastal altimetry. They are, in the chronological order in which they have appeared: (1) more detailed and coastal-specific data editing; (2) improvements in key correction fields; and (3) new schemes for radar echo analysis (retracking) optimizing the retrieval of the leading edge of the radar return signal (waveform). In more detail:

1. Historically, some research initiatives made progress by abandoning the systematic rejection of along-track data considered as suspect because of the water depth or their proximity to land. For example, in the ALBICOCCA project (Altimeter-Based Investigations in Corsica, Capraia, and Contiguous Area, and funded by Centre National d'Études Spatiales [CNES]), the data in shallower water and closer to land were examined and evaluated based on separate editing criteria for individual correction terms, range, and finally deriving the sea level anomalies (SLAs); the X-TRACK post-processing algorithm is derived from this approach and now routinely produces regional altimetry products that are largely distributed (Birol et al. 2016). Similarly, Feng and Vandemark (2011) showed that a significant volume of data could be reclaimed in the coastal zone by judicious application of standard data quality flags, especially the rain flag, and a revised wet troposphere range correction.
2. At the same time, improvements in the geophysical and propagation corrections that have to be applied to the altimeter range (see previous chapters, as well as the Radar Altimetry Tutorial at www.altimetry.info for a description of the altimetry measurement and its corrections) have significantly increased the accuracy of the basic retrieved data that were subjected to the detailed editing. The range correction that has had the most significant upgrade in the coastal zone is the wet tropospheric correction, which we discuss in Section 11.3.2; tidal models and dynamic atmospheric corrections have also had significant advances as we illustrate in 11.3.3.
3. Finally, retracking methods have been designed to accommodate nonstandard waveforms caused by proximity to land, such as the algorithm developed in the PEACHI project for Ka-band radars (Valladeau et al. 2015; see Section 11.4.1). Another specific approach is to restrict the tracking to the leading edge portion of the radar return, as carried out by the ALES algorithm (Passaro et al. 2014; see Section 11.4.2). This allows the range measurement to be retrieved using only the footprint radius of the radar reflection (i.e., neglecting the information in the tail of the waveform which in the coastal zone is usually the first to be corrupted by artifacts as we will illustrate in Section 11.4.2).

The other innovation in altimeter data analysis important for coastal studies is the retrieval of the height signal from wide rivers (and their estuaries) and flood plains. These developments, which are discussed in Chapter 14 of this book, allow an estimate of river discharges into the coastal ocean in some cases, especially for large rivers and during high discharges, when the river inputs have their greatest effects.

Coastal applications of altimeter data are now becoming widespread and are presented and discussed in Sections 11.5 and 11.6. We expect some applications to take advantage of a unique prerogative of coastal altimetry (i.e., that, being global, it provides measurements over remote stretches of the world's coasts where there are no other observing devices). In many cases, those measurements go back to 1991–1992 (i.e., the start of the European Remote Sensing [ERS]-1 and TOPEX/Poseidon [T/P] missions). In all coastal applications, two basic limitations of altimeter sampling remain: the wide spatial separation of the altimeter tracks and the long periods (usually 10 days or more) between repeat passes over the same track (or within a few kilometers for non-exact-repeat orbits). These limitations are important in the coastal domain, where temporal and spatial scales of physical processes decrease as land is approached and depths also decrease. The availability of data from multiple altimeters reduces the impact of these problems somewhat, but the fact remains that coastal altimetry will normally yield stronger benefits if it is integrated with *in situ* observations and models that it can complement synergistically.

Assimilation of altimeter data into coastal ocean circulation models may make the best use of the altimeter data in addressing shorter time and space scales. Altimeter data that are not assimilated are also useful in providing validation of the circulation models. In these ways, altimeter data constitute a unique component and powerful constraint within the hybrid monitoring systems needed to understand and predict the complex and rapidly changing coastal ocean environment.

In this chapter, we first examine the improvements in altimeter processing in the coastal zone following the order in which they appear in the altimeter processing chain, i.e., first improvements in retracking in Section 11.2, then improvements in corrections in Section 11.3, and then in Section 11.4 we describe the diverse data set now available for coastal altimetry. These are enabling applications: those that rely on observations alone are described in Section 11.5, while those that exploit the synergy with models are in Section 11.6.

11.2 DEALING WITH COASTAL WAVEFORMS

11.2.1 Pulse-Limited Waveforms

The Brown waveform model (Brown 1977) described in Chapter 1 of this book remains to date the fundamental model for the estimation of geophysical parameters from pulse-limited altimetry. As confirmed by the PISTACH project (Prototype Innovant de Système de Traitement pour les Applications Côtières et l'Hydrologie) (Mercier et al. 2010), in which a classification of Jason-1 waveforms was carried out, the Brown model is a good approximation for more than 90% of the waveforms over the open ocean, up to about 10–15 km from the coast. This is clearly seen in Figure 11.1 taken from Halimi et al. (2013), which shows the percentage of waveforms in three different classes as a function of distance from coast. Two questions then arise. The first question, which is central to coastal altimetry, is how we deal with the non-Brownian waveforms in the last 10–15 km. As one can see in the figure, these become more and more frequent (more than 50% at 5 km from the coastline) as one approaches the coast. The second question is whether that 7%–8% of non-Brownian waveforms away from the coast contain information that is not retrieved by standard processing, and if yes, how we can recover that information. (There is a hint that techniques that we develop for the coastal zone also may find applicability over the open ocean for those non-standard waveforms, as will be discussed later in the chapter.)

Departures of pulse-limited power waveforms from the Brown model are caused by variations of the surface height statistics and backscatter within the radar footprint. These non-homogeneities may occur over the open ocean due to intense fronts, sub-mesoscale activity, and natural or man-made slicks. In the coastal zone, they are also caused by the effects of shallow bathymetry, the sheltering of the sea induced by the coastal morphology, and the contamination by land entering the footprint. One particular case is the radar echo contamination caused by the very strong signal reflected from calm water in a marina or coastal recess, such as a sheltered bay.

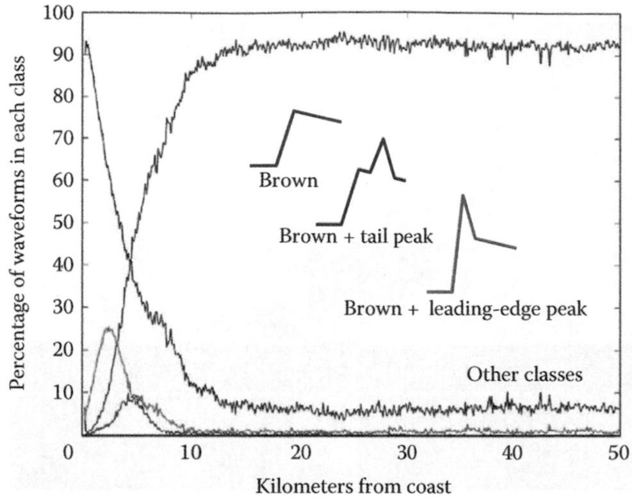

FIGURE 11.1 Percentages of observed classes of altimetric waveforms, as a function of distance from the coast, as found by the PISTACH project. (Reprinted from Halimi, A., et al., *IEEE Trans. Geosci. Rem. Sens.*, 1568–1577, 2013. With permission. © 2013 IEEE.)

Gómez-Enri et al. (2010) have studied the issue of contamination due to calm water with an example from overpasses of Envisat over a Mediterranean island; they confirmed that the signature of bright targets migrates following a hyperbolic law in the stack of sequential waveforms, which is expected on the basis of simple geometric considerations. As the altimeter extended footprint (i.e., the portion of the surface illuminated by the full waveform not just the leading edge) passes over the bright feature, the resulting peak in power moves from the tail of the waveform toward the leading edge and back, as is clearly visible in the example in Figure 11.2. Whether the peak reaches the leading edge or not depends on whether it is enough close to nadir and on its elevation. Targets at a height of several meters above sea level, such as the decks of large ships or flat icebergs, may generate signatures in the portion of the waveform preceding the leading edge (the "thermal noise" section); this concept is at the basis of studies identifying ships or icebergs in altimeter data (Tournadre et al. 2006, 2008). Similar considerations apply to dim features (i.e., areas of reduced backscatter such as those that may be caused by a low-lying island or sandbank). In this case, the waveform will show a power dip.

A comprehensive overview of retracking methods is in Gommenginger et al. (2011). Passaro et al. (2014) have reviewed the different approaches taken in literature for the processing of coastal pulse-limited waveforms that do not conform to the Brown model. There are four main approaches:

1. Algorithms based on some form of classification of the waveforms depending on their shape and the use of a different functional form for the fitting of each class. This approach, pioneered by Deng and Featherstone (2006), has then been used by Andersen (2010), Berry et al. (2010), and Yang et al. (2012).
2. Empirical models of the waveforms—for instance, using threshold values for the retracking in the presence of more peaky waveforms (Deng and Featherstone 2006; Hwang et al. 2006; Bao et al. 2009; Lee et al. 2010).
3. The processing of a stack of successive waveforms to detect the signature of bright targets as presented by Gómez-Enri et al. (2010) for the purpose of "cleaning" the stack prior to retracking (Quartly 2010).
4. The fitting of a portion of the waveform ("sub-waveform") containing the leading edge but excluding the tail section where most artifacts appear due to bright targets. This approach

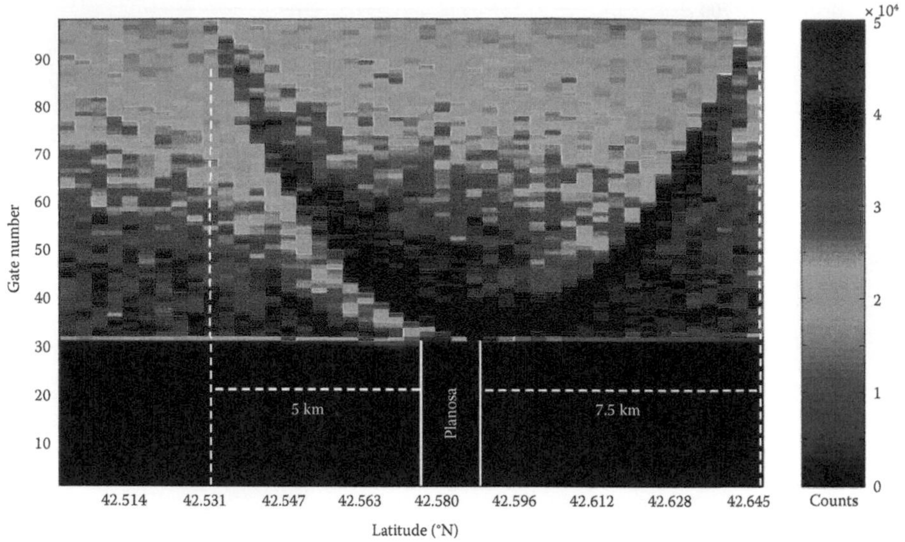

FIGURE 11.2 Envisat RA-2 power waveforms for orbital cycle 49, track 128 overpassing Pianosa Island in the Mediterranean Sea. Each column in the plot is a 20-Hz waveform (i.e., along the vertical axis are the receiving window power bins, also called gates). The horizontal axis represents the flight path of the satellite plotted by latitude; the color scale is in unit of power counts, corrected for automatic gain control variations. The vertical white lines mark the limits of the island. The bright hyperbolic feature is most likely due to a strong radar return from a sheltered bay on the northern side of the island. (Reprinted from Gómez-Enri, J., et al., *IEEE Geosci. Rem. Sens. Lett.*, 7, 474–478, 2010. With permission. © 2010 IEEE.)

was first presented in the PISTACH project for the so-called "RED" (for *reduced* portion of waveform) retracker (Mercier et al. 2009, 2010) and has since been adopted and developed by several authors (Guo et al. 2010; Yang et al. 2011; Idris and Deng 2012; Yang et al. 2012; Passaro et al. 2014).

The sub-waveform approach is emerging as the most promising one. This can be explained with the following intuitive considerations on the fitting process. As explained in detail in Chapter 1 of this book, properties of the leading edge yield the three main altimeter observables: range (derived from the leading edge position), significant wave height (from the leading edge rise time), and wind (from the maximum amplitude reached by the leading edge). A maximum likelihood estimator solving for these three parameters (MLE3) should in principle be able to give a good estimate of those parameters by fitting the leading edge portion only of the waveform, provided an independent estimate of a further variable in Brown's model, the off-nadir pointing angle or mispointing, is available from the onboard star trackers.* The tail of the waveform would normally not add information on the three main parameters and can be used instead to estimate mispointing in a four-parameter retracking scheme. In practice, for waveforms that conform well to the Brown model, such as those gathered in homogeneous open-ocean conditions where the noise is exclusively Raleigh, that is, speckle (plus a relatively small thermal noise), the use of the whole waveform in an MLE3 scheme helps constrain the fitting and improve the precision; reducing the fitting window under these "optimal" circumstances yields suboptimal estimates, as clearly illustrated by Passaro et al. (2014) with a Monte Carlo simulation. Conversely, in all those cases where the tail of the waveform is affected by artifacts due to non-homogeneities in the extended footprint (such as when bright targets

* Star trackers on board altimetric missions provide the precise pointing knowledge required for radar altimeter by pointing a great number of stars.

are present), fitting the whole waveform also has to accommodate the corrupted tail and therefore compromises the precision in the retrieval of the leading edge properties. In those latter cases, an approach that somehow reduces the fitting window to encompass only the leading edge option, or the leading edge plus a relatively small portion of the tail, may well be more robust. It must be noted that when the leading edge of the waveform is corrupted (which for instance may happen due to an off-nadir bright target at an elevation above the sea surface), then the sub-waveform approach is prone to significant errors.

In an Adaptive Leading-Edge Subwaveform (ALES) retracker, the retracking of each waveform is performed in two passes. A first pass looks at the rising portion of the waveform and provides a rough estimate of the significant wave height (SWH) from the slope of that portion. This estimate is then entered into an algorithm that selects the sub-waveform (i.e., sets the width of the fitting window over which a Brown fitting is performed in the second pass). Figure 11.3, taken from Passaro et al. (2014), shows three examples of Jason-2 waveforms in various conditions and the corresponding sub-waveform fit by the ALES retracker. As it can be seen, the width of the fitting window varies with the SWH; it can also be observed, in the middle panel referring to a coastal case, that the sub-waveform approach is not affected by a disturbance (likely from a bright target) in the tail portion of the waveform and therefore still achieves a good fit of the leading edge.

11.2.2 SAR Waveforms

SAR altimetry is intrinsically promising for coastal applications by virtue of the higher signal-to-noise ratio and along-track resolution; in SAR mode, a pulse repetition frequency higher than conventional altimetry allows slicing of the footprint in bands perpendicular to the flight direction by exploiting the Doppler frequency shift of the radar echoes from targets fore and aft of the satellite nadir point, resulting in a much higher along-track spatial resolution (300 m). However, this is only an advantage when the satellite crosses the coast at an angle that is close to normal to the coastline. When the ground track is more oblique to the coast, the SAR footprint (which extends in the across-track dimension and is essentially pulse-limited in that dimension) will encounter land and adversely impact the radar echo. Some experimental techniques have been proposed to account for all those non-optimal cases in which land or bright targets enter the footprint. Egido (2014) has used a coastal digital elevation model (DEM) and accurate geolocation of the echoes to reject those portions of the echoes that come from land. A sub-waveform approach has also been proposed for SAR waveforms by Thibaut et al. (2014). This is an area where research has only recently started, but more studies are expected to arise due to the availability of global SAR mode data from Sentinel-3 and the need to exploit in full the potential of this mission over the coastal zone.

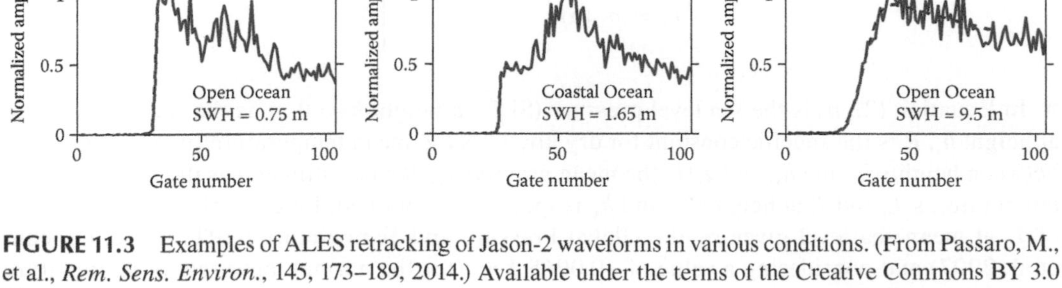

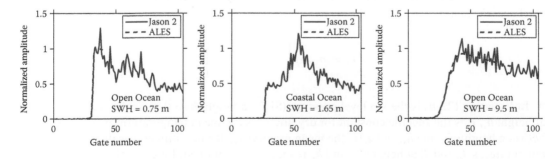

FIGURE 11.3 Examples of ALES retracking of Jason-2 waveforms in various conditions. (From Passaro, M., et al., *Rem. Sens. Environ.*, 145, 173–189, 2014.) Available under the terms of the Creative Commons BY 3.0 License.

11.3 IMPROVEMENTS IN RANGE AND GEOPHYSICAL CORRECTIONS

The accurate retrieval of sea surface height from satellite altimetry with centimeter-level accuracy requires the knowledge of all terms involved in the altimeter measurement system with similar or better accuracy, namely: satellite height above a reference ellipsoid from precise orbit determination (POD); altimeter range from dedicated retracking, including all instrument effects; and all range and geophysical corrections. As detailed in Chapter 1 of this book, a full set of range and geophysical corrections must be modeled and applied to altimetry: dry troposphere, wet troposphere, ionosphere, sea state bias, dynamic atmospheric correction, solid earth tide, ocean tide, load tide, and pole tide. The first four terms are the range corrections required to account for the interaction of the radar signal with the atmosphere and with the sea surface; the others refer to geophysical phenomena that need to be removed in order to separate them from the signal of interest, unless those phenomena are part of the very process that we aim to observe (as is, for instance, the case of the dynamic atmospheric contribution to storm surges).

The corrections with particular issues in the coastal regions are mainly the wet tropospheric correction (WTC), the sea state bias (SSB), and to somewhat lesser extent the dry tropospheric correction (DTC). Ocean tides may or may not be removed depending on the application; for example, models that simulate tides would retain this signal during data assimilation. Research is ongoing on these corrections and the main findings are detailed in the following subsection; the sole exception for this is the SSB, for which at the time of this writing we are not aware of any specific study to derive a solution for the coastal ocean—neither for conventional (pulse-limited) altimetry nor for SAR altimetry.

11.3.1 Dry Troposphere

The dry tropospheric correction (DTC) accounts for the path delay due to the dry neutral gases in the atmosphere. With an absolute value of about 2.3 m at sea level and a range of 0.2 m, the DTC is the largest range correction in satellite altimetry. However, using surface pressure from *in situ* observations or from an atmospheric model, such as those from the European Center for Medium-range Weather Forecasts (ECMWF), the DTC can be retrieved globally with an accuracy better than 1 cm, using the modified Saastamoinen model (Davis et al. 1985):

$$\Delta R_{dry} = -\frac{0.0022768\ p_s}{1 - 0.00266 cos\,(2\varphi) - 0.28 \cdot 10^{-6}\ h_s} \tag{11.1}$$

where p_s is the surface pressure in hPa, φ is the geodetic latitude, h_s is the surface height above the geoid (in meters), and the DTC, ΔR_{dry}, results in meters.

The DTC has an almost linear dependence on height of about 2.5 cm per each 100 m, directly related with the dependence of atmospheric pressure on height, which, according to Hopfield (1969) can be modeled as:

$$p_s = p_0\ exp\left[-\frac{g_m(h_s - h_0)}{RT_m}\right] \tag{11.2}$$

In Equation (2), p_0 is the sea level pressure (SLP) at height $h_0 = 0$, p_s is the surface pressure at height h_s, R is the specific constant for dry air, T_m is the mean temperature (in K) of the layer between heights h_0 and h_s, and g_m is the mean gravity; T_m can be estimated as the mean value of temperatures T_0 and T_s at heights h_0 and h_s, respectively, obtained, for example, from the values of T_0 at mean sea level given by the Global Pressure and Temperature (GPT) model (Boehm et al. 2007) and considering a value of $-0.0065\ Km^{-1}$ for the normal lapse rate of temperature with height.

The best sources of atmospheric pressure are the ECMWF Operational model (Miller et al. 2010) or the ECMWF ReAnalysis (ERA)-Interim model (Dee et al. 2011), provided as global grids at 0.125° × 0.125° (or better) and 0.75° × 0.75° spatial sampling, respectively, and 6-h intervals. The first model is not uniform, having undergone several updates. For this reason, for delayed-time products such as the geophysical data records (GDRs), ERA-Interim is the best model for the whole altimeter era, while ECMWF Operational is the most accurate model after 2004 (Fernandes et al. 2014; Legeais et al. 2014).

The DTC is provided in altimeter products as along-track interpolated values derived either from SLP or from surface pressure. In the first case, the derived DTCs are appropriate for use in coastal areas, having the same accuracy as in the open ocean. In the last case, in steep coastal areas, surrounded by land with elevations of, for example, 500–1000 m, errors in the DTC up to 5 cm or more (as shown in the example in Figure 11.4), increasing linearly as the coast is approached, may occur (Fernandes et al. 2014). These errors are systematic and will affect the derivation (e.g., of surface currents) in these regions. For a review of DTC errors in the context of inland water studies, in some aspects similar to those in the coastal regions, see for instance Fernandes et al. 2014.

11.3.2 Wet Troposphere

With an absolute value of less than 50 cm but with large space–time variability, the WTC is one of the major error sources, particularly in coastal altimetry.

In spite of the continuous progress in the modeling of the WTC by means of numerical weather models (NWMs) (e.g., Miller et al. 2010), an accurate enough retrieval of this effect can only be achieved through actual observations of the atmospheric water vapor content at the time and location of the altimetric measurements. For this purpose, dedicated microwave radiometers (MWRs) have been embedded in most altimetric missions.

Two main types of nadir-looking radiometers have been deployed in the altimetric satellites: two-band in ERS-1, ERS-2, Envisat, GEOSAT Follow-On (GFO), SARAL, and Sentinel-3, and three-band on T/P, Jason-1, Jason-2, and Jason-3. All of them have one band in the water vapor absorption line between 21 and 23.8 GHz plus one or two in "atmospheric window" channels, required to account for the effect of surface emissivity and for the cloud scattering (surface roughness is accounted for through the altimeter backscattering coefficient in a two-band configuration).

The algorithms used to retrieve the WTC from the measured brightness temperatures of the various MWR channels assume a surface ocean emissivity and are valid for ocean conditions, light rain, and

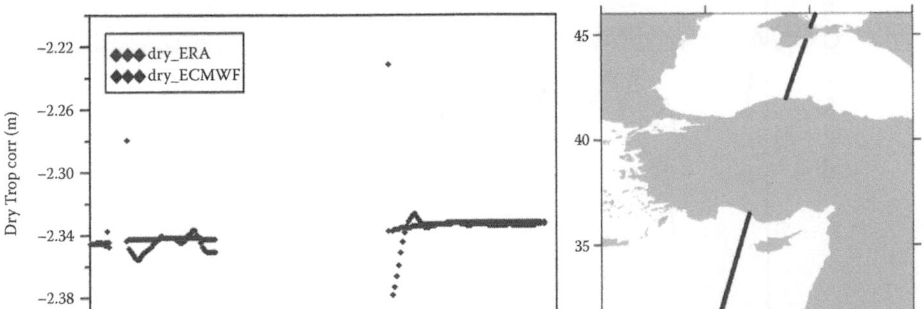

FIGURE 11.4 Illustration of DTC errors along Envisat cycle 12, pass 128 over the Mediterranean Sea and the Black Sea. The DTC from ERA-Interim is provided at sea level, while ECMWF Operational is provided at the height of model orography.

wind speed lower than 20 ms^{-1} (Obligis et al. 2006; Picard et al. 2015). Therefore, in the presence of surfaces with different emissivity, such as land or ice, the measurements lie outside the range of validity of the retrieval algorithm (Desportes et al. 2007). Thus, in spite of the high accuracy of MWR-derived WTC in open ocean, hampered by the contamination from land, ice, and rain in the radiometer footprint, the correction may be highly degraded, particularly in coastal and polar regions. Consequently, most current altimeter products fail to provide valid MWR-derived WTC in these regions.

It is worth noting that, with the contamination being directly related to the largest –3 dB width of the antenna pattern among the different channels, an instrument with a sharp spatial resolution, such as AltiKa (8 and 12 km at 23.8 and 37 GHz, respectively) (Valladeau et al. 2015), presents a very low contamination level, mainly occurring in the last 10 km approaching the coast (thin solid line in Figure 11.5). This raises the interest on higher frequency measurements (e.g., 89 GHz) offering a smaller footprint (see the following dedicated section).

In addition to contamination issues, noisy values caused by instrumental problems, jumps, and drifts may also occur in one or more channels, as illustrated by Scharroo et al. (2004).

In recent years, six main approaches have been proposed for correcting the altimeter measurements in the coastal regions, where the estimation from MWR measurements become invalid: (1) Basic extrapolation of the last valid measurement over the ocean; (2) a dynamically linked model (DLM) or composite approach; (3) the Land Contamination Algorithm (LCA); (4) the Mixed-Pixel Algorithm (MPA); (5) the GNSS (Global Navigation Satellite System) derived Path Delay (GPD) approach; and (6) variational method.

The first method is actually basic: The last valid brightness temperature (BT) measurement over ocean is extrapolated up to the coast. Validity in this context stands for non-contamination by land and is defined by a threshold on the land proportion in the field of view (FOV), estimated from a high-resolution land-sea mask and the theoretical half width of the FOV. This algorithm is applied on Envisat, Sentinel-3, and AltiKa (bold solid line on Figure 11.5) MWR.

The second method (DLM) was first used by Fernandes et al. (2003) and has been implemented in the European Space Agency's (ESA) COASTALT (COASTal ALTimetry) project on development of radar altimetry data processing in the oceanic coastal zone (COASTALT 2009; Obligis et al. 2011). A similar approach, the composite correction, has been used by Mercier et al. (2012) and is being implemented in

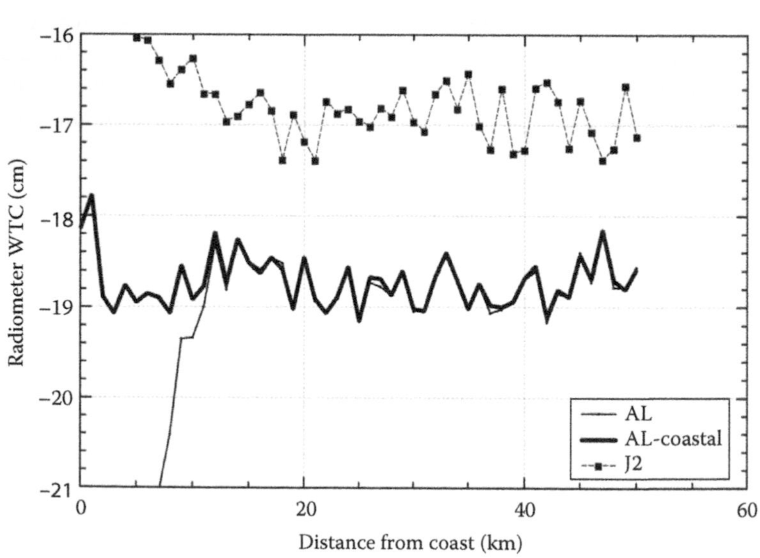

FIGURE 11.5 Wet Tropospheric Correction for AltiKa without any specific coastal processing (thin solid line) and with extrapolation of the last valid ocean brightness temperature (bold solid line). (From Valladeau, G., et al., *Marine Geodesy*, 38(Suppl 1), 124–142, 2015.)

the Archiving, Validation, and Interpretation of Satellite Data in Oceanography (AVISO) products. The composite correction is a conceptually simple method, which consists of replacing the MWR measurements near the coast (less than 50 km) by ECMWF model values. The ECMWF correction is shifted to the nearest valid radiometer measurement in the transition zone. Interpolation and detrending are also applied in complex cases (Mercier et al. 2010).

The Land Contamination Algorithm (LCA) was proposed around the end of the 1990s. This method (Ruf and Giampaolo 1998; Bennartz 1999; Desportes et al. 2007) is based on a correction of the land contamination of the BTs before applying the wet tropospheric correction retrieval algorithm. The correction is generally a function of the land fraction and land surface emissivity in the radiometer footprint. An example is shown in Figure 11.6.

The MPA has been developed at the Jet Propulsion Laboratory (JPL, USA). It is based on the existing open-ocean algorithm for the reference missions but extending to mixed land-ocean scenes, thus enabling retrievals in the coastal zone. It parameterizes log-linear coefficients as a function of the 18.7-GHz land fraction using a database of modeled coastal land BTs. The method requires an accurate land/sea mask and is directly applicable only to missions possessing three-band radiometers, including the 18.7-GHz channel. It has been successfully applied first to Jason-2 data (Brown 2010) and later to Jason 1.

The GPD Plus (GPD+) is the most recent version of the methods developed at the University of Porto to retrieve improved WTC, both for missions carrying an onboard MWR and for missions such as CryoSat-2, which does not possess such an instrument (Fernandes and Lázaro 2016). The GPD+ are wet path delays based on: (1) WTC from the onboard MWR measurements whenever they exist and are valid and (2) new WTC values estimated, by data combination through space–time objective analysis (OA) of all available observation in the vicinity of the estimation point, whenever the previous are considered invalid. Three types of observations have been considered: WTC from valid MWR values, WTC derived from scanning imaging MWR (Si-MWR) onboard various remote sensing missions, and GNSS-derived WTC from coastal and island stations. The underlying

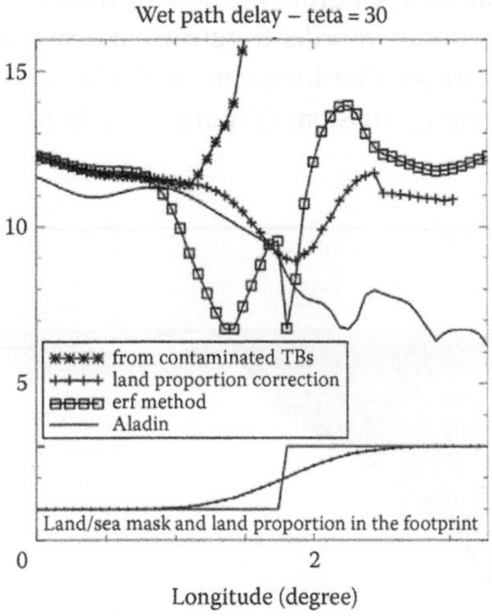

FIGURE 11.6 Wet path delays (in centimeters) on a coastal approach. The two solid lines at the bottom show the land/sea mask (land is on the right) and the land proportion in the radiometer footprint. The lines in the upper part of the figure are the path delays from the contaminated brightness temperatures (crossed solid line, with an unrealistic deviation on approaching the coast), from an analytical correction method ("erf method", line with squares, which also displays a prominent coastal artifact) and from the land contamination algorithm ("land proportion correction"); the latter is closest to the path delay predicted by the ALADIN model (solid line) in the coastal strip. (Reprinted from Desportes, C., et al., *IEEE Trans. Geosci. Rem. Sens.*, 45(7), 2139–2149, 2007. With permission. © 2007 IEEE.)

method, first developed for coastal altimetry in the scope of COASTALT, has been described in Fernandes et al. (2010, 2015) and Fernandes and Lázaro (2016). It improves data retrieval not only near the coast but also in open-ocean, by correcting the baseline MWR-derived WTC from other sources of error (ice and rain contamination and instrument malfunction).

In the scope of the Sea Level CCI project (http://www.esa-sealevel-cci.org) GPD+ WTC have been derived for the main altimetric missions (T/P, Jason-1, Jason-2, ERS-1, ERS-2, Envisat, CryoSat-2, and SARAL/AltiKa) and was later extended to GEOSAT Follow-On (Fernandes and Lázaro 2016). To ensure consistency and the long-term stability of the WTC, the large set of radiometers used in the GPD+ estimations have been inter-calibrated, using as reference the set of Special Sensor Microwave Imager (SSM/I) and SSMI/I Sounder (SSM/IS) on board the DMSP satellite series, due to their well-known stability and independent calibration (Wentz 2013).

The GPD+ corrections have been shown to reduce the sea level anomaly variance with respect to previous versions, to model-derived corrections, and to the AVISO Composite correction. They also have significant impacts in the estimation of regional sea level trends. Improvements are particularly significant in the coastal and polar regions, for T/P, and all ESA missions (Figure 11.7).

The last method developed and tested is based on a one-dimensional variational scheme (1DVAR) to retrieve the wet path delay for altimetry near coasts. This method, one output of which is shown in Figure 11.8, combines radiometric measurements, an *a priori* information on atmosphere (background data provided by ECMWF), and surface emissivity. Contrary to the previously described approaches, the 1DVAR scheme does not directly estimate an integrated content but will retrieve specific humidity and temperature profiles, from which the WTC is eventually computed. The combination of profile estimation and surface emissivity constraint makes this method very flexible and able to handle on-the-fly surface emissivities estimated from the model (e.g., from FASTEM-5, a fast microwave oceanic surface emissivity model over ocean), atlases (Karbou et al. 2005), or retrieved from coincident dedicated channel observations and consequently adapted to every atmospheric situation (standard or above upwelling regions) and surface type (ocean, land, sea ice, hydrology targets, and any transitions).

All these methods have in common handling the contamination on the measured brightness temperatures with coarse spatial resolution in order to retrieve a correction as clean as possible. Then, a new generation of instruments, with smaller footprints and additional atmospheric and surface information, is expected in future altimetry missions (Figure 11.9). In parallel to a classical low-frequency

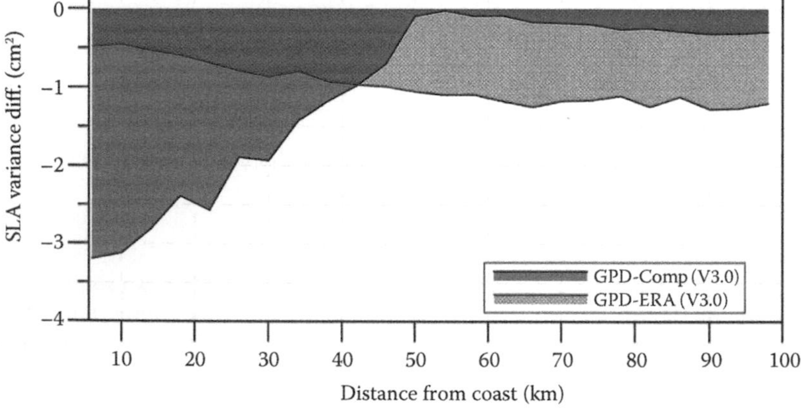

FIGURE 11.7 Variance difference between GPD+ and Composite WTC (blue) and ERA (orange), function of distance from coast, for Envisat MWR Reprocessing V3.0.

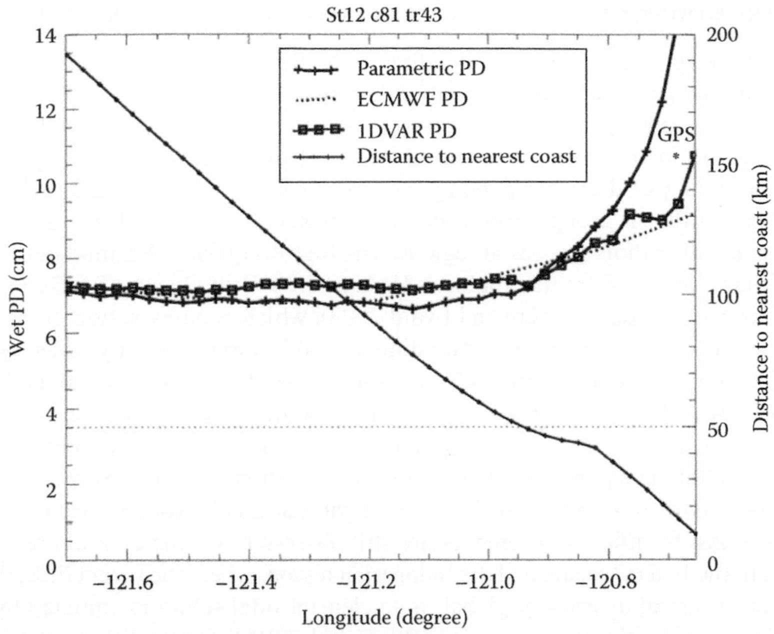

FIGURE 11.8 Wet Tropospheric Correction retrieved with a one-dimensional variational approach (squared solid line), compared with other methods, that is, a parametric approach and the ECMWF model, on a coastal approach. The wet path delay from a Global Positioning System (GPS) coastal station is also shown. (Reprinted from Desportes, C., et al., *IEEE Trans. Geosci. Rem. Sens.*, 48(3), 1001–1008, 2010. With permission. © 2010 IEEE.)

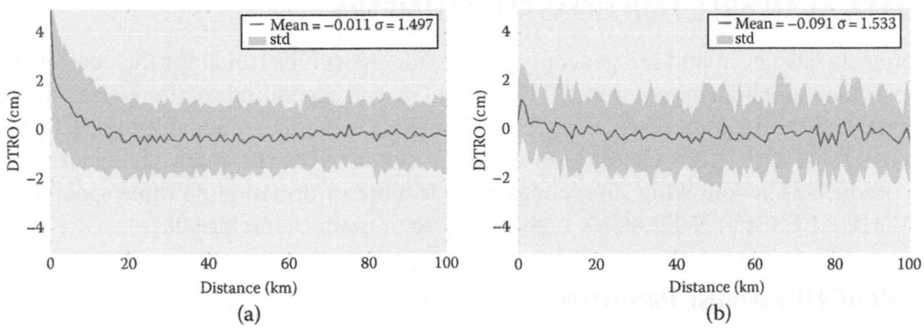

FIGURE 11.9 Difference between the retrieved Wet Tropospheric Correction (WTC) from a combination of (89 Hz + 157 GHz + 183.3 GHz) channels from Microwave Humidity Sounder (MHS) and the ECMWF WTC, as a function of along distance to coast over the Mediterranean Sea. (a) Neural network: The land contamination occurs shoreward of 20 km as shown by the increase in the correction difference; (b) 1DVAR approach: The land contamination is clearly reduced (Credit: ACCRA ESA study).

channel with a spatial resolution of about 30 km, the radiometers would embark a set of high-frequency channels, with footprints of about 10 km. Typical observation frequencies would be 50 GHz dedicated to surface emissivity and atmospheric temperature, 89 GHz dedicated to cloud content, and 183 GHz dedicated to both water vapor and cloud content.

In conclusion, the future of the WTC on coastal regions relies on a new generation of microwave radiometers and dedicated retrieval algorithms ready to take advantage of additional information on surface, atmospheric temperature and humidity, and cloud content.

11.3.3 Recent Improvements in Coastal Tides and Dynamic Atmospheric Correction

The standard altimetry corrections corresponding to the tides, to the changing atmospheric pressure (i.e., inverted barometer effect), and to the high-frequency atmospheric forcing are provided by global models. These geophysical effects tend to be much larger, complex, and associated with shorter wavenumbers in shelf and near-coastal waters than in deep waters. It places a critical demand on the accuracy of the corresponding numerical solutions (Andersen and Scharroo 2011). As a consequence, different groups worldwide have devoted substantial efforts to improving correction models and algorithms in coastal regions. The high-frequency dynamic ocean response to atmospheric forcing provided by the Toulouse Unstructured Grid Ocean (T-UGO) model hydrodynamic finite element model (Carrère and Lyard 2003), which is an evolution of the widely used MOG2D (Modèle aux Ondes de Gravité—two dimensions) is now commonly added to the inverted barometer correction. Considerable efforts have been oriented to improve spatial model resolution, numerical schemes, realism of the underlying hydrodynamic models and data assimilation techniques, better knowledge of the local near-coastal bathymetry and coastlines, developing regional models, and assembling independent data sets for the validation. Regular exercises of tidal model inter-comparisons (as in Stammer et al. 2014) highlight that much progress has been achieved in global tidal solutions. Significant differences are still observed in some coastal areas, such as the Amazonian shelf, the Indian Ocean, and the Indonesian region, where the recent FES2014 tidal atlas (the most recent release of a series of global finite element tidal solutions initiated by Le Provost et al. 1994) appears to be the most accurate (Birol et al. 2016). In parallel, modeling individual coastal regions has proven to offer additional progress for both the tidal and atmospherically forced high-frequency corrections (Roblou et al. 2011). Regional model corrections based on the T-UGO 2D hydrodynamic code are now available over an increasing number of coastal seas (Maraldi et al. 2007; Pairaud et al. 2008; Le Bars et al. 2010). This correction issue continues to be at the forefront of current research, and we can expect to see further important progress in the coming years.

11.4 DATA AVAILABLE FOR COASTAL ALTIMETRY

Several projects have been and are generating data with algorithms tuned for the coastal zone, and data are provided at a higher post rate (20 Hz or 40 Hz, corresponding to along-track distances of approximately 350 and 175 m, respectively), which makes them more amenable to coastal altimetry applications. Table 11.1 (from Cipollini et al. 2017) summarizes the main characteristics of the available products. The following subsections provide more information on three specific products (i.e., PEACHI, ALES, and X-TRACK), highlighting their main characteristics.

11.4.1 PEACHI Expertise Prototype

With the objective to ensure the complementarity but also the continuity with the Level-2 S-GDR products provided in the open ocean, the Prototype for Expertise on Altimetry for Coastal, Hydrology and Ice (PEACHI) project has been set up as an initiative of the French space agency, CNES. The PEACHI prototype is designed not only to serve coastal applications but is also considered as a reference to compute and provide dedicated algorithms over ice regions and for hydrological studies. The PEACHI project thus focuses on a handful of key algorithm improvements with regard to the operational Geophysical Data Record (GDR) products. The main objectives of the prototype are the following:

1. Validate and implement the existing algorithms before their application in the operational products.
2. Develop new algorithms linked to scientific objectives (coastal areas, surface hydrology, rain cells, continental and sea ice, etc.).
3. Ensure both complementarity and continuity with the altimeter measurements over the open ocean.

TABLE 11.1
Available Coastal Altimetry Products as of July 2017

ID	Produced by	Altimeter	Product level	Posting rate	Coverage	Download from	Comments
PISTACH	CLS CNES	j2	L2	20 Hz	Global	AVISO+	Experimental Jason-2 products for Hydrology and Coastal studies with specific processing. Will be discontinued at the end of 2016 in favor of PEACHI
PEACHI	CLS CNES	sa, (j2 to be added soon)	L2	40 Hz	Global	AVISO+ / ODES	Experimental SARAL/AltiKa products including dedicated retracking and corrections leading to more accurate products for coastal zones, hydrology and ice. From 2017 expected to generate also j2 products
XTRACK	LEGOS-CTOH	tx, j1, j2, gfo, en (sa to be added soon)	L2, L3	1 Hz 20 Hz (test)	23 regions covering the whole coastal ocean	CTOH AVISO+ / ODES	Specific processing using improved data screening and latest corrections available
ALES	NOC	j2, n1, (j1, j3 to be added soon)	L2	20 Hz	Global, <50 km from coast	PODAAC	Experimental products from the ALES processor included in SGDR-type files alongside the standard products and corrections.
SARvatore	ESA-ESRIN	c2 (SAR only)	L2	20 Hz	SAR mode regions	ESA G-POD	On-demand Processing service for the CryoSat-2 SAR mode data where the user can configure some processing parameters to meet specific requirements (for instance for the coastal zone)
COP	ESA	c2 (LRM/ PLRM)	L2	20 Hz	Global	ESA	Global products for CryoSat-2 from an Ocean processor (output is in PLRM over the SAR mode regions) - but no specific coastal processing
COSTA	DGFI-TUM	e2, en (other altimeters available on request)	L3	1 Hz 20 Hz	Mediterranean and North Sea (other regions available on request)	PANGAEA	Dedicated coastal altimetry sea level measurements based on enhanced ALES retracker

Note: The abbreviations used for the altimeters are e1: ERS-1 (1991-1996); tx: TOPEX (1992-2002); e2: ERS-2 (1995–2011); gfo: GEOSAT Follow-On-1 (2000–2008); j1: Jason-1 (2002–2013); n1: Envisat (2002–2012); j2: Jason-2 (2008–present); c2: CryoSat-2 (2010-present); sa: SARAL/AltiKa (2013–present); j3: Jason-3 (2016–present). For CryoSat-2 (c2), a further specification is added when data are only available from the Low-Resolution Mode and Pseudo-Low Resolution Mode (LRM/PLRM) or only from the Synthetic Aperture Radar (SAR) mode regions. The abbreviations used for product levels are L2: along-track data with corrections; L3: data projected on regularly spaced reference points along the nominal ground tracks of the satellite. From Cipollini et al. (2017), with additions, available under the terms of the Creative Commons Attribution 4.0 International License.

11.4.1.1 SARAL/AltiKa

The PEACHI Expertise Prototype was first initiated with the SARAL/AltiKa space mission launched in 2013. The main objective of the SARAL/AltiKa mission is to better observe the open ocean, but the mission is also targeting secondary objectives, such as the study of coastal dynamics. The prototype has been designed to complement SARAL/AltiKa processing software and the dissemination of the operational Level-2 products with some new or improved algorithms and therefore to analyze and improve processing dedicated to the Ka-band radar altimeter (Valladeau et al. 2016).

Over coastal regions, the prototype aims to provide end users with new waveform retrackers, a geometrical waveform classification, improved two-dimensional and new three-dimensional SSB, new tide models, better altimeter wind correction and a new wet troposphere correction (Valladeau et al. 2015).

Figure 11.10 displays the standard deviation of the altimeter range with regard to the distance from the coast. Compared to Jason-2, better performances were expected for SARAL/AltiKa close to the coasts, thanks to 40 Hz sampling and a smaller waveform footprint. These improvements are due to AltiKa's larger bandwidth, lower orbit, increased pulse repetition frequency (PRF), and reduced antenna beamwidth (Steunou et al. 2015). Real measurements confirm these expectations with an excellent behavior of SARAL/AltiKa closer to the coast in terms of standard deviations of the range, significant wave height, and sigma naught. Until 5.7 km off the coast, the standard deviation of the range remains stable for AltiKa (lower than 10 cm), while it increases for Jason-2 close to 10 km from the coast. Note that differences in the level of standard deviations in the open ocean originate in the temporal sampling of both altimeter missions (40 Hz for SARAL and 20 Hz for Jason-2).

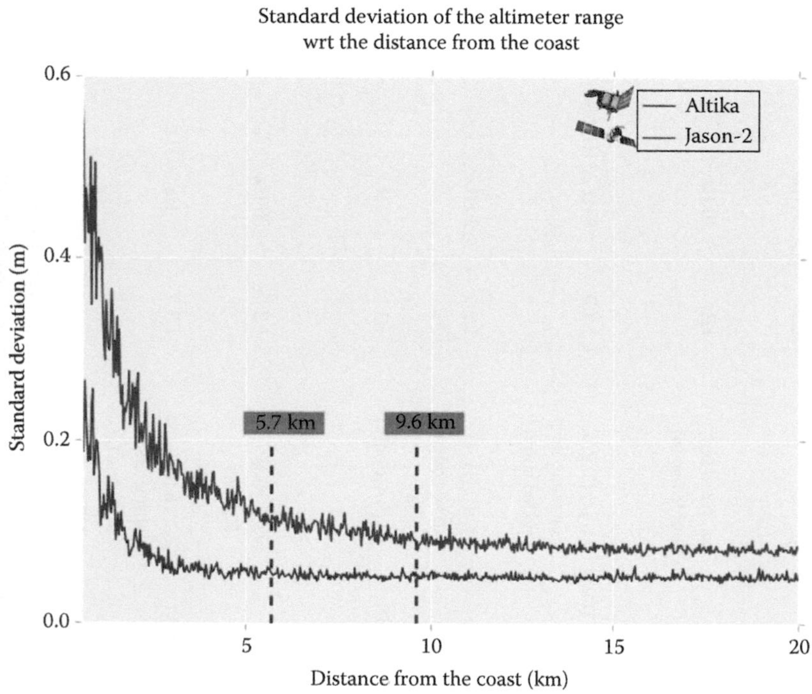

FIGURE 11.10 Performance of SARAL/AltiKa and Jason-2 altimeters near the coast: standard deviation of the altimeter range with respect to the distance from the coast. AltiKa (red curve) is remarkably more precise. (From Cipollini, P., et al., *Surv. Geophys.*, 38, 33, 2017.) Available under the terms of the Creative Commons Attribution 4.0 International License.

TABLE 11.2
List of PEACHI SLA Products Available and the First and Last Dates of the Time Series

Mission	Start	End
SARAL/AltiKa	03/14/2013 (1)	07/04/2016 (35)
SARAL/AltiKa drifting orbit	07/04/2016 (100)	12/26/2016 (104)
Jason-2	12/27/2014 (239)	10/02/2016 (303)
Jason-2 interleaved	10/13/2016 (305)	12/15/2016 (311)

Note: The corresponding cycle number is indicated in parentheses.

11.4.1.2 Implementation on Jason-2 and Jason-3

The PEACHI Expertise Prototype has been designed to serve multiple altimeter missions with a homogenous processing sequence and thus corresponding data sets provided to the end users. Therefore, the PEACHI prototype is also computed on-the-fly for the Jason-2 and Jason-3 missions, and the same panel of algorithms are computed as for SARAL/AltiKa for better cross-comparisons over coastal regions. On Jason-2, the prototype replaces the PISTACH Prototype released in 2009 (Dufau et al. 2011). Concerning Jason-3, the PEACHI-J3 experimental processing prototype is delivering delayed-time products to the expert users' community. PEACHI-J3 products are based on S-(I)GDR Jason-3 official products, enriched with specific algorithm outputs such as numerical retracking estimates, new wet tropospheric correction, and other useful corrections such as three-dimensional SSB correction and updated geophysical corrections.

11.4.1.3 Data Availability and Delivery Mode

PEACHI products are made from S-GDR altimeter products with specific processing (improvements of conventional satellite radar altimetry parameters lead to more accurate products over several surface types: open ocean, coastal regions). Constantly seeking to address its users' needs, AVISO+ proposes new features and products for its dissemination service called Online Data Extraction Service (ODES, http://odes.altimetry.cnes.fr) in order to provide users and applications with a wider range of altimetry-derived data and services. The summary of the products currently available is in Table 11.2.

11.4.2 ALES

11.4.2.1 ALES Data Set: Availability and Reliability

The ALES retracking algorithm (see Section 11.2.1) has already been implemented for Jason-1, Jason-2, ERS-2, Envisat, and SARAL/AltiKa, while the adaptation to Jason-3, CryoSat-2 Low Rate Mode, and ERS-1 is under development and validation.

Currently, Envisat and Jason-2 ALES data sets have been made freely available in the global coastal ocean (see Table 11.3), between 0 and 50 km from the coastline, distributed by the Physical Oceanography Distributed Active Archive Center (PO.DAAC) through the portal https://podaac.jpl.nasa.gov/data set/ALES_L2_OST_JASON2_V1. The data format follows the original Sensor Geophysical Data Record (SGDR): Range, Significant Wave Height (SWH), and Backscatter Coefficient (Sigma 0) estimated from ALES are available along every track together with the standard altimetry products. By combining the ALES Range estimates with the corrections needed to extract the sea level height (see Section 11.3), the user can analyze coastal sea level anomalies that are usually reliable up to 3 km from the coast (Passaro et al. 2014).

TABLE 11.3
List of ALES Products Available from PO.DAAC and the First and Last Dates of the Time Series

Mission	Start	End
Jason-2 GDR-D	07/21/2008 (2)	05/15/2015 (252)
Envisat-v2.1	05/14/2002 (6)	10/21/2010 (94)

Note: The corresponding cycle number is indicated in parentheses. Further products based on the ALES algorithm are now becoming available from the COastal Sea level Tailored ALES (COSTA) Project at https://doi.pangaea.de/10.1594/PANGAEA.871920 (see Table 11.1).

The ALES range has been validated by means of comparison with tide gauges in terms of correlations and root-mean-square (RMS) errors in different regions, including in bays with jagged coastlines such as the Gulf of Trieste (Passaro et al. 2014) and the Strait of Gibraltar (Gómez-Enri et al. 2016), showing consistent improvements. As an example, comparing Jason-1 track 161 with the Trieste tide gauge at the along-track location closest to the gauge (approximately 5 km from the coast), the ALES time series has a correlation coefficient of 0.93 with the time series extracted from the gauge at the times of the altimeter overpasses, as opposed to 0.60 for the standard product. Also, the SWH estimations have been validated against buoys in the German Bight: Improvements in terms of correlation are seen up to more than 20 km within the coast, while the noise of the high-rate SWH retrieval is steadily less than in the standard products (Passaro et al. 2015b).

The future objective of the ALES project is the distribution of averaged coastal sea level anomalies (i.e., a Level 3 product) by combining the ALES retracking with a new SSB correction, a post-processing based on outlier detection, and a set of up-to-date external geophysical corrections.

11.4.2.2 ALES Data Set Improves Coastal Sea Level Research

Figure 11.11 shows an example of the improvement that it is possible to obtain in terms of data quality and quantity with ALES compared to an altimetry data set that has been designed for the open ocean, such as the ESA Sea Level Climate Change Initiative (SL_cci) data set (http://www.esa-sealevel-cci.org/products). In the figure, each 1-Hz point of the North Sea/Baltic Sea intersection zone corresponds to the standard deviation (std) of the 2002–2010 sea level computed from Envisat in the specific location. The ALES product shows smooth variations of the std, with the exception of a few locations: No abrupt change is to be expected, given that consecutive points are spaced by roughly 7 km. This is not verified in the SL_cci, which is not tailored for coastal exploitation, and signs of corrupted estimations are evident in the unrealistically high std of several 1-Hz locations. The reliability of ALES in comparison to standard altimetry products, such as SL_cci and the Radar Altimetry Database System (RADS), is evident from the histogram in Figure 11.12, in which the ALES statistics of the RMS of the time series at each 1-Hz point of the same area of study show almost no outliers.

In the same region, the ALES Envisat product has been used to perform a sub-regional analysis of the sea level annual variability. The reliability of the data set has been verified by direct comparison to the annual signal estimated from the seven tide gauges (TGs) in the same area (Table 11.4). These differences demonstrate that the ALES estimates are more similar to the TG results than the corresponding values from standard altimetry products (SL_cci and RADS).

The availability of coastal estimates of SWH and backscatter coefficients is allowing the recomputation of the SSB correction in the coastal ocean (see Section 11.4.3), which is one of the main

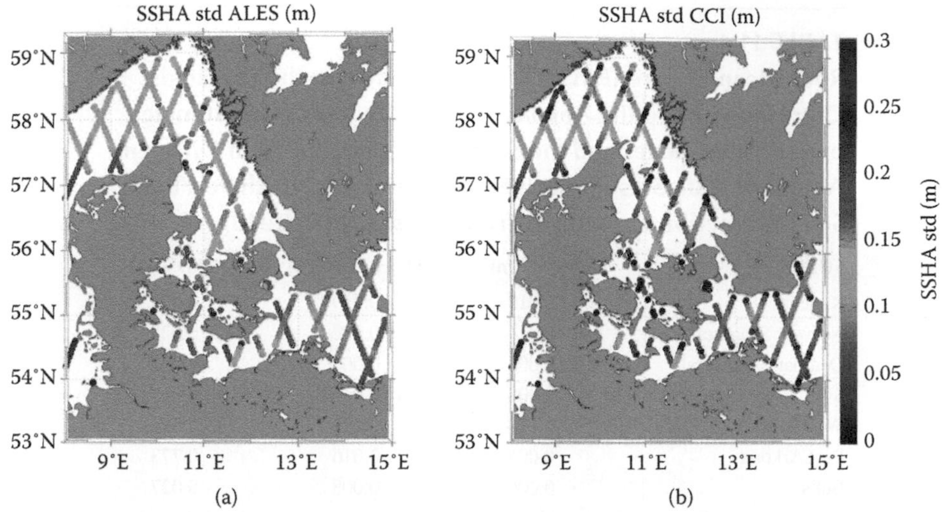

FIGURE 11.11 North Sea/Baltic Sea intersection: Comparison between (a) ALES reprocessed and (b) SL_cci data sets in terms of standard deviation of the Envisat SSHA time series (2002–2010) for each 1-Hz location. (Reprinted from Passaro, M., et al., *J. Geophys. Res. Oceans*, 120(4), 3061–3078, 2015a. With permission. © 2015 American Geophysical Union.)

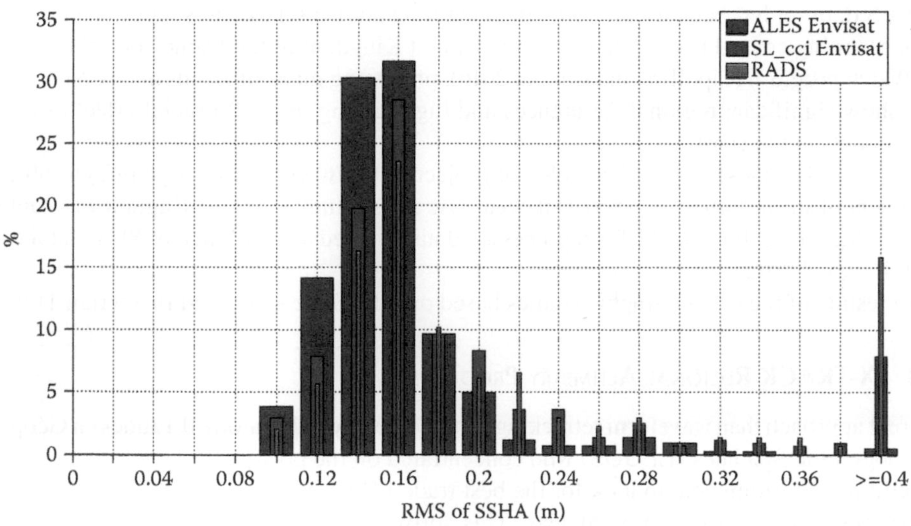

FIGURE 11.12 Histogram of the root mean square of SSHA computed at each 1-Hz location in the North Sea/Baltic Sea intersection from ALES reprocessed (red), Envisat SL_cii (blue), and Radar Altimetry Database System (RADS; green) data sets. Reprinted from Passaro, M., et al., *J. Geophys. Res. Oceans*, 120(4), 3061–3078, 2015a. With permission. © 2015 American Geophysical Union.)

sources of uncertainty in coastal altimetry. The SSB correction in the standard products is based on an empirical function of SWH and wind (derived from the backscatter coefficient estimated by the retracking). Preliminary studies along the coast of Spain have shown that the use of this relation with the ALES SWH and wind estimates leads to a reduction of the along-track coastal SLA uncertainty by up to 25% (Gómez-Enri et al. 2016).

TABLE 11.4
Root-Mean-Square (RMS) Difference between the Sinusoids Corresponding to the Annual Cycle of the Sea Level Estimated by the Tide Gauges and the Sinusoids Estimated from Different Altimetry Data Sets for Each Sub-Basin of the North Sea/Baltic Sea Intersection Zone (Data from Envisat 2002–2010)

SUBBASIN	ALES (m)	SL_cci Env (m)	RADS (m)
Kattegat	0.012	0.019	0.023
Norway Skagerrak	0.014	0.022	0.028
Denmark Skagerrak	0.008	0.008	0.012
Sweden Skagerrak	0.008	0.010	0.013
West Arkona	0.004	0.031	0.024
East Arkona	0.005	0.016	0.773
Belts	0.006	0.008	0.027

Source: Reprinted and reformatting from Passaro, M., et al., *J. Geophys. Res. Oceans*, 120(4), 3061–3078, 2015a. With permission. © 2015 American Geophysical Union.

11.4.2.3 Examples of Usage of the ALES Data Set

The ALES data set has been used to provide a sea level advice service to support agencies responsible for planning coastal flood defenses in the United Kingdom in the framework of the Sea Level Space Watch program (Cipollini and Calafat 2016), producing maps of trends and seasonal variability that show significant regional differences and highlighting the still crucial effect of the choice of tidal models in the coast.

ALES was also chosen in the ESA eSurge project (http://www.storm-surge.info) to observe the development of storm surges in the coastal ocean: As a case study, Cyclone Hagupit was captured by Jason-2 on May 12, 2014, and ALES reprocessed data revealed a clear bulge of SSHA at the storm's center (Harwood 2015).

More results of the oceanographic studies based on ALES are discussed in Section 11.6.

11.4.3 X-TRACK Regional Altimetry Products

A different approach than waveform retracking was adopted by Laboratoire d'Études en Géophysique et Océanographie Spatiales (LEGOS) who concentrated on the post-processing step and developed a data-editing algorithm able to look for the best trade-off between the quantity and the quality of altimetry observations over marginal seas. This software, called X-TRACK (Roblou et al. 2011; Birol et al. 2016), analyzes the behavior of the different altimeter corrective terms as a whole and then edits and recomputes the suspicious correction values with the aim of maximizing the number of useful near-shore sea level data finally available for the user. It is now routinely operated at the Center for Topographic studies of the Ocean and Hydrosphere (CTOH, French observation service dedicated to satellite altimetry studies), and a large part of the archive of altimetry data has been reanalyzed. X-TRACK regional products are freely distributed (http://ctoh.legos.obs-mip.fr/products/coastal-products/ and http://www.aviso.altimetry.fr/en/data/products/sea-surface-height-products/regional/x-track-sla.html), providing Level 3 along-track sea level anomaly (SLA) products from different altimetry missions for coastal and regional applications. Some products combine the observations from different missions that are on the same orbit. They are all computed in the same way, resulting in coherent multi-mission data sets, which is particularly important for climate research. Table 11.5 indicates the 1-Hz SLA products available and, for each, the first and last dates of the time series. Together with mono-mission products (T/P+ T/P interleaved, Envisat, GFO, Jason-1 + Jason-1 interleaved, and Jason-2) users can retrieve SLA for T/P + Jason-1 + Jason-2 and T/P interleaved + Jason-1 interleaved combined missions. The processing is done on a regional basis and, in parallel to the altimeter missions, the areas covered by X-TRACK regional products have progressively increased since 2007, and now include all the coastal oceans (Figure 11.13).

For each region, both raw and along-track low-pass filtered (using a 40-km cutoff frequency) 1-Hz SLA time series along the nominal ground tracks are available. They are provided together with: (1) an along-track mean sea surface height profile based on the data over the largest whole number

TABLE 11.5
List of X-TRACK SLA Products Available and the First and Last Dates of the Time Series

Mission	Start	End
TOPEX/Poseidon (T/P)	02/28/1993 (17)	08/12/2002 (364)
Jason-1 GDR-C	01/15/2002 (1)	01/27/2009 (259)
Jason-2 GDR-D	07/12/2008 (1)	05/26/2016 (290)
T/P+Jason-1+Jason-2	02/28/1993 (17 of T/P)	05/26/2016 (290 of Jason-2)
T/P interleaved	09/21/2002 (369)	10/09/2005 (481)
Jason-1 interleaved	02/10/2009 (262)	02/15/2012 (372)
Envisat-v2.1	10/01/2002 (10)	09/14/2010 (92)
GEOSAT Follow-On (GFO)	01/08/2000 (37)	09/08/2008 (222)
T/P interleaved+Jason-1 interleaved	09/21/2002 (369 of T/P)	02/15/2012 (372 of Jason-1)

Note: The corresponding cycle number is indicated in parentheses.

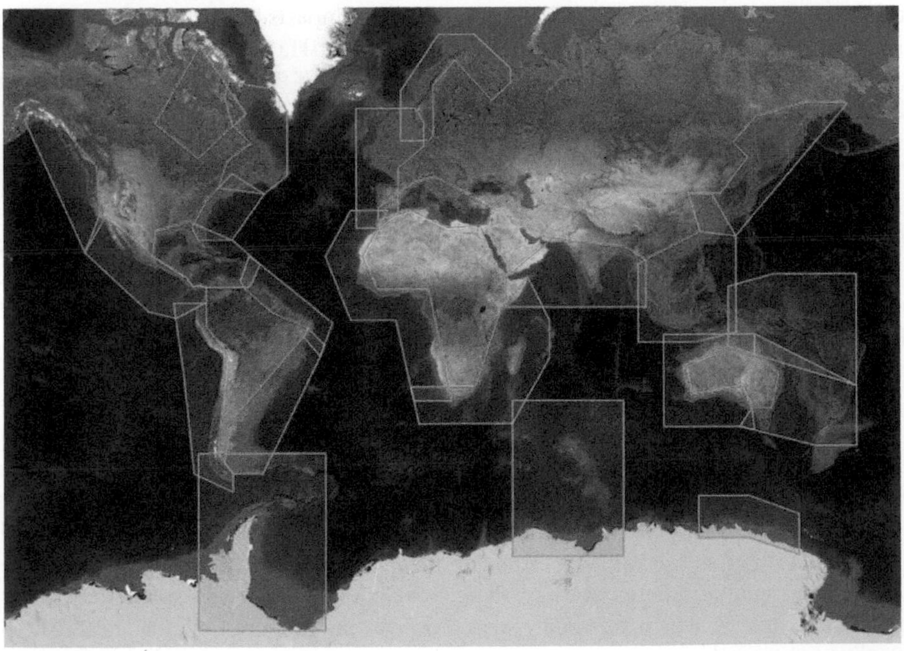

FIGURE 11.13 Map of X-TRACK regional products in 2016.

of years covered by the SLA data, (2) the tidal and dynamical atmospheric correction (DAC) used (applied to SLA data but provided for specific applications), (3) the distance to the nearest coast (in 2016, computed from the NOAA product at a spatial resolution of 0.04°, http://oceancolor.gsfc.nasa.gov/DOCS/DistFromCoast/), and (4) the mean dynamic topography (MDT).

The historical X-TRACK processing methodology is described in Vignudelli et al. (2005) and Roblou et al. (2011). It is continuously evolving and the editing strategy has been entirely revisited in the 2016 version. If the general principle remains the same, each altimetry corrective term is now edited in a different way in order to take into account its individual characteristics (i.e., the corresponding geophysical variations along the track; for more details, see Birol et al. 2016). It was shown to result in higher correlations between coastal altimeter and tide gauge sea level anomalies (by 15% in average) compared with the previous version of the algorithm. The suitability of the X-TRACK approach was shown in different regional studies, where compared to the operational AVISO processing algorithm, it results in a significant increase in the quantity of available data near coastlines (Vignudelli et al. 2005; Durand et al. 2008, Birol et al. 2010; Melet et al. 2010, Passaro et al. 2014; Jebri et al. 2016). For example, Birol et al. (2016) showed that along Western Africa, sea-level variations derived from X-TRACK data are observed closer to land (5 km) compared to AVISO (10 km). Sea-level statistics are more robust due to the larger and more stable data availability (Cipollini et al. 2016), and this leads to an improved observation of the near-shore ocean dynamics in different regions, in particular the boundary circulation (Durand et al. 2008, Birol et al. 2010; Bouffard et al. 2011; Liu et al. 2012; Jebri et al. 2016).

The improvement of the coastal altimetry SLA data has led the CTOH to also derive the empirical harmonic constants (amplitude and phase lags) for a number of tidal constituents by using harmonic analysis (Birol et al. 2016). The tidal constants database is computed using the 1-Hz SLA time series processed by X-TRACK, taking advantage of the T/P, Jason-1, and Jason-2 long time series. It is also distributed on a regional basis. The files hold the along-track estimates of amplitude and phase for a number of tidal constituents (see Table 11.3 of Birol et al. 2016) and two along-track error estimates associated with the tidal constants: (1) a formal error computed from the inverse method theory and (2) an error accounting for the ocean background signal associated to aliased tidal frequencies.

All these products are validated before distribution. Comparisons with Level 3 AVISO products and a number of tide gauge records are made for the 1-Hz SLA. Altimetry tidal constants are systematically compared to global model solutions. Many of the corresponding diagnostics are available on the CTOH web site. A new release of the products is proposed every year (including longer time series, new developments, new altimeter missions, etc.). A specific digital object identifier (DOI) has been created to identify X-TRACK products: 10.6096/CTOH_X-TRACK_2015_01. Finally, X-TRACK is not only a product but also a data service. It has been developed considering the input from users, and specific data sets can be computed on request for particular applications. This is important because no consensus exists yet concerning standard coastal altimetry processing and products—this is something on which the community of developers and users of coastal altimetry will hopefully converge in the next few years.

11.5 APPLICATIONS USING OBSERVATIONS ALONE

The oceanographic applications that are made possible or improved in coastal domains by the use of altimeter data are primarily those that require information about sea level and surface currents. A smaller number of applications use estimates of surface wave heights along altimeter tracks (for instance, Passaro et al. 2015b), while very few use altimeter estimates of wind speed. The fact that wind and wave conditions change more rapidly (in hours to a day) than circulation features may explain their lower rate of use along altimeter tracks, which are widely spaced over several hundred kilometers and 10–35 days. Meteorological buoys and atmospheric models provide wind estimates every 1–4 h; these winds, in turn, are used to force wave models. Those relying on frequent updates of wind and wave conditions, including forecasts, use the model fields. The sparse altimeter observations are more likely to be used to validate model wave and wind forecasts or to look at specific events in retrospective studies.

SSH and SLA fields over the ocean, however, are the unique contribution of altimeters. Coastal tide gauges, bottom pressure sensors, and inverted echo sounders can provide more rapid and complete time series at select points, but only altimeters provide data with enough coverage to create gridded time series of two-dimensional surface height fields on weekly to monthly timescales. Geostrophic currents, calculated from the height gradients, are also available over the global coastal ocean with coverage unobtainable from other sources, either as cross-over vector velocities, cross-track components of velocity, or as gridded fields.

Applications that combine altimeter data with numerical and statistical models are discussed in the next section. Here we consider applications that use observations without models, either from one or more altimeters by themselves or in combination with other satellite and *in situ* data sets. Our consideration of applications can also be divided spatially into those that consider the large-scale boundary currents (extending several hundred kilometers offshore) and others that are confined to the narrower region within several tens of kilometers of land—even extending onshore to quantify river discharge and coastal inundation.

Previous reviews of ocean remote sensing include coastal and fisheries applications in the large-scale boundary currents. Although most of these use sea surface temperature (SST) and ocean color, a few utilize altimeter data. Mitchum and Kilonsky (1995) compare altimeter measurements of sea level variability to tide gauge measurements, connecting inherently coastal observations to large-scale oceanography. Ikeda and Dobson (1995) discuss early retrievals of surface waves and winds from altimeters, although not in coastal contexts. Janssen (2000) and Atlas and Hoffman (2000) discuss altimeter wave and wind measurements over the open ocean. Garzoli and Goni (2000) provide altimeter-derived transport estimates in boundary currents (the Agulhas and Benguela). Oliveira et al. (2000), while looking at the movement of meddies (salty lenses of Mediterranean water), also use along-track altimeter data to characterize these large-scale eddies, which pass near the eastern boundary current next to the Iberian Peninsula. Reynolds et al. (2000) present an early example of assimilation of altimeter data into a large-scale ocean model that simulates the 1997 to 1998 El Niño, the topic of the next section. The more recent review of "Oceanography from Space" by Barale et al. (2010) describes applications of altimeter data for large-scale ocean circulation and sea level, MDT, the marine geoid, and planetary waves. Although coastal regions are not discussed per se, these discussions provide the background and underlying basis for coastal applications.

Vignudelli et al. (2011) present the first comprehensive review of the use of altimetry in coastal regions and discuss regional applications of altimeter-derived fields (primarily coastal currents and sea levels) in the northwest Mediterranean Sea; the Caspian Sea; Black Sea; Barents and White Seas; the coasts of North America; China's coastal seas; Australia; and in lakes in South America, East Africa, and Asia. For publications since 2011, the reader is directed to the literature databases maintained at NASA's Jet Propulsion Laboratory for the Ocean Surface Topography Science Team (search "sealevel.jpl.nasa.gov" under "Science" and "Literature Database") and for the Surface Water, Ocean Topography Science Team (search "swot.jpl.nasa.gov" under "Science" and "Publications" as well as under "Applications"). Presentations from a series of Coastal Altimetry Workshops held since February 2008 can also be found on the COASTALT Website (http://www.coastalt.eu).

Over the longest timescales, decadal studies of sea level rise from altimeter data depict global maps of the basins, showing the spatial structure of the rising seas (Nerem et al. 2010; Stammer et al. 2013; Ablain et al. 2015). For instance, the eastern North Pacific has a low value of increased sea level (some regions are even decreasing), while the western tropical Pacific is rising faster than the mean rate. Combinations of altimeter and tide gauge data have been used to extend these reconstructions, depicting the global patterns of sea level rise over the past 50 years (Church et al. 2004; Church and White 2006; Hamlington et al. 2011; Meyssignac et al. 2012). More regional studies of sea level rise/change have also been published for the Mediterranean Sea (Calafat et al. 2009), the Indian Ocean (Han et al. 2010), Europe (Feng et al. 2013), the Pearl River Delta (He et al. 2014),

Malaysia (Luu et al. 2015), and the Bay of Biscay (Marcos et al. 2007). An issue related to sea level rise is the vertical movement of the coastal land (Wöppelmann and Marcos 2016). Altimeter data are combined with tide gauges, Global Positioning System (GPS), and other data to investigate vertical land motion in a number of locations (see Wöppelmann and Marcos 2016), including Taiwan (Chang et al. 2012) and the coast of Turkey (Yildiz et al. 2013).

On seasonal timescales, new estimates of spatial and temporal variability in the seasonal cycles of coastal sea level have been published for the northwest Pacific by Feng et al. (2015) and the southwest Atlantic by Ruiz Etcheverry et al. (2016). A common finding on seasonal and shorter timescales is that tide gauges produce more energetic signals than standard altimeter fields. This problem has been addressed by the use of the reprocessed ALES coastal along-track data to derive seasonal cycles of sea level for the North Sea/Baltic Sea intersection zone (Passaro et al. 2015a) and the Indonesian Seas (Passaro et al. 2016, also using CryoSat-2 SAR data). In both regions, the use of ALES data reveals the higher amplitude of annual cycles of the sea level in the presence of coastal currents: the Norwegian Coastal Current in the Skagerrak Sea (Passaro et al. 2015a) and the Java Coastal Current south of Java (Passaro et al. 2016). In the first case, ALES data filled the gap between the higher annual amplitude (approximately 9 cm) registered by a coastal tide gauge and the lower amplitude in the open sea (approximately 4 cm), showing a gradient of annual amplitude that follows the bathymetry slope within 30 km of the coast. In the latter study, it was possible to detect the semiannual Kelvin Wave traveling from the Indian Ocean into the internal seas of Indonesia.

Moving from sea level to studies of regional ocean circulation, an overview of methods of estimating surface currents from satellite data, including altimetry in coastal regions, is given by Klemas (2012). Regional studies of shelf-slope circulation constitute some of the most common applications of coastal and boundary current altimetry, as represented by examples from the Bay of Biscay (Herbert et al. 2011), the northwest Mediterranean Sea (Bouffard et al. 2012), the eastern Gulf of Mexico and West Florida Shelf (Liu et al. 2012, 2014), the southern Bay of Bengal (Wijesekera et al. 2015), and the Agulhas Current (Krug and Tournadre 2012; Krug et al. 2014). In some regions—for example, the southwest Atlantic study of Strub et al. (2015)—fields of MDT are added to the SLA data to calculate absolute geostrophic circulation. The MDT fields may be global (Rio et al. 2014a) or specialized regional MDT fields, such as in the Mediterranean Sea (Rio et al. 2014b). Improvements in the tidal models used to correct the altimeter data are indicated by the fact that some of the aforementioned studies (e.g., Strub et al. 2015) succeed in using standard processed altimeter data over the wide continental shelf.

On synoptic scales of days to weeks, many coastal regions are affected by offshore eddies, as in the case of the meddies (Oliveira et al. 2000) and Agulhas eddies (Garzoli and Goni 2000) noted earlier. An altimetric survey of mesoscale eddies in the global ocean is reported in Chelton et al. (2011b) and is extended to examine the impact of eddies on satellite-derived fields of surface chlorophyll pigment concentrations in Chelton et al. (2011a). Mechanisms by which winds interact with eddies to create patterns of upwelling and downwelling are further explored in Gaube et al. (2015). These same mechanisms apply in both open and coastal ocean regions. One of the more comprehensive descriptions of surface and sub-surface characteristics of eddies in the ocean's major eastern boundary currents is that of Pegliasco et al. (2015), combining altimeter and sub-surface float data. Other descriptions of eddy characteristics and eddy kinetic energy are presented by Bouffard et al. (2012) in the northwest Mediterranean and Caballero et al. (2008) in the Bay of Biscay.

Investigations of the velocity fields next to the coast often use the specially processed data sets described in Section 11.5. This allows analyses to reach closer to the coast and also recovers some of the more energetic signals revealed by tide gauges in comparison to standard altimeter processing. Examples of the use of ALES data to resolve more energetic seasonal cycles of sea level and velocities are presented earlier. On the shorter synoptic scales, X-TRACK regional altimetry products have been used in a wide range of applications: coastal ocean circulation, mesoscale dynamics, hydrodynamic and tidal model validation, tides, development, or validation of new altimeter processing methods (Vignudelli et al. 2005; Durand et al. 2008; Birol et al. 2010; Melet et al. 2010; Bouffard et al. 2011; Liu et al. 2012; Passaro et al. 2014; Birol et al. 2016; Jebri et al. 2016). Figure 11.14 presents the

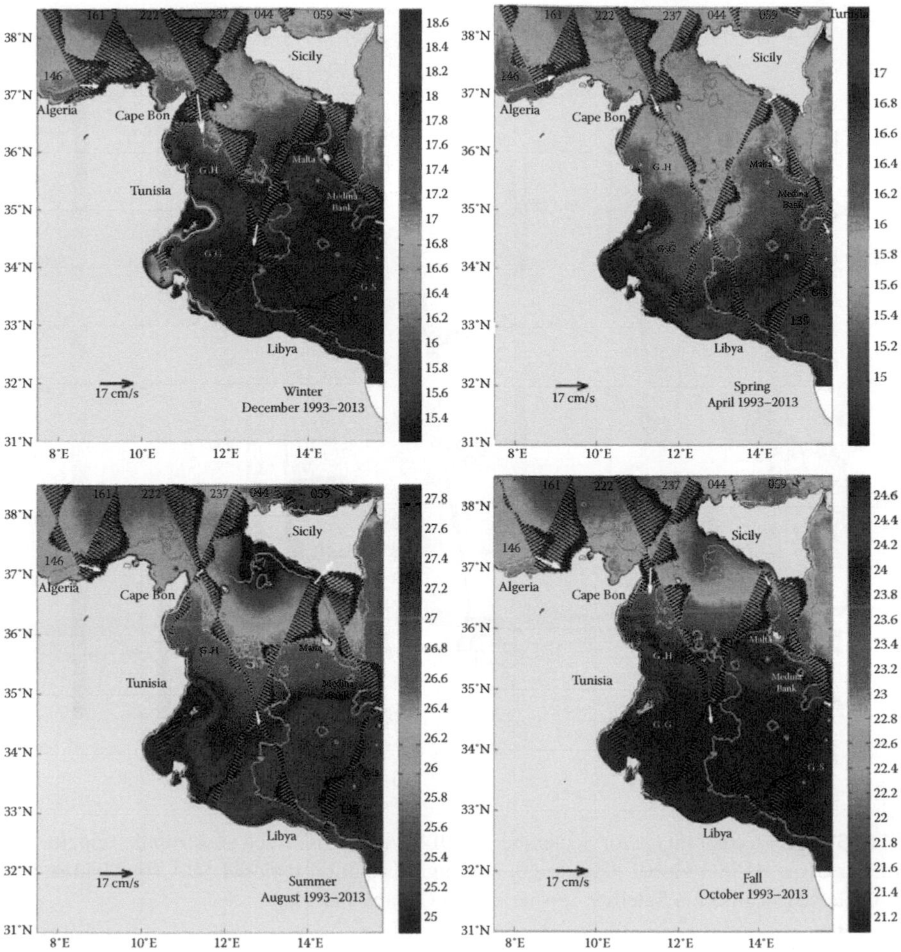

FIGURE 11.14 Maps of cross-track (black) and crossover (white) geostrophic currents for X-TRACK and SST over the climatological cycle. The 200-m isobath (gray solid line) is from the ETOPO2v1 global gridded database. SST data are from the German Aerospace Center (DLR) product with 1.1-km resolution. G.H. is the Gulf of Hammamet, G.G. is the Gulf of Gabes, and G.S is the Gulf of Sirte. (Reprinted from Jebri, F., et al., *J. Geophys. Res. Oceans*, 121(7), 4888–4909, 2016. With permission. © 2016 American Geophysical Union.)

example of the surface circulation over the central Mediterranean Sea observed in Jebri et al. (2016) from X-TRACK altimetry data and remotely sensed sea surface temperature (SST) observations. In this study, after a comparison between altimeter and *in situ* observations, the seasonal evolution of the surface circulation was analyzed in details. The combined SLA and SST data sets clearly depict different current regimes and bifurcations, which provides the basis for a new conceptual model of the seasonal circulation scheme for the Central Mediterranean. This analysis includes variations of the path and temporal behavior of the main regional circulation features (i.e., the Atlantic Tunisian Current, the Atlantic Ionian Stream, the Atlantic Libyan Current, and the Sidra Gyre), along with a new coastal current over the Gulf of Gabes (see Jebri et al. 2016 for more details).

In the Gulf of St. Lawrence, details of the circulation are also revealed by specialized processing, in this case the one adopted for the PEACHI data set. Figure 11.15 compares the cross-track velocities derived from standard processing with those from PEACHI data, with both overlain on satellite-derived surface chlorophyll concentrations. The greater detail in the PEACHI-derived velocities is consistent with advection of the surface chlorophyll field.

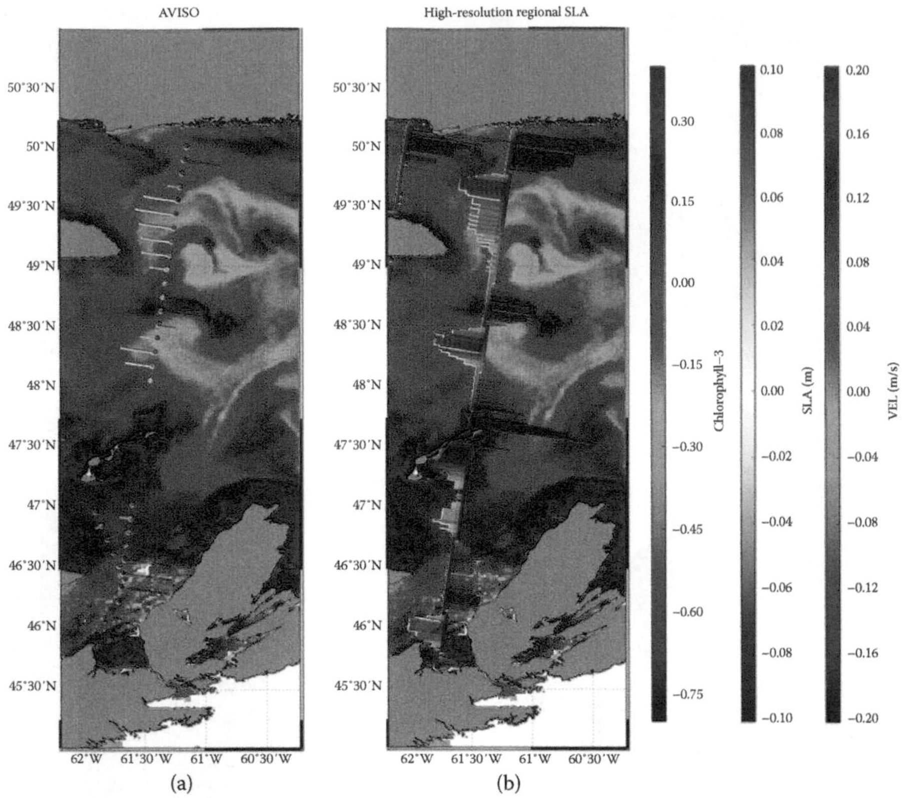

FIGURE 11.15 Ocean color image for September 8, 2014, with across-track geostrophic velocities computed from altimeter SLA. AVISO global data set (a), and high-resolution regional SLA (b). (Figure courtesy of C. Dufau, Collecte Localisation Satellite / Fisheries and Oceans Canada.)

Decreased SLA next to the coast is an indicator of wind-driven upwelling, as demonstrated along the Bonnie coast off southeast Australia (Cahill and Dufau 2015). In this study, PEACHI data allowed analysis of along-track SLA data close to the coast, which documented decreased coastal sea levels during a wind-driven upwelling event during November 2013, in agreement with colder satellite SST fields and coastal radar current fields showing surface flow to the northwest (Figure 11.16). Other studies have revealed low coastal sea levels associated with upwelling off Vietnam (Kuo et al. 2004) and in the coastal regions of the major eastern boundary currents (Strub et al. 2013).

Risien and Strub (2016) take a different approach to the improvement of sea-level fields along the U.S. West Coast. Following Saraceno et al. (2008), they blend tide gauge and altimeter data in the 75-km region next to the coast. This recaptures much of the greater energy observed with current meters over the coast, energy that is missing in geostrophic velocities calculated from standard gridded altimeter fields.

On the shortest timescales, papers by Madsen et al. (2007, 2015) have blended tide gauge and altimeter data to predict storm surge magnitudes in the complex region connecting the North Sea and Baltic Sea with multilinear regression models, anticipating more deterministic predictions using numerical circulation models. A similar statistical approach has been used to predict sea levels along the northwest coast of Britain by Cheng et al. (2012). Other very short timescale phenomena include tsunami wave elevations, which may be caught by opportunistic altimeter passes, such as reported by Troitskaya and Ermakov (2008). Another example of relatively short-term changes in sea level include the effects of expansion due to the heat content of eddies or other water masses. The influence of ocean surface heat content on the strength of tropical cyclones is now well

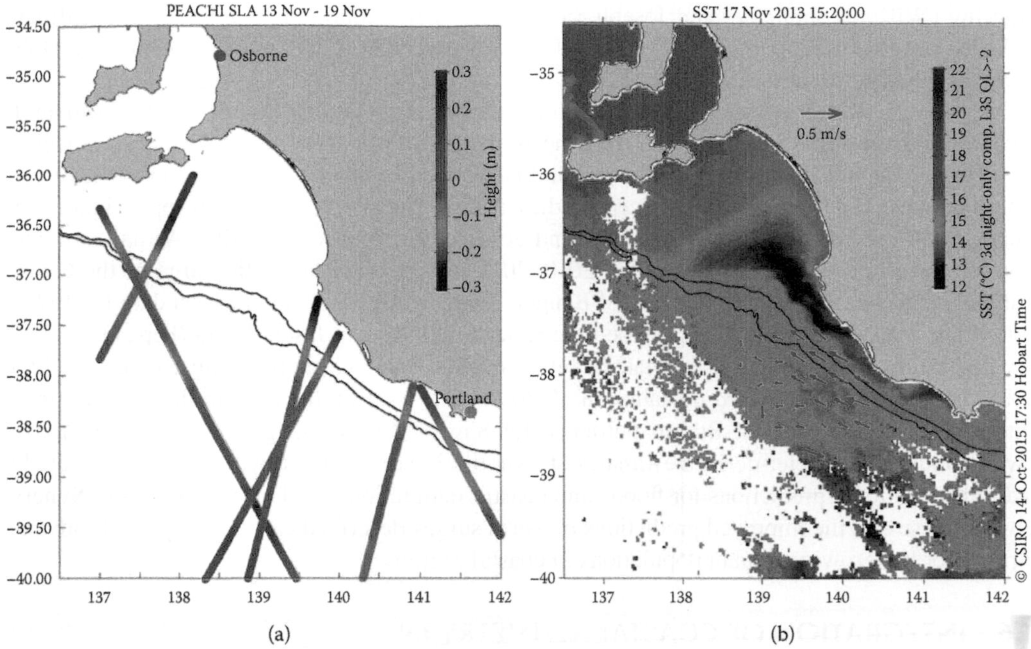

FIGURE 11.16 Along-track altimeter SLA (a) from November 13 to 19, 2013, compared to an SST image taken on November 17 (b). Altimeter data in (a) are SLA PEACHI products using Jason-2 and SARAL/AltiKa observations provided by CNES. Data in (b) are SST from the Australian Integrated Marine Observing System. Pink arrows display high-frequency (HF) radar velocities measured for the same day. Figure courtesy of M. Cahill, Commonwealth Scientific and Industrial Research Organisation (CSIRO).

documented and used in predictions of hurricane intensity (Halliwell et al. 2015) as happened in Hurricane Katrina (Scharroo et al. 2005). SLA is also useful as an indicator of ocean heat content due to El Niño signals in the California Current (Wilczak et al. 2007).

In addition to sea level and geostrophic velocities, estimates of significant wave height are derived from the slope of the leading edge of the return radar signal. These have been used in several studies to both evaluate the satellite estimates and infer properties of the coastal ocean wave field. In three tropical coastal regions, Hithin et al. (2015) found significant wave height estimates from the SARAL/AltiKa and Jason-2 altimeters compare well with *in situ* measurements. Comparing ERS-2, Envisat, Jason-1, and Jason-2 to significant wave heights at stations along the Indian coast, Jason-2 produced the best agreement with *in situ* measurements. Gridded, multi-mission data sets consistently overestimated the significant wave heights; the poorest agreement was found at the wave buoy closest to the coast. For significant wave heights less than 0.5 m, the satellite estimates were unreliable. Differences between the altimeters make it problematic to determine long-term changes in wind wave heights, as suggested by Dragani et al. (2010) along the southeast coast of South America.

For operational applications, along-track altimeter significant wave height estimates are routinely used by the U.S. NOAA NWS NCEP Ocean Prediction Center (OPC; http://www.opc.ncep.noaa.gov) for their open ocean analyzes and predictions. Data from all available altimeters (U.S. and international) are displayed by themselves and as overlays on satellite and forecast model fields (winds and waves). Near real-time data help forecasters to: determine if a specific weather system is behaving as models have predicted; observe conditions over data sparse areas; and correct large-scale biases in wave models. The altimeter data complement other data sources such as buoy winds and waves, scatterometer winds, satellite imagery and integrated displays with complimentary observations, imagery, and numerical model output. For coastal regions inshore of approximately 110 km (60 nm), the use of the wave data is more limited, due to its sparse sampling (William Schneider and Joseph Sienkiewicz, pers. commun., August 2016). Altimeter SLA data are also used operationally by the U.S. Navy and NOAA, by assimilation into the global ocean circulation forecast models.

One of the newest applications of altimeter data in the nearshore coastal ocean is to estimate the discharge of large rivers into the ocean from the altimetry-derived river heights, which is especially promising when looking forward to the availability of swath altimeter data. Most of the investigations conducted so far are proof of concept studies, using other sources of data to approximate what is expected from the swath altimeters (Durand et al. 2014; Yoon et al. 2016). Actual estimates, however, have been made by Papa et al. (2010, 2012) for the very large discharge of the Ganga-Brahmaputra River system into the Bay of Bengal. Using T/P, ERS-2, and Envisat data for 1993 to 2008 (Papa et al. 2010) and extending the time series to 2011 using Jason-2 data (Papa et al. 2012), they find errors in the discharge estimate of 17% for 1993–2011, within the requirements of 15% to 20%. The interannual standard deviation of 12,300 $m^3 s^{-1}$ is much larger than the data set uncertainties of 2,180 $m^3 s^{-1}$ for the Brahmaputra and 1,458 $m^3 s^{-1}$ for the Ganga River. Combinations of new methods for altimeter-derived estimates of river discharge and river hydrology are expected to lead to better model predictions for floods and coastal inundations (Pedinotti et al. 2014). Synergy between these and the improved predictions of storm surges described earlier will provide advance warning for the growing human populations in coastal regions.

11.6 INTEGRATION OF COASTAL ALTIMETRY IN COASTAL OBSERVING SYSTEMS

Hydrodynamic models of coastal oceanography augment observing systems by extending the spatial and temporal scope of data—notably, to sub-surface depths in the case of satellite observations and also beyond the duration of typical sustained time series platform deployments. Hydrodynamic models offer the ability to deliver long retrospective analyzes and, importantly, future predictions. When operated as real-time forecast systems, models assist decision-making for water quality, public health, maritime operations, and a host of other applications in coastal and littoral zones.

Reconciling model output and observations in applied coastal modeling entails model skill assessment and forecast verification by comparison to independent observations but also the formal merger of the two through data assimilation methods to derive estimates of the ocean state that are more accurate than models or observations alone. Data assimilation is common practice in Numerical Weather Prediction, and there has been substantial progress in the past decade in mesoscale ocean prediction spurred by Global Ocean Data Assimilation Experiment (GODAE) initiatives (Bell et al. 2015) in which the assimilation of satellite altimeter data features prominently, in the format of both along-track data and multi-mission blended, gridded products. But despite the progress detailed earlier in developing coastal-corrected along-track altimeter data products that extend valid data into shallow coastal seas and to within kilometers of the coast, the uptake of altimetry in mesoscale forecast systems has not been matched by comparable advances in coastal modeling.

This is not to say that altimetry has not been shown to usefully inform coastal circulation. Better-constrained mesoscale variability in boundary current regimes significantly enhances predictive skill in shelf seas. On the U.S. West Coast, assimilating altimeter data has a strong impact on the modeled transport of the California Current that flows along the shelf-break and also on SST variability adjacent to the coast (Moore et al. 2011). Other modeling systems such as those for the Gulf of Mexico (Barth et al. 2008) and East Australia (Oke et al. 2009) find the inclusion of satellite altimetry in deep water improves the fidelity of coastal currents.

In a comparison of seven real-time models (both local and global/basin) that span the U.S. East Coast Mid-Atlantic Bight, the model that assimilated coastal altimeter data was the most skillful (Wilkin and Hunter 2013), but that model also assimilated surface currents from coastal HF-radar systems and *in situ* data from a comprehensive coastal observatory. Other studies have considered the merits of assimilating HF-radar surface currents (Paduan and Shulman 2004; Oke et al. 2009), but no study has yet systematically explored the relative merits, and complementarities, of radar currents and coastal altimetry in models. As we approach the era of wide-swath coastal altimetry, a study that characterizes the value of a fully global coastal satellite altimeter observation system versus a distributed network of rapid sample repeat local coastal HF-radar systems is urgently required.

One impediment to more widespread adoption of coastal altimetry in conjunction with truly coastal models (notionally, inside the 100-m isobath) is reconciling sea level variability as typically defined in mesoscale altimetry with added dynamics that are in important in coastal waters. High-frequency sea level variability due to tides, coastal-trapped waves, and storm surge are important drivers of coastal inundation risk. GODAE operational models, and the aforementioned regional shelf modeling systems (Barth et al. 2008; Moore et al. 2011), do not include tides or the influence of the inverse barometer (IB) effect on coastal sea level; they assimilate altimeter data from which those dynamics have been removed by the application of a Dynamic Atmosphere Correction (DAC) and harmonic tidal model. In a coastal ocean model that explicitly resolves these high-frequency dynamics, the DAC and tide correction remove a signal that should be retained in sea level data assimilation. To do otherwise would either misrepresent the model-data misfit cost function that underlies data-assimilative algorithms or demand a thoughtful re-specification of the so-called representation-error in data assimilation.

Defining absolute sea level above datum requires a consistent definition from global to coastal regimes. Even when the resolution of MDT is locally enhanced for regional and coastal applications (e.g., Rio et al. 2007), a clear definition of the altimetric mean sea surface reference level may be absent, yet reporting and predicting sea level anomaly with respect to a local geodetic datum consistent with coastal tide gauges is vital to making such predictions valuable to users.

One realm where it has been demonstrated that sea level forecast skill improves with the use of coastal altimetry is storm surge prediction. Madsen et al. (2015) use an ensemble optimal interpolation (EnOI) approach to assimilate a blended altimeter and coastal tide gauge analysis into a three-dimensional baroclinic hydrodynamic model of the Baltic and North seas achieving a reduction in RMS error of on average 20% across all sub-regions of the domain. Using a four-dimensional variational approach (Courtier 1997), De Biasio et al. (2016) assimilated coastal altimetry in a two-dimensional shallow water equation storm surge model for the Adriatic Sea. The skill improvement with assimilation was less pronounced than in the Baltic and North Sea model, which they attribute to the rather small extent of the Adriatic Sea and therefore relatively extended periods with no passes of a satellite. However, it should be noted that Madsen et al. (2015) did not explore the impact of altimetry in the absence of supporting tide gauge data in their blended sea level input data, so studies of the value of coastal altimetry alone remain inconclusive.

A further consideration is that multidecadal scale coastal model downscaling simulations require open boundary conditions from a global ocean model. Ideally, these would use a consistent configuration through the period of interest, yet many global-scale GODAE systems have altered their sea level datum, altimeter data assimilation scheme, and altimeter data processing version over the course of ongoing model configuration development. The GLORYS global ocean reanalysis (Ferry et al. 2012) for the altimetric era is an example of a model-based analysis that brings consistency and stability to products that coastal ocean modelers might adopt as boundary conditions for long regional down-scaling reanalyses.

Beyond data assimilation and prediction, ocean models can also make an important contribution to observing systems themselves through quantitative analysis of the resolution and impact of observational strategies. For example, Le Hénaff et al. (2009) used the Representer Matrix Spectrum method to contrast nadir altimeters with wide-swath altimetry in a model of the Bay of Biscay, finding the latter significantly superior in its ability to constrain coastal circulation physics yet also highlighting the significant influence of adding complementary *in situ* vertical profile observations of velocity and temperature.

Efforts to build upon these experiences are being promoted by several communities of practice in coastal modeling and observing. The GODAE OceanView Coastal Ocean and Shelf Seas Task Team (COSS-TT), the Coastal Altimetry Workshop (CAW) community, and their inaugural joint Altimetry for Regional and Coastal Ocean Modeling (ARCOM) workshop in 2015, have collectively voiced the desire for an agreed "best" set of coastal-corrected sea level anomaly data that removes the necessity for specialist knowledge of altimetry on the part of model users, and also gridded multi-mission sea level and/or surface current products (Kourafalou et al. 2015) that are accurate globally for the many typical coastal physical oceanography applications.

11.7 CONCLUSIONS

This chapter has described the technical advances and the applications that are making coastal altimetry a reality. It should be apparent from what has been presented that this relatively new discipline is still largely at an experimental status, and that there is room for further improvements, but this challenge is naturally presented by the peculiarities of the coastal domain. Altimetry over the open ocean has long had a mature, fully operational status, and can often be used by itself for the description of ocean dynamics at meso- and large scales and at sea level; conversely, its extension into the coastal zone, where the spatial and temporal scales decrease, naturally calls for the integration of the altimetric measurements with *in situ* observations and models, in order to mitigate for the limitations of altimetric sampling. The scope for integrating altimetry in coastal observing systems becomes strikingly clear when one considers its global coverage of the coasts and the availability of long time series that allow a robust statistical description of the variability of the coastal sea level, SWH and wind fields. Moreover, altimeters provide absolute measurements of sea level (i.e., not affected by land movement) that complement tide gauges and can in principle help derive the vertical land movement and close the link between the open observations of sea level (one of the best indicators of the changes occurring in the climate system) and those at the coast.

As the impacts of global change in the oceans are felt most severely at the coast, it is not surprising that the research community has multiplied efforts to make altimetry more usable in this crucial domain. This chapter has clearly shown that several research products are available, from a number of research groups, covering different missions and regions. The coastal altimetry community is working hard to apply these products to a number of studies, for different processes. The intense activities of calibration and validation of the various coastal altimetry data sets will hopefully lead to a convergence of the many currently available data sets into a small number of well-documented products, facilitating their uptake by the user community, including the regional and coastal modelers.

The prospects for coastal altimetry are very good—not just by virtue of all these activities but also because the evolution of altimetry as a whole is moving toward systems that have intrinsically better coastal altimetry capabilities. SAR altimetry, convincingly demonstrated by CryoSat-2, is now available over the global ocean from Sentinel-3A and allows better signal-to-noise and higher resolution that, in turn, benefit in particular the coastal observations. Another quantum leap is expected in the next decade with the arrival of full two-dimensional altimetry from SWOT. In the meantime, technical and algorithmic developments from coastal altimetry are challenging our understanding of the short scales over the open ocean, too, and may result in a better quantification of the sub-mesoscales that are of great important for biophysical interactions and ocean/atmosphere exchange.

The coastal altimetry community (http://www.coastalt.eu/community) remains very lively, and this is demonstrated by the success of the Coastal Altimetry Workshop, the tenth edition of which (CAW-10) was held in Florence, Italy, in February 2017, with more than 100 participants. This is very promising to ensure that coastal altimetry makes a real impact for coastal oceanographers, modelers, and climate scientists.

ACKNOWLEDGMENT

We thank Madeleine Cahill, Claire Dufau, Jesus Gómez-Enri, Fatma Jebri and Bruno Picard for providing some of the figures.

REFERENCES

Ablain, M., Cazenave, A., Larnicol, G., Balmaseda, M., Cipollini, P., Faugère, Y., Fernandes, M. J., Henry, O., Johannessen, J. A., Knudsen, P., Andersen, O., Legeais, J., Meyssignac, B., Picot, N., Roca, M., Rudenko, S., Scharffenberg, M. G., Stammer, D., Timms, G., & Benveniste, J. (2015). Improved sea level record over the satellite altimetry era (1993–2010) from the Climate Change Initiative project. Ocean Science, 11(1), 67–82, https://doi.org/10.5194/os-11-67-2015.

Andersen, O. (2010). The DTU10 gravity field and mean sea surface. *Second international symposium of the gravity field of the Earth (IGFS2)*. Fairbanks, AK: University of Alaska, Fairbanks.

Andersen, O. B., & Scharroo, R. (2011). Range and geophysical corrections in coastal regions: And implications for mean sea surface determination. In *Coastal Altimetry* (eds. S. Vignudelli, A. Kostianoy, P. Cipollini, J. Benveniste), pp. 103–145. Berlin, Heidelberg: Springer.

Atlas, R., & Hoffman, R. N. (2000). The use of satellite surface wind data to improve weather analysis and forecasting at the NASA Data Assimilation Office. In *Satellite Oceanography and Society* (ed. D. Halpern). *Elsevier Oceanography Series*, Volume 63, pp. 57–78. Amsterdam: Elsevier Science.

Bao, L., Lu, Y., & Wang, Y. (2009). Improved retracking algorithm for oceanic altimeter waveforms. *Progress in Natural Science*, 19, 195–203. doi:10.1016/j.pnsc.2008.06.017.

Barale, V., Gower, J. F. R., & Alberotanza, L. (Eds.). (2010). *Oceanography from Space: Revisited*. Dordrecht, The Netherlands: Springer Science & Business Media.

Barth, A., Alvera-Azcárate, A., & Weisberg, R. H. (2008). Assimilation of high frequency radar currents in a nested model of the West Florida Shelf. *Journal of Geophysical Research*, 113, C08033. doi:10.1029/2007JC004585.

Bell, M. J., Schiller, A., Le Traon, P. Y., Smith, N. R., Dombrowsky, E., & Wilmer-Becker, K. 2015. An introduction to GODAE OceanView. *Journal of Operational Oceanography*, 8(suppl 1), s2–11. doi:10.1080/1755876X.2015.1022041

Bennartz, R. (1999). On the use of SSM/I measurements in coastal regions. *Journal of Atmospheric and Oceanic Technology*, 16(4), 417–431. doi:10.1175/1520-0426(1999)016<0417:OTUOSI>2.0.CO;2.

Berry, P., Freeman, J., & Smith, R. (2010). An enhanced ocean and coastal zone retracking technique for gravity field computation. In *Gravity, Geoid and Earth Observation* (ed. S. Mertikas), pp. 213–220. Berlin, Heidelberg: Springer-Verlag.

Birol, F., Cancet, M., & Estournel, C. (2010). Aspects of the seasonal variability of the Northern Current (NW Mediterranean Sea) observed by altimetry. *Journal of Marine Systems*, 81(4), 297–311. doi:10.1016/j.jmarsys.2010.01.005.

Birol F., Fuller N., Lyard F., Cancet M., Niño F., Delebecque C., Fleury S., et al. (2016). Coastal applications from nadir altimetry: Example of the X-TRACK regional products. *Advances in Space Research*, 59, 936–953. doi:10.1016/j.asr.2016.11.005.

Boehm, J., Heinkelmann, R., & Schuh, H. (2007). Short note: A global model of pressure and temperature for geodetic applications. *Journal of Geodesy*, 81, 679–683. doi:10.1007/s00190-007-0135-3.

Bouffard, J., Renault, L., Ruiz, S., Pascual, A., C. Dufau, C., & Tintore, J. (2012). Sub-surface small-scale eddy dynamics from multi-sensor observations and modeling. *Progress in Oceanography*, 106, 62–79. doi:10.1016/j.pocean.2012.06.007.

Bouffard, J., Roblou, L., Birol, F., Pascual, A., Fenoglio-Marc, L., Cancet, M., Morrow, R., & Ménard, Y. (2011). Introduction and assessment of improved coastal altimetry strategies: Case study over the Northwestern Mediterranean Sea. In *Coastal Altimetry* (eds. S. Vignudelli, A. Kostianoy, P. Cipollini, J. Benveniste), pp. 297–330. Berlin, Heidelberg: Springer.

Brown, G. (1977). The average impulse response of a rough surface and its applications. *IEEE Transactions on Antennas and Propagation*, 25, 67–74. doi:10.1109/TAP.1977.1141536.

Brown, S. (2010). A novel near-land radiometer wet path-delay retrieval algorithm: Application to the Jason-2/OSTM advanced microwave radiometer. *IEEE Transactions on Geoscience and Remote Sensing*, 48(4), 1986. doi:10.1109/TGRS.2009.2037220.

Caballero, A., Pascual, A., Dibarboure, G., & Espino, M. (2008). Sea level and Eddy Kinetic Energy variability in the Bay of Biscay, inferred from satellite altimeter data. *Journal of Marine Systems*, 72(1), 116–134.

Cahill, M., & Dufau, C. (2015). Monitoring Coastal Upwelling using Altimetry: A Feasibility Study. 9th Coastal Altimetry Workshop, Reston (USA), 18–19 October, 2015.

Calafat, F. M., Gomis D., & Marcos, M. (2009). Comparison of Mediterranean sea level fields for the period 1961–2000 as given by a data reconstruction and a 3D model. *Global Planetary Change*, 68, 175–184. doi:10.1016/j.gloplacha.2009.04.003.

Carrère, L., & Lyard, F. (2003). Modeling the barotropic response of the global ocean to atmospheric wind and pressure forcing-comparisons with observations. *Geophysical Research Letters*, 30(6), 1275. doi:10.1029/2002GL016473.

Chang, E. T. Y., Chao, B. F., Chiang, C.-C., & Hwang, C. (2012). Vertical crustal motion of active plate convergence in Taiwan derived from tide gauge, altimetry, and GPS data. *Tectonophysics*, 578, 98–106. doi:10.1016/j.tecto.2011.10.002.

Chelton, D. B., Gaube, P., Schlax, M. G., Early, J. J., & Samelson, R. M. (2011a). The influence of non-linear mesoscale eddies on near-surface oceanic chlorophyll, *Science*, 334, 328. doi:10.1126/science.1208897.

Chelton, D. B., Schlax, M. G., & Samelson, R. M. (2011b). Global observations of non-linear mesoscale eddies. *Progress in Oceanography*, 91, 167–216. doi:10.1016/j.pocean.2011.01.002.

Cheng, Y., Andersen, O. B., & Knudsen, P. (2012). Integrating non-tidal sea level data from altimetry and tide gauges for coastal sea level prediction. *Advances in Space Research*, 50(8), 1099–1106. doi:10.1016/j.asr.2011.11.016.

Church, J. A., & White, N. J. (2006). A 20th century acceleration in global sea-level rise. *Geophysical Research Letters*, 33, L01602. doi:1029/2005GL024826.

Church, J. A., White, N. J., Coleman, R., Lambeck, K., & Mitrovica, J. X. (2004). Estimates of the regional distribution of sea level rise over the 1950–2000 period. *Journal of Climate*, 17(13), 2609–2625. doi:10.1175/1520-0442.

Cipollini, P., & Calafat, F. (2016). Altimeter processing and data for Sea Level SpaceWatch Phase 2. Sea Level SpaceWatch Workshop, 22nd March 2016, National Oceanography Centre, Southampton. Available at http://www.satoc.eu/projects/sealevelsw/docs/SLSW2_finalWS_Altimeter_results.pdf. Accessed 24 July 2017.

Cipollini, P., Calafat, F. M., Jevrejeva, S., Melet, A., & Prandi, P. (2017). Monitoring sea level in the coastal zone with satellite altimetry and tide gauges. *Surveys in Geophysics*, 38, 33. doi:10.1007/s10712-016-9392-0.

COASTALT. (2009). Wet Tropospheric Corrections in Coastal Areas, COASTALT CCN2, Deliverable D2.1b v 1.2., 30/06/2009. Document code COASTALT-D21b-12.

Courtier, P. (1997). Dual formulation of four-dimensional variational assimilation. *Quarterly Journal of the Royal Meteorological Society*, 123(544), 2449–2461. doi:10.1002/qj.49712354414.

Davis, J. L., Herring, T. A., Shapiro, I. I., Rogers, A. E. E., & Elgered, G. (1985). Geodesy by radio interferometry: effects of atmospheric modelling errors on estimates of baseline length. *Radio Science*, 20(6), 1593–1607. doi:10.1029/RS020i006p01593.

De Biasio, F., Vignudelli, S., della Valle, A., Umgiesser, G., Bajo, M., & Zecchetto, S. (2016). Exploiting the potential of satellite microwave remote sensing to hindcast the storm surge in the Gulf of Venice. *IEEE Journal of Selected Topics in Applied Earth Observations and Remote Sensing*., *9*(11), 5089–5105. doi:10.1109/JSTARS.2016.2603235.

Dee, D. P., Uppala, S. M., Simmons, A. J., Berrisford, P., Poli, P., Kobayashi, S., Andrae, U., et al. (2011). The ERA-Interim reanalysis: Configuration and performance of the data assimilation system. *Quarterly Journal of the Royal Meteorological Society*, 137, 553–597. doi:10.1002/qj.828.

Deng, X., & Featherstone, W. (2006). A coastal retracking system for satellite radar altimeter waveforms: Application to ERS-2 around Australia. *Journal of Geophysical Research*, 111(C6), C06012. doi:10.1029/2005JC003039.

Desportes, C., Obligis, E., & Eymard, L. (2007). On the wet tropospheric correction for altimetry in coastal regions. *IEEE Transactions on Geoscience and Remote Sensing*, 45(7), 2139–2149. doi:10.1109/TGRS.2006.888967.

Desportes, C., Obligis, E., & Eymard, L. (2010). One-dimensional variational retrieval of the wet tropospheric correction for altimetry in coastal regions. *IEEE Transactions on Geoscience and Remote Sensing*, 48(3), 1001–1008. doi:10.1109/TGRS.2009.2031494.

Dragani, W. C., Martin, P. B., Simionato, C. G., & Campos, M. I. (2010). Are wind wave heights increasing in south-eastern south American continental shelf between 32° S and 40° S? *Continental Shelf Research*, 30(5), 481–490. doi:10.1016/j.csr.2010.01.002.

Dufau, C., Martin-Puig, C., & Moreno, L. (2011). User requirements in the coastal ocean for satellite altimetry. In *Coastal Altimetry* (eds. S. Vignudelli, A. Kostianoy, P. Cipollini, J. Benveniste), pp. 51–60. Berlin, Heidelberg: Springer.

Durand, F., Shankar, D., Birol, F., & Shenoi, S. S. C. (2008). An algorithm to estimate coastal currents from satellite altimetry: A case study for the East India. *Coastal Current. Journal of Oceanography*, 64, 831–845. doi:10.1007/s10872-008-0069-2.

Durand, M., Neal, J., Rodríguez, E., Andreadis, K. M., Smith, L. C., & Yoon, Y. (2014). Estimating reach-averaged discharge for the River Severn from measurements of river water surface elevation and slope. *Journal of Hydrology*, 511, 92–104. doi:10.1016/j.jhydrol.2013.12.050.

Egido, A. (2014). Geo-referencing of the Delay/Doppler Level-1 stack with application to coastal altimetry. 8th Coastal Altimetry Workshop, Konstanz (Germany), 23–24 October, 2014.

Feng, G., Jin, S., & Zhang, T. (2013). Coastal sea level changes in Europe from GPS, tide gauge, satellite altimetry and GRACE, 1993–2011. *Advances in Space Research*, 51(6), 1019–1028. doi:10.1016/j.asr.2012.09.011.

Feng, H., & Vandemark, D. (2011). Altimeter data evaluation in the coastal Gulf of Maine and mid-Atlantic Bight Regions. *Marine Geodesy*, 34(3–4), 340–363.

Feng, X., Tsimplis, M. N., Marcos, M., Calafat, F. M., Zheng, J., Jordà, G., & Cipollini, P. (2015). Spatial and temporal variations of the seasonal sea level cycle in the northwest Pacific. *Journal of Geophysical Research: Oceans*, 120(10), 7091–7112. doi:10.1002/2015JC011154.

Fernandes, M. J., Bastos, L., & Antunes, M. (2003). Coastal satellite altimetry—Methods for data recovery and validation. In *Gravity and Geoid 2002,* Proceedings of 3rd Meeting of the international Gravity and Geoid commission (ed. I. N. Tziavos), pp. 302–307. Thessaloniki: Ziti.

Fernandes M. J., & Lázaro, C. (2016). GPD+ Wet Tropospheric Corrections for CryoSat-2 and GFO altimetry missions. *Remote Sensing*, 8(10), 851. doi:10.3390/rs8100851.

Fernandes, M. J., Lázaro, C., Ablain, M., & Pires, N. (2015). Improved wet path delays for all ESA and reference altimetric missions. *Remote Sensing of Environment*, 169, 50–74. doi:10.1016/j.rse.2015.07.023.

Fernandes, M. J., Lazaro, C., Nunes, A. L., Pires, N., Bastos, L., & Mendes, V. B. (2010). GNSS-derived path delay: An approach to compute the wet tropospheric correction for coastal altimetry. *IEEE Geoscience and Remote Sensing Letters*, 7(3), 596–600.

Fernandes, M. J., Lázaro, C., Nunes, A. L., & Scharroo, R. (2014). Atmospheric corrections for altimetry studies over inland water. *Remote Sensing*, 6(6), 4952–4997.

Ferry, N., Parent, L., Garric, G., Bricaud, C., Testut, C.-E., Le Galloudec, O., Lellouche, J.-M., et al. (2012). GLORYS2V1 global ocean reanalysis of the altimetric era (1992–2009) at mesoscale. *Mercator Quarterly Newsletter* 44, 29–39.

Garzoli, S. L., & Goni, G. J. (2000). Combining altimeter observations and oceanographic data for ocean circulation and climate studies. In *Satellite Oceanography and Society* (ed. D. Halpern). Elsevier Oceanography Series, Volume 63, pp. 79–97. Amsterdam: Elsevier Science.

Gaube, P., Chelton, D. B., Samelson, R. M., Schlax, M. G., & O'Neill, L. W. (2015). Satellite observations of mesoscale eddy-induced Ekman pumping. *Journal of Physical Oceanography*, 45(1), 104–132. doi:10.1175/JPO-D-14-0032.1.

Gómez-Enri, J., Cipollini, P., Passaro, M., Vignudelli, S., Tejedor, B., & Coca, J. (2016). Coastal Altimetry Products in the Strait of Gibraltar. *IEEE Transactions on Geoscience and Remote Sensing*, 54 (9), 5455–5466. doi:10.1109/TGRS.2016.2565472.

Gómez-Enri, J., Vignudelli, S., Quartly, G. D., Gommenginger, C. P., Cipollini, P., Challenor, P. G., & Benveniste, J. (2010). Modeling ENVISAT RA-2 waveforms in the coastal zone: Case study of calm water contamination. *IEEE Geoscience and Remote Sensing Letters*, 7, 474–478. doi:10.1109/LGRS.2009.2039193.

Gommenginger, C., Thibaut, P., Fenoglio-Marc, L., Quartly, G., Deng, X., Gómez-Enri, J., Challenor, P., & Gao, Y. (2011). Retracking altimeter waveforms near the coasts. In *Coastal Altimetry* (eds. Vignudelli et al.), pp. 61–101. Berlin, Heidelberg: Springer.

Guo, J., Gao, Y., Hwang, C., & Sun, J. (2010). A multi-subwaveform parametric retracker of the radar satellite altimetric waveform and recovery of gravity anomalies over coastal oceans. *Science China Earth Sciences*, 610–616. doi:10.1007/s11430-009-0171-3.

Halimi, A., Mailhes, C., Tourneret, J. -Y., Thibaut, P., & Boy, F. (2013). Parameter estimation for peaky altimetric waveforms. *IEEE Transactions on Geoscience and Remote Sensing*, 1568–1577. doi:10.1109/TGRS.2012.2205697.

Halliwell, Jr., G. R., Gopalakrishnan, S., Marks, F., & Willey, D. (2015). Idealized study of ocean impacts on tropical cyclone intensity forecasts. *Monthly Weather Review*, 143(4), 1142–1165. doi:10.1175/MWR-D-14-00022.1.

Hamlington, B. D., Leben, R. R., Nerem, R. S., Han, W., & Kim, K. Y. (2011). Reconstructing sea level using cyclostationary empirical orthogonal functions. *Journal of Geophysical Research: Oceans*, 116(C12). doi:10.1029/2011JC007529.

Han, W., Meehl, G. A., Rajagopalan, B., Fasullo, J. T., Hu, A., Lin, J., Large, W. G., et al. (2010). Patterns of Indian Ocean sea-level change in a warming climate. *Nature Geoscience*, 3(8), 546–550. doi:10.1038/ngeo901.

Harwood, P. (2015). *DUE eSurge Final Report*. Ref: D430. Available at http://due.esrin.esa.int/files/20150310050027.pdf. Accessed 24 July 2017.

He, L., Li, G., Li, K., & Shu, Y. (2014). Estimation of regional sea level change in the Pearl River Delta from tide gauge and satellite altimetry data. *Estuarine, Coastal and Shelf Science*, 141, 69–77. doi:10.1016/j.ecss.2014.02.005.

Herbert, G., Ayoub, N., Marsaleix, P., & Lyard, F. (2011). Signature of the coastal circulation variability in altimetric data in the southern Bay of Biscay during winter and fall 2004. *Journal of Marine Systems*, 88(2), 139–158. doi:10.1016/j.jmarsys.2011.03.004.

Hithin, N. K., Remya, P. G., Nair, T. B., Harikumar, R., Kumar, R., & Nayak, S. (2015). Validation and intercomparison of SARAL/AltiKa and PISTACH-derived coastal wave heights using in-situ measurements. *IEEE Journal of Selected Topics in Applied Earth Observations and Remote Sensing*, 8(8), 4120–4129. doi:10.1109/JSTARS.2015.2418251.

Hopfield, H. S. (1969). Two-quartic tropospheric refractivity profile for correcting satellite data. *Journal of Geophysical Research*, 74(18), 4487–4499. doi:10.1029/JC074i018p04487.

Hwang, C., Guo, J., Deng, X., Hsu, H. -Y., & Liu, Y. (2006). Coastal gravity anomalies from retracked GEOSAT/GM altimetry: Improvement, limitation and the role of airborne gravity data. *Journal of Geodesy*, 80(4), 204–216. doi:10.1007/s00190-006-0052-x.

Idris, N., & Deng, X. (2012). The retracking technique on multi-peak and quasi-specular waveforms for Jason-1 and Jason-2 missions near the coast. *Marine Geodesy*, 217–237. doi:10.1080/01490419.2012.718679.

Ikeda, M., & Dobson, F.W. (1995). *Oceanographic Applications of Remote Sensing*. New York: CRC Press.

Janssen, P. (2000). ECMWF wave modeling and satellite altimeter wave data. In *Satellite Oceanography and Society* (ed. D. Halpern). Elsevier Oceanography Series, Volume 63, pp. 35–56. Amsterdam: Elsevier Science.

Jebri, F., Birol, F., Zakardjian, B., Bouffard, J., & Sammari, C. (2016). Exploiting coastal altimetry to improve the surface circulation scheme over the central Mediterranean Sea. *Journal of Geophysical Research: Oceans*, 121(7), 4888–4909. doi:10.1002/2016JC011961.

Karbou, F., Prigent, C., Eymard, L., & Pardo, J.R. (2005). Microwave land emissivity calculations using AMSU measurements. *IEEE Transactions on Geoscience and Remote Sensing*, 43(5), 948–959. doi:10.1109/TGRS.2004.837503.

Klemas, V. (2012). Remote Sensing of Coastal and Ocean Currents: An Overview. *Journal of Coastal Research*, 28(3), 576–586. https://doi.org/10.2112/JCOASTRES-D-11-00197.1

Kourafalou, V., De Mey, P., & Wilmer-Becker, K. (2015). Report of the 4th GODAE OceanView Coastal Ocean and Shelf Seas Task Team (COSS-TT) International Coordination Workshop (COSS-ICW4), Lisbon, Portugal, Sept. 2015. Available at https://www.godae-oceanview.org/files/download.php?m=documents&f=160204121952-ICWS4Lisbonreportfinal.pdf. Accessed 24 July 2017.

Krug, M., & Tournadre, J. (2012). Satellite observations of an annual cycle in the Agulhas Current. *Geophysical Research Letters*, 39(15), L15607. doi:10.1029/2012GL052335.

Krug, M., Tournadre, J., & Dufois, F. (2014). Interactions between the Agulhas Current and the eastern margin of the Agulhas Bank. *Continental Shelf Research*, 81, 67–79. doi:10.1016/j.csr.2014.02.020.

Kuo, N. J., Zheng, Q., & Ho, C. R. (2004). Response of Vietnam coastal upwelling to the 1997–1998 ENSO event observed by multisensor data. *Remote Sensing of Environment*, 89(1), 106–115. doi:10.1016/j.rse.2003.10.009.

Le Bars, Y., Lyard, F., Jeandel, C., & Dardengo, L. (2010). The AMANDES tidal model for the Amazon estuary and shelf. *Ocean Modelling*, 31(3), 132–149. doi:10.1016/j.ocemod.2009.11.001.

Lee, H., Shum, C. K., Emery, W., Calmant, S., Deng, X., Kuo, C.-Y., Roesler, C., & Yi, Y. (2010). Validation of Jason–2 altimeter data by waveform retracking over California coastal ocean. *Marine Geodesy*, 33, 304–316. doi:10.1080/01490419.2010.488982.

Legeais, J. F., Ablain, M., & Thao, S. (2014). Evaluation of wet troposphere path delays from atmospheric reanalyzes and radiometers and their impact on the altimeter sea level. *Ocean Science*, 10(6), 893–905. doi:10.5194/os-10-893-2014.

Le Hénaff, M., De Mey, P., & Marsaleix, P. (2009). Assessment of observational networks with the Representer Matrix Spectra method—Application to a 3D coastal model of the Bay of Biscay. *Ocean Dynamics*. 59(1), 3–20. doi:10.1007/s10236-008-0144-7.

Le Provost, C., Genco, M. L., Lyard, F., Vincent, P., & Canceil, P. (1994). Spectroscopy of the world ocean tides from a finite element hydrodynamic model. *Journal of Geophysical Research: Oceans*, 99(C12), 24777–24797. doi:10.1029/94JC01381.

Liu, Y., Weisberg, R. H., Vignudelli, S., & Mitchum, G. T. (2014). Evaluation of altimetry-derived surface current products using Lagrangian drifter trajectories in the eastern Gulf of Mexico. *Journal of Geophysical Research: Oceans*, 119(5), 2827–2842. doi:10.1002/2013JC009710.

Liu, Y., Weisberg, R. H., Vignudelli, S., Roblou, L., & Merz, C. R. (2012). Comparison of the X-TRACK altimetry estimated currents with moored ADCP and HF radar observations on the West Florida Shelf. *Advances in Space Research*, 50(8), 1085–1098. doi:10.1016/j.asr.2011.09.012.

Luu, Q. H., Tkalich, P., & Tay, T. W. (2015). Sea level trend and variability around Peninsular Malaysia. *Ocean Science*, 11(4), 617–628. doi:10.5194/osd-11-1519-2014.

Madsen, K. S., Høyer, J. L., Fu, W., & Donlon, C. (2015). Blending of satellite and tide gauge sea level observations and its assimilation in a storm surge model of the North Sea and Baltic Sea. *Journal of Geophysical Research: Oceans*, 120(9), 6405–6418. doi:10.1002/2015JC011070.

Madsen, K. S., Høyer, J. L., & Tscherning, C. C. (2007). Near-coastal satellite altimetry: Sea surface height variability in the North Sea–Baltic Sea area. *Geophysical Research Letters*, 34(14), L14601. doi:10.1029/2007GL029965.

Maraldi, C., Galton-Fenzi, B., Lyard, F., Testut, L., & Coleman, R. (2007). Barotropic tides of the southern Indian Ocean and the Amery Ice Shelf cavity. *Geophysical Research Letters*, 34(18), L18602. doi:10.1029/2007GL030900.

Marcos, M., Wöppelmann, G., Bosch, W., & Savcenko, R. (2007). Decadal sea level trends in the Bay of Biscay from tide gauges, GPS and TOPEX. *Journal of Marine Systems*, 68(3), 529–536. doi:10.1016/j.jmarsys.2007.02.006.

Melet, A., Gourdeau, L., & Verron, J. (2010). Variability in Solomon Sea circulation derived from altimeter sea level data. *Ocean Dynamics*, 60(4), 883–900. doi:10.1007/s10236-010-0302-6.

Mercier, F., Picot, N., Thibaut, P., Cazenave, A., Seyler, F., Kosuth, P., & Bronner, E. (2009). CNES/PISTACH project: An innovative approach to get better measurements over in-land water bodies from satellite altimetry. Early results. *EGU General Assembly Conference Abstracts*, 11, 11674.

Mercier, F., Picot, N., Guinle, T., Cazenave, A., Kosuth, P., & Seyler, F. (2012). The PISTACH project for coastal and hydrology altimetry: 2012 project status and activities. Presentation at 20 Years of Progress in Radar Altimetry Symposium, Venice, Italy.

Mercier, F., Rosmorduc, V., Carrere, L., & Thibaut, P. (2010). *Coastal and Hydrology Altimetry Product (PISTACH) Handbook*. CLS-DOS-NT-10-246, Issue 1.0. Toulouse: CNES.

Meyssignac, B., Becker, M., Llovel, W., & Cazenave, A. (2012). An assessment of two-dimensional past sea level reconstructions over 1950–2009 based on tide-gauge data and different input sea level grids. *Surveys in Geophysics*, 33(5), 945–972.

Miller, M., Buizza, R., Haseler, J., Hortal, M., Janssen, P., & Untch, A. (2010). Increased resolution in the ECMWF deterministic and ensemble prediction systems. *ECMWF Newsletter*, 124, 10–16.

Mitchum, G., and B. Kilonsky. (1995) Observations of tropical sea level variability from altimeters. In *Oceanographic Applications of Remote Sensing* (eds. M. Ikeda, F. W. Dobson), pp. 113–127. New York: CRC Press.

Moore, A. M., Arango, H. G., Broquet, G., Edward, C., Veneziani, M., Powell, B., Foley, D., Doyle, J., Costa, D., & Robinson, P. (2011). The Regional Ocean Modeling System (ROMS) 4-dimensional variational data assimilation systems. Part III: Observation impact and observation sensitivity in the California Current System. *Progress in Oceanography*, 91, 74–94. doi:10.1016/j.pocean.2011.05.005.

Nerem, R. S., Chambers, D. P., Hamlington, B. D., Leben, R. R., Mitchum, G. T., Phillips, T., & Willis, J. K. (2010). Observations of recent sea level change. Paper presented at the IPCC Workshop on Sea Level Rise and Ice Sheet Instabilities. Intergovernmental Panel on Climate Change, Malaysia. June 2010.

Obligis, E., Desportes, C., Eymard, L., Fernandes, M. J., Lázaro, C., & Nunes, A. L. (2011). Tropospheric corrections for coastal altimetry. In *Coastal altimetry* (eds. S. Vignudelli, A. Kostianoy, P. Cipollini, J. Benveniste), 147–176, Springer Berlin Heidelberg, doi:10.1007/978-3-642-12796-0_6.

Obligis, E., Eymard, L., Tran, N., Labroue, S., & Femenias, P. (2006). First three years of the microwave radiometer aboard Envisat: In-flight calibration, processing, and validation of the geophysical products. *Journal of Atmospheric and Oceanic Technology*, 23(6), 802–814. doi:10.1175/JTECH1878.1.

Oke, P. R., Sakov, P., & Schulz, E. (2009). A comparison of shelf observation platforms for assimilation in an eddy-resolving ocean model. *Dynamics of Atmospheres and Oceans*, 48(1), 121–142. doi:10.1016/j.dynatmoce.2009.04.002.

Oliveira, P. B., Serra, N., Fiúza, A., & Ambar, I. (2000). A Study of Mediterranean salt lenses using in situ and satellite observations. In *Satellite Oceanography and Society* (ed. D. Halpern), Elsevier Oceanography Series, Volume 63, pp. 125–148. Amsterdam: Elsevier Science.

Paduan, J. D., & Shulman, I. (2004). HF radar data assimilation in the Monterey Bay area. *Journal of Geophysical Research: Oceans*, 109(C7), C07S09. doi:10.1029/2003JC001949.

Pairaud, I. L., Lyard, F., Auclair, F., Letellier, T., & Marsaleix, P. (2008). Dynamics of the semi-diurnal and quarter-diurnal internal tides in the Bay of Biscay. Part 1: Barotropic tides. *Continental Shelf Research*, 28(10), 1294–1315. doi:10.1016/j.csr.2008.03.004.

Papa, F., Bala, S. K., Pandey, R. K., Durand, F., Gopalakrishna, V. V., Rahman, A., & Rossow, W. B. (2012). Ganga–Brahmaputra river discharge from Jason–2 radar altimetry: An update to the long–term satellite–derived estimates of continental freshwater forcing flux into the Bay of Bengal. *Journal of Geophysical Research: Oceans*, 117(C11). doi:10.1029/2012JC008158.

Papa, F., Durand, F., Rossow, W. B., Rahman, A., & Bala, S. K. (2010). Satellite altimeter-derived monthly discharge of the Ganga-Brahmaputra River and its seasonal to interannual variations from 1993 to 2008. *Journal of Geophysical Research: Oceans*, 115(C12). doi:10.1029/2009JC006075.

Passaro, M., Cipollini, P., Vignudelli, S., Quartly, G. D., & Snaith, H. M. (2014). ALES: A multi-mission adaptive subwaveform retracker for coastal and open ocean altimetry. *Remote Sensing of Environment*, 145, 173–189. doi:10.1016/j.rse.2014.02.008.

Passaro, M., Cipollini, P., & Benveniste, J. (2015a). Annual sea level variability of the coastal ocean: The Baltic Sea-North Sea transition zone. *Journal of Geophysical Research: Oceans*, 120(4), 3061–3078. doi:10.1002/2014JC010510.

Passaro, M., Dinardo, S., Quartly, G. D., Snaith, H. M., Benveniste, J., Cipollini, P., & Lucas, B. (2016). Cross-calibrating ALES Envisat and CryoSat-2 Delay–Doppler: A coastal altimetry study in the Indonesian Seas. *Advances in Space Research*, 58(3), 289–303. doi:10.1016/j.asr.2016.04.011.

Passaro, M., Fenoglio-Marc, L., & Cipollini, P. (2015b). Validation of significant wave height from improved satellite altimetry in the German Bight. *IEEE Transactions on Geoscience and Remote Sensing*, 53(4), 2146–2156.

Pedinotti, V., Boone, A., Ricci, S., Biancamaria, S., & Mognard, N. (2014). Assimilation of satellite data to optimize large-scale hydrological model parameters: A case study for the SWOT mission. *Hydrology and Earth System Sciences*, 18(11), 4485–4507. doi:10.5194/hess-18-4485-2014.

Pegliasco, C., Chaigneau, A., & Morrow, R. (2015). Main eddy vertical structures observed in the four major Eastern Boundary Upwelling Systems. *Journal of Geophysical Research: Oceans*, 120(9), 6008–6033. doi:10.1002/2015JC010950.

Picard, B., Frery, M. L., Obligis, E., Eymard, L., Steunou, N., & Picot, N. (2015). SARAL/AltiKa wet tropospheric correction: In-flight calibration, retrieval strategies and performances. *Marine Geodesy*, 38(suppl 1), 277–296. doi:10.1080/01490419.2015.1040903.

Quartly, G. D. (2010). Hyperbolic retracker: Removing bright target artifacts from altimetric waveform data. ESA SP-686, Living Planet Symposium 2010, Bergen, Norway, (28 June – 2 July, 2007). ESA Publication, SP-686. Noordwijkerhout, NL: ESA.

Reynolds, R. W, Behringer, D., Ji, M., Leetmaa, A., Maes, C., Vossepoel, F., & Xue Y. (2000). Analyzing the 1993–1998 interannual variability of NCEP model ocean simulations: The contribution of TOPEX/Poseidon observations. In *Satellite Oceanography and Society* (ed. D. Halpern). Elsevier Oceanography Series, Volume 63, pp. 299–308. Amsterdam: Elsevier Science.

Rio, M. H., Mulet, S., & Picot, N. (2014a). Beyond GOCE for the ocean circulation estimate: Synergetic use of altimetry, gravimetry, and in situ data provides new insight into geostrophic and Ekman currents. *Geophysical Research Letters*, 41(24), 8918–8925. doi:10.1002/2014GL061773.

Rio, M. H., Pascual, A., Poulain, P. M., Menna, M., Barceló-Llull, B., & Tintoré, J. (2014b). Computation of a new mean dynamic topography for the Mediterranean Sea from model outputs, altimeter measurements and oceanographic in-situ data. *Ocean Science*, 10, 731–744. doi:10.5194/os-10-731-201ence.

Rio, M.-H., Poulain, P. M., & Pascual, A., Mauri, E., Larnicol, G., & Santoleri, R. (2007). A mean dynamic topography of the Mediterranean Sea computed from altimetric data, in-situ measurements and a general circulation model. *Journal of Marine Systems*, 65, 484–508. doi:10.1016/j.jmarsys.2005.02.006.

Risien, C. M., & Strub, P. T. (2016). Blended sea level anomaly fields with enhanced coastal coverage along the US West Coast. *Scientific Data*, 3, 160013. doi:10.1038/sdata.2016.13.

Roblou, L., Lamouroux, J., Bouffard, J., Lyard, F., Le Hénaff, M., Lombard, A., Marsaleix, P., De Mey P., & Birol, F. (2011). Post-processing altimeter data towards coastal applications and integration into coastal models. In *Coastal Altimetry* (eds. S. Vignudelli, A. Kostianoy, P. Cipollini, J. Benveniste), pp. 217–246. Berlin, Heidelberg: Springer.

Ruf, C. S., & Giampaolo, J. C. (1998). Littoral antenna deconvolution for a microwave radiometer. *Proceedings of 1998 International Geoscience and Remote Sensing Symposium*, Seattle, WA, Cat. #98CH36174, 378–380. Doi:10.1109/IGARSS.1998.702911.

Ruiz Etcheverry, L. A., Saraceno, M., Piola, A. R., & Strub, P. T. (2016). Sea level anomaly on the Patagonian continental shelf: Trends, annual patterns and geostrophic flows. *Journal of Geophysical Research: Oceans*, 121(4), 2733–2754. doi:10.1002/2015JC011265.

Saraceno, M., Strub, P. T., & Kosro, P. M. (2008). Estimates of sea surface height and near-surface alongshore coastal currents from combinations of altimeters and tide gauges. *Journal of Geophysical Research: Oceans*, 113(C11). doi:10.1029/2008JC004756.

Scharroo, R., Lillibridge, J. L., Smith, W. H. F., & Schrama, E. J. O. (2004). Cross-calibration and long-term monitoring of the microwave radiometers of ERS, TOPEX, GFO, Jason, and Envisat. *Marine Geodesy*, 27(1–2), 279–297. doi:10.1080/01490410490465265.

Scharroo, R., Smith, W. H., & Lillibridge, J. L. (2005). Satellite altimetry and the intensification of Hurricane Katrina. *Eos, Transactions American Geophysical Union*, 86(40), 366–366. doi:10.1029/2005EO400004.

Stammer, D., Cazenave, A., Ponte, R. M., & Tamisiea, M. E. (2013). Causes for contemporary regional sea level changes. *Annual Review of Marine Science*, 5, 21–46.

Stammer, D., Ray, R. D., Andersen, O. B., Arbic, B. K., Bosch, W., Carrere, L., Cheng, Y., et al. (2014). Accuracy assessment of global barotropic ocean tide models. *Reviews of Geophysics*, 52, 243–282. doi:10.1002/2014RG000450.

Steunou, N., Desjonquères, J. D., Picot, N., Sengenes, P., Noubel, J., & Poisson, J. C. (2015). AltiKa altimeter: Instrument description and in flight performance. *Marine Geodesy*, 38(suppl 1), 22–42. doi:10.1080/01490419.2014.988835.

Strub, P. T., Combes, V., Shillington, F. A., & Pizarro, O. (2013). Currents and processes along the eastern boundaries. In *Ocean Circulation and Climate: A 21st Century Perspective* (eds. G. Siedler, S. M. Griffies, J. Gould, J. A. Church), pp. 339–384. Oxford: Academic Press.

Strub, P. T., James, C., Combes, V., Matano, R. P., Piola, A. R., Palma, E. D., Saraceno, M., Guerrero, R., Fenco, H., & Ruiz-Etcheverry, L. A. (2015). Altimeter-derived seasonal circulation on the southwest Atlantic shelf: 27°–43° S. *Journal of Geophysical Research: Oceans*, 120(5), 3391–3418. doi:10.1002/2015JC010769.

Thibaut, P., Aublanc, J., Moreau, T., Boy, F., & Picot, N. (2014). Delay/Doppler Waveform Processing in Coastal Zones. 8th Coastal Altimetry Workshop, Konstanz (Germany), 23–24 October, 2014.

Tournadre, J., Chapron, B., Reul, N., & Vandemark, D. C. (2006). A satellite altimeter model for ocean slick detection. *Journal of Geophysical Research*, 111, C04004. doi:10.1029/2005JC003109.

Tournadre, J., Whitmer, K., & Girard-Ardhuin, F. (2008). Iceberg detection in open water by altimeter waveform analysis. *Journal of Geophysical Research*, 113, C08040. doi:10.1029/2007JC004587.

Troitskaya, Y. I., & Ermakov, S. A. (2008). Manifestations of the Indian Ocean tsunami of 2004 in satellite nadir–viewing radar backscatter variations. *Geophysical Research Letters*, 33(4). L04607. doi:10.1029/2005GL024445.

Valladeau, G., Thibaut, P., Picard, B., Poisson, J. C., Tran, N., Picot, N., & Guillot, A. (2015). Using SARAL/AltiKa to improve Ka-band altimeter measurements for coastal zones, hydrology and ice: The PEACHI prototype. *Marine Geodesy*, 38(suppl 1), 124–142. doi:10.1080/01490419.2015.1020176.

Valladeau, G., Thibaut, P., Picot, N., Guillot, A., Boy, F., Le Gac, S., & the PEACHI team. (2016). *PEACHI Project: Improving Ka-Band Altimeter Measurements*. Prague: ESA Living Planet.

Vignudelli, S., Cipollini, P., Roblou, L., Lyard, F., Gasparini, G. P., Manzella, G., & Astraldi, M. (2005). Improved satellite altimetry in coastal systems: Case study of the Corsica Channel (Mediterranean Sea). *Geophysical Research Letters*, 32(7). L07608. doi:10.1029/2005GL022602.

Vignudelli, S., Kostianoy, A. G., Cipollini, P., & Benveniste J. (Eds). (2011). *Coastal Altimetry*. Berlin Heidelberg: Springer-Verlag, 578 p., doi:10.1007/978-3-642-12796-0.

Wentz, F. J. (2013). *SSM/I Version-7 Calibration Report*. Remote Sensing Systems Technical Report, 11012, 46. Santa Rosa, Calif.: Remote Sensing Systems.

Wijesekera, H. W., Jensen, T. G., Jarosz, E., Teague, W. J., Metzger, E. J., Wang, D. W., Jinadasa, S. U. P., Arulananthan, K., Centurioni, L. R., & Fernando, H. J. S. (2015). Southern Bay of Bengal currents and salinity intrusions during the northeast monsoon. *Journal of Geophysical Research: Oceans*, 120(10), 6897–6913. doi:10.1002/2015JC010744.

Wilczak, J. M., Leben, R. R., & McCollum, D. S. (2007). Upper-ocean thermal structure and heat content off the US West Coast during the 1997–1998 El Niño event based on AXBT and satellite altimetry data. *Progress in Oceanography*, 74(1), 48–70. doi:10.1016/j.pocean.2007.02.006.

Wilkin, J., & Hunter, E. (2013). An assessment of the skill of real-time models of middle Atlantic Bight continental shelf circulation. *Journal of Geophysical Research*, 118, 2919–2933. doi:10.1002/jgrc.20223223.

Wöppelmann, G., & Marcos, M. (2016). Vertical land motion as a key to understanding sea level change and variability. *Reviews of Geophysics*, 54(1), 64–92.

Yang, L., Lin, M., Liu, Q., & Pan, D. (2012). A coastal altimetry retracking strategy based on waveform classification and subwaveform extraction. *International Journal of Remote Sensing*, 33, 7806–7819.

Yang, Y., Hwang, C., Hsu, H., Dongchen, E., & Wang, H. (2011). A subwaveform threshold retracker for ERS-1 altimetry: A case study in the Antarctic Ocean. *Computers and Geosciences*, 41, 88–98.

Yildiz, H., Andersen, O. B., Simav, M., Aktug, B., & Ozdemir, S. (2013). Estimates of vertical land motion along the southwestern coasts of Turkey from coastal altimetry and tide gauge data. *Advances in Space Research*, 51(8), 1572–1580. doi:10.1016/j.asr.2012.11.011.

Yoon, Y., Garambois, P. A., Paiva, R. C., Durand, M., Roux, H., & Beighley, E. (2016). Improved error estimates of a discharge algorithm for remotely sensed river measurements: Test cases on Sacramento and Garonne Rivers. *Water Resources Research*, 52(1), 278–294. doi:10.1002/2015WR017319.

12 Monitoring Waves and Surface Winds by Satellite Altimetry

Applications

Saleh Abdalla and Peter A. E. M. Janssen

12.1 INTRODUCTION

Sea state forecasting started more than 60 years ago when there was an evident need for knowing the wave state for landing operations during World War II. The past seven decades have seen a significant development in ocean wave forecasting from simple manual techniques to sophisticated numerical wave models based on physical principles. For a concise review of these developments and the resulting progress in wave forecasting, see Janssen (2008). In particular, in the 1980s, development was rapid because of the need to improve wave modeling in rapidly varying circumstances and because of the availability of powerful computers needed to solve the energy balance equation. Furthermore, there was the prospect of the availability of wave data from satellites such as GEOSAT, European Remote Sensing (ERS) 1, TOPEX-Poseidon, and ERS-2. In contrast to the conventional observing systems that could only provide local information on the sea state, for the first time, altimeters would be able to provide wave height information on a global scale. As a consequence, a group of mainly European wave modelers started to develop a surface wave model from the first principles (i.e., a model that solves the energy balance equation for surface gravity waves). The source functions in the energy balance included an explicit representation of the wind input, nonlinear interactions, and dissipation by white capping (see Janssen 2004).

Early investigations into the quality of the WAve Model (WAM) results were based on a comparison with Seasat altimeter wave height data (Janssen et al. 1989; Bauer et al. 1992) and with GEOSAT data (Romeiser 1993). Generally, modeled wave heights, obtained by forcing the WAM model with European Center for Medium-Range Weather Forecasts (ECMWF) winds showed good agreement with observed wave heights, but there were also considerable differences. Romeiser (1993) found considerable regional and seasonal differences between modeled wave height and the GEOSAT data. During the Southern Hemisphere winter, WAM underestimated wave height by about 20% in large parts of the Southern Ocean and the tropical oceans. These discrepancies could be ascribed to shortcomings in the wave model physics and in the driving ECMWF wind fields, which at the end of the 1980s were too low in the Southern Hemisphere because the atmospheric model had a fairly low resolution (T_L106, corresponding to a spatial resolution of 190 km).

The shortcomings in the wave model physics were mainly ascribed to too much dissipation of swell and a not strong enough wind input source function. In November 1991, the next version of WAM, named WAM Cycle 4 (Janssen 1991, 2004; Komen et al. 1994), became part of the ECMWF wave prediction system. In addition, in September 1991, the horizontal and vertical resolutions of the ECMWF atmospheric general circulation model were doubled to produce a better representation of surface winds, in particular for the Southern Ocean.

Therefore, in late 1991, there was sufficient confidence in the quality of the ECMWF wind-wave forecasting system that it could be used for the validation of the first European Remote Sensing satellite (ERS-1) altimeter wind and wave products. The European Space Agency (ESA) launched ERS-1 in July 1991 and started dissemination of observed products in near real time

(NRT) soon thereafter. Comparison of ERS altimeter wind and wave products with corresponding ECMWF fields identified problems in the ERS wind speed and wave height retrieval algorithms (Hansen and Günther 1992; Janssen et al. 1997b), which, because of the strong interaction between the wave model community and ESA, resulted in correction of the operational retrieval algorithms.

Because of the promising validation results, altimeter wave height data have been assimilated in the United Kingdom Met Office (UKMO) and ECMWF global wave forecasting systems since mid-1993. The analysis scheme used by ECMWF is based on the optimum interpolation (OI) method, details of which are more fully described by Janssen et al. (1989) and Lionello et al. (1992). In general, the assimilation has led to an improved wave analysis as follows from comparisons with independent observations from buoys. In addition, Bauer and Staabs (1998) compared ECMWF wave analyses with TOPEX/Poseidon altimeter wave height data and they found an increase of correlation between the two after the ERS-1 wave data assimilation was switched on.

ERS-1 was the first satellite that provided NRT information on sea state parameters including the significant wave height (SWH) and wind speed from the altimeter, the wind vector from the scatterometer, and the two-dimensional wave spectra from the Synthetic Aperture Radar (SAR). The NRT availability of these sea state parameters is an important factor because, as will be clear later, these data provided a significant stimulus to the improvement of operational ocean wave forecasting, while, on the other hand, the need to have reliable wave predictions stimulated research in the development of altimeter wind and wave products. To demonstrate this point, some illustrative examples are discussed in Section 12.2, which also provides a brief description of the altimeter and its measurements.

Section 12.3 describes altimeter wind speed observations, their quality, and their importance for monitoring the quality of modeled surface wind. Nevertheless, the physics behind the retrieval algorithm is not fully understood yet. The wind speed observations follow from an empirical relation between the radar backscatter and wind speed. A well-known example is the "classical" Chelton-Wentz retrieval algorithm (MCW) of Witter and Chelton (1991), which is based on a few hundred collocations between wind speed observations from buoys and radar backscatter from GEOSAT. An improved version of this, based on much more collocations, was later developed by Abdalla (2007, 2012). However, the radar backscatter is inversely proportional to the mean square slope (MSS), and it is suggested hereafter that the MSS is a physically interesting quantity to observe as it is closely related to air-sea interaction. Alternatively, a number of researchers have made attempts at developing a new retrieval algorithm that includes a measure of the sea state, for example, the SWH (Lefèvre et al. 1994; Gourrion et al. 2002). A comparison of results from the Gourrion algorithm and the Abdalla algorithm will be presented. Finally, the wind speed retrieval algorithms are normally only applied to relatively modest wind speeds up to 20–25 ms^{-1}, but it is known that the radar backscatter even provides useful information in hurricane conditions (Quilfen et al. 2006).

Section 12.4 is dedicated to the altimeter SWH, which is four times the square root of the total wave energy. It is almost equivalent to the visual wave height that used to be reported by sailors. The quality and the importance of SWH, specifically for ocean wave data assimilation and model diagnostics, are also discussed. The assimilation of altimeter wave height data is of importance for the quality of the wave analysis and wave forecast, as judged by comparisons with *in situ* observations. In addition, it turns out that these data have been of great help in diagnosing problems in the formulation of the physics source functions of the wave model. The wave analysis and forecast are strongly constrained by wind forcing and, to a lesser extent, by errors in the wave model. Because over the past 15 years the quality of surface winds has improved quite considerably (see Janssen 2004) and also the wave model has improved (Bidlot et al. 2007), the impact of the assimilation of altimeter wave height data is now less pronounced than in the early days. Furthermore, we will show that altimeter wave height data are very useful in obtaining a global wave height climatology. In a first approach, one simply averages altimeter wave heights over large enough boxes over the oceans over a large enough period in time. A more promising approach is perhaps from Caires et al. (2005) who took ECMWF reanalysis data for wave height and corrected the wave height climatology by means of buoy and TOPEX altimeter wave height data.

In Section 12.5, we will summarize the applications of altimeter wind and wave observations. The impact of SWH data assimilation on the quality of the model analysis and forecasts will be highlighted using observations from various altimeters. The use of altimeter observations in assessing the model performance will be described. The triple collocation technique can be used for the estimation of random errors in the model as well as the observations. Furthermore, a simple technique to estimate the model effective resolution is briefly described. The section concludes with presenting some of the efforts toward the evaluation of the sea state climatology.

Section 12.6 presents a few developments in satellite sea state measurements. SAR or a delay-Doppler altimetry revolutionized the altimetry field. The SAR altimeter applies a coherent processing of groups of echoes resembling the acquisition from a synthetic aperture antenna—hence, the commonly used name SAR altimetry. Each along-track resolvable cell is viewed from more than one angle leading to much narrower footprints in the along-track direction (250–300 m compared to several kilometers in the case of conventional altimetry). This opened a gate for higher resolution measurements and further measurement capabilities near the coasts and ice edges. A new method of observing the spectral properties of surface gravity waves, based on the principles of Real-Aperture Radar (RAR) is another interesting development. Compared to the SAR (actual SAR not the SAR altimetry), the advantage of RAR is that the spectra are not distorted by velocity bunching effects. Nevertheless, due to speckle noise, there is still a limit of detection of the short waves, expected to be around 70 m, which is smaller by a factor of three than the azimuthal cut-off wavelength of a typical SAR.

Finally, we conclude with a summary of conclusions in Section 12.7.

12.2 THE ALTIMETER AND ITS OCEAN MEASUREMENTS

The radar altimeter is a nadir looking active microwave instrument. This instrument emits pulses, and by measuring the travel time of the return pulse, after extensive corrections for atmospheric delays, and accurate determination of the satellite orbit, for example, information on the mean sea level may be obtained. A good approximation of the backscattered return, which may be described by specular reflection, is inversely proportional to the MSS of the sea surface. Because, according to Cox and Munk (1954), mean square slope is closely related to the surface wind speed, the radar backscatter is a good measure for wind speed. Finally, the radar altimeter also provides a measure of the SWH through the distortion of the mean shape of the return pulse. The earlier return from the wave crests and the retarded return from the wave troughs lead to a deformation of the return pulse that can directly be related to the SWH. To determine the mean pulse shape, on the order of one thousand pulses are averaged, yielding one SWH measurement about every 6–7 km along the satellite track. For a Gaussian sea surface, the relation between pulse shape and the root mean square (RMS) of the sea surface displacement can be determined theoretically (although there are small corrections needed that are caused by deviations from normality; compare Janssen 2000; Gómez-Enri et al. 2007). This model has been confirmed by numerous comparisons with *in situ* measurements. Figure 12.1 shows typical waveform from averaging of about 100 individual returned signals from the ocean surface within one second. The shape of the waveform is used to infer various parameters. The typical accuracy of the SWH of the older generation of radar altimeters is thought to be the maximum of 0.5 m or 10% of wave height in the range of 1 to 20 m, while the wind speed accuracy is between 1.5 and 2 ms^{-1} (for wind speed in the range of 0 to 20 ms^{-1}). As will be discussed later based on a triple collocation technique, nowadays the accuracy of SWH is thought to be 6% of SWH, while the error in wind speed is around 1.0 ms^{-1}.

The radar altimeter is an important component of the payload of a number of satellites such as the Seasat, GEOSAT, ERS-1, ERS-2, TOPEX/Poseidon, GEOSAT Follow-On (GFO), Envisat, Jason-1, Jason-2, Jason-3, CryoSat-2, SARAL-AltiKa (or just SARAL), and Sentinel-3 series. For operational models, the data need to be available in NRT (i.e., within 3 h). The ERS-1/2, Envisat, Jason-1, Jason-2, Jason-3, CryoSat-2, SARAL, and Sentinel-3 series provide these fast delivery products. In general, two polar orbiting satellites give good coverage in about 6 h. During most of 2016 and

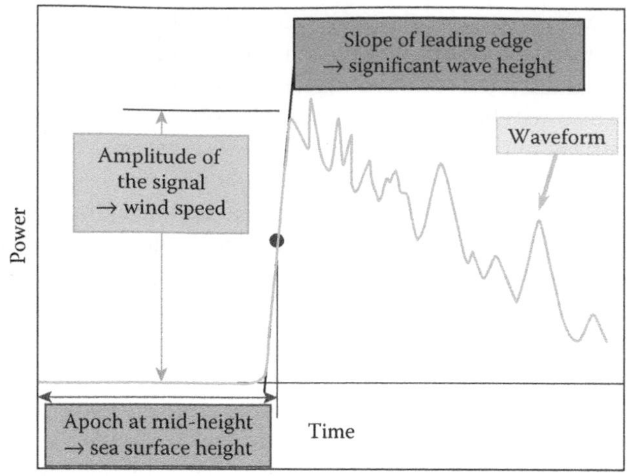

FIGURE 12.1 Information extracted from a radar echo reflected from the ocean surface (after averaging approximately 100 individual waveforms within approximately 1/20th of a second).

early 2017, there were six active altimetry missions; namely: Jason-2, Jason-3, SARAL, CryoSat-2, Sentinel-3A, and Hai Ying [HY] 2A; orbiting the globe, a record number of altimeters. All, except HY-2A, are providing altimeter data in NRT.

Most of the modern radar altimetry missions carry dual-frequency altimeters to estimate the impact of the atmosphere on the radar signal (ionospheric impact). The primary (or the only in the case of mono-frequency altimeters) electromagnetic wave frequency is usually the Ku band, which has a wavelength of about 2.5 cm (e.g., ERS-1, ERS-2, Envisat, Jason-1, Jason-2, Jason-3, and Sentinel-3). The exception is SARAL, which implements the Ka band with a wavelength of about 0.8 cm. The secondary frequency is mainly the C band with a wavelength of about 5.5 cm (e.g., TOPEX, Jason-1, Jason-2, Jason-3, and Sentinel-3). Envisat used the S band with a wavelength of about 9 cm as the secondary frequency.

Radar altimeters are invaluable instruments that are able to measure SWH and surface wind speed in addition to sea surface height and ice, land, and lake characteristics. Most altimetry missions accommodate a microwave radiometer instrument to measure the atmospheric humidity for the purpose of the determination of the impact of atmospheric humidity on the altimeter measurements. The water vapor causes a delay of the radar signal, which can cause errors in the order of tens of centimeters of sea surface height. Furthermore, it causes an attenuation that is reflected in wind speed inaccuracies. For longer electromagnetic wavelengths (e.g., Ku and C bands), this attenuation is negligible. However, for shorter wavelengths such as the Ka band, the impact is rather high. Additional products such as total column water vapor are also available from those microwave radiometers. Typical daily coverage of various altimeters is shown in Figure 12.2.

Before we start the discussion on the value of altimeter wind speed and wave height data for numerical wave prediction, it should be noted that, on the other hand, wind and wave model products have been quite useful in the validation and calibration of ESA's altimeter wind and wave product. The necessary calibration and validation of a satellite sensor requires large amounts of ground truth that should cover the full range of possible events. In particular, the number of reliable wave *in situ* measurements is very limited and, because of financial restrictions, dedicated field campaigns are possible only at a few sites. In contrast to that, model data are relatively cheap and provide global data sets for comparison. Thus, the combination of both *in situ* observations and model data seems to be an optimal calibration and validation (cal/val) data set. During the ERS-1, ERS-2, and Envisat cal/val campaigns, the altimeter-model comparisons have been very effective in identifying errors and problems in the altimeter processing and retrieval algorithms. A few examples are given now.

Just after the launch of ERS-1, the global mean altimeter wave height was about 1 m higher than computed by the model. The investigation of the detected bias led to the discovery of a small offset in

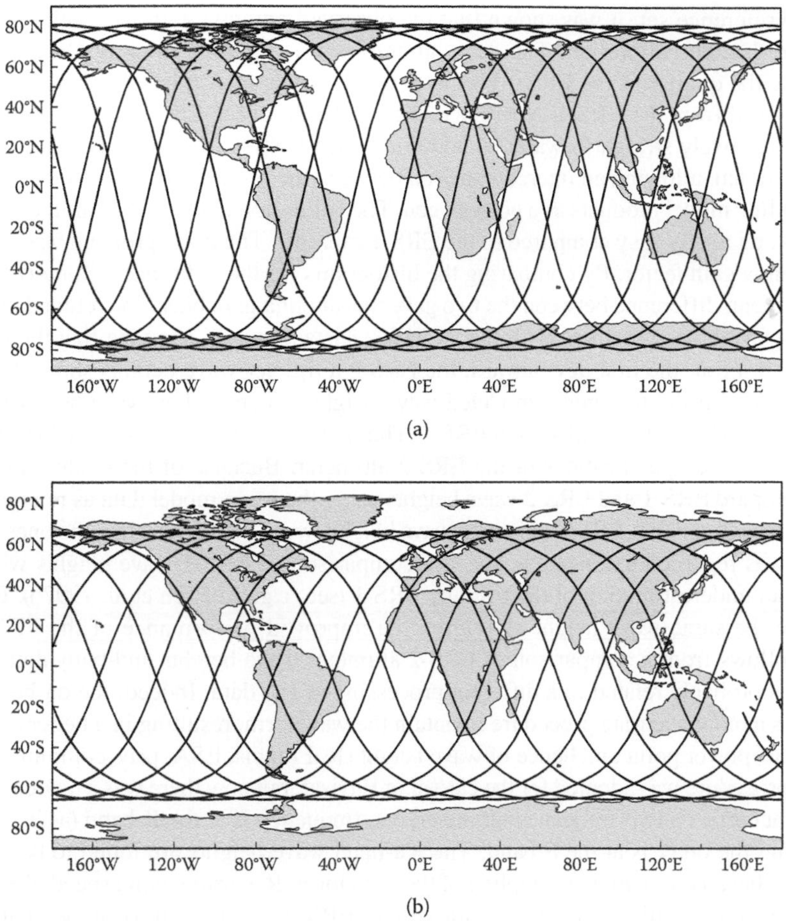

FIGURE 12.2 Typical daily coverage of ERS/Envisat/SARAL family (a) and TOPEX/Jason family (b).

the pre-launch instrument characterization data. When the processing algorithm was updated at all ground stations, the performance of the altimeter wave height was found to be satisfactory as follows from an almost zero wave height bias and a standard deviation of error of 0.5 m when compared with modeled wave height. The second example occurred during the operational phase of ERS-1. A bug was discovered in the processing algorithm that led to unrealistically shaped wave height distributions. This bug was removed at the beginning of 1994 and resulted not only in a much improved shape of the wave height histograms but also in a reduction of mean wave height of about 30 cm (Bauer and Staabs 1998).

For altimeter winds, a different approach needs to be followed because engineering and geophysical calibration cannot be separated as there is no absolute calibration of the backscatter against independent data from manmade targets or stable known targets of opportunity readily available. Later on, there were some ESA efforts to calibrate the Envisat RA-2 σ_0 using one transponder, but it takes a number of years before an accurate estimate of the absolute value of the radar backscatter is found. Once the absolute value is known, all altimeter missions can be corrected to provide an absolute σ_0, provided there is sufficient overlap in time among the different missions. For the initial data calibration, the system gain as determined by pre-launch instrument characterization was used, and for the initial geophysical calibration, algorithms from previous missions such as Seasat and GEOSAT were used. First comparisons with ECMWF winds uncovered several problems in the initial algorithm. The problems were solved in a couple of weeks, but differences of 20% remained. This difference corresponds to a small (0.8 dB) bias in antenna gain. After thorough validation of the ECMWF reference set, it was shown that the observed antenna gain was well within the error budget for pre-launch calibration. The data calibration was updated in early December 1991 and since that date the quality of the ERS-1 altimeter wind speeds reached an acceptable level.

Having learned from the ERS-1 experience, the validation of the ERS-2 altimeter wind and wave products was relatively straightforward. In addition, when ERS-2 was launched, the ERS-1 satellite was still operational, allowing an intercomparison between the products from ERS-1 and ERS-2 using the corresponding model products as a go-between. The validation of the ERS-2 altimeter wind speeds was, therefore, relatively easy compared to the ERS-1 exercise. The main problem was again to determine the antenna gain factor. By comparing the histograms for the radar backscatter σ_0 from the two satellites, the mean difference between the two gave the antenna gain bias. The retuned altimeter wind speeds gave a favorable agreement with the ECMWF surface winds, showing that the tuning procedure was sound. From the first day onward, the ERS-2 altimeter wave heights showed a remarkably good agreement with the first-guess modeled wave height, except at low wave height where ERS-2 had a somewhat higher cut-off value than ERS-1. The higher cut-off value is caused by the somewhat different instrumental specifications of the ERS-2 altimeter. Because of the tandem mission, it was possible to compare ERS-1 and ERS-2 wave heights using the wave model data as reference standard. As a result, it was found that ERS-2 altimeter wave heights were 8% higher than the ones from ERS-1. This change was regarded as favorable because comparison of ERS-1 wave heights with buoy data had revealed an underestimation of the truth by ERS-1 (see, e.g., Janssen et al. 1997b). Because both altimeters use the same wave height algorithm, the improved performance of the ERS-2 altimeter (which also follows from a comparison of ERS-2 altimeter wave heights and buoy data; see Janssen et al. 1997b) is probably related to a different processing of the data. Indeed, the on-board processor on ERS-2 uses a more accurate procedure to obtain the waveform, resulting in a better estimate of the slope at the half-power point and hence of wave height (R. Francis, ESA, priv. commun., May 1997).

A similar procedure was adopted for the cal/val of Envisat, but now there was a tandem mission with ERS-2. Envisat carries a two-frequency altimeter, one frequency is at the K band (as in case of ERS-1 and ERS-1) and the other is at the S band. The Ku-band wave heights are found to be of the highest quality, mainly because of a higher sampling of the waveform. Regarding wind speed, the antenna gain bias was obtained in a similar fashion as in the case of ERS-2, and by comparing with analyzed wind speed, the altimeter wind speed based on recalibrated σ_0 was found to be of high quality.

12.3 ALTIMETER SURFACE WIND SPEED

12.3.1 Principle of Wind Speed Measurement

To a good approximation, the strength of backscattered return (Figure 12.1), which may be described by specular reflection, is inversely proportional to MSS of the sea surface. As according to Cox and Munk (1954) mean square slope is closely related to the surface wind speed, the radar backscatter is a good measure for wind speed.

Although there is a considerable scatter in the relation between the radar backscatter σ_0 and the wind speed U_{10}, reflecting that perhaps the physics behind the altimeter wind speed retrieval is not fully understood, it should be pointed out that the altimeter wind speed product has nevertheless proven its value, as will described later. Researchers have often wondered whether there are not additional sea state parameters relevant in the relation between backscatter and wind speed. For a nadir-looking radar, the main scattering mechanism is specular reflection. Therefore, the backscatter is proportional to the joint probability density of slopes $p(\eta_x,\eta_y)$ of the surface η, where η_x and η_y are the slope components in two orthogonal directions. The radar backscatter $\sigma_0(\theta)$ at an incidence angle θ is then given by the classical result (Barrick 1968; Valenzuela 1978):

$$\sigma_0(\theta) = \frac{\pi |R(0)|^2}{\cos^4\theta}\, p(\eta_x,\eta_y)\ \Big|_{spec} \tag{12.1}$$

where $R(0)$ is the Fresnel reflection coefficient for normal incidence, and the probability density of slopes $p(\eta_x,\eta_y)$ is evaluated at the specular points. Accordingly, only surface facets normal to the direction of the incident radiation contribute to the backscattering. For small slopes, the probability density function (pdf) of the surface slope is given by a Gaussian distribution (Cox and Munk 1954)

$$p(\eta_x,\eta_y)=\frac{1}{2\pi\,\sigma_u\,\sigma_c}\exp\left[-\frac{1}{2}\left(\frac{\eta_x^2}{\sigma_u^2}+\frac{\eta_y^2}{\sigma_c^2}\right)\right] \tag{12.2}$$

where σ_u^2 and σ_c^2 are the slope variances in the along and cross directions. Combining (Equation 12.2) and (Equation 12.1) one finds that, for normal incidence ($\theta = 0$)

$$\sigma_0(\theta)=\frac{|R(0)|^2}{2\,\sigma_u\,\sigma_c} \tag{12.3}$$

which for an isotropic surface, with $\sigma_u^2 = \sigma_c^2 = s^2/2$ where s^2 is the total slope variance, becomes

$$\sigma_0(\theta)=\frac{|R(0)|^2}{s^2} \tag{12.4}$$

In order to apply (Equation 12.4), it should be realized that only a portion of the total mean square slopes of the ocean surface is included in s^2, namely only those ocean waves contribute whose wavenumber is smaller than a cut-off wavenumber, k_R, which is usually taken as one-third of the electromagnetic radiation wavenumber (Brown 1990). With $F(k,\varphi)$ the surface wave spectrum, where k is the wavenumber and φ the wave propagation direction, the slope variance s^2 therefore becomes

$$s^2=\int_0^{k_R}\int_0^{2\pi} k\,dk\,d\varphi\,k^2\,F(k,\varphi) \tag{12.5}$$

Equation 12.4 clearly shows how the radar backscatter return depends on the sea state through the MSS, s. Traditionally, however, the radar backscatter has been interpreted in terms of the surface wind speed, as Cox and Munk (1954) showed that there is a correlation between MSS and surface wind. Nevertheless, the relation between these two parameters is certainly not perfect. Several studies (Monaldo and Dobson 1989; Glazman and Pilorz 1990; Glazman and Greysukh 1993; Lefèvre et al. 1994; Hwang et al. 1998; Gourrion et al. 2002) suggest that in the presence of swell the radar backscatter may depend on both the local wind speed and the sea state. Therefore, a number of researchers have made attempts at developing a retrieval algorithm that includes a measure of the sea state. Most authors choose the SWH as measure of sea state because this parameter is readily available from the radar altimeter. Gourrion et al. (2002) have performed the most extensive study in this direction, and one of their main results shows the relation between TOPEX backscatter σ_0 and surface wind speed U_{10} from the National Aeronautics and Space Administration (NASA) Scatterometer (NSCAT) varies with respect to the SWH H_s (as obtained from TOPEX). Their results show that for low wind speeds in particular there is a dependence of the radar backscatter on SWH, and a two-dimensional model of the type:

$$U_{10}=f(\sigma_0,H_s) \tag{12.6}$$

and its inverse was developed using a neural network approach. Gourrion et al. (2002) presented clear evidence that, compared to the one-dimensional models such as the MCW retrieval algorithm,

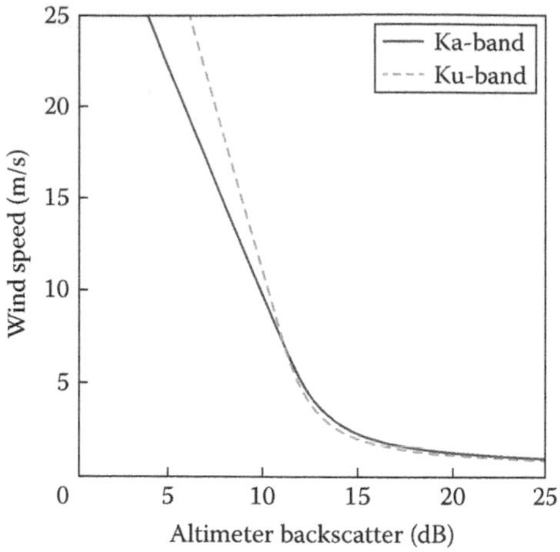

FIGURE 12.3 The empirical relation between wind speed and altimeter backscatter (for Ku and Ka bands).

the sea state-dependent algorithm (called from now on the Gourrion algorithm) performed better. Consequently, the Jason-1 and Jason-2 altimeter wind speed observations are obtained using this retrieval algorithm. The good performance of the Gourrion algorithm is reflected by the validation efforts at ECMWF where routinely the Jason-1/2 wind speed is compared with analyzed ECMWF wind speed. The typical value for the global scatter index is slightly more than 16% that, with a mean observed wind speed of 8 ms^{-1}, corresponds to a standard deviation of error of 1.25 ms^{-1}.

The empirical relations between surface wind speed and the altimeter backscatter σ_0 for the Ku band (as provided by Envisat) and the Ka band (as provided by SARAL) are shown in Figure 12.3. One should keep in mind that the stronger the wind is, the rougher the surface (i.e., the larger the MSS) is and therefore the smaller the backscatter is. Note that the relation between the MSS and backscatter in the case of the altimeter is the inverse of that of the scatterometer (e.g., De Chiara 2015).

12.3.2 Quality of Altimeter Wind Speed Observations

Surface wind speed product from altimeters is not assimilated at the ECMWF integrated forecasting system (IFS). Instead, it is used for independent verification of the model winds. For example, it is used to assess various model changes. Therefore, verification of this product and monitoring its quality are very important tasks. An automated system was set up to monitor the availability and the quality of the altimeter wind speed product in NRT. Currently, altimeter wind speed from Sentinel-3A, Jason-2, Jason-3, CryoSat-2, and SARAL are routinely received and monitored.

For the purpose of altimeter wind speed verification, an *in situ* wind speed measurement is only trusted if it is associated with an acceptable SWH value. Therefore, rejection of wave height in an *in situ* record invalidates the wind speed measurement in the same record. In fact, the same assumption is used for the quality control of altimeter data.

The verification of altimeter wind speeds is done against both IFS model analysis (Figure 12.4a) and *in situ* (buoy) measurements within a few hundreds of offshore buoys and platforms (Figure 12.4b). Although the *in situ* measurements are widely accepted as the "ground truth" (although there is an increasing evidence using the triple collocation technique that this is not true; see, for example, Janssen et al. 2007; Abdalla et al. 2011), the geographical coverage of those *in situ* stations is limited to offshore Europe and Northern America and parts of the tropics (denoted by the blue dots in the map shown in Figure 12.4c). On the other hand, the model comparison provides a real global assessment of altimeter wind speed.

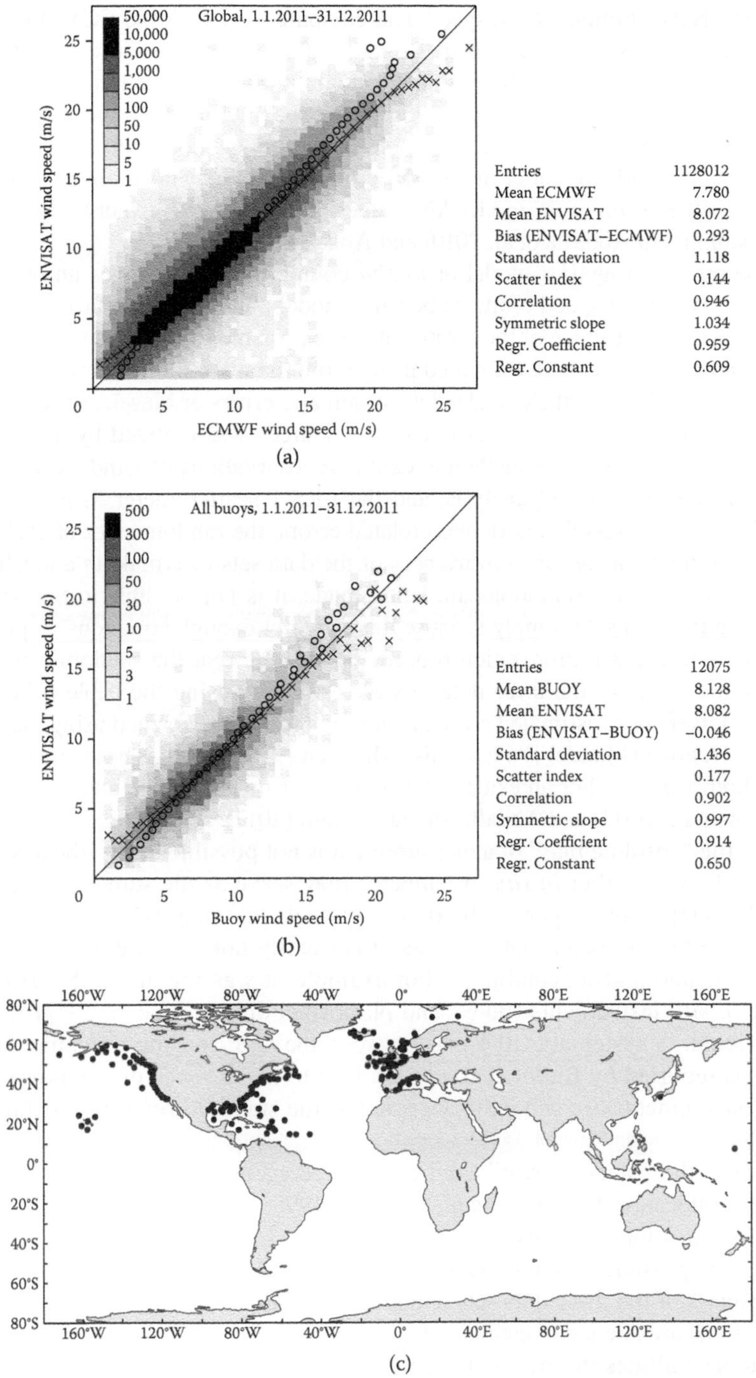

FIGURE 12.4 Comparison between (a) Envisat NRT wind speed and ECMWF IFS model AN (b) and in situ measurements for all data from January 1, 2011, to December 31, 2011. The number of collocations in each 0.5 m/s × 0.5 m/s two-dimensional bin is color-coded as in the legend. The "x" symbols are the means of the bins for given x-axis values (the model in the former and the in situ in the latter), while the "o" symbols are the means for given y-axis values (altimeter). Typical locations of in situ measurements (c) are also shown.

In general, the NRT altimeter wind speed data compare well with the ECMWF model analysis and the *in situ* observations as can be seen in the scatter plots shown in Figure 12.4. The scatter plots in Figure 12.4a and b represent two-dimensional histograms showing the number of observations in each two-dimensional bin of 0.5 m/s × 0.5 m/s of wind speed. Typically, the standard deviation of the difference (SDD) between altimeter winds and the model or the *in situ* winds is slightly above 1.0 ms^{-1}, which corresponds to scatter index (SI, defined as the SDD normalized by the mean of the reference, which is the *in situ* or the model AN) values of about 15%. Further verification results can be found, for example, in Abdalla et al. 2010) and Abdalla (2013a).

The direct comparison against model or *in situ* counterparts only gives an upper limit to the errors of the altimeter wind speed products because model and buoy data have their own errors as well. In general, with the absence of the "truth," it is not possible to estimate the wind speed errors. However, the random errors can be estimated using more than two data sources provided that their errors are not correlated. Nevertheless, absolute systematic errors or biases cannot be estimated.

The need for estimates of errors of different data sources was realized by Stoffelen (1998). He proposed to use a triple collocation method to calibrate observations of winds from a scatterometer using winds from buoys, a model analysis, and the ERS-1 scatterometer. It is straightforward to show that, with three data sets that have uncorrelated errors, the random error of each data type can be estimated from the variances and covariances of the data sets (Tokmakian and Challenor 1999). However, unless additional assumptions are being made, it is not possible to perform an absolute calibration among the data sets, simply because there are not enough equations. A possible way out of this dilemma is to use a minimization procedure. Assume that the random errors are not correlated and that the errors of the three data sets are estimated using the triple collocation method (Stoffelen 1998). Given these estimated errors, calibration is then performed using the neutral regression approach of Deming (Mandel 1964; see also Marsden 1999), which is based on the minimization of the error in both variants. For an extensive discussion of this approach and a number of applications see Janssen et al. (2007) and Abdalla and De Chiara (2017).

Quantifying the "absolute bias" of a measurement is not possible due to the absence of a standard reference. Buoys or other *in situ* instruments may serve as the standard reference because they are usually subject to proper calibration. Although this may be correct under controlled environmental conditions such as laboratories, it is usually not the case in the open ocean with harsh atmospheric and marine conditions. For example, it was found that SWH measurements from different *in situ* instruments (buoys and platforms) do not necessarily provide consistent results. In particular, a systematic 10% bias between the U.S. and the Canadian buoy networks was identified as reported by Bidlot et al. (2008). Due to the absence of any other options, however, *in situ* measurements are generally accepted as the standard reference as far as the bias is concerned.

The variance of the "random error" can be estimated even in the absence of an absolute truth using the triple collocation technique (e.g., Stoffelen 1998; Janssen et al. 2007). This technique can be summarized as follows: Given three independent estimates of the truth, T, with unknown random errors, it is possible to show that the error variance in each estimate can be found using the total covariances of the three data sets in addition to the "unknown" covariances of the errors. Further assumptions are needed to estimate the error covariances. The assumption of uncorrelated errors, for example, nullifies the error covariance terms. If this assumption is not correct, the error estimates will not be correct. It is also important to note that although the errors in two data sets may not be correlated directly, it may be possible to have a pseudo-correlation due to the nonlinear nature of both errors.

Abdalla et al. (2011) used the triple collocation method to estimate the random errors of Envisat, Jason-1, and Jason-2 wind speeds. In addition to the altimeters, the buoy measurements and the ECMWF IFS model fields are used. In order to guarantee a consistent model data set, output of the parallel ECMWF model run with the version corresponding to T_L1279 is used before the operational implementation of this model version (on January 26, 2010). The NRT altimeter wind speeds were collocated with the ECMWF IFS model forecasts (1-day) and buoy wind speeds globally over a year (August 2009 to July 2010). In order to alleviate the difference in scales among various data sources, the altimeter data are averaged along the track to produce super-observations with scales in between 75 and 100 km while the buoy data are averaged in time (every 5-hourly measurements), which results in scales of about 100 km. It is worthwhile mentioning that the model needs four to eight grid points (for the model version at the time, this corresponds to 65–120 km) to accurately represent atmospheric features, per Abdalla et al. (2013). The results are summarized in Table 12.1. The time series of the monthly wind speed errors of Jason-1, Jason-2, and Envisat altimeters are shown together with the buoy and model 1-day forecast (shown in Figure 12.5). Although there was no geographical discrimination applied while selecting the data, the applicability of the results is restricted to the areas where buoy measurements exist. This is mainly offshore along the coasts of North America and Europe (Figure 12.4c).

TABLE 12.1
The Absolute and Relative (SI) Wind Speed Random Errors (Standard Deviations) of Jason-2, Envisat, Jason-1, Model 1-Day Forecast, and Buoy as Estimated Using Triple Collocation Method for the Period from August 1, 2009, to July 31, 2010

	Jason-2		ENVISAT		Jason-1	
Number of Collocations	19,856		15,552		19,613	
	Absolute (ms^{-1})	SI (%)	Absolute (ms^{-1})	SI (%)	Absolute (ms^{-1})	SI (%)
Altimeter error	1.00	11.9	0.93	11.1	1.01	12.0
Model forecast error	0.97	11.5	0.94	11.2	0.97	11.6
Buoy error	1.15	13.6	1.14	13.7	1.17	13.8

Source: Table 2 of Abdalla, S., et al., *Mar. Geod.*, 34(3–4), 393–406. (Copyright © European Centre for Medium-Range Weather Forecasts, 2011. Reprinted with permission.)

Note: Results are valid for offshore along the coasts of Northern America and Europe.

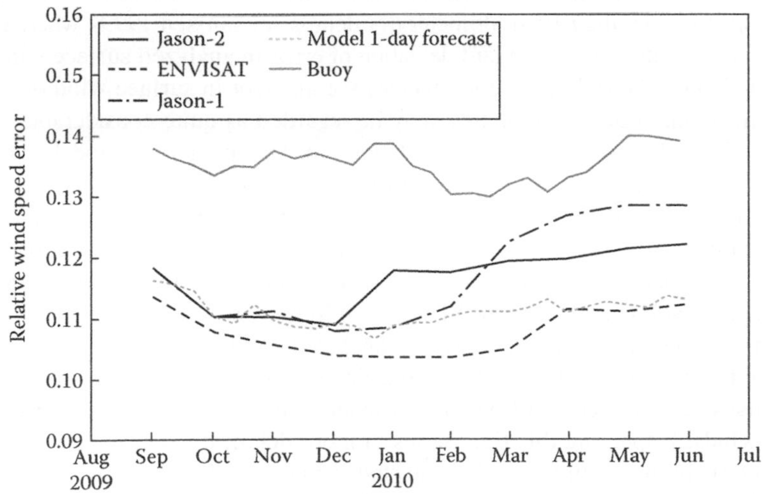

FIGURE 12.5 The time series of the monthly wind speed errors of Jason-1, Jason-2, and Envisat altimeters, buoys, and the model 1-day forecast. (Figure 7 of Abdalla, S., et al., *Mar. Geod.*, 34(3–4), 393–406. Copyright © European Centre for Medium-Range Weather Forecasts, 2011. Reprinted with permission.)

Abdalla and De Chiara (2017) used wind speed data from Meteorological Operational (Metop) Advanced Scatterometer (ASCAT)-A and B, NRT Jason-2, and ECMWF IFS model analysis and forecasts at lead times of 1–10 days to estimate wind speed random errors. The period covered is 2 years, from August 2013 to July 2015. This triple collocation study differs from previous studies that are limited to the locations of the buoy measurements (mainly offshore along North America and Europe) by being of a global nature. Furthermore, the use of the model forecasts at long lead times helps in tackling the impact of error correlations due to the fact that IFS assimilates scatterometer data. Assuming that, at long forecast time leads, the model wind speed errors lose their correlation with the ASCAT errors and the model retains enough predictability skill at those lead times, it is possible to estimate the wind speed error of Jason-2 to be 0.7 ms^{-1}. Note that this value is about 30% lower than that found by Abdalla et al. (2011) as listed in Table 12.1. The difference can be attributed to the fact that the collocations used by Abdalla and De Chiara (2017) cover the whole globe while the collocations used by Abdalla et al. (2011) are limited to the locations of the buoys. As will be shown later, in Figure 12.9, the altimeter wind speed quality in the Northern Hemisphere (NH), where most of the buoys are located, is not as good as in the other regions. Furthermore, buoys very close to the coast are usually rejected in order to avoid the adverse impact of the local small-scale features that are not resolvable by the model and to avoid the possible land contamination on satellite measurements. Nevertheless, some measurements may still be affected by the coastal activities.

12.3.3 Benefits of Altimeter Wind Speed Observations

Although there is a considerable scatter in the relation between the radar backscatter σ_0 and the wind speed U_{10}, reflecting that perhaps the physics behind the altimeter wind speed retrieval is not fully understood, it should be pointed out that the altimeter wind speed product has nevertheless proven its value. Three examples are mentioned. The first example concerns the validation of a new method of weather analysis, namely the variational method, which replaced the OI scheme in the ECMWF atmospheric forecasting system. Its static version, called 3DVAR, was introduced operationally by the end of January 1996 and had in general a positive impact on the forecast performance of the ECMWF forecast system (Andersson et al. 1998). When 3DVAR became operational, the new surface wind observations from the ERS-1 scatterometer were also presented to the data assimilation system. The scatterometer data had a positive impact on the quality of the surface wind speed analysis. This follows from a verification of analyzed wind speed in the Southern Hemisphere against altimeter wind speeds from the ERS-1 altimeter (for details see Janssen 1999), where the OI scheme without scatterometer data gave a standard deviation of error in analyzed surface wind of 1.99 ms^{-1} and the combination of 3DVAR and scatterometer gave an error in surface wind of only 1.77 ms^{-1}. The reduction in standard deviation of 10% may be regarded as quite considerable. This shows at once the value of the altimeter winds as a validation tool for operational model changes. Because these winds are not used in the analysis system, they may be considered as an independent source of information on the quality of the surface analysis, which makes them even more valuable.

Indeed, the altimeter winds have been used in assessing operational model performance and changes and the introduction of new data types such as surface winds from the Special Sensor Microwave Imager (SSMI) and from ASCAT on board Metop. As a second example, in Figure 12.6, a time series of the standard deviation of error of analyzed wind speed against the ERS-2 altimeter wind speed over the period May 1995 to October 2000 is presented. The area is the whole globe.

It is clear that over this period ECMWF has made considerable improvements in the quality of the surface winds. Inspecting this time series in more detail, it appears that at several instants fairly sudden jumps in the accuracy of the analyzed surface wind speed may be found. These can be related to changes in the operational weather forecasting system. For example, the transition around February 1996 is connected to the introduction of 3DVAR and ERS-2 scatterometer data; the jump in December 1997 is connected to the dynamic version of the variational analysis scheme, called 4DVAR, while the transition in July 1998 is connected to the two-way interaction of wind and

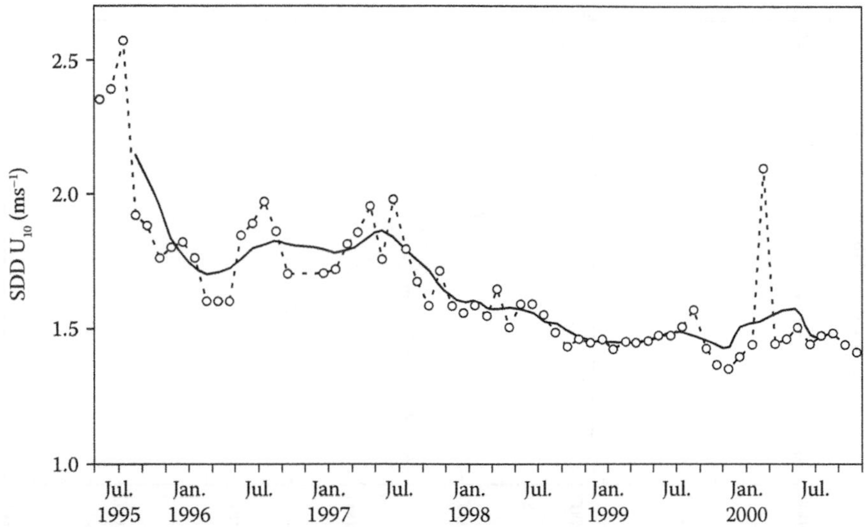

FIGURE 12.6 Standard deviation of the difference (SDD) of analyzed wind speed as obtained from the comparison between analyzed surface wind field and ERS-2 altimeter wind speed over the period May 1995 to October 2000. Area is the whole globe. The filled circles are monthly, data while the continuous line is a 6-month running average. (Figure 5.29 of Janssen 2004. Copyright © Cambridge University Press, 2004. Reprinted with permission.)

waves. A more detailed discussion is given in Janssen (2004). Note that at the end of the period, the quality of the altimeter wind speed worsened because of problems with the gyros on board ERS-2; these gyros help to determine the orientation of the satellite, and the radar backscatter σ_0 is sensitive to errors in the zenith angle. This is illustrated in Figure 12.6 by the significant outlier in February 2000.

As a third example, the operational introduction of ASCAT surface winds is briefly discussed. The ASCAT instrument is similar in design to the Active Microwave Instrument (AMI) scatterometers on board ERS-1 and ERS-2 from which data has been operationally assimilated at ECMWF since January 30, 1996. Triplets of radar backscatter from three antennas are combined to estimate surface vector winds over the global oceans. Both AMI and ASCAT operate at the C band (5.3 GHz) and have the same antenna geometry. Two main differences are a different range of incidence angle (optimized for ASCAT to enhance performance in wind direction) and the fact that ASCAT carries two sets of antennas (providing two swaths that double the coverage). ASCAT data have been monitored at ECMWF from the start of dissemination by the European Organization for the Exploitation of Meteorological Satellites (EUMETSAT) via the EUMETCast system on the January 31, 2007.

Surface winds are inverted from available (Level 1b) backscatter triplets on the basis of a modified version of the geophysical model function, CMOD5 (Hersbach et al. 2007). Resulting winds are collocated with operational short-range ECMWF forecast winds. The monitoring confirms that the ASCAT instrument is working well.

Besides a few short periods of data interruption, data volume is constant, and quality is found to be high and stable. The main concern is the availability of currently only one transponder out of a set of three, which prevents an absolute calibration of the Level 1B product. Although the calibration of the ASCAT instrument is still preliminary and non-optimal according to several groups that have worked with the data so far, it appears very stable. Instrument noise is low. Appropriate bias corrections, therefore, allow for the retrieval of a high-quality wind product. Especially the quality of ASCAT wind direction is found to be excellent, outperforming scatterometer data from ERS-2 for example. Assimilation experiments with the ASCAT data confirmed the high quality of the product as there were small but significant improvements in forecast performance, in particular in

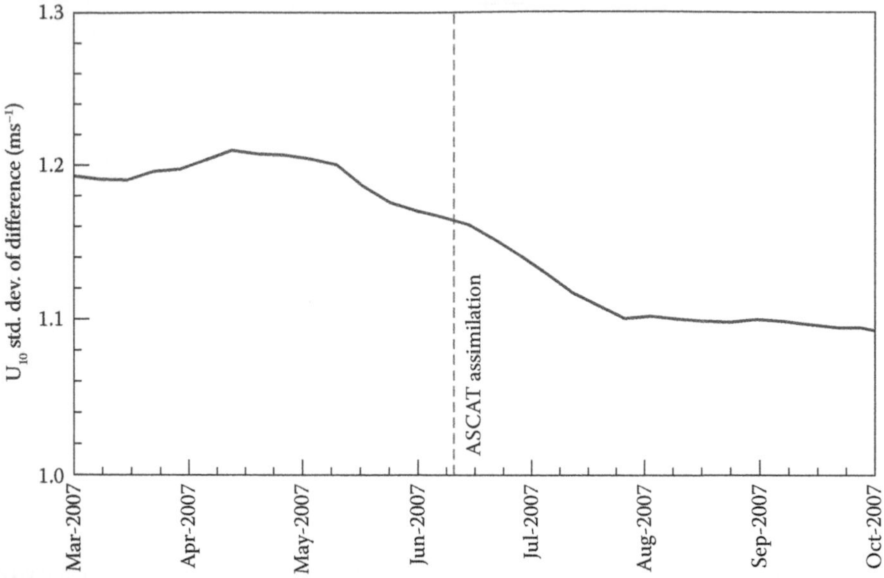

FIGURE 12.7 Time series of standard deviation of the wind speed difference (SDD) between Envisat altimeter and analyzed ECMWF wind speed. Note that on June 12, 2007, the assimilation of ASCAT data was introduced in operations giving a closer match (lower SDD) between altimeter and analyzed wind.

the Southern Hemisphere scores of 1000 and 500 hPa geopotential. On June 12, 2007, ECMWF was the first to present these data to the analysis scheme. Improvements of the analyzed surface winds were already evident after a few days, as is shown in Figure 12.7. Here, for different areas on the globe, time series of normalized standard deviation of error (SI) of the difference between Envisat altimeter wind speed and analyzed wind speed are shown for the period of May 11 until June 20, 2007. The assimilation of ASCAT data commenced on June 12, and the impact of these new observations on the quality of the surface analysis is clearly visible after a few days.

12.3.4 Altimeter Wind Speed Data and Problems

Nevertheless, from long-term monitoring of the quality of the altimeter winds, it was found that in particular in the Northern Hemisphere there is a clear underestimation of wind speed by the altimeter during the late spring and early summer, whereas in the remainder of the year the bias is much smaller (Janssen 1999). This problem is illustrated in Figure 12.8, which shows over the period of December 1996 until January 2000 a pronounced seasonal cycle in the bias between ERS-2 altimeter wind speed and analyzed wind speed for the Northern Hemisphere, while no such cycle is seen in the tropics and in the Southern Hemisphere. The wind speed climatologies in different areas of the globe are really different, in particular regarding the seasonal cycle in the mean wind speed. Areas such as the Southern Hemisphere and tropical oceans show hardly any seasonal variation around the annual mean wind speed (the amplitude of the seasonal cycle is less than 0.5 ms^{-1}), while in the Northern Hemisphere the amplitude of the seasonal cycle is much larger, about 1.5 ms^{-1}. The largest seasonal variations are found in the North Atlantic with an amplitude in the seasonal cycle of about 3 ms^{-1}. A similar remark applies to the seasonal variations in mean SWH. Therefore, it should not come as a big surprise that there are only seasonal variations in the Northern Hemisphere. One may argue that this is perhaps a problem with the model winds in the Northern Hemisphere, but a validation of altimeter wind speed against wind speeds from buoys in the Northern Hemisphere gives a similar picture (Janssen 1999). This is seen in Figure 12.9 that shows for July 1997 (which has according to Figure 12.8 the largest

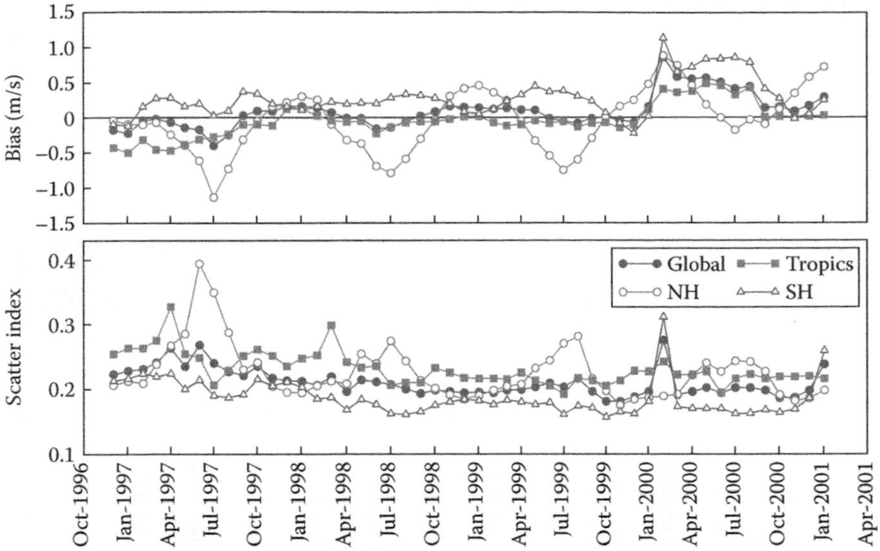

FIGURE 12.8 Comparison of ERS-2 altimeter wind speeds with analyzed model winds for the globe, the Northern Hemisphere, the tropics, and the Southern Hemisphere. Shown are monthly values of bias (alt-mod) and scatter index over the period of December 1996 until January 2001. Note the clear seasonal cycle in the bias for the Northern Hemisphere.

negative bias) a scatter diagram between the two. From the scatter diagram, it is immediately evident that the underestimation of wind speed by the altimeter mainly occurs in the wind speed range below 10 ms^{-1}. With a mean buoy wind of 6.4 ms^{-1}, a fairly large negative bias in the altimeter wind speed of −1.3 of ms^{-1} is found. This bias is a robust feature in the summertime. On average, one finds for June, July, and August 1997, a negative bias of −1 ms^{-1}, whereas by contrast in the winter of 1996 to 1997 the bias is only −0.06 ms^{-1}. Similar findings were reported by Chen et al. (2000) who compared TOPEX wind speeds with 6 years of buoy data from the Japan Meteorological Agency. These authors pointed out that the seasonal cycle in the altimeter bias depends on the choice of the retrieval algorithm. The MCW algorithm (which is the one used for the ERS-2 wind retrieval) gave a very similar seasonal cycle to the one displayed in Figure 12.8, but the Hwang et al. (1998) algorithm gave a seasonal cycle that is just the opposite of the seasonal cycle obtained with the MCW scheme.

This prompted Abdalla (2007, 2012) to refine the MCW algorithm by doing a tuning exercise based on much more data than used by Witter and Chelton (1991), namely 2 months of analyzed ECMWF wind fields and the buoy data for the same period. In particular, attention was paid to the low wind speed regime as in that range the largest discrepancies between altimeter winds and buoy data exists (compare Figure 12.9). The resulting algorithm may be regarded an improvement as follows from a validation of the Abdalla algorithm applied to Envisat radar backscatter against buoy data over a 1-year period, giving a better fit in the low wind speed range, a reduction in standard deviation of error of 5%, and a reduction in bias from −0.59 ms^{-1} to −0.13 ms^{-1}. A similar reduction in standard deviation of error was found when comparing the newly obtained altimeter winds against analyzed wind speed over a 2-year period from April 2003 until April 2005. There was also a reduction in the amplitude of the seasonal cycle in the bias, although the decrease was smaller than hoped for. For Envisat, using the MCW algorithm, the amplitude in the seasonal cycle was found to be about 0.5 ms^{-1}, and with the Abdalla algorithm an amplitude of 0.4 ms^{-1} was obtained. This new wind retrieval algorithm was introduced for Envisat in October 2005.

Before possible reasons for a seasonal cycle in the Northern Hemisphere altimeter bias are discussed, it is worthwhile to point out that other satellite instruments may have a similar problem.

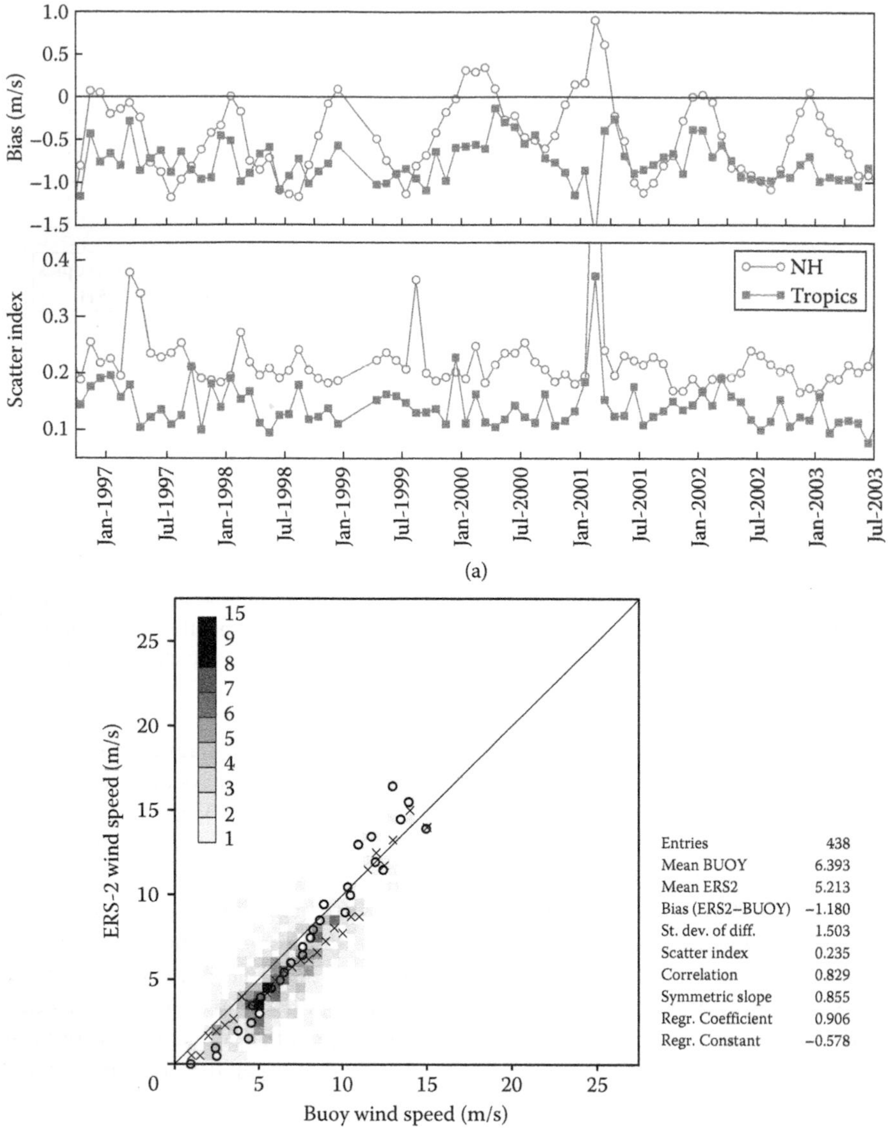

(b)

FIGURE 12.9 Comparison of altimeter wind speeds with buoy data as a time series (a) and the comparison for the Northern Hemispheric buoy data for July 1997 as a scatter plot (b), confirming the underestimation of winds by the altimeter in the summertime, in particular in the low wind speed range.

An example is the ERS-2 scatterometer for the North Atlantic area over the period January 2004 until June 2006 (Abdalla and Hersbach 2006), which has a bias that depends on the season. The amplitude of the seasonal cycle is however, somewhat smaller, about 0.3 ms^{-1}.

The following discussion will be in the context of the assumption that the radar backscatter σ_0 is caused by specular reflection from the short ocean waves (i.e., the amount of backscatter is related to the state of the short waves). We will discuss a number of additional effects that, so far, have not been considered in the standard backscatter model for operational wind speed retrieval (an exception is a form of sea state effects for Jason series), and we indicate how the additional physics may contribute to a better understanding of the seasonal cycle in the altimeter wind speed bias.

12.3.4.1 Effect of Slicks

One plausible reason for the seasonal cycle in the altimeter bias may be related to the presence of slicks, as it is known that slicks of natural origin play an important role in the damping of high frequency ocean waves (Alpers and Hühnerfuss 1982), and from Janssen et al. (1998), it is known that the effect of slicks is important in determining the radar backscatter. Thus, in the presence of slicks, the ocean surface appears to be smoother resulting in a larger backscatter than when slicks are absent. Hence, when a retrieval algorithm is used that does not take slicks into account, a lower wind speed will be retrieved because the radar backscatter is inversely proportional to the wind speed. Since slicks will be washed away at winds above 7 to 10 ms^{-1}; this underestimation of the strength of the wind should be most pronounced in the low wind speed range, which agrees with the findings from the validation of altimeter winds against buoy winds (compare Figure 12.9) and also against analyzed wind speed (not shown). In order to investigate whether slicks are relevant for the underestimation of the wind speed by the altimeter during the Northern Hemisphere summer, the spatial distribution of slicks is required as a function of the time in the year. Information on this was obtained from monthly mean observations of ocean color by SeaWiFs, which are displayed as global maps on the Web. Because ocean color is a measure of biological activity in the upper layer of the ocean and slicks are produced by plankton, it seems plausible to relate ocean color to the presence of slicks. In doing so, it was found that in the spring and summer of 1998, the main area of biological production was in the North Atlantic, where production peaks in May. By contrast, in the North Pacific, biological production is confined to coastal areas (e.g., near Japan); thus when the presence of slicks would be relevant for the bias problem in altimeter winds, one should expect that the main problem occurs in the North Atlantic in May 1998. This inference is, however, not supported by the comparison of altimeter winds against analyzed ECMWF winds. According to the present verification of altimeter winds, the main bias problem seems to occur in June, July, and August when there is, compared to May, relatively little biological activity. Also, when studying the verification statistics for the North Pacific and the North Atlantic separately, it follows that the underestimation of wind speed by the altimeter is similar in both ocean basins. Therefore, although effects of slicks are expected to be relevant in the retrieval of wind speed from an altimeter, it appears that an explanation of the bias problem in terms of a single cause such as the presence of slicks is not very likely.

12.3.4.2 Neutral versus 10-m Winds

Another interesting deviation from the standard backscatter model is related to the physics of how short waves are generated by wind. Since the contribution by Miles (1957), there is a growing belief that ocean waves are forced by the surface stress and not by the surface wind. As a consequence, instruments such as the altimeter and the scatterometer observe the surface stress. Rather than correlating radar backscatter, σ_0, with the wind speed at 10 m height, U_{10}, it may be more appropriate to correlate σ_0 with the equivalent neutral wind U_{10N} because this measure of surface wind is more closely connected to the surface stress. This approach has been followed, for example, in the development of the QuikSCAT geophysical model function, and at ECMWF the wave prediction model is forced by neutral winds.

The use of neutral winds, rather than 10-m winds, is in particular relevant for wind speeds up to 10 ms^{-1} and, therefore, may have some impact on the seasonal variations of the altimeter bias. By inspecting a number of ECMWF surface wind fields in the winter of 2006 and the summer of 2006, one may obtain information such as the fraction of unstable and stable cases and the systematic difference between U_{10} and U_{10N}. (For a similar exercise on the winter of 2001, see Abdalla and Hersbach 2006; Brown et al. 2006.) We restrict our attention to the Northern Hemisphere oceans. One then finds that in the winter time the fraction of unstable cases on average is 79%; therefore, the fraction of stable cases is 21%, while the average difference between U_{10} and U_{10N} is -0.24 ms^{-1}. In the summertime, there are more stable cases as the fraction of stable cases increases to 42%; consequently, the fraction of unstable cases is on average 58%, while the difference between 10-m wind and neutral wind is almost zero, about -0.02 ms^{-1}.

Hence, interpreting the radar backscatter in terms of the neutral wind rather than the 10-m wind speed will result in a reduction of the amplitude of the seasonal cycle of about (0.24 – 0.02)/2 = 0.11 ms^{-1}. Therefore, although theoretically it is expected that it is more appropriate to relate σ_0 with the neutral wind speed, it is not expected that the use of neutral winds will explain a large part of the seasonal cycle in the altimeter bias.

12.3.4.3 Sea State Effects

Although there is a reasonable correlation between σ_0 and surface wind speed U_{10}, researchers have often wondered whether there are not additional sea state parameters relevant in the relation between backscatter and wind speed—and for good reasons, as it was clear from Equation 12.4 and Equation 12.5. One may wonder to what extent the sea state information has resulted in a better wind speed retrieval. For this reason, Abdalla (2007, 2012) improved the pure wind speed algorithm of Chelton and Wentz by using a much larger data set (consisting of both ECMWF winds and *in situ* observations) for fitting the unknown coefficients, and this algorithm was introduced for Envisat in 2005. Global monitoring the Jason-1 Operational Sensor Data Record products at ECMWF suggests that the Jason Ku band σ_0 is about 0.4 dB higher than that of Envisat. Accordingly, the Jason-1 σ_0 values were reduced by this amount before applying the Abdalla algorithm. As can be seen in Figure 12.10, this algorithm is performing somewhat better than the Gourrion algorithm, which is based on backscatter and SWH as given by Equation 12.6. A reason for this perhaps counter-intuitive finding may be that the comparison of the results of the Gourrion algorithm with alternative algorithms was not quite fair because the older algorithms were trained on a much smaller data set than the sea state dependent algorithm. Another possible explanation is that SWH is just one parameter describing the sea state, and by itself, it may not be enough to account for the overall impact sea state.

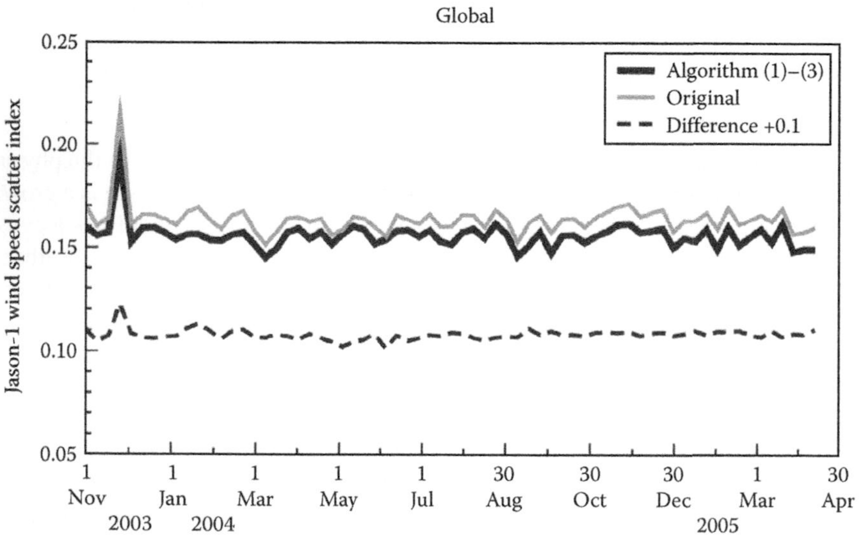

FIGURE 12.10 Time series of the wind speed scatter index between Jason RA and the European Center for Medium-Range Weather Forecasts (ECMWF) analyzed wind speed from the Gourrion algorithm (thin gray line) and the algorithm proposed by Abdalla (2012) (thick black line) over the period from November 1, 2003, until April 9, 2005, for the global oceans. The difference, Gourrion – Abdalla (dashed line), shifted by 0.1, is always above 0.1, indicating that the algorithm proposed by Abdalla is in slightly better agreement with the model. (Figure 9 of Abdalla, S., *Mar. Geod.,* 35(Suppl. 1), 276–298. Copyright © European Centre for Medium-Range Weather Forecasts, 2012. Reprinted with permission.)

12.3.5 Backscatter versus Mean Square Slope

It is concluded from the previous discussion that the addition of sea state information to the retrieval of wind speed from the radar backscatter gives only a very limited benefit. There have also been suggestions and attempts (compare Hwang et al. 1998; Gommenginger et al. 2003) to relate radar backscatter and mean wave period. However, this approach might only work for wind-seas as explained already by Hwang et al. (1998). This was confirmed by the validation work of Caires et al. (2005), who found that a meaningful retrieval of mean wave period could be obtained in only in 35% of the cases. The selection of these wind-sea cases was based on a criterion obtained from a wave prediction model. Nevertheless, following a neural network approach, Quilfen et al. (2005) generated an empirical algorithm that gave a much better agreement with buoy data than the Gommenginger algorithm. This new algorithm should perform better for swell cases.

On the other hand, it may perhaps be a good idea to follow the suggestion of the theoretical approach of the past (which is briefly summarized in Equations 12.1–12.5). Clearly, this approach suggests that the radar backscatter σ_0 is related to the MSS s^2 of the sea surface. The MSS is in itself a useful quantity to monitor as it measures the roughness of the sea surface, which is closely connected to the air-sea momentum transfer. Therefore, let us assume for the moment that the radar backscatter simply provides the MSS, an assumption that is supported by many studies in the past, see, for example, Valenzuela (1978), Apel (1994), and Freilich and Vanhoff (2003).

A problem with the MSS is, however, that this quantity is very difficult to model. This follows immediately from the definition of the slope as given in Equation 12.5 and the empirical knowledge that the short gravity waves follow the Phillips' wavenumber spectrum, which is proportional to k^{-4}. As a consequence, the integral in Equation 12.5 has a logarithmic divergence for large wavenumbers, which implies that the value of the MSS depends on detailed knowledge of the shape of the gravity-capillary wave spectrum. Fortunately, in the past two decades, we have seen considerable progress in our knowledge of the gravity-capillary wave spectrum, starting with the experimental investigations of Jähne and Riemer (1990) and Klinke and Jähne (1992), followed by semi-empirical model efforts of Apel (1994), Elfouhaily et al. (1997), and Liu et al. (2000). There is even a model based on the solution of a parameterization of the energy balance equation for short gravity-capillary waves, called the Viers-1 model (compare Janssen et al. 1998), which was applied with some success to retrieve wind vectors from the scatterometer.

The results of a simulation for the nadir radar backscatter as a function of surface wind speed based on the isotropic model (Equations 12.4–12.5) for the cross section at normal incidence are shown in Figure 12.11. The low-wavenumber part of the slope variance, s^2, is obtained from two-dimensional operational wave spectra produced by the ECMWF version of the WAM model (ECWAM) model while the high-wavenumber part of the slope variance is obtained from the VIERS-1 (Verification and Interpretation of ERS-1) gravity-capillary model of Janssen et al. (1998). This model basically describes the balance among wind input (including the effects of feedback of waves on the wind), three-wave interactions, and dissipation by white capping and molecular viscosity. The boundary condition for the energy flux at low wavenumbers is determined by the energy flux provided by the resolved waves in the WAM model. In order to check the realism of the short wave model, the MSS calculation was applied to the case of visible light, and a good agreement was found with the MSS observations of Cox and Munk (1954).

Note that, as discussed earlier, an absolute calibration of σ_0 is not readily available, but in a comparison with results from a simulation of the radar backscatter, it is highly desirable to have an absolute observed estimate. A dedicated long-term field campaign was designed for the Envisat RA-2 altimeter using a ground transponder developed at the European Space Research and Technology Centre (ESTEC). The campaign faced a lot of practical difficulties (Monica Roca, personal communications). The absolute radar cross-section accuracy is expected to be in the region of $\pm$ 0.2 dB. The results of this campaign indicate that the actual Envisat RA-2 altimeter backscatter has a negative bias of about 1.0 $\pm$ 0.2 dB (Pierdicca et al. 2013). Correcting for this bias, the global average

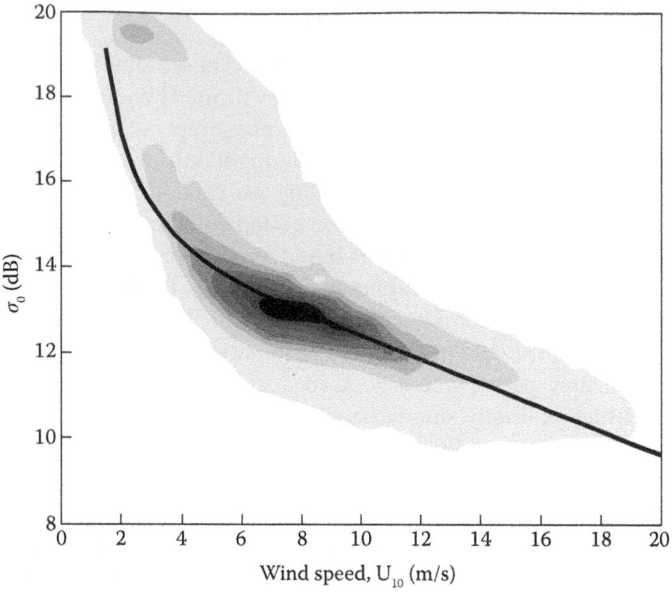

FIGURE 12.11 Two-dimensional histogram showing the relation between radar backscatter σ_0 and surface wind speed U_{10} obtained from a model of radar backscatter based on specular reflection. Here the slope is obtained from the ECWAM / VIERS-1 model. As a reference, the σ_0–U_{10} relation according to Abdalla is shown as the thick line.

radar backscatter from Envisat is 13.34 dB. Accordingly, in the Abdalla algorithm, appropriate corrections have been applied to agree with this global average of the absolute backscatter.

Note that for the σ_0 values reported in the Envisat products, a bias of –3.24 dB has already been applied to make the data consistent with those of ERS-2. Thus, the actual Envisat backscatter is 3.24 dB higher than the ones reported in the Envisat disseminated products.

In order to obtain the $\sigma_0 - U_{10}$ relation, MSSs for the Ku-band case ($k_R = 285$ rad/m) are obtained from the two-dimensional spectra and the radar backscatter then follows from Equation 12.4. It was found by trial and error that, in order to obtain optimal results, the nadir reflection coefficient R(0) must have the value $R(0) = 0.82$, which is about 5% higher than obtained from the estimates of the complex reflectivity of the ocean at 14 GHz, using realistic salinity and sea surface temperature (Apel 1994).

Using only wave model information to simulate σ_0, there is a fair agreement with the results of the Abdalla algorithm for wind speeds above 5 ms^{-1}. However, in the low wind speed range, the wave model seems to overestimate the backscatter. The reason for this is related to the discretization of the two-dimensional wave spectrum, which has coarse spectral resolution beyond 0.5 Hz. Hence, for low wind speed, the WAM model does not properly represent the local wind sea. Nevertheless, it can be concluded that a wave model provides a reasonable estimate of the nadir radar backscatter; therefore, if observations of radar back scatter are available, they could be assimilated by a variational analysis approach as is quite common nowadays at operational weather centers. Because of the high-frequency nature of the observed information, it is expected that this analysis will result in an improved specification of air-sea interaction.

12.3.6 Extreme Winds

Ku-band altimeter wind speeds are usually only regarded as reliable in the wind speed range below 20 to 25 ms^{-1}. One of the reasons for this is that the slope $\partial\sigma_0/\partial U_{10}$ seems to decrease for increasing surface wind speed; hence, σ_0 becomes less sensitive to changes in wind speed, and thus an increased accuracy in the determination of the radar backscatter is required. Another reason is that the Ku-band backscatter is seriously affected by rain contamination, which is often present in extreme events such as tropical cyclones and hurricanes. Furthermore, extreme events are rare; and therefore, it is difficult to develop proper empirical algorithms.

Nevertheless, Young (1993) addressed the special case of altimeter wind inversion at high winds by examining the rare events when the GEOSAT altimeter swath crossed over strong cyclones. Ground truth was obtained from a model for cyclone winds for six extreme events within the period of 1987 to 1989. Inputs of the cyclone model were parameters such as central pressure and the radius of maximum wind. Here, central pressure was obtained from analyzed weather maps, while the radius of maximum winds was obtained from the distance between minimum and maximum σ_0 as Young made sure that only those satellite tracks were considered that went right through the eye of the storm. The resulting model function, valid for winds above 20 ms^{-1}, relates GEOSAT Ku-band backscatter to wind speed using a simple linear model:

$$U_{10} = -6.4\,\sigma_0 + 72 \tag{12.7}$$

The truth as generated by the cyclone model is of questionable accuracy because analyzed central pressures at the end of the 1980s in the Southern Hemisphere were based on a manual analysis; and therefore, their quality may be fairly poor. Nevertheless, Gourrion et al. (2002) plotted TOPEX σ_0 as function of QuikSCAT (QSCAT) wind speed in clear sky conditions (as inferred from collocated TOPEX radiometer data), and the parameterization in Equation 12.7 was found to be a surprisingly good fit for wind speeds above 20 ms^{-1}. However, how accurate are QSCAT winds for extreme cases? Although the winds from SeaWinds using the QSCAT model function may be regarded an improvement over past scatterometer models (e.g., Donnelly et al. 1999), a comparison of QSCAT winds against 1 year of buoy data involving 114 buoys suggests that for winds larger than 15 ms^{-1} QSCAT winds overestimate the truth. At a wind speed of 20 ms^{-1}, the overestimation already amounts to 1 ms^{-1}. Therefore, in order to get a reliable estimation of the wind speed by the altimeter in extreme circumstances, more work is evidently required.

Finally, measurements at the Ku band are strongly affected by rain (see, e.g., Quartly 1997, 1998). Information from a radiometer is required to ensure that only observations are flagged as valid that are not affected by rain. However, in extreme circumstances such as those occurring near hurricanes, a considerable number of high wind speeds occur during rainfall, which are missed when observed with a Ku-band radar, with serious consequences as the location of the low may be completely missed. A way to circumvent this omission is by combining observations from a dual-frequency altimeter at the C band and Ku band (Quilfen et al. 2006) as is present on board the Jason series and the Sentinel-3 and Sentinel-6 series. As the C-band altimeter is almost insensitive to the effects of rain while the Ku band altimeter shows a considerable sensitivity to rain, there is the potential to simultaneously estimate wave height, wind speed, and rain rate during hurricane conditions. Quilfen et al. (2006) carefully selected Jason-1 altimeter tracks intersecting Tropical Cyclone Isabel using a systematic screening through the National Atmospheric Oceanic and Atmospheric Administration Hurricane Research Division (NOAA HRD) fields. A second selection was made to ensure that the Jason track was located in the vicinity of the eye of the hurricane. For the Ku band, winds were retrieved using the Gourrion algorithm, which was smoothly extended into the large wind speed regime (U_{10} greater than 20 ms^{-1}) by means of Young's (1993) algorithm (Equation 1.7). For the C band, a modification to the algorithm was needed in order to take account of a different sensitivity to foam near the ocean surface. Rain rates were derived using the classical Marshall and Palmer (1948) relation for the frequency-dependent attenuation coefficient of radiation by rain. Finally, SWH was derived from an analysis of the C-band waveform as in rainy conditions SWH from the Ku band may be seriously affected. Therefore, dual-frequency altimeters have the potential to provide valuable information for extreme events, but an independent validation of the measured information is highly desirable.

12.4 SIGNIFICANT WAVE HEIGHT

12.4.1 Principle of SWH Measurement

SWH is defined as the mean height of the highest one-third of the surface ocean waves. It is considered the most important altimeter product as far as the wave prediction is concerned. SWH is used for data assimilation to improve the model analysis and forecast. Therefore, the task of verification and monitoring of SWH is very important.

The temporal variation of the received returned altimeter signal is called the waveform. As shown in Figure 12.1, there is no received signal during the time when the signal is in its way to and back from the ocean surface. Then a sharp jump, termed here as the "leading edge," takes place in the received power as most of the signal arrives back within a short period of time. If the surface reflecting the signal is flat, the jump will be instantaneous. However, the existence of the ocean waves causes the inclination of the leading edge. The higher the SWH is, the more inclined (or the smaller the slope of) the leading edge. This relation between the slope of the waveform leading edge and the SWH is exploited to derive the altimetry SWH product.

SWH product is the most robust altimeter measurement. The errors in the SWH are mainly due to algorithms used for the waveform retracking as the individual waveforms are very noisy. The instrument characterization is important as incorrect values lead to additional errors.

12.4.2 Quality of Altimeter SWH Data

It is clear that it is of the utmost importance to assess the quality and accuracy of the altimeter wave height observations. The typical engineering requirements of the older generation of satellites are that the accuracy of the SWH observations should be better than the maximum of 0.5 m and 10% of wave height in the range of 1–20 m. The newer instruments, such as the RA-2 altimeter on board Envisat, have to satisfy the even more stringent requirement on accuracy of the maximum of 0.25 m and 5% of wave height (Figure 12.12).

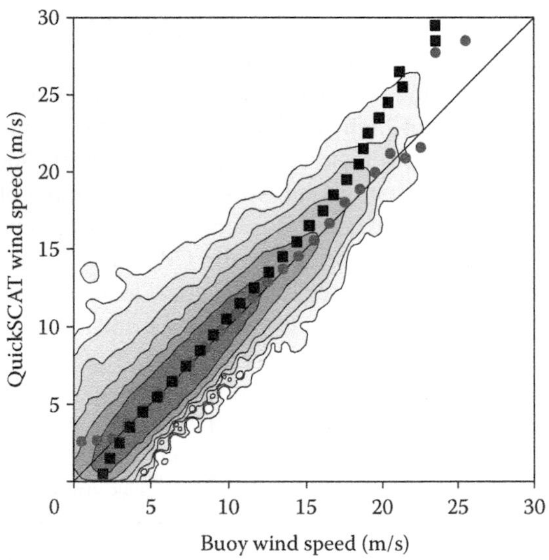

FIGURE 12.12 Comparison of QuikSCAT neutral winds with non-neutral wind speed observations from 114 buoys as received via World Meteorological Organization (WMO) Global Telecommunication System (GTS) over the year 2004. The number of collocations is 106,497. An average positive bias of 0.24 ms^{-1} for QuikSCAT is consistent with the average increase of 0.2 ms^{-1} due to stability effects. For winds above 10–15 ms^{-1}, however, QuikSCAT winds show an overestimation of wind speed. Circles (squares) represent binned averages for given buoy (QuikSCAT) wind speed.

How can one test whether these requirements have been satisfied? A first indication of the quality of the altimeter wave height product is to compare them with model wave height fields (e.g., Figure 12.13a) or with buoy observations (e.g., Figure 12.13b). Further verification results can be found, for example, in Abdalla (2010, 2013). However, this will only give an upper limit to the

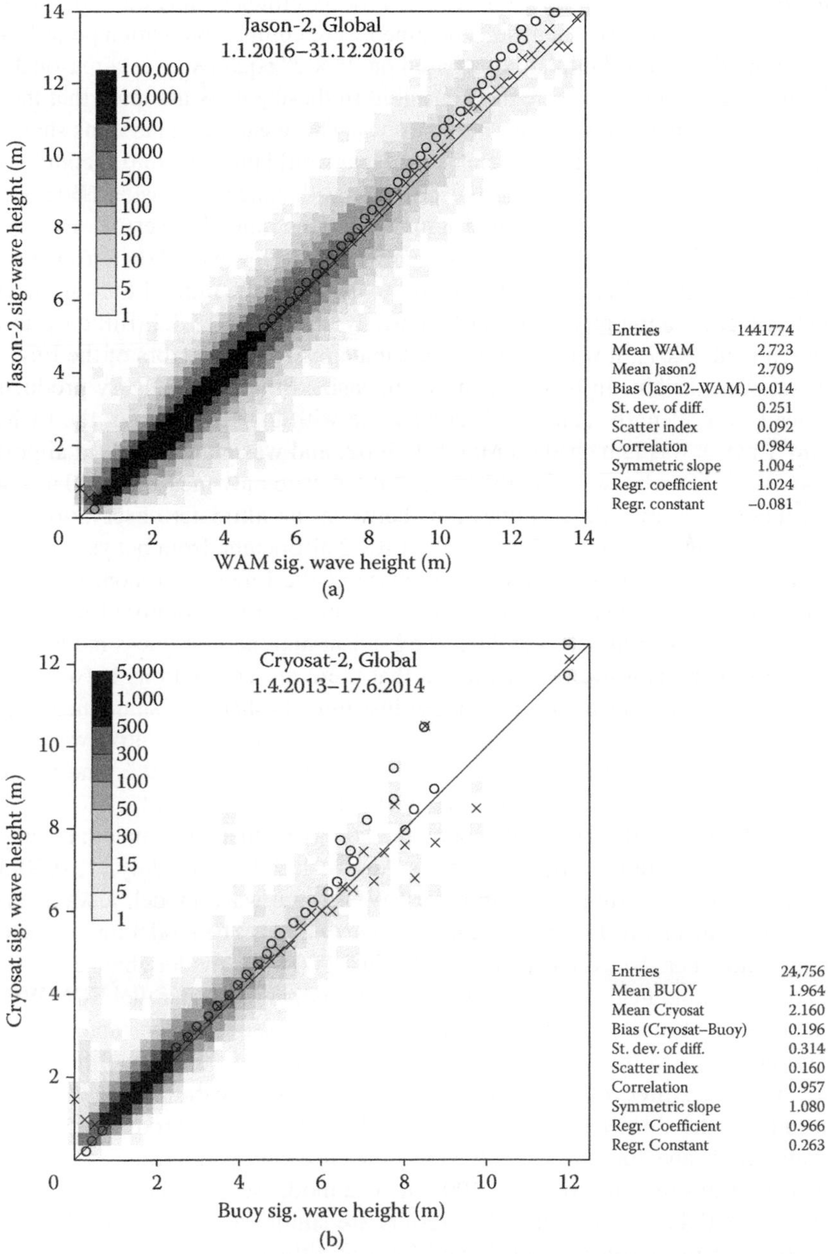

FIGURE 12.13 (a) Comparison between Jason-2 and ECMWF model first guess (FG) significant wave height (SWH) values during the whole year of 2016 over all global water surfaces. The number of collocations in each 0.25 m × 0.25 m two-dimensional bin is color-coded as in the legend. The "x" symbols are the means of the bins for given x-axis values (model), while the "o" symbols are the means for given y-axis values (Jason-2 Altimeter). (b) Comparison between CryoSat-2 near real time surface wave height (NRT SWH) and in situ measurements (buoys) for all data between April 1, 2013, and June 17, 2014.

errors of the altimeter wave height products as model and buoy data have errors as well. Initial work on the calibration of altimeter SWH used comparisons with buoys alone. Carter et al. (1992) used regression-based methods to compare GEOSAT to the U.S. buoy network and found that GEOSAT gave an estimate that was 13% lower compared to the buoys. Cotton and Carter (1994, 1995) used an alternative method. Instead of using individual crossings of the buoy positions by the altimeter, they looked at monthly means over 2° × 2° squares from the satellites compared with monthly means from buoys within those squares. The results obtained were comparable with a point-by-point calibration establishing that a SWH climatology based on 2° × 2° squares gives a reasonable approximation to the true wave climate. The regressions used in these papers assumed that the buoys had negligible error compared to the altimeter observations. These early comparisons showed that this was not the case and that better methods were needed that would take this into account. Cotton et al. (1997) and Challenor and Cotton (1997) used orthogonal distance regression (Boggs et al. 1987, 1989) to overcome this problem. This method assumes that the ratio of the error variances is known and, in altimeter calibration work, it has always been set to 1.0. However, if data from more than two sources are included in the calibration, it is possible to calculate the ratio of the variances directly.

Using the triple collocation method described briefly in Section 12.3.2 (full description can be found in Janssen et al. 2007), it was possible to estimate the random errors of the Envisat and the ERS-2 altimeter wave height. In this case, it is emphasized that fast delivery products are used that are averaged over a length scale that is compatible with the resolution of the ECMWF wave model (55 km). Envisat was launched on March 1, 2002, and was maneuvered in almost the same polar orbit as ERS-2, in such a way that the two satellites were only measuring 20 min apart. As a consequence, there are ample close collocations between the altimeter observations, giving five collocated data sets, namely from the Envisat and ERS-2 altimeters, from buoys, from model first-guess, and analysis. Note that during the initial phase of the Envisat mission, Envisat data were not assimilated because the quality of the new data needed to be monitored first, while ERS-2 wave height data were assimilated. Therefore, the Envisat and the first-guess model errors can be regarded as independent. However, it was found that there are correlations between Envisat and ERS-2 altimeter wave height errors because these instruments share the same measurement principle. When correlated errors are taken into account, it is found that the relative wave height errors (i.e., random error divided by the mean wave height) are 6%, 6.5%, 8%, 9%, and 5% for Envisat, ERS-2, buoys, first-guess, and wave analysis, respectively.

Abdalla et al. (2011) used the triple collocation method to estimate the random errors of Envisat, Jason-1, and Jason-2 SWH during the period from August 1, 2009, until July 31, 2010. In order to avoid the complications of assimilating altimeter SWH in the wave model, a wave model standalone hindcast run without any data assimilation was carried out. The model forced the operational ECMWF wind fields. For the period preceded the ECMWF IFS model change in resolution on January 26, 2010, the wind fields were obtained from the output of the parallel ECMWF model run with the same version of the model that was operationally implemented on January 26, 2010. The NRT SWH observations from each altimeter were collocated with the model hindcast and buoy independently producing three different data sets. In order to alleviate the difference in scales among various data sources, the altimeter data are averaged along the track to produce super-observations with scales between 75 and 100 km, while the buoy data are averaged in time every 5-hourly measurements, which results in scales of about 100 km. The model scale at the time is estimated to be around 65–120 km (Abdalla et al. 2013). The results are summarized in Table 12.2. The monthly wave height errors of the Jason-1, Jason-2, and Envisat altimeters are shown in Figure 12.14. It should be stressed that the validity of the results is restricted to the areas where buoy measurements exist, which is mainly offshore along the coasts of North America and Europe (Figure 12.4c).

An alternative approach was followed by Greenslade and Young (2004). They applied the Hollingsworth-Lönnberg method to obtain the random model and ERS-2 error from the structure of the spatial correlation function. Assuming uncorrelated observations errors, they found a 12% ERS-2 error, while the Australian wave model error was 24%.

TABLE 12.2
The Absolute and Relative (SI) Significant Wave Height Random Errors (Standard Deviations) of Jason-2, Envisat, Jason-1, Model Hindcast, and Buoy as Estimated Using Triple Collocation Method for the Period from August 1, 2009 to July 31, 2010

	Jason-2		ENVISAT		Jason-1	
Number of	13,920		11,005		13,281	
Collocations	Absolute (m)	SI (%)	Absolute (m)	SI (%)	Absolute (m)	SI (%)
Altimeter error	0.130	5.4	0.152	6.2	0.192	7.8
Model hindcast error	0.234	9.7	0.235	9.7	0.241	9.8
Buoy error	0.206	8.6	0.203	8.4	0.218	8.9

Source: Table 1 of Abdalla, S., et al., *Mar. Geod.*, 34(3–4), 393–406. Copyright © European Centre for Medium-Range Weather Forecasts, 2011. Reprinted with permission.
Note: Results are valid for offshore along the coasts of North America and Europe.

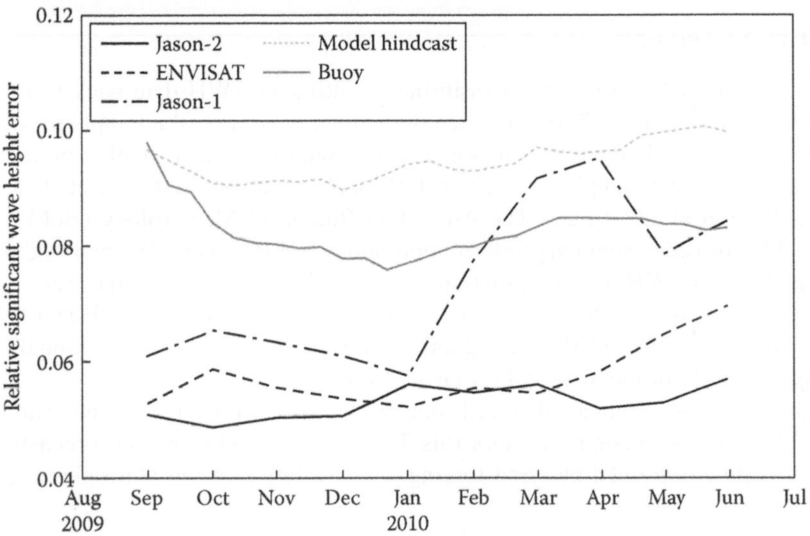

FIGURE 12.14 The time series of the monthly significant wave height errors of Jason-1, Jason-2, and Envisat altimeters, buoys, and the model hindcast. (Figure 4 of Abdalla, S., et al., *Mar. Geod.*, 34(3–4), 393–406. Copyright © European Centre for Medium-Range Weather Forecasts, 2011. Reprinted with permission.)

From the triple collocation studies as well as the other approaches presented so far, only bulk statistics have been produced. With a mean wave height of 2.5 m, one finds that the standard deviation of error is of the order of 15–16 cm; hence, it is concluded that, concerning wave height, the ERS-2, Envisat, Jason-1, and Jaosn-2 missions have comfortably satisfied the engineering requirements of 50 and 25 cm, respectively. On the other hand, there are biases: about −4% of wave height for ERS-2 and +2% for Envisat (Janssen et al. 2007). These biases are, however, relatively small in magnitude realizing that the buoy data, although of high quality, are not perfect either. Nevertheless, there may be problems with the method of retrieving SWH from the altimeter waveform, giving rise to systematic errors. For example, as pointed out earlier, it is assumed that the pdf of the sea surface elevation is Gaussian, but due to the fact that surface gravity waves have sharper crests and wider troughs, there is a positive skewness that gives small deviations from normality.

These deviations lead to the well-known sea-state bias in the altimeter range measurement (compare, e.g., Jackson 1979; Lipa and Barrick 1981; Hayne and Hancock 1982; Srokosz 1986), but they also give rise to small corrections to the SWH (Janssen 2000; Gómez-Enri et al. 2007). The sign of the correction depends, however, on the manner the SWH is obtained from the waveform. For ERS-2, SWH was obtained from the slope of the waveform at the half power point. In that event, Janssen (2000) has shown that finite skewness gives rise to an increase in SWH of at most 10% in very steep wave conditions, but on average the correction is about 4%. On the other hand, for Envisat, wave height is obtained as a result of a fitting procedure of the theoretical waveform to the waveform data. In that case, Gómez-Enri et al. (2007) found that a finite skewness will give a reduction in wave height of about 10 cm. Therefore, skewness effects could help to explain the biases seen in the ERS-2 and Envisat altimeter data.

Note finally that Gómez-Enri et al. (2007) also provide estimates of the skewness of the ocean surface from the fitting procedure as measured from space. However, remarkably, there are large areas (in particular in the tropics) where the skewness is found to be negative. This is odd as a straightforward estimate by Jackson (1979) shows that skewness should be positive. Clearly, more work is needed to understand this surprising result.

12.5 APPLICATIONS

12.5.1 Data Assimilation

The earliest documented efforts toward assimilating altimeter SWH date back to the late 1980s (Esteva 1988; Janssen et al. 1987, 1989). Since then, there were several attempts to achieve operational altimeter wave height data assimilation at operational numerical weather prediction (NWP) centers around the world. Examples are ECMWF (Lionello et al. 1992; Abdalla et al. 2004), Meteo-France (Skandrani et al. 2004), and The Australian Bureau of Meteorology (BoM) (Greenslade 2001). Table 12.3 provides summary information about NWP centers that run operational wave systems with altimeter SWH data assimilation as emerged from a recent e-mail survey carried out by J.-R. Bidlot among the members of the Expert Team on Waves and Coastal Hazards Forecasting Systems (ETWCH) of the World Meteorological Organisation (WMO) Joint Technical Commission for Oceanography and Marine Meteorology (JCOMM).

Wave data assimilation is not as advanced compared to what has been done in the area of weather forecasting. There are two main reasons for this. First, in contrast to weather forecasting, the wave forecast is strongly constrained by wind forcing. The evolution of the atmospheric conditions is

TABLE 12.3
Operational Wave Models that Assimilate Altimeter Significant Wave Height as of Early 2017

Organization	Country	Assimilation Technique	Start	Source/Reference
ECMWF	–	OI	15 Aug. 1993	Janssen et al. (1997a)
DWD	Germany	OI	2008	T. Bruns/Li et al. (2014)
MeteoFrance	France	OI	17 Mar 2011	L. Aouf/Lefèvre and Aouf (2012)
JMA	Japan	OI	24 Oct. 2012	N. Kohno/Kohno et al. (2011)

Note: DWD, Deutscher Wetterdienst; JMA, Japan Meteorological Agency.
Summary of an e-mail survey carried out by J.-R. Bidlot among members of the Expert Team on Waves and Coastal Hazards Forecasting Systems (ETWCH) of World Meteorological Organisation (WMO) Joint technical Commission for Oceanography and Marine Meteorology (JCOMM).

mainly controlled by the atmospheric initial state, whereas the initial wave field loses its influence after a relatively short time, ranging from a few hours to a number of days depending mainly on the basin size, the sea-state conditions, the wind strength, and on the atmospheric dynamical time scale. In theory, a perfect wave model driven by perfect winds would produce perfect wave fields after a certain time, whatever the initial state might have been. However, this is not the case for the atmospheric model for which chaotic behavior makes it very sensitive to its initial conditions. The second reason for the late introduction of wave data assimilation is that before the advent of satellites, only *in situ* wave data were available. In particular, the observations from wave buoys are of high quality, but the data coverage is limited to coastal areas in the Northern Hemisphere. Therefore, these data are of little use in global wave data assimilation. In practice, these *in situ* data are used for the control and monitoring of NWP and wave models.

The prospect of the advent of satellite data encouraged NWP centers to study the possibility of including wave data assimilation schemes in their operational wave forecast suites. For wave analysis, the wind fields are provided from the analysis of the atmospheric model. Satellite wave data are assimilated to improve the initial sea state used for the wave forecast. Verified against independent *in situ* wave measurements, the benefit of altimeter wave height data assimilation on wave height analysis can be clearly seen from Figure 12.15, where for the years between 1999 and 2007 we show time series of the scatter index (standard deviation of error normalized with mean observed value) of wave height as function of time for the ECMWF operational analysis (with data assimilation) and for a hindcast run with the wave model that did not assimilate satellite data. In the summers up to 2004, there were considerable differences between wave height simulations with and without data assimilation, suggesting that satellite data assimilation played an important role in the quality of the wave height analysis. However, in the spring of 2004 and in the spring of 2005, wave model improvements were introduced into operations that had a considerable positive impact on wave forecasting performance. As a consequence, because of these model improvements, the impact of the assimilation of altimeter and SAR data on the accuracy of the wave height analysis has diminished from then on.

Data assimilation also has a positive impact on the wave forecast as can be seen from Figure 12.16, but the size of the impact depends on the area. The left panels of Figure 12.16 show the impact over the whole globe, and judging by the scores on the scatter index and the absolute correlation, there is a modest impact up to 2.5 days in the forecast. The bias in wave height is a minor problem on a global scale. However, there are areas that have significant systematic errors.

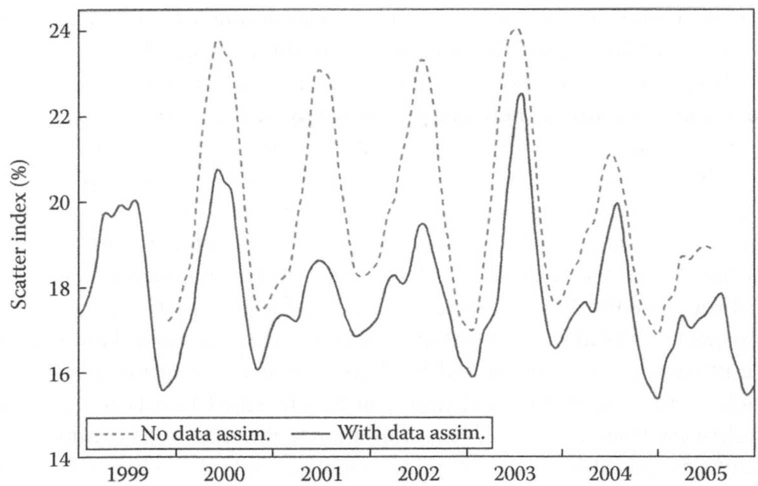

FIGURE 12.15 Impact of ERS-2 altimeter wave height data assimilation on ECMWF wave height analysis as compared to independent in situ wave height observations between 1999 and 2005. Shown is the three monthly mean scatter index.

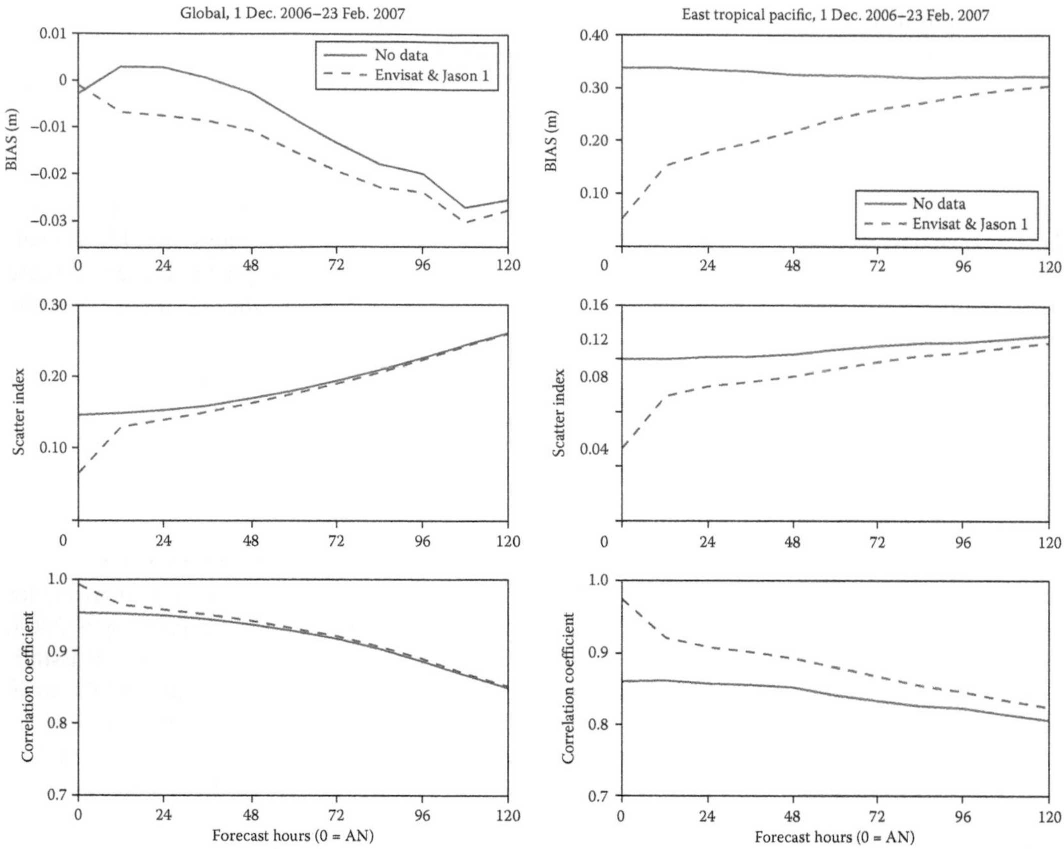

FIGURE 12.16 Impact of ERS-2 altimeter wave height data assimilation on ECMWF wave height forecast obtained by comparing wave height forecast against Envisat wave height observations. Period is about 3 months in the winter of 2006 to 2007. Left panels are for the global area and right panels are for the eastern Pacific.

One such area is the eastern tropical Pacific. The corresponding forecast scores are shown in the right panels of Figure 12.16. Clearly, forecast impact of data is much larger and longer lasting in areas where swell systems (which give a long memory to the forecast system because their lifetime is large) dominate and where there are significant systematic errors.

The degree of improvement in the wave height forecast achieved by assimilating wave data depends on a number of factors. We briefly discussed already the accuracy of the forecast surface winds and the quality of the wave model itself, realizing that a good wave forecasting system will show less impact of wave data on forecast skill. A second important factor is the quality of the wave data that are assimilated in the model fields. This issue will be discussed in the next section while here we briefly discuss the third important factor, namely the assimilation procedure.

When the prospect of global wave height data emerged, the first assimilation methods that were developed were obviously the simplest and the least expensive in terms of computer resources. Several approaches are conceivable, and they can be classified into two categories: sequential methods and multi-time level methods. The assimilation techniques most commonly used for operational applications are based on instantaneous sequential methods such as OI (e.g., Lionello et al. 1992) and successive corrections (e.g., Thomas 1988). Such methods are very attractive due to their low computational cost. However, the corrections are done at a local scale and at a one-time level. As from experience it is known that the main error source is the driving wind field, it would make sense to use the winds as the control variable in the analysis scheme (i.e., to modify the winds in such a way that an optimal agreement with the observations for wave height is obtained). For wind sea, updates to the wind field may be obtained in the context of a single time level approach. However, in case of swell, this approach does not work because swells have been generated by remote storms a number of days ago. The assimilation of altimeter wave height data represents an additional challenge because it only provides information on the integral over frequency and direction of the wave spectrum, whereas modern wave models are based on a spectral description. Applying the assimilation method results in a wave height increment that must be translated to a corresponding change in the local wave spectrum. For wind sea, this is fairly straightforward to do by using the evolution laws for wind generated waves, as obtained, for example, during the Joint North Sea Wave Project (JONSWAP) (Hasselmann et al. 1973) field campaign. However, for swell, it is assumed that during the transformation the MSS is invariant, which may be plausible, but in practice this assumption is hard to justify (Lionello et al. 1992; Greenslade 2001).

A number of multi-time level methods have been developed and tested in the past. One such method is based on the Kalman Filter (KF). The KF has the additional advantage in that it provides error statistics on the model variables. The KF propagates a forecast error covariance matrix (FECM) that gives further information on the model state. The problem of implementing these techniques arises from the dimension of FECM, which then has implications on the required number of model integrations. Some simplifications are required to reduce the cost of such methods (Voorrips 1998).

Another multi-time level method is based on the variational approach. Such methods are based on the minimization of a cost function and often use the adjoint technique in order to compute the gradient of the cost function. Multi-level time variational techniques take into account the history of the observations under the constraint of the wave model dynamics. A promising first step was reported by de las Heras et al. (1994) and Hersbach (1998), who managed to create the "true" wind forcing based on wave height observations alone.

These studies were using the adjoint of the WAM model, which was based either on coding the analytical adjoint or by means of an automatic procedure. However, the high cost of these methods has slowed down their introduction in the field of wave forecasting, although the variational approach is by now widely used in operational weather forecasting. Simplifications are always required for operational implementations. For example, simplifications in the tangent linear model as well as a reduction of its resolution would allow a significant reduction of its cost as it is necessary to carry out between 10 and 100 integrations of the adjoint of the tangent linear model to converge toward the optimum trajectory. Such an approach works well in the atmospheric context because the main interest is in an analysis of the large scales. However, for wave modeling, the main interest is in wave development during storms, which have smaller scales—say of the order of 500 km; and therefore, a simplification such as a reduction in resolution is probably counterproductive. Using an approximate tangent-linear model, Voorrips and de Valk (1997) compared results of the variational approach with an OI method. No advantage was found for the variational method, however, presumably because of a not optimally calibrated tangent-linear model.

Finally, there is a problem common to all assimilation methods that was by Greenslade and Young (2004). A key element in any data assimilation scheme is that knowledge is required of the spatial correlation of the model error. In most schemes, this is modeled by a simple exponential distribution involving the ratio of the distance between the two points of interest and a correlation length scale, where the correlation length scale is regarded to be independent of location and is adjusted in such a way that in some sense optimal results are achieved. This is an *ad hoc* approach, and Greenslade and Young (2004) show, using the Hollingsworth- Lönnberg (1986) method, how in a rational way the correlations in model error may be obtained by a comparison with superobbed altimeter wave height data (superobbing is required in order to avoid correlations that do exist between individual altimeter observations, see Janssen et al. 2007). It then turns out that the correlation length scale is a function of location on the globe, with large spatial scales (of the order of 700 km) in the tropics (where swells prevail), while in the extratropics, correlation scales are considerable smaller (of the order of 400 km) because wind seas, which have smaller scales, are an important component of the sea state in the storm tracks. Greenslade and Young (2005) have used these inhomogeneous model error correlations in the Australian wave analysis system, and an improved forecast skill was found when compared to the operational analysis system, with a fixed correlation length scale of 300 km. Preliminary tests at ECMWF, however, have only shown a modest impact of this promising change on the model correlation matrix; hence, more work is evidently needed.

Despite the aforementioned shortcomings, even today, most weather centers with wave modeling capabilities are using OI schemes or related schemes to assimilate SWH from a number of altimeters (see, e.g., Table 12.3). Admittedly, these schemes involve a number of strong assumptions to relate wave height increments to changes in the wave spectrum. Therefore, instruments that would provide spectral information such as SAR are ideally suited for wave assimilation because no assumption in the mapping from the observed information to spectral change is required. The introduction of SAR data assimilation has been tested in a number of studies (e.g., Breivik et al. 1998; Dunlap et al. 1998; Aouf et al. 2006b) and was introduced operationally at ECMWF in February 2003 (Abdalla et al. 2004). Forecast impact of the introduction of SAR data was similar to that of altimeter wave height data, which suggests that, although the wave height assimilation schemes involve a number of strong assumptions, the analysis results seem nevertheless realistic.

In closing, we note that, in the course of the last three decades, we have seen—coinciding with considerable improvements in the quality of the surface winds and the wave model—a gradual reduction of the size of the forecast impact by assimilating altimeter wave height data. Nevertheless, weather forecasting centers need to provide the best possible wind and wave forecast and need to deliver a forecast product with a consistent quality in time. A way to deliver a consistent product is to use observations to constrain the analysis in order to keep it on the right track, and for this reason, assimilation of altimeter wave height data will continue. In this context, there is a continuous confrontation between altimeter wave height data and their model counter parts, which can only benefit both. As an example, we mention some problems with the ECMWF wave prediction system that were discovered in the summer of 2003, when for a number of weeks ERS-2 could not deliver observations of the sea state because the on-board tape drives had failed, as it turned out indefinitely. This loss of data had serious consequences for the quality of the ECMWF wave analysis as is plainly clear from Figure 12.15, where the scatter index of the wave height analysis shows a maximum in the summer of 2003.

In April 2005, a new version of the dissipation source function was introduced, which used an alternative definition of the integral parameters, such as mean frequency, in the expression for the dissipation. The new definition is given in Bidlot et al. (2007); see also Janssen (2007). As can be seen in Figure 12.15, this change had a further beneficial impact on the accuracy of the wave height analysis in the Northern Hemisphere summer of 2005 and subsequent summers.

The first operational implementation of SWH assimilation in the global ECMWF IFS was realized on August 15, 1993. The history of ocean wave data assimilation in terms of instruments used is shown in Figure 12.17. The impact of altimeter SWH data assimilation can be assessed through comparison with independent *in situ* data, the model's own analysis, and wave and atmospheric data from other instruments.

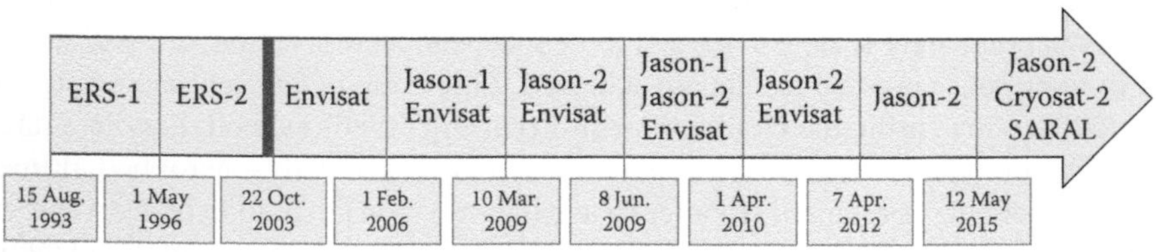

FIGURE 12.17 Timeline of altimeter surface wave height (SWH) assimilation at ECMWF. There is currently good resilience in altimeter SWH observations as data are being provided by three satellites. Jason-3 and Sentinel-3A data are expected to be added in 2017.

Figure 12.18 shows the mean difference between, on the one hand, the SWH analysis when Jason-3, CryoSat-2, and SARAL observations are assimilated and, on the other, the SWH analysis from a model run without any SWH data assimilation. Both are stand-alone wave model runs uncoupled with the atmosphere. Altimeter SWH data assimilation clearly affects the analysis, and detailed evaluation shows that it improves it. For example, the model is known to overestimate the SWH in the area in the eastern Pacific off Central America. The data assimilation corrects this overestimation by reducing the wave height.

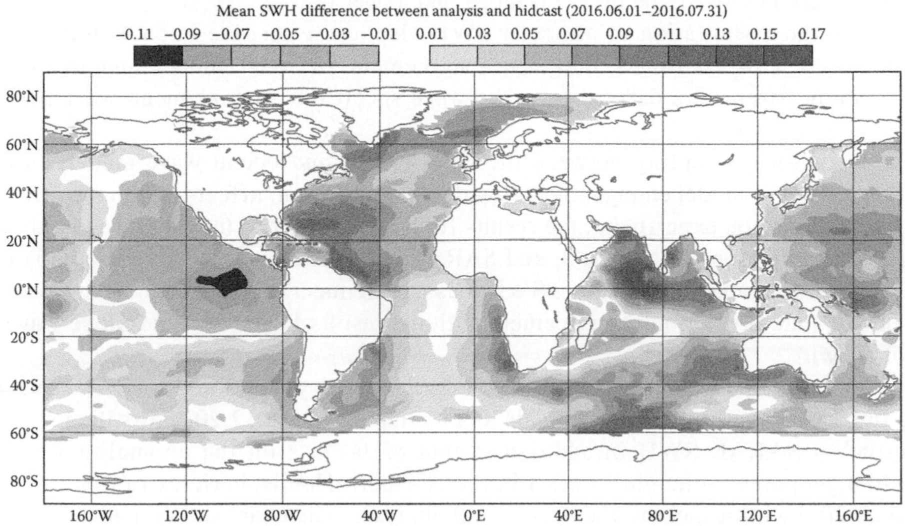

FIGURE 12.18 Mean impact, in June and July 2016, of assimilating Jason-3, CryoSat-2, and SARAL surface wave height (SWH) data on the SWH analysis, expressed as the difference in SWH between an ECWAM stand-alone model run at a resolution of 0.25° (IFS Cycle 42R1), assimilating data from the three satellites and another model run without any data assimilation.

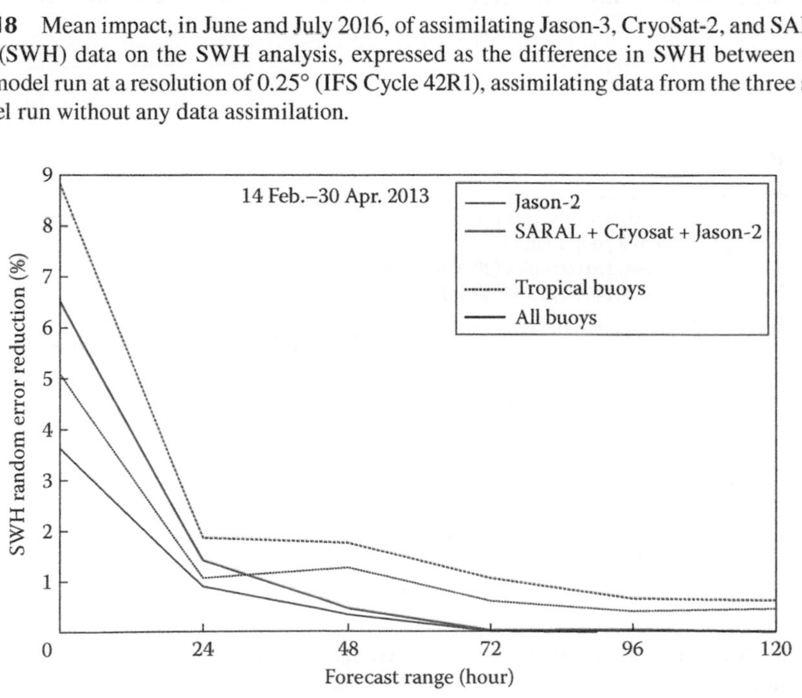

FIGURE 12.19 Impact of assimilating altimeter data on reducing the surface wave height (SWH) random error in an ECWAM stand-alone model run at a resolution of 0.25° (IFS Cycle 40r1) as verified against *in situ* buoy data, averaged over the period February 14 to April 30, 2013.

In situ wave data are not assimilated and can therefore be used as independent data for data assimilation impact assessment. Figure 12.19 shows the percentage by which SWH errors are reduced at analysis time and at various forecast ranges compared to *in situ* measurements from buoys and platforms. The assimilation of one satellite alone (Jason-2) reduces the error by about 3.5% (about 5% in the tropics) at analysis time, while assimilating SWH from three altimeters (Jason-2, CryoSat-2, and SARAL) reduces the error by 6.5% (about 9% in the tropics). The impact of data assimilation decreases with forecast range and vanishes after about 2 days in the extratropics (latitudes higher than 20°), which is usually dominated by active wave generation. In the tropics, which is dominated by swell, the impact is larger and longer lasting.

In general, the assimilation of altimeter SWH also has a positive impact on the predicted wave spectrum. This translates into better agreement between model and measured sea state-describing parameters derived from the wave spectrum, such as mean wave period (not shown).

The tight two-way coupling between the atmospheric and ocean wave models in the IFS means that any wave model change, including data assimilation, affects the atmospheric fields. In another assimilation experiment, the results from a full IFS run (coupled wave-atmospheric model runs) using Jason-2, CryoSat-2, and SARAL SWH measurements were compared with the results of only assimilating Jason-2 SWH measurements. The experiment showed a positive impact on sea state predictions in agreement with the results from the stand-alone wave model runs (not shown).

Furthermore, the additional altimeter SWH data from CryoSat-2 and SARAL have a small positive impact on some atmospheric fields. For example, Figure 12.20 shows the mean impact of assimilating SARAL SWH in addition to that of Jason-2 on the anomaly correlation of the 500 hPa geopotential height forecast in the Northern Hemispheric extra tropics (latitudes higher than 20°) with respect to the operational analysis. The chart shows a generally positive impact, although on most days the effect is not statistically significant at a confidence level of 95%.

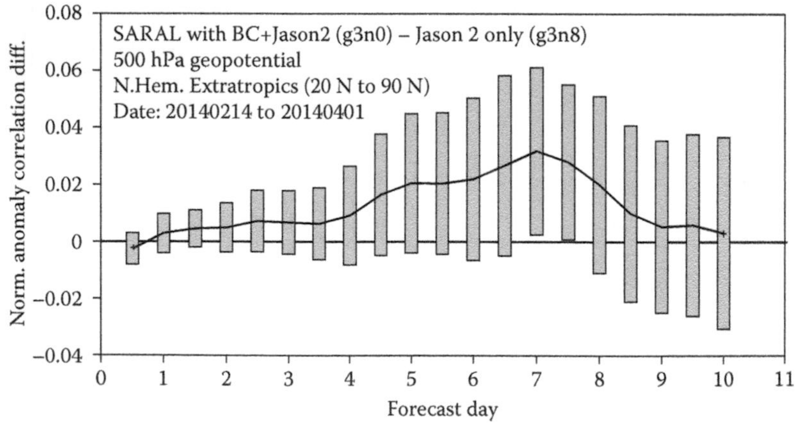

FIGURE 12.20 Mean impact of assimilating SARAL surface wave height (SWH) on the geopotential anomaly correlation at 500 hPa in the extratropical Northern Hemisphere, for forecasts produced by IFS Cycle 40R1 (atmosphere and waves) at the resolution TL511 (corresponding to a grid spacing of about 40 km) between February 14, 2014, and April 1, 2014. Vertical bars show 95% confidence intervals. (Figure 10a of Abdalla, S., *Mar. Geod.*, 38, 365–380. Copyright © European Centre for Medium-Range Weather Forecasts, 2015. Reprinted with permission.)

12.5.2 Estimation of Effective Model Resolution

Spectral analysis using discrete Fourier transform is an attractive tool to resolve data series into their underlying simple sinusoidal functions covering all possible scales. This concept can be used to reveal the ability of NWP models in resolving various scales by comparison with available theoretical and empirical (mainly from satellite data) spectra. The first step is to establish a reference against which the model spectra will be compared.

Theoretical, experimental, and observational studies show that atmospheric energy spectra follow a power law in the form of k^{-n} where k is the wavenumber (i.e., reciprocal of the horizontal scale). Theoretical studies (e.g., Lilly 1989) suggest that in the upper atmosphere the exponent n has a value of three at large scales (small wavenumbers) changing down to five/three at smaller scales. There is little known about the shape of the spectrum at the surface. Oceanic surface wind observations (e.g., scatterometers) show agreement with the upper atmosphere theory as far as the exponent at small scales is concerned (i.e., $n = 5/3$). However, it has been found by earlier studies that the value of n for larger scales varies between 2.4 and 2.6.

Surface wind speed product from the radar altimeter (RA-2) aboard the European Space Agency (ESA) Environmental Satellite (Envisat) has a relatively short sampling interval of about 7 km. A typical RA-2 measurement covers a footprint of a few kilometers (typically below 10 km). This resolution is enough to resolve scales in the order of a few tens of kilometers. This makes RA-2 surface wind speed measurements a very good candidate to study the properties of the atmospheric spectra at the ocean surface.

All continuous RA-2 data records of 7168-km length (1024 measurements sampled at 7 km) during a period of 1 year from August 1, 2010, to July 31, 2011, were extracted by Abdalla et al. (2013). There were 685 records in total. A fast Fourier transform (FFT) was used to compute the corresponding spectra after applying detrending (i.e., removing linear trends) and windowing (i.e., ensuring the periodicity). The average of the 685 spectra is plotted in Figure 12.21a. Comparing the average spectrum with the wavenumber power law, it is seen that the RA-2 spectrum follows a power law with $n = 2.5$ for large scales and $n = 5/3$ for smaller scales. The transition occurs at around 400-km wavelength. The spectral shape is in very good agreement with other (surface) wind spectra available in the literature from other instruments (e.g., Nastrom and Gage 1985;

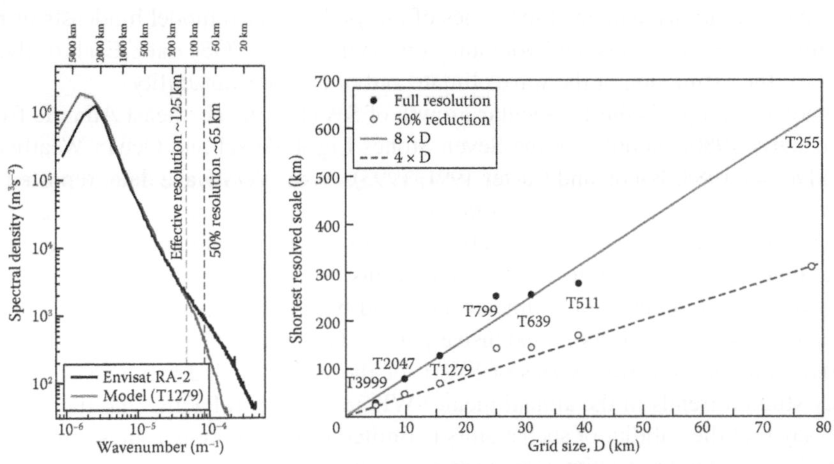

FIGURE 12.21 Spectra of Envisat altimeter and (a) model 10-m wind speed spectra and (b) effective model resolution of various IFS model configurations. The approximate grid spacings for each model are as follows: T_L3999: 5 km; T_L2047: 10 km; T_L1279: 16 km; T_L799: 25 km; T_L639: 31 km; T_L511: 39 km; and T_L255: 78 km. (Reproduced from Abdalla, S., et al., *ECMWF Newslett.*, 137, 19–22. Copyright © European Centre for Medium-Range Weather Forecasts, 2013. Reprinted with permission.)

Xu et al. 2011). Therefore, the mean RA-2 spectrum is used here as a reference against which the corresponding model spectra can be compared.

At the time of the study, the high-resolution operational forecasts (HRES) of ECMWF IFS used an NWP model with a spectral truncation of T_L1279, which corresponds to a horizontal grid resolution of 16 km (between January 26, 2010, and March 7, 2016); 560 ECMWF operational forecast 10-m wind speed fields over the ocean during the period from October 1 to October 10, 2011 were used to compute the model ocean-surface wind speed spectra. The spectra were averaged to produce the model wind speed spectrum shown in Figure 12.21a. The T_L1279 model spectrum coincides well with that of the RA-2 and only starts to deviate at a wavelength of about 120 km, that is, about eight times the model grid spacing (16 km). This means that the model is able to fully resolve the same structures as does RA-2 for all scales in excess of 120 km. This scale is termed as the effective model resolution.

The same procedure was carried out for other model resolutions (both coarser and finer). The deviation of the model spectrum from the reference altimeter spectrum is happening at more or less at eight times the model grid spacing. The results are summarized in Figure 12.21b.

Beyond the full spectrally resolved energy, it is sensible to accept scales with partial spectral energy content to be quite useful information. If we relax the strict definition of the effective resolution to what can be termed as effective useful resolution by requiring the presence of at least 50% of the variability of that scale (open circles in Figure 12.21b), the effective useful resolution of the current HRES configuration becomes three to five grid spacings. This needs to be taken into account when users of ECMWF products interpret the detailed information in the forecast fields.

12.5.3 Sea State Climatology

Design and planning activities of marine vessels, structures (e.g., harbors and offshore platforms), and activities (e.g., recreation and oil production) require a good knowledge of the wind and wave climate, including the most severe wind and wave conditions that are expected during the lifetime of vessels and structures. In the past and certainly in the pre-satellite era, this knowledge was often difficult to obtain as the amount of observations of the sea state is fairly small because it is mostly confined to the shipping routes in the Northern Hemisphere. Presently, there are four major sources of wind wave observations, namely, time series of buoys, long-term model hindcasts or reanalyses, satellite altimeter measurements, and voluntary observing ship (VOS) data. Each of these sources can be used for the estimation of the wave climate and of climate variability.

A well-known example is the 1% secular growth of SWH in the northeast Atlantic from the late 1960s to the early 1980s recorded at the Seven Stones Light Vessel and Ocean Weather Station L (Carter and Draper 1988; Bacon and Carter 1991, 1993). Also, VOS wave data, reported by marine officers worldwide since 1856, are an important source of wave data for wave height climatology. In the past, a large effort was devoted to correcting these observations of many biases and to minimizing observational errors inherent to visual observations. In addition, these observations have been validated against *in situ* buoy data, altimeter data, and wave model hindcasts. The strong point of the VOS data set is that it covers an extensive period. Gulev and Grigorieva (2004) developed a 120-year-long homogenized time series of SWH for the well-sampled locations of the major ship routes. They studied trends in the annual mean SWH in the North Pacific and the North Atlantic. Needless to say that the validity of such results is limited to the shipping routes only.

Satellite observations from altimeters cover the globe with quite homogeneous sampling and in the future are expected to become superior with respect to VOS observations. However, at present, these data cover a too short period to be of value for climate variability studies. Also, when using different satellites, intercalibration of the different instruments is required but, as discussed in the previous section, intercalibration procedures using triple collocation methods are available nowadays. Nevertheless, altimeter data already give valuable information on the climate as shown in Figure 12.22 by the example of maximum surface wind speed and SWH in the Black Sea obtained

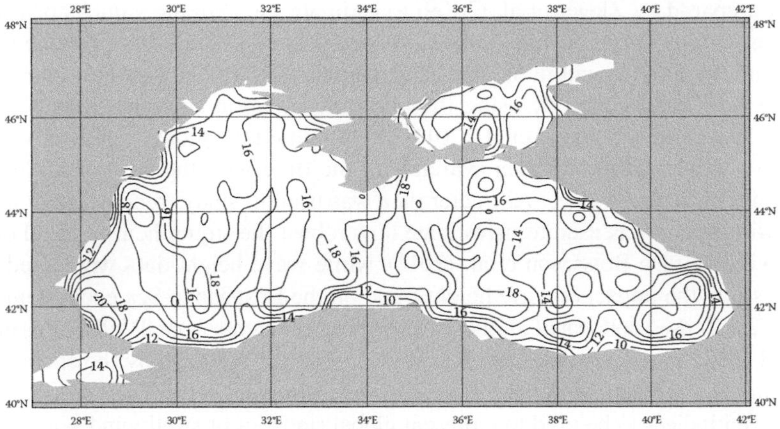

(a)

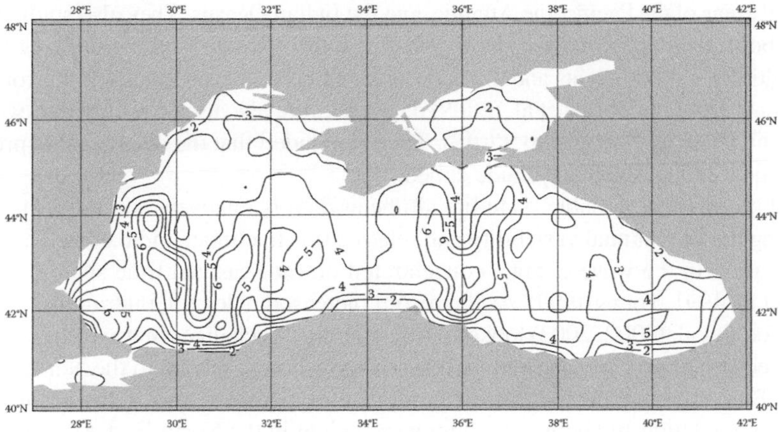

(b)

FIGURE 12.22 The (a) absolute maximum surface wind speed and (b) significant wave height over the Black Sea as measured by Envisat radar altimeter during its lifetime between 2002 and 2012. (Figures 2 and 3 of Abdalla, S., Wind and Wave data selection for climate studies in enclosed seas. In E. Özhan (ed.), *Global Congress on ICM: Lessons learned to address new challenges, MEDCOAST, Mediterranean Coastal Foundation*, Dalyan, Turkey, pp. 1347–1358, 2013b.)

from 10 years of Envisat altimeter data (Abdalla 2013b). Queffeulou and Bentamy (2007) studied the wave height climate in the Mediterranean based on 13 years of TOPEX-Poseidon data. However, the way altimeter onboard polar satellites sample the ocean (see Figure 12.2), there are many spatial gaps in the wave height climatology derived from the altimeter wave height data. Although this is less of an issue in the case of the Envisat family (approximately three passes with one degree in latitude), it is worse in the case of the TOPEX-Poseidon with gaps of about three times in size. Of course, this is a price to be paid in order to have more frequent sampling with 10-day repeat cycles for the latter versus 35-day cycles for the former. If the current number of altimeters in orbit (six of them) is maintained for few decades, this would eliminate this sampling shortcoming in the future, providing a proper inter-calibration among the altimeters is carried out (see, e.g., Zieger et al. 2009).

The estimation of extreme SWHs over the whole globe using altimeter data from GEOSAT, ERS-1, and TOPEX-Poseidon was studied by Alves and Young (2003). They examined various methods and distributions and highlighted some of the limitations of this approach including the space–time undersampling issue. Vinoth and Young (2011) used an extended database of 23-year altimeter data prepared by Zieger et al. (2009) to estimate the extreme values corresponding to a 100-year return period. They concluded that altimeter data can be used to produce extreme value wave heights and wind speed within 5% and 10%, respectively, of buoy measurements. The under-sampling issue can be alleviated by using the initial distribution approach rather than the peaks over threshold. Furthermore, Young et al. (2011, 2012) studied the global trends in wind speed and SWH using all available altimeter data available at the time with the second study concentrated on the trends in the extreme values. Altimeter data was used to study trends at regional scales. For example, the wind speed was used to investigate how robust the strengthening trend of the tropical Pacific trade winds by De Boisséson et al. (2014), while wave height data were used by Woo and Park (2017) to study the impact of wave height trend on the East/Japan Sea ecosystem.

Chen et al. (2002) used collocated altimeter SWH measurements from TOPEX-Poseidon and scatterometer wind speed measurements from NSCAT and QuikSCAT to study the spatial and seasonal pattern of dominant swell and wind wave regimes over the whole globe. They proposed two energy-related normalized indices to be used to represent global statistics of swell/wind sea probabilities and intensities. They found three well-defined, tongue-shaped zones of swell dominance located in the eastern tropical areas of the Pacific, the Atlantic, and the Indian Oceans. They also studied the seasonal variability of both swell and wind sea. Jiang and Chen (2013) confirmed the findings of Chen et al. (2002) using altimeter wave heights and the collocated radiometer wind speeds, both from Jason-1.

Altimeter data from the Zieger et al. (2009) database were also used to study the swell decay rate in the Southern Ocean by Young et al. (2013). They concluded that the decay rate is proportional to wavenumber squared and swell amplitude cubed.

The period covered by "reliable" satellite altimeter data is relatively short—some investigation related studying the interannual variability and relating this to climate indicators. The first attempts to look at variation in the wave climate concentrated on the seasonal time scale (Challenor et al. 1990; Carter et al. 1991; Young 1999). As the record increased in length, interannual scales could be considered. Woolf et al. (2002, 2003) looked at the interannual changes in wave climate in the North Atlantic referred to earlier. They showed that these changes were related to the North Atlantic oscillation (NAO). By comparing the patterns of variability from year to year in the altimeter data, they produced a simple relationship between winter wave height in the North-East Atlantic and the NAO index. Hindcasts produced from this relationship compared well with both GEOSAT altimetry and with *in situ* data from weather ships. Wimmer et al. (2006) looked at the estimation of extreme waves from altimeter data alone.

For studying changes in wave climate, wave hindcasts simulated by advanced wave models driven by reanalyzed winds are attractive. In particular, when this is combined with an assimilation of appropriately calibrated altimeter wave height data, an optimal, continuous product for wave height climatology is to be expected. This is the approach followed by Caires and Sterl (2005). Their starting point was wave fields obtained from ERA-40 (Uppala et al. 2005), which is a global reanalysis of meteorological and ocean-wave variables from 1957 to 2002. The data consist of 6-hourly fields on a $1.5° \times 1.5°$ latitude/longitude grid covering the whole globe. Initial validation of the data revealed a generally good description of variability and trends. However, because of the coarse resolution, underestimation of high wave height and wind speed events was noted as well. Therefore, 100-year return values of SWH obtained from the ERA-40 reanalysis need to be corrected for this underestimation. Caires and Sterl (2005) used the peaks-over-threshold method to estimate 100-year return values from buoy data and model data, and the model data were corrected using a linear regression based on a comparison with the extreme statistics from the buoy data. Buoy data, although of high quality, are limited to a restricted number of locations. In order to consolidate this linear calibration, a comparison between modeled and global independent TOPEX-Poseidon extreme statistics was performed as well. The resulting 100-year return SWH map is given in Figure 12.16. This plot shows that the most extreme wave conditions are in the storm track regions and that the highest return values occur in the North Atlantic. In agreement with operational experience, the highest average conditions are found in the Southern

Hemisphere storm track, but the most extreme conditions occur in the North Atlantic. Caires and Sterl (2005) also studied certain aspects of climate variability, but this aspect of the reanalysis is somewhat uncertain. Inhomogeneities in reanalysis winds could influence the patterns of climate variability in the wave hindcasts. These inhomogeneities are primarily present in the Southern Hemisphere (and to a lesser extent in the Northern Hemisphere) where due to the introduction of satellite data there has been a massive increase in the number of data used in the atmospheric analysis. Although the ERA-40 team took great care in intercalibration of the most important satellite data sources and in removing systematic differences between the model and the satellite data, there is of course no guarantee that the introduction of new data types might not lead to artificial trends. Despite this caveat, it seems that the Caires-Sterl approach is the most promising way forward to obtain a reliable wave height climatology. Clearly, altimeter wave height data play a vital role in the establishment of such a climatology, but these data should become available over a much longer period.

12.6 NEW DEVELOPMENTS

12.6.1 SAR (Delay-Doppler) Altimetry

Delay-Doppler, commonly known as SAR, altimetry can be considered a revolution in the field of altimetry. Coherent processing of several tens (usually 64) of echoes within each altimeter burst during a few milliseconds allows the narrowing of the altimeter footprint in the azimuthal direction (i.e., along the track) to about 250–300 m compared to a few kilometers in the case of conventional altimeters. The procedure implements the principles of SAR along the track (and hence the use of the name SAR altimetry) to exploit the slight frequency shifts, caused by the Doppler effect, in the forward- and aft-looking parts of the beam. This results in high-resolution measurements.

The CryoSat mission was pioneering in implementing delay-Doppler (SAR) altimetry. The Synthetic Aperture Interferometric Radar Altimeter (SIRAL) is the main payload of CryoSat. SIRAL was designed to measure ice-sheet elevation and sea-ice freeboard, which is the height of ice protruding from the water. Therefore, it operates in SAR mode over ice surfaces. In order to make use of its high-resolution capabilities, it operates in SAR mode near the coasts as well. Furthermore, for the sake of scientific exploration, SIRAL also operates in SAR mode over few ocean areas (boxes). Two of those boxes are shown in Figure 12.23a. Over the major part of ocean bodies, it operates as a conventional altimeter in what is known as the low bit rate mode (LRM).

The SAR mode over the few ocean boxes enabled scientists to develop processing techniques to exploit the high-resolution SAR mode for oceanographic measurements including SWH and wind speed. Figure 12.23b and c shows the comparison between CryoSat-2 SAR mode wind speed and SWH and their model counterparts in the northeastern Atlantic box. The comparison is clearly as good as with conventional altimetry if not slightly better.

The Sentinel-3 series of satellites, which is part of the space component of the European Commission (EC) Copernicus Global Monitoring for Environment and Security programmer, carry SAR altimeters (SRAL), which are able to operate in SAR mode. In fact, Sentinel-3A, the first in the series, which was launched on February 16, 2016, is the first satellite to provide 100% SAR-altimetry coverage. The decision to operate exclusively (except for one cycle during the commissioning phase) in SAR mode was motivated by the excellent results obtained from CryoSat mission SAR mode over the few ocean areas. The early results obtained from the first few cycles of Sentinel-3A proved that the SAR mode ocean products are of high quality. Figure 12.24 shows the comparison between Sentinel-3A wind speed and SWH against the model over the whole globe. The comparison is as good as those that involve conventional altimetry products.

Most of the future satellite altimetry missions will be operating in SAR mode. This includes the remaining satellites of the Sentinel-3 series (starting with Sentinel-3B, which will be launched

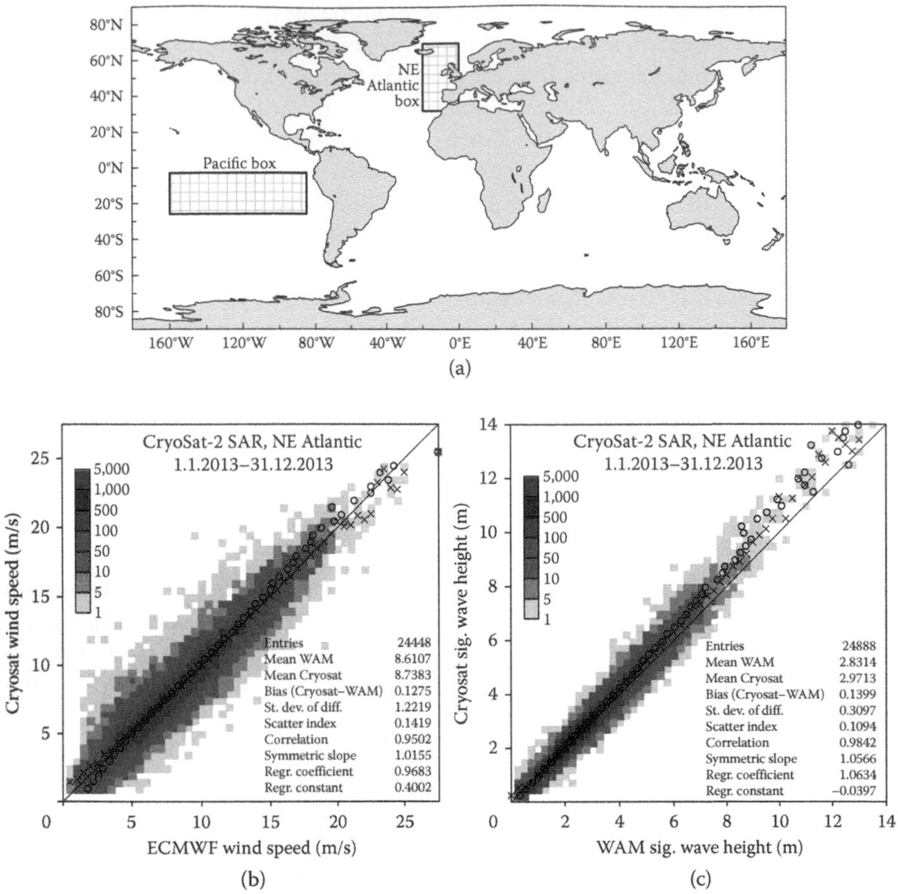

FIGURE 12.23 (a) Two of the open ocean regions where CryoSat-2 operates in SAR mode (note that the Pacific box is approximate as its boundaries have slightly changed over the years). Comparison between Cyosat-2 SAR mode and European Center for Medium-Range Weather Forecasts (ECMWF) model (b) surface wind speed and (c) significant wave height values during the whole year of 2013 in the NE Atlantic box. Refer to Figures 12.4 (wind speed) and 12.14 (wave height) for the meaning of the color coding and the "x" and "o" symbols.

toward the end of 2017) and the Sentinel-6 (also known as Jason-CS) series planned for 2020 and beyond. However, the latter is implementing a new technology of operation called the interleave mode, which will enable the altimeter to operate both in SAR and LRM modes at the same time.

12.6.2 CFOSAT

From the discussion given so far, it is evident that altimeter missions have proven their usefulness by measuring SWH and the surface wind speed over the oceans. A shortcoming of the altimeter missions is that they cannot provide information on the spectral distribution of wave energy with respect to frequency and propagation direction of the waves. SAR is currently the only technology for estimating the two-dimensional wave spectrum from a space-borne sensor. But SAR has a major drawback as it is well-known that the wavelike patterns visible in SAR images of the ocean surface may be considerably different from the actual ocean wave field. The motion of the ocean surface leads to Doppler misregistrations in the azimuth (flight direction) leading to a distorted image spectrum and even a cut-off in the azimuth direction. This so-called velocity bunching effect is proportional to the range-to-velocity ratio of the platform.

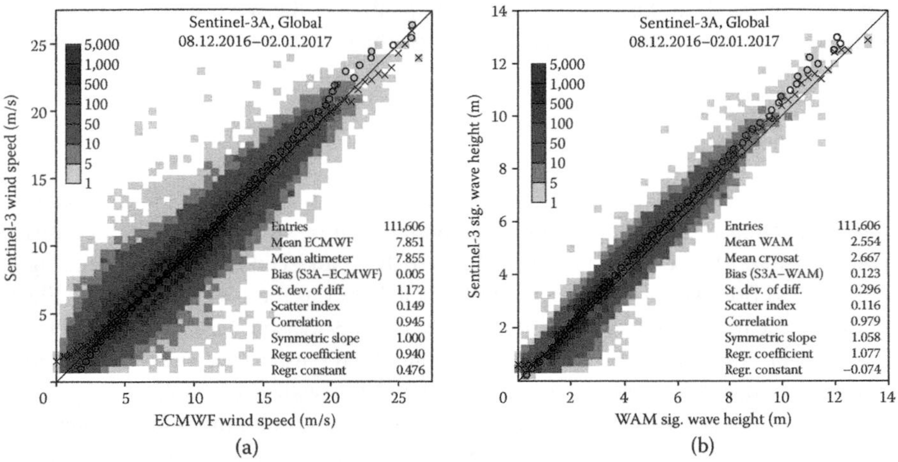

FIGURE 12.24 Comparison between Sentinel-3A SAR Mode and ECMWF model (a) surface wind speed and (b) significant wave height values during the period from December 8, 2016, to January 2, 2017. Refer to Figures 12.4 (wind speed) and 12.14 (wave height) for the meaning of the color coding and the "x" and "o" symbols.

For present and future missions, this ratio is high, giving a considerable limitation of the usefulness of SAR spectra over the oceans. Nevertheless, using model spectra as a first guess in the nonlinear inversion scheme of Hasselmann and Hasselmann (1991), assimilation of SAR spectra commenced at ECMWF in February 2003 (Abdalla et al. 2004) with some positive impact on forecast skill.

However, owing to the velocity bunching effect, there is only useful information from SAR for the long waves as the azimuthal cut-off wavelength is typically larger than 200 m. It is, therefore, highly desirable to explore new capabilities to avoid the limitations of an SAR. Such a capability is already known for some time as an RAR does not suffer from the velocity bunching effect but would usually require a very large antenna to achieve the required resolution, except when an intelligent method of scanning is applied. Such a possibility is offered with the satellite project CFOSAT (Chinese French Oceanic Satellite), which will embark on a Ku-band radar whose technology is derived from altimeter techniques but will use geometry and data processing appropriate for the estimation of directional spectra of long waves. The concept is derived from a development in the 1970s on board airplanes (Jackson and Walton 1985; Jackson et al. 1985; Hauser et al. 1992). The initial space-borne concept proposed under the name SWIMSAT with one beam pointing at 10° incidence (Hauser et al. 2001) has now evolved into a six-beam system, one pointing nadir and five pointing off-nadir between 2° and 10°. The five off-nadir beams, each having a footprint of approximately 18 km in diameter, will scan over 360° in azimuth. The principle is to measure the modulation of the radar signal due to the tilt of the long waves in each look direction. The Fourier analyzed modulation will provide the radial wave spectrum, while the azimuth scan will be used to determine the propagation direction. Estimates of the two-dimensional wave spectrum will be possible at a scale of 70 km × 70 km, for waves ranging from about 70 m to more than 400 m. The feasibility of such a concept was demonstrated in Hauser et al. (2001), and the impact of assimilating spectral data in wave forecast models was assessed in various recent studies using either simulations (Aouf et al. 2006a) or SAR observations (Aouf et al. 2006b). Results show that the assimilation of spectral information gives forecast skills, which are improved with respect to the case of assimilation of SWH only, and that this improvement is the most significant for mean wave direction and mean frequency. Also, the role of the cut-off wavelength was studied. For SWIMSAT, with a cut-off wavelength of 70 m, a larger impact on forecast skill was found than for an SAR (with a typical azimuthal cut-off wavelength of 200 m). But, these were results obtained with ECMWF wind fields from October 2000. In the course of time, the quality of the wind fields and the wave model have increased considerably, thus most likely reducing the impact of data assimilation as discussed already in Section 12.5.1.

12.7 CONCLUDING REMARKS

The last quarter of a century witnessed significant improvements in availability and quality of ocean wave and wind instrumental measurements. The constellation of satellite altimeters consisted of one (e.g., GEOSAT) before 1991 or at most two (e.g., ERS-1/ERS-2 and TOPEX-Poseidon) during the 1990s except for the period from May 1995 till June 1996 when there was three as both ERS satellites were in active operation. Recently, the number has risen to six (Sentinel-3A, Jason-2, Jason-3, CryoSat-2, SARAL-AltiKa, and HY-2A). NRT availability (within a few hours from sensing) necessary for operational wind and wave data assimilation has only been possible since 1991, with ERS-1 followed by ERS-2. Until early 2002, only one mission was able to deliver NRT data. Recently, all missions but one (HY-2A) are delivering NRT data with three missions (mainly Sentinel-3A and Jason-3 and, to a large extent, Jason-2) are operational in a sense that they are committed to providing almost full coverage within a few hours. The others (CryoSat-2 and SARAL-AltiKa) provide this service on the best effort basis.

At the same time, massive improvements in our ability to forecast ocean waves were attained. The reasons for these improvements have been discussed to some extent in this paper: vast improvements in atmospheric modeling resulting in accurate surface winds, a large increase in observations of the sea state and the atmosphere—notably provided by satellites, and the more realistic modeling of the dynamics of ocean waves, which is discussed in Janssen (2007).

In particular, altimeter sea state data have played an important role in the progress in ocean wave forecasting. On the one hand, altimeter sea state data can be used very effectively as a tool to diagnose problems in the atmospheric and ocean-wave model. On the other hand, despite the improvements in wave forecasting, altimeter wave height data are important in improving the wave height analysis. Furthermore, the wave height and wind speed data play a key role in the construction of a global wave height climatology, but altimeter observations over a much longer period are needed.

The retrieval of SWH from the altimeter waveform seems to be relatively well-understood, while triple collocation studies do suggest that altimeter wave height data are more accurate than expected from the engineering requirements. Although there is still discussion in the literature about how to obtain optimal benefits from the measurement of the radar backscatter, σ_0, the wind speed product derived from it has nevertheless proven its value. The altimeter wind speed has about the same accuracy as the wind speed from scatterometers. In addition, even for high wind speed, this product is useful. In particular, when considering dual-frequency altimeters, there is the potential to obtain wind speed, wave height, and rain rate at the same time.

Ocean waves and surface wind speed measurements using delay-Doppler (SAR) altimetry turned out to be of high quality. Most of future altimetry missions are expected to operate in that mode. There is the promise of obtaining two-dimensional wave spectra from SWIMSAT in a relatively straightforward manner. Finally, the SWOT mission, although not intended to provide data in NRT, promises to provide altimetry data across a swath of 120 km in high resolution and precision.

In conclusion, sea state parameters from altimeters have proven to be of tremendous value for ocean wave research and operational wave forecasting. And there is the promise for more valuable products to come.

REFERENCES

Abdalla, S. 2007. Ku-band radar altimeter surface wind speed algorithm. *ECMWF Tech. Memo*. 524, ECMWF, Reading, UK, 16 pp, https://www.ecmwf.int/search/elibrary/, last accessed: 14 July 2017.

Abdalla, S., 2012. Ku-Band Radar Altimeter Surface Wind Speed Algorithm. *Marine Geodesy*, **35**(Suppl. 1), 276–298.

Abdalla, S., 2013a. *Global validation of Envisat wind, wave and water vapour products from RA-2, MWR, ASAR and MERIS (2011–2012)*. ESA Cont. Report 21519/08/I-OL, ECMWF, Reading, 48 pp.

Abdalla, S., 2013b. Wind and Wave data selection for climate studies in enclosed seas. In E. Özhan (ed.), *Global Congress on ICM: Lessons learned to address new challenges*, MEDCOAST, Mediterranean Coastal Foundation, Dalyan, Turkey, pp. 1347–1358.

Abdalla, S., 2015. SARAL/AltiKa wind and wave products: Monitoring, validation and assimilation. *Mar. Geod.*, **38**, 365–380.

Abdalla, S., Bidlot, J. R., and Janssen, P., 2004. Assimilation of ERS and Envisat wave data at ECMWF, *Proc. 2004 Envisat & ERS Symposium*, 6–10 September, Salzburg, ESA Proc. SP-572 (CD) paper 403, 10 pp.

Abdalla, S. and De Chiara, G., 2017. Estimating random errors of scatterometer, altimeter and model wind speed data. *J. Select. Top. Appl. Earth Observ. Rem. Sens,* **10**(5), 2406–2414.

Abdalla, S. and Hersbach, H., 2006. *The technical support for global validation of ERS wind and wave products at ECMWF*. Final report for ESA contract 18212/04/I-OL, ECMWF, Reading, UK, 50 pp, https://www.ecmwf.int/search/elibrary/, last accessed: 14 July 2017.

Abdalla, S., Isaksen, L., Janssen, P. A. E. M., and Wedi, N., 2013. Effective spectral resolution of ECMWF atmospheric forecast models. *ECMWF Newslett.*, **137**, 19–22, ECMWF, Reading, UK, https://www.ecmwf.int/search/elibrary/, last accessed: 14 July 2017.

Abdalla, S., Janssen, P. A. E. M., and Bidlot, J.B., 2011. Altimeter near real time wind and wave products: Random error estimation. *Marine Geodesy*, **34**(3–4), 393–406.

Abdalla, S., Janssen, P. A. E. M., and Bidlot, J.-R., 2010. Jason-2 OGDR wind and wave products: Monitoring, validation and assimilation. *Marine Geodesy*, **33**(Suppl. 1), 239–255.

Alpers, W. and Hühnerfuss, H., 1989. The damping of ocean waves by surface films: A new look at an old problem. *J. Geophys. Res.*, **94**, 6251–6265.

Alves, J. H. G. M. and Young, I. R., 2003. On estimating extreme wave heights using combined GEOSAT, TOPEX/Poseidon and ERS-1 altimeter data. *Appl. Ocean Res.*, **25**, 167–186. doi: 10.1016/j.apor.2004.01.002.

Andersson, E., Haseler, J., Undén, P., Courtier, P., Kelly, G., Vasiljevic, D., Brankovic, C., et al., 1998. The ECMWF implementation of three-dimensional variational assimilation (3D-Var). III: Experimental results. *Q. J. Roy. Meteorol. Soc.*, **124**, 1831–1860.

Aouf, L., Lefèvre, J.-M., and Hauser, D., 2006a. Assimilation of directional wave spectra in the wave model WAM: An impact study from synthetic observations in preparation for the SWIMSAT Satellite Mission. *J. Atmos. Oceanic Technol.*, **23**, 448–463.

Aouf, L., Lefèvre, J.-M., Hauser, D., Chapron, B., and Collard, F. 2006b. The impact of using upgraded processing of ASAR level 2 wave products in the assimilation system. *Proceedings of SEASAR European Space Agency (ESA) Workshop*, Frascati, 23–26 January.

Apel, J. R., 1994. An improved model of the ocean surface wave vector spectrum and its effects on radar backscatter. *J. Geophys. Res.*, **99**, 16269–16291, SP-613, paper 51-6, 6pp.

Bacon, S. and Carter, D. J. T., 1991. Wave climate changes in the North Atlantic and North Sea. *Int. J. Climatol.*, **11**, 545–588.

Bacon, S. and Carter, D. J. T., 1993. A connection between mean wave height and atmospheric pressure gradient in the North Atlantic. *Int. J. Climatol.*, **13**, 423–436.

Barrick, D. 1968. Rough surface scattering based on the specular point theory. *IEEE Trans. Antennas Propagat.*, **16**(4), 449–454.

Bauer, E., Hasselmann, S., Hasselmann, K., and Graber, H. C., 1992. Validation and assimilation of Seasat altimeter wave heights using the WAM wave model. *J. Geophys. Res.* **C97**, 12671–12682.

Bauer, E. and Staabs, C., 1998. Statistical properties of global significant wave heights and their use for validation. *J. Geophys. Res.*, **103**, 1153–1166.

Bidlot, J.-R., Durrant, T., and Queffelou, P., 2008. Assessment of the systematic differences in wave observations from moorings. *Presentation at JCOMM Technical Workshop on Wave Measurements from Buoys*, New York, October JCOMM Tech. Report No. 47, 30 slides, http://www.wmo.int/pages/prog/amp/mmop/jcomm_reports.html, last accessed 14 July 2017.

Bidlot, J. R., Janssen, P., and Abdalla, S., 2007. A revised formulation of ocean wave dissipation and its model impact, *ECMWF Tech. Memo.* **509**, ECMWF, Reading, UK, 27 pp, https://www.ecmwf.int/search/elibrary/, last accessed: 14 July 2017.

Boggs, P. T., Byrd, R. H., and Schnabel, R. B., 1987. A stable and efficient algorithm for nonlinear orthogonal distance regression. *SIAM J. Sci. Statist. Comput.*, **8**(6), 1052–1078.

Boggs, P. T., Donaldson, J. R., Byrd, R. H., and Schnabel, R. B., 1989. Odrpack—Software for weighted orthogonal distance regression. *ACM Trans. Math. Softw.*, **15**(4), 348–364.

Breivik, L.-A., Reistad, M., Schyberg, H., Sunde, J., Krogstad, H. E., and Johnsen, H., 1998. Assimilation of ERS SAR wave spectra in an operational wave model. *J. Geophys. Res.* **103**, 7887–7900.

Brown, A. R., Beljaars, A. C. M., and Hersbach, H., 2006. Errors in parametrizations of convective boundary-layer turbulent momentum mixing. *Q. J. R. Meteorol. Soc.* **132**, 1859–1876.

Brown, G. S. 1990. Quasi-specular scattering from air–sea interface. In G. Geernaet and W. Plant (ed.), *Surface Waves and Fluxes*, Springer Dordrecht, the Netherlands, pp. 1–40.

Caires, S. and Sterl, A., 2005. 100-Year return value estimates for ocean wind speed and significant wave height from the ERA-40 data. *J. Clim.*, **18**, 1032–1048.

Caires, S., Sterl, A., and Gommenginger, C. P., 2005. Global ocean mean wave period data: Validation and description. *J. Geophys. Res.*, **110**, C02003. doi: 10.1029/2004JC002631.

Carter, D. J. T., Challenor, P. G., and Srokosz, M. A., 1992. An assessment of GEOSAT wave height and wind-speed measurements. *J. Geophys. Res.* **97**(C7), 11383–11392.

Carter, D. J. T. and Draper, L., 1988. Has the north-east Atlantic become rougher? *Nature*, **332**, 494.

Carter, D. J. T., Foale, S., and Webb, D. J., 1991. Variations in global wave climate throughout the year. *Int. J. Rem. Sens.*, **12**(8), 1687–1697.

Challenor, P. G. and Cotton, P. D., 1997. The SOC contribution to the ESA working group calibration and validation of ERS-2 FD measurements of significant wave height and wind speed. *Proceedings of the CEOS Wind and Wave Validation Workshop, ESTEC*, June 2–5, Noordwijk, ESA Publications, pp. 95–100.

Challenor, P. G., Foale, S., and Webb, D. J., 1990. Seasonal-changes in the global wave climate measured by the GEOSAT altimeter. *Int. J. Rem. Sens.*, **11**(12), 2205–2213.

Chen, G., Chapron, B., and Ezraty, R., 2002. A global view of swell and wind sea climate in the ocean by satellite altimeter and scatterometer. *J. Atmos. Oceanic Technol.*, **19**, 1849–1859.

Chen, G., Lin, H., and Ma, J., 2000. On the seasonal inconsistency of altimeter wind speed algorithms. *Int. J. Rem. Sens.* **21**, 2119–2125.

Cotton, P. D. and Carter, D. J. T., 1994. Cross calibration of TOPEX, Ers-1, and GEOSAT wave heights. *J. Geophys. Res.*, **99**(C12), 25025–25033.

Cotton, P. D. and Carter, D. J. T., 1995. Cross calibration of TOPEX, Ers-1, and GEOSAT wave heights (99, Pg 25025). *J. Geophys. Res.*, **100**(C4), 7095–7095.

Cotton, P. D., Challenor, P. G., and Carter, D. J. T., 1997. An assessment of the accuracy and reliability of GEOSAT, ERS-1, ERS-2 and TOPEX altimeter measurements of significant wave height and wind speed. *Proceedings of the CEOS Wind and Wave Validation Workshop*, ESTEC, June 2–5, Noordwijk, ESA Publications, pp. 81–94.

Cox, C. S. and Munk, W. H., 1954. Statistics of the sea surface derived from sun glitter. *J. Marine Res.*, **13**, 198–227.

De Boisséson, E., Balmaseda, M. A., Abdalla, S., Källén, E., and Janssen, P.A.E.M., 2014. How robust is the recent strengthening of the tropical Pacific trade winds? *Geophys. Res. Lett.*, **41**(12), 4398–4405. doi: 10.1002/2014GL060257.

De Chiara, G., 2015. Active techniques for wind observations: Scatterometer. *Proceedings of the Seminar on Use of Satellite Observations in Numerical Weather Prediction*, ECMWF, September 8–12, ECMWF, Reading, UK, 12 pp, https://www.ecmwf.int/search/elibrary/, last accessed: 14 July 2017.

De las Heras, M. M., Burgers, G., and Janssen, P. A. E. M., 1994. Variational wave data assimilation in a third generation wave model. *J. Atmos. Oceanic Technol.* **11**, 1350–1369.

Donnelly, W. J., Carswell, J. R., McIntosh, R. E., Chang, P. S., Wilkerson, J., Marks, F., and Blck, P. G., 1999. Revised ocean backscatter models at C and KU band under high-wind conditions. *J. Geophys. Res.* **104**, 11485–11497.

Dunlap, E. M., Olsen, R. B., Wilson, L., De Margerie, S., and Lalbeharry, R., 1998. The effect of assimilating ERS-1 fast delivery wave data into the North Atlantic WAM model. *J. Geophys. Res.* **103**, 7901–7915.

Elfouhaily, T., Chapron, B., Katsaros, K., and Vandemark, D., 1997. A unified directional spectrum for long and short wind-driven waves. *J. Geophys. Res.* **102**(C7), 15781–15796.

Esteva, D. C., 1988. Evaluation of preliminary experiments assimilating Seasat significant wave heights into a spectral wave model. *J. Geophys. Res.*, **93**, 14099–14106.

Freilich, M. H. and Vanhoff, B. A., 2003. The relationship between winds, surface roughness and radar backscatter at low incidence angles from TRMM precipitation radar measurements. *J. Atmos. Oceanic Technol.*, **20**, 549–562.

Glazman, R. and Greysukh, A., 1993. Satellite altimeter measurements of surface wind. *J. Geophys. Res.*, **98**, 2475–2483.

Glazman, R. and Pilorz, S., 1990. Effects of sea maturity on satellite altimeter measurements of surface wind. *J. Geophys. Res.*, **95**, 2857–2870.

Gómez-Enri, J., Gommenginger, C. P., Srokosz, M. A., and Challenor, P. G., 2007. Measuring global ocean wave skewness by retracking RA-2 envisat waveforms. *J. Atm. Oceanic Technol.*, **24**, 1102–1116.
Gommenginger, C. P., Srokosz, M. A., Challenor, P. G., and Cotton, P. D., 2003. Measuring ocean wave period with satellite altimeters: A simple empirical model. *Geophys. Res. Lett.* **30**, 2150. doi: 10:10.1029/2003GL017743.
Gourrion, J., Vandemark, D., Bailey, S., Chapron, B., Gommenginger, G. P., Challenor, P. G., and Srokosz, M. A., 2002. A two-parameter wind speed algorithm for Ku-band altimeters. *J. Atmos. Oceanic Technol.*, **19**, 2030–2048.
Greenslade, D. J. M., 2001. The assimilation of ERS-2 significant wave height data in the Australian region. *J. Mar. Sys.*, **28**, 141–160.
Greenslade, D. J. M. and Young, I. R., 2004. Background errors in a global wave model determined from altimeter data. *J. Geophys. Res.*, **109**, C09007. doi: 10.1029/2004JC002324.
Greenslade, D. J. M. and Young, I. R., 2005. The impact of inhomogeneous background errors on a global wave data assimilation system. *J. Atmos. Ocean Sci.*, **10**, 61–93.
Gulev, S. K. and Grigorieva, V., 2004. Last century changes in ocean wind wave height from global visual wave data. *Geophys. Res. Lett.*, **31**, L24302. doi: 10.1029/2004GL021040.
Hansen, B. and Günther, H., 1992. ERS-1 radar altimeter validation with the WAM model. *Proceedings of the ERS-1 Geophysical Validation Workshop*, European Space Agency, Paris, pp. 157–161.
Hasselmann, K., Barnett, T. P., Bouws, E., Carlson, H., Cartwright, D. E., Enke, K., Ewing, J. A., et al., 1973. Measurements of wind-wave growth and swell decay during the Joint North Sea Wave Project (JONSWAP). *Dtsch. Hydrogr. Z. Suppl. A.*, **8**(12), 95.
Hasselmann, K. and Hasselmann, S., 1991. On the nonlinear mapping of an ocean wave spectrum into an SAR image spectrum and its inversion. *J. Geophys. Res.*, **C96**, 10713–10729.
Hauser, D., Caudal, G., Rijckenberg, G. J., Vidal-Madjar, D., Laurent, G., and Lancelin, P., 1992. RESSAC: A new airborne FM/CW radar ocean wave spectrometer. *IEEE Trans. Geosci. Rem. Sens.*, **30**, 981–995.
Hauser, D., Soussi, E., Thouvenot, E., and Rey, L., 2001. SWIMSAT: A real-aperture radar to measure directional spectra of ocean waves from space—Main characteristics and performance simulation. *J. Atmos. Oceanic Technol.* **18**, 421–437.
Hayne, G. S. and Hancock, D. W., 1982. Sea state related altitude error in the Seasat radar altimeter. *J. Geophys. Res.* **87**, 3227–3231.
Hersbach, H., 1998. Application of the adjoint of the WAM model to inverse modeling. *J. Geophys. Res.*, **103**, 10469–10488.
Hersbach, H., Stoffelen, A., and de Haan, S., 2007. An improved C-band scatterometer ocean geophysical model function: CMOD5. *J. Geophys. Res.*, **112**, C03006.
Hollingsworth, A. and Lönnberg, P., 1986. The statistical structure of short-range forecasts errors as determined from radiosonde data: I. The wind field. *Tellus A*, **38**, 111–136.
Hwang, P.A., Teague, W.J., Jacobs, G.A., and Wang, D.W., 1998. A statistical comparison of wind speed, wave height and wave period derived from altimeters and ocean buoys in the Gulf of Mexico region, *J. Geophys. Res.*, **103**, 10451–10468.
Jackson, F. C., 1979. The reflection of impulses from a nonlinear random sea. *J. Geophys. Res.*, **84**, 4939–4943.
Jackson, F. C. and Walton, T. W., 1985. A comparison of *in situ* and airborne radar observations of ocean wave directionality. *J. Geophys. Res.* **90**, 1005–1018.
Jackson, F. C., Walton, T. W., and Baker, P. L., 1985. Aircraft and satellite measurement of ocean wave directional spectra using scanning-beam microwave radars. *J. Geophys. Res.* **90**, 987–1004.
Jähne, B. and Riemer, K. S., 1990. Two-dimensional wavenumber spectra of small-scale water surface waves. *J. Geophys. Res.* **95**, 11531–11546.
Janssen, P. A. E. M., 1991. Quasi-linear theory of wind wave generation applied to wave forecasting. *J. Phys. Oceanogr.*, **21**, 1631–1642.
Janssen, P. A. E. M., 1999. Wave modeling and altimeter wave height data. *ECMWF Tech. Memo.* 269, ECMWF, Reading, UK, 35 pp, https://www.ecmwf.int/search/elibrary/, last accessed: 14 July 2017.
Janssen, P. A. E. M., 2000. ECMWF wave modeling and satellite altimeter wave data. In D. Halpern (ed.), *Satellites, Oceanography and Society*, Elsevier Science BV, Amsterdam, the Netherlands, pp. 35–56.
Janssen, P. A. E. M., 2004. *The interaction of ocean waves and wind.* Cambridge University Press, Cambridge, UK, 300+viii pp.
Janssen, P. A. E. M., 2008. Progress in ocean wave forecasting. *J. Comput. Phys.*, 227, 3572–3594. doi: 10.1016/j.jcp.2007.04.029.
Janssen, P. A. E. M., Abdalla, S., Hersbach, H., and Bidlot, J.-R., 2007. Error estimation of Buoy, Satellite and Model Wave height data. *J. Atmos. Ocean. Technol.*, 1665–1677.

Janssen, P. A. E. M., Hansen, B., and Bidlot, J.-R., 1997a. Verification of the ECMWF wave forecasting system against buoy and altimeter data. *Weather Forecast.*, **12**, 763–784.

Janssen, P. A. E. M., Hansen, B., and Bidlot, J.-R., 1997b. Validation of ERS satellite wave products with the WAM model. *CEOS Wind and Wave Validation Workshop*, 3–5 June, ESTEC, The Netherlands, ESA WPP-147, pp. 101–108.

Janssen, P. A. E. M., Lionello, P., Reistad, M., and Hollingsworth, A., 1987. *A study of the feasibility of using sea and wind information from the ERS-1 satellite, part 2: Use of scatterometer and altimeter data in wave modeling and assimilation*. ECMWF report, ESA, Reading.

Janssen, P. A. E. M., Lionello, P., Reistad, M., and Hollingsworth, A., 1989. Hindcasts and data assimilation studies with the WAM model during the Seasat period. *J. Geophys. Res.*, **C94**, 973–993.

Janssen, P. A. E. M., Wallbrink, H., Calkoen, C. J., van Halsema, D., Oost, W. A., and Snoeij, P., 1998. VIERS-1 scatterometer model. *J. Geophys. Res.*, **103**(No. C4), 7807–7831.

Jiang, H. and Chen, G., 2013. A global view on the swell and wind sea climate by Jason-1 mission: A revisit. *J. Atmos. Oceanic Technol.*, **30**, 1833–1841.

Klinke, J. and Jähne, B., 1992. two-dimensional wavenumber spectra of short wind waves-results from wind-wave facilities and extrapolation to the ocean. *Proceedings of Conference on Optics of the Air-Sea Interface, Theory and Measurements*, San Diego, CA, USA, L. Estep (ed.), International Society for Optical Engineering, Bellingham, Washington, USA, 245–257.

Kohno, N., Miura, D. and Yoshita, K., 2011. The development of JMA wave data assimilation system. *Proceedings of the 12th International Workshop on Wave Hindcasting and Forecasting*, Kohala Coast, Hawaii, USA, 8 pp. http://www.waveworkshop.org/12thWaves/papers/full_paper_Kohno_et_al.pdf, last accessed: 14 July 2017.

Komen, G.J., Cavaleri, L., Donelan, M., Hasselmann, K., Hasselmann, S., and Janssen, P.A.E.M., 1994. *Dynamics and Modeling of Ocean Waves*. Cambridge University Press, Cambridge, UK, 532 p.

Lefèvre, J., Barckicke, J., and Menard, Y., 1994. A significant wave height dependent function for TOPEX/ Poseidon wind speed retrieval. *J. Geophys. Res.*, **99**, 25035–25046.

Lefèvre, J.-M. and Aouf, L., 2012. Latest developments in wave data assimilation. *Proceedings of ECMWF Workshop on Ocean Waves*, ECMWF, Reading, UK 175-188. http://www.ecmwf.int/sites/default/files/ elibrary/2012/10699-latest-developments-wave-data-assimilation.pdf last accessed 14 July 2017.

Li, X.-M., Lehner, S., and Bruns, T., 2014. Simultaneous measurements by advanced SAR and radar altimeter on potential improvement of ocean wave model assimilation. *IEEE Trans. Geosci. Rem. Sens.*, **52**(5), 2508–2518.

Lilly, D. K., 1989. Two-dimensional turbulence generated by energy sources at two scales. *J. Atmos. Sci.*, **46**, 2026–2030.

Lionello, P., Gunther, H., and Janssen, P.A.E.M., 1992. Assimilation of altimeter data in a global third generation model. *J. Geophys. Res.*, **C97**, 14453–14474.

Lipa, B. and Barrick, D. E., 1981. Ocean surface height-slope probability density function from Seasat altimeter echo. *J. Geophys. Res.*, **86**, 10921–10930.

Liu, Y., Su, M.-Y., Yan, X.-H., and Liu, W. T., 2000. The mean-square slope of ocean surface waves and its effect on radar backscatter. *J. Atmos. Oceanic Technol.*, **17**, 1092–1105.

Mandel, J., 1964. *The statistical analysis of experimental data.* Wiley-Interscience, NY, USA 410 pp.

Marsden, R. E., 1999. A proposal for a neutral regression. *J. Atmos. Oceanic Technol.*, **16**, 876–883.

Marshall, J. S. and Palmer, W. M., 1948. The distribution of raindrops with size. *J. Meteorol.*, **5**, 165–166.

Miles, J. W., 1957. On the generation of surface waves by shear flows. *J. Fluid Mech.* **3**, 185–204.

Monaldo, F. and Dobson, E., 1989. On using significant wave height and radar cross section to improve radar altimeter measurements. *J. Geophys. Res.*, **94**, 12699–12701.

Nastrom, G. D. and Gage, K. S., 1985. A climatology of atmospheric wavenumber spectra of wind and temperature observed by commercial aircraft. *J. Atmos. Sci.*, **42**, 950–960.

Pierdicca, N., Bignami, C., Roca, M., Féménias, P., Fascetti, M., Mazzetta, M., Loddo, C. N., Martini, A. and Pinori, S., 2013. Transponder calibration of the Envisat RA-2 altimeter Ku band sigma naught. *Adv. Space Res.*, **51**(8), 1478–1491.

Quartly, G. D., 1997. Achieving accurate altimetry across storms: Improved wind and wave estimates from C-band. *J. Atmos. Oceanic Technol.*, **14**, 705–715.

Quartly, G. D., 1998. Understanding the effects of rain on radar altimeter waveforms data: Part I: Theory. *J. Atmos. Oceanic Technol.*, **15**, 1361–1378.

Queffeulou, P. and Bentamy, A., 2007. Analysis of wave height variability using altimeter measurements: Application to the Mediterranean Sea. *J. Atmos. Oceanic Technol.*, **24**, 2078–2092.

Quilfen, Y., Chapron, B., Collard, F., and Serre, M., 2005. Calibration/validation of an altimeter wave period model and application to TOPEX/Poseidon and Jason-1 altimeters. *Marine Geodesy*, **27**, 535–550.

Quilfen, Y., Tournadre, J., and Chapron, B., 2006. Altimeter dual-frequency observations of surface winds, waves and rain rate in tropical cyclone Isabel. *J. Geophys. Res.*, **111**, C01004. doi: 10.1029/2005JC003068.

Romeiser, R., 1993. Global validation of the wave model WAM over a one year period using GEOSAT wave height data. *J. Geophys. Res.*, **C98**, 4713–4726.

Skandrani, C., Lefèvre, J.-M. and Queffeulou, P., 2004. Impact of multisatellite altimeter data assimilation on wave analysis and forecast. *Marine Geodesy* **27**, 1–23.

Srokosz, M. A., 1986. On the joint distribution of surface elevation and slope for a nonlinear random sea, with application to radar altimetry. *J. Geophys. Res.*, **91**, 995–1006.

Stoffelen, A., 1998. Error modeling and calibration: Toward the true surface wind speed. *J. Geophys. Res.*, **103**, 7755–7766.

Thomas, J. P., 1988. Retrieval of energy spectra from measured data for assimilation into a wave model. *Q. J. Roy. Meteorol. Soc.*, **144**, 781–800.

Tokmakian, R. and Challenor, P. G., 1999. On the joint estimation of model and satellite sea surface height anomaly errors. *Ocean Model.*, **1**, 39–52.

Uppala, S. M., KÅllberg, P. W., Simmons, A. J., Andrae, U., Bechtold, V. D. C., Fiorino, M., Gibson, J. K., et al., 2005. The ERA-40 re-analysis. *Q. J. Roy. Meteorol. Soc.*, **131**, 2961–3012. doi: 10.1256/ qj.04.176.

Valenzuela, G. R., 1978. Theories for the interaction of electromagnetic and oceanic waves—A review. *Bound. Layer Meteorol.*, **13**(1–4), 61–85.

Vinoth, J. and Young, I.R., 2011. Global estimates of extreme wind speed and wave height. *J. Climate*, **24**, 1647–1665.

Voorrips, A., 1998. Sequential data assimilation methods for ocean wave models. PhD thesis, Delft University of Technology.

Voorrips, A. and de Valk, C., 1997. *A comparison of two operational wave data assimilation methods*. Preprint 97-06, Royal Netherlands Meteorological Institute (KNMI), De Bilt, the Netherlands.

Wimmer, W., Challenor, P., and Retzler, C., 2006. Extreme wave heights in the north Atlantic from altimeter data. *Renew. Ener.*, **31**(2), 241–248.

Witter, D. L. and Chelton, D. B., 1991. A GEOSAT altimeter wind speed algorithm and a method for altimeter wind speed algorithm development. *J. Geophys. Res.* **96**, 8853–8860.

Woo, H. J. and Park, K. A., 2017. Long-term trend of satellite-observed significant wave height and impact on ecosystem in the East/Japan Sea. *Deep Sea Res. II Topic. Stud. Oceanogr*. doi: 10.1016/j.dsr2.2016.09.003.

Woolf, D. K., Challenor, P. G., and Cotton, P. D., 2002. Variability and predictability of the North Atlantic wave climate. *J. Geophys. Res.*, **107**(C10), 3145, 14 pp.

Woolf, D. K., Cotton, P. D., and Challenor, P. G., 2003. Measurements of the offshore wave climate around the British Isles by satellite altimeter. *Phil. Trans. Roy. Soc. Lond. A. Math. Phys. Eng. Sci.*, **361**(1802), 27–31.

Xu, Y., Fu, L. L., and Tulloch, R., 2011. The global characteristics of the wavenumber spectrum of ocean surface wind. *J. Phys. Oceanogr.*, **41**, 1576–1582.

Young, I.R., 1993. An estimate of the GEOSAT altimeter wind speed algorithm at high wind speeds. *J. Geophys. Res.*, **98**, 20275–20286.

Young, I. R., 1999. Seasonal variability of the global ocean wind and wave climate. *Int. J. Climatol.*, **19**, 931–950.

Young, I. R., Babanin, A., and Zieger, S., 2013. The decay rate of ocean swell observed by altimeter. *J. Phys. Oceanogr.*, **43**, 2322–2333.

Young, I. R., Vinoth, J., Zieger, S., and Babanin, A. V., 2012. Investigation of trends in extreme value wave height and wind speed. *J. Geophys. Res.*, **117**, C00J06. doi: 10.1029/2011JC007753.

Young, I. R., Zieger, S., and Babanin, A. V., 2011. Global trends in wind speed and wave height. *Science*, **332**(6028), 451–455. doi: 10.1126/science.1197219.

Zieger, S., Vinoth, J., and Young, I. R., 2009. Joint calibration of multiplatform altimeter measurements of wind speed and wave height over the past 20 years. *J. Atmos. Oceanic Technol.*, **26**, 2549–2564, doi: 10.1175/2009JTECHA1303.1

13 Tides and Satellite Altimetry

Richard D. Ray and Gary D. Egbert

13.1 INTRODUCTION

Like most fields of science over the past century or two, the subject of oceanic tides appears to have progressed through a few major phases that, in hindsight at least, can be readily identified. During each phase, certain problems were singled out as especially important, and certain experimental approaches, often reflecting new technological developments, were deemed especially fruitful (Cartwright 1999). Practical applications were often major drivers—and funders—of advancement, but not always.

Consider the problem of determining accurate co-tidal charts for the global ocean (Figures 13.3 and 13.4 are examples). Mid-twentieth century research mostly followed traditional paths of classical mathematical physics, with efforts devoted to constructing partial differential equations, for very idealized geometries, to be solved by analytical or semi-analytical means (e.g., Proudman and Doodson 1936; among many other papers). This body of work gave important insight into the behavior of tides in the ocean, especially the Atlantic, but it also came with serious limitations; see reviews by Proudman (1944), Doodson (1958), and Hansen (1962). By the 1960s and 1970s analytical methods had mostly given way to numerical methods (Hendershott 1977), which reached a temporary plateau with the work of Schwiderski (1980). Although the needs of navigation have always motivated much work on coastal tide problems, much of the older work on deep ocean tides was of a more basic nature without immediate practical application. The global tide posed a tempting, but difficult, mathematical challenge, sometimes with ties to fundamental problems in geophysical fluid dynamics (e.g., Proudman 1942). Yet the global tide has always had importance to a few subjects further afield, notably parts of geodesy, whole-earth geophysics, and astronomy (Lambeck 1977; Munk and MacDonald 1960). In the 1960s, for example, Walter Munk's encouragement of early deep-sea pressure recordings arose in part from his work on the classical problem of tidal friction and its implications for the histories of the Earth's rotation and the moon's orbit (Munk 1980).

Since 1992, the subject has entered what can aptly be called the "Age of Altimetry." The primary motivation for the work—to predict and remove tidal variability from altimeter records—has also been the source of its advances. TOPEX/Poseidon (T/P) and its successor missions Jason-1 through Jason-3, plus complementary satellite missions as described throughout this book, provided the tidal community with a wealth of new measurements over hitherto unexplored regions of the ocean.

Altimetry acts in essence as a tide gauge at every spot of the ocean, although one with an admittedly unusual and sometimes problematic sampling rate (see the following). It is almost a cliché to say a subject has been revolutionized by some new measurements or some new understanding. But in the case of tides and altimetry, it is no exaggeration. The problem of tidal friction, which preoccupied workers from Darwin and Jeffreys to Munk and Cartwright, has been solved, at least in broad outline. Part of that solution entails internal tides, for which altimetry has given us a fascinating new view, if only from the surface. Moreover, through internal tides and their possible implications for ocean mixing, the connections between tides and the rest of physical oceanography have never been stronger or more evident. We should not forget the (some might say) more mundane topic of tidal prediction because the needs for accurate prediction have been extremely demanding, stimulating much international effort, and indeed the near-global constraints from altimetry have culminated in ocean-tide models now capable of predicting tidal heights anywhere in the open ocean with an accuracy approaching 1 cm. Tidal predictions for coastal and polar seas remain deficient, calling for renewed efforts. New needs to predict internal tides raise unique problems with new levels of difficulty.

These topics are reviewed in the present chapter. Fortunately for us, we need not start from scratch because the late Christian Le Provost wrote a very thorough review of this subject eight years after the launch of T/P (Le Provost 2001). Because of the many advances that have occurred since 2001, we will concentrate in large part on those. We readily refer readers to Le Provost's review to fill in the many gaps we no doubt leave.

13.2 TIDAL ALIASING

When observing tides with a satellite altimeter, the problem of aliasing can scarcely be avoided. Signals nominally confined to narrow frequency bands near 1 and 2 cycles per day (cpd) are, after sampling with an altimeter, dispersed throughout the whole frequency spectrum in a seemingly random manner. In the worst cases, a tidal signal is aliased to the zero frequency (i.e., the temporal mean), where it is essentially unobservable, or to the seasonal cycle, where it becomes difficult to separate from large non-tidal variability. If we wish to minimize estimation noise, then the ocean's red background spectrum implies that aliasing into shorter periods is preferable to aliasing into longer periods.

Fortunately for the T/P-Jason series of satellites—and so far, for no others– considerations of tidal aliasing played an important role in orbit design. Most major tidal constituents in T/P data are aliased to relatively short periods, generally less than 70 days. Yet as explained below, it is difficult to avoid all cases of problematic aliasing, and even for T/P, the K_1 constituent was aliased uncomfortably close to the semiannual period. Large tabulations of alias periods for both major and minor constituents have been published (e.g., Andersen et al. 2006; Le Provost 2001; Wunsch 2015, 207) for each of the main satellite altimeter missions. Those tabulations need not be repeated here, but Table 13.1 does list alias periods for a few major constituents of each tidal band.

There is an underlying pattern to the tidal aliasing for any satellite, and to understand this it is helpful to return to basic aliasing concepts. For a satellite flying over a location that repeats every N days (N need not be an integer), all frequencies exceeding the Nyquist frequency of $0.5/N$ cpd are aliased to frequencies below $0.5/N$ in the familiar accordion-like folding pattern (e.g., Bendat and Piersol 1971, Figure 7.8). Note that the frequency interval $0.5/N$ is generally much narrower than the frequency range covered by any of the tidal bands, so there is repeated folding inside each band. This is illustrated in Figure 13.1 for each of the major satellite altimeter missions. From this diagram, the alias frequency of any constituent can be immediately read off the left-hand axis and its alias period from the right-hand axis. The T/P-Jason series has the shortest repeat period, so it has the least amount of folding within each tidal band. One sees how challenging it is to design an orbit with no important constituent falling into seasonal or longer periods. The orbit of Sentinel-3 is most unfortunate in this respect.

Figure 13.2 is an expanded view of the frequency interval near the K_1 tidal group. The location of the vertex of each V-shaped line in this interval is fundamental because it fixes the whole aliasing pattern according to each satellite's orbital precession rate. Each vertex marks a frequency that corresponds in inertial space to the precession rate (the lower abscissa in the figure), so any signal at this frequency is sampled at only a single phase; that is, it is frozen. For example, the vertex for a sun-synchronous orbit falls at the S_1 frequency; in the semidiurnal band, one finds a similar vertex falling on S_2. Moreover, any sun-synchronous satellite always aliases P_1 and K_1 into the annual cycle and π_1 and ψ_1 into the semiannual cycle. Similarly, a perfectly polar, non-precessing orbit could never observe K_1, and it would alias S_1 and ψ_1 into the annual cycle. In light of the tight cluster of tidal lines in the K_1 group and the orbit precession rates being inherently centered near K_1, it becomes difficult to find orbits that avoid problematic aliasing. For T/P, the only way

TABLE 13.1
Tidal Alias Periods (Days) for Selected Constituents

Tide	Frequency (°/h)	T/P-Jason (9.92 d)	Geosat (17.05 d)	Envisat (35.00 d)	Sentinel-3 (27.00 d)
Mf	1.09803	36.2	68.7	79.9	1147.0
O_1	13.94304	45.7	113.0	75.1	277.0
P_1	14.95893	88.9	4466.7	365.2	365.2
K_1	15.04107	173.2	175.4	365.2	365.2
N_2	28.43973	49.5	52.1	97.4	141.0
M_2	28.98410	62.1	317.1	94.5	157.5
S_2	30.00000	58.7	168.8	∞	∞
M_4	57.96821	31.1	158.6	135.1	78.8
MS_4	58.98410	1083.9	361.0	94.5	157.5

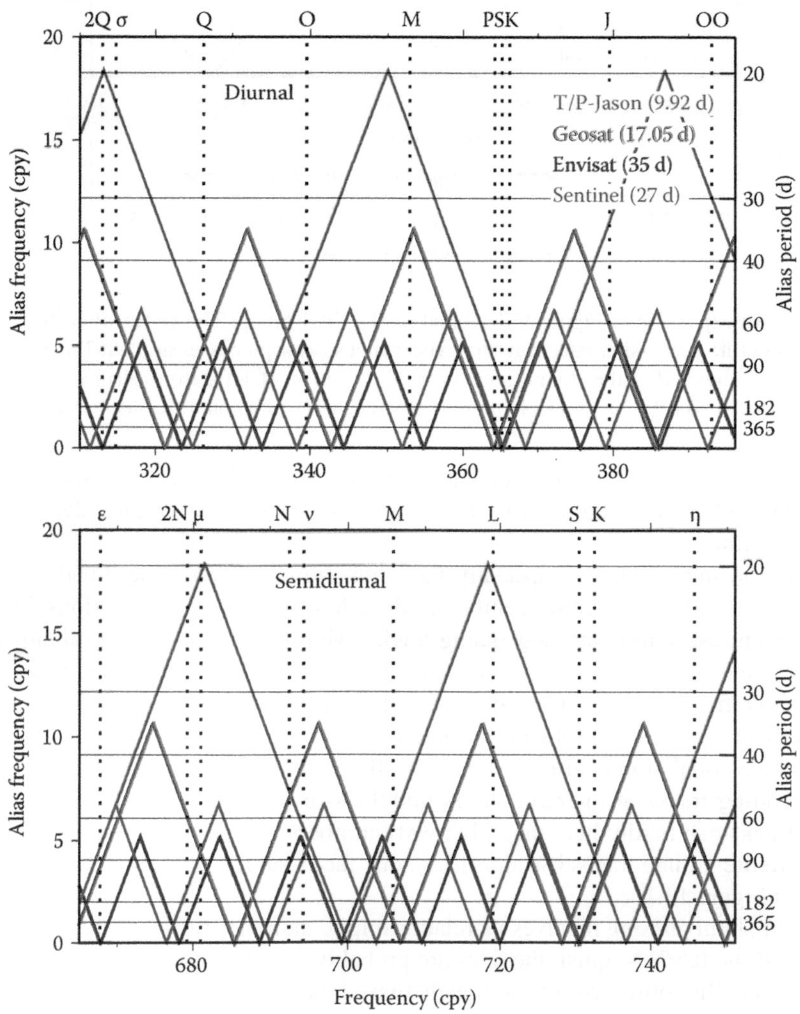

FIGURE 13.1 Aliasing, or frequency folding, across the diurnal and semidiurnal tidal bands, for the four major altimeter orbits. Major (and some minor) tidal constituents are labeled across the tops of each band. Alias frequencies for any satellite can be read off the left axis. All frequencies are in units of cycles per year.

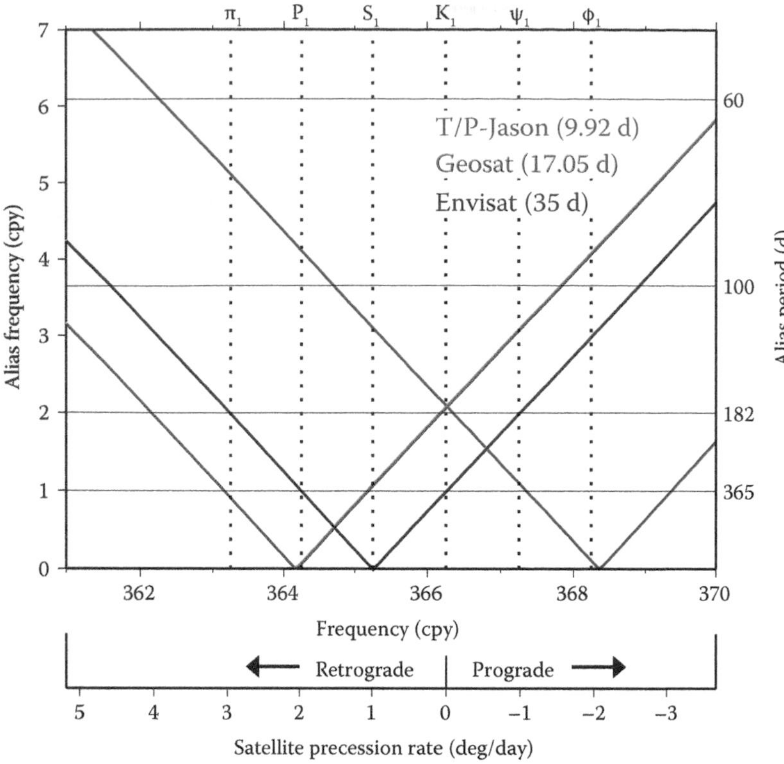

FIGURE 13.2 An expansion of Figure 13.1 around the K_1 tidal group. The vertex of each aliasing curve is defined by the satellite precession rate. Any sun-synchronous orbit precesses at 0.9867°/day, aliases S_1 to an infinite period, and aliases P_1 and K_1 to the annual period. The orbit of T/P-Jason sacrifices observations of the tiny φ_1 tide in order to alias P_1, S_1, and K_1 to periods shorter than semiannual, K_1 just barely.

to move the K_1 alias farther from the semiannual period would have been to speed up its precession rate. That could have been done by lowering its inclination or its altitude, but neither was desirable at the time.

Some tidal aliasing problems, especially those involving mapping the broad-scale barotropic tides, can be mitigated somewhat by combining the additional phase observations from neighboring tracks or from ascending and descending tracks. Moreover, badly aliased minor constituents can sometimes be estimated by inference from major constituents. This was the case for GEOSAT, where the nearly frozen P_1 could be reasonably inferred from the mapped K_1 (Cartwright and Ray 1990). But many aliasing problems are intractable. For example, for sun-synchronous orbits, the phase of the S_2 tidal argument is identical for all neighboring tracks, as well as for all ascending and descending tracks at intersections in middle to low latitudes. And combining data from neighboring tracks can hardly help map high wavenumbers. In particular, when mapping internal tides along satellite ground tracks (Section 13.5), one faces the full aliasing problems as laid out in Table 13.1 and Figure 13.1.

A final point about aliasing involves its relationship to stationarity. Of all the signals whose frequencies exceed the $0.5/N$ Nyquist, the tides are probably the easiest to deal with because they are stationary. The satellite-observed tide, although aliased, is still at a fixed frequency. Nonstationary signals are much more difficult to analyze. This is one reason—in addition to having extremely small amplitudes—that internal tides are difficult to study in altimetry because a significant component is likely to be nonstationary (Section 13.5.2). For example, consider seasonal-like modulations to M_2, generating energy within a small band ±1 cpy around the central M_2 line: In T/P-Jason data, this narrow band is distributed over a range of alias periods from 53 to 75 days, and thus much of it is surely buried in the background continuum.

13.3 BAROTROPIC TIDAL MODELS FOR AND FROM SATELLITE ALTIMETRY

For most users of satellite altimetry, the tides are a nuisance, a signal that must be removed from measurements in order to study smaller, usually subtler non-tidal signals. In many kinds of better sampled oceanographic measurements, a simple low-pass filter suffices as a de-tiding method, but the aliasing problem prohibits such simplicity for altimetry. Instead, at any location and time, the tidal height must be independently predicted as accurately as possible and subtracted from the measured sea-surface heights (SSHs). Global tide models have been devised to accomplish this, and predictions from one or more models are always included in project-released data records, although those users concentrating on small, near-coastal regions have always been advised to employ local tide models that can exploit local hydrodynamic knowledge and the most accurate, high-resolution bathymetry.

Any global tide model accurate enough to be used to "correct" satellite altimeter data, at least for the foreseeable future, must itself rely on historical altimeter measurements. Although much progress has been made in recent years in developing more realistic and accurate hydrodynamic models of tides (e.g., Green 2010; Ngodock et al. 2016), the stringent accuracy requirements of altimetry still require models constrained by global observations, for which altimetry itself is the best source.

The development of tidal models constrained by satellite altimeter data is thoroughly covered by Le Provost (2001), so only a brief summary is given here. Le Provost identified four main approaches: (1) direct empirical tidal analysis of altimetry; (2) empirical tidal analysis of altimeter residuals after first removing a prediction with an adopted prior model; (3) tidal analysis of altimetry (or residuals) in terms of some kind of spatial basis functions; and (4) use of inverse methods that solve the hydrodynamic equations via data assimilation. Approaches (1) and (2) are especially valuable in the deep, open ocean where tidal wavelengths are long enough to be well-sampled by satellite ground tracks. Moreover, there are always advantages to analyzing data with as few prior dynamical assumptions as possible. Empirical methods, however, tend to fail in shallow water if tidal wavelengths grow much shorter than satellite track spacings. Approach (3), based on global basis functions, has been most useful when the altimeter time series was short but subsequently has been found to be rather limited and inflexible, with less accurate results. It is being resurrected, in a fashion, for mapping baroclinic tides. Approach (4) is, of course, the most flexible because it allows a variable tradeoff between fitting data and satisfying dynamical equations of motion. It is the preferred approach for shallow seas. Its computational burden can be significant.

As might be suspected, approach (4) itself covers a wide range of different methods by which the tradeoff between measurements and dynamics can be managed. Many of these are reviewed by Egbert and Bennett (1996). "Nudging" methods date back at least to Schwiderski (1980), and more formal inverse methods based on conjugate gradient solutions were explored by Zahel (1995) and more recently by Taguchi et al. (2014). Representer methods were applied early on to the T/P data (Egbert et al. 1994; Le Provost et al. 1995), but they quickly grew unwieldy for large altimeter data sets. Subsequently, Egbert and Erofeeva (2002) found a number of approximations and simplifications that allowed the approach to be efficiently applied to large tidal problems. In this chapter, we often refer to the latest in the Egbert-Erofeeva series of inverse solutions, TPXO8. This solution assimilated over 20 years of altimetry data, with a series of higher resolution (1/30°) local coastal models nested into a 1/6° global base solution.

The fruit of these efforts is a number of very accurate global atlases of the barotropic tides, as depicted for example by the cotidal charts in Figures 13.3–13.5 for the largest semidiurnal constituent, M_2, and the largest diurnal constituent, K_1. Le Provost (2001) discusses some characteristics of these charts, as have many other authors. A notable feature of the K_1 constituent in Figure 13.5 is that it presents a nice depiction of the classic Antarctic Kelvin wave, a wave

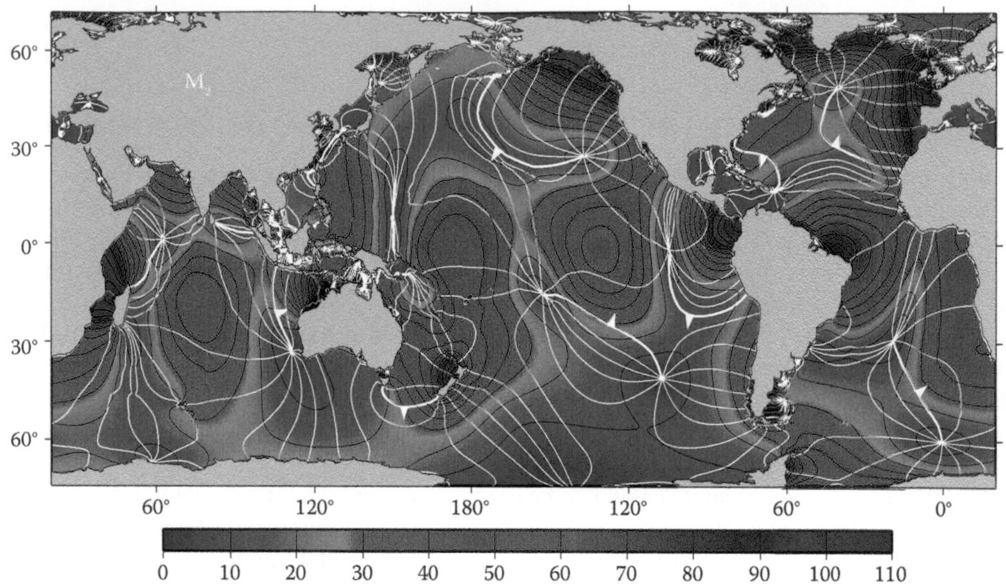

FIGURE 13.3 Cotidal chart of the principal semidiurnal constituent M_2 according to the global inverse solution TPXO8. Amplitudes in color are in centimeters; the color scale saturates in many shallow seas. Phase lines are drawn every 30°, or once per mean lunar hour; the thickest white lines mark high water when the mean moon passes Greenwich, with small arrows showing propagation direction. Independent bottom-pressure measurements in the deep ocean agree with this chart with a root-mean-square difference of 5.2 mm.

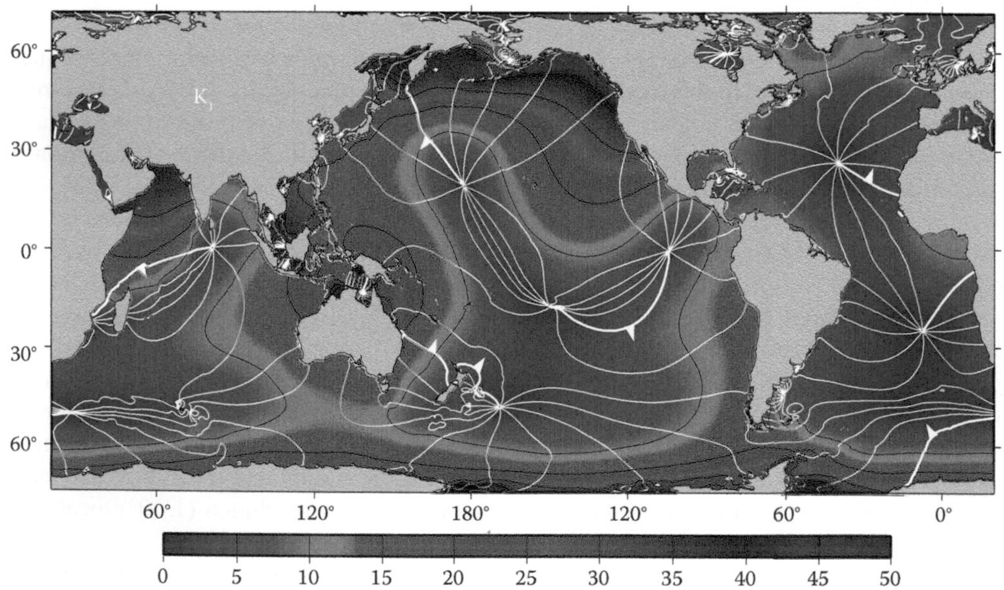

FIGURE 13.4 As in Figure 13.3, but for the declinational diurnal constituent K_1. Phase lines are drawn every 30°, or once every two sidereal hours. Independent bottom-pressure measurements in the deep ocean agree with this chart with a root-mean-square difference of 4.4 mm.

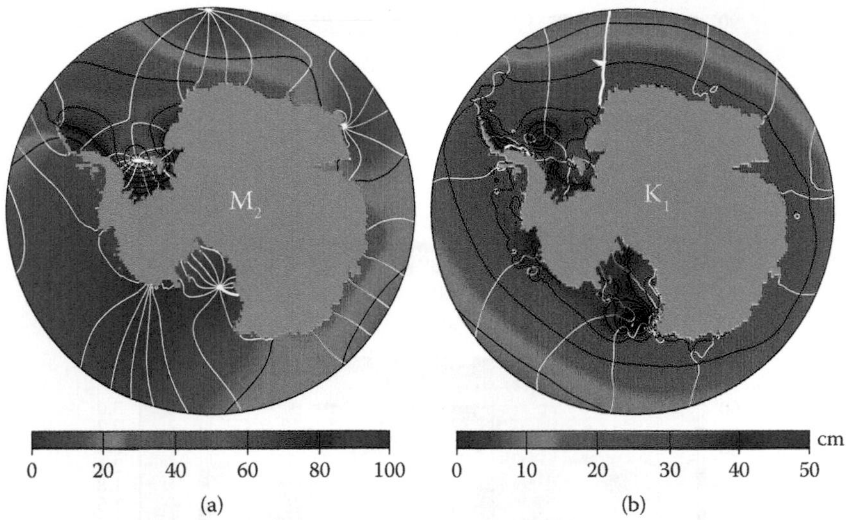

FIGURE 13.5 Cotidal charts of (a) M_2 and (b) K_1 in the oceans around Antarctica, including beneath Antarctic ice shelves, according to the global inverse model TPXO8. The M_2 amphidrome in the Ross Sea rotates clockwise. Unlike Figures 13.3 and 13.4, the charts here are less accurate, especially near Antarctica, owing to a spareness of high-quality data.

that completes a single circumpolar circuit in one tidal period, with amplitudes rising toward the coastline. This K_1 Kelvin wave represents the excitation of a particular oceanic normal mode with a period thought to be around 30 hours (Platzman et al. 1981; Müller 2007); the excitation is even stronger in diurnal constituents Q_1 and O_1, whose longer periods are closer to the modal period (Cartwright and Ray 1991).

How accurate are these altimeter-constrained barotropic models? The problem was recently addressed by Stammer et al. (2014) in the first comprehensive assessment of global tide models since early efforts by Andersen et al. (1995) and Shum et al. (1997). Among their results were maps of standard deviation computed from seven global models, reproduced here in Figure 13.6. These maps depict the degree of consistency among models, but they still have implications for accuracy. Models are very consistent in the open ocean where standard deviations are only a few millimeters. At least some of the models, however, must be relatively inaccurate in shallow seas and along certain coastlines, as well as in polar regions, which lack comparably good constraints. In the open ocean, the standard deviations are also inflated in a few regions where the tidal amplitudes themselves are large—for example, the North Pacific for K_1 (compare Figure 13.4)—but even there the standard deviations are generally sub-centimeters.

To assess the accuracies of barotropic models, Stammer et al. (2014) employed a variety of tests. Some of these tests entailed examining the reduction in variance in various kinds of independent measurements, such as independent altimeter data, but also in different types of satellite data such as ground-to-satellite laser ranging and satellite-to-satellite microwave ranging (GRACE data) in which the tidal models are used to compute gravitational forces (and resulting orbit perturbations) on the satellites. Some of the most useful tests were comparisons to independent tide measurements from coastal tide gauges and from seafloor pressure recorders. As an example, in Table 13.2, we employ the same test data against model TPXO8; the tabulated results are generally comparable to most of the other models examined by Stammer et al. (2014). In keeping with the suggestions from Figure 13.6, the model errors are very small in the deep ocean with no constituent exceeding a root-mean-square (RMS) difference of 0.6 cm (and some of those differences must surely arise from errors in the test data). Model errors grow larger in shallower water. With only 56 stations, the coastal assessment can hardly be definitive, and almost certainly there will be coastal locations

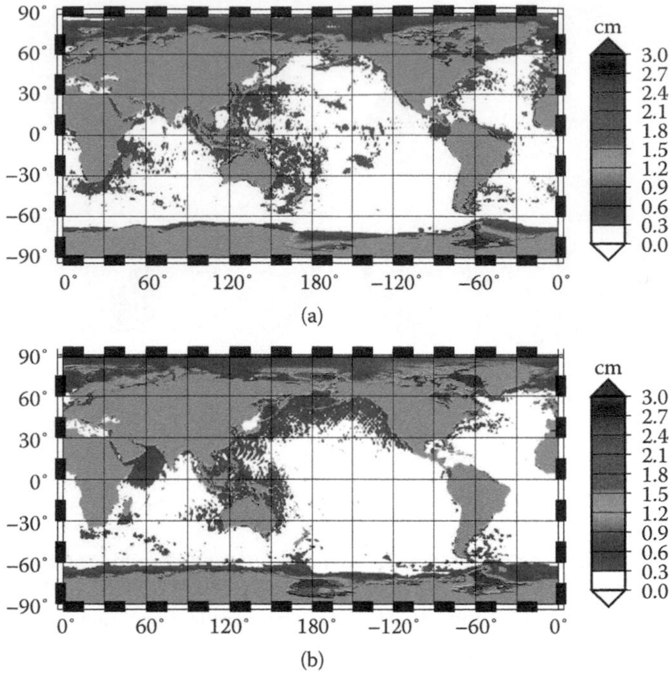

FIGURE 13.6 Standard deviations for (a) M_2 and (b) K_1 tides, computed for seven modern global tide models by Stammer et al. (2014). Throughout most of the deep ocean the seven models are consistent to better than 3 mm. They diverge in polar regions and in near-coastal waters (where values greatly saturate the scale bar limit of 3 cm). (Reproduced from Stammer, D., et al., *Rev. Geophys.*, 52, 243–282, 2014. With permission.)

where the model is significantly poorer than the listed RSS of 8 cm, but the main point is to show the general inflation of model errors near the coasts. Such errors must be kept in mind by altimeter users who rely on corrections from global models in such regions. But notwithstanding these larger errors, there has been considerable improvement over the past decade in the accuracies of global models in shallow-water regions, even though still more progress is needed (Ray et al. 2011).

Although comparisons of major constituents to tide-gauge estimates, such as those in Table 13.2, are a useful guide to possible errors in altimeter tide corrections, a more definitive assessment is by comparing a full tide prediction with measured "ground truth" sea-surface heights. This can test all constituents—not just a handful of major constituents—as well as the tide-prediction algorithms and software. Figure 13.7 summarizes the results of such tests with 12 seafloor pressure stations from the DART international tsunami network. The 12 stations are globally well distributed but were otherwise randomly selected; all are in deep water. For each pressure time series (each with 15-minute sampling, of duration 8–31 months), conversion to equivalent sea-surface heights was done by using a climatological model of ocean density; signals from atmospheric tides (primarily affecting constituents S_1 and S_2) were removed with a model (Ray and Ponte 2003); and a high-pass filter cutoff of approximately two days was applied to remove most non-tidal variability. The filtered time series were then compared with predicted heights from model TPXO8 based on prediction algorithms similar to those employed in generating T/P and Jason geophysical data records. In particular, 20 minor constituents were inferred from the TPXO8 major constituents, and standard adjustments for nodal and perigee modulations were applied. As Figure 13.7 shows, the tidal predictions are rarely worse than 2 cm RMS, and for several stations, they approach 1 cm. This figure (a similar figure based on tide model GOT4.10c is given by Egbert and Ray 2017) thus provides strong evidence for the statement that the best global tide models are now capable of predicting the deep ocean tide with an accuracy of about 1.5 cm. In fact, the tide prediction must be somewhat better than this because the

TABLE 13.2
Root-Mean-Square Differences (cm) of Tide Model TPXO8 against Tide Gauge Estimates

	Q_1	O_1	P_1	K_1	N_2	M_2	S_2	K_2	M_4	RSS
Deep ocean (151 stations)	0.15	0.31	0.18	0.44	0.20	0.52	0.34	0.15	0.07	0.90
Shelf seas (119 stations)	0.83	1.25	0.82	1.66	2.02	3.76	2.27	1.11	0.88	5.50
Open coastal (56 stations)	0.34	0.80	0.53	1.23	2.33	6.51	2.40	0.41	1.68	7.50

Note: RSS denotes root sum of squares across all listed constituents.

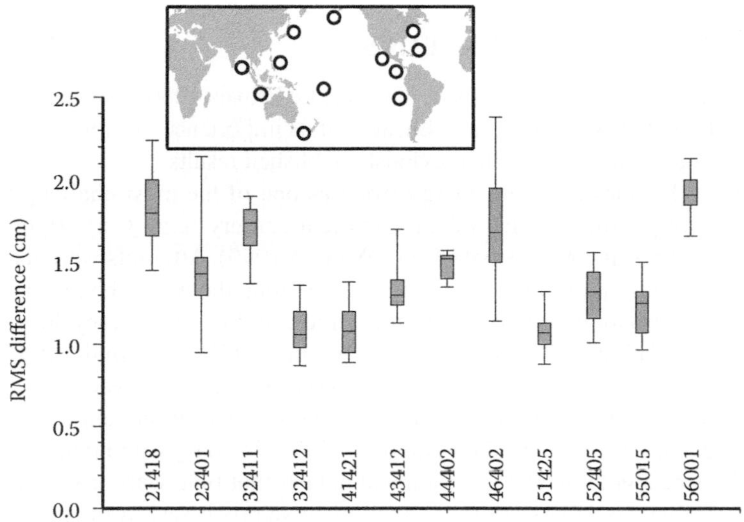

FIGURE 13.7 Tidal prediction error (cm) at 12 DART bottom-pressure stations, with predictions based on the TPXO8 global model. The prediction error was computed as the root-mean-square (RMS) difference between hourly observed and predicted tides, for each month of each time series. The "box and whisker" symbol marks the median, the upper and lower quartiles, and the minimum and maximum of the RMS errors over all months examined. The observed pressures were high-pass filtered to remove most non-tidal variability. Signals from atmospheric barometric tides were removed from the bottom-pressure data by means of an air-tide model. DART station numbers are listed at bottom, and their locations are given in the inset map. A similar diagram for model GOT4.10c is given by Egbert and Ray (2017).

filtered bottom-pressure data must still retain some amount of non-tidal variability caused by high-frequency winds and pressures.

As Figure 13.6 emphasizes, barotropic models are considerably more inaccurate in polar regions, especially in the Canadian archipelago and underneath Antarctic ice shelves. In these regions, the tide can be spatially complex, but high-quality observational constraints and adequate bathymetry are generally lacking. Satellite altimeter measurements of ice shelves are critically important for cryospheric and climate studies (e.g., Paolo et al. 2016), so there is clearly a need for better tide models in these locations. Padman et al. (2016) review some recent progress, and they discuss how better tide models are also needed for clarifying ice-ocean interactions.

An interesting additional complication sometimes arises in both polar regions and in some lower latitude marginal seas: a significant seasonal modulation in tidal constituents (e.g., Müller et al. 2014). At a few polar locations, the seasonal change is remarkably large; Godin (1986) reported amplitude changes of ±30% in James Bay, on the southern end of Hudson Bay.

In general, changes across most of the Arctic Ocean are smaller, 5 cm or less (Kagan et al. 2011), and are near zero in the deep interior basin. Seasonal changes significant enough to affect altimeter measurements have been observed in the Yellow and East China seas (Kang et al. 2002) and in the North Sea (Gräwe et al. 2014). In fact, seasonal modulations of M_2 are common enough in British waters that two additional constituents at frequencies ±1 cpy from M_2, often dubbed MA_2 and MB_2, are routinely introduced into their tidal analyses (e.g., Amin 1982). In polar seas, the likely explanation for seasonal modulations rests with changes in frictional damping between ice-covered and ice-free conditions (Müller et al. 2014; Prinsenberg 1988; St. Laurent et al. 2008). In non-polar shallow seas, the changes appear to be related to seasonal changes in stratification, a topic explored in some detail by Müller (2012). Standard tide corrections applied to altimetry have so far ignored seasonal modulations.

13.4 BAROTROPIC TIDAL ENERGETICS

One area where great progress has been made since the Le Provost (2001) review concerns the energetics and dissipation of the barotropic tide. Because of its importance, we review this topic in some detail and take the opportunity to update previously published results.

How and where the tides dissipate energy stood as one of the most challenging problems in physical oceanography through much of the twentieth century (e.g., Cartwright 1977; Jeffreys 1920; Munk 1997; Munk and MacDonald 1960; Wunsch 1990). Attempts were made to reconcile three global energy rates: (a) dissipation in shallow seas (long thought to be the main sink), (b) the rate of working by astronomical tide-generating forces, and (c) the energy loss implied by the moon's secular acceleration. As discussed by Cartwright (1977) and Munk (1997), early efforts to reconcile these rates featured repeating cycles of convergence and divergence. Jeffreys (1920) was thought to have solved the problem; Munk (1968) reviewed subsequent evidence showing he had not. A rough balance was achieved by Lambeck (1977) but only with large error uncertainties. In fact, available oceanographic and astronomical data at that time were insufficient to constrain accurately either the energy inputs or the acting dissipative processes. At the time of the launch of T/P, there was an approximate balance between (b) and (c)—see Cartwright and Ray (1991)—and none at all with (a).

Satellite geodesy, especially satellite laser ranging, provides direct constraints on the total tidal dissipation (e.g., Lambeck 1977; Platzman 1984). These totals eventually converged (e.g., Ray 1994) to the now well-determined global total of 3.7 TW (all constituents; 2.5 TW for M_2) for the Earth system as a whole (oceans, atmosphere, and solid Earth). However, before T/P, the "where" and "how" of dissipation remained unclear. Estimates of dissipation in shallow seas from *in situ* data—1.7 TW for M_2 by Miller (1966)—came up well short of the required total. Generation of internal tides and waves by interaction of tidal flows with bottom topography were often discussed as a likely additional energy sink (e.g., Cartwright 1977), but quantitative estimates of the significance of this process varied widely (e.g., Baines 1982; Morozov 1995; Munk 1997; Sjöberg and Stigebrandt 1992; Wunsch 1975).

The situation changed with the availability of precise T/P altimeter data and the resulting very accurate global charts of tidal elevations. These charts could be used to deduce—in fact, to map at somewhat broad scale—the barotropic tidal energy dissipation (Egbert and Ray 2000, 2001). It was found that a significant fraction (at least 25% to 30%) of the energy in the surface tide is lost in the deep ocean over rough and steep topographic features, essentially confirming the previously suggested importance of baroclinic conversion and internal tide generation in the tidal energy balance (e.g., Morozov 1995; Sjöberg and Stigebrandt 1992). These results have important implications for the broader field of oceanography as the substantial transfer of energy to the deep ocean internal wave field implies that the tides are likely a significant source of mechanical energy for vertical ocean mixing (Wunsch and Ferrari 2004) and may thus play an important role in maintenance of the meridional overturning circulation (Munk and Wunsch 1998).

In the approach of Egbert and Ray (2000, 2001), global maps of barotropic tidal energy dissipation are inferred from the balance between work done by tidal forces and the divergence of barotropic tidal energy flux

$$D = W - \nabla \cdot \mathbf{P} = \rho g \langle H \mathbf{u} \cdot \nabla(\eta_{EQ} + \eta_{SAL})\rangle - \rho g \langle \nabla \cdot H \mathbf{u}\eta \rangle. \quad (13.1)$$

Here $\mathbf{u}$ is the tidal velocity, H is water depth, η tidal elevation, and η_{EQ} and η_{SAL} are, respectively, the astronomical forcing, and self-attraction and loading, represented as equilibrium tidal heights. The brackets $\langle \cdot \rangle$ denote time averages over a tidal cycle. To employ (1), estimates of currents $\mathbf{u}$ are required, and these must be inferred indirectly from the altimetrically constrained elevation field η by using the frequency-domain linear shallow-water equations. In appropriately simplified form, these are

$$i\omega \mathbf{u} + f\hat{\mathbf{z}} \times \mathbf{u} = -g\nabla(\eta - \eta_{EQ} - \eta_{SAL}) - \mathbf{F} \quad (13.2)$$

$$i\omega\,\eta = -\nabla \cdot H\,\mathbf{u}, \quad (13.3)$$

where f is the Coriolis parameter, ω the tidal frequency, and $\mathbf{F}$ represents dissipative stresses.

Egbert and Ray (2000, 2001) used two different approaches to estimate $\mathbf{u}$. First, the variational data assimilation scheme of Egbert et al. (1994) was used to estimate both the η and $\mathbf{u}$ fields from the T/P data. Second, gridded tidal elevation fields η obtained by fitting T/P altimeter data were substituted into (2–3), and the currents $\mathbf{u}$ were estimated by weighted least squares (Ray 2001). To test possible sensitivity of the computed $\mathbf{u}$ to the assumed dissipation, different forms for $\mathbf{F}$ were tried, both linear and quadratic in velocity, with a range of friction parameters. Provided mass conservation was strongly enforced, similar dissipation maps were obtained, demonstrating the robustness of the observed dissipation patterns to details of how currents are inferred from the directly observed elevation field. These calculations were subsequently extended to construct dissipation maps for the eight largest tidal constituents (Egbert and Ray 2003), the result being that at least 1 TW (out of a total ocean dissipation of 3.5 TW) is dissipated in deep water over rough bottom topography. Similar efforts to map tidal dissipation have been reported by Taguchi et al. (2014), who reached similar conclusions concerning the importance of deep ocean dissipation.

An updated version of the map for M_2, computed from the 1/6° TPXO8 global solution, is shown in Figure 13.8a. This new dissipation map is broadly similar to those obtained previously. Not surprisingly, given the long time series of high-precision altimetric data now available, the new map is considerably cleaner, with fewer large patches of negative dissipation and much sharper resolution of some areas of intense deep ocean dissipation, especially in the Pacific and Indian oceans. The close correspondence among the spatial patterns of elevated open-ocean dissipation and the locations of island chains and arcs, shelf edges, and other steep bathymetric features leaves little doubt that the mapped dissipation in the open ocean represents energy transfer from the barotropic surface tide to internal tides and waves (e.g., Garrett and Kunze 2007). As discussed in Egbert and Ray (2001), this conclusion is further reinforced through comparison between the mapped dissipation and maps of conversion rates computed with simple linear theories for baroclinic wave generation (Baines 1982; Sjöberg and Stigebrandt 1992; Morozov 1995).

In Figure 13.8b, we carry this comparison one step further, plotting barotropic to baroclinic conversion rates computed from a high-resolution tide and wind-forced global run of the Hybrid Coordinate Ocean Model (HYCOM) summarized by Arbic et al. (2012). The results plotted are taken from Buijsman et al. (2016), a study focused on tidal dissipation in this three-dimensional stratified ocean model, and they represent total semidiurnal (as opposed to M_2) baroclinic conversion. In the deep ocean, there is very close correspondence between patterns of barotropic dissipation mapped from altimetry and baroclinic conversion computed from the model, with a close match even at the level of rather small-scale

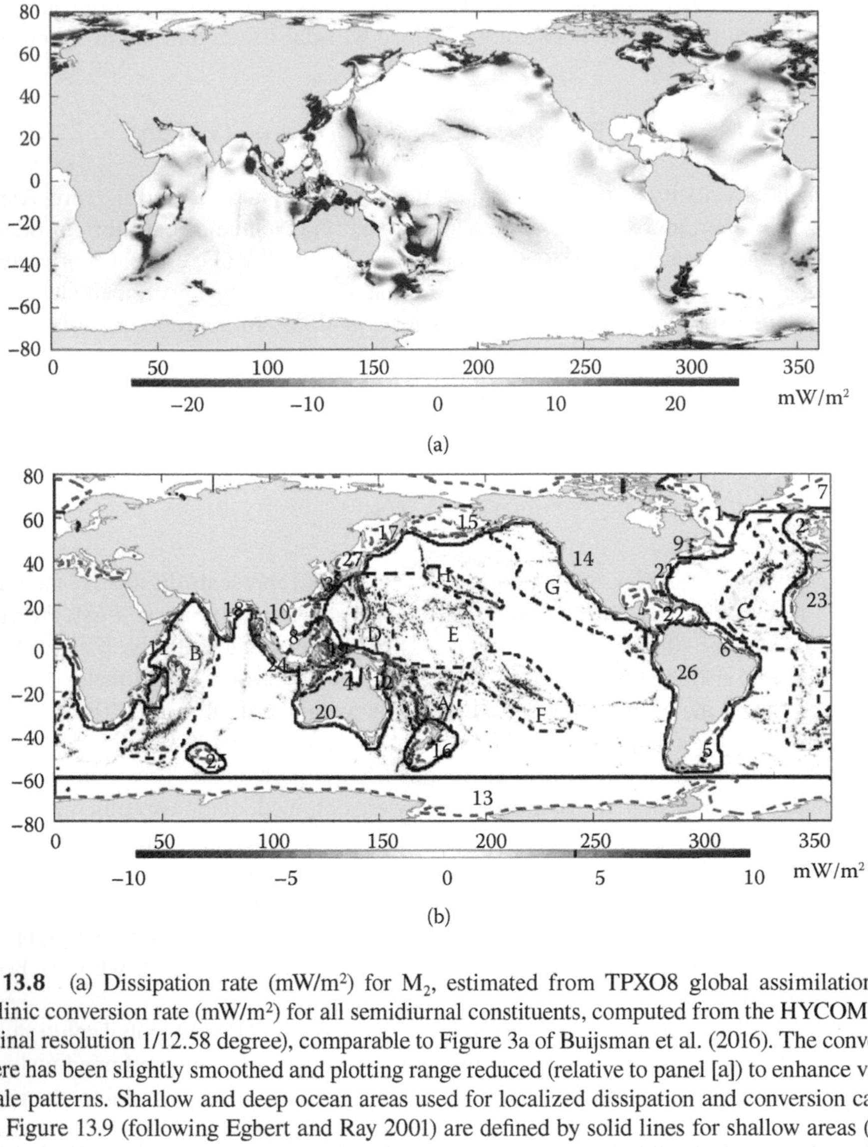

FIGURE 13.8 (a) Dissipation rate (mW/m^2) for M_2, estimated from TPXO8 global assimilation solution. (b) Baroclinic conversion rate (mW/m^2) for all semidiurnal constituents, computed from the HYCOM 18.5 solution (nominal resolution 1/12.58 degree), comparable to Figure 3a of Buijsman et al. (2016). The conversion rate plotted here has been slightly smoothed and plotting range reduced (relative to panel [a]) to enhance visibility of global scale patterns. Shallow and deep ocean areas used for localized dissipation and conversion calculations plotted in Figure 13.9 (following Egbert and Ray 2001) are defined by solid lines for shallow areas (numbered 1–27) and dashed lines for deep ocean areas (labeled A–I). The green dashed line is the more formally defined boundary between deep and shallow areas discussed in the text.

(500 km or so) details. The mapped dissipation (Figure 13.8a) is necessarily somewhat smoother, with more diffuse dissipation spread over larger areas. Some of this difference undoubtedly reflects resolution limits and blurring by the regularization implicit in the inversion. However, the resolution of the HYCOM simulation imposes limits on the representation of conversion by smaller scale (and often more diffuse) bottom topography such as abyssal hills. Indeed, only the first three baroclinic modes are well resolved by the 32 layers used for the HYCOM run (Arbic et al. 2012), and Buijsman et al. (2016) estimate that only 30% of the total baroclinic conversion is directly captured by this HYCOM run; the remaining conversion in the simulation was represented with a parameterized internal wave drag, comparable to that commonly used in barotropic models (e.g., Jayne and St. Laurent 2001). We consider it likely that some of the diffuse dissipation in the empirical maps is realistic.

To provide a more quantitative summary of dissipation patterns, Egbert and Ray (2001) integrated the altimetric estimates over a series of patches, corresponding to major shallow seas and continental shelf areas, as well as some of the major deep ocean bathymetric features. Terms in these surface integrals that depend on velocities can be reduced to line integrals around patch edges, making the integrals insensitive to details in the (likely poorly resolved) velocity fields in areas of complex bathymetry (Egbert and Ray 2001). We repeat this analysis here, using the same patches as in the original publication; these are shown for reference in Figure 13.8b. Results are shown graphically in Figure 13.9, where we compare results from the new TPXO8 dissipation map (solid line), with those obtained from TPXO4a (dashed line). Error bars (one sigma) for the older estimates (obtained as described in Egbert and Ray 2001) are also plotted. In all but a few cases, the new and old estimates agree within error bars.

In Figure 13.9, we also plot HYCOM baroclinic energy conversion, integrated over the same patches (red symbols). In the deep ocean areas (A–I, lower part of plot) magnitudes are correlated, but the HYCOM baroclinic conversion is almost always less than the estimated dissipation, in most cases by a factor of roughly two. Two notable exceptions are Hawaii, where the two are nearly the

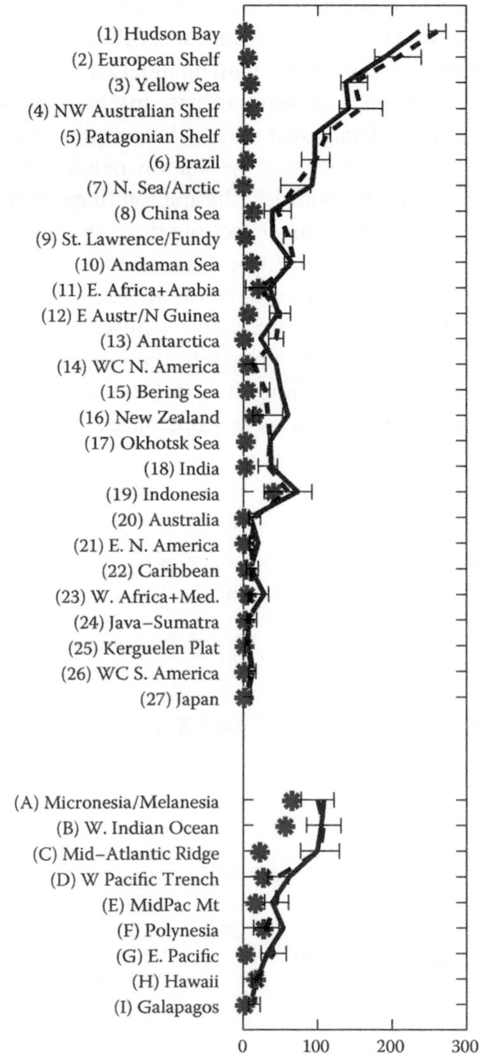

FIGURE 13.9 Energy dissipation rates (GW) for shallow and deep patches defined in Figure 13.8b computed from TPXO8 (solid line). Dashed line gives comparable results from TPXO4a, with error bars, from Egbert and Ray (2001). Symbols give semidiurnal baroclinic energy conversion from the HYCOM run (Buijsman et al. 2016) integrated over each patch.

same (17.2 versus 18.8 GW) and the Mid-Atlantic Ridge, where baroclinic conversion is less than a quarter of the mapped dissipation (22.3 versus 99.5 GW). These results are broadly consistent with the findings of Buijsman et al. (2016) that a significant fraction of the internal wave drag is not directly modeled, owing to limitations on vertical resolution. Internal waves generated at the Hawaiian Ridge, which is isolated and steep, are dominantly low mode (e.g., Zhao et al. 2010), allowing most of the conversion to be captured in the HYCOM runs. In contrast, internal tides generated over the expansive deep rough topography of the Mid-Atlantic Ridge are dominated by shorter wavelengths, with a spectral peak at Mode 5 (St. Laurent and Garrett 2002). Thus, much of the deep ocean energy conversion in the Atlantic is missed by the HYCOM simulation.

There is little HYCOM conversion in most of the areas considered as shallow seas or continental shelves by Egbert and Ray (1–27 in Figure 13.9), where tidal dissipation is expected to be dominated by bottom friction. However, there are a few exceptions, where the HYCOM numbers amount to a substantial fraction of the total mapped dissipation. The most significant of these are (19) Indonesia (40.4 of 72.8 GW); (11) E. Africa, Arabia (19.7/35.3); (8) China Sea (12.7/38.6); (16) New Zealand (14.4/59.5); and (10) Andaman Sea (10.4/64.4). In some cases, this is a result of the conservative definition of shallow areas in Egbert and Ray (2001), with boundaries drawn well out into the deep ocean to avoid areas of topographic complexity, where currents might be poorly defined. New Zealand, the Bering Sea, and the area around Madagascar are good examples of this. One can also clearly see baroclinic conversion at the shelf break in most of the shallow patches; a particularly clear example is off northeastern Brazil. Of course, in the empirical dissipation maps, it is impossible to separate this sink of energy from the (generally much stronger) bottom drag on the actual shelf.

It is enlightening to compare mapped barotropic dissipation and baroclinic conversion at higher resolution, focusing on the very complex area around Southeast Asia and Indonesia (Figure 13.10). In many respects, the two zoomed maps are remarkably similar, with many details in the modeled baroclinic conversion recovered at least qualitatively in the empirical dissipation maps. We again use integrals over small patches (defined in Figure 13.10b) to make more quantitative comparisons. These patches are divided into three groups (Table 13.3): large shallow seas along the continental

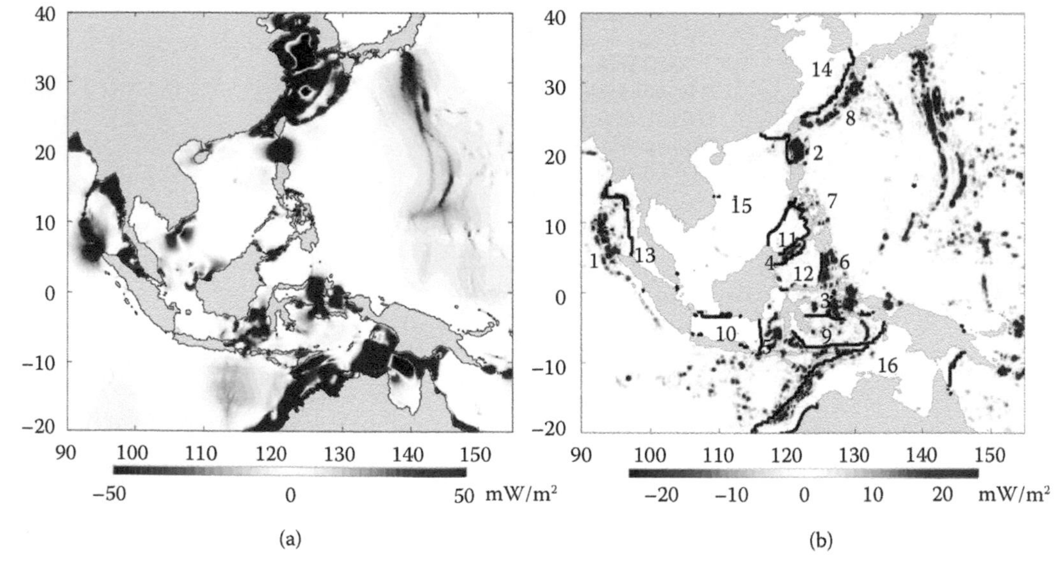

FIGURE 13.10 Zoom of Figure 13.8, showing (a) dissipation and (b) baroclinic conversion in Western Pacific/Indonesia region. Ocean patches used for local integrals given in Table 13.3 are identified in panel (b). Numbers 1–8 identify areas with submerged ridges or island chains that separate deeper well-defined seas (9–12). Some continental shelf areas are also identified (13–16).

shelf (13–16), small (and generally deep) Indonesian seas (9–12), and the island chains and submarine ridges that separate these seas from each other or from the Pacific or Indian oceans (1–8).

TPXO8 dissipation is strongest in the continental shelf areas, where there is very little baroclinic conversion. Except for northwest Australia, baroclinic conversion at the shelf break is negligible in comparison to the total dissipation estimated from TPXO8. There is also comparatively little dissipation, or baroclinic conversion, in the deep Indonesian seas. Unsurprisingly, the baroclinic conversion within this area of complex bathymetry is concentrated in narrow bands over the shoaling bathymetric features that define boundaries between individual seas. Most of the (non-shelf) dissipation is concentrated in these same narrow bands. As with the larger patches in the global compilation of Figure 13.9, amplitudes of modeled conversion and empirically mapped dissipation correlate, with the dissipation estimates generally higher by a factor of two or more.

A regional modeling study of tides in the Indonesian Archipelago by Nagai and Hibiya (2015) is also relevant to this discussion. Using the MITgcm with 100 vertical levels and 0.01-degree lateral resolution, they simulated the M_2 surface and internal tide. Almost all the baroclinic conversion in their high-resolution model occurs in the same island chains and narrow straits that dominate the TPXO8 dissipation map shown in Figure 13.10. Baroclinic conversion for four of the areas (3–6) listed in Table 13.3 can be computed for the model from results shown in Figure 11 in Nagai and Hibiya (2015). Agreement between modeled conversion and our estimates of dissipation is generally good: 26.8 versus 31.1 GW for Area 3, 10.6 versus 17.5 GW for Area 4, 15.7 versus 19.6 for Area 5, and 7.8 versus 2.7 for Area 6.

These results clearly demonstrate the significant improvement in resolution of barotropic dissipation that can be achieved with the long altimetric time series now available. More quantitative and spatially precise estimates of tidal inputs to the internal wave field are now possible, and the distinction between dissipation due to bottom friction and baroclinic conversion is at least somewhat clearer. For example, there is a clear separation between dissipation on the continental shelf within the Andaman Sea (Area 13, 44.3 GW) and at the Andaman-Nicobar Islands that define the edge of

TABLE 13.3
High-Resolution Estimates of Barotropic Dissipation (GW) for Indonesia and East Asia

Area	M_2 Diss. TPXO8	HYCOM[a] Conversion	K_1 Diss.TPXO8
(1) Andaman/Nicobar Islands	18.5	9.3	0.1
(2) Luzon Strait	18.8	11.7	15.4
(3) Malluccas	31.1	15.8	8.3
(4) Sibutu-Baslin Ridge	17.5	5.9	5.1
(5) Makassar-Timor	19.6	11.6	1.0
(6) Sangihe Islands	2.7	5.8	4.5
(7) Phillipines	5.7	1.1	1.5
(8) Riyuku arc	9.7	6.2	0.2
(9) Banda Sea	1.6	1.8	0.6
(10) Java Sea	5.5	0.0	4.6
(11) Sulu Sea	1.4	0.3	0.4
(12) Celebes Sea	2.2	–0.3	0.4
(13) Andaman Sea	44.3	1.2	0.7
(14) Yellow Sea	126.9	1.4	3.9
(15) China Sea	13.5	0.8	18.7
(16) NW Australian Shelf	137.7	6.5	20.0

[a] HYCOM baroclinic conversion represents the entire semidiurnal band.

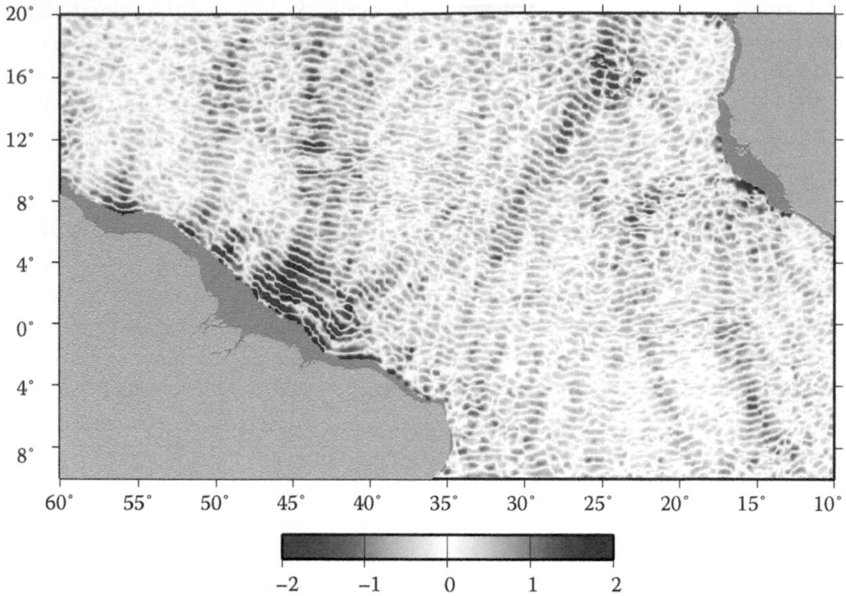

FIGURE 13.11 A snapshot of sea surface height (cm) of the M_2 internal tide (strictly, the component in-phase with the generating potential) from one of the earliest empirical analyses of along-track multi-mission satellite altimetry (Ray and Zaron 2016). Some of the largest waves, such as those off the Amazon shelf, saturate the 2-cm color scale.

the Indian Ocean (Area 1, 18.5 GW). Similarly, there is a clear separation between the Ryukyu arc (Area 8, 9.7 GW) and the Yellow Sea (Area 14, 126.9 GW).

We can, of course, extend our picture of tidal dissipation and baroclinic conversion by considering additional constituents. As a step in this direction, we include estimates for K_1 from the Southeast Asia and Indonesia patches in the last column of Table 13.3. There are four areas where baroclinic conversion for this diurnal constituent appears to be significant. The most prominent is the Luzon Strait, where the estimated dissipation is almost as large for K_1 as for M_2, with a total of 34.2 GW for these two constituents. Assuming that dissipation of other constituents of the same species can be scaled (with the square of relative forcing amplitude; Egbert and Ray (2003)), leads to a rough estimate of 49 GW for total tidal dissipation. In fact, internal tides in this area have been the focus of intense study over the past decade through the Internal Waves in Straits Experiment (IWISE; Alford et al. 2015). Extensive modeling in support of IWISE, combined with observations, provides estimates of baroclinic conversion by tidal flow over the sills of the Luzon Strait. A large scale far-field model resulted in a total baroclinic conversion (for eight dominant tidal constituents) of 24 GW. However, higher resolution near-field models increased this by a factor of 1.8, approaching our own rough estimate of total dissipation.

Other areas of high K_1 conversion include the Sibutu-Basilin Ridge (4) and the Sangihe Islands (6), which bound the Celebes Sea, and the Moluccas Islands, which divide the Indonesian Seas to the south from the Pacific. In the next section, we show that all of these areas, together with the Luzon Strait, are indeed among the largest sources of diurnal internal tides.

We now revisit the division between shallow and deep ocean tidal dissipation. In Egbert and Ray (2003), the boundary between the shallow and deep areas used for this calculation (shown in Figure 2 of Egbert and Ray 2001) was drawn conservatively (and informally) to avoid possible contamination of the deep ocean dissipation estimate by leakage from shallow seas. Here, we use a more formal definition of shallow areas: For each ocean point, a circle of radius 200 km is considered; if 20% of the area of the circle has a depth less than 300 m (including land), then that point is considered shallow.

Thus, essentially all points within roughly 150 km of land/continental shelf are included in the shallow zone, which is shown by the green dashed contour in Figure 13.8b. Most small islands and ridges are included in the deep zone. For M_2, 0.88 TW of dissipation (36.9% of the total) is classified as deep with this definition. Results for the four largest semidiurnal and diurnal constituents are given in Table 13.4. Summed over all eight TPXO8 constituents, total dissipation is 3.52 TW, with 1.28 TW (36.3%) in the deep ocean. Not surprisingly, given the tighter definition of shallow areas, these new estimates increase the fraction of deep ocean dissipation relative to the earlier total estimate of 29.3% of 3.51 TW given by Egbert and Ray (2003)). The increase is most notable in the diurnal constituents for which the deep fraction has increased dramatically from 11.3% to 39.3%. Examination of dissipation maps (not shown, but comparable to Figure 1 in Egbert and Ray 2003) reveals that much of the diurnal dissipation occurs in the North Pacific along the Aleutian Arc and Kuril Islands, in areas inside the shallow sea boundary used in the earlier calculations.

As these areas are well north of the critical latitudes for diurnal constituents where free gravity waves are not permitted, this dissipation cannot be interpreted as baroclinic conversion in the usual sense. Much of the energy lost from diurnal constituents along the North Pacific margin is most likely related to generation of continental shelf waves trapped to the steep topography along the margins of the Pacific basin. Both modeling and observational studies have shown that such waves are common and can have very large amplitudes in the North Pacific (Cherniawsky et al. 2001; Foreman et al. 2000; Kowalik and Polyikov 1999). Shelf waves, being associated with strong currents, may potentially dissipate significant energy. Furthermore, they are sensitive to stratification and, with their short wavelengths (hundreds of kilometers, comparable to low-mode internal waves), may interact nonlinearly with non-tidal ocean flows. Thus, one can anticipate that energy will leak from the surface tide (which TPXO8 represents) into smaller scale (and likely temporally variable/incoherent) ocean motions. Although less important than internal wave generation on a global scale, these processes are probably quite important in the North Pacific and perhaps deserve more attention.

It is also instructive to integrate the HYCOM baroclinic conversion over the shallow and deep ocean domains. Out of a global total of 460.7 GW, about 112 GW, or 24%, occur within the shallow area. Much of this dissipation occurs at the continental shelf break, although some reflects the inevitably approximate nature of the boundary. In any event, this suggests that some of the 1.56 TW of M_2 dissipation in shallow seas results from baroclinic conversion. Indeed, one of the clearer generators of internal tide, discussed in the next section, is the Amazon shelf break. Conservatively, 100 GW of the shallow dissipation is likely due to baroclinic conversion. Because the HYCOM estimates

TABLE 13.4
Total Energy Dissipation, Kinetic Energy (KE), and Potential Energy (PE) for Main Constituents of Model TPXO8, with Division between Shallow and Deep Seas

	Dissipation (TW)								
Tide	Total	Shallow	Deep	% Deep	PE (× 10^{15} J)	KE	%KE Shallow	%KE < 1000	Q
M_2	2.441	1.563	0.878	36.9	127.7	179.7	22.8	21.3	18
S_2	0.385	0.247	0.138	35.8	20.2	29.3	18.4	19.3	19
N_2	0.110	0.075	0.050	31.9	5.7	7.8	20.1	21.4	17
K_2	0.030	0.021	0.009	32.0	1.6	2.3	18.2	19.5	19
K_1	0.336	0.203	0.133	39.6	16.2	38.7	19.1	46.3	12
O_1	0.173	0.107	0.066	38.3	7.8	18.0	19.0	42.7	10
P_1	0.034	0.020	0.014	41.3	1.6	3.4	18.1	42.0	11
Q_1	0.007	0.004	0.003	41.1	0.3	0.6	22.0	36.0	9
Total	3.516	2.240	1.276	36.3	181.1	279.8			

are low by a factor of two or more, baroclinic conversion in the shallow patch is probably closer to 200 GW. The total baroclinic conversion could then be well over 1 TW for M_2 alone. On the other hand, some of the dissipation in deep water should undoubtedly be ascribed to bottom friction, or at least to very localized dissipative processes, so the total baroclinic conversion rate may still be somewhat less than 1 TW.

Table 13.4 also gives estimates of kinetic and potential energy, along with global specific dissipation Q^{-1}, for all major constituents. We divide the kinetic energy between shallow and deep areas, using two approaches. With the first, used also by Egbert and Ray (2003), we define shallow-ocean regions as those with a depth of 1000 m or less. With the second approach, we use the shallow/deep division developed for the dissipation calculations. For semidiurnal constituents, both definitions result in a similar fraction of kinetic energy in shallow water—on the order of 20%. For diurnal constituents, the two definitions produce dramatically different results—more than 40% for the first approach (as in Egbert and Ray 2003), less than half this for the second. This clearly demonstrates that more than 20% of the diurnal kinetic energy is concentrated along the edges of the continental shelves, at depths about 1000 m, consistent with an important role for trapped shelf waves in diurnal tidal energetics.

13.5 BAROCLINIC TIDES

Internal or baroclinic tides are internal gravity waves with tidal periodicity. They are generated by the interaction of barotropic tides with bottom topography—notably continental shelves, mid-ocean ridges, and seamounts—all locations highlighted in the previous section as sinks of barotropic tidal energy. When barotropic tidal currents impact such large topographic features, deep dense water is swept upward where buoyancy provides a restoring force to set up internal waves. Internal displacements are typically meters to tens of meters, or even larger near a few intense generation spots. Notable exceptions are places with weak barotropic tides, such as the Mediterranean Sea. General references for the subject are Wunsch (1975), Hendershott (1981), Vlasenko et al. (2005), and Garrett and Kunze (2007).

An internal tide has a small, but non-zero, vertical displacement at the ocean surface. The displacement is dependent on the underlying stratification but generally is of the order 10^{-3} times the maximum internal displacements (e.g., Apel 1987) and thus rarely more than a few centimeters. Modern satellite altimetry is capable of detecting these waves, even in the face of the severe aliasing discussed in Section 13.2. In fact, altimetry has become an important tool for studying internal tides, giving us a hitherto inconceivable global view—although only from the surface—while shedding light on internal-tide generation, propagation, and dissipation, as well as revealing new aspects about the oceanic medium through which the waves propagate (Ray and Zaron 2016; Zhao 2016; Zhao et al. 2012).

Measurements—both satellite and *in situ*—of open-ocean internal tides can sometimes lead to contradictory conclusions. Are internal tides stationary, or nonstationary, or (most likely) some combination of both? Because internal tides depend critically on the ocean's stratification and background currents, both of which vary on all time scales, they cannot be perfectly stationary. Variability is apparent in many *in situ* observations (e.g., Levine and Richman 1989); indeed, in some places, internal tides appear and disappear in seemingly erratic fashion (Magaard and McKee 1973). In this picture, internal tides are characterized by a fairly broad-band spectrum, peaking at the diurnal and/or semidiurnal periods but essentially nonstationary (Wunsch 1975). Yet the detection of internal tides in satellite altimetry (Ray and Mitchum 1996) relied on the waves being coherent in both time and space—phase-locked with the astronomical potential in order to overcome aliasing and sufficiently spatially coherent to be recognizable and separable from the barotropic tide. In fact, there had been previous indications, at least in some locations, that the internal tide could have a large stationary component (Hendry 1977; Sherwin 1988). Dushaw et al. (1995) detected a coherent internal-tide signal in long-range acoustic measurements.

Satellite altimetry detects primarily internal tides corresponding to the first baroclinic mode. This is expected on theoretical grounds because, for a given internal displacement, higher order modes are suppressed at the surface (e.g., Hendershott 1981; Wunsch 1975). First-mode baroclinic tides have wavelengths of the order 100 km to 200 km for semidiurnal constituents, twice that for diurnal constituents, corresponding to phase speeds around 3 m s^{-1}. Phase speeds for higher modes are slower and hence more sensitive to changes in background ocean currents and more incoherent in time. Relative to *in situ* measurements, which readily observe all modes, altimetric measurements of internal tides have a built-in modal filter that emphasizes stationarity. Nonetheless, the fact remains that standard altimeter analyses that rely on detection of a coherent wave must miss some fraction of the true tidal signal. We return to this topic in Section 13.5.2.

13.5.1 Stationary Baroclinic Tides

Over the past few years, concerted efforts have begun to map in fine spatial detail the SSH field of the stationary internal tides. Like the surface tide, part of the motivation is to develop tidal-prediction models so that unwanted tidal variability may be removed from satellite altimeter measurements. There is evidence that internal-tide signals have leaked into the current generation of gridded SSH anomaly products (Ray and Zaron 2016), although the leakage is likely too small to affect all but a few altimeter applications. More importantly, internal tide signals need to be removed in new high-resolution swath-mapping altimetry that aims to study the oceanic sub-mesoscale (Chavanne and Klein 2010; Durand et al. 2010; Fu and Ferrari 2008). There are, of course, many other reasons to map the internal-tide field, not the least being to support efforts to understand the sources and sinks of baroclinic tidal energy and the pathways of this energy as it propagates and dissipates throughout the ocean. But many of those kinds of studies do not require the very high prediction accuracy needed for "correcting" altimetry, so high-resolution numerical ocean models (e.g., Arbic et al. 2012) are often the tool of choice in such studies—they have ready access to the (model) ocean's interior—assuming they can be sufficiently validated with real observations, including from altimetry. In contrast, building prediction capabilities accurate enough to correct altimetry requires altimetry itself to provide the necessary data constraints.

The ground-track patterns of all current and historical altimeter missions—GEOSAT, T/P, ERS-1, Sentinel-3, and their successor missions—are now marginally dense enough to support direct mapping of the Mode-1 internal-tide field with standard gridding and interpolation methods. One approach is to determine tides along every repeating ground track and grid the resulting elevation fields (Ray and Zaron 2016). The approach is probably unsatisfactory for solar tides because so many historical missions were flown in sun-synchronous orbits, but initial efforts for lunar tides, especially for the dominant M_2 constituent, are promising—see Figure 13.11. The approach should also work after several years of data are collected from the future Surface Water and Ocean Topography (SWOT) wide-swath altimeter mission (Durand et al. 2010).

Figure 13.11 reveals tidal internal waves with amplitudes of a few millimeters or greater throughout the whole equatorial Atlantic. Even without the corresponding phase diagram, the largest amplitudes in the figure clearly delineate important generation regions, including the Amazon shelf break and the topography surrounding Cape Verde. Some of the beam-like patterns, evident also in global scale maps (e.g., Arbic et al. 2004; Ray and Zaron 2016; Zhao et al. 2016), are the manifestation of complex constructive and destructive wave interference from multiple generation sites (Rainville et al. 2010).

In contrast to a purely empirical approach, Dushaw et al. (2011) investigated an approach that requires the mapped field conform to theoretical frequency-wavenumber characteristics determined by the ocean's climatological stratification. These constraints act to suppress noise and to guide the interpolation between altimeter tracks. Using this method, Dushaw (2015) obtained global results solely from T/P altimetry. The mapped fields are notably smoother than empirical fields such as in Figure 13.11. Zhao et al. (2012, 2016) also invoke climatological dispersion relationships to map internal tide fields by localized, iterative plane-wave fitting.

Finally, we may anticipate that formal data assimilation schemes applied to numerical tidal models will also be developed because these methods should help suppress noise and overcome data limitations, including inadequate spatial resolution from satellite tracks spaced too widely apart. An early global effort, as yet unpublished, along these lines is by Egbert and Erofeeva (2014), who modified their barotropic codes to simulate reduced-gravity dynamics appropriate for Mode-1 waves while assimilating along-track tidal estimates. Coupling between modes can be handled by an additional force-like term involving gradients of seafloor topography. An example output for the K_1 constituent is shown in Figure 13.12. A notable feature of this figure is the large internal tide generated at Luzon Strait, undoubtedly a reflection of the large barotropic energy loss noted in Table 13.3. Because the phase speed of internal tides depends on the Coriolis parameter, the eastward trending wavetrains from Luzon Strait refract southward as the propagate, which is clearly seen in Figure 13.12. Zhao (2014) discusses these effects in more detail and shows filtered maps depicting refraction in both K_1 and O_1. Another notable feature of Figure 13.12 is a trapped standing wave in the Celebes Sea. How much this wave is arising from dynamics alone is not clear, although empirical maps suggest a somewhat similar pattern. This does emphasize the point that invoking dynamics in data assimilation involves more than mandating a preferred wavelength to fit and that the dynamics may be capable of yielding more physically realistic maps that empirical methods (or plane-wave fitting) fail to capture because of inadequate sampling.

On the other hand, one advantage of empirical methods that avoid making dynamical assumptions is that, if the data are adequate, such assumptions can be directly tested from the mapped results. Consider the case of the wavelength of internal tides. These wavelengths can be determined

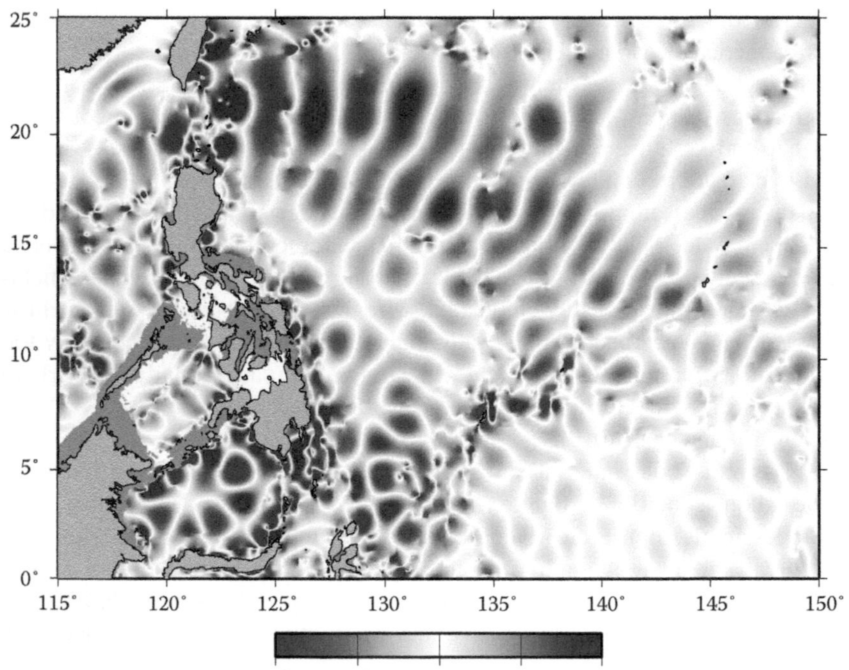

FIGURE 13.12 Sea surface height in-phase component (cm) of the K_1 Mode-1 internal tide, from a preliminary data-assimilation inverse solution, using the hydrodynamic model of Egbert and Erofeeva (2002) with reduced gravity appropriate for baroclinic Mode 1. Data from multiple altimeter missions were assimilated. Some of the largest waves saturate the 2-cm color scale. Note the eastward propagating wavetrain from the Luzon Strait bending southward owing to the change in phase speed with the Coriolis parameter. (See also Zhao 2014).

directly from empirical fields (such as Figure 13.11) and can then be compared against theoretical wavelengths determined by solving the vertical structure eigenvalue problem for a given ocean stratification. For example, Figure 13.13 shows a two-dimensional wavenumber spectrum for a 20° by 20° region northeast of New Zealand based on a strictly empirical multisatellite map (Ray and Zaron 2016). A strong circular peak representing Mode-1 propagation in nearly all directions is evident, as is a much fainter Mode-2 peak with propagation mainly toward the south. Both modal peaks are pronounced in the radial average spectrum (lower inset of figure). By computing such radial spectra in successive regions around the globe, a map of the empirically determined baroclinic M_2 wavelength can be constructed. Such a map is shown in Figure 13.14a, and it can be compared with the theoretical wavelengths based on the World Ocean Atlas (Locarnini et al. 2010) mean stratification (Figure 13.14b). Overall, the comparison is very good and suggests that adopting the theoretically determined wavelengths as a mapping constraint (as done by Dushaw et al. 2011; Zhao et al. 2016) is likely beneficial. But there are also minor differences in the two panels (e.g., the eastern Pacific shows wavelength differences of 10 km or more). Some of these differences occur in regions where historical hydrographic data may be lacking, such as west of Chile, so that differences may well stem from errors in the theoretical wavelengths. Differences in the vicinity of the Antarctic Circumpolar Current likely stem from errors in the altimeter-based results since these are suspect in regions of high mesoscale variability and/or small tidal amplitudes.

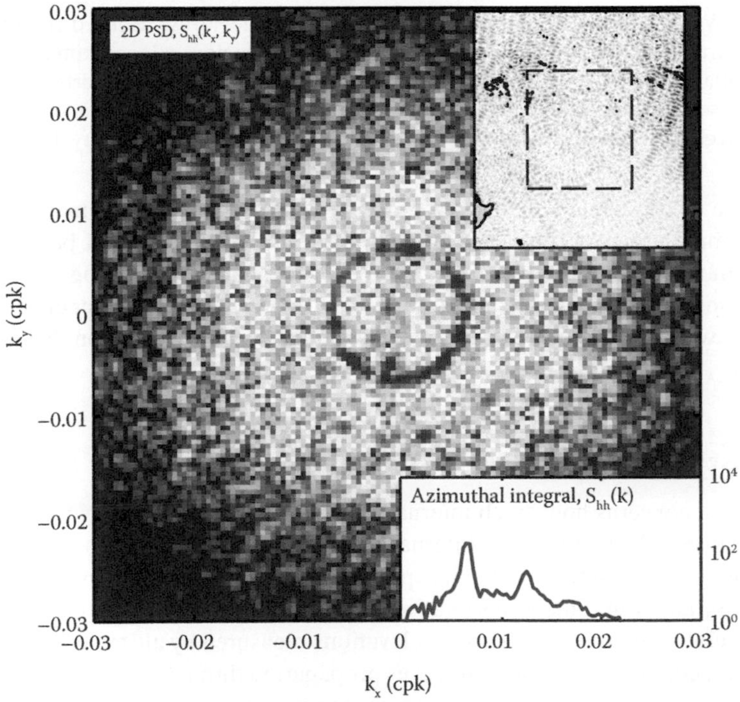

FIGURE 13.13 Two-dimensional internal-tide wavenumber spectrum for an ocean region roughly 2000 km square, northeast of New Zealand, based on the empirical internal-tide solution of Ray and Zaron (2016). The nearly circular pattern of energy corresponds to a wavenumber magnitude of $\sqrt{\left(k_x^2+k_y^2\right)}$ and indicates propagation in nearly all directions over the (relatively large) region. The azimuthal integral of the two-dimensional spectrum (inset, lower right) shows a pronounced peak for the first baroclinic mode with wavelength approximately 160 km. A secondary peak from a second mode occurs at approximately 80 km, with propagation mainly toward the south. (Reproduced from Ray, R. D., and Zaron, E. D., *J. Phys. Oceanogr.*, 46, 3–21, 2016. © American Meteorological Society. With permission.)

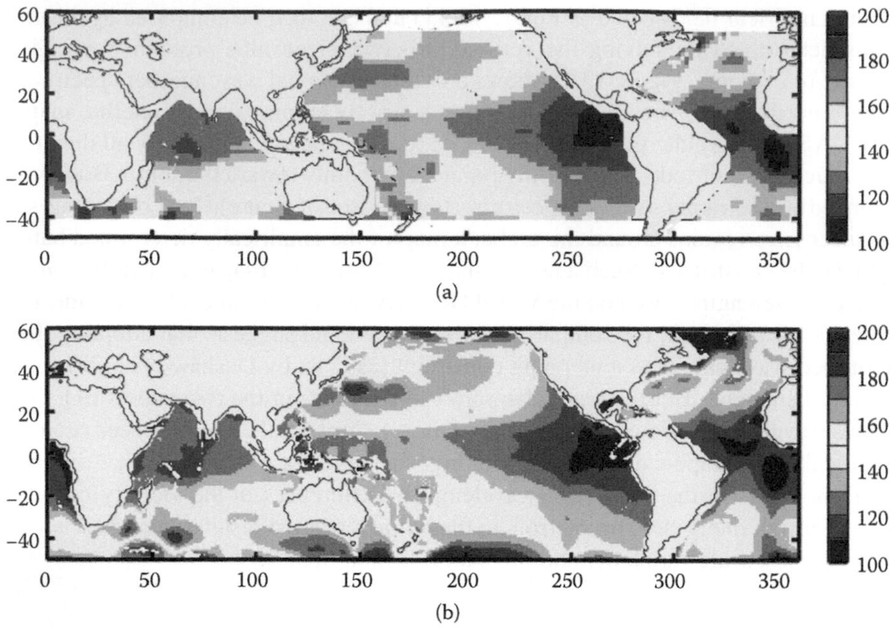

FIGURE 13.14 Wavelengths (km) of the M_2 first baroclinic mode. (a) As deduced from two-dimensional wavenumber spectra, as in Figure 13.13, from overlapping ocean regions of approximate 2000-km window sizes. (b) As calculated by solving the Sturm-Liouville eigenvalue problem for the vertical structure function, with mean ocean stratification defined by the World Ocean Atlas, subsequently smoothed with a 1000-km window. Reproduced from Ray and Zaron (2016) © American Meteorological Society. Used with permission.

Regarding use of baroclinic tide maps as a new type of correction to altimetry, testing work by the altimeter community has begun. Even though these early maps are still being refined, and no doubt will continue to be, they have already been shown capable of reducing the SSH variance in independent altimeter data (e.g., Ray and Zaron 2016, Figure 13.11). As happened with barotropic tide corrections, we can expect much future testing and comparison work on these new baroclinic tide corrections.

13.5.2 Nonstationary Baroclinic Tides

A critical question concerns how much internal-tide variability is missed by maps of the coherent tide, such as Figures 13.11 and 13.12. Internal tides can be temporally modulated by changes in generation, propagation, scattering, and dissipation. Past *in situ* observations suggest the variability can be substantial, even if these observations are sometimes skewed by high baroclinic modes that contribute relatively little to the surface elevations measured by altimeters. Numerical experiments with idealized, low-mode, internal waves propagating through variable media are useful for highlighting various mechanisms that may come into play (e.g., Dunphy and Lamb 2014; Ponte and Klein 2015). For example, Ponte and Klein explored how a low-mode plane-wave internal tide propagates through a mesoscale eddy field of increasingly energetic turbulence. When eddy kinetic energies are weak, internal tides can pass through the eddy field with little scattering, but with more vigorous eddies, corresponding to RMS velocities around 50 cm s^{-1}, the internal tide becomes almost completely incoherent. Based on such experiments, one might expect that western boundary currents and strong equatorial jets may completely destroy internal tide coherence. See also work by Kelly et al. (2016) on numerical experiments regarding interactions of internal tides with the Gulf Stream.

In fact, empirical maps such as Figure 13.11 do resolve very little coherent signal in regions of strong mesoscale variability. Although an obvious explanation for this is the difficulty of extracting a tiny tidal signal from a noisy red background when the tidal signal is aliased to long periods, numerical experiments such as those of Kelly et al. (2016) suggest that, in fact, there may be no coherent tidal signal to extract in these places.

Especially relevant to exploring this problem is the recent appearance of a number of global eddy-resolving general circulation models, forced now by the astronomical tidal potential in addition to standard wind and buoyancy forcing. (Combined circulation-tidal models have, of course, been used for many years in regional coastal ocean modeling.) Notable efforts include the HYCOM model, as summarized by Arbic et al. (2012) and discussed earlier, and the Max Planck Institute Ocean Model (MPI-OM), described by Müller et al. (2010); both models have now been used to investigate properties of the global internal tide (Müller 2013; Shriver et al. 2012, 2014). Regarding the incoherence problem, Shriver et al. (2014) find that the internal-tide SSH variability tends to increase with increasing mean amplitude, but when normalized by the mean amplitude the variability decreases. Thus, perhaps as expected, the internal tide is relatively more stationary near strong generation regions and grows more nonstationary with distance. The variability is comparable to the mean amplitude in regions of weak tides, such as the tropical Pacific and eastern Indian Ocean.

Real ocean observations are often more difficult to interpret. The most valuable data on tidal surface heights, of course, come from long time series of tide-gauge measurements, but these point measurements are often ambiguous without knowledge of the spatial variability. Probably the most in-depth study of a tide-gauge time series is the work of Colosi and Munk (2006) on the century-long Honolulu gauge. The variability in the harmonic "constants," manifested in the sea-level spectrum by a pronounced cusp around the tidal line(s), was interpreted in part as caused by the surface signature of the varying internal tide. They deduced an amplitude of the M_2 surface tide of 16.6 cm and of the internal tide of 1.8 cm, with their combined sum phase-modulated by varying propagation times from a remote generation point. (Intriguingly, Colosi and Munk also attribute a slight secular increase in the Honolulu M_2 amplitude over the course of the twentieth century to a 28-degree rotation of the internal tide vector caused by ocean warming.) Colosi and Munk built their analysis on earlier work by Mitchum and Chiswell (2000), who found that variable M_2 amplitudes are correlated with mean sea level at a number of tide gauges along the Hawaiian Ridge. The correlation could be positive or negative, depending on distance to an assumed source point (depending on depth to thermocline), while the M_2 modulations were found to be coherent for long distances along the Hawaiian Ridge. Mitchum and Chiswell could not directly determine the mean internal tide, considering it subsumed at any one tide gauge by the surface tide, but the variable internal tide was found to be smaller than the mean internal tide as estimated from nearby echo-sounder data as well as from altimetry. (As an aside, we note that in older tide-gauge data, any cusps in tidal spectra could well be caused by time-keeping errors, although this should not occur in modern data tied by satellite links to stable clocks; Zaron and Jay (2014) have made one of the most careful studies of time-keeping (and other) errors at two dozen open-ocean tide gauges in the Pacific. Time-keeping errors appear evident in Hilo data, less so in Honolulu data.)

In summary, both general circulation models and tide-gauge data suggest significant variability in the surface signature of internal tides. This variability is likely to be weaker than the mean internal tide, at least near strong generation spots such as the Hawaiian Ridge. It now seems that satellite altimetry can also shed light on this question, even though the tidal signals are badly aliased. We thus close this section with some very recent work based on altimetry.

Zaron (2015) has made an original study of SSH differences in dual-satellite (Jason-2 and CryoSat-2) crossover data. These crossover data occur with variable time lags; a few are almost simultaneous with less than a few hours separation. By binning the mean square SSH differences (after some critical corrections) as a function of time lag and analyzing the resulting lagged time series, Zaron was able to estimate a mean global variance of the nonstationary semidiurnal tide. He found it to be approximately 30% of the mean internal tidal variance.

In another approach to analyzing altimeter data, Ray and Zaron (2011) examined along-track wavenumber spectra along multiple repeated passes. As shown in Figure 13.15, there is often a pronounced bump in these spectra around the expected theoretical wavenumber for Mode-1 tides, and sometimes for higher modes as well, although any bump is necessarily broadened by the change in Coriolis parameter and ocean stratification over the length of the satellite track. (Part of the tidal energy is also likely mapped into lower wavenumbers, depending on track orientation.) These spectra are the result of removing (via a model) only the barotropic tide from the satellite data—the black curve in Figure 13.15. If tidal harmonics are also estimated point-by-point along the satellite track, and the tidal signal then predicted and removed before the spectra are computed, any residual bump—the red curve in Figure 13.15—will be due to the nonstationary internal tide. As the figure shows, for this particular track across the northwest Pacific, most of the Mode-1 bump is removed, implying a mostly stationary wave. Much of the Mode-2 bump, which is much smaller, remains. The variance in the two modal bands is approximately 1.19 cm^2 and 0.26 cm^2 before de-tiding and 0.23 cm^2 and 0.12 cm^2 after. Ray and Zaron (2011) also showed an example of a track off the Amazon Shelf for which most of the Mode-1 spectral bump was nonstationary.

Zaron (2017) has followed up this initial wavenumber spectrum study by applying the same method systematically to track segments all around the global ocean. His results are summarized in Figure 13.16. In many respects consistent with the HYCOM study of Shriver et al. (2014), the figure shows that nonstationary tides can be larger than stationary tides, notably so in locations where the overall internal-tide variance is fairly weak, including throughout the tropics in nearly all longitudes except the eastern Atlantic. The tide is also predominantly nonstationary throughout much of the Indian Ocean with the exception of the western region near the Mascarene Ridge, which is a known generator of very large internal tides (e.g., Morozov

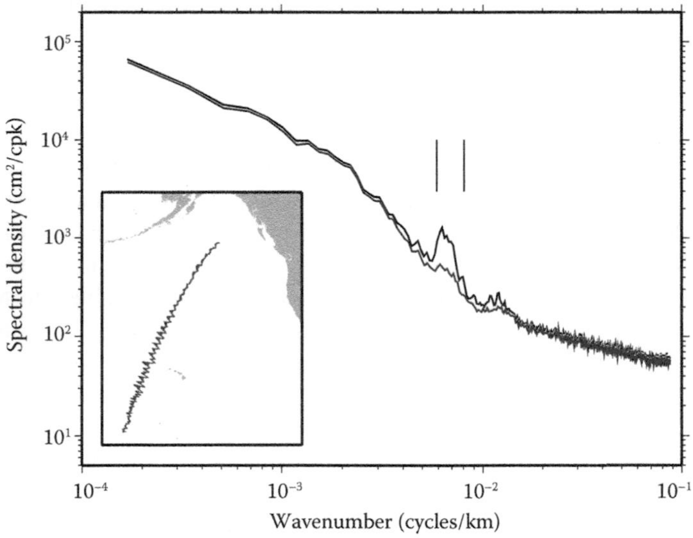

FIGURE 13.15 Wavenumber spectrum for a T/P-Jason track in the North Pacific. The black curve shows the 17-year average wavenumber spectrum of sea surface height after a standard model is used to remove the barotropic tide. The red curve shows the same spectrum after estimating and removing residual tides (relative to model) point by point along the track. Vertical lines denote the expected wavenumber range of Mode-1 semidiurnal internal waves along this track, accounting for variation of the Coriolis parameter and the mean ocean stratification; this range nicely delineates the observed peak in the spectrum. A second peak, corresponding to Mode 2, is evident at approximately twice the Mode-1 wavenumber. The de-tided spectrum still contains elevated variance at both the first- and second-mode wavenumbers, which is attributed to the nonstationary internal tide. (Reproduced from Ray, R. D., and Zaron, E. D., *Geophys. Res. Lett.*, 38, L17609, 2011. With permission.)

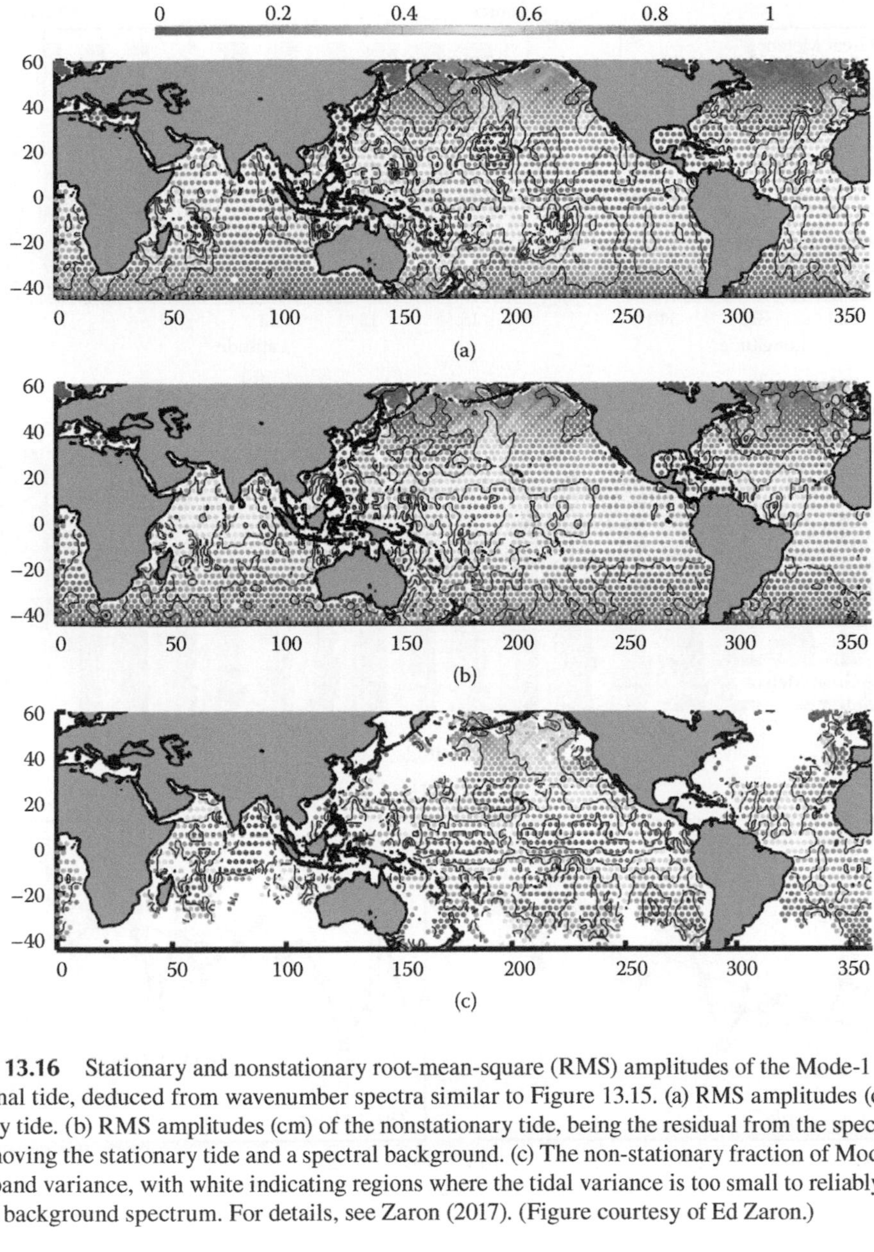

FIGURE 13.16 Stationary and nonstationary root-mean-square (RMS) amplitudes of the Mode-1 semidiurnal internal tide, deduced from wavenumber spectra similar to Figure 13.15. (a) RMS amplitudes (cm) of the stationary tide. (b) RMS amplitudes (cm) of the nonstationary tide, being the residual from the spectral hump after removing the stationary tide and a spectral background. (c) The non-stationary fraction of Mode-1 semidiurnal band variance, with white indicating regions where the tidal variance is too small to reliably estimate from the background spectrum. For details, see Zaron (2017). (Figure courtesy of Ed Zaron.)

and Vlasenko 1996) and where amplitudes of the coherent component measured by altimetry can reach 5 cm. Note that regions of high mesoscale variability, which are places where nonstationarity might also be expected, are blanked out in the figure because the overall internal-tide variance could not be determined from the background spectrum; the altimeter spectra there are evidently overwhelmed by the mesoscale. Aside from these whited-out regions, most of the South Pacific, the South Atlantic, and the North Pacific are dominated by the stationary tide. The northeast Atlantic, also a region with multiple sources of large internal tides (e.g., Pairaud et al. 2010; Pingree and New 1995), appears mostly nonstationary in the altimeter data. Note that in some places, perhaps the tropical western Pacific, it is possible that part of the nonstationary variance is arising from variability in Mode-2 diurnal tides rather than (or in addition to) Mode-1 semidiurnal tides.

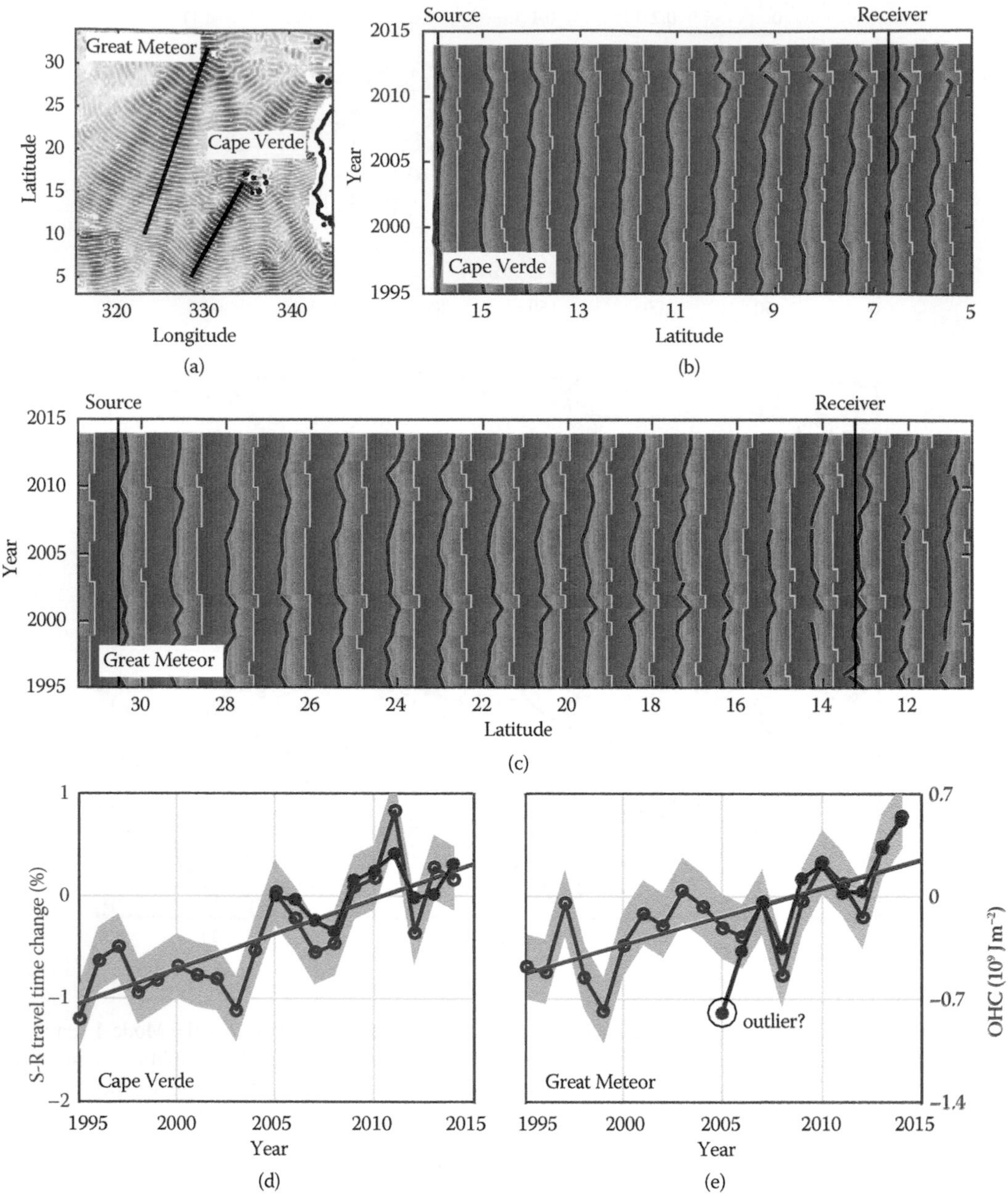

FIGURE 13.17 Variability in the propagation speed of internal tides and its relation to ocean heat content, from analysis by Zhao (2016) of satellite altimeter tidal measurements in Atlantic Ocean. (a) Location map showing filtered, southward-propagating internal tidal waves, as analyzed along two paths, with sources at Great Meteor Seamount and Cape Verde. (b,c) Time-latitude contours of tidal phase along each path. Red lines delineate co-phase lines, with wavelength approximately 150 km in space. The tilting of these lines implies the propagation speed increases slightly between 1995 and 2014. (d,e) The "source-to-receiver" travel time change rate (blue line) and Argo-measured ocean heat content changes (red line). The outlier in (e) is possibly due to the small number of Argo floats in 2005. (Reproduced from Zhao, Z., *Geophys. Res. Lett.*, 43, 9157–9164, 2016. With permission.)

As a global average, Zaron (2017) found about 44% of the internal tide variance is nonstationary, somewhat higher than the 30% found in his 2015 study; the differences are not unexpected given the very different approaches and the limitations inherent in each. In any event, we can conclude that nonstationarity is significant, which at a minimum has some obvious implications for constructing internal-tide correction models for altimeter data. Significant nonstationarity also has important implications for discussions about energetics of the internal tide and the relationships with dissipation and mixing. Any constraints on these quantities arising from altimeter studies must attempt to account for nonstationarity. Results such as those shown in Figure 13.16 could provide a starting point. In fact, the variances depicted in this figure could provide a first-order adjustment to energy terms for any calculations based on analysis of the coherent tide.

There is one great benefit to nonstationarity of internal tides. Assuming the tidal variability can be sufficiently well determined in both time and space, we might use the results to learn about changes in the ocean interior. In fact, exciting new results along these lines have been obtained by Zhao (2016), who examined the tidal change in phase every two years along three long internal-tide beams in the Pacific and Atlantic oceans; his Atlantic results are reproduced in Figure 13.17. By considering the perturbation in the ocean's mean climatological stratification necessary to explain a given change in the tidal phase, Zhao could determine changes in ocean heat content averaged over the path. Comparisons with the same quantity determined from Argo float data are remarkably good (Figure 13.17d, e) and show that altimetry can potentially extend the Argo heat-content time series back until at least 1995, if not before. These results will no doubt garner considerable attention. Although Zhao's results require analyzing tidal phase modulations along long internal tide beams, it is likely that the approach can be simplified, and perturbations can be mapped systematically across ocean regions. Complications can be expected from variable background currents and salinity and from tradeoffs between temporal resolution and aliasing. We anticipate a large number of follow- up studies in the future as this potential new tool for probing the ocean interior is investigated, extended, and put to use.

13.6 OUTSTANDING ISSUES

Most applications of satellite altimetry require the large tidal signals in the data be removed as an initial step in data processing. Because of aliasing, this de-tiding can be accomplished only with accurate tidal models. The best global barotropic models now appear capable of predicting deep ocean tidal heights with an accuracy approaching 1 cm. Tests against seafloor pressure data, shown in Figure 13.7, reveal residuals of 1 to 2 cm, but this is based on integrating the residual pressure variance across a wide frequency band of 0.5 to 48 cpd; some part of that variance is surely non-tidal. A large number of complementary tests were made on seven recent global models by Stammer et al. (2014).

Tidal prediction in the polar seas is a remaining deficiency in current global models. Tides in the Canadian archipelago and Baffin Bay, along the Siberian shelf, and beneath the Antarctic ice shelves are complex and poorly constrained by accurate observations. Required bathymetric data for hydrodynamic constraints are also generally inadequate. Under the ice shelves, both knowledge of ice thickness and ice grounding zones are needed but are not well known. However, it seems likely that progress can be made even with existing data because new altimetry from CryoSat-2 and Sentinel-3 are accumulating as we write and are waiting to be properly utilized. Future data from the Ice, Cloud, and Land Elevation Satellite (ICESat-2) and SWOT (below latitude 78°) are anticipated to be useful for model improvements in polar regions.

The ability of global tide models to provide accurate height predictions in shallow seas and near-coastal waters has been improving in recent years (Lyard et al. 2006; Ray et al. 2011). But as altimeter measurements and applications are extended ever closer to the coasts, prediction inaccuracies of global models will undoubtedly be revealed, which is one reason we continue to recommend use of local tide models for local applications. Inadequate knowledge of bathymetry is, and will continue to be, one of the primary roadblocks to progress. There is also a dearth of publicly available, high-quality tide-gauge data along some coastlines; such data are needed for testing as well as assimilation. We anticipate future SWOT data will add useful high-resolution measurements. But many complicating factors affect tides in shallow waters, as well as in polar seas, such as the possible presence of significant nonlinear compound constituents and the seasonal modulations mentioned earlier. To address such problems requires long time series of dense, high-quality data, far more than any one space mission can provide. Although recent progress has been clear, we expect tidal predictions in these regions will remain for some time far less accurate than those now achievable in the deep ocean.

We have devoted an entire section of this chapter (Section 13.4) to a discussion of barotropic tidal energetics, a classic problem in the history of geophysics and one which satellite altimetry has recently played a key role in addressing. We have, however, said almost nothing about baroclinic tidal energetics. The topic is exceedingly important because of its connections to the problem of ocean mixing. Determining how much baroclinic energy gets deposited where in the ocean, by what mechanism, and how it affects the diapycnal diffusivity of the deep ocean (see Waterhouse et al. 2014) are all critical problems to be addressed in coming years. There has indeed been some important work with satellite altimetry on the problem of baroclinic tidal energetics—for example, mapping of energy fluxes and flux divergences (e.g., Zhao et al. 2016). The problem is more difficult than the corresponding barotropic problem for a number of reasons, not the least being the smallness of the (aliased) signal that an altimeter must extract from a noisy background and deciphering what that small signal implies about the underlying water column (e.g., Wunsch 2013). Perhaps even more problematic is accounting for the evidently significant nonstationary part of internal tides, even for the first baroclinic mode, as depicted in Figure 13.16 (also Shriver et al. 2014). But efforts such as those behind Figure 13.16 mark important progress and suggest ways that nonstationarity can be partially accounted for.

As we have seen, the time-variable part of internal tides is also a signal to exploit. The possibility of using current and historical satellite altimetry to investigate changes in the ocean interior, and especially changes in ocean heat content, is an exciting development, with the promise of new insights and possibly new surprises in years to come. The final chapter on tides and satellite altimetry has not yet been written.

ACKNOWLEDGMENTS

We thank Edward Zaron for many discussions and for providing several figures, some as yet unpublished. We thank Maarten Buijsman for providing the global grid of baroclinic conversion values used for Figures 13.8b and 13.10b. Philip Woodworth and Detlef Stammer gave us useful comments on an earlier version of this chapter. Support from NASA through the Ocean Surface Topography program is gratefully acknowledged.

REFERENCES

Alford, M. H., T. Peacock, J. A. MacKinnon, J. D. Nash, M. C. Buijsman, L. R. Centuroni, et al. 2015. The formation and fate of internal waves in the South China Sea. *Nature, 521*, 65–69.

Amin, M. 1982. On analysis and prediction of tides on the west coast of Great Britain. *Geophys. J. R. Astr. Soc., 68,* 57–78.

Andersen, O. B., G. D. Egbert, S. Y. Erofeeva, and R. D. Ray. 2006. Mapping nonlinear shallow-water tides: A look at the past and future. *Ocean Dyn., 56,* 416–429.

Andersen, O. B., P. L. Woodworth, and R. A. Flather. 1995. Intercomparison of recent global ocean tide models. *J. Geophys. Res., 100,* 25261–25282.

Apel, J. R. 1987. *Principles of Ocean Physics*. Academic Press, London.

Arbic, B. K., S. T. Garner, R. W. Hallberg, and H. L. Simmons. 2004. The accuracy of surface elevations in forward global barotropic and baroclinic tide models. *Deep-Sea Res. II, 51*(25), 3069–3101.

Arbic, B. K., J. G. Richman, J. F. Shriver, P. Timko, E. J. Metzger, and A. J. Wallcraft. 2012. Global modeling of internal tides within an eddying ocean general circulation model. *Oceanography, 25*, 20–29.

Baines, P. G. 1982. On internal tide generation models. *Deep-Sea Res., 29,* 307–338.

Bendat, J. S. and A. G. Piersol. 1971. *Random Data: Analysis and Measurement Procedures.* Wiley-Interscience, New York.

Buijsman, M. C., J. K. Ansong, B. K. Arbic, J. G. Richman, J. F. Shriver, P. G. Timko, A. J. Wallcraft, C. B. Whalen, and Z. Zhao. 2016. Impact of parameterized internal wave drag on the semidiurnal energy balance in a global ocean circulation model. *J. Phys. Oceanogr., 46*, 1399–1419.

Cartwright, D. E. 1977. Oceanic tides. *Rep. Prog. Phys., 40*, 665–708.

Cartwright, D. E. 1999. *Tides: A Scientific History*. Cambridge University Press, Cambridge, 292 pp.

Cartwright, D. E. and R. D. Ray. 1990. Oceanic tides from Geosat altimetry. *J. Geophys. Res., 95*, 3069–3090.

Cartwright, D. E. and R. D. Ray. 1991. Energetics of global ocean tides from Geosat altimetry. *J. Geophys. Res., 96*, 16897–16912.

Chavanne, C. P. and P. Klein. 2010. Can oceanic sub-mesoscale processes be observed with satellite altimetry? *Geophys. Res. Lett., 37,* L22602.

Cherniawsky, J. Y., M. G. G. Foreman, W. Crawford, and R. F. Henry. 2001. Ocean tides from TOPEX/Poseidon sea level data. *J. Atmos. Oceanic Tech., 18,* 649–664.

Colosi, J. A. and W. Munk. 2006. Tales of the venerable Honolulu tide gauge. *J. Phys. Oceanogr., 36*, 967–996.

Doodson, A. T. 1958. Oceanic tides. *Adv. Geophys., 5*, 117–152.

Dunphy, M. and K. G. Lamb. 2014. Focusing and vertical mode scattering of the first mode internal tide by mesoscale eddy interaction. *J. Geophys. Res.,* 119, 523–536.

Durand, M., L.-L. Fu, D. P. Lettenmaier, D. E. Alsdorf, E. Rodriguez, and D. Esteban-Fernandez. 2010. The Surface Water and Ocean Topography mission: Observing terrestrial surface water and oceanic sub-mesoscale eddies, *Proc. IEEE, 98*, 766–779.

Dushaw, B. D. 2015. *An empirical model for mode-1 internal tides derived from satellite altimetry: Computing accurate tidal predictions at arbitrary points over the world oceans.* Tech. Rep., Applied Physics Laboratory, University of Washington, Seattle, 107 p.

Dushaw, B. D., B. D. Cornuelle, P. F. Worcester, B. M. Howe, D. S. Luther. 1995. Barotropic and baroclinic tides in the central North Pacific Ocean determined from long-range reciprocal acoustic transmissions. *J. Phys. Oceanogr., 25,* 631–647.

Dushaw, B. D., P. F. Worcester, M. A. Dzieciuch. 2011. On the predictability of mode-1 internal tides. *Deep-Sea Res., 58,* 677–698.

Egbert, G. D., and A. F. Bennett. 1996. Data assimilation methods for ocean tides. In *Modern Approaches to Data Assimilation in Ocean Modeling*. P. Malanotte-Rizzoli, ed. Elsevier, Amsterdam, pp. 147–179.

Egbert, G. D., A. F. Bennett, and M. G. Foreman. 1994. TOPEX/POSEIDON tides estimated using a global inverse model. *J. Geophys. Res., 99,* 24821–24852.

Egbert, G. D., and S. Y. Erofeeva. 2002. Efficient inverse modeling of barotropic ocean tides. *J. Atmos. Oceanic Tech., 19*, 183–204.

Egbert, G. D., and S. Y. Erofeeva. 2014. Mapping M2 internal tides using a data-assimilative reduced gravity model. In *AGU Fall Meeting*, San Francisco, CA, December 2014.

Egbert, G. D., and R. D. Ray. 2000. Significant dissipation of tidal energy in the deep ocean inferred from satellite altimeter data. *Nature, 405*, 775–778.

Egbert, G. D., and R. D. Ray. 2001. Estimates of M2 tidal energy dissipation from Topex/Poseidon altimeter data. *J. Geophy. Res., 106*, 22475–22502.

Egbert, G. D., and R. D. Ray. 2003. Semidiurnal and diurnal tidal dissipation from Topex/Poseidon altimetry. *Geophys. Res. Lett., 30*(17), 1907.

Egbert, G. D., and R. D. Ray. 2017. Tidal prediction. In *The Sea*. K. Brink et al., eds. Vol. 17, in press.

Foreman, M. G. G., W. R. Crawford, J. Y. Cherniawsky, R. F. Henry, and M. R. Tarbotton. 2000. A high- resolution assimilating tidal model for the northeast Pacific Ocean. *J. Geophys. Res., 105*, 28629–28651.

Fu, L.-L. and R. Ferrari. 2008. Observing oceanic sub-mesoscale processes from space. *Eos, 89*, 488–489.

Garrett, C. and E. Kunze. 2007. Internal tide generation in the deep ocean. *Annu. Rev. Fluid Mech., 39*, 57–87.

Godin, G. 1986. Modification by an ice cover of the tide in James Bay and Hudson Bay. *Arctic, 39,* 65–67.

Gräwe, U., H. Burchard, M. Müller, and H. M. Schuttelaars. 2014. Seasonal variability in M2 and M4 tidal constituents and its implications for the coastal residual sediment transport. *Geophys. Res. Lett., 41,* 5563–5570.

Green, J. A. M. 2010. Ocean tides and resonance. *Ocean Dyn., 60,* 1243–1253.

Hansen, W. 1962. Tides. In *The Sea, Vol. 1: Physical Oceanography*. M. N. Hill, ed. Cambridge: Harvard University Press, pp. 764–801.

Hendershott, M. C. 1977. Numerical models of ocean tides. In *The Sea, Vol. 6: Marine Modeling.* I. N. McCave, J. J. O'Brien, and J. H. Steele, eds. Cambridge: Harvard University Press, pp. 47–95.

Hendershott, M. C. 1981. Long waves and ocean tides. In *Evolution of Physical Oceanography*. B. A. Warren and C. Wunsch, eds. Cambridge: MIT Press, pp. 292–341.

Hendry, R. M. 1977. Observations of the semidiurnal internal tide in the western North Atlantic Ocean. *Phil. Trans. R. Soc. London,* A286, 1–24.

Jayne, S. R. and L. C. St. Laurent. 2001. Parameterizing tidal dissipation over rough topography. *Geophys. Res. Lett, 28*, 811–814.

Jeffreys, H., 1920. Tidal friction in shallow seas. *Proc. R. Soc. London, Ser. A, 221*, 239–264.

Kagan, B. A., E. V. Sofina, and A. A. Timofeev. 2011. Modeling of the M2 surface and internal tides and their seasonal variability in the Arctic Ocean. *J. Marine Res., 69,* 245–276.

Kang, S. K., M. G. G. Foreman, H.-J. Lie, J.-H. Lee, J. Cherniawsky, and K. D. Yum. 2002. Two-layer tidal modeling of the Yellow and East China Seas with application to seasonal variability of the M2 tide. *J. Geophys. Res.,* 107(C3), 3020.

Kowalik, Z. and I. Polyakov. 1999. Diurnal tides over Kashevarov Bank, Okhotsk Sea. *J. Geophy. Res., 104,* 5361–5380.

Kelly, S. M., P. F. Lermusiaux, T. F. Duda, and P. J. Haley. 2016. A coupled-mode shallow water model for tidal analysis: Internal-tide reflection and refraction by the Gulf Stream. *J. Phys. Oceanogr., 46,* 3661–3679.

Lambeck, K. 1977. Tidal dissipation in the oceans: Astronomical, geophysical and oceanographic consequences. *Phil. Trans. R. Soc. London A, 287*, 545–594.

Le Provost, C. 2001. Ocean tides. In *Satellite Altimetry and Earth Sciences: A Handbook of Techniques and Applications*. L.-L. Fu and A. Cazenave, eds. Academic Press, San Diego, CA, pp. 267–304.

Le Provost, C., A. F. Bennett, and D. E. Cartwright. 1995. Ocean tides for and from TOPEX/POSEIDON. *Science, 267*, 639–642.

Levine, M. D. and J. G. Richman. 1989. Extracting the internal tide from data: Methods and observations from the Mixed Layer Dynamics Experiment. *J. Geophys. Res., 94*, 8125–8134.

Locarnini, R. A., A. V. Mishonov, J. I. Antonov, T. P. Boyer, H. E. Garcia, O. Maranova, et al. 2010. *Temperature: Vol. 1, World Ocean Atlas 2009*. NOAA NESDIS 68, Silver Spring, 184 pp.

Lyard, F., F. Lefevre, T. Letellier, and O. Francis. 2006. Modelling the global ocean tides: Modern insights from FES2004. *Ocean Dyn., 56*, 394–415.

Magaard, L. and W. D. McKee. 1973. Semi-diurnal tidal currents at 'Site D.' *Deep-Sea Res.,* 20, 997–1009.

Miller, G. R. 1966. The flux of tidal energy out of the deep ocean. *J. Geophy. Res., 71*, 2485–2489.

Mitchum, G. T. and S. M. Chiswell. 2000. Coherence of internal tide modulations along the Hawaiian Ridge. *J. Geophys. Res., 105*, 28653–28661.

Morozov, E. G., 1995. Semidiurnal internal wave global field. *Deep-Sea Res., 42*, 135–148.

Morozov, E. G. and V. I. Vlasenko. 1996. Extreme tidal internal waves near the Mascarene ridge. *J. Marine Syst., 9,* 203–210.

Müller, M. 2007. *A large spectrum of free oscillations of the world ocean including the full ocean loading and self-attraction effects*. PhD thesis, Universität Hamburg, 101 pp.

Müller, M. 2012. The influence of changing stratification conditions on barotropic tidal transport and its implications for seasonal and secular changes of tides. *Cont. Shelf Res., 47*, 107–118.

Müller, M. 2013. On the space- and time-dependence of barotropic-to-baroclinic tidal energy conversion. *Ocean Modelling, 72*, 242–252.

Müller, M., J. Y. Cherniawsky, M. G. G. Foreman, and J.-S. von Storch. 2014. Seasonal variation of the M2 tide. *Ocean Dyn., 64*, 159–177.

Müller, M., H. Haak, J. H. Jungclaus, J. Sündermann, and M. Thomas. 2010. The effect of ocean tides on a climate model simulation. *Ocean Modelling, 35*, 304–313.

Munk, W. H. 1968. Once again – Tidal friction. *Q. J. R. Astr. Soc., 9,* 352–375.

Munk, W. H. 1980. Affairs of the sea. *Ann. Rev. Earth Planet Sci., 8,* 1–16.

Munk, W. H. 1997. Once again: Once again – Tidal friction. *Prog. Oceanogr., 40,* 7–36.

Munk, W. H. and G. J. F. MacDonald. 1960. *The Rotation of the Earth: A Geophysical Discussion*. Cambridge University Press, Cambridge, 323 pp.

Munk, W. H. and C. Wunsch. 1998. Abyssal recipes II: Energetics of tidal and wind mixing. *Deep-Sea Res., 45*, 1977–2010.

Nagai, T., and T. Hibiya. 2015. Internal tides and associated vertical mixing in the Indonesian Archipelago. *J. Geophys. Res.-Oceans, 120,* 3373–3390.

Ngodock, H. E., I. Souopgui, A. J. Wallcraft, J. G. Richman, J. F. Shriver, and B. K. Arbic. 2016. On improving the accuracy of the M2 barotropic tides embedded in a high-resolution global ocean circulation model. *Ocean Modelling, 97,* 16–26.

Padman, L., M. R. Siegfried, and H. A. Fricker. 2016. Ocean tide influences on the Antarctic and Greenland ice sheets. *Rev. Geophys., under review.*

Paolo, F. S., H. A. Fricker, and L. Padman. 2016. Constructing improved decadal records of Antarctic ice shelf height change from multiple satellite radar altimeters. *Rem. Sens. Environ., 177,* 192–205.

Pingree, R. D., and A. L. New. 1995. Structure, seasonal development and sunglint spatial coherence of the internal tide on the Celtic and Armorican shelves and in the Bay of Biscay. *Deep-Sea Res., 42,* 245–284.

Pairaud, I. L., F. Auclair, P. Marsaleix, F. Lyard, and A. Pichon. 2010. Dynamics of the semi-diurnal and quarter-diurnal internal tides in the Bay of Biscay. Part 2: Baroclinic tides. *Cont. Shelf Res., 30,* 253–269.

Platzman, G. W. 1984. Planetary energy balance for tidal dissipation, *Rev. Geophys. Space Phys., 22,* 73–84.

Platzman, G. W., G. A. Curtis, K. S. Hansen, R. D. Slater. 1981. Normal modes of the world ocean. Part II: Description of modes in the period range 8 to 80 hours. *J. Phys. Oceanogr., 11,* 579–603.

Ponte, A. L. and P. Klein. 2015. Incoherent signature of internal tides on sea level in idealized numerical simulations. *Geophys. Res. Lett., 42,* 1520–1526.

Prinsenberg, S. J. 1988. Damping and phase advance of the tide in western Hudson Bay by the annual ice cover. *J. Phys. Oceanogr., 18,* 1744–1751.

Proudman, J. 1942. On Laplace's differential equations for the tides. *Proc. R. Soc. London A, 179,* 261–288.

Proudman, J. 1944. The tides of the Atlantic Ocean (George Darwin Lecture). *Mon. Not. R. Astr. Soc., 104,* 244–256.

Proudman, J. and A. T. Doodson. 1936. Tides in oceans bounded by meridians. *Phil. Trans. R. Soc. Lond. A, 235,* 273–342.

Rainville, L., T. M. S. Johnston, G. S. Carter, M. A. Merrifield, R. Pinkel, P. F. Worcester, et al. 2010. Interference pattern and propagation of the M2 internal tide south of the Hawaiian Ridge. *J. Phys. Oceanogr., 40,* 311–325.

Ray, R. D. 1994. Tidal energy dissipation: Observations from astronomy, geodesy, and oceanography. In *The Oceans.* S. K. Majumdar et al., eds. Pennsylvania Academy of Science, pp. 171–185.

Ray, R. D. 2001. Inversion of oceanic tidal currents from measured elevations. *J. Mar. Syst., 28,* 1–18.

Ray, R.D., G. D. Egbert, and S. Y. Erofeeva. 2011. Tide predictions in shelf and coastal waters: Status and prospects. In *Coastal Altimetry.* S. Vignudelli, A. Kostianoy, P. Cipollini, and J. Benveniste, eds. Berlin: Springer-Verlag, pp. 191–216.

Ray, R. D. and G. T. Mitchum. 1996. Surface manifestation of internal tides generated near Hawaii. *Geophys. Res. Lett., 23,* 2101–2104.

Ray, R. D. and E. D. Zaron. 2011. Nonstationary internal tides observed with satellite altimetry. *Geophys. Res. Lett., 38,* L17609.

Ray, R. D. and E. D. Zaron. 2016. M2 internal tides and their observed wavenumber spectra from satellite altimetry. *J. Phys. Oceanogr., 46,* 3–21.

Schwiderski, E. W. 1980. On charting global ocean tides. *Rev. Geophys. Space Phys., 18,* 243–268.

Sherwin, T. J. 1988. Analysis of an internal tide observed on the Malin Shelf, north of Ireland. *J. Phys. Oceanogr., 18,* 1035–1050.

Shriver, J. F., B. K. Arbic, J. G. Richman, R. D. Ray, E. J. Metzger, A. J. Wallcraft, and P. G. Timko. 2012. An evaluation of the barotropic and internal tides in a high-resolution global ocean circulation model. *J. Geophys. Res., 117,* C10024.

Shriver, J. F., J. G. Richman, and B. K. Arbic. 2014. How stationary are the internal tides in a high-resolution global ocean circulation model? *J. Geophys. Res., 119,* 2769–2787.

Shum, C. K., et al. 1997. Accuracy assessment of recent ocean tide models. *J. Geophys. Res., 102,* 25,173–25, 194.

Sjöberg, B. and A. Stigebrandt. 1992. Computations of the geographical distribution of the energy flux to mixing processes via internal tides and the associated vertical circulation in the ocean, *Deep-Sea Res.,* 39, 269–291.

Stammer, D., R. D. Ray, O. B. Andersen, B. K. Arbic, W. Bosch, L. Carrere, et al. 2014. Accuracy assessment of global barotropic ocean tide models. *Rev. Geophys., 52,* 243–282.

St. Laurent, L. C. and C. Garrett. 2002. The role of internal tides in mixing the deep ocean. *J. Phys. Oceanogr., 32,* 2882–2899.

St. Laurent, P., F. J. Saucier, and J.-F. Dumais. 2008. On the modification of tides in a seasonally ice-covered sea. *J. Geophys. Res., 113,* C11014.

Taguchi, E., D. Stammer, and W. Zahel. 2014. Inferring deep ocean tidal energy dissipation from the global high-resolution data-assimilative HAMTIDE model. *J. Geophys. Res., 119*, 4573–4592.

Vlasenko, V., N. Stashchuk, and K. Hutter. 2005. *Baroclinic Tides: Theoretical Modeling and Observational Evidence*. Cambridge University Press, Cambridge, 351 pp.

Waterhouse, A. F., J. A. MacKinnon, J. D. Nash, M. H. Alford, E. Kunze, H. L. Simmons, et al. 2014. Global patterns of diapycnal mixing from measurements of the turbulent dissipation rate. *J. Phys. Oceanogr., 44*, 1854–1872.

Wunsch, C. 1975. Internal tides in the ocean. *Rev. Geophys. Space Phys., 13*, 167–182.

Wunsch, C. 1990. Comment on R. N. Stewart's "Physical Oceanography to the end of the twentieth century." In *Quo Vadimus: Geophysics for the Next Generation.* Geophys. Monogr. Ser., Vol. 60. G. D. Garland and J. Apel, eds. American Geophysical Union, Washington, DC, p. 69.

Wunsch, C. 2013. Baroclinic motions and energetics as measured by altimeters. *J. Atmos. Oceanic Tech., 30,* 140–150.

Wunsch, C. 2015. *Modern Observational Physical Oceanography: Understanding the Global Ocean.* Princeton University Press.

Wunsch, C. and R. Ferrari. 2004. Vertical mixing, energy, and the general circulation of the oceans. *Annu. Rev. Fluid Mech., 36*, 281–314.

Zahel, W. 1995. Assimilating ocean tide determined data into global tidal models. *J. Mar. Syst.*, 6, 3–13.

Zaron, E. D. 2015. Nonstationary internal tides observed using dual-satellite altimetry. *J. Phys. Oceanogr., 45*, 2239–2246.

Zaron, E. D. 2017. Mapping the nonstationary internal tide with satellite altimetry. *J. Geophys. Res. Oceans,* 122, 539–554

Zaron, E. D. and D. A. Jay. 2014. An analysis of secular changes in tides at open-ocean sites in the Pacific. *J. Phys. Oceanogr., 44,* 1704–1726.

Zhao, Z. 2014. Internal tide radiation from the Luzon Strait. *J. Geophys. Res.-Oceans, 119*, 5434–5448.

Zhao, Z. 2016. Internal tide oceanic tomography. *Geophys. Res. Lett., 43*, 9157–9164.

Zhao, Z., M. H. Alford, and J. B. Girton. 2012. Mapping low-mode internal tides from multisatellite altimetry. *Oceanography, 25(2)*, 42–51.

Zhao Z., M. H Alford, J. B. Girton, L. Rainville, H. L. Simmons. 2016. Global observations of open-ocean mode-1 M2 internal tides. *J. Phys. Oceanogr., 46,* 1657–1684.

Zhao, Z., M. H. Alford, J. A. MacKinnon, and R. Pinkel. 2010. Long-range propagation of the semidiurnal internal tide from the Hawaiian Ridge. *J. Phys. Oceanogr., 40*, 713–736.

14 Hydrological Applications of Satellite Altimetry

Rivers, Lakes, Man-Made Reservoirs, Inundated Areas

Jean-François Cretaux, Karina Nielsen, Fréderic Frappart, Fabrice Papa, Stéphane Calmant, and Jérôme Benveniste

14.1 INTRODUCTION

Lakes and rivers, which play an essential role in supplying freshwater for basic human and economic needs, are also highly linked to the continental water cycle and are impacted by climate changes. The major part of the world's demand on water relies on continental surface waters (rivers, lakes, and artificial reservoirs) and less on underground aquifer and seawater desalinization. It is consequently crucial to quantify changes in the continental water cycle and to describe in an accurate manner the quantity of freshwater available on the Earth as well as its variability on timescales ranging from sub-seasonal (actually we even need hourly/daily for some applications) to interdecadal. Water is not homogeneously distributed on Earth and does not coincide with the domestic, industrial, and agricultural needs. At global scale, river discharge is predominant in equatorial and temperate climate regions. Lakes are also irregularly distributed across continents with high concentration in the Northern Hemisphere and in boreal regions (Bousquet et al. 2006; Downing et al. 2006).

Lake, reservoir, and wetland distribution has a great importance for large-scale study of the environment, biodiversity (Dudgeon et al. 2006), water resources inventory for agriculture needs, industrial activities, the economy and human welfare, and also for studying the impact of climate change on the water cycle (Rast and Straskraba 2000; Huttunen et al. 2003; Dudgeon et al. 2006; Tranvik et al. 2009; Downing 2010; Seekell et al. 2014). However, until recently, inland water distribution was not accounted for in global climate models (GCMs) and was ignored in global estimates of ecosystem processes (Downing 2010). It is now well recognized that continental waterbodies play a central role in the exchange of carbon and methane with the atmosphere (Seekell et al. 2014).

A lake acts as an integrator of climate change (Adrian et al. 2009; Schindler 2009; Williamson et al. 2009), with different response times from seasonal to interdecadal, depending on its morphology (Mason et al. 1994). Any change in rainfall and air temperature has direct and indirect consequences on lake water storage (Robertson and Ragotzkie 1990; De Wit and Stankiewicz 2006). Therefore, an adequate network of lakes worldwide can provide a large amount of information on climate change impact on surface water. For example, polar and alpine lakes are among the most sensitive to climate change and can be considered as suitable proxies. As deeply demonstrated in Williamson et al. (2009, pp. 2279), "Lakes and reservoirs are simultaneously serving as sentinels, integrators, and regulators of climate change."

For rivers, the hydrological cycle is closely linked to climate variability on the one hand and to water uses for irrigation and hydroelectric power generation on the other hand. Climate varies at different timescales due to various forcing factors. Impacts on rivers mainly result from changes in rainfall over the river watershed. To characterize river hydrological regimes and fluctuations, the current variables used are discharge trends and periodicities. River discharge is an Essential Climate Variable (GCOS 2011).

The quantity of water available for human needs (agriculture, industry, tourism, etc.) is very low in comparison to the total water content on the Earth. Furthermore, more than 260 rivers are transboundary rivers that are draining over a total of 145 countries (Wolf et al. 1999; Sood and Mathukumalli 2011). Sharing the water resources is therefore a crucial question, potentially with some conflicting international situations such as in the case of construction of dams or overdevelopment of irrigation in an upstream country. Moreover, population increase and consequently increase in freshwater demand aggravates transboundary issues (Gleditsch and Hegre 2000). River discharge is the most informative variable but is hard to measure directly and preferentially. It is water level that is measured and then converted into discharge using empirical laws called rating curves (Manning 1891). The current quantitative knowledge of river discharge worldwide is limited due to the drastic reduction of *in situ* observations (e.g., Global Runoff Data Center). A challenge is, therefore, to develop alternative methodologies to measure river discharge mostly from satellite remote sensing techniques.

Although an improved description of the components of the global water cycle is now recognized as being of major importance, the global distribution and spatiotemporal variations of continental water extent and volume are still poorly known (Alsdorf et al. 2007; Papa et al. 2010b). Until recently, our knowledge of the spatiotemporal variations of continental waters relied on sparse *in situ* observations and hydrological models. *In situ* gauge measurements help quantify discharge into river channels but provide comparatively little information about spatial dynamics of terrestrial water in floodplains and wetlands or aquifers. The lack of spatially complete measurements of inundation/wetland locations, sizes, water volume changes, and hydrologic models prevents proper partitioning of precipitation (minus evapotranspiration). In addition, information is limited to continental-to-global scales (Alsdorf and Lettenmaier 2003; Decharme et al. 2012; Getirana and Peters-Lidar 2012; Yamazaki et al. 2012; Emery et al. 2016).

Several basic questions related to the surface water budget are still open, for example: How much freshwater is stored at the surface/near surface of continents? What are the spatial and temporal dynamics in terrestrial surface water storage?

14.1.1 Past, Present, and Future of Satellite Altimetry

Satellite altimetry has been extensively used over the last two decades to calculate the water height variations over the Earth's lakes, rivers, reservoirs, and floodplains.

The first use of satellite altimetry for hydrology was with the GEOSAT mission launched in 1985 (Morris and Gill 1994a). However, the measurement error budget was quite high, especially for the orbit. Thanks to substantial progress in orbit determination from geodetic systems, the U.S.-French, TOPEX/Poseidon (T/P) satellite, launched in 1992, is considered as the first sensor producing water height accurate enough to be used in surface water studies. Since then, altimetry systems have significantly evolved.

The first generation of radar altimeters (European Remote Sensing [ERS] 1, T/P, ERS-2, Envisat, Jason-1, Jason-2, GEOSAT Follow On [GFO]) used a pulse-limited mode of operation in the Ku band (13.6 GHz), called low resolution mode (LRM). The last mission relying only on the pulse-limited mode is Jason-3, launched in the beginning of 2016. It is the follow-on mission of Jason-2, with the same orbital and instrumental characteristics. In 2020, the European Space Agency (ESA) will launch the "Jason Continuity of Service" (Jason-CS/Sentinel-6A), followed by Sentinel-6B a few years later, which both will continue the TOPEX/Jason-series observation on the same "TOPEX" orbit. This generation of missions will also operate in the Ku band in LRM mode but also with a new mode called delay-Doppler or Synthetic Aperture Radar (SAR) mode, exploiting the Doppler frequency change with the speed of the spacecraft and firing 18,000 pulses per second instead of 2000, enabling new processing algorithms to reduce the resolution down to 350 m and significantly improving the precision. The SAR mode was first designed for the CryoSat mission, to improve sea ice monitoring. CryoSat-2 was launched in 2010. With the very positive benefits this new technique offered to all domains of altimetry, ocean, cryosphere and inland water, this mode was selected as the radar altimeter on Sentinel-3, launched in 2016.

Although not being specifically designed for hydrology, the SAR mode has led to many studies using CryoSat-2 data showing considerable benefits of SAR mode altimetry for continental water monitoring. The advantages of satellite altimetry are well known: almost global coverage, spatial and temporal coherence of data, archive of past data, durability, and continuity of data. In terms of precision or temporal sampling, a measurement carried out *in situ* in optimal conditions will intrinsically be preferable to any altimetry measurement. However, in many regions of the Earth, the increasing scarcity of operating ground networks for the monitoring of lakes and rivers levels contrasts greatly with the increasing need to understand the linkage between surface water resources and climate change on one hand, and increasing societal need for freshwater resources on another hand.

This has progressively led space agencies to include hydrological objectives in the definition of new missions. This is the reason why the recent missions, CryoSat-2 and Sentinel-3 launched by ESA in 2010 and 2016, and SARAL/AltiKa launched in 2013 by CNES (Centre National d'Études Spatiales) and ISRO (Indian Space Research Organization) have explored new concepts of satellite measurements. SARAL/AltiKa operated in the Ka band (35.75 GHz) allowing a significant reduction of the footprint by a factor of two to three and of the ionospheric effect on the range measurement. This has allowed calculating water height of narrow lakes and rivers with decimeter accuracy, as shown in Arsen et al. (2015) for lakes and Frappart et al. (2015b) for the rivers.

In 2003, the National Aeronautics and Space Administration (NASA) launched the ICESat (Ice, Cloud, and Land Elevation Satellite) mission that for the first time carried a laser altimeter, called the Geoscience Laser Altimeter System, or GLAS (Zwally et al. 2003). On a near-polar orbit with a 91-day repeat cycle, the GLAS altimeter has provided very high accuracy elevation of the Earth's surface thanks to the very small footprint of the laser echo (70 m). Data provided by the National Snow & Ice Data Center (NSIDC) were "ready to use" without any corrections to apply, except for geoid undulations. The ICESat data have widely been used to calculate water level variations over the lakes in the Tibetan Plateau, allowing coverage of dozens of lakes that were not surveyed by any other instruments (Wang et al. 2013; Zhang et al. 2013; Cretaux et al. 2016) and over the Amazon river (Hall et al. 2012).

In April 2010, ESA launched the CryoSat-2 mission with a new radar instrumentation initially designed to study sea ice and polar ice sheets. It is placed on a 369-day polar repeat orbit leading to an intertrack spacing of only 7.5 km at the Equator and less than 6 km around 40° of latitude. CryoSat-2 carries two antennas to enable cross-track interferometry. The switching among the three modes, LRM, SAR, and SARIn (SAR Interferometric mode) is based on a geographical mask that scientists can request to be changed to target a specific location in a specific mode. The SARIn mode has the capability to very precisely geolocate the reflecting point within the radar footprint. Although the temporal spacing between two repeat measurements is not optimal for hydrology, it is of high interest for lakes that can be covered by several tracks. Moreover, the dense coverage allows sampling of a very large number of small lakes on Earth (Kleinherenbrink et al. 2015). It is also being used for roughly estimating the mean slope on a large number of rivers worldwide. CryoSat-2 filled the 2-year gap from 2011 to 2013 at the end of the Envisat mission until the launch of the SARAL/AltiKa mission (Cretaux et al. 2016).

More interesting for hydrology is the Sentinel-3A mission launched by ESA in February 2016. It will be followed 18 months later by the Sentinel-3B mission to be placed in an interleaved orbit. The constellation of these two satellites will furthermore increase the Earth coverage. Both Sentinel-3 satellites carry a Ku-band altimeter that works either in LRM or in SAR mode, the latter being programmed for the operational mission to work over the ocean as well as overall continents. In SAR mode, the large circular footprint is sliced in the cross-track direction into bands that should allow a much better selection of the water body and a reduction of contamination of the echo from the surrounding terrain compared to LRM measurements.

In the 2020s decade, the Jason-CS/Sentinel-6 and the Sentinel-3A and B will constitute a constellation providing a dense coverage of the Earth's surface allowing also mission inter-comparison. The Jason-CS/Sentinel-6 will have a new feature: the simultaneous operation of both the LRM and SAR mode. This will permit an accurate link to the time series started with the LRM missions and extend them with the SAR mode missions.

In 2021, the Surface Water and Ocean Topography (SWOT) mission will be launched for a 3-year lifetime. It is currently under joint development by NASA, CNES, the Canadian Space Agency (CSA), and the United Kingdom Space Agency (UKSA). SWOT will carry a Ka-Band Radar Interferometer (KaRin) providing water elevation in two dimensions on two swaths of 50 km each on both sides of the ground track of the satellite. These two swaths will be separated by a band of 20 km without observations (Rodriguez 2015). Then, SWOT will also embark a Ku-band nadir altimeter working in LRM, inherited from the Jason technology. SWOT will have a 20.86-day repeat cycle. The requirement of the mission is to provide water height on all lakes wider than 250 m by 250 m and along rivers wider than 100 m. The expected accuracy is 10 cm for each 1 km^2 of lake and river surface and 1.7 cm/km in slope for rivers wider than 100 m every 10 km length.

SWOT's main objectives for hydrology will be to study hydrological processes and understand the role and impact of water storage change on the global water cycle and to study the sensitivity of the continental water cycles to human activities (irrigation, transboundary water management) and extreme events (droughts and floods). It will allow understanding of the dynamics of floodplains and their role in the terrestrial carbon budget. It will also provide a global inventory of the spatial distribution of freshwater on the continents and their spatiotemporal variability. After the 3-year lifetime of the SWOT mission, the discharge of rivers worldwide will be estimated and the connectivity among rivers, lakes, and floodplains will be modeled. SWOT will allow filling our gap in knowledge about lake interactions with climate processes. The SWOT data will also be assimilated into GCM models (Biancamaria et al. 2010; Bates et al. 2014). SWOT will provide to water managers and scientists an independent, reliable, and accurate measurement system of diverse variables controlling the flow and the water storage changes at basin scale (Cretaux et al. 2015).

14.1.2 Short History of Past Applications of Satellite Altimetry on Surface Water

After an experiment on Skylab in 1973, the first space-borne radar altimeter mission, GEOS 3, was launched in 1975 followed by Seasat in 1978 (Cudlip et al. 1992) and GEOSAT in 1985 (Koblinsky et al. 1993, Morris and Gill 1994a). It demonstrated the potential of this technique for monitoring surface waters.

Progress in orbit determination (accuracy reduced from metric to centimetric in the radial direction) using Global Positioning System (GPS), the Doppler Orbitography and Radiopositioning Integrated by Satellite (DORIS) system, and Satellite Laser Ranging (SLR) allowed the more extensive use of this technique for the monitoring of lakes and rivers. T/P was the precursor of this new discipline. In the 1990s, two articles still considered as basic references were published to describe the use of altimetry for the monitoring of lakes, with data collected over the American Great Lakes (Birkett 1995) and over the Amazon Basin for rivers (Birkett 1998).

Since then, a large number of publications followed for lakes (Birkett et al. 1999; Mercier et al. 2002; Coe and Birkett 2005; Hwang et al. 2005; Medina et al. 2008; Lee et al. 2011; Phan et al. 2011; Singh et al. 2012; Song et al. 2013; Cretaux et al. 2015; Kleinherenbrink et al. 2015), rivers (Alsdorf et al. 2001; Berry et al. 2005; Leon et al. 2006; Silva et al. 2010; and many others), floodplains (Pandey et al. 2014), and calibration of large-scale hydrological models (Getirana 2010; Emery et al. 2016).

If many of these earlier studies were principally dedicated to demonstrating that satellite altimetry provides accurate and useful data to study lakes rivers and floodplains, over the last 10 years, it has also been widely used in synergy with other space sensors for determination of derived products, such as water storage changes in lakes and reservoirs (Gao et al. 2012; Duan and Bastiaanssen 2013; Sima and Tajrishy 2013; Cretaux et al. 2016), river discharge (Papa et al. 2010b, 2012; Tarpanelli et al. 2015; Frappart et al. 2015b; Paris et al. 2016), or volume variations of floodplains (Frappart et al. 2008, 2011, 2012, 2015b; Papa et al. 2015).

14.1.3 Objectives of This Chapter

After more than 20 years of using satellite altimetry for hydrology, the number of published studies has exponentially increased. Based on research to improve data processing, or exploration of new applications, satellite altimetry became more than a promising tool. Today, the data are assimilated into models describing the hydrological processes at basin scale (Paiva et al. 2013a, 2013b; Tourian et al. 2017). Satellite altimetry has also been widely used to understand the impact of climate change at regional scale: for example, the Tibetan Plateau (Cretaux et al. 2016) or in East Africa (Becker et al. 2014). Used in synergy with a global database for surface water extent dynamics, it has also been used to measure and interpret the water volume changes over large basins: for example, the Amazon or the Niger river basins (Frappart et al. 2012).

With the development of SWOT and operational systems such as the Sentinel-3 mission, this technique definitively entered into a new era where continental water is now fully considered as the primary objective together with oceanography. Global databases delivering products from satellite altimetry for hydrology have also contributed to the development of applications.

In Section 14.2 of this chapter, we will give some basics of the interpretation of satellite altimetry measurements; in Section 14.3, we will explain how to use altimetry for hydrology; and in Section 14.4, we will show some examples of applications.

14.2 SATELLITE ALTIMETRY: MEASUREMENT AND INTERPRETATION

14.2.1 General Principle of Satellite Altimetry (see also Chapter 1 of this book)

Satellite radar altimeters are designed to measure the two-way travel time of propagation of short pulses emitted on board and reflected by the Earth's surface. This time measurement is converted into a distance measurement between the satellite and the reflected surface, so-called *range*. The range, R, between the satellite and the reflecting surface is thus given by the following equation:

$$R = c\,\frac{\Delta t}{2} \tag{14.1}$$

with c the speed of light and Δt the two-way travel time of the radar pulse between the satellite and the ground reflector.

The height of the reflecting surface, H, with respect to a fixed reference (see as follows) is given by the following equation:

$$H = a - R + \Sigma C_p + \Sigma C_g \tag{14.2}$$

where **a** is the height of the satellite above the reference ellipsoid. The terms ΣC_p and ΣC_g correspond to the sum of the corrections that must be applied in order to determine a precise estimation of H.

It is not the purpose of this chapter to enter into details of these different techniques, but it is easy to see that precise altimetry is only possible because of the very high accuracy (1–2 cm in the radial direction) of the positioning of the satellites thanks to these systems (see Chapter 1 for more detail about the corrections).

These corrections are essentially (for hydrology) of two types:

- Corrections of propagation through the atmosphere (ΣC_p), which is slower than the speed of light used in Equation 14.1 because of the presence of ions and water in the ionosphere and the troposphere.
- Geophysical corrections (ΣC_g) that aim to correct for the crustal deformation due to pole and solid earth tides at the time of measurement with respect to a reference Earth.

Because H is provided in the geodetic frame of the orbitography systems, it is an ellipsoidal height and it can be corrected from the local undulation of the geoid N in order to be converted into an orthometric height, one that is useful for hydrology:

$$h = H - N \tag{14.3}$$

In hydrology, h is the geoidal altitude of the lake or river, and this is the variable that is then used by the hydrologists.

The electromagnetic bias, which is related to the sea state and therefore to the amplitude of waves at the ocean surface, generally is not taken into account in hydrology, except for some large lakes, such as the Caspian Sea or the Great Lakes of North America.

Echoes are acquired due to a tracking system placed on board the satellite known as "trackers." Waveforms consist in the power distribution of accumulated echoes. In order to determine the time elapsed between the signal emission and its return, waveforms are modeled using a theoretical predefined analytic function. After accumulation of the echoes, oceanic waveforms have standard "ocean-like" or "Brown-like" shapes, where initial thermal noise is followed by a sharp rise called leading edge (LE) and a gently end-sloping plateau mixed with noise known as the trailing edge (TE). Over continents, a huge diversity of waveforms exists. Many of them are strongly contaminated by noise from multiple land returns, especially in rapidly changing topography. However, because the K band used in satellite altimetry reflects fully on water, the water body in the beam footprint will return much more energy than non-waterbodies do, and waveforms useful for continental hydrology (lakes, rivers, and reservoirs) are generally expected to possess at least one distinctive leading edge from radar beam interaction with water surface.

The purpose of the onboard tracker is to interpret received waveforms, adapt to changes in the pulse shape, adjust the reception window, and gain attenuation so the next incoming echoes are properly captured. Several tracking algorithms have been implemented in the successive missions. They are briefly reviewed hereafter.

The Envisat tracking algorithm does not perform a precise estimate of range or other parameters as was the case on earlier altimeters (ERS or T/P). Its primary function was to maintain LE of a waveform around the nominal tracking point within the analysis window thanks to the resolution selection logic algorithm (RSL). This approach increases the robustness of the tracker.

Most continental tracking anomalies are found over steep topography regions where altitude rate (in m/s) combined with noise artifacts exceed the agility of the tracker. Anomalies also occur when uncommon echoes are acquired during several tracking cycles, leading to incorrect reception window shift. When a tracking anomaly is identified, the altimeter performs a sweep looking for an appropriate target, and during this phase, important science measurements are not acquired.

Launched just before Envisat, the Jason-1 Poseidon-2 altimeter was designed for oceanography. The mission primary objective impacted the conception of the tracker. The corresponding algorithm fits a model of ocean return echoes to an average return. This concept is frequently inappropriate for specular echoes originating from ice and lakes because there is no generic model of inland echoes analogous to the Brown model for the open ocean (Brown 1977) that may be used for this purpose.

In the Poseidon-2 range loop, the reception window is separated into three sub-windows. The control algorithm will try to keep the amplitude of the trailing edge in the third window at a constant level while the α–β tracker (second order filter closed loop) will adjust the reception window position.

This concept is particularly well adapted for open ocean surfaces and has given excellent results in these conditions. However, over continental surfaces, specular or quasi-specular echoes (from lakes and rivers), or uncommon echoes with front contamination, will always cause a tracking anomaly. Consequently, Jason-1 provided measurements over the few biggest lakes and reservoirs only and almost none for rivers.

14.2.2 Jason-2 DIODE/DEM Tracker

In the nominal tracking mode, the Jason-2 Poseidon-3 altimeter uses the "median tracker" concept, which again ensures the independence of shape pulse power weighting similar to the Free Model Tracker (FMT) design on board Envisat. However, the combination of a digital elevation model (DEM) with DORIS immediate onboard determination (DIODE) in an open-loop tracking was tested for Jason-2. The DIODE system is onboard real time orbit determination software (Jayles et al. 2010). This mode has the ability to replace the standard tracking closed-loop. The reception window position is planned directly by the altimeter, by combining the altitude information provided by DORIS/DIODE on the one hand and the altitude derived from the onboard DEM on the other. Depending on the quality and sampling of this fragmentary DEM, the reception window position can be estimated with a precision of few meters. The combined use of DIODE data and the DEM altitude ensures target tracking abilities independent of the echoes previously returned whenever the target was still in the altimeter sight. This mode can be extremely useful for tracking upcoming areas of special interest, such as rivers and lakes, since it ensures a good tracking in areas where conventional trackers fail. As the results obtained with DIODE/DEM on Jason-2 were very satisfactory, it has been implemented on the Jason-3, SARAL/AltiKa, and Sentinel-3 missions.

14.2.3 Review of the Different Modes (LRM, SAR, and SARIn)

As mentioned earlier, radar altimeters function in different modes: LRM, SAR, and SARIn. In LRM, the radar footprint covers several tens of square kilometers. Therefore, an object remains in the footprint before and after the satellite is at the zenith of this object. If the reflecting surface is homogeneous within the footprint, the reflecting energy decreases with distance to the footprint center. But, frequently, the reflected energy from a water body located at the edge of the footprint could be higher than the ones from the footprint center (e.g., a small river or a small lake surrounded by a dense forest). The tracking system on board the satellite will center the reception window on this energy peak, and the range will then be estimated. It will induce a geometrical error that may be corrected "a posteriori" as described in Section 14.3.3.

The SAR mode is based on the principle of a variation of reflected power from an angular distance with respect to the nadir. Instead of emitting the radar pulse at a low rate (1–2 kHz) such as in LRM for decorrelating the individual echoes, the pulses are emitted at a higher frequency (approximately 10–20 kHz) in SAR mode to take advantage of the fact that echoes will be highly correlated. The SAR mode allows slicing the footprint in bands within the footprint, thereby exploiting the Doppler frequency shift, for a much higher spatial resolution. Moreover, it will increase the signal to noise ratio compared to LRM.

For many years, considerable efforts have been devoted to enhancing the capabilities of radar altimetry. The SARIn mode on board the CryoSat-2 mission came with several major improvements. First of all, the instrument on board CryoSat-2 has two antennas, 1.1 m apart. The delay in the signal arrival time at both antennas allows the angle of arrival to be calculated. Hence, the altimeter can solve the location of the reflecting facet both along-track and across-track leading to a full three-dimensional positioning of the reflecting water body (Bercher and Calmant 2013). The geometric correction that needs to be applied due to this shift with respect to the footprint center is therefore easily estimated because the relative position of the reflecting point and the satellite is known (see Section 14.3.5).

14.2.4 Review of the Geophysical Corrections

14.2.4.1 Dry Tropospheric Correction

A dry tropospheric correction (DTC) must be applied to the ranges in order to account for the delay of propagation of the radar pulse through the atmosphere. The DTC takes into account air density and is calculated using the atmospheric pressure through the following standard relationship (Saastamoinen 1972):

$$\Delta R_{dry}(m) = -2.277\ P_{atm}(h)(1 + 0.0026 \cos 2\varphi) \tag{14.4}$$

where P_{atm} is the atmospheric pressure at the level of the lake and ϕ is the latitude of the point (in order to take into account the flatness of the Earth). Because it is practically not possible to measure atmospheric pressure everywhere, using local instrumentation, the DTC is derived from atmospheric reanalysis (e.g., European Center for Medium Range and Weather Forecast [ECMWF]; National Center for Environmental Prediction [NCEP]). Note that this can lead to some errors over the continents if the local altitude is not properly taken into account in Equation 14.4 (Cretaux et al. 2009).

14.2.4.2 Wet Tropospheric Correction

The wet tropospheric correction (WTC) is related to the water vapor contained in the atmosphere. This correction can be calculated using two approaches: by means of a bi- or tri-frequency onboard radiometer or by means of global meteorological models as done for the DTC. Radiometers measure the instantaneous brightness temperatures at the nadir, a quantity dependent on the atmospheric water vapor content. This measurement has the obvious advantage of time coincidence with the radar altimeter measurements. However, radiometers do not operate properly over the continental areas because the measurement is polluted by the inhomogeneity of the soil emissivity, except for very large lakes. To avoid this problem, a WTC based on the ECMWF reanalysis is used. The validation of meteorological data is difficult. Therefore, there are no published studies allowing the estimation of the accuracy of this correction on rivers. Over lakes, however, using local GPS precise positioning near Lake Issykkul in Central Asia, Cretaux et al. (2009, 2011) has shown that residual errors of a few centimeters can still affect the estimation of this correction and therefore the lake level estimate.

14.2.4.3 Ionospheric Correction

The ionospheric correction is due to the interaction of the electromagnetic pulse with the free electrons (measured in terms of total electron content, TEC) in the high atmosphere. Their density, $n_e(z)$, varies spatially and also temporarily mainly due to the solar activity. The delay of propagation in the ionosphere varies with the frequency of the electromagnetic signal and the TEC and is calculated using the following equation:

$$\Delta R_{ion}(f) = \frac{40.3\ 10^6}{f^2} \int_0^a n_e(z)\,dz \tag{14.5}$$

where a is the altitude of the satellite above the reference ellipsoid, and f is the frequency of the EM wave.

This estimation of the ionospheric delay is fully based on the measurement of the TEC (through $\int_0^a n_e(z)dz$). The TEC can be deduced from global models such as the global ionospheric model (GIM) distributed by the Jet Propulsion Laboratory (JPL), and the precision in terms of ΔR_{ion} is at the centimeter level in the Ku band and a few millimeters in the Ka band (SARAL/AltiKa).

14.2.4.4 Sea State Bias Correction

The sea state bias (SSB), which is composed of an instrumental bias (INB) and an electromagnetic bias (EMB), principally depends on significant wave height (SWH) and wind velocity. Over the ocean, this effect can reach a few centimeters, especially for high SWH and wind speed (Gaspar et al. 2002). It is not considered over continental waterbodies, except for the very large ones where waves can develop. Actually, the fact that no SSB has to take into account lakes and rivers is an important point in the determination of the systematic biases (see Section 14.2.5).

14.2.4.5 Tidal Corrections

Tides affect both the water surface and the height of the bed of waterbodies.

The liquid tide modifies the water level in estuaries, mostly propagating from the ocean. For example, the tide still reaches a few centimeters at Obidos station, more than 900 km from the mouth of the Amazon. However, there is no global tidal model in estuaries that would enable to correct *a priori* water level in rivers for this effect.

The solid tide is linked to the lunar and solar gravitation effect—essentially depending on their position in the same manner as the oceanic tides. The vertical displacements of the Earth crust linked to this effect are on the order of 20 cm in general. They are well modeled, such as in the Preliminary Reference Earth Model (PREM) (Dziewonski and Anderson 1981), and are available in the Geophysical Data Records (GDRs) of radar altimetry missions.

The polar tide is related to the flatness of the Earth and to the polar motion. The vertical displacements at the Earth's surface linked to the polar tide are at the centimeter level and are directly given in the GDRs. This correction can, therefore, be applied when calculating water height on lakes or rivers.

14.2.5 Review of the Biases and Their Determination

The range estimation is affected by several errors, whatever the satellite altimetry mission, and whatever the retracking algorithm used. These errors can be separated into three parts:

- Systematic errors
- Geographically correlated errors
- Variable error

The variable error is determined from the dispersion between *in situ* time series and altimetric time series. This error is the one generally considered to characterize the quality of the hydrological products inferred from satellite altimetry. The systematic error, or bias, can be estimated (and thus corrected) by the mean error value deduced from the comparison of the altimetric height series with independent height values generally obtained using GPS. The geographically correlated error is seen as a dispersion of the systematic error with respect to its mean. The determination of the constant error at a given site, which is the sum of the systematic errors and the geographically correlated error, is essential in order to:

- Provide long time series deduced from measurements of different time series at a crossover point.
- Calculate the river slope water line between two virtual stations using two different time series.

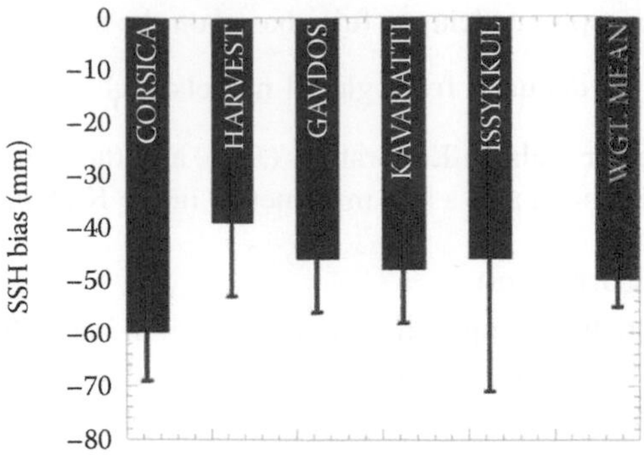

FIGURE 14.1 Example of typical waveforms, from left to right: specular, ocean-like (Brown model), contaminated, and noisy.

Moreover, linking time series from different missions requires accurate monitoring of biases and drifts for each parameter contributing to the final estimate of the water level.

Over the last 15 years, several groups have developed calibration/validation (C/V) sites to measure the absolute altimeter bias of each satellite. Such calibration activities have been performed in the framework of "post launch" research in order to quantify system performances for each mission. They have been repeated at different sites worldwide: Harvest platform in California (Haines et al. 2010); Corsica site—Cape Senetosa and Aspretto (Bonnefond et al. 2010); Gavdos Island (Mertikas et al. 2010) and Ibiza (Martinez-Benjamin et al. 2004) in the Mediterranean; and Bass Strait in Australia (Watson et al. 2011).

Calibration of radar altimeters is based on the principle of estimating the sea surface height (SSH) with altimetry at some comparison point using independent measurements of the sea level, including tide gauges, GPS surveys, moorings, or gravity/leveling profiles.

Shum et al. (2003) and Cretaux et al. (2009) pointed out that the variety of calibration sites for altimetry had to be enlarged in order to have more global distribution and more robust assessment of the altimetry system and to check if specific conditions lead to different estimation of absolute bias of the instruments. Actually, the only studies dedicated to calculating instrumental biases over rivers are Calmant et al. (2012) for Envisat and *Ice-1* retracker and Seyler et al. (2009) for Jason-2 with *Ice-1* and *Ice-3*. For lakes, some estimations have been proposed in Shum et al. (2003), Cheng et al. (2010), and Cretaux et al. (2009, 2011, 2013) for most altimetry missions and for several retrackers. The altimeter biases of Saral/AltiKa mission has been estimated over several cal/val sites including the Lake Issykkul and provide very similar results. It is illustrated by figure 14.1

14.3 SATELLITE ALTIMETRY FOR HYDROLOGY: SOME BASICS

14.3.1 Review of the Retracking and Height Retrieval

The electromagnetic (EM) wave emitted by a radar altimeter at any of the commonly used frequencies (S, C, Ku, and Ka) illuminates an area of several tenths of square kilometers.

Footprint information for each altimetry LRM mission based on the technical characteristics of the antennas is given in Table 14.1.

To obtain the highest possible accuracy on range measurements, waveforms are down linked to the Earth and the interpretation of these data together with meaningful geophysical quantities is performed on the ground. This is called waveform retracking. This approach grants flexibility in data reprocessing with dedicated algorithms.

The altimeter footprint on land often contains different surface types: calm water, rougher water, and various types of land, leading to a variety of radar echo shapes. Over larger lakes and rivers,

TABLE 14.1
Antenna Aperture and Footprint Diameter for Each Altimetry LRM Mission Based on Their Technical Characteristics

Altimeter	Altitude (km)	Wavelength (m)/ Frequency band	Antenna Diameter (m)	Antenna Aperture (°)	Footprint Diameter (km)
T/P	1336	0.022 (Ku)	1.5	1.1	~26
		0.0566 (C)		2.64	~62
Jason-1, 2, 3	1336	0.022 (Ku)	1.2	1.28	~30
		0.0566 (C)		3.4	~79
ERS-1 & 2	785	0.0217 (Ku)	1.2	1.3	~18
Envisat	782–800	0.0217 (Ku)	1.2	1.3	~18
		0.0938 (S)		5.5	~77
Saral/AltiKa	814	0.0084 (Ka)	1	0.59	~8
GFO	800	0.0222 (Ku)	1.1	1.41	~20

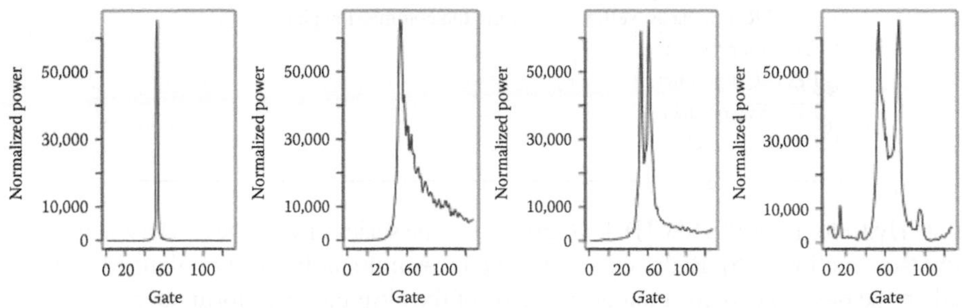

FIGURE 14.2 Estimation of the absolute bias of the SARAL/AltiKa altimeter from different experiments.

ocean-like echoes are often observed in the central part, while specular waveforms are returned from closer to the shore, where the water is calmer. However, due to the inhomogeneous surface, the echo is often a mixture of several reflectors causing the waveforms to be multipeaked (see Figure 14.2).

With the launch of Envisat, range estimates derived from four different retracking algorithms (Ocean, Offset Center of Gravity [OCOG] or *Ice-1, Ice-2*, and *Sea Ice*) were available in the GDR for the first time. Over the Great Lakes, ranges obtained using the ocean retracker provide the best results. Over rivers, it is generally preferred to use ranges computed with the Ice-1 retracking algorithm (Frappart et al. 2006a; Silva et al. 2010). Recently, studies performed on lakes and rivers about 200 m wide showed that water level based on the Sea Ice retracking algorithm exhibit similar precisions to the ones computed with Ice-1 (Sulistioadi et al. 2015; Biancamaria et al. 2017).

To generalize the use of altimetry for hydrology, Jason-2 and Jason-3 GDR contain Ice-1 retracked ranges while SARAL/AltiKa GDR contain ranges retracked using the same four algorithms as Envisat. Efforts to provide homogeneous radar altimetry data sets over the whole period of observations were also undertaken, starting with the reprocessing of T/P data using the Envisat algorithms in the framework of the Comtribution de l'Altimetrie Spatiale à l'Hydrologie (CASH) project (Seyler et al. 2005). Recently, reprocessing of ERS data was performed by ESA, for ERS-1 and 2, in the framework of the Reprocessing of Altimeter Products for ERs (REAPER) project (Femenias et al. 2015), and by the Centre for Topographic Studies of the Oceans (CTOH) for ERS-2 (Frappart et al. 2016). Other initiatives applied different retracking algorithms and provided dedicated processing for hydrology products. CNES made available Ice-3 retracked ranges in the Jason-2 GDR in the framework of the coastal and hydrology products (Prototype Innovant de Système de Traitement pour les Applications

TABLE 14.2
Retracking Available for All the LRM Altimetry Missions Put into Orbit after 1990

	Ocean	Ice-1	Ice-2	Sea Ice
T/P	Yes	Yes[a]	Yes[a]	Yes[a]
Jason-1	Yes	No	No	No
Jason-2	Yes	Yes + Ice-3[b]	Yes	Yes
Jason-3	Yes	√	Yes	Yes
ERS-1	Yes[c]	Yes[c]	Yes[c]	Yes[c]
ERS-2	Yes[c]	Yes[c,d]	Yes[c,d]	Yes[c]
Envisat	Yes	Yes	Yes	Yes
Saral/AltiKa	Yes	Yes	Yes	Yes
GFO	Yes	No	No	No
Sentinel-3A	Yes	Yes	Yes	Yes

Note: Sentinel-3 is a 100% SAR mode mission during its operation; there is an LRM mode as well, tested during the commission phase.
[a] CASH reprocessing.
[b] PISTACH product.
[c] REAPER product.
[d] CTOH product.

Côtières et l'Hydrologie, or PISTACH). ESA provides time series of water levels based on an automatic processing (see Section 14.5). Radar echoes from a measurement inside a predefined water mask are processed on the basis of classification of the shape of the altimeter waveform (Berry et al. 2005). The retracking algorithms available for each altimetry LRM mission are summarized in Table 14.2.

The following sections give a brief overview of aforementioned retracking algorithms.

14.3.1.1 Ocean and Ice-2 Retrackers

The Ocean and Ice-2 retrackers are based on the physical modeling of the waveforms. Over ocean-like surfaces, the altimeter echo can be described as the sum of the reflections from elementary surface facets distributed around the mean surface ordered by their arrival time or range (Brown 1977). This radar echo (or waveform W) results from the convolution of the impulse response shape (I) with the antenna pattern function (Ant) and the point distribution function (pdf):

$$W(t) = I * pdf * Ant \tag{14.6}$$

Over land and ice sheets, the EM waves emitted by the altimeter are likely to penetrate most of the underneath surfaces. The resulting radar response can be composed of surface and volume echoes. Over ice sheets, it is the sum of surface and snowpack layers echoes. Over land, the resulting waveform can be the sum of echoes from the canopy, ground layers, and possibly snow layers depending on the wavelength of THE EM wave, the nature and the water content of the soil, and the vegetation. For very smooth or very wet surfaces, the waveforms are almost specular. The echo shape is also affected by the surface topography slope and curvature. In these cases, it is necessary to take into account the scattering distribution ($fscat$) that describes the vertical profile of the reflecting surfaces to model the waveform (Legrésy et al. 2005):

$$W(t) = I * pdf * fscat * Ant \tag{14.7}$$

The analytical expression of the different terms can be found in Legrésy et al. (2005).

14.3.1.2 OCOG Retracker

The OCOG retracker (Wingham et al. 1986) determines the date of the Leading Edge Point (LEP) by a weighted sum of the power received in each of the receiving window power bins:

$$LEP = \sum_{i=1+n_1}^{N-n_2} iP_i^2(t) \Big/ \sum_{i=1+n_1}^{N-n_2} P_i^2(t) - \frac{1}{2} \cdot \left(\sum_{i=1+n_1}^{N-n_2} P_i^2(t)\right)^2 \Big/ \sum_{i=1+n_1}^{N-n_2} P_i^4(t) \quad (14.8)$$

Here P_i is the power of the i'th gate and N is the total number of gates; n_1 and n_2 are the numbers of bins—also called gates—that are affected by noise. The values of n_1 and n_2 are determined.

The OCOG retracker is robust and will always provide a height estimate. The OCOG amplitude is also used as input to other retrackers such as the threshold and improved threshold retrackers. In the GDRs, the OCOG retracker is also named *Ice-1*.

The *Ice-3* retracker (only proposed in the PISTACH project) is derived from the *Ice-1* retracker except that it is applied to a smaller window selected around the main leading edge of the waveform (–10, +20 samples). This processing is intended to provide better performances than Ice-1 in the case of small peaks present at the beginning of the waveform (PISTACH Handbook 2009).

14.3.1.3 Sea Ice Retracker

As no model currently describes sea ice waveforms, no multi-parameter fit of the radar echo can be retrieved. A straightforward threshold technique was developed to retrack data over sea ice (Laxon 1994). The sea ice waveform amplitude (A) is first identified by finding the maximum value of the echo:

$$A = \max_{n \in N}(P_i) \quad (14.9)$$

The tracking offset is then determined by finding the point on the waveform, using linear interpolation, where the radar echo is greater than a threshold corresponding to half the waveform amplitude (Laxon 1994).

14.3.1.4 Threshold Retracker

Although this algorithm is not used by any agency to provide alternate ranges in the GDRs, we briefly present the threshold algorithm because it is used by some research groups to compute their own ranges from the waveforms instead of using the ranges proposed in the GDRs. The threshold retracker (Davis 1997) was developed to obtain elevation changes of the ice sheet. The algorithm seeks to identify the decimal range gate on the leading edge that corresponds to a given threshold level. The threshold level is often determined as a certain percentage of the OCOG amplitude or the maximum power amplitude of the waveform. Based on analysis of waveforms over the ice sheet, Davis (1997) suggested threshold levels of 10%–20% for waveforms dominated by volume scattering and 50% for those dominated by surface scattering. For inland water different threshold values have been suggested, for example: 10% (Schwatke et al. 2015) and 50% (Kuo and Kao 2011).

The threshold level T_h is defined as

$$T_h = T_N + q(A - T_N). \quad (14.10)$$

where T_N is the thermal noise observed on the very first gates, and A is the amplitude of the waveform (e.g., based on the amplitude of the maximum waveform power); q is the threshold percentage.

The decimal retracking gate, G_r, which corresponds to the threshold level, is estimated by a linear interpolation of the power at the gates before and after on the threshold level, T_h.

The philosophy behind the improved threshold retrackers is to consider the waveform as being composed of a number of sub-waveforms each with a leading edge and a corresponding retracking point. This results in one or more height estimates for each waveform, where only one of these heights is related to the water surface at nadir. The optimal sub-waveform and the corresponding retracking point may be determined directly (Lee et al. 2008) or from external data (Hwang et al. 2006).

Gommenginger et al. (2011) and Villadsen et al. (2015) use the standard deviation of the power differences to estimate the threshold values, which ensures that the threshold values are scaled according to the individual waveforms.

A modified version of the improved threshold retracker has recently been used to obtain water levels of rivers and lakes. Villadsen et al. (2015) used the primary sub-waveform (Jain et al. 2015) to determine the water level, which is defined as the first sub-waveform where the retracking point is larger than 25% of the maximum power. To ensure more stable height estimates, two additional gates are added to the sub-waveform. This version is referred to as the narrow primary peak (NPP) retracker. Schwatke et al. (2015) used the first sub-waveform to determine the water level. Boergens et al. (2016) used the sum of the power in each sub-waveform as a weight in order to detect the most likely sub-waveform that originates from the reflection of the water surface. The sub-waveform with the most power and highest amplitude is defined as the most likely water return.

Considering multiple waveforms simultaneously might help to identify the persistent peak, which is related to water surface at nadir from the peaks detected by the modified threshold algorithm. This approach is used for the Multiple Waveform Persistent Peak (MWaPP) retracker (Villadsen et al. 2016), which is developed for inland water SAR waveforms. Here, the information from a given and four neighboring waveforms is used simultaneously in the retracking procedure (Figure 14.3).

14.3.2 Hooking Effect

If the reflecting surface was homogeneous in the footprint, the backscattered energy received by the sensor would decrease with the incidence angle and the distance to the center of the illuminated disk. Contrary to the ocean, the land surfaces are very heterogeneous. The surface encompassed in the altimeter footprint can be composed of elements with very different dielectrical properties. As a consequence, backscatter energy from off-nadir elements can be significantly greater than contribution from elements at nadir to the sensor (e.g., a small water body—lake or river—surrounded by forest). The onboard tracking system will then progressively center the receiving window to the dominant power peak, and the range will then be estimated from this peak. As the most power-reflecting element of the scene is not at nadir, the range measurement is a measurement of sloping distance, which increases as the satellite moves toward or away from the object. Off-nadir reflections (i.e., hooking effect) are responsible for hyperbolic profiles present in both the received power and the range, as presented in Figure 14.4, with the example of SARAL/AltiKa measurements in the region of Lake Mamori in the Amazon Basin. They can be modeled at first order with parabolas as follows (see the scheme in Figure 14.5):

$$H_0 = H_i + \frac{1}{2\rho_0}\left(1+\left(\frac{\partial a}{\partial s}\right)^2\right)ds_i \tag{14.11}$$

where s is the along-track coordinate, H_0 is the altimeter height at nadir, ρ_0 the altimeter range at nadir corrected from the geophysical effects, H_i the altitude of the satellite at s_i the coordinates of the slant measurements along the altimeter track, $\partial a/\partial s$ the rate of altitude variation of the satellite along the orbital segment, and ds the along track difference between two along track measurements.

The altimeter height at nadir is obtained by estimating the summit of the parabola representing the actual water level:

$$H_0 = as_0^2 + bs_0 + c \tag{14.12}$$

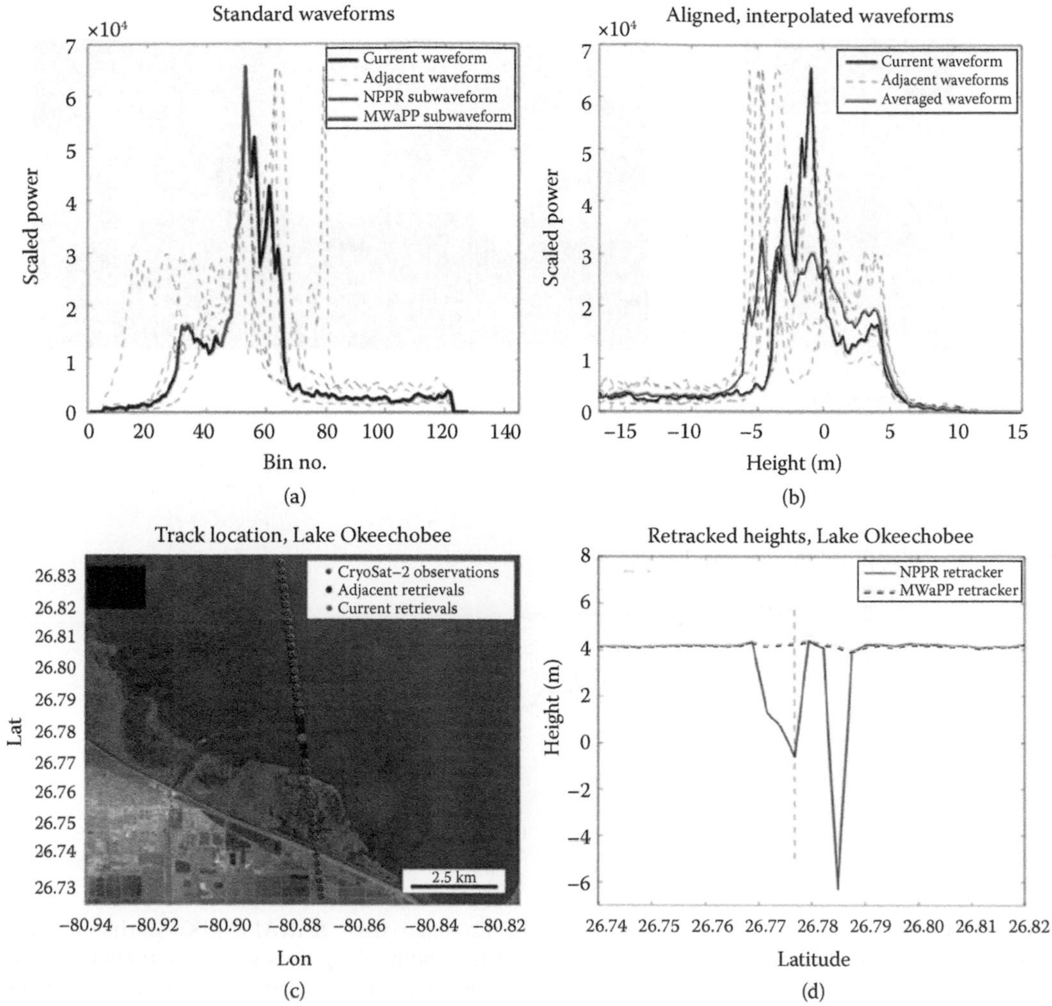

FIGURE 14.3 Illustration of the Multiple Waveform Persistent Peak (MWaPP) method. (Adapted from Villadsen, H., et al., *J. Hydrol.*, 537, 234–247, 2016.) (a) The current (solid line) and the four adjacent waveforms along with the sub-waveforms and retracking points obtained with the narrow primary peak (NPP) retracker and the MWaPP retracker. (b) The aligned current, adjacent, and averaged waveforms. (c) The location of the five considered waveforms. (d) The estimated height based on the MWaPP retracker and the NPPR retracker. Note, that large off-nadir returns are damped in the average waveform, while the peak related to nadir (the persistent peak) is enhanced.

where s_0 is the location of the nadir along the altimeter track; a, b, and c are the coefficients of the parabola estimating through a least-square fitting of the altimeter data affected by hooking.

$$s_0 = -\frac{b}{2a} \text{ and } H_0 = c - \frac{b^2}{4a} \tag{14.13}$$

If slant measurements are considered as an artifact to be corrected to obtain an accurate measurement of water stage, they enhance the number of detectable small waterbodies. For very small lakes and narrow rivers that can be viewed as bright targets, it is essential to adjust the parabolas to all relevant measurements. This allows estimating water levels more accurately instead of if only the

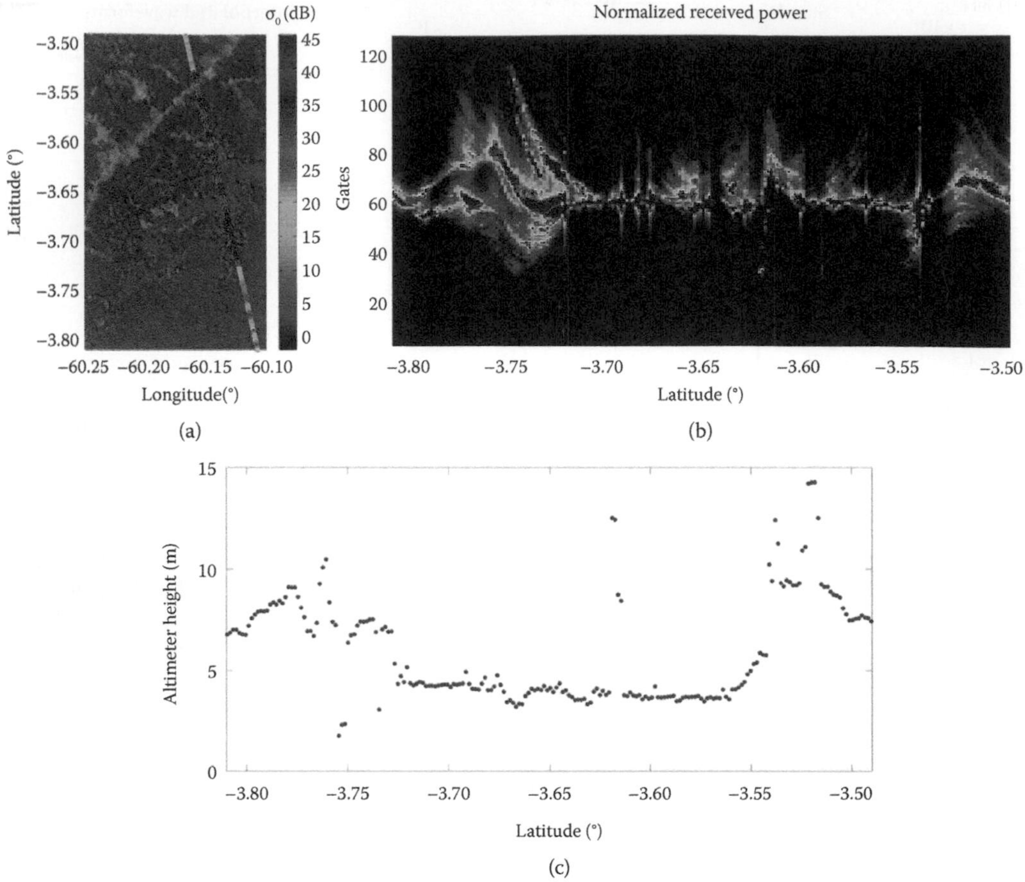

FIGURE 14.4 (a) Backscattering coefficients at the Ka band from SARAL/AltiKa during Cycle 9 (12/24/2014) in Lake Mamori area (Amazon Basin) and corresponding along-track profiles of (b) radar echoes (or waveforms) and (c) altimeter heights. The received power is represented relative to the reception gate for all waveforms. Distinct parabolas can be seen between gates 40 and 70 at latitudes around 3.73°S, 3.65°S, and 3.62°S. The vertex of each of the parabolas corresponds to the lake itself and two small tributaries. A parabola corresponding to the lake in the altimeter height can also be observed.

measurements at nadir were used (if it even exists). Practically, thanks to this artifact, hundreds of time series of water stages for small rivers were built, especially in the case of rivers flowing under a dense canopy, such as in the Amazon or in the Congo basins. Figure 14.6 illustrates such a case with an example of ERS/Envisat/SARAL altimetry tracks crossing the Rio Pardo, a minor tributary of the Amazon that is just 30 m wide during the low water flow period.

14.3.3 Geoid Gradient Correction over Lakes

Considering Equation 14.3, the ellipsoid height calculated using satellite altimetry needs to be corrected for the geoid undulation. This can be done using existing available global geoid models, such as EIGEN2008. In such case, the geoid correction is calculated by interpolating the model to the position of the altimetry measurement. However, the geoid models do not have a spatial resolution that is high enough to capture the small wavelength undulation of a lake surface due to gravity. In many cases, such as the lakes Issykkul, Tanganyika, or Baikal, the local geoid may reach several meters with important vertical gradients that can be accurately measured using the altimetry data themselves. When compared to the corresponding gradient inferred from global geoid models, the discrepancy may reach several decimeters.

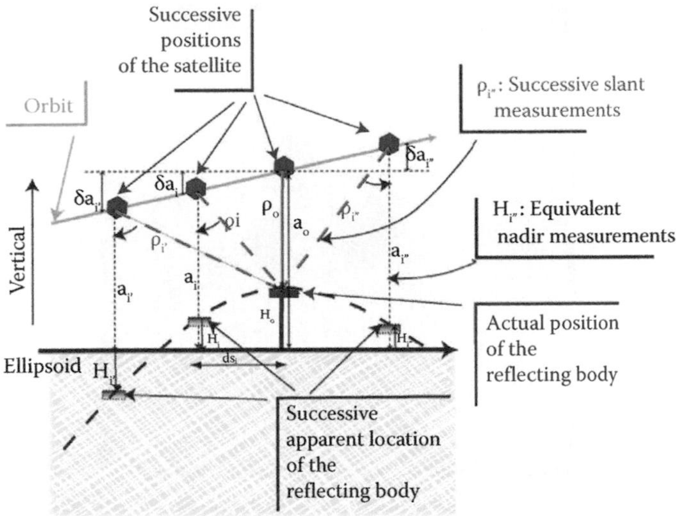

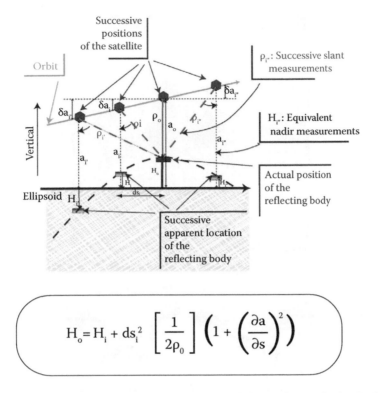

$$H_o = H_i + ds_i^2 \left[\frac{1}{2\rho_0}\right]\left(1 + \left(\frac{\partial a}{\partial s}\right)^2\right)$$

FIGURE 14.5 Diagram of the slant measurement. The true height H_o of the reflecting body (lake, narrow river, etc.) may be deduced from successive slant H_i measurements from either side of its actual position by means of a correction that is a quadratic function of the distance to the nadir ds_i (a parabolic shape in height profiles H_i [ds_i]). *Copyright, Calmant et al. 2017, Principes de l'altimétrie satellitaire radar pour les applications continentales, ISTE Edition, Télédétection pour l'observation des surfaces continentales*

In order to estimate this gradient along the altimeter tracks, it is necessary to stack all profiles from each cycle and average them along a so-called average track; this method, which has been described in Birkett (1995) and Cretaux et al. (2016), is called the "repeat track technique." It results in a mean vertical profile along the average track that then is used in Equation 14.3 for the geoid correction. It is also illustrated in Figures 14.7 and 14.8.

14.3.4 Selection and Editing of Measurements for Hydrology

A virtual station is defined as a given intersection of an orbit ground track and a water body (i.e., lake, reservoir river channel, floodplain, or wetland). There, the variations of the altimeter height from one cycle to the next can be associated with changes in water level. Altimetry-derived water stages were first obtained by averaging each cycle of data in a rectangular window at the crossing (Koblinsky et al. 1993; Morris and Gill 1994a, 1994b). Over rivers and small lakes, due to the small number of available data at cross-sections, the average was replaced by the median to limit the impact of outliers (Frappart et al. 2006a).

As the altimeter footprint with its diameter of several kilometers generally encompasses an inhomogeneous surface over land, the precise determination of the height is likely to be altered by two effects:

1. The radar echo measured by the antenna is composed of returns from both water and non-water reflectors with different elevations. These non-water reflectors can be vegetation, bare soil—eventually wet rivers, and lake banks (sandbanks that appear at low water stage).

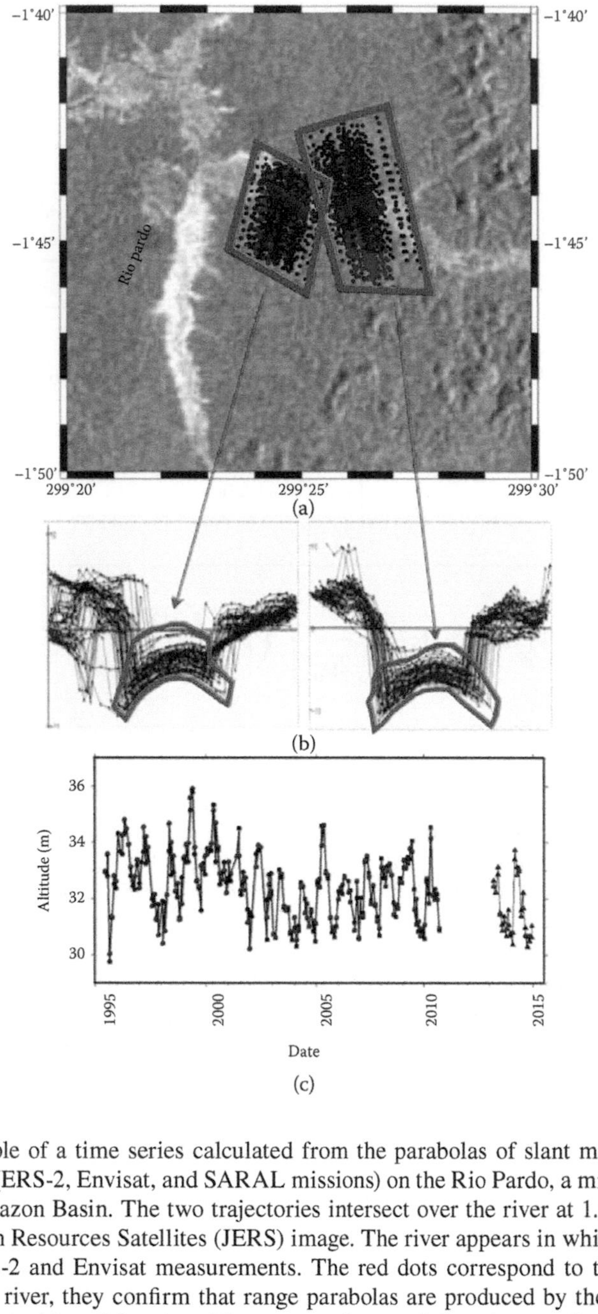

FIGURE 14.6 Example of a time series calculated from the parabolas of slant measurements at an orbit intersection at 35 days (ERS-2, Envisat, and SARAL missions) on the Rio Pardo, a minor tributary of the Rio Negro basin in the Amazon Basin. The two trajectories intersect over the river at 1.5-day intervals. (a) The map is a Japanese Earth Resources Satellites (JERS) image. The river appears in white. The dark blue points correspond to the ERS-2 and Envisat measurements. The red dots correspond to the vertex of parabolas. Located just above the river, they confirm that range parabolas are produced by the river. (b) Parabolas in altimetry profiles of both sides of the river. The green polygons define the set of measurements located in the top figure. (c) Altitude time series of water levels after correction of parabolas. Assuming that the water level has not changed between two passages of the same cycle, we get an overestimation of measurement error, which is 40 cm for ERS-2 and 25 cm for Envisat (Silva et al. 2010). Copyright, Calmant et al. 2017, Principes de l'altimétrie satellitaire radar pour les applications continentales, ISTE Edition, Télédétection pour l'observation des surfaces continentales

2. The surface that reflects the energy peak that dominates the radar echo is not at the nadir of the antenna, causing an underestimation of the altimeter height (due to an overestimate of the range).

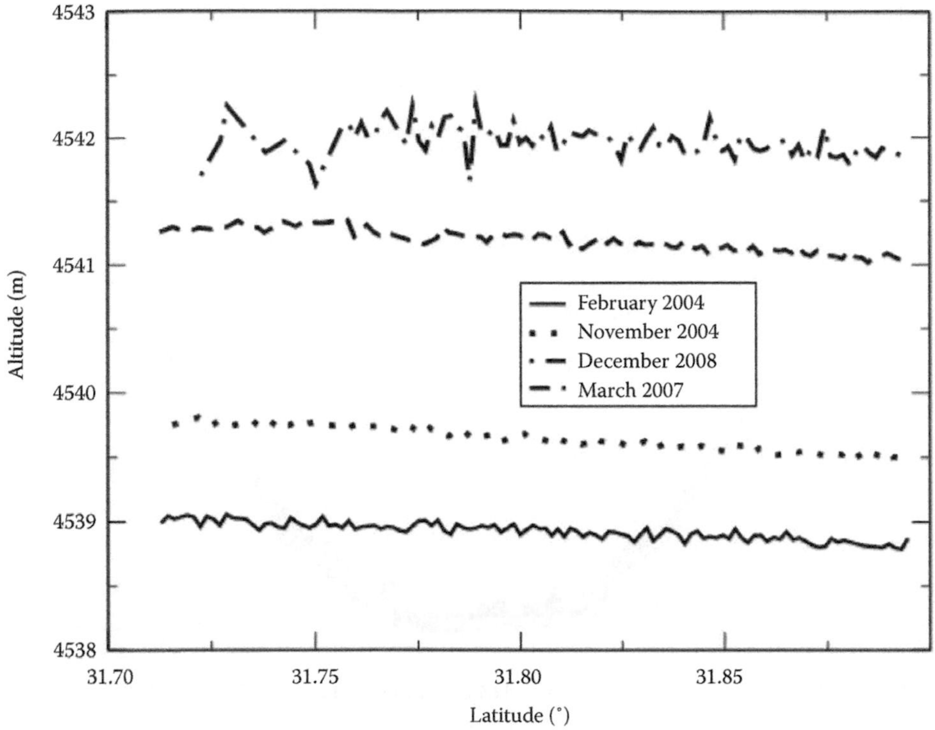

FIGURE 14.7 An example of a vertical profile of an ICESat track crossing Lake Ziling in the Tibetan Plateau. Four different cycles are represented. The average difference between each track is due to hydrological variation of the lake, and the slope observed on each cycle corresponds to the vertical gradient. After correction of this slope, the average water height of the lake for each pass can be calculated.

To tackle these problems, a refined selection of the valid altimetry data needs to be performed, especially over rivers and wetlands to accurately estimate the water levels. Several software have been developed to enable visual selection of valid altimetry data to build virtual stations, such as the Virtual Altimetry Station (VALS) software (Santos da Silva et al. 2010) and the Multi-Mission Altimetry Processing Software (MAPS, Frappart et al. 2015b). Data processing in MAPS is composed of three main steps, which are presented in the following detailed example:

1. A coarse delineation of the virtual stations is performed using Google Earth. The polyline is saved in a KML file that is then loaded into MAPS. Data located within the polyline are extracted from a database storing the altimetry measurements contained in the GDRs. The height is computed using Equations 14.2 and 14.3 where all flagged values (either on the orbit, the range, or the corrections) are discarded (Figure 14.9).
2. A refined selection of the valid altimetry data. Profiles of altimeter heights are displayed for each cycle (Figure 14.9). Outliers and measurements over non-water surfaces are eliminated based on visual inspection. Parabolic profiles caused by off-nadir reflections (i.e., hooking effect) can be corrected using (Equations 14.12 and 14.13) (see Section 14.3.3 on hooking effect) (Figure 14.9 and 14.10).
3. The computation of the water level time series. The altimetry-based water level is computed for each cycle using the median of the selected altimetry heights, along with their respective deviation (i.e., mean absolute deviations) after correction for off-nadir effect. This process is repeated for each cycle to construct a water level time series at the virtual station (Figure 14.11).

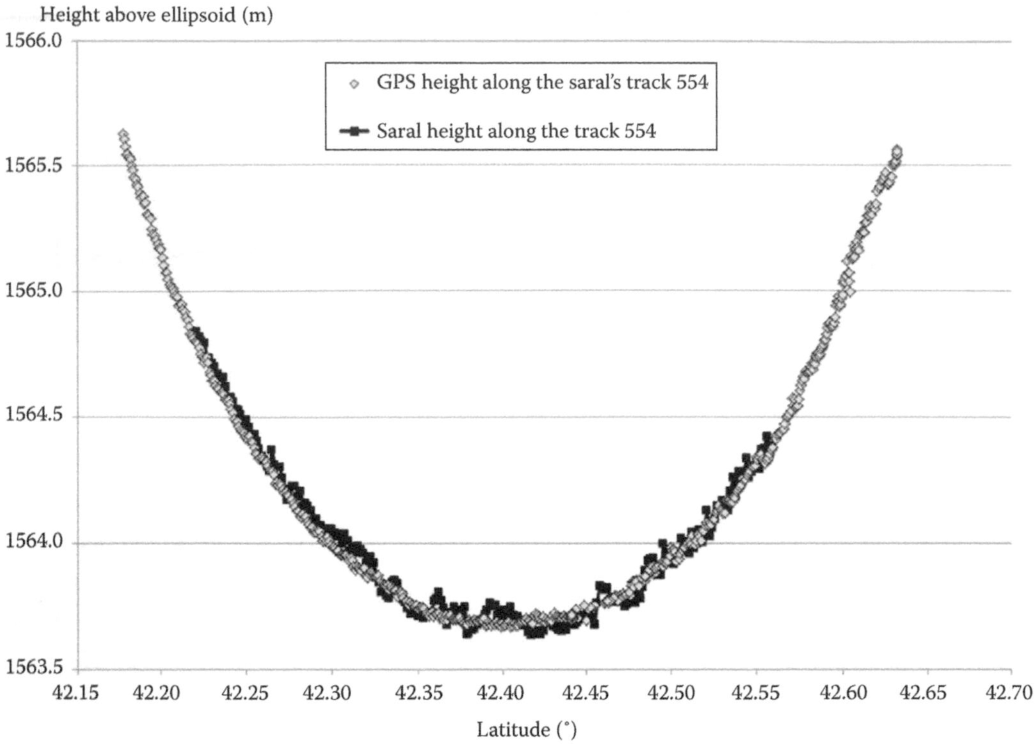

FIGURE 14.8 A vertical profile along the SARAL/AltiKa track N° 554 with altimeter measurements and with GPS survey (done during field work over this lake in July 2014). From this curve, we may see that a north/south profile along Lake Issykkul can present approximately 2 m of gradient, which was not accurately detected using the global models: Along this profile, we have measured more than 30 cm of errors comparing the aforementioned values and the EGM2008 model.

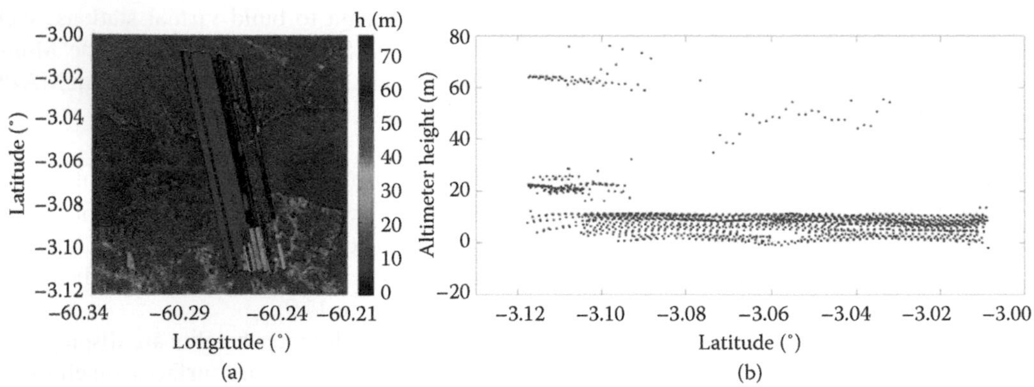

FIGURE 14.9 Example of a coarse delineation of a virtual station for SARAL/AltiKa track 0149 on the Negro River, one of the major tributaries of the Amazon River, at the outlet of the basin, close to Manaus. SARAL/AltiKa-based altimeter heights are presented in (a) geographical representation and (b) as along-track profiles. Copyright: Villadsen et al. 2016.

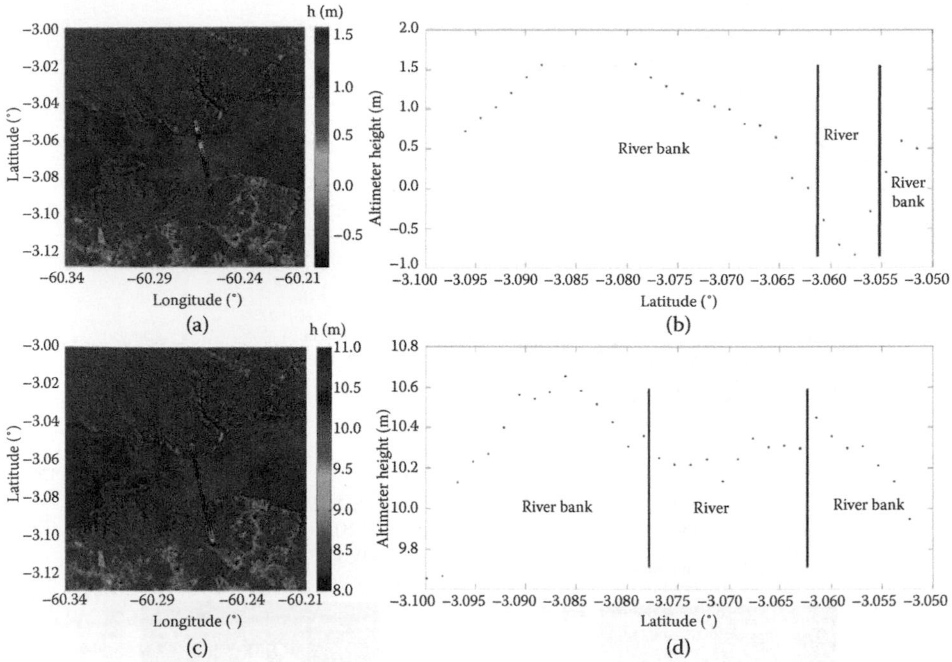

FIGURE 14.10 (a–d) Same as Figure 14.9 for two SARAL/AltiKa cycles (008 on11/19/2013 and 013 on05/13/2013 during low and high waters, respectively). Important changes both in height and width can be observed between the two cycles.

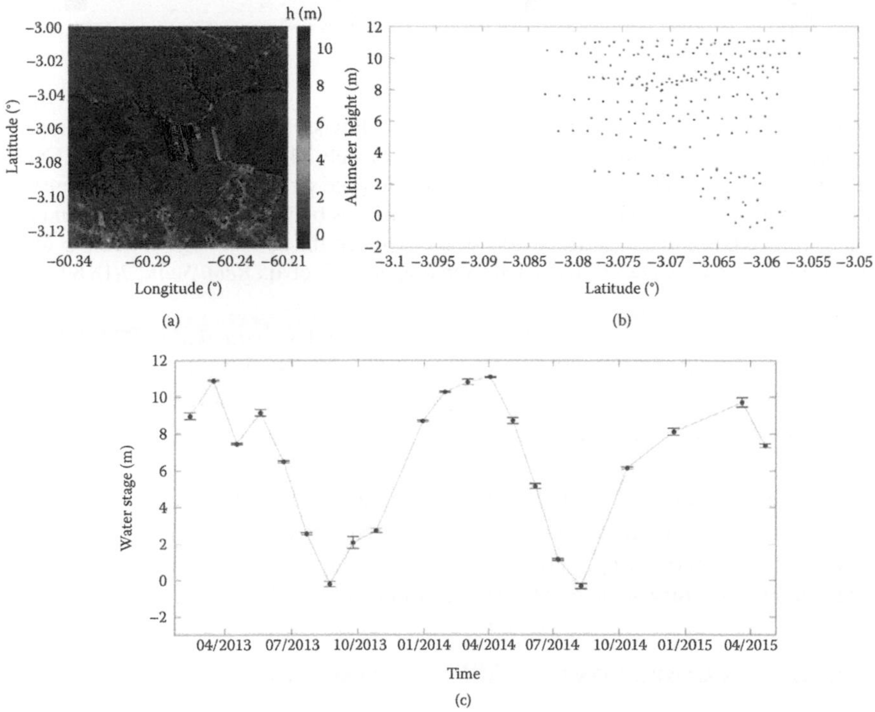

FIGURE 14.11 Same as Figure 14.10 for (a) and (b) panels for 27 SARAL/AltiKa cycles from March 2013 to September 2015. (c) Time series of water stage based on this selection.

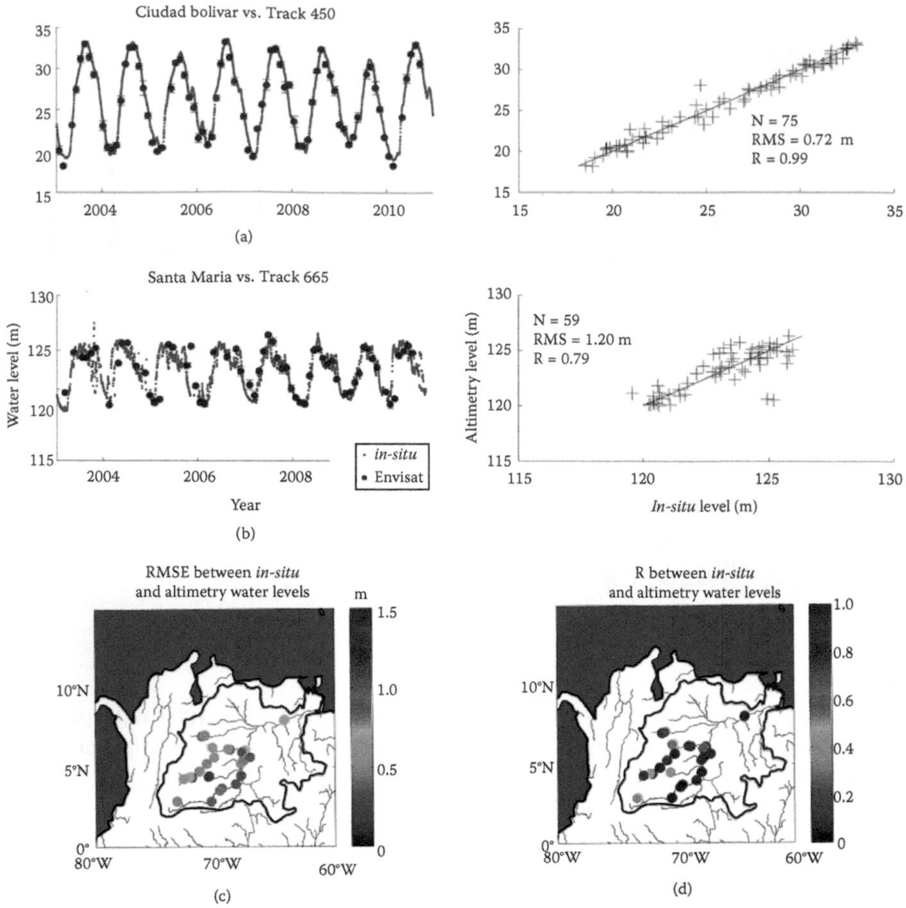

FIGURE 14.12 Two examples of water levels derived from Envisat radar altimetry and compared with in situ gauge records over 2003 to 2010. (a) Track 450 at approximately 74 km upstream from the Ciudad Bolivar gauge; (b) Track 665 for the Santa Maria gauge. In both cases, time series of water levels are presented on the left and the corresponding scatter plots on the right. Comparisons between altimetry-based and in situ water levels over the Orinoco Basin in 28 virtual station locations materialized with dots, in terms of (c) root-mean-square error (RMSE) (d) and range (R). (Adapted from Frappart, F., et al., *Rem. Sens.*, 7(1), 89–110, 2015a.)

Over rivers, the accuracy between altimetry-derived and *in situ* water stages has increased from approximately 1 m in early studies (Koblinsky et al. 1993; Birkett 1998) to 0.2 m to 0.5 m (Figure 14.12) in more recent ones (Berry et al. 2005; Frappart et al. 2006a; Silva et al. 2010; Papa et al. 2012) for the following reasons:

- The use of high-frequency data (10, 18, 20, 40 Hz) instead of 1-Hz data (approximately 7 km of sampling along the track) as commonly used over open ocean
- The availability of retracking algorithms (e.g., Ice-1) that are better suited for inland water
- The development of software (e.g., MAPS) that allows a refined selection of valid altimetry data

14.3.5 Cross-Track Correction for SARIn and Snagging

Snagging is the phenomenon that occurs when the dominating reflection originates from an off-nadir target located in the across-track direction. It is typically seen for inhomogeneous surfaces such as inland water, coastal areas, and sea ice. Figure 14.13 illustrates the principle. The range, *R*,

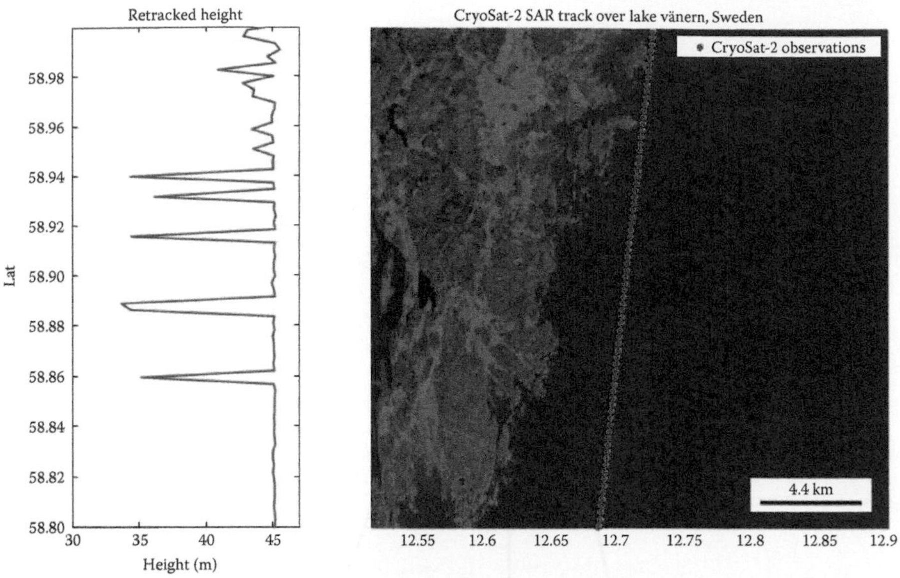

FIGURE 14.13 An example of snagging (adapted from Villadsen et al., 2016). (a) Height variations along a track over Lake Vänern (the location of the track is displayed on the right panel). (b) The location of CryoSat-2 observations.

at nadir is wrongly estimated too large, resulting in an underestimation of the water level. Correcting for snagging is generally more challenging because no distinct height profile can be identified in the along-track direction, such as the parabolic profile seen in the hooking effect (see Section 14.3.3).

The two antennas on board CryoSat-2, which both are operational in SARIn mode, offer the possibility to locate the origin of the reflecting target in the footprint. When the position is known, it is possible to correct for the range error as described in Armitage and Davidson (2014). Figure 14.14 describes the geometry of the satellite and the surface. Here M is the location of a bright target and S is the nadir point. R_M is the range, to the target, measured by the satellite, R is the range to nadir, and dh_p is the range error. θ is the angle of the antenna boresight direction, OP, and the direction of the bright target, OM.

If the surface is assumed to be horizontal, the range error is given by

$$dh_p \simeq 1.113.R_m \frac{\left[\dfrac{\phi}{k_0 B} - \chi\right]^2}{2}, \tag{14.14}$$

where k_0 is the carrier wave number and B is the distance among the two antennas; χ the baseline roll angle and ϕ is the phase difference that arises due to the slightly different travel time of the signal from the bright target to the two antennas.

The correct range R is obtained by subtracting the range correction from the measured range.

$$R = R_m - dh_p \tag{14.15}$$

The capability of the SARIn mode offers a unique possibility to obtain improved height estimates in complex areas such as inland water, where snagging causes problems.

Figure 14.15 displays the height and positions of two tracks crossing Lake Tangra Yumco, which is located on the Tibetan Plateau. The left plot displays the nadir (blue crosses) and corrected positions (orange dots) of the observations. It is worth noticing that the actual position of the measurements can be several kilometers from nadir. The right plots show the corrected (orange dots) and

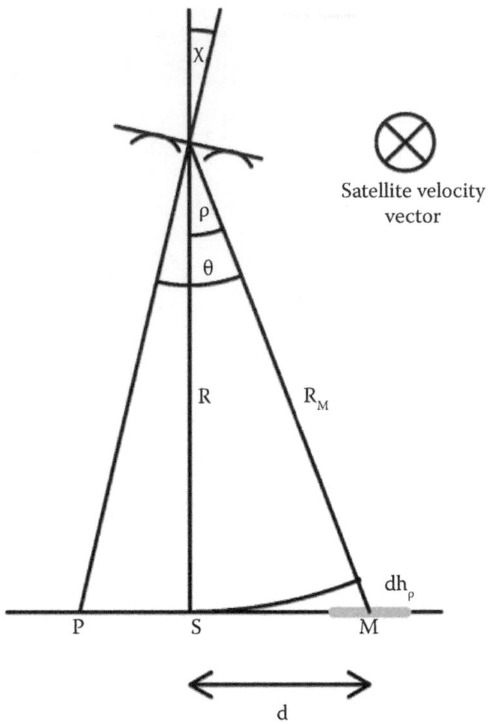

FIGURE 14.14 The geometry of SARIn, where M is the location of a bright target, S is the nadir point, R_M is the range—to the target—measured by the satellite, and R is the range to nadir; dh_ρ is the range error and θ is the angle among the antenna boresight direction, OP, and the direction of the bright target, OM.

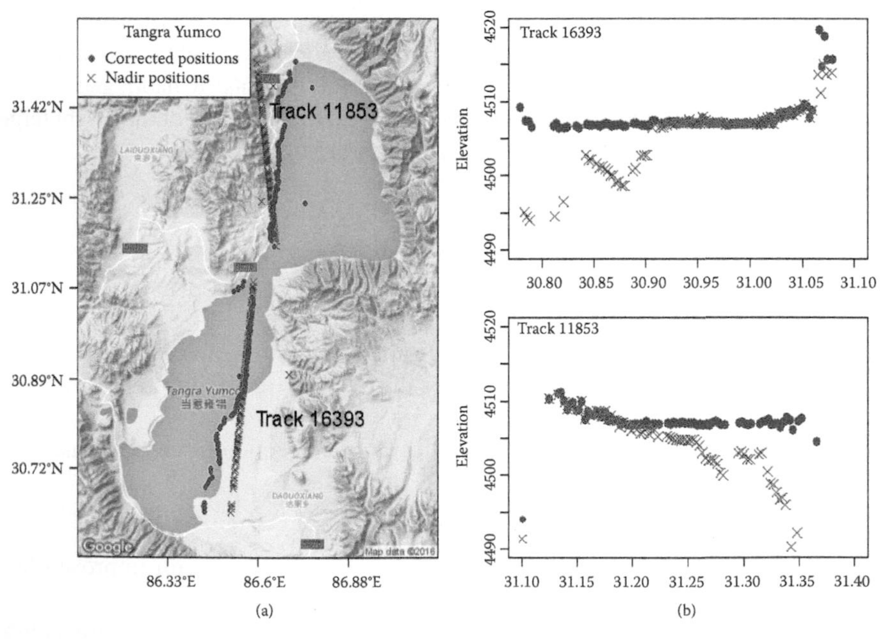

FIGURE 14.15 An example of range correction. (a) Displays the position of the observations at nadir and the relocated positions. The right plots (b) show the water level as a function of latitude before and after the range correction dh_ρ is applied.

uncorrected (blue crosses) heights. After correcting for the range error, the water level can be determined more accurately. This demonstrates an encouraging feature with SARIn, namely that tracks passing in the vicinity of a water body may also contribute with additional height measurements. This was demonstrated in Abulaitijiang et al. (2015) for fjords in Greenland, where observations located as far as 13 km from the fjords provide valid SSHs.

14.4 EXAMPLES OF APPLICATION

14.4.1 Use of Altimetry over Lakes and Reservoirs

14.4.1.1 Regional Survey of Lakes (Tibetan Plateau Lakes)

The Tibetan Plateau is the highest and most extensive highland in the world and is generally considered as the Third Pole. It is one of the most sensitive regions of the Earth to climate changes (Kang et al. 2010; Huang et al. 2011).

The climate system of the Tibetan Plateau is characterized by high spatiotemporal variability of rainfalls with very dry conditions in the northwest (50 mm/year of rainfall on average) to more humid conditions in the southeast (700 mm/year on average). Moreover, it is influenced by a synergy of Indian and East Asian monsoon and westerly jet (Li et al. 2007). More than 60% of rainfall occurs between June and September (Kang et al. 2010).

Several studies have been published to investigate the impact of climate change on the Tibetan Plateau, particularly in terms of water resources. It has been widely recognized and admitted that the temperature increased over the last decades, by 0.36°C/decade from 1964 to 2007 (Wang et al. 2008) and by 0.16°C/decade from 1955 to 1996 (Liu and Chen 2000), values significantly higher than the global mean. In contrast to the rather uniform air temperature increase over the Tibetan Plateau, long-term changes in precipitation represent a regional pattern of large interannual variability (Kang et al. 2010). The region became drier over the last 30 years in its northeastern and western parts (Kang et al. 2010; Huang et al. 2011) with some acceleration (Huang et al. 2011). However, it became much wetter in the eastern and central parts (Kang et al. 2010). Significant warming over the last 30 years has been observed over the Tibetan Plateau with pronounced effects in winter (Huang et al. 2011).

Changes of climate conditions over the Tibetan Plateau have also been extensively studied in recent years from a hydrological point of view. Climate change has led some lakes to shrink (those fed by precipitation and river runoff as observed by Huang et al. (2011), such as over the source of the Yellow River). In other regions, lakes have expanded (those fed by glaciers and permafrost melting such as in the western and central Tibetan Plateau; (see Liu et al. 2009; Huang et al. 2011). However, it has been shown by Huang et al. (2011) that although precipitation is decreasing in a given region, the lake abundance and size may increase. This indicates that a changing lake water storage depends on the direct effect from precipitation and/or temperature changes at the regional scale but also on the indirect effect through glaciers and snow melt as well as on permafrost degradation, which increases soil moisture and thickening of the active layer (Kang et al. 2010; Huang et al. 2011; Zhang et al. 2011b). Indeed, even in the case when the general trend of precipitation over the Tibetan Plateau is positive, it represents only a few millimeters of increase per year (Zhang et al. 2011b), which cannot explain the rate of lake level changes that is often decimeters per year. The impacts of climate change on lakes in highlands such as the Tibetan Plateau are indeed of different natures—direct consequences of rainfalls and temperature changes (through evaporation over open water surfaces) to a small part but, more significantly, indirect ones through glaciers and snow melt and permafrost degradation, increasing soil moisture and active layer thickening (Kang et al. 2010; Huang et al. 2011; Zhang et al. 2011b).

However, very few regions over the Tibetan Plateau exhibit weather and hydrological data sets of more than four to five decades (Huang et al. 2011). Moreover, *in situ* gauges for different parameters such as precipitation, glacier retreat, or lake level variations are still sparsely spread over the Tibetan Plateau, although large efforts have been made over recent years by the Chinese academy of science. Consequently, satellite data offer a very useful tool for the monitoring of lakes'

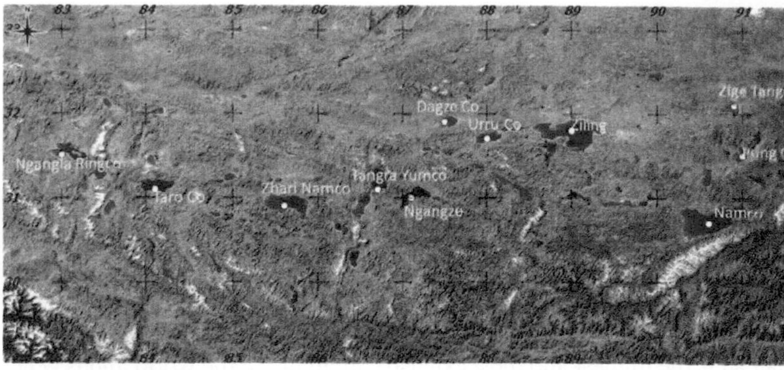

FIGURE 14.16 Map of the central Tibetan Plateau where several big lakes are monitored using satellite altimetry.

environment changes over the Tibetan Plateau, particularly for essential climate variables such as lake level, surface, and storage changes (see Figure 14.16).

Satellite altimetry (radar and laser) have been widely used for that purpose. Many studies have been focused on a specific area: an individual or limited group of lakes (Zhang et al. 2013) such as the Nam-Co and the Ziling (Zhang et al. 2011a, 2011b), the Ngoring-Co (Lee et al. 2011), the Ngangze and La'nga (Hwang et al. 2005), or the Kokonor* (Lee et al. 2011; Zhang et al. 2011b), while others have investigated global lake level changes related to climate change, using satellite laser altimetry and/or satellite imageries as the principal source of information (Liu et al. 2009; Zhang et al. 2010, 2011b; Huang et al. 2011; Phan et al. 2011; Lei et al. 2013; Song et al. 2013; Kleinherenbrink et al. 2015). Generally, authors try to explain lake level changes in terms of global warming and determine which component of the water balance is the most influential.

Recent studies have assumed that lakes of the Tibetan Plateau are in a phase of expansion as a consequence of climate change observed over many years. However, most of these studies are based on short time spans (Phan et al. 2011; Zhang et al. 2011b; Song et al. 2013; Wu et al. 2014). It was demonstrated from about 100 lakes monitored between 2003 and 2009 by ICESat data that the water level of a majority of lakes has increased on average by approximately 20 cm/year. Song et al. (2013) calculated water storage changes of a large number of lakes using ICESat and showed a general increase in water storage over the decade 2000–2011 of approximately 6.79 km^3/year.

However, information about lake level changes is still limited because of the short duration of *in situ* and satellite measurements or the very limited number of studied lakes. It is thus necessary to rely on lake water balance modeling under the influence of climate changes.

Recently, Cretaux et al. (2016) showed that water level time series can be extended to more than 20 years, and then the interpretation of lake storage changes is quite different. The time response to any small change in climate condition depends on many factors including lake morphology (Mason et al. 1994). In such a situation, the continuity of satellite altimetry missions since the mid-1990s is essential, and we can see in Figure 14.17a through c that the fluctuations of some large lakes over the Tibetan Plateau are observed at a decadal time interval. Some lakes present long-term expansion (Ziling), while others present interannual oscillations (Zhari-Namco) or are in equilibrium (Namco). The time series presented in Figure 14.17 is extracted from the Hydroweb Web site (http://hydroweb.theia-land.fr/).

* Lake Kokonor (Mongolian name) is also called *Qinghai* (Chinese name) in many articles.

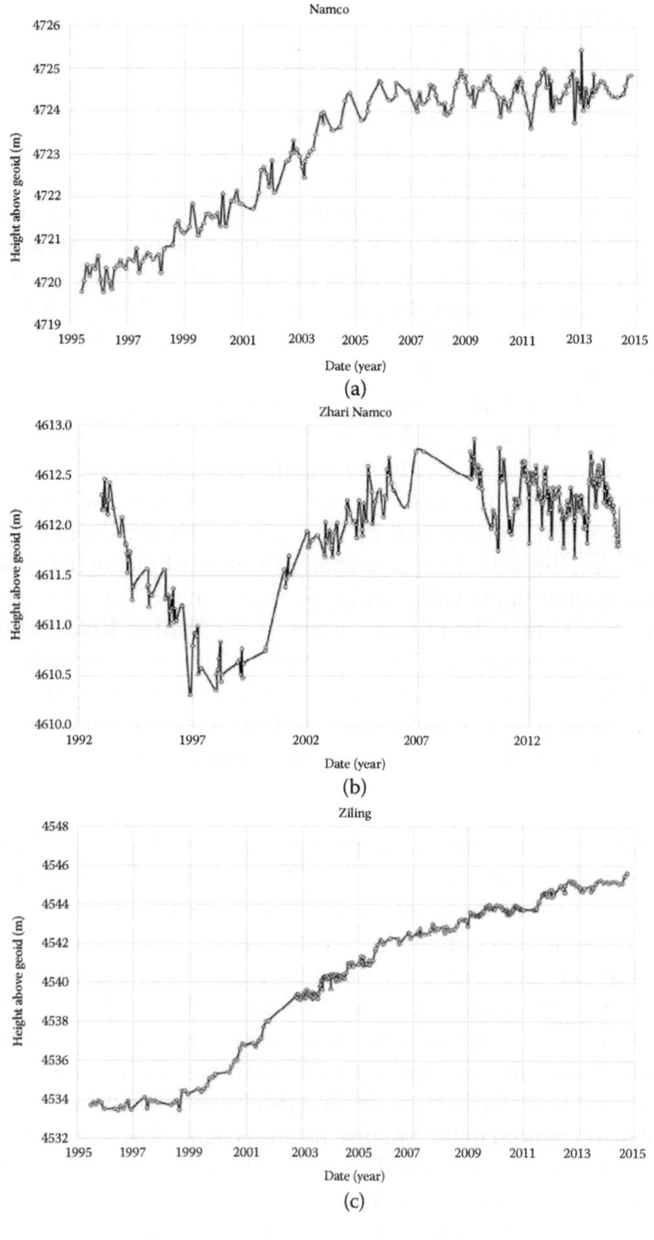

FIGURE 14.17 Water height from multi-satellite measurements of the Tibetan lakes (a) Namco (b) Zhari-Namco, and Ziling (c).

14.4.1.2 Case Study on Lakes Using SAR/SARIn

The performance of the new altimetry modes SAR and SARIn is evaluated by comparison with in situ data. To investigate the potential improvement from these new modes, the estimated water levels are compared with results obtained from conventional altimetry such as Envisat. The following lakes are considered: Arresø (Denmark), Okeechobee (Florida, United States), and Vänern (Sweden)—where SAR mode data is available, and Great Salt Lake (Utah, United States) and Walker Lake (Nevada, United States)—where SARIn mode data is available. These represent lakes of different sizes and locations in different surroundings. The settings of the lakes are introduced in the following section.

14.4.1.2.1 Lakes in SAR and SARIn Modes

Arresø is the largest lake located in Denmark. Its surface area is 40 km^2 and its average depth approximately 3 m (Wikipedia). It is fed by several small streams and drains into the Roskilde Fjord.

Okeechobee is the seventh largest freshwater lake in the United States with a surface area of 1900 km^2. Relative to its size, the lake is quite shallow, with an average depth of 2.7 m. The primary inflow is from Kissimmee River, Fisheating Creek, and Taylor Creek, whereas the primary outflow is through the Everglades, Caloosatchee River, and St. Lucie River. The lake is primarily surrounded by wetlands. Okeechobee Lake is used as water supply for agriculture, domestic use, and industry. It is further used for enhancement of fish and wildlife and for recreation.

Vänern is the largest lake in Sweden with a surface area of 5650 km^2 and an average depth of 27 m. The primary in- and outflow is Klarälven and Göta älv, respectively. The coastline is complex, and the lake contains several small islands.

The Great Salt Lake in Utah (United States) is a saline terminal lake, with no outlet besides evaporation. It has a surface area of 4400 km^2 and an average depth of 4.9 m. Hence, the lake is very shallow relative to its size, which makes it sensitive to changes in water inflow, precipitation, and evaporation. The water inflow originates from three major rivers: the Jordan, Weber, and Bear. The Great Salt Lake is divided into a northern and southern part by the causeway for the Lucin Cutoff. Due to a larger inflow in the southern part, the water level is generally 15 cm to 61 cm higher in this part.

The Walker Lake, located in the Great Basin, Nevada (United States), is also a terminal lake, with a surface area of 130 km^2. The primary inflow is the Walker River, which is an important source of water for irrigation in Nevada. Water diversions of the river have reduced its flow such that the level in Walker Lake dropped approximately 55 m between 1882 and 2016 (USGS database).

The satellite tracks from both CryoSat-2 (blue) and Envisat (red) are shown in Figure 14.18 for the five lakes: Arresø, Okeechobee, Vänern, Great Salt Lake, and Walker Lake.

14.4.1.2.2 Data

The CryoSat-2 data used in this analysis are the ESA Level 1b Baseline C data product in SAR and SARIn modes, which contain waveforms with 256 gates. The waveforms are retracked by a modified version of the improved threshold retracker with an 80% threshold value (Villadsen et al. 2015). The water levels with respect to the EGM2008 geoid (Pavlis et al. 2012) are derived from Equation 14.3. The Envisat data used is the Level 2 product RA-2 GDRs, which contains 18-Hz range measurements. Here, the water levels based on the Ice-1 retracker are derived. Finally, lake masks from the Global Lakes and Wetlands Database (Lehner and Döll 2004) and the Danish Geodata Agency (sdfe.dk) are used to retain only the measurements related to the lakes.

Gauge data for Vänern is available from the Swedish Meteorological and Hydrological Institute (SMHI). These data are referenced to the Swedish height system "Rikets höjdsystem 1900" (RH 00). Data from Okeechobee, Great Salt Lake, and Walker are obtained from the National Water Information System, waterdata.usgs.gov/nwis, and are relative to NGVD 1929. Gauge data from Arresø was provided by the Danish Nature Agency. These data are references to the Danish height system DVR90.

14.4.1.2.3 Method

Waveforms related to inland water might be influenced by the surrounding land in which case the waveform can display multiple peaks or be distorted. These waveforms may therefore lead to noisy and even erroneous water level estimates. This should be taken into account when deriving the water level time series. One approach, as described in Nielsen et al. (2015), is to assume that an observation follows a mixed distribution consisting of a Gaussian and a Cauchy distribution. The Cauchy distribution has a heavier tail compared to the Gaussian, thus erroneous observations will have a smaller influence on the estimated mean value.

The water level is varying with time, thus measurements taken over a short time span tend to be similar. To exploit this information, the final water level time series can be constructed by using a state-spaced model composed of two parts: a process model that describes the true height variations

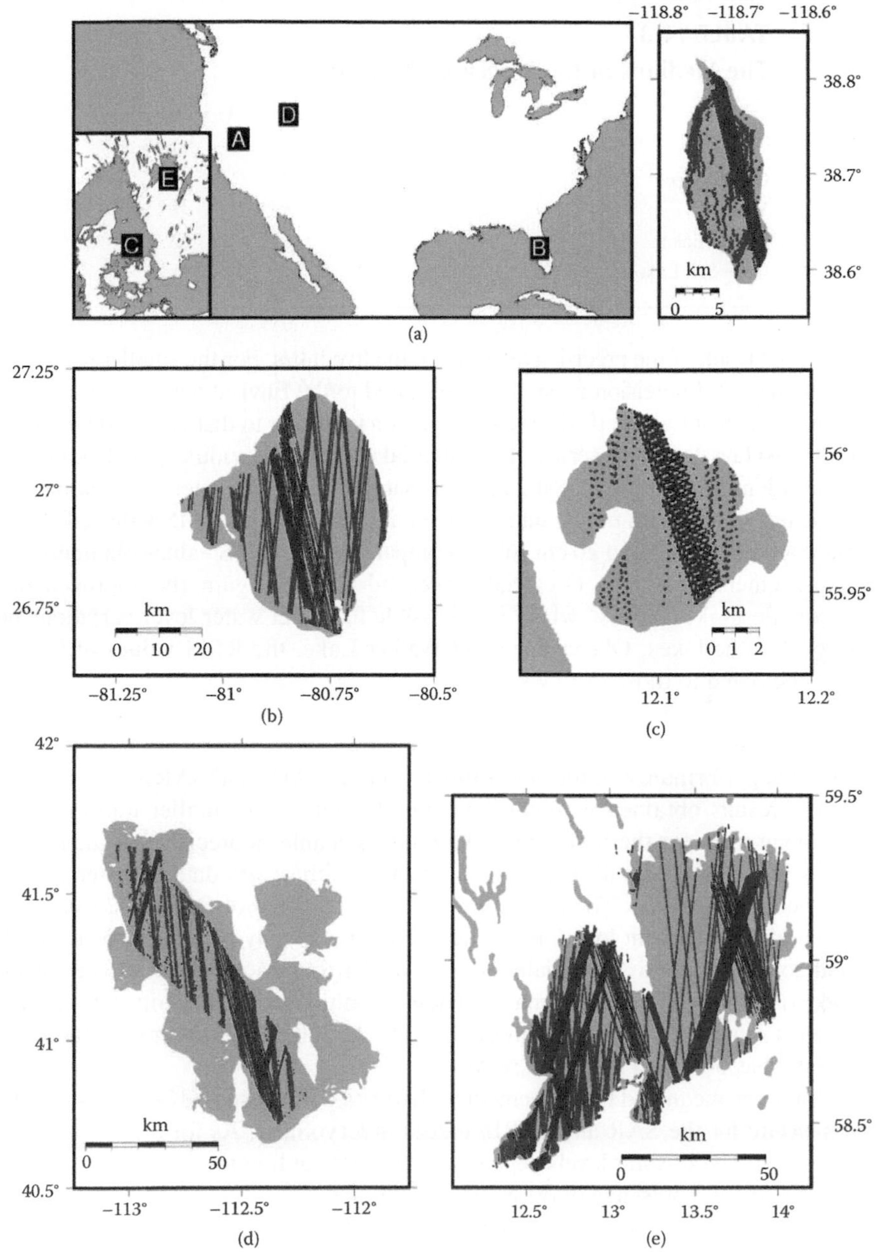

FIGURE 14.18 Study area: (a) Walker Lake, (b) Lake Okeechobee, (c) Arresø, (d) Great Salt Lake, and (e) Vänern. The ground location of CryoSat-2 and Envisat tracks are indicated with blue and red, respectively.

with time and an observation model that describes the distribution of the data (Nielsen et al. 2015). The methodology to derive times series has been collected in an "R" package, "tsHydro," which is freely available from https://github.com/cavios/tshydro.

14.4.1.2.4 Results

Evaluation of the CryoSat-2 SAR and SARIn modes is performed by computation of precision and agreements with *in situ* data. Here, the precision is defined as the standard deviation of the along-track mean value. Hence, when several tracks are considered, the precision shows some spread around the median.

TABLE 14.3
The Medians of the Precision Distribution

Lake	CryoSat-2 (cm)	Envisat (cm)
Arresø	2	7
Okeechobee	1	1
Vänern	1	1
Walker Lake	4	7
Great Salt Lake	1	1

Table 14.3 shows the result of the precision analysis for the five lakes. For the smaller lakes, Arresø and Walker Lake, an improved precision is observed compared to the Envisat results. For the larger lakes, Okeechobee, the Great Salt Lake and Vänern, the precision is similar to that obtained with Envisat.

Figure 14.19 displays the time series of the five lakes. The individual panels contain the raw retracked observation from both CryoSat-2 and Envisat, the modeled water level estimates, and the gauge data. The agreement with *in situ* data is shown in Table 14.4 as RMS values. For CryoSat-2, the RMS values are between 2 and 20 cm. When compared to the RMS values obtained for Envisat, there is an improvement for Arresø, Great Salt Lake, and Vänern. Again, the improvement is much better for the smallest lake, Arresø, where it is possible to detect water level variations below the decimeter level. For the lakes, Okeechobee and Walker Lake, the RMS values of CryoSat-2 are slightly higher compared to those of Envisat.

14.4.1.2.5 Discussion

In this analysis, the performance of the new altimetry modes, SAR and SARIn, has been evaluated and compared to results obtained with conventional altimetry. The smaller footprint of approximately 300 m along-track for the SAR and SARIn modes enables a precision of the mean of just a few centimeters, even for the small lakes. The agreement with *in situ* data is better or comparable to results obtained with Envisat. The long repeat period (369 days) of CryoSat-2 causes the tracks to cross a given lake at different locations. Hence, some tracks may be located near the shoreline, where the land contamination is potentially larger. The drifting track pattern also entails potential residual geoid error in water level estimates. These conditions may explain why the agreement with *in situ* data is sometimes better for Envisat. All in all, with the new altimetry modes SAR and SARIn, CryoSat-2 has provided improved results.

The precision of the mean and the agreement with *in situ* data presented here should not be taken as a general measure for the SAR and SARIn modes of CryoSat-2. As for conventional altimetry, the quality of the estimated water levels depends on the surrounding topography, the complexity of the lake shoreline, and the presence of off-nadir waterbodies.

Snagging, which is a well-known problem (see Section 14.3.6), has to a large degree been solved with the SARIn mode, where it is possible to correct for the off-nadir range error. However, the problem of snagging is still present in the SAR mode data, which is also the altimeter mode of Sentinel-3. Hence, erroneous observations due to off-nadir signal will still be present. However, the recently developed SAR mode retracker MWaPP (Villadsen et al. 2016) has demonstrated promising results with respect to snagging.

The water level of rivers and lakes is an important climatic parameter because it represents the hydrological balance of the surrounding area. Closed-basin lakes are especially sensitive to changes in precipitation and evaporation and therefore serve as markers of the regional climate. The Global Climate Observing System (GCOS) has proposed a prioritized list of almost 80 lakes that need to be monitored to assess the state of the global and regional climate. A part of these lakes is monitored using *in situ* measurements, but for the majority, satellite altimetry in synergy with optical or SAR images provide water height, extent, and volume of these lakes (Cretaux et al. 2016).

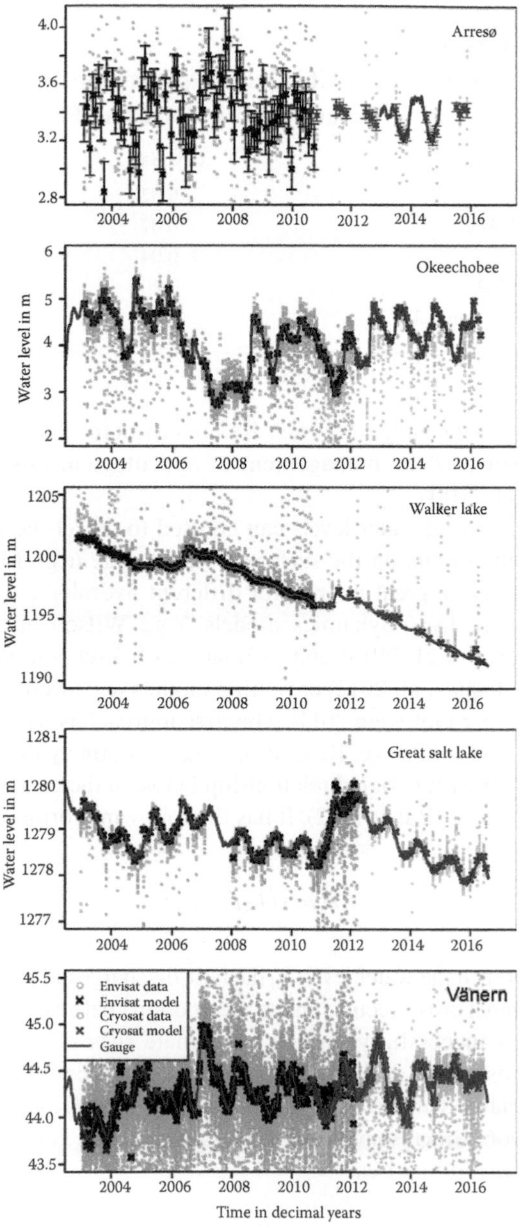

FIGURE 14.19 Time series of water levels for Arresø, Okeechobee, Vänern, Great Salt Lake, and Walker Lake. The individual time series display gauges data (red), modeled water levels based on CryoSat-2 (blue), and Envisat (black), and the raw retracked along-track water level measurements of CryoSat-2 (light blue) and Envisat (gray). Copyright Villadsen 2016, Satellite altimetry for land hydrology, DTU, Research Ph.D. thesis – Annual report

14.4.2 Use of Satellite Altimetry over Rivers

Altimetry also allows long-term observations of river level variations (e.g., Birkett 1998; Frappart et al. 2006a; Silva et al. 2010; Biancamaria et al. 2011; Michailovsky et al. 2012), especially when combining measurements from several missions. This technique can now be used as an external and independent source of information to monitor ungauged basins and

TABLE 14.4
Differences with In Situ Data, Expressed in Terms of RMS (in meters) Values for CryoSat-2 and Envisat

Lake	CryoSat-2	Envisat
Arresø	0.02	
Okeechobee	0.07	0.04
Vänern	0.04	0.09
Walker Lake	0.20	0.13
The Great Salt Lake	0.07	0.09

to cross-check existing *in situ* gauge records. Radar altimetry provides an independent data set that can be used to support the management of hydrological observation networks in the Amazon Basin (Silva et al. 2014).

Time series of altimetry-based water levels can be used to detect the signature of climate variability and extreme climatic events on the surface water storage (e.g., Maheu et al. 2003; Becker et al. 2014). They also present a great interest for different hydrological applications that include calibration and validation of hydrodynamics models (e.g., Wilson et al. 2007; Getirana 2010; Yamazaki et al. 2012; Paiva et al. 2013) and estimation of discharge using either rating curves (e.g., Kouraev et al. 2004; Papa et al. 2012) or routing models (e.g., León et al. 2006; Hossain et al. 2014; Michailovsky and Bauer-Gottwein 2014). The first approach is based on the use of a relationship between water stage and discharge. River discharge is routinely determined from water level *in situ* measurements through a functional relationship between the two quantities known as stage-discharge rating or rating curve (Rantz 1982). It has the following form:

$$Q(t) = \alpha\left(h(t) - h_0\right)^{\beta} \tag{14.16}$$

where Q is the river discharge, h the water level, h_0 the null-discharge elevation, and α and β are related to the geometry of the channel cross-section and to the friction coefficient modulating the discharge.

This approach has been successfully applied to estimate river discharge using altimetry-based water levels when it was possible to derive the rating curve over a common period of observation. It allows deriving river discharge with an accuracy better than 20% (Kouraev et al. 2004; Leon et al. 2006; Zakharova et al. 2006; Birkinshaw et al. 2010; Getirana and Peters-Lidard 2012). The disadvantage of this method is that it requires:

1. A common period of availability between flow data and water height data.
2. The two data sets need to be collected at a sufficiently short interval in order to be able to make the hypothesis that there was no major hydrological change (e.g., confluence with a tributary, phase difference between time series, etc.).

Leon et al. (2006) raised the issue of proximity in a calculation of rating curves in the Negro River Basin. This basin is one of the Amazon's main tributaries, but its upstream section is rarely gauged. In Leon et al. (2006), flow was routed applying a hydrodynamic model (a simplified version of Mushkingum Cunge, in this case, Cunge 1969) using all the possible flow data available upstream and downstream of the section being studied.

A record at each virtual station can then be extracted from the output of the model and the rating curve is calculated between the altimetric series and the series of flow models (Figure 14.20). The second approach consists of calculating flow using a rain-discharge model. This method estimates

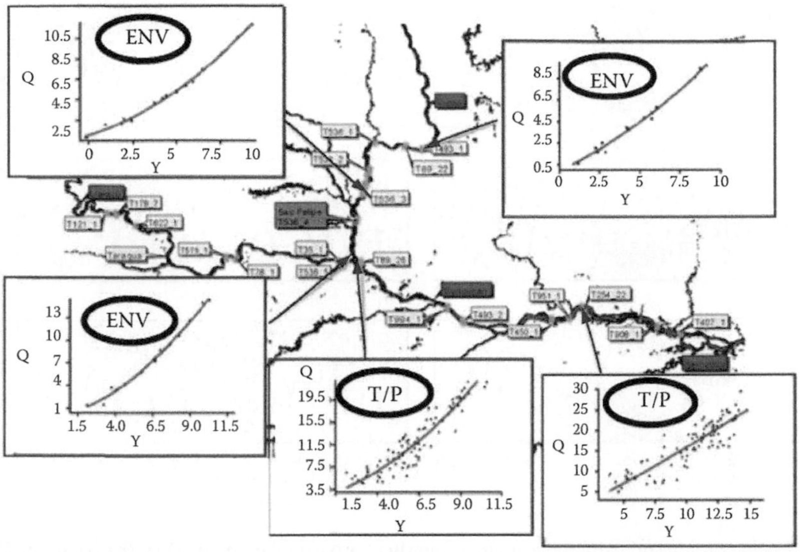

FIGURE 14.20 Rating curve on the Rio Negro using Q flows (× 10^4 m^3s) obtained through the spread of flows between measurement sites and a set of heights taken at the T/P and Envisat virtual stations. Heights are converted into depth Y (m) by subtracting from a reference depth Y_0 so that $Q(Y_0) = 0$. Here, we can note the increase in quality among the three Envisat (ENV) series and the two T/P series—more populated but significantly noisier. Copyright, Calmant et al. 2017, Principes de l'altimétrie satellitaire radar pour les applications continentales, ISTE Edition, Télédétection pour l'observation des surfaces continentales

the percentage of rain captured by the vegetation, the percentage that evaporates, the percentage that seeps into the ground, and the percentage that reaches the river through streams or underground flow. The first study of this type was carried out by Getirana et al. (2009) in Amazonian sub-basins. The methodology was then extended to cover the whole Amazon Basin (Getirana and Peters-Lidard 2012; Paris et al. 2016). These two studies are very similar in terms of the data used, the flow model, and the database from virtual stations (Silva et al. 2010). The advantage of the methodology put forward by Paris et al. (2016) is that in addition to the flow itself, it provides an uncertainty on the value based on uncertainties in altimetry data, in flow models, and the quality of the adjustment of the relationship to the height-flow pairs.

Based on a high density of altimetry virtual stations (approximately 600) and some assumptions on the interactions between surface and ground waters, the topography of the groundwater table was estimated during a low water period in the alluvial plain of central Amazon (Pfeffer et al. 2014). These first maps offer a way to monitor changes in groundwater. Following the drought in 2005, Pfeffer et al. (2014) observed an abrupt drop in low water level in most of the study zone. Later on, the water level gradually rose from north to south and returned to its average value between 2007 and 2008. This result suggests an important "memory effect" in the groundwater (Figure 14.21).

The synergy between satellite altimetry and imagery allows the development of new applications of remote sensing for land hydrology (as presented in Section 14.4.4: Use of altimetry over floodplains). Measurements of river velocities were derived from multispectral images coupled with altimetry-based water levels allow the determination of river discharges (e.g., Tarpanelli et al. 2015).

14.4.3 Use of Satellite Altimetry over Floodplains

Recently, some efforts have been undertaken to quantify the surface freshwater storage in floodplains and wetlands and its variations at seasonal to interannual time scales, using multiple satellite observations. In general, the approach combines satellite-derived surface water extent observations

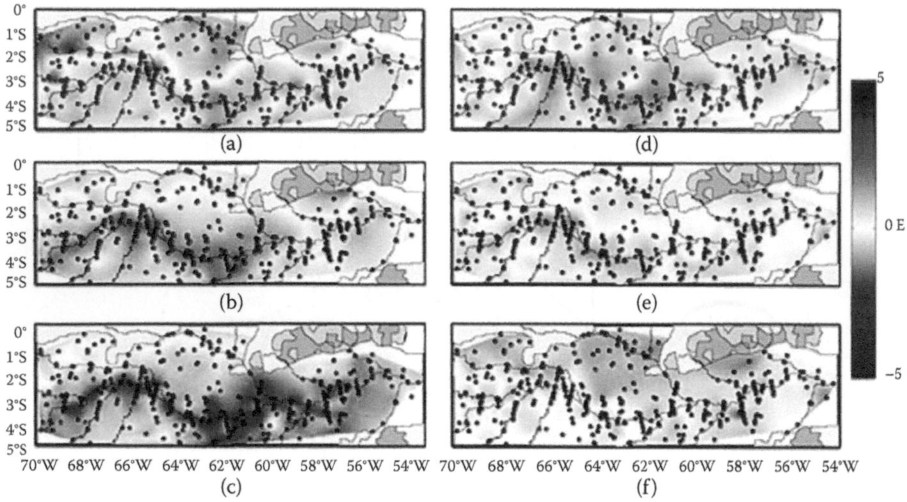

FIGURE 14.21 Deviation (m) from the average groundwater base level in (a) 2003, (b) 2004, (c) 2005, (d) 2006, (e) 2007, and (f) 2008. (From Pfeffer, J., et al., *Geophys. Res. Lett.*, 41, 1981–1987, 2014.) The locations of the 593 Envisat virtual stations used are materialized with black dots.

and radar altimeter-derived height variation of rivers, wetlands, and flood inundations. As a pioneer work, Frappart et al. (2006b) combined ERS-2, Envisat, and T/P satellite water level time series with surface water extent derived from satellite imagery data from the SPOT-4 Vegetation instrument to estimate surface water volume change in the Mekong River. However, the use of visible/near-infrared observations to detect the variations in surface water limits the development of the methodology to regions with low cloud cover and/or low vegetation cover.

A more robust methodology was then derived using a multi-satellite technique (the Global Inundation Extent from Multi-Satellite, called hereafter GIEMS; Prigent et al. 2007, 2016; Papa et al. 2010a) that captures the extent of episodic and seasonal inundation, wetlands, rivers, lakes, and irrigated agriculture at the global scale. The technique uses a complementary suite of satellite observations covering a large wavelength range, mainly based on passive microwave emissivities, which offer the ability to monitor the variations of water extent under clouds and under densely vegetated areas (Prigent et al. 2007; Papa et al. 2010a).

The combination of GIEMS with radar altimeter-derived height variation of rivers, wetlands, and flood inundations to estimate surface water volume change was first developed and applied over the Rio Negro, a sub-basin of the Amazon (Frappart et al. 2008, 2011), and was also tested with success over a boreal environment in the Ob River basin (Frappart et al. 2010). More recently, it was developed over the Orinoco floodplains (Frappart et al. 2015a) or to estimate the amount of freshwater store in the Ganges-Brahmaputra floodplains, rivers, and wetlands (Papa et al. 2015).

Here, we present in more detail the use of satellite altimetry over floodplains and wetlands along with satellite-derived estimates of surface water extent as illustrated in Frappart et al. (2012). Using continuous water level observations derived from Envisat radar altimeter between 2003 and 2007 along with GIEMS observations, Frappart et al. (2012) provides a monthly map of water level for the entire Amazon Basin along with variations of surface water storage, highlighting the exceptional drought of 2005.

For the very first time, a continuous mapping of surface water levels and surface water volumes, as well as their temporal dynamics at interannual time scale, were presented for the Amazon River, the largest drainage basin on Earth. First, monthly surface water level maps are obtained by combining GIEMS surface water extent maps (Prigent et al. 2007, 2012, 15.22a; Papa et al. 2010b) with

534 altimetry-derived water levels in the Amazon Basin (Silva et al. 2012). The location of Envisat RA-2 altimetry stations is shown in Figure 14.22b.

Using a bilinear interpolation (Frappart et al. 2011), monthly maps of surface water levels with a spatial resolution of 0.25° and referenced to EGM2008 geoid are obtained for the period 2003–2007 when all data sets overlapped. The error on these estimates is lower than 10% (Frappart et al. 2008, 2011). A map of minimum water levels was estimated for the entire observation period using a hypsometric approach to take into account the difference in altitude between the river and the floodplain.

Focusing on the signature of the 2005 drought on Amazon surface water, the map of anomaly of minimum water levels for 2005 (Figure 14.23) shows that the whole wetland complex of the Central Amazon exhibits large negative values, with the greatest anomalies registered for the Purus

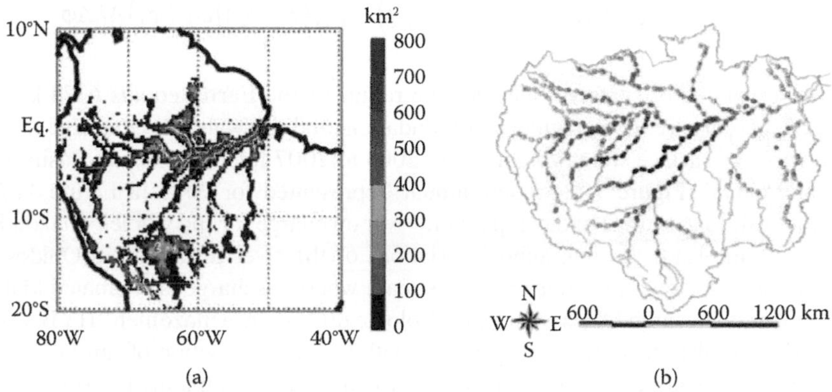

FIGURE 14.22 (a) Spatial distribution over the Amazon Basin of the annual maximum surface water extent averaged over 1993–2007, derived from the global inundation extent from multi-satellites (GIEMS); (b) the location of Envisat RA-2 altimetry stations over the Amazon.

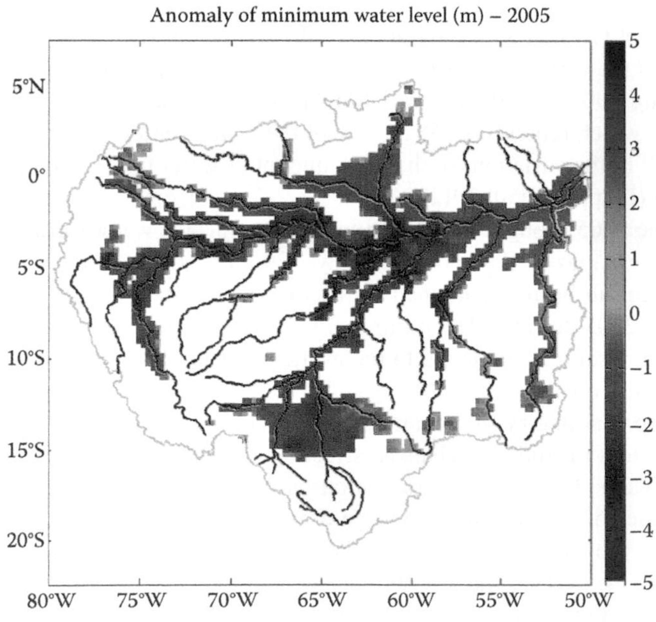

FIGURE 14.23 Map of the anomaly of water level for 2005 over the Amazon (2003–2007 reference period). Copyright, Frappart et al. 2012.

(64.9°–61°W and 2°–4.5°S), Madeira (between 55.67°–59.9°W and 1.25°–5.25°S), and Mamiraua (between 64.67°–67.4°W and 1.4–3.1°S) wetlands.

These minima derived from radar altimetry are consistent with anomalies (computed on longer time periods) of levels estimated from *in situ* gauge records: –2.4 m at Tabatinga (69.9°W, 4.25°S), −4.8 m in Iquitos (72.28°W, 3.43°S), between 2 and 5 m on several locations along the Amazonas (Peru) and its major tributaries, and along the Solimões and its southern tributaries, –4 m at Manaus (60.04°W, 3.15°S) at the mouth of the Negro River.

Second, surface water volume variations for the Amazon River can also be estimated using the surface water levels maps. At basin scale, the time-variations of surface water volume are simply computed as (Frappart et al. 2011):

$$V_{SW}(t) = R_e^2 \sum_{j \in S} P(\lambda_j, \varphi_j, t)\left(h(\lambda_j, \varphi_j, t) - h_{min}(\lambda_j, \varphi_j)\right)\cos(\varphi_j)\Delta\lambda\Delta\varphi \tag{14.17}$$

where V_{SW} is the volume of surface water; R_e (the radius of the Earth) equals 6378 km; $P(\lambda_j,\varphi_j,t)$, $h(\lambda_j,\varphi_j,t)$, and $h_{min}(\lambda_j,\varphi_j)$ are the percentage of inundation, and the water level at time t.

The time series of surface water volume from 2003 to 2007 for the Amazon Basin was decomposed into interannual (Figure 14.24a) and annual (represented for 2005 in Figure 14.24b) terms using a 13-month sliding average and compared to river discharge for the whole Amazon Basin. The surface water volume leads the interannual variations of the river discharge in Obidos (55.68°W, 1.92°S), the last station along the Amazon main stream where discharge is estimated (data obtained from Environmental Research Observatory), Hydrologie du Bassin Amazonien (HYBAM) R = 0.93 with R the linear correlation coefficient with a 1-month lag. The reduction of rainfall over southern Amazonia since 2002 caused a decrease in water stored in the floodplains up to the minimum of 2005, also observed on stream flow. The annual cycle of surface water storage for 2005 was close to or above the mean from February to June 2005, peaking in May with a value around $+\sigma$ (one standard deviation or STD). Then, it became significantly below the mean (values lower than $-\sigma$) from July to December (Figure 14.24b). These results are also in good agreement with what was observed in river discharge at Obidos.

The combination of GIEMS with Envisat-derived water levels also helps to provide the first pluriannual estimates of the variations of surface water storage in a large basin at a monthly time scale. It reveals that during 2003 to 2007, the variations in surface water reservoirs varied from 800 to 1,000 km^3 per year, which represents 155 to 20% of the water volume that flew out of the Amazon Basin and about half of the variations of the total amount of water in the Amazon Basin as detected using GRACE data (Frappart et al. 2012).

The impact of the 2005 drought was quantified for the surface water storage and the total water storage for the whole Amazon Basin (respectively, 129 and 245 km^3 below the 2003–2007 average), and for the three sub-basins mentioned earlier for which different hydrological behaviors were observed during the 2005 drought. The minimum volume of water stored in the Amazon was 71% lower for the surface reservoir compared to the average during 2003–2007 and 29% lower for the total hydrological reservoirs.

In conclusion, radar altimetry in combination with satellite-derived estimates of surface water extent is a powerful tool to study the changes affecting the hydrological cycle in large river basins covered with floodplains. It is a newly observation-based technique that helps to better understand the complex dynamics of surface water in large drainage basins (i.e., backwater effects, Amazon flood-pulse linked to the strong seasonality of the rainfall, or time residence of water in the floodplains). For instance, the surface water level maps give unique and valuable spatial information on the time evolution of floodplain reservoirs during the hydrological cycle in response to rainfall forcing caused by interannual and long-term variability. They permit direct identification of the regions most severely affected by exceptionally low water levels during extreme events such as droughts (or major floods).

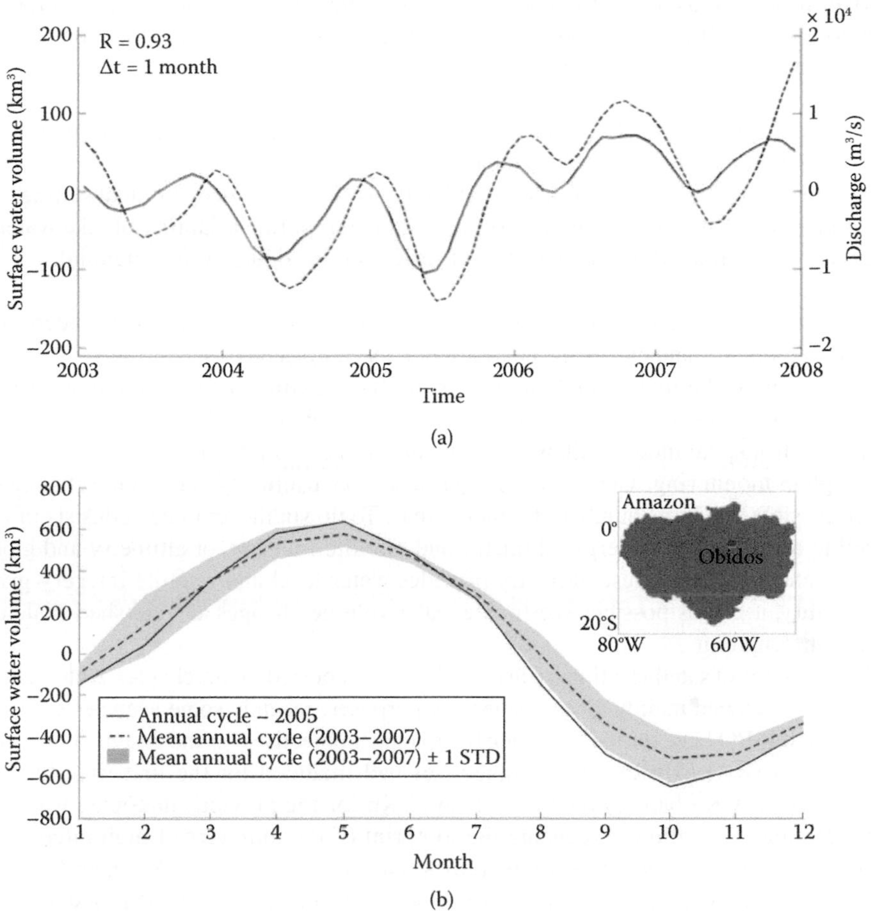

FIGURE 14.24 (a) Interannual variations of surface water volume of the Amazon (black) and discharge at Obidos (dotted blue) between 2003 and 2007. (b) Annual cycle of surface water volume of the Amazon for 2005 (blue) and average (dotted black) ± standard (gray area). Copyright, Frappart et al. 2012.

The estimated spatial and temporal patterns of surface water storage are also key parameters as they can be used to separate the individual contribution of GRACE-derived total water storage, such as isolating the variations of subsurface water storage (groundwater plus soil moisture). For instance, over the Amazon, Frappart et al. (2017) disaggregate for the first time the GRACE-derived total water storage over the Amazon Basin for the period of 2003–2007. Bringing together all these hydrological variables with estimates of river discharge, rainfall, and surface evaporation will improve our knowledge of different water storage components and their contributions toward the terrestrial water budget over the Amazon Basin.

These new and unique surface water storage data sets will soon be developed on the global scale and will play a key role in the validation of future hydrology-oriented satellite missions such as the SWOT mission dedicated to surface hydrology.

14.5 CONCLUSIONS AND PERSPECTIVES

As seen earlier through some cases studies, satellite altimetry has allowed the addressing of a large number of questions in hydrology. Some issues may be of a scientific nature, such the impact of climate change on continental waters. Some may have societal applications, such as the determination of large river discharge as a tool for a more efficient water management or for the monitoring of the water storage in artificial reservoirs. Concerning the questions related to the monitoring of lake water storage, it is clear from past literature that:

- Lakes are responding to climate change (at seasonal, interannual, long-term timescales) in their direct vicinity as well as at a global scale.
- They present a large variety of behaviors that depend on their geographical setting.
- In contrast to what can be observed over rivers, the temporal variability of lake water level is generally dominated by a long temporal scale—decadal and even centennial.

In order to analyze the cause and the consequences of lake level variability, we need to perform long-term measurements and ensure the continuity of space missions.

Concerning rivers, the most useful variable is discharge, which is not directly accessible from space. However, as has been shown in some examples, a combination of *in situ* data and use of rating curves, or of hydrological models, allows deriving discharge along rivers.

For floodplain monitoring, satellite altimetry is also very useful. Water volume change is for the first time accessible in a large number of remote areas. To do so, the approach consists of using different satellite techniques in synergy: altimetry and satellite imagery, or altimetry and global scale database of water extent. Because altimetry provides water level and satellite imagery gives water extent variability, it is thus possible to estimate water volume changes at watershed scale using the two types of information.

Many applications of satellite altimetry in hydrology are reported in this chapter. Although this technique has been developed mainly for oceanography purposes, recently some agencies such as CNES, NASA, ESA, and ISRO have started to include hydrological objectives for the design of new altimeters.

In 2013, the SARAL/AltiKa mission objectives had included for the first time the survey of continental waters. A Ka-band altimeter (instead of Ku for the previous missions) was carried on board this satellite. This allowed reducing the footprint of the altimeter of high interest for small waterbodies. It has been shown by several studies that this choice of the Ka band has drastically reduced errors when compared to *in situ* measurements (Arsen et al. 2015; Schwatke et al. 2015).

The Sentinel-3A mission has also been designed to monitor continental waters. It carries a SAR altimeter similar to the one on board CryoSat-2. The decision to operate in SAR mode over the land was actively supported by the user community relying on the very promising results of the CryoSat-2 SAR mode. The gain in term of footprint is two orders of magnitude with respect to an LRM altimeter. Moreover the signal to noise ratio was also improved with respect to LRM. One of the characteristics of the SAR mode is indeed to allow focusing the measurements on the reflecting target, much more precisely than in LRM, therefore eliminating many polluted sources generally affecting small waterbodies. In the past, many potential waterbodies were not observed because the LRM measurements were too noisy or even not acquired by the tracking function of the altimeter.

Moreover, the orbit of Sentinel-3 is a fully new one. Although this alters the possibility of having a long-term monitoring in the unique orbit of the ERS-2, Envisat, and SARAL/AltiKa type missions, this new 27-day repeat orbit allows making a survey of new lakes, reservoirs, and virtual stations along rivers. Sentinel-3A will be complemented in 2018 by Sentinel-3B, which will be placed in an interleave orbit, allowing to double the coverage of the Earth's surface. These two missions, along with Sentinel-1 and Sentinel-2, built and launched by ESA, are the first of an ambitious European program, Copernicus, to maintain an operational satellite constellation for Earth observation over several decades.

One of the main evolutions that are expected in space hydrology comes from the SWOT mission. The requirements of this mission are to observe all rivers wider than 100 m and all lakes larger than 250 m by 250 m on a repeat cycle of 21 days. Due to the SWOT orbital configuration, in fact, the majority of lakes and rivers will be observed twice or more time during the 1-day cycle. The whole continental surface will be covered because the inclination of the orbit will be 77.6°. The KaRIN instrument will also measure water area and discriminate them from non-water surfaces. This will constitute a unique instrument for three-dimensional mapping of floodplains, lakes, and rivers worldwide. The required precision is 10 cm for surfaces of 1 km^2, 20 cm for surfaces of 250 m by 250 m, and 45 cm for surfaces of 1 ha. It is required to estimate discharge every 10 km for all rivers wider than 100 m with a slope expected precision of 1.7 cm/km. Thanks to this instrument, two components of the global water cycle will be estimated: the global discharge of rivers to the ocean and the total water storage changes in lakes and floodplains.

Some key hydrological questions are driving the SWOT mission development:

- What are the spatiotemporal scales of hydrological processes controlling the distribution of Earth's water storage and their dynamic through the mass transport cross the continents?
- What are the impacts of human activities on continental waters, for example, through dams' constructions or water intake for irrigation?
- What is the sensitivity at different scales (regional to global) of continental water mass distribution to climate, to the presence of wetland, or in the case of extreme droughts?
- How can water extent data be assimilated in models describing floodplains' hydrological processes?
- What is of interest and how can the SWOT data be used in combination with other missions such as Soil Moisture Active Passive (SMAP), GRACE Follow-On (GRACE-FO), Global Precipitation Measurement (GPM), Sentinel-6, and so on, for the quantification of the global water cycle and for closing the continental water budget?
- How can the SWOT data, along with historical remote sensing data, be relied upon for establishment of long-term chronicles usable for climate change studies?

Over the last 10 years, several groups worldwide have set up lake and river databases with water height inferred from satellite altimetry. These databases deliver their products in a free mode of access and are very useful in many regions where no *in situ* data are available. The current existing databases are the following:

Hydroweb (http://hydroweb.theia-land.fr) was developed at Legos (France). It produces and delivers river level data from satellite altimetry on more than 1000 virtual stations spread over about 20 river basins worldwide. It also provides lake level, surface extent, and volume variations for about 160 lakes. Some of the products are delivered in NRT.

G-REALM (www.pecad.fas.usda.gov/cropexplorer) has been developed by the United States Department of Agriculture (USDA) and is focused on lake and reservoir level variations using the T/P, Jason-1, Jason-2, and Envisat satellites. It produces and distributes archived and NRT water level parameters on a network of more about 200 lakes.

River & Lake (tethys.eaprs.cse.dmu.ac.uk/RiverLake) has been developed by De Monfort University (DMU, Leicester, UK) for ESA. It processes altimetry data from ERS-2, Envisat, and Jason-2 on lakes and rivers using NRT fully automated data processing.

Hidrosat (hidrosat.ana.gov.br), developed by the Agencia Nacional de Aguas (ANA, Brasilia, Brazil) to process height over rivers, lakes, and reservoirs in South America (essentially over the Amazon Basin) using Envisat and Jason-2 data.

DAHITI (dahiti.dgfi.tum.de) developed by the Deutsches Geodätisches Forschungsinstitut (DFG) of Munich University (Technical University of Munich, Germany). It processes ESA and CNES/NASA missions on lakes and rivers worldwide. It is based on multi-satellite data processing.

AltWater (Altimetry for Inland Water, http://altwater.dtu.space/) developed at DTU Space, National Space Institute, in Denmark. It currently provides water level time series for a few dozens of lakes using CryoSat-2 data.

Considering the variety of methodologies developed by the different teams involved in the databases listed to produce time series of water level on lakes and rivers, the products may differ significantly from one product to another one. Some comparison of the products of these databases with *in situ* measurements can be found in Schwatke et al. (2015) and Ričko et al. (2012).

REFERENCES

Abulaitijiang, A., O. B. Andersen, and L. Stenseng. 2015. Coastal sea level from inland CryoSat-2 interferometric SAR altimetry. *Geophys. Res. Lett.* 42, 1841–1847. http://dx.doi.org/10.1002/2015GL063131.

Adrian, R., C.M. O'Reilly, H. Zagarese, et al. 2009. Lakes as sentinels of climate change. *Limnol. Oceanogr.* 54(6), 2283–229. http://dx.doi.org/10.4319/lo.2009.54.6_part_2.2283.

Alsdorf, D.E., C.M. Birkett, T. Dunne, J. Melack, and L. Hess. 2001. Water level changes in a large Amazon lake measured with spaceborne radar interferometry and altimetry. *Geophys. Res. Lett.* 28(14), 2671–2674. http://dx.doi.org/10.1029/2001GL012962.

Alsdorf, D.E, and D.P. Lettenmaier. 2003. Tracking fresh water from space. *Science* 301, 1491–1494. http://dx.doi.org/10.1126/science.1089802.

Alsdorf, D.E., E. Rodriguez, and D.P. Lettenmeier. 2007. Measuring surface water from space. *Rev. Geophys.* 45, RG2002. http://dx.doi.org/10.1029/2006RG000197.

Armitage, T.W., and M.W. Davidson. 2014. Using the interferometric capabilities of the ESA CryoSat-2 mission to improve the accuracy of sea ice freeboard retrievals. *IEEE Trans. Geosci. Remote Sens.* 52, 529–536. http://dx.doi.org/10.1109/TGRS.2013.2242082.

Arsen, A., J.F. Cretaux, and R. Abarca-Del-Rio. 2015. Use of SARAL/AltiKa over mountainous lakes, intercomparison with Envisat mission *J. of Adv. Space Res.* The SARAL/ALtiKa satellite Altimetry Mission, 38, 534–548. http://dx.doi.org/10.1080/01490419.2014.1002590.

Bates, P.D., J.C. Neal, D. Alsdorf, and G.J.P. Schumann. 2014. Observing global surface water flood dynamics. *Surv. Geophys.* 35(3), 839–852. http://dx.doi.org/10.1007/s10712-013-9269-4.

Becker, M., J.S. Silva, S. Calmant, V. Robinet, and F. Seyler. 2014. Water level fluctuations in the Congo Basin derived from Envisat satellite altimetry. *Remote Sens.* 6(10), 9340–9358. http://dx.doi.org/10.3390/rs6109340.

Bercher, N., and S. Calmant. 2013. A review of CryoSat-2/SIRAL applications for the monitoring of river water levels. Proceedings of the ESA Living Planet Symposium 2013, L. Ouwehand Ed., *ESA special publication SP-722*.

Berry, P.A.M., J.D. Garlick, J.A. Freeman, and E.L. Mathers. 2005. Global inland water monitoring from multi-mission altimetry, *Geophys. Res. Lett.* 32, L16401. http://dx.doi.org/10.1029/2005GL022814.

Biancamaria, S., K.M. Andreadis, M. Durand, et al. 2010. Preliminary characterization of SWOT hydrology error budget and global capabilities. *IEEE JSTARS* Special Issue on Microwave Remote Sensing for Land Hydrology Research and Applications, 3(1), 6–19. http://dx.doi.org/10.1109/JSTARS.2009.2034614.

Biancamaria, S., F. Frappart, A.-S. Leleu, et al. (in revision). 2017. Satellite altimetry water elevations performance over a 200 m wide river: Evaluation over the Garonne River. Adv. *Space Res.* http://dx.doi.org/10.1016/j.asr.2016.10.008.

Biancamaria, S., F. Hossain, and D.P. Lettenmaier. 2011. Forecasting transboundary river water elevations from space. *Geophys. Res. Lett.* 38, L11401. http://dx.doi.org/10.1029/2011GL047290.

Birkett, C.M. 1995. The contribution of TOPEX/POSEIDON to the global monitoring of climatically sensitive lakes. *J. Geophys. Res.* 100(C12), 25179–25204. http://dx.doi.org/10.1029/95JC02125.

Birkett, C.M. 1998. Contribution of the TOPEX NASA Radar Altimeter to the global monitoring of large rivers and wetlands. *Water Resour. Res.* 34(5), 1223–1239. http://dx.doi.org/10.1029/98WR00124.

Birkett, C.M., R. Murtugudde, and T. Allan. 1999. Indian Ocean climate event brings floods to East Africa's lakes and the Sudd Marsh. *Geophys. Res. Lett.* 26, 1031–1034. http://dx.doi.org/10.1029/1999GL900165.

Birkinshaw, S.J., G.M. O'Donneell, P. Moore, C.G. Kilsby, H.J. Fowler, and P.A.M. Berry. 2010. Using satellite altimetry data to augment flow estimation techniques on the Mekong River. *Hydrol. Processes* 24, 3811–3825. http://dx.doi.org/10.1002/hyp.7811.

Boergens, E., D. Dettmering, C. Schwatke, C. and F. Seitz. 2016. Treating the hooking effect in satellite altimetry data: A case study along the Mekong River and its tributaries. *Remote Sens.* 8, 91. http://dx.doi.org/10.3390/rs8020091.

Bonnefond, P., P. Exertier, O. Laurain, et al. 2010. Absolute Calibration of Jason-1 and Jason-2 Altimeters in Corsica during the formation flight phase. *Mar. Geod.* 33(S1), 80–90. http://dx.doi.org/10.1080/01490419.2010.487790.

Bousquet, P., P. Ciais, J.B. Miller, et al. 2006. Contribution of anthropogenic and natural sources to atmospheric methane variability. *Nature* 443, 439–443. http://dx.doi.org/10.1038/nature05132.

Brown, G.S. 1977. The average impulse response of a rough surface and its application. *IEEE Trans. Antenna Propag.* 25(1), 67–73. http://doi.org/10.1109/TAP.1977.1141536.

Calmant, S., J. Santos da Silva, D. Medeiros Moreira, et al. 2012. Detection of Envisat RA2 / ICE-1 retracked radar altimetry bias over the Amazon Basin Rivers using GPS. *Adv. Space Res.* 51(8), 1551–1564. http://dx.doi.org/10.1016/j.asr.2012.07.033.

Cheng, K.-C., C-Y. Kuo, H.-Z. Tseng, et al. 2010. Lake surface height calibration of Jason-1 and Jason-2 over the Great Lakes. *Mar. Geod.* 33(S1), 186–203.

Coastal and Hydrology Altimetry product (PISTACH) handbook, 2009, 64 p.

Coe, M.T., and C.M. Birkett. 2005. Water Resources in the Lake Chad BasIn: Prediction of river discharge and lake height from satellite radar altimetry. *Water Resour. Res.* 40(10). http://dx.doi.org/10.1029/2003WR002543.

Cretaux, J.-F., R. Abarca Del Rio, M. Berge-Nguyen, et al. 2016. Lake volume monitoring from space. *Surv. Geophys.* 37, 269–305. http://dx.doi.org/10.1007/s10712-016-9362-6.

Crétaux, J.-F., M. Bergé-Nguyen, S. Calmant, et al. 2013. Calibration of envisat radar altimeter over Lake Issykkul. *J. Adv. Space Res.* 51(8), 1523–1541. http://dx.doi.org/10.1016/j.asr.2012.06.039.

Cretaux, J.-F., S. Biancamaria, A. Arsen, M. Bergé-Nguyen, and M. Becker. 2015. Global surveys of reservoirs and lakes from satellites and regional application to the Syrdarya river basin. *Environ. Res. Lett.* 10(1), AN: 015002. http://dx.doi.org/10.1088/1748-9326/10/1/015002.

Cretaux, J.F., S. Calmant, V. Romanovski, et al. 2009. An absolute calibration site for radar altimeters in the continental domain: Lake Issykkul in Central Asia. *J. Geod.* 83(8), 723–735. http://dx.doi.org/10.1007/s00190-008-0289-7.

Cretaux, J.F., S. Calmant, V. Romanovski, et al. 2011. Absolute calibration of Jason radar altimeters from GPS kinematic campaigns over Lake Issykkul. *Mar. Geod.* 34(3–4), 291–318. http://dx.doi.org/10.1080/01490419.2011.585110.

Cudlip, W., J.K. Ridley, and C.G. Rapley. 1992. The use of satellite radar altimetry for monitoring wetlands. In: *Remote sensing and global change*. Proceedings of the 16th Annual Conference of Remote Sensing Society, London, UK, pp. 207–216.

Cunge, J.A. 1969. On the subject of a flood propagation computation method (Mushkingum method). *Hydraulic Res.* 7(2), 205–230. http://dx.doi.org/10.1080/00221686909500264.

Davis, C.H. 1997. A robust threshold retracking algorithm for measuring ice-sheet surface elevation change from satellite radar altimeters. *IEEE Trans. Geosci. Remote Sens.* 35, 974–979. https://doi.org/10.1109/36.602540.

Decharme, B., R. Alkama, F. Papa, S. Faroux, H. Douville, and C. Prigent. 2012. Global off-line evaluation of the ISBA-TRIP flood model. *Clim. Dyn.* 38, 7/8, 1389-1412, doi:10.1007/s00382011-1054-9

De Wit, M., and J. Stankiewicz. 2006. Changes in surface water supply across Africa with predicted climate change. *Science* 311(5769), 1917–1921. http://dx.doi.org/10.1126/science.1119929.

Downing, J.A. 2010. Emerging global role of small lakes and ponds: Little things mean a lot. *Limnetica* 29(1), 9–24.

Downing, J.A., Y.T. Prairie, J.J. Cole, et al. 2006. The global abundance and size distribution of lakes, ponds, and impoundments. *Limnol. Oceanogr.* 51, 2388–2397. http://dx.doi.org/10.4319/lo.2006.51.5.2388.

Duan, Z., and W.G.M. Bastiaanssen. 2013. Estimating water volume variations in lakes and reservoirs from four operational satellite altimetry databases and satellite imagery data. *Remote Sens. Env.* 134, 403–16. http://dx.doi.org/10.1016/j.rse.2013.03.010.

Dudgeon, D., A.H. Arthington, M.O. Gessner, et al. 2006. Freshwater biodiversity: importance, threats, status and conservation challenges. *Biol. Rev.* 81, 163–182. http://dx.doi.org/10.1017/S1464793105006950.

Dziewonski, A.M., and D.L. Anderson. 1981. Preliminary reference earth model. *Phys. Earth Planet. Inter.* 25(4), 297–356. http://dx.doi.org/10.1016/0031-9201(81)90046-7.

Emery, C., S. Biancamaria, A. Boone, et al. 2016. Temporal variance-based sensitivity analysis of the river-routing component of the large-scale hydrological model ISBA-TRIP: Application on the Amazon Basin. *J. Hydrometeorol.* http://dx.doi.org/10.1175/JHM-D-16-0050.1.

Femenias, P., S. Baker, D.J. Brockley, et al. 2015. Reprocessing of the ERS-1 and ERS-2 altimetry missions. The REAPER project. Abstracts Book. Ocean Surface Topography Science Team Meeting, Lake Constance, Germany, 2014.

Frappart, F., S. Calmant, M. Cauhopé, F. Seyler, and A. Cazenave. 2006a. Preliminary results of ENVISAT RA-2 derived water levels validation over the Amazon basin. *Remote Sens. Environ.* 100(2), 252–264. http://dx.doi.org/10.1016/j.rse.2005.10.027.

Frappart, F., K. Dominh, J. Lhermitte, G. Ramilllien, A. Cazenave, A. and T. LeToan. 2006b. Water volume change in the lower Mekong Basin from satellite altimetry and imagery data. *Geophys. J. Int.* 167, 570–584. http://dx.doi.org/10.1111/j.1365-246X.2006.03184.x.

Frappart, F., B. Legrésy, F. Niño, et al. 2016. An ERS-2 altimetry reprocessing compatible with ENVISAT for long-term land and ice sheets studies. *Remote Sens. Environ.* 184, 558–581. http://dx.doi.org/10.1016/j.rse.2016.07.037.

Frappart, F., F. Papa, A. Guentner, et al. 2017. The spatio-temporal variability of groundwater storage in the Amazon River Basin. *Adv. Water. Resour.* Submitted.

Frappart, F., F. Papa, A. Guntner, et al. 2010. Interannual variations of the terrestrial water storage in the lower Ob' Basin from a multisatellite approach. *Hydrol. Earth Syst. Sci.* 14(12), 2443–2453. http://dx.doi.org/105194/hess-14-2443-2010.

Frappart, F., F. Papa, A. Güntner, et al. 2011. Satellite-based estimates of groundwater storage variations in large drainage basins with extensive floodplains. *Remote Sens. Environ.* 115, 1588–1594. http://dx.doi.org/10.1016/j.rse.2011.02.003.

Frappart, F., F. Papa, Y. Malbeteau, et al. 2015a. Surface freshwater storage variations in the Orinoco floodplains using multi-satellite observations. *Remote Sens.* 7(1), 89–110. http://dx.doi.org/10.3390/rs70100089.

Frappart F., F. Papa, V. Marieu, et al. 2015b. Preliminary assessment of SARAL/AltiKa observations over the Ganges-Brahmaputra and Irrawaddy Rivers. *Mar. Geod.* 38(Supl), 568–580. http://dx.doi.org/10.1080/01490419.2014.990591.

Frappart, F., F. Papa, J. Santos da Silva, et al. 2012. Surface freshwater storage and dynamics in the Amazon basin during the 2005 exceptional drought. *Environ. Res. Lett.* 7, 044010 (7 p). http://dx.doi.org/10.1088/1748-9326/7/4/044010.

Gao, H., C.M. Birkett, and D.P. Lettenmeir. 2012. Global monitoring of large reservoir storage from satellite remote sensing. *Water Resour. Res.* 48, W09504. http://dx.doi.org/10.1029/2012WR012063.

Gaspar, P., S. Labroue, F. Ogor, G. Lafitte, L. Marchal, and M. Rafanel. 2002. Improving nonparametric estimates of the sea state bias in radar altimeter measurements of sea level. *J. Atmos. & Oceanic Technol.* 19, 1690–1707. http://dx.doi.org/10.1175/1520-0426(2002)019<1690:INEOTS>2.0.CO;2.

GCOS. 2011. Systematic observation requirements for satellite-based data products for climate (2011 update)—Supplemental details to the satellite-based component of the "Implementation plan for the global observing system for climate in support of the UNFCCC (2010 update)." GCOS-154 (WMO, December 2011).

Getirana, A. 2010. Integrating spatial altimetry data into the automatic calibration of hydrological models. *J. Hydrol.* 387(3–4), 244–255. http://dx.doi.org/10.1016/j.jhydrol.2010.04.013.

Getirana, A.C.V., M-P. Bonnet, S. Calmant, E. Roux, O.C. Rotunno Filho, and J.W. Mansur, 2009. Hydrological monitoring of poorly gauged basins based on rainfall–runoff modeling and spatial altimetry. *J. Hydrol.* 379, 205–219. http://dx.doi.org/10.1016/j.jhydrol.2009.09.049.

Getirana, A.C.V., and C. Peters-Lidard. 2012. Water discharge estimates from large radar altimetry datasets in the Amazon basin. *Hydrol. Earth Syst. Sci. Discuss.* 9, 7591–7611. http://dx.doi.org/10.5194/hessd-9-7591-2012.

Getirana, A.C.V., and C. Peters-Lidard. 2013. Water discharge estimates from large radar altimetry datasets in the Amazon basin. *Hydrol. Earth Syst. Sci.* 17, 923–933. http://dx.doi.org/10.5194/hess-17-923-2013.

Gleditsch, N.P., and H. Hegre. 2000. Shared rivers and interstate conflict. *Polit. Geogr.* 19(8), 971–96. http://dx.doi.org/10.1016/S0962-6298(00)00038-X.

Gommenginger, C., P. Thibaut, L. Fenoglio-Marc, et al. 2011. Retracking altimeter waveforms near the coasts. In Vignudelli, S., Kostianoy, A.G., Cipollini, P., Benveniste, J. (Eds.), *Coastal altimetry.* Springer, Berlin, pp. 61–101.

Haines, B.J., S.D. Desai, and G.H. Born. 2010. The harvest experiment: Calibration of the climate data record from TOPEX/Poseidon, Jason-1 and the Ocean Surface Topography Mission. *Mar. Geod.* 33(S1), 91–113. http://dx.doi.org/10.1080/01490419.2010.491028.

Hall, A., G. Schumann, J. Bamber, P. Bates. and M. Trigg. 2012. Geodetic corrections to Amazon River water level gauges using ICESat altimetry. *WRR* 48, W06602. http://dx.doi.org/10.1029/2011WR10895.

Hossain, F., A.H.M. Siddique-E-Akbor., L.C. Mazumdar, et al. 2014. Proof of concept of an operational altimeter-based forecasting system for transboundary flow. *IEEE J. Sel. Topics Applied Earth Observ. Remote Sens.* 7(2), 587–601. http://dx.doi.org/10.1109/JSTARS.2013.2283402.

Huang, L., J. Liu, Q. Shao, and R. Liu. 2011. Changing inland lakes responding to climate warming in northern Tibetan Plateau. *Clim. Change* 24, 479–502. http://dx.doi.org/10.1007/s10584-011-0032-x.

Huttunen, J.T., J. Alm, A. Liikanen, et al. 2003. Fluxes of methane, carbon dioxide and nitrous oxide in boreal lakes and potential anthropogenic effects on the aquatic greenhouse gas emissions. *Chemosphere* 52(3), 609–621.

Hwang, C., J. Guo, J.X. Deng, H.-Y. Hsu, and Y. Liu, 2006. Coastal gravity anomalies from retracked GEOSAT/GM altimetry: Improvement, limitation and the role of airborne gravity data. *J. Geod. Springer*, 80, 204–216. http://dx.doi.org/10.1007/s00190-006-0052-x.

Hwang, C., M.-F. Peng, J. Ning, J. Luo, and C.-H. Sui. 2005. Lake level variations in China from TOPEX/Poseidon altimetry: data quality assessment and links to precipitation and ENSO. *Geophys. J. Int.* 161, 1–11. http://dx.doi.org/10.1111/j.1365-246X.2005.02518.x.

Jayles, C., J.-P. Chauveau, and F. Rozo. 2010. DORIS/Jason-2: Better than 10 cm on-board orbits available for near-real-time altimetry. *Adv. Space Res.* 46(12), 1497–1512. http://dx.doi.org/10.1016/j.asr.2010.04.030.

Kang, S., Y. Xu, Q. You, W. Flügel, N. Pepin, and T. Yao. 2010. Review of climate and cryospheric change in the Tibetan Plateau. *Environ. Res. Lett.* 5, 015101. http://dx.doi.org/10.1088/1748-9326/5/1/015101.

Kleinherenbrink, M., R.C. Lindenbergh, and P.G. Ditmar. 2015. Monitoring of lake level changes on the Tibetan Plateau and Tian Shan by retracking Cryosat SARIn waveforms. *J. Hydrol.* 521, 119–131. http://dx.doi.org/10.1016/j.jhydrol.2014.11.063.

Koblinsky, C. J., R. T. Clarke, A. C. Brenner, and H. Frey. 1993. Measurement of river level variations with satellite altimetry. *Water Resour. Res.* 29(6), 1839–1848. http://dx.doi.org/10.1029/93WR00542.

Kouraev, A., E.A. Zakharov, O. Samain, N.M. Mognard-Campbell, and A. Cazenave. 2004. Ob' river discharge from Topex/Poseidon satellite altimetry. *Remote Sens. Environ.* 93, 238–245. http://dx.doi.org/10.1016/j.rse.2004.07.007.

Kuo, C.-Y., and H.-C. Kao. 2011. Retracked Jason-2 altimetry over small water bodies: Case study of Bajhang River, Taiwan. *Mar. Geod.* 34(3–4), 382–392. http://dx.doi.org/10.1080/01490419.2011.584830.

Laxon, S. 1994. Sea ice altimeter processing scheme at the EODC. *Int. J. Remote Sens.* 15(4), 915–924. http://dx.doi.org/10.1080/01431169408954124.

Lee, H., C.-K. Shum, Y. Yi, A. Braun, A. and C.-Y. Kuo. 2008. Laurentia crustal motion observed using TOPEX/POSEIDON radar altimetry over land. *J. Geodyn.* Elsevier, 46, 182–193. http://dx.doi.org/10.1016/j.jog.2008.05.001.

Lee, H., C.-K. Shum, K.-H. Tseng, J.-Y. Guo, and C.-Y. Kuo. 2011. Present day lake level variation from Envisat altimetry over the North eastern Qinghai-Tibetan Plateau: Links with precipitation and temperature. *Terr. Atmos. Ocean. Sci.* 22(2). http://dx.doi.org/10.3319/TAO.2010.08.09.01(TibXS).

Legrésy, B., F. Papa, F. Rémy, F., G. Vinay, M. van den Bosch, M., and O.Z Zanife. 2005. ENVISAT radar altimeter measurements over continental surfaces and ice caps using the ICE-2 retracking algorithm. *Remote Sens. Environ.* 95, 150–163. http://dx.doi.org/10.1016/j.rse.2004.11.018.

Lehner, B., and P. Döll. 2004. Development and validation of a global database of lakes, reservoirs and wetlands. *J. Hydrol.* 296(1), 1–22. http://dx.doi.org/10.1016/j.jhydrol.2004.03.028.

Lei, Y., T. Yao, B.W. Bird, K. Yang, J. Zhai, and Y. Sheng. 2013. Coherent lake growth on the central Tibetan Plateau since the 1970s: Characterization and attribution. *J. Hydrol.* 483, 61–67. http://dx.doi.org/10.1016/j.jhydrol.2013.01.003.

León, J.G., S. Calmant, F. Seyler, et al. 2006. Estimation of stage-discharge rating curves and mean water depths from radar altimetry data and hydrological modeling in the upper Negro River basin. *J. Hydrol.* 328(3–4), 481–496. http://dx.doi.org/10.1016/j.jhydrol.2005.12.006.

Leon, J.G., F. Seyler, S. Calmant, M-P. Bonnet, and M. Cauhope. 2006. Hydrological parameter estimation for ungauged basin based on satellite altimeter data and discharge modeling. A simulation for the Caqueta River (Amazonian Basin, Colombia). *Hydrol. Earth Syst. Sci.* 3(5), 3023–3059.

Li, X.Y., H.Y. Xu, Y.L. Sun, D.S. Zhang, and Z.P. Yang. 2007. Lake-level change and water Balance analysis at Lake Qinghai, West China during recent decades. *Water Resour. Manage.* 21, 1505–1516. http://dx.doi.org/10.1007/s11269-006-9096-1.

Liu, J., S. Wang, S. Yu, D. Yang, and L. Zhang. 2009. Climate warming and growth of high-elevation inland lakes on the Tibetan Plateau. *Global Planet. Change* 67, 209–217. http://dx.doi.org/10.1016/j.gloplacha.2009.03.010.

Liu, X.D., and B.D. Chen. 2000. Climatic warming in the Tibetan Plateau during recent decades. *Int. J. Climatol.* 20(14), 1729–1742. http://dx.doi.org/10.1002/1097-0088(20001130)20:14<1729::AID-JOC556>3.0.CO;2-Y.

Maheu, C., A. Cazenave, and C.R. Mechoso. 2003. Water level fluctuations in the Plata Basin (South America) from Topex/Poseidon satellite altimetry. *Geophys. Res. Lett.* 30, 1143. http://dx.doi.org/10.1029/2002GL016033.

Manning, R. 1891. On the flow of water in open channels and pipes. *Trans. Civil Eng. Ireland* 20, 161–207.

Martinez-Benjamin, J.J., M. Martinez-Garcia, S. Gonzalez Lopez, et al. 2004. Ibiza absolute calibration experiment: Survey and preliminary results. *Mar. Geod.* 27(S3–4), 657–681. http://dx.doi.org/10.1080/01490410490883342.

Mason, I.M., M.A.J. Guzkowska, C.G. Rapley, and F.A. Street-Perrot. 1994. The response of lake levels and areas to climate change. *Clim. Change* 27, 161–197. http://dx.doi.org/10.1007/BF01093590.

Medina, C., J. Gomez-Enri, J. Alonso, and P. Villares. 2008. Water level fluctuations derived from Envisat Radar altimetry (RA-2) and in situ measurements in a subtropical water body: Lake Izabal (Guatemala). *RSE*. http://dx.doi.org/10.1016/J.rse.2008.05.001.

Mercier, F., A. Cazenave, and C. Maheu. 2002. Interannual lake level fluctuations in Africa from Topex-Poseidon: Connections with ocean-atmosphere interactions over the Indian Ocean. *Global Planet. Change* 32, 141–163. http://dx.doi.org/10.1016/S0921-8181(01)00139-4.

Mertikas, S.P., R.T. Ioannides, I.N. Tziavos, et al. 2010. Statistical models and latest results in the determination of the absolute bias for the radar altimeters of Jason satellites using the Gavdos facility. *Mar. Geod.* 33(S1), 114–149. http://dx.doi.org/10.1080/01490419.2010.488973.

Michailovsky, C.I., and P. Bauer-Gottwein. 2014. Operational reservoir inflow forecasting with radar altimetry: the Zambezi case study. *Hydrol. Earth Syst. Sci.* 18(3), 997–1007. http://dx.doi.org/10.5194/hess-18-997-2014.

Michailovsky, C.I., S. McEnnis, P.A.M. Berry, R. Smith, and P. Bauer-Gottwein. 2012. River monitoring from satellite radar altimetry in the Zambezi River basin. *Hydrol. Earth Syst. Sci.* 16, 2181–2192. http://dx.doi.org/10.5194/hess-16-2181-2012.

Morris, C.S., and S.G. Gill. 1994a. Variation of Great Lakes water levels derived from GEOSAT altimetry. *Wat. Resour. Res.* 30(4), 1009–1017. http://dx.doi.org/10.1029/94WR00064.

Morris, C.S., and S.K. Gill. 1994b. Evaluation of the TOPEX/POSEIDON altimeter system over the Great Lakes. *J. Geophys. Res.* 99(C12), 24527–24539. http://dx.doi.org/10.1029/94JC01642.

Nielsen, K., L. Stenseng, O.B. Andersen, H. Villadsen, and P. Knudsen. 2015. Validation of CryoSat-2 SAR mode based lake levels. *Remote Sens. Environ.* 171, 162–170. http://dx.doi.org/10.1016/j.rse.2015.10.023.

Paiva, R.C.D., D.C. Buarque, and W. Colischonn, et al. 2013a. Large-scale hydrologic and hydrodynamic modeling of the Amazon River basin. *Water Resour. Res.* 49(3), 1216–1243. http://dx.doi.org/10.1002/wrcr.20067.

Paiva, R.C.D., W. Collischonn, M.-P. Bonnet, et al. 2013b. Assimilating in situ and radar altimetry data into a large-scale hydrologic-hydrodynamic model for streamflow forecast in the Amazon. *Hydrol. Earth Syst. Sci.* 17, 2929–2946. http://dx.doi.org/10.5194/hess-17-2929-2013.

Pandey, R.K., J.-F. Cretaux, M. Bergé-Nguyen, et al. 2014. Water level estimation by remote sensing for 2008 Flooding of the Kosi River. *Int. J. of Remote Sens.* 35(2), 424–440. http://dx.doi.org/10.1080/01431161.2013.870678.

Papa, F., S.K. Bala, R.K. Pandey, F. Durand, A. Rahman, and W. B. Rossow. 2012. Ganga-Brahmaputra river discharge from Jason-2 radar altimetry: An update to the long-term satellite-derived estimates of continental freshwater forcing flux into the Bay of Bengal. *J. Geophys. Res.* 117, C11021. http://dx.doi.org/10.1029/2012JC008158.

Papa, F., F. Durand, W.B. Rossow, A. Rahman, and S.K. Bala. 2010a. Seasonal and interannual variations of the Ganges-Brahmaputra river discharge, 1993–2008 from satellite altimeters. *J. Geophys. Res.* 115, C12013. http://dx.doi.org/10.1029/2009JC006075.

Papa, F., F. Frappart, Y. Malbeteau, et al. 2015. Satellite-derived surface and sub-surface water storage in the Ganges-Brahmaputra River Basin. *J. Hydrol. Regional Stud.* 4, 15–35. http://dx.doi.org/10.1016/j.ejrh.2015.03.004.

Papa, F., C. Prigent, F. Aires, C. Jimenez, W.B. Rossow, and E. Matthews. 2010b. Interannual variability of surface water extent at global scale, 1993–2004. *J. Geophys. Res.* 115, D12111. http://dx.doi.org/10.1029/2009JD012674.

Paris, A., J. Santos da Silva, R. Dias de Paiva, et al. 2016. Global determination of rating curves in the Amazon Basin. *Water Resour. Res.*

Pavlis, N.K., S.A. Holmes, S.C. Kenyon, and J.K. Factor. 2012. The development and evaluation of the Earth Gravitational Model 2008 (EGM2008). *J. Geophys. Res. Solid Earth (1978–2012)*, 117(B4). http://dx.doi.org/10.1029/2011JB008916.

Pfeffer, J., F.F. Seyler, M.-P. Bonnet, et al. 2014. Low-water maps of the groundwater table in the central Amazon by satellite altimetry. *Geophys. Res. Lett.* 41, 1981–1987. http://dx.doi.org/10.1002/2013GL059134.

Phan, V.H., R. Lindenbergh, and M. Menenti. 2011. ICESat derived elevation changes of Tibetan lakes between 2003 and 2009. *Int. J. Appl. Earth Observ. Geoinf. http://dx.doi.org/*10.1016/j.jag.2011.09.015.

Prigent, C., D.P. Lettenmaier, F. Aires, and F. Papa. 2016. Toward a high-resolution monitoring of continental surface water extent and dynamics, at global scale: From GIEMS (Global Inundation Extent from Multi-Satellites) to SWOT (Surface Water and Ocean Topography). *Surv. Geophys.* 37(2), 339–355. http://dx.doi.org/10.1007/s10712-015-9339-x.

Prigent C., F. Papa, F. Aires, G. Jimenez, WB. Rossow, and E. Mattews. 2012. Changes in land surface water dynamics since the 1990s and relation to population pressure. *Geophys. Res. Lett.* 39, L08403. http://dx.doi.org/10.1029/2012/2012GRL051276.

Prigent, C., F. Papa, F. Aires, W.B. Rossow, and E. Matthews. 2007. Global inundation dynamics inferred from multiple satellite observations, 1993–2000. *J. Geophys. Res.* 112, D12107. http://dx.doi.org/10.1029/2006JD007847

Rantz, S.E. 1982. Measurement and computation of streamflow: Volume 2. Computation of discharge. Water Supply Paper 2175, U. S. Geol. Surv., pp. 285–631.

Rast, W., and M. Straskraba. 2000. *Lakes and Reservoirs, Similarities, Differences and Importance. Short Series on Planning and Management of Lakes and Reservoirs, UNEP-IETC (International Environment Technological Center) /ILEC (International Lake Environment Committee Foundation)*, Vol. 1, 24 p. Available at http://www.ilec.or.jp/en/pubs/p2/lake-resvr.

Ričko, M., C.M. Birkett, J.A. Carton, and J.-F. Cretaux. 2012. Intercomparison and validation of continental water level products derived from satellite radar altimetry. *J. Appl. Remote Sens.* 6, Art No.: 061710. http://dx.doi.org/10.1117/1.JRS.6.061710.

Robertson, D.M., and R.A. Ragotzkie. 1990. Changes in the thermal structure of moderate to large sized lakes in response to changes in air temperature. *Aquat. Sci.* 52(4), 360–380. http://dx.doi.org/10.1007/BF00879763.

Rodriguez, E. 2015. Surface Water and Ocean Topography project. Science requirement document, release February 2015, JPL D-61923.

Saastamoinen, J. 1972. Atmospheric correction for the troposphere and stratosphere in radio ranging of satellites. *Geophys. Monogr.* 15, American Geophysical Union, Washington DC. http://dx.doi.org/10.1029/GM015p0247.

Schindler, D.W. 2009. Lakes as sentinels and integrators for the effects of climate change on watersheds, airsheds, and landscapes. *Limnol. Oceanogr.* 54(6), 2349–2358. http://dx.doi.org/10.4319/lo.2009.54.6_part_2.2349.

Schwatke, C., D. Dettmering, W. Bosch, and F. Seitz. 2015. DAHITI–an innovative approach for estimating water level time series over inland waters using multimission satellite altimetry. *Hydrol. Earth Syst. Sci.* Copernicus GmbH, 19, 4345–4364. http://dx.doi.org/10.5194/hess-19-4345-2015.

Seekell, D.A., J.A. Carr, C. Gudasz, and J. Karlsson. 2014. Upscaling carbon dioxide emissions from lakes. *Geophys. Res. Lett.* 41(21), 7555. http://dx.doi.org/10.1002/2014GL061824.

Seyler, F., M.-P. Bonnet, S. Calmant, et al. 2005. «CASH « Contribution de l'Altimétrie Spatiale à l'hydrologie, Rapport intermédiaire d'opération d'une recherche financée par le Ministère de la Recherche, Octobre 2005, Décision d'aide n: 04 T 131.

Seyler, F., S. Calmant, J. da Silva, N. Filizola, G. Cochonneau, M.-P. Bonnet, A.-C. Zoppas Costi. 2009. Inundation risk in large tropical basins and potential survey from radar altimetry: Example in the Amazon Basin. *Mar. Geod.* 32(3), 303–319. http://dx.doi.org/10.1080/01490410903094809.

Shum, C.K., Y. Yi, K. Cheng, C. Kuo, A. Buran, S. Calmant, and D. Chambers. 2003. Calibration of Jason-1 altimeter over Lake Erie. *Mar. Geod.* 26(3–4), 335–354.

Silva, J., S. Calmant, F. Seyler, O. Corrêa Rotunno Filho, G. Cochonneau, and W.J. Mansur. 2010. Water levels in the Amazon basin derived from the ERS 2 and ENVISAT radar altimetry missions. *Remote Sens. Environ.* 114, 2160–2181. http://dx.doi.org/10.1016/j.rse.2010.04.020.

Silva, J., S. Calmant, F. Seyler, D.M. Moreira, D. Oliveira, and A. Monteiro. 2014. Radar altimetry aids managing gauge networks. *Water Resour. Manag.* 28, 587–603. http://dx.doi.org/10.1007/s11269-013-0484-z.

Silva, J.S., F. Seyler, S. Calmant, et al. 2012. Water level dynamics of Amazon wetlands at the watershed scale by satellite altimetry. *Int. J. of Remote Sens.* 33(11), 3323–3353. http://dx.doi.org/10.1080/01431161.2010.531914.

Sima, S., and M. Tajrishy. 2013. Using satellite data to extract volume-area elevation relationships for Urmia Lake, Iran. *J. Great Lakes Res.* 39(1), 90–99. http://dx.doi.org/10.1016/j.jglr.2012.2.013.

Singh, A., F. Seitz, and C. Schwatke. 2012. Interannual water storage changes in the Aral Sea from multi-mission satellite altimetry, optical remote sensing, and GRACE satellite gravimetry. *Remote Sens. Environ.* 123, 187–195. http://dx.doi.org/10.1016/j.rse.2012.01.001.

Song, C., B. Huang, and L. Ke. 2013. Modeling and analysis of lake water storage changes on the Tibetan Plateau using multi-mission satellite data. *Remote Sens. Environ.* 135, 25–35. http://dx.doi.org/10.1016/j.rse.2013.03.013.

Sood, A., and B.K.P. Mathukumalli. 2011. Managing international river basins: Reviewing India–Bangladesh transboundary water issues, *Int. J. River Basin Manage.* 9, 43–52.

Sulistioadi, Y.B., K.-H. Tseng, C.K. Shum, et al. 2015. Satellite radar altimetry for monitoring small rivers and lakes in Indonesia. *Hydrol. Earth Syst. Sci.* 19, 341–359. http://dx.doi.org/10.5194/hess-19-341-2015.

Tarpanelli, A., L. Brocca, S. Barbetta, M. Faruolo, T. Lacava, and T. Moramarco, 2015. Coupling MODIS and Radar Altimetry Data for Discharge Estimation in Poorly Gauged River Basins. *IEEE Trans. Geosci. Remote Sens.* 8(1), 141–149. http://dx.doi.org/10.1109/JSTARS.2014.2320582.

Tourian, M., C. Schwatke, and N. Sneeuw. 2017. River discharge estimation at daily resolution from satellite altimetry over an entire river basin. *J. Hydrol.* 546, 230–247. http://dx.doi.org/10.1016/j.hydrol.2017.01.009.

Tranvik, L.J., J.A. Downing, J.B. Cotner, et al. 2009. Lakes and reservoirs as regulators of carbon cycling and climate. *Limnol. Oceanogr.* 54(6 part 2), 2298–2314.

Villadsen, H., O.B. Andersen, L. Stenseng, K. Nielsen, and P. Knudsen. 2015. Cryosat-2 altimetry for river level monitoring—Evaluation in the Ganges-Brahmaputra river basin. *Remote Sens. Environ. http://dx.doi.org/*10.1016/j.rse.2015.05.025.

Villadsen, H., X. Deng, O.B. Andersen, L. Stenseng, N. Nielsen, and P. Knudsen. 2016. Improved inland water levels from SAR altimetry using novel empirical and physical retrackers. *J. Hydrol.*, 537, 234–247. http://dx.doi.org/10.1016/j.jhydrol.2016.03.051.

Wang, B., Q. Bao, B. Hoskins, G. Wu, and Y. Liu. 2008. Tibetan Plateau warming and precipitation change in East Asia. *Geophys. Res. Lett.* 35, L14702. http://dx.doi.org/10.1029/2008GL034330.

Wang, X., P. Gong, P., Y. Zhao, et al. 2013. Water-level changes in China's large lakes determined from ICESat/GLAS data. *Remote Sens. Environ.* 132, 131–144. http://dx.doi.org/10.1016/j.rse.2013.01.005.

Watson, C., R.N. White, J. Church, et al. 2011. Absolute calibration in Bass Strait, Australia: TOPEX, Jason-1 and OSTM/Jason-2. *Mar. Geod.* 34, 242–260. http://dx.doi.org/10.1080/01490419.2011.584834.

Williamson, C.E., J.E. Saros, W.F. Vincent, and J.-P. Smol. 2009. Lakes and reservoirs as sentinels, integrators, and regulators of climate change. *Limnol. Oceanogr.* 54(6), 2273. http://dx.doi.org/10.4319/lo.2009.54.6_part_2.2273.

Wilson, M.D., P.D. Bates, D.E. Alsdorf, et al. 2007. Modeling large-scale inundation of Amazonian seasonally flooded wetlands. *Geophys. Res. Lett.* 34(15), L15404. http://dx.doi.org/10.1029/2007GL030156.

Wingham, D., C. Rapley, and H. Griffiths. 1986. New techniques in satellite altimeter tracking systems. ESA Proceedings of the 1986 International Geoscience and Remote Sensing Symposium(IGARSS'86) on Remote Sensing: Today's Solutions for Tomorrow's Information Needs 1986, 3.

Wolf, A., J. Nathrius, J. Danielson, B. Ward, and J. Pender. 1999. International river basins of the world. *Int. J. Water Resour. Dev.* 15, 387–427. http://dx.doi.org/10.1080/07900629948682.

Wu, Y., H. Zheng, B. Zhang, and D. Chen. 2014. Long-term changes of lake level and water budget in the Nam Co Lake Basin, Central Tibetan Plateau. *J. Hydrometeorol.* 15, 1312–1322. http://dx.doi.org/10.1175/JHM-D-13-093.1.

Yamazaki, D., H. Lee, D. E. Alsdorf, et al. 2012. Analysis of the water level dynamics simulated by a global river model: A case study in the Amazon River. *Water Res. Res.* 48, W09508. http://dx.doi.org/10.1029/2012WR011869.

Zakharova, E.A., A.V. Kouraev, A. Cazenave, and F. Seyler. 2006. Amazon River discharge estimated from TOPEX/Poseidon altimetry. *CR Geosci.* 338, 188–196. http://dx.doi.org/10.1016/j.crte.2005.10.003.

Zhang, B., Y. Wu, L. Zhu, J. Wang, J. Li, and D. Chen. D. 2011a. Estimation and trend detection of water storage at Nam Co Lake, central Tibetan Plateau. *J. Hydrol.* 405, 161–170. http://dx.doi.org/10.1016/j.jhydrol.2011.05.018.

Zhang, G., H. Xie, S. Kang, D. Yi, and S. Ackley. 2011b. Monitoring lake level changes on the Tibetan Plateau using ICESat altimetry. *Remote Sens. Environ.* 115, 1733–1742. http://dx.doi.org/10.1016/j.rse.2011.03.005.

Zhang, G., H. Xie, T. Yao, and S. Kang. 2013. Water balance estimates of ten greatest lakes in China using ICESat and Landsat data. *Chin. Sci. Bull.* 1–15. http://dx.doi.org/10.1007/s11434-013-5818-y.

Zhang, G., H. Xie, and M. Zhu. 2010. Water level changes of two Tibetan lakes Nam Co and Selin Co from ICESat Altimetry data, 2010. Second ITA International Conference on Geoscience and Remote Sensing, 978-1-4244-8515-4/10/$26.00 ©2010 IEEE GRS2010.

Zwally, H.J.R., C. Schutz, J. Bentley, T. Bufton, J. Herring, J. Minster, and R.Y. Spinhirne. 2003. *GLAS/* ICESat L1B Global Elevation Data. Version 33: GLA 06, Boulder and Colorado USA.

15 Applications of Satellite Altimetry to Study the Antarctic Ice Sheet

Frédérique Remy, Anthony Memin, and Isabella Velicogna

15.1 INTRODUCTION

About 76% of the total freshwater on Earth is stored in a frozen state inside the polar ice sheets, Greenland and Antarctica. These two ice sheets play a critical role in the Earth's climate. Both constitute the world's glacial archives, revealing past climate conditions, and are sensitive indicators of current climate fluctuations. The Greenland Ice Sheet holds a volume of ice equivalent to a 7-m sea level rise; Antarctica holds a volume of ice equivalent to a 56-m sea level rise. At present, these ice sheets contribute sooner, faster, and more significantly to sea level than anticipated from climate models (IPCC 2013). There is potential for an increase in sea level of more than one meter by the end of this century and several more meters beyond the next century (IPCC 2013).

Because of the expected major contribution of the ice sheets to future sea level rise, monitoring present-day ice sheet evolution is imperative. It is equally crucial to understand the causes of ice mass change in order to model and predict future spatiotemporal evolutions and, in turn, impacts on sea level variations.

Surface topography is one of the most pertinent parameters for monitoring an ice sheet but also for constraining ice flow models. Ice sheet surface elevation is observed using radar or laser altimeters on aircraft and satellites; the distance between the instrument and the surface is measured together with precise information on the location of the altimeter. Precision radar altimetry at high latitudes started in 1991 with the launch of the European Remote Sensing (ERS) 1 mission by the European Space Agency (ESA). This was the first mission that used onboard specialized instruments dedicated to study the poles. With an inclined orbit of 81.6°, the ERS-1 mission covered almost the whole of Greenland in the Northern Hemisphere and more than 75% of Antarctica in the Southern Hemisphere. In addition, radar altimetry, which primarily measures distances using radar waves, also provides information about the snowpack that is related to climate forcing. ERS-1 was followed by ERS-2 in 1995 and Envisat in 2003, providing a continuous radar altimetry record until 2010. In 2002, the National Aeronautics Space Administration (NASA) launched ICESat-1 (Ice, Cloud, and Land Elevation Satellite), a laser altimeter capable of higher vertical precision than radar altimeters but hampered by clouds, pointing errors, and limited duration of the lasers (Abdalati et al. 2010). NASA Operation IceBridge was launched in 2009 to bridge the gap with the ICESat-2 laser altimeter to be launched in late 2018 with a new photon counter technology. Meanwhile, ESA launched CryoSat-2 in 2010 operating in the Ku band with an interferometric, higher resolution mode (SIRAL-SAR Interferometric Radar Altimeter) along the coasts of the ice sheets. A follow-on mission to CryoSat-2 is planned in the Ka band for improved capabilities.

In this chapter, we focus on Antarctica and first discuss a few general properties of this ice sheet. We show why ice sheet topography is one of the most pertinent parameters for studying the dynamics of the ice sheet. Then we discuss specific aspects of the altimetry technique applied for studying land ice. We finally address various areas to which radar and laser altimetry have made notable contributions: surface climatology, ice dynamics, and monitoring of ice volume variations.

15.2 THE ANTARCTICA ICE SHEET

15.2.1 General Characteristics

Antarctica (Figure 15.1) has a surface of 14 million km² and an average glacial thickness of 2000 m and sometimes exceeding 4000 m. It represents 90% of terrestrial ice and contains the equivalent of 56 m mean sea level. The ice sheet in East Antarctica, namely between 0 and 180°E, is relatively stable, but it appears that in recent decades glaciers in West Antarctica have accelerated, thinned, and lost an increasing amount of ice (Sutterley et al. 2015). The Transantarctic Mountain Range separates the continent into two parts, east and west. Western Antarctica is smaller, with lower surface elevation, and has the majority of its bedrock below sea level. East Antarctica, however, includes large sectors with a bedrock below sea level—referred to as marine-based ice sheets—that hold far more ice together than all the ice in West Antarctica (Fretwell et al. 2013).

Antarctica is the coldest, the highest (on average), the driest, and the windiest continent on the planet. The temperature decreases from the coast going inward from −15°C to −60°C on average, with record freezing temperatures approaching −90°C. The air is very dry at this temperature, and it snows less here than it rains in the Sahara. The cold and dense air from the center of the continent hurls down the slopes, creating strong and persistent katabatic winds. At Dumont d'Urville, the average speed of katabatic winds is 40 km/h, with a maximum three or four times this value. These winds sculpt the surface of the snow from the centimeter scale (microroughness) to the meter scale (sastrugi). They move significant amounts of snow, making direct measurement of snowfall accumulation challenging and remote sensing difficult. In fact, the majority of sensors used in the polar regions operate in the microwave range, with a wavelength typically ranging from a couple of millimeters to 10 cm, of the same order of magnitude as the surface microroughness. The extended size of the continent and its challenging environment explain why spaceborne observations are

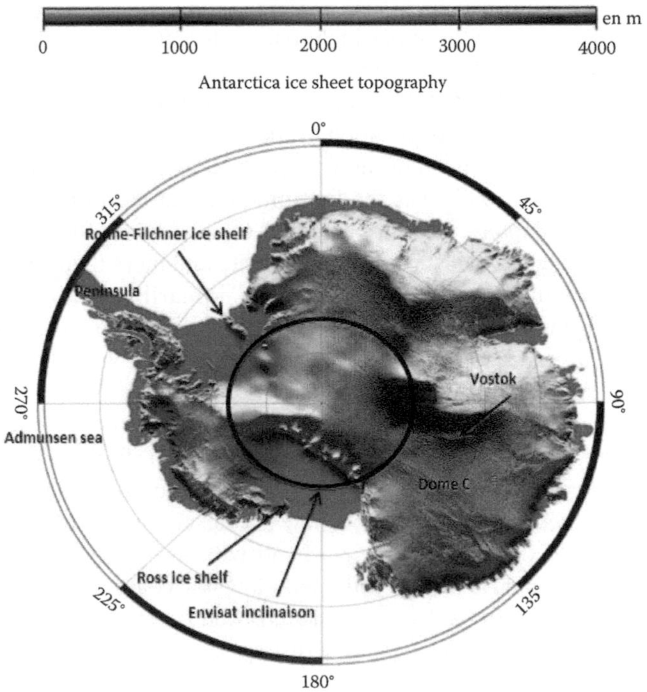

FIGURE 15.1 The topography of Antarctica obtained from the geodetic orbit of ERS-1. The central circle indicates the limit of the altimeter visibility. The names of key locations discussed in this chapter are indicated. *In situ* data were used to extrapolate the topography beyond ERS-1's reach.

essential to study the Antarctic ice sheet. The interaction between electromagnetic waves at radar frequencies and the rough surface is complex. *In situ* measurements allowing us to describe this interaction, at least in statistical terms, remain essential for understanding and modeling it.

Snowfall is light in Antarctica. Expressed in water equivalent, it varies from a couple of centimeters per year in the center of the continent to a few tens of centimeters near the coast. The average rate of snowfall is 17 cm/year versus a global average of 1 m/y (Van de Berg et al, 2006). Integrated over Antarctica, snowfalls represent 2200 billion tons per year, corresponding to a sea level variation of 6 mm if distributed uniformly on the ocean surface. Slight imbalance in snowfall can induce significant variations in sea level. Throughout time, snowfalls accumulate and form successive layers that sink and gradually densify into ice. As a result of its own weight, the newly formed ice flows downstream in the downslope direction of the surface. The flow velocity is low, of the order of decimeters per year in the center of the continent, to several 100 meters per year near the coast (Rignot et al. 2011b). Flow velocity reaches kilometers per year for the fastest glaciers in the Amundsen Sea Embayment sector of West Antarctica. Thus, it takes thousands of years for the ice to reach the ocean, where it melts from below and breaks into icebergs. Due to this long period of residence, polar ice sheets are the "Planet Earth's glacial archives." For instance, the ice core at Dome C allowed retrieving climatic terrestrial history 800,000 years back in time, hence, over several glacial cycles (EPICA Community Members 2004).

The volume of Antarctica is controlled by the balance between mass gains (i.e., precipitation) and mass losses (i.e., sublimation), evaporation, drift and melt of snow, and ice discharge by the outlet glaciers. The behavior of the ice sheet is thus determined by different meteorological, dynamic, and mechanical processes that depend on climate at different timescales (Figure 15.2). At the large scale, the profile of the ice sheet surface is quasi-parabolic. In the center, the slope is small, less than a few meters per kilometer, and increases toward the coast. The atmospheric processes—snowfall, sublimation, evaporation, snow drift, and snow/ice melting at the surface—react instantaneously to climate forcing. For example, warmer air stores more humidity, heating will thus create an excess precipitation. Antarctica is also sensitive to boundary conditions at sea level, to the stability of ice shelves that react on longer timescales. Nevertheless, the flow of glaciers in the Amundsen Sea Embayment of West Antarctica has been accelerating on a decadal timescale. Lastly, snow is an excellent insulator. A "wave" of temperature will take an excessive amount of time before substantially changing the temperature of the ice and in turn the ice velocity. A variation in the volume of ice will change the way in which it depresses the bedrock, which will in turn adjust by isostasy.

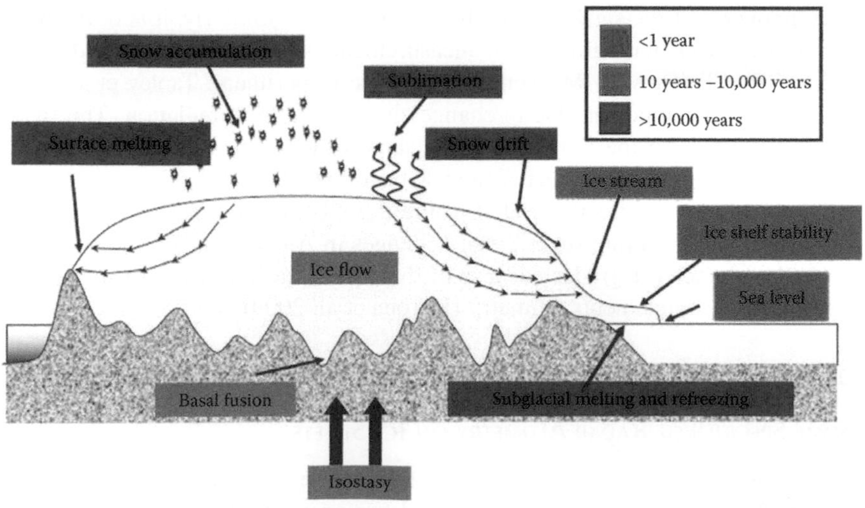

FIGURE 15.2 The functioning of a polar ice sheet and the reaction time of the different components.

This process acts slowly compared to climate. The ice sheet thus lives and reacts over a broad range of timescales, from seasonal to major glacial cycles, and from decadal to annual and even daily fluctuations (Ritz et al. 2001). The ice sheet mass balance (i.e., ice gain or ice loss) at a given point and given epoch is extremely difficult to predict. It has to be measured with a variety of complementary instruments and observed within the context of the global climate system.

15.2.2 How to Observe the Evolution of the Ice Sheet

To date ice cores, constrain the discharge of ice and numerical models, measure the mass balance, and predict the effects of climate warming on the ice sheet evolution, numerous observations are necessary. The laws of mechanics, the physical processes that allow ice to deform or slide, the role of coastal and basal boundary conditions, and the effect of longitudinal constraints on ice flow are topics of importance that are still the subject of intense research. For example, the spatiotemporal distribution of snowfall is not well constrained by observations. Indeed, the snowpack depends on the size of snow grains, the density of snow, and the snow stratification, which are all subject to climate forcing and difficult to measure. The ultimate question is the state of mass balance of the ice sheet. Does it grow because of a surplus in snowfall accumulation due to warmer air that holds more humidity? Or does it lose mass because of the acceleration of its outlet glaciers and enhanced discharge of ice into the Southern Ocean?

Boarded on satellites, numerous sensors are used to understand the different unknown parameters of the ice sheet. These instruments operate in the microwave, infrared, and visible part of the spectrum and also include instruments that measure the time-variable gravity field of the Earth. The snow is cold, dry, and not dense, so microwave signals penetrate and emit information about the snowpack. For example, scatterometer data allows us to measure the direction of katabatic winds through surface anisotropy that they create. Radar interferometry is important to measure the speed of outlet glaciers, which—combined with airborne measurements of ice thickness using depth sounders—provides estimates of the glacier ice discharge along the periphery of the continent (Rignot et al. 2011a). Optical or near infrared sensors are sensitive to albedo, the size of snow grains, and the roughness of the surface.

Altimeters are useful sensors because they measure the surface topography resulting from meteorological and dynamic processes. Moreover, successive observations allow us to estimate the ice sheet mass balance. Altimeters make it possible to estimate the surface elevation and therefore the volume of the ice sheet and its variations. It is, however, not possible to distinguish a precipitation deficit from a loss of mass through ice discharge. For that purpose, it is necessary to couple the altimeter measurements with observations of ice flow and changes in ice flow with time in order to identify the components of the signal caused by ice dynamics. Similarly, it is of interest to couple the altimeter measurements with gravimetric measurements such as those recorded by the satellite gravity mission GRACE (Gravity Recovery and Climate Experiment; Tapley et al. 2004) to constrain the mass balance component due to changes in snowfall accumulation. The twin satellites of the GRACE mission orbit the Earth about the same time as the Envisat altimeter. "Tom and Jerry" measure the inter-satellite distance using a K-band ranging system. The variation in distance between the two satellites is directly related to the mass change on Earth. Given that, GRACE appropriately helps in determining surface-mass changes in Antarctica. Near the coast, the performance of the radar altimeter degrades because of the higher surface slope. In such places, optical stereography favorably complements altimetry (Korona et al. 2009).

15.3 POLAR ALTIMETRY

15.3.1 Some Specifics of Radar Altimetry on Ice Sheets

Here, we discuss specifics of measurements of polar ice sheets by radar altimeters. First among these is the satellite orbit. The altimeters dedicated to oceanography (e.g., the TOPEX/Poseidon and Jason series) have an inclination of 66° that covers only southern Greenland and misses most of Antarctica. The European satellites ERS-1 (1991–2000), ERS-2 (1995–2003), and Envisat (2003–2010) have an inclination of 81.6° and orbit around the Earth in 35 days. This geometry has allowed repeat monitoring of the ice sheets for nearly 25 years.

The second specific is the slope error. Altimetry does not measure the distance between the satellite and the sub-satellite point (the nadir) but the smallest distance between the satellite and the ground. If the surface is inclined, the point of impact, where the measurement is taken, is shifted upward. The resulting error depends on the square of the slope; therefore, it is rapidly prohibitive as surface slope increases. Poor orbit repetition and the presence of a perpendicular slope across the track cause errors in height measurement.

A third specific is associated with the pronounced topography of ice sheets compared to the relatively smooth ocean surface for which radar altimeters are optimized. The tracking loop of the altimeter is not able to monitor topographic fluctuations from the length of the satellite ground track to such an extent that altimetric waveform is no longer centered in the receiving window. The data must therefore be processed to estimate the height of the surface. This process, named *retracking*, benefits from various algorithms dedicated to polar ice sheets. One algorithm, named *Ice-1*, calculates the center of gravity of the waveform and searches for the rising edge position for which the energy is proportional to the center of gravity. Another, named *Ice-2*, adjusts the rising edge by an error function and uses its center as a corresponding point to the surface. The advantage of *Ice-2* is that, in addition to estimating the height, it estimates the width of the leading edge, the slope of the trailing edge, and total backscatter.

The fourth specific, most critical issue is the depth of penetration of the electromagnetic waves into the dry and cold snow. The penetration depth depends on the sensor frequency and the dielectric properties of the medium. The so-called "volume" echo affects the waveform so that the surface of the ice sheet appears farther away than its actual distance from the satellite. This additional echo is represented in Figure 15.3. The error of the distance measurement depends on the relationship between the intensity of the surface echo and that of the volume, with the two echoes being significantly variable. The first is sensitive to the surface roughness that varies with wind, and the second depends on snowpack on a seasonal timescale. This variability creates fluctuations in measured height. Moreover, the properties of snowpack change with time and create a bias on the height measurements that yields an artificial variation in volume (Remy and Parouty 2009).

Since March 2013, the Centre National d'Etudes Spatiales (CNES) and the Indian Space Research Organization (ISRO) have put on the same Envisat orbit, inclined at 81.6°, and the SARAL

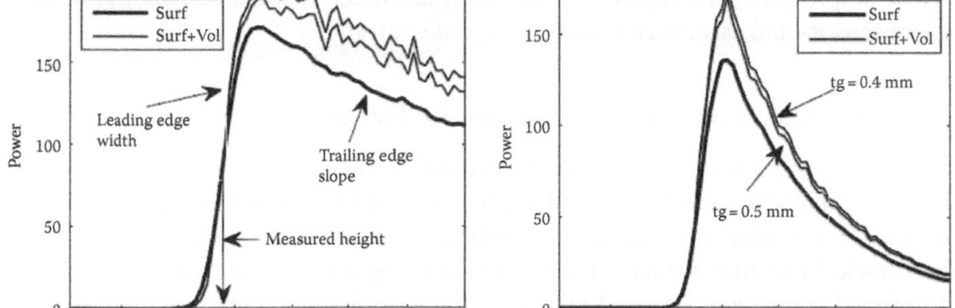

FIGURE 15.3 Simulated altimetric waveform at the Ku (13.6 GHz) (a) and Ka band (36 GHz) (b) and the parameters estimated using the "*Ice-2*" retracking algorithm. The surface slope is 1 m/km. The volume echoes (Vol) were simulated by using two snow grain sizes: tg = 0.4 and 0.5 mm. The surface echoes correspond to the "Surf" curves.

satellite (Satellite with ARgos and ALtika, or "simple" in Indian), which has the AltiKa altimeter on board. This altimeter works at the Ka band (35.75 GHz) instead of the Ku band (13.6 GHz) used for prior altimeters (see Chapter 1). At the Ka band, the ionospheric effect causing propagation delays is lower. Thus, there is no need for another instrument to correct for the induced error (C band "5.6 GHz" for the Jason series and S band "3.6 GHz" for Envisat). Moreover, the Ka band antenna is smaller for the same resolution. The theory of electromagnetic scattering suggests that the volume echo should be less at the Ka band than at the Ku band due to weaker wave penetration (Figure 15.3). The first results show that the volume echo at the Ka band is stronger than at the Ku band, but it only affects the first tens of centimeters beneath the surface, making its effect on height restitution less significant (Remy et al. 2015).

As mentioned earlier, the CryoSat-2 satellite, launched in 2010, carries SIRAL, a new type of delay-Doppler radar altimeter (Raney 1998). Over the interior part of the ice sheets, SIRAL operates as a conventional pulse limited radar system (low resolution mode, LRM). In more complex high-sloping terrain, the system operated in an "Interferometric Synthetic Aperture Radar" (SIN) mode using a novel antenna. These new features allow the satellite to monitor changes in complex topography and the high relief marginal areas of the ice sheets.

15.3.2 Characteristics of Laser Altimetry on Ice Sheets

The NASA's ICESat-1 mission was launched in 2003 into a near-circular 600-km altitude orbit (Schenk and Csatho 2012). The main instrument on ICESat-1 was the Geoscience Laser Altimeter System (GLAS), which consisted of three individual and identical 1064-nm Nd-YAG lasers (L1, L2, and L3). After the premature failure of the L1 laser, the ICESat-1 mission was adjusted to acquire data in discrete campaigns. This decision limited the total number of surface elevation measurements compared to the initial design but extended the overall mission lifetime to meet the ice sheet elevation change detection objectives (Abdalati et al. 2010). Overall, the mission fired nearly two billion shots with a footprint of about 40 m before the failure of its final laser on October 11, 2009 (Schenk and Csatho 2012). The dominant error in the ICESAt measurements is the 10- to 20-m error in footprint location that produces errors of 15–30 cm on a one-degree surface. The track spacing was 30 km at 70 degrees during the three 33-day campaigns per year. The decay in laser energy during the mission also affected the height accuracy. The measurements were still affected by the unknown surface slope due to the uncertainty in laser pointing.

The follow-on to the ICESat mission (ICESat-2) is planned for launch in 2018 (Abdalati et al. 2010). It will carry a multi-beam laser altimetry based on photon counter technology that will solve the surface slope problem. In contrast with the radar altimeter, there is no penetration depth issue with the laser sensors and associated spurious seasonal variability.

15.3.3 Methodology for Constructing Height Time Series

Two major sources of errors complicate the process for constructing height time series. The first is a slope perpendicular to the satellite track that affects the results obtained from one passage of the satellite over the same point on the ground to another. A track from one cycle varies by ± 1 km from the nominal track. Over that distance, the height varies by a couple of m/km, making its impact on height retrieval significant. To avoid this error, the variation in height is estimated by using the crossing points of the satellite's ascending and descending tracks. At the crossover point of the tracks, the slope is identical and cancels out by subtraction. In order to consider a maximum number of points and better sample the ice sheet surface, other algorithms use all the data along the track after estimating the surface slope along the track. The data set is 20 times denser and allows us to obtain continuous profiles.

The second source of error for evaluating a time series is the variation of the electromagnetic wave penetration into snow. When the backscatter increases because of added echo volume, we find that the height of the surface is lower, the leading edge stretches out, and the trailing edge slope rises (Figure 15.3). We therefore apply an empirical correction while adjusting the height fluctuations and those from retracking parameters (see previous section).

When using altimetric series, the error due to the poor temporal repetition of the satellite ground track and the presence of surface slope must be corrected. Flament and Rémy (2012) have used a biquadratic form of the spatial unknowns. To reduce height variations due to penetration into the snowpack, we use the three classical parameters of the waveform (backscatter, trailing edge slope, and leading edge width) and calculate the sensitivity of the estimated height for each parameter. Regarding the temporal variations, we estimate the trend but also the seasonal variations, so we add three more parameters: time, cosine, and sine for a year-long time interval. We obtain a system of equations with 12 unknowns. This equation is better constrained when the temporal series is long. Remy et al. (2014) show that after about 50 cycles, the error on the estimated height trend is less than a few percent to the rate of snow accumulation. The Envisat satellite flew for 8 years, offering 85 repetitive cycles, which represents a large quantity of data. The inversion of these parameters is thus robust and provides estimates of the temporal trend in height and backscatter, their seasonal variations, and an estimation of the surface slope at a kilometric scale.

15.4 CONTRIBUTION OF ALTIMETRY TO STUDYING ANTARCTIC CLIMATE

Altimetry offers complementary observations to radiometers or scatterometers for polar climate studies. To correctly interpret the height variations and detect noise sources that affect the estimated trends, it is important to understand the "physics of measurement." The waveform is controlled by the intensity of the volume and surface echo. The penetration in a medium controls the stretching of the leading edge and the lifting of the trailing edge (Figure 15.3). The comparison between altimetric radar data at the Ku band and the laser altimeter data from the ICESat altimeter indicates an average penetration depth of 2 m to 5 m (Michel et al. 2014) principally controlled by volume scattering from snow grains. Larger grains increase the loss by scattering, which results in a reduction of the penetration depth. On a seasonal scale, the variations in waveform parameters come from the densification of the snowpack and the growth of snow grains. The surface down to several meters depth contribute to the volume echo, but after reprocessing, the error induced on height is only about a meter.

The successive snow layers vary throughout the seasons, and snowpack is thus stratified. This stratification induces dielectric discontinuities at each stratum. The importance of the dielectric variability on the echo depends on the altimeter frequency. At the S band (3.2 GHz), the variability of the dielectric constant dominates the shape of the echo, but this is insignificant at the Ka band (35.75 GHz) and comparable to snow grain scattering at the Ku band (13.6 GHz) (Remy et al. 2015).

As on the ocean, wind creates several scales of roughness on snow. The microroughness plays a role on the backscatter that is also affected by the surface slope *via* the antenna gain. Sastrugi also stretch out the leading edge of the altimeter. The scale of roughness to which the radar is sensitive also depends on the radar wavelength of the instrument.

At the seasonal scale, we observe one of the mysterious phenomena of altimetry in Antarctica. The spatial distribution of the maximum backscatter date exhibits a bimodal distribution well marked geographically (Figure 15.4). The maximum occurs in the beginning of April, at the end of the austral summer at the S band, or in the beginning of October, at the end of the austral winter at the Ka band. This difference remains unexplained at present but reflects changes in scattering mechanisms at the different frequencies.

15.5 MAIN SURFACE CHARACTERISTICS OF THE ANTARCTIC ICE SHEET

In this section, we summarize average characteristics of the ice sheet, assuming in first approximation that it is in a state of mass balance.

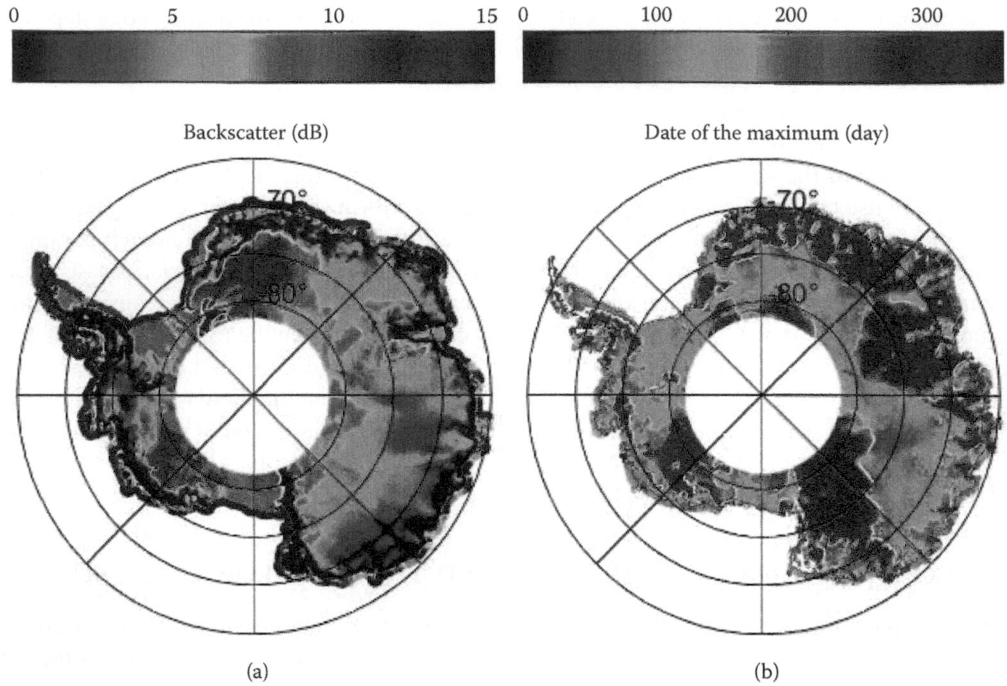

FIGURE 15.4 The backscatter coefficient (a) varies by 15 dB (or a factor of 30 in power). The spatial distribution of the maximum backscatter date (day of the year) shows a bimodal distribution with a marked geographic transition (b).

15.5.1 Surface Topography

From a kilometer to hundreds of kilometers or even at a continental scale, the ice sheet is entirely shaped by current and past climatological or dynamical processes (Figure 15.1). The topography deduced from ERS-1 geodetic orbit provides interesting information. First, at the continental scale, the profile of the polar ice sheet is quasi-parabolic, like a plastic body on a nonuniform base. This quasi-parabolic shape reflects the balance between accumulation by snow precipitation and loss by ice discharge. At a smaller scale, a few hundreds of kilometers, a mathematical transformation allows us to reveal stretched structures that are barely visible to the naked eye. This transformation requires us to calculate the surface curve in the perpendicular direction of the slope y, in other words the term $\partial h^2/\partial y \partial y$, where h is the surface elevation of the ice sheet. The curve perpendicular to the slope (Figure 15.5) shows the discharge of ice into different basins. This curve is mathematically related to the convergence and divergence of the discharge, $1/l(x)\ \partial l/dx$ where l(x) is the width of the channel.

Intuitively, two parallel channels along the x-direction receive the same amount of snow and produce the same outflow. If one discharges more slowly, it will be slightly thicker and vice versa. This irregularity in height is controlled by the bedrock topography near the coast, where they find their origin upstream and reveal the largest drainage basins. We find a pattern of rapid and slow channels all around the coast separated by 250 km.

On a scale of a few tens of kilometers, we observe features directly related to ice dynamics. For instance, the observed smooth and flat zone a couple of hundreds of kilometers long and a few dozen kilometers wide near Vostok (see Figure 15.1) reflects a subglacial lake. Vostok Lake is the largest subglacial lake in Antarctica, but there are others. In fact, the bottom of the ice sheet, near the bedrock, is isolated from atmospheric cold by thousands of meters of ice and heated by the geothermal flow, so ice melts from below. According to the shape of the bedrock, the resulting water is

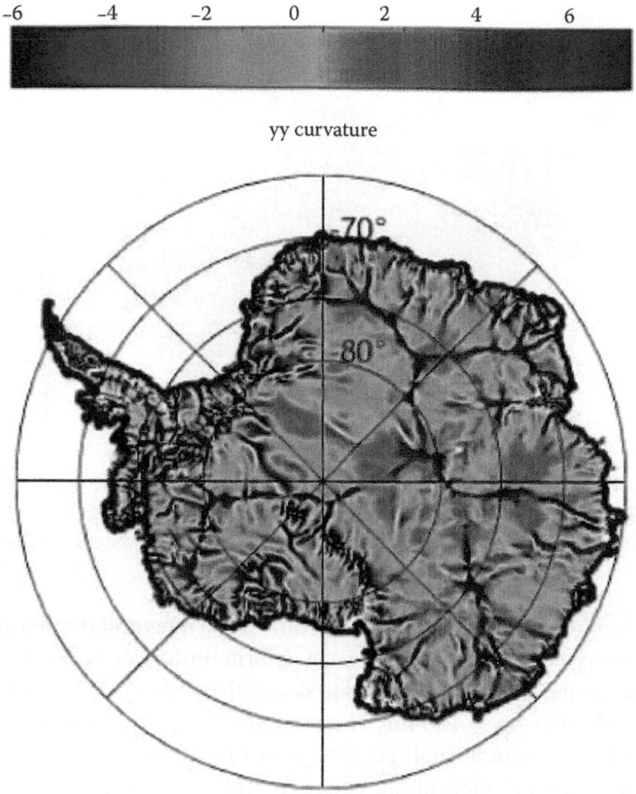

FIGURE 15.5 Curve at a scale of 100 km, expressed in cm/km², perpendicular to the slope. Extremely coherent stretched structures appear, they show the discharge and allow us to visualize the different basins.

discharged and stored to form a lake. Vostok Lake is the most famous subglacial lake in Antarctica. Its existence was suspected in the 1960s because echoes obtained from low-frequency radars used to estimate ice thickness had an unusual intensity. Its existence was confirmed with ERS-1 when it was possible to identify the subglacial lake signature on the ice sheet surface. Because ice slides over water on the subglacial lake, it creates a flat and smooth zone at the surface.

Since then, several hundreds of subglacial lakes have been identified by echoes from low-frequency radars or by their laser altimetric signature at the surface (Smith et al. 2009). The "emptying" and "filling" of these lakes has been detected and monitored (Fricker et al. 2007).

Lastly, the most typical feature of the surface is an undulation with a wavelength 10 to 20 km long and an amplitude of a few meters that affects the entire continent (Figure 15.1). These undulations are weak at the center of the ice sheet and more pronounced near the coast. They are due to the effect of the bedrock on ice flow.

15.5.2 Constraints for Numerical Models

The topography of the surface is a fundamental constraint for ice sheet models. The direction of the largest slope drives the direction of flow; the angle between the two directions of the largest slope is responsible for flow convergence or divergence; and the intensity of the slope controls the basal constraint that drives the flow intensity or driving stress. Thus, assuming the ice sheet is close to equilibrium, a simple model with a reliable topography makes it possible to deduce the velocity of the ice flow. The model assumes that ice is incompressible and that, between two channels of ice discharges, the outgoing ice mass is equal to the mass entering by precipitation. The topography provides the direction

FIGURE 15.6 Balance velocity of the Antarctic Ice Sheet on a logarithmic scale. The majority of the ice sheet velocity is less than 10 m/y. Velocity is several orders of magnitude larger along outlet glaciers, which represent a small percentage of the coast but drain the majority of the ice sheet into the southern ocean.

and convergence of the flow. Knowledge of snow accumulation rates and the density of the ice allow us to reconstruct the velocity field. Antarctica does not deform uniformly with time (Testut et al. 2003). More than 80% of the ice sheet is subject to motion slower than 10 m/y. In several places, however, the outlet glaciers reach velocities up to 100 m/y (Figure 15.6). In fact, the ice sheet is drained by a few major glaciers that only represent a small percentage of the Antarctic coastline. This illustrates the need to monitor specific outlet glaciers because they control the volume of the entire ice sheet.

The models used by glaciologists are complex (Ritz et al. 2001) and take into account the physics and boundary conditions of the ice flow. Numerous satellite and *in situ* observations are necessary to parameterize the different processes and force the current initial and past conditions. Surface topography is used as a boundary condition. The models help us better constrain the dating of ice cores (Parrenin et al. 2004), especially away from a dome, because of the horizontal flow between the dome and the ice core location. Precise topography and density measurements allow us to date layers of ice cores and compare them with the chronology deduced from core drilling.

15.6 TEMPORAL VARIATIONS

15.6.1 Different Times and Signals

Figure 15.2 suggests that Antarctica is never in perfect equilibrium because the dynamics and climatic processes that control its volume change are numerous and react on different timescales. We have seen that Antarctica responds rapidly to fluctuations in accumulation rates. We expect an increase in accumulation in the future due to warmer air conditions. The heat wave from the beginning of the Holocene, 15,000 years ago, has not reached the base of the ice sheet where ice deforms. The central part of the Antarctic ice sheet discharges in a glacial rhythm and may still respond to an increase in snowfall. The time it takes for the ice sheet to reach a state of mass balance with accumulation is the time of relaxation. This time depends on ice velocity, surface slope, and snow density (Remy et al. 2002). It varies from a couple hundred of years at the coast to hundreds of thousands of years in the center. In other words, the surface of the ice sheet cannot be balanced between two fluctuations in accumulation rate for which the variability is estimated at the 10% level on a timescale of 30 years. The surface of the ice sheet also randomly varies during snowfalls. We calculate that the probability of having an artificial decrease in Antarctica volume yielding a 0.5 mm/y increase of sea level is 12%.

15.6.2 Lake Drainage

When assessing time-changing data obtained from remote sensing measurements, we deal with outliers to clean the data. Outliers are isolated data that significantly deviate from the average variability of the time series. Classically, we remove outliers that have a difference from the mean larger than 3 σ (σ being the standard deviation). Sometimes, outliers reflect an anomaly that corresponds to a geophysical signal. For example, if we look at the height maps derived from Envisat data, we expect height variations to be weak and smoothly continuous. Flament et al. (2014) detected and isolated rapid variations up to 1 m from these time series (Figure 15.7). These anomalies are aligned and spread along the surface in a coherent pattern. They reflect the draining of a subglacial lake and a waterfall that is 400-km long.

The impact of the lake draining on the surface creates a hole so deep that an altimeter has difficulty probing its center elevation. Using images from the High Resolution Stereoscopic (HRS) sensor on SPOT-5, we find that the surface may cave by 70 m over 200 km². One lake lost 6.4 km³ of water. The water then flowed to other cavities, raising the surface and so on for more than 400 km (Figure 15.7). The draining and discharge under the ice took 2 years. The water is trapped under the ice sheet until it reaches the ocean.

Such lake drainage has had limited to no influence on ice flow despite the change in basal lubrication it produces, but it provides information about the flow of subglacial water beneath the ice sheet, with ramifications about the age of the subglacial water contained in the lakes and the interpretation of ice layers and ice cores.

15.6.3 Firn Compaction

The density of the ice sheet surface layer varies due to the compression of low-density snow into glacier ice. New snow is compacted by thermal and wind stresses, and firn is further compacted by

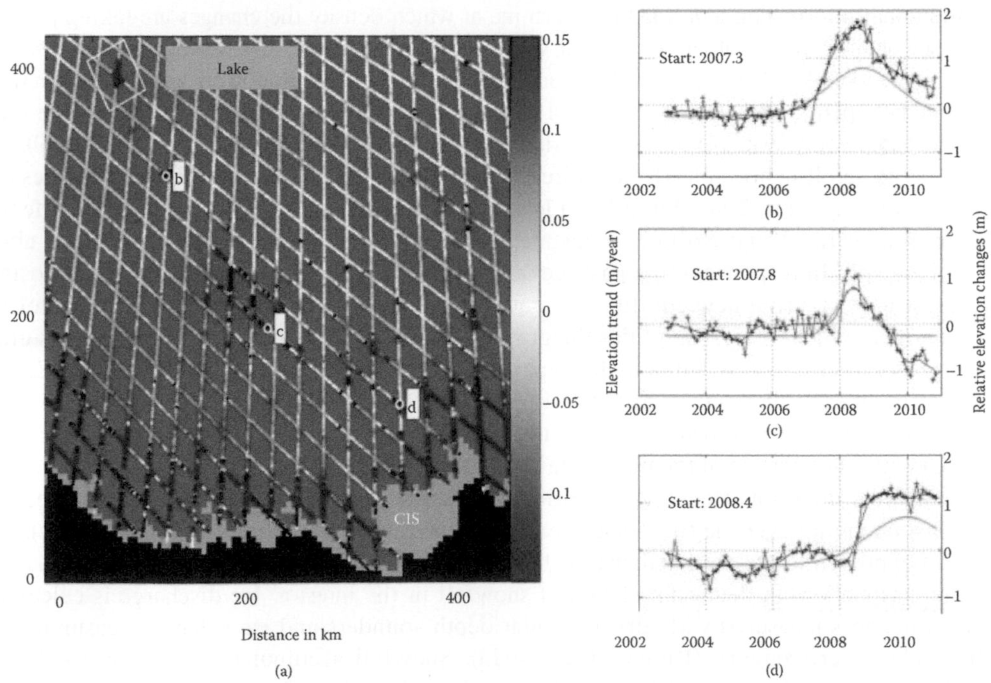

FIGURE 15.7 Varying height abnormalities appear well aligned on the ice's discharge lines (a). We see that the wave at point (b) arrives in the beginning of 2007, at (c) in mid-2008, and at (d) at the end of 2008 (right). (From Flament, T., et al., *Cryosphere*, 8, 1–15, 2014. With permission.)

the weight of the overlying snow and firn. Firn becomes glacier ice once the air-filled pores among the grains of snow completely close off, which occurs approximately at a density of 830 kg/m^3 (Van de Berg et al, 2006). The equivalent ice thickness of a column of firn and snow is calculated using the height and density of the snow layer (r_{snow}, 100–550 kg/m^3), the height and density of the firn layer (r_{firn}, 550–830 kg/m^3), and the density of glacier ice (ρ_{ice}, 917 kg/m^3) (Ligtenberg et al. 2011).

In the absence of significant melting, the rate of compaction depends on the temperature, snow accumulation rate, and wind speed (van den Broeke 2006; Ligtenberg et al. 2011). In Antarctica, firn-layer thickness variations can be the same order of magnitude as the elevation changes (Helsen et al. 2008). For glaciers in temperate regions, densification is rapid, and the firn-layer thickness is shallow (40–60 m). For the cold and dry interiors of ice sheets, densification is slow and the firn-layer thickness exceeds 100 m (van den Broeke 2006). The rate and spatial variability of firn compaction is particularly important for calculating ice sheet mass balance from repeat altimetry measurements of surface elevation.

Measurements of snow densification rates in the dry snow area of Greenland over 2004 to 2011 along a 500-km-long portion of the Expeditions Glaciologiques Internationales au Groenland (EGIG) line have been used to estimate firn compaction and its dependence on temperature and snow accumulation (Morris and Wingham 2014). A follow-on study determined that 7 years of data are needed to detect the small trend in ice sheet mass balance at the location of the EGIG line (Morris and Wingham 2015).

15.6.4 Present-Day Mass Balance

Since the 1990s, numerous altimeters have been used to estimate volume changes of the ice sheet. The series of ERS-1, ERS-2, and Envisat provided 20 years of continuous and homogeneous altimeter data since 1992. ICESat-1 provided precision measurements between 2003 and 2009. CryoSat-2 has been providing ongoing measurements since 2010. Unfortunately, altimeters only measure the change in volume. Additional information is needed to determine at which density the changes are taking place to convert the volume changes into mass changes, which refers to the firn compaction problem discussed in the previous section. In addition, radar altimeters are affected by a penetration error that varies in time, and they do not operate well along steep coastal regions. Data from laser altimetry do not have the same limitations but are only available over short periods between 2003 and 2009 (Pritchard et al. 2009).

The gravity satellite mission GRACE directly measures the spatiotemporal mass changes and has been operating since 2002, but it has a low spatial resolution (300–400 km) and is affected by uncertainties in glacial-isostatic adjustment following the Pleistocene deglaciation about 18,000 years ago. In response to the past redistribution of melting ice, the Earth is still isostatically adjusting to reach a new equilibrium state. This adjustment redistributes the mass of the Earth's mantle, which linearly modifies the gravity field of the Earth over Antarctica. This gravity change is difficult to assess due to discrepancies between models of deglaciation and Earth's internal structure (Whitehouse et al. 2012; Ivins et al. 2013). The effect of the Glacial-Isostatic Adjustment (GIA) in Antarctica is difficult to correct and contributes to the largest uncertainty in GRACE-based estimates of the mass balance (Barletta et al. 2013; Velicogna and Wahr 2013). This uncertainty does not affect the estimation of temporal changes in ice mass balance because it is a constant signal over the timescales considered in satellite applications (i.e., decades).

A third approach to the mass balance is the comparison of the ice discharge from the glaciers along the periphery with the accumulation of snowfall in the interior. Ice discharge is calculated from ice thickness measured with airborne radar depth sounders and ice velocity measured from satellite radar interferometry (Rignot et al. 2011a). Snowfall accumulation is reconstructed by regional atmospheric climate models forced by reanalysis data along the periphery (van den Broeke et al. 2006). The mass balance is the difference between these two large numbers affected by uncertainties. This approach is less precise than the gravity measurements but offers details about the partitioning of mass balance between ice dynamics and surface mass balance that the other

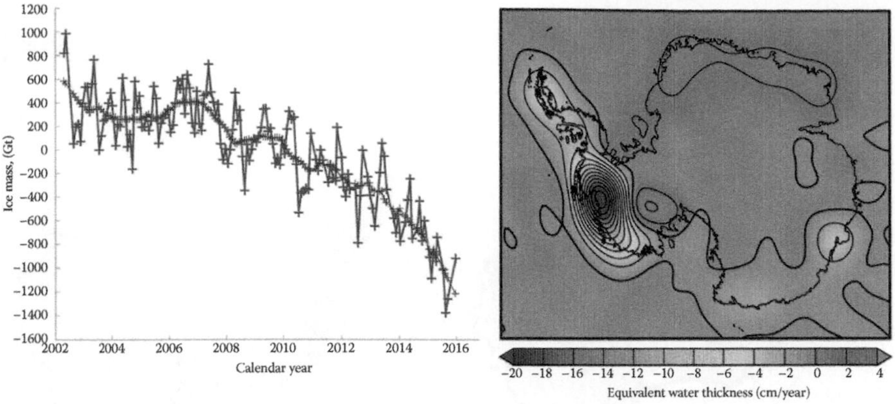

(a) (b)

FIGURE 15.8 Time series of ice mass, in Gt (Gigatonne), for the Antarctic ice sheet estimated from GRACE are shown in blue. Also shown, in red, are data filtered for the seasonal dependence using a 13-month filter. The GRACE data have corrected Glacial Isostatic Adjustment (GIA) using the Ivins et al. (2013) model (a). Antarctica's linear trend in mass, or average mass loss, from 2002 to 2015 from GRACE, in cm/year of water (b).

techniques cannot provide. Altimetry helps identify regions of change, in particular over outlet glaciers. Ultimately, these three measurement techniques are used in combination to better resolve the mass balance and understand the physical processes that control it.

A recent study gathering the scientific community used these multi-geodetic techniques and estimated that from 1992 to 2011 Antarctica lost 71 ± 53 Gt/year, contributing to an increase in sea level of 0.2 mm/year (Shepherd et al. 2012). The eastern part of the Antarctic ice sheet appears to be still in a state of mass balance. The combination of methods suggests a possible mass gain of 14 ± 43 Gt/y. The most important mass changes are concentrated in the western part where 85 ± 40 Gt of mass is lost every year. This loss of mass increased to 141 ± 9 Gt/y between January 2003 and June 2014 (Sutterley et al. 2015), leading to a mass balance for the entire ice sheet of −92 ± 10 Gt/y (Velicogna and Wahr 2013). This mass balance estimate agrees with that estimated from altimetry data of −84 ± 22 Gt/y between January 2003 and December 2013 (Martin-Espagnol 2015) and an estimate for 2002 to 2015 of 95 ± 50 Gt/y using GRACE data (Forsberg et al. 2016) (Figure 15.8). Changes in mass, estimated for 2002 to 2015 using GRACE observations are larger for the Greenland ice sheet. Forsberg et al. (2017) estimated a loss of 264 ± 25 Gt/y. The record shows significant acceleration in mass loss rate, in particular during summer 2012. Mass decrease is still continuing.

On average, linear trends in surface elevation estimated from altimetry range between ± 0.15 m/y (Figure 15.9). Comparing the trends for the ERS-2 period (1995–2003) and for Envisat (2003–2010) reveals changes in signs in several places in East Antarctica. Those observed along the coast of Queen Maud Land (see Figure 15.1) reflect a multi-year change in snowfall (Mémin et al. 2015). Interannual changes in ice sheet accumulation are quasi-periodic with a period of 4–6 years leading to anomalies that propagate eastward and circle Antarctica in 9–10 years. The spatiotemporal properties of the anomalies are correlated with that expected from the Antarctic circumpolar wave, known to affect several key climate parameters (White and Peterson 1996; White and Cherry 1998; White 2000; Fischer et al. 2004; Autret et al. 2013).

15.6.5 Acceleration of Outlet Glaciers

As mentioned in Section 15.6.3, we observe a continuous mass loss from glaciers in the western part of the Antarctic ice sheet that is increasing with time (Flament and Rémy 2012; Mouginot

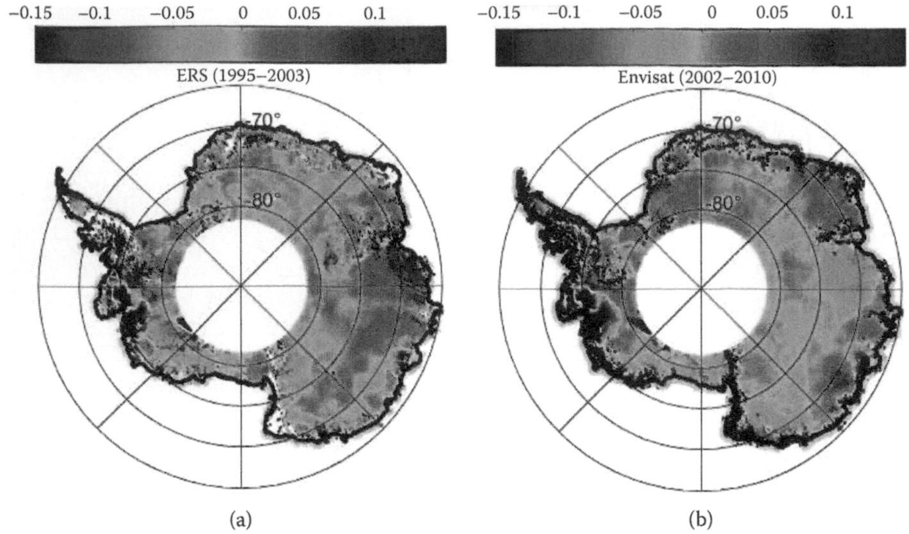

FIGURE 15.9 Changes in Antarctic height (m/y) measured during the ERS-2 (a) and Envisat (b) period. We notice a systematic mass loss from glaciers west of the Amundsen Sea and opposite signals in the eastern side.

et al. 2014). For example, the loss in volume from Pine Island Glacier has increased from 7 km^3/year between 2002 and 2006 to 48 km^3/year between 2006 and 2010. The thinning propagates upstream and is currently detected 300 km from the coast. This sector of Antarctica could have catastrophic consequences for sea level because it contains enough ice to raise sea level by 1.2 m and could entrain the collapse of the rest of West Antarctica for a potential sea level rise of 3 m.

The rapid changes in the dynamics of glaciers (e.g., flowing into the Amundsen Sea Embayment sector of West Antarctica) lead to a different view of the ice sheet evolution from that driven by uniform velocities of the ice sheet interior. In Greenland, the acceleration of glaciers is affected by the increase in surface melt, the warmer ocean temperature, and the disappearance of floating sections or ice tongues. Melting at the surface is not a significant driver in Antarctica; the dominant forcing is the intrusion of subsurface, warm, salty waters beneath the ice shelves. Up to now, numerical ice sheet models have focused on changes over long periods of time. These models are now evolving toward shorter timescales and must include processes taking place at the grounding line, where the glacier ice meets with the ocean (Figure 15.2). What happens at the grounding line depends in part on the bedrock slope (Durand et al. 2009). It is also important to understand how ice interacts with the ocean, which brings issues about the shape of the sea floor in front of the glaciers and beneath ice shelves, ocean circulation in front of the glaciers and its connections with larger scale circulation, and a detailed knowledge of the bed topography over the entire domain. Among these, detailed surface topography will provide important observational constraints on the rates of ice sheet thinning, glacier thinning, ice shelf melting, and migrating drainage boundaries in areas of most significant changes. Hence, the trend is for more precise elevation data, higher spatial resolution, and higher temporal revisit to examine details at the glacier per glacier level.

15.7 SUMMARY AND PERSPECTIVE

Radar altimetry for polar ice sheets started in 1991 with the launch of ERS-1. Over the last 25 years, altimeter satellites allowed significant improvement in understanding the dynamics of the ice sheets. Providing elevation time series, however, requires special attention due to the penetration of radar signals in snow and the lack of long and dense time series of measurements. With about 50 repetitive cycles, most errors are significantly reduced. Envisat's 85 cycles have allowed us to perform robust inversions, providing the seasonal signal and trends over different time intervals. The mission detected subglacial lakes and coherent structures with low frequencies linked to ice flow and offered new constraints for ice flow models. In combination with gravity measurements, ice velocity measurements from radar interferometry and reconstruction of surface mass balance from regional atmospheric climate models, radar altimetry has helped constrain estimates of the mass balance of the Antarctic ice sheet and determine the partitioning of the changes between surface/atmospheric processes and changes in ice dynamics. With new instruments such as SARAL, Sentinel-3, and others, we will be able to extend the time series started in 1991 and develop a better understanding of the physics of the measurements but also obtain data with better spatial resolution. The ICESat-2 laser altimetry will be launched in late 2018, with multiple beams allowing a direct determination of the surface slope across the track, denser time series, and continuous data for several years. The GRACE satellite mission will have a successor, GRACE Follow-On (GRACE-FO) to be launched in 2018, which will improve the precision of gravity measurements. The NASA Indian Space Agency (ISRO), NISAR, radar interferometry mission to be launched in 2020 will complement the existing series of SAR satellites including RADARSAT-2, TerraSAR-X, and Sentinel-1a/b (Kumar et al. 2016). Collectively, these instruments and others will significantly improve our current knowledge of the mass balance of ice sheets.

REFERENCES

Abdalati, et al. 2010. The ICESat-2 laser altimetry mission. *Proc. IEEE*, 98, 735–751.

Autret, G., Rémy, F., Roques, S., 2013. Multi-scale analysis of Antarctic surface temperature series by empirical mode decomposition. *J. Atmos. Ocean. Technol.* 30, 649–654. doi: 10.1175/JTECH-D-11-00050.1.

Barletta, V.R., Sørensen, L.S., Forsberg, R., 2013. Scatter of mass changes estimates at basin scale for Greenland and Antarctica. *Cryosphere* 7, 1411–1432. doi: 10.5194/tc-7-1411-2013.

Durand, G., Gagliardini, O., de Fleurian, B., Zwinger, T., Le Meur, E., 2009. Marine ice sheet dynamics: Hysteresis and neutral equilibrium. *J. Geophys. Res.* 114, 1–10.

EPICA Community Members, 2004. Eight glacial cycles from an Antarctic ice core. *Nature* 429: 623–628.

Fischer, H., Traufetter, F., Oerter, H., Weller, R., Miller, H., 2004. Prevalence of the Antarctic Circumpolar Wave over the last two millennia recorded in Dronning Maud Land ice. *Geophys. Res. Lett.* 31, L08202. doi: 10.1029/2003GL019186.

Flament, T., Berthier, E., Remy, F., 2014. Cascading water underneath the Antarctic ice sheet (Wilkes Land) observed using altimetry and digital elevation models. *Cryosphere* 8, 1–15.

Flament, T., Rémy, F., 2012. Dynamics thinning of Antarctic glaciers from along-track repeat radar altimetry. *J. Glaciology*, 58(211), 830–840.

Fretwell, P, Pritchard, H.D., Vaughan, D.G., Bamber, J.L., Barrand, N.E., Bell, R., Bianchi, C., et al., 2013. Bedmap2: Improved ice bed, surface and thickness data sets for Antarctica. *Cryosphere* 7, 375–393.

Fricker, H.A., T. Scambos, R. Bindschadler, L. Padman. 2007. An active subglacial water system in West Antarctica mapped from space. *Science* 315(5818), 1544–1548.

Forsberg, R., Simonsen, S.B. Sørensen, L. S., Nilsson, J, 2016. Mass balance of Greenland and the Canadian Ice Caps from combined altimetry and GRACE inversion., abstract from ESA Living Planet Symposium 2016, Prague, Czech Republic.

Forsberg, R., Simonsen, S.B. Sørensen, L.S., Nilsson, J, 2016. Mass balance of Greenland and the Canadian Ice Caps from combined altimetry and GRACE inversion., abstract from ESA Living Planet Symposium 2016. Prague, Czech Republic.

Forsberg, R., Sorensen, L., Simonsen, S., 2017. Greenland and Antarctica ice sheet mass changes and effects on global sea level. *Surv. Geophys.*, 38(1), 89–104.

Helsen, M.M., van den Broeke, M.R., van de Wal, R.S., van de Berg, W.J., van Meijgaard, E, Davis, C.H. and I. Goodwin, 2008. Elevation changes in Antarctica mainly determined by accumulation variability, *Science*, 320 (5883), 1626–1629.

IPCC, 2013, Intergovernmental Panel on Climate Change, Climate Change, The Physical Science Basis.

Ivins, E.R., James, T.S., Wahr, J., Schrama, J.O., Landerer, F.W., Simon, K.M., 2013. Antarctic contribution to sea level rise observed by GRACE with improved GIA correction. *J. Geophys. Res.* 118, 31268, 3126. doi: 10.1002/jgrb.50208.

Korona, J., Berthier, E., Bernard, M., Rémy, F., Thouvenot, E., 2009. SPIRIT: SPOT 5 stereoscopic survey of Polar ice: Reference images and topographies during the international Polar year. *ISPRS J. Photogram. Rem. Sens.*, 64, 204–212.

Kumar, R, Rosen, P., Misra, T., 2016. NASA-ISRO synthetic aperture radar: Science and applications. Proceedings of the SPIE 9881, Earth Observing Missions and Sensors: Development, Implementation, and Characterization IV, 988103, May 2, 2016. doi: 10.1117/12.2228027.

Ligtenberg, S. R. M., Helsen, M. M. and M. R. van den Broeke, 2011, An improved semi-empirical model for the densification of Antarctic firn, The cryosphere, 5, 809–819.

Martín-Espagnol, A., Zammit-Mangion, A., Clarke, P.J., Flament, T., Helm, V., King, M.A., Luthcke, S.B., et al. 2015. Spatial and temporal Antarctic Ice Sheet mass trends, glacio-isostatic adjustment, and surface processes from a joint inversion of satellite altimeter, gravity, and GPS data. *J. Geophys. Res.* 121, 182–200. doi: 10.1002/2015JF003550.

Michel, A., Flament, T., Remy, F., 2014. Study of the penetration bias of Envisat Altimeter observations over Antarctica in comparison to ICESat Observations. *Rem. Sens.*, 6, 9412–9434.

Mémin, A., Flament, T., Alizier, B., Watson, C., R, s, F., 2015. Detection of the Antarctic Circumpolar Wave in a combined analysis of satellite gravimetry and altimetry data. *Earth Planet. Sci. Lett.*, 422, 150–156.

Morris, E.M. and D. J Wingham, 2014, Densification of polar snow: Measurements, modeling, and implications for altimetry, *J. Geophys. Res.*, 119 (2), 349–365.

Morris, E. M., & Wingham, D. J., 2015, Uncertainty in mass-balance trends derived from altimetry: a case study along the EGIG line, central Greenland. *J. Glaciol.*, 61 (226), 345–356.

Mouginot, J., Rignot, E., Scheuchl B., 2014. Sustained increase in ice discharge from the Amundsen Sea Embayment, West Antarctica. *Geophys. Res. Lett.*, 41(5), 1576–1584.

Parrenin, F., Rémy, F., Ritz, C., Siegert, M., Jouzel, J., 2004. New modelling of the Vostok ice flow line and the implication fort the glaciological chronology of the Vostok ice core. *J. Geophys. Res.*, 109, D20102.

Pritchard, H.D., R.J. Arthern, D.G. Vaughan, and L.A. Edwards, 2009, 'Extensive dynamic thinning on the margins of the Greenland and Antarctic ice sheets', *Nature*, 461, 971–975.

Raney, R. K., 1998, The delay Doppler radar altimeter, *IEEE Trans. Geosci. Remote Sens.*, 36, 1578–1588.

Remy, F., Flament, T., Michel, A., Blumstein, D., 2015. Envisat and SARAL/AltiKa observations of the Antarctic ice sheet: A comparison between the Ku-band and the Ka-band. *Marine Geodesy*, 38(1), 510–521, 2015.

Remy, F., Flament, T., Michel, A., Verron, J., 2014. Ice sheet survey over Antarctica with satellite altimetry: ERS-2, EnviSat, SARAL/AltiKa, the key importance of continuous observations along the same repeat orbit. *Int. J. Rem. Sens.*, 35(14), 5497–5512.

Remy, F., Testut, L., Legrésy, B., 2002. Random fluctuations of snow accumulation over Antarctica and its relation with sea level change. *Clim. Dyn.*, 19, 267–276.

Rignot, E., Mouginot, J., Scheuchl, B., 2011a. Ice flow of the Antarctic Ice Sheet. *Science*, 333, 1427–1430. doi: 10.1126/science.1208336.

Rignot, E., Velicogna, I., van den Broeke, M., Monaghan, A., 2011b. Acceleration of the contribution of the Greenland and Antarctic ice sheets to sea level rise. *Geophys. Res. Lett.* 38, L05503. doi: 10.1029/2011GL046583.

Ritz, C., Rommelaere, V., Dumas, C., 2001. Modeling the evolution of Antarctic ice sheet over the last 420 000 years: Implications for altitude changes in the Vostok region. *J. Geophys. Res.*, 106(D23), 31943–31964.

Schek T. and Csatho, B., 2012, A new methodology for detecting ice sheet surface elevation changes from laser altimetry data. *IEEE Trans. Geosci. Remote Sens.* 50(9), 1–25.

Shepherd, A., Ivins, E.R., Geruo, A., Barletta, V.R., Bentley, M.J., Bettadpur, S., Briggs, K. H., et al. A reconciled estimate of ice-sheet mass balance. *Science*, 338, 1183–1189.

Smith, B.E., Fricker, H.A., Joughin, I.R., Tulaczyk, S., 2009. An inventory of active subglacial lakes in Antarctica detected by ICESat (2003–2008). *J. Glaciol.*, 55(192), 573–595.

Sutterley, T.C., Velicogna, I., Rignot, E., Mouginot, J., Flament, T., van den Broeke, M., van Wessem, J.M., Reijmer, C. H., 2015. Mass loss of the Amundsen Sea Embayment of West Antarctica from four independent techniques. *Geophys. Res. Lett.*, 41(23), 8421–8428.

Tapley, B.D., Bettadpur, S., Watkins, M., Reigber, C., 2004. The gravity recovery and climate experiment: Mission overview and early results. *Geophys. Res. Lett.*, 31, L09607., doi: 10.1029/2004GL019920.

Testut, L., Coleman, R., Remy, F., Legrésy, B., 2003. Precise drainage pattern of Antarctica derived from high resolution topography. *Ann. Glaciol.*, 37, 337–343.

Van de Berg WJ, van den Broeke MR, Reijmer CH, (2006). Reassessment of the Antarctic surface mass balance using calibrated output of a regional atmospheric climate model. *J. Geophys. Res.-Atmos.* 111(D11): D11104.

Velicogna, I., Wahr, J., 2013. Time-variable gravity observations of ice sheet mass balance: Precision and limitations of the GRACE satellite data. *Geophys. Res. Lett.*, 40. 3055–3063, doi: 10.1002/grl.50527.
White, B.W., 2000. Influence of the Antarctic Circumpolar Wave on Australian precipitation from 1958 to 1997. *J. Clim.*, 13, 2125–2141.
White, B.W., Cherry, N.J., 1998. Influence of the Antarctic Circumpolar Wave upon New Zealand temperature and precipitation during autumn–winter. *J. Clim.*, 12, 960–976.
White, B.W., Peterson, R.G., 1996. An Antarctic circumpolar wave in surface pressure, wind, temperature and sea-ice extent. *Nature*, 380, 699–702. doi: 10.1038/380699a0.
Whitehouse, P.L., Bentley, M.J., Milne, G.A., King, M.A., Thomas, I.D., 2012. A new glacial isostatic adjustment model for Antarctica: Calibrated and tested using observations of relative sea-level change and present-day uplift rates. *Geophys. J. Int.*, 190, 1464–1482. doi: 10.1111/j.1365-246X.2012.05557.x.

16 Advances in Imaging Small-Scale Seafloor and Sub-Seafloor Tectonic Fabric Using Satellite Altimetry

R. Dietmar Müller, Kara J. Matthews, and David T. Sandwell

Key words: tectonics, microplates, fracture zone, passive margin, sedimentary basin, fault, mid-ocean ridge

16.1 INTRODUCTION

In this chapter, we first review advances in satellite altimetry for imaging small-scale structures on the seafloor, focusing on oceanic microplates and fracture zones, followed by imaging fault systems on stretched continental crust. The two examples we focus on are on the Falkland Plateau and the Lord Howe Rise, both stretched, submerged continental plateaus where satellite altimetry data now allow the mapping of detailed fault fabrics. We also review the utility of satellite altimetry data for understanding the timing and nature of major tectonic reorganizations in the ocean basins.

16.2 SATELLITE-DERIVED GRAVITY FOR TECTONIC MAPPING

16.2.1 Brief History

There are two approaches for mapping the topography and thus the tectonics of the deep ocean. Shipboard surveys, using multibeam sonar technology, provide the highest resolution (approximately 100 m) and accuracy (approximately 10 m) but to date only about 11% of the deep ocean has been mapped with this technology. Another 5.5% of the seafloor has been mapped by older, single-beam sonars that offer only about 1000 m spatial resolution. These older data are critical for filling large coverage gaps in the most remote ocean areas. The topography of the remaining 83% of the seafloor is mapped at low resolution (approximately 6000 m) from satellite-derived gravity. This is possible because the topography of the seafloor represents a large contrast in density between the seawater and the rock. Topographic highs associated with, for example, ridges or seamounts have an extra mass relative to the surrounding seafloor. This extra mass produces a local increase in the pull of gravity or a gravity anomaly. Moreover, seafloor topography that is buried by lower density sediments will also produce gravity anomalies because the sediments have a lower density than the basement rock. Thus, an accurate mapping of the gravity field can be used to infer seafloor topography even if it is buried by sediment.

Satellite altimeters can provide high spatial resolution maps of the ocean surface topography, which to a first approximation is an equipotential surface called the geoid. As will be discussed, the geoid maps can be converted to maps of gravity anomaly and thus can reveal the tectonics of the seafloor. The accuracy and resolution of these altimeter-derived maps depend on three factors: spatial track density, altimeter range precision, and diverse track orientation. Over the past 40 years, there have been a number of altimeter missions having the ability to map seafloor tectonics but only a few missions have the dense track coverage needed for mapping at high spatial resolution. Next, we review the significant milestones in tectonic mapping from altimetry.

Seasat: The first such maps were provided by the Seasat altimeter mission. Although the satellite failed after just 90 days in orbit, the gravity maps constructed by Haxby et al. (1983) revolutionized our understanding of deep ocean tectonics. In particular, the Seasat gravity maps revealed large, previously unknown seamount chains in the southern oceans as well as a more accurate mapping of the plate boundaries.

ERS-1 and GEOSAT: In mid-1995, better altimeter coverage became available from the European Remote Sensing (ERS) 1 altimeter (European Space Agency) and the GEOSAT altimeter (U.S. Navy). The GEOSAT satellite, launched in 1985, collected high-precision sea surface height measurements in a non-repeat orbit for 1.5 years and continued in a repeating orbit for another 3 years. These data remained classified until the ERS-1 mission completed a 1-year geodetic mapping phase. The combined ERS-1 and GEOSAT mappings provided a major improvement in gravity field accuracy and resolution (Andersen and Knudsen 1998; Cazenave et al. 1996; Hwang and Parsons 1996; Sandwell and Smith 1997; Tapley and Kim 2001) that prompted a second revolution in our understanding of the tectonics of the deep oceans (e.g., Müller et al. 1997; Sandwell et al. 2006; Wessel 2001). The next significant improvement came from a more precise analysis of the raw radar waveforms from ERS-1 and GEOSAT (Maus et al. 1998; Sandwell and Smith 2005). The two-pass waveform retracking approach led to a factor of 1.5 improvement in the accuracy of the marine gravity field that revealed the segmentation of the seafloor spreading ridges as well as a partial mapping of the abyssal fabric on the slow spreading ridges where abyssal relief can exceed 1000 m. In addition, the improved gravity models also revealed thousands of previously uncharted seamounts (Wessel 2001).

Jason-1 and CryoSat-2: Over the past 6 years, two new satellite altimeters have provided a dramatically improved view of the tectonics of the deep ocean floor (Sandwell et al. 2014). Most of the new information comes from the CryoSat-2 satellite launched by the European Space Agency in 2010. CryoSat-2 has routinely collected altimetry data over ice, land, and oceans since July 2010 (Wingham et al. 2006). The satellite has a long, 369-day repeat cycle resulting in an average ground track spacing of 3.5 km at the equator. To date, it has completed more than six geodetic mappings of the ocean surface. These data are augmented by a complete 14-month geodetic mapping of the ocean surface by Jason-1 from its lower inclination orbit of 66° that compliments the higher inclination orbit CryoSat-2 (88°); the lower inclination provides more accurate recovery of the east-west sea surface slope, especially at low latitudes. The most recent global marine gravity anomaly map (Figure 16.1) based on all of these geodetic mission data with two-pass retracking for optimal range precision (Garcia et al. 2014) has an accuracy that is two times better than the maps derived from GEOSAT and ERS-1.

16.2.2 Methodology and Limitations

Radar altimeters measure the topography of the ocean surface, which is nearly an equipotential surface of the Earth called the geoid. Geoid height, N, is the deviation in the shape of the Earth from an ellipsoidal shape defined by the WGS84 ellipsoid. Typical variations in geoid height are ± 33 m whereas variations in ocean surface height due to currents, and eddies are ± 1 m. At the horizontal length scales of interest for seafloor mapping of less than 50 km, geoid height variations dominate the sub-mesoscale oceanographic signals except in areas of high mesoscale variability, such as the eastward extension of the Gulf Stream. In these areas, we rely on temporal averaging to reduce the oceanographic noise. We enhance these small-spatial scale signals in geoid by taking first and second horizontal derivatives. This also provides us with gravity anomaly and vertical gravity gradient, respectively. Let $\Phi(x,y,z)$ be the anomalous gravitational potential as a function of a local Cartesian coordinate system where z is up. Variations in potential cause in bumps in the geoid height through Brun's formula (Heiskanen and Moritz 1967)

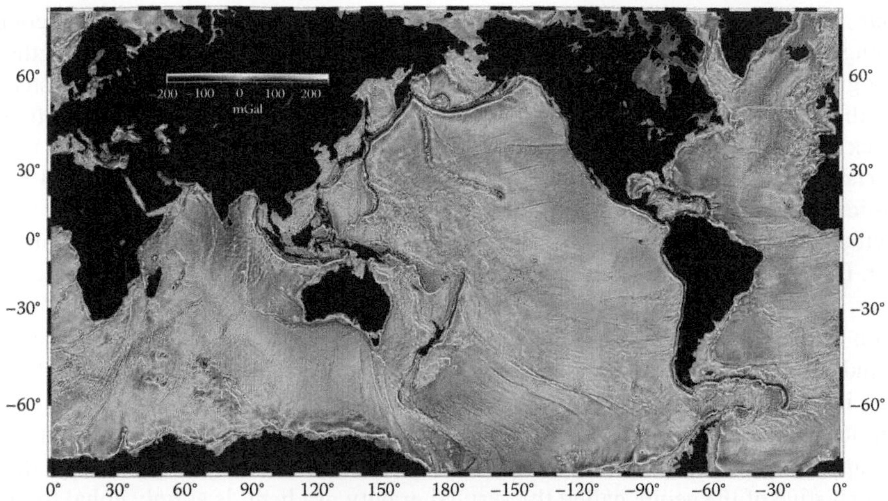

FIGURE 16.1 Gravity anomaly in milligal from GEOSAT, ERS-1, Envisat, Jason-1, and Cryosat-2 radar altimetry. Grid is available at: ftp://topex.ucsd.edu/pub/global_grav_1min

$$N(x,y) = \frac{\Phi(x,y,0)}{g}$$

where g is the average local acceleration of gravity (approximately 9.8 ms^2). A positive potential anomaly produces a positive geoid height because for the ocean surface to remain at an equipotential it must be displaced away from the center of the Earth to a lower potential to offset the local high in the potential. For the areas on and above the ocean surface, the anomalous potential satisfies Laplace's equation.

$$\frac{\partial^2 \Phi}{\partial x^2} + \frac{\partial^2 \Phi}{\partial y^2} + \frac{\partial^2 \Phi}{\partial z^2} = 0$$

The gravity anomaly is the first vertical derivative of the potential anomaly, so $\Delta g = -\frac{\partial \phi}{\partial z}$ and the vertical gravity gradient (VGG) is $\frac{\partial \Delta g}{\partial z} = -\frac{\partial^2 \phi}{\partial z^2}$. Therefore, using Laplace's equation and Brun's formula, it is clear that the VGG is the curvature of the ocean surface

$$\frac{\partial \Delta g}{\partial z} = g\left(\frac{\partial^2 N}{\partial x^2} + \frac{\partial^2 N}{\partial y^2}\right)$$

There is a similar relationship between gravity anomaly and first horizontal derivatives of the geoid that is more easily derived in the Fourier transform domain (Haxby et al. 1983). Note that both the gravity anomaly and the VGG require that the geoid or sea surface height is resolved in both horizontal dimensions. This requires complete altimeter coverage of the surface. Moreover, the errors in the gravity and VGG are smaller when the altimeter tracks are more orthogonal.

For our investigation of seafloor tectonics, we use Laplace's equation in spherical coordinates at wavelengths longer than 20 km and in Cartesian coordinates at shorter wavelengths to construct the gravity anomaly and vertical gravity gradient. Because we are not interested in constructing the geoid, we perform the first derivative directly on the raw altimeter profiles. These along-track slopes are combined in a biharmonic spline analysis to construct east and north grids of sea surface slope (e.g., Sandwell and Smith 2009). As discussed in Olgiati et al. (1995), there are significant computational and accuracy benefits to constructing gravity from along-track slopes rather than first constructing a geoid model and then differentiating to obtain gravity. Moreover, factors that affect the absolute height accuracy of altimetric sea level—such as radial orbit error, ionosphere/troposphere delays and deep ocean tides (Chelton et al. 2001)—have correlation scales long enough that they yield negligible error (1 microradian) in along-track slope (Sandwell and Smith 2009). For gravity recovery at length scales less than 50 km, the remaining important error sources are coastal tide model errors, sub-mesoscale variability, and atmospheric fronts with sharp variations in wet tropospheric delay. We rely on averaging over several mapping cycles to reduce the noise of these sources. As shown previously, the second vertical derivative of the geoid, called the vertical gravity gradient, is simply equal to the curvature of the ocean surface through Laplace's equation. Therefore, computation of VGG is a local operation that does not suffer from the coastal Fourier edge effects that occur when computing vertical gravity from the sea surface slope.

We emphasize that taking the first derivative of the geoid to form gravity anomaly enhances the short wavelength tectonic signals of interest. Moreover, the second derivative needed to form the VGG further enhances the shortest wavelengths, but it also enhances the short wavelength altimeter noise. The error budget for recovery of gravity anomaly and especially VGG is dominated by the range precision of the radar measurement. The main source of environmental noise is the roughness of the ocean surface due to ocean waves. Also note that the amplitude of the gravity signal decreases exponentially with decreasing wavelength for wavelengths smaller than the mean ocean depth. Therefore, a factor of two improvement in the spatial resolution requires a factor of four reduction in altimeter noise. This can either be achieved by improved radar technology or by repeating the measurement 16 times.

16.2.3 Improved Radar Technology—Current and Future

The most important contribution of the new altimeters (Jason-1 and CryoSat-2) is related to a 1.25 times improvement in range precision with respect to the older altimeters (ERS-1 and GEOSAT) (Garcia et al. 2014). The newer altimeters have pulse repetition frequencies (PRFs) of 1950 Hz and 2060 Hz, respectively, while the older instruments were technologically limited to lower values of 1020 Hz. Theoretically, this approximate doubling of PRF should result in a square root of two improvement in range precision; the actual improvement is somewhat smaller (1.25) perhaps reflecting the onset of echo correlation at the 2 kHz PRF. Nevertheless, this improvement in range precision maps directly into an improvement in gravity field accuracy.

CryoSat-2 was also operated in a new Synthetic Aperture Radar (SAR) mode over very limited areas of the oceans. This mode has a much higher PRF of 18.2 kHz, and the highly correlated echoes are summed coherently in bursts of 64 pulses to form a long synthetic aperture. This enhances along-track resolution in the form of a set of narrow beams distributed in the along-track direction (Wingham et al. 2006). Unlike the conventional pulse-width limited geometry, the resulting echo waveforms have useful information in both the leading and trailing edges. This, together with an increase in the effective number of independent samples resulting from the SAR technique, reduces the height noise by a factor of approximately 1.4 compared to conventional altimeters. Comparison of height noise performance indeed shows this expected improvement for CryoSat-2's SAR, but similar gains for pulse-width limited echoes are obtained by a two-pass processing scheme in which the slowly varying ocean wave height is first estimated and smoothed and then excluded from the estimation process in the second pass (Garcia et al. 2014).

There are two technological developments that will further improve the altimeter range precision. The first is a new shorter wavelength altimeter called AltiKa aboard the SARAL spacecraft launched by the Indian Space Research Organization (ISRO) and French space agency, Centre National d'Etudes Spatiales (CNES), in 2013. Recent studies show that the range precision of AltiKa is two times better than Ku-band altimeters operated in non-SAR mode (e.g., Jason-1 and Cryosat-2 low resolution mode, or LRM) (Smith 2015; Zhang and Sandwell 2016). In July 2016, the orbit of SARAL/AltiKa was changed from the 35-day repeat trackline of ERS and Envisat to a new non-repeat geodetic mission. There will be another major improvement in the accuracy of the gravity field if it survives for 1.3 years to obtain complete coverage. The second technology is a planned swath altimeter mission called Surface Water and Ocean Topography (SWOT) scheduled for launch in 2021. Engineering estimates of range precision are up to a factor of five times better than the current altimeters, so one can expect another significant boost in resolution in 2025.

The remainder of this chapter is focused on the imaging of small-scale tectonics in the deep oceans and continental margins using mainly maps of VGG, especially in areas with enigmatic, poorly understood seafloor tectonic fabric such as seen in the Java Sea (Figure 16.2). The curved lineaments shown here were recently discussed by Zahirovic et al. (2016), who proposed that they may reflect successive generations of volcanic arcs that are now curved because of oroclinal bending in a compressional episode driven by the rotation of Borneo. We also note that marine gravity has many other applications, such as inertial guidance of moving platforms, planning shipboard surveys, petroleum exploration, estimating the strength of the lithosphere, and searching for meteorite impacts on the ocean floor. The gravity is also used for predicting seafloor bathymetry in uncharted areas.

16.3 OCEANIC MICROPLATES

Oceanic microplates are small and independently rotating rigid plates that form at spreading centers, either between two plates or at a triple junction, due to ridges propagating into, and capturing fragments of, existing oceanic lithosphere. They proceed to grow due to accretion at the new spreading center. Based on our knowledge from well-mapped microplates, they range in size from only a couple of thousand to more than 1.2×10^6 km^2 in area and are active for less than a few million years to around 10 million years. Although microplates are short-lived, transient features, they are an important record of spreading center reorganization and periods of asymmetric lithospheric accretion. Microplate formation at spreading ridges has almost exclusively been studied in the Pacific, particularly in the eastern Pacific, for more than 40 years. Here, there are several actively forming microplates (e.g., Easter, Juan Fernandez, Galapagos and North Galapagos) and many extinct Cenozoic ones that have since been detached from an active spreading ridge (e.g., Mathematician, Bauer, Friday, Selkirk and Hudson) (Figure 16.3), enabling their formation and associated seafloor structures to be studied in depth. Yet the recent discovery of a microplate in the Indian Ocean, based on advances in seafloor imaging using satellite altimetry, now enables the study of microplate formation outside of the Pacific to gain a broader understanding of where and how these structures develop (Matthews et al. 2016).

16.3.1 Models for Microplate Formation

In the wake of the pioneering work of Hey and colleagues on propagating rifts (e.g., Hey 1977; Hey et al. 1980), the origins and mechanisms of microplate formation received considerable attention, particularly in the 1980s and 1990s, as ship-derived geophysical data sets enabling mapping along the East Pacific Rise expanded (see Searle 2013, for a review). The pivotal "edge-driven" model of microplate formation, presented by Schouten et al. (1993), focuses on

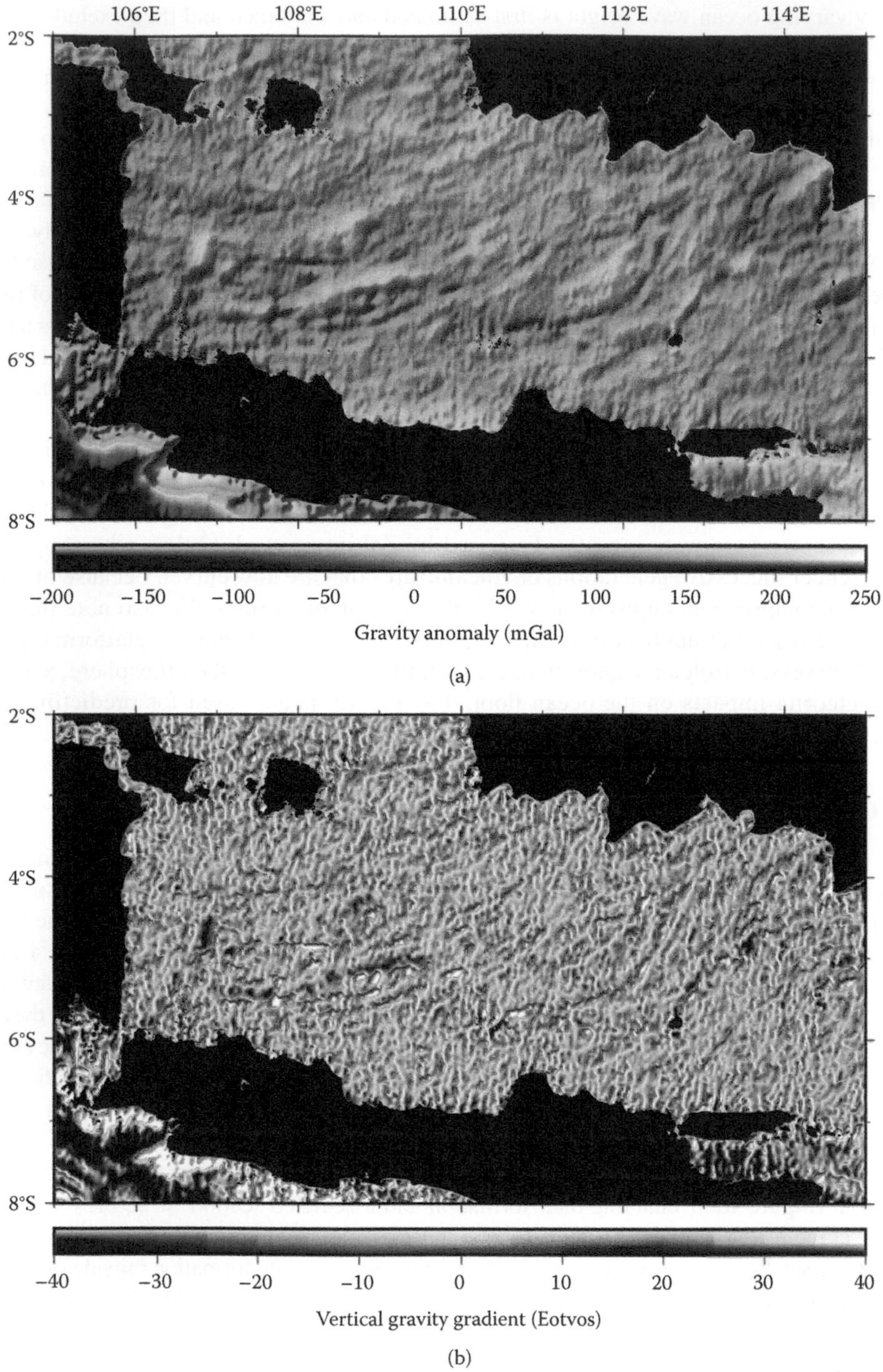

FIGURE 16.2 (a) Gravity anomaly of the Java Sea. (b) Vertical gravity gradient (VGG) of Java Sea reveals small-scale anomalies related to basement structure.

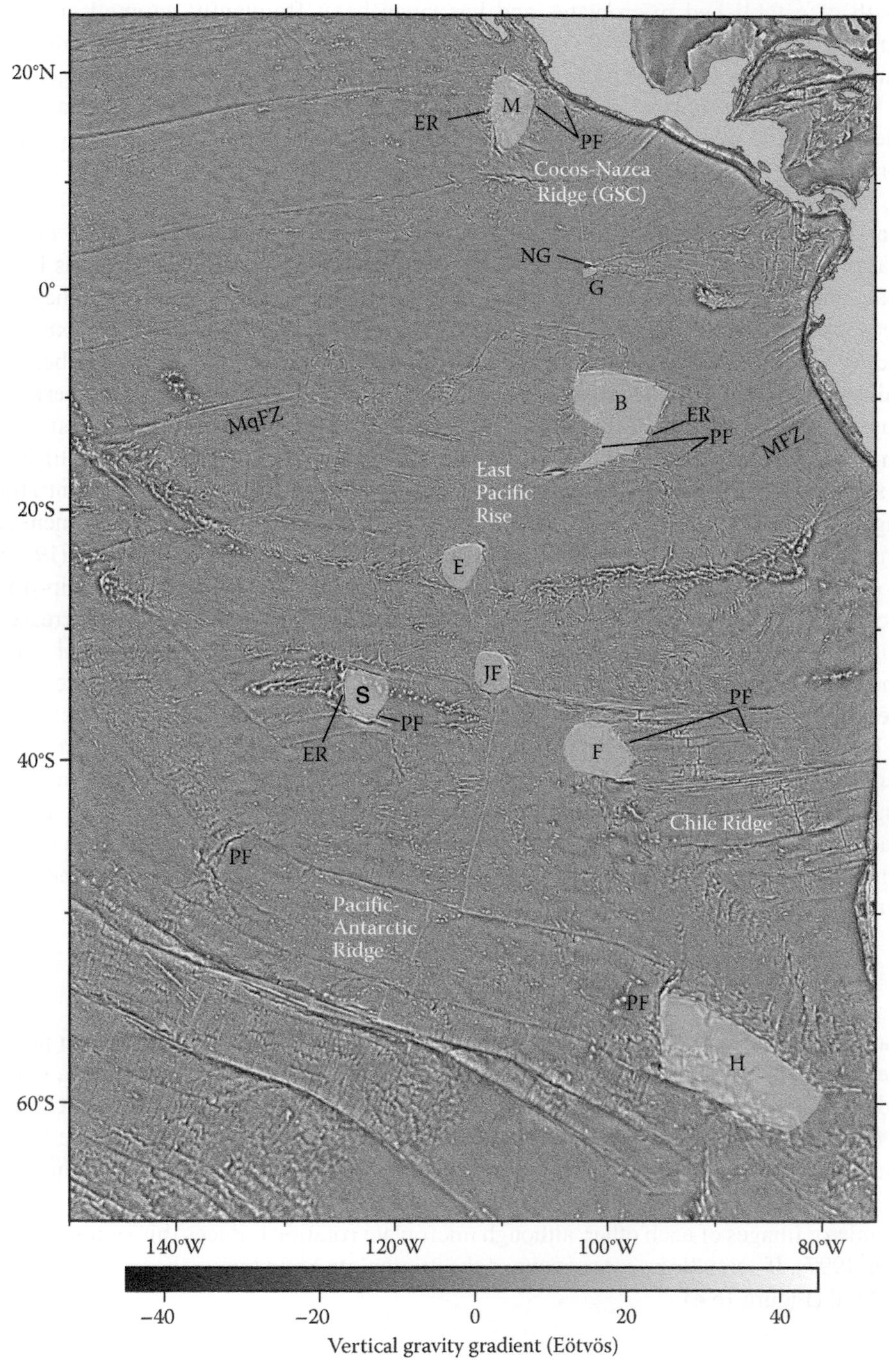

FIGURE 16.3 Pacific vertical gravity gradient map and microplate vertical gravity gradient (VGG) map (Sandwell et al. 2014) of the eastern Pacific Ocean basin showing active (yellow) and extinct microplates (green). The Pacific–Antarctic Ridge and East Pacific Rise produce positive VGG anomalies, while the Chile Ridge produces a negative VGG anomaly. Examples of the pseudo faults (PF) and extinct ridges (ER) associated with the extinct microplates are highlighted. B, Bauer Microplate; E, Easter Microplate; F, Friday Microplate; G, Galapagos Microplate; GSC, Galapagos Spreading Centre; H, Hudson Microplate; JF, Juan Fernandez Microplate; M, Mathematician Microplate; MFZ, Mendaña Fracture Zone; MqFZ, Marquesas Fracture Zone; NG, North Galapagos Microplate; S, Selkirk Microplate.

the growth of established microplates and has since been frequently adopted and modified. According to Schouten et al.'s (1993) "edge-driven" model of microplate formation, rotation is driven by the drag of the bounding plates and is analogous to the motion of a gear or roller bearing between two pieces of wood that are sliding past each other. An important aspect of the model is that the microplate has two axes of rotation located at each of its margins with the neighboring plates, and these are no-slip boundaries. Furthermore, the microplate is assumed to be rigid, with deformation confined to its edges.

Bird and Naar (1994) presented a model for microplate formation based on their investigation of the Easter and Juan Fernandez microplates. In their model, microplate formation is linked with rift propagation from transform faults, possibly at intratransform spreading centers. The microplate forms between the existing spreading ridge and the propagating ridge. One spreading axis eventually dies, leading to microplate detachment. Bird and Naar (1994) recognized that a number of factors may facilitate rift propagation from the transform fault. For instance, warmer lithospheric temperatures may reduce the effective lithospheric thickness difference across the intratransform region; warm and young lithosphere, such as at an intratransform spreading center, is more likely to concentrate stress and crack; and finally, excess relief due to the excess melt supply at an intratransform spreading center may be a driving force for ridge propagation (Phipps Morgan and Parmentier 1985).

Eakins (2002) built on the models of Schouten et al. (1993) and Bird and Naar (1994), based on studying formation of the Bauer and Hudson microplates from high-resolution ship-track data. In his model for microplate initiation and growth, microplates form when coupling increases across a free slip boundary, such as a transform. This increased coupling leads to tearing of one or both plates and the initiation of a propagator. As the propagator grows, the microplate core rotates as it is dragged by the adjacent plates. This initial increase in shear coupling across a previously free-slipping boundary could be caused by plate motion changes that either lead to transtension (Bird and Naar 1994) or transpression. According to Eakins (2002), propagation from an intratransform spreading center is not necessary but may facilitate the process and localize the rifting. Four drivers and/or facilitators of microplate formation are currently recognized that can occur in combination with each other. These include plate reorganizations, hotspot activity, fast spreading, and triple junction migration (Hey 2004).

16.3.2 Associated Seafloor Structures

Microplates are associated with oblique seafloor structures rather than those parallel or perpendicular to the direction of spreading prior to ridge propagation (Hey 2004). For example, due to their independent rotation, fan-shaped patterns of isochrons and rotated abyssal hills are characteristic features (Eakins 2002; Hey 2004; Hey et al. 1988; Mammerickx et al. 1988). Ridge propagation during microplate formation produces a pair of pseudo faults, which are linear boundaries between the young lithosphere accreted at the new ridge and the older preexisting lithosphere. Pseudo faults are often mirror images of each other, although microplate rotation reduces this symmetry (Tebbens and Cande 1997). If spreading ceases at one of the microplate boundaries, then an extinct ridge will be left behind (Figure 16.4).

16.3.3 Recent Advances in Mapping the Structure and History of Microplates Using Satellite Altimetry

Seafloor structures that are produced during microplate formation have typically been mapped using ship-track data sets, yet the larger structures such as pseudo faults, extinct ridges, and sometimes abyssal hills can also be illuminated in satellite altimetry-derived gravity maps of the seafloor, which provide a proxy for seafloor topography. In particular, VGG data better resolve short-wavelength linear tectonic structures compared to free-air gravity data, and the recent near-global VGG

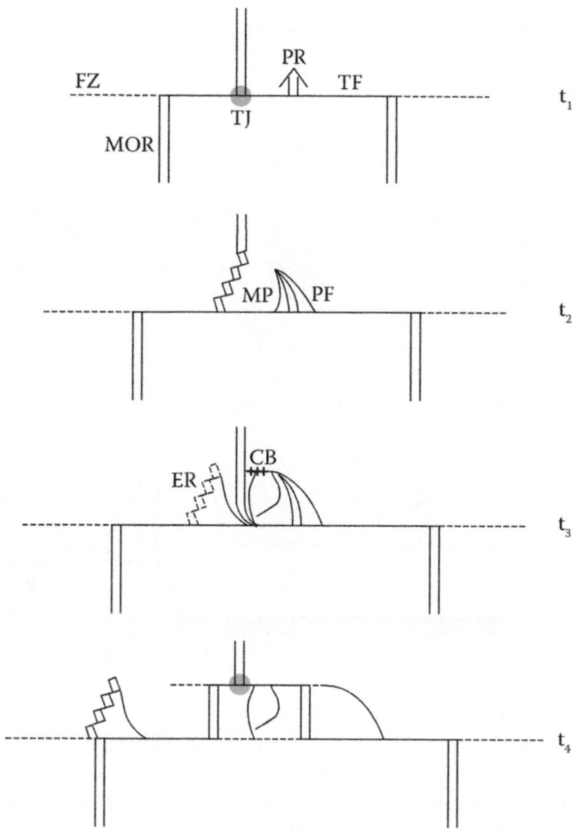

FIGURE 16.4 Microplate formation and triple junction migration. Bird et al.'s (1999) model for microplate formation associated with triple junction migration as has been observed in the Pacific. The model begins with a new ridge propagating from a transform fault at the ridge-transform-transform triple junction (t_1). The microplate forms between the propagating ridge and the preexisting triple junction ridge (t_{2-3}). Once the transform fault that forms the southern boundary of the microplate "freezes," the microplate detaches (t_4). In this particular model, the triple junction migrates northward and the microplate attaches to the southern plate at the triple junction. However, the ridge can propagate into any of the three plates at the triple junction. The triple junction (highlighted pink) is only present during t_1 and t_4.

data set of Sandwell et al. (2014) can resolve features as small as 6 km in width. Improvements in satellite data not only reveal new structures on the seafloor but also confirm the identification of features that were previously only poorly resolved and thus uncertain.

16.3.3.1 Indian Ocean

The mapping of seafloor tectonic fabric in the Indian Ocean using the VGG data set of Sandwell et al. (2014) recently led to the discovery of the approximately 47-Ma Mammerickx Microplate (Figure 16.5), an extinct microplate west of the Ninetyeast Ridge, centered on approximately 83°E and 21.5°S (Matthews et al. 2016). Prior to this study, remnants of a system of dual spreading and ridge rotation associated with Pacific-style microplate formation had not yet been clearly identified in the Indian or Atlantic oceans. The Mammerickx Microplate is bounded in the north by an extinct ridge and in the south by a pseudo fault, indicating that microplate formation occurred during a southward relocation of the Antarctic-Indian spreading ridge that isolated a fragment of the Antarctic Plate (Figure 16.5b). The conjugate pseudo fault to that which forms the southern boundary of the microplate is located on the Antarctic Plate, centered on 65°E and 41°S to the north of

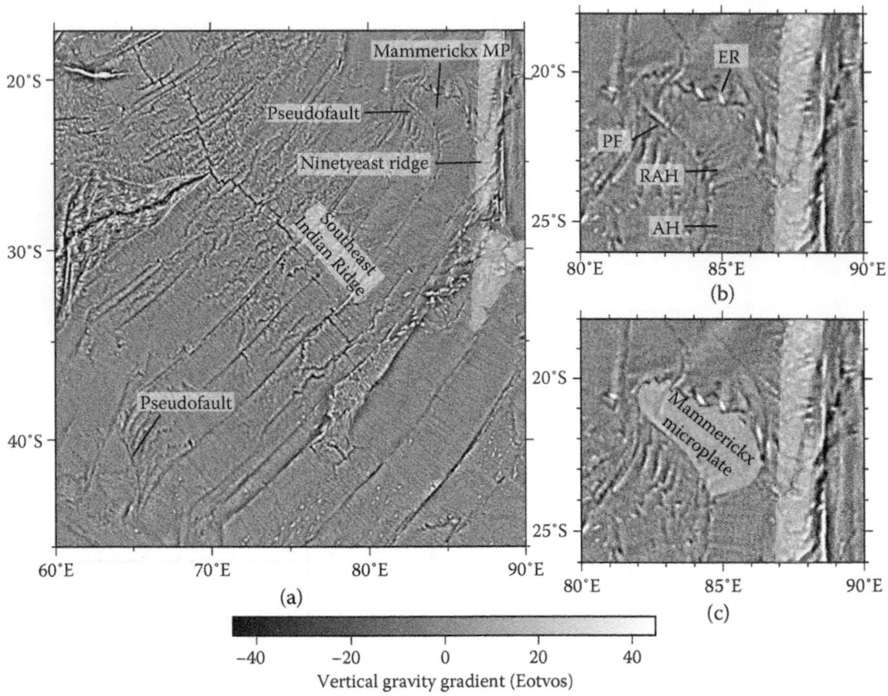

FIGURE 16.5 (a–c) Mammerickx Microplate, Indian Ocean. Vertical gravity gradient (VGG) maps (Sandwell et al. 2014) of the Indian Ocean showing the Mammerickx Microplate and associated structures. Large igneous provinces are shaded orange (Whittaker et al. 2015). AH, abyssal hills; ER, extinct ridge; MP, microplate; PF, pseudo fault; RAH, rotated abyssal hills.

the Kerguelen Plateau (Figure 16.5a). These pseudo faults were produced during rapid (likely less than 2 million years) westward ridge propagation from a developing transform fault (Matthews et al. 2016). Independent microplate rotation is indicated by a slight asymmetry in the pseudo fault pair, as well as counterclockwise rotated abyssal hill fabric that formed during rotation and southward growth of the dying spreading ridge, which was also confirmed by ship-track bathymetry data. Matthews et al. (2016) provided a detailed review of the potential driving mechanisms for the formation of the microplate and proposed that it was likely triggered by a plate reorganization, namely India-Eurasia collision, and fast spreading and hotspot activity, which had been present for tens of millions of years in the region, facilitated the event by contributing to the production of warm, thin, and weak lithosphere. The use of satellite gravity data, with the highest resolution yet published (Sandwell et al. 2014), to map microplate structures is methodologically significant (Goff and Cochran 1996) as typically these structures have been investigated using expensive ship-track data.

16.3.3.2 Pacific Ocean

The central Pacific region is characterized by extremely fast full spreading rates between the Pacific and Nazca plates, which rose from about 10 cm/year to more than 17 cm/year in the mid-late Cenozoic between magnetic chrons 18 (40 Ma) and 5 (11 Ma) north of the Chile Fracture zone (Wright et al. 2016). Such fast spreading rates are known to facilitate ridge propagation and microplate formation as they produce a wide region of hot, relatively thin lithosphere along the spreading ridge and enhance stress changes along the ridge (Hey 2004; Hey et al. 1985). The spreading asymmetry analysis by Wright et al. (2016) indicates large asymmetries in the spreading corridor east of the Juan Fernandez Microplate (Figure 16.6), with about 55%–60% of the crust generated at the mid-ocean ridge left on the Nazca Plate. In this region, the VGG data set of Sandwell et al. (2014)

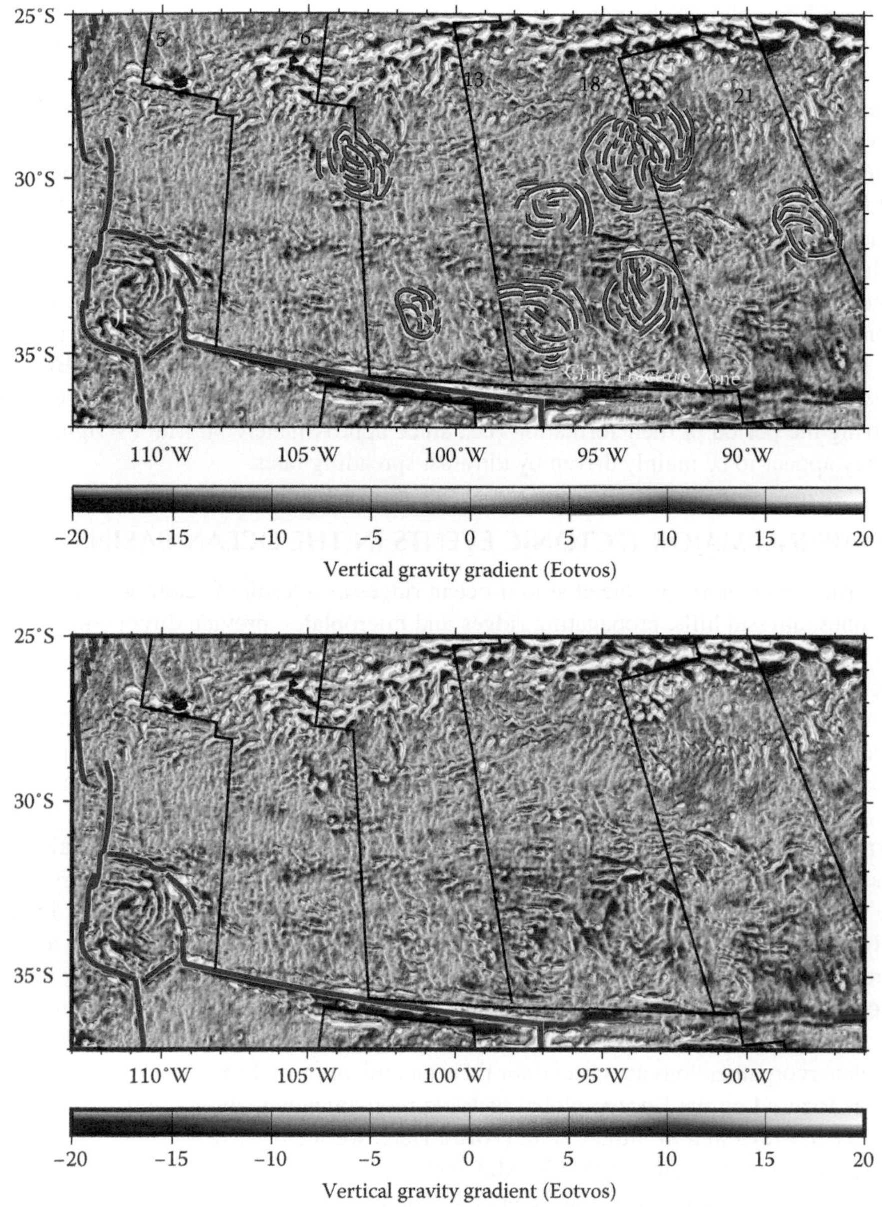

FIGURE 16.6 Paleo-Microplates in the Central Pacific Ocean. Vertical gravity gradient (VGG) map (Sandwell et al. 2014) of the spreading corridor east of the Juan Fernandez (JF) microplate and north of the Chile Fracture Zone without (top) and with (bottom) interpretation of microplate tectonic fabric. Current plate boundaries are outlined in red. VGG anomalies, outlined in magenta, illustrate the location of nine paleo-microplates. Their fabric has been interpreted by following the troughs of associated negative VGG anomalies. Schematic seafloor isochrons are shown for Chrons 5 (11 Ma), 6 (20 Ma), 13 (33 Ma), 18 (39 Ma), and 21 (48 Ma) (Wright et al. 2016) to provide rough age estimates for the tectonic stages when these microplates were active. Note the curved seafloor fabric indicative of microplate rotation and the narrow, partly curved VGG highs, likely corresponding to compressional ridges, as mapped around the Easter Microplate (Rusby and Searle 1993).

reveals previously unmapped paleo-microplates and confirms the existence of several paleo-microplates that were identified by Matthews et al. (2011) using an older version of the VGG data (Sandwell and Smith 2009). The extinct microplates are clearly visible through the curved lineations caused by their rotation, driven by the drag of the Nazca and Pacific plates, akin to that proposed for the Easter Microplate (Schouten et al. 1993) (Figure 16.6). After their formation and a period of rotation between a propagating and a receding ridge tip (Hey 2004), they were all eventually accreted to the Nazca Plate. The identification of these paleo-microplates is not only based on the curved seafloor fabric they have left behind due to their rotation history about stage poles proximal to their center but also on curved compressional ridges, which are expressed as narrow VGG highs (Figure 16.6). Compressional brittle deformation of the crust within microplates has been well documented for the Easter Microplate (Rusby and Searle 1993). Here, narrow ridges, up to 1 km high, have been shown to be associated with thin-skinned thrusting in response to local changes in the stress field (Rusby and Searle 1993). These features are preserved on old ocean crust and stand out among the background abyssal hill fabric (Figure 16.6). The apparent ubiquity of paleo-microplates in this spreading corridor may indicate that they were not driven by a major plate reorganization, as has been inferred for the formation for the Easter and Juan Fernandez microplates (Bird and Naar 1994), because there is no evidence for frequent plate reorganizations between the Nazca and Pacific Plates during the period of their formation (i.e., since approximately 50 Ma) (Wright et al. 2016). Instead they appear to be mainly driven by ultrafast spreading rates.

16.4 MAPPING MAJOR TECTONIC EVENTS IN THE OCEAN BASINS

Tectonic structures that are produced at mid-ocean ridges as a result of seafloor spreading, such as fracture zones, abyssal hills, propagating ridges and microplates, provide direct constraints on the direction and/or speed of relative plate motions. When combined with crustal age data, these structures allow for the construction of robust plate reconstruction models. Fracture zones and abyssal hills, the most pervasive structures on the seafloor, are particularly valuable for charting the motion of the plates. Their orientation records the direction of plate motion, and their morphologies and spacing can indicate the nature of the spreading regime, such as whether it was ultra-fast, fast, intermediate, slow, or ultra-slow (e.g., Sandwell and Smith 2009).

Patterns in the spatial distribution of structures that originate at spreading ridges, along with the occurrence of fracture zone bends and sudden changes in fracture zone morphology, also reveal evidence for plate reorganizations, including reorientations, relocations, or changes in segmentation of divergent plate boundaries and plate motion changes. Although continents are also sensitive to plate reorganizations, compared to the oceanic realms, they are more affected by deformation, tectonic overprinting, and erosional processes, all of which make it more difficult to unambiguously attribute continental structures and events to plate reorganizations. Therefore, seafloor fabric indicators of plate reorganizations are crucial for their identification and analysis. An enormous volume of work has focused on the Eocene global tectonic reorganization, about approximately 50 Ma in age, and recently reviewed in Müller et al. (2016). Here, we focus particularly on the controversial mid-Cretaceous (approximately 105–100 Ma) reorganization and review to what extent satellite gravity data, combined with other marine geophysical and geological data sets, are contributing to an improved understanding of the structural expression, age, and regional occurrence of this event.

16.4.1 The Enigmatic Mid-Cretaceous Tectonic Event

A mid-Cretaceous (approximately 100 Ma) spreading reorganization in the Indian Ocean was originally proposed by Powell et al. (1988). The now well-mapped associated seafloor fabric illustrates that this event produced one of the most prominent suites of fracture zone bends on Earth, in the Wharton Basin west of Australia, that are now well-resolved in satellite free-air and vertical gravity gradient data sets (Figure 16.7). These fracture zone bends express an approximately

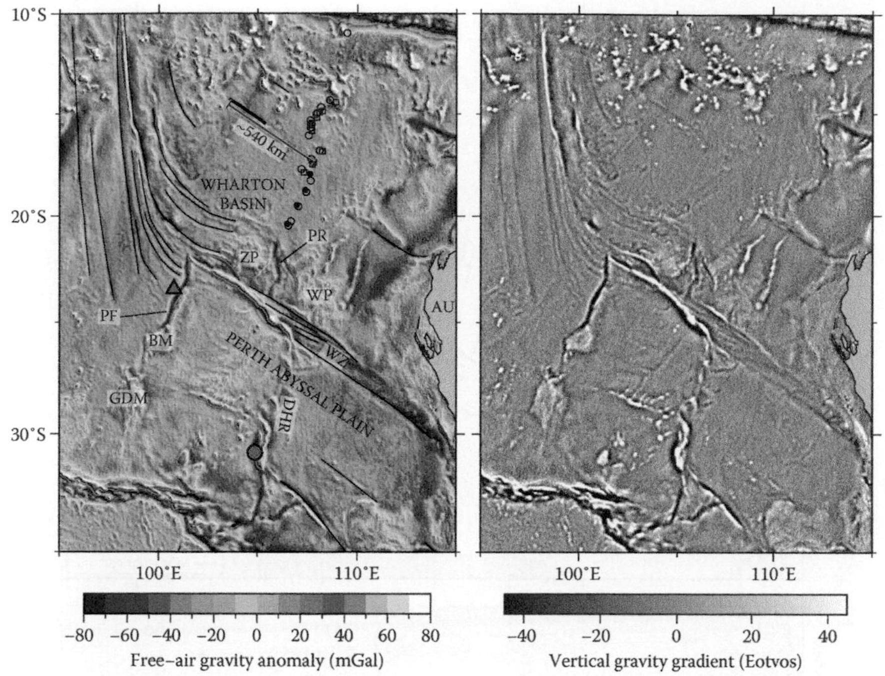

FIGURE 16.7 Wharton Basin Gravity Anomaly Maps. Free-air gravity (a) and vertical gravity gradient (VGG; b) maps of the Wharton Basin and Perth Abyssal Plain to the west of Australia (AU) (Sandwell et al. 2014). Fracture zones are represented by thin black lines, although a single fracture zone in the Wharton Basin that is discussed in the text is drawn as a thick black line (Wessel et al. 2015). M0 magnetic anomaly picks in the Wharton Basin are from Müller et al. (1998) (squares), Ségoufin et al. (2004) (hollow circles), and Gibbons et al. (2013) (filled circles). DSDP Site 256 is a purple triangle, and dredge site 5 from Whittaker et al. (2016) is a yellow circle. BM, Batavia Microcontinent; DHR, Dirck Hartog Ridge; GDM, Gulden Draak Microcontinent; PF, pseudo fault; PR, extinct propagating ridge feature; WP, Wallaby Plateau; WZ, Wallaby-Zenith Fracture Zone; ZP, Zenith Plateau.

50° clockwise change in the direction of spreading between Australia and India. Other events reported to have been concomitant with formation of the Wharton Basin fracture zone bends include a minor change in the direction of spreading between Africa and Antarctica, which produced broad fracture zone bends to the southwest of Madagascar and off the Queen Maud Land margin of Antarctica (Bernard et al. 2005), and a change in the direction of spreading between Antarctica and India that produced fracture zone bends in the Enderby Basin and the Bay of Bengal (Rotstein et al. 2001). Veevers (2000) reviewed the mid-Cretaceous event in a wider spatial context and showed that a series of tectonic and stratigraphic regime changes in Australia, New Zealand, and the Pacific Rim, and bends in Pacific hotspot trails that indicate a swerve of the Pacific Plate, also occurred in the mid-Cretaceous at about 100 Ma. Most of this work predated the use of detailed satellite altimetry-derived gravity anomalies to map seafloor structures associated with this event.

Matthews et al. (2012) produced the first detailed interpretation of changes in seafloor fabric that occurred from approximately 110–90 Ma, based on satellite gravity data. They concluded that the reorganization was likely global in scale and initiated at approximately 105–100 Ma. This was based on the widespread occurrence of prominent fracture zone bends (Wharton Basin, Enderby Basin, Bay of Bengal, Weddell Sea, and South Atlantic) and fracture zone terminations (Central and South Atlantic) in the Indian and Atlantic oceans (Figure 16.8) and hotspot trail bends in the Pacific, as noted by Veevers (2000). Unfortunately, seafloor structures that form at approximately 105–100 Ma are difficult to date as they coincide with the Cretaceous Normal Superchron (CNS),

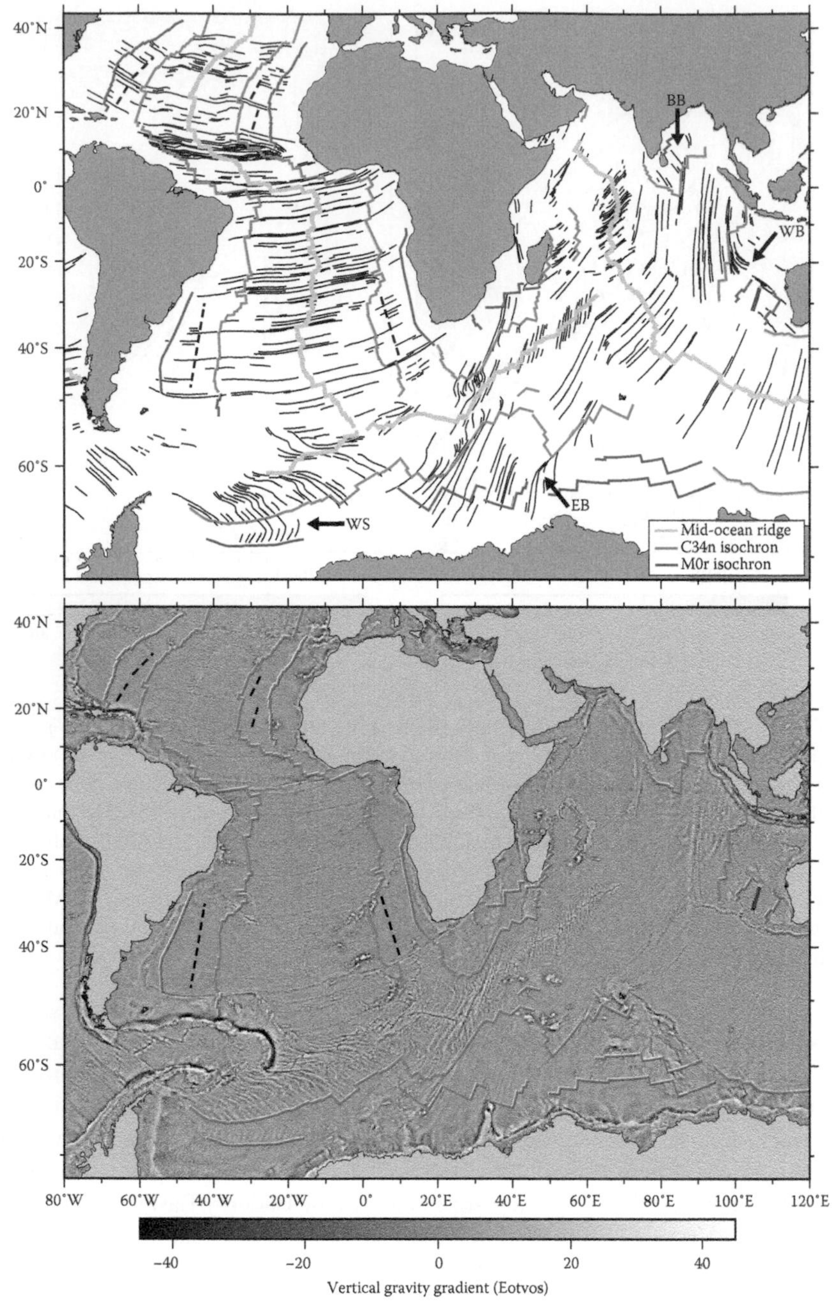

FIGURE 16.8 Tectonic Fabric of The Atlantic and Indian Oceans. Seafloor fabric map of the Atlantic and Indian oceans showing fracture zones (black lines) and mid-ocean ridges (bold gray lines) (a). Arrows point to the fracture zone bends in the Weddell Sea (WS), Enderby Basin (EB), Bay of Bengal (BB), and Wharton Basin (WB), and a dashed line demarcates the change in seafloor fabric of the Central and South Atlantic, from rougher to smoother (Matthews et al. 2012). Seafloor that formed during the Cretaceous Normal Superchron is bounded by the C34n (83 Ma, young end) and M0r (120.6 Ma, young end) isochrons of Seton et al. (in prep). Fracture zones are from Wessel et al. (2015), and mid-ocean ridges are from Sandwell and Smith (2009). The extinct Dirck Hartog Ridge in the Perth Abyssal Plain, to the southeast of the Wharton Basin fracture zone bends, is plotted as an orange line (Whittaker et al. 2016). A vertical gravity gradient (VGG) map of the region (Sandwell et al. 2014) is also provided (b), in which the aforementioned seafloor tectonic structures can be identified.

a prolonged period (approximately 37 million years) of stable magnetic field polarity. Matthews et al. (2012) therefore used absolute ages from ocean drilling data where available (Wharton Basin and Central Atlantic), along with extrapolations of published pre- and post-CNS spreading rates to estimate the ages of the seafloor structures they identified in the Indian and Atlantic oceans within CNS seafloor. They assigned an age of 104 Ma to the spreading reorganization that produced the major fracture zone bends in the Wharton Basin, 98 Ma to the Antarctica-Indian spreading reorganization, 105 Ma to fracture zone bends in the Weddell Sea, and 101 and 102–96 Ma to increases in spreading rate in the Central and South Atlantic, respectively, that they identified from a decrease in seafloor roughness (see Matthews et al. 2012, Appendix B). Of the different ocean regions, their estimate for the Wharton Basin reorganization was most well constrained as they were able to use four different dating approaches, including the use of a dated Deep Sea Drilling Project dredge sample in the location of the fracture zone bends (Davies et al. 1974).

To obtain tighter constraints on the tectonic evolution of the eastern Indian Ocean during the mid-Cretaceous, Desa and Ramana (2016) recently reanalyzed ship-track magnetic anomalies in the Wharton Basin, Bay of Bengal, and Enderby Basin, focusing on the identification of reversals within the CNS that have been dated at 92 and 108 Ma (Granot et al. 2012). These internal markers in the CNS were mapped in these regions and led Desa and Ramana (2016) to propose an age of 102 Ma for the spreading reorganizations at the Indian-Australian and Indian-Antarctic ridges.

Despite mounting work in support of a major plate reorganization at about 100 Ma, it was recently proposed by Talwani et al. (2016) that the change in motion of the Indian plate may have initiated in the Early Cretaceous, as early as 118 Ma. This was based on their estimates of the age of the fracture zone bends in the Enderby Basin, the eruption of the Rajmahal and Sylhet traps from the Kerguelen Plume, and the proposal that the reorganization coincided with a northward ridge jump in the Eastern Enderby Basin (Talwani et al. 2016). This ridge jump event isolated the continental Elan Bank and was previously described by Gaina et al. (2007) and Gibbons et al. (2013), although these authors did not link the ridge jump with formation of the fracture zone bends.

Dating the Enderby Basin fracture zone bends is difficult as they formed during the CNS; furthermore, there are no absolute age constraints nearby—for instance, from seafloor dredging or drilling. Talwani et al. (2016) measured the distance between the bend in the Kerguelen Fracture Zone and their new M0 magnetic anomaly pick to the south and, using a spreading rate of 40 mm/y, they arrived at an estimated age of 116.2 Ma for the bend, an age similar to that of the ridge jump and eruption of the Rajmahal and Sylhet traps. However, Talwani et al.'s (2016) interpretation of the M0 magnetic anomaly is greater than 130 km north of the M0 picks of Gaina et al. (2007), specifically their Model 2, which assumes the ridge jump affected the Eastern Enderby Basin and is consistent with the work of Gibbons et al. (2013), who identified a pseudo fault in the Eastern Enderby Basin (Figure 16.9). Furthermore, their M0 magnetic anomaly interpretation in the Western Enderby Basin (east of the Gunnerus Ridge) is 100 and 140 km north of the interpretations of Eagles and König (2008) and Marks and Tikku (2001), respectively, which are all derived from the same ship data.

16.4.2 Insights from Combining Satellite Altimetry with Geological and Geophysical Ship Data

Detailed seafloor fabric observations from satellite altimetry, paired with other marine geophysical data to support a major reorganization in the Indian Ocean, have recently grown to include a major mid-Cretaceous ridge jump (Gibbons et al. 2013; Watson et al. 2016; Williams et al. 2013) in the Perth Abyssal Plain east of the major Wharton Basin fracture zone bends (Figure 16.9). The greater than 500-km-long, roughly north-south trending Dirck Hartog Ridge is an extinct spreading ridge that formed during a westward jump in the Australian-Indian spreading center (Watson et al. 2016). Although the structure has the appearance of an extinct ridge in VGG maps, its classification has been confirmed by ship-track magnetic anomaly interpretations (Williams et al. 2013) and detailed

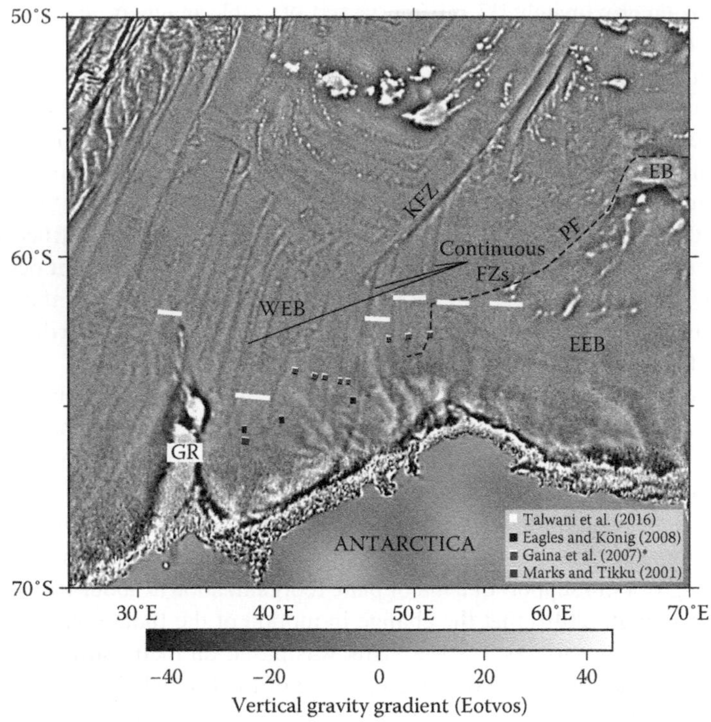

FIGURE 16.9 Vertical gravity gradient of the Enderby basin. Vertical gravity gradient (VGG) map (Sandwell et al. 2014) of the Western (WEB) and Eastern Enderby Basin (EEB) showing the M0r (120.6 Ma, young end) magnetic anomaly identifications of Talwani et al. (2016), Eagles and König (2008), Gaina et al. (2007), and Marks and Tikku (2001). The pseudo fault (PF) identified in the study of Gibbons et al. (2013), which formed during the northward ridge jump that isolated the Elan Bank (EB), is shown as a dashed line. To the west of this boundary, fracture zones are continuous (examples of this are identified in the map), suggesting that spreading was continuous in this corridor and unaffected by the ridge jump. GR, Gunnerus Ridge; KFZ, Kerguelen Fracture Zone.

analysis of swath bathymetry (Watson et al. 2016). An age of 102.3 ± 2.5 Ma was obtained from $^{40}Ar/^{39}Ar$ dating of a plagioclase crystal in a gabbro sample from the Dirck Hartog Ridge (Whittaker et al. 2016). The ridge jump event resulted in the calving off of two continental fragments (Batavia and Gulden Draak microcontinents) from the Indian passive margin, both of which align with the pseudo fault that formed during ridge propagating (Whittaker et al. 2016) (Figure 16.7). These combined data suggest that a spreading reorganization at approximately 105 Ma caused a ridge jump and subsequent microcontinent formation at approximately 104–101 Ma, immediately prior to the formation of the curved Wharton Basin fracture zones clearly visible in the VGG image (Whittaker et al. 2016) (Figure 16.7). These timings are consistent with the results of Desa and Ramana (2016), who proposed an age of 102 Ma for the ridge jump based on their identification of CNS internal time markers in ship-track magnetic anomaly profiles.

To understand the timing and nature of the Indian Ocean reorganization, one needs to consider events at all of its plate boundaries as a major reorganization at one boundary will affect its other boundaries. Talwani et al. (2016) suggested that the Wharton Basin fracture zone bends (Australian-Indian spreading ridge) could be as old as 117 Ma, forming some time during the period 100–117 Ma. In the Wharton Basin, to the northeast of the fracture zone bends, at least 540 km of quite smooth seafloor was produced between the time of anomaly M0 and the change in seafloor spreading (Figure 16.7). This observation is supported by interpretation of seafloor fabric to the north of the Zenith Plateau, where a well-resolved fracture zone can be identified and the seafloor fabric shows no sign of ridge extinction or ridge jumps (Matthews et al. 2012). The production of this amount of seafloor and considering a variety of published spreading rates for just prior to M0 (3.5–5.5 cm/year) places an upper limit on the age on the reorganization of approximately 105.2–110.8 Ma (based on the timescale of Gee and Kent 2007). In contrast, Talwani et al. (2016) proposed that measuring the distance between the suite of fracture zone bends and the M0 anomalies may result in an estimate as old as 117 Ma for the reorganization. However, the seafloor to the east of the fracture zone bends, which Talwani et al. (2016) relied upon for their estimates, has been affected by a complex tectonic history including a major ridge jump (Gibbons et al. 2013; Watson et al. 2016; Whittaker et al. 2013; Williams et al. 2013) and possibly aborted ridge propagation in the region between the Zenith and Wallaby plateaus (see VGG image and interpretation in Figure 16.7); therefore, it is not reliable to measure the distance between the bends and the M0 picks in this location immediately to the east of the fracture zone bends. Talwani et al.'s (2016) interpretation of the Wharton Basin fracture zone bends also did not take into account the seafloor age of 101 ± 1 Ma obtained from DSDP Site 256 (Davies et al. 1974) that directly overlaps the fracture zone bends, which is in excellent agreement with the more recently acquired date for the Dirck Hartog Ridge (Whittaker et al. 2016).

In conclusion, it is impossible, in terms of tectonic evolution, for an extinct Dirck Hartog Ridge, with an age of 102.3 ± 2.5 Ma (Whittaker et al. 2016), to be located east of Wharton basin fracture zone bends, if those bends formed from a reorganization at approximately 118–117 Ma. A variety of other plate motion changes, as well as Australia-Antarctica and India-Madagascar breakup, occurred in the mid-Cretaceous and coincide with the final breakup between Africa and South America in the equatorial Atlantic (Somoza and Zaffarana 2008), lending additional support to an approximately 100 Ma age of the major fracture zone bends between India and Australia. Finally, an onset of India's northward motion at 118 Ma, instead of around 100 Ma, is also irreconcilable with models for the separation of India from Madagascar. The breakup and spreading history of India and Madagascar is intimately linked to India's change in motion (Gibbons et al. 2013). Their separation initiated around approximately 100 Ma, following the approximately 100 Ma reorganization, and culminated in seafloor spreading from south to north from 94 to 84 Ma (Gibbons et al. 2013). An interpretation of VGG seafloor images supplemented with marine geological and geophysical ship data allows us to robustly map the tectonic history of the ocean basins, even within the CNS.

16.4.3 What Caused the 100 Ma Event?

Several authors have speculated about what drove a major plate reorganization at approximately 105–100 Ma. Zahirovic et al. (2016) recently proposed that commencement of Indian Plate subduction following a ridge subduction event at its northern margin may have triggered the spreading reorganization recorded in the Wharton Basin fracture zones. In this scenario, the Neotethyan spreading ridge that formed the northern margin of the Indian Plate since approximately 155 Ma was subducted at the intraoceanic Kohistan-Ladakh subduction zone from approximately 110–90 Ma, producing a significant phase of magmatic accretion from approximately 105–99 Ma, high-temperature granulite facies metamorphism at approximately 95 Ma, and a hot mantle source at the Kohistan Arc. This ridge subduction event would have changed the balance of plate-driving forces acting on the Indian Plate; in particular, it strengthened northward slab pull, to which the Australian-Indian and Antarctic-Indian spreading ridges responded by clockwise rotation, as reflected in the major fracture zone bends.

16.5 MAPPING SUB-SEAFLOOR TECTONIC FABRIC

Marine gravity anomalies derived from satellite altimetry have been used to help delineate the boundary between continental and oceanic crust (COB) (Pawlowski 2008) as well as regional fault systems on stretched continental crust (McGrane et al. 2001; Trung et al. 2004). However, the most recent satellite-derived vertical gravity gradient grid (Sandwell et al. 2014) allows the detailed mapping of relatively closely-spaced faults (less than 10 km), in contrast to most previous applications. Here, we focus on two examples where the new generation of gravity data can be used to map regional fault systems at scales that were previously achieved only by relatively closely spaced seismic reflection profiles.

16.5.1 North Falkland Basin

Our first example is the North Falkland Basin, which is embedded in the northern portion of the Falkland Plateau (Figure 16.8). The basin is located north of the Falkland Islands (Figure 16.10). Strain rate inversion of its stratigraphy has shown that the basin formed during a single phase or rifting close to the Jurassic-Cretaceous boundary, lasting from about 150–125 Ma (Jones et al. 2004). Its structural fabric has been mapped based on seismic reflection data (Richards and Hillier 2000) (Figure 16.11), an interpretation that was recently refined by Lohr and Underhill (2015). This is a good area to ground-truth the VGG data for the interpretation of buried structures as it is

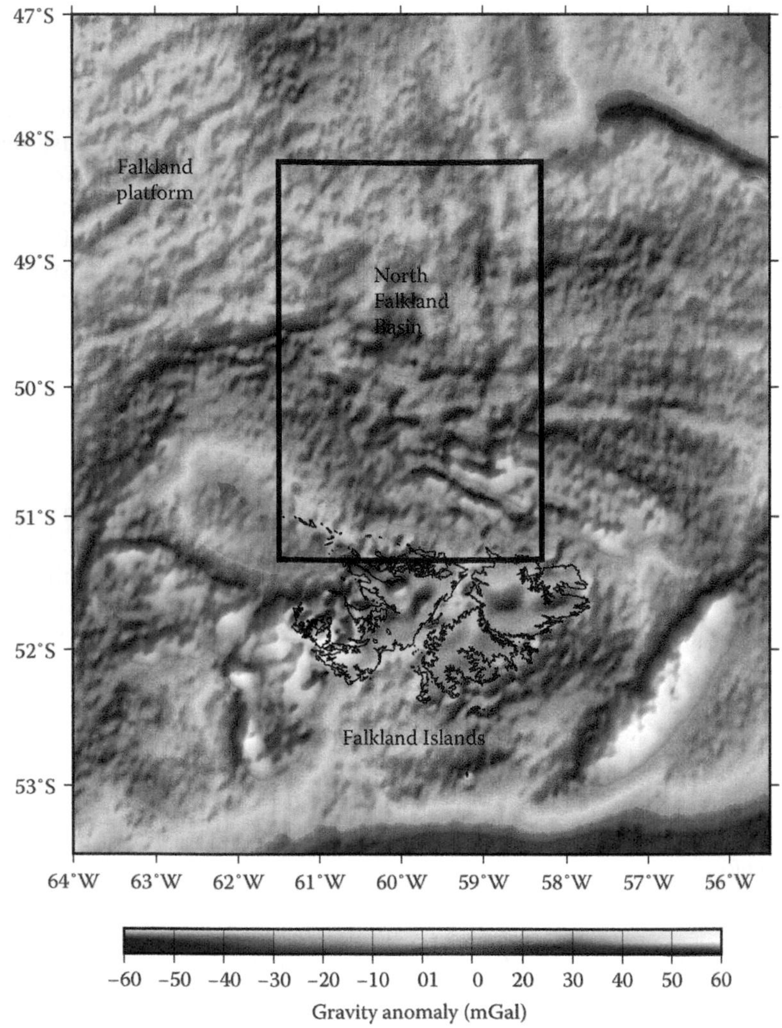

FIGURE 16.10 Free-Air Gravity Anomalies of The Falkland Plateau. Free-air gravity anomaly map of the Falkland Plateau. Black box outlines the region shown in Figure 16.11.

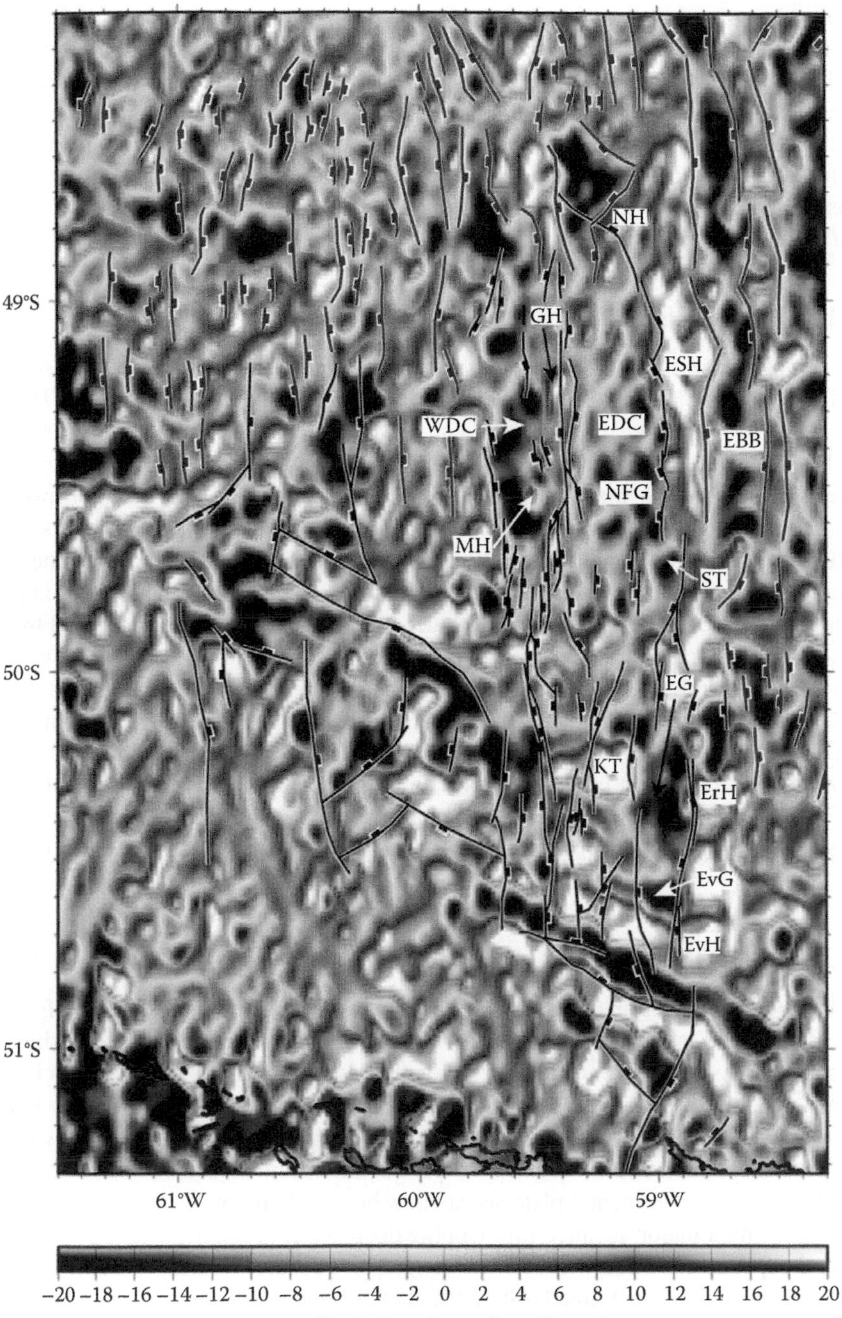

FIGURE 16.11 Vertical Gravity Gradient and Structural Fabric of The North Falkland Basin. Vertical gravity gradient (VGG) map (Sandwell et al. 2014) of the North and South Falkland Basins (see Figure 16.10 for location). Black fault interpretations are from Richards and Hillier (2000), and faults shown in red are interpreted from the VGG, with filled boxes indicating the dip direction. Note good correlation between linear, positive VGG anomalies and elevated blocks bounded by normal faults interpreted by Richards and Hillier (2000). NFB, North Falkland Basin; ESH, Eastern Structural High; EBB, Eastern Backarc-Basin; KT, Keppel Terrace; EG, Ernest Graben; ErH Ernest High; EvH, Evie High; EvG, Evie Graben; EDC, Eastern Depo-Centre; IGH, Intra-Graben High; MH, Minke High; WDC, Western Depo-Centre; ST, Sanson Terrace; NH, Northern High.

quite well mapped. The fault interpretation of Richards and Hillier (2000) is overlain over the VGG on Figure 16.11 and illustrates an excellent correlation between normal faults interpreted to bound structural basement highs and the edges of positive VGG anomalies, illustrating that an interpretation based just on the VGG data would provide a fairly similar result. In Figure 16.9, we have extended the interpretation of Richards and Hillier (2000) to the north, east, and west, following the assumption that VGG highs correspond to structural basement highs and are bounded by normal faults. This example demonstrates the power of the new VGG data to interpret buried basement fault fabrics in frontier areas with poor seismic data coverage.

16.5.2 Lord Howe Rise

Our next example represents a deep water frontier area, namely the Lord Howe Rise, a stretched, submerged continental plateau with poor seismic data coverage (Bache et al. 2014). The most comprehensive regional structural interpretation can be found in Collot et al. (2009) and is shown in Figure 16.10. It outlines major troughs and highs in the region, as well as some regional transcurrent faults offsetting the main structural fabric oriented north-south to northwest-southeast (Figure 16.12). The Lord Howe Rise falls into the category of wide, asymmetric rifts, with a width of about 200 km, while the conjugate Australian margin is extremely narrow. Brune et al. (2014) recently suggested that such hyperextended crust typical for wide, asymmetric rifts is produced by steady-state rift migration. Rift migration in a succession of phases is accomplished by sequential, oceanward-younging, upper crustal faults, and is balanced through lower crustal flow. This process results in a dense network of faults dissecting a "basin and range" basement fabric as is clearly seen in the VGG image for the Lord Howe Rise (Figure 16.12), where VGG highs outline horst blocks while VGG lows delineate basement troughs. We have used the VGG image, following the insights gained from the North Falkland Basin interpretation, to interpret the distributed fault fabric of this wide rift. This is the first regional structural map of this region, which may form a useful framework for future seismic imaging as well as scientific drilling by the Integrated Ocean Drilling Program in 2017.

16.6 CONCLUSIONS AND FUTURE OUTLOOK

The current generation of altimetry data have yielded a new marine gravity anomaly grid with an improvement in signal-to-noise ratio by a factor of two; this advance has been critical in imaging small-scale seafloor and sub-seafloor features that were previously too noisy to interpret. The new data have enabled the identification of extinct spreading ridges buried deeply beneath thick sediments (Sandwell et al. 2014) as well as previously unknown paleo-microplates in the Indian (Matthews et al. 2016) and Pacific oceans. Deeply buried basement fault fabrics along continental margins and submerged continental plateaus can now be clearly mapped in vertical gravity gradient images, representing a major advance for mapping deep-water frontier regions where seismic data are sparse and expensive to obtain. The future merging of the latest marine satellite-derived gravity grid with the Gravity Field and Steady-State Ocean Circulation Explorer (GOCE) gravity field on continents should result in improved continental fit reconstructions (e.g., Braitenberg 2015).

ACKNOWLEDGMENTS

This work was supported by Australian Research Council (ARC) grants DP130101946 and IH130200012. K. Matthews acknowledges funding from the European Research Council grant agreement n. 639003 DEEP TIME. D. Sandwell was supported by the National Geospatial-Intelligence Agency (HM0177-13-1-0008) and NASA SWOT program (NNX16AH64G). We thank Lauren Harrington for her help with digitizing and editing figures.

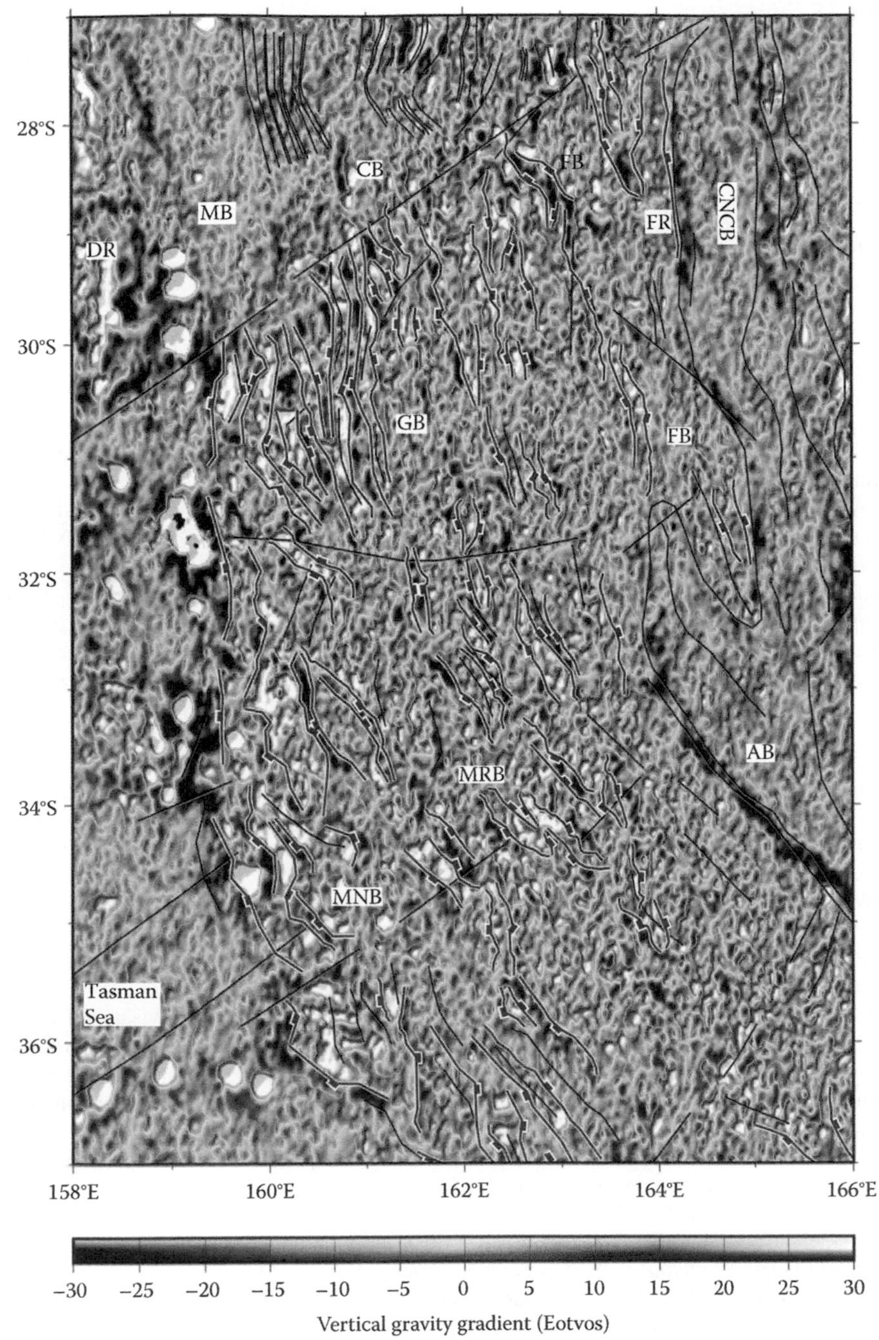

FIGURE 16.12 Vertical Gravity Gradient and Structural Fabric Of The Lord Howe Rise. Vertical gravity gradient (VGG) map (Sandwell et al. 2014) of the Lord Howe Rise, a stretched, submerged continental plateau east of Australia. The Lord Howe Rise stretches from the eastern edge of the Tasman Sea in the west to the edge of the Fairway Basin (FB) and Aotea Basin (AE) to the east. Tectonic lineaments shown as black lines are taken from Collot et al. (2009), while red lines correspond to normal faults interpreted from the VGG image, with filled boxes indicating the inferred dip direction. DR, Dampier Ridge; MB, Middleton Basin; CP, Capel Basin; FB, Faust Basin; GB, Gower Basin; MRB, Moore Basin; MNB, Monowai Basin; FR, Fairway Ridge; CNCB, Central New Caledonia Basin.

REFERENCES

Andersen, O.B., and P. Knudsen. 1998. Global marine gravity field from the ERS-1 and GEOSAT geodetic mission altimetry. *Journal of Geophysical Research: Oceans* 103 (C4): 8129–8137.

Bache, F., N. Mortimer, and R. Sutherland, et al. 2014. Seismic stratigraphic record of transition from Mesozoic subduction to continental breakup in the Zealandia sector of eastern Gondwana. *Gondwana Research* 26 (3): 1060–1078.

Bernard, A., M. Munschy, Y. Rotstein, and D. Sauter. 2005. Refined spreading history at the Southwest Indian Ridge for the last 96 Ma, with the aid of satellite gravity data. *Geophysical Journal International* 162: 765–778.

Bird, R.T., and D.F. Naar. 1994. Intratransform origins of mid-ocean ridge microplates. *Geology* 22 (11): 987–990.

Bird, R.T., S.F. Tebbens, M.C. Kleinrock, and D.F. Naar. 1999. Episodic triple-junction migration by rift propagation and microplates. *Geology* 27 (10): 911–914.

Braitenberg, C. 2015. Exploration of tectonic structures with GOCE in Africa and across-continents. *International Journal of Applied Earth Observation and Geoinformation* 35: 88–95.

Brune, S., C. Heine, M. Pérez-Gussinyé, and S.V. Sobolev. 2014. Rift migration explains continental margin asymmetry and crustal hyper-extension. *Nature Communications* 5: 1–9.

Cazenave, A., P. Schaeffer, M. Berge, C. Brossier, K. Dominh, and M.C. Gennero. 1996. High-resolution mean sea surface computed with altimeter data of ERS-1 (Geodetic Mission) and TOPEX-POSEIDON. *Geophysical Journal International* 125 (3): 696–704.

Chelton, D.B., J.C. Ries, B.J. Haines, L.-L. Fu, and P.S. Callahan. 2001. Satellite altimetry. *International Geophysics* 69: 1–ii.

Collot, J., R. Herzer, Y. Lafoy, and L. Geli. 2009. Mesozoic history of the Fairway-Aotea Basin: Implications for the early stages of Gondwana fragmentation. *Geochemistry, Geophysics, Geosystems* 10 (12), 1–24.

Davies, T.A., B.P. Luyendyk, K.S. Rodolfo, et al., eds. 1974. *Leg XXVI, Initial reports of the Deep Sea Drilling Project.* Vol. 26. Washington, DC: US Govn't Printing Office.

Desa, M.A., and M.V. Ramana. 2016. Middle cretaceous geomagnetic field anomalies in the eastern Indian Ocean and their implication to the tectonic evolution of the Bay of Bengal. *Marine Geology* 382: 111–121.

Eagles, G., and M. König. 2008. A model of plate kinematics in Gondwana breakup. *Geophysical Journal International* 173 (2): 703–717.

Eakins, B.W. 2002. *Structure and Development of Oceanic Rifted Margins, Earth Sciences.* University of California, San Diego, CA.

Gaina, C., R.D. Müller, B. Brown, T. Ishihara, and S. Ivanov. 2007. Breakup and early seafloor spreading between India and Antarctica. *Journal of Geophysics International* 170 (1): 151–170.

Garcia, E.S., D.T. Sandwell, and W.H.F. Smith. 2014. Retracking CryoSat-2, Envisat and Jason-1 radar altimetry waveforms for improved gravity field recovery. *Geophysical Journal International* 196 (3): 1402–1422. doi: 10.1093/gji/ggt469.

Gee, J.S., and D.V. Kent. 2007. Source of Oceanic magnetic anomalies and the geomagnetic polarity timescale. *Treatise Geophys* 5: 455–507.

Gibbons, A.D., J.M. Whittaker, and R.D. Müller. 2013. The breakup of East Gondwana: Assimilating constraints from Cretaceous ocean basins around India into a best-fit tectonic model. *Journal of Geophysical Research: Solid Earth* 118 (3): 808–822.

Goff, J.A., and J.R. Cochran. 1996. The Bauer scarp ridge jump: A complex tectonic sequence revealed in satellite altimetry. *Earth and Planetary Science Letters* 141 (1): 21–33.

Granot, R., J. Dyment, and Y. Gallet. 2012. Geomagnetic field variability during the Cretaceous Normal Superchron. *Nature Geoscience* 5 (3): 220–223.

Haxby, W.F., G.D. Karner, J.L. LaBrecque, and J.K. Weissel. 1983. Digital images of combined oceanic and continental data sets and their use in tectonic studies. *Eos, Transactions American Geophysical Union* 64 (52): 995–1004.

Heiskanen, W.A., Moritz, H., 1967. Physical geodesy. Bulletin Géodésique, 86, 491–492.

Hey, R.N. 1977. A new class of "pseudo faults" and their bearing on plate tectonics: A propagating rift model. *Earth and Planetary Science Letters* 37 (2): 321–325.

Hey, R.N. 2004. Propagating rifts and microplates at mid-ocean ridges. In *Encyclopedia of Geology*, edited by R.C. Selley, R. Cocks, and I. Plimer. London: Academic Press, pp. 396–405.

Hey, R.N., F.K. Duennebier, and W.J. Morgan. 1980. Propagating rifts on midocean ridges. *Journal of Geophysical Research* 85 (B7): 3647–3658.

Hey, R.N., H.W. Menard, T.M. Atwater, and D.W. Caress. 1988. Changes in direction of seafloor spreading revisited. *Journal of Geophysical Research* 93 (B4): 2803–2811.

Hey, R.N., D.F. Naar, M.C. Kleinrock, W.J. Phipps Morgan, E. Morales, and J.-G. Schilling. 1985. Microplate tectonics along a superfast seafloor spreading system near Easter Island. *Nature* 317: 320–325.

Hwang, C., and B. Parsons. 1996. An optimal procedure for deriving marine gravity from multi-satellite altimetry. *Geophysical Journal International* 125 (3): 705–718.

Jones, S.M., N. White, P. Faulkner, and P. Bellingham. 2004. Animated models of extensional basins and passive margins. *Geochemistry, Geophysics, Geosystems* 5 (8), 1—38.

Lohr, T., and J.R. Underhill. 2015. Role of rift transection and punctuated subsidence in the development of the North Falkland Basin. *Petroleum Geoscience* 21 (2–3): 85–110.

Mammerickx, J., D.F. Naar, and R.L. Tyce. 1988. The mathematician paleoplate. *Journal of Geophysical Research* 93 (B4): 3025–3040.

Marks, K.M., and A.A. Tikku. 2001. Cretaceous reconstructions of East Antarctica, Africa and Madagascar. *Earth & Planetary Science Letters* 186 (3–4): 479–495.

Matthews, K.J., R.D. Müller, and D.T. Sandwell. 2016. Oceanic microplate formation records the onset of India–Eurasia collision. *Earth and Planetary Science Letters* 433: 204–214.

Matthews, K.J., R.D. Müller, P. Wessel, and J.M. Whittaker. 2011. The tectonic fabric of the ocean basins. *Journal of Geophysical Research* 116 (B12): B12109.

Matthews, K.J., M. Seton, and R.D. Müller. 2012. A global-scale plate reorganization event at 105–100 Ma. *Earth & Planetary Science Letters* 355: 283–298.

Maus, S., C.M. Green, and J.D. Fairhead. 1998. Improved ocean-geoid resolution from retracked ERS-1 satellite altimeter waveforms. *Geophysical Journal International* 134 (1): 243–253.

McGrane, K., P.W. Readman, and B.M. O'Reilly. 2001. Interpretation of transverse gravity lineaments in the Rockall Basin. *Geological Society, London, Special Publications* 188 (1): 393–399.

Müller, R.D., D. Mihut, and S. Baldwin. 1998. A new kinematic model for the formation and evolution of the Northwest and West Australian margin. In *The Sedimentary Basins of Western Australia 2*, edited by P.G. Purcell and R.R. Purcell. Perth: Petroleum Exploration Society of Australia, 55–72.

Müller, R.D., W.R. Roest, J.-Y. Royer, L.M. Gahagan, and J.G. Sclater. 1997. Digital isochrons of the world's ocean floor. *Journal of Geophysical Research: Solid Earth* 102 (B2): 3211–3214.

Müller, R.D., M. Seton, S. Zahirovic, et al. 2016. Ocean basin evolution and global-scale plate reorganization events since Pangea breakup. *Annual Review of Earth and Planetary Sciences* 44: 107–138.

Olgiati, A., G. Balmino, M. Sarrailh, and C.M Green. 1995. Gravity anomalies from satellite altimetry: Comparison between computation via geoid heights and via deflections of the vertical. *Bulletin Géodésique* 69 (4): 252–260.

Pawlowski, R. 2008. The use of gravity anomaly data for offshore continental margin demarcation. *The Leading Edge* 27 (6): 722–727.

Phipps Morgan, J., and E.M. Parmentier. 1985. Causes and rate-limiting mechanisms of ridge propagation: A fracture mechanics model. *Journal of Geophysical Research* 90 (B10): 8603–8612.

Powell, C., S.R. Roots, and J.J. Veevers. 1988. Pre-breakup continental extension in East Gondwanaland and the early opening of the eastern Indian Ocean. *Tectonophysics* 155: 261–283.

Richards, P.C., and B.V. Hillier. 2000. Post-drilling analysis of the North Falkland Basin – Part 1: Tectono-stratigraphic framework. *Journal of Petroleum Geology* 23 (3): 253–272.

Rotstein, Y., M. Munschy, and A. Bernard. 2001. The Kerguelen Province revisited: Additional constraints on the early development of the Southeast Indian Ocean. *Marine Geophysical Researches* 22 (2): 81–100.

Rusby, R.I., and R.C. Searle. 1993. Intraplate thrusting near the Easter microplate. *Geology* 21 (4): 311–314.

Sandwell, D.T., R.D. Müller, W.H.F. Smith, E. Garcia, and R. Francis. 2014. New global marine gravity model from CryoSat-2 and Jason-1 reveals buried tectonic structure. *Science* 346 (6205): 65–67.

Sandwell, D.T., and W.H.F. Smith. 1997. Marine gravity anomaly from GEOSAT and ERS 1 satellite altimetry. *Journal of Geophysical Research: Solid Earth* 102 (B5): 10039–10054.

Sandwell, D.T., and W.H.F. Smith. 2005. Retracking ERS-1 altimeter waveforms for optimal gravity field recovery. *Geophysical Journal International* 163 (1): 79–89.

Sandwell, D.T., and W.H.F. Smith. 2009. Global marine gravity from retracked GEOSAT and ERS-1 altimetry: Ridge segmentation versus spreading rate. *J. Geophys. Res.* 114 (B1): B01411.

Sandwell, D.T., W.H.F. Smith, S. Gille, et al. 2006. Bathymetry from space: Rationale and requirements for a new, high-resolution altimetric mission. *Comptes Rendus Geoscience* 338 (14): 1049–1062.

Schouten, H., K.D. Klitgord, and D.G. Gallo. 1993. Edge-driven microplate kinematics. *Journal of Geophysical Research: Solid Earth* 98 (B4): 6689–6701.

Searle, R. 2013. *Mid-ocean ridges*. Cambridge University Press, Cambridge, 330 pp.

Ségoufin, J., P. Munschy, P. Bouysse, et al. 2004. *Map of the Indian Ocean (1:20 000 000), sheet 1: "Physiography," sheet 2: "Structural map."* Commission for the Geological Map of the World, Paris.

Smith, W.H.F. 2015. Resolution of seamount geoid anomalies achieved by the SARAL/AltiKa and Envisat RA2 satellite radar altimeters. *Marine Geodesy* 38 (Suppl. 1): 644–671.

Somoza, R., and C.B. Zaffarana. 2008. Mid-cretaceous polar standstill of South America, motion of the Atlantic hotspots and the birth of the Andean cordillera. *Earth and Planetary Science Letters* 271 (1): 267–277.

Talwani, M., M.A. Desa, M. Ismaiel, and K.S. Krishna. 2016. The Tectonic origin of the Bay of Bengal and Bangladesh. *Journal of Geophysical Research: Solid Earth* 121 (7): 4836–4851.

Tapley, B.D., and M.-C. Kim. 2001. Applications to geodesy. In *Satellite Altimetry and Earth Sciences*, edited by L.-L. Fu and A. Cazenave. New York: Academic Press, 371–406.

Tebbens, S.F., and S.C. Cande. 1997. Southeast Pacific tectonic evolution from early oligocene to present. *Journal of Geophysical Research* 102 (B6): 12061–12084.

Trung, N.N., S.-M. Lee, and B.C. Que. 2004. Satellite gravity anomalies and their correlation with the major tectonic features in the South China Sea. *Gondwana Research* 7(2): 407–424.

Veevers, J.J. 2000. Change of tectono-stratigraphic regime in the Australian plate during the 99 Ma (mid-Cretaceous) and 43 Ma (mid-Eocene) swerves of the Pacific. *Geology* 28 (1): 47–50.

Watson, S.J., J.M. Whittaker, J.A. Halpin, et al. 2016. Tectonic drivers and the influence of the Kerguelen plume on seafloor spreading during formation of the early Indian Ocean. *Gondwana Research* 35: 97–114.

Wessel, P. 2001. Global distribution of seamounts inferred from gridded GEOSAT/ERS-1 altimetry. *Journal of Geophysical Research: Solid Earth* 106 (B9): 19431–19441.

Wessel, P., K.J. Matthews, R.D. Müller, et al. 2015. Semiautomatic fracture zone tracking. *Geochemistry, Geophysics, Geosystems* 16 (7): 2462–2472.

Whittaker, J.M., Williams, S.E., Müller, R.D., 2013. Revised tectonic evolution of the Eastern Indian Ocean. Geochemistry, Geophysics, Geosystems 14, 1891–1909.

Whittaker, J.M., J.C. Afonso, S. Masterton, et al. 2015. Long-term interaction between mid-ocean ridges and mantle plumes. *Nature Geoscience* 8 (6): 479–483.

Whittaker, J.M., S.E Williams, J.A. Halpin, et al. 2016. Eastern Indian Ocean microcontinent formation driven by plate motion changes. *Earth and Planetary Science Letters* 454: 203–212.

Williams, S.E., J.M. Whittaker, R. Granot, and D.R. Müller. 2013. Early India-Australia spreading history revealed by newly detected Mesozoic magnetic anomalies in the Perth Abyssal Plain. *Journal of Geophysical Research: Solid Earth* 118 (7): 3275–3284.

Wingham, D.J., C.R. Francis, S. Baker, et al. 2006. CryoSat: A mission to determine the fluctuations in Earth's land and marine ice fields. *Advances in Space Research* 37 (4): 841–871.

Wright, N.M., M. Seton, S.E. Williams, and R.D. Müller. 2016. The late cretaceous to recent tectonic history of the Pacific Ocean basin. *Earth-Science Reviews* 154: 138–173.

Zahirovic, S., K.J. Matthews, N. Flament, et al. 2016. Tectonic evolution and deep mantle structure of the eastern Tethys since the latest Jurassic. *Earth Science Reviews* 162: 293–337.

Zhang, S., and D.T. Sandwell. 2016. Retracking of SARAL/AltiKa radar altimetry waveforms for optimal gravity field recovery. *Marine Geodesy*, 40 (1), 40–56.

17 Ocean Modeling and Data Assimilation in the Context of Satellite Altimetry

Detlef Stammer and Stephen M. Griffies

17.1 INTRODUCTION

The ocean is a forced-dissipative fluid, with forcing active predominantly at the boundaries and molecular dissipation active throughout the water column. The ocean domain is bounded by complex land-sea boundaries, and motions are constrained by planetary rotation and density stratification. In response to forces, the ocean flow exhibits a rich variety of dynamical phenomena, including boundary currents, boundary layers, large-scale gyres and jets, linear and nonlinear waves, quasi-coherent vortices, quasi-geostrophic turbulence, and three-dimensional turbulence. The governing thermo-hydrodynamical ocean equations are nonlinear, admitting turbulent processes that impart a cascade of mechanical energy and tracer variance across these scales. Together, these phenomena cover temporal scales ranging from less than an hour (e.g., surface waves) to hundreds of years (deep circulation and mixing processes), with spatial scales ranging from a few millimeters (small-scale turbulence and molecular diffusion) to the circumference of the Earth (global circulation).

17.1.1 Observational Oceanography and Ocean Circulation Modeling

Observing the vast scale range of ocean phenomena remains a huge scientific and engineering challenge. Satellite remote sensing is able to measure the near-global surface ocean within just a few days, thereby providing ocean measurements unprecedented in their space–time resolution by conventional *in situ* technology. Satellite altimetry provides information about the dynamically important sea surface elevation, which on the rotating Earth offers a means to determine the near-surface geostrophic circulation (see Section 17.3.8), including its mesoscale eddy component (e.g., Stammer 1998; Griffies et al. 2015), as well as the movement of heat and salt within the ocean interior (e.g., Peyser et al. 2016). However, to bring satellite observations into the context of the time-varying, full-depth ocean circulation requires a dynamical framework to extrapolate surface information into the ocean interior. Ocean modeling provides such a framework and furthermore helps to interpret and correct satellite altimeter observations. Consequently, ocean models are essential for using satellite altimetry to study and to predict the ocean.

Ocean circulation modeling is a maturing field of computational fluid dynamics. It provides a dynamical description of the time-varying ocean circulation, including a means to quantify the relation among the movement of seawater, forces acting on the ocean, and dissipation and mixing processes affecting the flow. With increasing integrity of numerical methods and subgrid scale parameterizations, and with enhancements in computer power allowing for refinements to grid resolution, ocean general circulation models (OGCMs) have become an essential tool for exploring dynamical interactions within the ocean. For example, ocean models are used to investigate mechanistic hypotheses for ocean phenomena; to consider future scenarios of the ocean as part of the fully coupled Earth system, such as those associated with human-induced climate warming; and to nowcast and forecast ocean conditions on weekly to decadal timescales. Ocean models are tested against observations to assess their fidelity as a scientific tool. Such tests have been ongoing for decades (e.g., Stammer et al. 1996), with tests now extended to climate models using a broad suite of observations (e.g., Flato et al. 2013; Danabasoglu et al. 2014; Griffies et al. 2014, 2016). Data from various observing networks have thus become indispensable for model assessments. For example, Argo floats offer a means to evaluate simulations of interior ocean properties such as temperature and salinity (Riser et al. 2016). As a complement to these *in situ* measurements, satellite remote sensing offers a view of global surface ocean properties, such as sea surface temperature, surface salinity, and surface height. Offering a physical understanding of observations provides a mandate for many ongoing efforts in ocean and climate modeling, including ocean data assimilation.

17.1.2 The Need for Ocean Data Assimilation

Ocean models deviate from observations due to model uncertainties (e.g., subgrid scale parameterizations), initialization errors, and boundary forcing errors. Hence, there is an ongoing need to improve ocean models and to keep them consistent with the observed time-evolving ocean state. Ocean data assimilation (ODA) helps to accomplish this task by constraining ocean simulations with available observations. The resulting data-assimilated product allows for the study of changing ocean circulation, transport, and tracer fields. Furthermore, dynamically consistent "smoothers" allow for uncertain model parameters to be adjusted so the model more closely reflects the observed ocean. Parameters can also be used to improve process-oriented forward models, including coupled climate and Earth system models. Satellite altimetry provides an essential constraint for all uses of data assimilation, including climate applications such as ocean reanalysis (this chapter) and operational oceanography (Chapter 19). Ultimately, ODA aims to improve the skill of climate predictions by improving forecast models and by providing accurate descriptions of the present climate state as initial conditions for coupled climate models in support of climate services. ODA also provides a means to evaluate ocean models in comparison to observational measurements.

17.1.3 Aims for This Chapter

In the remainder of this chapter, we introduce elements of ocean modeling and ODA, both with a focus on sea level. Our presentation assumes that readers have some exposure to ocean fluid dynamics, though the material is reasonably self-contained and can thus be used as a starting point for further study. The remainder of this chapter consists of the following sections. We introduce the dynamical framework for ocean circulation models in Section 17.2. In Section 17.3, we introduce various theoretical frameworks to use for understanding sea level evolution and spatial structure. In Section 17.4, we summarize elements of data assimilation, with an emphasis on ocean climate studies. In Section 17.5, we present recent examples of ocean modeling and data assimilation in the context of satellite altimetry. We conclude with an outlook in Section 17.6.

17.2 OCEAN GENERAL CIRCULATION MODELS

OGCMs provide numerical solutions for the equations describing the mechanical and thermodynamical evolution of seawater on a rotating sphere. We here summarize salient features of the ocean equations and ocean circulation models, with specific focus on aspects related to sea level. The discussion is terse, with far more details available in Griffies (2004), Vallis (2006), and Griffies and Adcroft (2008).

17.2.1 The Hydrostatic Primitive Equations

The large-scale ocean circulation is generally well approximated by motion of a stably stratified layer of fluid on a rapidly rotating sphere, with lateral scales of motion far larger than vertical.

The lateral to vertical scale separation allows us to assume the fluid is relatively shallow, in which case the vertical momentum equation is well approximated by the hydrostatic balance (Vallis 2006). To formulate the underlying governing equations, we consider a continuum fluid parcel of mass density ρ, volume δV, mass $\delta M = \rho\delta V$, and center of mass velocity

$$\upsilon = \boldsymbol{u} + \hat{z}w = (u, v, w), \tag{17.1}$$

with the velocity defined in a reference frame rotating with the spherical planet. The linear momentum of each parcel, $\upsilon\delta M$, evolves according to Newton's Second Law. Forces affecting the large-scale ocean circulation arise from the Coriolis effect (rotating reference frame), gravity, pressure, and friction. The thermo-hydrodynamical ocean equations result from coupling the momentum budget to the First Law of thermodynamics, which determines the evolution of enthalpy (heat). The hydrostatic ocean primitive equations can be written in the following manner*:

$$\text{horizontal momentum } \rho\left(\frac{D}{Dt} + \boldsymbol{f}\wedge\right)\boldsymbol{u} = -\nabla_z p + \rho F \tag{17.2a}$$

$$\text{hydrostatic balance } \frac{\partial p}{\partial z} = -\rho\, g \tag{17.2b}$$

$$\text{mass continuity } \frac{D\rho}{Dt} = -\rho\nabla\cdot\upsilon \tag{17.2c}$$

$$\text{tracer conservation (e.g., heat and salt) } \rho\left(\frac{DC}{Dt}\right) = -\nabla\cdot\mathbf{D} \tag{17.2d}$$

$$\text{equation of state } \rho = \rho(s, \Theta, p). \tag{17.2e}$$

Equation (17.2a) provides the budget for horizontal linear momentum, with

$$\frac{D}{Dt} = \frac{\partial}{\partial t} + \upsilon\cdot\nabla \tag{17.3}$$

the time derivative following a material fluid parcel. The material time derivative has two terms: $\partial/\partial t$ is the Eulerian time tendency computed at a fixed point in space, and $\upsilon\cdot\nabla$ is the advective term arising from fluid motion. Advection of momentum is a nonlinear term giving rise to the complexity and richness of turbulent fluid motion. For a hydrostatic fluid, the Coriolis force takes the form $-\rho f\, \hat{\mathbf{z}} \wedge \boldsymbol{u}$, with Coriolis parameter

$$f = 2\Omega \sin\phi, \tag{17.4}$$

where $\hat{\mathbf{z}}$ is the local vertical direction oriented perpendicular to a surface of constant geopotential, $\Omega \approx 7.29 \times 10^{-5}\,\text{s}^{-1}$ is the rotational rate of the Earth, and ϕ is the latitude. Linear momentum is also affected by the downgradient horizontal pressure force, $-\nabla_z p$. Irreversible exchanges of momentum between parcels (i.e., momentum mixing), and between parcels and the ocean boundaries, are parameterized by the friction operator ρF, with Laplacian and/or biharmonic operators most commonly used for the ocean interior (e.g., Smagorinsky 1993; Griffies and Hallberg 2000; Large et al. 2001; Chapter 17 of Griffies 2004; Fox-Kemper and Menemenlis 2008; Jochum et al. 2008).

* "Primitive" here refers to the choice to represent the momentum budget in terms of the velocity field rather than the alternative vorticity and divergence.

Equation (17.2b) is the vertical momentum equation as approximated by the inviscid hydrostatic balance, with p the hydrostatic pressure and g the gravitational acceleration. We assume that the gravitational acceleration is constant in space and time, which is sufficient for many applications. However, space–time variations of gravity are important when considering tidal motions (e.g., Arbic et al. 2010) as well as changes to the static equilibrium sea level as occur with land ice melt (Mitrovica et al. 2001; Kopp et al. 2010).

Furthermore, Vinogradova et al. (2015) showed that self-attraction and loading (SAL) effects caused by the redistribution of mass within the land-atmosphere-ocean system can have a measurable impact on sea level. The mass continuity equation (17.2c) arises from the conservation of mass for the fluid parcel, $D(\delta M)/Dt = 0$, as well as the kinematic result that the infinitesimal parcel volume is materially modified according to the velocity divergence. The concentration of a material tracer, C, represents the mass of trace constituent per mass of the seawater parcel

$$C = \left(\frac{\text{mass of tracer in parcel}}{\text{mass of seawater in parcel}} \right). \tag{17.5}$$

Additionally, the heat content (potential enthalpy) per mass of a fluid parcel is given by the heat capacity times the conservative temperature, Θ. This heat content evolves in the same mathematical form as other conservative material tracers such as salt (McDougall 2003). Consequently, we can consider conservative temperature as the "concentration" of heat.

Although the parcel mass is a material constant, the parcel tracer content and heat content are generally modified by mixing or stirring in the presence of concentration gradients. In this context, mixing is an irreversible exchange of fluid properties among parcels (e.g., diffusion), whereas stirring is a reversible rearrangement of parcels (e.g., advection). The convergence of the tracer flux vector D incorporates subgrid scale mixing and stirring processes in the tracer equation (17.2d). Common means to parameterize subgrid processes involve diffusive mixing across density surfaces in the ocean interior (diapycnal diffusion, as reviewed in MacKinnon et al. 2013); mixing across geopotential surfaces in the well-mixed surface boundary layer (e.g., Large et al. 1994); diffusive mixing along neutral directions in the interior (Solomon 1971; Redi 1982; McDougall et al. 2014); and eddy-induced advection in the ocean interior (e.g., Gent and McWilliams 1990; Gent et al. 1995; Griffies 1998; Fox-Kemper et al. 2013). Given knowledge of the conservative temperature (Θ), absolute salinity (S), and pressure (p), we make use of an empirically determined equation of state (Equation 17.2e) to diagnose the *in situ* density (IOC et al. 2010).

17.2.2 Flux-Form Ocean Equations

To formulate the discrete equations of an ocean model, it is useful to transform the material parcel equations into Eulerian flux-form equations. The flux-form provides a framework for numerical methods that conserve mass and linear momentum according to fluxes across grid cell boundaries. In contrast, discretizations based on the material form, also known as the "advective" form, generally lead to spurious sources of scalars and momentum. Spurious scalar sources (e.g., mass, heat, salt and carbon) are unacceptable for climate simulations. Replacing the material time derivative by the Eulerian time derivative and advection via Equation (17.3) transforms the continuity Equation (17.2c) into

$$\frac{\partial \rho}{\partial t} = -\nabla \cdot (\upsilon \rho) \cdot \tag{17.6}$$

Likewise, combining the tracer equation (17.2d) and continuity equation (17.6) leads to the Eulerian flux-form tracer equation

$$\frac{\partial(\rho C)}{\partial t} = -\nabla\cdot(\rho C \upsilon + \mathbf{D})\cdot \tag{17.7}$$

Notably, there are no subgrid scale terms on the right-hand side of the mass continuity equation (17.6). This result follows because we formulated the equations for a mass conserving fluid parcel and made use of its center of mass velocity, υ. Operationally, this compatibility between mass and tracer budgets is ensured so long as the subgrid scale tracer flux, D, vanishes in the presence of a spatially constant tracer concentration, in which case the tracer equation (17.7) reduces to mass continuity (17.6).

17.2.3 Basics of Finite Volume Discrete Ocean Equations

Numerical solutions to the ocean equations are commonly obtained by approximating the differential operators in space and time through discrete difference operators involving finite length or finite volume elements. For example, a forward approximation to the Eulerian time derivative of a tracer is given by

$$\frac{\partial C}{\partial t} \approx \frac{C(t+\Delta t)-C(t)}{\Delta t}, \tag{17.8}$$

where Δt is the discrete time step. Similar approximations are made for spatial derivatives. In turn, the model domain is split into discrete domains or grid cells, with the discrete variables representing the mean value of the continuous variable over the cell. There are various ways to arrange the suite of model variables horizontally, following from the atmospheric model work of Arakawa and Lamb (1977) (see Haidvogel and Beckmann [1999] for an oceanographic focus). Discretization is also applied in the vertical, dividing the ocean according to depth, potential density, bottom topography, or other choices (Griffies et al. 2000).

We can schematically write the discrete ocean equations in the form of an algebraic equation that updates the ocean state vector, $x(t)$, given the ocean state at the previous time step

$$x(t + \Delta t) = \mathrm{A}[x(t), t] + \mathrm{F}(t). \tag{17.9}$$

The ocean state vector includes temperature, salinity, sea level, velocity, and pressure at each grid point (i.e., the suite of model fields sufficient to specify the discrete ocean state). The time-stepping operator, A, evolves the ocean state forward from one time step to the next, starting from an initial condition. Forcing is provided by F, which includes mechanical boundary conditions (stresses at the air-sea, ice-sea, and Earth-sea boundaries) and buoyancy boundary conditions (heat and freshwater fluxes, including river runoff). As part of the boundary conditions, we need information about the bottom topography (sea floor and coastlines).

When discretizing in time, we must keep the time step, Δt, small enough to ensure that information transmitted by waves, advection, and diffusion does not propagate fully across a grid cell within a single time step. This constraint is the essence of the Courant-Friedrichs-Levy (CFL) condition (Durran 1999). In Section 17.2.7, we see the CFL condition in action when discussing linear wave modes in the ocean.

17.2.4 Oceanic Boussinesq Approximation

The oceanic Boussinesq approximation is based on the observation that dynamically relevant density changes (i.e., changes impacting horizontal pressure gradients) are quite small in the ocean, thus motivating an asymptotic expansion around a global mean density (see Section 9.3 of Griffies and Adcroft 2008). Operationally, the Boussinesq approximation replaces nearly all occurrences of the *in situ* density in the primitive equations with a constant Boussinesq reference density, ρ_o. The key exceptions are the hydrostatic balance and the equation of state, where the full density is retained. Notably, the oceanic Boussinesq approximation does *not* assume a linear equation of state, which contrasts to the common use of the Boussinesq approximation in other areas of fluid dynamics (e.g., Chandrasekhar 1961).

When making the Boussinesq approximation, the mass continuity equation (17.2c) reduces to volume conservation, so that the Boussinesq velocity has zero divergence

$$\nabla \cdot \upsilon = 0 \tag{17.10}$$

A divergent-free velocity filters out all acoustic modes (e.g., sound waves and Lamb waves), which is useful because acoustic modes are not relevant for the large-scale circulation. The Boussinesq approximation is based on the scaling $|\upsilon^d| \ll |\upsilon|$, where υ is the prognostic velocity appearing in the Boussinesq momentum equations, and υ^d is a divergent velocity field that balances material changes in density through the continuity equation. That is, the divergent velocity field satisfies

$$\frac{D\rho}{Dt} = -\rho \nabla \cdot \upsilon^d, \tag{17.11}$$

where, to leading order, the material time derivative only involves the non-divergent velocity. It is in this manner that the oceanic Boussinesq approximation admits material density changes from thermohaline effects (i.e., changes in temperature and salinity), which in turn impact the large-scale circulation. In Section 17.3.1, we see how the Boussinesq approximation impacts sea level equations. Although Boussinesq models have kinematics based on volume conservation, this property does *not* mean the total volume of the Boussinesq ocean remains constant. Indeed, Boussinesq models generally allow for the transfer of volume across the ocean boundary through rivers, precipitation, and evaporation. It is only when making the assumption of a constant *total* volume that Boussinesq models preclude boundary volume fluxes, in which case they must introduce virtual salt fluxes (see Section 17.2.5).

The geopotential vertical coordinate is arguably the simplest choice for discretizing the vertical direction. This coordinate thus formed the basis for the first ocean circulation model of Bryan (1969) and many subsequent models. The geopotential coordinate is naturally suited for Boussinesq kinematics because the seawater volume between two geopotentials remains constant for a Boussinesq fluid. Consequently, most large-scale ocean models make the oceanic Boussinesq approximation. Nevertheless, there are methods for incorporating non-Boussinesq effects into hydrostatic ocean models through use of pressure as the vertical coordinate rather than geopotential (Huang et al. 2001; DeSzoeke and Samelson 2002; Marshall et al. 2004; Griffies and Greatbatch 2012). These pressure-based methods are being considered for some ocean climate models being developed today.

17.2.5 Ocean Responses to Virtual Salt Fluxes versus Real Water Fluxes

Some Boussinesq ocean models are formulated with a globally constant volume, with the ocean boundary impermeable to water fluxes. A globally constant volume is fundamental to the rigid lid algorithm of Bryan (1969), whereas free surface algorithms offer a natural means for water fluxes to cross the surface boundary and thus to change the ocean volume (e.g., Griffies et al. 2001). Even so, some free surface models maintain a constant total volume (e.g., see Table 1 in Griffies et al. 2014).

For models based on a globally constant volume, it is necessary to introduce virtual salt fluxes to parameterize impacts on density that would otherwise arise from boundary water fluxes. For example, rain added to the ocean dilutes the surface salinity. This dilution is parameterized in a virtual salt flux model through the removal of salt. That is, virtual salt flux models transfer salt rather than water across the surface boundary (Huang 1993; Griffies et al. 2001; Yin et al. 2010b). Either process alters the ocean density and hence the horizontal and vertical density gradients.

Perturbations to density interfaces induce linear wave responses that propagate as baroclinic gravity and Rossby waves. The term "baroclinic" refers to the non-alignment of pressure and density isolines, which in turn generate vorticity (see Chapter 4 of Vallis 2006). Stammer (2008) studied the response of sea level to meltwater perturbations in the North Atlantic, with meltwater parameterized by virtual salt fluxes. The associated density perturbations propagated as baroclinic waves traveling around the planet on multidecadal timescales.

Besides altering the density field, water crossing the ocean directly changes the sea level (volume per area of an ocean column) and the bottom pressure (mass per area of an ocean column). Information about these material changes propagates as barotropic waves (fluctuations of the full ocean column). Barotropic waves travel roughly 100 times faster than baroclinic waves and can thus span the global ocean in a few days. Lorbacher et al. (2010) examined the timescales and propagation pathways for such barotropic wave processes arising from high latitude melt events. They found that barotropic sea level signals of Greenland melt are felt in the tropical west Pacific within a week. This result contrasts to the virtual salt flux models, which do not exhibit such barotropic signals from melt events; they only carry baroclinic signals such as those considered by Stammer (2008). Correspondingly, virtual salt flux models cannot be used to study changes in sea level associated with melting land ice (Kopp et al. 2010).

Another limitation of virtual salt flux models arises from different responses of the Atlantic Meridional Overturning Circulation (AMOC) to meltwater pulses. As shown by Yin et al. (2010b), virtual salt flux models exaggerate their freshening effect relative to the response seen in real water flux models. The reason for the different responses arises from the use of a constant reference salinity in the virtual salt flux models. As changes to the Atlantic overturning are important for regional sea level changes (Yin et al. 2009; Lorbacher et al. 2010; Goddard et al. 2015), it is useful to remove unnecessary assumptions, such as virtual tracer fluxes, when considering model responses to climate change associated with melt events.

Making use of real water flux boundary conditions allows for the ocean mass to change according to the net evaporation, precipitation, and runoff. A surface mass flux in turn is associated with a barotropic Goldsbrough-Stommel circulation (Goldsbrough 1933; Stommel 1957; Huang and Schmitt 1993), which has an associated imprint in sea level on the order of 1 cm (Liu et al. 2016).

17.2.6 Impacts from Changes to the Gravitational Geopotential

Inhomogeneities in mass distributions of the solid Earth cause the Earth's gravity field to be non-ellipsoidal, thus also causing the sea surface to deviate from a reference ellipsoid. These inhomogeneities evolve over geological timescales (millions of years), in which case they are assumed fixed for ocean circulation modeling. However, there is a response of the solid Earth to mass redistributions associated with melting land ice, notably polar ice sheets over Greenland and Antarctica.

Changes to the Earth gravity field and rotation vector alter the static equilibrium sea level (Farrell and Clark 1976; Mitrovica et al. 2001), which defines the surface of a resting ocean. As land ice melts, changes to the static equilibrium sea level will emerge from among changes in dynamical sea level (see Kopp et al. 2010; Slangen et al. 2012, 2014). Such static equilibrium changes are computed offline and added to climate model results. Ultimately, climate models need to incorporate both effects on sea level, such as in Kuhlmann et al. (2011) and Vinogradova et al. (2015). Additionally, gravity changes have an impact on ice sheet dynamics by altering grounding lines (Gomez et al. 2010), motivating the coupling of gravity models to ocean and ice sheet models.

17.2.7 Fast and Slow Dynamics

Linear oceanic wave motions exhibit a range of timescales. When developing an economical time-stepping algorithm for the numerical primitive equations, it is essential to decompose the wave dynamics into fast and slow motions. Ideally, this timescale decomposition will allow one to numerically time step the fast waves using small time steps, required for numerical stability according to the CFL constraint (e.g., Durran 1999), while the slow motions (waves, advection, and diffusion) can utilize longer time steps. However, fast and slow modes couple in the ocean, so any putative split is not complete, thus requiring careful considerations (see, e.g., Killworth et al. 1991; Griffies et al. 2001; Chapter 12 of Griffies 2004; Shchepetkin and McWilliams 2005; Hallberg and Adcroft 2009).

Acoustic modes are irrelevant for the general circulation and are in effect filtered through the hydrostatic approximation or are fully filtered through the Boussinesq approximation (DeSzoeke and Samelson 2002; Griffies and Adcroft 2008). Gravity waves, both baroclinic and barotropic, were introduced in Section 17.2.5. A barotropic gravity wave rapidly carries information about mass and volume perturbations, which in turn affects sea level. In contrast, baroclinic gravity waves carry information about changes to interior ocean density interfaces. Barotropic gravity waves travel at speeds $\sqrt{gH}$ (H denoting the ocean depth), which in the deep ocean can be 50–100 times faster than baroclinic waves (e.g., 100 m s^{-1} vs. 1–2 m s^{-1}). To split between barotropic (also called external) and baroclinic (also called internal) motions, we may attempt a formal eigenmode decomposition (e.g., Chapter 6 of Gill 1982). However, this decomposition only works when the flow is close to linear, which is not the case in the real ocean. Furthermore, with topography, eigenmodes are strongly coupled (Hallberg and Rhines 1996).

A practical means to partially split between barotropic and baroclinic modes is to depth integrate the primitive equations. The depth-averaged motions well approximate the barotropic mode, and depth-dependent deviations approximate baroclinic modes. The numerics associated with these "split-explicit" methods concern details of the time-stepping algorithm, particularly for the fast depth-integrated motions, as well as determining what portion of the dynamics to place in the fast versus slow equations. We expose some of the issues by formulating the depth-integrated kinematics, which are based on a budget for the mass per horizontal area in a column of seawater

$$\frac{\partial}{\partial t}\left(\int_{-H}^{\eta} \rho\, dz\right) = -\nabla \cdot U^{\rho} + Q_m. \tag{17.12}$$

That is, the column mass changes according to the convergence of mass transported horizontally by the currents

$$U^{\rho} = \int_{-H}^{\eta} \rho \mathbf{u}\, dz, \tag{17.13}$$

and from mass crossing the ocean free surface, Q_{m}, through precipitation, evaporation, sea ice melt/form, and river runoff. Combining this mass budget (a kinematical balance) with the hydrostatic balance (a dynamical balance) renders a prognostic equation for the pressure difference

$$\frac{1}{g}\frac{\partial (p_b - p_a)}{\partial t} = -\nabla \cdot U^{\rho} + Q_m, \tag{17.14}$$

where

$$p_b = p_a + g\int_{-H}^{\eta} \rho\, dz \tag{17.15}$$

is the hydrostatic pressure at the ocean bottom, and p_a is the pressure applied on the top of the ocean from the atmosphere, sea ice, or ice shelf. A similar formulation follows for the horizontal momentum equation, though with some subtleties related to extracting the fast portion of the depth-integrated dynamics from the slow portion (see Section 12.2 of Griffies [2004] for details and references). The result of this development is the depth-integrated momentum equation

$$\left(\frac{\partial}{\partial t} + f\hat{z}\wedge\right)U^{\rho} = -\left(\frac{p_b - p_a}{\rho_0 g}\right)\nabla\left(p_b + \rho_0\Phi_b\right) + \boldsymbol{G}, \tag{17.16}$$

where $\Phi_b = -g\,H$ is the geopotential at the ocean bottom. The term $\boldsymbol{G}$ contains the vertical integral of nonlinear terms, friction, and the depth-integrated slow portion of the pressure gradient.

To help understand the free linear modes of the depth-integrated system, we drop all forces and nonlinear terms to render the shallow water system

$$\frac{\partial U^{\rho}}{\partial t} = -H\nabla p_b \tag{17.17a}$$

$$\frac{\partial p_b}{\partial t} = -g\nabla \cdot U^{\rho}, \tag{17.17b}$$

where the bottom at $z = -H$ is assumed to be flat. This linear system admits non-dispersive external gravity waves with phase speed

$$C = \sqrt{gH}. \tag{17.18}$$

These waves transmit information about changes in the bottom pressure or, equivalently, changes in the mass per area of a fluid column. Furthermore, assuming the ocean has a constant density, bottom pressure takes the form $p_b = \rho_o\, g\, (H + \eta)$, so that bottom pressure waves are equivalent to sea surface waves.

The essential features of a split-explicit algorithm involve time-stepping the depth-integrated mass budget (17.14) and momentum budget (17.16), making use of small time steps to stably resolve barotropic gravity waves. The slower dynamics are approximated by the full velocity field with the depth-averaged velocity removed. The resulting depth-dependent motions have timescales dominated by baroclinic gravity waves and advection. The slow dynamics can be integrated with a longer time step than the fast motions, which is important because the slow dynamics are three-dimensional and so more expensive computationally. There are many details required to bring these ideas into a working algorithm. The interested reader can find further discussion in Chapter 12 of Griffies (2004) and Section 11 of Griffies and Adcroft (2008), along with even more specialized and detailed discussions in Killworth et al. (1991), Griffies et al. (2001), Shchepetkin and McWilliams (2005), and Hallberg and Adcroft (2009).

17.3 SEA LEVEL TENDENCIES AND SPATIAL PATTERNS

The upper ocean is typically characterized by breaking surface gravity waves (e.g., Cavaleri et al. 2012), in which case there is no mathematically smooth ocean "surface." Nonetheless, for large-scale hydrostatic modeling, and for large-scale observational oceanography, we define the upper ocean interface as a smooth, non-overturning, permeable, free boundary surface

$$\text{ocean free surface: } z = \eta(x, y, t). \tag{17.19}$$

The ocean free surface provides our mathematical representation of sea level. Furthermore, the effects of turbulent wave breaking, which are inherently non-hydrostatic, are incorporated into parameterizations of air-sea boundary fluxes and upper ocean wave induced mixing. We discussed one form of sea level evolution when considering the fast and slow modes in Section 17.2.7. Here, we derive further kinematic expressions based on the mass continuity equation. We then make the hydrostatic approximation, which connects changes in sea level to changes in pressure at the ocean top and bottom boundaries. Notably, we here ignore changes in the land-sea boundaries (i.e., the ocean bottom at $z = -H(x, y)$ is static). We also assume the geopotential takes the standard form, $\Phi = gz$. Far more details on material from this section can be found in Griffies and Greatbatch (2012) and Griffies et al. (2014).

17.3.1 Sea Level Tendencies and Mass Continuity

We derive a kinematic expression for sea level evolution by integrating the mass continuity equation (17.2c) over the full ocean depth and making use of surface and bottom kinematic boundary conditions. The resulting sea level tendency is given by

$$\frac{\partial \eta}{\partial t} = \frac{Q_m}{\rho(\eta)} - \nabla \cdot U - \int_{-H}^{\eta} \frac{1}{\rho} \frac{D\rho}{Dt} dz, \tag{17.20}$$

with $U = \int_{-H}^{\eta} \mathbf{u}\, dz$ being the vertically integrated horizontal velocity. Equation (17.20) partitions sea level evolution into a boundary mass flux, Q_m, which is the mass per time per horizontal area of water crossing the ocean surface, the convergence of vertically integrated horizontal ocean currents, and material time changes of the *in situ* density.

Taking the global area mean of Equation 17.20 renders an evolution equation for global mean sea level, $\bar{\eta}$

$$\frac{\partial \bar{\eta}}{\partial t} = \overline{\left(\frac{Q_m}{\rho(\eta)}\right)} - \frac{V}{S}\left\langle \frac{1}{\rho} \frac{D\rho}{Dt} \right\rangle, \tag{17.21}$$

where V is the total seawater volume, S the ocean surface area, < > is the global volume mean operator, and $\overline{(\,)}$ is the ocean area mean operator. Evolution of global mean sea level is thus affected by surface boundary mass fluxes and material density changes. In this framework, ocean currents are seen to redistribute ocean volume yet not to alter global mean sea level. Hence, as emphasized by Griffies and Greatbatch (2012), Equation (17.20) provides a useful analysis framework to study how physical processes impact the evolution of global mean sea level.

17.3.2 Non-Boussinesq Steric Effect and the Boussinesq Sea Level

The *non-Boussinesq steric effect* refers to the sea level tendency associated with material density changes in Equation (17.20)

$$\left(\frac{\partial \eta}{\partial t}\right)^{non-Bouss\ steric} = -\int_{-H}^{\eta} \frac{1}{\rho} \frac{D\rho}{Dt} dz. \tag{17.22}$$

This term is where buoyancy forcing (e.g., boundary heat fluxes), mixing, and subgrid scale processes directly impact sea level.

Notably, the non-Boussinesq steric effect is absent in Boussinesq fluids (recall discussion in Section 17.2.4). That is, Boussinesq kinematics is based on conserving volume, not mass. Integrating $\nabla \cdot \upsilon = 0$ over a seawater column, and using kinematic boundary conditions at the ocean surface and bottom, leads to the Boussinesq sea level equation

$$\left(\frac{\partial \eta}{\partial t}\right)^{bouss} = \frac{Q_m}{\rho_0} - \nabla \cdot U. \tag{17.23}$$

As discussed by Losch et al. (2004) and Griffies and Greatbatch (2012), Boussinesq and non-Boussinesq fluids capture very similar large-scale patterns of dynamical sea level.* Examples of

* Dynamic sea level refers to the sea level normalized to have zero area mean. This component of sea level responds directly to dynamical ocean processes. See more discussion of dynamic sea level in Section H7 of Griffies et al. (2016).

such patterns include sea level depressions under anomalously dense fluid columns (e.g., a dense mesoscale eddy). However, as discussed in Section 17.3.3, Boussinesq fluids require an adjustment of their prognostic sea level to capture the global mean of the non-Boussinesq fluid.

17.3.3 Evolution of Global Mean Sea Level

Greatbatch (1994) noted that a Boussinesq fluid will not alter its prognostic sea level under a uniform heating, thus exposing a limitation of Boussinesq models for simulating the observed global thermosteric sea level rise associated with anthropogenic warming (Church et al. 2013). The problem arises from dropping the non-Boussinesq steric term (17.22) as it appears in the sea level equation (17.20). The non-Boussinesq steric term is where heating directly impacts sea level. With the aim of accounting for this missing physical process in Boussinesq models, Greatbatch (1994) introduced a time-dependent global adjustment to the Boussinesq sea level.

To help in understanding the basics, we here summarize elements of Section 4.5 and Appendix D of Griffies and Greatbatch (2012), whereby we formulate the evolution equation for global mean sea level in both non-Boussinesq and Boussinesq fluids.

17.3.3.1 Mass Conserving Non-Boussinesq Fluids

The global mean density of liquid seawater, $\langle\rho\rangle$, is the ratio of the total seawater mass, M, to its volume, V

$$\langle\rho\rangle = \frac{M}{V}. \tag{17.24}$$

It follows that the time tendency of the total seawater mass is given by

$$\frac{\partial M}{\partial t} = V\left(\frac{\partial\langle\rho\rangle}{\partial t}\right) + \langle\rho\rangle\left(\frac{\partial V}{\partial t}\right). \tag{17.25}$$

The total seawater mass changes if there is a nonzero mass flux across the ocean surface

$$\frac{\partial M}{\partial t} = S\overline{Q_m}, \tag{17.26}$$

where $\overline{Q_m}$ is the total mass per horizontal area per time of material crossing the ocean boundaries (i.e., mean precipitation plus runoff minus evaporation), and S is the total ocean surface area. Additionally, the total seawater volume

$$V = \iint dx\,dy \int_{-H}^{\eta} dz \tag{17.27}$$

has a time tendency arising from changes to the area mean sea level*

$$\frac{\partial V}{\partial t} = S\left(\frac{\partial\overline{\eta}}{\partial t}\right). \tag{17.28}$$

Bringing these results together leads to an evolution equation for the global mean sea level[3]

$$\frac{\partial\overline{\eta}}{\partial t} = \frac{\overline{Q_m}}{\langle\rho\rangle} - \left(\frac{V}{S}\right)\frac{1}{\langle\rho\rangle}\left(\frac{\partial\langle\rho\rangle}{\partial t}\right) \tag{17.29}$$

* Recall that we assume the lateral ocean boundaries to be fixed in time (i.e., there is no wetting and drying).

This equation is structurally similar to Equation (17.21) based on the non-Boussinesq steric form. Setting Equation (17.29) equal to Equation (17.21) leads to the identity

$$\overline{\left(\frac{Q_m}{\rho(\eta)}\right)} - \left(\frac{V}{S}\right)\left\langle\frac{1}{\rho}\frac{D\rho}{Dt}\right\rangle = \frac{\overline{Q_m}}{\langle\rho\rangle} - \left(\frac{V}{S}\right)\frac{1}{\langle\rho\rangle}\left(\frac{\partial\langle\rho\rangle}{\partial t}\right). \tag{17.30}$$

17.3.3.2 Volume-Conserving Boussinesq Fluids

A Boussinesq fluid conserves volume rather than mass, with this property leading to some unfamiliar and unphysical behavior. For example, consider the case of zero boundary fluxes of volume, so that $\partial V/\partial t = 0$. With $M = V\langle\rho\rangle$ and with the total volume constant, the Boussinesq seawater mass picks up a source associated with changes in global mean density: $\partial M/\partial t = V\,\partial\langle\rho\rangle/\partial t$. For the case of surface warming with a positive thermal expansion coefficient, maintaining a constant Boussinesq volume means there must be a decrease in ocean mass (and bottom pressure) as density decreases. This mass change is physically spurious because it does not appear in the real ocean, although it does appear in a Boussinesq ocean model. When interested in the mass distribution of a Boussinesq ocean model, such as needed for angular momentum (Bryan 1997), bottom pressure (Ponte 1999), or static equilibrium perturbations (Kopp et al. 2010), one must account for spurious mass changes. To account for the missing global steric effect in a Boussinesq ocean model, Greatbatch (1994) introduced a diagnosed sea level tendency according to

$$\left(\frac{\partial\eta}{\partial t}\right)^{bouss}_{diag} = \left(\frac{\partial\eta}{\partial t}\right)^{bouss} + \left(\frac{\partial\overline{\eta}}{\partial t}\right)^{non-basic\ steric} \tag{17.31}$$

where $(\partial\eta/\partial t)^{bouss}$ is given by the prognostic equation for Boussinesq sea level Equation (17.23), and the global mean of the non-Boussinesq steric term is set to

$$\left(\frac{\partial\overline{\eta}}{\partial t}\right)^{non-bouss\ steric} = -\frac{V}{S}\left\langle\frac{1}{\rho}\frac{D\rho}{Dt}\right\rangle \tag{17.32a}$$

$$= -\frac{V}{S}\left(\frac{1}{\langle\rho\rangle}\frac{\partial\langle\rho\rangle}{\partial t}\right). \tag{17.32b}$$

As seen from Equation (17.30), the approximation going from Equation (17.32a) to (17.32b) is appropriate when making the Boussinesq approximation, in which case

$$\overline{\left(\frac{Q_m}{\rho(\eta)}\right)} \approx \frac{\overline{Q_m}}{\langle\rho\rangle} \approx \frac{1}{\rho_0}\overline{Q_m} \tag{17.33}$$

The global steric effect on the right-hand side of Equation (17.32b) involves the time tendency of the global mean *in situ* density $\langle\rho\rangle$. This time tendency is easily diagnosed from model output (e.g., see Section H9.5 of Griffies et al. 2016). Adjusting the Boussinesq sea level with the global steric effect ensures that the diagnosed global mean sea level emulates sea level in the non-Boussinesq model. In particular, it incorporates global thermosteric sea level rise in global warming experiments.

17.3.4 Sea Level Tendencies and the Hydrostatic Balance

We now deduce relations for sea level tendencies based on the hydrostatic balance (17.2b). These relations hold for *both* Boussinesq and non-Boussinesq ocean models as they rely only on the hydrostatic approximation.

Integrating the hydrostatic balance from the ocean bottom at $z = -H(x, y)$ to the surface at $z = \eta(x, y, t)$ leads to the bottom pressure equation (17.15). From that equation, we see that the bottom pressure, p_b, equals the sum of the pressure applied at the sea surface, p_a, plus the weight per horizontal area of seawater in the liquid ocean column. This balance holds instantaneously. Consequently, adding mass to the ocean surface instantaneously increases bottom pressure, no matter how deep the ocean. Such instantaneous signal propagation results from assuming the hydrostatic fluid, in which vertically propagating acoustic waves formally have an infinite phase speed.

Taking the time derivative of the bottom pressure equation (17.15) renders

$$\frac{\partial(p_b - p_a)}{\partial t} = g\rho(\eta)\frac{\partial \eta}{\partial t} + g\int_{-H}^{\eta} \frac{\partial \rho}{\partial t} dz. \tag{17.34}$$

where $\rho(\eta) = \rho(z = \eta)$ is density at the ocean free surface. This equation represents a diagnostic balance among three time tendencies: changes in the mass of seawater in an ocean column (left-hand side) are balanced by changes in the sea level and depth-integrated changes in density. Following Gill and Niiler (1973), we rearrange Equation (17.34) to yield a diagnostic expression for the sea level tendency

$$\frac{\partial \eta}{\partial t} = \underbrace{\left(\frac{1}{g\rho(\eta)}\right)\frac{\partial(p_b - p_a)}{\partial t}}_{\textit{mass tendency}} \underbrace{- \frac{1}{\rho(\eta)}\int_{-H}^{\eta} \frac{\partial \rho}{\partial t} dz}_{\textit{local steric tendency}}. \tag{17.35}$$

This decomposition connects changes in ocean volume to changes in ocean mass and changes in ocean density. We refer to the density changes as the *local steric effect* to distinguish it from the global steric effect described in Section 17.3.3 (see in particular Equation 17.32b).

The decomposition (17.35) provides the basis for various diagnostic analyses of regional sea level changes. In particular, it allows us to attribute sea level changes to changes in salinity (halosteric effects), temperature (thermosteric effects), and mass (bottom pressure changes). Recent studies using this decomposition are Lowe and Gregory (2006), Landerer et al. (2007a, 2007b), Yin et al. (2009, 2010a), Pardaens et al. (2011), Griffies et al. (2014), and Landerer et al. (2015). Furthermore, as previously noted, the decomposition (17.35) holds for both a Boussinesq or non-Boussinesq ocean model. In particular, the local steric term gives rise to a positive sea level anomaly for a fluid column with a negative density anomaly (e.g., warm core mesoscale eddy).

It has recently become feasible to diagnose each of the three terms in the decomposition (17.35) using observationally based measurements. Namely, the sea level tendency is measured by satellite altimetry (e.g., this book); the mass tendency is measured by the gravity field (e.g., GRACE); and the density (or *local steric*) term is measured by *in situ* temperature and salinity (e.g., Argo). We discuss facets of this balance in the following.

17.3.5 Sea Level Tendencies due to Mass Changes

The hydrostatic expression for the bottom pressure (17.15) indicates that the pressure difference $p_b - p_a$ changes when mass per area within a seawater column changes. Furthermore, the column mass budget is given by Equation 17.12, which allows us to write the equivalent expressions for sea level change arising from mass changes

$$\left(\frac{\partial \eta}{\partial t}\right)^{\textit{mass changes}} = \underbrace{\left(\frac{1}{g\rho(\eta)}\right)\frac{\partial(p_b - p_a)}{\partial t}}_{\textit{mass tendency}} = \underbrace{\frac{-\nabla \cdot U^{\rho} + Q_m}{\rho(\eta)}}_{\textit{mass convergence}} \tag{17.36}$$

Mass converging to a column causes the column to increase its thickness and thus to raise the sea level. Signals of mass changes propagate through barotropic wave processes, which rapidly transmit mass induced sea level changes around the World Ocean (e.g., see Section 17.2.6, Ponte 2006a and Lorbacher et al. 2012).

17.3.6 Sea Level Tendencies due to Local Steric Changes

The second term on the right-hand side of Equation (17.35) arises from local depth-integrated density changes, which we refer to as the *local steric* effect

$$\left(\frac{\partial \eta}{\partial t}\right)^{\text{local steric}} = -\frac{1}{\rho(\eta)} \int_{-H}^{\eta} \frac{\partial \rho}{\partial t} dz. \tag{17.37}$$

This local steric effect is distinguished from the non-Boussinesq steric effect discussed in Section 17.3.1. As density in the column decreases, such as when a fluid column warms or freshens, then the column expands and sea level rises. The local steric term in Equation (17.37) thus arises from changes in temperature and salinity (and pressure).* In many regions, such as the Atlantic, the ocean is both warming and getting saltier, so that the *thermosteric* (temperature-induced) sea level rise is partially compensated by *halosteric* (salinity-induced) sea level fall (e.g., Figure 17.6; see also Yin et al. 2010a). Finally, we note that changes in steric sea level propagate throughout the World Ocean on a baroclinic timescale and are therefore far slower than the barotropic signals that transmit mass changes (Bryan 1996; Hsieh and Bryan 1996; Stammer 2008; Lorbacher et al. 2012).

17.3.7 Sea Level Changes due to Applied Surface Loading

A changing atmospheric surface pressure and/or sea ice loading leads to changes in the sea surface height (Ponte 1993; Wunsch and Stammer 1997; Ponte 2006b). Fluctuations in the applied surface pressure in turn force barotropic ocean fluctuations (e.g., Arbic 2005). The isostatic sea level response to changes in applied pressure is the *inverse barometer*. A 1-mbar change in applied pressure approximately corresponds to a 1-cm inverse barometer sea level.

To help understand the inverse barometer sea level, consider a case where the applied surface pressure is modified, yet the ocean bottom pressure and ocean density are unchanged, thus leaving currents unchanged. We can realize this situation so long as the sea level adjusts to provide exact compensation for changes in the applied pressure. Making use of Equation (17.35) renders

$$\left(\frac{\partial \eta}{\partial t}\right)^{\text{inverse barometer}} = -\left(\frac{1}{g\rho(\eta)}\right)\frac{\partial p_a'}{\partial t}. \tag{17.38}$$

We here introduce the anomalous applied pressure

$$p_a' = p_a - \overline{p}_a, \tag{17.39}$$

where $\overline{p}_a$ is the ocean area mean applied pressure. As reviewed in Appendix C of Griffies and Greatbatch (2012), such inverse barometer responses of sea level are commonly realized under sea ice and under atmospheric pressure loading. Although the sea surface height changes, the "effective sea level"

* Pressure-induced changes are generally subdominant, so that the local steric effect is generally determined by changes in temperature and salinity.

$$\eta' = \eta + \left(\frac{p_a'}{g\overline{\rho}}\right) \tag{17.40}$$

remains close to constant, where we introduced the area mean surface density, $\overline{\rho}$. For example, the sea surface height is depressed if the applied pressure increases. If the depressed sea surface height maintains an inverse barometer response, then the effective sea level remains unchanged. The effective sea level was introduced by Campin et al. (2008) in the context of ocean/sea-ice modeling, and it is commonly used for reporting sea level in climate models (see Section H7 in Griffies et al. 2016).

17.3.8 Sea Level Gradients and Ocean Circulation

From the hydrostatic balance (17.2b), we can write the pressure at an arbitrary depth z in the form

$$p(z) = p_a + g\int_z^{\eta} \rho\, dz', \tag{17.41}$$

which then leads to the horizontal pressure gradient

$$\nabla_z p(z) = \nabla p_a + g\rho(\eta)\nabla\eta + g\int_z^{\eta} \nabla_z \rho\, dz' \tag{17.42a}$$

$$\approx g\overline{\rho}\nabla\eta' + g\int_z^{\eta} \nabla_z \rho\, dz' \tag{17.42b}$$

In the last step, we made use of the effective sea level (Equation 17.40) discussed in relation to the inverse barometer. Equation 17.42b is the basis for approximate dynamical expressions considered in the following.

17.3.8.1 Surface Ocean

A particularly simple relation between sea level and surface ocean currents occurs when gradients in sea level balance the Coriolis force created by surface currents

$$g\nabla\eta' = -f\,\hat{z}\wedge\mathbf{u}, \tag{17.43}$$

where $\mathbf{u}$ is the surface horizontal velocity. This *geostrophic balance* forms the basis for how surface geostrophic currents are diagnosed from sea level measurements (Wunsch and Stammer 1998). A slight generalization is found by including the turbulent momentum flux τ^s through the ocean surface boundary, in which case the sea level gradient takes the form

$$g\nabla\hat{\eta} = -f\,\hat{z}\wedge\mathbf{u} + \frac{\tau^s}{\rho_o h_E}, \tag{17.44}$$

where h_E is the Ekman depth over which the boundary stresses penetrate the upper ocean (e.g., see Section 2.12 of Vallis 2006). As noted by Lowe and Gregory (2006), surface currents in balance with surface wind stresses tend to flow parallel to the sea level gradient, whereas geostrophically balanced surface currents are aligned with surfaces of constant sea level. For equatorial regions, where the Coriolis force vanishes, the east-west sea level tilt is in balance with the surface easterly wind stress.

17.3.8.2 Full Ocean Column

Vertically integrating the linearized form of the horizontal momentum budget (17.2a) in the absence of horizontal friction leads to

$$(g\rho_o H)\nabla\eta' = \tau^s + Q_m\mathbf{u}_m - \tau^b - (\partial_t + f\hat{z}\wedge)U^\rho - \mathrm{B}\cdot \tag{17.45}$$

In this equation, τ^s and τ^b are the turbulent boundary momentum fluxes at the surface and bottom. Q_m $\mathbf{u}_m$ is the horizontal advective momentum flux associated with surface boundary fluxes of mass, with $\mathbf{u}_m$ the horizontal momentum per mass of material crossing the ocean surface.* Finally,

$$B = g\int_{-H}^{\eta} dz \int_{z}^{\eta} \nabla_z \rho \, dz' \tag{17.46}$$

is a horizontal pressure gradient arising from horizontal density gradients throughout the ocean column. Lowe and Gregory (2006) employed the steady state version of the balance (17.45) while ignoring boundary terms (see their Equation 17.7),

$$(g\rho_o \mathrm{H})\nabla\eta' \approx -f\hat{z}\wedge U^{\rho} - \mathrm{B} \tag{17.47}$$

to help interpret the sea level patterns in their climate model simulations.

17.3.8.3 Barotropic Geostrophic Balance and Transport through a Section

As seen by Equation 17.45, sea level gradients balance many terms, including surface fluxes, internal pressure gradients, and vertically integrated transport. Dropping all terms except the Coriolis leads to a geostrophic balance for the vertically integrated flow, whereby Equation (17.45) reduces to

$$(g\rho_o H)\nabla\eta' = f\hat{z}\wedge U^{\rho} \cdot \tag{17.48}$$

This equation is equivalent to

$$U^{\rho} = -\left(\frac{g\rho_o H}{f}\right)\hat{z}\wedge\nabla\eta', \tag{17.49}$$

implying that, for a constant-depth ocean and constant Coriolis parameter (f-plane approximation), the effective sea level is the stream function for the vertically integrated flow. Following Wunsch and Stammer (1998), we use Equation (17.48) to see how much vertically integrated transport is associated with a sea level deviation. The meridional transport between two longitudes x_1 and x_2 is given by

$$\int_{x_1}^{x_2} dx \, \mathrm{V}^{\rho} = \frac{g\rho_o H}{f}\left[\eta(x_2) - \eta(x_1)\right], \tag{17.50}$$

where we assumed a flat ocean bottom. The horizontal distance drops out from the right-hand side, so that the meridional geostrophic transport depends only on the sea level difference across the zonal section and not on the length of the section. As an example of Equation 17.50, assume the ocean depth is $H = 4000$ m and set $f = 7.3 \times 10^{-5}\ \mathrm{s}^{-1}$ (30° latitude). Each centimeter of sea level difference corresponds to a mass transport of roughly $6 \times 10^9\ \mathrm{kg\ s}^{-1}$ (which equals six mass Sverdrups).

17.4 OCEAN DATA ASSIMILATION

By making use of a broad suite of mathematical and computational tools, ODA aims to provide optimal descriptions of the time-varying ocean circulation. Hence, ODA supports studies of

* In ocean models, u_m is generally taken as the surface ocean horizontal velocity.

ocean dynamics, in particular by allowing for the estimate of unobservable quantities. ODA gains much from and offers much to satellite altimetry. In this section, we borrow from Stammer et al. (2016), who review data assimilation approaches and applications to ocean climate problems (see also Wunsch and Heimbach 2013). Optimal here means that all available information about data, the model, and uncertainties are used in a mathematically rigorous way. In practice, this is often not realized through simplifications, making results suboptimal at best.

Many ODA methods were developed in numerical weather prediction (NWP) to help initialize atmospheric weather forecasts (e.g., Bouttier and Courtier 1999). Others were advocated specifically for estimating the time-varying ocean circulation (Wunsch 1996). Most ODA approaches are variants of least squares methods used to combine models with observational based data, with least squares optimal if the errors are Gaussian. Ideally, the ODA solutions encompass estimates of dynamically consistent state fields, along with model parameters such as mixing coefficients and sub-grid scale closure. Additionally, and critically, optimal solutions provide error estimates for the ocean state and parameters. However, in practice, certain aspirations remain unmet.

17.4.1 Elements of Ocean Data Assimilation

Measurements from the ocean do not generally correspond to fields carried in an ocean model. Additionally, all measurements have errors. Consequently, we must map between ocean measurements and the model ocean state, and we do so with a functional relation of the form*

$$\mathbf{E}\mathbf{x} + \mathbf{n} = \mathbf{y}. \tag{17.51}$$

This equation assumes a linear relation between observations and model state variables, where $\mathbf{E}$ is the "observational" matrix that maps the model state, $\mathbf{x}$, to the ocean measurement, $\mathbf{y}$. The vector $\mathbf{n}$ represents measurement uncertainties, including data noise. In some cases, altimetric sea level data cannot be assimilated directly but instead needs to be projected onto vertical temperature and salinity profiles. ODA methods are used to estimate the time-dependent ocean state, $\mathbf{x}^{\sim}(t)$, along with the uncertain model parameters. Optimal estimates are found through minimizing an objective or cost function, J. The cost function is a weighted norm of the vector of differences between observations and their model equivalents. We write the cost function as

$$J = \sum_t \left((\mathbf{y} - \mathbf{E}\tilde{\mathbf{x}})^T \mathbf{R} (\mathbf{y} - \mathbf{E}\tilde{\mathbf{x}}) \right) + \mathbf{x}_0^{\mathrm{T}} \mathbf{Q} \mathbf{x}_0 + \mathbf{u}^{\mathrm{T}} \mathbf{P} \mathbf{u}, \tag{17.52}$$

where T refers to an algebraic transpose, and the sum extends over the time considered for the assimilation. Here $\mathbf{x}^{\sim}(t)$ is the estimated ocean state at time t; $\mathbf{x}_0$ is an adjustment of the initial ocean state; and $\mathbf{u}$ denotes here and in the following assimilation section an adjustment to the model parameters. During the optimization process, optimal values for both the initial state and model parameters are sought, so long as they remain within prior error bounds. The matrices $\mathbf{R(t)}$, $\mathbf{Q}$, and $\mathbf{P}$ are inverses of error covariance matrices for the observations, model, and model parameters including initial conditions, respectively. Covariances play a central role in determining the optimal solution since they select individual solutions from the manifold of possible solutions.

Data assimilation methods can be separated into filtering methods and smoothing methods. Within these two categories, some methods are simplified—computationally efficient but also not optimal any more (e.g., optimal interpolation)—whereas others are more rigorous yet computationally intensive (e.g., Kalman filters, adjoint models). Ocean data assimilation schemes vary in the way the individual data assimilation components are defined and the extent to which optimal values

* We drop time dependence from each of the vectors in equation (51) for notational brevity.

are subject to constraints (e.g., Wunsch 1996). Some schemes are also influenced by the sophistication of the forward model. These details impact how assimilation schemes incorporate observations, whether a solution is sought for a constrained or unconstrained optimization problem, and in the level of accuracy used to describe *a priori* error estimates of observations and the model dynamics. Furthermore, different assimilation methods make different compromises between the fidelity and range of temporal and spatial scales to be represented and the degree of dynamical consistency sought in the solution. Given the substantial differences in methods, there are many differences in the resulting solutions. The chosen method is generally dependent on the scientific or operational application.

Simplified filtering methods are relatively straightforward to set up and are computationally inexpensive. In Figure 17.1, we illustrate how filtering methods sequentially estimate the ocean state at discrete points in time. They work by statistically merging observations available at the analysis steps with the forecast model state (also called the background state). As a result of previous assimilation cycles, the model state contains information from past observations. As part of the filtering method, the analysis increment aims to correct the model state. Notably, this step violates conservation principles (e.g., conservation of mass, heat and salt) because it introduces sources/sinks to the ocean dynamical equations. Analysis increments may also introduce temporal discontinuities in the ocean state evolution. These discontinuities can be ameliorated via "incremental analysis updating" (IAU; Bloom et al. 1996). Nonetheless, IAU corrections remain dynamically unbalanced. Consequently, temporal evolution of the ocean state realized through filtering methods is not generally consistent dynamically. Smoothing methods (e.g., adjoint models, see as follows) estimate values of uncertain model parameters while ingesting observations. They do so by making use of both past and future observations (see Figure 17.1), which then leads to an optimal assimilation of the observations and improvements to the model. That is, smoother methods use observations from the future and the past to constrain the ocean circulation in a retrospective analysis. In contrast to filtering methods, smoothers do not change the prognostic model state at analysis times. Rather, smoothers estimate an ocean state by changing model-independent parameters (as opposed to elements of the prognostic state) such that the simulated state optimally matches the observed ocean state over an extended time period (years to several decades). The resulting ocean state evolution is dynamically self-consistent because it evolves according to the conservation of mass, heat, salt, and momentum over the estimation period (i.e., there are no added source/sink terms in the prognostic equations). More precisely, the optimal ocean state evolution is typically found by running the forward model using optimally estimated model parameters, such as initial conditions, surface forcing, and subgrid scale mixing parameters. Smoothers, such as four-dimensional variational/adjoints, are generally quite demanding computationally and algorithmically. The payoff is that they provide a

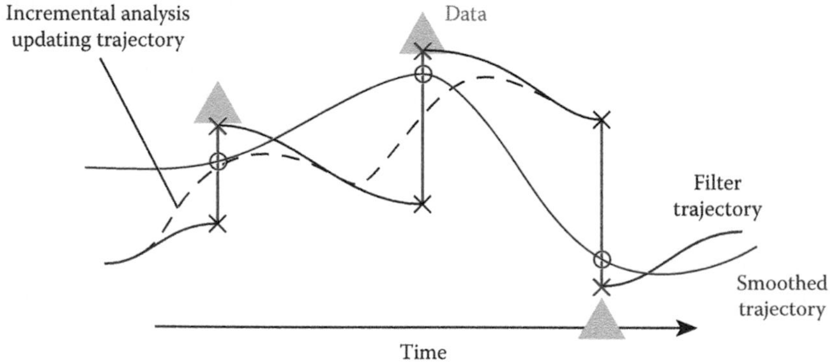

FIGURE 17.1 Schematic of the difference between the two general methods used for assimilation: filters and smoothers, as they produce an estimated state. (From Stammer, D., et al., *Ann. Rev. Mar. Sci.*, 8, 491–518, 2016. With permission.)

dynamically self-consistent ocean state while exploiting information contained in the data. Hence, by remaining dynamically self-consistent, the optimal ocean state can be used to study physical aspects of a data constrained evolving ocean.

17.4.2 Some Details for Filtering Methods

Filtering for ODA includes three major approaches: optimal interpolation (OI), three-dimensional variational assimilation (3DVAR), and Kalman filtering (KF; Kalman 1960) in various forms. The OI and 3D-VAR approaches are approximations of the Kalman filter. OI is the simplest form of an optimal least-squares minimum variance estimator (e.g., Gandin 1963). For each observation, a correction of the model is defined based on the difference between observation and model (called the "innovation"). Interpolated values are calculated from a linear combination of the innovations weighted by the inverse of the sum of the estimated observation error variance and the background error variance at observation points. OI provides an optimal instantaneous estimate for a particular set of temporally constant weights. However, the OI solution is suboptimal over the entire measurement period because there is no optimization over time (e.g., Fukumori 2002), and the covariance matrices are correspondingly static.

The KF is also a minimum variance estimator that was developed to improve predictions, and it provides useful estimates of uncertainties. It evolves the model state error covariance matrix according to the underlying dynamics and the assumed error covariance matrix of the numerical model. The mathematical formulation for obtaining an improved state and error estimate at synoptic times is then given by

$$\tilde{\mathbf{x}}(\mathbf{t},-)+\mathbf{K}(\mathbf{t})[\mathbf{y}(\mathbf{t})-\mathbf{E}(\mathbf{t})\tilde{\mathbf{x}}(\mathbf{t},-)] \quad (17.53)$$

$$\mathbf{K}(\mathbf{t})=\mathbf{P}(\mathbf{t},-)\mathbf{E}(\mathbf{t})^{T}[E(\mathbf{t})\mathbf{P}(\mathbf{t},-)\mathbf{E}(\mathbf{t})^{T}+\mathbf{R}(\mathbf{t})]^{-1} \quad (17.54)$$

with updated uncertainty

$$\mathbf{P}(\mathbf{t})=\mathbf{P}(\mathbf{t},-)-\mathbf{K}(\mathbf{t})\mathbf{E}(\mathbf{t})\mathbf{P}(\mathbf{t},-). \quad (17.55)$$

$\tilde{\mathbf{x}}$ **(t, –)** is the prediction at the same time step as simulated by the model forward run before making use of information available in the new observations; **K(t)** is the Kalman gain matrix, determining the increment by which the simulation needs to be corrected through the new observations. Finally, **P(t)** is the error covariance of the estimated solution, which declines through time while stepping the model forward.

In practice, evolving the model state error covariance matrix involves a large computational burden. Hence, the complete KF is not feasible for assimilating observations into global ocean models. Consequently, approximations to the KF have been devised, such as the partitioned KF, which solves the larger estimation problem by partitioning it into a series of smaller calculations (Fukumori 2002). An extended KF (EKF; Gelb 1974) was developed for weakly nonlinear problems under the tangent-linear approximation. For stronger nonlinear problems, Evensen (1994) proposed a different extension of the KF, called the ensemble KF (EnKF), to estimate the model forecast error covariance matrix by means of a limited number of Monte Carlo simulations from a set of parallel analyses. In contrast to other realizations of the linear KF, the EnKF is suitable for high-resolution global eddy-permitting assimilation. Another variant is the Singular Evolutive Extended Kalman (SEEK) filter and its interpolated variant the Singular Evolutive Interpolated Kalman (SEIK) filter developed by Pham et al. (1998).

In contrast to the KF, the 3D-VAR formalism leads to a maximum likelihood estimator, which treats elements in the cost function, *J*, independently in time and seeks an approximate solution through iterative minimization (e.g., Derber and Rosati 1989; Courtier et al. 1998). Its implementation requires the existence of the adjoint of the observation operators but not of the full GCM; 3D-VAR eliminates the need to split the analysis domain into subsections and provides a more general framework for including complex (including nonlinear) constraints in the cost function such as nonlinear observation operators, dynamical balance constraints, and physically motivated conservation relationships (Ricci et al. 2005; Weaver et al. 2005).

17.4.3 Smoother Methods

The development of two major smoother approaches was essential for the application of smoothing methods to ocean data assimilation: the optimal Rauch-Tung-Striebel (RTS, Rauch et al. 1965) smoother and the adjoint method. These methods have different algorithmic properties but are equivalent as long as a problem is truly linear and assumptions about data and model dynamic constraint errors are the same (e.g., Bennett 2002; Lee et al. 2009). Comprehensive mathematical expositions of original smoother formulations are provided by Bouttier and Courtier (1999) and Wunsch (1996).

The optimal RTS smoother is a minimum variance estimator and thus a recursive algorithm that seeks estimates of the state vector and associated uncertainty at each point in time based on all observations both before and after (e.g., Cohn and Dinovitzer, 1994).

$$\tilde{\mathbf{x}}(t,+)=\tilde{\mathbf{x}}(t)+\mathbf{L}(t+1)\left[\tilde{\mathbf{x}}(t+1,+)-\tilde{\mathbf{x}}(t+1,-1)\right] \tag{17.56a}$$

$$\mathbf{L}(t+1)=\mathbf{P}(t)\,\mathbf{A}(t)^{\mathrm{T}}\,\mathbf{P}(t+1,-)^{-1} \tag{17.56b}$$

$$\tilde{\mathbf{u}}(t,+)=\tilde{\mathbf{u}}(t)+\mathbf{M}(t+1)\left[\tilde{\mathbf{x}}(t+1,+)-\tilde{\mathbf{x}}(t+1,-)\right] \tag{17.56c}$$

$$\mathbf{M}(t+1)=\mathbf{Q}(t)\,\Gamma(t)^{\mathrm{T}}\,\mathbf{P}(t+1,-)^{-1} \tag{17.56d}$$

$$\mathbf{P}(t,+)=\mathbf{P}(t)+\mathbf{L}(t+1)\,[\mathbf{P}(t+1,+)-\mathbf{P}(t+1,-)\,]\,\mathbf{L}(t+1)^{\mathrm{T}} \tag{17.56e}$$

$$\mathbf{Q}(t,+)=\mathbf{Q}(t)+\mathbf{M}(t+1)\,[\mathbf{P}(t+1,+)\mathbf{P}(t+1,-)]\,\mathbf{M}(t+1)^{\mathrm{T}}. \tag{17.56f}$$

The operators **L** and **M** are defined in the aforementioned equations and $\tilde{\mathbf{x}}(t,1)$ is the estimated state that uses information from the future to correct the model state at time step t. The RTS smoother uses the Kalman filter in a forward sweep and then successively improves the prior estimates driven by the model increments $\left(\tilde{\mathbf{x}}(t+1,+)-\tilde{\mathbf{x}}(t+1,-)\right)$. We note that no data are involved in the RTS smoother because all information has been used in the filter. However, we need to store the matrix $\mathbf{P}(t)$ at each time step. The use of information from the future thereby leads to uncertainties that are smaller than those associated with filtered results (Fukumori 2002). It is complementary to the KF in that it acts to "smooth" the filtered results by estimating model parameters required to reduce the temporal discontinuities that result from the sequential input of data.

In contrast, the "whole domain" adjoint or Lagrange multiplier approach estimates the ocean state in an iterative way by changing model parameters (i.e., initial conditions, boundary forcing, mixing coefficients, etc.) and using observations that are distributed in time over the entire model run (e.g., Sasaki 1970; Talagrand and Courtier 1987; Thacker and Long 1988).

$$\frac{1}{2}\frac{\partial J}{\partial\tilde{\mathbf{u}}(t,+)}=\mathbf{Q}(t)^{-1}\tilde{\mathbf{u}}(t,+)+\mathbf{\Gamma}^{T}\boldsymbol{\mu}(t+1)=0 \tag{17.57a}$$

$$\frac{1}{2}\frac{\partial J}{\partial\boldsymbol{\mu}(t)}=\tilde{\mathbf{x}}(t,+)-\mathbf{A}\tilde{\mathbf{x}}(t-1,+)-\mathbf{B}\mathbf{q}(t-1)-\mathbf{\Gamma}\tilde{\mathbf{u}}(t-1,+)=0 \tag{17.57b}$$

$$\frac{1}{2}\frac{\partial J}{\partial \tilde{\mathbf{x}}(0,+)} = \mathbf{P}(0)^{-1}\left(\mathbf{x}(0,+) - \tilde{\mathbf{x}}(0)\right) + \mathbf{A}^T\mu(1) = 0 \quad (17.57c)$$

$$\frac{1}{2}\frac{\partial J}{\partial \tilde{\mathbf{x}}(t,+)} = -\mathbf{E}(t)^T\mathbf{R}(t)^{-1}\left[\mathbf{y}(t) - \mathbf{E}(t)\tilde{\mathbf{x}}(t,+)\right] - \mu(t) + \mathbf{A}^T\mu(t+1) = 0 \quad (17.57d)$$

$$\frac{1}{2}\frac{\partial J}{\partial \tilde{\mathbf{x}}(t_f)} = -\mathbf{E}(t_f)^T R\left(t_f\right)^{-1}\left[\mathbf{y}(t_f) - \mathbf{E}(t_f)\tilde{\mathbf{x}}(t_f)\right] - \mu(t_f) = 0. \quad (17.57e)$$

Where $\mathbf{Bq}(t)$ is the external forcing of the forward model and $\mathbf{\Gamma u}^{\sim}(\mathbf{t})$ represents the unknowns. Equation 17.57d is the adjoint model $\mu(t) = \mathbf{AT}\ \mu(t+1) - \mathbf{E}(t)^{\mathrm{T}}\ \mathbf{R}(t)^{-1} \cdot \mathbf{E}(t)\ \mathbf{x}^{\sim}(t,+) - \mathbf{y}(t)$, in which the model-data differences appear as a source term. The method is based on the assumption that model equations are correct ("strong constraint" formalism). The approach can deal with weakly nonlinear problems. However, it might fail for turbulent flow (i.e., highly nonlinear systems [Tanguay et al. 1995]).

Because the number of equations is equal to the number of unknowns, with a large computer we could solve this system in one step. However, in practice, the solution has to be obtained iteratively by running the model forward in time over the entire modeling period, which can be as long as 50 years or longer. The model-data differences are stored during this run and serve as input for the adjoint model, which is run backward in time. The adjoint provides a sensitivity measure of the cost function J with respect to model parameters. These sensitivities are then used to correct model parameters in an iterative loop until a minimum of the cost function is obtained.

17.5 APPLICATIONS WITH RESPECT TO ALTIMETRY

There are numerous literature examples of ocean modeling and assimilation approaches to oceanographic problems. Here we provide representative examples that focus on altimetry related applications. Among these studies, ocean models are compared with altimeter measures to evaluate model quality (e.g., Stammer et al. 1996; Griffies and Treguier 2013). Additionally, comparisons help to interpret altimeter measures dynamically or to help identify measurement errors. In general, data assimilation studies help improve models by constraining with *in situ* and satellite observations, and using estimates of the changing circulation to study quantities such as ocean currents, transports, and sea level. Some studies are concerned with improving model parameters (e.g., Ferreira et al. 2005; Menemenlis et al. 2005; Stammer 2005; Liu et al. 2012), others with providing initial conditions for climate prediction systems (Pohlmann et al. 2009; Polkova et al. 2014; Marotzke et al. 2016).

17.5.1 Process Modeling

An example state of the art simulation with a rich mesoscale eddy field is shown in Figure 17.2, which includes various near-surface flow parameters simulated by the Coral Triangle Regional Ocean Modeling System (CT-ROMS) model. CT-ROMS was designed to investigate the patterns of vulnerability of marine ecosystems in the Coral Triangle and how these patterns will change in response to climate forcing through the end of the twenty-first century (see http://www.ctroms.ucar.edu for details). Besides sea surface height, the system provides simulations of many parameters that can be observed by satellites, such as temperature, salinity, chlorophyll, and surface currents. All those fields can be compared against observations to further improve the model simulations.

Figure 17.3 illustrates how numerical simulations can help bring altimeter information into a dynamical framework. Here, we see the mean vertical velocity (w) at 25 m depth as well as its standard deviation (taken from Figure 10 of Koch et al. 2010). The spatial pattern of the standard

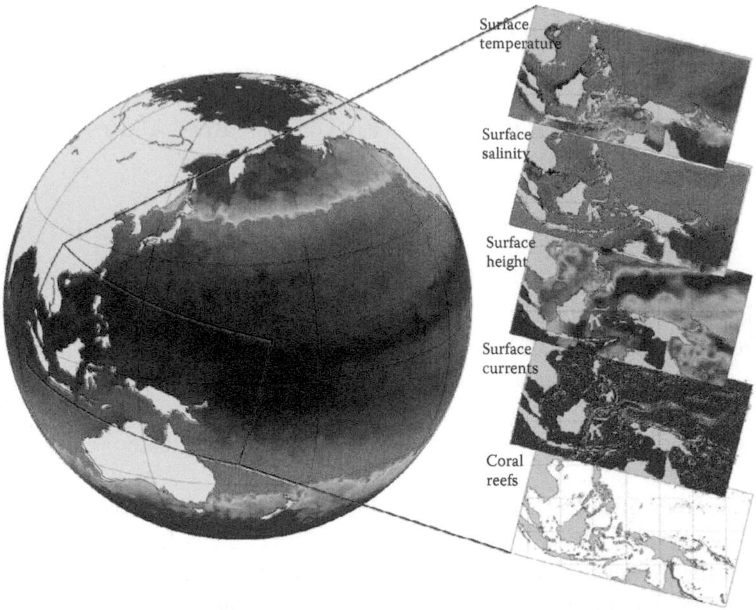

FIGURE 17.2 The Regional Ocean Modeling System (ROMS) is widely used for applications from the basin to coastal and estuarine scales http://www.myroms.org. CT-ROMS was designed to investigate the patterns of vulnerability of marine ecosystems in the Coral Triangle and how these patterns will change in response to climate forcing through the end of the 21st Century. (Taken from http://www.ctroms.ucar.edu/ct-roms.html.) Shown is a snap shot of the surface temperature field. In the inset simultaneous fields of sea surface temperature, sea surface salinity, sea surface height, surface currents and the location of coral reefs are displaced for the box marked in black.

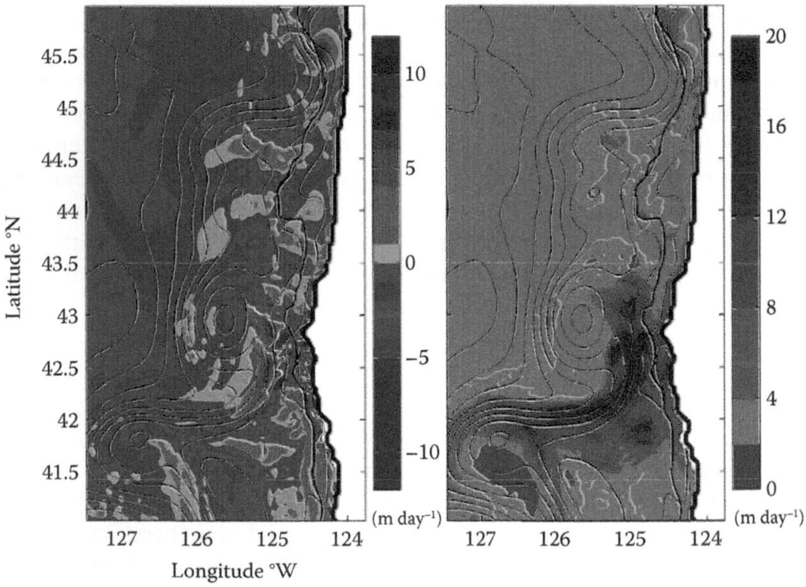

FIGURE 17.3 (a) Mean vertical velocity at 25 m depth. (b) Standard deviation of vertical velocity. In both panels, the contour shows the sea surface height (SSH). Data have been averaged over 26 July to 21 August. Contour intervals for SSH are 0.03 m. The solid black line shows the 200 m isobath. (From Koch, A. O., et al., *J. Geophys. Res.*, 115, 2010. With permission.)

deviations shows relatively large values strongly concentrated in the region of the separated jet marked by strong gradients in sea level. That region extends along the jet from a location within about 60 km of the coast near Cape Blanco (43N). The behavior in that location is characterized by concentrated mean upwelling velocities offshore in the jet, with downwelling velocities immediately inshore.

Model-altimetry comparisons were performed in terms of statistical quantities such as eddy variability of sea surface height or eddy kinetic energy defined as $EKE = 1/2(u'^2 + v'^2)$, where (u', v') represent the anomalies of the zonal and meridional flow component from a time average. Satellite altimetry is essential for providing the observational information required to further improve models in their representation of the eddy component. An example of such a comparison (Figure 17.4) shows EKE computed from altimeter data and similar fields provided by a hierarchy of model runs as described by Biri et al. (2016). In this case, altimetry is used as the truth to identify the fidelity of model simulations as a function of spatial resolution using nominal grid sizes of 16, 8, and 4 km. The figure reveals that a horizontal model grid spacing close to 4 km is required to approach the observed spatial EKE pattern and amplitudes. The improvement for grids finer than 8 km is especially obvious in the so-called Northwest Corner of the North Atlantic and along the Azores front.

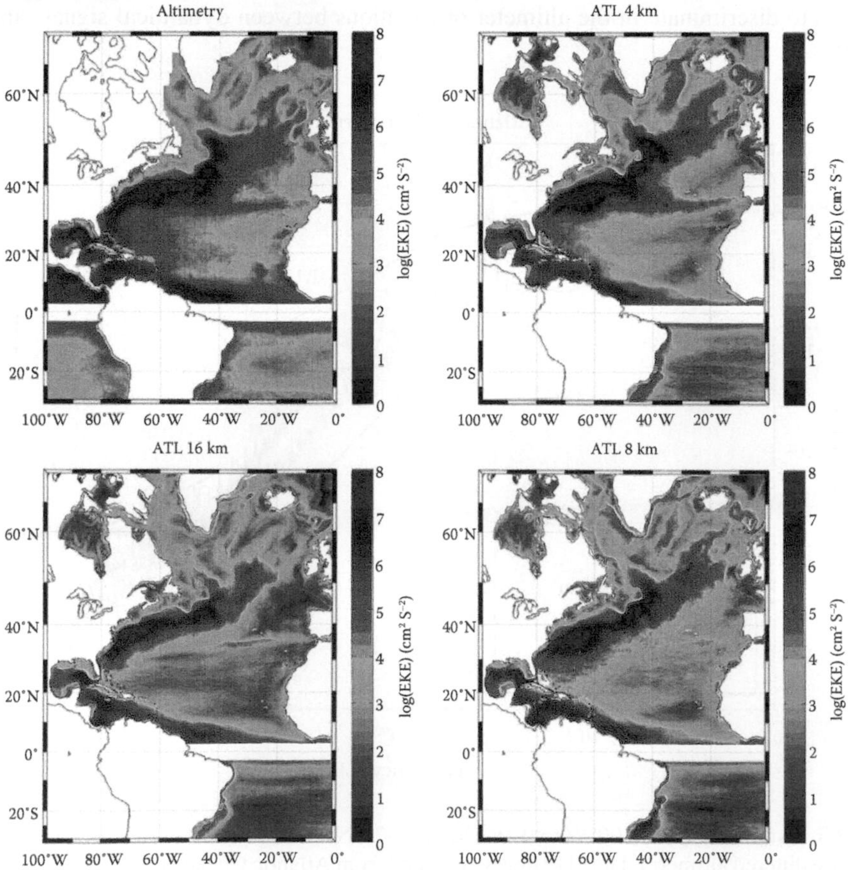

FIGURE 17.4 Eddy Kinetic Energy per mass ($cm^2\ s^{-2}$ in logarithmic scale) inferred from satellite altimetry and from a numerical simulation of the Atlantic Ocean (ATL) using horizontal grid spacings indicated on top of each panel. (From Biri, S., et al., *J. Geophys. Res.*, 121, 4157–4177, 2016. With permission.)

Altimeter data are not error-free, and a comparison needs to take their uncertainties into account, as suggested by Figure 17.5. The figure shows basin-averaged SSH frequency spectra from altimetry and from the same hierarchy of model simulations used in the previous figure (see Biri et al. [2016] for details). This study aims to separate, through a model-data intercomparison, dynamical signals from altimeter noise. The figure reveals that on periods longer than 100 days the shape of the spectra simulated by the model hierarchy is similar to the altimetric result, albeit resulting in slightly lower energy levels. However, on periods shorter than about 180 days, all model spectra follow a substantially steeper spectral slope close to f^{-2}.

Since the model results are available as daily averages, they resolve frequencies down to a Nyquist cut-off of 1–2 days and therefore can be used to reveal problems in the altimeter data arising from aliasing high-frequency motions. The respective frequency spectrum derived from the 10-day re-sampled 4 km model data indeed more closely follows the altimeter spectrum (gray curve), indicating that some fraction of the altimetric spectra likely displays too high an energy level at the high-frequency end simply due to aliasing of unresolved energy, with a severity that depends on the geographic position. Respective aliasing effects can be expected to impact the altimetric estimate especially in high latitudes, and along all frontal axes, where fast barotropic motions are pronounced. Enhanced aliasing might also appear close to the equator where, again, enhanced fast motions exist. The aliasing of tides and internal waves in altimetric frequency spectra is a general and unresolved problem in its own. Other data errors are likely to contribute on high-frequencies as well (Xu and Fu 2012) and model-data comparisons will play an increasingly important role in the future to discriminate in the altimeter observations between dynamical signals and residual uncertainties.

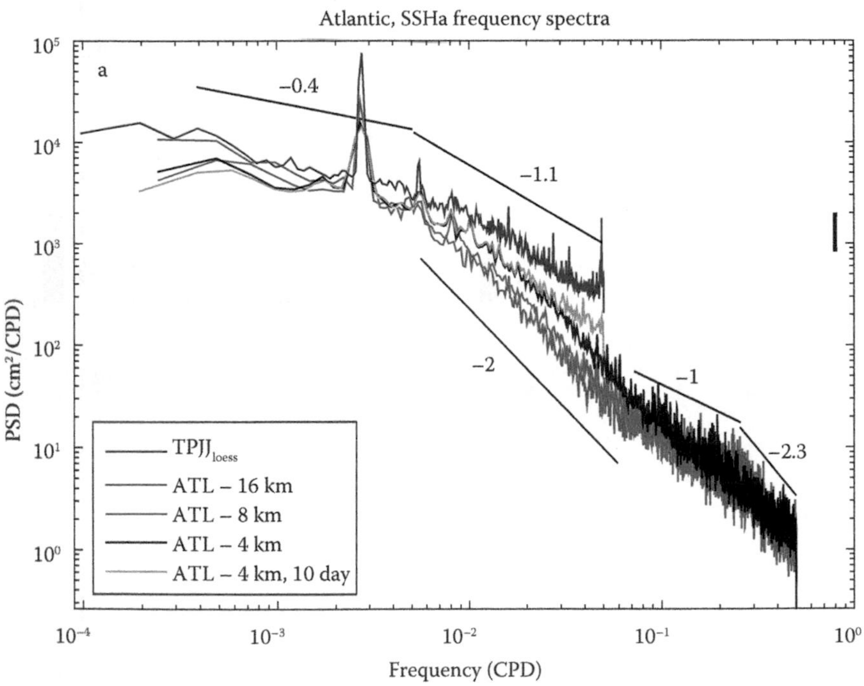

FIGURE 17.5 Atlantic Ocean (100°W to 0°W and 30°S–70°N) averaged sea surface height frequency spectra from Loess-filtered altimetry data (TPJJloess - red) and from Atlantic Ocean numerical simulations (called ATL) using spatial resolutions as indicated in the inset (blue, magenta, and black lines), including 10 day re-sampled data from the 4 km model (grey line). The vertical line on the right side of the spectra corresponds to the 95% confidence interval. No dealiasing correction was applied to the altimeter data. (From Biri, S., et al., *J. Geophys. Res.*, 121, 4157–4177, 2016. With permission.)

17.5.2 Assimilation of Sea Level into Models

Historically, ocean observations are very sparse, making it difficult to extract ocean climate signals from the limited observations extending more than a few years into the past. This problem is further exacerbated in studies extending back in time for a few decades, preceding the altimeter

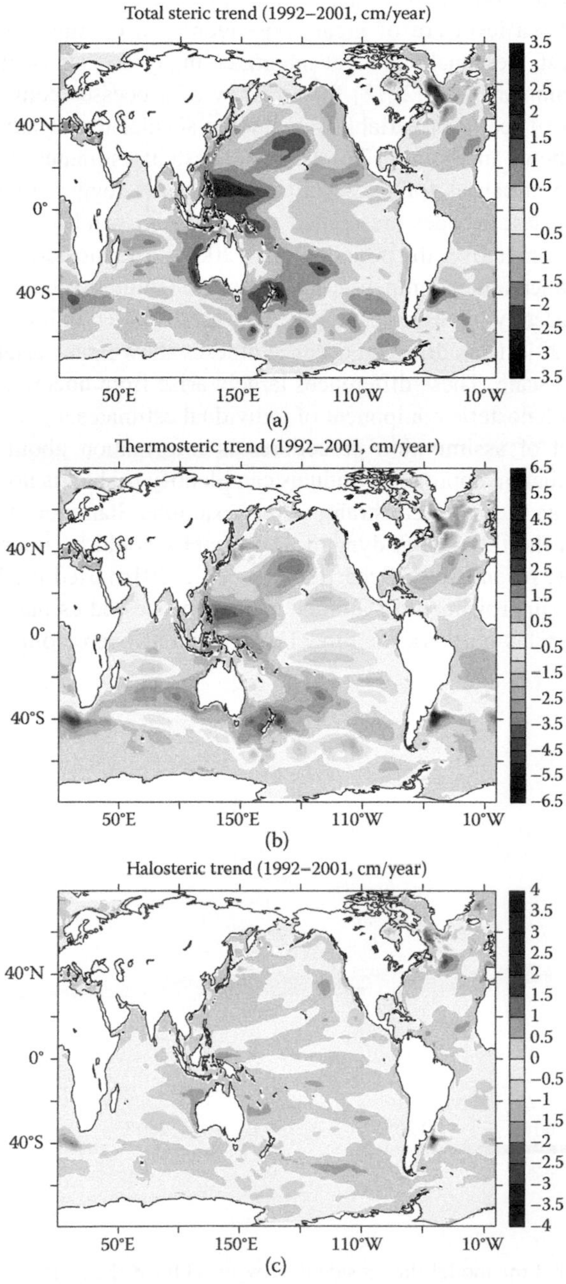

FIGURE 17.6 (a) Model steric SSH trend estimated over 1992–2001 of the 50-year run from the model potential density field. The global mean trend of 1.5 mm/year was removed from the steric trend. The corresponding thermosteric (b) and halosteric (c) sea level change include the globally averaged trends of 1.8 mm/year and –0.3 mm/year, respectively. All fields are in cm/year. (From Köhl, A., and Stammer, D., *J. Phys. Oceanogr.*, 38, 1913–1930, 2008. With permission.)

and Argo eras. Therefore, many ocean syntheses aim to assimilate all available ocean data sets to develop a quantitative understanding of regional sea level variability and trends (e.g., Wunsch et al. 2007; Carton and Giese 2008; Köhl and Stammer 2008; Balmaseda et al. 2013; Piecuch and Ponte 2014). Besides altimetry, data sets assimilated typically include Argo, XBT, hydrography, SST and salinity, gravity as well as surface flux fields. For details regarding all applied data constraints see, for example, Wunsch et al. (2007) and Köhl and Stammer (2008).

Changes in regional sea level are of great consequence to coastal regions and therefore are of major societal interest. Regional sea level integrates many aspects of the ocean state and the climate system (see Stammer et al. [2013] for a review of processes contributing to regional sea level change). Analyses of sea level variability including secular trends are provided by Köhl and Stammer (2008) and shown in Figure 17.6. These consistent estimates of the ocean circulation enable the identification of causes for sea level trends. Results suggest that much of the regional sea level trends over the last decades are related to changes in wind forcing. The figure suggests that most of the sea level trend over the period 1992 to 2001 is thermosteric (temperature related) in nature, with a much smaller, albeit compensating, halosteric (salinity-related) effect. Similar studies were performed more recently by Storto et al. (2015), who showed that linear trends in steric height over the period 1993 to 2010 from different ocean syntheses show large variations among individual products on the regional scale. These differences largely arise from uncertainties in the deep ocean and discrepancies in the halosteric component of individual estimates.

An important output of assimilation procedures is information about model-data residuals. Depending on the assimilation approach, residuals can identify problems not only in the underlying model but also in the data used as a constraint. As an example, Stammer et al. (2007) showed that model-data residuals of the time-mean dynamic topography could be interpreted as errors in the EGM96 geoid. In a more recent study, Scharffenberg et al. (2016) used the ECCO state estimation to document the quality-improvement of altimeter data reprocessed as part of the ESA CCI effort (Ablain et al. 2015). For this purpose, the model data residuals were computed relative to various

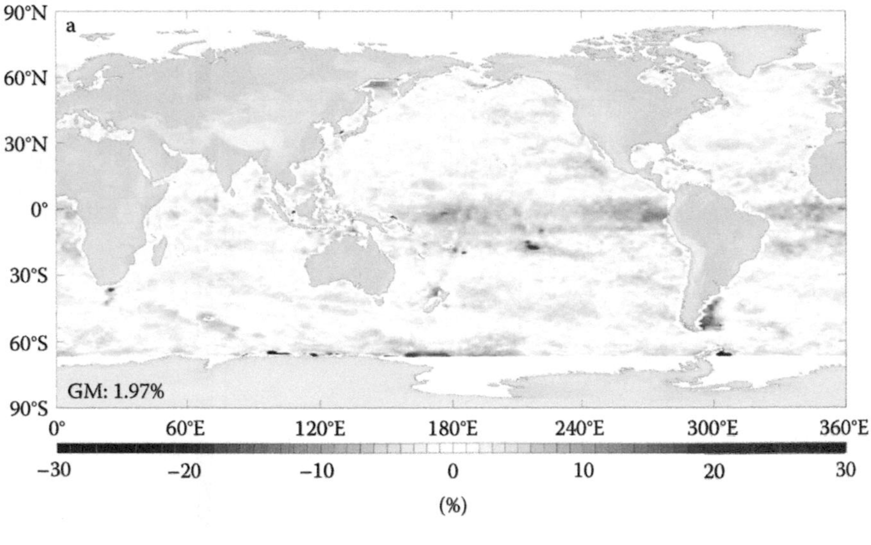

FIGURE 17.7 Reduction of the model-data residual between TOPEX data and the GECCO ocean synthesis obtained through the reprocessing of the TOPEX data through the ESA CCI project. Shown is the root-mean-square (RMS) model-data residual reduction as positive percentages, archived after reprocessing the altimeter data. The global mean (GM) percentage improvement is given for latitudes between 66°S and 66°N. The figure illustrates how data assimilation efforts can be used to test the consistency of observations with the information available from other observations and the dynamics embedded in a circulation model. (From Scharffenberg, M., et al., *Surveys Geophys.*, 2016. With permission.)

altimeter data sets using assimilation results from various altimeter data products as constraints. As illustrated in Figure 17.7, the system was able to identify regions where the data quality of the new altimeter product computed in the framework of the ESA SL CCI (Ablain et al. 2015) improved upon the reprocessing. The figures display percentage changes in the RMS model-data residuals using the old and the new altimeter data products. Positive RMS ratios in the figure show areas of data improvements, whereas negative RMS ratios show areas of data degradation through the reprocessing relative to the model dynamics and the information embedded in the other data set used as constraints during the assimilation process. Improvements of up to 30% can be seen in all equatorial regions, on the Argentine shelf, and in parts of the Antarctic Circumpolar Current of the Southern Ocean. In most other parts of the ocean, improvements of up to 10% are evident. Similar analyses have been used by numerical weather prediction centers to monitor the quality of the observing system. Ocean assimilation efforts are also used to monitor and improve the ocean observing system, including remote sensing products.

Figure 17.8 shows the SST field as simulated by the coupled GFDL-CM2.6 climate model (Delworth et al., 2012; Griffies et al., 2015). As coupled climate models refine their ocean and atmospheric resolution, we expect to improve simulations of ocean and atmosphere flows through coupled data assimilation. This figure illustrates what happens after one passes through an important oceanic threshold that allows for ocean mesoscale eddies to form (see also Chapter 11 of this book). These eddies fill the ocean interior with a sea of geostrophic turbulence. The production of these eddies occurs through baroclinic instability, much as occurs in the atmosphere as seen through mid-latitude cyclones and anticyclones (e.g., Vallis, 2006). It is a challenge to confront these simulations with observations. The estimate of kinetic energy in the surface flow provided by satellite altimetry is especially useful to evaluate eddying models (Delworth et al., 2012; Griffies et al., 2015), and will remain so into the future.

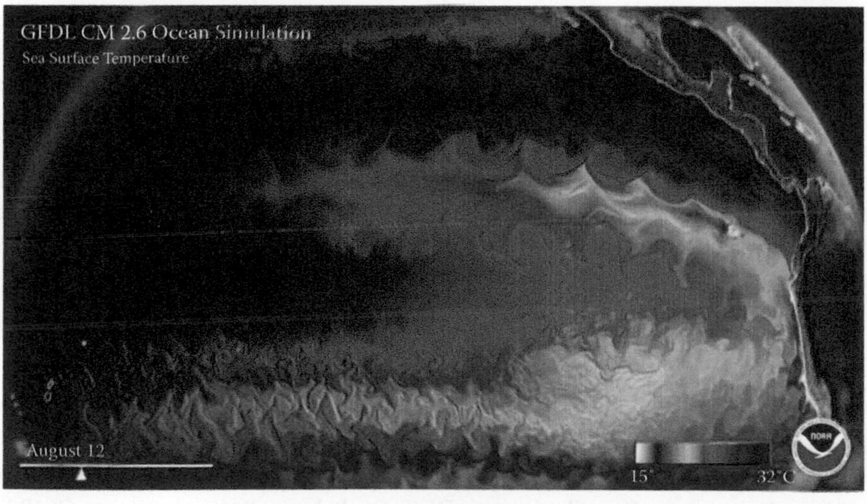

FIGURE 17.8 Daily mean sea surface temperature (SST) in the tropical Pacific from the coupled climate model GFDL-CM2.6, shown here as an example of a future aim for ocean data assimilation. The ocean component in this model has a nominal resolution of roughly 10 km in low latitudes and 5 km in high latitudes. We see the large meanders associated with tropical instability waves, as well as the wake downstream from the Galapagos Islands in the eastern equatorial Pacific. Also note the poleward decreasing length scale of mesoscale eddy fluctuations, reflecting the dependence on the Coriolis parameter of the Rossby radius of deformation so that the Rossby radius generally decreases poleward. Animation of the daily mean SST is available at https://vimeo.com/27076776, and details of the model configuration and simulation features are documented in Delworth et al. (2012) and Griffies et al. (2015).

17.6 SUMMARY AND CONCLUDING REMARKS

Over the past 20 years, ocean models have improved considerably and now entail an important, indispensable means for studying the ocean circulation and its role in the coupled climate system. Coincident with this improvement, we have now 25 years of high-quality altimetry data. Altimetry was the first essential observational element to test the realism of eddy variability in ocean models. As we have discussed in this chapter, this work will continue and will certainly lead to further improvements in ocean circulation models. Data assimilation can be considered one branch in this context. In addition, model-data intercomparison efforts provide valuable information about model skill.

We are now entering an era where roles are exchanged, whereby models become increasingly important to test altimeter data quality and to help correct biases in altimeter data. Additionally, we see how models can be useful to identify aliasing problems associated with high-frequency oceanic motions. The entire data noise problem is an aspect that received little attention in the past. Dedicated studies are urgently needed to place increasing focus on the dynamical content available from altimeter data records. This work will include aspects of eddy-mean flow interactions, eddy-wind interaction, and the dissipation of eddy energy.

In essence, circulation models are an integral element of satellite altimetry and of the entire climate observing system. This role for models holds especially for data assimilation. It is through data assimilation that we combine observations with ocean dynamics to produce a quantitative picture of the ocean circulation. ODA is the ultimate means for synthesizing ocean and climate observations into a dynamically consistent description of the time-varying ocean and climate state. We propose that more emphasis in ODA exercises needs to be placed on understanding the time mean ocean state, thereby increasing use of precise geoid information and thus offering feedback to the geodesy community.

We anticipate enhancements of model improvements (ocean and especially coupled climate models) through parameter estimations. Altimetry will continue to play an essential role in this context. At the same time, increasing emphasis will be placed on interdisciplinary approaches (either the ocean alone or in a fully coupled climate system). Interdisciplinary approaches highlight the role of physical-biological or physical-chemical interactions. Altimetry will then be seen as one element of the full ocean climate state as indicated in Figure 17.2.

To some extent, this central role for altimetry has already materialized in climate prediction efforts, in which coupled climate models are being initialized by the actual ocean state. This approach is used in seasonal forecasts (Kirtman et al. 2013), and it is now pursued for decadal timescales and semi-operational systems for decadal climate predictions (Marotzke et al. 2016). Information in the ocean state constitutes the essential memory determining forecast skill. Altimetry, combined with *in situ* hydrography from Argo, is part of the foundation for climate model initialization.

ACKNOWLEDGMENTS

This work is supported in part through the European NaClim project as well as the German national RACE and MiKLIP projects. S.M.G. acknowledges ongoing support from the NOAA Geophysical Fluid Dynamics Laboratory. Contributions to DFG funded excellence initiative CLISAP of the Universita¨ t Hamburg. We thank Rui Ponte, Tony Rosati, Carl Wunsch, and Jianjun Yin for useful comments on this chapter.

REFERENCES

Ablain, M., Cazenave, A., Larnicol, G., Balmaseda, M., Cipollini, P., and Faugére, Y. Improved sea level record over the satellite altimetry era (19932010) from the Climate Change Initiative project, *Ocean Science*, 11, 67–82, 2015. http://doi.org/10.5194/os-11-67-2015.

Arakawa, A., and Lamb, V. R. The UCLA general circulation model, in *Methods in Computational Physics: General Circulation Models of the Atmosphere*, edited by J. Chang, vol. 17, pp. 174–265, Academic Press, Amsterdam. 1977.

Arbic, B. Atmospheric forcing of the oceanic semidiurnal tide, *Geophysical Research Letters*, 32, L02610, 2005.

Arbic, B., Wallcraft, A., and Metzger, E. Concurrent simulation of the eddying general circulation and tides in a global ocean model, *Ocean Modelling*, 32, 175–187, 2010.

Balmaseda, M., Mogensen, K., and Weaver, A. Evaluation of the ECMWF ocean reanalysis system ORAS4, *Quarterly Journal of the Royal Meteorological Society*, 139, 1132–1161, 2013.

Bennett, A. *Inverse Modeling of the Ocean and Atmosphere*, Cambridge University Press, Cambridge, UK, 2002.

Biri, S., Serra, N., Scharffenberg, M., and Stammer, D. Atlantic sea surface height and velocity spectra inferred from altimetry and a hierarchy of models, *Journal of Geophysical Research*, 121, 4157–4177, 2016.

Bloom, S., Takacs, L., Da Silva, A., and Ledvina, D. Data assimilation using incremental analysis updates, *Monthly Weather Review*, 124, 1256–1271, 1996.

Bouttier, F., and Courtier, P. *Data Assimilation Concepts and Methods*, ECMWF Meteorological Training Course Lecture Series, pp. 1–59, 1999.

Bryan, F. O. The axial angular momentum balance of a global ocean general circulation model, *Dynamics of Atmospheres and Oceans*, 25, 191–216, 1997.

Bryan, K. A numerical method for the study of the circulation of the World Ocean, *Journal of Computational Physics*, 4, 347–376, 1969.

Bryan, K. The steric component of sea level rise associated with enhanced greenhouse warming: A model study, *Climate Dynamics*, 12, 545–555, 1996.

Campin, J.-M., Marshall, J., and Ferreira, D. Sea ice-ocean coupling using a rescaled vertical coordinate $z*$, *Ocean Modelling*, 24, 1–14, 2008.

Carton, J., and Giese, B. A reanalysis of ocean climate using Simple Ocean Data Assimilation (SODA), *Monthly Weather Review*, 136, 2999–3017, 2008.

Cavaleri, L., Fox-Kemper, B., and Hemer, M. Wind waves in the coupled climate system, *Bulletin of the American Meteorological Society*, 93, 1651–1661, 2012.

Chandrasekhar, S. *Hydrodynamic and Hydromagnetic Stability*, 654 pp., Dover Publications, New York, 1961.

Church, J. A., Clark, P. U., Cazenave, A., Gregory, J. M., Jevrejeva, S., Levermann, A., Merrifield, M. A., et al. *Sea level change, in Climate Change 2013: The Physical Science Basis. Contribution of Working Group I to the Fifth Assessment Report of the Intergovernmental Panel on Climate Change*, pp. 1137–1216, Cambridge University Press, Cambridge, UK, 2013.

Cohn, M. Z., and Dinovitzer, A. S. (1994). Application of structural optimization. *Journal of Structural Engineering*, 120(2), 617–650,

Courtier, P., Andersson, E., Heckley, W., Vasiljevic, D., Hamrud, M., Hollingsworth, A., Rabier, F., et al. The ECMWF implementation of three-dimensional variational assimilation (3D-Var). I: formulation, *Quarterly Journal of the Royal Meteorological Society*, 124, 1783–1807, 1998.

Danabasoglu, G., Yeager, S. G., Bailey, D., Behrens, E., Bentsen, M., Bi, D., Biastoch, A., Böning, C. W., et al. North Atlantic simulations in coordinated ocean-ice reference experiments phase II (CORE-II). Part I: Mean states, *Ocean Modelling*, 73, 76–107, 2014, http://www.sciencedirect.com/science/article/pii/S1463500313001868.

Delworth, T. L., Rosati, A., Anderson, W., Adcroft, A. J., Balaji, V., Benson, R., Dixon, K., et al. Simulated climate and climate change in the GFDL CM2.5 high-resolution coupled climate model, *Journal of Climate*, 25, 2755–2781, 2012.

Derber, J., and Rosati, A. A global oceanic data assimilation system, *Journal of Physical Oceanography*, 19, 1333–13 347, 1989.

DeSzoeke, R. A., and Samelson, R. M. The duality between the Boussinesq and Non-Boussinesq Hydrostatic Equations of Motion, *Journal of Physical Oceanography*, 32, 2194–2203, 2002.

Durran, D. R. *Numerical Methods for Wave Equations in Geophysical Fluid Dynamics*, 470 pp., Springer Verlag, Berlin, 1999.

Evensen, G. Sequential data assimilation with a nonlinear quasi-geostrophic model using Monte Carlo methods to forecast error statistics, *Journal of Geophysical Research–Oceans*, 99, 10143–10162, 1994.

Farrell, W., and Clark, J. On postglacial sea level, *Geophysical Journal of the Royal Astronomical Society*, 46, 646–667, 1976.

Ferreira, D., Marshall, J., and Heimbach, P. Estimating eddy stresses by fitting dynamics to observations using a residual-mean ocean circulation model and its adjoint, *Journal of Physical Oceanography*, 35, 1891–1910, 2005.

Flato, G., Marotzke, J., Abiodun, B., Braconnot, P., Chou, S. C., Collins, W., Cox, P., et al. *Evaluation of Climate Models, in Climate Change 2013: The Physical Science Basis. Contribution of Working Group I to the Fifth Assessment Report of the Intergovernmental Panel on Climate Change*, pp. 741–866, Cambridge University Press, Cambridge, UK, 2013.

Fox-Kemper, B., Lumpkin, R., and Bryan, F. Lateral transport in the ocean interior, in *Ocean Circulation and Climate, 2nd Edition: A 21st Century Perspective*, edited by G. Siedler, S. M. Griffies, J. Gould, and J. Church, vol. 103 of International Geophysics Series, pp. 185–209, Academic Press, Amsterdam. 2013.

Fox-Kemper, B., and Menemenlis, D. Can large eddy simulation techniques improve mesoscale rich ocean models?, in *Ocean Modeling in an Eddying Regime*, edited by M. Hecht and H. Hasumi, vol. 177 of Geophysical Monograph, pp. 319–338, American Geophysical Union, Washington, D.C. 2008.

Fukumori, I. A partitioned Kalman filter and smoother, *Monthly Weather Review*, 130, 1370–1383, 2002.

Gandin, L. Objektivnyi Analyz Meteorologicheskikh Polei, 1963, edited by A. Gelb, *Applied Optimal Estimation*, MIT Press, Cambridge, MA, 1974.

Gelb, A, 1974. *Applied Optimal Estimation*. Cambridge, MA: MIT Press.

Gent, P. R., and McWilliams, J. C. Isopycnal mixing in ocean circulation models, *Journal of Physical Oceanography*, 20, 150–155, 1990.

Gent, P. R., Willebrand, J., McDougall, T. J., and McWilliams, J. C. Parameterizing eddy-induced tracer transports in ocean circulation models, *Journal of Physical Oceanography*, 25, 463–474, 1995.

Gill, A. *Atmosphere-Ocean Dynamics*, vol. 30 of International Geophysics Series, 662 + xv pp., Academic Press, London, 1982.

Gill, A. E., and Niiler, P. The theory of the seasonal variability in the ocean, *Deep-Sea Research*, 20, 141–177, 1973.

Goddard, P., Yin, J., Griffies, S. M., and Zhang, S. An extreme event of sea level rise along the Northeast Coast of North America in 2009–2010, *Nature Communications*, 6, 6346–6355, 2015.

Goldsbrough, G. Ocean currents produced by evaporation and precipitation, *Proceedings of the Royal Society of London*, A141, 512–517, 1933.

Gomez, N., Mitrovica, J. X., Huybers, P., and Clark, P. U. Sea level as a stabilizing factor for marine-ice-sheet grounding lines, *Nature Geoscience*, 3, 850–853, 2010.

Greatbatch, R. J. A note on the representation of steric sea level in models that conserve volume rather than mass, *Journal of Geophysical Research*, 99, 12767–12771, 1994.

Griffies, S. M. The Gent-McWilliams skew-flux, *Journal of Physical Oceanography*, 28, 831–841, 1998.

Griffies, S. M. *Fundamentals of Ocean Climate Models*, 518+xxxiv pp., Princeton University Press, Princeton, NJ, 2004.

Griffies, S. M., and Adcroft, A. J. Formulating the equations for ocean models, in *Ocean Modeling in an Eddying Regime*, edited by M. Hecht and H. Hasumi, vol. 177 of Geophysical Monograph, pp. 281–317, American Geophysical Union, Washington, D.C. 2008.

Griffies, S. M., Böning, C. W., Bryan, F. O., Chassignet, E. P., Gerdes, R., Hasumi, H., Hirst, A., Treguier, A.-M., and Webb, D. Developments in ocean climate modelling, *Ocean Modelling*, 2, 123–192, 2000.

Griffies, S. M., Danabasoglu, G., Durack, P. J., Adcroft, A. J., Balaji, V., Böning, C. W., Chassignet, E. P., et al. OMIP contribution to CMIP6: Experimental and diagnostic protocol for the physical component of the Ocean Model Intercomparison Project, *Geoscientific Model Development*, 9, 3231–3296, 2016.

Griffies, S. M., and Greatbatch, R. J. Physical processes that impact the evolution of global mean sea level in ocean climate models, *Ocean Modelling*, 51, 37–72, 2012.

Griffies, S. M., and Hallberg, R. W. Biharmonic friction with a Smagorinsky viscosity for use in large-scale eddy-permitting ocean models, *Monthly Weather Review*, 128, 2935–2946, 2000.

Griffies, S. M., Pacanowski, R., Schmidt, M., and Balaji, V. Tracer conservation with an explicit free surface method for *z*-coordinate ocean models, *Monthly Weather Review*, 129, 1081–1098, 2001.

Griffies, S. M., and Treguier, A.-M. Ocean models and ocean modeling, in *Ocean Circulation and Climate, 2nd Edition: A 21st Century Perspective*, edited by G. Siedler, S. M. Griffies, J. Gould, and J. Church, vol. 103 of International Geophysics Series, pp. 521–552, Academic Press, Amsterdam. 2013.

Griffies, S. M., Winton, M., Anderson, W. G., Benson, R., Delworth, T. L., Dufour, C., Dunne, J. P., et al. Impacts on ocean heat from transient mesoscale eddies in a hierarchy of climate models, *Journal of Climate*, 28, 952–977, 2015.

Griffies, S. M., Yin, J., Durack, P. J., Goddard, P., Bates, S., Behrens, E., Bentsen, M., et al. An assessment of global and regional sea level for years 1993–2007 in a suite of interannual CORE-II simulations, *Ocean Modelling*, 78, 35–89, 2014.

Haidvogel, D. B., and Beckmann, A. *Numerical Ocean Circulation Modeling*, Imperial College Press, London, 1999.

Hallberg, R., and Adcroft, A. Reconciling estimates of the free surface height in Lagrangian vertical coordinate ocean models with mode-split time stepping, *Ocean Modelling*, 29, 15–26, 2009.

Hallberg, R., and Rhines, P. Buoyancy-driven circulation in an ocean basin with isopycnals intersecting the sloping boundary, *Journal of Physical Oceanography*, 26, 913–940, 1996.

Hsieh, W., and Bryan, K. Redistribution of sea level rise associated with enhanced green-house warming: A simple model study, *Climate Dynamics*, 12, 535–544, 1996.

Huang, R. X. Real freshwater flux as a natural boundary condition for the salinity balance and thermohaline circulation forced by evaporation and precipitation, *Journal of Physical Oceanography*, 23, 2428–2446, 1993.

Huang, R. X., Jin, X., and Zhang, X. An oceanic general circulation model in pressure coordinates, *Advances in Atmospheric Physics*, 18, 1–22, 2001.

Huang, R. X., and Schmitt, R. W. Goldsbrough-Stommel circulation of the World Oceans, *Journal of Physical Oceanography*, 23, 1277–1284, 1993.

IOC, SCOR, and IAPSO: The international thermodynamic equation of seawater-2010: *Calculation and use of thermodynamic properties, Intergovernmental Oceanographic Commission, Manuals and Guides No. 56*, 196pp, UNESCO, 2010. available from http://www.TEOS-10.org.

Jochum, M., Danabasoglu, G., Holland, M., Kwon, Y.-O., and Large, W. Ocean viscosity and climate, *Journal of Geophysical Research*, 114, C06017, 2008. doi:10.1029/2007JC004 515.

Kalman, R. A new approach to linear filtering and prediction problems, *Journal of Fluids Engineering*, 82, 35–45, 1960.

Killworth, P. D., Stainforth, D., Webb, D. J., and Paterson, S. M. The development of a free-surface Bryan-Cox-Semtner ocean model, *Journal of Physical Oceanography*, 21, 1333–1348, 1991.

Kirtman, B. P., Stockdale, T., and Burgman, R. The ocean's role in modeling and predicting seasonal-to-interannual climate variations, in *Ocean Circulation and Climate, 2nd Edition: A 21st century perspective*, edited by G. Siedler, S. M. Griffies, J. Gould, and J. Church, vol. 103 of International Geophysics Series, pp. 625–644, Academic Press, Amsterdam. 2013.

Koch, A. O., Kurapov, A. L., and Allen, J. S. Nearsurface dynamics of a separated jet in the coastal transition zone off Oregon, *Journal of Geophysical Research*, 115, C08020, doi:10.1029/2009JC005704. 2010.

Köhl, A., and Stammer, D. Variability of the meridional overturning in the North Atlantic from 50-year GECCO state estimation, *Journal of Physical Oceanography*, 38, 1913–1930, 2008.

Kopp, R. E., Mitrovica, J. X., Griffies, S. M., Yin, J., Hay, C. C., and Stouffer, R. J. The impact of Greenland melt on regional sea level: A preliminary comparison of dynamic and static equilibrium effects, *Climatic Change Letters*, 103, 619–625, 2010.

Kuhlmann, J., Dobslaw, H., and Thomas, M. Improved modeling of sea level patterns by incorporating self-attraction and loading, *Journal of Geophysical Research*, 116, C11036, doi:10.1029/2011JC007399, 2011.

Landerer, F., Jungclaus, J., and Marotzke, J. Regional dynamic and steric sea level change in response to the IPCC-A1B Scenario, *Journal of Physical Oceanography*, 37, 296–312, 2007a.

Landerer, F., Jungclaus, J., and Marotzke, J. Ocean bottom pressure changes lead to a decreasing length-of-day in a warming climate, *Geophysical Research Letters*, 34, L06307, 2007b. doi:10.1029/2006GL029 106.

Landerer, F., Wiese, D. N., Bentel, K., Boening, C., and Watkins, M. North Atlantic meridional overturning circulation variations from GRACE ocean bottom pressure anomalies, *Geophysical Research Letters*, 42, 8114–8121, 2015.

Large, W., McWilliams, J., and Doney, S. Oceanic vertical mixing: A review and a model with a nonlocal boundary layer parameterization, *Reviews of Geophysics*, 32, 363–403, 1994.

Large, W. G., Danabasoglu, G., McWilliams, J. C., Gent, P. R., and Bryan, F. O. Equatorial circulation of a global ocean climate model with anisotropic horizontal viscosity, *Journal of Physical Oceanography*, 31, 518–536, 2001.

Lee, T., Awaki, T., Balmaseda, M., Greiner, E., and Stammer, D. Ocean state estimation for climate research, *Oceanography*, 22, 160–167, 2009.

Liu, C., Köhl, A., and Stammer, D. Adjoint-based estimation of eddy-induced tracer mixing parameters in the global ocean, *Journal of Physical Oceanography*, 42, 11861206, 2012.

Liu, X., Köhl, A., and Stammer, D. On the dynamical response of the ocean to volume flux forcing, *Journal of Geophysical Research*, in press, 2017.

Lorbacher, K., Dengg, J., Böning, C., and Biastoch, A. Regional patterns of sea level change related to interannual variability and multidecadal trends in the Atlantic Meridional Overturning Circulation, *Journal of Physical Oceanography*, 23, 4243–4254, 2010.

Lorbacher, K., Marsland, S. J., Church, J. A., Griffies, S. M., and Stammer, D. Rapid barotropic sea-level rise from ice-sheet melting scenarios, *Journal of Geophysical Research*, 117, C06003, 2012.

Losch, M., Adcroft, A., and Campin, J.-M. How sensitive are coarse general circulation models to fundamental approximations in the equations of motion?, *Journal of Physical Oceanography*, 34, 306–319, 2004.
Lowe, J. A., and Gregory, J. M. Understanding projections of sea level rise in a Hadley Centre coupled climate model, *Journal of Geophysical Research: Oceans*, 111, n/a–n/a, 2006. http://dx.doi.org/10.1029/2005JC003421.
MacKinnon, J., St. Laurent, L., and Garabato, A. N. Diapycnal mixing processes in the ocean interior, in *Ocean Circulation and Climate, 2nd Edition: A 21st century perspective*, edited by G. Siedler, S. M. Griffies, J. Gould, and J. Church, vol. 103 of International Geophysics Series, pp. 159–183, Academic Press, Amsterdam. 2013.
Marotzke, J., Müller, W., Vamborg, F., Becker, P., Cubasch, U., Feldmann, H., Kaspar, F., et al. MiKlip a national research project on decadal climate prediction, *Bulletin of the American Meteorological Society*, 2016, 97, 2379–2394, 10.1175/BAMS- D-15-00184.1.
Marshall, J., Adcroft, A., Campin, J.-M., Hill, C., and White, A. Atmosphere-ocean modeling exploiting fluid isomorphisms, *Monthly Weather Review*, 132, 2882–2894, 2004.
McDougall, T. J. Potential enthalpy: A conservative oceanic variable for evaluating heat content and heat fluxes, *Journal of Physical Oceanography*, 33, 945–963, 2003.
McDougall, T. J., Groeskamp, S., and Griffies, S. M. On geometric aspects of interior ocean mixing, *Journal of Physical Oceanography*, 44, 2164–2175, 2014.
Menemenlis, D., Fukumori, I., and Lee, T. Using Green's functions to calibrate an ocean general circulation model, *Monthly Weather Review*, 133, 1224–1240, 2005.
Mitrovica, J. X., Tamisiea, M. E., Davis, J. L., and Milne, G. A. Recent mass balance of polar ice sheets inferred from patterns of global sea-level change, *Nature*, 409, 1026–1029, 2001.
Pardaens, A., Gregory, J., and Lowe, J. A model study of factors influencing projected changes in regional sea level over the twenty-first century, *Climate Dynamics*, 2011. 10.1029/2011GL047678.
Peyser, C., Yin, J., Landerer, F., and Cole, J. E. Pacific sea level rise patterns and global surface temperature variability, *Geophysical Research Letters*, 43, 2016. doi : 10.1002/2016GL069401.
Pham, D., Verron, J., and Roubaud, M. A singular evolutive extended Kalman filter for data assimilation in oceanography, *Journal of Marine Systems*, 16, 323–340, 1998.
Piecuch, C., and Ponte, R. Mechanisms of global mean steric sea level change, *Journal of Climate*, 27, 824–834, 2014.
Pohlmann, H., Jungclaus, J., Marotzke, J., Köhl, A., and Stammer, D. Initializing decadal climate predictions with the GECCO oceanic synthesis: Effects on the North Atlantic, *Journal of Climate*, 22, 3926–3938, 2009.
Polkova, I., Köhl, A., and Stammer, D. Impact of different initialization procedures on the predictive skill of an ocean-atmosphere coupled model, *Climate Dynamics*, 42, 31513169, 2014.
Ponte, R. M. Variability in a homogeneous global ocean forced by barometric pressure, *Dynamics of Atmospheres and Oceans*, 18, 209–234, 1993.
Ponte, R. M. A preliminary model study of the large-scale seasonal cycle in bottom pressure over the global ocean, *Journal of Geophysical Research*, 104, 1289–1300, 1999.
Ponte, R. M. Oceanic response to surface loading effects neglected in volume-conserving models, *Journal of Physical Oceanography*, 36, 426–434, 2006a.
Ponte, R. M. Low-frequency sea level variability and the inverted barometer effect, *Journal of Atmospheric and Oceanic Technology*, 23, 619–629, 2006b.
Rauch, H., Striebel, C., and Tung, F. Maximum likelihood estimates of linear dynamic systems, *AIAA Journal*, 3, 1445–1450, 1965.
Redi, M. H. Oceanic isopycnal mixing by coordinate rotation, *Journal of Physical Oceanography*, 12, 1154–1158, 1982.
Ricci, S., Weaver, A., Vialard, L., and Rogel, P. Incorporating temperature-salinity constraints in the background-error covariance of variational ocean data assimilation, *Monthly Weather Review*, 133, 317–338, 2005.
Riser, S., Freeland, H., Roemmich, D., Wijffels, S., Troisi, A., Belbeoch, M., Gilbert, D., et al. Fifteen years of ocean observations with the global Argo array, *Nature Climate Change*, 6, 145–153, 2016.
Sasaki, Y. Some basic formalism in numerical variational analysis, *Monthly Weather Review*, 98, 875–883, 1970.
Scharffenberg, M., Köhl, A., and Stammer, D. Testing the quality of sea level data using the GECCO adjoint assimilation approach, *Surveys in Geophysics*, 38, 349–383, doi:10.1007/s10712-016-9401-3, 2016.
Shchepetkin, A., and McWilliams, J. The regional oceanic modeling system (ROMS): A split-explicit, free-surface, topography-following-coordinate oceanic model, *Ocean Modelling*, 9, 347–404, 2005.
Slangen, A., Katsman, C., van de Wal, R., Vermeersen, L., and Riva, R. Towards regional projections of twenty-first century sea-level change based on IPCC SRES scenarios, *Climate Dynamics*, 2012. doi:10.1007/s00 382–011–1057–6.

Slangen, A., Carson, M., Katsman, C., van de Wal, R., Hoehl, A., Vermeersen, L., and Stammer, D. Projecting twenty-first century regional sea-level changes, *Climatic Change*, 124, 317–332, 2014.

Smagorinsky, J. Some historical remarks on the use of nonlinear viscosities, in *Large Eddy Simulation of Complex Engineering and Geophysical Flows*, edited by B. Galperin and S. A. Orszag, pp. 3–36, Cambridge University Press, Cambridge, UK, 1993.

Solomon, H. On the representation of isentropic mixing in ocean models, *Journal of Physical Oceanography*, 1, 233–234, 1971.

Stammer, D. On eddy characteristic, eddy transports, and mean flow properties, *Journal of Physical Oceanography*, 28, 727–739, 1998.

Stammer, D. Adjusting internal model errors through ocean state estimation, *Journal of Physical Oceanography*, 35, 1143–1153, 2005.

Stammer, D. Response of the global ocean to Greenland and Antarctic ice melting, *Journal of Geophysical Research*, 113, doi:10.1029/2006JC004079, 2008.

Stammer, D., Köhl, A., and Wunsch, C. Impact of the GRACE geoid on ocean circulation estimates, *Journal of Atmospheric and Oceanic Technology*, 24, 1464–1478, 2007.

Stammer, D., Balmaseda, M., Heimbach, P., Köhl, A., and Weaver, A. Ocean data assimilation in support of climate applications: Status and perspectives, *Annual Reviews of Marine Science*, 8, 491–518, 2016.

Stammer, D., Cazenave, A., Ponte, R. M., and Tamisiea, M. E. Causes for contemporary regional sea level changes, *Annual Reviews of Marine Science*, 5, 21–46, 2013.

Stammer, D., Tokmakian, R., Semtner, A., and Wunsch, C. How well does a 1/4o global circulation model simulate large-scale oceanic observations?, *Journal of Geophysical Research*, 101, C11, 25 779–25 811, 1996.

Stommel, H. A survey of ocean current theory, *Deep Sea Research*, 4, 149–184, 1957.

Storto, A., Masina, S., Balmaseda, M., Guinehut, S., Xue, Y., Szekely, T., Fukumori, I., et al. Steric sea level variability (1993–2010) in an ensemble of ocean reanalyses and objective analyses, *Climate Dynamics*, 2015. https://doi.org/10.1007/s00382-015-2554-9.

Talagrand, O., and Courtier, P. Variational assimilation of meteorological observations with the adjoint vorticity equation I: Theory, *Quarterly Journal of the Royal Meteorological Society*, 113, 1311–1328, 1987.

Tanguay, M., Bartello, P., and Gauthier, P. Four-dimensional data assimilation with a wide range of scales, *Tellus A*, 47, 974–997, 1995.

Thacker, W., and Long, R. Fitting dynamics to data, *Journal of Geophysical Research Oceans*, 93, 1227–1240, 1988.

Vallis, G. K. *Atmospheric and Oceanic Fluid Dynamics: Fundamentals and Large-scale Circulation*, Cambridge University Press, Cambridge, 1st edn., 745 + xxv pp., 2006.

Vinogradova, N. T., Ponte, R. M., Quinn, K. J., Tamisiea, M. E., Campin, J.-M., and Davis, J. L. Dynamic adjustment of the ocean circulation to self-attraction and loading effects, *Journal of Physical Oceanography*, 45, 678–689, 2015.

Weaver, A., Deltel, C., Machu, E., Ricci, S., and Daget, N. A multivariate balance operator for variational ocean data assimilation, *Quarterly Journal of the Royal Meteorological Society*, 131, 3605–3625, 2005.

Wunsch, C. *The Ocean Circulation Inverse Problem*, Cambridge University Press, Cambridge, 442 + xiv pp., 1996.

Wunsch, C., and Heimbach, P. Dynamically and kinematically consistent global ocean circulation and ice state estimates, in *Ocean Circulation and Climate, 2nd Edition: A 21st Century Perspective*, edited by G. Siedler, S. M. Griffies, J. Gould, and J. Church, vol. 103 of International Geophysics Series, pp. 553–579, Academic Press, Amsterdam, 2013.

Wunsch, C., Ponte, R., and Heimbach, P. Decadal trends in sea level patterns: 1992–2004, *Journal of Climate*, 20, 5889–5911, 2007.

Wunsch, C., and Stammer, D. Atmospheric loading and the oceanic "inverted barometer" effect, *Reviews of Geophysics*, 35, 79–107, 1997.

Wunsch, C., and Stammer, D. Satellite altimetry, the marine geoid, and the oceanic general circulation, *Annual Reviews of Earth Planetary Science*, 26, 219–253, 1998.

Xu, Y., and Fu: The effects of altimeter instrument noise on the estimation of the wavenumber spectrum of sea surface height, *Journal of Physical Oceanography*, 42, 2229–2233, 2012.

Yin, J., Griffies, S. M., and Stouffer, R. Spatial variability of sea-level rise in 21st century projections, *Journal of Climate*, 23, 4585–4607, 2010a.

Yin, J., Schlesinger, M., and Stouffer, R. Model projections of rapid sea-level rise on the northeast coast of the United States, *Nature Geosciences*, 2, 262–266, 2009.

Yin, J., Stouffer, R., Spelman, M. J., and Griffies, S. M. Evaluating the uncertainty induced by the virtual salt flux assumption in climate simulations and future projections, *Journal of Climate*, 23, 80–96, 2010b.

18 Use of Satellite Altimetry for Operational Oceanography

Pierre-Yves Le Traon, Gérald Dibarboure, Gregg Jacobs, Matt Martin, Elisabeth Rémy, and Andreas Schiller

18.1 INTRODUCTION

In this chapter, operational oceanography will be addressed from the ocean analysis and forecasting perspective.* Operational oceanography can be used to improve safety of life at sea, wealth creation, and security and protection of the marine environment. The Global Ocean Data Assimilation Experiment (GODAE) and its follow up the GODAE OceanView international program (Smith and Lefebvre 1997; International GODAE Steering Team 2000; Bell et al. 2009, 2015) have had a major impact on the development of operational oceanography capabilities. Modeling and data assimilation systems have been developed on global, regional, and coastal scales. They now operationally assimilate *in situ* and satellite data to provide regular and systematic reference information on the physical state, variability, and dynamics of the ocean and marine ecosystems. This capacity encompasses the description of the current situation (analysis), the prediction of the situation a few days ahead (forecast), and the provision of consistent retrospective data records in the past (reanalysis). Products serve a wide range of applications and downstream services (e.g., Bell et al. 2009).

Satellite altimetry is a critical observing system required for operational oceanography (e.g., Le Traon 2013; Le Traon et al. 2015). Global ocean forecasting is not possible without it, and no other observing system can enable global ocean forecasting in the absence of altimetry. It provides global, real time, all-weather sea level measurements with high space and time resolution. Sea level is an integral of the ocean interior properties and is a strong constraint for inferring the four-dimensional ocean circulation through data assimilation. Only altimetry can constrain, in particular, the four-dimensional mesoscale circulation in ocean models that is required for most operational oceanography applications. High resolution from multiple altimeters is required to adequately represent ocean eddies and associated currents in models.

All global operational oceanography systems rely on the near-real time availability of high quality multiple altimeter data sets. This requires an adequate altimeter constellation, near-real time high-quality multiple altimeter data processing capabilities and efficient altimeter data assimilation techniques.

This chapter reviews the development of global operational oceanography and the unique contribution of satellite altimetry to it. Section 18.2 summarizes the history of operational oceanography development and describes the operational oceanography infrastructure and its applications and users. Section 18.3 focuses on the essential role of satellite altimetry and the close relationships with the development of operational oceanography. General requirements for the satellite altimetry constellation are examined. Synergies with other satellite and *in situ* (Argo) observations are also addressed. Use and impact of satellite altimetry for operational oceanography is detailed in Section 18.4. This covers near real time data processing issues including the merging of multiple missions, use of mean dynamic topography (MDT), assimilation in ocean forecasting models, and impact of multiple altimeter data assimilation in ocean forecasting models.

* Role of satellite altimetry for wave analysis and forecasting is addressed in a specific chapter (Abdalla et al., this issue).

The last section deals with future prospects; we discuss the evolution of altimeter technology and the future of the altimeter constellation as well as the long-term perspective on the evolution of operational oceanography and its future requirements for altimeter observations. A final synthesis is given in the conclusion.

18.2 OPERATIONAL OCEANOGRAPHY

18.2.1 History of Development

Today's operational ocean forecasting systems are underpinned by three key components: the global *in situ* and remote sensing (physical and biogeochemical) ocean observing system, hydrodynamic numerical models, and data assimilation techniques.

Since the end of the 1970s with the launch of Seasat (Born et al. 1979), and thanks to major technological and scientific advances, satellite remote sensing has demonstrated its unique capability to provide near real time measurements of sea level anomalies (SLAs), sea surface temperature (SST), and ocean color. Through many years of development, these key observations have, for the first time, allowed us to observe the upper ocean at mesoscale and in near real time, which enabled eddy-permitting ocean forecasting applications (Fu and Cazenave 2001).

Because the ocean is electromagnetically opaque, remote sensing provides information only near the water surface. The increasing availability of remotely sensed data has been complemented by the realization of the network of more than 3500 Argo profiling floats that report temperature and salinity profiles to a 2000-m depth in a timely fashion. Argo floats have transformed the *in situ* ocean measurement network in the new millennia. This allows continuous monitoring of the temperature, salinity, and velocity of the upper ocean, with all data being relayed and made publicly available within hours of collection.

Over the last 20 years, the global ocean observing system (*in situ* and remote sensing) has been progressively implemented and led to a revolution in the amount of data available for climate research and forecasting applications. This progress in the global ocean observing system was accompanied by advances in supercomputing technology, allowing the development and operational implementation of eddy-permitting (order 10 km horizontal grid size and less) basin-scale ocean circulation models. The 1990s saw the emergence of the first large-scale eddy-resolving models (Semtner and Chervin 1992) and the first ocean-atmosphere coupled climate change projections (IPCC First Assessment Report 1990).

In the mid-1990s, the research community and operational agencies saw an emerging opportunity for near real time ocean forecasts similar to those produced in numerical weather prediction (NWP): combining numerical models and observations via data assimilation in order to provide ocean prediction products on various space and time scales. This development was facilitated through an international framework provided by GODAE that aimed at advancing ocean data assimilation by synthesizing satellite and *in situ* observations with state-of-the art models of global ocean circulation. In the past few years, ocean forecasting has matured to a stage where many nations have implemented global and basin-scale ocean analyses and short-term forecast systems that provide routine products to the oceanographic community serving a variety of applications in areas such as marine environmental monitoring and management, ocean climate, defense, and industry applications. The successor of GODAE, called GODAE OceanView, continues the tradition of GODAE and is a key platform for international collaboration on operational oceanography albeit with a broader scope than GODAE. Bell et al. (2009) and Schiller (2011) and references therein provide detailed accounts of the history of operational oceanography.

18.2.2 Operational Oceanography Infrastructure

The flow diagram (Figure 18.1) captures the main functional components and sources of inputs initially developed during GODAE and required by any state-of-the-art ocean forecasting system. These are: the data and product servers, the assimilation centers, and the product servers. These capture many of the interactions required to ensure or enhance the quality of the systems and their outputs. The measurement network and data assembly and processing centers provide the main inputs to the

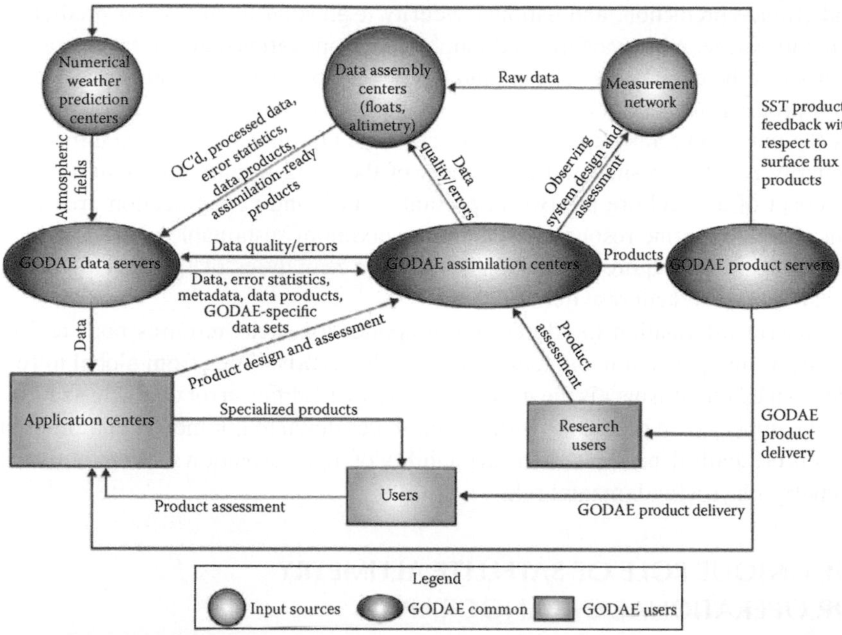

FIGURE 18.1 Functional components of operational ocean forecasting systems developed during the Global Ocean Data Assimilation Experiment (GODAE). (Adapted from Schiller, A., and Brassington, G. B., *Operational Oceanography in the 21st Century,* Springer, Dordrecht, The Netherlands, p. 745, 2011.)

assimilation centers (top center and right of Figure 18.1). Significant progress has been achieved in the capabilities of ocean prediction systems and data and product servers (see left of middle row of Figure 18.1), often underpinned by concepts and technologies that enable the observed data to be discovered, visualized, downloaded, intercompared, and analyzed all over the world. Progress in ocean data assimilation (the central item in Figure 18.1) has been pivotal in the synthesis of global and regional ocean forecasting systems and observations. Most forecasting centers now operate systems with 1/10° or finer horizontal grid spacing, have a global capability, make use of community ocean models, and assimilate *in situ* profile data, altimeter data, some form of surface temperature data, and even biogeochemical data. Product assessments and interactions with research users (lower right area of Figure 18.1) have been key activities since the inception of operational ocean forecasting systems. Nowadays, observing system design aids the implementation of new observing systems and is used to illustrate the complementarity of various observing system components such as SST, altimeter, and profile data for mesoscale prediction. Ecological and biogeochemical applications are an emerging area of research in ocean forecasting. The lower left part of Figure 18.1 depicts the information flows to application centers (also known as downstream services) and users.

18.2.3 Applications and Users

Over the past 10 years, there has been increased attention to the development of products and downstream services and the demonstration of their utility for applications. A comprehensive overview about the growing number of applications can be found in a special issue about GODAE OceanView in the *Journal of Operational Oceanography* (2015, Vol. 8).

Marine and coastal environmental monitoring, environmental regulations (e.g., marine strategy in Europe), maritime transport and ship routing, marine safety, search and rescue operations, pollution forecasting, and combating oil spill and chemical marine pollution events are important application areas. Weather and extreme event forecasting (e.g., tropical cyclones and storm surges), seasonal and climate prediction, and national security (e.g., acoustic and sonar prediction applications, search and rescue, mine warfare, and amphibious applications) also rely on precise description and forecast of the ocean state. The oil and gas and deep sea mining industries use operational oceanography products to monitor environmental conditions of production or extraction sites as well as to assess the environmental impact of their activities. Optimization and monitoring of marine renewable energy production sites and assessment of the multi-trophic productivity and the environmental impact of aquaculture are two important and expanding application areas. Sustainable management of living marine resources, achieving maximum sustainable yield, and rebuilding of overexploited stocks also require regular and long-term reanalyses of the ocean state. The same holds for ocean and ecosystem research.

Providing useful information to address these application areas requires both real-time high-resolution analysis and prediction and long-term reanalysis capabilities from global to regional and coastal scales. An essential aspect is the need to precisely (including error characterization) describe and forecast the upper ocean processes with spatial scales down to kilometers and timescales down to hours. This is critically dependent on the availability of high-resolution satellite observations, and satellite altimetry plays a fundamental role.

18.3 THE UNIQUE ROLE OF SATELLITE ALTIMETRY FOR OPERATIONAL OCEANOGRAPHY

18.3.1 The Close Relationships between Operational Oceanography and Satellite Altimetry

There are strong links between satellite altimetry and operational oceanography. Satellite altimeters provide all weather observations and unique observations at high space and time resolution to partly resolve the mesoscale variability and coastal variability. Altimeter sea level observations are also the only remotely sensed information that reflects the ocean state far below the surface. Sea level is an integral part of the density structure of the ocean interior and temperature, and salinity variations at all depths contribute to variations in sea level. These characteristics explain the unique and fundamental role of satellite altimetry for data assimilation and operational oceanography.

Because of the critical reliance on altimetry, the development of operational oceanography and operational altimetry occurred simultaneously. At the end of the 1990s, the satellite altimetry community was keen to develop further the operational use of altimetry, and this required an integrated approach merging satellite and *in situ* observations with models. GODAE was thus set up in 1997 (Smith and Lefebvre 1997), and its main demonstration (2002–2008) was phased with the Jason-1 and Envisat altimeter missions. Since then, GODAE and GODAE OceanView have been maintaining strong links with satellite altimetry communities, and major progress has been made to develop and optimize the use of satellite altimeter data for operational oceanography.

18.3.2 Synergies with Other Satellite and *In Situ* Observations

Ocean analysis and forecasting models are strongly dependent on the availability of altimeter data from multiple satellites. High-resolution altimetry is mandatory to constrain the mesoscale circulation. At least three to four altimeters are required (see discussion that follows). Most operational model resolutions are now at least 1/12° and 1/36° at global and regional scales, respectively. Coastal ocean analysis and forecasting systems use resolution of a few kilometers or less (e.g., Kourafalou et al. 2015; Stanev et al. 2016). This poses even stronger requirements for the altimeter constellation. SST observations are also essential observations to be assimilated in ocean models. The Group for High Resolution Temperature (GHRSST, e.g., Donlon et al. 2007) now offers a suite of tailored global high-resolution SST products (based on microwave and infrared sensors), in near real time, on a daily basis, to support operational forecast systems. Other important satellite observations for operational oceanography include sea ice, surface waves, sea surface salinity, ocean color radiometry, and winds (see Le Traon et al. 2015 for a review).

Argo and the other elements of the global *in situ* observing system (e.g., tropical moorings) are, on the other hand, mandatory to constrain large-scale temperature and salinity fields that are poorly constrained by satellite observations. Although Argo does not resolve the mesoscale, the joint use of Argo and altimetry through data assimilation techniques can provide a good representation of mesoscale temperature and salinity fields (e.g., Le Traon et al. 2001). Guinehut et al. (2012) have quantified how the joint use of altimetry, SST, and Argo observations improves the reconstruction of three-dimensional mesoscale temperature and salinity fields. The capability of Argo to complement altimeter observations to constrain modeling and assimilation systems is a major asset and explains why these two observing systems now provide the backbone of the global observations for operational oceanography (e.g., Le Traon 2013). A review of the role of altimeter and Argo observation data to constrain global ocean models is given by Oke et al. (2009). Turpin et al. (2016) demonstrate the essential role of Argo observations to constrain together with altimeter data a global data assimilation system.

18.3.3 General Requirements/Constellation

The need for long-term observations, continuity, and high resolution as well as operational constraints (e.g., ensuring that a given parameter continues to be monitored even when a satellite fails) requires international cooperation and the development of virtual constellations as promoted by the Committee on Earth Observing Satellites (CEOS) (e.g., Escudier and Fellous 2008).

Le Traon et al. (2006) have defined the main priorities for altimeter missions in the context of the European Copernicus Marine Service. Their Tables 18.1 and 18.2 give general requirements for different applications of altimetry and characteristics of altimeter missions.

TABLE 18.1
User Requirements for Different Applications of Altimetry

	Application Area	Accuracy[a]	Spatial Resolution	Revisit Time	Priority
1	Climate applications and reference mission	1 cm[a]	300–500 km	10–20 days	High
2	Ocean nowcasting/forecasting for mesoscale applications	3 cm[a]	50–100 km	7–15 days	High
3	Coastal/local	3 cm[a]	10 km	1 day	Low[b]

[a] For the given resolution;
[b] Limited by feasibility (but strong requirement).

TABLE 18.2
Altimeter Mission Characteristics

Class	Orbit	Mission Characteristics	Revisit Interval	Track Separation at the Equator
A	Non-sun synchronous	High accuracy for climate applications and to reference other missions	10–20 days	150–300 km
B	Polar	Medium-class accuracy	20–35 days	80–150 km

Table 18.2 categorizes altimeter missions according to the accuracy and precision as well as sampling characteristics of the overall system. Class A missions are those such as the Jason series that provide very precise observations required for climate change monitoring. Class B missions are those with lower accuracy and precision that provide higher spatial resolution for the mesoscale field. The main operational oceanography requirements for satellite altimetry can be summarized as follows:

- Need to maintain a long time series of a high accuracy altimeter system (Jason series) to serve as a reference mission and for climate applications. It requires one Class A altimeter with an adequate overlap between successive missions to allow a precise intercalibration.
- The main requirement for medium- to high-resolution altimetry is to fly three Class B altimeters in addition to the Jason series (Class A). Most operational oceanography applications (e.g., marine security and pollution monitoring) require high-resolution surface currents that cannot be adequately reproduced without a high-resolution altimeter system. Pascual et al. (2006, 2009) showed that at least three, but preferably four, altimeter missions are needed for near real time monitoring of the mesoscale circulation. Such a scenario would also provide an improved operational reliability, and it would enhance the spatial and temporal sampling for monitoring and forecasting significant wave height (see Abdalla et al. this volume).

These requirements are based on the use of conventional nadir-viewing altimeter instruments. In parallel, there is a need to develop and test innovative instrumentation (e.g., along-track Synthetic Aperture Radar [SAR] processing, wide swath altimetry with the Surface Water Ocean Topography [SWOT] mission) to better answer existing and future operational oceanography requirements for high to very high resolution (e.g., mesoscale to sub-mesoscale and coastal dynamics) (see discussion in Sections 18.5 and 18.6).

18.4 USE AND IMPACT OF SATELLITE ALTIMETRY FOR OPERATIONAL OCEANOGRAPHY

This section starts with a review of the altimeter constellation over the past 25 years and analyzes how it addressed the main operational oceanography requirements (see earlier). It then provides an overview of near real time multiple altimeter data processing focusing on data and products required for modeling and data assimilation. Data assimilation issues are then discussed and a summary of impact of altimeter data assimilation for ocean analysis and forecasting is given.

18.4.1 Evolution of the Altimeter Constellation over the Last 25 Years

Figure 18.2 shows an overview of the radar altimetry timeline. From 1992 up to now, there were always at least two altimeters flying simultaneously. Over the last 15 years, three to four altimeters have been flying simultaneously.

Constellation efficiency is not simply the number of satellites but rather is influenced by several additional factors (see Dibarboure and Lambin 2015). One needs to take into account the time needed before an altimeter becomes operational (e.g., after the initial calibration phase), that some altimeters (e.g., Jason series) are set for a couple of months on the same track at almost the same time as their predecessor for calibration purposes, and that some altimeter missions experience failures that degrade data service.

Finally, some altimeter systems are not optimized for oceanography (e.g., geodetic missions such as CryoSat-2). These missions do not have instrumentation to provide high-accuracy sea surface height (SSH), and a significant portion of the observations become redundant with other missions where ground tracks are very close together.

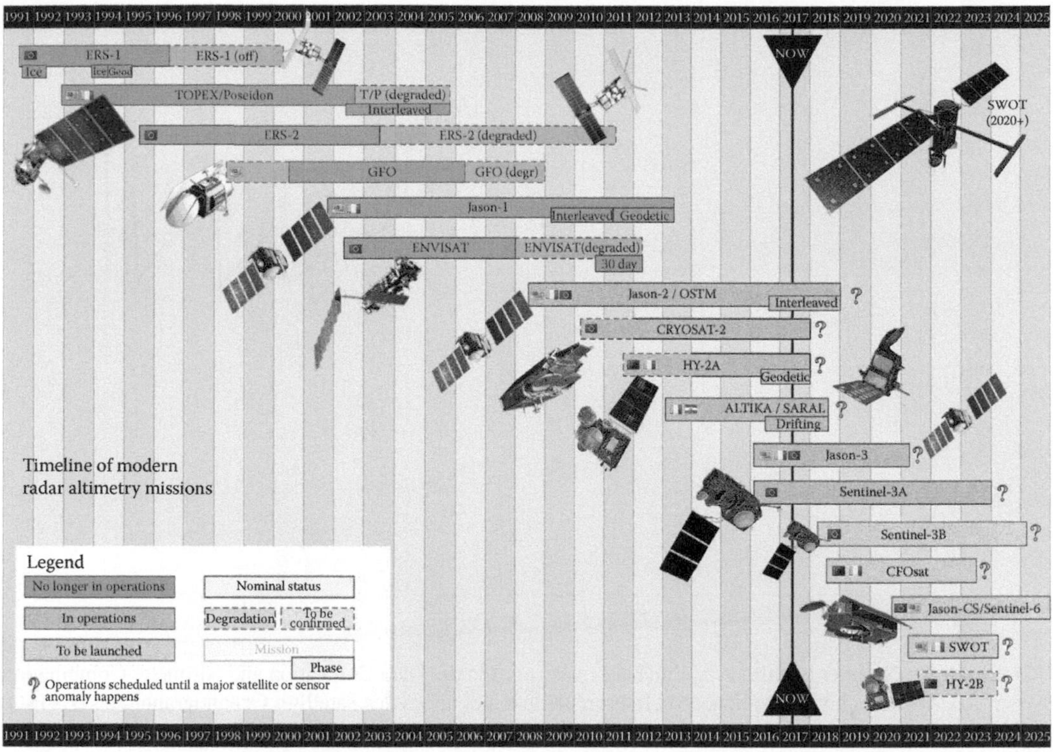

FIGURE 18.2 Overview of the timeline of altimetry missions.

Figure 18.3 shows the number of altimeters used by the multi-mission system AVISO/DUACS, which was computed as the monthly ratio of the number of measurements actually usable divided by the theoretical number of measurements expected for each sensor. This provides a good estimation of the number of altimeter systems readily available for operational oceanography. This shows that except for the 2003 to 2006 time period, this number was always below three and if geodetic missions are not taken into account closer to two. The minimum requirement for operational oceanography (at least three to four missions) was thus met only during a limited time period despite outstanding efforts made by space agencies and data centers to ensure that new missions are ingested in the operational data stream as soon as possible and to maintain high quality and timeliness despite the aging of altimetry missions (see Dibarboure and Lambin 2015; Le Traon et al. 2015).

As discussed in Section 18.5, perspectives for the coming years are (much) better and most likely three to four altimeters flying simultaneously will be available over the next decade. Requirements also evolve, however, and observation capabilities are not following and lag behind the evolution of modeling and data assimilation system resolution.

Beyond these changes in the altimeter constellation, it is important to note that several improvements highly beneficial for operational oceanography were also achieved over the last 25 years. Le Traon et al. (2015) review the main improvements: more precise orbits, improved algorithms, better mean sea surfaces (MSSs) and MDTs (see as follows) and reduced noise level with the use of Ka-band (SARAL/AltiKa) and delay-Doppler altimetry (so-called SAR Mode, or SAR, of CryoSat-2 and Sentinel-3).

18.4.2 Multiple Altimeter Data Processing for Operational Oceanography

Altimeter data processing includes different steps: Level 0 and Level 1 (from telemetry to calibrated sensor measurements), Level 2 (from sensor measurements to geophysical SSH data), Level 2P (homogenization of standards and geophysical corrections between satellite missions), Level 3 (intercalibration,

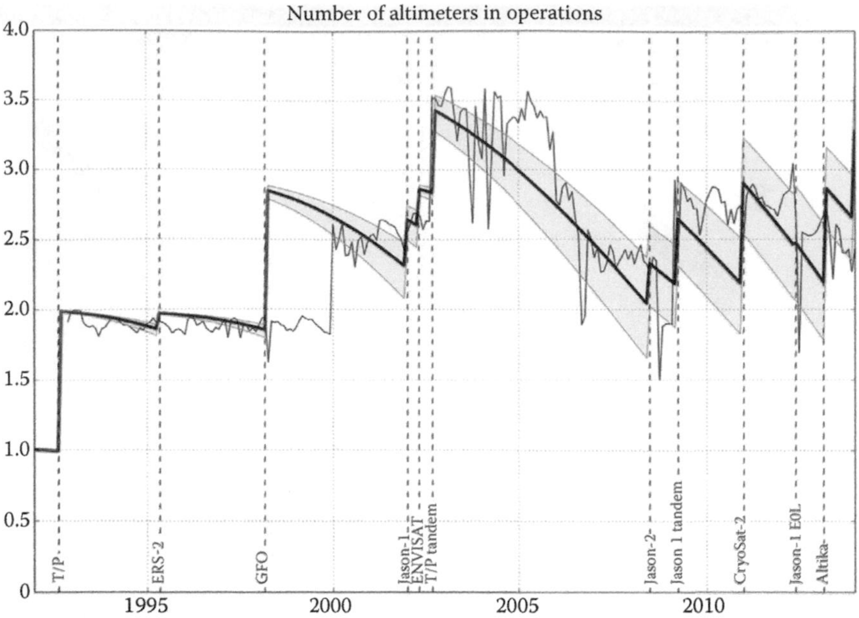

FIGURE 18.3 Number of altimeters (thin blue curve) used in the Data Unification and Altimeter Combination System (DUACS)/Archivage, Validation et Interprétation des données des Satellites Océanographiques (AVISO) system and computed as the monthly ratio of the number of actual measurements divided by the theoretical number of measurements for one altimeter. The thick black line is the model prediction (will be removed—new figure to be produced including extension up to 2016). (Derived from Dibarboure, G., and Lambin, *J., Mar. Geod.*, 38, 684–703, 2016. With permission.)

large-scale error reduction, along-track gridding, and extraction of consistent SLA data between missions), and Level 4 (merging of different sensors and mapping of SLA on regular space/time grids).

All processing ranging from Level 0 to Level 2 is carried out by satellite agencies as part of their altimeter ground segment activities. In contrast, the assembly of multiple Level 2 data flows from different missions and the generation of Level 2P to Level 3 and Level 4 products is usually carried out by specific data processing centers or thematic assembly centers. The role of these data processing centers is, in particular, to provide modeling and data assimilation centers with the real time and delayed mode data sets required for data assimilation. This also includes uncertainty estimates that are critical to an effective use of data in modeling and data assimilation systems. An active link between the assembly centers and the data assimilation centers is also needed to organize feedback on data quality control performed at the level of data assimilation centers (e.g., comparing an observation with a model forecast).

In theory, it is better for advanced assimilation schemes to use raw satellite data (Level 2) because the data error structure is generally more easily described. In practice, however, high level altimeter data processing (e.g., Level 3 cross-calibration, see as follows) is needed before along-track altimeter data are assimilated into ocean models (the cross-calibration process cannot be easily done within the assimilation systems). To that extent, the multi-mission altimeter products used today by most of the modeling and data assimilation centers are Level 3 SLA data. In most cases, precise MDTs are also required so that models can be constrained by the absolute dynamic topography (ADT). In other situations, the Level 3 SLA is used to construct sub-surface temperature and salinity anomalies to which climatological mean temperature and salinity are added.

The following sections give an overview of the main multi-mission algorithms and high-level products. Readers may refer to Dibarboure et al. (2011 a,b) and Pujol et al. (2016) for more in-depth presentations of the algorithms. More details about higher-level products are also given in AVISO (2016) or Copernicus Marine Environment Monitoring Service (CMEMS) (2016).

18.4.2.1 Level 2 Data Assembly and Homogenization

The main altimetry product delivered by the satellite agencies is the Level 2 (L2), also known as Geophysical Data Record (GDR). There are three types of L2 products with increasing latency and accuracy:

- Fast delivery products delivered in 3 to 6h. This class of product is sometimes called Operational GDR (OGDR), or near real time (NRT).
- Intermediate products delivered in 24–48 h. This class of product is sometimes called Interim GDR (IGDR) or slow time critical (STC).
- Delayed time products delivered in 30 to 60 days or more. This class of product is sometimes called GDR or non-time-critical (NTC).

Operational real time systems use the first two categories, and the third type is more commonly used in reanalyses and offline validation. The main difference between OGDR and IGDR products is that the former use slightly less accurate onboard orbit determination and atmospheric model forecasts, whereas the latter benefit from a more precise orbit determination computed in the ground segment as well as environmental corrections based on atmospheric model analyses.

Standards and processing used for GDR-class products can be heterogeneous because they are selected by different agencies, or different subsets of altimetry experts. This is emphasized in NRT due to the additional pressure stemming from the short latency. Yet homogeneity is a strong requirement for most altimetry users, especially when the goal is to merge multiple sensors in a global mapping or ocean model assimilation scheme.

Therefore GDR data sets may be complemented with ancillary data sets ranging from updated instrumental corrections, orbit solutions, geophysical corrections (e.g., tidal models, model-based ionosphere, or atmospheric corrections), to reference surfaces (e.g., MSS). To illustrate, systems such as CMEMS/SL-TAC, AVISO, ALPS, or NOAA/RADS operationally acquire and apply up to 30 different types of ancillary data. The enhanced and homogenized GDR product is generally known as Level 2P.

18.4.2.2 Intercalibration, Orbit, and Large-Scale Error Correction

There are still residual discrepancies among altimetry products. Differences may originate in residual sensor biases (geographically correlated errors) or geophysical discrepancies (e.g., residual tidal signatures or imperfect atmospheric corrections). The multi-mission cross-calibration process ensures that data sets from all satellites provide consistent and accurate information (e.g., Figure 18.4). Although the sequence can be rather complex, two algorithms account for the bulk of the error minimization: the crossover error reduction (hereafter, CER) and the long-wavelength error reduction (hereafter LER).

The CER is a two-step cross-calibration process. The first step is a global crossover minimization performed on the so-called reference mission. TOPEX/Poseidon (T/P was the first reference mission, then the Jason series replaced it. The crossover minimization performed on the reference mission is derived from Tai (1988), and it uses a one and two cycle per revolution analytical model. The second CER step is a dual crossover minimization of each other satellite derived from Le Traon and Ogor (1998). Using the CER process, the orbit precision of any satellite can be improved almost to the level of the reference mission. This process is particularly efficient in the open ocean and at latitudes lower than 66° (TOPEX/Jason latitude limit) but is less accurate poleward of 66° due to the lack of a predefined analytical model.

The LER is then used as a second cross-calibration process. It is derived from Le Traon et al. (1998) and Ducet et al. (2000). An optimal interpolation process is used to estimate and remove local biases between neighboring ground tracks from each satellite. The algorithm requires a

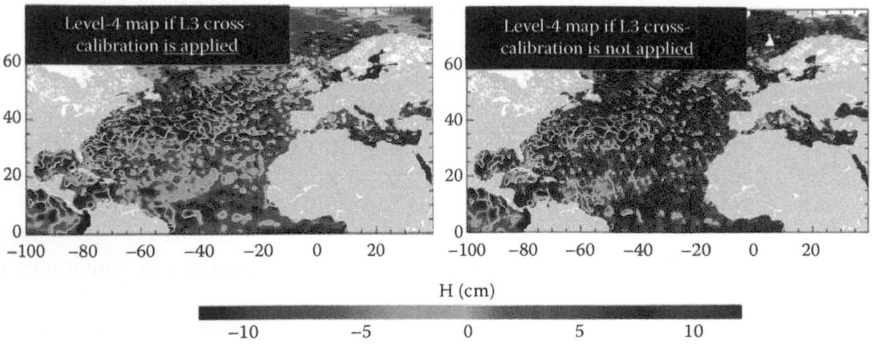

FIGURE 18.4 Impact of the Level 3 multi-mission cross-calibration on the multi-altimeter merging (Level 4). Example of a Jason-2 plus CryoSat-2 map.

good statistical description of the problem to be solved: sea level variability that should be ignored because it is not long-wavelength error (LWE), localized correlation scales, mission-specific noise, and variance of the LWE to be removed.

18.4.2.3 Calculation of Sea Level Anomalies

The final step of the Level 3 processing is to compute along-track SLAs from the SSH (i.e., using an MSS).

For altimetry missions using a repeat orbit, such as TOPEX/Jason or European Remote Sensing (ERS)/Envisat/AltiKa, the MSS model can be local (i.e., an average of 20 years of SSH on this charted track). The averaging process filters out the dynamic part of the SSH, leaving a mean surface of sub-centimetric precision. When the time series is shorter (e.g., GEOSAT Follow-On [GFO] or HY-2A), the challenge is twofold: One needs to separate the mean surface from SLA, and also to correct for interannual signals.

For uncharted tracks (e.g., Sentinel-3 launched in early 2016) and geodetic missions (e.g., CryoSat-2), it is not possible to compute such a mean profile. The approach is then to build a gridded MSS model using an optimal interpolation of the mean topography from all altimetry measurements available over a 20-year period. The challenge is to properly separate mesoscale anomalies from the mean topography. This is particularly challenging and critical for geodetic altimeters. It is important to highlight that the errors associated with geodetic missions are quite problematic for assimilation systems: Any error in the MSS gridded reference will create a similar SLA error that is (1) constant in time, (2) with a spatial structure, and (3) common to all altimeters using this gridded reference. These properties violate a fundamental assumption of operational assimilation systems. Because the additional observations are still quite valuable, some assimilation systems attempt to mitigate this issue through bias correction techniques (see Section 18.4.3).

The analysis of MSS models older than Jason-1 GM (geodetic mission) predicted an error ranging from 1 to 3 cm root mean square (RMS), that is, an error level high enough to limit the ability of Jason-1 GM to observe small mesoscale features over half of the oceans (Dibarboure et al. 2012). The analysis of Jason-1 GM data and the comparison with other phases confirmed this order of magnitude. It showed that errors from these imperfect MSS models were more than 10%–15% of the SLA variance for wavelengths ranging from 30 to 150 km (e.g., Dufau et al. 2016).

Faugere et al. (pers. commun., 2016) report that the 400-day geodetic data set collected by Jason-1 GM and the 4 years of CryoSat-2 were instrumental in improving the latest generations of MSS models (e.g., DTU15 and CNES15). This massive addition of geodetic data significantly reduced the error of older models: The difference among green, red/blue, and black spectra in Figure 18.5 shows that the global MSS error variance has been reduced by up to 50% for wavelengths ranging from 25 to 100 km. The remaining error in recent models is still significant: The black spectrum

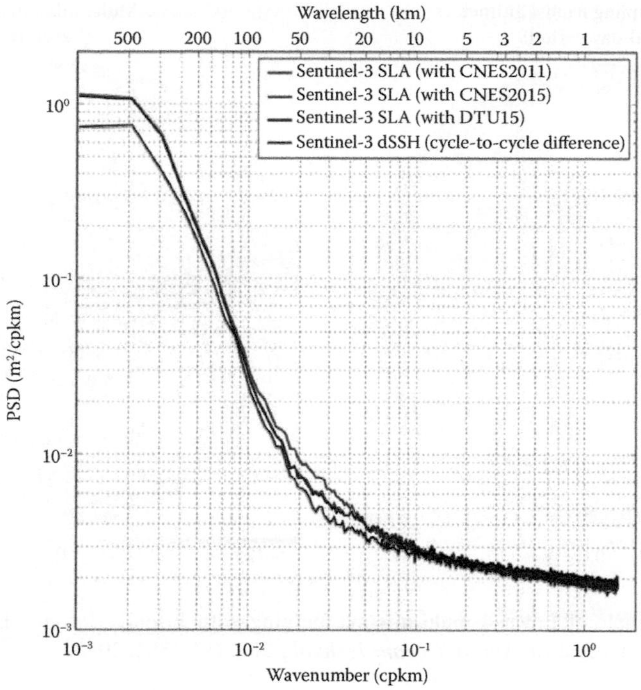

FIGURE 18.5 First global assessment of the error of recent mean sea surface (MSS) models using Sentinel-3 (Faugère et al., personal communication). The green spectrum is the mean global along-track wavenumber spectrum of Sentinel-3 (based on the CNES2011MSS model). The red and blue spectra are based on sea level anomaly (SLA) using 2015 MSS models. The black spectrum is the spectrum of half the differences between SLA from subsequent cycles separated by 27 days. When the SLA content is entirely decorrelated, the black spectrum is superimposed with the mean SLA spectrum. The discrepancy between red/blue spectra and the black spectra from 10 to 100 km is very clear, and it is likely due to residual errors in 2015 MSS models.

in Figure 18.5 is computed from SSH differences between two 27-day cycles of Sentinel-3. The difference between the red/blue and black spectra for wavelengths ranging from 10 to 100 km is substantial, and it mainly originates in the error of 2015 MSS models.

18.4.2.4 Strengthening the Links with Data Assimilation Systems

A recent and still ongoing evolution of operational model assimilation is the transition from standard global Level 3 altimetry products to a new type of model-tailored altimetry product. Typical changes include *ad hoc* filtering, subsampling, and physical content or references: Some models may want to have a different type of processing for tides or atmospheric correction (e.g., use a barotropic model correction of the SLA) or MDT that is more consistent with the model bathymetry and mean currents. Dufau (2014), Dobricic et al. (2012), and Benkiran et al. (2017) illustrate how such an *ad hoc* Level 3 production interacts with changes in the modeling and data assimilation systems.

18.4.2.5 Mapping SLA on Regular Space and Time Grids

Arguably the most convenient and widely used altimetry product is the gridded maps (Level 4) of SLA merging Level 3 altimeter products from all altimeters in operations (e.g., Figure 18.6). Such high-level data products can be directly used for applications (e.g., marine safety or offshore applications). They are also useful to intercompare with data assimilation systems (e.g., statistical vs. dynamical interpolation), and they complement products derived through modeling and data assimilation systems.

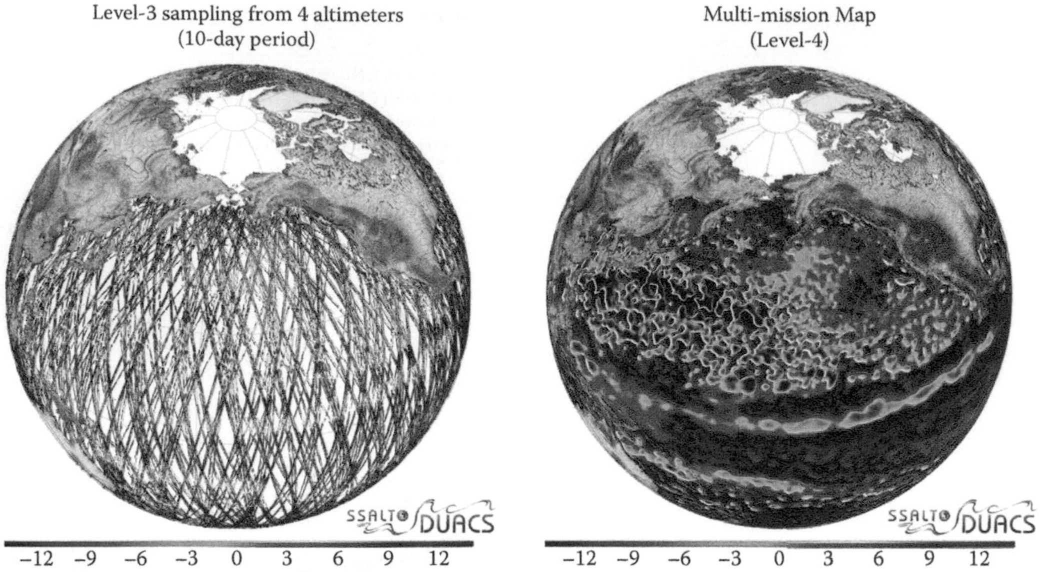

FIGURE 18.6 Example of Level 4 multi-mission merging with Jason-2, Jason-1, Envisat, and CryoSat. (From Dibarboure, G. F., et al., *J. Atmos. Ocean. Technol.*, 31, 1337–1362, 2012.)

The AVISO/DUACS multi-mission mapping process is based on an optimal interpolation derived from Ducet et al. (2000). The gridding process requires *a priori* knowledge of the covariance of sea level and measurement errors. Sea level covariance functions and error budgets are derived from an analysis of T/P, Jason-1, and ERS data (Jacobs et al. 2001; Le Traon et al. 2003). The critical parameters used in the mapping are the error characterization (including representation error) and signal covariance and propagation models as they largely control the scales observed on multi-mission maps (Pujol et al. 2016) within the limit of the sampling capability of the constellation, as highlighted by Chelton et al. (2011).

18.4.2.6 Geoid and Mean Dynamic Topography

Accurate knowledge of the marine geoid is a fundamental element for the full exploitation of altimetry for oceanographic applications and, in particular, for assimilation into operational ocean forecasting systems. SSH measured by an altimeter is the sea level above the ellipsoid, which differs from the ADT by the value of the geoid height. The ADT is usually obtained by estimating an MDT and adding it to the altimetric SLAs. The MDT is obtained, at spatial scales where the geoid is known with sufficient accuracy, as the difference between an altimeter MSS height (MSSH) and a geoid model.

Thanks to the recent dedicated GRACE (Gravity Recovery and Climate Experiment) and GOCE (Gravity Field and Steady-State Ocean Circulation Explorer) space gravity missions, the knowledge of the geoid at scales of around 100 to 150 km has greatly improved in the past years, so that the ocean MDT is now resolved at those scales with centimeter accuracy. However, the true ocean MDT over a given period (e.g., 10 to 20 years) contains scales shorter than 100 to 150 km that are not resolved in geoid models based on remote sensing. To compute higher resolution MDT, space gravity data can be combined with altimetry and oceanographic *in situ* measurements such as hydrological profiles from the Argo array and velocity measurement from drifting buoys (e.g., Maximenko et al. 2009; Rio et al. 2014). This approach was used by Rio et al. (2014) to compute the CNES-CLS13 MDT, which is used in several global data assimilation systems (see next section).

18.4.3 Assimilation in Ocean Forecasting Models

Operational ocean forecasting systems assimilate a variety of data with the aim of obtaining a good estimate of both the large-scale and mesoscale ocean state in order to produce forecasts and reanalysis. The assimilated data provide complementary information (e.g., Oke et al. 2008; Lellouche et al. 2013; Lea et al. 2014): SST data provide high-resolution information about temperature in the mixed layer, *in situ* profiles of temperature and salinity such as those from Argo provide information about the large-scale density structure in the ocean, and the satellite altimeter sea level data provide information about the mesoscale ocean dynamics and can be used to infer corrections to the sub-surface density and velocity. Many of the systems also assimilate sea-ice concentration observations (Hebert et al. 2015; Posey et al. 2015). These various observations are combined with a previous model forecast using data assimilation techniques including both variational (3DVAR: e.g., Usui et al. 2006; Balmaseda et al. 2013; Cummings and Smedstad 2013; Waters et al. 2014; or 4DVAR: e.g., Kurapov et al. 2011; Ngodock et al. 2016; Usui et al. 2015) and ensemble-based methods (Ensemble Optimal Interpolation: Oke et al. 2013b; Ensemble Kalman Filter: Sakov et al. 2012; and a fixed-basis implementation of the Singular Evolutive Extended Kalman [SEEK] filter: Lellouche et al. 2013). An overview of these operational data assimilation systems is provided in Martin et al. (2015).

The innovation (observation minus model forecast) contains information about sea level model forecast errors due to various ocean processes. Most data assimilation systems assume that these sea level errors can be split into errors due to baroclinic and barotropic processes. Determining how much of the signal to use to correct the baroclinic or barotropic part of the model is usually done through statistical estimates from previous model runs. The baroclinic part of the signal can be used to infer corrections to the sub-surface density structure in the model, either through dynamical balance relationships (such as those described in Weaver et al. 2005; or Cooper and Haines 1996), through statistical estimates from an ensemble of model runs (Oke et al. 2008) or statistics based on historical observations (Fox et al. 2002). The statistical relations between the sea level and sub-surface structure are among the most critical components in the assimilation system. Without these, any sea level change to the model would propagate away as a barotropic wave. The barotropic part of the observation can be used to correct the model sea level without making density corrections. Both baroclinic and barotropic parts can also be used to infer corrections to the model's velocity structure through the geostrophic relationship (away from the equator). The altimeter data therefore can be used to provide corrections to the model's three-dimensional density structure and dynamics.

The relation between sea level observations and sub-surface density structure has been shown to be strong at mid- to low latitudes where the largest errors are often due to misplacement of the depth of the thermocline (Carnes et al. 1990). At high latitudes, however, where the stratification is often weak and most of the error in sea level is due to barotropic processes, it is not always possible to produce a change in sea level of the required magnitude by adjusting the sub-surface density. Most GODAE OceanView (GOV) systems therefore only make barotropic changes where the stratification is weak or do not assimilate the altimeter data at all there.

The GOV systems mainly assimilate along-track Level 3 altimeter SLA data from various satellites (see previous section). These SLA data are anomalies from a long timescale (20-year) mean of the SSH observed from the satellites. A variety of approaches exist for relating the observed SLA to the model state, and all these require information on the mean ocean state. The dependence on mean ocean state introduces challenges in all the methods. One approach uses historical correlations between steric height variations and sub-surface temperature and salinity (e.g., as observed by Argo). Given the altimetry observed SLA, temperature and salinity anomalies are estimated, and a climatological mean temperature and salinity is added. This synthetic profile is then provided to the assimilation system as an observation (Rowley and Mask 2014). Alternatively, the correlations between SLA and sub-surface density may be embedded directly into the assimilation so that the construction of a synthetic profile is not necessary. In this case, the model SSH must be compared to the observed SLA.

To compare the altimetry SLA observations to the model sea level, an MDT is required: The observed SLA data are added to this MDT before comparing to the model sea level in order to provide the data assimilation system with the innovations. The data assimilation then uses this information together with the innovations from the other observing networks and, with the observation errors and model background errors, produces a set of assimilation increments on the model grid that are then added into the model, often using an incremental analysis updating (IAU) scheme.

The choice of an MDT is an important part of the design of the scheme to assimilate altimeter data. The average SSH from the models is often quite different from the MDT generated from observations. This can be due to model errors—for instance, the average path of the Gulf Stream may not be accurately represented in the models, and the observed MDT also contains errors, particularly at small spatial scales (the larger scales are now well observed from GOCE and GRACE). If the observed MDT was used directly, the assimilation of the sea level observations (SLA plus MDT) would be attempting to correct for these biases as well as the anomalies, and data assimilation schemes are generally not designed to deal with biases of this nature.

This explains why some data assimilation systems use an MDT from a free model run, and others use the mean from a previous reanalysis run that only assimilated temperature and salinity data. These approaches assume that the SSH observations are not biased compared to the underlying model and the assimilation will correct the location of the mesoscale ocean features, but this approach will usually not enable the data assimilation to correct the large-scale structure of the model dynamics (e.g., the average Gulf Stream path).

Given the large improvements in MDT estimations in past years, derived in particular from GOCE and GRACE (see previous section), it is, however, much more preferable to use observed MDTs (e.g., Stammer et al. 2007; Haines et al. 2011). Such improved MDTs strongly improve the data assimilation system performance (both for analysis and forecasts) (Haines et al. 2011). This requires, however, improved data assimilation schemes that take into account model and observation biases. Lea et al. (2008) proposed to use the observation-minus-forecast (i.e., innovation) values to update an estimate of the bias due to errors in the MDT (and remove it) during the data assimilation process, and this observation bias correction was implemented in the Forecasting Ocean Assimilation Model (FOAM) system (Waters et al. 2014). Mercator Ocean uses an iterative process to improve the CNES-CLS MDT, taking into account high-resolution analysis innovations. In addition, a 3DVAR bias correction scheme is applied to correct the large-scale, slowly evolving error of the model (Lellouche et al. 2013).

The assimilation of altimeter data is also challenging in shallow waters. The relation between the observed SLA and underlying density structure is much weaker. In addition, other processes produce significant discrepancies among observations and models. These include tides that may require very high model resolution to accurately predict them as well as atmospheric forcing errors. Because of the ambiguity in the error source (density, tides, or wind forcing), most of the GOV systems do not attempt to assimilate the altimeter data in shallow waters. Development versions of some of the global systems are now beginning to represent tides (e.g., the Naval Research Laboratory (NRL) Hybrid Coordinate Ocean Model (HYCOM) system), and assimilation of altimeter data in these systems becomes more challenging. The application of altimeter data to correcting tides in a global model is conducted as an analysis separate from the regular assimilation of mesoscale information (Buijsman et al. 2015). Regional systems have been developed that assimilate altimeter data into ocean models in the presence of tides but also do not normally attempt to correct the tidal signal (e.g., Kurapov et al., 2011; Rowley and Mask 2014; Benkiran et al. 2017).

18.4.4 Impact of Multiple Altimeter Data Assimilation in Ocean Forecasting Models

GOV systems have been used to investigate the impact of altimeter data within their data assimilation and forecasting framework in a number of global and regional studies summarized in Oke et al. (2015a, 2015b). Most of these studies demonstrate the impact of the altimeter data in the context of the other observing systems using observing system evaluations (OSEs) whereby an experiment is run assimilating all available data and a parallel run of the system is carried out assimilating all the data except for the data type to be investigated. The difference between the two runs then shows the impact of the withheld data in the context of all the other data, and the two runs can be assessed by comparing the outputs with assimilated and independent data. A complementary approach for estimating the influence of the altimeter observations on the analysis is the computation of the so-called "degrees of freedom for signal" (DFS) that represents the equivalent number of independent observations that constrain the model

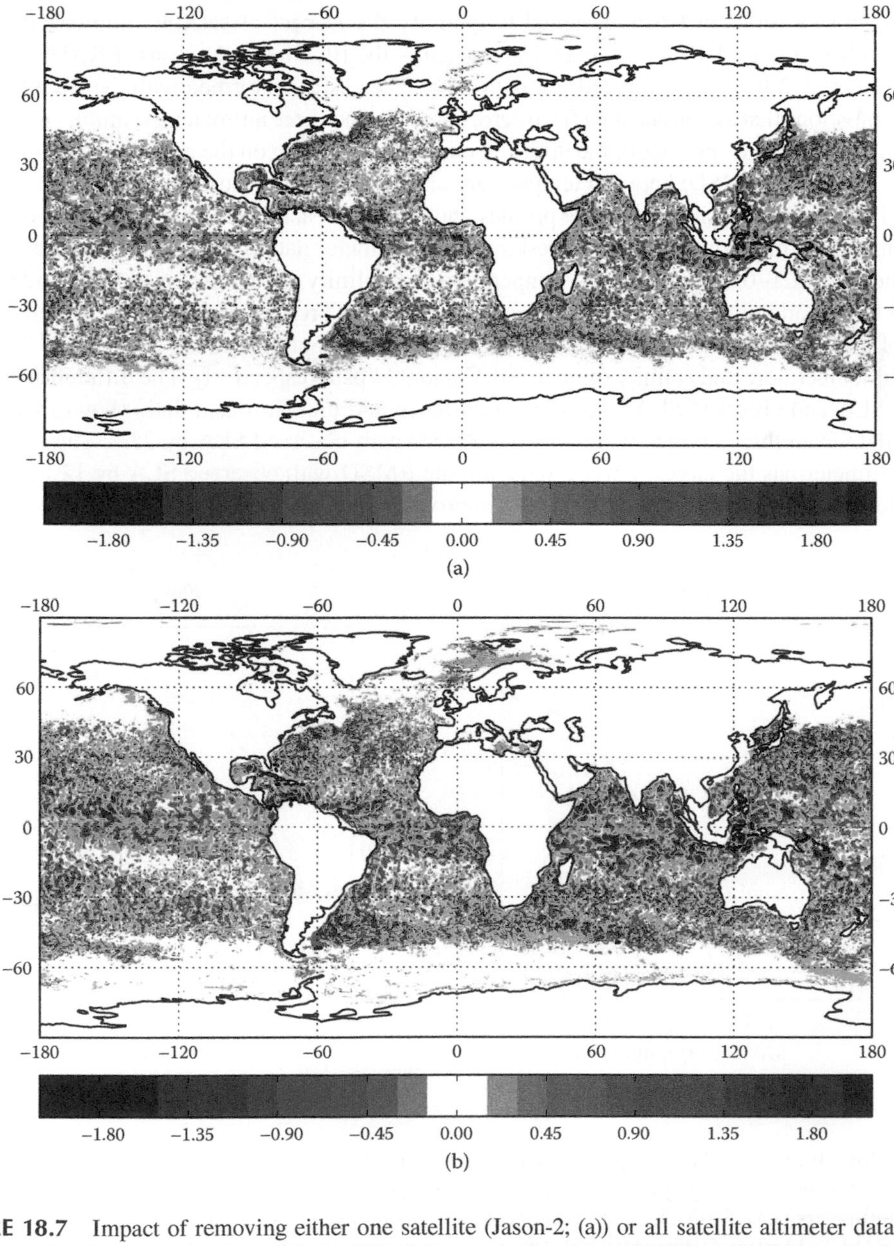

FIGURE 18.7 Impact of removing either one satellite (Jason-2; (a)) or all satellite altimeter data (b) from the Forecasting Ocean Assimilation Model (FOAM) system after 1 month while assimilating all other ocean-observing systems: map of the temperature difference (°C) at 109.7-m depth, calculated from daily average fields from the last day of the experiments. (Taken from Lea, D. J., et al., *Q. J. Roy. Meteorol. Soc.*, 140, 2037–2049, 2014. With permission.)

analysis at the observation point (Cardinali et al. 2004; Oke et al. 2008). DFS quantifies the influence of observations on analyses without having to run dedicated experiments withholding some data.*

Some OSE results are presented here to illustrate the major impact and role of multiple altimeters for ocean analysis and prediction systems. Removing one altimeter (Jason-2) and all altimeters (Jason-2, Jason-1, and Envisat) in the FOAM global data assimilation system while assimilating all other ocean observing systems has a major impact on sea level analysis and forecasts. It also has a large impact on temperature fields. Figure 18.7 shows daily averaged temperature differences at 110-meter depth after 1 month between the runs with and without altimeter data assimilation (one altimeter removed in (a) and all altimeters removed in (b)) (Lea et al. 2014).

Results from a series of OSEs designed to quantify the impact of satellite altimeter data on a regional, eddy-resolving 1/10° ocean reanalysis system—the Bluelink Reanalysis (BRAN; version 3) system (Oke et al. 2013b) are presented in Figure 18.8. The OSEs performed with the BRAN system include simulations that assimilate data from zero, one, two, and three altimeters, spanning a 12-month period. Figure 18.8 shows profiles of the 90th percentile (i.e., focusing on the more extreme impacts) of the RMS difference (RMSD) between temperature and salinity in the Australian region for different OSEs. All experiments assimilate *in situ* profiles and SST, and each OSE assimilates SLA data from a different number of altimeters. The greatest impact of altimeter data is made with the addition of the first altimeter (1ALT–0ALT), where the temperature and salinity change by up to 1.5°C and 0.4 psu, respectively. The addition of the second altimeter (2ALT–1ALT) results in changes of temperature and salinity of up to 1°C and 0.18 psu, respectively; and the addition of the third altimeter (3ALT–2ALT) changes of temperature and salinity of up to 0.8°C and 0.14 psu, respectively. The difference between analyzed SLA and observed SLA (including assimilated and unassimilated data) shows that assimilation of data from three altimeters reduces the RMSD with observed SLA by 23%. The addition of the first altimeter has the largest impact, reducing the RMSD with observed SLA by 12% and, with the addition of the second and third altimeters, improving SLA by a further 8% and 3%, respectively.

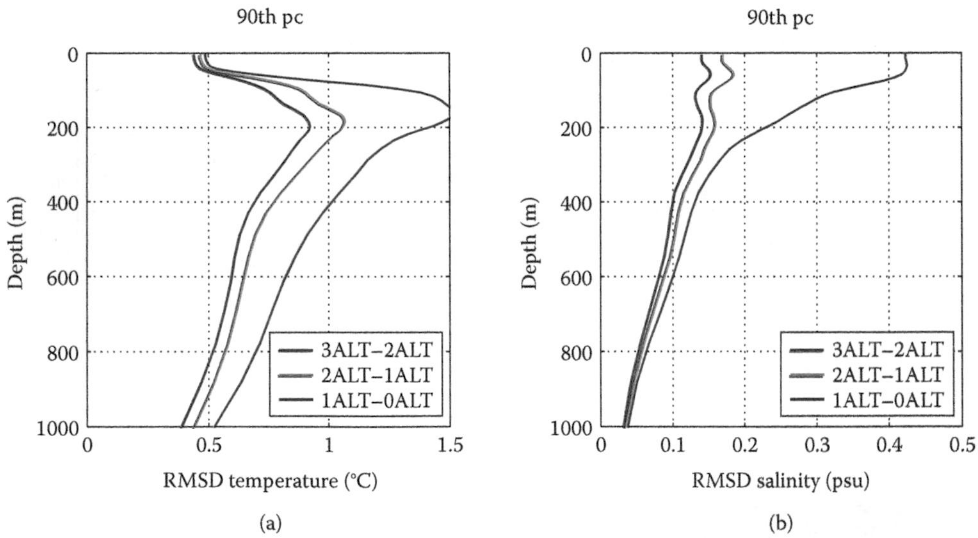

FIGURE 18.8 Profiles of the 90th percentile of the root-mean-squared difference for temperature (a) and salinity (b) between observing system evaluations performed using the Bluelink system with data from one and zero altimeters (1ALT–0ALT; blue), two and one altimeters (2ALT–1 ALT; green), and three and two altimeters (3ALT–2ALT; red) (Courtesy of P. Oke).

* DFS calculation may require, however, an additional model run (e.g. Oke et al. 2008).

Altimeter OSEs have also been carried out by Mercator Ocean with a North Atlantic 1/12° resolution model (Remy et al. 2017) over a 1-year time period. The assimilation code is based on statistical data assimilation procedure, inherited from SEEK. Satellite SST maps, Level 3 along track SLA and *in situ* temperature and salinity profiles, are assimilated every week. The model equivalent of the SLA is computed by removing from the model SSH a synthetic MDT based on the CNES/CLS 2009 MDT. The system is described in detail in Lellouche et al. (2013). The impact of assimilating different altimeter missions (Jason-2, Jason-1 Geodetic Mission, and CryoSat-2) is analyzed by comparing analyses and short-term forecasts (less than 7 days) with Jason-2 along-track data. The first altimeter data set assimilated has the largest impact. Figure 18.9 shows errors for a given configuration (E_X) expressed as a percentage of variance reduction compared to the best configuration, that is, analyzes when all altimeter data sets (Jason-2, Jason-1 Geodetic Mission, and CryoSat-2) are assimilated, that is, E_X = 100 (analysis or forecast error variance—error variance J2 + J1G + C2 analysis)/(error variance J2 + J1G + C2 analysis). Compared to the three-altimeter configuration, assimilating only two altimeters increases the forecast errors by 10%–20%, depending on which altimeter is withheld, and assimilating only one altimeter increases the forecast error by more than 30%.

Similar results are demonstrated by Jacobs et al. (2014) in which a 1.5-year-long OSE is conducted in the western Pacific at 3-km resolution using permutations of the T/P, Jason-1, Envisat, and GFO altimeters. The spatial correlation of the various OSEs to the nature run indicates that sea level correlation increases rapidly with the first satellite and asymptotically reaches a value of 0.99 with all four satellites. However, mesoscale features may be misplaced 50 km or more, and the spatial correlation is only slightly degraded. A more stringent metric is the spatial correlation of properties related to the fronts of eddies such as the frontogenesis forcing and surface divergence. The spatial correlation of these properties increases relatively linearly with the number of satellites, and with four satellites, these reach peak spatial correlations of 0.59 and 0.57, respectively (Figure 18.10). These levels indicate that prediction of frontal positions is only beginning to show skill with four satellites.

These OSE studies demonstrate the value of satellite altimeter data for constraining an eddy-resolving ocean model not only for sea level but also for ocean state variables not directly measured. It shows that the addition of the first altimeter has the largest impact but that there are quantitative

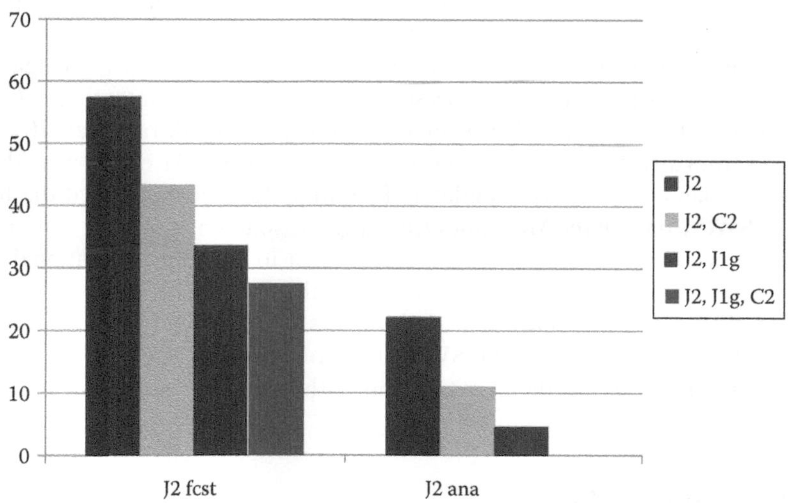

FIGURE 18.9 Observing system evaluation (OSE) carried out with the 1/12° North Atlantic Mercator Ocean data assimilation system for the 2012 year (Remy et al. 2017). Impact of assimilating different altimeter missions (Jason-2, Jason-1 Geodetic Mission, CryoSat-2) is analyzed by comparing analyses and short-term forecasts (less than 7 days) with Jason-2 along-track data. Errors for a given configuration (E_x) are displayed in percentage of variance reduction compared to the best configuration (i.e., analyses when all altimeter data sets are assimilated).

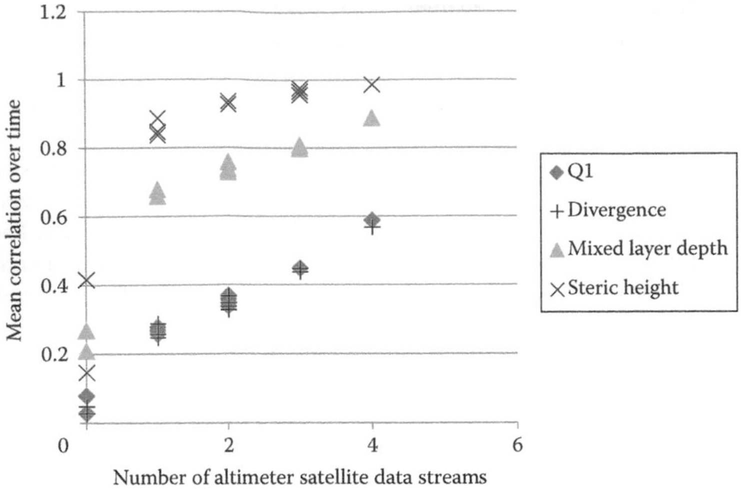

FIGURE 18.10 The correlation coefficients from observing system evaluations averaged over 1.5 years of ocean parameters as a function of number of satellite altimeters assimilated (Jacobs et al. 2014). The steric height correlation increases rapidly with just one data stream, and the marginal improvement of additional altimeters is small and decreases with increasing altimeters. The Mixed Layer Depth (MLD) marginal improvement is relatively constant from one to four altimeters. Frontal forcing Q_1 and surface divergence (vertical velocity) show a nearly linear increase in correlation as the number of altimeters assimilated increases and thus show a constant marginal improvement.

improvements seen by the addition of a second and third altimeter. As model resolution increases, the requirements for higher resolution will become even stronger.

The impact of future observations can also be assessed using OSSEs. OSSEs typically use two different models. One model is used to perform a "truth" run—and it is treated as if it is the real ocean. The truth run is sampled in a manner that mimics either an existing or future observing system—yielding synthetic observations. The synthetic observations are assimilated into the second model, and the model performance is evaluated by comparing it against the truth run. Mercator Ocean has performed first OSSEs of SWOT observations using a 1/12° regional model of the Iberian Biscay region that includes tidal forcing (Benkiran et al. 2017). The truth run was derived from a 1/36° model run over the same region. SWOT errors were derived using the Jet Propulsion Jab (JPL) SWOT Simulator. This first study demonstrated the feasibility of assimilation of SWOT data in the Mercator Ocean high-resolution models. It also quantified the very significant improvement of SWOT data with respect to existing conventional nadir altimeters to constrain ocean models (Figure 18.11). The results are (much) better with SWOT (with a nadir altimeter) than with three conventional altimeters. The system is able to sustain the right level of mesoscale activity in spite of the SWOT time repetitivity (21 days). SWOT swaths help the assimilation to create less artificial extrapolation than within the gaps of the conventional altimeters.

18.5 FUTURE PROSPECTS

18.5.1 Evolution of Operational Oceanography and New Challenges

The marine environment plays an increasingly important role in shaping economies and infrastructures and touches upon many aspects of our lives, including food supplies, energy resources, national security, and recreational activities. The recent OECD report "The Ocean Economy in 2030" notes that the global ocean economy could double in size by 2030, reaching a gross value

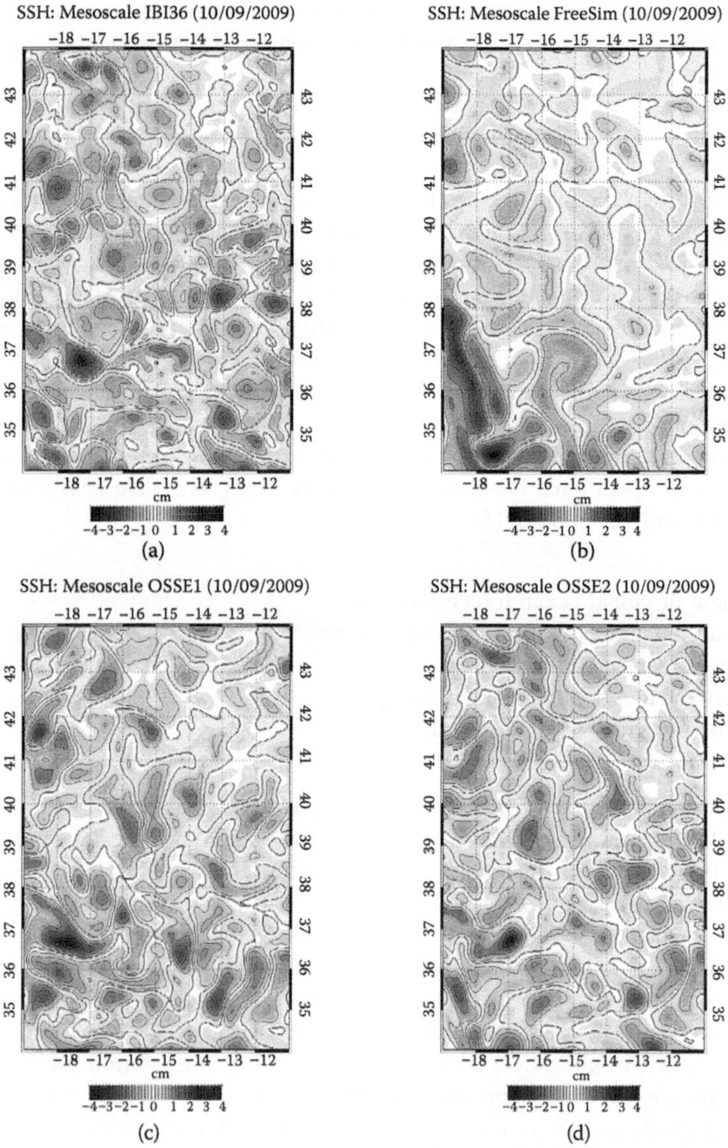

FIGURE 18.11 Observing System Simulation Experiment (OSSE) to assess the impact of SWOT data assimilation with the Ireland-Biscay-Iberian (IBI) 1/12° Mercator Ocean regional data assimilation system (Benkiran et al. 2017). For a given day in the year 2009, high pass filtered (wavelengths less than 200 km) maps are given for the (a) nature run (« truth ») obtained from the 1/36° model; (b) free run from the 1/12° model; (c) OSSE2 with assimilation of Jason-2, Envisat, and Jason-1 GM data; and (d) OSSE1 with assimilation of SWOT data.

added of around $3 trillion USD. As a consequence, it is almost inevitable that pressures on the ocean's natural assets will increase in the coming years and that this will add to the major stresses the oceans face and will face due to climate change.

More than ever, there is a need to continuously observe and monitor the oceans. This is imperative to understand and predict the evolution of our weather and climate. This is also essential for a better and sustainable management of our oceans and seas in support of the development of the blue economy. Operational oceanography plays a major role there. The capacity of operational oceanography to provide relevant responses to current and future applications remains, however, very dependent on advances in research and development and improvement of operational oceanography capabilities.

Despite significant progress over the last two decades, many scientific challenges and opportunities remain to be tackled in operational oceanography, from the observing system to modeling, data assimilation, and product dissemination.

In situ and remote sensing data are now routinely assimilated in global and regional ocean models to provide an integrated description of the ocean state, allowing for forecasts of the ocean's mesoscale of up to 10 days. However, despite this progress, some underlying problems remain to be solved with the specific nature of these depending on the individual forecasting system. An example is the assimilation of altimeter data that works well with many systems when there is a strong ocean thermocline (i.e., in low to mid-latitudes), but that is often less effective in polar waters where the thermocline is weaker or nonexistent or due to seasonal ice cover or shallow salinity stratification masking the thermocline signal.

The subsequent paragraphs highlight some of the future scientific challenges within each of five key research areas in operational oceanography (a more comprehensive analysis can be found in Schiller et al. 2015, and She et al. 2016):

- Development of global ocean forecast systems
- Intercomparison and validation of these systems
- Observing system evaluation (OSE)
- Marine ecosystem assessment and prediction
- Short- to medium-range coupled prediction

Improved representation of sub-mesoscale ocean processes (scales of 1–10 km or less) could improve air-sea fluxes and weather forecasts, reduce biases in the forecasts of temperatures, and improve the predictions of near-surface currents. Not surprisingly, this is an active area of research using field studies and large eddy simulations to develop improved parameterizations. Coupled atmosphere-wave-ocean models will allow more complete representation of sub-mesoscale processes, such as fronts and Langmuir circulation, provided the models have sufficient spatial and temporal resolution to resolve them. Improving the model resolution also involves improving the atmospheric forcing and bathymetry—a challenge over the next years.

In ocean models, one might expect improvements in fidelity at least until the next scale of ocean processes (perhaps sub-mesoscale and internal waves) starts to fall within the models' scope and new predictability challenges arise. The predictive capability of the forecast systems could however already be limited by our ability to initialize (constrain) errors with the observational coverage. Constellations of three to four satellite altimeters resolve a useful fraction of sea level variance, but (much) higher space/time resolution will be required to constrain future global and regional data assimilation systems. This can be achieved through a combination of swath altimetry with conventional altimetry (see the following).

There is a strong potential for extending the predictive range of ocean forecasts. As shown by Metzger et al. (2014), the predictive skill of ocean forecasts extends beyond 14 days in many areas dominated by mesoscale ocean circulation. The ocean forecast skill extends beyond the atmosphere in large part because of the difference in the fluids and stratification involved. Although ocean features are of much smaller scale than the equivalent atmospheric features, the timescales are much greater. As forecasting centers move toward global coupled systems such as the Earth System Prediction Capability (Theurich et al. 2016), the ocean becomes the long-term controller of the energy cycle and enables longer-term forecasts in the Earth system.

Because of the computational expense of resolving the ocean mesoscale, most of the operational oceanography community has been slower to implement ensemble prediction systems than the weather prediction and seasonal forecasting communities (moreover, one needs to use an ensemble of forcing from boundary conditions). It is now computationally feasible to develop global and regional ensemble prediction systems, and these will offer valuable estimates of and insights into the spread of uncertainty and levels of predictability that can be achieved by these systems.

Another set of issues arises from the emergence of multiple estimate approaches, a challenge also well-known to NWP. Through the expansion of operational oceanography systems, there may be several estimates at the same location from different global and regional systems, from ensembles, as well as from observational estimates. It is thus important to be able to identify the different processes represented by each system and the specific benefits of regional and coastal systems. Multi-model ensembles have shown the potential to provide substantial gains over using single system products; however, the limits and advantages of this approach need to be better understood. As the resolution of the systems cross the boundary between eddy-permitting and eddy-resolving, issues such as the representation of sub-grid scale parameterizations need to be addressed. The construction of a multi-parameter multi-model ensemble that can reflect the range of data assimilation and modeling approaches in operational oceanography systems is clearly a challenge for the years to come.

With all the advancements in operational ocean forecasting incorporating altimeter observations, there remain significant problems to be addressed. The diversity in data assimilation approaches reflects both a broad energetic science area in which new ideas are being proposed and explored as well as an area in which there exists large uncertainty in the solution to the problem. There are different assimilation cycles from daily to several days, different time periods of data used, and different methods for relating the SLA to sub-surface density. The area of operational altimeter observation assimilation into ocean forecast models will remain a highly active area for many years to come.

Observing system decision-makers now and in the future need quantitative feedback on the state of the system. To try to meet this need, NRT OSEs are being developed. Operational centers perform equivalent OSEs, withholding the same observations from their respective forecast systems, while at the same time, using a setup that is identical to their own operational system. By performing the OSEs in NRT, the results will be relevant to the present-day observing system and facilitate identification of potential "gaps". If multiple operational ocean forecasting systems are used for the NRT OSEs, results can be intercompared and the most robust findings identified and disseminated to interested parties. The hope is that community NRT OSEs will become an integral and routine part of all operational centers' activities—providing a regular, relevant, and robust evaluation of the current state of the observing system. The first examples of NRT OSEs were performed by Lea et al. (2014) using FOAM, the Met Office's ocean assimilation and forecasting system. This study demonstrated the strength of this approach and took steps toward the development of Observation Impact Statements (OISs; see Lea 2012). Development of OSSE capabilities is required, in parallel, to contribute to the design of future ocean observing systems.

The use of outputs from operational systems is progressively expanding to include biogeochemical applications. Operational oceanography products are likely to have sufficient skill in the not-too-distant future to provide key information to many applications related to living marine resources, from regular monitoring to statutory advice on ecosystems. Despite the growing amount of information, the outputs are still not extensively used by potential end users, suggesting more research is required to make them fit for purpose. The capacity to deliver "integrated biophysical products" based on syntheses between observations and model-derived information is likely to yield more systematic benefits to users if the quality and reliability of the products is scientifically proven.

As a first step toward coupled physical-biogeochemical data assimilation, physical data assimilation products were used to drive biogeochemical models (Brasseur et al. 2009). The analysis of biogeochemical tracer distributions revealed, however, the risk of a degradation of modeled biogeochemical distributions (e.g., chlorophyll-a, nutrients and sea-to-air carbon dioxide fluxes) by physical data assimilation most notably in the equatorial Pacific (e.g., Gehlen et al. 2015). Understanding the causes of spurious vertical fluxes and identifying a solution are critical to the development of biogeochemical applications of physical data assimilation and remains a priority for the operational oceanography community.

The availability of forecast systems of the ocean state provides the opportunity to develop high-resolution coupled ocean-atmosphere prediction systems for the short to medium range. Making progress in this field is a grand challenge due to the complexity of coupled infrastructure, coupled modeling, observational requirements (including experimental campaigns), and the resulting need for more diverse teams of scientific experts. The groups active in coupled prediction research are pursuing a wide range of applications including global weather forecast systems and predictions of hurricanes, tropical cyclones and typhoons, extratropical storms, high-latitude weather and sea ice, as well as coastal upwelling, sea breezes, and sea fog. First systems run in coupled prediction mode are described by Brassington et al. (2015). A related challenge is the development of coupled data assimilation, which must correctly handle the different temporal and spatial scales across the ocean, wave, sea-ice, and atmospheric environments. There is a need to explore new approaches to address coupled model biases and to optimize the weighting of coupled covariances. Further progress will require community-established benchmarks, test cases, targeted observation campaigns, and coordination across the existing international teams.

Finally, a key long-term challenge is the sustainability and expansion of the physical and biogeochemistry Global Ocean Observing System (GOOS), which is the foundation for operational oceanography and climate applications. The ocean is, in particular, still critically under sampled for biological and biogeochemical properties. The lack of data is the main obstacle to the implementation of operational systems suitable for the routine and accurate monitoring of the ocean biogeochemical state. Research can contribute to the specification of essential physical and biological/biogeochemical observations, to identifying the best sampling strategies, and to formulating recommendations to improve the observing capacity needed to sustain fully fledged integration of biogeochemistry into operational systems. There is a clear need for a roadmap regarding future ocean color missions needed to consolidate the space component in the next 20 years, combining conventional low-Earth orbit and geostationary orbit missions. New *in situ* observing programs such as BioGeoChemical Argo autonomous floats offer major opportunities to improve synergies with present and future ocean color satellite missions. A first solid set of recommendations was provided by the OceanObs09 conference (Claustre et al. 2010) and recently led to a design and implementation plan for BioGeoChemical Argo (BioGeoChemical Argo Task Team 2016). New opportunities also arise in regional seas (e.g., high-frequency radars and gliders) and the advent of "intelligent" new sensors and new remote sensing technologies both from space and *in situ*.

18.5.2 Evolution of Altimetry Technology and Impact on Operational Oceanography

Radar altimetry is going through two major transitions originating in changes of the programmatic landscape and in the development of new altimetry technologies. The former provides a solid background for coverage and constellation robustness improvement, and the latter paves the way for a substantial improvement in the precision and accuracy of recent and future altimetry measurements. In this context, wide-swath altimeter/interferometer technology to be demonstrated with the SWOT mission might be a game changer.

18.5.2.1 Improving Coverage and Robustness

Probably the most important evolution of altimetry for operational oceanography is the transition of uncoordinated programs and one-shot concepts (e.g., AltiKa for Ka-band altimetry or CryoSat-2 for SAR-mode) to sustained multi-agency operational programs (e.g., Europe's Copernicus with Sentinel-3 series, U.S./Europe with the Jason series). The altimeter community has been extremely lucky to benefit from past missions with exceptional lifespans —for example, 11 years for GFO and Jason-1 or 13 years for TOPEX/Poseidon. However, the design life for these satellites was a 3-year nominal mission. The concept of virtual altimetry constellation developed by CEOS (see Escudier and Fellous 2008) made it possible for space and environmental agencies to coordinate their efforts in sustained programs where altimetry missions are thought and developed as a series of two to four missions (e.g., Jason-3 and Jason-CS, or Sentinel-3A and Sentinel-3C) with a vision for follow-on programs. Moreover, these operational programs are complemented by challenging technology demonstrators such as wide-swath altimetry with SWOT and the combination of altimetry and wind/wave scatterometry with China France Oceanography SATellite (CFOsat).

This paradigm change not only makes it possible for operational oceanography to consistently rely on ocean topography in the coming years, but it also substantially strengthens the constellation robustness and coverage. The analysis of Dibarboure and Lambin (2015) predicts that the probability of having two to four altimeters in operation (the minimum required for delayed time and real time monitoring of mesoscale) is larger than 80% until 2024 thanks to the combination of operational satellites and so-called contributing missions.

In this context, the wide-swath mission outlined by Fu et al. (2012) provides a breakthrough in coverage thanks to its two-dimensional topography images (also combined with two-dimensional images of radar backscatter that might give precious insights on the nature of additional parameters related to SSH derivatives, sea state, and ocean currents). With a 130-km swath, SWOT could demonstrate an observing capability of up to four altimeters in Level 4 products or ocean model assimilation (Pujol et al. 2012; Benkiran et al. 2017).

18.5.2.2 Improving Precision and Accuracy

In addition to sampling improvements, new generations of ocean topography payloads (altimeter, precise orbit determination, microwave radiometer) pave the way for more precise and more accurate products.

Thanks to its Ka band and high pulse repeat frequency, AltiKa has a noise level that is substantially better than Jason-class altimeters, which were already a major improvement from older altimetry missions such as GEOSAT or ERS-1. Yet AltiKa still uses a so-called low resolution mode (LRM) altimeter with a circular footprint of the order of 7 km. In comparison, CryoSat-2 demonstrated the value of a new altimeter technology: the delay-Doppler mode, also known as SAR mode (SARM) featuring a thin stripe-shaped footprint as small as 300 m in the along-track direction. The benefits extend to a lower noise floor, less corruption from rain cells, and rapid sea state changes (otherwise misinterpreted by LRM altimeters; see Dibarboure et al. 2014), and better coastal stability. Sentinel-3A recently provided the first global coverage in SARM, giving new insights on the benefit of this new technology. The main benefit for operational oceanography is the ability to observe and to assimilate wavelengths smaller than 80 to 100 km that might be difficult to trust with LRM altimeters, as shown by Dufau et al. (2016).

Because this technology is relatively new, the design of new climate-focused mission Jason-CS (Sentinel 6) may benefit from further improvements where LRM and SARM are activated concurrently using so-called interleaved technology (as opposed to the short bursts of Sentinel-class altimeters) as presented by Cullen and Francis (2013). In addition to retro-compatibility with the older LRM technology to ensure a consistent climate record, the interleaved mode makes it possible to further reduce the noise level to observe smaller scales.

Similarly, the SWOT mission intends to provide a breakthrough in observability thanks to its new Ka-band radar interferometer technology. The NASA/CNES/CSA SWOT mission is a wide-swath altimetry mission using Ka-band SAR-interferometric technology, aiming to achieve an order of magnitude improvement in both spatial resolution and SSH precision. SWOT should be launched in 2021. SWOT aims to provide global maps of oceanic sea level and geostrophic currents (up to 78°N and S) on a 1- or 2-km grid with full coverage every 21 days. SWOT will provide an effective resolution of 15 km for ocean dynamics, compared to the 150-km resolution available today from gridded conventional nadir altimeter data. Although SWOT is a research-oriented mission, its new approach to altimetry has the potential to provide valuable fine-scale data sets for the next generation of ocean models and marine applications. In addition to the two-dimensional imaging capability discussed earlier, SWOT should feature an unprecedented noise level, allowing it to observe wavelengths as small as 15 km (resp. 30 km) over 68 % (resp. 95%) of the oceans. This will be an asset in observing small mesoscale to investigate the link with sub-mesoscale, internal waves, and internal tides (see Fu et al. 2012). In turn, this may emphasize the need to develop new algorithms to separate balanced dynamics from internal tides (e.g., Arbic et al. 2012; Ponte and Klein 2015) as additional processing steps in Level 2 or Level 3 altimetry processors or new Level 4 interpolators to account for the imbalance between the high resolution/coverage of SWOT images and its sparse temporal sampling (e.g., Ubelmann et al. 2016).

18.6 CONCLUSION

Satellite altimetry has had a major impact on the development of operational oceanography, and it remains the most important satellite observing system required for ocean analysis and forecasting. Operational oceanography systems are heavily dependent on the availability of near real time altimeter data from multiple missions. Over the past 20 years, there has been major progress in the development of near real time multi-mission high-quality altimeter products. There have been many improvements (e.g., improved algorithms, timeliness, new SLA products, and new MDTs). Improvements of models and data assimilation techniques have also resulted in a better use of multiple altimeter missions jointly with *in situ* observations.

The impact of the altimeter constellation on the performance of modeling and data assimilation systems is well evidenced, and there is a clear benefit of having three to four coordinated altimeters flying simultaneously (i.e., the minimum requirement for operational oceanography). Analysis of the satellite altimetry constellation over the past 25 years showed, however, that this minimum requirement was met only during a limited time period. Review of future missions for the next decade shows clear progress. The altimeter constellation will be consolidated, in particular, with the development of the European Copernicus Sentinel long-term program. It will be essential, however, to strengthen international collaboration to optimize and make the best use of altimeter missions from a growing number of space agencies. Efforts to ensure homogenized and intercalibrated data sets from multiple missions should also be pursued.

New swath techniques with a potentially very large impact for operational oceanography will also be demonstrated with the SWOT mission. This is essential. Operational oceanography now uses high-resolution models with data assimilation (e.g., less than 10 km at global scale, a few kilometers at regional scales, a few 100 meters at coastal scales) and will likely move to even higher resolution in the next decade (e.g., a few kilometers at the global scale). This poses much stronger requirements for an altimeter constellation. Observation capabilities now critically lag behind modeling capabilities. Swath techniques will likely play a major role in the coming decades and planning for operational swath missions should start as soon as possible.

REFERENCES

Arbic, B. K., J. G. Richman, J. F. Shriver, P. G. Timko, E. J. Metzger, and A. J. Wallcraft. 2012. Global modeling of internal tides within an eddying ocean general circulation model. *Oceanography* 25(2): 20–29. doi: 10.5670/oceanog.2012.38.

AVISO. 2016. *SSALTO/DUACS User Handbook: (M)SLA and (M)ADT Near-Real Time and Delayed Time Products. CNES document* SALP-MU-P-EA-21065-CLS, latest version available on AVISO website: http://www.aviso.oceanobs.com/ en/data/tools/aviso-user-handbooks/index.html (accessed August, 2016).

Balmaseda, M. A., K. Mogensen, and A. T. Weaver. 2013. Evaluation of the ECMWF ocean reanalysis system ORAS4. *Q. J. Roy. Meteorol. Soc.* 139:1132–1161. doi: 10.1002/qj.2063.

Bell, M. J., M. Lefebvre, P. Y. Le Traon, N. Smith, and K. Wilmer-Becker. 2009. The global ocean data assimilation experiment. *Oceanography* 22(3): 14–21.

Bell, M. J., A. Schiller, P. Y. Le Traon, N. R. Smith, E. Dombrowsky, and K. Wilmer-Becker. 2015. An introduction to GODAE OceanView. *J. Oper. Oceanogr.* 8(suppl. 1): s2–s11. 10.1080/1755876X.2015.1022041.

Benkiran, M., E. Remy, E. Greiner, Y. Drillet, and P. Y. Le Traon. 2017. An Observing System Simulation Experiment to evaluate the impact of SWOT in a regional data assimilation system. *Remote Sensing of the Environment* (in press).

Biogeochemical Argo Task Team. 2016. *The Rationale, Design and Implementation Plan for Biogeochemical Argo*. http://doi.org/10.13155/46601

Born, G. H., J. A. Dunne, and D. B. Lame. 1979. Seasat mission overview. *Science* 204(4400):1405–1406.

Brasseur, P., et al. 2009. Integrating biogeochemistry and ecology into ocean data assimilation systems. *Oceanography* 22(3): 206–215.

Brassington, G. B., et al. 2015. Progress and challenges in short- to medium-range coupled prediction. *J. Oper. Oceanogr.* 8(Suppl 2): s239–s258.

Buijsman, M. C., B. K. Arbic, J. A. M. Green, R. W. Helber, J. G. Richman, J. F. Shriver, P. G. Timko, and A. Wallcraft. 2015. Optimizing internal wave drag in a forward barotropic model with semidiurnal tides. *Ocean Model.* 85: 42–55. doi: 10.1016/j.ocemod.2014.11.003.

Cardinali, C., S. Pezzulli, and E. Andersson. 2004. Influence-matrix diagnostic of a data assimilation system. *Q. J. Roy. Meteorol. Soc.* 130: 2767–2786. doi: 10.1256/qj.03.205.

Carnes, M. R., J. L. Mitchell, and P. W. Witt. 1990. Synthetic temperature profiles derived from GEOSAT altimetry: Comparison with air-dropped expendable bathythermograph profiles. *J. Geophys. Res. Oceans* 95(C10): 17979–17992.

Chelton, D. B., M. G. Schlax, and R. M. Samelson. 2011. Global observations of nonlinear mesoscale eddies. *Prog. Oceanogr.* 91: 167–216.

Claustre, H., et al. 2010. Bio-Optical Profiling Floats as New Observational Tools for Biogeochemical and Ecosystem Studies: Potential Synergies with Ocean Color Remote Sensing. *S.l: ESA Publication WPP-306. Proceedings of OceanObs'09: Sustained Ocean Observations and Information for Society*. Venice, Italy, 21–25 September, 2009, Eds. J. Hall, D. E. Harrison, D. Stammer.

CMEMS. 2016. *Copernicus Marine Environment Monitoring Service—Sea Level Thematic Assembly Center Product User Manual*. Available online: http://marine.copernicus.eu/documents/PUM/CMEMS-SL-PUM-008-017-036.pdf (accessed August 2016).

Cooper, M., and K. Haines. 1996. Altimetric assimilation with water property conservation. *J. Geophys. Res.* 101(C1): 1059–1077.

Cullen, R., and R. Francis. 2013. The Jason-CS Ocean Surface Topography—Mission Payload Design and Development. *Presented at the 2013 GODAE symposium*. Available online: https://www.godae.org/~godae-data/Symposium/GOV-posters/S2.1-03-Cullen.pdf (accessed November 2016).

Cummings, J. A., and O. M. Smedstad. 2013. Variational data assimilation for the global ocean. In *Data Assimilation for Atmospheric, Oceanic and Hydrologic Applications*. II. Eds. Park S. K., Xu L DOI 10.1007/978-3-642-35088-713, © Springer-Verlag Berlin Heidelberg.

Dibarboure, G., F. Boy, J. D. Desjonqueres, S. Labroue, Y. Lasne, N. Picot, and P. Thibaut. 2014. Investigating short-wavelength correlated errors on low-resolution mode altimetry. *J. Atmos. Oceanic Technol.* 31(6): 1337–1362.

Dibarboure, G., and J. Lambin. 2015. Monitoring the ocean surface topography virtual constellation: Lessons learned from the contribution of SARAL/AltiKa. *Mar. Geod.* 38(suppl. 1): 684–703.

Dibarboure, G., M. I. Pujol, F. Briol, P.-Y. Le Traon, G. Larnicol, N. Picot, F. Mertz, and M. Ablain. 2011a. Jason-2 in DUACS: First tandem results and impact on processing and products. *Mar. Geod.* 34(3–4): 214–241: doi: 10.1080/01490419.2011.584826.

Dibarboure, G., M. I. Pujol, F. Briol, P.-Y. Le Traon, G. Larnicol, N. Picot, F. Mertz, and M. Ablain. 2011b. Jason-2 in DUACS: Updated system description, first tandem results and impact on processing and products. *Mar. Geod.* 34(3–4): 214–241.

Dibarboure, G., et al. 2012. Finding desirable orbit options for the "Extension of Life" phase of Jason-1. *Mar. Geod.* 35:(suppl. 1). http://dx.doi.org/10.1080/01490419.2012.717854

Dobricic, S., C. Dufau, P. Oddo, N. Pinardi, M. I. Pujol, and M. H. Rio. 2012. Assimilation of SLA along track observations in the Mediterranean with an oceanographic model forced by atmospheric pressure. *Ocean Sci.* 8(5): 787–795.

Donlon, C., et al. 2007. The global ocean data assimilation experiment high-resolution Sea Surface Temperature Pilot Project. *Bull. Am. Meteorol. Soc.* 88(8): 1197–1213. doi: 10.1175/BAMS-88-8-1197.

Ducet, N., P. Y. Le Traon, and G. Reverdin. 2000. Global high resolution mapping of ocean circulation from the combination of TOPEX/POSEIDON and ERS-1/2. *J. Geophys. Res. Oceans* 105(C8): 19477–19498.

Dufau, C. 2014. The TAPAS initiative, an efficient link between MFCs and SL TAC. Presented at the 2014 MyOcean Science Days Conference. Available online: http://lgge.osug.fr/meom/Projets/MYOCEAN/Events/MSD2014/Agenda/S3-1_Dufau_MSD2014.pdf (accessed September 2014).

Dufau, C., M. Orsztynowicz, G. Dibarboure, R. Morrow, and P.Y. Le Traon. 2016. Mesoscale resolution capability of altimetry: Present and future. *J. Geophys. Res. Oceans 06/2016*. doi: 10.1002/2015JC010904.

Escudier, P., and J. L. Fellous. 2008. The next 15 years of Satellite Altimetry: Ocean Surface Topography Constellation User Requirements Document. Report of the Ocean Surface Topography Virtual Constellation (OST VC) at the 27th Committee on Earth Observation Satellites (CEOS) SIT Meeting Produced and distributed by Eumetsat, Germany. October 2009. http://ceos.org/images/OST/SatelliteAltimetryReport_2009-10.pdf.

Fox, D. N., W. J. Teague, C. N. Barron, M. R. Carnes, and C. M. Lee. 2002. The Modular Ocean Data Assimilation System (MODAS). *J. Atmos. Oceanic Technol.* 19(2): 240–252.

Fu, L. L., D. Alsdorf, R. Morrow, E. Rodriguez, and N. Mognard. 2012. SWOT: The surface wand ocean topography mission wide—Swath altimetric measurement of water elevation on earth. *JPL Publication 12-05*. Available online: http://swot.jpl.nasa.gov/files/SWOT_MSD_final-3-26-12.pdf (accessed February 2012).

Fu, L. L., and A. Cazenave. 2001. Satellite altimetry and earth sciences. In *A Handbook of Techniques and Applications.* International Geophysics Series, Volume 69, Academic Press, San Diego, CA, p. 463.

Gehlen, M., et al. 2015. Building the capacity for forecasting marine biogeochemistry and ecosystems: Recent advances and future developments. *J. Oper. Oceanogr.* 8 (S1): s168–s187.

Guinehut, S., A. L. Dhomps, G. Larnicol, and P. Y. Le Traon. 2012. High resolution 3D temperature and salinity fields derived from in-situ and satellite observations. *Ocean Sci.* 8: 845–857. doi: 10.5194/os-8-845-2012.

Haines, K., Johannessen, J. A., Knudsen, P., Lea, D., Rio, M.-H., Bertino, L., Davidson, F., and Hernandez, F., 2011. An ocean modelling and assimilation guide to using GOCE geoid products, *Ocean Sci.* 7: 151–164. doi: 10.5194/os-7-151-2011.

Hebert, D. A., R. A. Allard, E. J. Metzger, P. G. Posey, R. H. Preller, A. J. Wallcraft, M. W. Phelps, and O. M. Smedstad. 2015. Short-term sea ice forecasting: An assessment of ice concentration and ice drift forecasts using the U.S. Navy's Arctic Cap Nowcast/Forecast System. *J. Geophys. Res. Oceans* 120. 1–19, doi: 10.1002/2015JC011283.

International GODAE Steering Team. 2000. *The Global Ocean Data Assimilation Experiment Strategic Plan.* GODAE Report No. 6. December, 2000.

IPCC First Assessment Report. 1990. *Scientific Assessment of Climate Change—Report of Working Group I.* J. T. Houghton J. T., G. J. Jenkins, J. J. Ephraums (eds.), Cambridge University Press, UK, p. 365.

Jacobs, G. A., C. N. Barron, and R. C. Rhodes. 2001. Mesoscale characteristics. *J. Geophys. Res. Oceans* 106(C9): 19581–19595.

Jacobs, G. A., J. G. Richman, J. D. Doyle, P. L. Spence, B. P. Bartels, C. N. Barron, R. W. Helber, and F. L. Bub. 2014. Simulating conditional deterministic predictability within ocean frontogenesis. *Ocean Model.* 78: 1–16.

Kourafalou, V. H, et al. 2015. Coastal ocean forecasting: Science foundation and user benefits. *J. Oper. Oceanogr.* 8(suppl. 1): s147–s167. doi: 10.1080/1755876X.2015.1022348.

Kurapov, A. L., D. Foley, P. T. Strub, G. D. Egbert, and J. S. Allen. 2011. Variational assimilation of satellite observations in a coastal ocean model off Oregon. *J. Geophys. Res.* 116: C05006. doi: 10.1029/2010JC006909.

Lea, D. J. 2012. *Observation Impact Statements for Operational Ocean Forecasting*. Met Office Forecasting R&D Technical Report no. 568.

Lea, D. J., J.-P. Drecourt, K. Haines, and M. J. Martin. 2008. Ocean altimeter assimilation with observational- and model-bias correction. *Q. J. Roy. Meteorol. Soc.* 134: 1761–1774.

Lea, D. J., Martin, M. J., and P. R. Oke. 2014. Demonstrating the complementarity of observations in an operational ocean forecasting system. *Q. J. Roy. Meteorol. Soc.* 140: 2037–2049. doi: 10.1002/qj.2281.

Lellouche, J. M., O. Le Galloudec, M. Drévillon, C. Régnier, E. Greiner, G. Garric, N. Ferry, C. Desportes, C. E. Testut, and C. Bricaud. 2013. Evaluation of global monitoring and forecasting systems at Mercator Océan. *Ocean Sci.* 9: 57–81. doi: 10.5194/os-9-57-2013.

Le Traon, P. Y. 2013. From satellite altimetry to Argo and operational oceanography: Three revolutions in oceanography. *Ocean Sci.* 9(5): 901–915. doi: 10.5194/os-9-901-2013.

Le Traon, P. Y., et al. 2015. Use of satellite observations for operational oceanography: recent achievements and future prospects. *J. Oper. Oceanogr.* 8(supp. 1): s12–s27. doi: 10.1080/1755876X.2015.1022050.

Le Traon, P. Y., M. Rienecker, N. Smith, P. Bahurel, M. Bell, H. Hurlburt, and P. Dandin. 2001. *Operational Oceanography and Prediction—A GODAE Perspective, in Observing the Oceans in the 21st Century*, edited by C.J. Koblinsky and N.R. Smith. GODAE project office, Bureau of Meteorology, 529–545.

Le Traon, P.Y., Faugère Y., Hernandez F., Dorandeu J., Mertz F., and M. Ablain. 2003. Can we merge GEOSAT Follow-On with TOPEX/POSEIDON and ERS-2 for an improved description of the ocean circulation? *J. Atmos. Oceanic Technol.* 20: 889–895.

Le Traon, P. Y., J. Johannessen, I. Robinson, and O. Trieschmann. 2006. *Report from the Working Group on Space Infrastructure for the GMES Marine Core Service*. GMES Fast Track Marine Core Service Strategic Implementation Plan.

Le Traon, P.Y., F. Nadal, and N. Ducet. 1998. An improved mapping method of multisatellite altimeter data. *J. Atmos. Oceanic Technol.* 15: 522–533.

Le Traon, P.Y., and F. Ogor. 1998. ERS-1/2 orbit improvement using TOPEX/POSEIDON: The 2 cm challenge. *J. Geophys. Res. Oceans* 103: 8045–8057.

Martin, M. J., et al. 2015. Status and future of data assimilation in operational oceanography. *J. Oper. Oceanogr.* 8(suppl. 1): s28–s48.

Maximenko, N., P. Niiler, M.-H. Rio, O. Melnichenko, L. Centurioni, D. Chambers, V. Zlotnicki, and B. Galperin. 2009. Mean dynamic topography of the ocean derived from satellite and drifting buoy data using three different techniques. *J. Atmos. Oceanic Technol.* vol 26, pp. 1910–1919. doi: http://dx.doi.org10.1175/2009JTECHO672.1.

Metzger, E. J., et al. 2014: US Navy Operational Global Ocean and Arctic Ice Prediction Systems. *Oceanography* 27: 3. doi:10.5670/oceanog.2014.66.

Ngodock, H., M. Carrier, I. Souopgui, S. Smith, P. Martin, P. Muscarella, and G. Jacobs. 2016. On the direct assimilation of along-track sea-surface height observations into a free-surface ocean model using a weak constraints four-dimensional variational (4D-Var) method. *Q. J. R. Meteorol. Soc.* 142: 1160–1170. doi: 10.1002/qj.2721.

Oke, P. R., M. Balmaseda, M. Benkiran, J. A. Cummings, E. Dombrowsky, Y. Fujii, S. Guinehut, G. Larnicol, P.-Y. Le Traon, and M. J. Martin. 2009. Observing system evaluations using GODAE systems. *Oceanography* 22(3): 144–153. doi: 10.5670/oceanog.2009.72.

Oke, P. R., G. B. Brassington, D. A. Griffin, and A. Schiller. 2008. The Bluelink Ocean Data Assimilation System (BODAS). *Ocean Model.* 20: 46–70. doi: 10.1016/ j.ocemod.2007.11.002.

Oke, P. R., et al. 2015a. Assessing the impact of observations on ocean forecasts and reanalyses: Part 1, Global studies. *J. Oper. Oceanogr.* 8(suppl. 1): s49–s62. doi: 10.1080/1755876X.2015.1022067.

Oke, P. R., et al. 2015b. Assessing the impact of observations on ocean forecasts and reanalyses: Part 2, Regional applications. *J. Oper. Oceanogr.* 8(suppl. 1): s63–s79. doi: http://dx.doi.org10.1080/1755876X.2015.1022080.

Oke, P. R., P. Sakov, M. L. Cahill, J. R. Dunn, R. Fiedler, D. A. Griffin, J. V. Mansbridge, K. R. Ridgway, and A. Schiller. 2013b. Towards a dynamically balanced eddy-resolving ocean re-analysis: BRAN3. *Ocean Model.* 67: 52. doi: 10.1016/j.ocemod.2013.03.008.

Pascual, A., C. Boone, G. Larnicol, and P.Y. Le Traon. 2009. On the quality of real-time altimeter gridded fields: Comparison with in situ data. *J. Atmos. Oceanic Technol.* 26: 556–569.

Pascual, A., Y. Faugere, G. Larnicol, and P. Y. Le Traon. 2006. Improved description of the ocean mesoscale variability by combining four satellite altimeters. *Geophys. Res. Lett.* 33(2): Art. No. L02611.

Ponte, A. L., and P. Klein. 2015. Incoherent signature of internal tides on sea level in idealized numerical simulations. *Geophys. Res. Lett.* 42(5): 1520–1526.

Posey, P. G., et al. 2015. Improving Arctic sea ice edge forecasts by assimilating high horizontal resolution sea ice concentration data into the US Navy's ice forecast systems. *The Cryosphere* 9: 1735–1745.

Pujol, M. I., Dibarboure, G., Le Traon, P. Y., and P. Klein. 2012. Using high-resolution altimetry to observe mesoscale signals. *J. Atmos. Oceanic Technol.* 29(9): 1409–1416.

Pujol, M.-I., Y. Faugère, G. Taburet, S. Dupuy, C. Pelloquin, M. Ablain, and N. Picot. 2016. DUACS DT2014: The new multi-mission altimeter dataset reprocessed over 20 years. *Ocean Sci.* 12: 1067–1090. doi: 10.5194/os-12-1067-2016.

Remy, E., P. Y. Le Traon, and S. Verrier. 2017. Impact of the altimeter constellation on a high resolution ocean analysis and forecasting system (submitted).

Rio, M. H., S. Mulet, and N. Picot. 2014. Beyond GOCE for the ocean circulation estimate: Synergetic use of altimetry, gravimetry, and in situ data provides new insight into geostrophic and Ekman currents. *Geophys. Res. Lett.* 41(24): 8918–8925.

Rowley, C., and A. Mask. 2014: Regional and coastal prediction with the relocatable ocean nowcast/forecast system. *Oceanography* 27(3): 44–55. doi: 10.5670/oceanog.2014.67.

Sakov, P, F. Counillon, L. Bertino, K. A. Lisaeter, P. R. Oke, and A. Korablev. 2012. TOPAZ4: An ocean-sea ice data assimilation system for the North Atlantic and Arctic. *Ocean Sci.* 8: 633–656. doi: 10.5194/os-8-633-2012.

Schiller, A. 2011. Ocean Forecasting in the 21st Century. From the early days to tomorrow's challenges. In: *Operational Oceanography in the 21st Century*. Eds. Andreas S., Gary B. B. Springer, Dordrecht, The Netherlands, pp. 3–28.

Schiller, A., et al. 2015. Synthesis of new scientific challenges for GODAE OceanView. *J. Oper. Oceanogr.* 8(suppl. 2): s259–s271. doi: 10.1080/1755876X.2015.1049901.

Schiller, A., and G. B. Brassington (eds.). 2011. *Operational Oceanography in the 21st Century*. Springer. Dordrecht, The Netherlands. p. 745.

Semtner, A. J., and R. M. Chervin. 1992. Ocean general circulation from a global eddy resolving model. *J. Geophys. Res.* 97: 5493–5550.

She, J., et al. 2016. Developing European operational oceanography for Blue Growth, climate change adaptation and mitigation, and ecosystem-based management. *Ocean Sci.* 12: 953–976. doi: 10.5194/os-12-953-2016.

Smith, N., and M. Lefebvre. 1997. Monitoring the oceans in the 2000s: An integrated approach. The Global Ocean Data Assimilation Experiment (GODAE). International Symposium. Biarritz. Theurich, G., C. DeLuca, T. Campbell, F. Liu, K. Saint, M. Vertenstein, J. Chen, et al. 2015. The earth system prediction suite: Toward a coordinated US modeling capability. *Bull. Am. Meteorol. Soc.* 93(4): 485–498.

Stammer, D., A. Köhl, and C. Wunsch. 2007. Impact of accurate geoid fields on estimates of the ocean circulation. *J. Atmos. Oceanic Technol.* 24(8): S1464–1478. doi: 10.1175/JTECH2044.1.

Stanev, E. V., J. Schulz-Stellenfleth, J. Staneva, S. Grayek, S. Grashorn, A. Behrens, W. Koch, and J. Pein. 2016. Ocean forecasting for the German Bight: From regional to coastal scales. *Ocean Sci.* 12: 1105–1136. doi: 10.5194/os-12-1105-2016.

Tai, C. K. 1988. GEOSAT crossover analysis in the tropical Pacific 1. Constrained sinusoidal crossover adjustment. *J. Geophys. Res.* 93(C9): 10621–10629. doi: 10.1029/JC093iC09p10621.

Theurich, G., et al. 2016. The Earth System prediction suite: Toward a coordinated US modeling capability. *Bull. Am. Meteorol. Soc.* 97: 1229–1247. doi: 10.1175/BAMS-D-14-00164.1.

Turpin, V., E. Remy, and P. Y. Le Traon. 2016. How essential are Argo observations to constrain a global ocean data assimilation system? *Ocean Sci.* 12: 257–274. doi: 10.5194/os-12-257-2016.

Ubelmann, C., B. Cornuelle, and L. L. Fu. 2016. Dynamic mapping of along-track ocean altimetry: Method and performance from Observing System Simulation Experiments. *J. Atmos. Oceanic Technol.* 32: 177–184. doi: 10.1175/JTECH-D-14-00152.1.

Usui, N., Y. Fujii, K. Sakamoto, and M. Kamachi. 2015. Development of a four-dimensional variational assimilation system for coastal data assimilation around Japan. *Mon. Weather Rev* Vol 143, pp. 3874–3892. doi: 10.1175/MWR-D-14-00326.1.

Usui, N., S. Ishizaki, Y. Fujii, H. Tsujino, T. Yasuda, and M. Kamachi. 2006. Meteorological Research Institute Multivariate Ocean Variational Estimation (MOVE) System: Some early results. *Adv. Space Res.* 37: 806–822.

Waters, J., D. J. Lea, M. J. Martin, I. Mirouze, A. Weaver, and J. While. 2014. Implementing a variational data assimilation system in an operational 1/4 degree global ocean model. *Q. J. R. Meteorol. Soc.* 141(87): 333–349. doi: 10.1002/qj.2388.

Weaver, A. T., C. Deltel, E. Machu, S. Ricci, and N. Daget. 2005. A multivariate balance operator for variational ocean data assimilation. *Q.J.R. Meteorol. Soc.*, 131: 3605–3625. doi: 10.1256/qj.05.119.

Index

J

K

L

M

N

O

P

Q

R

S

X

Y

Z

Printed and bound by CPI Group (UK) Ltd, Croydon, CR0 4YY
01/11/2024
01782601-0014